The following curriculum standards were taken from the *Principles and Standards fo*... by the National Council of Teachers of Mathematics, 2000.

MATHEMATICAL PROCESS STANDARDS

Problem Solving
STANDARD

Instructional programs from prekindergarten through grade 12 should enable all students to—
- Build new mathematical knowledge through problem solving
- Solve problems that arise in mathematics and in other contexts
- Apply and adapt a variety of appropriate strategies to solve problems
- Monitor and reflect on the process of mathematical problem solving

Communication
STANDARD

Instructional programs from prekindergarten through grade 12 should enable all students to—
- Organize and consolidate their mathematical thinking through communication
- Communicate their mathematical thinking coherently and clearly to peers, teachers, and others
- Analyze and evaluate the mathematical thinking and strategies of others
- Use the language of mathematics to express mathematical ideas precisely

Representation
STANDARD

Instructional programs from prekindergarten through grade 12 should enable all students to—
- Create and use representations to organize, record, and communicate mathematical ideas
- Select, apply, and translate among mathematical representations to solve problems
- Use representations to model and interpret physical, social, and mathematical phenomena

Reasoning and Proof
STANDARD

Instructional programs from prekindergarten through grade 12 should enable all students to—
- Recognize reasoning and proof as fundamental aspects of mathematics
- Make and investigate mathematical conjectures
- Develop and evaluate mathematical arguments and proofs
- Select and use various types of reasoning and methods of proof

Connections
STANDARD

Instructional programs from prekindergarten through grade 12 should enable all students to—
- Recognize and use connections among mathematical ideas
- Understand how mathematical ideas interconnect and build on one another to produce a coherent whole
- Recognize and apply mathematics in contexts outside of mathematics

PreK–2 Standards
NUMBER AND OPERATIONS
In prekindergarten through grade 2 all students should—

Understand numbers, ways of representing numbers, relationships among numbers, and number systems and should—
- Count with understanding and recognize "how many" in sets of objects;
- Use multiple models to develop initial understandings of place value and the base-ten number system;
- Develop understanding of the relative position and magnitude of whole numbers and of ordinal and cardinal numbers and their connections;
- Develop a sense of whole numbers and represent and use them in flexible ways, including relating, composing, and decomposing numbers;
- Connect number words and numerals to the quantities they represent, using various physical models and representations;
- Understand and represent commonly used fractions, such as $\frac{1}{4}$, $\frac{1}{3}$, and $\frac{1}{2}$

Understand meanings of operations and how they relate to one another and should—
- Understand various meanings of addition and subtraction of whole numbers and the relationship between the two operations;
- Understand the effects of adding and subtracting whole numbers;
- Understand situations that entail multiplication and division, such as equal groupings of objects and sharing equally

Compute fluently and make reasonable estimates and should—
- Develop and use strategies for whole-number computations, with a focus on addition and subtraction;
- Develop fluency with basic number combinations for addition and subtraction;
- Use a variety of methods and tools to compute, including objects, mental computation, estimation, paper and pencil, and calculators

ALGEBRA
In prekindergarten through grade 2 all students should—

Understand patterns, relations, and functions and should—
- Sort, classify, and order objects by size, number, and other properties;
- Recognize, describe, and extend patterns such as sequences of sounds and shapes or simple numeric patterns and translate from one representation to another;
- Analyze how both repeating and growing patterns are generated

Represent and analyze mathematical situations and structures using algebraic symbols and should—
- Illustrate general principles and properties of operations, such as commutativity, using specific numbers;
- Use concrete, pictorial, and verbal representations to develop an understanding of invented and conventional symbolic notations

Use mathematical models to represent and understand quantitative relationships and should—
- Model situations that involve the addition and subtraction of whole numbers, using objects, pictures, and symbols

Analyze change in various contexts and should—
- Describe qualitative change, such as a student's growing taller;
- Describe quantitative change, such as a student's growing two inches in one year

PreK–2 Standards
GEOMETRY
In prekindergarten through grade 2 all students should—

Analyze characteristics and properties of two- and three-dimensional geometric shapes and develop mathematical arguments about geometric relationships and should—
- Recognize, name, build, draw, compare, and sort two- and three-dimensional shapes;
- Describe attributes and parts of two- and three-dimensional shapes;
- Investigate and predict the results of putting together and taking apart two- and three-dimensional shapes

Specify locations and describe spatial relationships using coordinate geometry and other representational systems and should—
- Describe, name, and interpret relative positions in space and apply ideas about relative position;
- Describe, name, and interpret direction and distance in navigating space and apply ideas about direction and distance;
- Find and name locations with simple relationships such as "near to" and in coordinate systems such as maps

Apply transformations and use symmetry to analyze mathematical situations and should—
- Recognize and apply slides, flips, and turns;
- Recognize and create shapes that have symmetry

Use visualization, spatial reasoning, and geometric modeling to solve problems and should—
- Create mental images of geometric shapes using spatial memory and spatial visualization;
- Recognize and represent shapes from different perspectives;
- Relate ideas in geometry to ideas in number and measurement;
- Recognize geometric shapes and structures in the environment and specify their location

MEASUREMENT
In prekindergarten through grade 2 all students should—

Understand measurable attributes of objects and the units, systems, and processes of measurement and should—
- Recognize the attributes of length, volume, weight, area, and time;
- Compare and order objects according to these attributes;
- Understand how to measure using the nonstandard and standard units;
- Select an appropriate unit and tool for the attribute being measured

Apply appropriate techniques, tools, and formulas to determine measurements and should—
- Measure with multiple copies of units of the same size, such as paper clips laid end to end;
- Use repetition of a single unit to measure something larger than the unit, for instance, measuring the length of a room with a single meterstick;
- Use tools to measure;
- Develop common referents for measures to make comparisons and estimates

NINTH EDITION

A PROBLEM SOLVING APPROACH

MATHEMATI

FOR ELEMENTARY SCHOOL TEACH

NINTH EDITION

A PROBLEM SOLVING APPROACH TO MATHEMATICS FOR ELEMENTARY SCHOOL TEACHERS

Rick Billstein
University of Montana

Shlomo Libeskind
University of Oregon

Johnny W. Lott
University of Montana

Boston San Francisco New York
London Toronto Sydney Tokyo Singapore Madrid
Mexico City Munich Paris Cape Town Hong Kong Montreal

Publisher: Greg Tobin
Acquisitions Editor: Carter Fenton
Project Editor: Katie Nopper
Editorial Assistant: Rachel Monaghan
Managing Editor: Karen Wernholm
Senior Production Supervisor: Kathleen A. Manley
Senior Designer: Dennis Schaefer
Photo Researcher: Beth Anderson
Digital Assets Manager: Marianne Groth
Media Producer: Sharon Smith
Software Development: Mary Durnwald and Janet McHugh
Senior Marketing Manager: Becky Anderson
Marketing Coordinator: Maureen McLaughlin
Senior Author Support/Technology Specialist: Joe Vetere
Senior Prepress Supervisor: Caroline Fell
Rights and Permissions Advisor: Dana Weightman
Senior Manufacturing Buyer: Evelyn Beaton
Cover and Text Design: Leslie Haimes
Production Coordination, Composition, Illustrations: Pre-Press Company, Inc.

Cover art and activity reprinted with permission of Sterling Publishing Company, Inc. New York, New York, from Ivan Moscovich's *Mastermind Collection*: *The Hinged Square and Other Puzzles*, p. 50, copyright © 2004 by Ivan Moscovich.

School book pages used with permission from Scott Foresman–Addison Wesley Mathematics, copyright © 2005 Pearson Education, Inc.; Scott Foresman–Addison Wesley Middle School Math Course Level 1-3 © 2002 by Pearson Education, Inc., publishing as Pearson Prentice Hall; McDougal Littell Math Thematics, copyright © 1999 McDougal Littell, Inc.; and Connected Mathematics, Measurement: Filling and Wrapping, copyright © 2004 Michigan State University, Glenda Lappan, James T. Fey, William M. Fitzgerald, Susan N. Friel, and Elizabeth Defanis Phillips, published by Pearson Education, Inc., publishing as Pearson Prentice Hall.

NAEP questions from the National Assessment of Educational Progress (NAEP).

TIMSS questions from Trends in International Mathematics and Science Study (TIMSS).

Excerpts from NCTM Standards reprinted with permission from *Principles and Standards for School Mathematics,* copyright 2000 by the National Council of Teachers of Mathematics (NCTM). All rights reserved. NCTM does not endorse the content or validity of these alignments.

For permission to use additional copyrighted material, grateful acknowledgment is made to the copyright holders on page 1031, which is hereby made part of this copyright page.

Many of the designations used by manufacturers and sellers to distinguish their products are claimed as trademarks. Where those designations appear in this book, and Addison-Wesley was aware of a trademark claim, the designations have been printed in initial caps or all caps.

The Geometer's Sketchpad and Dynamic Geometry are registered trademarks of Key Curriculum Press. Sketchpad is a trademark of Key Curriculum Press.

Library of Congress Cataloging-in-Publication Data
Billstein, Rick.
A problem solving approach to mathematics for elementary school teachers / Rick Billstein, Shlomo Libeskind, Johnny W. Lott.—9th ed.
p. cm.
ISBN 0-321-33179-6—ISBN 0-321-33130-3 (IE)
1. Mathematics—Study and teaching (Elementary) 2. Problem solving—Study and teaching (Elementary) I. Libeskind, Shlomo. II. Lott, Johnny W., 1944- III. Title.
QA135.6.B55 2006
372.7—dc22

2005050914

Copyright © 2007 Pearson Education, Inc. All rights reserved. No part of this publication may be reproduced, stored in a retrieval system, or transmitted, in any form or by any means, electronic, mechanical, photocopying, recording, or otherwise, without the prior written permission of the publisher. Printed in the United States of America. For information on obtaining permission for use of material in this work, please submit a written request to Pearson Education, Inc., Rights and Contracts Department, 75 Arlington Street, Suite 300, Boston, MA 02116, fax your request to 617-848-7047, or e-mail at http://www.pearsoned.com/legal/permissions.htm.

1 2 3 4 5 6 7 8 9 10—DOW—10 09 08 07 06

To my mother Esther as she continues to fight all the hard knocks and challenges that life now gives her each day. There could be no better role model. —RWB

In memory of my parents Mendel and Genia Libeskind ז״ל. —SL

To Ouidamai, my mother, and Carolyn, my wife — always inspirations. —JWL

Contents

Preface

The ninth edition of *A Problem Solving Approach to Mathematics for Elementary School Teachers* is designed to meet the education needs of prospective elementary teachers who will be the future's high-quality teachers. This edition continues to be heavily skill-based, but students will now benefit from additional emphasis on active and collaborative learning. A few new features have been included in order to highlight current issues in the field of education. The content has been revised and updated to better prepare students for when they will be instructors in their own classrooms.

Standards of the NCTM

We focus on the National Council of Teachers of Mathematics (NCTM) publication, *Principles and Standards of School Mathematics* (2000) (hereafter referred to as *Principles and Standards.*)

We also emphasize the need for the teaching of mathematics to include:

Improved!
- Algebraic thinking—there is an increase in the focus on algebraic thinking woven throughout the text in this edition;

Improved!
- Updated data analysis and statistical thinking—this material has been expanded to include more content on populations, sampling, and surveys;
- Mathematical reasoning;
- Emphasis on the use of geometric utilities;
- Conjecturing, inventing, and problem solving; and
- Connecting mathematics, its ideas, and its applications.

Our goals remain:

- To present appropriate mathematics in an intellectually honest and mathematically correct manner.
- To use problem solving as an integral part of mathematics.
- To approach mathematics in a sequence that instills confidence AND challenges students.
- To provide opportunities for alternate forms of teaching and learning.
- To provide communication problems to develop writing skills and allow students to practice explanation.
- To encourage the integration of technology tools.
- To provide core mathematics for prospective elementary teachers in a way so that they are challenged to determine why mathematics is done as it is.
- To provide core mathematics that allows instructors to use methods integrated with content.

The ninth edition allows instructors a variety of approaches to teaching, encourages discussion and collaboration among students and with their instructors, and allows for the integration of projects into the curriculum. Most importantly, it promotes discovery and active learning for the students and the teacher.

Features

We continue to incorporate various study aids and features that facilitate learning.

New!

- **Research Notes:** Notes have been added in order to highlight current research projects in mathematics or mathematics education as it relates to the content.

New!

- **Problem sets:** 20% of the exercises have been revised. We have retained the six different types of problems: (1) ongoing assessment, (2) communication, (3) open-ended, (4) cooperative learning, (5) technology, and (6) review. New exercises have also been added from the Trends in International Mathematics and Science Study (TIMSS) and the National Assessment of Educational Progress (NAEP).
- **Historical Notes** add context and humanize the mathematics.
- **Brain Teasers** provide a different avenue for problem solving. They are solved in the Instructor's Edition, and may be assigned or used by the teacher to challenge students.
- **Definitions** are either set off in text or presented as key terms in the margin for quick review.
- **Key Terms** are presented in the margins for quick review.
- ***Updated*** **School Book Pages** are included to show how the mathematics is actually introduced to the K-8 student. Students are now asked to complete many of the activities on the student pages so they can see what is expected in elementary schools.
- **Laboratory Activities** are integrated throughout the book to provide hands-on learning exercises. Answers are in the Instructor's Edition. A separate activities book is also available as a supplement.
- **Now Try This** activities appear throughout each chapter, and are intended to help students become actively involved in their learning, to facilitate the development and improvement of their critical thinking and problem-solving skills, and to stimulate both in-class and out-of-class discussion. Answers are now in the back of the text.
- **Problem-solving strategies** are often highlighted in italics, and indicated by
- **Relevant quotes** from the *Principles and Standards* are incorporated throughout the text, and are marked by the standards icon .
- **Questions from the Classroom** present questions posed from actual classrooms; the questions are involved enough to help instructors decide where to place emphasis in a course. Answers are in the Instructor's Edition.
- **Chapter Outlines** at the end of each chapter help students review the chapter.
- **Chapter Reviews** at the end of each chapter have been rewritten to allow students to test themselves more effectively.
- **Optional sections,** as well as problems based on these sections, are marked with an asterisk (*). **More difficult problems** are marked with a star (★). **Problems numbered in color** have answers at the back of the book.
- **Selected Bibliographies** have been updated and revised, and appear at the end of each chapter.
- **Cartoons** teach or emphasize important material, and add levity.
- **Communication Problems** require students to explain or justify their answers.
- **Full color** has been used for pedagogical reasons and to help students better visualize concepts. Figures are modern, attractive, and easy to follow. All of the pages taken from elementary mathematics texts are presented in full color.
- **Relevant and realistic problems** are more accessible and appealing to students of diverse backgrounds.

- **Technology Corners** include use of spreadsheets, both graphing and scientific calculators, The Geometer's Sketchpad, and computer activities. Answers are now in the back of the text.

Content Highlights

The ninth edition has new content added, and existing content has been re-organized in many chapters.

Chapter 1:

The reorganization of this chapter puts mathematics and problem solving first, followed by an expanded section on exploration with patterns. New problems and new student pages are other additions, as is a new section on Fibonacci sequences. In the algebra section, there is increased work with solving equations using pan balances.

Chapter 2:

New problems and student pages are included, as are new models to better explain computations involving whole numbers. Work with algebraic reasoning in this chapter has been expanded.

Chapter 3:

New Historical Notes, Research Notes, and Now Try This problems are presented in this chapter. There are also many new examples, problems in the Ongoing Assessments, and student pages. The improved exposition includes much more on algebraic thinking, with examples and problems on algebraic computations analogous to their respective arithmetic computations. In addition, greater emphasis has been placed on the division algorithm, both in the text and in Ongoing Assessments.

Chapter 4:

This chapter contains many new problems and greater emphasis on algebraic thinking, including functions and absolute value functions. New Research Notes and Now Try This problems are also included.

Chapter 5:

This chapter has a new discussion of fractions relating parts to the whole. Models for computations involving fractions are another addition. The section on proportional reasoning has been expanded to now include a discussion of how the relationship between two ratios in a proportion is multiplicative rather than additive and why this is so. There is a new discussion of when two quantities vary proportionally and a discussion of the constant of proportionality.

Chapter 6:

In this chapter, an increased emphasis on the use of number lines to order decimals incorporates both fractions and decimals and further ties them together. The reordered sections introduce real numbers before applications of decimals as percents. Each section contains many new problems and references, and geometric sequences are incorporated as a tool used to approach repeating decimals. Fraction bars or percent bars help in considering percentage problems.

Chapter 7:

New problems are included.

Chapter 8:

This chapter contains references to the American Statistical Association's latest draft document on recommended data analysis and probability to be studied in grades K-12. An examination of categorical and numerical data and different treatments of these data is included, with graphical material reorganized to follow these different treatments. Following the ASA's recommendations, the language is updated, and the fact that mode is not really a measure of central tendency is explained. The distinction between continuous and discrete data is made. Mean absolute deviation is introduced as an alternative to standard deviation, and the need to report mean with standard deviation or mean absolute deviation is acknowledged. Additional work on abuses of statistics, more examples, and more problems are also present in this chapter.

Chapter 9:

New Historical Notes and examples focusing on the variety of approaches to solving a problem are included in this chapter. Greater emphasis is placed on algebraic thinking, and new Questions from the Classroom have been added.

Chapter 10:

This chapter contains more challenging problems, more work tying slope and similar triangles together, and the addition of applications of coordinate geometry.

Chapter 11:

More challenging problems are presented in this chapter. The topic of measurement is slightly reorganized with more alternative solutions in converting measures given.

Chapter 12:

This chapter now places greater emphasis on transformations with coordinate geometry.

Appendices

Appendices in this edition include:

I. Using a Spreadsheet

II. Graphing Calculators

III. Using a Geometry Drawing Utility (based on The Geometer's Sketchpad)

Appendix III has been completely revised and includes 11 new Geometer's Sketchpad (GSP 4.04) activities. These lab activities are designed to be used with the various geometry sections of the book. The appropriate text sections are listed in each of the labs.

Calculator Usage

As stated in *Principles and Standards*, coverage of calculators is necessary and timely. The use of the graphing calculator is presented, where relevant, in the Technology online exercises and import TestGen tests for added flexibility. All student work is tracked in MathXL's online gradebook. Students can take chapter tests in MathXL and receive personalized study plans based on their test results. The study plan diagnoses weaknesses and links students directly to tutorial exercises for the objectives they need to study and retest. Students can also access supplemental animations and video clips directly from selected exercises. MathXL is available to qualified adopters. For more information, visit our Web site at www.mathxl.com, or contact your Addison-Wesley sales representative.

MyMathLab

MyMathLab is a series of text-specific, easily customizable online courses for Addison-Wesley textbooks in mathematics and statistics. Powered by CourseCompass™ (Pearson Education's online teaching and learning environment) and MathXL® (our online homework, tutorial, and assessment system), MyMathLab gives you the tools you need to deliver all or a portion of your course online, whether your students are in a lab setting or working from home. MyMathLab provides a rich and flexible set of course materials, featuring free-response exercises that are algorithmically generated for unlimited practice and mastery. Students can also use online tools, such as video lectures, animations, and a multimedia textbook, to independently improve their understanding and performance. Instructors can use MyMathLab's homework and test managers to select and assign online exercises correlated directly to the textbook, and they can also create and assign their own online exercises and import TestGen tests for added flexibility. MyMathLab's online gradebook—designed specifically for mathematics and statistics—automatically tracks students' homework and test results and gives the instructor control over how to calculate final grades. Instructors can also add offline (paper-and-pencil) grades to the gradebook. MyMathLab is available to qualified adopters. For more information, visit our Web site at www.mymathlab.com or contact your Addison-Wesley sales representative.

Acknowledgments

For past editions of this book, many noted and illustrious mathematics educators and mathematicians have served as reviewers. To honor the work of the past as well as to honor the reviewers of this edition, we list all but place asterisks by this edition's reviewers.

Leon J. Ablon
*Paul Ache
G.L. Alexanderson
Haldon Anderson
Bernadette Antkoviak
Richard Avery
Sue H. Baker
Jane Barnard
Joann Becker
Cindy Bernlohr
James Bierden
Jackie Blagg
Jim Boone
Sue Boren
Barbara Britton
Beverly R. Broomell
Jane Buerger
Maurice Burke
David Bush
Laura Cameron
Louis J. Chatterley
Phyllis Chinn
Donald J. Dessart
Ronald Dettmers
Jackie Dewar
Amy Edwards
Lauri Edwards
Margaret Ehringer
Albert Filano
Marjorie Fitting
Michael Flom
Martha Gady
Dwight Galster
Sandy Geiger
Glenadine Gibb
Don Gilmore
Diane Ginsbach
Elizabeth Gray
Jerrold Grossman
Alice Guckin
*Jennifer Hegeman
Joan Henn
Boyd Henry
*Linda Hintzman
Alan Hoffer
E. John Hornsby, Jr.
Judith E. Jacobs
Donald James
*Thomas R. Jay
Jerry Johnson
Wilburn C. Jones
Robert Kalin
Sarah Kennedy
Steven D. Kerr
Leland Knauf
Margret F. Kothmann
Kathryn E. Lenz
Hester Lewellen
Ralph A. Liguori
Don Loftsgaarden
Sharon Louvier
Stanley Lukawecki
Judith Merlau
Barbara Moses
Cynthia Naples
Charles Nelson
Glenn Nelson
Kathy Nickell
Dale Oliver
*Mark Oursland
Linda Padilla
Dennis Parker
Clyde Paul
Keith Peck
Barbara Pence
Glen L. Pfeifer
Debra Pharo
Jack Porter
Edward Rathnell
Sandra Rucker
Jennifer Rutherford
Helen R. Santiz
*Sherry Scarborough
Jane Schielack
Barbara Shabell
M. Geralda Shaefer
Nancy Shell
Wade H. Sherard
Gwen Shufelt
Julie Sliva
Ron Smit
Joe K. Smith
William Sparks
Virginia Strawderman
Mary M. Sullivan
Viji Sundar
Sharon Taylor
*Jo Temple
C. Ralph Verno
Hubert Voltz
John Wagner
Edward Wallace
Virginia Warfield
Lettie Watford
Mark F. Weiner
Grayson Wheatley
Jim Williamson
Ken Yoder
Jerry L. Young
Deborah Zopf

An Introduction to Problem Solving

PRELIMINARY PROBLEM

When a book is printed, pages are passed through a printing press and then folded to form a book. To see how this works start out with a simple book made from one $8\frac{1}{2} \times 11$-in. sheet of paper. Fold the page in half lengthwise to form a book and number the pages 1 through 4. When you open your sheet of paper, the numbers 2 and 3 are on one side of the paper and the numbers 1 and 4 are on the other side. The sum of the numbers on both sides of the sheet equals 5 and the sum of all the page numbers is 10. If two sheets of paper are used to build an 8-page book and the pages are numbered, predict the sum of the numbers on each page and the sum of all of the page numbers. Build your book to see if you were correct. Try the same thing with 3 pages.

Suppose you are building a 100-page book, how many sheets will you need? What is the sum of the two page numbers that occur on the same side of the sheet? What is the sum of all the page numbers in this book?

Problem solving has long been recognized as one of the hallmarks of mathematics. The greatest goal of learning mathematics is to have people become good problem solvers. We do not mean doing exercises that are routine practice for skill building. What does *problem solving* mean? George Polya (1887–1985), one of the great mathematicians and teachers of the twentieth century, pointed out that "solving a problem means finding a way out of difficulty, a way around an obstacle, attaining an aim which was not immediately attainable." (Polya 1981, p. ix)

In *Principles and Standards of School Mathematics* (NCTM 2000), we find the following:

Problem solving means engaging in a task for which the solution method is not known in advance. In order to find a solution, students must draw on their knowledge, and through this process, they will often develop new mathematical understandings. Solving problems is not only a goal of learning mathematics but also a major means of doing so. Students should have frequent opportunities to formulate, grapple with, and solve complex problems that require a significant amount of effort and should then be encouraged to reflect on their thinking. (p. 52)

Further, we find that

Instructional programs from prekindergarten through grade 12 should enable all students to

- *build new mathematical knowledge through problem solving;*
- *solve problems that arise in mathematics and in other contexts;*
- *apply and adapt a variety of appropriate strategies to solve problems;*
- *monitor and reflect on the process of mathematical problem solving.* (p. 52)

Students learn mathematics as a result of solving problems. Exercises, or practice problems, serve a purpose in learning mathematics, but problem solving must be a focus of school mathematics. Your mathematical experience often determines whether situations are *problems* or *exercises*. In the "Fox Trot" cartoon, the word *problems* appears to be a challenge to Paige, but not to Jason. For him, they are exercises.

FOXTROT © Bill Amend, Reprinted with permission of UNIVERSAL PRESS SYNDICATE. All rights reserved.

RESEARCH NOTE A reasonable amount of tension and discomfort improves students' problem-solving performance. The motivation is the release of tension after the problem has been solved. If the tension is not present, the problem is either an *exercise* or the students are "generally unwilling to attack the problem in a serious way" (Bloom and Broder 1950; McLeod 1985).

Worthwhile, interesting problems, not just routine word problems, must be a part of elementary students' mathematical experience. To engage students in worthwhile tasks, problems should be introduced in a familiar context. Good mathematical problem solving occurs when all of the following are present:

1. Students are presented with a situation that they understand but do not know how to proceed directly to a solution.
2. Students are interested in finding the solution and attempt to do so.
3. Students are required to use mathematical ideas to solve the problem.

In this text, you will have many opportunities to solve problems. Each chapter opens with a problem that can be solved by using the concepts developed in the chapter. A hint for the solution to the problem is given at the end of each chapter. Throughout the text, there are numerous problems solved using a four-step process and others solved using other formats.

Working with other students to solve problems can enhance your problem-solving ability and communication skills. We encourage *cooperative learning* and encourage students to work in groups whenever possible. To encourage group work and help identify when cooperative learning could be useful, we identify activities that involve tasks where it might be helpful to have several people gathering data, or the problems might be such that group discussions could lead to strategies for solving the problem.

RESEARCH NOTE Students solving problems in small groups use cognitive behaviors and processes that are very similar to those of expert mathematical problem solvers (Artz and Armour-Thomas 1992).

Throughout the book, we remind you of definitions of mathematical terms. However, you do not come to this section or the book with a *tabula rasa*, or blank slate. You have mathematical experience from school and from daily life. We ask you to build on what you know, to correct any misconceptions you have, and to challenge yourself to learn as we try to challenge you. Most of all, we ask you to think about why you do what you do in mathematics. Having you understand and question mathematics will help make you a better learner and consequently a better teacher.

1-1 Mathematics and Problem Solving

George Polya (1887–1985) described the experience of problem solving in his book *How to Solve It* (p. v):

> *A great discovery solves a great problem, but there is a grain of discovery in the solution of any problem. Your problem may be modest; but if it challenges your curiosity and brings into play your inventive facilities, and if you solve it by your own means, you may experience the tension and enjoy the triumph of discovery.*

To solve a problem, we must first understand both the task and the given information. Next, it is helpful to determine a strategy to accomplish the task. Once we arrive at a solution, we should determine whether the solution makes sense and is reasonable. This process of problem solving can be described using a four-step process similar to the one developed by George Polya.

▶ **HISTORICAL NOTE**

George Polya (1887–1985) was born in Hungary and received his Ph.D. from the University of Budapest. He moved to the United States in 1940, and after a brief stay at Brown University he joined the faculty at Stanford University and taught there until—and after—his retirement. In addition to being a preeminent mathematician, he focused on the vital importance of mathematics education. At Stanford, he published 10 books, including *How To Solve It* (1945), which has been translated into 15 languages. ◀

Four-Step Problem-Solving Process

1. **Understanding the problem**
   a. Can you state the problem in your own words?
   b. What are you trying to find or do?
   c. What are the unknowns?
   d. What information do you obtain from the problem?
   e. What information, if any, is missing or not needed?
2. **Devising a plan**
   The following list of strategies, although not exhaustive, is very useful:
   a. Look for a pattern.
   b. Examine related problems and determine if the same technique applied to them can be applied to the current problem.
   c. Examine a simpler or special case of the problem to gain insight into the solution of the original problem.
   d. Make a table or list.
   e. Make a diagram.
   f. Write an equation.
   g. Use guess and check.
   h. Work backward.
   i. Identify a subgoal.
   j. Use indirect reasoning.
   k. Use direct reasoning.
3. **Carrying out the plan**
   a. Implement the strategy or strategies in step 2 and perform any necessary actions or computations.
   b. Check each step of the plan as you proceed. This may be intuitive checking or a formal proof of each step.
   c. Keep an accurate record of your work.
4. **Looking back**
   a. Check the results in the original problem. (In some cases, this will require a proof.)
   b. Interpret the solution in terms of the original problem. Does your answer make sense? Is it reasonable? Does it answer the question that was asked?
   c. Determine whether there is another method of finding the solution.
   d. If possible, determine other related or more general problems for which the techniques will work.

Avoiding Mind Sets

If problems are approached in only one way, a mind set may be formed. For example, consider the following. Spell the word *spot* three times out loud. "S-P-O-T! S-P-O-T! S-P-O-T!" Now answer the question, "What do you do when you come to a green light?" Write an answer. If you answered "Stop," you may be guilty of having formed a mind set. You do not stop at a *green* light.

Consider the following problem: "A shepherd had 36 sheep. All but 10 died. How many lived?" Did you answer "10"? If you did, you are catching on and are ready to try some problems. If you did not answer "10," then you did not understand the question. *Understanding the problem* is the first step in the four-step problem-solving process. Using the four-step process does not guarantee a solution to a problem, but it does provide a systematic means of attacking problems.

Strategies for Problem Solving

To aid in problem solving, *Principles and Standards* provides the following guidance.

Instead of teaching problem solving separately, teachers should embed problems in the mathematics-content curriculum. When teachers integrate problem solving into the context of mathematical situations, students recognize the usefulness of strategies. Teachers should choose specific problems because they are likely to prompt particular strategies and allow for the development of certain mathematical ideas. (p. 119)

RESEARCH NOTE Problem-solving ability develops slowly over time, perhaps because the many understandings and skills needed for problem solving develop at different rates. A key element in developing problem-solving skills is multiple, continuous experience in solving problems with different contexts and at different levels of ability (Kantowski 1981).

Strategies are tools that might be used to discover or construct the means to achieve a goal. For each strategy described next, an example is given that can be solved with that strategy. Often, problems can be solved in more than one way as seen in the cartoon on page 6. You may devise a different strategy to solve the sample problems. There is no one best strategy to use.

HISTORICAL NOTE

Carl Gauss (1777–1855) is regarded as the greatest mathematician of the nineteenth century and one of the greatest mathematicians of all time. Born to humble parents in Brunswick, Germany, he was an infant prodigy who, it is said, at age 3 corrected an arithmetic error in his father's bookkeeping. Gauss claimed that he could figure before he could talk.

Gauss made contributions in the areas of astronomy, geodesy, and electricity. After Gauss's death, the King of Hanover ordered a commemorative medal prepared in his honor. On the medal was an inscription referring to Gauss as the "Prince of Mathematics," a title that stayed with his name.

NONTRADITIONAL SOLUTION

Strategy: Look for a Pattern

PROBLEM SOLVING **Gauss's Problem**

When Carl Gauss was a child, his teacher required the students to find the sum of the first 100 natural numbers. The teacher expected this problem to keep the class occupied for some time. Gauss gave the answer almost immediately. Can you?

natural numbers

Understanding the Problem The **natural numbers** are 1, 2, 3, 4, ... Thus, the problem is to find the sum $1 + 2 + 3 + 4 + \ldots + 100$.

Devising a Plan The strategy *look for a pattern* is useful here. One version of the story about young Gauss reports that he listed the numbers as shown in Figure 1-1. To discover the original sum, Gauss then divided the sum in Figure 1-1 by 2.

FIGURE 1-1

$$\begin{array}{r} 1 + \ \ 2 + \ \ 3 + \ \ 4 + \ \ 5 + \ldots + \ 98 + \ 99 + 100 \\ \underline{100 + \ 99 + \ 98 + \ 97 + \ 96 + \ldots + \ \ 3 + \ \ 2 + \ \ 1} \\ 101 + 101 + 101 + 101 + 101 + \ldots + 101 + 101 + 101 \end{array}$$

Carrying Out the Plan There are 100 sums of 101. Thus, the total can be found by computing $\frac{100 \cdot 101}{2}$, or 5050.

Looking Back The method is mathematically correct because addition can be performed in any order, and multiplication is repeated addition. Also, the sum in each pair is always 101 because when we move from any pair to the next, we add 1 to the top and subtract 1 from the bottom, which does not change the sum, for example, $2 + 99 = (1 + 1) + (100 - 1) = 1 + 100$, $3 + 98 = (2 + 1) + (99 - 1) = 2 + 99 = 101$, and so on.

A more general problem is to find the sum of the first n natural numbers, $1 + 2 + 3 + 4 + 5 + 6 + \ldots + n$, where n is any natural number. We use the same plan as before and notice the relationship in Figure 1-2. There are n sums of $n + 1$ for a total of $n(n + 1)$. Therefore, the sum $1 + 2 + 3 + \ldots + n$ is given by $\frac{n(n+1)}{2}$.

FIGURE 1-2

$$\begin{array}{r} 1 + \quad 2 + \quad 3 + \quad 4 + \ldots + n \\ n + (n-1) + (n-2) + (n-3) + \ldots + 1 \\ \hline (n+1) + (n+1) + (n+1) + (n+1) + \ldots (n+1) \end{array}$$

A different strategy for finding the sum $1 + 2 + 3 + \ldots + n$ involves the strategy of *making a diagram* and thinking of the sum geometrically as a stack of blocks. To find the sum, consider the stack in Figure 1-3(a) and a stack of the same size placed differently, as in Figure 1-3(b). The total number of blocks in the stack in Figure 1-3(b) is $n(n + 1)$, which is twice the desired sum. Thus the desired sum is $n(n + 1)/2$.

FIGURE 1-3

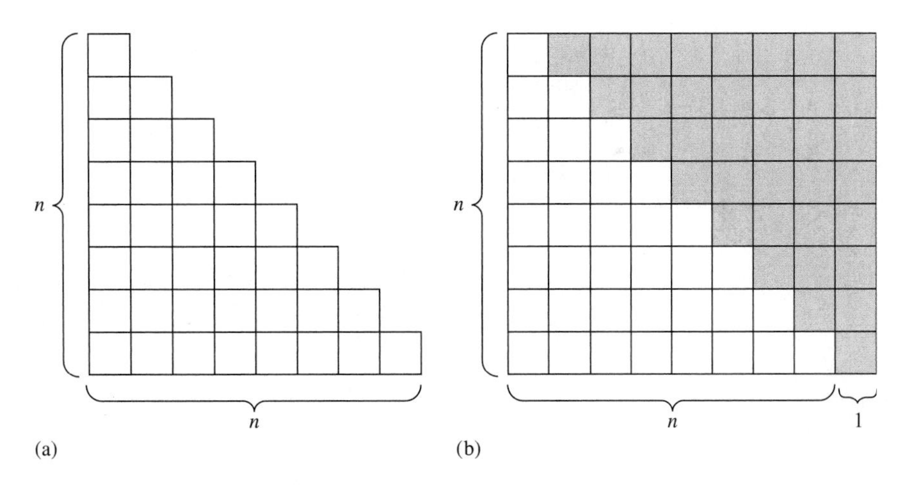

REMARK The sum $1 + 2 + 3 + 4 + 5 + \ldots + n = \frac{n(n + 1)}{2}$ will be further examined in the next section when we discuss arithmetic sequences.

NOW TRY THIS 1-1

One cut on a log produces two pieces, two cuts produce three pieces, and three cuts produce four pieces. How many pieces are produced by 10 cuts? Assume the cuts are made in the same manner as the first three cuts. How many pieces are produced by n cuts?

Strategy: Examine a Related Problem

PROBLEM SOLVING **Sums of Even Natural Numbers**

Find the sum of the even natural numbers less than or equal to 100. Devise a strategy for finding that sum and generalize the result.

Understanding the Problem The problem is to find the sum of the even natural numbers $2 + 4 + 6 + 8 + \ldots + 100$

Devising a Plan Recognizing that the sum can be separated into two simpler parts *related* to Gauss's original problem helps us devise a plan. Consider the following:

$$\begin{aligned} 2 + 4 + 6 + 8 + \ldots + 100 &= 2 \cdot 1 + 2 \cdot 2 + 2 \cdot 3 + 2 \cdot 4 + \ldots + 2 \cdot 50 \\ &= 2(1 + 2 + 3 + 4 + \ldots + 50) \end{aligned}$$

Thus, we can use Gauss's method to find the sum of the first 50 natural numbers and then double that.

Carrying Out the Plan We carry out the plan as follows:

$$\begin{aligned} 2 + 4 + 6 + 8 + \ldots + 100 &= 2(1 + 2 + 3 + 4 + \ldots + 50) \\ &= 2 \cdot [50(50 + 1)/2] \\ &= 2550 \end{aligned}$$

Thus, the sum is 2550.

Looking Back A different way to consider this problem is to realize that there are 25 sums of 102, as shown in Figure 1-4.

FIGURE 1-4

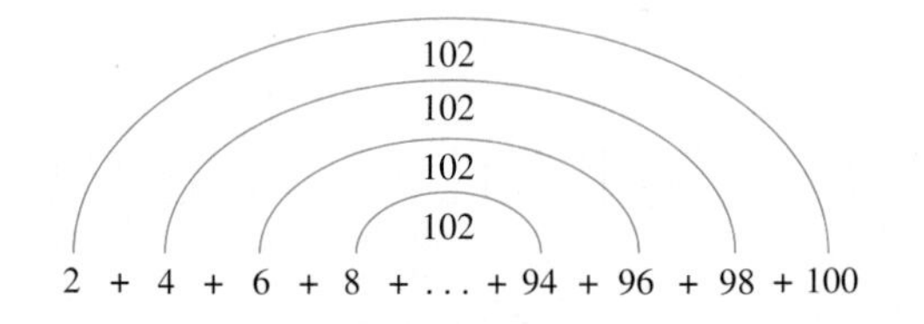

Thus, the sum is $25 \cdot 102$, or 2550.

NOW TRY THIS 1-2

Find the sum of the odd natural numbers less than 100.

Strategy: Examine a Simpler Case

One strategy for solving a complex problem is to *examine a simpler case* of the problem and then consider other parts of the complex problem. An example is shown on the student page on page 9.

SAMPLE SCHOOL BOOK PAGE:

SOLVING A SIMPLER PROBLEM

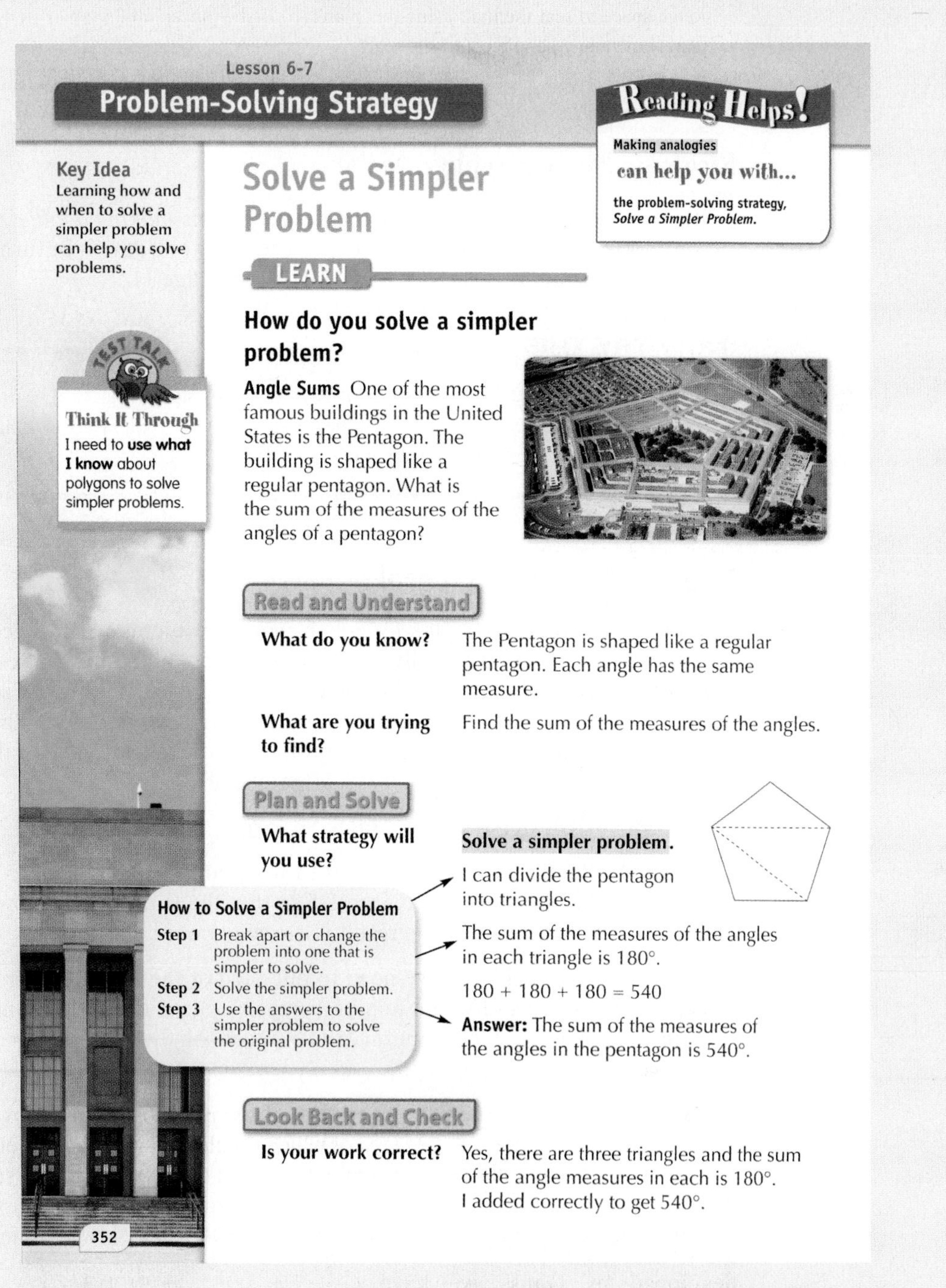
Lesson 6-7

Problem-Solving Strategy

Reading Helps!

Making analogies can help you with... the problem-solving strategy, *Solve a Simpler Problem.*

Key Idea
Learning how and when to solve a simpler problem can help you solve problems.

Solve a Simpler Problem

LEARN

How do you solve a simpler problem?

Angle Sums One of the most famous buildings in the United States is the Pentagon. The building is shaped like a regular pentagon. What is the sum of the measures of the angles of a pentagon?

TEST TALK
Think It Through
I need to **use what I know** about polygons to solve simpler problems.

Read and Understand

What do you know?	The Pentagon is shaped like a regular pentagon. Each angle has the same measure.
What are you trying to find?	Find the sum of the measures of the angles.

Plan and Solve

What strategy will you use? **Solve a simpler problem.**

I can divide the pentagon into triangles.

The sum of the measures of the angles in each triangle is 180°.

180 + 180 + 180 = 540

Answer: The sum of the measures of the angles in the pentagon is 540°.

How to Solve a Simpler Problem
Step 1 Break apart or change the problem into one that is simpler to solve.
Step 2 Solve the simpler problem.
Step 3 Use the answers to the simpler problem to solve the original problem.

Look Back and Check

Is your work correct? Yes, there are three triangles and the sum of the angle measures in each is 180°. I added correctly to get 540°.

352

Source: Scott Foresman-Addison Wesley, Grade 5, 2005 (p. 352).

NOW TRY THIS 1-3

Nikki is setting up tables for a noon luncheon in the gym. She has 25 small square tables that hold 1 person to a side. She plans to put 25 of these tables in a row to make one long rectangular table that is only one table wide. (a) If 60 people will be attending the lunch, will she have enough space to seat them all using this plan? (b) If she placed all 25 tables in the form of a big square, how many people could she seat?

Strategy: Make a Table

A strategy that is often used in elementary school is *making a table*. A table can be used to look for patterns that emerge in the problem, which in turn can lead to a solution. An example of this strategy is shown on page 11.

NOW TRY THIS 1-4

Molly and Karly started a new job the same day. After they start work, Molly is to visit the home office every 15 days and Karly is to visit the home office every 18 days. How many days will it be before they both visit the home office the same day?

Strategy: Identify a Subgoal

As we attempt to devise a plan for solving some problems, it may become apparent that the problem could be solved if the solution to a somewhat easier or more familiar related problem could be found. In such a case, finding the solution to the easier problem may become a subgoal of the primary goal of solving the original problem. The following Magic Square problem shows an example of this.

PROBLEM SOLVING **A Magic Square**

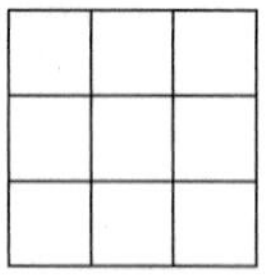

Arrange the numbers 1 through 9 into a square subdivided into nine smaller squares like the one shown in Figure 1-5 so that the sum of every row, column, and main diagonal is the same. (The result is a *magic square*.)

FIGURE 1-5

Understanding the Problem We need to put each of the nine numbers $1, 2, 3, \ldots, 9$ in the small squares, a different number in each square, so that the sums of the numbers in each row, in each column, and in each of the two major diagonals are the same.

Devising a Plan If we knew the fixed sum of the numbers in each row, column, and diagonal, we would have a better idea of which numbers can appear together in a single row, column, or diagonal. Thus our *subgoal* is to find that fixed sum. The sum of the nine numbers, $1 + 2 + 3 + \ldots + 9$, equals 3 times the sum in one row (why?). Consequently, the fixed sum is obtained by dividing $1 + 2 + 3 + \ldots + 9$ by 3. Using the process developed by Gauss, we have $(1 + 2 + 3 + \ldots + 9) \div 3 = \left(\dfrac{9 \cdot 10}{2}\right) \div 3$, or $45 \div 3 = 15$, so the sum in each row, column, and diagonal must be 15. Next, we need to decide what numbers could occupy the various squares. The number in the center space

SAMPLE SCHOOL BOOK PAGE:

MAKING A TABLE

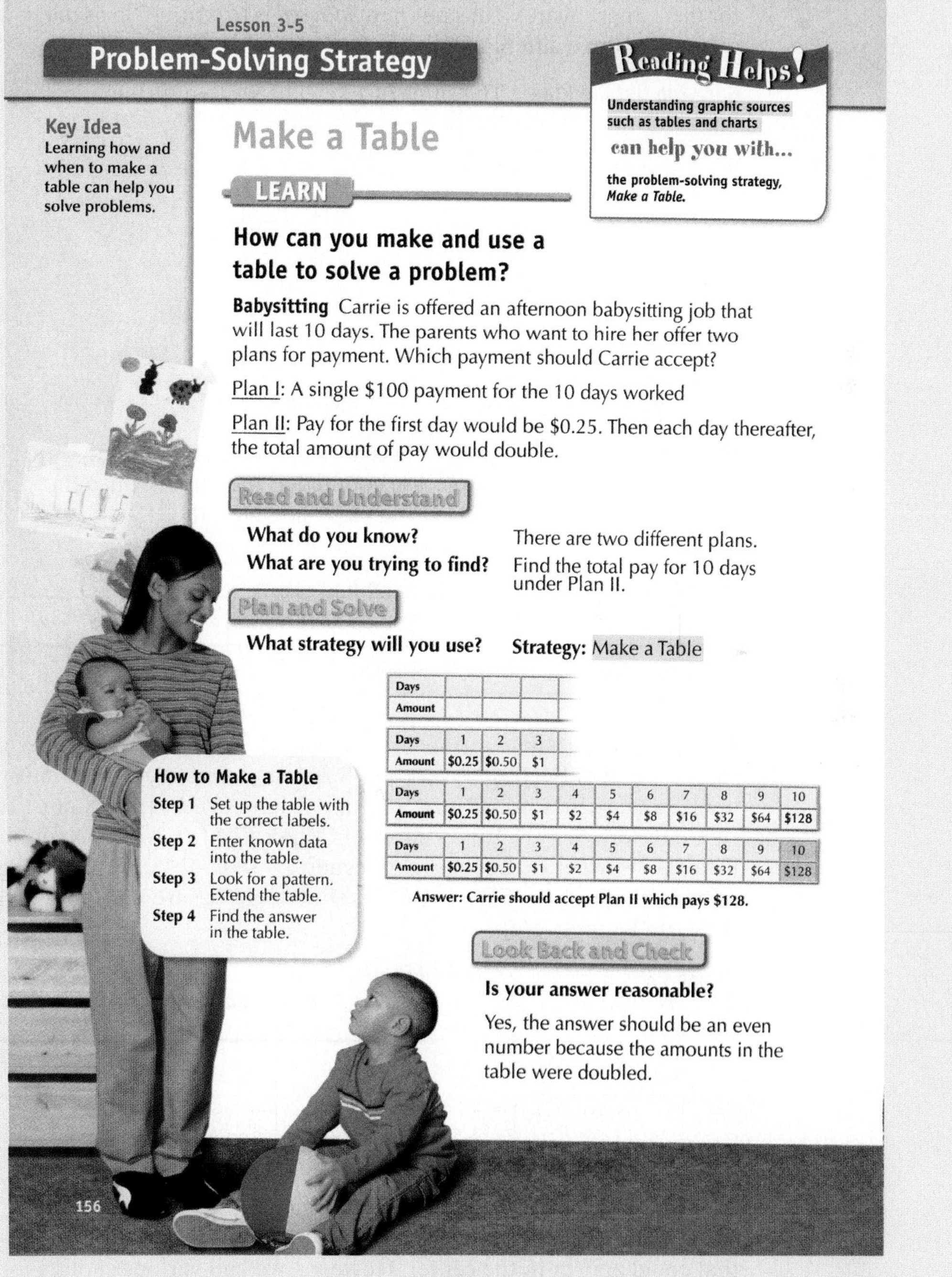

Lesson 3-5

Problem-Solving Strategy

Key Idea
Learning how and when to make a table can help you solve problems.

Make a Table

Reading Helps!
Understanding graphic sources such as tables and charts **can help you with...** the problem-solving strategy, *Make a Table.*

LEARN

How can you make and use a table to solve a problem?

Babysitting Carrie is offered an afternoon babysitting job that will last 10 days. The parents who want to hire her offer two plans for payment. Which payment should Carrie accept?

Plan I: A single $100 payment for the 10 days worked

Plan II: Pay for the first day would be $0.25. Then each day thereafter, the total amount of pay would double.

Read and Understand

What do you know? There are two different plans.

What are you trying to find? Find the total pay for 10 days under Plan II.

Plan and Solve

What strategy will you use? **Strategy:** Make a Table

Days			
Amount			

Days	1	2	3
Amount	$0.25	$0.50	$1

Days	1	2	3	4	5	6	7	8	9	10
Amount	$0.25	$0.50	$1	$2	$4	$8	$16	$32	$64	$128

Days	1	2	3	4	5	6	7	8	9	10
Amount	$0.25	$0.50	$1	$2	$4	$8	$16	$32	$64	$128

Answer: Carrie should accept Plan II which pays $128.

How to Make a Table

Step 1 Set up the table with the correct labels.
Step 2 Enter known data into the table.
Step 3 Look for a pattern. Extend the table.
Step 4 Find the answer in the table.

Look Back and Check

Is your answer reasonable?

Yes, the answer should be an even number because the amounts in the table were doubled.

156

Source: Scott Foresman-Addison Wesley, Grade 6, 2005 (p. 156).

will appear in four sums, each adding to 15 (two diagonals, the second row, and the second column). Each number in the corners will appear in three sums of 15. (Do you see why?) If we write 15 as a sum of three different numbers 1 through 9 in all possible ways, we could then count how many sums contain each of the numbers 1 through 9. The numbers that appear in at least four sums are candidates for placement in the center square, whereas the numbers that appear in at least three sums are candidates for the corner squares. Thus our new *subgoal* is to write 15 in as many ways as possible as a sum of three different numbers from the set $\{1, 2, 3, \ldots, 9\}$.

Carrying Out the Plan The sums of 15 can be written systematically as follows:

$$9 + 5 + 1$$
$$9 + 4 + 2$$
$$8 + 6 + 1$$
$$8 + 5 + 2$$
$$8 + 4 + 3$$
$$7 + 6 + 2$$
$$7 + 5 + 3$$
$$6 + 5 + 4$$

Notice that the order in each sum is not important. (Do you see why?) Hence, $1 + 5 + 9$ and $5 + 1 + 9$, for example, are counted as the same. Notice that 1 appears in only two sums, 2 in three sums, 3 in two sums, and so on. Table 1-1 summarizes this pattern.

Table 1-1

Number	1	2	3	4	5	6	7	8	9
Number of Sums Containing the Number	2	3	2	3	4	3	2	3	2

The only number that appears in four sums is 5; hence, 5 must be in the center of the square. (Why?) Because 2, 4, 6, and 8 appear 3 times each, they must go in the corners. Suppose we choose 2 for the upper left corner. Then 8 must be in the lower right corner. (Why?) This is shown in Figure 1-6(a). Now we could place 6 in the lower left corner or upper right corner. If we choose the upper right corner, we obtain the result in Figure 1-6(b). The magic square can now be completed, as shown in Figure 1-6(c).

FIGURE 1-6

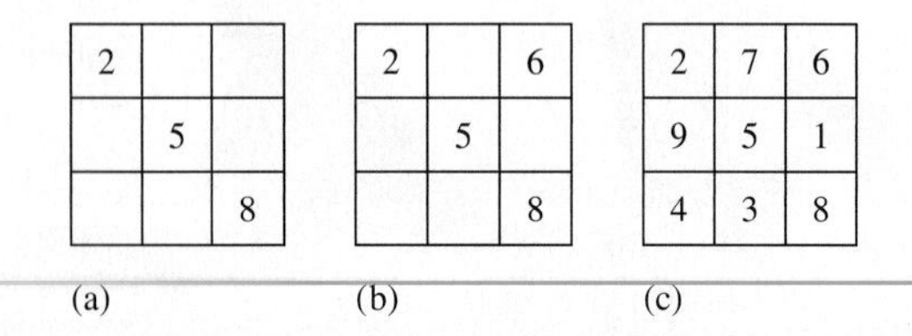

Looking Back We have seen that 5 was the only number among the given numbers that could appear in the center. However, we had various choices for a corner, and hence it seems that the magic square we found is not the only one possible. Can you find all the others?

Another way to see that 5 could be in the center square is to consider the sums $1 + 9, 2 + 8, 3 + 7, 4 + 6$, as shown in Figure 1-7. We could add 5 to each to obtain 15.

FIGURE 1-7

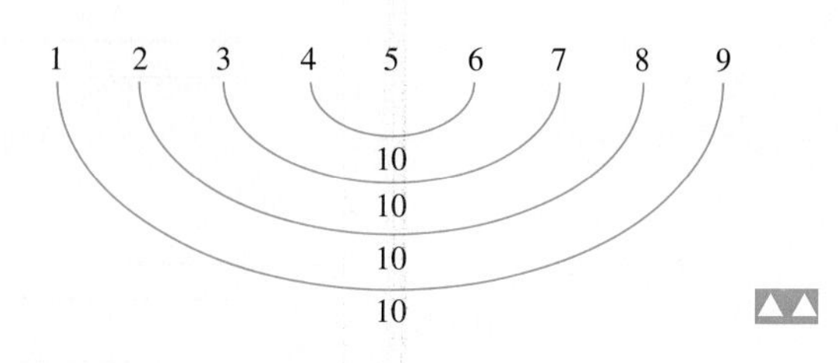

NOW TRY THIS 1-5

Five friends decided to give a party and split the costs equally. Al spent \$4.75 on invitations, Betty spent \$12 for drinks and \$5.25 on vegetables, Carl spent \$24 for pizza, Dani spent \$6 on paper plates and napkins, and Ellen spent \$13 on decorations. Determine who owes money to whom and how the money can be paid.

PROBLEM SOLVING **50-m Race Problem**

Bill and Jim ran a 50-meter race. Jim was at the 45-meter mark when Bill crossed the finish line.

a. To make the race closer Jim started 5 m ahead of Bill, who lined up at the starting line. If each runner runs the same speed as in the first race, who will win the race?
b. If Jim starts at the starting line and Bill starts 5 m behind, who will win the race?

Understanding the Problem When Bill and Jim run a 50-meter race, Bill wins by 5 meters. If Bill starts at the starting line and Jim is given a 5-meter head start, we are to determine who will win the race. If Jim starts at the starting line and Bill starts 5 m behind, we are to determine who will win.

Devising a Plan A strategy to determine the winner under each condition is to *draw a diagram*. A diagram for the original 50-m race is given in Figure 1-8(a). In this case, Bill wins by 5 m. In the next race, Jim is given a 5-m head start and hence we have the diagram in Figure 1-8(b). In the last race, because Bill starts 5 m behind, we use Figure 1-8(a) and move Bill back 5 m, as shown in Figure 1-8(c). From the diagram we can determine the results in each case.

Carrying Out the Plan From Figure 1-8(b) we see that if Jim is given a 5-m head start, then the race will end in a tie. If Bill starts 5 m behind Jim, then at 45 m they will be tied. Because Bill is faster than Jim, Bill will cover the last 5 m faster than Jim and win the race.

Looking Back The diagrams show the solution makes sense and is appropriate. Other problems can be investigated involving racing and handicaps. For example, if Bill and Jim run on a 50-m oval track, how many laps will it take for Bill to lead Jim by one full lap. (Assume the same speeds as earlier.)

FIGURE 1-8

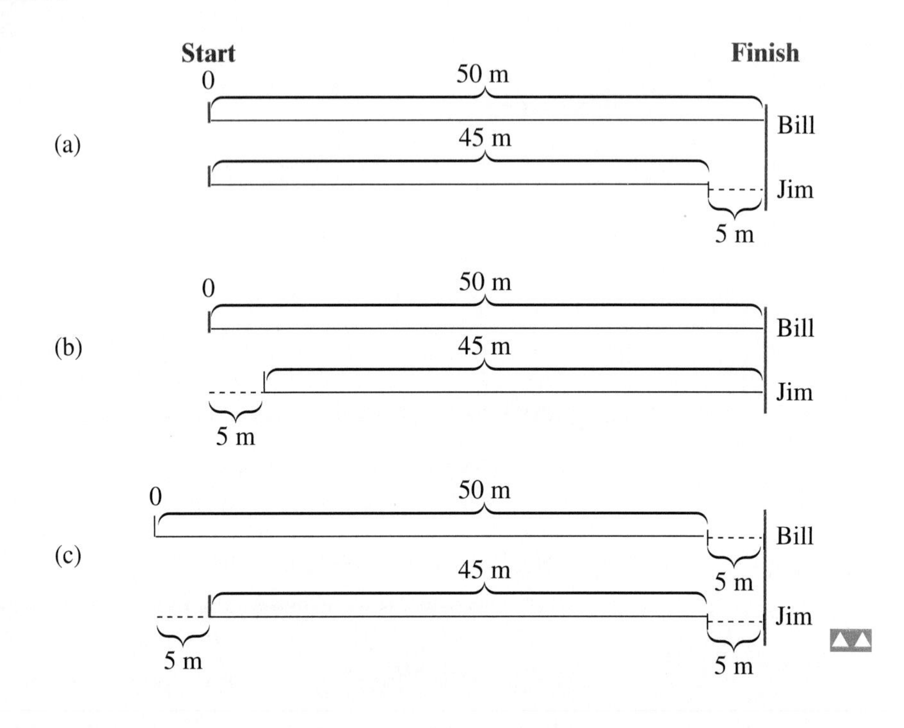

NOW TRY THIS 1-6

An elevator stopped at the middle floor of a building. It then moved up 4 floors and stopped. It then moved down 6 floors, and then moved up 10 floors and stopped. The elevator was now 3 floors from the top floor. How many floors does the building have?

Strategy: Guess and Check

In the strategy of *guess and check*, we first guess at a solution using as reasonable a guess as possible. Then we check to see whether the guess is correct. If not, the next step is to learn as much as possible about the solution based on the guess before making the next guess. This strategy can be regarded as a form of trial and error, where the information about the error helps us choose what trial to make next.

RESEARCH NOTE Students in Grades 1–3 rely primarily on the *guess-and-check* strategy when faced with a mathematical problem. As students enter Grades 6–12 this tendency decreases. Older students benefit more from the observed "errors" after a guess when formulating a new "trial" (Lester 1975).

The guess-and-check strategy is often used when a student does not know how to solve the problem more efficiently or if the student does not yet have the tools to solve the problem in a faster way. The guess-and-check strategy is demonstrated on the student page on page 15.

SAMPLE SCHOOL BOOK PAGE:

GUESS AND CHECK

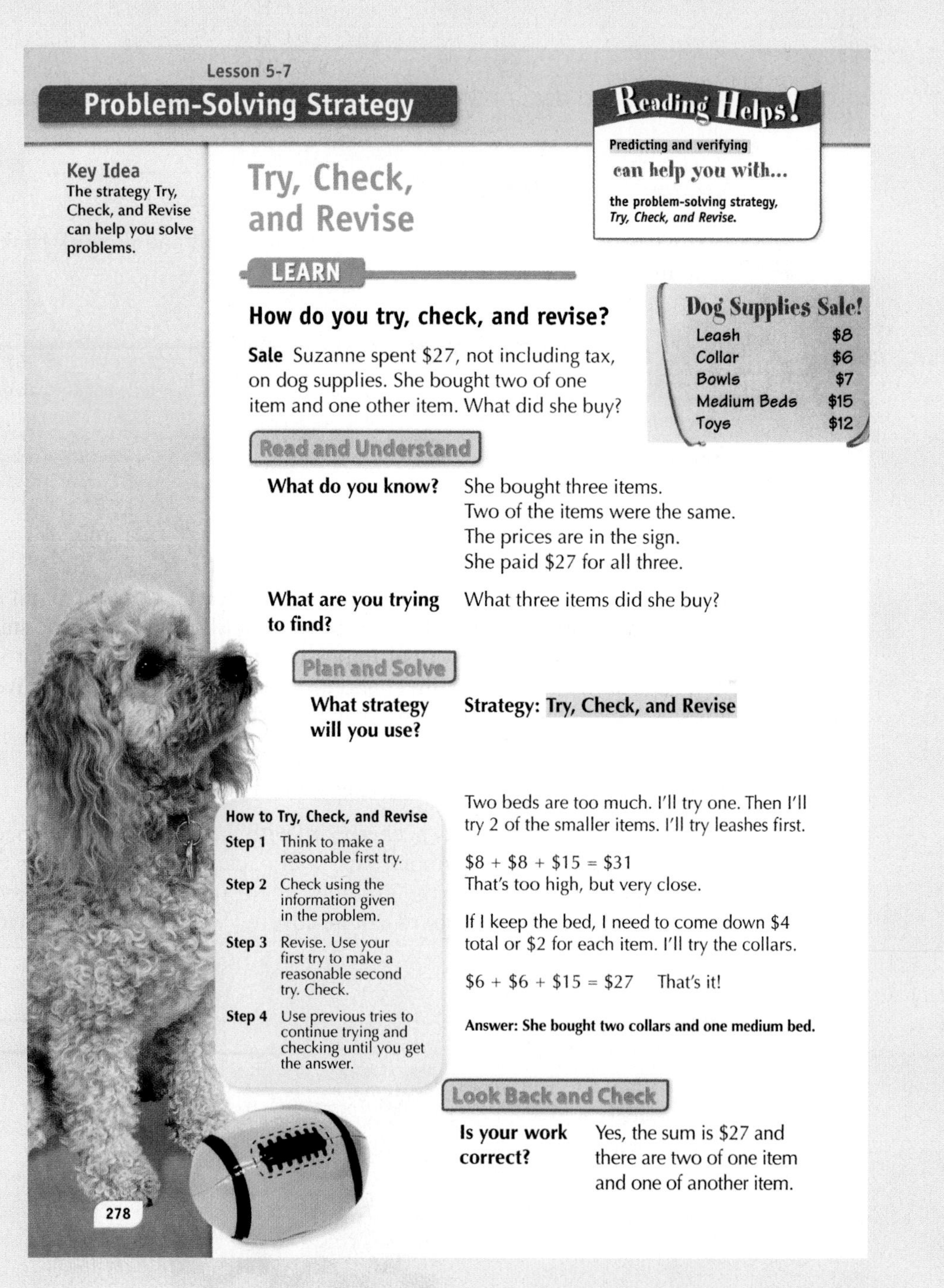
Lesson 5-7

Problem-Solving Strategy

Reading Helps!

Predicting and verifying can help you with...

the problem-solving strategy, *Try, Check, and Revise.*

Key Idea

The strategy Try, Check, and Revise can help you solve problems.

Try, Check, and Revise

LEARN

How do you try, check, and revise?

Sale Suzanne spent $27, not including tax, on dog supplies. She bought two of one item and one other item. What did she buy?

Dog Supplies Sale!	
Leash	$8
Collar	$6
Bowls	$7
Medium Beds	$15
Toys	$12

Read and Understand

What do you know?	She bought three items. Two of the items were the same. The prices are in the sign. She paid $27 for all three.
What are you trying to find?	What three items did she buy?

Plan and Solve

What strategy will you use?	**Strategy: Try, Check, and Revise**

How to Try, Check, and Revise

Step 1 Think to make a reasonable first try.

Step 2 Check using the information given in the problem.

Step 3 Revise. Use your first try to make a reasonable second try. Check.

Step 4 Use previous tries to continue trying and checking until you get the answer.

Two beds are too much. I'll try one. Then I'll try 2 of the smaller items. I'll try leashes first.

$8 + $8 + $15 = $31
That's too high, but very close.

If I keep the bed, I need to come down $4 total or $2 for each item. I'll try the collars.

$6 + $6 + $15 = $27 That's it!

Answer: She bought two collars and one medium bed.

Look Back and Check

Is your work correct?	Yes, the sum is $27 and there are two of one item and one of another item.

278

Source: Scott Foresman-Addison Wesley, Grade 4, 2005 (p. 278).

NOW TRY THIS 1-7

A cryptarithm is a collection of words where each unique letter represents a unique number. Find the digits that can be substituted in the following:

$$\begin{array}{r} \text{SUN} \\ +\ \text{FUN} \\ \hline \text{SWIM} \end{array}$$

Strategy: Work Backward

In some problems, it is easier to start with the result and to work backward. This is demonstrated on the student page on page 18.

NOW TRY THIS 1-8

Linda has an 80 average (mean) on her 11 math tests. Her teacher tells her she can drop her single low score of 50. What is her new average?

Strategy: Use Indirect Reasoning

To show that a statement is true, it is sometimes easier to show that it is impossible for the statement to be false. This can be done by showing that if the statement were false, something contradictory or impossible would follow. This approach is useful when it is difficult to start a direct argument and when negating the given statement gives us something tangible with which to work. An example follows.

PROBLEM SOLVING **Checkerboard Problem**

In Figure 1-9, we are given a checkerboard with the two squares on opposite corners removed and a set of dominoes such that each domino can cover two adjacent squares on the board. Can the dominoes be arranged in such a way that all the remaining squares on the board can be covered with no dominoes hanging off the board? If not, why not?

FIGURE 1-9

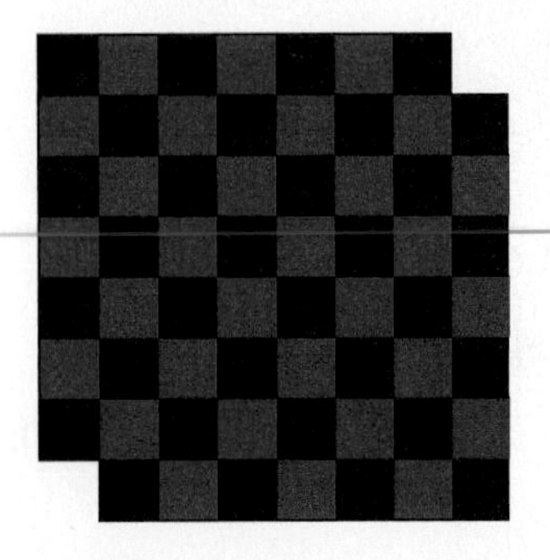

SAMPLE SCHOOL BOOK PAGE:

WORK BACKWARD

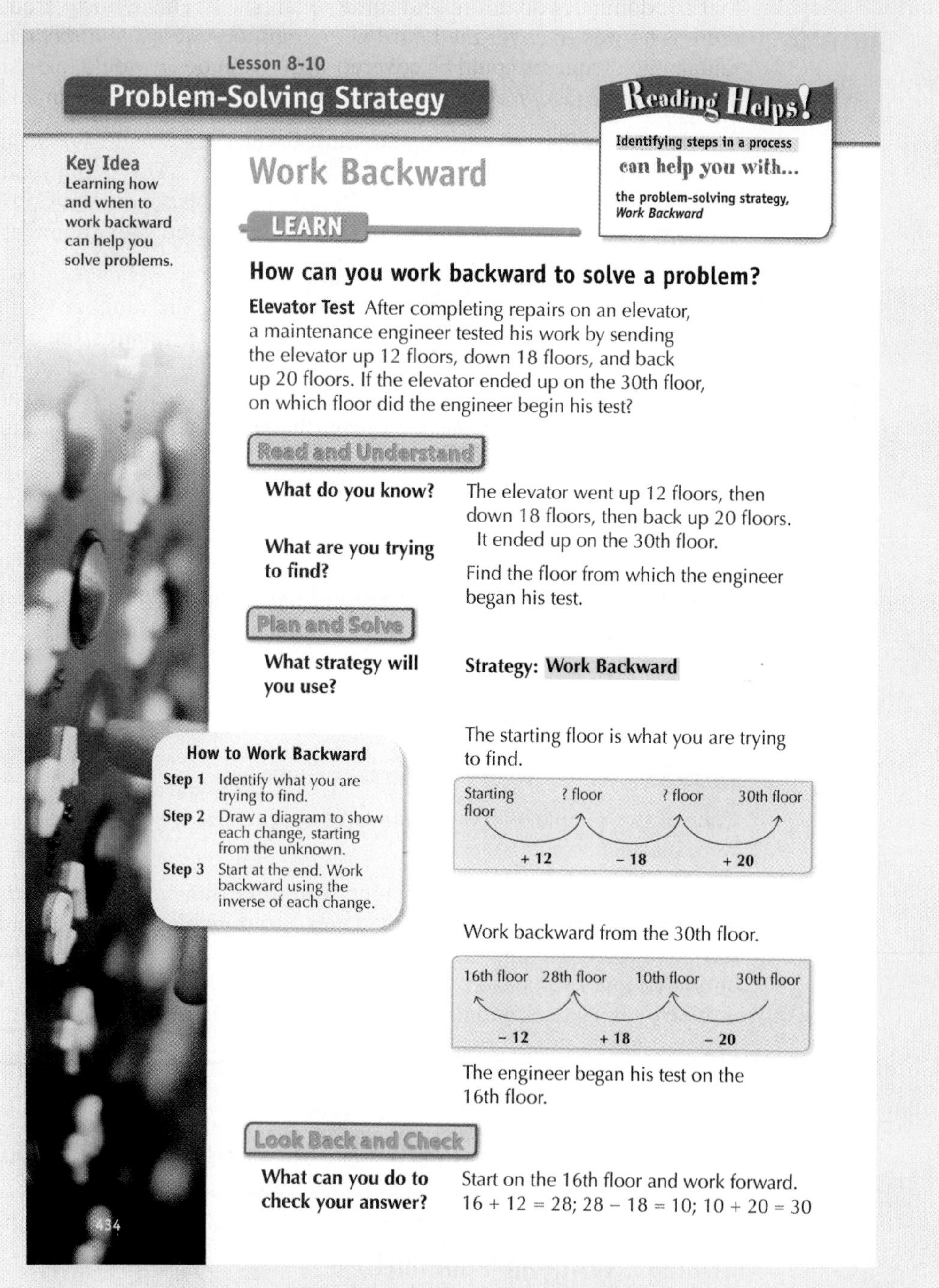
Lesson 8-10

Problem-Solving Strategy

Reading Helps!

Identifying steps in a process can help you with...

the problem-solving strategy, *Work Backward*

Key Idea
Learning how and when to work backward can help you solve problems.

Work Backward

LEARN

How can you work backward to solve a problem?

Elevator Test After completing repairs on an elevator, a maintenance engineer tested his work by sending the elevator up 12 floors, down 18 floors, and back up 20 floors. If the elevator ended up on the 30th floor, on which floor did the engineer begin his test?

Read and Understand

What do you know? The elevator went up 12 floors, then down 18 floors, then back up 20 floors. It ended up on the 30th floor.

What are you trying to find? Find the floor from which the engineer began his test.

Plan and Solve

What strategy will you use? **Strategy: Work Backward**

How to Work Backward
Step 1 Identify what you are trying to find.
Step 2 Draw a diagram to show each change, starting from the unknown.
Step 3 Start at the end. Work backward using the inverse of each change.

The starting floor is what you are trying to find.

Work backward from the 30th floor.

The engineer began his test on the 16th floor.

Look Back and Check

What can you do to check your answer? Start on the 16th floor and work forward. 16 + 12 = 28; 28 − 18 = 10; 10 + 20 = 30

434

Source: Scott Foresman-Addison Wesley, Grade 6, 2005 (p. 434).

Understanding the Problem Two red spaces on opposite corners were removed from the checkerboard in Figure 1-9. We are asked whether it is possible to cover the remaining 62 squares with dominoes the size of 2 squares.

Devising a Plan If we try to cover the board in Figure 1-9 with dominoes, we will find that the dominoes do not fit and some squares will remain uncovered. To show that there is no way to cover the board with dominoes, we use *indirect reasoning*. If the remaining 62 squares could be covered with dominoes, it would take 31 dominoes to accomplish the task. We want to show that this implies something impossible.

Carrying Out the Plan Each domino must cover 1 black and 1 red square. Hence, 31 dominoes would cover 31 red and 31 black squares. This is impossible, however, because the board in Figure 1-9 has 30 red and 32 black squares. Consequently, our assumption that the board in Figure 1-9 can be covered with dominoes is wrong.

Looking Back The counting of black and red squares implies that if we remove any number of squares from a checkerboard so that the number of remaining red squares differs from the number of remaining black squares, the board cannot be covered with dominoes. (Do you see why?) We could also investigate what happens when two squares of the same color are removed from an 8-by-7 board and other-sized boards. Also, is it always possible to cover the remaining board if two squares of opposite colors are removed?

NOW TRY THIS 1-9

Al, Bob, Carl, and Dan participate in either swimming, baseball, basketball, or tennis. Bob plays baseball. Al can't swim. Carl plays basketball. In what sports does each person participate?

Strategy: Use Direct Reasoning

PROBLEM SOLVING **Checker Games**

Each of two people won three games of checkers. Is it possible that only five games were played?

Solution We know that each person won three games. *Reasoning directly,* we see that if the people each won three games and they played each other, there would have had to be six games played. Thus, they could not play each other in all games and have three wins each. Could they in fact each win three games while playing only five total games and not have played each other? The answer is no and the situation is impossible.

NOW TRY THIS 1-10

Two U.S. coins have a total value of 35¢. One of the coins is not a quarter. What are the two coins?

Strategy: Write an Equation

A problem-solving strategy used in algebraic thinking is *writing an equation*. Section 1-3 will be devoted to algebra and writing equations.

ONGOING ASSESSMENT 1-1

1. Use the approach in Gauss's Problem to find the following sums (do not use formulas):
   a. $1 + 2 + 3 + 4 + \ldots + 99$
   b. $1 + 3 + 5 + 7 + \ldots + 1001$
2. Find the sum of $36 + 37 + 38 + 39 + \ldots + 146 + 147$.
3. How many ways can you make change for a \$50 bill using \$5, \$10, and \$20 bills?
4. Cookies are sold singly or in packages of 2 or 6. How many ways can you buy a dozen cookies?
5. The sign says you are leaving Missoula, Butte is 120 mi away, and Bozeman is 200 mi away. There is a rest stop halfway between Butte and Bozeman. How far is the rest stop from Missoula?
6. Alababa, Bubba, Cory, and Dandy are in a horse race. Bubba is the slowest, Cory is faster than Alababa but slower than Dandy. Name the finishing order of the horses.
7. Frankie and Johnny began reading a novel on the same day. Frankie reads eight pages a day and Johnny reads five pages a day. If Frankie is on page 72, what page is Johnny on?
8. How many squares are in the following figure?

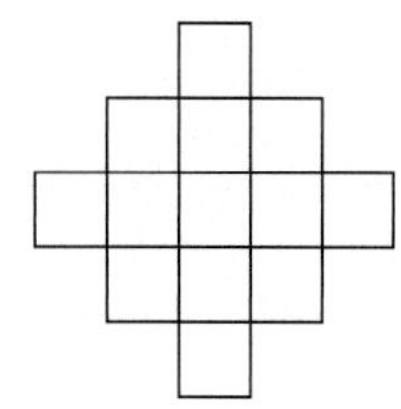

9. What is the largest sum of money—all in coins and no silver dollars—that you could have in your pocket without being able to give change for a dollar, a half-dollar, a quarter, a dime, or a nickel?
10. **a.** Without computing each sum find which is greater, O or E, and by how much.

$$O = 1 + 3 + 5 + 7 + \ldots + 97$$
$$E = 2 + 4 + 6 + 8 + \ldots + 98$$

   b. If $P = 1 + 3 + 5 + 7 + \ldots + 99$, which is greater, E or P, and by how much?
11. a. Place the digits 1, 2, 4, 5, and 7 in the following boxes so that in (i) the greatest product is obtained and in (ii) the greatest quotient is obtained:

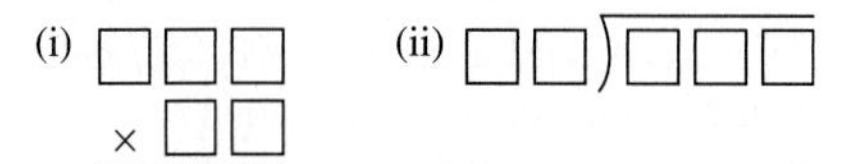

   b. Use the same digits as in (a) to obtain (i) the least product and (ii) the least quotient.
12. How many terms are there in the following sequence? 1, 8, 15, 22, . . . , 113
13. Suppose you could spend \$10 every minute, night and day. How much could you spend in a year? (Assume there are 365 days in a year.)
14. Refer to the following pattern and answer questions (a) through (c):

$$1 = 2^1 - 1$$
$$1 + 2 = 2^2 - 1$$
$$1 + 2 + 2^2 = 2^3 - 1$$
$$1 + 2 + 2^2 + 2^3 = 2^4 - 1$$
$$1 + 2 + 2^2 + 2^3 + 2^4 = 2^5 - 1$$

   a. Write a simpler expression for
$$1 + 2 + 2^2 + 2^3 + 2^4 + 2^5.$$
   Justify your answer.
   b. Write a simpler expression for the sum in the nth row in the pattern.
   c. Use a spreadsheet or a calculator to check your answer in (b) for $n = 15$.
15. How many different four-digit numbers have the same digits as 1993?
16. A compass and a ruler together cost \$4. The compass costs 90¢ more than the ruler. How much does the compass cost?
17. Kathy stood on the middle rung of a ladder. She climbed up three rungs, moved down five rungs, and then climbed up seven rungs. Then she climbed up the remaining six rungs to the top of the ladder. How many rungs are there in the whole ladder?
18. Same-sized cubes are glued together to form a staircaselike sequence of solids as shown:

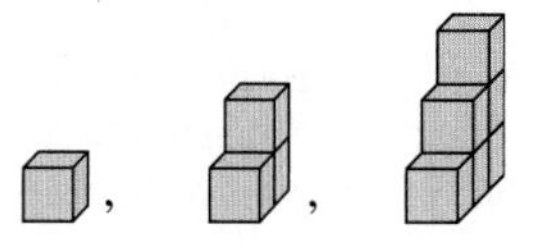

All of the unglued faces of the cubes need to be painted. How many squares will need to be painted in (a) the 100th solid? **(b)** the nth solid?
19. Marc goes to the store with exactly \$1.00 in change. He has at least one of each coin less than a half-dollar coin, but he does not have a half-dollar coin.
   a. What is the least number of coins he could have?
   b. What is the greatest number of coins he could have?
20. A farmer needs to fence a rectangular piece of land. She wants the length of the field to be 80 ft longer than the width. If she has 1080 ft of fencing material, what should the length and the width of the field be?
21. Find a 3-by-3 magic square using the numbers 3, 5, 7, 9, 11, 13, 15, 17, and 19.

22. Find the following sums:
    a. $1 + 6 + 11 + 16 + 21 + \ldots + 1001$
    b. $3 + 7 + 11 + 15 + 19 + \ldots + 403$
23. Eight marbles look alike, but one is slightly heavier than the others. Using a balance scale, explain how you can determine the heavier one in exactly
    a. three weighings **b.** two weighings
24. **a.** Find the sum of all the numbers in the following array.

1	2	3	4	5	6	...	100
2	4	6	8	10	12	...	200
3	6	9	12	15	18	...	300
			⋮				
100	200	300	400	500	600	...	$100 \cdot 100$

b. Generalize part (a) to a similar array in which each row has n numbers and there are n rows.
25. The 14 digits of a credit card are written in the boxes shown. If the sum of any three consecutive digits is 20, what is the value of A?

A	7										7		4

26. Recall the song "The Twelve Days of Christmas":

On the first day of Christmas my true love gave to me a partridge in a pear tree.
On the second day of Christmas my true love gave to me two turtle doves and a partridge in a pear tree.
On the third day of Christmas my true love gave to me three French hens, two turtle doves, and a partridge in a pear tree.

This pattern continues for 9 more days. After 12 days,
a. which gifts did my true love give the most? (Yes, you will have to remember the song.)
b. How many total gifts did my true love give to me?
27. Use the strategy of *indirect reasoning* to justify each of the following:
    a. If the product of two positive numbers is greater than 82, then at least one of the numbers is greater than 9.
    b. If the product of two positive numbers is greater than 81, then at least one of the numbers is greater than 9.
28. a. Using the existing lines on the checkerboard shown, how many squares are there?

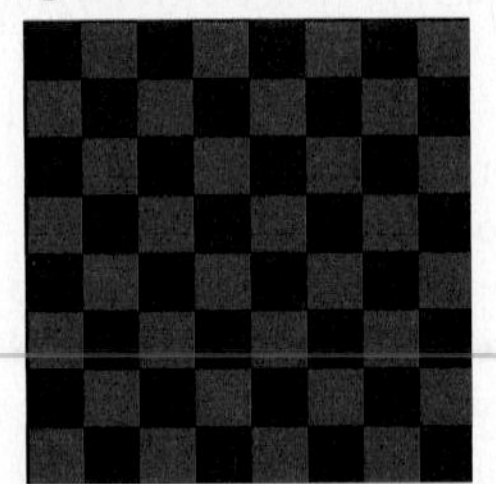

b. If the number of rows and columns of the checkerboard is doubled, is the number of squares doubled? Justify your answer.

Communication

29. Why is teaching problem solving an important part of mathematics?
30. Explain why it is impossible to have a 3-by-3 magic square with numbers 1, 3, 4, 5, 6, 7, 8, 9, and 10.

Open-Ended

31. Use exactly four 4s and any mathematical symbols to create the numbers 1 to 20 inclusive; for example, $4/4 + 4/4 = 2$ and $4 \times 4 + 4 - \sqrt{4} = 18$.
32. Choose a problem-solving strategy and make up a problem that would use this strategy. Write the solution using Polya's four-step approach.
33. Explain when you might use the *guess-and-check* strategy.

Cooperative Learning

34. Have each person in your group work the following problem: If 8 people shake hands with one another, how many handshakes take place?
a. Compare your strategies for working the problem. How are they the same? How are they different?
b. Find as many ways as possible to do the problem.
c. Generalize the solution for n people.
35. The distance around the world is approximately 40,000 km. Approximately how many people of average size in your group would it take to stretch around the world if they were holding hands?
36. Work in pairs on the following version of a game called NIM. A nongraphing calculator is needed for each pair.
a. Player 1 presses [1] and [+] or [2] and [+]. Player 2 does the same. The players take turns until the target number of 21 is reached. The first player to make the display read 21 is the winner. Determine a strategy for deciding who always wins.
b. Try a game of NIM using the digits 1, 2, 3, and 4, with a target number of 104. The first player to reach 104 wins. What is the winning strategy?
c. Try a game of NIM using the digits 3, 5, and 7, with a target number of 73. The first player to exceed 73 loses. What is the winning strategy?
d. Now play Reverse NIM with the keys [1] and [2]. Instead of [+], use [−]. Put 21 on the display. Let the target number be 0. Determine a strategy for winning Reverse NIM.
e. Try Reverse NIM using the digits 1, 2, and 3 and starting with 24 on the display. The target number is 0. What is the winning strategy?
f. Try Reverse NIM using the digits 3, 5, and 7 and starting with 73 on the display. The first player to display a negative number loses. What is the winning strategy?

Third International Mathematics and Science Study (TIMSS) Questions

4	11	6
9		5
8	3	10

The rule for the table is that numbers in each row and column must add up to the same number. What number goes in the center of the table?

a. 1
b. 2
c. 7
d. 12

TIMSS 2003, Grade 4

A car has a fuel tank that holds 45 L of fuel. The car consumes 8.5 L of fuel for each 100 km driven. A trip of 350 km was started with a full tank of fuel. How much remained in the tank at the end of the trip?

a. 15.25 L
b. 16.25 L
c. 24.75 L
d. 29.75 L

TIMSS 2003, Grade 8

Ten women are fishing all in a row in a boat. One seat in the center of the boat is empty. The five women in the front of the boat want to change seats with the five women in the back of the boat. A person can move from her seat to the next empty seat or she can step over one person without capsizing the boat. What is the minimum number of moves needed for the five women in front to change places with the five in back?

LABORATORY ACTIVITY

Place a half-dollar, a quarter, and a nickel in position *A* as shown in Figure 1-10. Try to move these coins, one at a time, to position *C*. At no time may a larger coin be placed on a smaller coin. Coins may be placed in position *B*. How many moves does it take to get them to position *C*? Now add a penny to the pile and see how many moves are required. This is a simple case of the famous Tower of Hanoi problem, in which ancient Brahman priests were required to move a pile of 64 disks of decreasing size, after which the world would end. How long would it take at a rate of one move per second?

FIGURE 1-10

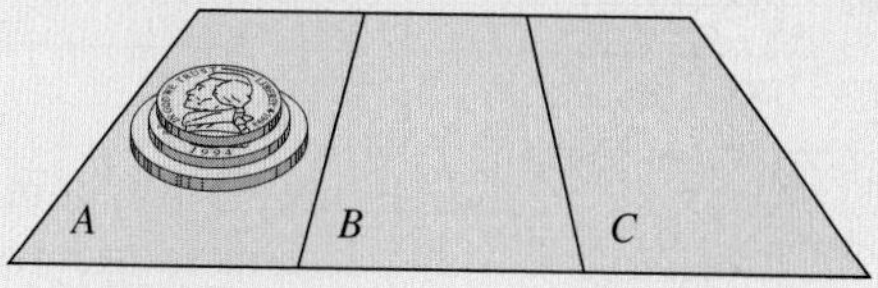

A B C

1-2 Explorations with Patterns

Mathematics has been described as the study of patterns. Patterns are everywhere—in wallpaper, tiles, traffic, and even television schedules. Police investigators study case files to find the *modus operandi,* or pattern of operation, when a series of crimes is committed. Scientists look for patterns in order to isolate variables so that they can reach valid conclusions in their research. In *Principles and Standards*, we find the following,

. . . students should investigate numerical and geometric patterns and express them mathematically in words or in symbols. They should analyze the structure of the pattern and how it grows or changes, organize this information systematically, and use their analysis to develop generalizations about the mathematical relationships in the pattern. (p. 159)

Patterns can be surprising. Consider Example 1-1.

EXAMPLE 1-1

a. Describe any patterns seen in the following:

$$1 + 0 \cdot 9 = 1$$
$$2 + 1 \cdot 9 = 11$$
$$3 + 12 \cdot 9 = 111$$
$$4 + 123 \cdot 9 = 1111$$
$$5 + 1234 \cdot 9 = 11111$$

b. Do the patterns continue? Why or why not?

Solution **a.** There are several possible patterns. For example, the numbers on the far left are natural numbers; that is, numbers from the set {1, 2, 3, 4, 5, ...} The pattern starts with 1 and continues to the next greater natural number in each successive line. The numbers "in the middle" are products of two numbers, the second of which is 9. The first number in the first product is 0; after that the first number is formed using natural numbers and including one more in each successive line. The resulting numbers on the right are formed using 1s and include an additional 1 in each successive line.

b. The pattern in the complete equation appears to continue for a number of cases, but it does not continue in general; for example,

$$13 + 123456789101112 \cdot 9 = 1{,}111{,}111{,}101{,}910{,}021.$$

This pattern breaks down when the pattern of digits in the number being multiplied by 9 contains previously used digits.

As seen in Example 1-1, determining a pattern on the basis of a few cases is not reliable. For all patterns found, we should explain why we think they either do or do not continue.

Patterns do not always have to be numerical, as shown in Now Try This 1-11.

NOW TRY THIS 1-11

a. Find three more terms to continue a pattern:
o, Δ, Δ, o, Δ, Δ, o ___, ___, ___

b. Describe the pattern found in words.

Inductive Reasoning

Inductive reasoning

conjecture

counterexample

Scientists make observations and propose general laws based on these observations and patterns seen. Statisticians use patterns when they form conclusions based on collected data. This process, **inductive reasoning**, is the method of making generalizations based on observations and patterns. Although inductive reasoning may lead to new discoveries, its weakness is that conclusions are drawn only from the collected evidence. If not all cases have been checked, the possibility exists that another case will prove the conclusion false. Inductive reasoning may lead to a **conjecture**, a statement thought to be true but not yet proved true or false. For example, considering only that $0^2 = 0$ and that $1^2 = 1$, a conjecture might be that *every number squared is equal to itself.* When we find an example that contradicts the conjecture, we provide a **counterexample** and prove the conjecture false in general. To show that the preceding conjecture is false, it is enough to exhibit only one counterexample, for example, $2^2 = 4$. Sometimes finding a counterexample is difficult, but the lack of a counterexample does not automatically make a conjecture true.

RESEARCH NOTE

In a study of the understanding of mathematical proofs, it was found that 80% of 11th grade, students did not understand the concept of a counterexample, and more than 70% of the students could not tell the difference between inductive and deductive reasoning, which included being unaware that inductive reasoning does not prove anything (Williams 1980).

Next, consider a pattern that does work and helps solve a problem. How can you find the sum of three consecutive natural numbers without performing the addition? Several examples are given here. Look for a pattern in these examples.

$$14 + 15 + 16 \qquad (\mathbf{45})$$
$$19 + 20 + 21 \qquad (\mathbf{60})$$
$$99 + 100 + 101 \qquad (\mathbf{300})$$

After studying the sums, a pattern of multiplying the middle number by 3 is apparent. Other numbers could be tried to see if a *counterexample* can be found. The pattern suggests other mathematical questions that could be considered. For example,

1. Does this work for any three consecutive natural numbers?
2. How can you find the sum of any odd number of consecutive natural numbers?
3. What happens if there is an even number of consecutive natural numbers?

A proof for (1) showing that the sum of three consecutive natural numbers is equal to 3 times the middle number is given next.

Proof

Let n be the first of three consecutive natural numbers. Then the three numbers are n, $n + 1$, and $n + 2$. The sum of these three numbers is $n + (n + 1) + (n + 2) = 3n + 3 = 3(n + 1)$. Therefore, the sum of the three consecutive natural numbers is 3 times the middle number.

The following discussion illustrates the danger of making a conjecture based on a few cases. In Figure 1-11, we choose points on a circle and connect them to form distinct, nonoverlapping regions. In this figure, 2 points determine 2 regions, 3 points determine 4 regions, and 4 points determine 8 regions. What is the maximum number of regions that would be determined by 10 points?

FIGURE 1-11

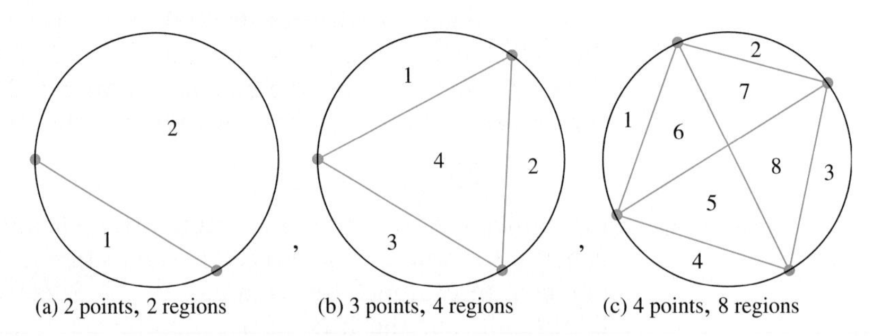

(a) 2 points, 2 regions, (b) 3 points, 4 regions, (c) 4 points, 8 regions

The data from Figure 1-11 are recorded in Table 1-2. It appears that each time we increase the number of points by 1, we double the number of regions. If this were true, then for 5 points we would have 2 times the number of regions with 4 points, or $2 \cdot 8 = 16$, and so on. If we base our conjecture on this pattern, we believe that for 10 points, we would have 512 regions. (Why?)

Table 1-2

Number of Points	2	3	4	5	6	...	10
Maximum Number of Regions	2	4	8				?

FIGURE 1-12

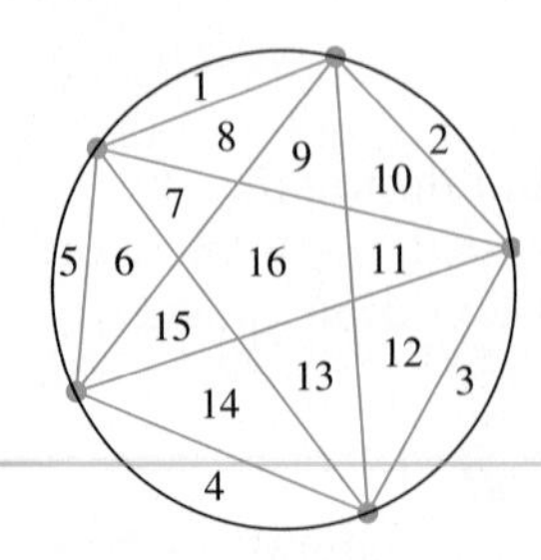

An initial check for this conjecture is to see whether we obtain 16 regions for 5 points. We obtain a figure similar to that in Figure 1-12, and our guess of 16 regions is confirmed. For 6 points, the pattern predicts that the number of regions will be 32. Choose the points so that they are neither symmetrically arranged nor equally spaced and count the regions carefully. You should obtain 31 regions and not 32 regions as predicted. No matter how the points are located on the circle, the guess of 32 regions is not correct. The counterexample tells us that the doubling pattern is not correct; note that it does not tell us whether or not there are 512 regions with 10 points, but only that the pattern is not what we conjectured.

A natural-looking pattern of 2, 4, 8, 16, . . . is suggested in this example, but the pattern does not continue, as shown when actual pictures are drawn. If we look at only the first four terms of the sequence 2, 4, 8, 16 without context, the doubling pattern is logical. In the context of counting the number of regions of a circle, however, the pattern is incorrect.

NOW TRY THIS 1-12

A *prime number* is a natural number with exactly two distinct positive numbers that divide it; for example, 2, 3, 5, 7, 11, 13 are primes. One day Amy makes a *conjecture* that the formula, $y = x^2 + x + 11$ will produce only prime numbers if you substitute the natural numbers, 1, 2, 3, 4, 5, . . . for x. She shows her work so far in Table 1-3 for $x = 1, 2, 3, 4$.

Table 1-3

x	1	2	3	4
y	13	17	23	31

a. What type of reasoning is Amy using?
b. Try the next several numbers and see if they seem to work.
c. Can you find a counterexample to show that Amy's conjecture is false?

Arithmetic Sequences

sequence

A **sequence** is an ordered arrangement of numbers, figures, or objects. A sequence has items or terms identified as *1st, 2nd, 3rd,* and so on. Often, sequences can be classified by their properties. For example, what property do the following first three sequences have that the fourth does not?

a. 1, 2, 3, 4, 5, 6, . . .
b. 0, 5, 10, 15, 20, 25, . . .
c. 2, 6, 10, 14, 18, 22, . . .
d. 1, 11, 111, 1111, 11111, 111111, . . .

arithmetic sequence

difference

In each of the first three sequences, each term—starting from the second—is obtained from the preceding one by adding a fixed number. In other words, the difference between consecutive numbers in the sequences is always the same. In part (d), the difference between the first two terms is $11 - 1$, or 10; the second difference is $111 - 11$, or 100; the third difference is $1111 - 111$, or 1000; and so on. Sequences such as the first three are arithmetic sequences. An **arithmetic sequence** is one in which each successive term is obtained from the previous term by the addition or subtraction of a fixed number, the **difference**. The difference in part (c) is 4 because 4 is the fixed number that is added each time to obtain the next number.

Arithmetic sequences can be generated from objects, as shown in Example 1-2.

EXAMPLE 1-2

Find a pattern in the number of matchsticks required to continue the pattern shown in Figure 1-13.

FIGURE 1-13

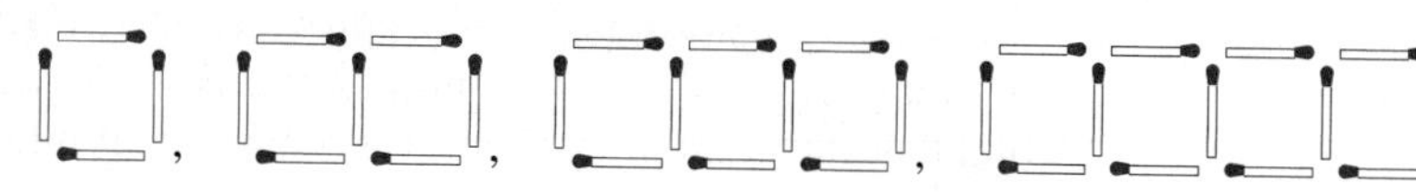

Solution Assume the matchsticks are arranged so that each figure has one more square on the right than the preceding figure. Note that the addition of a square to an arrangement requires the addition of three matchsticks each time. Thus, the numerical pattern obtained is 4, 7, 10, 13, 16, 19, . . . , an arithmetic sequence with a difference of 3. ▲

REMARK An informal description of an arithmetic sequence is one that can be described as an "add d" pattern, where d is the common difference. In Example 1-2, $d = 3$.

In the language of children, the pattern in Example 1-2 is "add 3," meaning that 3 must be added to any term after the first to find the next. This is an example of a **recursive pattern**. In a recursive pattern, after one or more consecutive terms are given to start, each successive term of the sequence is obtained from the previous term(s). For example, 3, 6, 9, . . . is another "add 3" sequence starting with 3, and 1, 2, 3, 5, 8, 13, . . . is a recursive pattern in which the next term (from the third on) is obtained by adding the two previous terms.

recursive pattern

A recursive pattern is typically used in a spreadsheet, as seen in Table 1-4 where column A tracks the order of the terms; the headers for the columns are A, B, etc. The first entry in the B column (in the B1 cell) is 4; and to find the term in the B2 cell, we use the number in the B1 cell and add 3. Once the B2 cell entry is found, the pattern is continued using the *Fill Down* command. In spreadsheet language, the formula = B1 + 3 finds any term after the first by adding 3 to the previous term. The formula, based on a recursive pattern, is a **recursive formula**. (For more explicit directions on using a spreadsheet, see Appendix I.)

recursive formula

Table 1-4

	A	B
1	1	4
2	2	7
3	3	10
4	4	13
5	5	16
6	6	19
7	7	22
8	8	25
9	9	28
10		
11		
12		
13		

If you want to find the number of matchsticks in the 100th figure in Example 1-2, you can use the spreadsheet or you can find a different type of general rule for finding the number of matchsticks when given the number of the term. The problem-solving strategy of *making a table* is again helpful here.

The spreadsheet in Table 1-4 provides an easy way to form a table. Column A gives the numbers of the terms and column B gives the terms of the sequence. If one is building such a table without a spreadsheet, it might look like Table 1-5. An *ellipsis,*

Table 1-5

Number of Term	Term
1	4
2	$7 = 4 + 3 = 4 + 1 \cdot 3$
3	$10 = (4 + 3) + 3 = 4 + 2 \cdot 3$
4	$13 = (4 + 2 \cdot 3) + 3 = 4 + 3 \cdot 3$
.	.
.	.
.	.
n	$4 + (n - 1)3$

denoted by three dots, indicates that the sequence continues in the same manner. Notice that each term is a sum of 4 and a certain number of 3s. We see that the number of 3s is 1 less than the number of the term. This pattern should continue, since the first term is $4 + 0 \cdot 3$ and each time we increase the number of the term by 1, we add one more 3. Thus, it seems that the 100th term is $4 + (100 - 1)3$, and, in general, the nth term, a_n, is $4 + (n - 1)3$.

***n*th term**

REMARK The ***n*th term**, usually denoted by a_n, of the sequence is $4 + (n - 1)3$, which could be written as $3n + 1$.

Still a different approach to finding the number of matchsticks in the 100th term of Figure 1-13 might be as follows: If the matchstick figure has 100 squares, we could find the total number of matchsticks by adding the number of horizontal and vertical sticks. There are $2 \cdot 100$ placed horizontally. (Why?) Notice that in the first figure, there are 2 matchsticks placed vertically; in the second, 3; and in the third, 4. In the 100th figure, there should be $100 + 1$ vertical matchsticks. Altogether there will be $2 \cdot 100 + (100 + 1)$, or 301 matchsticks in the 100th figure. Similarly, in the nth figure, there would be $2n$ horizontal and $(n + 1)$ vertical matchsticks, for a total of $3n + 1$. This discussion is summarized in Table 1-6.

Table 1-6

Number of Term	Number of Matchsticks Horizontally	Number of Matchsticks Vertically	Total
1	2	2	4
2	4	3	7
3	6	4	10
4	8	5	13
.	.	.	.
.	.	.	.
.	.	.	.
100	200	101	301
.	.	.	.
.	.	.	.
.	.	.	.
n	$2n$	$n + 1$	$2n + (n + 1) = 3n + 1$

If we were given the value of the term, we could use the formula for the nth term in Table 1-6 to *work backward* to find the number of the term. For example, given the term 1798, we know that $3n + 1 = 1798$. Therefore, $3n = 1797$ and $n = 599$. Consequently, 1798 is the 599th term. We could obtain the same answer by solving $4 + (n - 1) \cdot 3 = 1798$ for n.

In the matchstick problem, we found the nth term of a sequence. If the nth term of a sequence is given, we can find any term of the sequence, as shown in Example 1-3.

EXAMPLE 1-3

Find the first four terms of a sequence the nth term of which is given by the following and determine whether the sequence is arithmetic:

a. $a_n = 4n + 3$
b. $a_n = n^2 - 1$

Solution **a.** To find the first term, substitute $n = 1$ in the formula $4n + 3$ in order to obtain $4 \cdot 1 + 3$, or 7. Similarly, substituting $n = 2, 3, 4$, we obtain, respectively, $4 \cdot 2 + 3$, or 11; $4 \cdot 3 + 3$, or 15; and $4 \cdot 4 + 3$, or 19. Hence, the first four terms of the sequence are 7, 11, 15, and 19; this sequence is arithmetic with a difference of 4.

b. Substituting $n = 1, 2, 3, 4$ in the formula $n^2 - 1$, we obtain, respectively, $1^2 - 1$, or 0; $2^2 - 1$, or 3; $3^2 - 1$, or 8; and $4^2 - 1$, or 15. Thus, the first four terms of the sequence are 0, 3, 8, and 15. This sequence is not arithmetic, since it has no common difference.

NOW TRY THIS 1-13

In an arithmetic sequence with the 2nd term 11 and the 5th term 23, find the 100th term.

Generalizing Arithmetic Sequences

To generalize our work with arithmetic sequences, suppose the first term in an arithmetic sequence is a_1 and the difference is d. The strategy of *making a table* can be used to investigate the general term for the sequence $a_1, a_1 + d, a_1 + 2d, a_1 + 3d, \ldots$, as shown in Table 1-7. *The nth term of any sequence with first term a_1 and difference d is given by $a_n = a_1 + (n - 1)d$.* For example, in the arithmetic sequence 5, 9, 13, 17,

Table 1-7

Number of Term	Term
1	a_1
2	$a_1 + d$
3	$a_1 + 2d$
4	$a_1 + 3d$
5	$a_1 + 4d$
.	.
.	.
.	.
n	$a_1 + (n - 1)d$

21, 25, . . . , the first term is 5 and the difference is 4. Thus, the nth term is given by $a_1 + (n - 1)d = 5 + (n - 1)4$. Simplifying algebraically, we obtain $5 + (n - 1)4 = 5 + 4n - 4 = 4n + 1$. Check to see if $4n + 1$ generates the sequence 5, 9, 13, 17, 21, ...

REMARK The nth term of any arithmetic sequence with first term a_1 and difference d is $a_n = a_1 + d(n - 1)$, where d is a natural number.

EXAMPLE 1-4

The diagrams in Figure 1-14 show the molecular structure of alkanes, a class of hydrocarbons. C represents a carbon atom and H a hydrogen atom. A connecting segment shows a chemical bond. (Remark: CH_4 stands for C_1H_4.)

FIGURE 1-14

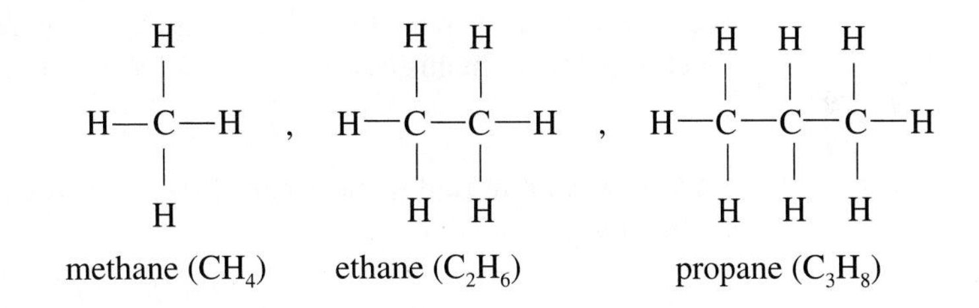

a. Hectane is an alkane with 100 carbon atoms. How many hydrogen atoms does it have?
b. Write a general rule for alkanes C_nH_m showing the relationship between m and n.

Solution **a.** To determine the relationship between the number of carbon and hydrogen atoms, we study the drawing of the alkanes and disregard the extreme left and right hydrogen atoms in each. We can see that for every carbon atom, there are two hydrogen atoms. Therefore, there are twice as many hydrogen atoms as carbon atoms plus the two hydrogen atoms at the extremes. For example, when there are 3 carbon atoms, there are $(2 \cdot 3) + 2$, or 8, hydrogen atoms. This notion is summarized in Table 1-8. If we extend the table for 4 carbon atoms, we get $(2 \cdot 4) + 2$, or 10, hydrogen atoms. For 100 carbon atoms, there are $(2 \cdot 100) + 2$, or 202, hydrogen atoms.

Table 1-8

No. of Carbon Atoms	No. of Hydrogen Atoms
1	4
2	6
3	8
.	.
.	.
.	.
100	?
.	.
.	.
.	.
n	m

b. In general, for n carbon atoms there would be n hydrogen atoms attached above, n attached below, and 2 attached on the sides. Hence, the total number of hydrogen atoms would be $2n + 2$. Because the number of hydrogen atoms was designated by m, it follows that $m = 2n + 2$.

EXAMPLE 1-5

A theater is set up in such a way that there are 20 seats in the first row and 4 additional seats in each consecutive row. The last row has 144 seats. How many rows are there in the theater?

Solution Because there are 4 additional seats in each consecutive row, the number of seats in the rows forms an arithmetic sequence. The first term, a_1, of the sequence is 20 and the difference, d, is 4. The last term in the sequence is 144. A computerized spreadsheet could easily be used to count the number of terms in the sequence, 20, 24, 28, . . . , 144. However, without technology, we count the terms as follows: In an arithmetic sequence, $a_n = a_1 + (n - 1)d$, where a_1 is the first term, d is the difference, and n is the number of the term. In this case, $a_1 = 20$ and $d = 4$. Therefore,

$$a_n = a_1 + (n - 1)d = 20 + (n - 1)4$$

We now want to find the number of the term when $a_n = 20 + (n - 1)4$ is equal to 144. Therefore,

$$\begin{aligned} 20 + (n - 1)4 &= 144 \\ (n - 1)4 &= 124 \\ n - 1 &= 31 \\ n &= 32. \end{aligned}$$

This tells us that when $n = 32$, the value of the term is 144. This implies that there are 32 rows in the theater. Instead of using a formula, we could have made a table and looked for a pattern.

Fibonacci Sequence

The popular book *The DaVinci Code* brought renewed interest to one of the most famous sequences of all time, the ***Fibonacci sequence***. The Fibonacci sequence is hinted at in the following cartoon.

FIBONACCI LEARNS TO COUNT.

The Fibonacci sequence is as follows:

$$1, 1, 2, 3, 5, 8, 13, 21, 34, 55, 89, 144, \ldots$$

The sequence is named after the Italian Leonardo de Pisa, better known by the nickname Fibonacci. This sequence is not *arithmetic* as there is no fixed difference, d. Before reading on, determine a rule for generating this sequence.

The standard mathematical way to represent a Fibonacci number is an F followed by a subscript indicating its place in the sequence. For example, F_1 represents the first term, F_2 represents the second term, F_3 represents the third term, and in general F_n represents the nth term. If we want to indicate the Fibonacci numbers that come after F_n, we write them as F_{n+1}, F_{n+2}, and so on. The number that comes before F_n is F_{n-1}. With this notation the rule for generating the Fibonacci sequence can be written as

$$F_n = F_{n-1} + F_{n-2}, \quad \text{for } n = 3, 4, 5, \ldots$$

Notice that this rule cannot be applied to the first two Fibonacci numbers. Because $F_1 = 1$ and $F_2 = 1$, then $F_3 = 1 + 1 = 2$. The *seeds* $F_1 = 1$ and $F_2 = 1$ and the rule $F_n = F_{n-1} + F_{n-2}$ give another example of a *recursive* definition because the rule in the sequence defines a number using previous numbers in the same sequence. Using the seeds and the rule, we can find any Fibonacci number. To find F_{100} with what we know right now we would have to know F_{98} and F_{99}.

NOW TRY THIS 1-14

a. Add the first three Fibonacci numbers.
b. Add the first four Fibonacci numbers.
c. Add the first five Fibonacci numbers.
d. Add the first six Fibonacci numbers.
e. What pattern is there in the sum and any of the remaining numbers in the Fibonacci sequence?
f. Write a rule for your pattern in part (e) using the notation for Fibonacci numbers.

▶ **HISTORICAL NOTE**

Leonardo Fibonacci (ca. 1125–1250 BCE) was born in Pisa, Italy, around 1125. His real family name was Bonaccio but he preferred the nickname Fibonacci, derived from the Latin for *filius Bonacci* meaning "son of Bonacci." In his travels, Leonardo learned arithmetic and algebra from the Moors. The Moors used the Hindu-Arabic number system that we use today. Around the year 1200, Fibonacci returned to Italy and dedicated himself to bringing the Hindu-Arabic system to Europe. In his first book, *Liber Abaci* (1202), he described the workings of the Hindu-Arabic system. One of the problems in his book was the now-famous rabbit problem, whose solution is the sequence 1, 1, 2, 3, 5, 8, 13, 21, . . . , which became known as the *Fibonacci sequence.* ◀

Geometric Sequences

geometric sequence

ratio

A child has 2 biological parents, 4 grandparents, 8 great grandparents, 16 great-great grandparents, and so on. The number of ancestors form the **geometric sequence** 2, 4, 8, 16, 32, ... Each successive term of a geometric sequence is obtained from its predecessor by multiplying by a fixed nonzero number, the **ratio**. In this example, both the first term and the ratio are 2. (The ratio is 2 because each person has two parents.) To find the nth term, a_n, examine the pattern in Table 1-9.

Table 1-9

Number of Term	Term
1	$2 = 2^1$
2	$4 = 2 \cdot 2 = 2^2$
3	$8 = (2 \cdot 2) \cdot 2 = 2^3$
4	$16 = (2 \cdot 2 \cdot 2) \cdot 2 = 2^4$
5	$32 = (2 \cdot 2 \cdot 2 \cdot 2) \cdot 2 = 2^5$
.	.
.	.
.	.

The pattern of the terms in Table 1-9 can be generalized as follows.

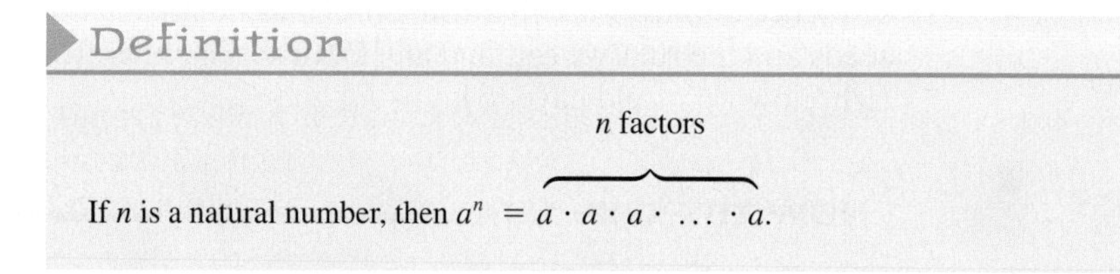

Definition

If n is a natural number, then $a^n = \overbrace{a \cdot a \cdot a \cdot \ldots \cdot a}^{n \text{ factors}}$.

When the given term is written as a power of 2, the number of the term is the **exponent**. Following this pattern, the 10th term, a_{10}, is 2^{10}, or 1024, the 100th term, a_{100}, is 2^{100}, and the nth term, a_n, is 2^n. Thus, the number of ancestors in the nth previous generation is 2^n.

exponent

Table 1-10

Number of Term	Term
1	a_1
2	a_1r
3	a_1r^2
4	a_1r^3
5	a_1r^4
.	.
.	.
.	.
n	a_1r^{n-1}

Finding the nth Term for a Geometric Sequence

It is possible to find the nth term, a_n, of any geometric sequence when given the first term and the ratio. If the first term is a_1 and the ratio is r, then the terms are as listed in Table 1-10. Notice that the second term is a_1r, the third term is a_1r^2, and the fourth term is a_1r^3. The power of r in each term is 1 less than the number of the term. This pattern continues since we multiply by r to get the next term. Thus, the nth term, a_n, is a_1r^{n-1}. For $n = 1$, we have $a_1r^{1-1} = a_1r^0$. Because the first term is a_1, $a_1r^0 = a_1$. When $a \neq 0$, for all numbers $r \neq 0$, then $r^0 = 1$, as discussed in Chapter 6. For the geometric sequence 3, 12, 48, 192, . . . , the first term is 3 and the ratio is 4, and so the nth term, a_n, is given by $a_n = a_1r^{n-1} = 3 \cdot 4^{n-1}$.

NOW TRY THIS 1-15

a. Two bacteria are in a dish. The number of bacteria triples every hour. Following this pattern, find the number of bacteria in the dish after 10 hours and after n hours.

b. Suppose that instead of increasing geometrically as in part (a), the number of bacteria increases arithmetically by 3 each hour. Compare the growth after 10 hours and after n hours. Comment on the difference in growth of a geometric sequence versus an arithmetic sequence.

REMARK The nth term of any geometric sequence with first term a_1 and a ratio r, is $a_n = a_1 \cdot r^{n-1}$, when n is a natural number and $r \neq 0$.

Other Sequences

figurate numbers

triangular numbers

Figurate numbers provide examples of sequences that are neither arithmetic nor geometric. Such numbers can be represented by dots arranged in the shape of certain geometric figures. The number 1 is the beginning of most patterns involving figurate numbers. The array in Figure 1-15 represents the first four terms of the sequence of **triangular numbers**.

FIGURE 1-15

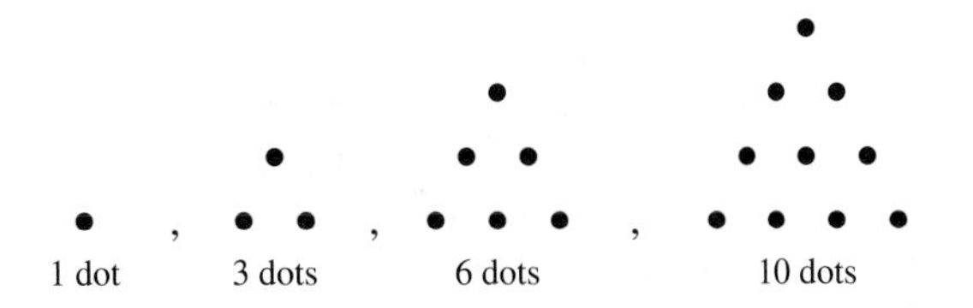

The triangular numbers can be written numerically as 1, 3, 6, 10, 15, . . . The sequence 1, 3, 6, 10, 15, . . . is not an arithmetic sequence because there is no common difference, as Figure 1-16 shows. It is not a geometric sequence because there is no common ratio. It is not a Fibonacci sequence.

FIGURE 1-16

1 3 6 10 15

(First difference) 2 3 4 5

However, the sequence of differences 2, 3, 4, 5, . . . is an arithmetic sequence with difference 1, as Figure 1-17 shows.

FIGURE 1-17

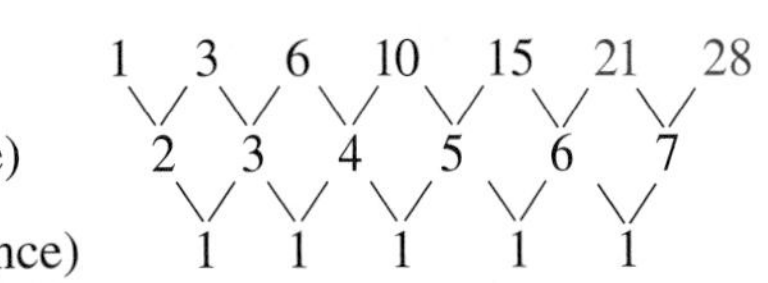

(First difference)

(Second difference)

The next successive terms for the original sequence are shown in color in Figure 1-17.

Table 1-11 suggests a pattern for finding the next terms and the nth term for the triangular numbers. The second term is obtained from the first term by adding 2; the third term is obtained from the second term by adding 3; and so on. In general, because the nth triangular number has n dots in the nth row, it is equal to the sum of the dots in the previous triangular number (the $(n - 1)$st one) plus the n dots in the nth row. Following this pattern, the 10th term is $1 + 2 + 3 + 4 + 5 + 6 + 7 + 8 + 9 + 10$, or 55, and the nth term, a_n is $1 + 2 + 3 + 4 + 5 + \ldots + (n - 1) + n$. This problem is similar to Gauss's problem in Section 1-1. Because of the work done in Section 1-1, we know that

$$a_n = \frac{n(n + 1)}{2}.$$

Table 1-11

Number of Term	Term
1	1
2	3 = 1 + 2
3	6 = 1 + 2 + 3
4	10 = 1 + 2 + 3 + 4
5	15 = 1 + 2 + 3 + 4 + 5
.	.
.	.
.	.
10	55 = 1 + 2 + 3 + 4 + 5 + 6 + 7 + 8 + 9 + 10

Next consider the first four *square numbers* in Figure 1-18. These square numbers, 1, 4, 9, 16, . . . , can be written as $1^2, 2^2, 3^2, 4^2$, and so on. The number of dots in the 10th array is 10^2; the number of dots in the 100th array is 100^2; and the number of dots in the nth array is n^2. The sequence of square numbers is neither arithmetic nor geometric.

FIGURE 1-18

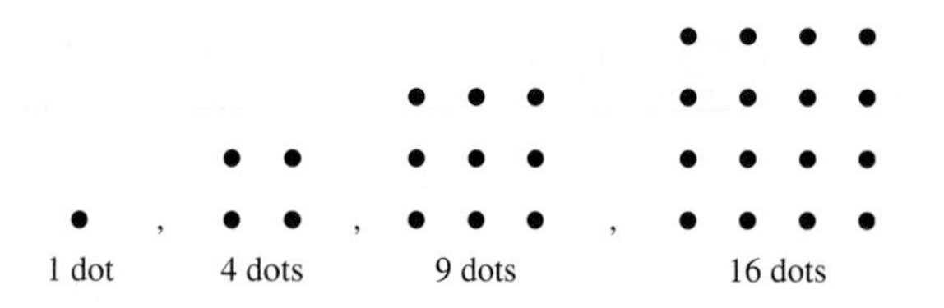

EXAMPLE 1-6

Assuming that the pattern discovered continues, find the seventh term in each of the following sequences:

a. 5, 6, 14, 29, 51, 80, . . .
b. 2, 3, 9, 23, 48, 87, . . .

Solution **a.** Following is the sequence of first differences:

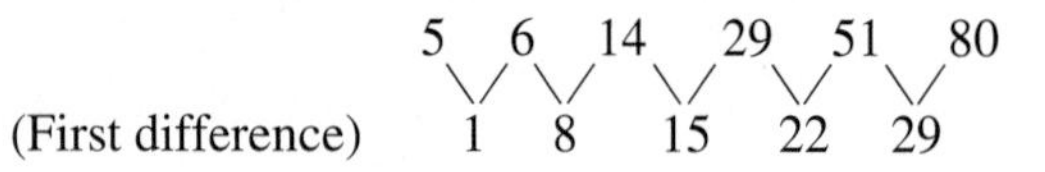

To discover a pattern for the original sequence, we try to find a pattern for the sequence of differences 1, 8, 15, 22, 29, ... This sequence is an arithmetic sequence with fixed difference 7:

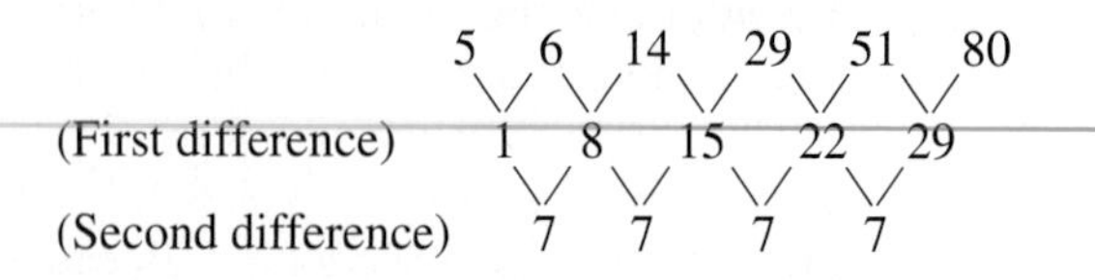

Thus the sixth term in the first difference row is 29 + 7, or 36, and the seventh term in the original sequence is 80 + 36, or 116. What number follows 116?

b. Because the second difference is not a fixed number, we go on to the third difference:

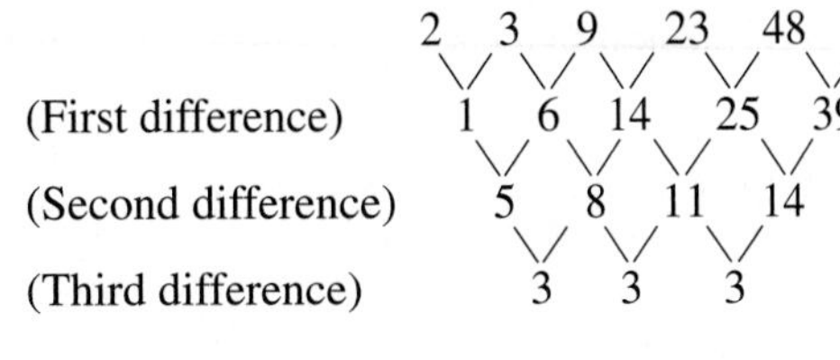

The third difference is a fixed number; therefore, the second difference is an arithmetic sequence. The fifth term in the second-difference sequence is 14 + 3, or 17; the sixth term in the first-difference sequence is 39 + 17, or 56; and the seventh term in the original sequence is 87 + 56, or 143. ▲

NOW TRY THIS 1-16

The first three figures of arrays of sticks are shown in Figure 1-19.

FIGURE 1-19

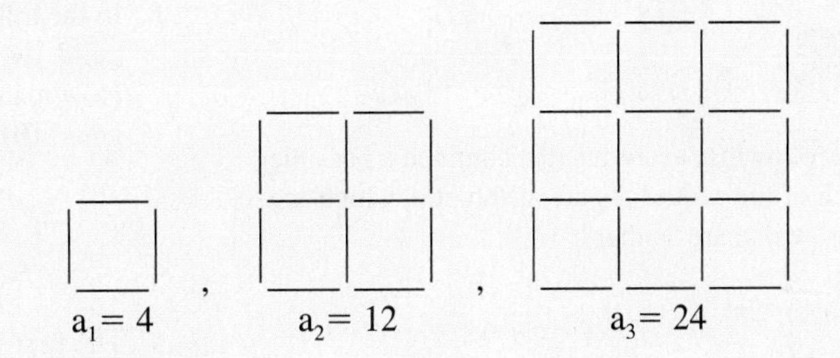

a. Draw the next array of sticks.
b. Build a table showing the term number and the number of sticks for $n = 1, 2, 3, 4$.
c. Use differences to predict the number of sticks for $n = 5, 6, 7$.
d. Is finding differences the best way to determine how many sticks there are for a_{100}? Tell how you would find a_{100} and a_n.

When asked to find a pattern for a given sequence, you first look for some easily recognizable pattern and determine whether the sequence is arithmetic or geometric. If a pattern is unclear, taking successive differences may help. *It is possible that none of the methods described reveal a pattern.*

BRAIN TEASER

Female bees are born from fertilized eggs, and male bees are born from unfertilized eggs. This means that a male bee has only a mother, whereas a female bee has a mother and a father. If the ancestry of a male bee is traced back 10 generations, how many bees are there in all 10 generations? (*Hint:* The Fibonacci sequence might be helpful.)

ONGOING ASSESSMENT 1-2

1. For each of the following sequences of figures, determine a possible pattern and draw the next figure according to that pattern:

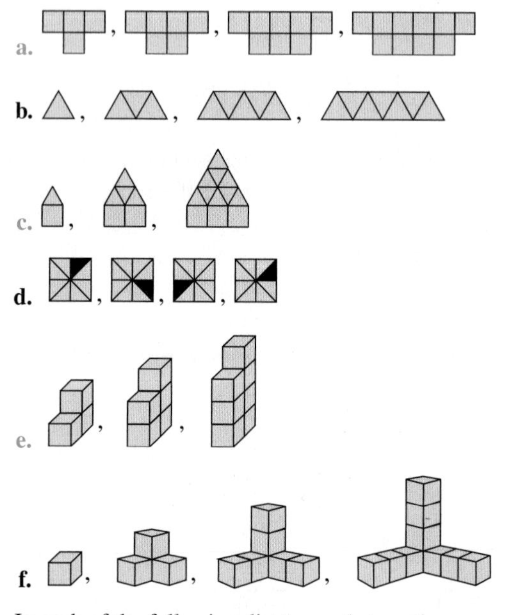

2. In each of the following, list terms that continue a possible pattern. Which of the sequences are arithmetic, which are geometric, and which are neither?
   a. 1, 3, 5, 7, 9
   b. 0, 50, 100, 150, 200
   c. 3, 6, 12, 24, 48
   d. 10, 100, 1,000, 10,000, 100,000
   e. 9, 13, 17, 21, 25, 29
   f. 1, 8, 27, 64, 125
3. Find the 100th term and the *n*th term for each of the sequences in problem 2.
4. Use a traditional clock face to determine the next three terms in the following sequence:

$$1, 6, 11, 4, 9, \ldots$$

5. Observe the following pattern:

$$1 + 3 = 2^2,$$
$$1 + 3 + 5 = 3^2,$$
$$1 + 3 + 5 + 7 = 4^2$$

   a. State a generalization based on this pattern.
   b. Based on the generalization in (a), find

$$1 + 3 + 5 + 7 + \ldots + 35.$$

6. In the pattern, 8, 16, 14, 10, . . . , the sum of digits can be used to create the next number. In this case, each succeeding number is double the sum of the digits in the previous number.
   a. Find the next three numbers in the sequence described.
   b. Find the next three numbers in the sequence 4, 16, 49, 169, 256, _____, _____, _____. Describe the rule you used.
   c. Find the next three numbers in the sequence 4, 16, 37, 58, 89, 145, 42, 20, _____, _____, _____. Describe the rule you used.
   d. What will happen if the sequence in part (c) is continued indefinitely?
7. The following geometric arrays suggest a sequence of numbers:

2, 6, 12, 20

   a. Find the next three terms.
   b. Find the 100th term.
   c. Find the *n*th term.
8. In the following pattern, one hexagon takes 6 toothpicks to build, two hexagons take 11 toothpicks to build, and so on. How many toothpicks would it take to build (a) 10 hexagons? **(b)** *n* hexagons?

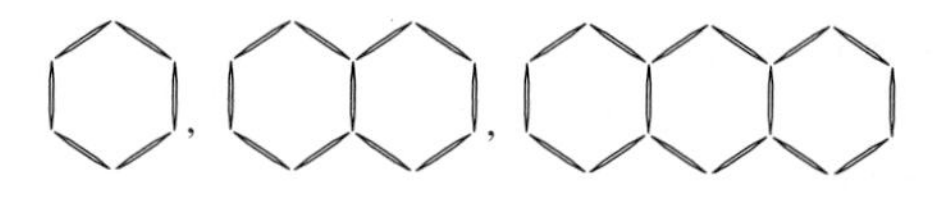

9. The first windmill takes 5 matchstick squares to build, the second takes 9 to build, and the third takes 13 to build, as shown. How many matchstick squares will it take to build (a) the 10th windmill? **(b)** the *n*th windmill? (c) How many matchsticks will it take to build the *n*th windmill?

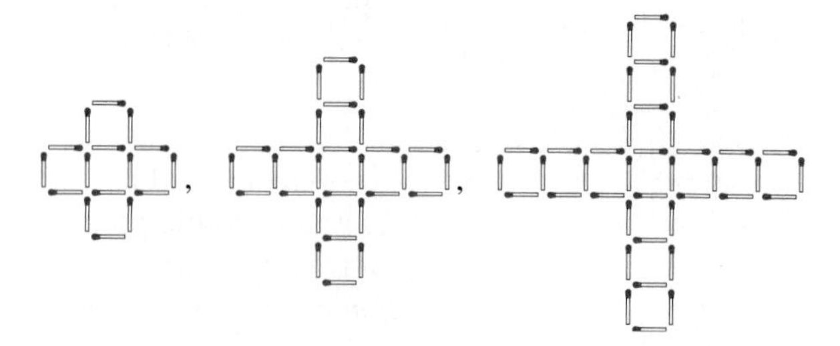

10. Each of the following figures is made of small triangles like the first one in the sequence. (The second figure is made of four small triangles.) Make a conjecture concerning the number of small triangles needed to make (a) the 100th figure and **(b)** the *n*th figure.

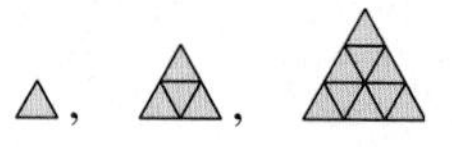

11. In the following sequence, the figures are made of cubes that are glued together. If the exposed surface needs to be painted, how many squares will be painted in (a) the 10th figure? **(b)** the *n*th figure?

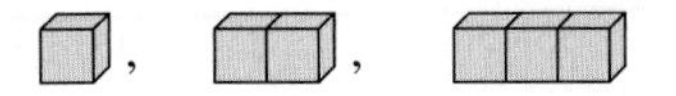

12. The school population for a certain school was predicted to increase by 50 students per year for the next 10 years. If the current enrollment is 700 students, what will the enrollment be after 10 years?

13. A tank contains 15,360 L of water. At the end of each day, half of the water is removed and not replaced. How much water is left in the tank after 10 days?

14. Joe's annual income has been increasing each year by the same amount. The first year his income was \$24,000, and the ninth year his income was \$31,680. In which year was his income \$45,120?

15. The first difference of a sequence is 2, 4, 6, 8, ... Find the first six terms of the original sequence in each of the following cases:
   a. The first term of the original sequence is 3.
   b. The sum of the first two terms of the original sequence is 10.
   c. The fifth term of the original sequence is 35.

16. List the next three terms to continue a pattern in each of the following. (Finding differences may be helpful.)
   a. 5, 6, 14, 32, 64, 115, 191
   b. 0, 2, 6, 12, 20, 30, 42

17. How many terms are there in each of the following sequences?
   a. 51, 52, 53, 54, . . . , 151 **b.** $1, 2, 2^2, 2^3, \ldots, 2^{60}$
   c. 10, 20, 30, 40, . . . , 2000 d. 9, 13, 17, 21, 25, . . . , 353
   e. 1, 2, 4, 8, 16, 32, . . . , 1024

18. Find the first five terms in each of the following:
   a. $a_n = n^2 + 2$ **b.** $a_n = 5n - 1$
   c. $a_n = 10^n - 1$ **d.** $a_n = 3n + 2$

19. Find a counterexample for each of the following:
   a. If x is a natural number, then $(x + 5)/5 = x + 1$
   b. If x is a natural number, then $(x + 4)^2 = x + 16$

20. Assume the following pattern of square tile figures continues and answer the questions that follow.

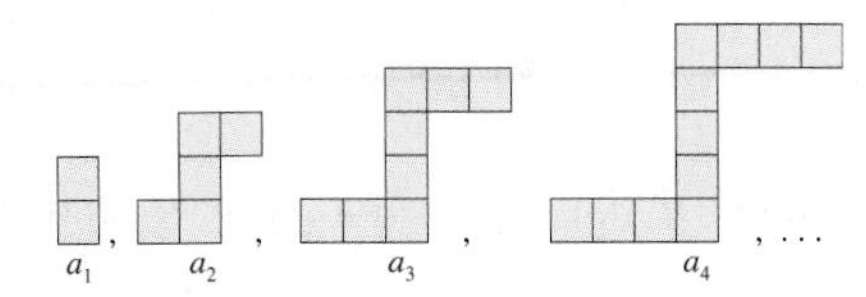

   a. How many square tiles are there in the sixth figure, a_6?
   b. What is the formula for a_n, where a_n is the number of square tiles in the *n*th figure?
   c. Is there a figure that has exactly 449 square tiles? If so, which one?

21. Assume that the following pattern of square tile figures continues and answer the questions that follow.

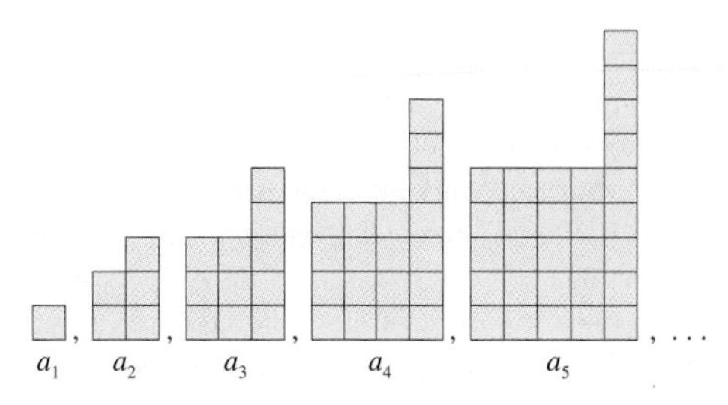

   a. How many square tiles are the in the sixth figure, a_6?
   b. What is the formula for a_n, where a_n is the number of square tiles in the *n*th figure?
   c. Is there a figure that has exactly 1259 square tiles? If so, which one?

22. Write consecutive odd numbers in triangular form as shown.

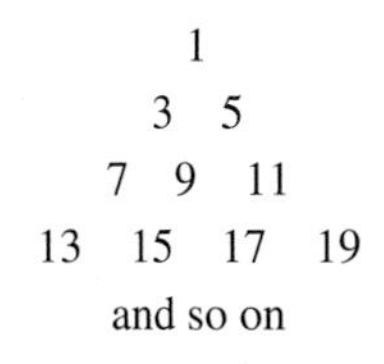

1
3 5
7 9 11
13 15 17 19
and so on

   a. Find the sum of each of the first five horizontal rows.
   b. What patterns do you notice?

23. Find the third, fourth, and fifth terms in the sequence if $a_1 = 2$, $a_2 = 5$, and $a_n = 2a_{n-1} - a_{n-2}$.

24. Consider the following sequences:

$$300, 500, 700, 900, 1100, 1300, \ldots$$
$$2, 4, 8, 16, 32, 64, \ldots$$

Find the number of the term in which the geometric sequence becomes greater than the arithmetic sequence.

25. Cut a piece of paper into five pieces. Take any one of the pieces and cut it into five pieces, and so on.
   a. What number of pieces can be obtained in this way?
   b. What is the number of pieces obtained in the *n*th cut?

26. The sequence 32, *a*, *b*, *c*, 512 is a geometric sequence. Find *a*, *b*, *c*.

27. Find the sum of the first 43 terms of an arithmetic sequence in which the 11th term is 83 and the 62nd term is 440.

Communication

28. Explain how the two sequences in each part are the same and how they are different.
   a. 2, 4, 6, 8, 10, . . . and 2, 4, 8, 16, 32, . . .
   b. 2, 4, 6, 8, 10, . . . and 3, 5, 7, 9, 11, . . .
   c. 5, 10, 15, 20, 25, . . . and 50, 100, 150, 200, 250, . . .

29. Study the following examples below:

$$1/2 + 1/3 = 5/6,\ 1/3 + 1/4 = 7/12,\ 1/5 + 1/7 = 12/35$$

a. Discuss the pattern you see and use inductive reasoning to come up with a rule for adding *unit fractions*, that is, fractions with numerators of 1.
b. Does your rule work all of the time? Why?

30. **a.** Explain the difference between *inductive* and *deductive* reasoning.
b. Give two examples of how inductive reasoning might be used in everyday life. Is a conclusion based on inductive reasoning certain?

31. a. If a fixed number is added to each term of an arithmetic sequence, is the resulting sequence an arithmetic sequence? Justify the answer.
b. If each term of an arithmetic sequence is multiplied by a fixed number, will the resulting sequence always be an arithmetic sequence? Justify the answer.
c. If the corresponding terms of two arithmetic sequences are added, is the resulting sequence arithmetic?

Open-Ended

32. Patterns can be used to count the number of dots on the Chinese checkerboard; two patterns are shown here. Determine several other patterns to count the dots.

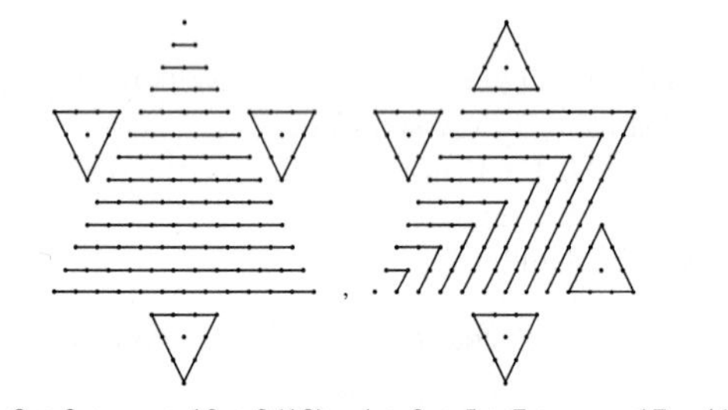

$1 + 2 + 3 + \ldots + 13 + 3(10)$ $\quad 1 + 3 + 5 + 7 + \ldots + 17 + 4(10)$

33. Make up a pattern involving figurate numbers and find a formula for the 100th term. Describe the pattern and how to find the 100th term.

Cooperative Learning

34. The following pattern is called *Pascal's triangle*. It was named for the mathematician Blaise Pascal (1623–1662).

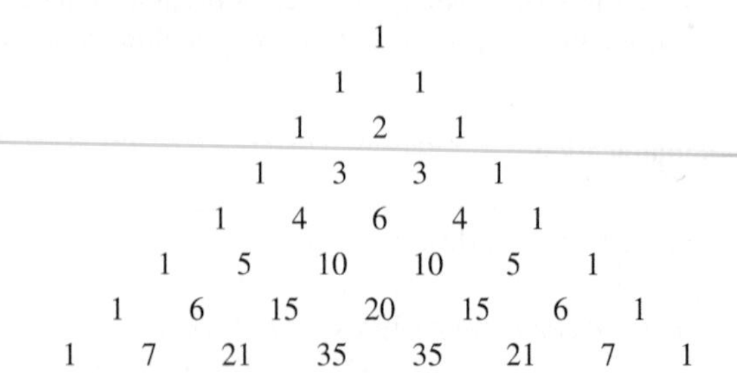

a. Have each person in the group find four different patterns in the triangle and then share them with the rest of the group.
b. Add the numbers in each horizontal row. Discuss the pattern that occurs.
c. Use what you have learned in part (b) to find the sum in the 16th row.
d. What is the sum of the numbers in the nth row?

35. If the following pattern continued indefinitely, the resulting figure would be called the *Sierpinski triangle*, or *Sierpinski gasket.*

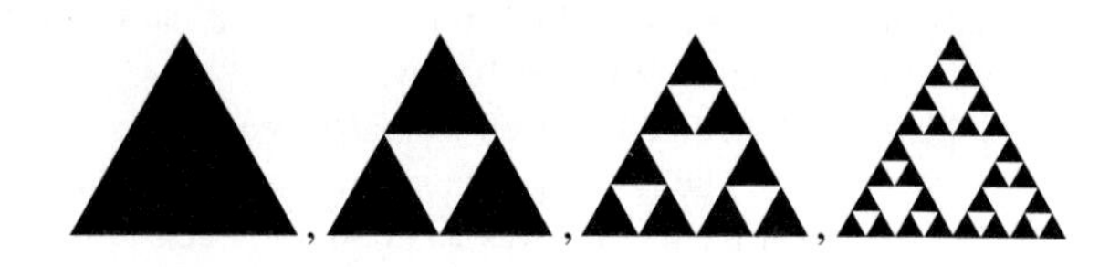

In a group, determine each of the following. Discuss different counting strategies.
a. How many black triangles would be in the fifth figure?
b. How many white triangles would be in the fifth figure?
c. If the pattern is continued for n figures, how many black triangles will there be?
d. If the pattern is continued for n figures, how many white triangles will there be?

Review Problems

36. In a baseball league consisting of 10 teams, each team plays each of the other teams twice. How many games will be played?

37. How many ways can you make change for 40¢ using only nickels, dimes, and quarters?

38. Nine dots are arranged as shown. Is it possible to connect all nine dots with line segments if you cannot retrace any part of a segment except a single point and you are not allowed to remove the pencil from the paper? Show your drawing if it can be done.

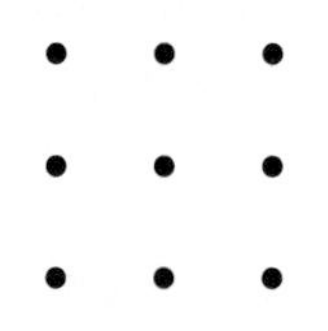

Third International Mathematics and Science Study (TIMSS) Questions

The numbers in the sequence 7, 11, 15, 19, 23, . . . increase by four. The numbers in the sequence 1, 10, 19, 28, 37, . . . increase by nine. The number 19 is in both sequences. If the two sequences are continued, what is the next number that is in BOTH the first and the second sequences?

TIMSS 2003, Grade 8

The three figures below are divided into small congruent triangles.

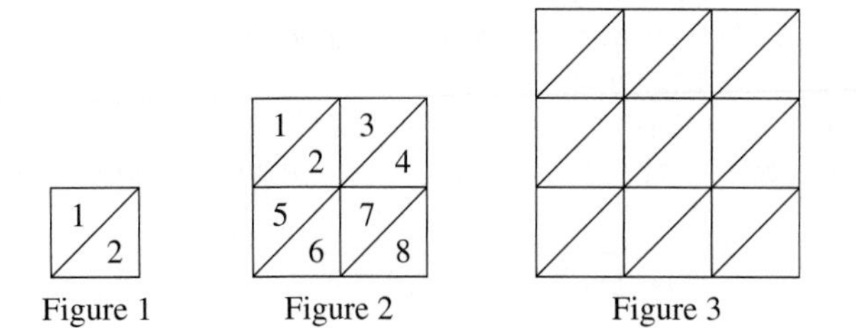

Figure 1 Figure 2 Figure 3

a. Complete the table below. First, fill in how many small triangles make up Figure 3. Then, find the number of small triangles that would be needed for the fourth figure if the sequence of figures is extended.

Figure	Number of Small Triangles
1	2
2	8
3	
4	

b. The sequence of figures is extended to the 7th figure. How many small triangles would be needed for Figure 7?

c. The sequence of figures is extended to the 50th figure. Explain a way to find the number of small triangles in the 50th figure that does not involve drawing it and counting the number of triangles.

TIMSS, Grade 8

1-3 Algebraic Thinking

One of the problem-solving strategies is *write an equation*. This strategy is part of the bigger picture of algebraic thinking. In *Principles and Standards*, we find the following:

Instructional programs from prekindergarten through grade 12 should enable all students to

- *understand patterns, relations, and functions;*
- *represent and analyze mathematical situations and structures using algebraic symbols;*
- *use mathematical models to represent and understand quantitative relationships;*
- *analyze change in various contexts.* (p. 37)

Because algebraic thinking is so important in mathematics at all levels, from the early grades on, we chose to include a separate section on it. Because patterns were included in the previous section, we concentrate on other features of algebraic thinking in this section.

For PreK–2, we find that *Principles and Standards* recommends that students be able to

- *use concrete, pictorial, and verbal representations to develop an understanding of invented and conventional symbolic notations;*
- *model situations that involve the addition and subtraction of whole numbers, using objects, pictures, and symbols.* (p. 90)

For Grades 3–5, we find the following:

- *represent and analyze patterns and functions, using words, tables, and graphs.*
- *represent the idea of a variable as an unknown quantity using a letter or a symbol;*
- *express mathematical relationships using equations;*
- *investigate how a change in one variable relates to a change in a second variable;*
- *identify and describe situations with constant or varying rates of change and compare them.* (p. 158)

For Grades 6–8, we have:

- *identify functions as linear or nonlinear and contrast their properties from tables, graphs, or equations;*
- *develop an initial conceptual understanding of different uses of variables;*
- *explore relationships between symbolic expressions and graphs of lines, paying particular attention to the meaning of intercept and slope;*
- *use symbolic algebra to represent situations and to solve problems, especially those that involve linear relationships;*
- *recognize and generate equivalent forms for simple algebraic expressions and solve linear equations.* (p. 222)

Developing Algebraic Thinking Skills

variable

One of the big ideas of algebraic thinking is the concept of **variable**. Understanding the concept of a variable is fundamental to mathematics. However, "variable" can mean many things in mathematics. It may stand for a missing element or an unknown, as in $x + 2 = 5$.

In this situation, many possible numbers could be placed in the sentence, but there is exactly one number that makes the sentence true. Here the unknown is 3.

In a different situation, a variable can represent more than one thing. For example, in a group of children, we could say that their heights vary with their ages. If h represents height and a represents age, then both h and a can have different values for different children in the group. Here a variable represents a changing quantity.

Variables can also be used in generalizations of patterns as we saw in Section 1-2.

To apply algebra in solving problems, we frequently need to translate given information into a mathematical expression involving variables designated by letters or words. In all such examples, we may name the variables as we choose.

RESEARCH NOTE

Students may have difficulty if they view algebra as generalized arithmetic. Arithmetic and algebra use the same symbols and signs but interpret them differently. Some misconceptions identified in research studies include:

1. Letters are used differently; for example, $6m$ is confused with 6 m, which means 6 meters (Booth 1988).
2. Arithmetic and algebra treat juxtaposition of two symbols differently; for example, "$8y$" denotes a multiplication and "87" means an addition, $80 + 7$. Another example is that students think "$3x = 34$" implies $x = 4$ (Matz 1982).

EXAMPLE 1-7

In each of the following, translate the given information into a symbolic expression involving quantities designated by letters:

a. One weekend, a store sold twice as many CDs as DVDs and 25 fewer tapes than CDs. If the store sold d DVDs, how many tapes and CDs did it sell?
b. French fries have about 12 calories apiece. A hamburger has about 600 calories. Akiva is on a diet of 2000 calories per day. If he ate f french fries and one hamburger, how many more calories can he consume that day?

Solution **a.** Because d DVDs were sold, twice as many CDs as DVDs implies $2d$ CDs. Thus, 25 fewer tapes than CDs implies $2d - 25$ tapes.

b. First, find how many calories Akiva consumed eating f french fries and one hamburger. Then, to find how many more calories he can consume, subtract this expression from 2000.

1 french fry	12 calories
f french fries	$12f$ calories

Therefore, the number of calories in f french fries and one hamburger is

$$600 + 12f.$$

The number of calories left for the day is $2000 - (600 + 12f)$, or $2000 - 600 - 12f$, or $1400 - 12f$. ▲

▶**HISTORICAL NOTE**

The word *algebra* comes from the Arabic book *Al-jabr wa'l muqabalah,* written by Mohammed al-Khowârizmî (ca. 825 CE). Algebra was introduced in Europe in the thirteenth and fourteenth centuries by Leonardo di Pisa (also called Fibonacci). Algebra was occasionally referred to as *Ars Magna*, or "the great art." Both Diophantus (ca. 250 CE) and François Viète (1540–1603) have been called "fathers of algebra." Little is known about Diophantus, a Greek, except that he is supposed to have lived to be 84 years old and that he wrote *Arithmetica*, a treatise originally in 13 books. Viète was a French lawyer who devoted his leisure time to mathematics. Not liking the word *algebra*, he referred to the subject as "the analytic art." ◀

Algebraic thinking is frequently needed to understand how simple number tricks work, as seen in Example 1-8.

EXAMPLE 1-8

A teacher instructed her class as follows:

Take any number and add 15 to it. Now multiply that sum by 4. Next subtract 8 and divide the difference by 4. Now subtract 12 from the quotient and tell me the answer, I will tell you the original number.

Analyze the instructions to see how the teacher was able to determine the original number.

Solution Translate the information into an algebraic form.

Instructions	Discussion	Symbols
Take any number.	Since any number is used, we need a variable to represent the number. Let n be that variable.	n
Add 15 to it.	We are told to add 15 to "it." "It" refers to the variable n.	$n + 15$
Multiply that sum by 4.	We are told to multiply "that sum" by 4. "That sum" is $n + 15$.	$4(n + 15)$
Subtract 8.	We are told to subtract 8 from the product.	$4(n + 15) - 8$
Divide the difference by 4.	The difference is $4(n + 15) - 8$. Divide it by 4.	$\frac{4(n + 15) - 8}{4}$
Subtract 12 from the quotient and tell me the answer.	We are told to subtract 12 from the quotient.	$\frac{4(n + 15) - 8}{4} - 12$

Translating what the teacher told the class to do results in the algebraic expression $\frac{4(n + 15) - 8}{4} - 12$. We are also told that we have to tell the teacher the answer obtained and she then produces the original number. Let's use the strategy of *working backward* to see if we can determine what happens. Suppose we tell the teacher that our final result is r. Think about how r was obtained. Just before we told the teacher "r," we had subtracted 12. To reverse that operation, we could add 12 to obtain $r + 12$. Prior to that we had divided by 4. To reverse that, we could multiply by 4 to obtain $4r + 48$. To get that result, we had subtracted 8, so that now we add 8 to obtain $4r + 56$. Just previous to that we had multiplied by 4 so now we divide $4r + 56$ by 4 to obtain $r + 14$. The first operation had been to add 15 so now we subtract 15 from $r + 14$ to get $r - 1$. Thus, the teacher knows when we tell her that our final result is r, it is 1 more than the number with which we started, or the number with which we started, n, is the result minus 1.

This can be shown as follows:

$$\begin{aligned}\frac{4(n + 15) - 8}{4} - 12 &= \frac{4n + 60 - 8 - 4 \cdot 12}{4} \\ &= \frac{4n + 4}{4} \\ &= \frac{4(n + 1)}{4} \\ &= n + 1.\end{aligned}$$

EXAMPLE 1-9

Figure 1-20 shows a sequence of figures containing small square tiles. Some of the tiles are shaded. Notice that the first figure has one shaded tile. The second figure has $2 \cdot 2$ or 2^2 shaded tiles. The third figure has $3 \cdot 3$ or 3^2 shaded tiles. Answer the following.

a. How many shaded tiles are there in the nth figure?
b. How many white tiles are there in the nth figure?

FIGURE 1-20

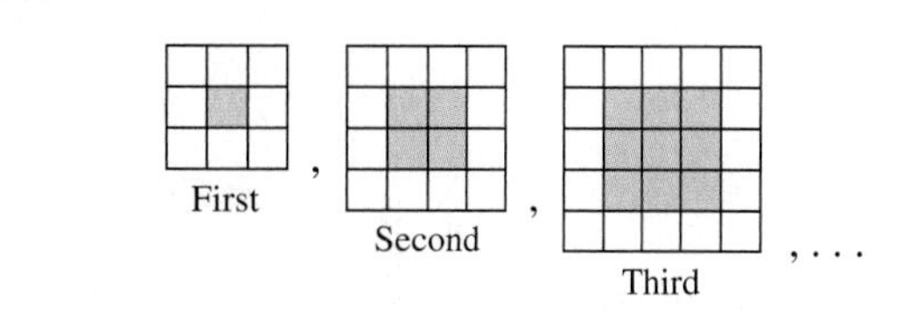

Solution **a.** The shaded tiles have sides with increasing lengths 1, 2, 3, and so on. In the nth figure, the length of a side of the shaded tiles would be n. Hence the nth figure has n^2 shaded tiles.

b. One way to think about the number of white tiles is to recognize that the number of white tiles on a side is 2 more than n, or $n + 2$. The number of white tiles could be 4 times $(n + 2)$, less any overlapping counting. In this case, each corner tile would be duplicated so that there are 4 white tiles overcounted giving us $4(n + 2) - 4$, or $4n + 4$ white tiles.

NOW TRY THIS 1-17

There are many other ways to count the tiles in Example 1-9. Describe two other ways of determining the number of white tiles in the nth term.

Algebraic thinking can occur in different ways. One example that uses pictures is seen in Example 1-10.

EXAMPLE 1-10

At a local farmer's market, three purchases were made for the prices shown in Figure 1-21. What is the cost of each object?

FIGURE 1-21

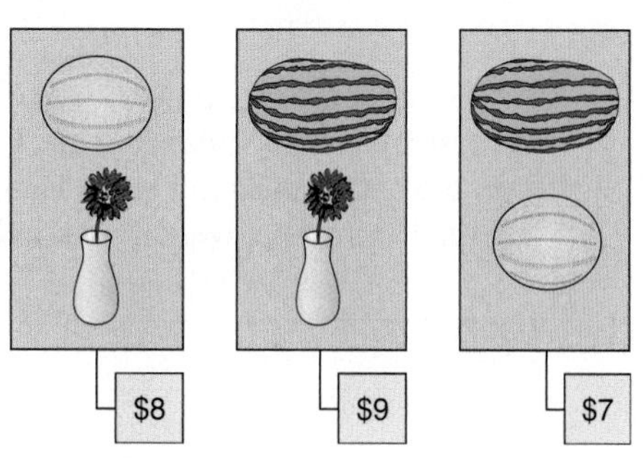

Solution Approaches to this problem may vary. For example, if the objects in the first two purchases are put together, the total cost would be \$8 + \$9, or \$17. That cost would be for two vases and one each of the cantaloupe and watermelon, as in Figure 1-22.

FIGURE 1-22

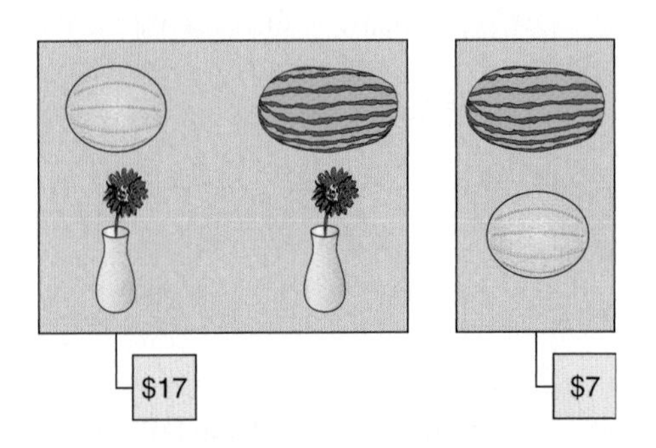

Now if the cantaloupe and watermelon are taken away from that total, then according to the cost of those two objects from the tag on the right, the cost should be reduced to \$10 for two vases. That means each of the two vases costs \$5. This in turn tells us that the cantaloupe costs \$8 − \$5, or \$3, and the watermelon costs \$9 − \$5, or \$4.

The solution in Example 1-10 could involve the strategy of *writing an equation*. But first we need a basic knowledge of solving equations.

Properties of Equations

Variables are usually associated with *equations*. When variables are thought of as unknowns, we can consider expressions such as $w + c = 7$. The *equal* sign indicates that the values on both sides of the equation are the same even though they do not look the same. Students sometimes mistakenly tend to think of an equal sign as an operator button on a calculator. As they read the problem they think of the "=" sign as meaning "give me the answer." (See Research Note below.)

RESEARCH NOTE

Van de Walle points out that students tend to see an equation such as $3x + 7 = 5 + 9$ as having two separate sides with things to do rather than two names for the same thing. The equal sign is often viewed as a symbol used to separate a problem from its answer. He points out that students fail to see $5 + 2$ as another name for 7 (Van de Walle 2001).

To solve equations, we need several properties of equality. Children discover many of these by using a balance scale. For example, consider two weights of amounts a and b on the balances, as in Figure 1-23(a). If the balance is level, then $a = b$. When we add an equal amount of weight c to both sides, the balance is still level, as in Figure 1-23(b).

FIGURE 1-23

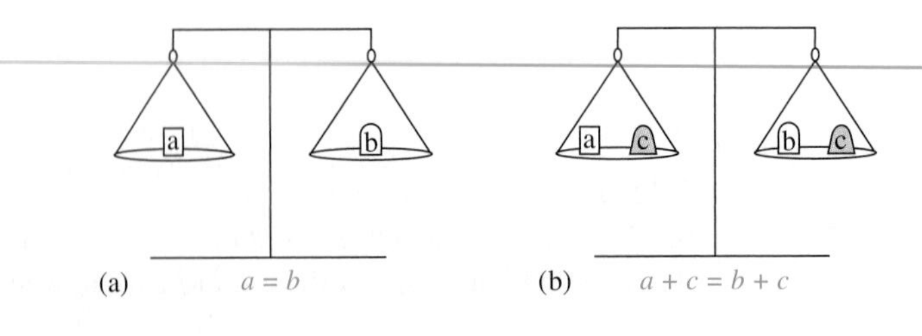

This demonstrates that *if* $a = b$*, then* $a + c = b + c$*, which is the Addition Property of Equality*.

Similarly, if the scale is balanced with amounts a and b, as in Figure 1-24(a), and we put additional a's on one side and an equal number of b's on the other side, the scale remains level, as in Figure 1-24(b).

FIGURE 1-24

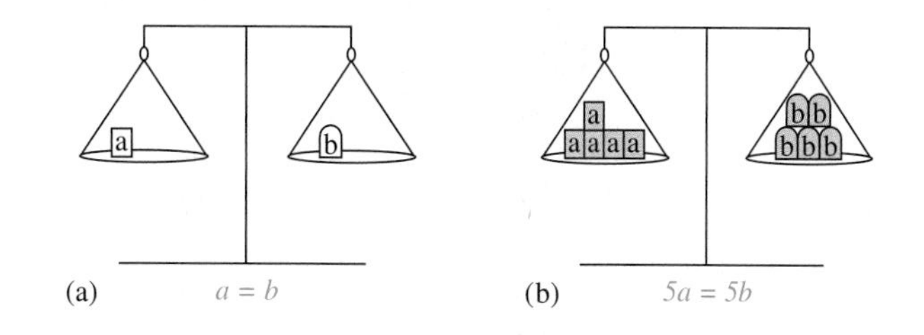

Figure 1-24 suggests that *if* c *is any real number and* $a = b$*, then* $ac = bc$, which is the *Multiplication Property of Equality*. These properties are summarized next.

Properties

The Addition Property of Equality For any numbers a, b, and c, if $a = b$, then $a + c = b + c$.
The Multiplication Property of Equality For any numbers a, b, and c, if $a = b$, then $ac = bc$.

The properties imply that we may add the same number to both sides of an equation or multiply both sides of the equation by the same number without affecting the equality. A new statement results from reversing the order of the *if* and *then* parts of the Addition Property of Equality. The new statement is the *converse* of the original statement. In the case of the addition property, the converse is a true statement. The converse of the Multiplication Property of Equality is also true when $c \neq 0$. These properties are summarized next.

Cancellation Properties of Equality

1. For any numbers a, b, and c, if $a + c = b + c$, then $a = b$.
2. For any numbers a, b, and c, with $c \neq 0$, if $ac = bc$, then $a = b$.

Substitution Property

Equality is not affected if we substitute a number for its equal. This property is referred to as the **Substitution Property**. Examples of substitution follow:

1. If $a + b = c + d$ and $d = 5$, then $a + b = c + 5$.
2. If $a + b = c + d$, if $b = e$, and if $d = f$, then $a + e = c + f$.

Solving Equations

Part of algebraic thinking involves operations on numbers and other elements represented by symbols. Finding solutions to equations is one part of algebra.

The balance-scale model is an excellent way to help understand the basic concepts used to solve equations and inequalities. We first explore equations (balance on the scale) and inequalities (tilt the scale). For example, consider Figure 1-25. If we release the pan on the left, what will happen? Upon release, the scale will tilt and we have an *inequality*, $2 \cdot 3 < 3 + (2 \cdot 2)$.

FIGURE 1-25

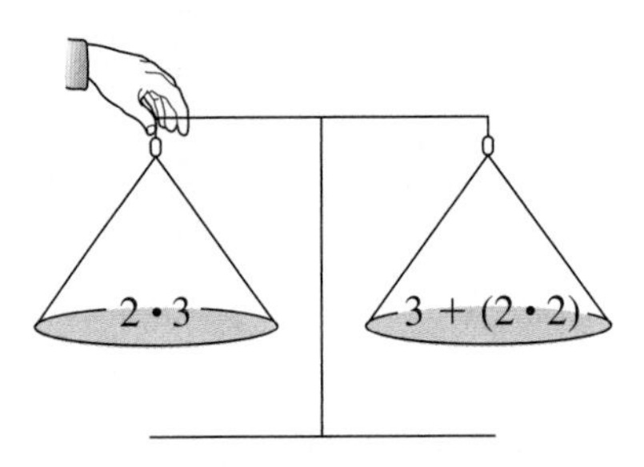

Next consider Figure 1-26. If we release the pan, then the sides will balance and we have the *equality* $2 \cdot 3 = (1 + 1) + 4$.

FIGURE 1-26

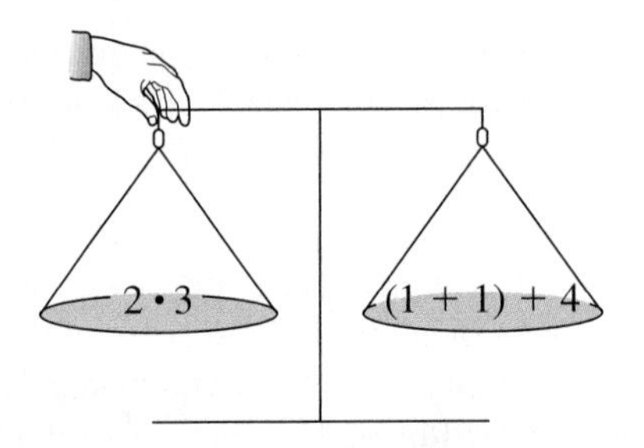

A balance scale can also be used to reinforce the idea of a replacement set for a variable. Name some solutions in Figure 1-27 that will keep the scale balanced. For example, $3 \cdot 2$ balances $2 \cdot 3$, $3 \cdot 6$ balances $2 \cdot 9$, and so on. Do you see any patterns in the numbers that will balance the scale?

FIGURE 1-27

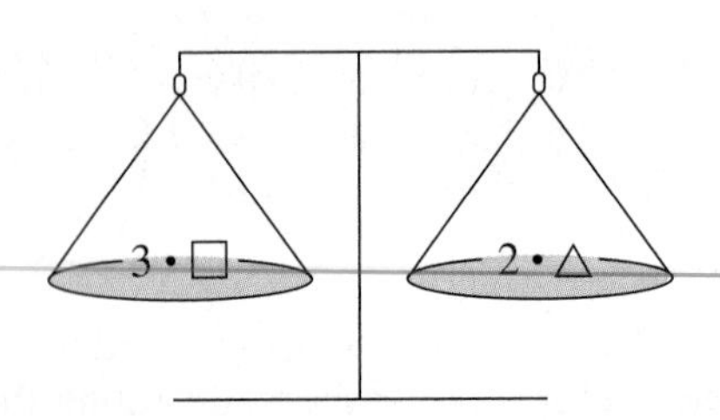

Other types of balance-scale problems can also help get ready for algebra. Work through Now Try This 1-18 before proceding.

NOW TRY THIS 1-18

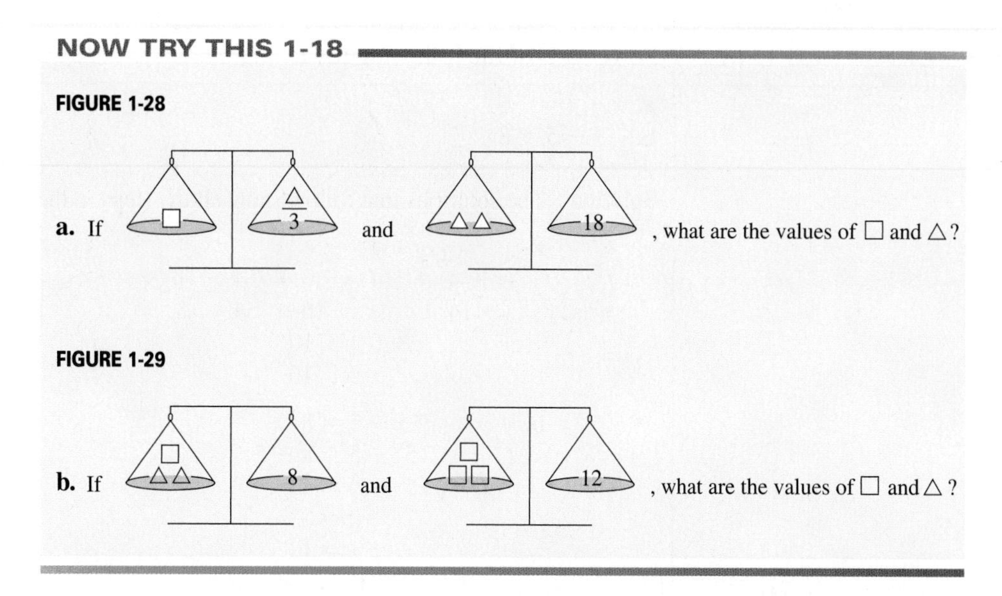

FIGURE 1-28

a. If [scale: □ balances △ and 3] and [scale: △△ balances 18], what are the values of □ and △?

FIGURE 1-29

b. If [scale: □△△ balances 8] and [scale: □□□ balances 12], what are the values of □ and △?

To solve equations, we may use the properties of equality developed earlier. Consider $3x - 14 = 1$. Put the equal expressions on the opposite pans of the balance scale. Because the expressions are equal, the pans should be level, as in Figure 1-30.

FIGURE 1-30

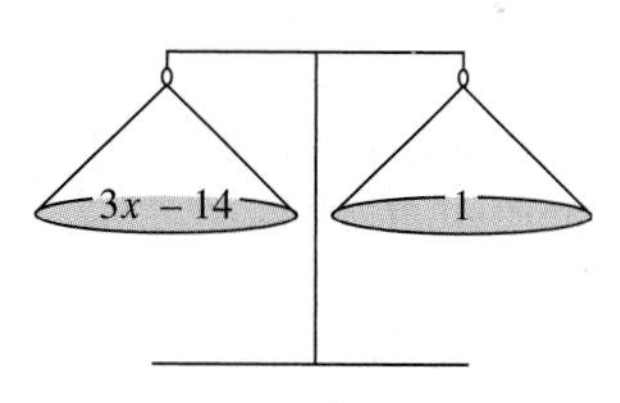

To solve for x, we use the properties of equality to manipulate the expressions on the scale so that after each step, the scale remains level and, at the final step, only an x remains on one side of the scale. The number on the other side of the scale represents the solution to the original equation. To find x in the equation of Figure 1-30, consider the scales pictured in successive steps in Figure 1-31. In Figure 1-31, each successive scale represents an equation that is equivalent to the original equation; that is, each has the same solution as the original. The last scale shows $x = 5$. To check that 5 is the correct solution, we substitute 5 for x in the original equation. Because $3 \cdot 5 - 14 = 1$ is a true statement, 5 is the solution to the original equation.

FIGURE 1-31

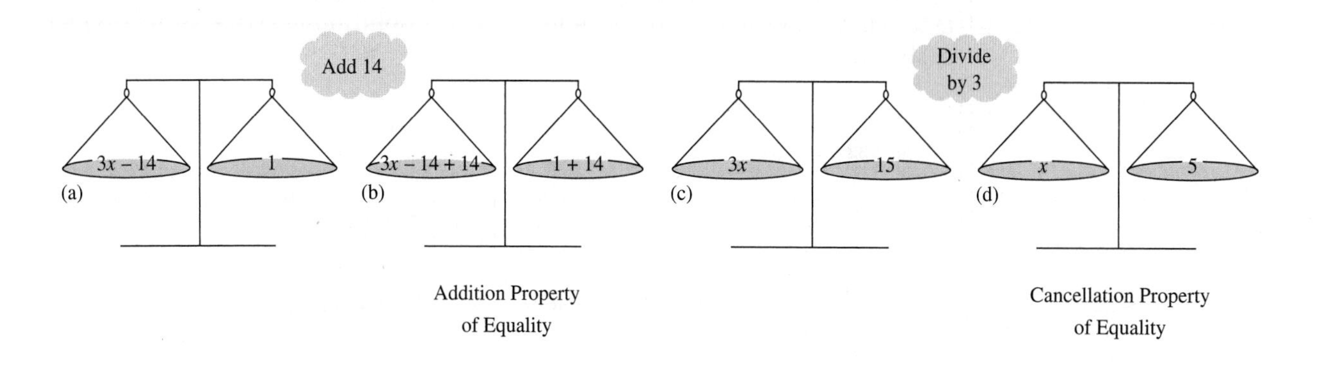

EXAMPLE 1-11

Solve each of the following for x:

a. $x + 4 = {}^{-}6$
b. ${}^{-}x - 5 = 8$

Solution The solutions that follow show all the steps in the process:

a.

$$\begin{aligned} x + 4 &= {}^{-}6 \\ (x + 4) + {}^{-}4 &= {}^{-}6 + {}^{-}4 \\ x + (4 + {}^{-}4) &= {}^{-}6 + {}^{-}4 \\ x + 0 &= {}^{-}10 \\ x &= {}^{-}10 \end{aligned}$$

b.

$$\begin{aligned} {}^{-}x - 5 &= 8 \\ ({}^{-}x + {}^{-}5) + 5 &= 8 + 5 \\ {}^{-}x + ({}^{-}5 + 5) &= 13 \\ {}^{-}x + 0 &= 13 \\ {}^{-}x &= 13 \\ ({}^{-}x)({}^{-}1) &= 13({}^{-}1) \\ x &= {}^{-}13 \end{aligned}$$

▲

▶ **HISTORICAL NOTE**

Mary Fairfax Somerville (1780–1872) was born in Scotland of upper-class parents. Her introduction to algebra came at about age 13 while reading a ladies' fashion magazine that contained some puzzles. Although not allowed to study mathematics formally, at age 27, widowed and with two children, she bought and studied a set of mathematics books. In her autobiography, she wrote, "I was sometimes annoyed when in the midst of a difficult problem someone would enter and say, 'I have come to spend a few hours with you.'" Shortly before her death, she wrote, "I am now in my ninety-second year, . . . , I am extremely deaf, and my memory of ordinary events, and especially of the names of people, is failing, but not for mathematical and scientific subjects. I am still able to read books on the higher algebra for four or five hours in the morning and even to solve the problems. Sometimes I find them difficult, but my old obstinacy remains, for if I do not succeed today, I attack them again tomorrow." ◀

Application Problems

The simple model in Figure 1-32 demonstrates a method for solving application problems. Formulate the problem as a mathematical problem, solve the mathematical problem, and then interpret the solution in terms of the original problem.

FIGURE 1-32

Application Problem	→	Mathematical Model
		↓
Original Problem Interpretation	←	Mathematical Solution

An example of this model at the third-grade level appears in Figure 1-33.

FIGURE 1-33

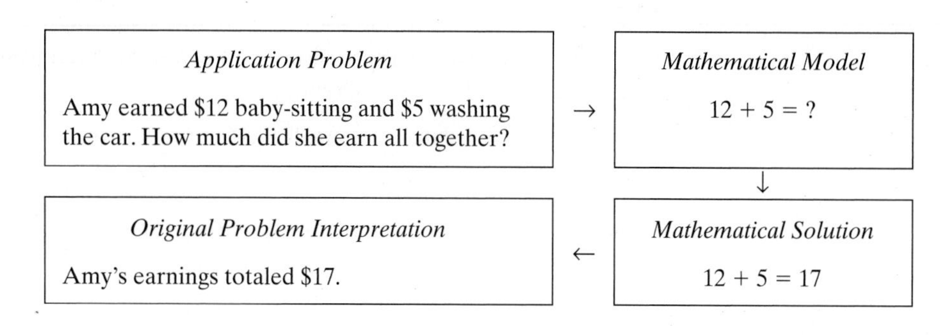

We can apply Polya's four-step problem-solving process to solving word problems in which the use of algebraic thinking is appropriate. In Understanding the Problem, we identify what is given and what is to be found. In Devising a Plan, we assign letters to the unknown quantities and translate the information in the problem into a model involving equations. In Carrying Out the Plan, we solve the equations or inequalities. In Looking Back, we interpret and check the solution in terms of the original problem.

In the following problems, we demonstrate Polya's four-step problem-solving process.

PROBLEM SOLVING **Overdue Books**

Bruno has five books overdue at the library. The fine for overdue books is 10¢ a day per book. He remembers that he checked out an astronomy book a week before he checked out four novels. If his total fine was $8.70, how long was each book overdue?

Understanding the Problem Bruno has five books overdue. He checked out an astronomy book seven days earlier than the four novels, so the astronomy book is overdue seven days more than the novels. The fine per day for each book is 10¢, and the total fine was $8.70. We need to find out how many days each book is overdue.

Devising a Plan Let d be the number of days that each of the four novels is overdue. The astronomy book is overdue seven days longer; that is, $d + 7$ days. To *write an equation* for d, we express the total fine in two ways. The total fine is $8.70. This fine in cents equals the fine for the astronomy book plus the fine for the four novels.

$$\text{Fine for each of the novels} = \underbrace{\text{fine per day}}_{10}\ \underbrace{\text{times}}_{\cdot}\ \underbrace{\text{the number of overdue days.}}_{d}$$

$$\text{Fine for the four novels} = \underbrace{\text{1 day's fine for four novels}}_{4 \cdot 10}\ \underbrace{\text{times}}_{\cdot}\ \underbrace{\text{number of overdue days.}}_{d}$$

$$= (4 \cdot 10)d$$

$$= 40d$$

$$\text{Fine for the astronomy book} = \underbrace{\text{Fine per day}}_{10}\ \underbrace{\text{times}}_{\cdot}\ \underbrace{\text{the number of overdue days.}}_{(d+7)}$$
$$= 10 \cdot (d + 7)$$

Because each of the expressions is in cents, we need to write the total fine of \$8.70 as 870¢ to produce the following:

$$\begin{array}{ccccc} \text{Fine for the four novels} & + & \text{Fine for the astronomy book} & = & \text{Total fine.} \\ 40\,d & + & 10(d+7) & = & 870 \end{array}$$

Carrying Out the Plan Solve the equation for d.

$$\begin{aligned} 40d + 10(d + 7) &= 870 \\ 40d + 10d + 70 &= 870 \\ 50d + 70 &= 870 \\ 50d &= 870 - 70 \\ 50d &= 800 \\ d &= 16 \end{aligned}$$

Thus, each of the four novels was 16 days overdue, and the astronomy book was overdue $d + 7$, or 23, days.

Looking Back To check the answer, follow the original information. Each of the four novels was 16 days overdue, and the astronomy book was 23 days overdue. Because the fine was 10¢ per day per book, the fine for each of the novels was $16 \cdot 10$¢, or 160¢. Hence, the fine for all four novels was $4 \cdot 160$¢, or 640¢. The fine for the astronomy book was $23 \cdot 10$¢, or 230¢. Consequently, the total fine was 640¢ + 230¢, or 870¢, which agrees with the given information of \$8.70 as the total fine.

The problem can also be solved without algebra. One way is to notice that the astronomy book was overdue for 7 days for a fine of 70¢ before the other four books were overdue. Thus, 870¢ − 70¢, or 800¢, is the fine for the five books. Therefore, the fine for one book is 800¢/5, or 160¢. Because the fine is 10¢ per day, each book was overdue 160/10, or 16, days. The astronomy book was checked out a week earlier and hence was overdue for 23 days.

PROBLEM SOLVING **Newspaper Delivery**

In a small town, three children deliver all the newspapers. Abby delivers 3 times as many papers as Bob, and Connie delivers 13 more than Abby. If the three children delivered a total of 496 papers, how many papers does each deliver?

Understanding the Problem The problem asks for the number of papers that each child delivers. It gives information that compares the number of papers that each child delivers as well as the total number of papers delivered in the town.

Devising a Plan Let a, b, and c be the number of papers delivered by Abby, Bob, and Connie, respectively. We translate the given information into *equations* as follows:

Abby delivers 3 times as many papers as Bob: $a = 3b$.

Connie delivers 13 more papers than Abby: $c = a + 13$.

Total delivery is 496: $a + b + c = 496$.

To reduce the number of variables, substitute $3b$ for a in the second and third equations:

$$c = a + 13 \quad \text{becomes} \quad c = 3b + 13.$$
$$a + b + c = 496 \quad \text{becomes} \quad 3b + b + c = 496.$$

Next, make an equation in one variable, b, by substituting $3b + 13$ for c in the equation $3b + b + c = 496$, solve for b, and then find a and c.

Carrying Out the Plan

$$\begin{aligned} 3b + b + 3b + 13 &= 496 \\ 7b + 13 &= 496 \\ 7b &= 483 \\ b &= 69 \end{aligned}$$

Thus, $a = 3b = 3 \cdot 69 = 207$. Also, $c = a + 13 = 207 + 13 = 220$. So, Abby delivers 207 papers, Bob delivers 69 papers, and Connie delivers 220 papers.

Looking Back To check the answers, follow the original information, using $a = 207$, $b = 69$, and $c = 220$. The information in the first sentence, "Abby delivers 3 times as many papers as Bob" checks, since $207 = 3 \cdot 69$. The second sentence, "Connie delivers 13 more papers than Abby" is true because $220 = 207 + 13$. The information on the total delivery checks, since $207 + 69 + 220 = 496$.

ONGOING ASSESSMENT 1-3

1. In the following, write an expression in terms of the given variable that represents the indicated quantity. For example, the distance d traveled at a constant speed of 60 mph during t hr could be written as $d = 60t$.
   a. The cost of having a plumber spend h hr at your house if the plumber charges \$20 for coming to the house and \$25 per hour for labor
   b. The amount of money in cents in a jar containing d dimes and some nickels and quarters, if there are 3 times as many nickels as dimes and twice as many quarters as nickels
   c. The sum of three consecutive integers if the least integer is x
   d. The amount of bacteria after n min if the initial amount of bacteria is q and the amount of bacteria doubles every minute. (*Hint:* The answer should contain q as well as n.)
   e. The temperature after t hr if the initial temperature is 40°F and each hour it drops by 3°F
   f. Pawel's total salary after 3 yr if the first year his salary was s dollars, the second year it was \$5000 higher, and the third year it was twice as much as the second year
   g. The sum of three consecutive odd natural numbers if the least is x
   h. The sum of three consecutive natural numbers if the middle is m
2. Pluto, once thought to be the planet farthest from the Sun in our solar system, was also the smallest. The following table gives the weight, P, on Pluto for a given weight, E, on Earth measured in the same units.

Weight on Earth (*E*)	1	2	3	4	5	10	100
Weight on Pluto (*P*)	0.04	0.08	0.12	0.16	0.2	0.4	4

Based on the information given in the table, answer each of the following:
   a. Find an equation for P in terms of E.
   b. Express E in terms of P.
   c. What is Debbie's weight on Pluto if her weight on Earth is 135 lb?
   d. Find the weight on Earth of an object that weighs 100 lb on Pluto.

3. a. If

and

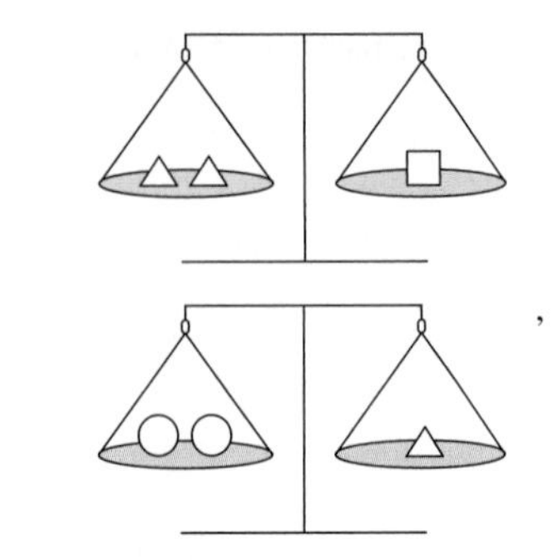

,

i. which shape weighs the most? Tell why.
ii. which shape weighs the least? Tell why.

b. If

and

and

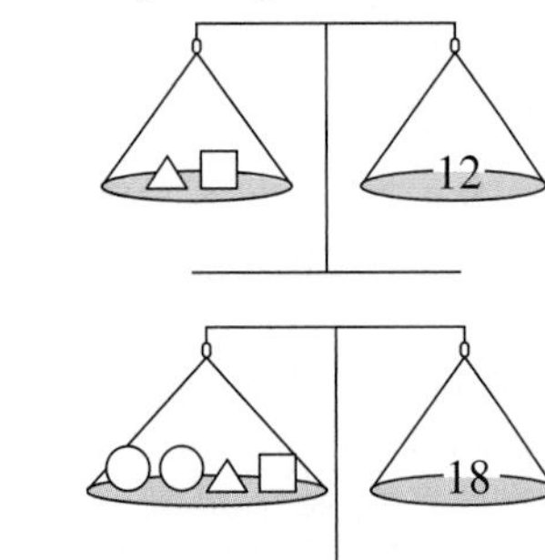

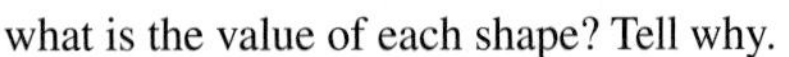

,

what is the value of each shape? Tell why.

4. Ryan is building matchstick square sequences, as shown. He used 67 matchsticks to form the last figure in his sequence. How many matchsticks will he use for the entire project?

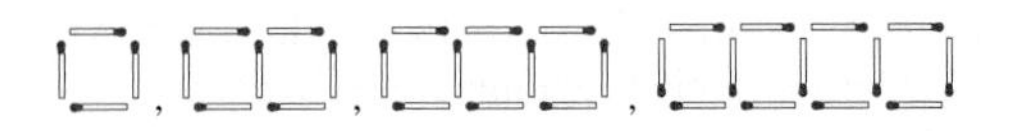

5. The formula for converting degrees Celsius (C) to degrees Fahrenheit (F) is $F = \left(\frac{9}{5}\right)C + 32$. Samantha reads that the temperature is 32°C in Spain. What is the Fahrenheit temperature?

6. To convert a temperature on the Celsius (C) scale to the Kelvin (K) scale, you need to add 273.15 to the Celsius temperature.
 a. Write a formula for K in terms of C.
 b. Write a formula for converting from the Kelvin scale to the Celsius scale.

7. Write an equation relating the variables described in each of the following situations:
 a. The pay, P, for t hr if you are paid \$8 an hour
 b. The pay, P, for t hr if you are paid \$15 for the first hour and \$10 for each additional hour
 c. The total pay, P, for a visit and t hr of gardening if you are paid \$20 for the visit and \$10 for each hour of gardening
 d. The total cost, C, of membership in a health club that charges a \$300 initiation fee and \$4 for each of n days attended
 e. The cost, C, of renting a midsized car for 1 day of driving m mi if the rent is \$30 per day plus 35¢ per mile.

8. Solve each of the following:
 a. $3 - x = -15$
 b. $-x - 3 = 15$
 c. $3x - 5 = 16$
 d. $2x + 5 = 3x - 4$

9. A teacher instructed her class as follows: *Take any number, multiply it by 3, add 49, and divide the result by 7. Subtract 7 from the quotient, divide the new result by 3, and tell me your answer. I will tell you the original number.* To determine each student's original number, the teacher multiplied each answer by 7. Explain how the teacher was able to tell each student's original number.

10. David has 3 times as much money as Rick. Together, they have \$400. How much does each have?

11. Some bowling leagues use handicaps to increase a team's chance of winning. Use the formula $H = 0.8\ (200 - A)$, where H is the handicap and A is the bowler's average, to answer the following questions:
 a. Betty has an average of 120. What is her handicap?
 b. Al has a 30 handicap. What is his average?

12. Draw pictures as in Example 1-10 to find a solution to the following:
 - Two silk butterflies and a silk rose cost \$18.
 - One silk butterfly and a silk rose cost \$11.

 What is the cost of each?

13. For a particular event, 812 tickets were sold for a total of \$1912. If students paid \$2 per ticket and nonstudents paid \$3 per ticket, how many student tickets were sold?

14. A man left an estate of \$64,000 to three children. The eldest child received 3 times as much as the youngest. The middle child received \$14,000 more than the youngest. How much did each child receive?

Communication

15. Students were asked to write an algebraic expression for the sum of three consecutive natural numbers. One student wrote $x + (x + 1) + (x + 2) = 3x + 3$. Another wrote $(x - 1) + x + (x + 1) = 3x$. Explain who is correct and why.

16. Explain how to solve the equation $3x + 5 = 5x - 3$ using a balance scale.

Open-Ended

17. In Example 1-8, a teacher instructed her class to take any number and perform a series of computations using that number. The teacher was able to tell each student's original number by subtracting 1 from the student's answer. Create similar instructions for students so that the

teacher needs to do only the following to obtain the student's original number:
a. Add 1 to the answer.
b. Multiply the answer by 2.
c. Multiply the answer by itself.

18. Think of a mathematical relationship that can be modeled using an equation. Identify the variables and write an equation that models the relationship.

Cooperative Learning

19. Examine several elementary school textbooks for Grades 1 through 4 and report on which algebraic concepts are introduced in each and how they are introduced.

Review Problems

20. List the terms to continue a possible pattern in the following sequences:
a. 7, 14, 21, 28, . . .
b. 4, 1, 8, 1, 12, . . .

21. Find the nth term for the following arithmetic sequence: 12, 32, 52, 72, . . .

22. Find how many terms are in the following arithmetic sequence:
6, 10, 14, 18, . . . , 86

23. In how many ways can you make change for $0.21?

Third International Mathematics and Science Study (TIMSS) Questions

Ali had 50 apples. He sold some and then had 20 left. Which of these is a number sentence that shows this?

a. $\square - 20 = 50$
b. $20 - \square = 50$
c. $\square - 50 = 20$
d. $50 - \square = 20$

TIMSS 2003, Grade 4

The objects on the scale make it balance exactly. On the left pan there is a 1 kg weight (mass) and half a brick. On the right pan there is one brick.

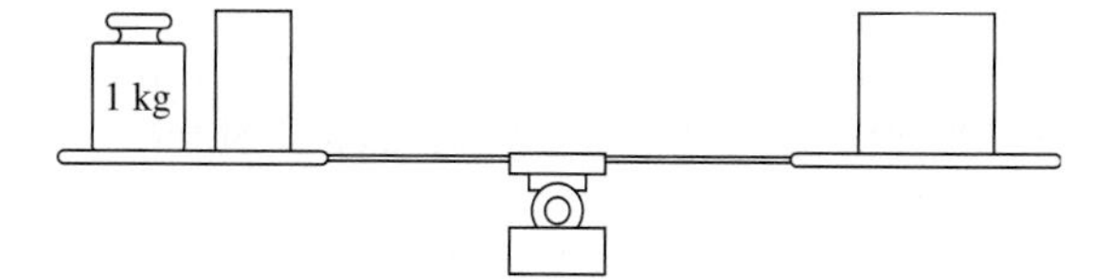

What is the weight (mass) of one brick?
a. 0.5 kg
b. 1 kg
c. 2 kg
d. 3 kg

TIMSS 2003, Grade 8

BRAIN TEASER

Find the next row in the following pattern and explain your pattern:

1
1 1
2 1
1 2 1 1
1 1 1 2 2 1

1-4 *Logic: An Introduction

statement

Logic is a tool used in mathematical thinking and problem solving. It is essential for reasoning. In logic, a **statement** *is a sentence that is either true or false, but not both.* The following expressions are not statements because their truth values cannot be determined without more information.

1. She has blue eyes.
2. $x + 7 = 18$.
3. $2y + 7 > 1$.
4. $2 + 3$
5. How are you?
6. Look out!
7. Carter was the best president.

Expressions (**1**), (**2**), and (**3**) become statements if, for (**1**), "she" is identified, and for (**2**) and (**3**), values are assigned to x and y, respectively. However, an expression involving he or she or x or y may already be a statement. For example, "If he is over 210 cm tall, then he is over 2 m tall" and "$2(x + y) = 2x + 2y$" are both statements because they are true no matter who *he* is or what the numerical values of x and y are.

Negation and Quantifiers

negation

From a given statement, it is possible to create a new statement by forming a **negation**. The negation of a statement is *a statement with the opposite truth value of the given statement.* If a statement is true, its negation is false, and if a statement is false, its negation is true. Consider the statement "It is snowing now." The negation of this statement may be stated simply as "It is not snowing now."

EXAMPLE 1-12

Negate each of the following statements:

a. $2 + 3 = 5$
b. A hexagon has six sides.

Solution **a.** $2 + 3 \neq 5$
b. A hexagon does not have six sides.

Sentences like "The shirt is blue" and "The shirt is green" are statements if put in context. However, they are not negations of each other. A statement and its negation must have opposite truth values. If the shirt is actually red, then both of the statements are false and, hence, cannot be negations of each other. However, the statements "The shirt is blue" and "The shirt is not blue" are negations of each other because they have opposite truth values no matter what color the shirt really is.

quantifiers

Some statements involve **quantifiers** and are more complicated to negate. Quantifiers include words such as *all, some, every*, and *there exists*.

- The quantifiers *all, every*, and *no* refer to each and every element in a set and are called *universal quantifiers*.
- The quantifiers *some* and *there exists at least one* refer to one or more, or possibly all, of the elements in a set and are called *existential quantifiers*.
- *All, every*, and *each* have the same mathematical meaning. Similarly, *some* and *there exists at least one* all have the same meaning.

Consider the following statement involving the existential quantifier *some* and known to be true: "Some professors at Paxson University have blue eyes." This means that at least one professor at Paxson University has blue eyes. It does not rule out the possibilities that all the Paxson professors have blue eyes or that some of the Paxson professors do not have blue eyes. Because the negation of a true statement is false, neither "Some professors at Paxson University do not have blue eyes" nor "All professors at Paxson have blue eyes" are negations of the original statement. One possible negation of the original statement is "No professors at Paxson University have blue eyes."

To discover if one statement is a negation of another, we use arguments similar to the preceding one to determine if they have opposite truth values in all possible cases.

General forms of quantified statements with their negations follow:

Statement	**Negation**
Some a are b.	No a is b.
Some a are not b.	All a are b.
All a are b.	Some a are not b.
No a is b.	Some a are b.

EXAMPLE 1-13

Negate each of the following regardless of its truth value:

a. All students like hamburgers.
b. Some people like mathematics.
c. There exists a natural number x such that $3x = 6$.
d. For all natural numbers, $3x = 3x$.

Solution **a.** Some students do not like hamburgers.
b. No people like mathematics.
c. For all natural numbers x, $3x \neq 6$.
d. There exists a natural number x such that $3x \neq 3x$.

Truth Tables

There is a symbolic system defined to help in the study of logic. If p represents a statement, the negation of the statement p is denoted by $\sim p$ and is read "not p." **Truth tables** are often used to show all possible true-false patterns for statements. Table 1-12 summarizes the truth tables for p and $\sim p$.

truth table

compound statement

From two given statements, it is possible to create a new, **compound statement** by using a connective such as *and*. For example, "It is snowing" and "The ski run is open" together with *and* give "It is snowing and the ski run is open." Other compound statements can be obtained by using the connective *or*. For example, "It is snowing or the ski run is open." The symbols $\wedge$ and $\vee$ are used to represent the connectives *and* and *or*, respectively. For example, if p represents "It is snowing" and q represents "The ski run is open," then "It is snowing and the ski run is open" is denoted by $p \wedge q$. Similarly, "It is snowing or the ski run is open" is denoted by $p \vee q$.

Table 1-12

Statement p	Negation $\sim p$
T	F
F	T

▶ HISTORICAL NOTE

George Boole (1815–1864), born in Lincoln, England, is called "the father of logic." At age 15, he began a teaching career, and while teaching arithmetic he studied advanced mathematics and physics. In 1849, he was appointed professor at Queens College in Cork, Ireland. He continued to teach there until his death in 1864. In his work he employed symbols to represent concepts and developed a system of algebraic manipulations to accompany the symbols. His work was a marriage of logic and mathematics. Many of Boole's ideas, such as Boolean algebra, have applications in computer science and in the design of telephone switching devices. ◀

The truth value of any compound statement, such as $p \wedge q$, is defined using the truth value of each of the simple statements. Because each of the statements p and q may be either true or false, there are four distinct possibilities for the truth value of $p \wedge q$, as shown in Table 1-13. The compound statement $p \wedge q$ is the **conjunction** of p and q and is defined to be true if, and only if, both p and q are true. Otherwise, it is false.

conjunction

disjunction

The compound statement $p \vee q$—that is, *p or q*—is a **disjunction**. In everyday language, *or* is not always interpreted in the same way. In logic, we use an *inclusive or*. The statement "I will go to a movie or I will read a book" means I will either go to a movie, or read a book, or do both. Hence, in logic, p or q, symbolized $p \vee q$, is defined to be false if both p and q are false and true in all other cases. This is summarized in Table 1-14.

Table 1-13

p	q	Conjunction $p \wedge q$
T	T	T
T	F	F
F	T	F
F	F	F

Table 1-14

p	q	Disjunction $p \vee q$
T	T	T
T	F	T
F	T	T
F	F	F

EXAMPLE 1-14

Classify each of the following as true or false.

$$p: 2 + 3 = 5 \qquad q: 2 \cdot 3 = 6 \qquad r: 5 + 3 = 9$$

a. $p \wedge q$ **c.** $\sim p \vee r$ **e.** $\sim(p \wedge q)$
b. $q \vee r$ **d.** $\sim p \wedge \sim q$ **f.** $(p \wedge q) \vee \sim r$

Solution **a.** p is true and q is true, so $p \wedge q$ is true.
b. q is true and r is false, so $q \vee r$ is true.
c. $\sim p$ is false and r is false, so $\sim p \vee r$ is false.
d. $\sim p$ is false and $\sim q$ is false, so $\sim p \wedge \sim q$ is false.
e. $p \wedge q$ is true so $\sim(p \wedge q)$ is false.
f. $p \wedge q$ is true and $\sim r$ is true, so $(p \wedge q) \vee \sim r$ is true.

Truth tables are used not only to summarize the truth values of compound statements; they also are used to determine if two statements are logically equivalent. Two statements are **logically equivalent** if, and only if, they have the same truth values. If p and q are logically equivalent, we write $p \equiv q$.

logically equivalent

EXAMPLE 1-15

Show that $\sim(p \wedge q) \equiv \sim p \vee \sim q$.

Solution Two statements are logically equivalent if they have the same truth values. Truth tables for these statements are given in Table 1-15 and Table 1-16.

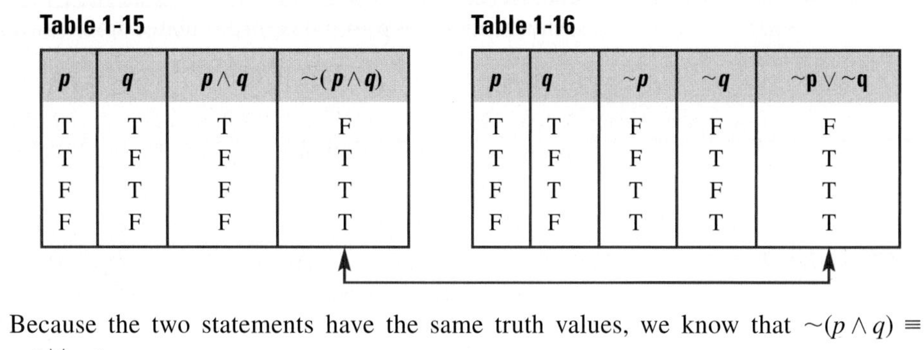

Table 1-15

p	q	$p \wedge q$	$\sim(p \wedge q)$
T	T	T	F
T	F	F	T
F	T	F	T
F	F	F	T

Table 1-16

p	q	$\sim p$	$\sim q$	$\sim p \vee \sim q$
T	T	F	F	F
T	F	F	T	T
F	T	T	F	T
F	F	T	T	T

Because the two statements have the same truth values, we know that $\sim(p \wedge q) \equiv \sim p \vee \sim q$.

Conditionals and Biconditionals

conditionals
implications
hypothesis
conclusion

Statements expressed in the form "if p, then q" are **conditionals**, or **implications**, and are denoted by $p \rightarrow q$. Such statements also can be read "p implies q." The "if" part of a conditional is the **hypothesis** of the implication and the "then" part is the **conclusion**. Many types of statements can be put in "if-then" form. An example follows:

Statement: All equilateral triangles have acute angles.

If-then form: If a triangle is equilateral, then it has acute angles.

An implication may also be thought of as a promise. Suppose Betty makes the promise "If I get a raise, then I will take you to dinner." If Betty keeps her promise, the implication is true; if Betty breaks her promise, the implication is false. Consider the following four possibilities:

	p	q	
(1)	T	T	Betty gets the raise; she takes you to dinner.
(2)	T	F	Betty gets the raise; she does not take you to dinner.
(3)	F	T	Betty does not get the raise; she takes you to dinner.
(4)	F	F	Betty does not get the raise; she does not take you to dinner.

Table 1-17

p	q	Implication $p \rightarrow q$
T	T	T
T	F	F
F	T	T
F	F	T

The only case in which Betty breaks her promise is when she gets her raise and fails to take you to dinner, case (2). If she does not get the raise, she can either take you to dinner or not without breaking her promise. The definition of implication is summarized in Table 1-17. Observe that the only case for which the implication is false is when p is true and q is false.

An implication can be worded in several equivalent ways, as follows:

1. If the sun shines, then the swimming pool is open. (If p, then q.)
2. If the sun shines, the swimming pool is open. (If p, q.)
3. The swimming pool is open if the sun shines. (q if p.)
4. The sun is shining implies the swimming pool is open. (p implies q.)
5. The sun is shining only if the pool is open. (p only if q.)
6. The sun's shining is a sufficient condition for the swimming pool to be open. (p is a sufficient condition for q.)
7. The swimming pool's being open is a necessary condition for the sun to be shining. (q is a necessary condition for p.)

Any implication $p \rightarrow q$ has three related implication statements, as follows:

Statement:	If p, then q.	$p \rightarrow q$
Converse:	If q, then p.	$q \rightarrow p$
Inverse:	If not p, then not q.	$\sim p \rightarrow \sim q$
Contrapositive:	If not q, then not p.	$\sim q \rightarrow \sim p$

EXAMPLE 1-16

Write the converse, the inverse, and the contrapositive for the following statement:
If I am in San Francisco, then I am in California.

Solution *Converse:* If I am in California, then I am in San Francisco.
Inverse: If I am not in San Francisco, then I am not in California.
Contrapositive: If I am not in California, then I am not in San Francisco. ▲

Example 1-16 can be used to show that if an implication is true, its converse and inverse are not necessarily true. However, the contrapositive is true. Let's check these observations on the following: *If a number is a natural number, the number is not 0.* We check the truth of the converse, inverse, and contrapositive.

Inverse: *If a number is not a natural number, then it is 0.* This is false, since $^-6$ is not a natural number but it also is not 0.

Converse: *If a number is not 0, then it is a natural number.* This is false, since $^-6$ is not 0 but neither is it a natural number.

Contrapositive: *If a number is 0, then it is not a natural number.* This is true.

The contrapositive of the last statement is the original statement. Hence, the preceding discussion suggests that if $p \rightarrow q$ is true, its contrapositive $\sim q \rightarrow \sim p$ is also true, and if the contrapositive is true, the original statement must be true. It follows that a statement and its contrapositive cannot have opposite truth values. We summarize this in the following property.

Property

Equivalence of a statement and its contrapositive The implication $p \rightarrow q$ and its contrapositive $\sim q \rightarrow \sim p$ are logically equivalent.

EXAMPLE 1-17

Use truth tables to show that $p \rightarrow q \equiv \sim q \rightarrow \sim p$.

Solution Truth tables for these statements are given in Table 1-18 and Table 1-19.

Table 1-18

p	q	$p \rightarrow q$
T	T	T
T	F	F
F	T	T
F	F	T

Table 1-19

p	q	$\sim q$	$\sim p$	$\sim q \rightarrow \sim p$
T	T	F	F	T
T	F	T	F	F
F	T	F	T	T
F	F	T	T	T

Because the two statements have the same truth values, we know that $p \to q \equiv {\sim}q \to {\sim}p$.

NOW TRY THIS 1-19

Use truth tables to determine if $p \to q \equiv {\sim}p \lor q$.

Connecting a statement and its converse with the connective *and* gives $(p \to q) \land (q \to p)$. This compound statement can be written as $p \leftrightarrow q$ and usually is read "***p* if and only if *q*.**" The statement "p if and only if q" is a **biconditional**. Build a truth table to determine when a biconditional statement is true.

if and only if ◆ biconditional

Valid Reasoning

valid reasoning

In problem solving, the reasoning is said to be **valid** if the conclusion follows unavoidably from true hypotheses. Note that if the hypotheses are false, then regardless of the truth value of the conclusion, the truth of the conditional statement forming the argument is true and the argument is valid. Thus, in all arguments in this section we assume the hypotheses are true. Consider the following example:

Hypotheses: All roses are red.
This flower is a rose.

Conclusion: Therefore, this flower is red.

The statement "All roses are red" can be written as the implication "If a flower is a rose, then it is red" and pictured with the **Euler diagram** in Figure 1-34(a).

Euler diagram

FIGURE 1-34

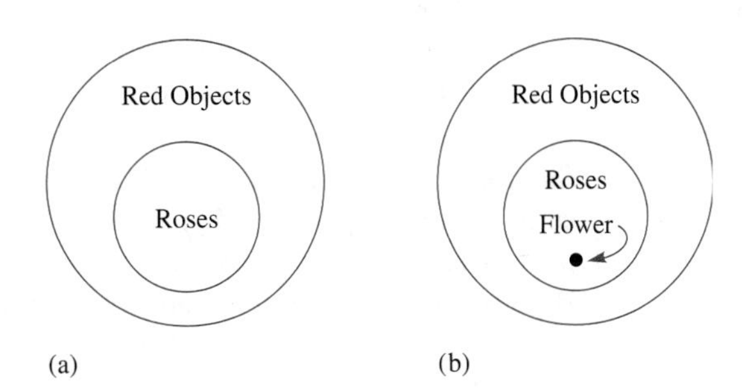

The information "This flower is a rose" implies that this flower must belong to the circle containing roses, as pictured in Figure 1-34(b). This flower also must belong to the circle containing red objects. Thus, the reasoning is valid because it is impossible to draw a picture that satisfies the hypotheses and contradicts the conclusion.

Consider the following argument with true hypotheses:

Hypotheses: All elementary school teachers are mathematically literate.
Some mathematically literate people are not children.

Conclusion: Therefore, no elementary school teacher is a child.

Let E be the set of elementary school teachers, M be the set of mathematically literate people, and C be the set of children. Then the statement "All elementary school

23. Consider the poem:

For want of a nail, the shoe was lost.
For want of a shoe, the horse was lost.
For want of a horse, the rider was lost.
For want of a rider, the battle was lost.
For want of a battle, the war was lost.
Therefore, for want of a nail, the war was lost.

a. Write each line as an *if . . . then . . .* statement.
b. Does the conclusion follow logically. Why?

24. Most students today use Internet search engines such as Yahoo or Google. Efficient use of a search engine requires some knowledge of the connectives AND, OR, and NOT. One common type of advanced search is called a *Boolean search* (see HISTORICAL NOTE). With a Boolean search you can increase the accuracy of your search by specifying relationships among the keywords and phrases. The operator AND tells the search engine to search for all documents that contain both words, for example, "sports AND baseball." Go online and explore the connectives AND, OR, and NOT. Explain your findings.

Open-Ended

25. Give two examples from mathematics for each of the following:
a. A statement and its converse are true.
b. A statement is true, but its converse is false.
c. An "if and only if" true statement.
d. An "if and only if" false statement.

Cooperative Learning

26. Each person in a group makes five statements similar to the ones in Examples 1-17 through 1-20 but concerning mathematical objects, each with a valid or invalid conclusion. The statements should be as varied as possible. Each group member exchanges statements with another person—not revealing which are valid and which are not—and determines which of the other person's statements are valid and which are not. The two group members compare their answers and discuss any discrepancies.

27. Discuss the paradox arising from the following:
a. This textbook is 1000 pages long.
b. The author of this textbook is Dante.
c. The statements (a), (b), and (c) are all false.

HINT FOR SOLVING THE PRELIMINARY PROBLEM

Predict your results for using two and three sheets of paper and see if you were correct. Build a table starting with one sheet of paper, then two sheets, three sheets, and so on, and see if you notice any patterns. Keep track of the number of sheets of paper, the number of pages in the booklet, the pairs of numbers on the same side of the sheet of paper, the sum of the two numbers on the same side of the paper, and the sum of all the page numbers. Gauss's Problem in Section 1-1 will help you find the sum of the page numbers.

Next try to generalize the problem if n represents the number of sheets of paper; p, the number of pages in the booklet; s, the sum of the two numbers on one side of the paper; and T, the sum of all the page numbers in the booklet.

QUESTIONS FROM THE CLASSROOM

1. A student claims she checked that $n^{50} > 2^n$ for $n = 1, 2, 3, \ldots, 50$. Hence, she claims that $n^{50} > 2^n$ should be true for all values of n. How do you respond?

2. A student says she read that Thomas Robert Malthus (1766–1834), a renowned British economist and demographer, claimed that the increase of population will take place, if unchecked, in a geometric sequence, whereas the supply of food will increase in only an arithmetic sequence. This theory implies that population increases faster than food production. The student is wondering why. How do you respond?

3. A student claims that the sequence 6, 6, 6, 6, 6, . . . never changes so it is neither arithmetic nor geometric. How do you respond?

4. A student claims that two terms are enough to determine any sequence. For example, 3, 6, . . . means the sequence would be 3, 6, 9, 12, 15, ... What is your response?

5. A student claims that algebra cannot be taught to students before Grade 8. How do you respond?

6. A student says that if the hypothesis is false, an argument cannot be valid. How do you respond?

7. A student asks why he just can't make "random guesses" rather than "intelligent guesses" when using the guess-and-check problem-solving strategy. How do you respond?

CHAPTER OUTLINE

I. Problem solving
 A. Problem solving can be guided by the following four-step process:
  1. Understanding the problem
  2. Devising a plan
  3. Carrying out the plan
  4. Looking back
 B. Important problem-solving strategies include the following:
  1. Look for a pattern.
  2. Make a table.
  3. Examine a simpler or special case of the problem to gain insight into the solution of the original problem.
  4. Identify a subgoal.
  5. Examine related problems and determine if the same technique can be applied.
  6. Work backward.
  7. Write an equation.
  8. Draw a diagram.
  9. Guess and check.
  10. Use indirect reasoning.
  11. Use direct reasoning.
 C. Beware of mind sets!

II. Mathematical patterns
 A. Patterns are an important part of problem solving.
 B. Patterns are used in **inductive reasoning** to form conjectures. Inductive reasoning is the method of making generalizations based on observations and patterns. A **conjecture** is a statement that is thought to be true but that has not yet been proved to be true or false.
 C. A **sequence** is a group of terms in a definite order.
  1. **Arithmetic sequence**: Each successive term is obtained from the previous one by the addition of a fixed number called the **difference**. The nth term, a_n, is given by $a_n = a_1 + (n - 1)d$, where a_1 is the first term and d is the difference.
  2. **Geometric sequence**: Each successive term is obtained from its predecessor by multiplying it by a fixed, nonzero number called the **ratio**. The nth term, a_n, is given by a_1r^{n-1}, where a_1 is the first term and r is the ratio.
  3. $a^n = \underbrace{a \cdot a \cdot a \cdot a \cdot a \cdot \ldots \cdot a}_{n \text{ factors}}$, where $n \neq 0$.
  4. $a^0 = 1$, where a is a natural number.
  5. Finding differences for a sequence is one technique for finding the next terms.

III. Properties of equality
 A. **Addition property:** For numbers a, b, and c, if $a = b$, then $a + c = b + c$.
 B. **Multiplication property:** For numbers a, b, and c, if $a = b$, then $ac = bc$.
 C. **Cancellation properties:** For numbers a, b, and c,
  1. if $a + c = b + c$, then $a = b$.
  2. if $c \neq 0$, and $ac = bc$, then $a = b$.
 D. Equality is not affected if we substitute a number for its equal.

*IV. Logic
 A. A **statement** is a sentence that is either true or false but not both.
 B. The **negation** of a statement is a statement with the opposite truth value of the given statement. The negation of p is denoted by $\sim p$.
 C. The **compound statement** $p \wedge q$ is the **conjunction** of p and q and is defined to be true if, and only if, both p and q are true.
 D. The compound statement $p \vee q$ is the **disjunction** of p and q and is true if either p or q or both are true.
 E. Statements of the form "if p, then q" are **conditionals** or **implications** and are false only if p is true and q is false.
 F. Given the conditional $p \rightarrow q$, the following can be found:
  1. **Converse:** $q \rightarrow p$
  2. **Inverse:** $\sim p \rightarrow \sim q$
  3. **Contrapositive:** $\sim q \rightarrow \sim p$
 G. If $p \rightarrow q$ is true, the converse and the inverse are not necessarily true, but the contrapositive is true.
 H. Two statements are **logically equivalent** if, and only if, they have the same truth value. An implication and its contrapositive are logically equivalent.
 I. The statement "$p \rightarrow q$ and $q \rightarrow p$" is written $p \leftrightarrow q$, a **biconditional**, and referred to as "p if and only if q."
 J. Laws to determine the validity of arguments include the **law of detachment,** ***modus tollens***, and the **chain rule**.

CHAPTER REVIEW

1. List three more terms that complete a pattern in each of the following:
 a. 0, 1, 3, 6, 10, ______, ______, ______
 b. 52, 47, 42, 37, ______, ______, ______
 c. 6400, 3200, 1600, 800, ______, ______, ______
 d. 1, 2, 3, 5, 8, 13, ______, ______, ______
 e. 2, 5, 8, 11, 14, ______, ______, ______
 f. 1, 4, 16, 64, ______, ______, ______
 g. 0, 4, 8, 12, ______, ______, ______
 h. 1, 8, 27, 64, ______, ______, ______
2. Classify each sequence in problem 1 as arithmetic, geometric, or neither.

3. Find a possible nth term in each of the following:
   a. 5, 8, 11, 14, . . .
   b. 0, 7, 26, 63, . . .
   c. 3, 9, 27, 81, 243, . . .
4. Find the first five terms of the sequences whose nth term is given as follows:
   a. $3n - 2$ **b.** $n^2 + n$ c. $4n - 1$
5. Find the following sums:
   a. $2 + 4 + 6 + 8 + 10 + \ldots + 200$
   b. $51 + 52 + 53 + 54 + \ldots + 151$
6. Produce a counterexample, if possible, to disprove each of the following:
   a. If two odd numbers are added, then the sum is odd.
   b. If a number is odd, then it ends in a 1 or a 3.
   c. If two even numbers are added, then the sum is even.
7. Complete the following magic square; that is, complete the square so that the sum in each row, column, and diagonal is the same.

16	3	2	13
	10		
9		7	12
4		14	

8. How many people can be seated at 12 square tables lined up end to end if each table individually holds four persons?
9. A shirt and a tie sell for \$9.50. The shirt costs \$5.50 more than the tie. What is the cost of the tie?
10. If fence posts are to be placed in a row 5 m apart, how many posts are needed for 100 m of fence?
11. A total of 129 players entered a single-elimination handball tournament. In the first round of play, the top-seeded player received a bye and the remaining 128 players played in 64 matches. Thus, 65 players entered the second round of play. How many matches must be played to determine the tournament champion?
12. **a.** Use patterns to predict the next two lines.

$$3 = \frac{3 \cdot 2}{2}$$

$$3 + 6 = \frac{6 \cdot 3}{2}$$

$$3 + 6 + 9 = \frac{9 \cdot 4}{2}$$

$$3 + 6 + 9 + 12 = \frac{12 \cdot 5}{2}$$

   b. Show that this pattern works in general for adding consecutive multiples of 3.
13. If a complete turn of a car tire moves a car forward 6 ft, how many turns of the tire occur before the tire goes off its 50,000-mi warranty?
14. The members of Mrs. Grant's class are standing in a circle; they are evenly spaced and are numbered in order. The student with number 7 is standing directly across from the student with number 17. How many students are in the class?
15. A carpenter has three large boxes. Inside each large box are two medium-sized boxes. Inside each medium-sized box are five small boxes. How many boxes are there altogether?
16. How many triangles are there in the following figure? Explain your reasoning.

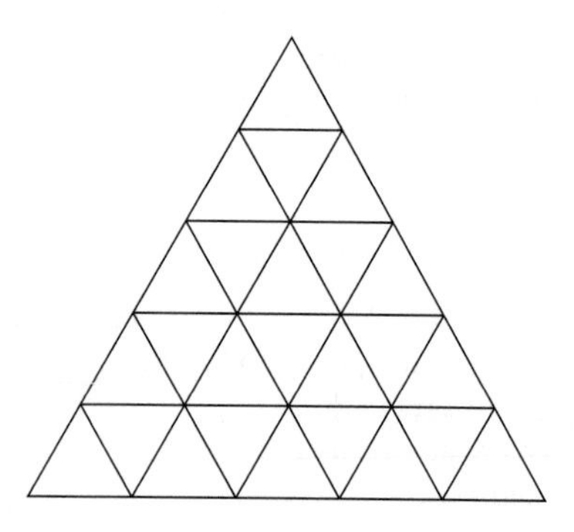

17. Mary left her home and averaged 16 km/hr riding her bicycle on an uphill trip to Larry's house. On the return trip over the same route, she averaged 20 km/hr. If it took 4 hr to make the return trip, how much cycling time did the entire trip take?
18. Use finite differences to find the next term in the following pattern:

5, 15, 37, 77, 141, ____

19. An ant farm can hold 100,000 ants. If the farm held 1500 ants on the first day, 3000 ants on the second day, 6000 ants on the third day, and so on in this manner, in how many days will the farm be full?
20. Toma's team entered a mathematics contest where teams of students compete by answering questions that are worth either 3 points or 5 points. No partial credit was given. Toma's team scored 44 points on 12 questions. How many 5-point questions did the team answer correctly?
21. There are three baskets sitting next to each other on a high shelf so that you cannot see the contents of any basket. Under the first basket is a sign that says APPLES. Under the second basket is a sign that says ORANGES, and under the third basket is a sign that says APPLES AND ORANGES. Each basket is incorrectly labeled. One basket contains all apples, one all oranges, and one a combination of apples and oranges. Is it possible to reach up on the shelf and without looking into any of the baskets select one piece of fruit and on the basis of knowing what that piece of fruit is correctly label all three baskets? Explain your reasoning.

22. Solve the following for x if possible:
 a. $3x + 5 = 17$
 b. $4 - 2x = 0$
 c. $3(x + 1) = 3x$
23. Three pieces of wood are needed for a project. They are to be cut from a 90-cm-long piece of wood. The longest piece is to be 3 times as long as the middle-sized piece and the shortest piece is to be 10 cm shorter than the middle-sized piece. Can this be done with two cuts? If so, tell how.
24. I am thinking of a number. If I double it, square the result, then divide by 2 and add 8, I get 40. What is my number?
***25.** Explain the difference between the following two statements:
 a. All students passed the final.
 b. Some students passed the final.
***26.** Which of the following are statements?
 a. The moon is inhabited.
 b. $3 + 5 = 8$.
 c. $x + 7 = 15$.
 d. Some women have Ph.Ds in mathematics.
***27.** Negate each of the following:
 a. Some women smoke.
 b. $3 + 5 = 8$.
 c. All heavy-metal rock is loud.
 d. Beethoven wrote only classical music.
***28.** Write the converse, inverse, and contrapositive of the following: If we have a rock concert, someone will faint.
***29.** Use truth tables to show that $p \rightarrow \sim q \equiv q \rightarrow \sim p$.
*30. Find valid conclusions for the following true hypotheses:
 a. All Americans love Mom and apple pie.
 Joe Czernyu is an American.
 b. Steel eventually rusts.
 The Statue of Liberty has a steel structure.
 c. Albertina passed Math 100 or Albertina dropped out.
 Albertina did not drop out.
*31. Write the following argument symbolically and then determine its validity:
 If you are fair-skinned, you will sunburn.
 If you sunburn, you will not go to the dance.
 If you do not go to the dance, your parents will want to know why you didn't go to the dance.
 Your parents do not want to know why you didn't go to the dance.
 Therefore, you are not fair-skinned.
***32.** State whether the conclusion is true or false in each case and tell why.
 a. If Bob scores at least 80 on the final, he will pass the course.
 Bob did not pass the course.
 Therefore, Bob did not score at least 80 on the final.
 b. If you build it, they will come.
 You build it.
 Therefore, they will come.

SELECTED BIBLIOGRAPHY

Ameis, J. "Stories Invite Children to Solve Mathematical Problems." *Teaching Children Mathematics* 8 (January 2002): 260–264.

Artz, S., and E. Armour-Thomas. "Development of a Cognitive-Metacognitive Framework for Protocol Analysis of Mathematical Problem Solving in Small Groups." *Cognition and Instruction* 9 (1992): 137–175.

Bishop, J., A. Otto, and C. Lubinski. "Promoting Algebraic Reasoning Using Student Thinking." *Mathematics Teaching in the Middle School* 6 (May 2001): 508–514.

Bloom, B., and L. Broder. *Problem Solving Processes of College Students*. Chicago, IL: University of Chicago Press, 1950.

Booth, L. "Children's Difficulties in Beginning Algebra." In *The Ideas of Algebra, K-12*, edited by A. Coxford and A. Shulte. Reston, VA: NCTM, 1988.

Bradley, E. "Is Algebra in the Cards?" *Mathematics Teaching in the Middle School* 2 (May 1997): 398–403.

Buschman, L. "Becoming a Problem-Solver." *Teaching Children Mathematics* 9 (October 2002): 98–103.

———. "Children Who Enjoy Problem-Solving." *Teaching Children Mathematics* 9 (May 2003): 539–544.

Chappell, M., and M. Strutchens. "Creating Connections: Promoting Algebraic Thinking with Concrete Models." *Mathematics Teaching in the Middle School* 7 (September 2001): 20–25.

Clement, L., and J. Bernhard. "A Problem-Solving Alternative to Using Key Words." *Mathematics Teaching in the Middle School* 10 (March 2005): 360–365.

Curcio, F. "Exploring Patterns in Nonroutine Problems." *Mathematics Teaching in the Middle School* 2 (February 1997): 262–269.

Day, R., and G. Jones. "Building Bridges to Algebraic Thinking." *Mathematics Teaching in the Middle School* 2 (February 1997): 208–212.

Dugdale, S., J. Matthews, and S. Guerro. "The Art of Posing Problems and Guiding Investigations." *Mathematics Teaching in the Middle School* 10 (October 2004): 140–147.

Femiano, R. "Algebraic Problem Solving in the Primary Grades." *Teaching Children Mathematics* 9 (April 2003): 444–449.

Ferrini-Mundy, J., G. Lappan, and E. Phillips. "Experiences with Patterning." *Teaching Children Mathematics* 3 (February 1997): 262–268.

Ferrucci, B., B. Yeap, and J. Carter. "A Modeling Approach for Enhancing Problem-Solving in the Middle Grades."

Mathematics Teaching in the Middle School 8 (May 2003): 470–475.

Fouche, K. "Algebra for Everyone: Start Early." *Mathematics Teaching in the Middle School* 2 (February 1997): 226–229.

Hylton-Lindsay, A. "Problem-Solving, Patterns, Probability, Pascal, and Palindromes." *Mathematics Teaching in the Middle School* 8 (February 2003): 288–293.

Hoosain, E., and R. Chance. "Problem-Solving Strategies of First Graders." *Teaching Children Mathematics* 10 (May 2004): 474–479.

Kantowski, M. "Problem Solving." In *Mathematics Education Research: Implications for the 80s*, edited by E. Fennema. Alexandria, VA: ASCD, 1981.

Koirala, H., and P. Goodwin. "Teaching Algebra in the Middle Grades Using Mathmagic." *Mathematics Teaching in the Middle School* 5 (May 2000): 562–566.

Krebs, A. "Studying Students' Reasoning in Writing Generalizations." *Mathematics Teaching in the Middle School* 10 (February 2005): 284–287.

Lannin, J. "Developing Algebraic Reasoning Through Generalization." *Mathematics Teaching in the Middle School* 8 (March 2003): 342–348.

Lester, F. "Developmental Aspects of Children's Ability to Understand Mathematical Proof." *Journal for Research in Mathematics Education* 6 (1975): 14–25.

Lubinski, C., and A. Otto. "Meaningful Mathematical Representation and Early Algebraic Reasoning." *Teaching Children Mathematics* 9 (October 2002): 76–80.

McLeod, D. "Affective Issues in Research on Teaching Mathematical Problem Solving." In *Teaching and Learning Mathematical Problem Solving: Multiple Research Perspectives*, edited by E. Silver. Hillsdale, NJ: LEA, 1985.

Maida, P. "Using Algebra Without Realizing It." *Mathematics Teaching in the Middle School* 9 (May 2004): 484–488.

Mann, R. "Balancing Act: The Truth Behind the Equals Sign." *Teaching Children Mathematics* 11 (September 2004): 65–69.

Martinez-Cruz, A., and E. Barger. "Adding a la Gauss." *Mathematics Teaching in the Middle School* 10 (October 2004): 152–155.

Matz, M. "Towards a Process Model for High School Algebra Errors." In *Intelligent Tutoring Systems*, edited by D. Sleeman and J. Brown. New York: Academic Press, 1982.

Mikusa, M. G. "Problem Solving Is More Than Solving Problems." *Mathematics Teaching in the Middle School* 4 (November 1998): 20–25.

Moran, G. "X-tending the Fibonacci Sequence." *Mathematics Teaching in the Middle School* 7 (April 2002): 452–454.

Novak, J., and J. Jacobs. "Sorting Symbol Strings." *Mathematics Teaching in the Middle School* 10 (March 2005): 334–338.

Olson, J. "Guess the Weight!" *Teaching Children Mathematics* 5 (October 1998): 104.

Olson, M., and J. Olson. "Puzzling Prices." *Teaching Children Mathematics* 7 (April 2001): 474.

Ploger, D. "Spreadsheets, Patterns, and Algebraic Thinking." *Teaching Children Mathematics* 3 (February 1997): 330–334.

Polya, G. *How to Solve It.* Princeton, NJ: Princeton University Press, 1957.

———. *Mathematical Discovery, Combined Edition.* New York: John Wiley & Sons, Inc., 1981.

Reid, D. "Describing Reasoning in Early Elementary School Mathematics." *Teaching Children Mathematics* 9 (December 2002): 234–237.

Rubenstein, R. "Building Explicit and Recursive Forms of Patterns with the Function Game." *Mathematics Teaching in the Middle School* 7 (April 2002): 426–431.

Siegel, M. "The Sum of Cubes: An Activity Review and Conjecture." *Mathematics Teaching in the Middle School* 10 (March 2005): 356–359.

Strutchens, M. "Multicultural Literature as a Context for Problem Solving: Children and Parents Learning Together." *Teaching Children Mathematics* 8 (April 2002): 448–454.

Swarthout, M. "Great Garlands." *Teaching Children Mathematics* 8 (December 2001): 217.

Thornton, S. "New Approaches to Algebra: Have We Missed the Point?" *Mathematics Teaching in the Middle School* 6 (March 2001): 388–392.

Usiskin, Z. "Doing Algebra in Grades K–4." *Teaching Children Mathematics* 3 (February 1997): 346–356.

Van de Walle, J. *Elementary and Middle School Mathematics: Teaching Developmentally*. New York: Addison Wesley Longman, 2001.

Van Reeuwijk, M., and M. Wijers. "Students' Construction of Formulas in Context." *Mathematics Teaching in the Middle School* 2 (February 1997): 230–236.

Verzoni, K. "Turning Students into Problem Solvers." *Mathematics Teaching in the Middle School* 3 (October 1997): 102–107.

Williams, E. "An Investigation of Senior High School Students' Understanding of Mathematical Proof." *Journal for Research in Mathematics Education* 11 (May 1980): 165–166.

Yolles, A. "Using Friday Puzzles to Discover Arithmetic Sequences." *Mathematics Teaching in the Middle School* 9 (November 2003): 180–185.

Sets, Whole Numbers, and Functions

PRELIMINARY PROBLEM

A baseball manager examined his roster and noticed the following.

- Every outfielder was a switch hitter.
- A third of the infielders were switch hitters.
- Half of all the switch hitters were outfielders.
- There are 12 infielders and 8 outfielders and no person played both positions.

How many switch hitters are neither infielders nor outfielders?

Georg Cantor, in the years from 1871 through 1884, created *set theory*, a new area of mathematics. His theories had a profound effect on research and mathematics teaching.

▶ HISTORICAL NOTE

Georg Cantor (1845–1918) was born in St. Petersburg, Russia. His family moved to Frankfurt when he was 11. Against his father's advice, Cantor pursued a career in mathematics and obtained his doctorate in Berlin at age 22. Most of his academic career was spent at the University of Halle. His hope of becoming a professor at the University of Berlin did not materialize because his work gained little recognition during his lifetime.

However, after his death Cantor's work was praised as an "astonishing product of mathematical thought, one of the most beautiful realizations of human activity."◀

The languages of set theory and basic set operations clarify and unify many mathematical concepts and are useful for teachers in understanding the mathematics covered in elementary school. Sets and relations between sets form a basis to teach children the *concept* of whole numbers and the concept of "less than" as well as addition, subtraction, and multiplication of whole numbers. After introducing set notation, relations between sets, set operations, and their properties, we will show how some of these concepts can be used to understand the concept of a whole number, operations on whole numbers, and the properties of these operations. We also use the concept of a set to define relations and functions.

In NCTM's *Principles and Standards for School Mathematics* (2000), we find:

Instructional programs from prekindergarten through grade 12 should enable all students to –

- *understand numbers, ways of representing numbers, relationships among numbers, and number systems;*
- *understand meanings of operations and how they relate to one another;*
- *compute fluently and make reasonable estimates.* (p. 32)

Further we find in the *Standards* that

During the primary grades, students should encounter a variety of meanings for addition and subtraction and subtraction of whole numbers.

In grades 3–5, helping students develop meaning for whole-number multiplication and division should become the central focus. (p. 34)

The concept of a function is a fundamental unifying concept in all of mathematics. The Algebra Standard lists "Understand Patterns, Relations, and Functions" in each grade level:

Algebraic concepts can evolve and continue to develop during prekindergarten through grade 2. They will be manifested through work classification, patterns and relations, operations with whole numbers, exploration of function, and step-by-step processes. (p. 91)

For Grades 3–5 the *Standards* point out that

Students should investigate numerical and geometric patterns and express them mathematically in words or symbols. They should analyze the structure of the pattern and how it grows or changes, organize this information systematically, and use their analysis to develop generalizations about the mathematical relationship in the pattern. (p. 159)

For Grades 6–8 the *Standards* point out that

The study of patterns and relationships in the middle grades should focus on patterns that relate to linear functions, which arise when there is a constant rate of change. Students should solve problems in which they use tables, graphs, words, and symbolic expressions to represent and examine functions and patterns of change. (p. 223)

2-1 Describing Sets

sets
elements ◆ members

A **set** is understood to be any collection of objects. Individual objects in a set are **elements**, or **members**, of the set. For example, each letter is an element of the set of letters in the English language.

We use braces to enclose the elements of a set and label the set with a capital letter for easy reference. The set of lowercase letters of the English alphabet can be written as

$$A = \{a, b, c, d, e, f, g, h, i, j, k, l, m, n, o, p, q, r, s, t, u, v, w, x, y, z\}.$$

The order in which the elements are written makes no difference, and *each element is listed only once*. For example, the set of letters in the word *book* could be written as $\{b, o, k\}$, $\{o, b, k\}$, or $\{k, o, b\}$.

We symbolize an element belonging to a set by using the symbol $\in$. For example, $b \in A$. If an element does not belong to a set, we use the symbol $\notin$. For example, the fact that A does not contain the Greek letter α (alpha) is written as $\alpha \notin A$.

REMARK In mathematics, the same letter, one lowercase and the other uppercase cannot be freely interchanged. For example, in the set A mentioned earlier we have $b \in A$ but $B \notin A$.

well defined

For a given set to be useful in mathematics, it must be **well defined**; that is, if we are given a set and some particular object, then we must be able to tell whether the object does or does not belong to the set. For example, the set of all citizens of Pasadena, California, who ate rice on January 1, 2003, is well defined. We may not know if a particular resident of Pasadena ate rice or not, but that resident either belongs or does not belong to the set. On the other hand, the set of all tall people is not well defined because we do not know which people qualify as "tall."

We may use sets to define mathematical terms. For example, the set N of *natural numbers* is defined by the following:

$$N = \{1, 2, 3, 4, \ldots\}.$$

Notice the use of *ellipsis* in the above set—three dots that indicate that the sequence continues in the same manner indefinitely.

When each element of the set is listed, as in $C = \{1, 2, 3, 4\}$, this is called the **listing method**. At other times **set-builder notation** is used. An example of set-builder notation is

listing method
set-builder notation

$$C = \{x \mid x \in N \text{ and } x < 5\}.$$

This notation is read as follows:

C	$=$	$\{$	x	$\mid$	$x \in N$	and	$x < 5\}$
Set C	is equal to	the set of	all elements x	such that	x is a natural number	and	x is less than 5

When the individual elements of a set are not known or they are too numerous to list, set-builder notation is used. For example, the set of decimals between 0 and 1 can be written as

$$D = \{x \mid x \text{ is a decimal between 0 and 1}\}.$$

This is read "D is the set of all elements x such that x is a decimal between 0 and 1." It would be impossible to list all the elements of D. Hence the set-builder notation is indispensable here.

EXAMPLE 2-1

Write the following sets using set-builder notation:

a. $\{2, 4, 6, 8, 10, \ldots\}$
b. $\{1, 3, 5, 7, \ldots\}$

Solution **a.** $\{x \mid x$ is an even natural number$\}$. Or because every even natural number can be written as 2 times some natural number, this set can be written as $\{x \mid x = 2n$, and $n \in N\}$ or, in a somewhat simpler form, as $\{2n \mid n \in N\}$.

b. $\{x \mid x$ is an odd natural number$\}$. Or because every odd natural number can be written as some even number minus 1, this set can be written as $\{x \mid x = 2n - 1$, and $n \in N\}$ or $\{2n - 1 \mid n \in N\}$.

▲

EXAMPLE 2-2

Each of the following sets is described in set-builder notation. Write each of the sets by listing its elements.

a. $A = \{2k + 1 \mid k = 3, 4, 5\}$
b. $B = \{a^2 + b^2 \mid a = 2 \text{ or } 3, \text{ and } b = 2, 3, \text{ or } 4\}$

Table 2-1

k	$2k + 1$
3	$2 \cdot 3 + 1 = 7$
4	$2 \cdot 4 + 1 = 9$
5	$2 \cdot 5 + 1 = 11$

Solution **a.** We substitute $k = 3, 4, 5$ in $2k + 1$ and obtain the corresponding values shown in Table 2-1. Thus, $A = \{7, 9, 11\}$.

b. Here $a = 2$ or 3 and $b = 2, 3$, or 4. Table 2-2 shows all possible combinations of a and b and the corresponding values of $a^2 + b^2$. Thus, $B = \{8, 13, 20, 18, 25\}$. Notice that 13 appears twice in the table but only once in the set. Why?

Table 2-2

$a \backslash b$	2	3	4
2	$2^2 + 2^2 = 8$	$2^2 + 3^2 = 13$	$2^2 + 4^2 = 20$
3	$3^2 + 2^2 = 13$	$3^2 + 3^2 = 18$	$3^2 + 4^2 = 25$

Definition of Equal Sets

Two sets are **equal** if and only if they contain exactly the same elements.

The order in which the elements are listed does not matter. If A and B are equal, written $A = B$, then every element of A is an element of B, and every element of B is an element of A. If A does not equal B, we write $A \neq B$. Consider sets $D = \{1, 2, 3\}$, $E = \{2, 5, 1\}$, and $F = \{1, 2, 5\}$. Sets D and E are not equal; sets E and F are equal.

One-to-One Correspondence

one-to-one correspondence

Consider the set of people $P = \{\text{Tomas, Dick, Mari}\}$ and the set of swimming lanes $S = \{1, 2, 3\}$. Suppose each person in P is to swim in a lane numbered 1, 2, or 3 so that no two people swim in the same lane. Such a person-lane pairing is a **one-to-one correspondence**. One way to exhibit a one-to-one correspondence is Tomas $\leftrightarrow$ 1, Dick $\leftrightarrow$ 2, and Mari $\leftrightarrow$ 3, as shown in Figure 2-1.

FIGURE 2-1

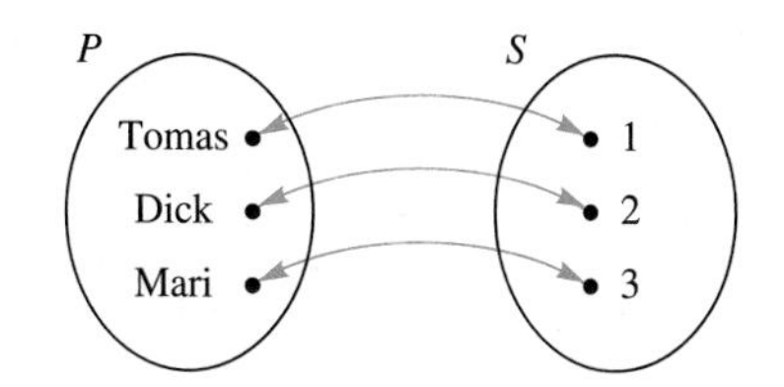

Other possible one-to-one correspondences exist between the sets P and S. There are several schemes for exhibiting them. For example, all six possible one-to-one correspondences between sets P and S can be listed as follows:

1. Tomas $\leftrightarrow$ 1
Dick $\leftrightarrow$ 2
Mari $\leftrightarrow$ 3

2. Tomas $\leftrightarrow$ 1
Dick $\leftrightarrow$ 3
Mari $\leftrightarrow$ 2

3. Tomas $\leftrightarrow$ 2
Dick $\leftrightarrow$ 1
Mari $\leftrightarrow$ 3

4. Tomas $\leftrightarrow$ 2
Dick $\leftrightarrow$ 3
Mari $\leftrightarrow$ 1

5. Tomas $\leftrightarrow$ 3
Dick $\leftrightarrow$ 1
Mari $\leftrightarrow$ 2

6. Tomas $\leftrightarrow$ 3
Dick $\leftrightarrow$ 2
Mari $\leftrightarrow$ 1

Notice that the diagram in (**1**) as well as Figure 2-1 represent a single one-to-one correspondence between the sets P and S. The correspondence Tomas $\leftrightarrow$ 1 can also be a one-to-one correspondence but between two different sets, namely the sets $\{\text{Tomas}\}$ and $\{1\}$.

Definition of One-to-One Correspondence

If the elements of sets P and S can be paired so that for each element of P there is exactly one element of S and for each element of S there is exactly one element of P, then the two sets P and S are said to be in **one-to-one correspondence**.

NOW TRY THIS 2-1

Consider a set of four people {A, B, C, D} and a set of four swimming lanes {1, 2, 3, 4}.

a. Exhibit all the one-to-one correspondences between the two sets.

b. How many such one-to-one correspondences are there?

c. Find the number of one-to-one correspondences between two sets with five elements each and explain your reasoning.

Table 2-3

1	2	3
Tomas	Dick	Mari
Tomas	Mari	Dick
Dick	Tomas	Mari
Dick	Mari	Tomas
Mari	Tomas	Dick
Mari	Dick	Tomas

Another method of demonstrating a one-to-one correspondence is to use a table, such as Table 2-3, where the lane numbers are listed across the top of the table and the possible pairings of swimmers to lanes are listed in the table.

We can also use a tree diagram to list the possible one-to-one correspondences, as Figure 2-2 shows. To read the tree diagram and see the one-to-one correspondence, we follow each branch. The person occupying a specific lane in a correspondence is listed below the lane number. For example, the top branch gives the pairing (Tomas, 1), (Dick, 2), and (Mari, 3).

FIGURE 2-2

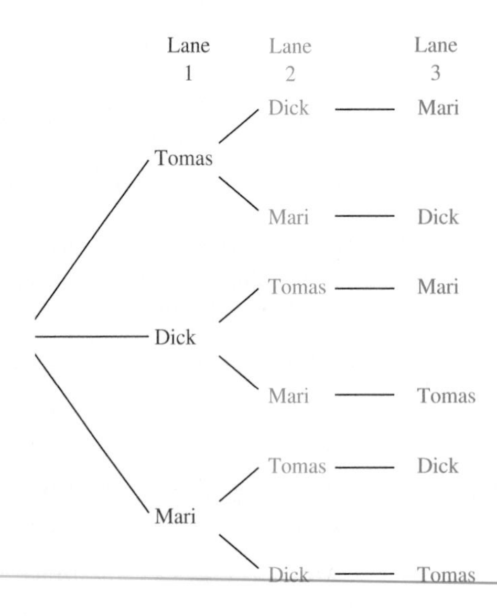

Observe in Figure 2-2 that in assigning a swimmer to lane 1 we have a choice of three people: Tomas, Dick, or Mari. If we put Tomas in lane 1, then he cannot be in lane 2, and hence the second lane must be occupied by either Dick or Mari. In the same way, we see that if Dick is in lane 1, then there are two choices for lane 2: Tomas or Mari. Similarly, if Mari is in lane 1, then again there are two choices for the second lane: Tomas or Dick. Thus, for each of the three ways we can fill the first lane, there are two subsequent ways to fill the second lane, and hence there are $2 + 2 + 2$, or $3 \cdot 2$, or 6 ways to arrange the swimmers in the first two lanes. Notice that for each arrangement of the swimmers in the first two lanes, there remains only one possible swimmer to fill the third lane. That is, if Mari fills the first lane and Dick fills the second, then Tomas must be in the third. Thus, the total number of arrangements for the three swimmers is equal to $3 \cdot 2$, or 6.

Similar reasoning can be used to find how many ice-cream arrangements are possible on a two-scoop cone if 10 flavors are offered. If we count chocolate and vanilla (chocolate on bottom and vanilla on top) different from vanilla and chocolate (vanilla on bottom and chocolate on top) and allow two scoops to be of the same flavor, we can proceed as follows. There are 10 choices for the first scoop and for each of these 10 choices there are 10 subsequent choices for the second scoop. Thus, the total number of arrangements is $10 \cdot 10$, or 100.

The counting argument used to find the number of possible one-to-one correspondences between the set of swimmers and the set of lanes and the previous problem about ice-cream-scoop arrangements are examples of the Fundamental Counting Principle.

Property

Fundamental Counting Principle If event M can occur in m ways and, after it has occurred, event N can occur in n ways, then event M followed by event N can occur in $m \cdot n$ ways.

NOW TRY THIS 2-2

Extend the Fundamental Counting Principle to any number of events.

Equivalent Sets

Suppose a room contains 20 chairs and one student is sitting in each chair with no one standing. There is a one-to-one correspondence between the set of chairs and the set of students in the room. In this case, the set of chairs and the set of students are **equivalent sets**.

equivalent sets

Definition of Equivalent Sets

Two sets A and B are **equivalent**, written $A \sim B$, if and only if there exists a one-to-one correspondence between the sets.

The term *equivalent* should not be confused with *equal*. The difference should be made clear by Example 2-3.

EXAMPLE 2-3

Let

$$A = \{p, q, r, s\}, \quad B = \{a, b, c\}, \quad C = \{x, y, z\}, \quad \text{and} \quad D = \{b, a, c\}.$$

Compare the sets, using the terms *equal* and *equivalent*.

Solution Each set is both equivalent to and equal to itself.

Sets A and B are not equivalent ($A \nsim B$) and not equal ($A \neq B$).
Sets A and C are not equivalent ($A \nsim C$) and not equal ($A \neq C$).
Sets A and D are not equivalent ($A \nsim D$) and not equal ($A \neq D$).
Sets B and C are equivalent ($B \sim C$) but not equal ($B \neq C$).
Sets B and D are equivalent ($B \sim D$) and equal ($B = D$).
Sets C and D are equivalent ($C \sim D$) but not equal ($C \neq D$). ▲

NOW TRY THIS 2-3

a. If two sets are equivalent, are they necessarily equal? Explain why or why not.
b. If two sets are equal, are they necessarily equivalent? Explain why or why not.

Cardinal Numbers

The concept of one-to-one correspondence can be used to introduce the notion of two sets having the same number of elements. Suppose a child knows how to count only to 3. The child might still tell that there are as many fingers on the left hand as on the right hand by matching the fingers on one hand with the fingers on the other hand. Naturally placing the fingers so that the left thumb touches the right thumb, the left index finger touches the right index finger, and so on, exhibits a one-to-one correspondence between the fingers of the two hands. Similarly, without counting, children realize that if every student in a class sits in a chair and no chairs are empty, there are as many chairs as students and vice versa.

One-to-one correspondence between sets is often used to introduce the concept of a number as follows. (In elementary school, the approach is similar but without the abstract notation.) The five sets $\{a, b\}$, $\{p, q\}$, $\{x, y\}$, $\{b, a\}$, and $\{*, \#\}$ are equivalent to one another and share the property of "twoness." These sets have the same cardinal number, namely, 2. The **cardinal number** of a set X, denoted $n(X)$, indicates the number of elements in the set X. If $D = \{a, b\}$, the cardinal number of D is 2, and we write $n(D) = 2$. If A is equivalent to B, then A and B have the same cardinal number; that is, $n(A) = n(B)$.

cardinal number

A set that contains no elements has cardinal number 0 and is an **empty**, or **null, set**. The empty set is designated by the symbol $\varnothing$ or $\{\ \}$. Three examples of sets with no elements are the following:

empty set ◆ null set

$$C = \{x \mid x \text{ was a state of the United States before 1200CE}\}$$
$$D = \{x \mid x \text{ is a natural number less than } 1\}$$
$$E = \{x \mid x \text{ is a natural number and } x^2 = 10\}$$

REMARK The empty set is often incorrectly recorded as $\{\varnothing\}$. This set is not empty but contains one element. Likewise, $\{0\}$ does not represent the empty set. Why?

finite set A set is a **finite set** if the cardinal number of the set is zero or a natural number. For example, the set of letters in the English alphabet is a finite set because it contains exactly 26 elements. Another way to think of this is that the set of letters in the English alphabet can be put into a one-to-one correspondence with the set $\{1, 2, 3, \ldots, 26\}$. The set of natural numbers N and the set of whole numbers, $W = \{0, 1, 2, 3, 4, \ldots\}$, are examples of **infinite sets**, sets that are not finite. The set of natural numbers with 0 is the set of whole numbers, W.

infinite set

The following "Peanuts" cartoon demonstrates how a lack of understanding of set theory concepts can lead to frustration.

NOW TRY THIS 2-4

Determine whether the set of natural numbers N is equivalent to the set of whole numbers $W = \{0, 1, 2, 3, 4, \ldots\}$.

More About Sets

universal set ◆ universe The **universal set**, or the **universe**, denoted U, is the set that contains all elements being considered in a given discussion. For this reason, you should know what the universal set is in any given problem. Suppose $U = \{x \mid x \text{ is a person living in California}\}$ and $F = \{x \mid x \text{ is a female living in California}\}$. The universal set and set F can be represented by a diagram, as in Figure 2-3(a). The universal set is usually indicated by a large rectangle, and particular sets are indicated by geometric figures inside the rectangle, as shown in Figure 2-3(a). This figure is an example of a **Venn diagram**, named after the Englishman John Venn (1834–1923) who used such diagrams to illustrate ideas in logic. The set of elements in the universe that are not in F is the set of males living in California and is the **complement** of F. It is represented by the shaded region in Figure 2-3(b).

Venn diagram

complement

FIGURE 2-3

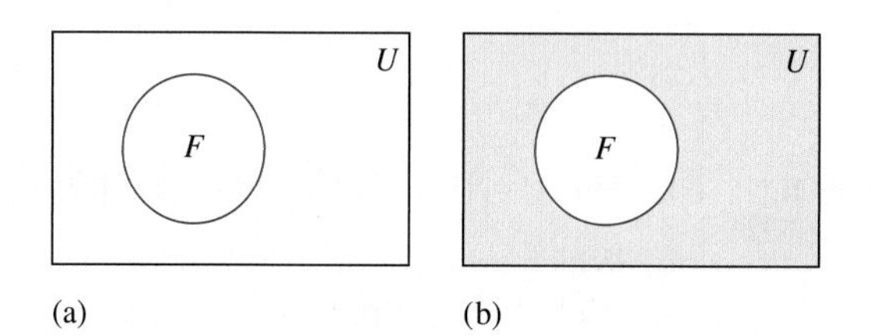

Definition of Set Complement

The **complement** of a set F, written $\overline{F}$, is the set of all elements in the universal set U that are not in F; that is, $\overline{F} = \{x \mid x \in U \text{ and } x \notin F\}$.

EXAMPLE 2-4

a. If $U = \{a, b, c, d\}$ and $B = \{c, d\}$, find (i) $\overline{B}$; (ii) $\overline{U}$; (iii) $\overline{\varnothing}$.
b. If $U = \{x \mid x \text{ is an animal in the zoo}\}$ and $S = \{x \mid x \text{ is a snake in the zoo}\}$, describe $\overline{S}$.
c. If $U = N$, $E = \{2, 4, 6, 8, \ldots\}$, and $O = \{1, 3, 5, 7, \ldots\}$, find (i) $\overline{E}$; (ii) $\overline{O}$.

Solution **a.** (i) $\overline{B} = \{a, b\}$; (ii) $\overline{U} = \varnothing$; (iii) $\overline{\varnothing} = U$

b. Because the individual animals in the zoo are not known, $\overline{S}$ must be described using set-builder notation:

$$\overline{S} = \{x \mid x \text{ is a zoo animal that is not a snake}\}.$$

c. (i) $\overline{E} = O$; (ii) $\overline{O} = E$

Subsets

Consider the sets $A = \{1, 2, 3, 4, 5, 6\}$ and $B = \{2, 4, 6\}$. All the elements of B are contained in A and we say that B is a **subset** of A. We write $B \subseteq A$. In general, we have the following definition.

subset

Definition of Subset

B is a **subset** of A, written $B \subseteq A$, if and only if every element of B is an element of A.

This definition allows B to be equal to A. The definition is written with the phrase "if and only if," which means "if B is a subset of A, then every element of B is an element of A, and if every element of B is an element of A, then B is a subset of A." *If both $A \subseteq B$ and $B \subseteq A$, then $A = B$.*

proper subset

If B is a subset of A and B is not equal to A, then B is a **proper subset** of A, written $B \subset A$. This means that every element of B is contained in A and there is at least one element of A that is not in B. To indicate this, sometimes a Venn diagram like the one shown in Figure 2-4 is used.

FIGURE 2-4

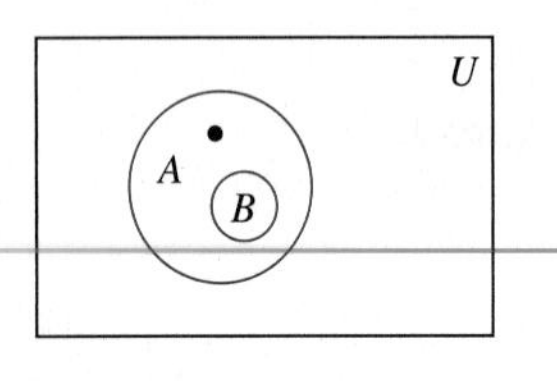

EXAMPLE 2-5

Given $A = \{1, 2, 3, 4, 5\}$, $B = \{1, 3\}$, $P = \{x \mid x = 2^n - 1, \text{ and } n \in N\}$:

a. Which sets are subsets of each other?
b. Which sets are proper subsets of each other?

Solution **a.** Because $2^1 - 1 = 1, 2^2 - 1 = 3, 2^3 - 1 = 7, 2^4 - 1 = 15$, and $2^5 - 1 = 31$, and so on, $P = \{1, 3, 7, 15, 31, \ldots\}$. Thus, $B \subseteq P$. Also $B \subseteq A$, $A \subseteq A$, $B \subseteq B$ and $P \subseteq P$.
b. $B \subset A$ and $B \subset P$

NOW TRY THIS 2-5

a. Suppose $A \subset B$. Can we always conclude that $A \subseteq B$?
b. If $A \subseteq B$, does it follow that $A \subset B$?

REMARK Notice the similarity of the subset symbols $\subset$ and $\subseteq$ to the inequality symbols $<$ and $\leq$. If $A \subseteq B$, then $n(A) \leq n(B)$, but the converse is not true. (Why?)

When a set A is not a subset of another set B, we write $A \not\subseteq B$. To show that $A \not\subseteq B$, we must find at least one element of A that is not in B. If $A = \{1, 3, 5\}$ and $B = \{1, 2, 3\}$, then A is not a subset of B because 5 is an element of A but not of B. Likewise, $B \not\subseteq A$ because 2 belongs to B but not to A.

Figures 2-4 and 2-5 show three relationships between two sets. In Figure 2-4, $B \subset A$. In Figure 2-5(a), A and B have no elements in common; in Figure 2-5(b), A and B have at least one element in common.

FIGURE 2-5

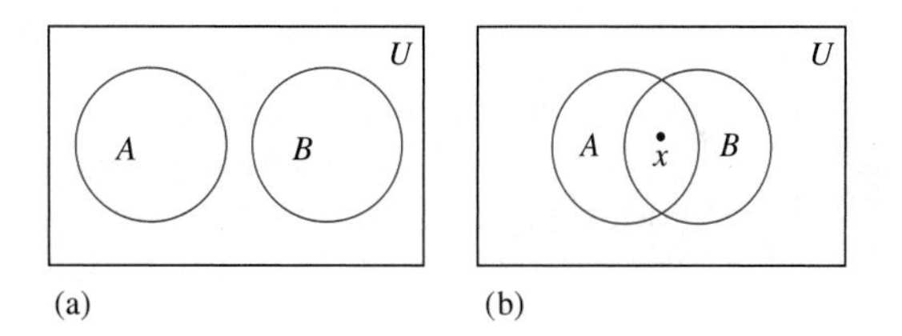

It is not obvious how the empty set fits the definition of a subset because no elements in the empty set are elements of another set. To investigate this problem, we use the strategies of *indirect reasoning* and *looking at a special case*.

For the set $\{1, 2\}$, either $\varnothing \subseteq \{1, 2\}$ or $\varnothing \not\subseteq \{1, 2\}$. Suppose $\varnothing \not\subseteq \{1, 2\}$. Then there must be some element in $\varnothing$ that is not in $\{1, 2\}$. Because the empty set has no elements, there cannot be an element in the empty set that is not in $\{1, 2\}$. Consequently, $\varnothing \not\subseteq \{1, 2\}$ is false. Therefore, the only other possibility, $\varnothing \subseteq \{1, 2\}$, is true.

The same reasoning can be applied in the case of the empty set and any other set. In particular, note that *the empty set is a subset of itself and a proper subset of any set other than itself.*

Subsets and elements of sets are often confused. We say that $2 \in \{1, 2, 3\}$. But because 2 is not a set, we cannot substitute the symbol $\subseteq$ for $\in$. However, $\{2\} \subseteq \{1, 2, 3\}$ and $\{2\} \subset \{1, 2, 3\}$. Notice that the symbol $\in$ cannot be used between $\{2\}$ and $\{1, 2, 3\}$.

Inequalities: An Application of Set Concepts

The notion of a proper subset and the concept of one-to-one correspondence can be used to define the concept of "less than" among natural numbers. The set $\{a, b, c\}$

has fewer elements than the set $\{w, x, y, z\}$ because when we try to pair the elements of the two sets, as in

$$\begin{array}{c}\{a, b, c\} \\ \updownarrow\ \updownarrow\ \updownarrow \\ \{x, y, z, w\},\end{array}$$

we see that there is an element of the second set that is not paired with an element of the first set. The set $\{a, b, c\}$ is equivalent to a proper subset of the set $\{x, y, z, w\}$.

less than

greater than

In general, *if A and B are finite sets, A has fewer elements than B if A is equivalent to a proper subset of B.* We say that $n(A)$ is **less than** $n(B)$ and write $n(A) < n(B)$. We say that a is **greater than** b, written $a > b$, if and only if $b < a$. Defining the concept of "less than or equal to" in a similar way is explored in the Ongoing Assessment.

We have just seen that if A and B are finite sets and $A \subset B$, then A has fewer elements than B and it is not possible to find a one-to-one correspondence between the sets. Consequently, A and B are not equivalent. However, when both sets are infinite and $A \subset B$, the sets could be equivalent. For example, consider the set N of counting numbers and the set E of even counting numbers. We have $E \subset N$, but it is still possible to find a one-to-one correspondence between the sets. To do so, we correspond each number in set N to a number in set E that is twice as great. That is, $n \in N$ corresponds to $2n \in E$, as shown next.

$$\begin{array}{l} N = \{1, 2, 3, 4, \ 5, \ldots, n, \ldots\}. \\ \qquad\ \ \updownarrow\ \updownarrow\ \updownarrow\ \updownarrow\ \ \updownarrow \qquad\ \ \updownarrow \\ E = \{2, 4, 6, 8, 10, \ldots, 2n, \ldots\}. \end{array}$$

Notice that in the correspondence, every element of N corresponds to a unique element in E and, conversely, every element of E corresponds to a unique element in N. For example, 11 in N corresponds to $2 \cdot 11$, or 22, in E. And 100 in E corresponds to $100 \div 2$, or 50, in N. Thus, $N \sim E$; that is, N and E are equivalent.

PROBLEM SOLVING **Passing a Senate Measure**

A committee of senators consists of Abel, Baro, Carni, and Davis. Suppose each member of the committee has one vote and a simple majority is needed to either pass or reject any measure. A measure that is neither passed nor rejected is considered to be blocked and will be voted on again. Determine the number of ways a measure could be passed or rejected and the number of ways a measure could be blocked.

Understanding the Problem We are asked to determine how many ways the committee of four could pass or reject a proposal and how many ways the committee of four could block a proposal. To pass or reject a proposal requires a winning coalition, that is, a group of senators who can pass or reject the proposal, regardless of what the others do. To block a proposal, there must be a blocking coalition, that is, a group who can prevent any proposal from passing but who cannot reject the measure.

Devising a Plan To solve the problem, we can *make a list* of subsets of the set of senators. Any subset of the set of senators with three or four members will form a winning coalition. Any subset of the set of senators with exactly two members will form a blocking coalition.

Carrying Out the Plan We list all subsets of the set $S = \{\text{Abel, Baro, Carni, Davis}\}$ that have at least three elements and all subsets that have exactly two elements. For ease, we identify the members as follows: A—Abel, B—Baro, C—Carni, D—Davis. All the subsets are given next:

$\varnothing$	$\{A\}$	$\{A, B\}$	$\{A, B, C\}$	$\{A, B, C, D\}$
	$\{B\}$	$\{A, C\}$	$\{A, B, D\}$	
	$\{C\}$	$\{A, D\}$	$\{A, C, D\}$	
	$\{D\}$	$\{B, C\}$	$\{B, C, D\}$	
		$\{B, D\}$		
		$\{C, D\}$		

There are five subsets with at least three members that can form a winning coalition and pass or reject a measure and six subsets with exactly two members that can block a measure.

Looking Back Other questions that might be considered include the following:

1. How many minimal winning coalitions are there? In other words, how many subsets are there of which no proper subset could pass a measure?
2. Devise a method to solve this problem without listing all subsets.
3. In "Carrying Out the Plan," 16 subsets of $\{A, B, C, D\}$ are listed. Use that result to systematically list all the subsets of a committee of five senators. Can you find the number of subsets of the five-member committee without actually counting the subsets?

NOW TRY THIS 2-6

Suppose a committee of U.S. senators consists of five members.

a. Compare the number of winning coalitions having exactly four members with the number of senators on the committee. What is the reason for the result?

b. Compare the number of winning coalitions having exactly three members with the number of subsets of the committee having exactly two members. What is the reason for the result?

Number of Subsets of a Set

How many subsets can be made from a set containing n elements? To obtain a general formula, we use the strategy of *trying simpler cases* first.

1. If $P = \{a\}$, then P has two subsets, $\varnothing$ and $\{a\}$.
2. If $Q = \{a, b\}$, then Q has four subsets, $\varnothing$, $\{a\}$, $\{b\}$, and $\{a, b\}$.
3. If $R = \{a, b, c\}$, then R has eight subsets, $\varnothing$, $\{a\}$, $\{b\}$, $\{c\}$, $\{a, b\}$, $\{a, c\}$, $\{b, c\}$, and $\{a, b, c\}$.

An alternative strategy for listing the number of subsets of a given set is to use a tree diagram. For example, tree diagrams for the subsets of $Q = \{a, b\}$ and $R = \{a, b, c\}$ are given in Figure 2-6.

FIGURE 2-6

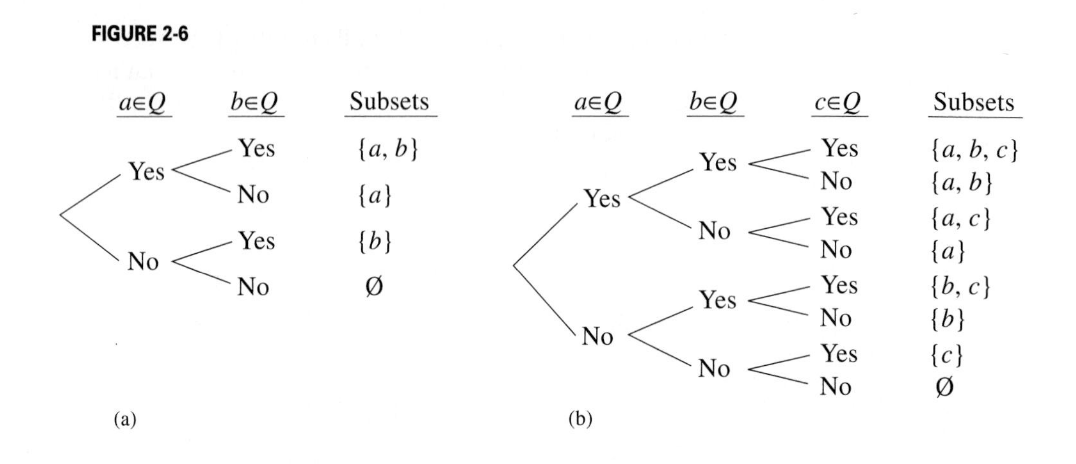

Using the information from these cases, we *make a table and search for a pattern*, as in Table 2-4.

Table 2-4

Number of Elements	Number of Subsets
1	2, or 2^1
2	4, or 2^2
3	8, or 2^3
.	.
.	.
.	.

Table 2-4 suggests that for four elements, there are 2^4, or 16, subsets. Is this guess correct? If $E = \{a, b, c, d\}$, then all the subsets of $D = \{a, b, c\}$ are also subsets of E. Eight new subsets are also formed by adjoining the element d to each of the eight subsets of D. The eight new subsets are $\{d\}$, $\{a, d\}$, $\{b, d\}$, $\{c, d\}$, $\{a, b, d\}$, $\{a, c, d\}$, $\{b, c, d\}$, and $\{a, b, c, d\}$. Thus, there are twice as many subsets of set E (with four elements) as there are of set D (with three elements). Consequently, there are $2 \cdot 8$, or 2^4, subsets of a set with four elements. In a similar way, there are $2 \cdot 2^4$, or 2^5, subsets of a set with five elements. Because including one more element in a finite set doubles the number of possible subsets of the new set, a set with six elements will have $2 \cdot 2^5$, or 2^6, subsets and so on. In each case, the number of elements and the power of 2 used to obtain the number of subsets are equal. *Therefore, if there are n elements in a set,* 2^n *subsets can be formed.* If we apply this result to the empty set—that is when $n = 0$—then we have $2^0 = 1$ because the empty set has only one subset—itself.

NOW TRY THIS 2-7

a. How many proper subsets does a set with four elements have?

b. How many proper subsets does a set with n elements have?

ONGOING ASSESSMENT 2-1

1. Write the following sets using the listing method or using set-builder notation:
   a. The set of letters in the word *mathematics*
   b. The set of states in the continental United States
   c. The set of natural numbers greater than 20
   d. The set of states in the United States that border the Pacific Ocean
2. Rewrite the following using mathematical symbols:
   a. P is equal to the set whose elements are a, b, c, and d.
   b. The set consisting of the elements 1 and 2 is a proper subset of $\{1, 2, 3, 4\}$.
   c. The set consisting of the elements 0 and 1 is not a subset of $\{1, 2, 3, 4\}$.
   d. 0 is not an element of the empty set.
   e. The set whose only element is 0 is not equal to the empty set.
3. Which of the following pairs of sets can be placed in one-to-one correspondence?
   a. $\{1, 2, 3, 4, 5\}$ and $\{m, n, o, p, q\}$
   b. $\{m, a, t, h\}$ and $\{f, u, n\}$
   c. $\{a, b, c, d, e, f, \ldots, m\}$ and $\{1, 2, 3, 4, 5, 6, \ldots, 13\}$
   d. $\{x \mid x$ is a letter in the word *mathematics*$\}$ and $\{1, 2, 3, 4, \ldots, 11\}$
   e. $\{\bigcirc, \Delta\}$ and $\{2\}$
4. How many one-to-one correspondences are there between two sets with
   a. 5 elements each?
   b. 6 elements each?
   c. n elements each?
5. How many one-to-one correspondences are there between the sets $\{x, y, z, u, v\}$ and $\{1, 2, 3, 4, 5\}$ if in each correspondence
   a. x must correspond to 5?
   b. x must correspond to 5 and y to 1?
   c. x, y, and z must correspond to odd numbers?
6. Which of the following represent equal sets?

   $A = \{a, b, c, d\}$ $B = \{x, y, z, w\}$
   $C = \{c, d, a, b\}$ $D = \{x \mid 1 \le x \le 4, x \in N\}$
   $E = \varnothing$ $F = \{\varnothing\}$
   $G = \{0\}$ $H = \{\ \}$
   $I = \{x \mid x = 2n + 1, n \in W\}$, where $W = \{0, 1, 2, 3, \ldots\}$
   $J = \{x \mid x = 2n - 1, n \in N\}$
7. Find the cardinal number of each of the following sets:
   a. $\{101, 102, 103, \ldots, 1100\}$
   b. $\{1, 3, 5, \ldots, 1001\}$
   c. $\{1, 2, 4, 8, 16, \ldots, 1024\}$
   d. $\{x \mid x = k^2, k = 1, 2, 3, \ldots, 100\}$
   e. $\{\{1, 2\}, \{3, 4\}, \{5, 6\}\}$
   f. $\{i + j \mid i \in \{1, 2, 3\}$ and $j \in \{1, 2, 3\}\}$
8. If U is the set of all college students and A is the set of all college students with a straight-A average, describe $\overline{A}$.
9. Suppose B is a proper subset of C.
   a. If $n(C) = 8$, what is the maximum number of elements in B?
   b. What is the least possible number of elements in B?
10. Suppose C is a subset of D and D is a subset of C.
    a. If $n(C) = 5$, find $n(D)$.
    b. What other relationship exists between sets C and D?
11. Indicate which symbol, $\in$ or $\notin$, makes each of the following statements true:
    a. 0 ___ $\varnothing$ **b.** $\{1\}$ ___ $\{1, 2\}$
    c. $\varnothing$ ___ $\varnothing$ **d.** $\{1, 2\}$ ___ $\{1, 2\}$
    e. 1024 ___ $\{x \mid x = 2^n$ and $n \in N\}$
    f. 3002 ___ $\{x \mid x = 3n - 1$ and $n \in N\}$
12. Indicate which symbol, $\subseteq$ or $\nsubseteq$, makes each part of problem 11 true.
13. Answer each of the following. If your answer is *no*, tell why.
    a. If $A = B$, can we always conclude that $A \subseteq B$?
    b. If $A \subseteq B$, can we always conclude that $A \subset B$?
    c. If $A \subset B$, can we always conclude that $A \subseteq B$?
    d. If $A \subseteq B$, can we always conclude that $A = B$?
14. Use the definition of *less than* to show each of the following:
    a. $2 < 4$ **b.** $3 < 100$ c. $0 < 3$
15. On a certain senate committee there are seven senators: Abel, Brooke, Cox, Dean, Eggers, Funk, and Gage. Three of these members are to be appointed to a subcommittee. How many possible subcommittees are there?
16. How many two-digit numbers can be formed if the first digit cannot be 0 and no digit can be repeated?

Communication

17. Explain the difference between a well-defined set and one that is not. Give examples.
18. Which of the following sets are not well defined? Explain.
    a. The set of wealthy schoolteachers
    b. The set of great books
    c. The set of natural numbers greater than 100
    d. The set of subsets of $\{1, 2, 3, 4, 5, 6\}$
    e. The set $\{x \mid x \neq x$ and $x \in N\}$
19. Is $\varnothing$ a proper subset of every nonempty set? Explain your reasoning.
20. Explain why $\{\varnothing\}$ has $\varnothing$ as an element and also as a subset.
21. Tell how you would show that $A \nsubseteq B$.
22. Explain why every set is a subset of itself.
23. Define *less than or equal to* in a way similar to the definition of *less than*.

Open-Ended

24. a. Give three examples of sets A and B and a universal set U such that $A \subset B$; find $\overline{A}$ and $\overline{B}$.
    b. Based on your observations, conjecture a relationship between $\overline{B}$ and $\overline{A}$.
    c. Justify your conjecture in (b) using a Venn diagram.

25. Find an infinite set A such that
    a. $\overline{A}$ is finite.
    b. $\overline{A}$ is infinite.
26. Describe two sets from real-life situations such that it is clear from using one-to-one correspondence, and not from counting, that one set has fewer elements than the other.

Cooperative Learning

27. a. Use a calculator if necessary to estimate the time in years it would take a computer to list all the subsets of $\{1, 2, 3, \ldots, 64\}$. Assume the fastest computer can list one subset in approximately 1 microsecond (one-millionth of a second).
    b. Estimate the time in years it would take the computer to exhibit all the one-to-one correspondences between the sets $\{1, 2, 3, \ldots, 64\}$ and $\{65, 66, 67, \ldots, 128\}$.

BRAIN TEASER

Mr. Gonzales's and Ms. Chan's seventh-grade classes in Paxson Middle School have 24 and 25 students, respectively. Linda, a student in Mr. Gonzales's class, claims that the number of school committees that could be formed to contain at least one student from each class is greater than the number of people in the world. Assuming that a committee can have up to 49 students, find the number of committees and determine if Linda is right.

2-2 Other Set Operations and Their Properties

Finding the complement of a set is an operation that acts on only one set at a time. In this section, we consider operations that act on two sets at a time.

FIGURE 2-7

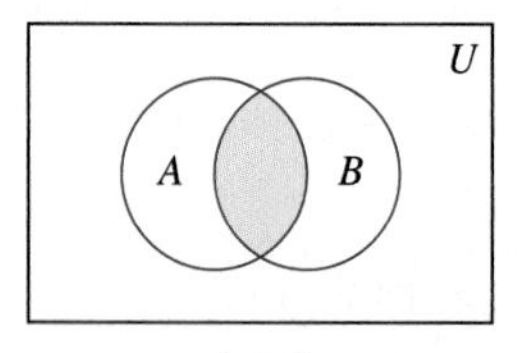

$A \cap B$

Set Intersection

Suppose that during the fall quarter, a college wants to mail a survey to all its students who are enrolled in both art and biology classes. To do this, the school officials must identify those students who are taking both classes. If A and B are the sets of students taking art courses and the set of students taking biology courses, respectively during the fall quarter, then the desired set of students includes those common to A and B, or the **intersection** of A and B. The intersection of sets A and B is the shaded region in Figure 2-7.

intersection

Definition of Set Intersection

The **intersection** of two sets A and B, written $A \cap B$, is the set of all elements common to both A and B. $A \cap B = \{x \mid x \in A \text{ and } x \in B\}$.

The key word in the definition of *intersection* is *and*. In everyday language, as in mathematics, *and* implies that both conditions must be met. In the example, the desired set is the set of those students enrolled in both art and biology.

disjoint sets

If sets such as A and B have no elements in common, they are **disjoint sets**. In other words, two sets A and B are disjoint if, and only if, $A \cap B = \varnothing$. For example, the set of males taking biology and the set of females taking biology are disjoint. The Venn diagram in Figure 2-8 implies that sets A and B are disjoint.

FIGURE 2-8

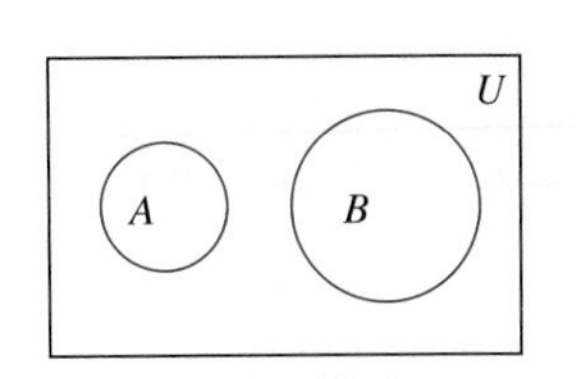

EXAMPLE 2-6

Find $A \cap B$ in each of the following:

a. $A = \{1, 2, 3, 4\}, B = \{3, 4, 5, 6\}$
b. $A = \{0, 2, 4, 6, \dots\}, B = \{1, 3, 5, 7, \dots\}$
c. $A = \{2, 4, 6, 8, \dots\}, B = \{1, 2, 3, 4, \dots\}$

Solution **a.** $A \cap B = \{3, 4\}$
b. $A \cap B = \varnothing$; therefore A and B are disjoint.
c. $A \cap B = A$ because all the elements of A are also in B.

FIGURE 2-9

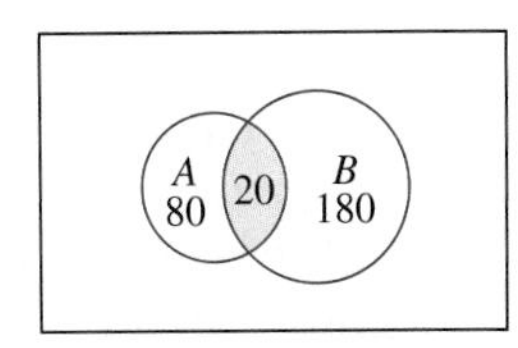

If A represents all students enrolled in art classes and B all students enrolled in biology classes, we may use a Venn diagram, taking into account that some students are enrolled in both subjects. If we know that 100 students are enrolled in art and 200 in biology and that 20 of these students are enrolled in both art and biology, then $100 - 20$, or 80, students are enrolled in art but not in biology and $200 - 20$, or 180, are enrolled in biology but not art. We can record this information as in Figure 2-9. Notice that the total number of students in set A is 100 and the total in set B is 200.

Set Union

union

If A is the set of students taking art courses during the fall quarter and B is the set of students taking biology courses during the fall quarter, then the set of students taking art or biology or both during the fall quarter is the **union** of sets A and B. The union of sets A and B is pictured in Figure 2-10.

FIGURE 2-10

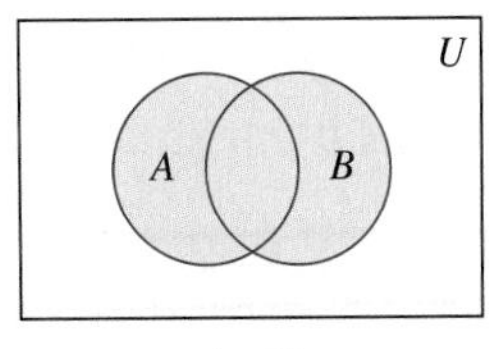

$A \cup B$

Definition of Set Union

The **union** of two sets A and B, written $A \cup B$, is the set of all elements in A or in B, $A \cup B = \{x \mid x \in A \text{ or } x \in B\}$.

The key word in the definition of *union* is *or*. In mathematics, *or* usually means "one or the other or both." This is known as the *inclusive or*.

EXAMPLE 2-7

Find $A \cup B$ for each of the following:

a. $A = \{1, 2, 3, 4\}, B = \{3, 4, 5, 6\}$ **b.** $A = \{0, 2, 4, 6, \dots\}, B = \{1, 3, 5, 7, \dots\}$
c. $A = \{2, 4, 6, 8, \dots\}, B = \{1, 2, 3, 4, \dots\}$

Solution **a.** $A \cup B = \{1, 2, 3, 4, 5, 6\}$
b. $A \cup B = \{0, 1, 2, 3, 4, \dots\}$
c. Because every element of A is already in B we have $A \cup B = B$.

NOW TRY THIS 2-8

Notice that in Figure 2-9, $n(A \cup B) = 80 + 20 + 180 = 280$, but $n(A) + n(B) = 100 + 200 = 300$; hence in general $n(A \cup B) \neq n(A) + n(B)$. Use the concept of intersection of sets to write a formula for $n(A \cup B)$.

Set Difference

complement of A relative to B ◆ **set difference**

If A is the set of students taking art classes during the fall quarter and B is the set of students taking biology classes, then the set of all students taking biology but not art is called the **complement of A relative to B**, or the **set difference** of B and A.

Definition of Relative Complement

The **complement of A relative to B**, written $B - A$, is the set of all elements in B that are not in A; $B - A = \{x \mid x \in B \text{ and } x \notin A\}$.

REMARK Note that $B - A$ is not read as "B minus A." *Minus* is an operation on numbers and *set difference* is an operation on sets.

A Venn diagram representing $B - A$ is shown in Figure 2-11(a). The shaded region represents all the elements that are in B but not in A. A Venn diagram for $B \cap \overline{A}$ is given in Figure 2-11(b). The shaded region represents all the elements that are in B and in $\overline{A}$. Notice that $B \cap \overline{A} = B - A$ because $B \cap \overline{A}$ is, by definition of intersection and complement, the set of all elements in B and not in A.

FIGURE 2-11

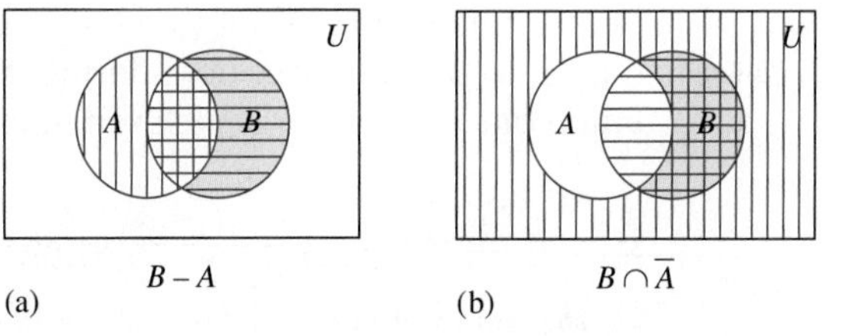

EXAMPLE 2-8

If $U = \{a, b, c, d, e, f, g\}$, $A = \{d, e, f\}$, $B = \{a, b, c, d, e, f\}$, and $C = \{a, b, c,\}$, find each of the following:

a. $A - B$
b. $B - A$
c. $B - C$
d. $C - B$

Solution **a.** $A - B = \varnothing$
b. $B - A = \{a, b, c\}$
c. $B - C = \{d, e, f\}$
d. $C - B = \varnothing$

▲

Properties of Set Operations

commutative property of set union

commutative property of set intersection

Because the order of elements in a set is not important, $A \cup B$ is equal to $B \cup A$. This is the **commutative property of set union**. It does not matter in which order we write the sets when the union of two sets is involved. Similarly, $A \cap B = B \cap A$. This is the **commutative property of set intersection**.

NOW TRY THIS 2-9

Use Venn diagrams and other means to find whether grouping is important when the same operation is involved. For example, is it always true that $A \cap (B \cap C) = (A \cap B) \cap C$? Similar questions should be investigated involving union and set difference.

REMARK The property $A \cap (B \cap C) = (A \cap B) \cap C$ is the *associative property of set intersection.* Similarly, $A \cup (B \cup C) = (A \cup B) \cup C$ is the *associative property of set union.*

EXAMPLE 2-9

Is grouping important when two different set operations are involved? For example, is it true that $A \cap (B \cup C) = (A \cap B) \cup C$?

Solution To investigate this, we let $A = \{a, b, c, d\}$, $B = \{c, d, e\}$, and $C = \{d, e, f, g\}$. Then

$$\begin{aligned} A \cap (B \cup C) &= \{a, b, c, d\} \cap (\{c, d, e\} \cup \{d, e, f, g\}) \\ &= \{a, b, c, d\} \cap \{c, d, e, f, g\} \\ &= \{c, d\} \\ (A \cap B) \cup C &= (\{a, b, c, d\} \cap \{c, d, e\}) \cup \{d, e, f, g\} \\ &= \{c, d\} \cup \{d, e, f, g\} \\ &= \{c, d, e, f, g\} \end{aligned}$$

In this case, $A \cap (B \cup C) \neq (A \cap B) \cup C$. So we have found a counterexample, that is, an example illustrating that the general statement is not always true.

FIGURE 2-12

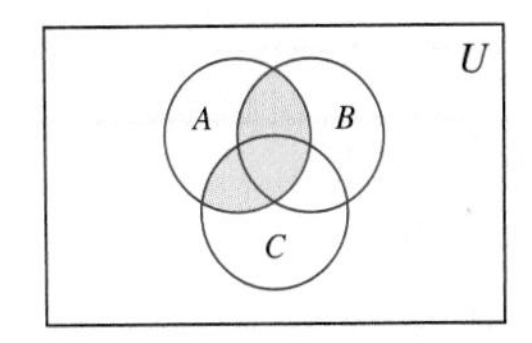

To discover an expression that is equal to $A \cap (B \cup C)$, consider the Venn diagram for $A \cap (B \cup C)$ shown by the shaded region in Figure 2-12. In the figure, $A \cap C$ and $A \cap B$ are subsets of the shaded region. The union of $A \cap C$ and $A \cap B$ is the entire shaded region. Thus, $A \cap (B \cup C) = (A \cap B) \cup (A \cap C)$. This property is stated formally next.

Property

Distributive property of set intersection over union For all sets A, B, and C,

$$A \cap (B \cup C) = (A \cap B) \cup (A \cap C).$$

NOW TRY THIS 2-10

If on both sides of the equation in the distributive property of set intersection over union the symbol $\cap$ is replaced by $\cup$ and the symbol $\cup$ is replaced by $\cap$, is the new property true? Explain why. What should this property be called?

EXAMPLE 2-10

Use set notation to describe the shaded portions of the Venn diagrams in Figure 2-13.

FIGURE 2-13

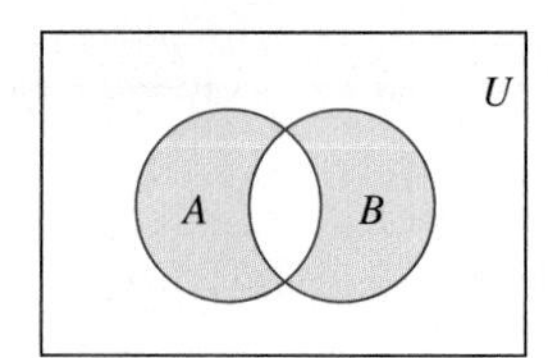

(a)

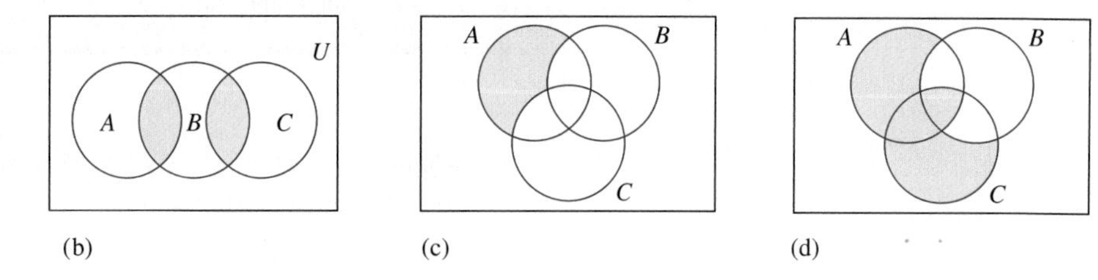

(b) (c) (d)

Solution The solutions can be described in many different, but equivalent, forms. The following are possible answers:

a. $(A \cup B) - (A \cap B)$, $(A \cup B) \cap \overline{(A \cap B)}$, or $(A - B) \cup (B - A)$
b. $(A \cap B) \cup (B \cap C)$, or $B \cap (A \cup C)$
c. $(A - B) - C$, or $A - (B \cup C)$, or $(A - (A \cap B)) - (A \cap C)$
d. $((A \cup C) - B)) \cup (A \cap B \cap C)$, or $(A - (B \cup C)) \cup (C - (A \cup B)) \cup (A \cap C)$

Using Venn Diagrams as a Problem-Solving Tool

Venn diagrams can be used as a problem-solving tool for modeling information, as shown in the following examples.

EXAMPLE 2-11

Suppose M is the set of all students taking mathematics and E is the set of all students taking English. Identify the students described by each region in Figure 2-14.

FIGURE 2-14

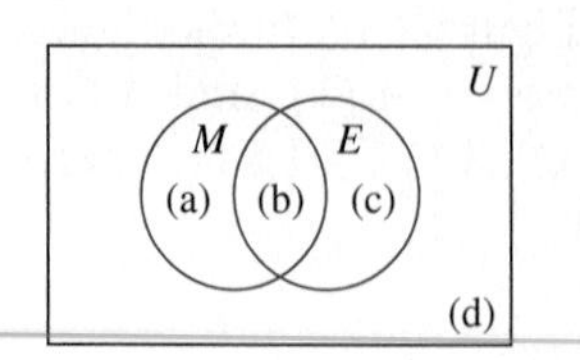

Solution

Region (a) contains all students taking mathematics but not English.
Region (b) contains all students taking both mathematics and English.
Region (c) contains all students taking English but not mathematics.
Region (d) contains all students taking neither mathematics nor English.

NOW TRY THIS 2-11

The following student page is another example of the use of Venn diagrams in modeling information. Answer questions 48 through 56 on the student page.

VENN DIAGRAMS

Extension

Venn Diagrams

This Venn diagram shows the number of sixth graders who take music, art, or drama at a school.

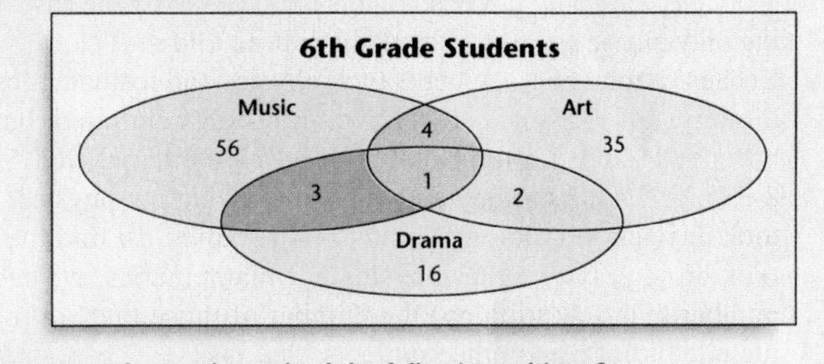

How many students take each of the following subjects?

48. music
49. art
50. drama
51. both music and art
52. both music and drama
53. both art and drama
54. all three subjects
55. music and art, but not drama
56. at least one of the three subjects

Source: McDougal Littell Middle Grades MATH Thematics, Book 1, 2005 (p. 169).

EXAMPLE 2-12

In a survey of 110 college freshmen that investigated their high school backgrounds, the following information was gathered:

25 took physics
45 took biology
48 took mathematics
10 took physics and mathematics
8 took biology and mathematics
6 took physics and biology
5 took all three subjects

a. How many students took biology but neither physics nor mathematics?
b. How many took physics, biology, or mathematics?
c. How many did not take any of the three subjects?

Solution To solve this problem, we *build a model* using sets. Because there are three distinct subjects, we should use three circles. The maximum number of regions of

FIGURE 2-15

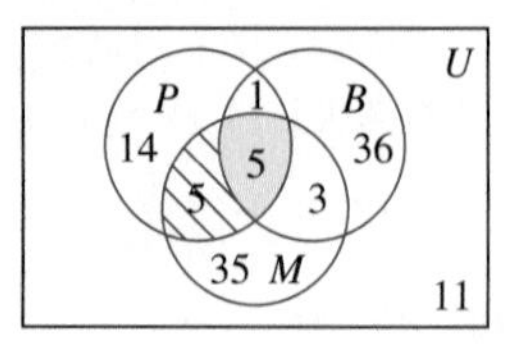

a Venn diagram determined by three circles is eight. In Figure 2-15, P is the set of students taking physics, B is the set taking biology, and M is the set taking mathematics. The shaded region represents the five students who took all three subjects. The lined region represents the students who took physics and mathematics, but who did not take biology.

In part (a) we are asked for the number of students in the subset of B that has no element in common with either P or M. That is, $B - (P \cup M)$. In part (b) we are asked for the number of elements in $P \cup B \cup M$. Finally, in part (c) we are asked for the number of students in $\overline{P \cup B \cup M}$, or $U - (P \cup B \cup M)$. Our strategy is to find the number of students in each of the eight nonoverlapping regions.

One mindset to beware of in this problem is thinking that the 25 who took physics, for example, took only physics. That is not necessarily the case. If those students had been taking only physics, then we should have been told so.

a. Because a total of 10 students took physics and mathematics and 5 of those also took biology, $10 - 5$, or 5, students took physics and math but not biology. Similarly, because 8 students took biology and mathematics and 5 took all three subjects, $8 - 5$, or 3, took biology and mathematics but not physics. Also $6 - 5$, or 1, student took physics and biology but not mathematics. To find the number of students who took biology but neither physics nor mathematics, we subtract from 45 (the total number that took biology) the number of those that are in the distinct regions that include biology and other subjects, that is, $1 + 5 + 3$, or 9. Because $45 - 9 = 36$, we know that 36 students took biology but neither physics nor mathematics.

b. To find the number of students in all the distinct regions in P, M, or B, we proceed as follows. The number of students who took physics but neither mathematics nor biology is $25 - (1 + 5 + 5)$, or 14. The number of students who took mathematics but neither physics nor biology is $48 - (5 + 5 + 3)$, or 35. Hence the number of students who took mathematics, physics, or biology is $35 + 14 + 36 + 3 + 5 + 5 + 1$, or 99.

c. Because the total number of students is 110, the number that did not take any of the three subjects is $110 - 99$, or 11.

Cartesian Products

Cartesian product

Another way to produce a set from two given sets is by forming the **Cartesian product**. This formation pairs the elements of one set with the elements of another set in a specific way. Suppose a person has three pairs of pants, $P =$ {blue, white, green}, and two shirts, $S =$ {blue, red}. According to the Fundamental Counting Principle, there are $3 \cdot 2$, or 6, possible different pant-and-shirt pairs, as shown in Figure 2-16.

FIGURE 2-16

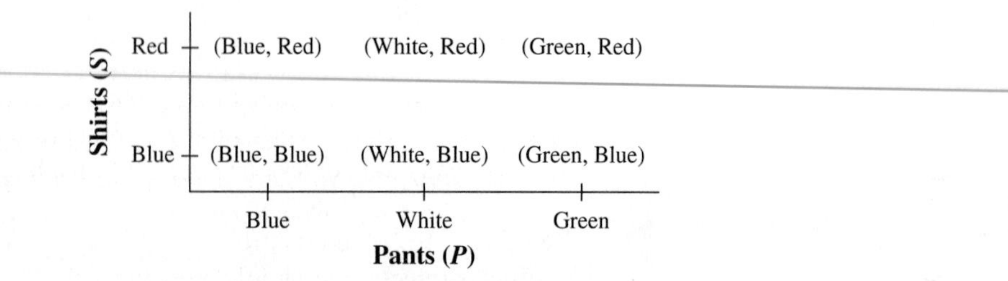

The pairs of pants and shirts form a set of all possible pairs in which the first member of the pair is an element of set P and the second member is an element of set S. The set of all possible pairs is given in Figure 2-16. Because the first component in each pair represents pants and the second component in each pair represents shirts, the order in which the components are written is important. Thus (green, blue) represents green pants and a blue shirt, whereas (blue, green) represents blue pants and a green shirt. Therefore, the two pairs represent different outfits. Because the order in each pair is important, the pairs are **ordered pairs**. The positions that the ordered pairs occupy within the set of outfits is immaterial. Only the order of the **components** within each pair is significant.

ordered pairs

components

equality for ordered pairs

The pant-and-shirt pairs suggest the following definition of **equality for ordered pairs**: $(x, y) = (m, n)$ *if, and only if, the first components are equal and the second components are equal.* A set consisting of ordered pairs such as the ones in the pants-and-shirt example is the Cartesian product of the set of pants and the set of shirts. A formal definition follows.

Definition of Cartesian Product

For any sets A and B, the **Cartesian product** of A and B, written $A \times B$, is the set of all ordered pairs such that the first component of each pair is an element of A and the second component of each pair is an element of B.

$$A \times B = \{(x, y) \mid x \in A \text{ and } y \in B\}$$

REMARK $A \times B$ is commonly read as "A cross B" and should never be read "A times B."

EXAMPLE 2-13

If $A = \{a, b, c\}$ and $B = \{1, 2, 3\}$, find each of the following:

a. $A \times B$ **b.** $B \times A$ **c.** $A \times A$

Solution **a.** $A \times B = \{(a, 1), (a, 2), (a, 3), (b, 1), (b, 2), (b, 3), (c, 1), (c, 2), (c, 3)\}$
b. $B \times A = \{(1, a), (1, b), (1, c), (2, a), (2, b), (2, c), (3, a), (3, b), (3, c)\}$
c. $A \times A = \{(a, a), (a, b), (a, c), (b, a), (b, b), (b, c), (c, a), (c, b), (c, c)\}$

It is possible to form a Cartesian product involving the null set. Suppose $A = \{1, 2\}$. Because there are no elements in $\varnothing$, no ordered pairs (x, y) with $x \in A$ and $y \in \varnothing$ are possible, so $A \times \varnothing = \varnothing$. This is true for all sets A. Similarly, $\varnothing \times A = \varnothing$ for all sets A. There is an analogy between the last equation and the multiplication fact that $0 \cdot a = 0$, where a is a natural number. In Section 2-4 we use the concept of Cartesian product to define multiplication of natural numbers.

ONGOING ASSESSMENT 2-2

1. a. Write $\{4, 5, 6, 7, 8, 9\}$ using set-builder notation.
   b. Write $\{x \mid x = 5n, \text{ where } n = 3, 6, 9\}$ using the listing method.
2. If $N = \{1, 2, 3, 4, \ldots\}$, $A = \{x \mid x = 2n - 1, n \in N\}$, $B = \{x \mid x = 2n, n \in N\}$, and $C = \{x \mid x = 2n + 1, n = 0, \text{ or } n \in N\}$, find the simplest possible expression for each of the following:
   a. $A \cup C$ b. $A \cap C$
   c. $A \cup B$ d. $A \cap B$
3. Use a Venn diagram or any other approach to decide whether the following pairs of sets are equal.
   a. $A \cap B$ and $B \cap A$
   b. $A \cup B$ and $B \cup A$
   c. $A \cup (B \cup C)$ and $(A \cup B) \cup C$
   d. $A \cup \varnothing$ and A
   e. $A \cup A$ and $A \cup \varnothing$
   f. $(A \cap A)$ and $(A \cap \varnothing)$
4. Tell whether each of the following is true or false for all sets A, B, or C. If false, give a counterexample.
   a. $A \cup \varnothing = A$
   b. $A - B = B - A$
   c. $A \cup A = A$
   d. $\overline{A \cap B} = \overline{A} \cap \overline{B}$
   e. $(A \cup B) \cup C = A \cup (B \cup C)$
   f. $(A \cup B) - A = B$
   g. $(A - B) \cup A = (A - B) \cup (B - A)$
5. If $B \subseteq A$, find a simpler expression for each of the following:
   a. $A \cap B$ b. $A \cup B$ c. $B - A$ d. $B \cap \overline{A}$
6. For each of the following, shade the portion of the Venn diagram that illustrates the set:
   a. $A \cup B$ b. $A \cap \overline{B}$
   c. $\overline{A \cap B}$ d. $(A \cap B) \cup (A \cap C)$
   e. $A \cap B$ f. $(A \cup B) \cap \overline{C}$
   g. $(A \cap B) \cup C$ h. $(\overline{A} \cap B) \cup C$

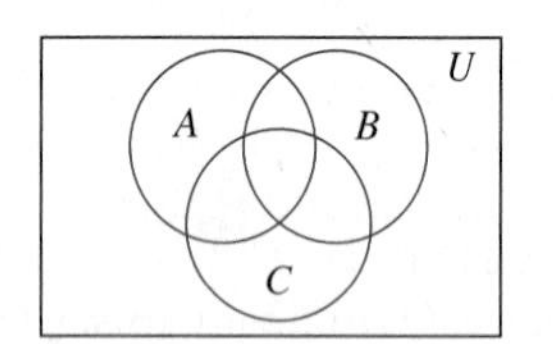

7. If S is a subset of universe U, find each of the following:
   a. $S \cup \overline{S}$ b. $\varnothing \cup S$ c. $\overline{U}$
   d. $\overline{\varnothing}$ e. $S \cap \overline{S}$ f. $\varnothing \cap S$
8. For each of the following conditions, find $A - B$:
   a. $A \cap B = \varnothing$ b. $B = U$
   c. $A = B$ d. $A \subseteq B$
9. a. Give two examples of sets A and B for which $A - B = \varnothing$. Show that in each example $A \subseteq B$.
   b. If for sets A and B we know that $A - B = \varnothing$, is it necessarily true that $A \subseteq B$? Justify your answer.
10. Use set notation to identify each of the following shaded regions:

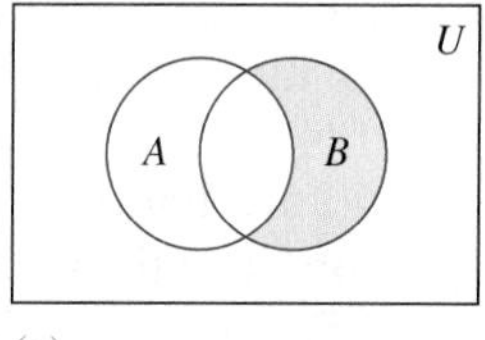

(a)

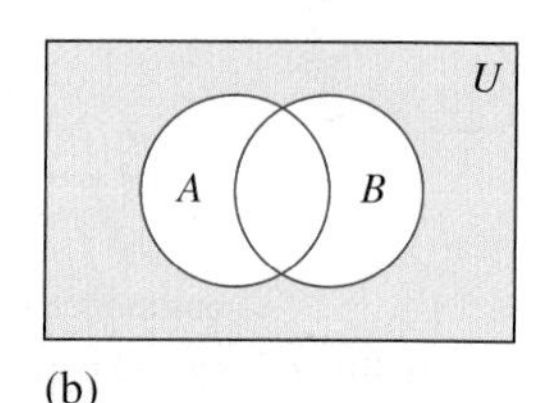

(b)

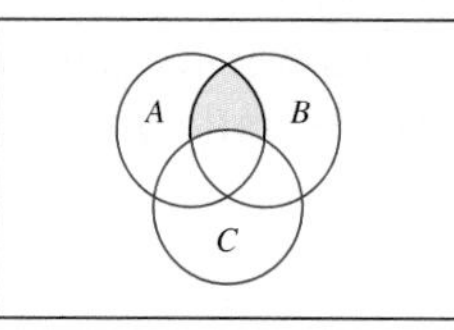

(c)

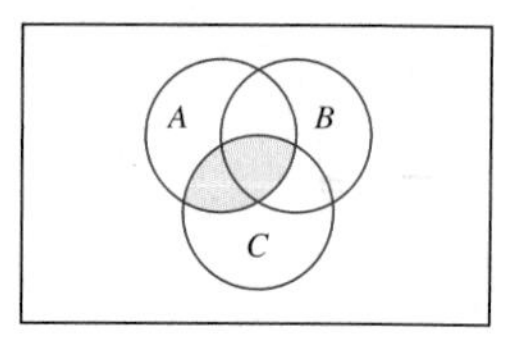

(d)

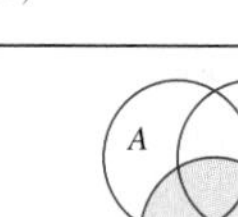

(e)

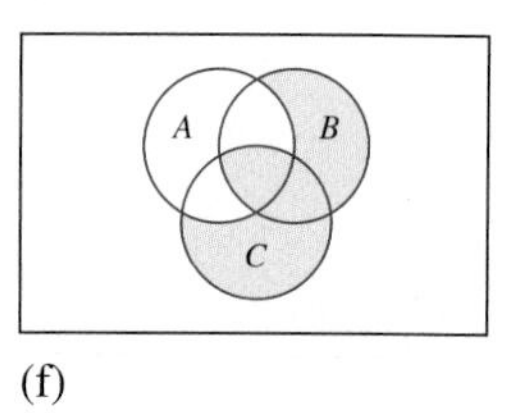

(f)

11. In the following, shade the portion of the diagram that represents the given sets:

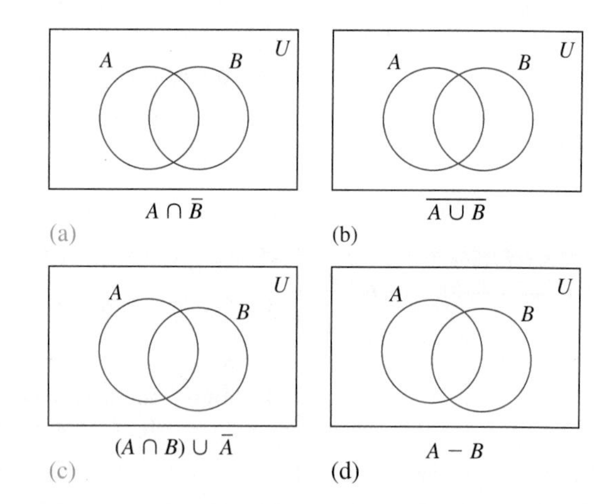

12. Use Venn diagrams to determine if each of the following is true:
    a. $A \cup (B \cap C) = (A \cup B) \cap C$
    b. $A \cap (B \cup C) = (A \cap B) \cup C$
    c. $A - (B - C) = (A - B) - C$
13. For each of the following pairs of sets, explain which is a subset of the other. If neither is a subset of the other, explain why.
    a. $A \cap B$ and $A \cap B \cap C$ b. $A \cup B$ and $A \cup B \cup C$
    c. $A \cup B$ and $(A \cup B) \cap C$ d. $A - B$ and $B - A$

14. a. If A has three elements and B has two elements, what is the greatest number of elements possible in (i) $A \cup B$? (ii) $A \cap B$? (iii) $B - A$? (iv) $A - B$?
    b. If A has n elements and B has m elements, what is the greatest number of elements possible in (i) $A \cup B$? (ii) $A \cap B$? (iii) $B - A$? (iv) $A - B$?
15. a. If $n(A \cup B) = 22$, $n(A \cap B) = 8$, and $n(B) = 12$, find $n(A)$.
    b. If $n(A) = 8$, $n(B) = 14$, and $n(A \cap B) = 5$, find $n(A \cup B)$.
16. If $n(A) = 4$, $n(B) = 5$, and $n(C) = 6$, what is the greatest and least number of elements in
    a. $A \cup B \cup C$? **b.** $A \cap B \cap C$?
17. The equation $\overline{A \cup B} = \overline{A} \cap \overline{B}$ and a similar equation for $\overline{A \cap B}$ are referred to as *DeMorgan's Laws* in honor of the famous British mathematician who first discovered them.
    a. Use Venn diagrams to show that $\overline{A \cup B} = \overline{A} \cap \overline{B}$.
    b. Discover an equation similar to the one in part (a) involving $\overline{A \cap B}$, $\overline{A}$, and $\overline{B}$. Use Venn diagrams to show that the equation holds.
    c. Verify the equations in (a) and (b) for specific sets.
18. Given that the universe is the set of all humans, $B = \{x \mid x$ is a college basketball player$\}$, and $S = \{x \mid x$ is a college student more than 200 cm tall$\}$, describe each of the following in words:
    a. $B \cap S$ **b.** $\overline{S}$ c. $B \cup S$
    d. $\overline{B \cup S}$ e. $\overline{B} \cap S$ **f.** $B \cap \overline{S}$
19. Suppose P is the set of all eighth-grade students at the Paxson School, with B the set of all students in the band and C the set of all students in the choir. Identify in words the students described by each region of the following figure:

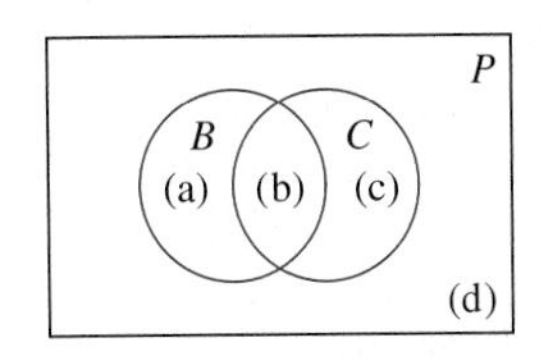

20. Fill in the Venn diagram with the appropriate numbers based on the following information:

$n(A) = 26$ $n(B \cap C) = 12$
$n(B) = 32$ $n(A \cap C) = 8$
$n(C) = 23$ $n(A \cap B \cap C) = 3$
$n(A \cap B) = 10$ $n(U) = 65$

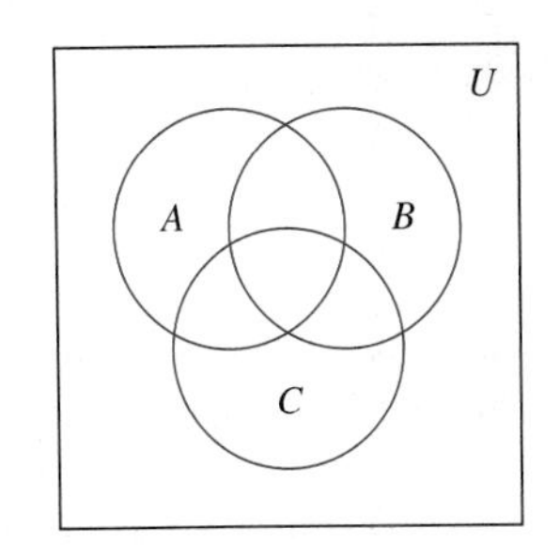

21. Of the eighth graders at the Paxson School, 7 played basketball, 9 played volleyball, 10 played soccer, 1 played basketball and volleyball only, 1 played basketball and soccer only, 2 played volleyball and soccer only, and 2 played volleyball, basketball, and soccer. How many played one or more of the three sports?
22. In a fraternity with 30 members, 18 take mathematics, 5 take both mathematics and biology, and 8 take neither mathematics nor biology. How many take biology but not mathematics?
23. Write the letters in the appropriate sections of the following Venn diagram using the following information:

    Set A contains the letters in the word *Iowa*.
    Set B contains the letters in the word *Hawaii*.
    Set C contains the letters in the word *Ohio*.

    The universal set U contains the letters in the word *Washington*.

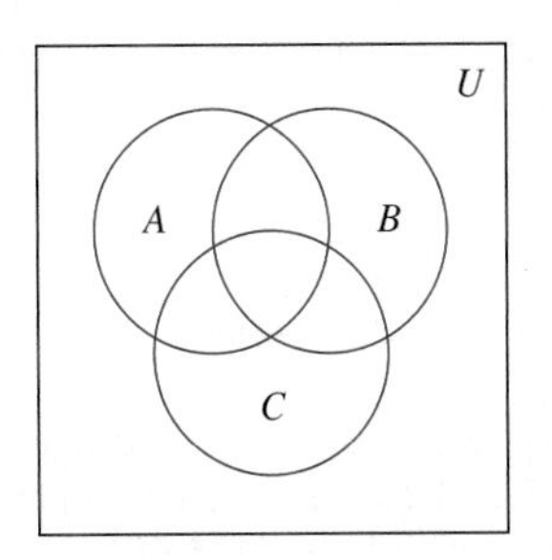

24. When three sets A, B, and C intersect as in the diagram above, eight nonoverlapping regions are created. Describe each of the regions in words and using set notation.
25. In Paul's bicycle shop, 40 bicycles are inspected. If 20 needed new tires and 30 needed gear repairs, answer the following:
    a. What is the greatest number of bikes that could have needed both?
    b. What is the least number of bikes that could have needed both?
    c. What is the greatest number of bikes that could have needed neither?
26. A pollster interviewed 500 university seniors who owned credit cards. She reported that 240 owned Goldcard, 290 had Supercard, and 270 had Thriftcard. Of those seniors, the report said that 80 owned only a Goldcard and a Supercard, 70 owned only a Goldcard and a Thriftcard, 60 owned only a Supercard and a Thriftcard, and 50 owned all three cards. When the report was submitted for publication in the local campus newspaper, the editor refused to publish it, claiming the poll was not accurate. Was the editor right? Why or why not?

27. The Red Cross looks for three types of antigens in blood tests: A, B, and Rh. When the antigen A or B is present, it is listed, but if both these antigens are absent, the blood is type O. If the Rh antigen is present, the blood is positive; otherwise, it is negative. If a laboratory technician reports the following results after testing the blood samples of 100 people, how many were classified as O negative? Explain your reasoning.

Number of Samples	Antigen in Blood
40	A
18	B
82	Rh
5	A and B
31	A and Rh
11	B and Rh
4	A, B, and Rh

28. Classify the following as true or false. If false, give a counterexample. Assume that A and B are finite sets.
   a. If $n(A) = n(B)$, then $A = B$.
   b. If $A \sim B$, then $A \cup B$ is not equivalent to B.
   c. If $A - B = \varnothing$, then $A = B$.
   d. If $B - A = \varnothing$, then $B \subseteq A$.
   e. If $A \subset B$, then $n(A) < n(B)$.
   f. If $n(A) < n(B)$, then $A \subset B$.

29. On the first day of tryouts for Little League, 128 boys of ages 10 (T), 11 (E), and 12 (W) showed up. They were asked what positions besides pitcher they wanted to play: infield (I), outfield (O), or catcher (C). The results are shown in the following table:

	I	*O*	*C*	Totals
10 (*T*)	28	14	12	54
11 (*E*)	18	20	8	46
12 (*W*)	10	12	6	28
Totals	56	46	26	128

Tell what each of the following means in words along with the number of boys indicated in each part:
   a. $I \cap W$ **b.** $C \cap (T \cup E)$
   c. $(I \cup O) \cap T$ **d.** $(T \cup E) \cap O$

30. Three announcers each tried to predict the winners of Sunday's professional football games. The only team not picked that is playing Sunday was the Giants. The choices for each person were as follows:

Phyllis: Cowboys, Steelers, Vikings, Bills
Paula: Steelers, Packers, Cowboys, Redskins
Rashid: Redskins, Vikings, Jets, Cowboys

If the only teams playing Sunday are those just mentioned, which teams will play which other teams?

31. Let $A = \{x, y\}$, $B = \{a, b, c\}$, and $C = \{0\}$. Find each of the following:
   a. $A \times B$
   b. $B \times A$
   c. $B \times \varnothing$
   d. $(A \cup B) \times C$
   e. $A \cup (B \times C)$

32. Tell whether each of the following is true or false and tell why.
   a. $(2, 5) = (1 + 1, 7 - 2)$
   b. $\{2, 5\} = \{5, 2\}$
   c. $(2, 5) = (5, 2)$
   d. $(2, 5) = \{2, 5\}$

33. For each of the following, the Cartesian product $C \times D$ is given by the following sets. Find C and D.
   a. $\{(a, b), (a, c), (a, d), (a, e)\}$
   b. $\{(1, 1), (1, 2), (1, 3), (2, 1), (2, 2), (2, 3)\}$
   c. $\{(0, 1), (0, 0), (1, 1), (1, 0)\}$

34. Answer each of the following:
   a. If A has five elements and B has four elements, how many elements are in $A \times B$?
   b. If A has m elements and B has n elements, how many elements are in $A \times B$?
   c. If A has m elements, B has n elements, and C has p elements, how many elements are in $(A \times B) \times C$?

35. If there are six teams in the Alpha league and five teams in the Beta league and if each team from one league plays each team from the other league exactly once, how many games are played?

36. José has four pairs of slacks, five shirts, and three sweaters. From how many combinations can he choose if he chooses a pair of slacks, a shirt, and a sweater each day?

Communication

37. Answer each of the following and justify your answer:
   a. If $a \in A \cap B$, is it true that $a \in A \cup B$?
   b. If $a \in A \cup B$, is it true that $a \in A \cap B$?

38. The primary colors are red, blue, and yellow. If each is considered a set, write an explanation of what we would expect to get with the intersection of each of these sets from the regions pictured in the following figure:

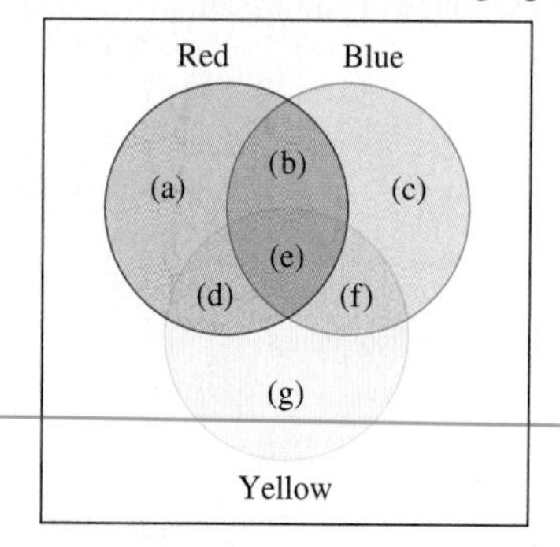

39. Is the operation of forming Cartesian products commutative? Explain why or why not.

40. If A and B are sets, is it always true that $n(A - B) = n(A) - n(B)$? Explain.

Open-Ended

41. Make up and solve a story problem concerning specific sets A, B, and C for which $n(A \cup B \cup C)$ is known and it is required to find $n(A)$, $n(B)$, and $n(C)$.

42. Describe a real-life situation that can be represented by each of the following:

a. $A \cap \overline{B}$

b. $A \cap B \cap C$

c. $A - (B \cup C)$

Cooperative Learning

43. Use set operations like union, intersection, complement, and set difference to describe the shaded region in the following figure in as many ways as possible. Compare your expressions with those of other groups to see which has the most. What is the total number of different expressions found by all the groups? Which expressions appeared in all the groups?

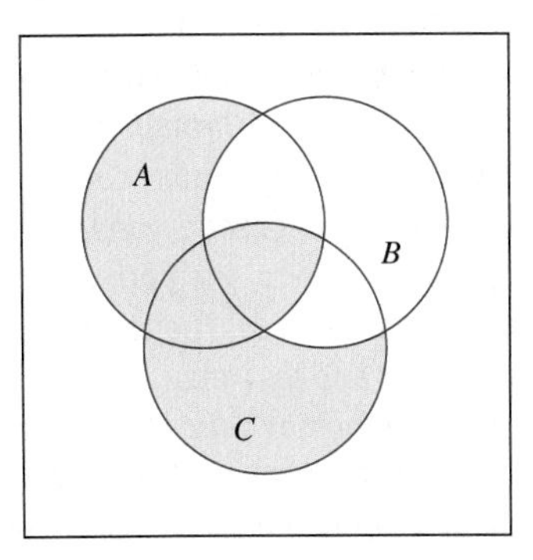

Review Problems

44. Find the number of elements in the following sets:

a. $\{x \mid x$ is a letter in *commonsense*$\}$

b. The set of letters appearing in the word *committee*

45. If $A = \{1, 2, 3, 4\}$ and $B = \{1, 2, 3, 4, 5\}$, answer the following questions:

a. How many subsets of A do not contain the element 1?

b. How many subsets of A contain the element 1?

c. How many subsets of A contain either the element 1 or 2?

d. How many subsets of A contain neither the element 1 nor 2?

e. How many subsets of B contain the element 5 and how many do not?

f. If all the subsets of A are known, how can all the subsets of B be listed systematically? How many subsets of B are there?

46. a. Which of the following sets are equal?

b. Which sets are proper subsets of the other sets?

$$A = \{2, 4, 6, 8, 10, \ldots\}$$
$$B = \{x \mid x = 2n + 2 \text{ and } n = 0, 1, 2, 3, 4, \ldots\}$$
$$C = \{x \mid x = 4n \text{ and } n \in N\}$$

47. Give examples from real life for each of the following:

a. A one-to-one correspondence between two sets

b. A correspondence between two sets that is not one-to-one

LABORATORY ACTIVITY

A set of attribute blocks consists of 32 blocks. Each block is identified by its own shape, size, and color. The four shapes in a set are square, triangle, rhombus, and circle; the four colors are red, yellow, blue, and green; the two sizes are large and small. In addition to the blocks, each set contains a group of 20 cards. Ten of the cards specify one of the attributes of the blocks (for example, red, large, square). The other 10 cards are negation cards and specify the lack of an attribute (for example, not green, not circle). Many set-type problems can be studied with these blocks. For example, let A be the set of all green blocks and B be the set of all large blocks. Using the set of all blocks as the universal set, describe elements in each set listed here to determine which are equal:

1. $A \cup B; B \cup A$

2. $\overline{A \cap B}; \overline{A} \cap \overline{B}$

3. $\overline{A \cap B}; \overline{A} \cup \overline{B}$

4. $A - B; A \cap \overline{B}$

2-3 Addition and Subtraction of Whole Numbers

In Section 2-1 we saw that the concept of one-to-one correspondence between sets can be used to introduce children to the concept of a number. The *Principles and Standards* in "Number and Operation" for Grades PreK–2 point out all students on that level should

- *understand meanings of operations and how they relate to one another;*
- *understand various meanings of addition and subtraction of whole numbers and the relationship between the two operations;*
- *understand the effects of adding and subtracting whole numbers;*
- *understand situations that entail multiplication and division, such as equal groupings of objects and sharing equally.* (p. 78)

In addition the *Principles and Standards* point out that

Counting is a foundation for students' early work with number. Young children are motivated to count everything from the treats they eat to the stairs they climb, and through their repeated experience with the counting process, they learn many fundamental number concepts. They can associate number words with small collections of objects and gradually learn to count and keep track of objects in larger groups. They can establish one-to-one correspondence by moving, touching, or pointing to objects as they say the number words. They should learn that counting objects in a different order does not alter the result, and they may notice that the next whole number in the counting sequence is one more than the number just named. Children should learn that the last number named represents the last object as well as the total number of objects in the collection. They often solve addition and subtraction problems by counting concrete objects, and many children invent problem-solving strategies based on counting strategies (Ginsburg, Klein, and Starkey 1998; Siegler 1996). (p. 79)

In the following "Peanuts" cartoon, it seems that Lucy's little brother has not yet learned to associate number words with a collection of objects. He will soon learn that this set of fingers can be put into one-to-one correspondence with many sets of objects that can be counted. He will associate the word *three* not only with Lucy's three upheld fingers but with other sets of objects with this same cardinal number.

When zero is joined with the set of natural numbers, $N = \{1, 2, 3, 4, 5, \dots\}$, we have the set of whole numbers, denoted $W = \{0, 1, 2, 3, 4, 5, \dots\}$. In this section, we provide a variety of models for teaching computational skills involving whole numbers and allow you to revisit mathematics at a level you must know in order to be competent teachers.

▶HISTORICAL NOTE Historians think that the word *zero* originated from the Hindu word *sūnya,* which means "void." Then *sūnya* was translated into the Arabic *sifr,* which when translated to Latin became *zephirum,* from which the word *zero* was derived. ◀

Addition of Whole Numbers

Children encounter addition in preschool years by combining objects and wanting to know how many objects there are in the combined set. They may "count on" as suggested by Carpenter and Moser, or they may count the objects to find the cardinal number of the combined set. A set model is one way to represent addition of whole numbers.

▶ RESEARCH NOTE Simple addition and subtraction problems can be better understood when students solve "joining" and "take-away" problems by directly modeling the situation or by using counting strategies such as counting on or counting back (Carpenter and Moser 1984). ◀

Set Model

Suppose Jane has 4 blocks in one pile and 3 in another. If she combines the two groups of blocks, how many blocks are there in the combined group? Figure 2-17 shows the solution as it might appear in an elementary school text. The combined set of blocks is the union of the disjoint sets of 4 blocks and 3 blocks. After the sets have been combined, children count the blocks to determine that there are 7 blocks in all. Note the importance of the sets being disjoint or having no elements in common. If the sets have common elements, then an incorrect conclusion can be drawn.

FIGURE 2-17

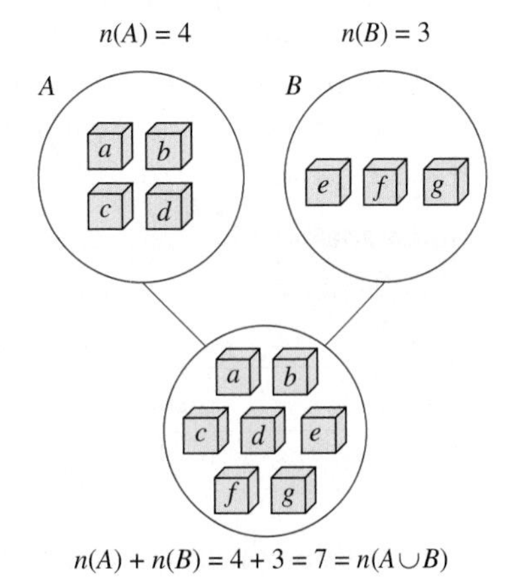

Using these ideas we can use set terminology to define addition formally.

Definition of Addition of Whole Numbers

Let A and B be two disjoint finite sets. If $n(A) = a$ and $n(B) = b$, then $a + b = n(A \cup B)$.

addends ◆ sum

The numbers a and b in $(a + b)$ are the **addends** and $(a + b)$ is the **sum**.

NOW TRY THIS 2-12

If the sets in the preceding definition of addition of whole numbers are not disjoint, explain why the definition is incorrect.

▶ HISTORICAL NOTE The symbol "+" first appeared in a 1417 manuscript and was a short way of writing the Latin word *et,* which means "and." However, Johann Widmann wrote a book in 1498 that made use of the + and − symbols for addition and subtraction. The word *minus* means "less" in Latin. First written as an *m,* it was later shortened to a horizontal bar. ◀

Number-Line Model

For some problems, the set model for addition that we have developed may not be the appropriate model. For example, consider the following questions:

1. Josh has 4 feet of red ribbon and 3 feet of white ribbon. How many feet of ribbon does he have altogether?
2. One day, Gail drank 4 ounces of orange juice in the morning and 3 ounces at lunch time. If she drank no other orange juice that day, how many ounces of orange juice did she drink for the entire day?

A *number line* can be used to model whole-number addition. Any line marked with two fundamental points, one representing 0 and the other representing 1, can be turned into a number line. The points representing 0 and 1 mark the ends of a *unit segment.* Other points can be marked and labeled, as shown in Figure 2-18. Any two consecutive points on the number line in Figure 2-18 mark the ends of a segment that has the same length as the unit segment.

FIGURE 2-18

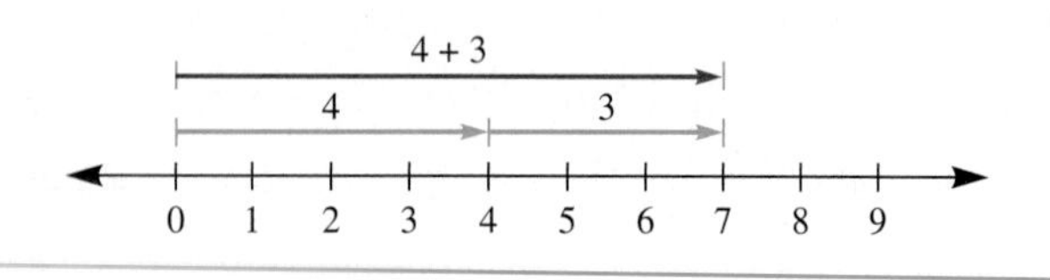

Addition problems can be modeled using directed arrows on the number line. For example, the sum of 4 + 3 is shown in Figure 2-18. Arrows representing the addends, 4 and 3, are combined into one arrow representing the sum. Figure 2-18 poses an inherent problem for students. If an arrow starting at 0 and ending at

3 represents 3, why should an arrow starting at 4 and ending at 7 represent 3? Students need to understand that the sum of any two directed arrows can be thought of as placing the endpoint of the first directed arrow at 0 and then joining to it the directed arrow for the second number with no gaps nor overlaps. The sum of the numbers can then be read. We have depicted the addends as arrows (or vectors) above the number line, but students typically concatenate the arrows directly on the line.

greater than ◆ less than

In Section 2-1 we used the concept of a set and the concept of a one-to-one correspondence to define *greater-than* relations. A number line can also be used to describe **greater-than** and **less-than** relations in the set of whole numbers. For example, in Figure 2-18, notice that 4 is to the left of 7 on the number line. We say, "four is less than seven," and we write $4 < 7$. We can also say "seven is greater than four" and write $7 > 4$. Since 4 is to the left of 7, there is a natural number that can be added to 4 to get 7, namely, 3. Thus $4 < 7$ because $4 + 3 = 7$. We can generalize this discussion to form the following definition of *less than*.

Definition of Less Than

For any whole numbers a and b, a is **less than** b, written $a < b$, if and only if there exists a natural number k such that $a + k = b$.

greater than or equal to ◆ less than or equal to

Sometimes equality is combined with the inequalities, greater than and less than, to give the relations **greater than or equal to** and **less than or equal to**, denoted $\geq$ and $\leq$. Thus, $a \leq b$ means $a < b$ or $a = b$. The emphasis with respect to these symbols is on the *or*. Observe that "$3 < 5$ or $3 = 5$" is a true statement, so $3 \leq 5$ is true. Both $5 \geq 3$ and $3 \geq 3$ are true statements.

Whole-Number Addition Properties

Any time two whole numbers are added, we are guaranteed that a unique whole number will be obtained. This property is sometimes referred to as the *closure property of addition of whole numbers*. We say that "the set of whole numbers is closed under addition."

Property

Closure property of addition of whole numbers If a and b are whole numbers, then $a + b$ is a whole number.

REMARK Note that the closure property implies that the sum of two whole numbers *exists* and that the sum is a *unique* whole number; for example, $5 + 2$ is the unique whole number 7.

NOW TRY THIS 2-13

Determine whether each of the following sets is closed under addition; that is, when you add any two elements from the set, is the sum always an element of the set?

a. $E = \{2, 4, 6, 8, 10, \ldots\}$
b. $F = \{1, 3, 5, 7, 9\}$

Figure 2-19(a) shows two additions. Pictured above the number line is $3 + 5$ and below the number line is $5 + 3$. The sums are the same. Figure 2-19(b) shows the same sums obtained with colored rods with the result being the same. Both illustrations in Figure 2-19 demonstrate the idea that two whole numbers can be added in either order. This property is true in general and is the *commutative property of addition of whole numbers*. We say that "addition of whole numbers is commutative." The word *commutative* is derived from *commute*, which means "to interchange."

FIGURE 2-19

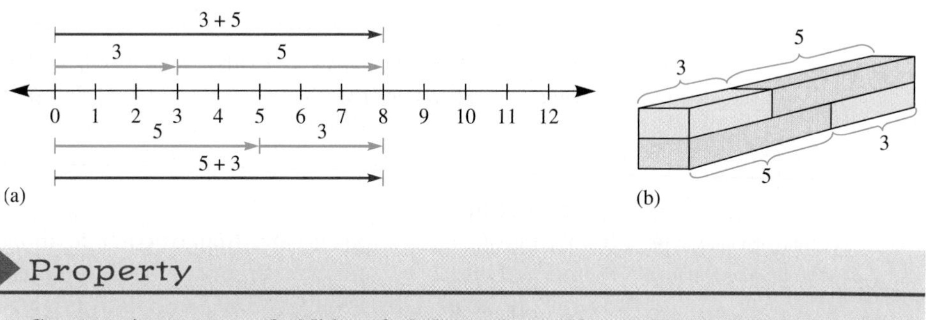

Property

Commutative property of addition of whole numbers If a and b are any whole numbers, then $a + b = b + a$.

The commutative property of addition of whole numbers is not obvious to many young children. They may be able to find the sum $9 + 2$ and not be able to find the sum $2 + 9$. Using *counting on*, $9 + 2$ can be computed by starting at 9 and then counting on two more as "ten" and "eleven." To compute $2 + 9$, the *counting on* is more involved. Students need to understand that $2 + 9$ is another name for $9 + 2$.

Another property of addition is demonstrated when we select the order in which to add three or more numbers. For example, we could compute $24 + 8 + 2$ by grouping the 24 and the 8 together: $(24 + 8) + 2 = 32 + 2 = 34$. (The parentheses indicate that the first two numbers are grouped together.) We might also recognize that it is easy to add any number to 10 and compute it as $24 + (8 + 2) = 24 + 10 = 34$. This example illustrates the *associative property of addition of whole numbers*. The word *associative* is derived from the word *associate*, which means "to unite."

Property

Associative property of addition of whole numbers If a, b, and c are whole numbers, then $(a + b) + c = a + (b + c)$.

When several numbers are being added, the parentheses are usually omitted, since the grouping does not alter the result.

RESEARCH NOTE

Students tend not to view the commutativity and associativity as distinct properties of the set of whole numbers, but rather as permissions to combine numbers in any order (Resnick 1992).

Another property of addition of whole numbers operates when one addend is 0. In Figure 2-20, set A has 5 blocks and set B has 0 blocks. The union of sets A and B has only 5 blocks.

FIGURE 2-20

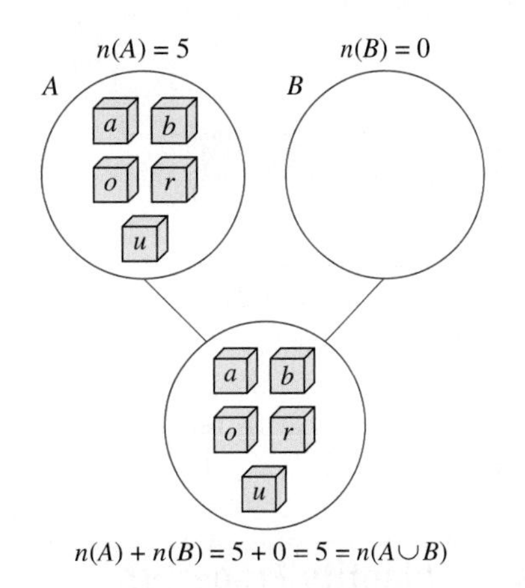

This example illustrates the following property of whole numbers.

Property

additive identity

Identity property of addition of whole numbers There is a unique whole number 0, the **additive identity**, such that for any whole number a, $a + 0 = a = 0 + a$.

Notice how the associative and identity properties are introduced on the Grade 3 student page on p. 102.

EXAMPLE 2-14

Which properties justify each of the following?

a. $5 + 7 = 7 + 5$
b. $1001 + 733$ is a unique whole number.
c. $(3 + 5) + 7 = (5 + 3) + 7$
d. $(8 + 5) + 2 = 2 + (8 + 5) = (2 + 8) + 5$
e. $(10 + 5) + (10 + 3) = (10 + 10) + (5 + 3)$

Solution
a. Commutative property of addition
b. Closure property of addition
c. Commutative property of addition
d. Commutative and associative properties of addition
e. Commutative and associative properties of addition combined

SAMPLE SCHOOL BOOK PAGE:

PROPERTIES OF ADDITION

What is the Associative Property?

The **Associative (grouping) Property of Addition** says that you can group addends in any way and the sum will be the same.

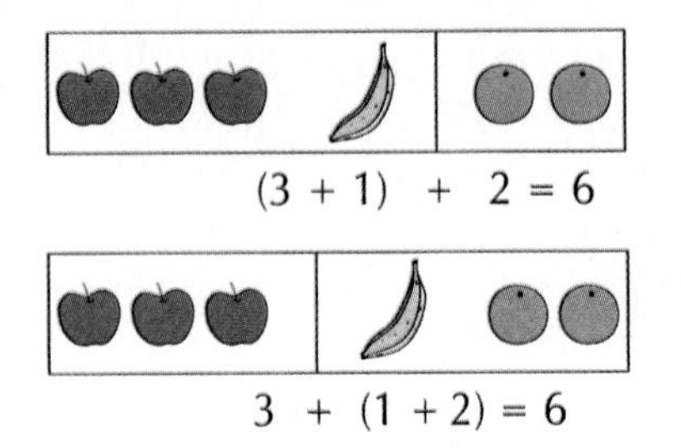

(3 + 1) + 2 = 6

3 + (1 + 2) = 6

Grouping symbols, like parentheses, ( ), show which numbers to add first.

So (3 + 1) + 2 = 3 + (1 + 2).

Talk About It

4. Evan says, "You can rewrite (5 + 3) + 2 as 8 + 2." Do you agree? Explain.

What is the Identity Property?

The **Identity (zero) Property of Addition** says that the sum of any number and zero is that same number.

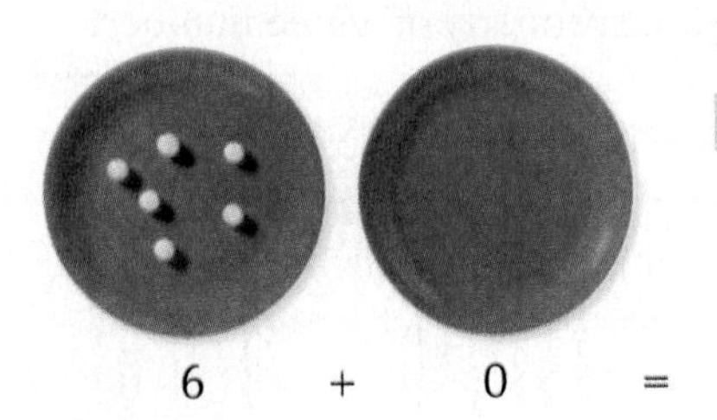

6 + 0 = 6

Talk About It

5. How could you use the Identity Property of Addition to find 536 + 0?

CHECK

For another example, see Set 2-1 on p. 116.

Find each sum.

1. 2 + (5 + 3) **2.** 3 + (1 + 4) **3.** 3 + 2 + 6

Write each missing number.

4. 5 + 2 = 2 + ___ **5.** ___ + 4 = 4 **6.** (1 + 4) + 3 = ___ + (4 + 3)

7. Number Sense What property of addition is shown in the following number sentence? Explain. 4 + (5 + 2) = (5 + 2) + 4

Source: Scott Foresman–Addison Wesley Math, Grade 3, 2005 (p. 67).

Mastering Basic Addition Facts

Certain mathematical facts are *basic addition facts*. Basic addition facts are those involving a single digit plus a single digit. In the following "Dennis The Menace" cartoon, it seems that Dennis has not mastered the basic addition facts.

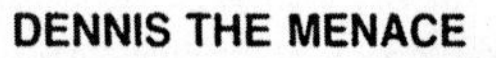

9-12
"MAKE UP YOUR MIND. FIRST YOU TELL ME 3 PLUS 3 IS SIX, AND NOW YOU SAY 4 PLUS 2 IS SIX!"

RESEARCH NOTE Students trying to master basic addition facts should be given experiences with derived fact strategies. For example, $5 + 6$ can be transformed into $(5 + 5) + 1$, which could be solved by finding the sum of the easier double $5 + 5 = 10$ and 1. Because this strategy builds on a student's number sense and meaningful relationships between basic combinations, it improves fact recall and provides a fallback mechanism for students (Fuson 1992; Steinberg 1985).

One method of learning the basic facts is to organize them according to different strategies, listed as follows:

1. *Counting On.* The strategy of counting on from the greater of the addends can be used any time we need to add whole numbers, but it is inefficient. It is usually used when one addend is 1, 2, or 3. For example, in the cartoon Dennis could have computed $4 + 2$ by starting at 4 and then counting on 5, 6. Likewise, $2 + 4$ would be computed by starting at 4 and then counting on 5, 6.
2. *Doubles.* The next strategy involves the use of *doubles*. Doubles such as $3 + 3$ in the cartoon receive special attention with students. After doubles are mastered, *doubles + 1* and *doubles + 2* can be learned easily. For example, if a student knows $6 + 6 = 12$, then $6 + 7$ is $(6 + 6) + 1$, or 1 more than the double of 6, or 13. Likewise, $7 + 9$ is $(7 + 7) + 2$ or 2 more than the double of 7, or 16.
3. *Making 10.* Another strategy is that of *making 10* and then adding any leftover. For example, we could think of $8 + 5$ as shown in Figure 2-21. Notice that we are really using the associative property of addition.

FIGURE 2-21

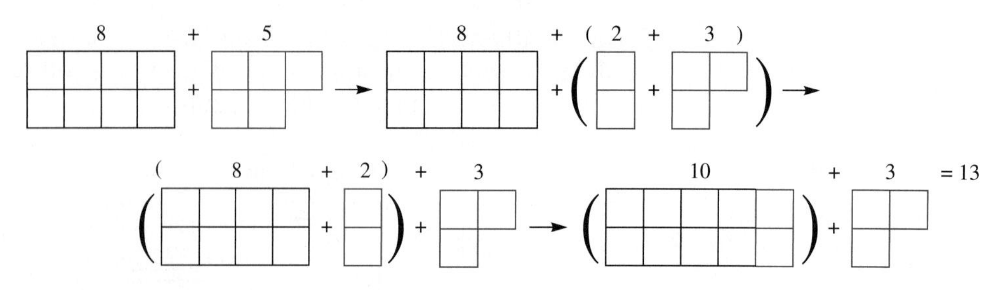

4. *Counting Back.* The strategy of *counting back* is usually used when one number is 1 or 2 less than 10. For example, because 9 is 1 less than 10 then $9 + 7$ is 1 less than $10 + 7$, or 16. In symbols, this is $9 + 7 = (10 + 7) - 1 = 16$. Also, $8 + 7 = (10 + 7) - 2 = 17 - 2 = 15$.

Many basic facts might be classified under more than one strategy. For example, we could find $9 + 8$ using *making 10* as $9 + (1 + 7) = (9 + 1) + 7 = 10 + 7 = 17$ or using a *double plus 1* as $(8 + 8) + 1$.

Subtraction of Whole Numbers

Subtraction of whole numbers can be modeled in several ways including the *set (take-away)* model, the *missing-addend* model, the *comparison* model, and the *number-line* model.

Take-Away Model

inverse operations

In elementary school, operations that "undo" each other are called **inverse operations**. Subtraction is the inverse operation for addition. It is sometimes hard for students to understand this inverse relationship between the two operations as seen in the following cartoon.

B.C.

One way to think about subtraction is this: Instead of imagining a second set of objects as being joined to a first set (as in addition), consider the second set as being *taken away* from a first set. For example, suppose we have 8 blocks and take away 3 of them, as shown in Figure 2-22. We record this process as $8 - 3 = 5$.

FIGURE 2-22

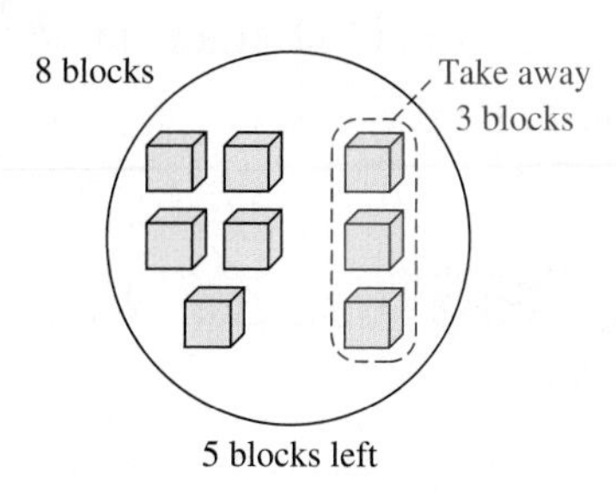

NOW TRY THIS 2-14

Recall how addition of whole numbers was defined using the concept of union of two disjoint sets. Similarly, write a definition of subtraction of whole numbers using the concept of set difference.

Missing-Addend Model

A second model for subtraction, the *missing-addend* model relates subtraction and addition. Recall that in Figure 2-22, $8 - 3$ is pictured as 8 blocks "take away" 3 blocks. The number of blocks left is the number $8 - 3$, or 5. This can also be thought of as the number of blocks that could be added to 3 blocks in order to get 8 blocks, that is,

$$\boxed{8 - 3} + 3 = 8.$$

missing addend The number $8 - 3$, or 5, is the **missing addend** in the equation

$$\boxed{} + 3 = 8.$$

The missing-addend model gives elementary school students an opportunity to begin algebraic thinking. An unknown is a major part of the problem of trying to decide the difference of 8 minus 3.

fact family Refer to the student page on p. 106 to see how Grade 3 students are shown how addition and subtraction are related using a **fact family**. Answer the *Talk About It* questions on the bottom of the student page.

Cashiers often use the missing-addend model. For example, if the bill for a movie is \$8 and you pay \$10, the cashier might calculate the change by saying "8 and 2 is 10." This idea can be generalized: For any whole numbers a and b, such that $a \geq b$, $a - b$ is the unique whole number such that $(a - b) + b = a$. That is, $a - b$ is the unique solution of the equation $\boxed{} + b = a$. The definition can be written more formally as follows:

Definition of Subtraction of Whole Numbers

For any whole numbers a and b such that $a \geq b$, $a - b$ is the unique whole number c such that $a = b + c$.

Comparison Model

A third way to consider subtraction is by using a *comparison* model. Suppose Juan has 8 blocks and Susan has 3 blocks and we want to know how many more blocks

SAMPLE SCHOOL BOOK PAGE:

RELATING ADDITION AND SUBTRACTION

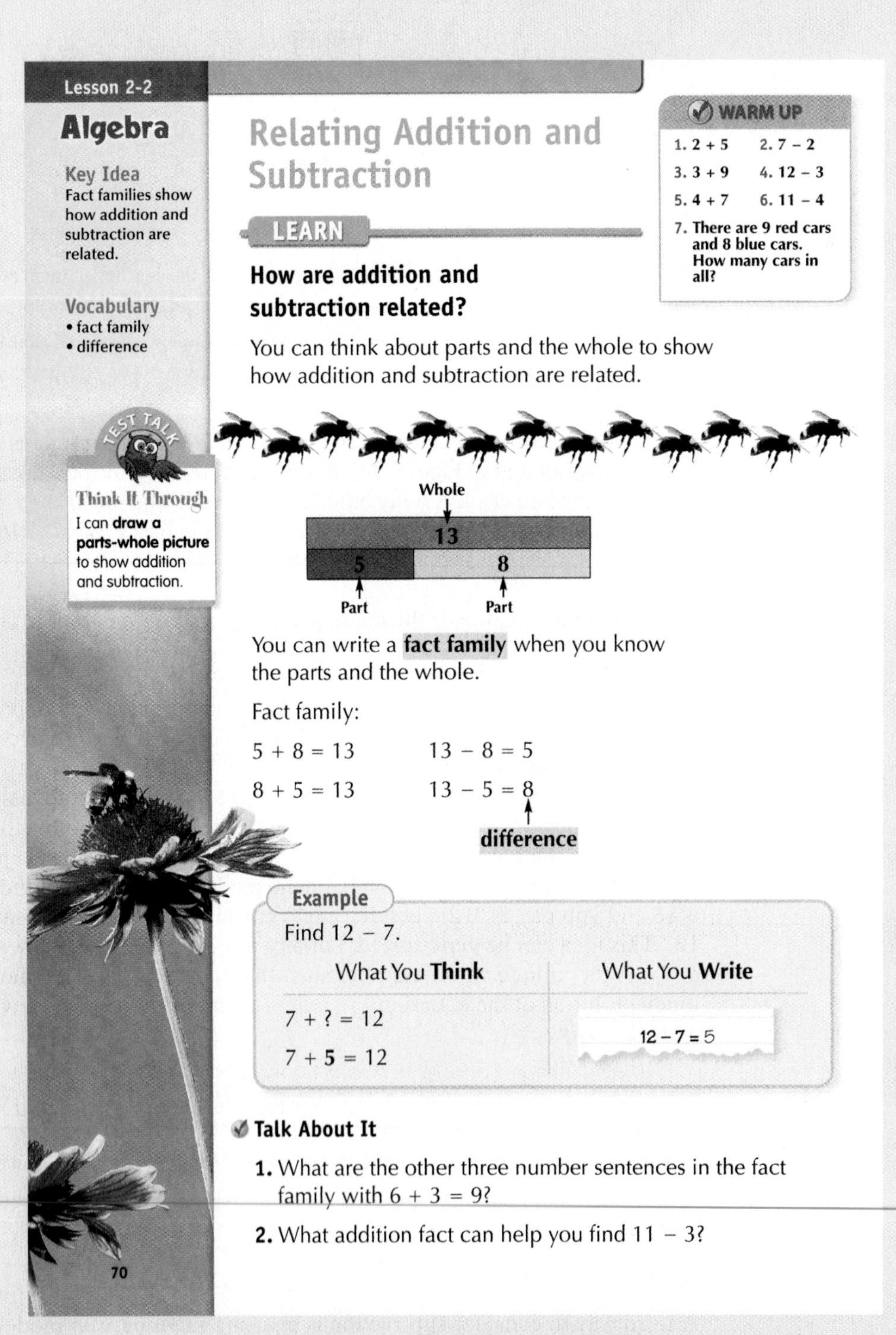

Lesson 2-2

Algebra

Key Idea
Fact families show how addition and subtraction are related.

Vocabulary
- fact family
- difference

TEST TALK

Think It Through
I can **draw a parts-whole picture** to show addition and subtraction.

Relating Addition and Subtraction

WARM UP

1. 2 + 5
2. 7 − 2
3. 3 + 9
4. 12 − 3
5. 4 + 7
6. 11 − 4
7. There are 9 red cars and 8 blue cars. How many cars in all?

LEARN

How are addition and subtraction related?

You can think about parts and the whole to show how addition and subtraction are related.

Whole
13
5 | 8
Part Part

You can write a **fact family** when you know the parts and the whole.

Fact family:

5 + 8 = 13   13 − 8 = 5

8 + 5 = 13   13 − 5 = 8

difference

Example

Find 12 − 7.

What You **Think**	What You **Write**
7 + ? = 12 7 + **5** = 12	12 − 7 = 5

Talk About It

1. What are the other three number sentences in the fact family with 6 + 3 = 9?
2. What addition fact can help you find 11 − 3?

70

Source: Scott Foresman–Addison Wesley Math, Grade 3, 2005 (p. 70).

Juan has than Susan. We can pair Susan's blocks with some of Juan's blocks, as shown in Figure 2-23, and determine that Juan has 5 more blocks than Susan. We also write this as $8 - 3 = 5$.

FIGURE 2-23

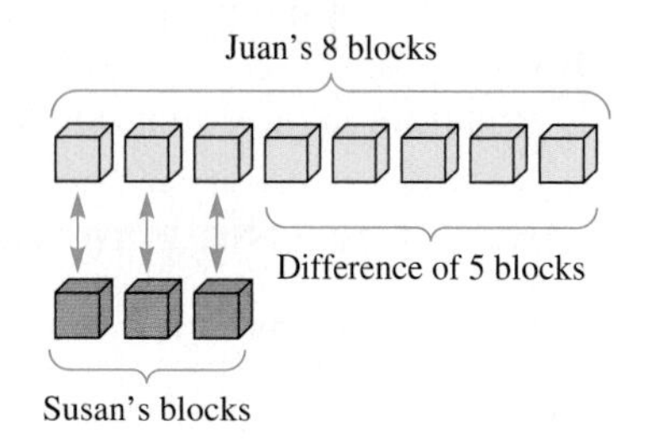

Number-Line Model

Subtraction of whole numbers can also be modeled on a number line, as suggested in Figure 2-24, where it is shown that $5 - 3 = 2$.

FIGURE 2-24

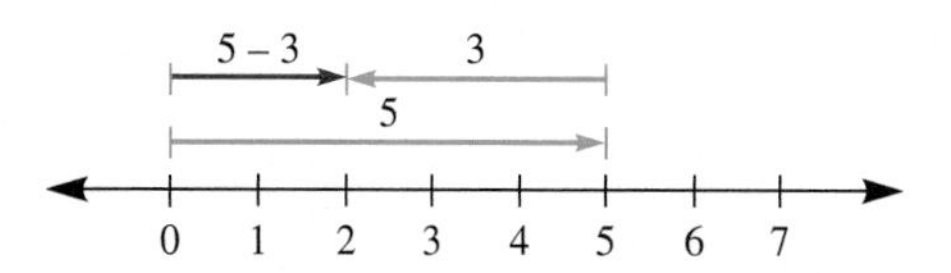

The following four problems illustrate why each of the four models for subtraction should be considered. In all four problems the answer is 5 but each can be thought of using a different model.

1. *Take-away model.* Al had $9 and spent $4, how much did he have left?
2. *Missing-addend model.* Al has read 4 chapters of a 9-chapter book. How many chapters does he have left to read?
3. *Comparison model.* Al has 9 books and Betty has 4 books. How many more books does Al have than Betty?
4. *Number-line model.* Al completed a 9-mile hike in two days. He hiked 4 miles on the second day. How far did he hike on the first day?

Properties of Subtraction

In an attempt to find $3 - 5$, we use the definition of subtraction: $3 - 5 = c$ if there is a solution to $c + 5 = 3$. Since there is no whole number c that satisfies the equation, $3 - 5$ is not meaningful in the set of whole numbers. In general, it can be shown that if $a < b$, then $a - b$ is not meaningful in the set of whole numbers.

NOW TRY THIS 2-15

Which of the following properties hold for subtraction of whole numbers? Explain.

a. Closure property
b. Associative property
c. Commutative property
d. Identity property

Introductory Algebra Using Whole Number Addition and Subtraction

In Chapter 1 we defined a *statement* as a sentence that could be classified as true or false. If variables were involved, then in many cases the sentence could not be classified as a statement. Sentences such as $9 + 5 = x$ and $12 - y = 4$ can become statements when x and y are given values. For example, if $x = 10$, then $9 + 5 = x$ is false. If $y = 8$, then $12 - y = 4$ is true. If the value that is used makes the equation true, it is a **solution** to the equation.

solution

NOW TRY THIS 2-16

Find the solution for each of the following where x is a whole number:

a. $x + 8 = 13$ **b.** $15 - x = 8$ **c.** $x > 9$ and $x < 11$

ONGOING ASSESSMENT 2-3

1. Give an example to show why, in the definition of addition, sets A and B must be disjoint.
2. For which of the following is it true that $n(A) + n(B) = n(A \cup B)$?
   a. $A = \{a, b, c\}$, $B = \{d, e\}$
   b. $A = \{a, b, c\}$, $B = \{b, c\}$
   c. $A = \{a, b, c\}$, $B = \varnothing$
3. If $n(A) = 3$, $n(B) = 5$, and $n(A \cup B) = 6$, what do you know about $n(A \cap B)$?
4. If $n(A) = 3$ and $n(A \cup B) = 6$,
   a. What are the possible values for $n(B)$?
   b. If $A \cap B = \varnothing$, what are the possible values of $n(B)$?
5. Explain whether the following given sets are closed under addition:
   a. $B = \{0\}$
   b. $T = \{0, 3, 6, 9, 12, \ldots\}$
   c. $N = \{1, 2, 3, 4, 5, \ldots\}$
   d. $V = \{3, 5, 7\}$
   e. $\{x \mid x \in W \text{ and } x > 10\}$
6. Each of the following is an example of one of the properties for addition of whole numbers. Identify the property illustrated.
   a. $6 + 3 = 3 + 6$
   b. $(6 + 3) + 5 = 6 + (3 + 5)$
   c. $(6 + 3) + 5 = (3 + 6) + 5$
   d. $5 + 0 = 5$
   e. $5 + 0 = 0 + 5$
   f. $(a + c) + d = a + (c + d)$
7. In the definition of *less than*, can the natural number k be replaced by the whole number k? Why or why not?
8. a. Recall how we have defined *less-than* and *greater-than* relations and give a similar definition using the concept of subtraction for each of the following:
   (i) $a < b$ (ii) $a > b$
   b. Use subtraction to define $a \geq b$.
9. Find the next three terms in each of the following arithmetic sequences:
   a. 8, 13, 18, 23, 28, ____, ____, ____
   b. 98, 91, 84, 77, 70, 63, ____, ____, ____
10. If A, B, and C each stand for a different single digit from 1 to 9, answer the following if

$$A + B = C.$$

   a. What is the greatest digit that C could be? Why?
   b. What is the greatest digit that A could be? Why?
   c. What is the smallest digit that C could be? Why?
   d. If A, B, and C are even, what number(s) could C be? Why?
   e. If C is 5 more than A, what number(s) could B be? Why?
   f. If A is 3 times B, what number(s) could C be? Why?
   g. If A is odd and A is 5 more than B, what number(s) could C be? Why?
11. If A, B, C, and D each stand for a different single digit from 1 to 9, answer each of the following if

$$\begin{array}{r} A \\ +\,B \\ \hline CD \end{array}$$

   a. What is the value of C? Why?
   b. Can D be 1? Why?
   c. If D is 7, what values can A be?
   d. If A is 6 greater than B, what is the value of D?
12. Assuming the same pattern continues, find the total of the terms in the 50th row in the following figure:

1	1st row
$1 - 1$	2nd row
$1 - 1 + 1$	3rd row
$1 - 1 + 1 - 1$	4th row
$1 - 1 + 1 - 1 + 1$	5th row

13. Make each of the following a magic square:

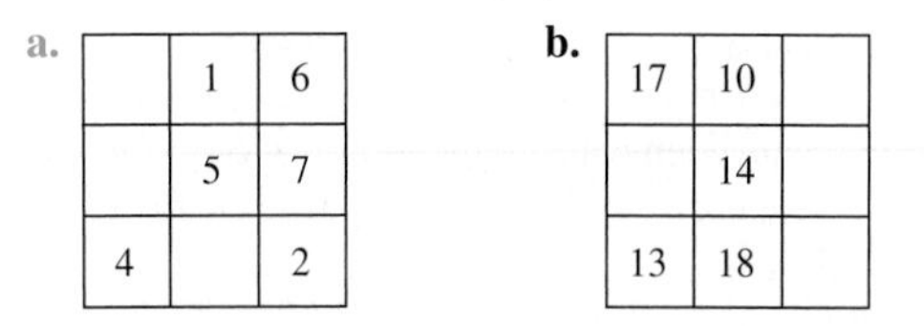

14. a. A domino set contains all number pairs from double-0 to double-6, with each number pair occurring only once; that is, the following domino counts as 2-4 and 4-2. How many dominoes are in the set?

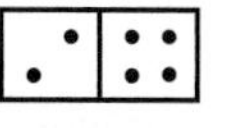

b. When considering the sum of all dots on a single domino in an ordinary set of dominoes, explain how the commutative property might be important.

15. a. At a volleyball game, the players stood in a row ordered by height. If Kent is shorter than Mischa, Sally is taller than Mischa, and Vera is taller than Sally, who is the tallest and who is the shortest?

b. Write possible heights for the players in part (a).

16. Make a calculator count backward from 27 to 0 as follows (use a constant operation if possible):

a. By 1s **b.** By 3s **c.** By 9s

17. If a calculator is made to count by 2s starting at 2, what is the thirteenth number in the sequence?

18. Rewrite each of the following subtraction problems as an equivalent addition problem:

a. $9 - 7 = x$

b. $x - 6 = 3$

c. $9 - x = 2$

19. Refer to the student page on page 106 to recall the definition of a *fact family*.

a. Write the fact family for $8 + 3 = 11$.

b. Write the fact family for $13 - 8 = 5$.

20. What conditions, if any, must be placed on *a*, *b*, and *c* in each of the following to make sure that the result is a whole number?

a. $a - b$ **b.** $a - (b - c)$

21. Show that each of the following is true. Give a property of addition to justify each step.

a. $a + (b + c) = c + (a + b)$

b. $a + (b + c) = (c + b) + a$

22. Illustrate $8 - 5 = 3$ using each of the following models.

a. Take-away **b.** Missing addend

c. Comparison **d.** Number line

23. Find the solution for each of the following:

a. $3 + (4 + 7) = (3 + x) + 7$

b. $8 + 0 = x$

c. $5 + 8 = 8 + x$

d. $x + 8 = 12 + 5$

e. $12 - x = x + 6$

f. $9 - x - 6 = 1$

g. $3 + x = x + 3$

h. $15 - x = x - 7$

i. $14 - x = 7 - x$

Communication

24. In Figure 2-18, arrows were used to represent numbers in completing an addition. Explain whether you think an arrow starting at 0 and ending at 3 represents the same number as an arrow starting at 4 and ending at 7. How would you explain this to students?

25. When subtraction and addition appear in an expression without parentheses, it is agreed that the operations are performed in order of their appearance from left to right. Taking this into account, answer the following:

a. Use an appropriate model for subtraction to explain why

$$a - b - c = a - c - b.$$

assuming that all expressions are meaningful.

b. Use an appropriate model for subtraction to explain why

$$a - b - c = a - (b + c).$$

26. Explain whether it is important for elementary students to learn more than one model for performing the operations of addition and subtraction.

27. Do elementary students still have to learn their basic facts when the calculator is part of the curriculum? Why or why not?

28. Explain how the following model can be used to illustrate each of the following addition and subtraction facts:

a. $9 + 4 = 13$ **b.** $4 + 9 = 13$

c. $4 = 13 - 9$ **d.** $9 = 13 - 4$

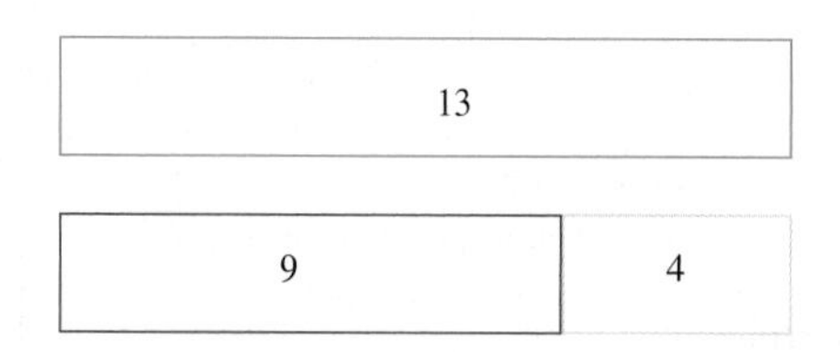

29. How are addition and subtraction related? Explain.

30. Why is 0 not an identity for subtraction? Explain.

Open-Ended

31. Describe any model not in this text that you might use to teach addition to students.

32. Suppose $A \subseteq B$. If $n(A) = a$ and $n(B) = b$, then $b - a$ could be defined as $n(B - A)$. Choose two sets A and B and illustrate this definition.

Cooperative Learning

33. Discuss with your group each of the following. Use the basic addition fact table shown.

+	0	1	2	3	4	5	6	7	8	9
0	0	1	2	3	4	5	6	7	8	9
1	1	2	3	4	5	6	7	8	9	10
2	2	3	4	5	6	7	8	9	10	11
3	3	4	5	6	7	8	9	10	11	12
4	4	5	6	7	8	9	10	11	12	13
5	5	6	7	8	9	10	11	12	13	14
6	6	7	8	9	10	11	12	13	14	15
7	7	8	9	10	11	12	13	14	15	16
8	8	9	10	11	12	13	14	15	16	17
9	9	10	11	12	13	14	15	16	17	18

a. How does the table show the closure property?
b. How does the table show the commutative property?
c. How does the table show the identity property?
d. How do the addition properties help students learn their basic facts?

34. Suppose that a number system used only four symbols, a, b, c, and d, and the operation Δ and the system operated as shown in the table. Discuss with your group each of the following:

Δ	a	b	c	d
a	a	b	c	d
b	b	c	d	a
c	c	d	a	b
d	d	a	b	c

a. Is the system closed? Why?
b. Is the system commutative? Why?
c. Does the system have an identity? If so, what is it?
d. Is the system associative? Why?

35. Have each person in your group choose a different grade textbook and report on when and how subtraction of whole numbers is introduced. Compare with different ways subtraction is introduced in this section.

Review Problems

36. Determine whether the following sets are equivalent. Justify your answers.
a. $\{2, 4, 6, 8, \ldots, 1000\}$ and $\{3, 6, 9, 12, 15, \ldots, 1500\}$
b. $\{1, 4, 7, 10, 13, \ldots, 2998\}$ and $\{1, 2, 3, 4, \ldots, 1000\}$
c. $\{0, 1, 2, 3, 4, \ldots\}$ and $\{1, 2, 3, 4, 5, \ldots\}$
d. $S = \{x \mid x = 2n, n \in N\}$ and $W = \{x \mid x = 4n, n \in N\}$

37. Decide which of the following are always true. If true, justify the statement, and if false, provide a counterexample.
a. $A - (B \cup C) = (A - B) \cap (A - C)$
b. If $A \cup B = B$, then $A \subseteq B$.
c. If $A \cap B = \varnothing$, then $A \cup B = U$.
d. If $A \not\subseteq B$ and $B \not\subseteq C$, then $A \not\subseteq C$.

38. a. How many one-to-one correspondences are possible between $A = \{a, b, c\}$ and $B = \{1, 2, 3\}$?
b. How many elements are there in $A \times B$?

39. Classify each of the following as true or false. If true, justify the statement, and if false, provide a counterexample.
a. If $A \cup B = A \cup C$, then $B = C$.
b. If $A \cap B = A \cap C$, then $B = C$.
c. If $A \cap B = \varnothing$ and $B \cap C = \varnothing$, then $A \cap C = \varnothing$.

40. A committee of senators has six members. A two-thirds vote is needed to carry any proposal. How many winning coalitions are there?

41. Oakridge has a population of 4800 and only one movie theater. One week the movie *Minority Report* was shown, and 3100 people went to see it. The following week, the movie *Gosford Park* was shown, and 2200 residents went to see it.
a. What is the greatest number of townspeople that could have seen both movies? Justify your answer.
b. What is the least number of people that could have seen both movies? Justify your answer.

42. If $U = \{a, b, c, d\}, A = \{a, b, c\}, B = \{b, c\}$, and $C = \{d\}$, find each of the following:
a. $A \cup \overline{B}$
b. $\overline{A \cap B}$
c. $A \cap \varnothing$
d. $B \cap C$
e. $B - A$

Third International Mathematics and Science Study (TIMSS) Questions

Ali had 50 apples. He sold some and then had 20 left. Which of these is a number sentence that shows this?

a. $\square - 20 = 50$
b. $20 - \square = 50$
c. $\square - 50 = 20$
d. $50 - \square = 20$

TIMSS 2003, Grade 4

4	11	6
9		5
8	3	10

The rule for the table is that numbers in each row and column must add up to the same number. What number goes in the center of the table?

a. 1
b. 2
c. 7
d. 12

TIMSS 2003, Grade 4

BRAIN TEASER

Use Figure 2-25 to design an *unmagic square*. That is, use each of the digits 1, 2, 3, 4, 5, 6, 7, 8, and 9 exactly once so that every column, row, and diagonal adds to a different sum.

FIGURE 2-25

2-4 Multiplication and Division of Whole Numbers

Multiplication of Whole Numbers

In this section, we explore the kind of problems that Grampa is having in the "Peanuts" cartoon. Why do you think he would have more troubles with "9 times 8" rather than "3 times 4"? If multiplication facts are only memorized, they may be forgotten. If students have models to develop the basic facts when needed, then all of the basic facts can be determined even if not automatically recalled.

In this section, we use three models to discuss multiplication: the *repeated-addition* model, the *array* model, and the *Cartesian-product* model. We also investigate the distributive property of multiplication over addition.

The *Principles and Standards* lists the following expectations under "Number and Operations Standards" for Grades 3–5:

- *understand various meanings of multiplication and division;*
- *understand the effects of multiplying and dividing whole numbers;*
- *identify and use relationships between operations, such as division as the inverse of multiplication, to solve problems;*
- *understand and use properties of operations, such as the distributivity of multiplication over addition.* (p. 148)

RESEARCH NOTE Students learning multiplication as a conceptual operation need exposure to a variety of models (for example, array and area). Access only to the "multiplication as repeated addition" and the term *times* leads to basic misunderstandings of multiplication that complicate future extensions to decimals and fractions (Bell et al. 1989; English and Halford 1995).

Repeated-Addition Model

FIGURE 2-26

Suppose we have a classroom with 5 columns of 4 chairs each, as shown in Figure 2-26. How many chairs are there altogether? We can think of this as combining 5 sets of 4 objects into a single set.

The 5 columns of 4 suggest the following addition:

$$\underbrace{4 + 4 + 4 + 4 + 4}_{\text{five 4s}} = 20$$

We write $4 + 4 + 4 + 4 + 4$ as $5 \cdot 4$ and say "5 times 4" or "5 multiplied by 4." The advantage of the multiplication notation is evident when the number of addends is great. Thus, if we have 25 columns of 4 chairs each, we can find the total number of chairs by adding 25 fours, or $25 \cdot 4$.

HISTORICAL NOTE

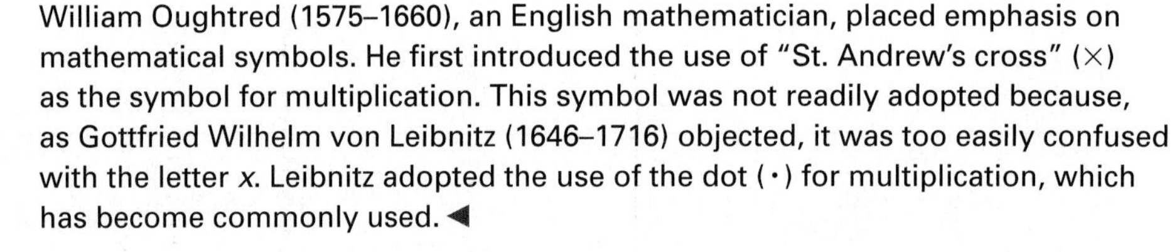

William Oughtred (1575–1660), an English mathematician, placed emphasis on mathematical symbols. He first introduced the use of "St. Andrew's cross" (×) as the symbol for multiplication. This symbol was not readily adopted because, as Gottfried Wilhelm von Leibnitz (1646–1716) objected, it was too easily confused with the letter x. Leibnitz adopted the use of the dot (·) for multiplication, which has become commonly used.

The *repeated-addition* model can be illustrated in several ways, including the use of a number line and the use of arrays. For example, using colored rods of length 4, we could show that the combined length of five of the rods can be found by joining the rods end-to-end, as in Figure 2-27(a). Figure 2-27(b) shows the process using arrows on a number line.

FIGURE 2-27

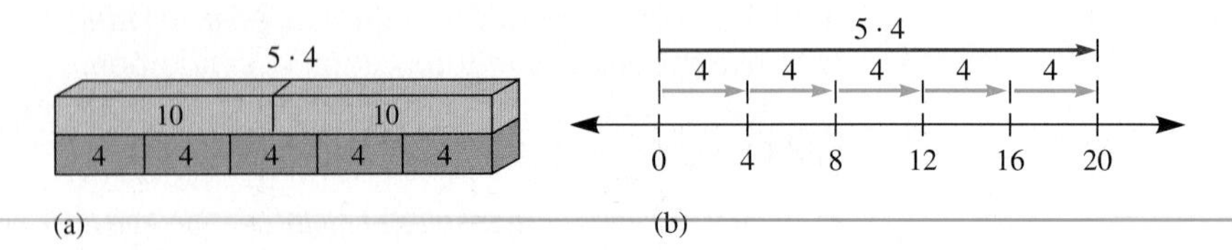

The constant feature on a calculator can help relate multiplication to addition. Students can find products on the calculator without using the × key. For example, if a calculator has the *constant feature*, then 5×3 can be found by pressing [+] [3] [=] [=] [=] [=] [=]. Each press of the equal sign will add 3 to the display. (Some calculators will work differently.)

The Array and Area Models

Another representation of repeated addition that is useful in exploring multiplication of whole numbers is an *array*. In Figure 2-28(a), we cross sticks to create intersection points, thus forming an array of points. The number of points on a single vertical stick is 4 and there are 5 sticks, forming a total of $4 \cdot 5$ points in the array. In Figure 2-28(b), the array is shown as a 4-by-5 grid. The number of unit squares required to fill in the grid is 20. These models motivate the following definition of multiplication of whole numbers.

FIGURE 2-28

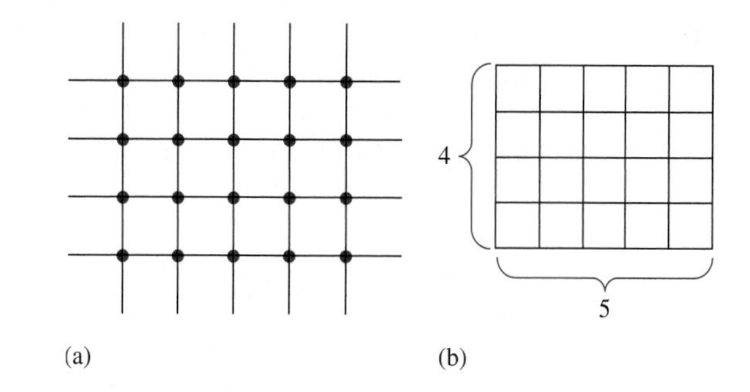

Definition of Multiplication of Whole Numbers

For any whole numbers a and $n \neq 0$,

$$n \cdot a = \underbrace{a + a + a + \cdots + a}_{n \text{ terms}}.$$

If $n = 0$, then $0 \cdot a = 0$.

Cartesian-Product Model

The *Cartesian-product* model offers another way to discuss multiplication. Suppose you can order a soyburger on light or dark bread with one condiment: mustard, mayonnaise, or horseradish. To show the number of different soyburger orders that a waiter could write for the cook, we use a *tree diagram*. The ways of writing the order are listed in Figure 2-29, where the bread is chosen from the set $B = \{\text{light, dark}\}$ and the condiment is chosen from the set $C = \{\text{mustard, mayonnaise, horseradish}\}$.

FIGURE 2-29

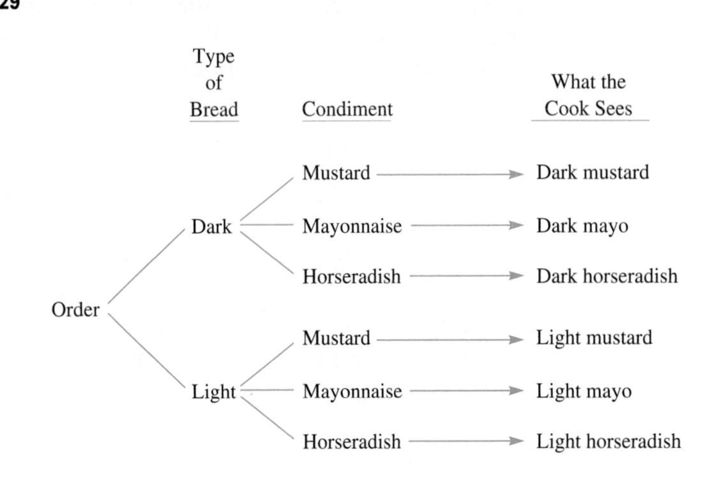

Each order can be written as an ordered pair, for example, (dark, mustard). The set of ordered pairs forms the Cartesian product $B \times C$. The Fundamental Counting Principle tells us that the number of ordered pairs in $B \times C$ is $2 \cdot 3$.

The preceding discussion demonstrates how multiplication of whole numbers can be defined in terms of Cartesian products. Thus, an alternative definition of multiplication of whole numbers follows.

Alternative Definition of Multiplication of Whole Numbers

For finite sets A and B, if $n(A) = a$ and $n(B) = b$, then $a \cdot b = n(A \times B)$.

product ◆ factors

REMARK In this definition, sets A and B do not have to be disjoint. The expression $a \cdot b$, or simply ab, is the **product** of a and b, and a and b are **factors**. Also, note that $A \times B$ indicates the Cartesian product, not multiplication. We multiply numbers, not sets.

NOW TRY THIS 2-17

How would you use the repeated-addition definition of multiplication to explain to a child unfamiliar with both the Fundamental Counting Principle and the concept of cross product that the number of possible outfits consisting of a shirt and pants combination—given 6 shirts and 5 pairs of pants—is $6 \cdot 5$?

The following four problems illustrate why each of the four models for multiplication are useful. In all four problems the answer is 15 but each can be thought of using a different model. Work through each problem using the suggested model.

1. *Repeated-addition model* One piece of gum costs 5¢, how much do three pieces cost?
2. *Number-line model* If Al walks 5 mph for 3 hr, how far has he walked?

3. *Area model* If a carpet is 5 ft by 3 ft, what is the area of the carpet?
4. *Cartesian-product model* Al has five shirts and three pairs of pants, how many different shirt-pant combinations are possible?

Properties of Whole-Number Multiplication

The set of whole numbers is *closed* under multiplication. That is, if we multiply any two whole numbers, the result is a whole number. This property is referred to as the *closure property of multiplication of whole numbers*. Multiplication on the set of whole numbers, like addition, has the commutative, associative, and identity properties.

Properties of Multiplication of Whole Numbers

Closure property of multiplication of whole numbers For whole numbers a and b, $a \cdot b$ is a unique whole number.

Commutative property of multiplication of whole numbers For whole numbers a and b, $a \cdot b = b \cdot a$.

Associative property of multiplication of whole numbers For whole numbers a, b, and c, $(a \cdot b) \cdot c = a \cdot (b \cdot c)$.

Identity property of multiplication of whole numbers There is a unique whole number 1 such that for any whole number a, $a \cdot 1 = a = 1 \cdot a$.

Zero multiplication property of whole numbers For whole number a, $a \cdot 0 = 0 = 0 \cdot a$.

FIGURE 2-30

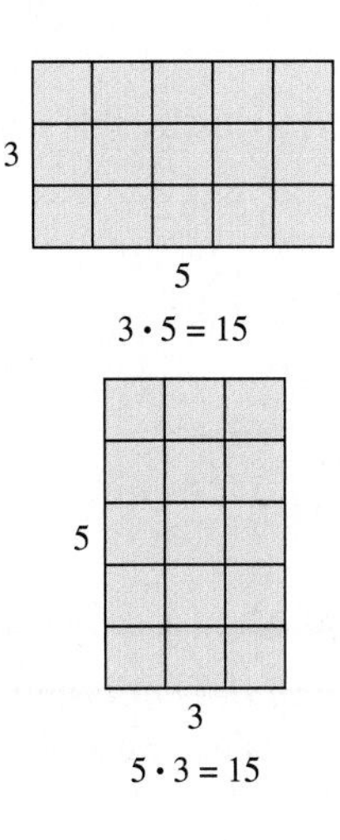

The *commutative property of multiplication of whole numbers* is illustrated easily by building a 3-by-5 grid and then turning it sideways, as shown in Figure 2-30. We see that the number of 1×1 squares present in either case is 15, that is, $3 \cdot 5 = 15 = 5 \cdot 3$. The commutative property can be verified by recalling that $n(A \times B) = n(B \times A)$.

The *associative property of multiplication of whole numbers* can be illustrated as follows. Suppose $a = 3$, $b = 5$, and $c = 4$. In Figure 2-31(a), we see a picture of $3 \cdot (5 \cdot 4)$ blocks. In Figure 2-31(b), we see the same blocks, this time arranged as $(3 \cdot 5) \cdot 4$. Because both sets of blocks in Figure 2-31(a) and (b) compress to the set shown in Figure 2-31(c), we see that $3 \cdot (5 \cdot 4) = (3 \cdot 5) \cdot 4$. The associative property is useful in computations such as the following:

$$3 \cdot 40 = 3 \cdot (4 \cdot 10) = (3 \cdot 4) \cdot 10 = 12 \cdot 10 = 120.$$

FIGURE 2-31

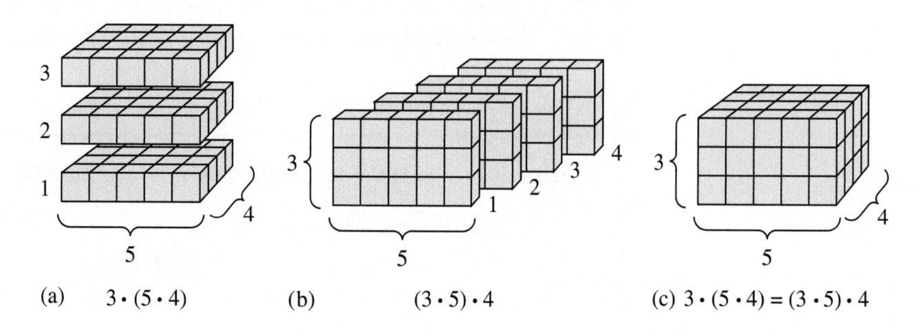

The *multiplicative identity for whole numbers* is 1. For example, $3 \cdot 1 = 1 + 1 + 1 = 3$. In general, for any whole number a,

$$a \cdot 1 = \underbrace{1 + 1 + 1 + \ldots + 1}_{a \text{ terms}} = a..$$

Thus, $a \cdot 1 = a$, which, along with the commutative property for multiplication, implies that $a \cdot 1 = a = 1 \cdot a$. Cartesian products can also be used to show that $a \cdot 1 = a = 1 \cdot a$.

Next, consider multiplication involving 0. For example $6 \cdot 0 = 0 + 0 + 0 + 0 + 0 + 0 = 0$. Thus we see that multiplying 0 by 6 yields a product of 0 and, by commutativity, $0 \cdot 6 = 0$. This is an example of the *zero multiplication property*. This property can also be verified by using the definition of multiplication in terms of Cartesian products.

The Distributive Property of Multiplication over Addition

The next property we investigate is the basis for understanding multiplication algorithms for whole numbers. The area of the large rectangle in Figure 2-32 equals the sum of the areas of the two smaller rectangles and hence $5 \cdot (3 + 4) = 5 \cdot 3 + 5 \cdot 4$.

FIGURE 2-32

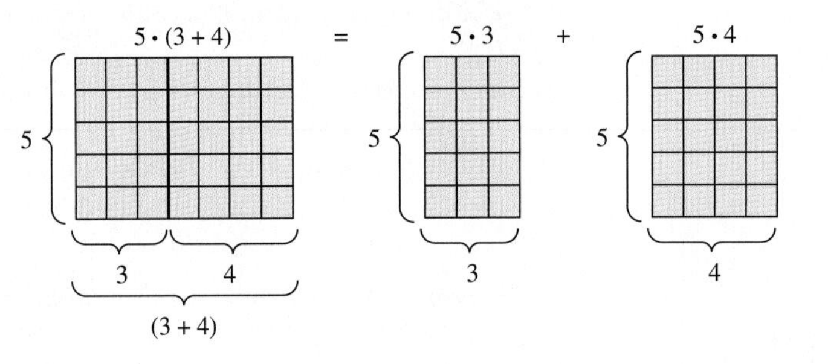

The properties of addition and multiplication also can be used to justify this result:

$$5 \cdot (3 + 4) = \underbrace{(3 + 4) + (3 + 4) + (3 + 4) + (3 + 4) + (3 + 4)}_{\text{Five terms}}$$

Definition of multiplication

$$= (3 + 3 + 3 + 3 + 3) + (4 + 4 + 4 + 4 + 4)$$

Commutative and associative properties of addition

$$= 5 \cdot 3 + 5 \cdot 4 \quad \text{Definition of multiplication}$$

This example illustrates the *distributive property of multiplication over addition* for whole numbers. Because in algebra it is customary to write $a \cdot b$ as ab, we state the distributive property of multiplication over addition as follows:

Property

Distributive property of multiplication over addition for whole numbers For any whole numbers a, b, and c,

$$a(b + c) = ab + ac.$$

REMARK Because the commutative property of multiplication of whole numbers holds, the distributive property of multiplication over addition can be rewritten as $(b + c)a = ba + ca$.

The distributive property can be generalized to any finite number of terms. For example, $a(b + c + d) = ab + ac + ad$.

The distributive property can be written as

$$ab + ac = a(b + c).$$

This is commonly referred to as *factoring*. Thus, the factors of $ab + ac$ are a and $(b + c)$.

Students find the distributive property of multiplication over addition useful when doing mental mathematics. For example, $13 \cdot 7 = (10 + 3) \cdot 7 = 10 \cdot 7 + 3 \cdot 7 = 70 + 21 = 91$. The distributive property of multiplication over addition is important in the study of algebra and in developing algorithms for arithmetic operations, as will be seen in Chapter 3. For example, it is used to combine like terms when we work with variables, as in $3x + 5x = (3 + 5)x = 8x$ or $3ab + 2b = (3a + 2)b$.

EXAMPLE 2-15

a. Use an area model to show that $(x + y)(z + w) = xz + xw + yz + yw$.

b. Use the distributive property of multiplication over addition to justify the result in part (a).

FIGURE 2-33

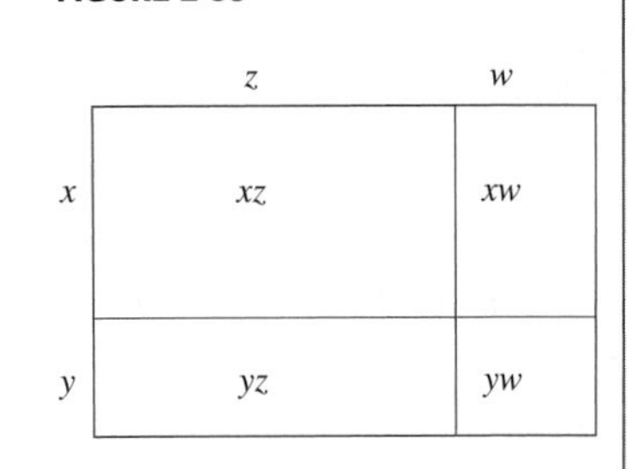

Solution **a.** Consider the rectangle in Figure 2-33 whose width is $x + y$ and whose length is $z + w$. The area of the entire rectangle is $(x + y)(z + w)$. If we divide the rectangle into smaller rectangles as shown, we notice that the sum of the areas of the four smaller rectangles is $xz + xw + yz + yw$. Because the area of the original rectangle equals the sum of the areas of the smaller rectangles, the result follows.

b. To apply the distributive property of multiplication over addition, we think about $x + y$ as one number and proceed as follows:

$(x + y)(z + w) = (x + y)z + (x + y)w$	The distributive property of multiplication over addition
$= xz + yz + xw + yw$	The distributive property of multiplication over addition
$= xz + xw + yz + yw$	The commutative property of addition

The properties of whole-number multiplication can reduce the 100 basic multiplication facts involving numbers 0–9 that students have to learn. For example, 19 facts involve multiplication by 0 and 17 more have a factor of 1. Therefore, knowing the zero multiplication property and the identity multiplication property allows students to know 36 facts. Next, 8 facts are doubles, such as $5 \cdot 5$, that students seem to know and that leaves 56 facts. The commutative property cuts this number in half because if students know $7 \cdot 9$, then by the commutative property they know $9 \cdot 7$. This leaves 28 facts that students can learn or use the associative and distributive properties to figure out. For example, $6 \cdot 5$ can be thought of as $(5 + 1) \cdot 5 = 5 \cdot 5 + 1 \cdot 5$, or 30.

Order of Operations

Difficulties involving the order of arithmetic operations sometimes arise. For example, many students will treat $2 + 3 \cdot 6$ as $(2 + 3)6$, while others will treat it as $2 + (3 \cdot 6)$. In the first case, the value is 30; in the second case, the value is 20. To avoid confusion, mathematicians agree that when no parentheses are present, multiplications are performed *before* additions. Thus, $2 + 3 \cdot 6 = 2 + 18 = 20$. This order of operations is not built into calculators that display an incorrect answer of 30.

Division of Whole Numbers

We discuss division using three models: the *set (partition)* model, the *missing-factor* model, and the *repeated-subtraction* model.

Set (Partition) Model

Suppose we have 18 cookies and want to give an equal number of cookies to each of three friends: Bob, Dean, and Charlie. How many should each person receive? If we draw a picture, we can see that we can divide (or partition) the 18 cookies into three sets, with an equal number of cookies in each set. Figure 2-34 shows that each friend received 6 cookies.

FIGURE 2-34

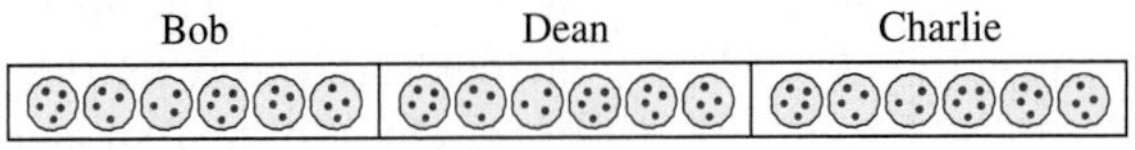

Bob Dean Charlie

The answer may be symbolized as $18 \div 3 = 6$. Thus, $18 \div 3$ is the number of cookies in each of three disjoint sets whose union has 18 cookies. In this approach to division, we partition a set into a number of equivalent subsets.

Missing-Factor Model

Another strategy for dividing 18 cookies among three friends is to use the *missing-factor* model. If each friend receives c cookies, then the three friends receive $3c$, or 18, cookies. Hence, $3c = 18$. Since $3 \cdot 6 = 18$, then $c = 6$. We have answered the division computation by using multiplication. This leads us to the following definition of division of whole numbers.

Definition of Division of Whole Numbers

For any whole numbers a and b, with $b \neq 0$, $a \div b = c$ if and only if c is the unique whole number such that $b \cdot c = a$.

dividend ◆ divisor ◆ quotient

REMARK The number a is the **dividend**, b is the **divisor**, and c is the **quotient**. Note that $a \div b$ can also be written as $\frac{a}{b}$ or $b\overline{)a}$.

Repeated-Subtraction Model

Suppose we have 18 cookies and want to package them in cookie boxes that hold 6 cookies each. How many boxes are needed? We could reason that if one box is filled, then we would have $18 - 6$ (or 12) cookies left. If one more box is filled, then there are $12 - 6$ (or 6) cookies left. Finally, we could place the last 6 cookies in a third box. This discussion can be summarized by writing $18 - 6 - 6 - 6 = 0$. We have found by repeated subtraction that $18 \div 6 = 3$. Treating division as repeated subtraction works well as long as there is no remainder. If there is a remainder, then fractions will arise, as seen in Chapter 6.

Calculators can be used to show that division of whole numbers can be thought of as repeated subtraction. For example, consider $135 \div 15$. If the calculator has a constant key, [K], press [1] [5] [−] [K] [1] [3] [5] [=] . . . and then count how many times you must press the [=] key in order to make the display read 0. Calculators with a different constant feature may require a different sequence of entries. For example, if the calculator has an automatic constant, we can press [1] [3] [5] [−] [1] [5] [=] and then count the number of times we press the [=] key to make the display read [0].

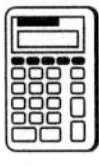

The Division Algorithm

Just as subtraction of whole numbers is not closed, division of whole numbers is not closed. For example, to find $27 \div 5$, we look for a whole number c such that $5c = 27$.

Table 2-5 shows several products of whole numbers times 5. Since 27 is between 25 and 30, there is no whole number c such that $5c = 27$. Because no whole number c satisfies this equation, we see that $27 \div 5$ has no meaning in the set of whole numbers, and the set of whole numbers is not closed under division.

Table 2-5

$5 \cdot 1$	$5 \cdot 2$	$5 \cdot 3$	$5 \cdot 4$	$5 \cdot 5$	$5 \cdot 6$
5	10	15	20	25	30

Even though the set of whole numbers is not closed under division, the operation of division is meaningful with whole numbers. For example, if 27 apples were to be divided among five students, each student would receive 5 apples and 2 apples would remain. The number 2 is the **remainder**. Thus, 27 contains five 5s with a remainder of 2. Observe that the remainder is a whole number less than 5. This operation is illustrated in Figure 2-35. The concept illustrated is the **division algorithm**. (In Chapter 6, we discuss how to "complete" the division using fractions.)

remainder

division algorithm

FIGURE 2-35

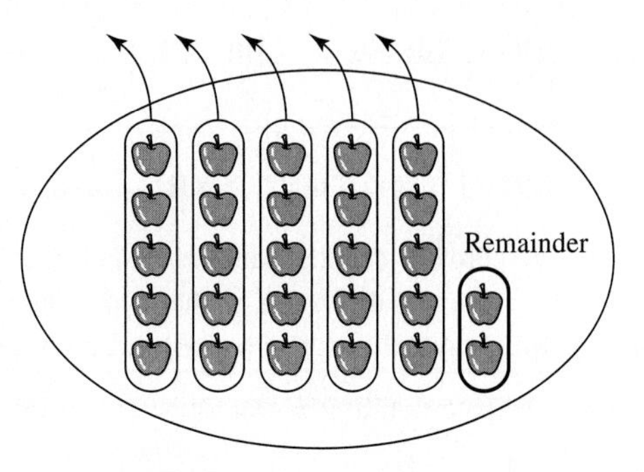

Definition of Division Algorithm

Given any whole numbers a and b with $b \neq 0$, there exist unique whole numbers q (quotient) and r (remainder) such that

$$a = bq + r \quad \text{with } 0 \leq r < b.$$

When a is "divided" by b and the remainder is 0, we say that a is *divisible* by b or that b is a *divisor* of a or that b *divides* a. By the *division algorithm*, a is divisible by b if $a = bq$ for a unique whole number q. Thus, 63 is divisible by 9 because $63 = 9 \cdot 7$. Notice that 63 is also divisible by 7.

Relating Multiplication and Division

In Section 2-3, we saw that subtraction and addition were related as inverse operations and we looked at fact families for both operations. In a similar way, division and multiplication are related. Division is the inverse of multiplication. We can again see this by looking at fact families as shown on the Grade 3 student page on page 121. Notice that question 1 in the *Talk About It* has students think of division as a *repeated-subtraction* model by skip counting backwards from a starting point. Question 2 has students think of division using the *missing-factor* model.

Next consider how the four operations of addition, subtraction, multiplication, and division are related. This is shown in Figure 2-36. Note that addition and subtraction are inverses of each other as are multiplication and division. Also note that multiplication is repeated addition, and division is repeated subtraction.

FIGURE 2-36

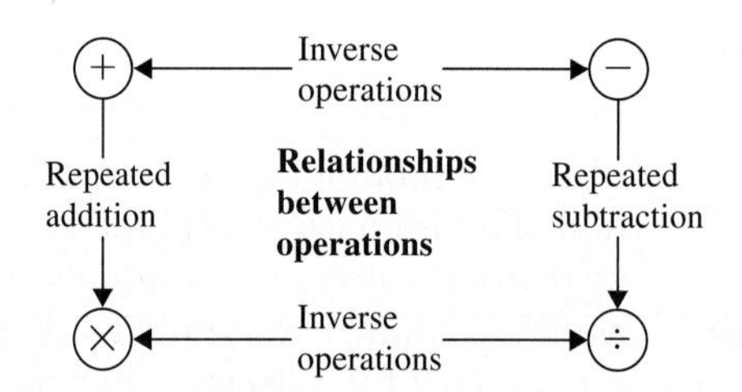

In Section 2-3, we have seen that the set of whole numbers is closed under addition and that addition is commutative and associative and has an identity. Conversely, subtraction did not have these properties. In this section, we have seen that multiplication has some of the same properties that hold for addition. Does it follow that division has some of the same properties as subtraction? Investigate this in Now Try This 2-18.

NOW TRY THIS 2-18

a. Provide counterexamples to show that the set of whole numbers is not closed under division and that division is neither commutative nor associative.

b. Why is 1 not the identity for division?

SAMPLE SCHOOL BOOK PAGE:

RELATING MULTIPLICATION AND DIVISION

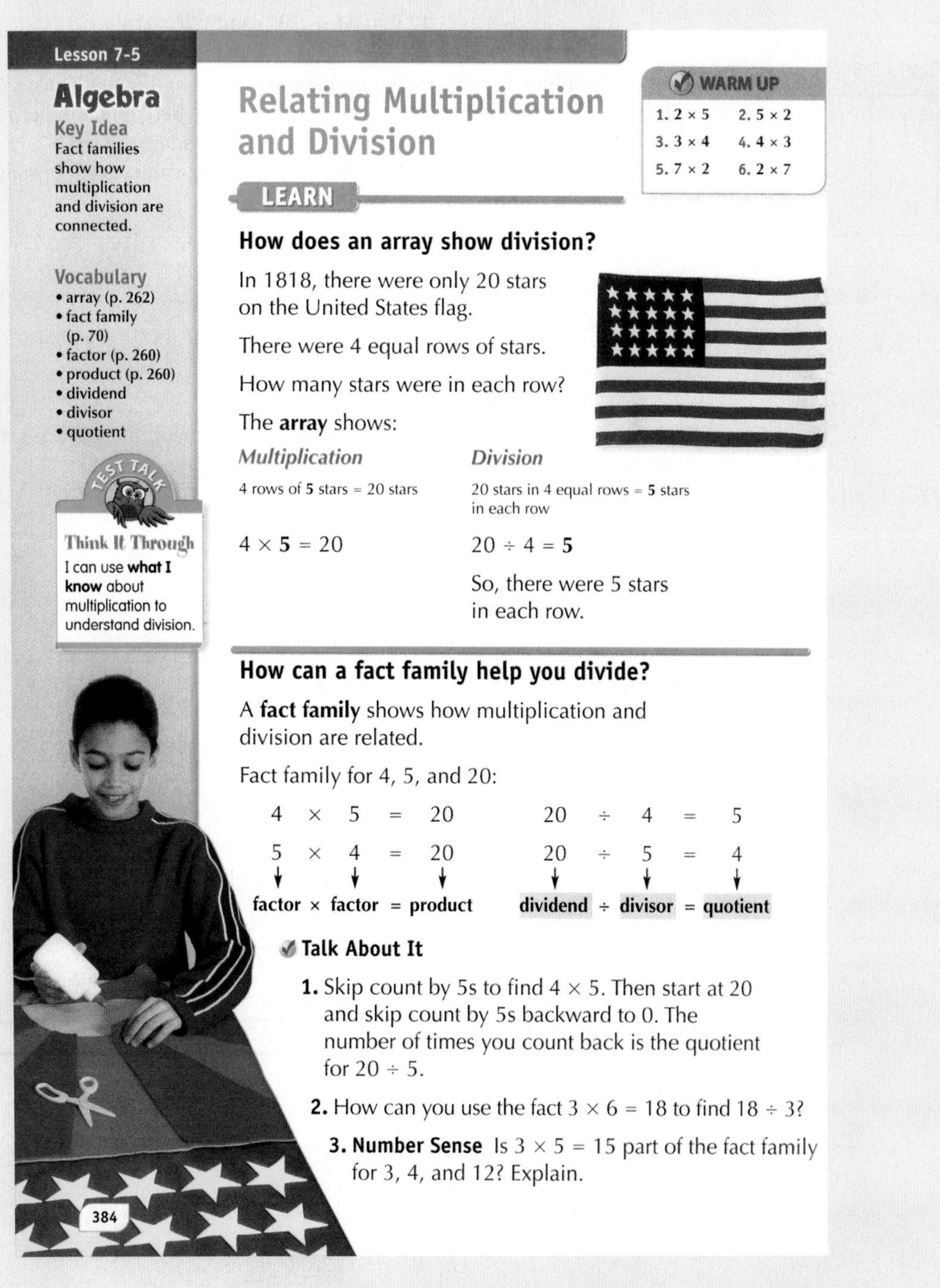

Lesson 7-5

Algebra

Key Idea
Fact families show how multiplication and division are connected.

Vocabulary
- array (p. 262)
- fact family (p. 70)
- factor (p. 260)
- product (p. 260)
- dividend
- divisor
- quotient

TEST TALK

Think It Through
I can use **what I know** about multiplication to understand division.

Relating Multiplication and Division

WARM UP

1. 2×5	2. 5×2
3. 3×4	4. 4×3
5. 7×2	6. 2×7

LEARN

How does an array show division?

In 1818, there were only 20 stars on the United States flag.

There were 4 equal rows of stars.

How many stars were in each row?

The **array** shows:

Multiplication	*Division*
4 rows of **5** stars = 20 stars	20 stars in 4 equal rows = **5** stars in each row
$4 \times \mathbf{5} = 20$	$20 \div 4 = \mathbf{5}$
	So, there were 5 stars in each row.

How can a fact family help you divide?

A **fact family** shows how multiplication and division are related.

Fact family for 4, 5, and 20:

4	×	5	=	20	20	÷	4	=	5
5	×	4	=	20	20	÷	5	=	4
factor	×	**factor**	=	**product**	**dividend**	÷	**divisor**	=	**quotient**

Talk About It

1. Skip count by 5s to find 4×5. Then start at 20 and skip count by 5s backward to 0. The number of times you count back is the quotient for $20 \div 5$.
2. How can you use the fact $3 \times 6 = 18$ to find $18 \div 3$?
3. **Number Sense** Is $3 \times 5 = 15$ part of the fact family for 3, 4, and 12? Explain.

384

Source: Scott Foresman–Addison Wesley Math, Grade 3, 2005 (p. 384).

EXAMPLE 2-16

If 123 is divided by a number and the remainder is 13, what are the possible divisors?

Solution If 123 is divided by b, then from the division algorithm we have

$$123 = bq + 13 \quad \text{and} \quad b > 13.$$

Table 2-6

1	110
2	55
5	22
10	11

Using the definition of subtraction, we have $bq = 123 - 13$, and hence $110 = bq$. Now we are looking for two numbers whose product is 110, where one number is greater than 13. Table 2-6 shows the pairs of whole numbers whose product is 110.

We see that 110, 55, and 22 are the only possible values for b because each is greater than 13.

NOW TRY THIS 2-19

When the marching band was placed in rows of 5, one member was left over. When the members were placed in rows of 6, there was still one member left over. However, when they were placed in rows of 7, nobody was left over. What is the smallest number of members that could have been in the band?

Division by 0 and 1

The whole numbers 0 and 1 deserve special attention with respect to division of whole numbers as seen in the following cartoon.

SAMPLE SCHOOL BOOK PAGE:

DIVISION BY 0 AND 1

Lesson 7-10

Key Idea
Thinking of related multiplication facts can help you understand division rules for 0 and 1.

Dividing with 0 and 1

WARM UP
1. 0 × 3 2. 8 × 1
3. 2 × 0 4. 1 × 9

LEARN

What are the division rules for 0 and 1?

Example A

	What You **Think**	What You **Write**
Divide a number by 1. 4 ÷ 1 =	1 times what number = 4? 1 × 4 = 4 So, 4 ÷ 1 = 4.	4 ÷ 1 = 4 or 1)4 (quotient 4)

Rule: When any number is divided by 1, the quotient is that number.

Example B

Divide a number by itself. 7 ÷ 7 =	7 times what number = 7? 7 × 1 = 7 So, 7 ÷ 7 = 1.	7 ÷ 7 = 1 or 7)7 (quotient 1)

Rule: When any number (except 0) is divided by itself, the quotient is 1.

Example C

Divide zero by a number. 0 ÷ 2 =	2 times what number = 0? 2 × 0 = 0 So, 0 ÷ 2 = 0.	0 ÷ 2 = 0 or 2)0 (quotient 0)

Rule: When zero is divided by a number (except 0) the quotient is 0.

Example D

Divide a number by zero. 3 ÷ 0 =	0 times what number = 3? There is no number that works, so, 3 ÷ 0 cannot be done.	3 ÷ 0 cannot be done.

Rule: You cannot divide a number by 0.

Talk About It

1. How can you tell without dividing that 427 ÷ 1 = 427?

4 ÷ 1 = 4 or 1)4

396

Source: *Scott Foresman–Addison Wesley Math, Grade 3, 2005* (p. 396).

Before reading on, try to find the values of the following three expressions:

1. $3 \div 0$ **2.** $0 \div 3$ **3.** $0 \div 0$

Consider the following explanations:

1. By definition, $3 \div 0 = c$ if there is a unique number c such that $0 \cdot c = 3$. Since the zero property of multiplication states that $0 \cdot c = 0$ for any whole number c, there is no whole number c such that $0 \cdot c = 3$. Thus, $3 \div 0$ is undefined because there is no answer to the equivalent multiplication problem.
2. By definition, $0 \div 3 = c$ if there exists a unique number such that $3 \cdot c = 0$. Because any number times 0 is 0, and in particular $3 \cdot 0 = 0$, then $c = 0$ and $0 \div 3 = 0$. Note that $c = 0$ is the only number that satisfies $3 \cdot c = 0$.
3. By definition, $0 \div 0 = c$ if there is a unique whole number c such that $0 \cdot c = 0$. Notice that for *any* c, $0 \cdot c = 0$. According to the definition of division, c must be unique. Since there is no *unique* number c such that $0 \cdot c = 0$, it follows that $0 \div 0$ is indeterminate, or undefined.

Division involving 0 may be summarized as follows. Let n be any nonzero whole number. Then,

1. $n \div 0$ is undefined;
2. $0 \div n = 0$;
3. $0 \div 0$ is indeterminate, or undefined.

Recall that $n \cdot 1 = n$ for any whole number n. Thus, by the definition of division, $n \div 1 = n$. For example, $3 \div 1 = 3$, $1 \div 1 = 1$, and $0 \div 1 = 0$. A Grade 3 discussion of division by 0 and 1 can be found on the student page on the previous page.

ONGOING ASSESSMENT 2-4

1. For each of the following, find, if possible, the whole numbers that make the equations true:
   a. $3 \cdot \square = 15$
   b. $18 = 6 + 3 \cdot \square$
   c. $\square \cdot (5 + 6) = \square \cdot 5 + \square \cdot 6$
2. In terms of set theory, the product na could be thought of as the number of elements in the union of n sets with a elements in each. If this were the case, what must be true about the sets?
3. Determine if the following sets are closed under multiplication:
   a. $\{0, 1\}$
   b. $\{0\}$
   c. $\{2, 4, 6, 8, 10, \ldots\}$
   d. $\{1, 3, 5, 7, 9, \ldots\}$
   e. $\{1, 4, 7, 10, 13, \ldots\}$
   f. $\{0, 1, 2\}$
4. a. If 5 is removed from the set of whole numbers, is the set closed with respect to addition? Explain.
   b. If 5 is removed from the set of whole numbers, is the set closed with respect to multiplication? Explain.
   c. Answer the same questions as (a) and (b) if 6 is removed from the set of whole numbers.
5. Rename each of the following using the distributive property of multiplication over addition so that there are no parentheses in the final answer:
   a. $(a + b)(c + d)$
   b. $3(x + y + 5)$
   c. $\square(\Delta + \bigcirc)$
   d. $(x + y)(x + y + z)$
   e. $a(b + c) - ac$
   f. $x(y + 1) - x$
6. Place parentheses, if needed, to make each of the following equations true:
   a. $4 + 3 \cdot 2 = 14$
   b. $9 \div 3 + 1 = 4$
   c. $5 + 4 + 9 \div 3 = 6$
   d. $3 + 6 - 2 \div 1 = 7$

7. The generalized distributive property for three terms states that for any whole numbers a, b, c, and d, $a(b + c + d) = ab + ac + ad$. Justify this property using the distributive property for two terms.

8. Using the distributive property of multiplication over addition, we can factor as in $x^2 + xy = x(x + y)$. Use the distributive property to factor each of the following:
   a. $xy + y^2$
   b. $47 \cdot 99 + 47$
   c. $xy + x$
   d. $(x + 1)y + (x + 1)$
   e. $a^2b + ab^2$

9. For each of the following, find whole numbers to make the statement true, if possible:
   a. $18 \div 3 = \square$
   b. $\square \div 76 = 0$
   c. $28 \div \square = 7$

10. A sporting goods store has designs for 6 shirts, 4 pairs of pants, and 3 vests. How many different shirt-pant-vest outfits are possible?

11. Which property is illustrated in each of the following:
   a. $6(5 \cdot 4) = (6 \cdot 5)4$
   b. $6(5 \cdot 4) = 6(4 \cdot 5)$
   c. $6(5 \cdot 4) = (5 \cdot 4)6$
   d. $1(5 \cdot 4) = 5 \cdot 4$
   e. $(3 + 4) \cdot 0 = 0$
   f. $(3 + 4)(5 + 6) = (3 + 4)5 + (3 + 4)6$

12. Students are overheard making the following statements. What properties justify their statements?
   a. I know that $9 \cdot 7$ is either 63 or 69 and I know they can't both be right.
   b. I know if I remember what $7 \cdot 9$ is, then I also know what $9 \cdot 7$ is.
   c. I know that $9 \cdot 0$ is 0 because I know that any number times 0 is 0.
   d. To find $9 \cdot 6$ I just remember that $9 \cdot 5$ is 45 and so $9 \cdot 6$ is just 9 more than 45, or 54.
   e. Any number times 1 is the same as we started, so $9 \cdot 1$ is 9.

13. The product $6 \cdot 14$ can be found by thinking of the problem as $6(10 + 4) = (6 \cdot 10) + (6 \cdot 4) = 60 + 24 = 84$.
   a. What property is being used?
   b. Use this technique to mentally compute $12 \cdot 32$.

14. The distributive property of multiplication over subtraction is

$$a(b - c) = ab - ac.$$

Use this property to find each of the following:
   a. $9(10 - 2)$
   b. $20(8 - 3)$

15. Show that $(a + b)^2 = a^2 + 2ab + b^2$ using
   a. The distributive property of multiplication over addition.
   b. An area model.

16. Show that if $b > c$, then $a(b - c) = ab - ac$ using:
   a. an area a model suggested by the given figure (express the shaded area in two different ways).

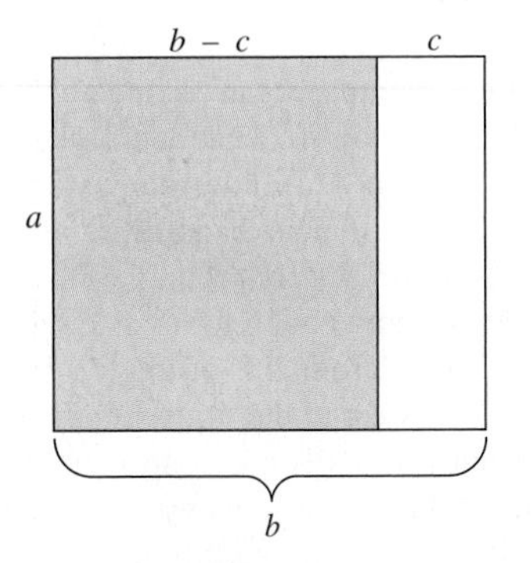

   b. the definition of subtraction in terms of addition and the distributive property of multiplication over addition.
   c. Name the property you have justified in parts (a) and (b).

17. If a and b are whole numbers with $a > b$, use the rectangles in the figure to explain why $(a + b)^2 - (a - b)^2 = 4ab$.

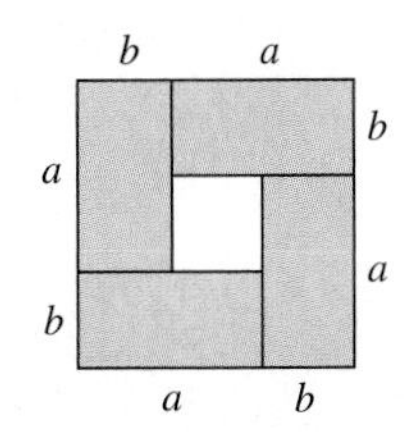

18. Show that the left-hand side of the equation is equal to the right-hand side and give a reason for every step.
   a. $(ab)c = (ca)b$
   b. $(a + b)c = c(b + a)$

19. Factor each of the following:
   a. $xy - y^2$
   b. $xy - y$
   c. $47 \cdot 101 - 47$
   d. $(x + 1)y - (x + 1)$
   e. $ab^2 - ba^2$

20. Rewrite each of the following division problems as a multiplication problem:
   a. $40 \div 8 = 5$
   b. $326 \div 2 = x$
   c. $48 \div x = 16$
   d. $x \div 5 = 17$

21. Think of a number. Multiply it by 2. Add 2. Divide by 2. Subtract 1. How does the result compare with your original number? Will this work all the time? Explain your answer.

22. Show that, in general, each of the following is false if a, b, and c are whole numbers:
   a. $(a \div b) \div c = a \div (b \div c)$
   b. $a \div (b + c) = (a \div b) + (a \div c)$

23. Suppose c is a divisor of a and of b. Show that $(a + b) \div c = (a \div c) + (b \div c)$ using
   a. a model.
   b. the definition of division in terms of multiplication and the distributive property of multiplication over addition.

24. Find the solution set for each of the following:
    a. $5x + 2 = 22$
    b. $3x + 7 = x + 13$
    c. $3(x + 4) = 18$

25. Millie and Samantha began saving money at the same time. Millie plans to save \$3 a month, and Samantha plans to save \$5 a month. After how many months will Samantha have exactly \$10 more than Millie?

26. String art is formed by connecting evenly spaced nails on the vertical and horizontal axes by line segments. Connect the nail farthest from the origin on the vertical axis with the nail closest to the origin on the horizontal axis. Continue until all nails are connected, as shown in the figure that follows. How many intersection points are created with 10 nails on each axis?

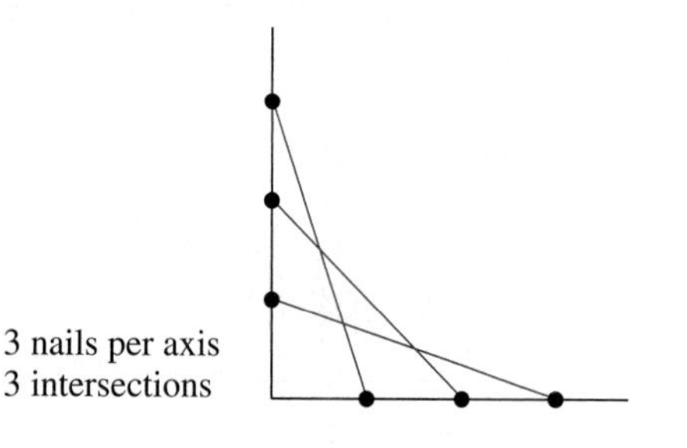

27. There were 17 sandwiches for 7 people on a picnic. How many whole sandwiches were there for each person if they were divided equally? How many were left over?

28. a. Find all pairs of whole numbers whose product is 36.
    b. Plot the points found in (a) on a grid.
    c. Compare the pattern shape formed by the points to the graph of the pattern shape that could be found using all pairs of whole numbers whose sum is 36.

29. A new model of car is available in 4 exterior colors and 3 interior colors. Use a tree diagram and specific colors to show how many color schemes are possible for the car.

30. Students were divided into 8 teams with 9 on each team. Later the same students were divided into teams with 6 on each team. How many teams were there then?

31. To find $7 \div 5$ on the calculator, press [7] [÷] [5] [=], which yields 1.4. To find the whole-number remainder, ignore the decimal portion of 1.4, multiply $5 \cdot 1$, and subtract this product from 7. The result is the remainder. Use a calculator to find the whole-number remainder for each of the following divisions:
    a. $28 \div 5$ b. $32 \div 10$
    c. $29 \div 3$ **d.** $41 \div 7$
    e. $49{,}382 \div 14$

32. Jonah has a large collection of marbles. He notices that if he borrows 5 marbles from a friend, he can arrange the marbles in rows of 13 each. What is the remainder when he divides his original number of marbles by 13?

33. Is it possible to find a whole number less than 100 that when divided by 10 leaves remainder 4 and when divided by 47 leaves remainder 17?

34. A number leaves remainder 6 when divided by 10. What is the remainder when the number is divided by 5? Justify your reasoning.

35. In the following problems, use only the designated number keys on the calculator. You may use any function keys.
    a. Use the keys [1], [9], and [7] exactly once each in any order and use any operations available to write as many of the whole numbers as possible from 1 to 20. For example, $9 - 7 - 1 = 1$ and $1 \cdot 9 - 7 = 2$.
    b. Use the [4] key as many times as desired with any operations to display 13.
    c. Use the [2] key three times with any operations to display 24.
    d. Use the [1] key five times with any operations to display 100.

36. In each of the following, tell what computation must be done last:
    a. $5(16 - 7) - 18$ **b.** $54/(10 - 5 + 4)$
    c. $(14 - 3) + (24 \cdot 2)$ **d.** $21{,}045/345 + 8$

37. Is $x \div x$ always equal to 1? Explain your answer.

38. Is $x \cdot x$ ever equal to x? Explain your answer.

39. Describe all pairs of whole numbers whose sum and product are the same.

40. Write an algebraic expression for each of the following:
    a. Width of a rectangle whose area is A and length is l
    b. Feet, f, in yards
    c. Hours, h, in minutes
    d. Days, d, in weeks

41. Find infinitely many whole numbers that leave remainder 3 upon division by 5.

42. The operation ⊙ is defined on the set $S = \{a, b, c\}$, as shown in the following table. For example, $a \odot b = b$ and $b \odot a = b$.

⊙	a	b	c
a	a	b	c
b	b	a	c
c	c	c	c

    a. Is S closed with respect to ⊙?
    b. Is ⊙ commutative on S?
    c. Is there an identity for ⊙ on S? If yes, what is it?
    d. Is ⊙ associative on S?

Communication

43. Suppose a student argued that $0 \div 0 = 0$ because "nothing divided by nothing" is "nothing." How would you help that person?

44. Sue claims the following is true by the distributive law, where a and b are whole numbers:

$$3(ab) = (3a)(3b)$$

How might you help her?

45. Can 0 be the identity for multiplication? Explain why or why not.

46. Suppose you forgot the product of $9 \cdot 7$. Give several ways that you could find the product using different multiplication facts and properties.

Open-Ended

47. Describe a real-life situation that could be represented by the expression $5 + 8 \cdot 6$.

48. How would you explain to a child that an even number has the form $2q$ and an odd number has the form $2q + 1$, where q is a whole number?

Cooperative Learning

49. Multiplication facts that most children have memorized can be stated in the table that is partially filled:

×	1	2	3	4	5	6	7	8	9
1									
2									
3									
4				16					
5							35		
6									
7									
8									72
9									81

a. Fill out the table of multiplication facts. Find as many patterns as you can. List all the patterns that your group discovered and explain why some of those patterns occur in the table.

b. How can the multiplication table be used to solve division problems?

c. Consider the odd number 35 shown in the multiplication table. Consider all the numbers that surround it. Note that they are all even. Does this happen for all odd numbers in the table? Explain why or why not.

Review Problems

50. Give a set that is not closed under addition.

51. Are the whole numbers commutative under subtraction? If not, give a counterexample.

52. Explain why if a, b, and c are whole numbers, and $c \neq 0$, then $a < b$ implies $ac < bc$.

Third International Mathematics and Science Study (TIMSS) Questions

In Toshi's class there are twice as many girls as boys. There are 8 boys in the class. What is the total number of boys and girls in the class?

a. 12
b. 16
c. 20
d. 24

TIMSS 2003, Grade 4

A piece of rope 204 cm long is cut into 4 equal pieces. Which of these gives the length of each piece in centimeters?

a. $204 + 4$
b. 204×4
c. $204 - 4$
d. $204 \div 4$

TIMSS 2003, Grade 4

LABORATORY ACTIVITY

Enter a positive whole number less than 20 on the calculator. If the number is even, divide it by 2; if it is odd, multiply it by 3 and add 1. Next, use the number on the display. Follow the given directions. Repeat the process.

1. Will the display eventually reach 1?
2. Which number less than 20 takes the most steps before reaching 1?
3. Do even or odd numbers reach 1 more quickly?
4. Investigate what happens with numbers greater than 20.

2-5 Functions

The *Principles and Standards* point out that the concept of a function is central in all of mathematics and in particular in algebra.

By viewing algebra as a strand in the curriculum from prekindergarten on, teachers can help students build a solid foundation of understanding and experience as a preparation for more sophisticated work in algebra in the middle grades and high school. For example, systematic experience with patterns can build up to an understanding of the idea of function (Erick Smith forthcoming), and experience with numbers and their properties lays a foundation for later work with symbols and algebraic expressions. By learning that situations often can be described using mathematics, students can begin to form elementary notions of mathematical modeling. (p. 37)

Also the *Principles and Standards* points out that

As they progress from preschool through high school, students should develop a repertoire of many types of functions. (p. 38)

In this section, we will explore functions as *rules*, *machines*, *equations*, *arrow diagrams*, *tables*, *ordered pairs*, and *graphs*.

Functions as Rules

The following is an example of a game called "guess my rule," often used to introduce the concept of a function.

When Tom said 2, Noah said 5. When Dick said 4, Noah said 7. When Mary said 10, Noah said 13. When Liz said 6, what did Noah say? What is Noah's rule?

The answer to the first question may be 9, and the rule could be "Take the original number and add 3"; that is, for any number n, Noah's answer is $n + 3$.

EXAMPLE 2-17

Guess the teacher's rule for the following responses:

a.

You	Teacher
1	3
0	0
4	12
10	30

b.

You	Teacher
2	5
3	7
5	11
10	21

c.

You	Teacher
2	0
4	0
7	1
21	1

Solution **a.** The teacher's rule could be "Multiply the given number n by 3," that is, $3n$.
b. The teacher's rule could be "Double the original number n and add 1," that is, $2n + 1$.
c. The teacher's rule could be "If the number n is even, answer 0; if the number is odd, answer 1." Another possible rule is "If the number is less than 5, answer 0; if greater than or equal to 5, answer 1."

▶ **HISTORICAL NOTE**

The Babylonians (ca. 2000 BCE) probably had a working idea of what a function was. To them, it was a table or a correspondence. René Descartes (1637), Gottfried Wilhelm von Leibnitz (1692), Johann Bernoulli (1718), Leonhard Euler (1750), Joseph Louis Lagrange (1800), and Jean Joseph Fourier (1822) were among the mathematicians contributing to the notion of a function. Leonhard Euler in 1734 may have been the first to use the notation $f(x)$. ◀

Functions as Machines

Another way to prepare students for the concept of a function is by using a "function machine." The following student page shows an example of a function machine. What goes in the machine is referred to as *input* and what comes out as *output*. Thus, on the student page, if the input to the first function machine is 8, the output is 16. Note that the output here is denoted by y. In later grades, a special notation for the output is used. For any input element x, the output is denoted $f(x)$, read "f of x." For the function machine pictured on the student page, when the input is 17, the output could be written $f(17)$. Because the output is 34, we have $f(17) = 34$. Because the machine works according to the rule "double it," $f(x) = 2x$ and $f(17) = 2 \cdot 17 = 34$. Do problems 1 through 6 on the student page.

REMARK On most graphing calculators, the function notation used is Y_1, Y_2, Y_3, . . . , and so on. Here, Y_1 acts like y as described on the student page or like $f(x)$ if the function rule is written in terms of x.

EXAMPLE 2-18

Consider the function machine in Figure 2-37. What will happen if the numbers 0, 1, 3, and 6 are entered?

Solution If the numbers output are denoted by $f(x)$, the corresponding values can be described using Table 2-7.

SAMPLE SCHOOL BOOK PAGE:

ALGEBRA

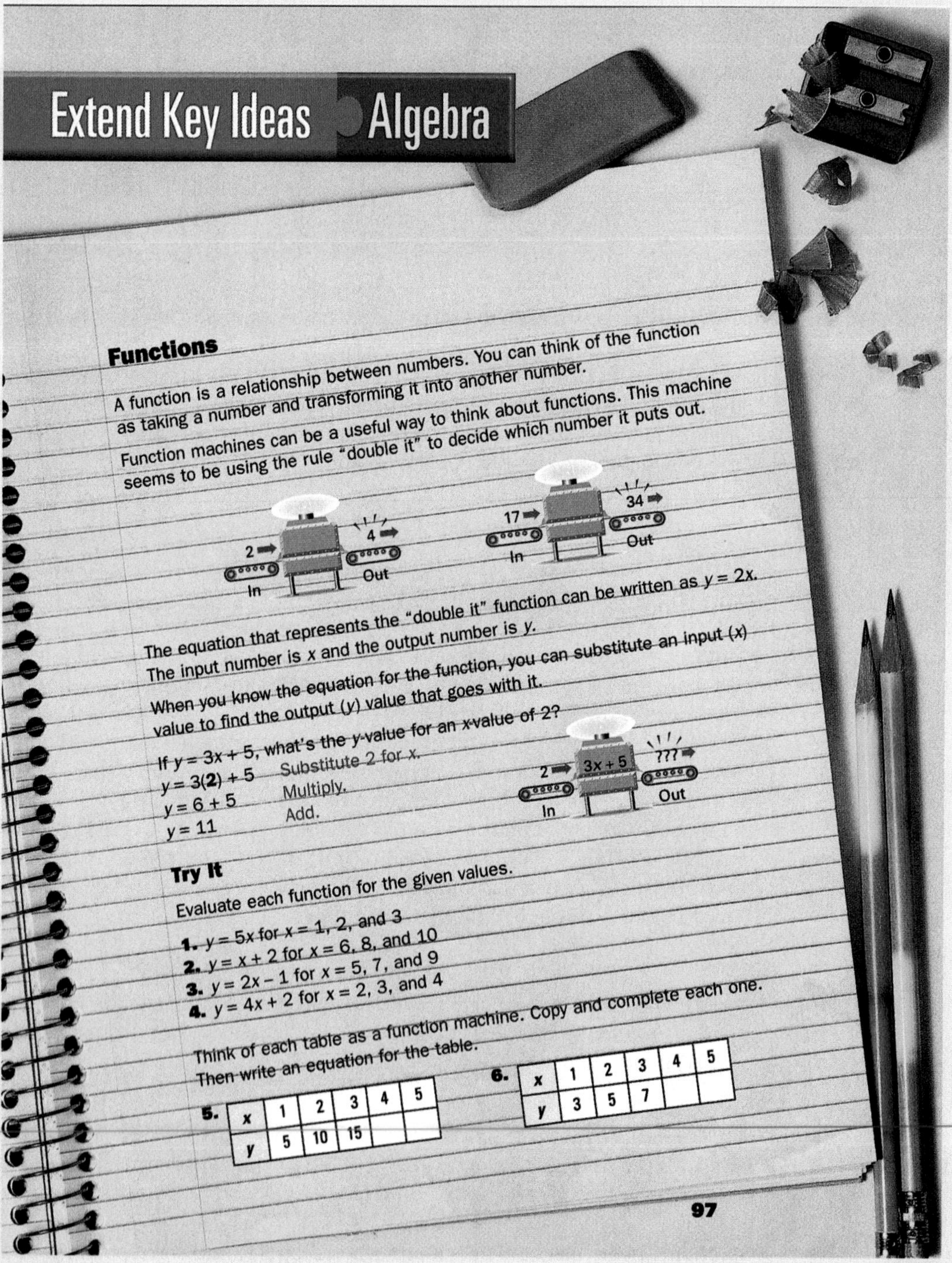

Extend Key Ideas Algebra

Functions

A function is a relationship between numbers. You can think of the function as taking a number and transforming it into another number.

Function machines can be a useful way to think about functions. This machine seems to be using the rule "double it" to decide which number it puts out.

2 → In → 4 Out

17 → In → 34 Out

The equation that represents the "double it" function can be written as $y = 2x$. The input number is x and the output number is y.

When you know the equation for the function, you can substitute an input (x) value to find the output (y) value that goes with it.

If $y = 3x + 5$, what's the y-value for an x-value of 2?

$y = 3(\mathbf{2}) + 5$ Substitute 2 for x.

$y = 6 + 5$ Multiply.

$y = 11$ Add.

2 → In → $3x + 5$ → ??? Out

Try It

Evaluate each function for the given values.

1. $y = 5x$ for $x = 1, 2,$ and 3
2. $y = x + 2$ for $x = 6, 8,$ and 10
3. $y = 2x - 1$ for $x = 5, 7,$ and 9
4. $y = 4x + 2$ for $x = 2, 3,$ and 4

Think of each table as a function machine. Copy and complete each one. Then write an equation for the table.

5.

x	1	2	3	4	5
y	5	10	15		

6.

x	1	2	3	4	5
y	3	5	7		

97

Source: Scott Foresman-Addison Wesley Middle School Math Course 2, 2002 (p. 97).

FIGURE 2-37

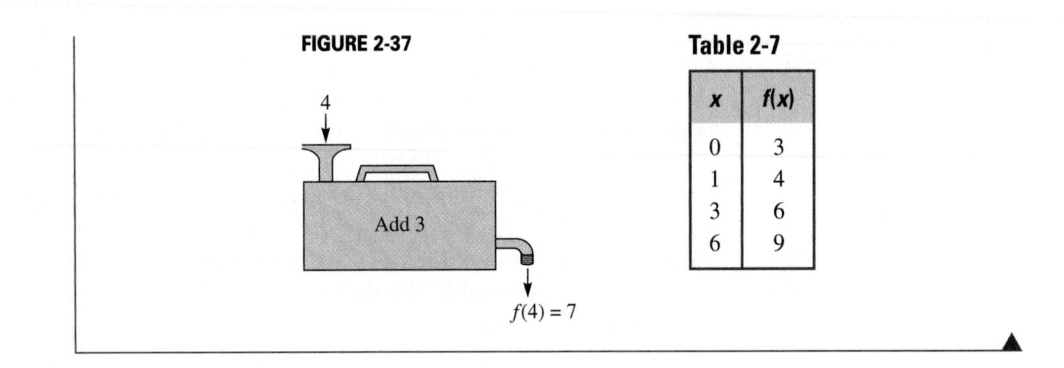

Table 2-7

x	$f(x)$
0	3
1	4
3	6
6	9

Functions as Equations

We can write an equation to depict the rule in Example 2-18 as follows. If the input is x, the output is $x + 3$; that is, $f(x) = x + 3$. The output values can be obtained by substituting the values 0, 1, 3, 4, and 6 for x in $f(x) = x + 3$, as shown:

$$
\begin{aligned}
f(0) &= 0 + 3 = 3 \\
f(1) &= 1 + 3 = 4 \\
f(3) &= 3 + 3 = 6 \\
f(4) &= 4 + 3 = 7 \\
f(6) &= 6 + 3 = 9
\end{aligned}
$$

In many applications, both the inputs and the outputs of a function machine are numbers. However, inputs and outputs can be any objects. For example, consider a particular candy machine that accepts only 25¢, 50¢, and 75¢ and outputs one of three types of candy with costs of 25¢, 50¢, and 75¢, respectively. A function machine associates *exactly one output with each input*. If you enter some element x as input and obtain $f(x)$ as output, then every time you enter the same x as input, you will obtain the same $f(x)$ as output. The idea of a function machine associating exactly one output with each input according to some rule leads to the following definition.

Definition of Function

A **function** from set A to set B is a correspondence from A to B in which each element of A is paired with one, and only one, element of B.

domain
range

The set A in the previous definition is the set of all allowable inputs and is the **domain** of the function. The set B is any set that includes all the possible outputs. The set of all outputs is the **range** of the function. Set B in the definition is any set that includes the range and can be the range itself. The distinction is made for convenience sake, since sometimes the range is not easy to find. For example, consider corresponding to each student at a university the student's Social Security number. This is a function from the set of all students to the set W of whole numbers. The range in this case is all the Social Security numbers of students who are enrolled at the university. The range is a proper subset of the set W. Normally, *if no domain is given to describe a function, then the domain is assumed to contain all elements for which the rule is meaningful.*

▶ **RESEARCH NOTE**

Students frequently have trouble with the language of functions (for example, *image, domain, range*, and *one-to-one*), which subsequently impacts their abilities to work with graphical representations of functions (Moakovits et al. 1988). ◀

A calculator contains many functions. Suppose a student enters [9] [×] [K] on the calculator that has a constant key, [K]. The student then presses [0] and hands the calculator to another student. The other student is to determine the rule by entering various numbers followed by the [=] key. Machines with an automatic constant feature can also be used.

Other buttons on a calculator are function buttons. For example, the [π] button always displays an approximation for π, such as 3.1415927; the [+/−] button either displays a negative sign in front of a number or removes an existing negative sign; and the [x^2] and [$\sqrt{\ }$] buttons square numbers and take the square root of numbers, respectively.

FIGURE 2-38

x

output is less than x

$f(x)$

Are all input-output machines function machines? Consider the machine in Figure 2-38. For any natural-number input x, the machine outputs a number that is less than x. If, for example, you input the number 10, the machine may output 9, since 9 is less than 10. If you input 10 again, the machine may output 3, since 3 is less than 10. Such a machine is not a function machine because the same input may give different outputs.

Functions as Arrow Diagrams

Arrow diagrams can be used to examine whether a correspondence represents a function. This representation is normally used when sets A and B are finite sets with few elements. The following example shows how arrow diagrams can be used to examine both functions and nonfunctions.

EXAMPLE 2-19

Which, if any, of the parts of Figure 2-39 exhibits a function from A to B? If a correspondence is a function from A to B, find the range of the function.

FIGURE 2-39

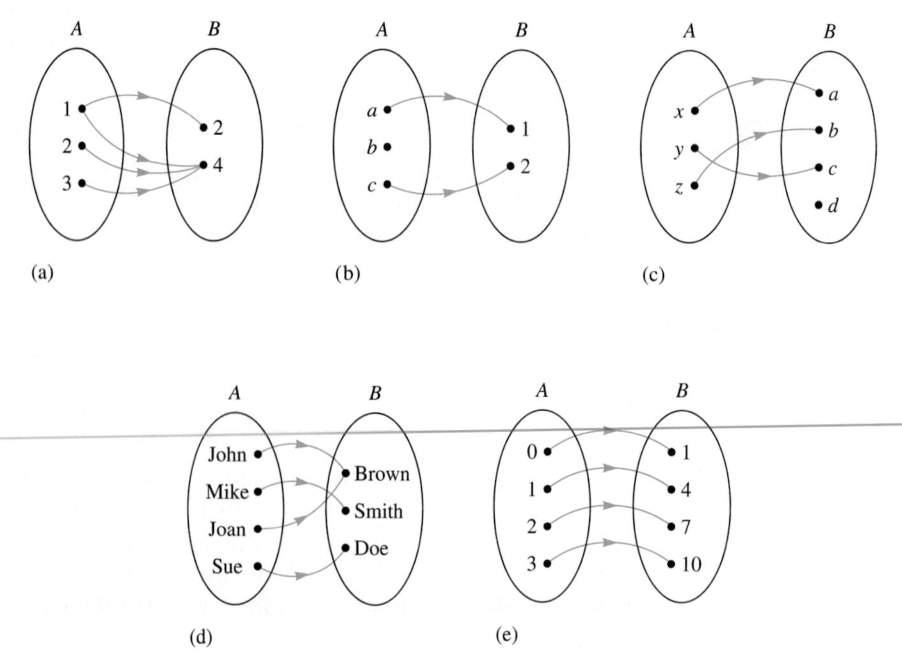

Solution **a.** Figure 2-39(a) does not define a function from A to B, since the element 1 is paired with both 2 and 4.
b. Figure 2-39(b) does not define a function from A to B, since the element b is not paired with any element of B. (It is a function from a subset of A to B.)
c. Figure 2-39(c) does define a function from A to B, since there is one and only one arrow leaving each element of A. The fact that d, an element of B, is not paired with any element in the domain does not violate the definition. The range is $\{a, b, c\}$ and does not include d because d is not an output of this function, as no element of A is paired with d.
d. Figure 2-39(d) illustrates a function, since there is only one arrow leaving each element in A. It does not matter that an element of set B, Brown, has two arrows pointing to it. The range is {Brown, Smith, Doe}.
e. Figure 2-39(e) illustrates a function whose range is $\{1, 4, 7, 10\}$.

Figure 2-39(e) also illustrates a one-to-one correspondence between A and B. In fact, any one-to-one correspondence between A and B defines a function from A to B as well as a function from B to A.

NOW TRY THIS 2-20

Determine which of the following are functions from the set of natural numbers to $\{0, 1\}$. Justify your reasoning.

a. For every natural-number input, the output is 0.
b. For every natural-number input, the output is 0 if the input is an even number and the output is 1 if the input is an odd number.

Functions as Tables and Ordered Pairs

Table 2-8

Amount of Advertising (in \$1000s)	Amount of Sales (in \$1000s)
0	1
1	3
2	6
3	8
4	10

Another useful way to describe a function is with a table. Consider the information in Table 2-8 relating the amount spent on advertising and the resulting sales in a given month for a small business. Note that for *Amount of Advertising* and *Amount of Sales*, the information is actually given in thousands of dollars. We could talk about a function between the amount of dollars in *Advertising* and the amount of dollars in *Sales*, or we could simplify a defined function as follows: If $A = \{0, 1, 2, 3, 4\}$ and $S = \{1, 3, 6, 8, 10\}$, the table describes a function from A to S, where A represents thousands of dollars in advertising and S represents thousands of dollars in sales.

The function could be given using ordered pairs. When 0 is the input and 1 is the output, that is recorded as the ordered pair $(0, 1)$. Similarly, the information in the second row is recorded as $(1, 3)$ and the rest of the information as $(2, 6)$, $(3, 8)$, and $(4, 10)$. The first component in the ordered pair is always an element in the domain and the second is the corresponding output.

EXAMPLE 2-20

Which of the following sets of ordered pairs represent functions? If a set represents a function, give its domain and range. If it does not, explain why.

a. $\{(1, 2), (1, 3), (2, 3), (3, 4)\}$ **b.** $\{(1, 2), (2, 3), (3, 4), (4, 5)\}$

c. $\{(1, 0), (2, 0), (3, 0), (4, 4)\}$ **d.** $\{(a, b) \mid a \in N, b = 2a\}$

Solution **a.** This is not a function because the input 1 has two different outputs.

b. This is a function with domain $\{1, 2, 3, 4\}$ and range $\{2, 3, 4, 5\}$.

c. This is a function with domain $\{1, 2, 3, 4\}$ and range $\{0, 4\}$. The output 0 appears more than once, but this does not contradict the definition of a function in that each input corresponds to only one output.

d. This is a function with domain N and range E, the set of all even natural numbers.

Functions as Graphs

Perhaps one of the most widely recognized representations of a function is as a graph. Graphs, as visual representations of functions, appear in newspapers and books and on television. To graph the function created from Table 2-8, consider the set of ordered pairs $\{(0, 1), (1, 3), (2, 6), (4, 10)\}$ and match each ordered pair to a point on the grid in Figure 2-40. We use the horizontal scale for the inputs and the vertical scale for the outputs and mark the point corresponding to $(0, 1)$ by starting at 0 on the horizontal scale and going up 1 unit on the vertical scale. To mark the point that corresponds to $(1, 3)$, we start at 0 and move 1 unit horizontally and then 3 units vertically. Marking the point that corresponds to an ordered pair is referred to as **graphing** the ordered pair. The set of all points that correspond to all the ordered pairs is the graph of the function.

graphing

FIGURE 2-40

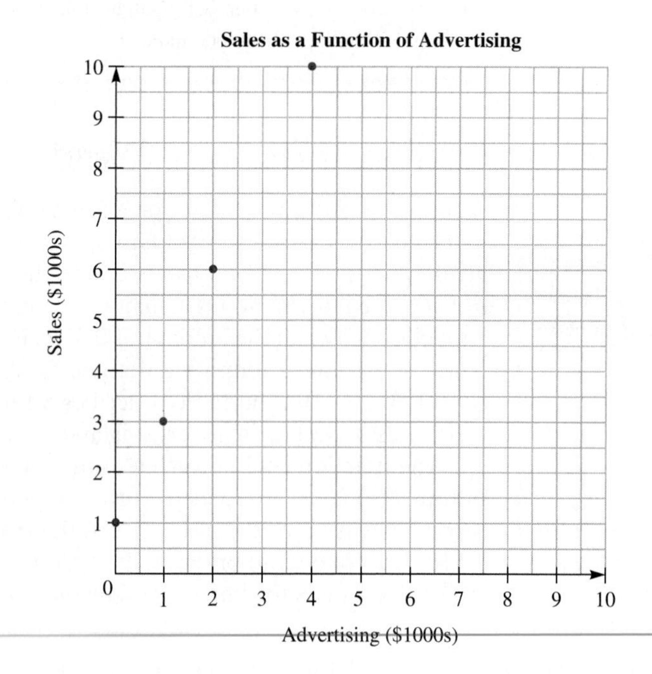

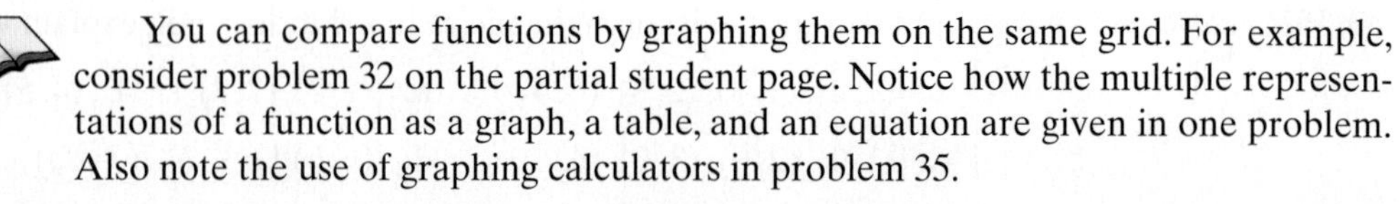

You can compare functions by graphing them on the same grid. For example, consider problem 32 on the partial student page. Notice how the multiple representations of a function as a graph, a table, and an equation are given in one problem. Also note the use of graphing calculators in problem 35.

SAMPLE SCHOOL BOOK PAGE:

USING GRAPHS

Make a table of values for each equation when $x = -1$, $x = 0$, and $x = 1$. Then graph each equation in a coordinate plane.

29. $y = x + 3$ 30. $y = x - 4$ 31. $y = 2 - x$

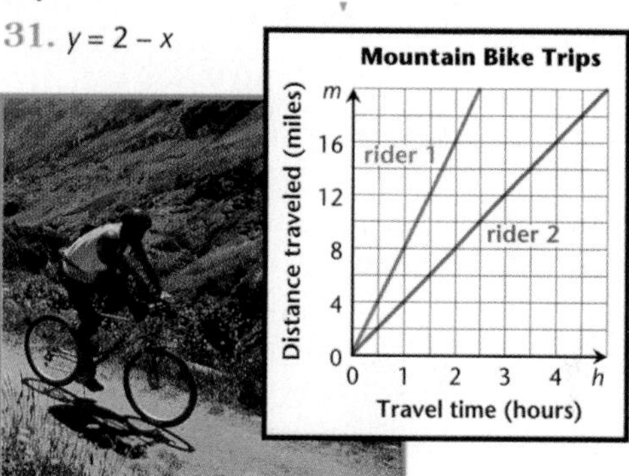

32. **Interpreting Data** The graph models two mountain bike trips. Riders with equal skills are riding on different trails.
    a. **Writing** Who is traveling on the easier trail? Explain your thinking.
    b. How far does each rider travel in 2 hours? Is this consistent with your answer to part (a)?
    c. Make a table of values for each trip.
    d. Write an equation for each trip.

33. **Challenge** Describe how you can use the graph in Exercise 32 to find the rate at which each rider is traveling.

34. **Bowling Costs** Some bowling alleys charge by the hour. Rashida compared the costs per hour for two people to bowl together at two different bowling alleys.

Bowling Alley A:	Bowling Alley B:
\$4 for 2 pairs of shoes plus \$17 per hour	\$2 for 2 pairs of shoes plus \$18 per hour

    a. Let x = the bowling time (in hours). Let y = the total charge (in dollars). Use them to model each option with an equation.
    b. Graph your equations from part (a) on one coordinate grid.
    c. When is Bowling Alley A less expensive?
    d. When are the charges the same for both bowling alleys?

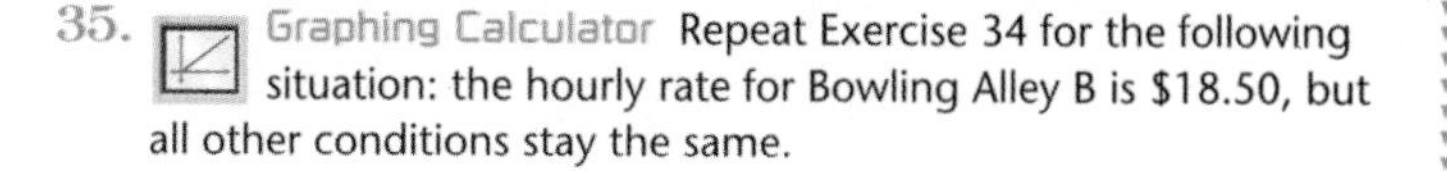

35. Graphing Calculator Repeat Exercise 34 for the following situation: the hourly rate for Bowling Alley B is \$18.50, but all other conditions stay the same.

Source: Middle Grades MATH Thematics, McDougal Littell, Grade 7, 2005 (p. 127).

NOW TRY THIS 2-21

Answer questions 32–34 on the student page.

EXAMPLE 2-21

Explain why a telephone company would not set rates for telephone calls as depicted on the graph in Figure 2-41.

FIGURE 2-41

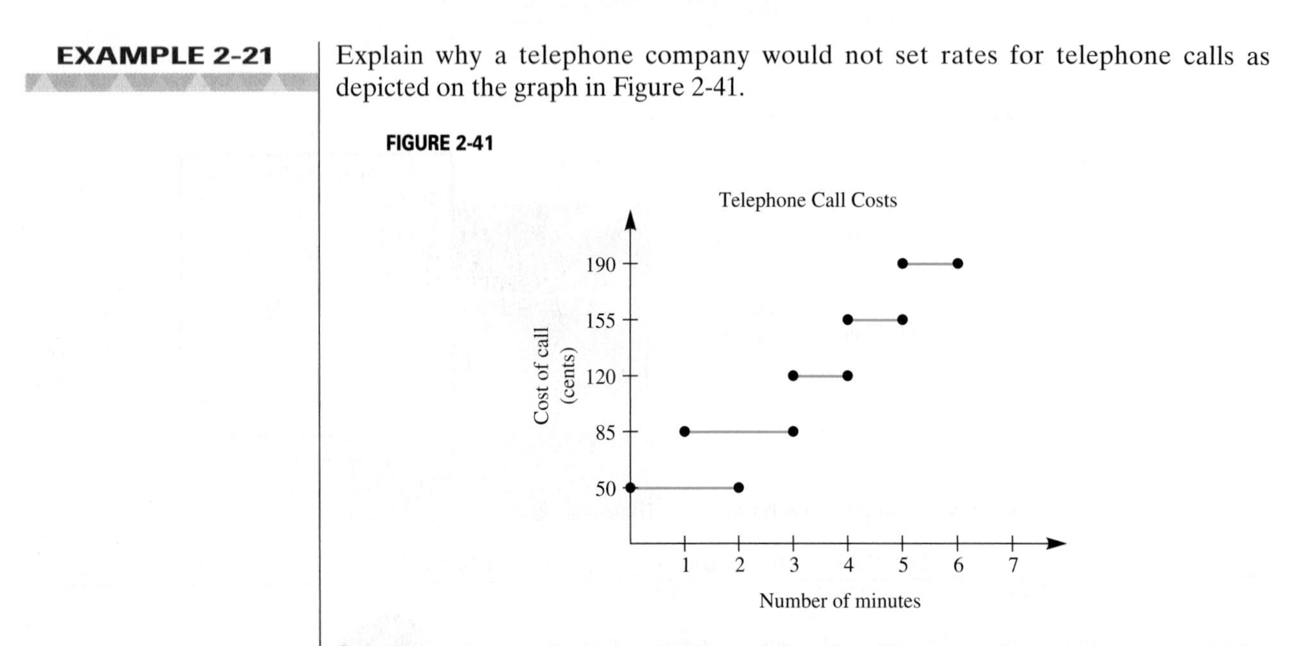

Solution The graph does not depict a function. For example, a customer could be charged either \$0.50 or \$0.85 for a 2-minute call.

Suppose you join a video rental club where your cost per rental is \$5. We have seen that one way to describe a function is by writing an equation. Based on the information in Table 2-9, the equation relating the number of videos rented to cost is $C = n \cdot 5$, where n is the number of videos rented.

Table 2-9

Number of Videos Rented	Cost in Dollars
1	$1 \cdot 5 = 5$
2	$2 \cdot 5 = 10$
3	$3 \cdot 5 = 15$
4	$4 \cdot 5 = 20$
5	$5 \cdot 5 = 25$
.	.
.	.
.	.
n	$n \cdot 5$

This could also be written as $f(n) = 5n$, where $f(n)$ is the cost of the rental in dollars. If we restrict the number of rentals to the first five natural numbers, the function can be described as the set of ordered pairs $\{(1, 5), (2, 10), (3, 15), (4, 20), (5, 25)\}$. Figure 2-42 shows the graph of the function. The graph consists of five points that are not connected. In graphing the function in Figure 2-42, we assume the domain to be the set of natural numbers.

FIGURE 2-42

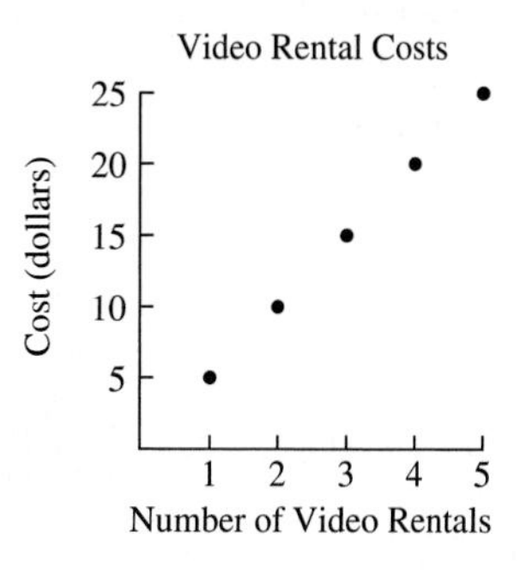

NOW TRY THIS 2-22

The graph in Figure 2-42 shows only points that are not connected. Do you think the points in the graph should be connected? Why or why not?

Sequences as Functions

Arithmetic, geometric, and other sequences introduced in Chapter 1 can be thought of as functions whose inputs are natural numbers and whose outputs are the terms of a particular sequence. For example, the arithmetic sequence 2, 4, 6, 8, . . . , whose nth term is $2n$ can be described as a function from N (natural numbers) to the set E (even natural numbers) using the rule $f(n) = 2n$, where n is a natural number and $f(n)$ stands for the value of the nth term.

EXAMPLE 2-22

If $f(n)$ denotes the nth term of a sequence, find $f(n)$ in terms of n for each of the following:

a. An arithmetic sequence whose first term is 3 and whose difference is 3
b. A geometric sequence whose first term is 3 and whose ratio is 3
c. The sequence: $1, 1 + 2, 1 + 2 + 3, 1 + 2 + 3 + 4, \ldots$

Solution **a.** The first term is 3, the second term is $3 + 3$, or $2 \cdot 3$, the third is $2 \cdot 3 + 3$, or $3 \cdot 3$, and the fourth term is $3 \cdot 3 + 3$, or $4 \cdot 3$, the nth term is $n \cdot 3$, and hence $f(n) = n \cdot 3 = 3n$, where n is a natural number.

b. The first term is 3, the second $3 \cdot 3$, or 3^2, the third $3 \cdot 3^2$, or 3^3, and so on. Hence, the nth term is 3^n and therefore $f(n) = 3^n$, where n is a natural number.

c. The nth term is $1 + 2 + 3 + 4 + \cdots + n$. In Chapter 1 we saw that this sum equals $\frac{n(n+1)}{2}$, so the function is $f(n) = \frac{n(n+1)}{2}$, where n is a natural number.

Composition of Functions

Consider the function machines in Figure 2-43. If 2 is entered in the top machine, then $f(2) = 2 + 4 = 6$. Six is then entered in the second machine and $g(6) = 2 \cdot 6 = 12$. The functions in Figure 2-43 illustrate the **composition of two functions**. In the composition of two functions, the range of the first function becomes the domain of the second function.

composition of two functions

FIGURE 2-43

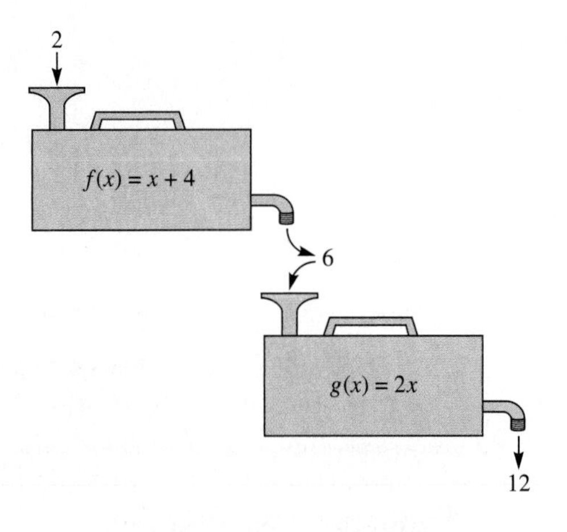

If the first function f is followed by a second function g, as in Figure 2-43, symbolize the composition of the functions as $g \circ f$. If we input 3 in the function machines of Figure 2-43 then the output is symbolized by $(g \circ f)(3)$. Because f acts first on 3, to compute $(g \circ f)(3)$ we find $f(3) = 3 + 4 = 7$ and then $g(7) = 2 \circ 7 = 14$. Hence, $(g \circ f)(3) = 14$ and $(g \circ f)(3) = g(f(3))$. Also note that $(g \circ f)(x) = g(f(x)) = 2 \cdot f(x) = 2(x + 4)$ and hence $g(f(3)) = 2(3 + 4) = 14$.

EXAMPLE 2-23

If $f(x) = 2x + 3$ and $g(x) = x - 3$, find the following:

a. $(f \circ g)(3)$ **b.** $(g \circ f)(3)$ **c.** $(f \circ g)(x)$ **d.** $(g \circ f)(x)$

Solution
a. $(f \circ g)(3) = f(0) = 2 \cdot 0 + 3 = 3$
b. $(g \circ f)(3) = g(9) = 9 - 3 = 6$
c. $(f \circ g)(x) = f(g(x)) = 2 \cdot g(x) + 3 = 2(x - 3) + 3 = 2x - 6 + 3 = 2x - 3$
d. $(g \circ f)(x) = g(f(x)) = f(x) - 3 = (2x + 3) - 3 = 2x$

REMARK Example 2-23 shows that composition of functions is not commutative, since $(f \circ g)(3) \neq (g \circ f)(3)$.

We have seen that a function can be represented in a variety of ways. Pictures of sets with arrows and function machines are used mostly as pedagogical devices in learning the concept of a function. The most common representations are a table, an equation, or a graph. Depending on the situation, one representation may be more useful than another. For example, if the domain of a function has many elements, a table is not a convenient representation. In later chapters, we learn how to graph certain kinds of equations. Graphing calculators are capable of graphing most functions given by equations with specified domains.

TECHNOLOGY CORNER

A sketch of the function $y = 2x + 1$ for x between 0 and 5 is shown in Figure 2-44. Use a graphing calculator to sketch the graphs of $y = 2x + b$ for three choices of b. What do the graphs seem to have in common? Why?

FIGURE 2-44

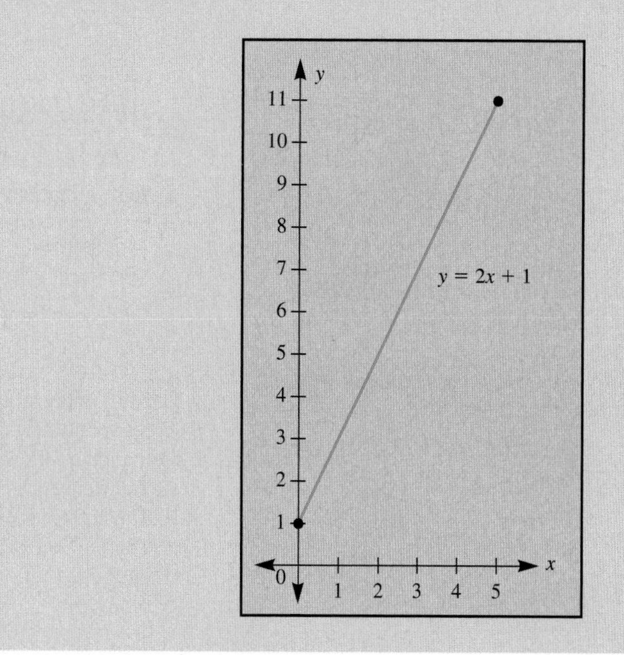

ONGOING ASSESSMENT 2-5

1. The following sets of ordered pairs are functions. Give a rule that could describe each function.
   a. $\{(2, 4), (3, 6), (9, 18), (12, 24)\}$
   b. $\{(5, 3), (7, 5), (11, 9), (14, 12)\}$
   c. $\{(2, 8), (5, 11), (7, 13), (4, 10)\}$
   d. $\{(2, 5), (3, 10), (4, 17), (5, 26)\}$
2. Which of the following are functions from the set $\{1, 2, 3\}$ to the set $\{a, b, c, d\}$? If the set of ordered pairs is not a function, explain why not.
   a. $\{(1, a), (2, b), (3, c), (1, d)\}$ **b.** $\{(1, c), (3, d)\}$
   c. $\{(1, a), (2, b), (3, a)\}$ **d.** $\{(1, a), (1, b), (1, c)\}$
3. a. Draw an arrow diagram of a function with domain $\{1, 2, 3, 4, 5\}$ and range $\{a, b\}$.
   b. How many possible functions are there in part (a)?
4. Suppose $f(x) = 2x + 1$ and the domain is $\{0, 1, 2, 3, 4\}$. Describe the function in the following ways:
   a. Draw an arrow diagram involving two sets.
   b. Use ordered pairs. c. Make a table.
   d. Draw a graph to depict the function.
5. Determine which of the following are functions from $W = \{0, 1, 2, 3, \ldots\}$ or a subset of W to W. If your answer is that it is not a function, explain why not.
   a. $f(x) = 2$ for all $x \in W$
   b. $f(x) = 0$ if $x \in \{0, 1, 2, 3\}$, and $f(x) = 3$ if $x \notin \{0, 1, 2, 3\}$
   c. $f(x) = x$
   d. $f(x) = 0$ for all $x \in W$ and $f(x) = 1$ if $x \in \{3, 4, 5, 6, \ldots\}$
   e. $f(x)$ is the sum of the digits in x for all $x \in W$.

6. **a.** Make an arrow diagram for each of the following:
(i) Rule: "is the double of."

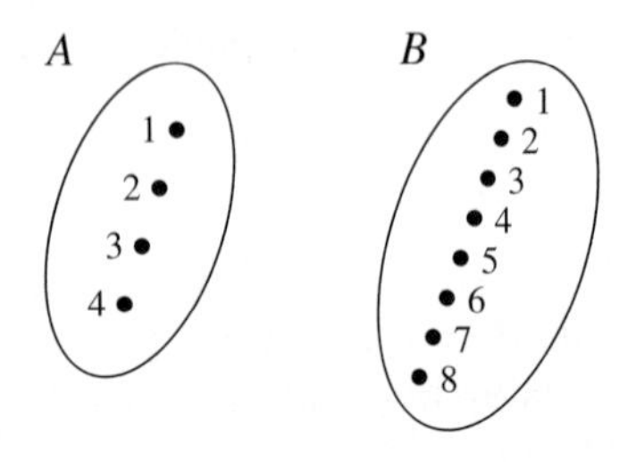

(ii) Rule: "is greater than."

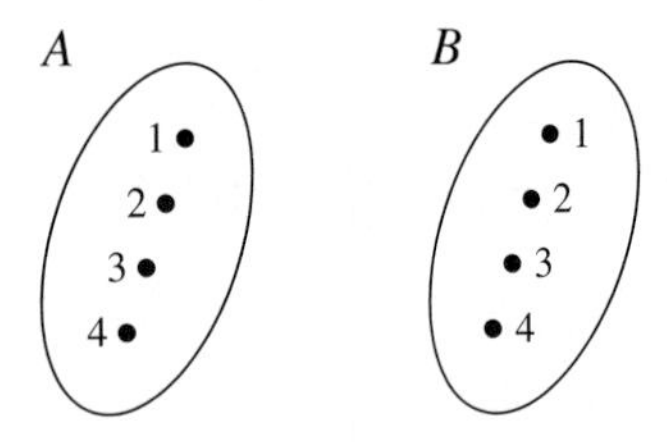

b. Which, if any, of the parts in (a) exhibits a function from A to B? If there is a function, tell why and find the range of the function.

7. Given the following arrow diagrams for functions from A to B, give a possible rule for the function:

a.

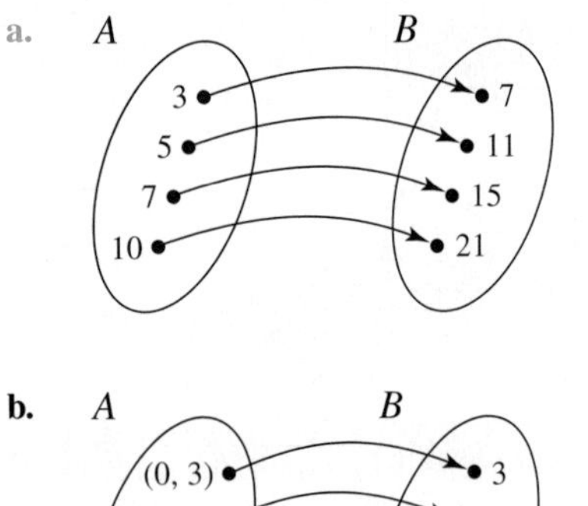

b.

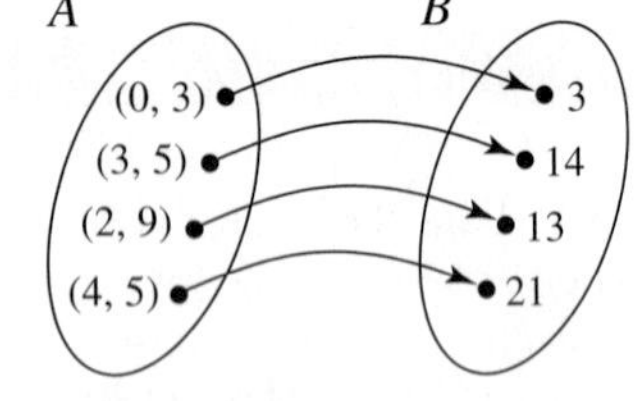

8. The dosage of a certain drug is related to the weight of a child as follows: 50 mg of the drug and an additional 15 mg for each 2 lb or fraction of 2 lb of body weight above 30 lb. Sketch the graph of the dosage as a function of the weight of a child for children who weigh between 20 and 40 lb.

9. According to wildlife experts, the rate at which crickets chirp is a function of the temperature; specifically, $C = T - 40$, where C is the number of chirps every 15 sec and T is the temperature in degrees Fahrenheit.

a. How many chirps does the cricket make per second if the temperature is 70°F?

b. What is the temperature if the cricket chirps 40 times in 1 min?

10. If taxi fares are \$3.50 for the first half mile and \$0.75 for each additional quarter mile, answer the following:

a. What is the fare for a 2-mi trip?

b. Write a rule for computing the fare for an n-mile trip by taxi.

11. For each of the following, guess what might be Latifah's rule. In each case, if n is your input and $L(n)$ is Latifah's answer, express $L(n)$ in terms of n.

a.

You	Latifah
3	8
4	11
5	14
10	29

b.

You	Latifah
0	1
3	10
5	26
8	65

c.

You	Latifah
6	42
0	0
8	72
2	6

12. The following graph shows arithmetic achievement-test scores for students of a sixth-grade class. From the graph, estimate the following:

a. The frequency of the score made most often

b. The highest score obtained

c. The number of boys who would have had to score 54 on the test in order to match the number of girls who scored 54.

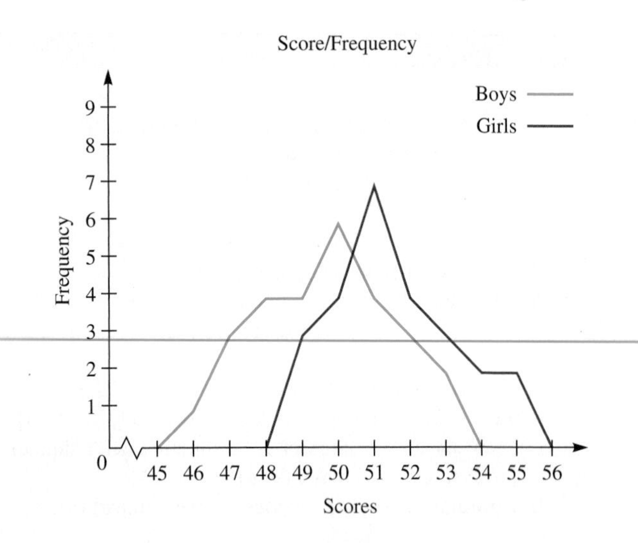

13. The *Principles and Standards* for Grades 6–8 points out that "in their study of algebra, middle grades students should encounter questions that focus on quantities that change." (p. 229). It poses the following problem along with graphs.

    ChitChat charges \$0.30 a minute for cellular phone calls. The cost per minute does not change, but the total cost changes as the telephone is used.

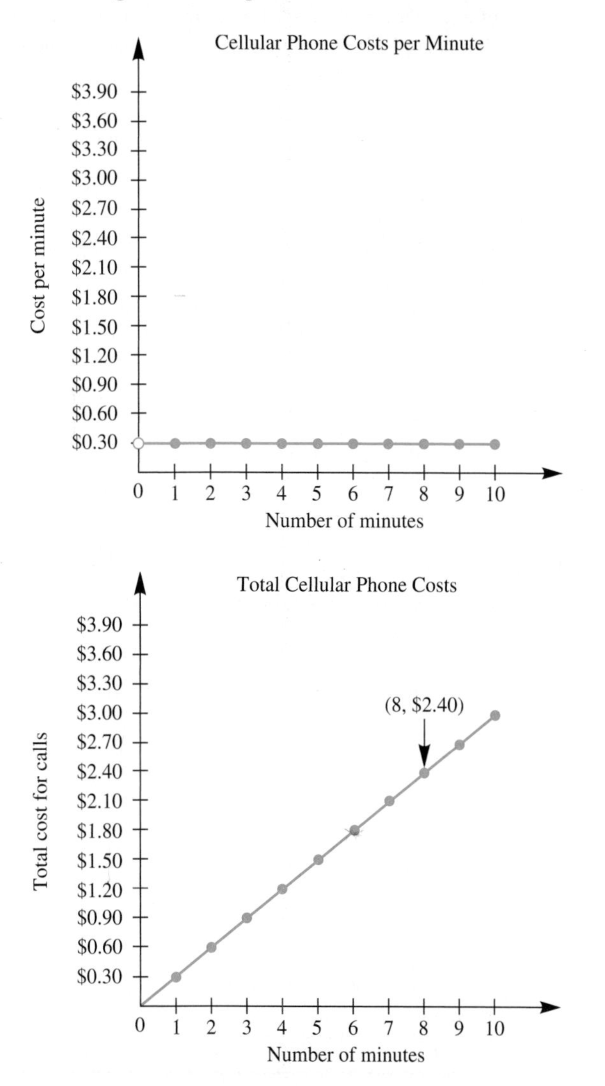

    a. When the number of minutes is 6, what do the values of the corresponding point on each graph represent?
    b. What kind of assumption about the charges needs to be made to allow the connection of the points on each graph? Explain.
    c. If time in minutes is t and the total cost for calls is c, write c as a function of t for each graph.

14. In the *Principles and Standards* for Grades 6–8, the "Algebra" section (p. 229) poses the following problem. Quick-Talk advertises monthly cellular phone service for \$0.50 a minute for the first 60 minutes but only \$0.10 a minute for each minute thereafter. Quick-Talk also charges for the exact amount of time used. Answer the following:
    a. Graph two graphs similar to the ones in problem 13, one graph showing the cost per minute as a function of number of minutes and the other showing the total cost for calls as a function of number of minutes up to 100 minutes.
    b. If you connect the points in the second graph in part (a), what kind of assumption needs to be made about the way the telephone company charges phone calls?
    c. Why does the total cost for calls consist of two line segments? Why is one part steeper than the other?
    d. The function representing the total cost for calls as a function of number of minutes talked can be represented by two equations. Write these equations.

15. For each of the following sequences, find a possible function whose domain is the set of natural numbers and whose outputs are the terms of the sequence.
    a. 3, 8, 13, 18, 23, . . .
    b. 3, 9, 27, 81, 243, . . .
    c. 2, 4, 6, 8, 10, . . .

16. Consider two function machines that are placed as shown. Find the final output for each of the following inputs:
    a. 5 **b.** 3 c. 10

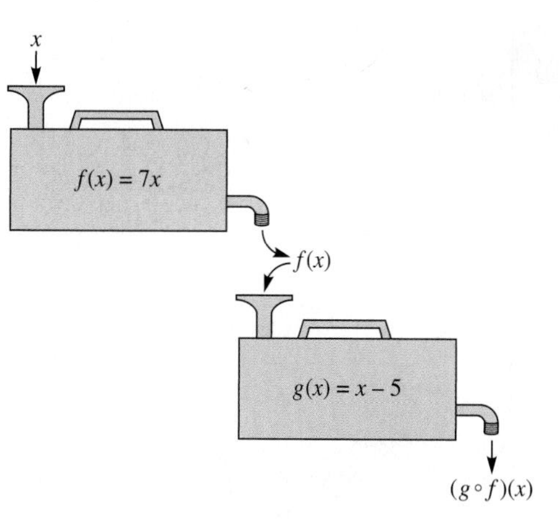

17. Let $t(n)$ represent the nth term of a sequence for $n \in N$. Answer the following:
    a. If $t(n) = 4n - 3$, determine which of the following are values of the function:
    (i) 1 (ii) 385 (iii) 389 (iv) 392
    b. If $t(n) = n^2$, determine which of the following are values of the function:
    (i) 1 (ii) 4 (iii) 9 (iv) 10 (v) 900
    c. If $t(n) = n(n + 1)$, determine which of the following are in the range of the function:
    (i) 2 (ii) 12 (iii) 2550 (iv) 2600

18. Consider a function machine that accepts inputs as ordered pairs. Suppose the components of the ordered pairs are

natural numbers and the first component is the length of the rectangle and the second is its width. The following machine computes the perimeter (the distance around a figure) of the rectangle. Thus, for a rectangle whose length, l, is 3 and whose width, w, is 2, the input is (3, 2) and the output is $2 \cdot 3 + 2 \cdot 2$, or 10. Answer each of the following:

a. For each of the following inputs, find the corresponding output: (1, 7), (2, 6), (6, 2), (5, 5).
b. Find the set of all the inputs for which the output is 20.
c. What is the domain and the range of the function?

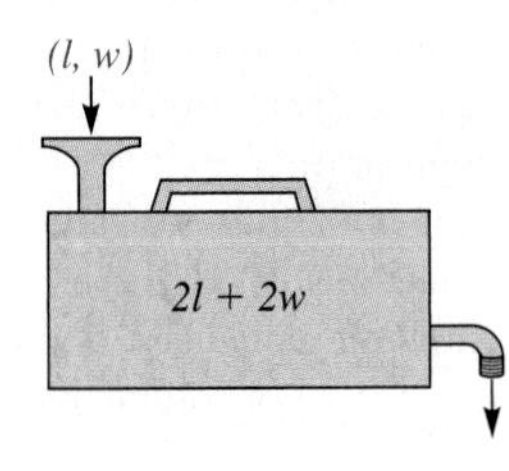

19. The following graph shows the relationship between the number of cars on a certain road at different times between 5:00 A.M. and 9:00 A.M.:

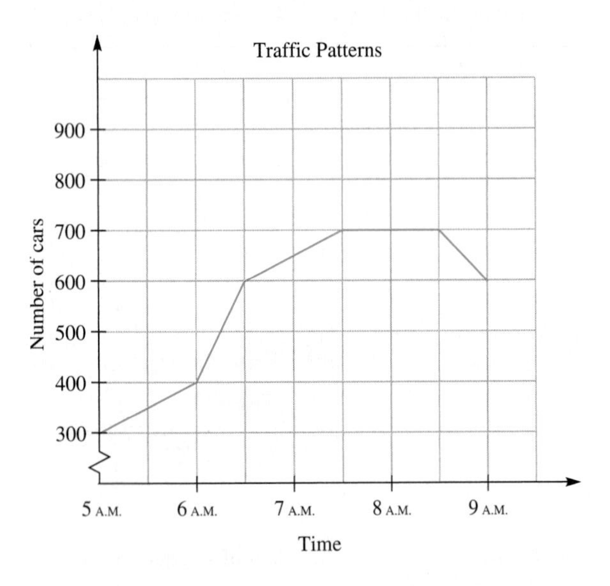

a. What was the increase in the number of cars on the road between 6:30 A.M. and 7:00 A.M.?
b. During which half hour was the increase in the number of cars the greatest?
c. What was the increase in the number of cars between 8:00 A.M. and 8:30 A.M.?
d. During which half hour(s) did the number of cars decrease? By how much?
e. The graph for this problem is composed of segments rather than just points as in Figure 2-42. Why do you think segments are used here instead of just points?

20. A health club charges a one-time initiation fee of \$100 plus a membership fee of \$40 per month.
a. Write an expression for the cost function $C(x)$ that gives the total cost for membership at the health club for x months.
b. Draw the graph of the function in (a).
c. The health club decided to give its members an option of a higher initiation fee but a lower monthly membership charge. If the initiation fee is \$300 and the monthly membership fee is \$30, use a different color and draw on the same set of axes the cost graph under this plan.
d. Determine after how many months the second plan is less expensive for the member.

21. A ball is thrown straight up. We know its height H in feet after t sec is given by the function $H(t) = 128t - 16t^2$.
a. Find $H(2)$, $H(6)$, $H(3)$, and $H(5)$. Why are some of the outputs equal?
b. Graph the function and from the graph find at what instant the ball is at its highest point. What is its height at that instant?
c. How long will it take the ball to hit the ground?
d. What is the domain of H?
e. What is the range of H?

22. A rectangular plot is bounded on one side by a straight river and on the other sides by a fence. Suppose 900 yd of fence are available and the length of the side of the rectangle parallel to the river is denoted by x.
a. Find an expression for the area $A(x)$ in terms of x.
b. Graph $A(x)$.
c. Use the graph in (b) or your calculator to estimate the length and width of the rectangle for which the area will be the largest.

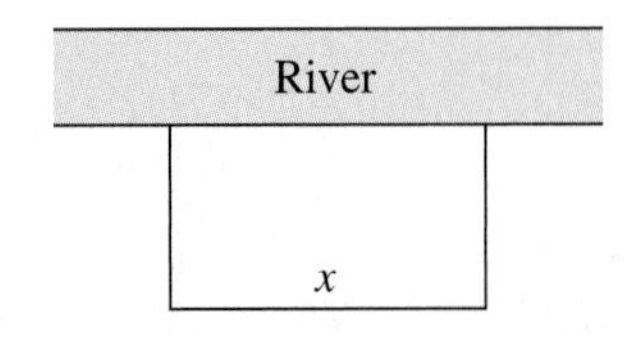

23. For each of the following sequences of matchstick figures, let $S(n)$ be the function giving the total number of matchsticks in the nth figure.
a. For each of the following, find the total number of matchsticks in the fourth figure.
b. For each of the following, find as simple a formula as possible for $S(n)$ in terms of n.

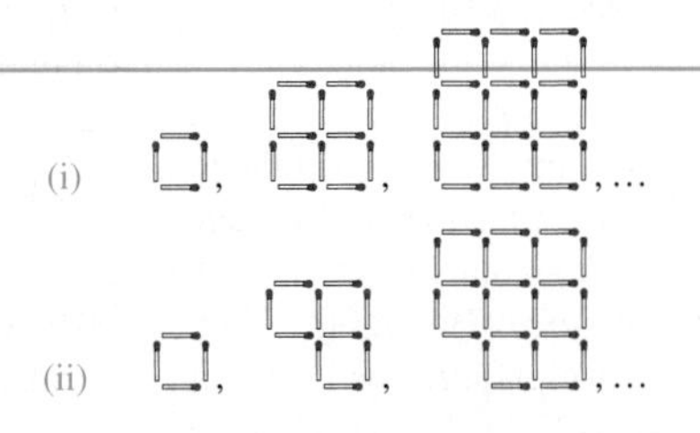

24. Assume the pattern continues for each of the following sequences of square tile figures. Let $S(n)$ be the function giving the total number of tiles in the nth figure. For each of the following find a formula for $S(n)$ in terms of n. Each square is divided into four squares in the subsequent figure.

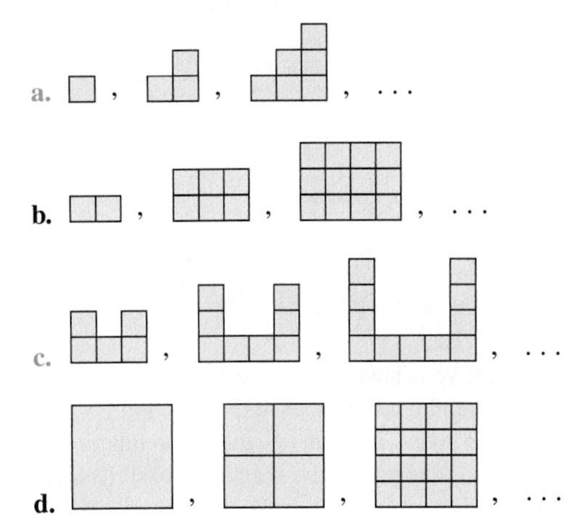

25. A function can be represented as a set of ordered pairs where the set of all the first components is the domain and the set of all the second components is the range. Is the converse also true? That is, is every set of ordered pairs a function whose domain is the set of first components and whose range is the set of second components? Justify your answer.
26. Suppose each point in the figure represents a child on a playground, the letters represent their names, and an arrow going from I to J means that I "is the sister of" J.

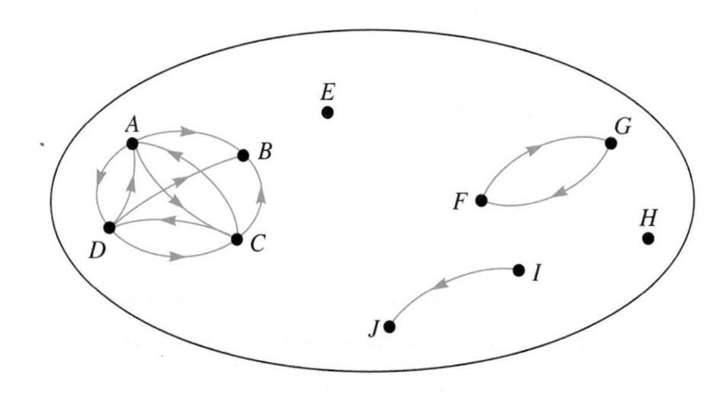

 a. Based on the information in the figure, who are definitely girls and who are definitely boys?
 b. Suppose we write "A is the sister of B" as an ordered pair (A, B). Based on the information in the diagram, write the set of all such ordered pairs.
 c. Is the set of all ordered pairs in (b) a function with the domain equal to the set of all first components of the ordered pairs and with the range equal to the set of all second components?

27. A generalization of the concept of a function is the concept of a *relation. Given two sets A and B, a relation from A to B is any set of ordered pairs in which the first components are from A and the second are from B.* Which of the following are functions and which are relations but not functions from the set of first components of the ordered pairs to the set of second components?
 a. {(Montana, Helena), (Oregon, Salem), (Illinois, Springfield), (Arkansas, Little Rock)}
 b. {(Pennsylvania, Philadelphia), (New York, Albany), (New York, Niagara Falls), (Florida, Ft. Lauderdale)}
 c. $\{(x, y) \mid x$ resides in Birmingham, Alabama, and x is the mother of y, where y is a U.S. resident$\}$
 d. $\{(1, 1), (2, 4), (3, 9), (4, 16)\}$
 e. $\{(x, y) \mid$ where x and y are natural numbers and $x + y$ is an even number$\}$
28. a. Is the rule "has a mother" a function whose domain is the set of all people?
 b. Is the relation "has a brother" a function from the set of all boys to the set of all boys?

The following definitions are needed to answer problems 29 and 30.

A relation on a set X. This is a relation from X to X. A relation on X may have one or more of the following properties: *The reflexive property.* A relation on a set X is reflexive if and only if for all a in X, a is related to a, that is (a, a) is in the set of ordered pairs. *The symmetric property.* A relation on a set X is symmetric if and only if for all elements a and b in X, whenever a is related to b, then b is also related to a; that is, if (a, b) is in the set of ordered pairs, so is (b, a).

The transitive property. A relation on a set X is transitive if and only if for all elements a, b, and c in X, whenever a is related to b and b is related to c, then a is related to c; that is, if (a, b) and (b, c) are in the set of ordered pairs, then (a, c) is also in the same set (a, b, and c do not have to be different).

An **equivalence relation** on a set X is any relation on X that satisfies the reflexive, symmetric, and transitive properties.

29. Tell whether each of the following is reflexive, symmetric, or transitive on the set of all people. Which are equivalence relations?
 a. "Is a parent of"
 b. "Is the same age as"
 c. "Has the same last name as"
 d. "Is the same height as"
 e. "Is married to"
 f. "Lives within 10 mi of"
 g. "Is older than"
30. Tell whether each of the following is reflexive, symmetric, or transitive on the set of subsets of a nonempty set. Which are equivalence relations?
 a. "Is equal to"
 b. "Is a proper subset of"
 c. "Is not equal to"
 d. "Has the same cardinal number as"

Communication

31. Does the diagram define a function from A to B? Why or why not?

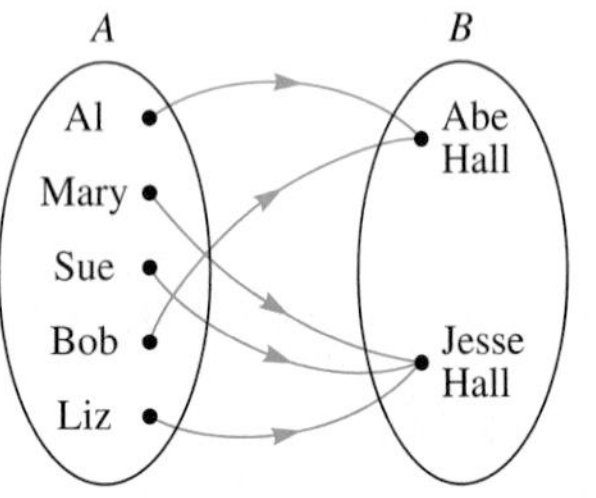

32. Is a one-to-one correspondence a function? Explain your answer and give an example.
33. Which of the following are functions from A to B? If your answer is "not a function," explain why not.
    a. A is the set of mathematics faculty at the university. B is the set of all mathematics classes. To each mathematics faculty member, we associate the class that person is teaching during a given term.
    b. A is the set of mathematics classes at the university. B is the set of mathematics faculty. To each mathematics class, we associate the teacher who is teaching the class.
    c. A is the set of all U.S. senators and B is the set of all Senate committees. We associate each senator to the committee of which the senator is chairperson.
34. When a boat is put in the water, its hull is partially above and partially below the water. The part below the water is the "draft" and the part above the water is the "freeboard." The following table shows the relationship between the draft and freeboard for a 50-cm-deep boat:

Draft	Freeboard
5	45
10	40
15	35
20	30
30	20

    a. Graph the freeboard as a function of the draft. Is it meaningful to connect the points graphed? Explain.
    b. If d stands for the draft in centimeters and $f(d)$ for the freeboard, write an equation expressing $f(d)$ in terms of d.

Open-Ended

35. Examine several newspapers and magazines and describe at least three examples of functions that appear. What is the domain and range of each function?
36. Give at least three examples of functions from A to B where neither A nor B is a set of numbers.
37. Draw a sequence of matchstick figures and describe the pattern in words. Find as simple an expression as possible for $S(n)$, the total number of matchsticks in the nth figure.
38. A function whose output is always the same regardless of the input is a *constant function*. Give several examples of constant functions from real life.
39. A function whose output is the same as its input is an *identity function*. Give several concrete examples of identity functions.

Cooperative Learning

40. Each person in a group picks a natural number and uses it as an input in the following function machine:

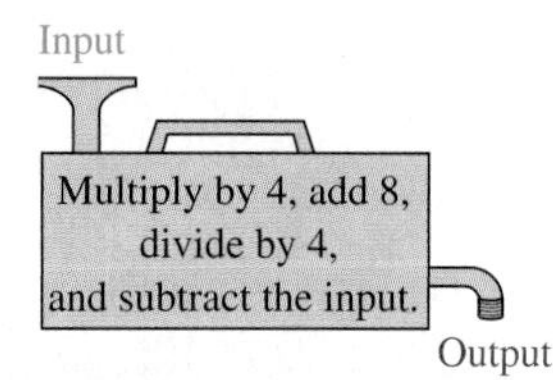

    a. Compare your answers. Based on the answers, make a conjecture about the range of the function.
    b. Based on your answer in (a), graph the function.
    c. Write the function in the simplest possible way using $f(x)$ notation.
    d. Justify your conjecture in (a).
    e. Make up similar function machines and try different inputs in your group.
    f. Devise a function machine in which the machine performs several operations, but the output is always the same as the input. Exchange your answer with someone in the group and check that the other person's function machine performs as required.
    g. Write the functions the group came up with in the simplest way using $f(x)$ notation and graph them.
41. In your group of four, work through the following. You will need a metric tape or meter stick.
    a. Place your math book on a desk and measure the distance (nearest tenth of a centimeter) from the floor to the top of the book. Record the distance.
    b. Place a second math book on top of the first and measure the distance (nearest tenth of a centimeter) from the floor to the top of the second book. Record the distance.
    c. Continue this procedure for all four of your math books and complete the following table and graph:

Number of Books	Distance from Floor
1	
2	
3	
4	
.	
.	
.	

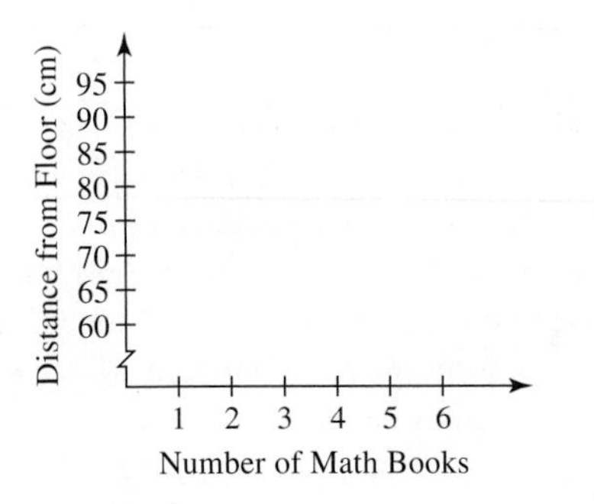

d. Without measuring, what is the distance from the floor with 0 books? 5 books?

e. Write a rule or function for $d(x)$, where $d(x)$ is the distance above the floor to the top of the stack of books and x is the number of books.

f. Suppose the distance from the floor to the ceiling is 2.5 m. If you stack the books as described above, how many books would be needed to reach the ceiling?

g. The function $h(x) = 3.4x + 70$ represents the height of another stack of x math books (in tenths of a centimeter) on a cabinet. What does the function tell you about the height of the cabinet? What does it tell you about the width of each book?

h. Suppose that a table with a stack of similar math books (more than 10) is 200 cm high. If the top math book is removed, the height is 197 cm. If a second book is removed, the height is 194 cm. What is the height if 5 books are removed?

i. Write a function $h(x)$ for the height of the stack after x books are removed.

Review Problems

42. Rewrite each of the following in equivalent form using only multiplication or addition operations:
    a. $109 - 11 = 98$
    b. $x - y = z$
    c. $60 \div 4 = 15$
    d. $x \div 3 = y$
    e. $(x - 3) \div 5 = 10$
43. Is the number of elements in $A \cup B$ always equal to the sum of the number of elements in A and the number of elements in B? Why or why not?
44. Shade the portion of the Venn diagram that illustrates the set $(B - C) \cup (A \cap B)$.

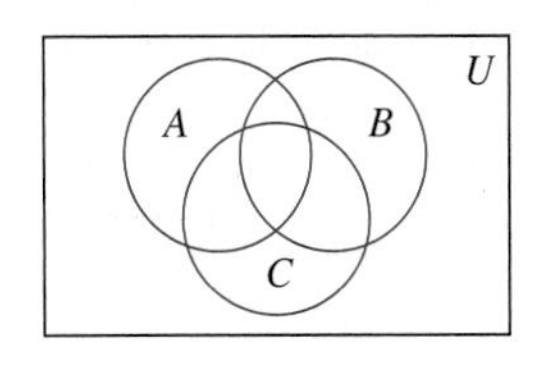

Third International Mathematics and Science Study (TIMSS) Questions

A number machine takes a number and operates on it. When the Input Number is 5, the Output Number is 9, as shown below.

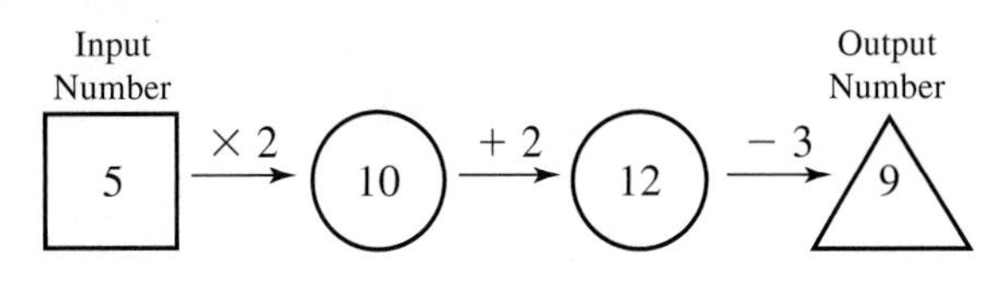

When the Input Number is 7, which of these is the Output Number?

a. 11
b. 13
c. 14
d. 25

TIMSS 2003, Grade 4

The graph represents the distance and time of a hike taken by Joshua and Liam.

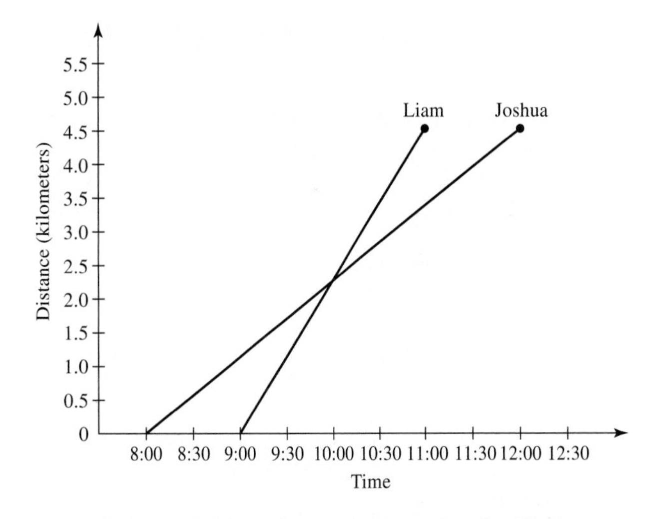

If they both started from the same place and walked in the same direction, at what time did they meet?

a. 8:00
b. 8:30
c. 9:00
d. 10:00
e. 11:00

TIMSS 2003, Grade 8

Only 10 rooms were vacant in the Village Hotel. Eleven men went into the hotel at the same time, each wanting a separate room. The clerk, settling the argument, said, "I'll tell you what I'll do. I'll put two men in Room 1 with the understanding that I will come back and get one of them a few minutes later." The men agreed to this. The clerk continued, "I will put the rest of you men in rooms as follows: the third man in Room 2, the fourth man in Room 3, the fifth man in Room 4, the sixth man in Room 5, the seventh man in Room 6, the eighth man in Room 7, the ninth man in Room 8, and the tenth man in Room 9." Then the clerk went back and got the extra man he had left in Room 1 and put him in Room 10. Everybody was happy. What is wrong with this plan?

HINT FOR SOLVING THE PRELIMINARY PROBLEM

Draw a Venn diagram for the problem with switch hitters, infielders, and outfielders all represented based on the information given. Next, fill in the sections of the diagram with numbers representing what we know about each group. Continue filling in the sections until all the information is used. Based on the Venn diagram, answer the question.

QUESTIONS FROM THE CLASSROOM

1. A student argues that $\{\emptyset\}$ is the proper notation for the empty set. What is your response?
2. A student asks, "If $A = \{a, b, c\}$ and $B = \{b, c, d\}$, why isn't it true that $A \cup B = \{a, b, c, b, c, d\}$?" What is your response?
3. A student says that she can show that if $A \cap B = A \cap C$, then it is not necessarily true that $B = C$; but she thinks that whenever $A \cap B = A \cap C$ and $A \cup B = A \cup C$, then $B = C$. What is your response?
4. A student claims that a finite set of numbers is any set that has a greatest element. Do you agree?
5. A student claims that the complement bar can be broken over the operation of intersections; that is, $\overline{A \cap B} = \overline{A} \cap \overline{B}$. What is your response?
6. A student claims that $\overline{A} \cap \overline{B}$ includes all elements that are not in A. What is your response?
7. A student asks whether a formula and a function are the same. What is your response?
8. A student states that either $A \subseteq B$ or $B \subseteq A$. Is the student correct?
9. A student is asked to find all one-to-one correspondences between two given sets. He finds the Cartesian product of the sets and claims that his answer is correct because it includes all possible pairings between the elements of the sets. How do you respond?
10. A student argues that adding two sets A and B, or $A + B$, and taking the union of two sets, $A \cup B$, is the same thing. How do you respond?
11. A student argues that $A = \{1, \{1\}\}$ has only one element. How do you respond?
12. A student asks "Does $2(3 \cdot 4)$ equal $(2 \cdot 3)(2 \cdot 4)$?" Is there a distributive property of multiplication over multiplication?
13. The division algorithm, $a = bq + r$, holds for $a \geq b$; $a, b, q, r \in W$, $r < b$; and $b \neq 0$. Does the division algorithm hold when $a < b$?
14. Can we define $0 \div 0$ as 1? Why or why not?
15. **a.** A student claims that for all whole numbers $(a \cdot b) \div b = a$. How do you respond?
    b. The student in part (a) claims that $0 \div 0 = 0$. The student's reasoning is, "If $a = 0$ and $b = 0$ are substituted in the equation in part (a), the result is $0 \cdot 0 \div 0 = 0$. But because $0 \cdot 0 = 0$, it follows that $0 \div 0 = 0$." How do you respond?
16. A student asks if division on the set of whole numbers is distributive over subtraction. How do you respond?
17. A student says that 0 is the identity for subtraction. How do you respond?
18. A student claims that on the following number line, the arrow doesn't really represent 3 because the end of the arrow does not start at 0. How do you respond?

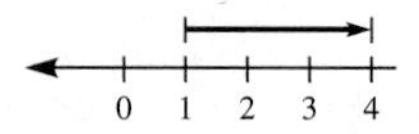

19. A student asks whether a function from A to B is related to the cross product $A \times B$. How do you respond?

20. A student claims that the following machine does not represent a function machine because it accepts two inputs at once rather than a single input. How do you respond?

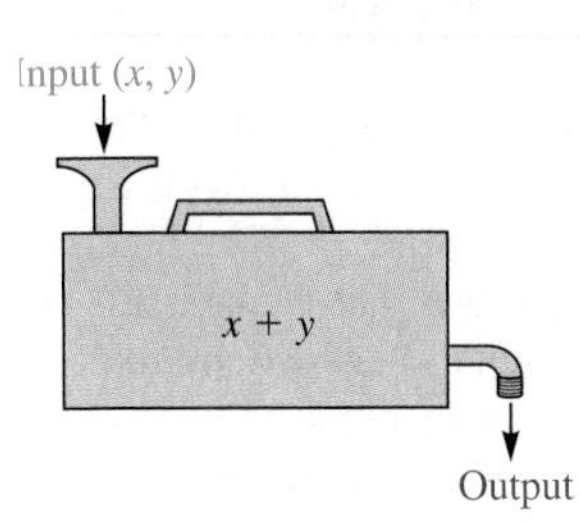

21. A student asks, "If every sequence is a function, is it also true that every function is a sequence?" How do you respond?

22. A student claims that the following does not represent a function, since all the values of x correspond to the same number.

x	0	1	2	3	4	5
y	1	1	1	1	1	1

How do you respond?

CHAPTER OUTLINE

I. Set definitions and notation
- **A.** A **set** can be described as any collection of objects.
- **B.** Sets should be **well defined** so that an object either does or does not belong to the set.
- **C.** An **element** is any **member** of a set.
- **D.** Sets can be specified by either **listing** all the elements or using **set-builder notation.**
- **E.** The **empty set**, written $\varnothing$, contains no elements.
- **F.** The **universal set** contains all the elements being discussed.

II. Relationships and operations on sets
- **A.** Two sets are **equal** if and only if they have exactly the same elements.
- **B.** Two sets A and B are in **one-to-one correspondence** if and only if each element of A can be paired with exactly one element of B and each element of B can be paired with exactly one element of A.
- **C.** Two sets A and B are **equivalent** if and only if their elements can be placed into one-to-one correspondence (written $A \sim B$).
- **D.** Set A is a **subset** of set B if and only if every element of A is an element of B (written $A \subseteq B$).
- **E.** Set A is a **proper subset** of set B if and only if every element of A is an element of B and there is at least one element of B that is not in A (written $A \subset B$).
- **F.** A set containing n elements has 2^n subsets.
- **G.** The **union** of two sets A and B is the set of all elements in A, in B, or in both A and B (written $A \cup B$).
- **H.** The **intersection** of two sets A and B is the set of all elements belonging to both A and B (written $A \cap B$).
- **I.** The **cardinal number** of a finite set S, $n(S)$, indicates the number of elements in the set.
- **J.** A set is **finite** if the number of elements in the set is zero or a natural number. Otherwise, the set is **infinite**.
- **K.** Two sets A and B are **disjoint** if they have no elements in common.
- **L.** The **complement** of a set A is the set consisting of the elements of the universal set not in A (written $\overline{A}$).
- **M.** The **complement of set A relative to set B** (set difference) is the set of all elements in B that are not in A (written $B - A$).
- **N.** The **Cartesian product** of sets A and B, written $A \times B$, is the set of all ordered pairs such that the first element in each pair is from A and the second element of each pair is from B.
- **O.** Properties of set operations
  1. Commutative property of set union
  2. Commutative property of set intersection
  3. Distributive property of set intersection over union

III. Whole numbers
- **A.** The set of **whole numbers** W is $\{0, 1, 2, 3, \ldots\}$.
- **B.** The basic operations for whole numbers are addition, subtraction, multiplication, and division.
  1. *Addition:* If $n(A) = a$ and $n(B) = b$, where $A \cap B = \varnothing$, then $a + b = n(A \cup B)$. The numbers a and b are **addends** and $a + b$ is the **sum.**
  2. *Subtraction:* If a and b are any whole numbers, then $a - b$ is the unique whole number c such that $a = b + c$.
  3. *Multiplication:* If a and b are any whole numbers, and $a \neq 0$, then
  $$ab = \underbrace{b + b + b + \ldots + b}_{a \text{ terms}}$$
  where a and b are **factors** and ab is the **product**.
  4. *Multiplication:* If A and B are sets such that $n(A) = a$ and $n(B) = b$, then $ab = n(A \times B)$.
  5. *Division:* If a and b are any whole numbers with $b \neq 0$, $a \div b$ is the unique whole number c such that $bc = a$. The number a is the **dividend**, b is the **divisor**, and c is the **quotient**.

6. **Division algorithm:** Given any whole numbers a and b, with $b \neq 0$, there exist unique whole numbers q and r such that $a = bq + r$, with $0 \leq r < b$.

C. Properties of addition and multiplication of whole numbers
1. *Closure:* If $a, b \in W$, then $a + b \in W$ and $ab \in W$.
2. *Commutative:* If $a, b \in W$, then $a + b = b + a$ and $ab = ba$.
3. *Associative:* If $a, b, c \in W$, then $(a + b) + c = a + (b + c)$ and $a(bc) = (ab)c$.
4. *Identity:* 0 is the unique identity element for addition of whole numbers; 1 is the unique identity element for multiplication.
5. *Distributive property of multiplication over addition:* If $a, b, c \in W$, then $a(b + c) = ab + ac$.
6. *Zero multiplication property:* For any whole number a, $a \cdot 0 = 0 = 0 \cdot a$.

D. Relations on whole numbers
1. $a < b$ if and only if there is a natural number c such that $a + c = b$.
2. $a > b$ if and only if $b < a$.

IV. Functions

A. A **function** from set A to B is a correspondence in which each element $a \in A$ is paired with one, and only one, element $b \in B$. If the function is denoted by f, we write $f(a) = b$. The element $a \in A$ is the input and $f(a)$ is the output. A is the **domain** of the function. B is any set containing all the outputs. The set of all the outputs is the **range** of the function.

B. A function can be represented by a rule, a table, an equation, an arrow diagram, a function machine, a set of ordered pairs, or a graph.

C. A sequence is a function whose domain is N, the set of natural numbers.

CHAPTER REVIEW

1. List all the subsets of $\{m, a, t, h\}$.
2. Let
$U = \{u, n, i, v, e, r, s, a, l\}$,
$A = \{r, a, v, e\}$, $\quad C = \{l, i, n, e\}$,
$B = \{a, r, e\}$, $\quad D = \{s, a, l, e\}$.
Find each of the following:
a. $A \cup B$ b. $C \cap D$
c. $\overline{D}$ d. $A \cap \overline{D}$
e. $\overline{B \cup C}$ f. $(B \cup C) \cap D$
g. $(\overline{A} \cup B) \cap (C \cap \overline{D})$ h. $(C \cap D) \cap A$
i. $n(\overline{C})$ j. $n(C \times D)$
3. Indicate the following sets by shading the figure:

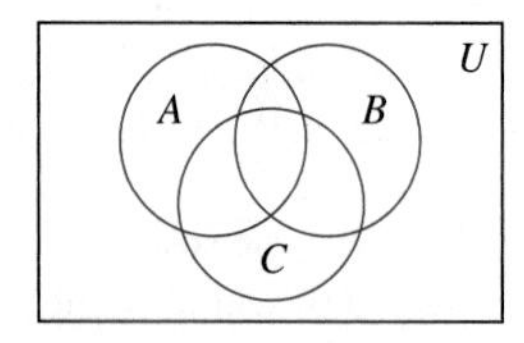

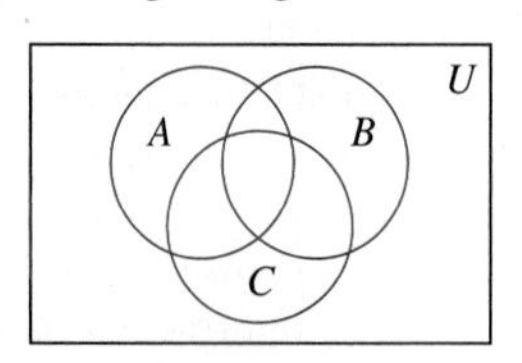

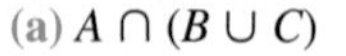

(a) $A \cap (B \cup C)$ (b) $\overline{(A \cup B)} \cap C$

4. Suppose you are playing a word game with seven distinct letters. How many seven-letter words can there be?
5. a. Show one possible one-to-one correspondence between sets D and E if $D = \{t, h, e\}$ and $E = \{e, n, d\}$.
b. How many one-to-one correspondences between sets D and E are possible?
6. Use a Venn diagram to determine whether $A \cap (B \cup C) = (A \cap B) \cup C$ for all sets A, B, and C.
7. According to a student survey, 16 students liked history, 19 liked English, 18 liked mathematics, 8 liked mathematics and English, 5 liked history and English, 7 liked history and mathematics, 3 liked all three subjects, and every student liked at least one of the subjects. Draw a Venn diagram describing this information and answer the following questions:
a. How many students were in the survey?
b. How many students liked only mathematics?
c. How many students liked English and mathematics but not history?
8. Describe, using symbols, the shaded portion in each of the following figures:

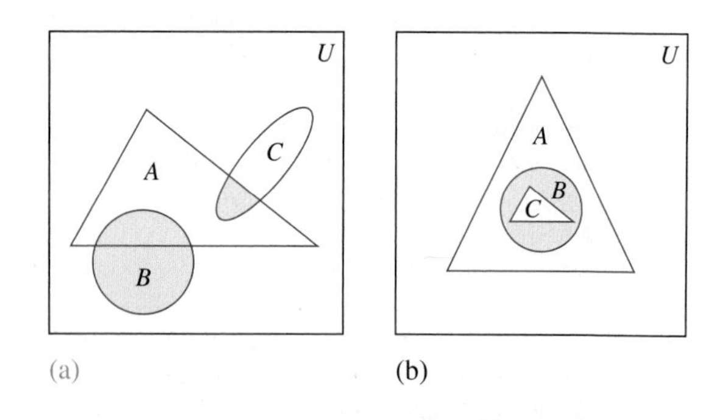

(a) (b)

9. Classify each of the following as true or false. If false, tell why.
a. For all sets A and B, either $A \subseteq B$ or $B \subseteq A$.
b. The empty set is a proper subset of every set.
c. For all sets A and B, if $A \sim B$, then $A = B$.
d. The set $\{5, 10, 15, 20, \ldots\}$ is a finite set.
e. No set is equivalent to a proper subset of itself.
f. If A is an infinite set and $B \subseteq A$, then B also is an infinite set.

g. For all finite sets A and B, if $A \cap B \neq \varnothing$, then $n(A \cup B) \neq n(A) + n(B)$.

h. If A and B are sets such that $A \cap B = \varnothing$, then $A = \varnothing$ or $B = \varnothing$.

10. Use Venn diagrams to decide whether each of the following is always true.
   a. $n(A \cup B) = n(A - B) + n(B - A) + n(A \cap B)$
   b. $n(A \cup B) = n(A - B) + n(B) = n(B - A) + n(A)$

11. Suppose P and Q are equivalent sets and $n(P) = 17$.
   a. What is the minimum number of elements in $P \cup Q$?
   b. What is the maximum number of elements in $P \cup Q$?
   c. What is the minimum number of elements in $P \cap Q$?
   d. What is the maximum number of elements in $P \cap Q$?

12. Case Eastern Junior College awarded 26 varsity letters in crew, 15 in swimming, and 16 in soccer. If awards went to 46 students and only 2 lettered in all sports, how many students lettered in two of the three sports?

13. Consider the set of northwestern states or provinces {Montana, Washington, Idaho, Oregon, Alaska, British Columbia, Alberta}. If a person chooses one element, show that in three *yes* or *no* questions, we can determine the element.

14. For each of the following, identify the properties of the operation(s) for whole numbers illustrated:
   a. $3 \cdot (a + b) = 3 \cdot a + 3 \cdot b$
   b. $2 + a = a + 2$
   c. $16 \cdot 1 = 1 \cdot 16 = 16$
   d. $6 \cdot (12 + 3) = 6 \cdot 12 + 6 \cdot 3$
   e. $3 \cdot (a \cdot 2) = 3 \cdot (2 \cdot a)$
   f. $3 \cdot (2 \cdot a) = (3 \cdot 2) \cdot a$

15. Using the definitions of less than or greater than, prove that each of the following inequalities is true:
   a. $3 < 13$ **b.** $12 > 9$

16. For each of the following, find all possible replacements to make the following statements true for whole numbers:
   a. $4 \cdot \square - 37 < 27$
   b. $398 = \square \cdot 37 + 28$
   c. $\square \cdot (3 + 4) = \square \cdot 3 + \square \cdot 4$
   d. $42 - \square \geq 16$

17. Use the distributive property of multiplication and addition facts, if possible, to rename each of the following:
   a. $3a + 7a + 5a$
   b. $3x^2 + 7x^2 - 5x^2$
   c. $x(a + b + y)$
   d. $(x + 5)3 + (x + 5)y$

18. How many 12-oz cans of juice would it take to give 60 people one 8-oz serving each?

19. Heidi has a brown pair and a gray pair of slacks; a brown blouse, a yellow blouse, and a white blouse; and a blue sweater and a white sweater. How many different outfits does she have if each outfit she wears consists of slacks, a blouse, and a sweater?

20. I am thinking of a whole number. If I divide it by 13, then multiply the answer by 12, then subtract 20, and then add 89, I end up with 93. What was my original number?

21. A ski resort offers a weekend ski package for \$80 per person or \$6000 for a group of 80 people. Which would be the less expensive option for a group of 80?

22. Josi has a job in which she works 30 hr/wk and gets paid \$5/hr. If she works more than 30 hr in a week, she receives \$8/hr for each hour over 30 hr. If she worked 38 hr this week, how much did she earn?

23. In a television game show, there are five questions to answer. Each question is worth twice as much as the previous question. If the last question was worth \$6400, what was the first question worth?

24. a. Think of a number.
   Add 17.
   Double the result.
   Subtract 4.
   Double the result.
   Add 20.
   Divide by 4.
   Subtract 20.

   Your answer will be your original number. Explain how this trick works.

   b. Fill in three more steps that will take you back to your original number.
   Think of a number.
   Add 18.
   Multiply by 4.
   Subtract 7.
   .
   .
   .

   c. Make up a series of instructions such that you will always get back to your original number.

25. Which of the following sets of ordered pairs are functions from the set of first components to the set of second components?
   a. $\{(a, b), (c, d), (e, a), (f, g)\}$
   b. $\{(a, b), (a, c), (b, b), (b, c)\}$
   c. $\{(a, b), (b, a)\}$

26. Given the following function rules and the domains, find the associated ranges:
   a. $f(x) = x + 3$; domain $= \{0, 1, 2, 3\}$
   b. $f(x) = 3x - 1$; domain $= \{5, 10, 15, 20\}$
   c. $f(x) = x^2$; domain $= \{0, 1, 2, 3, 4\}$
   d. $f(x) = x^2 + 3x + 5$; domain $= \{0, 1, 2\}$

27. Which of the following correspondences from A to B describe a function? If a correspondence is a function, find its range. Justify your answers.
   a. A is the set of college students, and B is the set of majors. To each college student corresponds his or her major.

b. A is the set of books in the library, and B is the set N of natural numbers. To each book corresponds the number of pages in the book.

c. $A = \{(a, b) \mid a \in N \text{ and } b \in N\}$, and $B = N$. To each element of A corresponds the number $4a + 2b$.

d. $A = N$ and $B = N$. If x is even, then $f(x) = 0$ and if x is odd, then $f(x) = 1$.

e. $A = N$ and $B = N$. To each natural number corresponds the sum of its digits.

28. A health club charges an initiation fee of \$200 that gives 1 month of free membership, and then charges \$55 per month.

a. If $C(x)$ is the total cost of membership in the club for x months, express $C(x)$ in terms of x.

b. Graph $C(x)$ for the first 12 months.

c. Use the graph in (b) to find when the total cost of membership in the club will exceed \$600.

d. When will the total cost of membership exceed \$6000?

29. If $f(x) = 3x - 2$, find

a. $f(6)/f(2)$ **b.** $f(2) + f(3)$

c. $f(2) \cdot f(4)$ **d.** $f(5) - f(3)$

e. $f(2 + 3)$ **f.** $f(2 \cdot 4)$

30. If the rule for the function is $f(x) = 4x - 5$ and $f(x) = 15$ is the output, what is the input?

31. Which of the following graphs represent functions? Tell why.

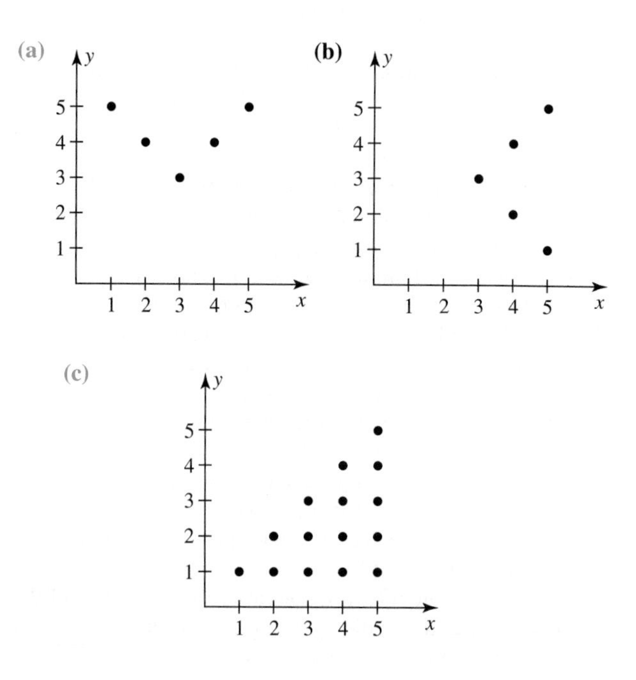

SELECTED BIBLIOGRAPHY

Bell, A., B. Greer, C. Mangan, and L. Grimison. "Children's Performance on Multiplicative Word Problems: Elements of a Descriptive Theory." *Journal for Research in Mathematics Education* 1989, 20(5): 434–449.

Billings, E., and M. Schultz McClure. "Mailing a Publication: Exploring Linear and Step Functions in a Real-World Context." *Mathematics Teaching in the Middle School* 10 (March 2005): 349–355.

Cantlon, D. "Kids + Conjecture = Mathematics Power." *Teaching Children Mathematics* 5 (October 1998): 108–112.

Carpenter, T., and J. Moser. "The Acquisition of Addition and Subtraction Concepts in Grades One Through Three." *Journal for Research in Mathematics Education* 15 (May 1984): 179–202.

Crouse, R., and A. Alison. "Tips for Beginners: The Human Coordinate System." *Mathematics Teacher* 84 (February 1991): 108–109.

English, L., and G. Halford. *Mathematics Education: Models and Processes.* Mahwah, NJ: LEA, 1995.

Fleege, P., and D. Thompson. "From Habitat to Legs: Using Science-Themed Counting Books to Foster Connections." *Teaching Children Mathematics* 7 (October 2000): 74–78.

Fuson, K. "Research on Learning and Teaching Addition and Subtraction of Whole Numbers." In *Handbook of Research on Mathematics Teaching and Learning,* edited by D. Grouws. New York: MacMillan, 1992.

Ginsburg, H., A. Klein, and P. Starkey. "The Development of Children's Mathematical Thinking: Connecting Research with Practice." In *Child Psychology in Practice,* edited by Irving E. Sigel and K. Ann Renninger, pp. 401–476, vol. 4 of *Handbook of Child Psychology,* edited by William Damon. New York: John Wiley & Sons, 1998.

Hines, E. "Exploring Functions with Dynamic Physical Models." *Mathematics Teaching in the Middle School* 7 (January 2002): 274–278.

Huinker, D. "Calculators as Learning Tools for Young Children's Explorations of Numbers." *Teaching Children Mathematics* 8 (February 2002): 316–321.

Johnston, A. "Introducing Function and Its Notation." *Mathematics Teacher* 80 (October 1987): 558–560.

Kline, K. "Kindergarten Is More Than Counting." *Teaching Children Mathematics* 5 (October 1998): 84–87.

Koirala, H., and P. Goodwin. "Teaching Algebra in the Middle Grades Using Mathmagic." *Mathematics Teaching in the Middle School* 5 (May 2000): 563–566.

Lambdin, D., R. Lynch, and H. McDaniel. "Algebra in the Middle Grades." *Mathematics Teaching in the Middle School* 6 (November 2000): 195–198.

Marlovits, Z., B. Eylon, and M. Bruckheimer. "Difficulties Students have with the Function Concept." In *The Ideas of Algebra, K-12,* edited by A. Coxford and A. Shulte. Reston VA: NCTM, 1988.

Miller, C., and T. Veenstra. "Fibonacci: Beautiful Patterns, Beautiful Mathematics." *Mathematics Teaching in the Middle School* 7 (January 2002): 298–300, 304–305.

Muller, G. "Menu of Problems," Problems 9 and 12. *Mathematics Teaching in the Middle School* 5 (January 2000): 306–307.

Resnick, L. "From Protoquantities to Operators: Building Mathematical Competence on a Foundation of Everyday Knowledge." In *Analysis of Arithmetic for Mathematics Teaching,* edited by D. Leinhardt, R. Putnam, and R. Hattrup. Hillsdale, NJ: LEA, 1992.

Rubenstein, R. "Building Explicit and Recursive Forms of Patterns with the Function Game." *Mathematics Teaching in the Middle School* 7 (April 2002): 426–431.

———. "The Function Game." *Mathematics Teaching in the Middle School* 2 (November-December 1996): 74–78.

Sakshaug, L., and K. Wohlhuter. "Responses to the Which Graph Is Which Problem." *Teaching Children Mathematics* 7 (February 2001): 352–353.

Sand, M. "A Function Is a Mail Carrier." *Mathematics Teacher* 89 (September 1996): 468–469.

Schneider, S., and C. Thompson. "Incredible Equations Develop Incredible Number Sense." *Teaching Children Mathematics* 7 (November 2000): 146–148, 165–168.

Schulman, L., and R. Eston. "A Problem Worth Revisiting." *Teaching Children Mathematics* 5 (October 1998): 72–77.

Shealy, B. "Becoming Flexible with Functions: Investigating United States Population Growth." *Mathematics Teacher* 89 (May 1996): 414–418.

Siegler, R. *Emerging Minds: The Process of Change in Children's Thinking.* New York: Oxford University Press, 1996.

Sisul, J. "Fostering Flexibility with Numbers in the Primary Grades." *Teaching Children Mathematics* 9 (December 2002): 2002–204.

Smith, E. "Patterns, Functions, and Algebra." In *A Research Companion to NCTM's Standards,* edited by J. Kirkpatrick, W. G. Martin, and D. S. Schifter. Reston, VA: NCTM, 2000.

Steinberg, R. "Instruction on Derived Facts Strategies in Addition and Subtraction." *Journal for Research in Mathematics Education* 1985, 16 (5): 337–355.

White, J., ed. "Counting Sheep" cartoon by Blair. *Mathematics Teaching in the Middle School* 5 (March 2000): 445.

Numeration Systems and Whole-Number Computation

PRELIMINARY PROBLEM

Noah and Ben are math enthusiasts; they claim that they found a new, quick method for squaring a number that ends in 5. To calculate $35 \cdot 35$, they multiply the first digit, 3, by (first digit + 1), that is, $3 \cdot 4$, or 12. They then append 25 and get 1225 as the answer.

a. Explain why this method for squaring a two-digit number ending in 5 works.
b. Explain why a similar approach for squaring a three-digit number works. For which three-digit number is the approach practical? Explain.

The NCTM's *Principles and Standards* Number and Operations Standard is reprinted here.

> *Instructional programs from prekindergarten through grade 12 should enable all students to*
>
> - *understand numbers, ways of representing numbers, relationships among numbers, and number systems;*
> - *understand meanings of operations and how they relate to one another;*
> - *compute fluently and make reasonable estimates.* (p. 32)

Further, *Principles and Standards* states:

> *The Number and Operations Standard describes deep and fundamental understanding of, and proficiency with, counting, numbers, and arithmetic, as well as an understanding of number systems and their structures. The concepts and algorithms of elementary arithmetic are part of number and operations, as are the properties and characteristics of the classes of numbers that form the beginnings of number theory.* (p. 32)

Also, in *Principles and Standards* we find:

> *In these* Standards, *understanding number and operations, developing number sense, and gaining fluency in arithmetic computation form the core of mathematics education for the elementary grades. As they progress from prekindergarten through grade 12, students should attain a rich understanding of numbers—what they are; how they are represented with objects, numerals, or on number lines; how they are related to one another; how numbers are embedded in systems that have structures and properties; and how to use numbers and operations to solve problems.* (p. 32)

Number and Operation Expectations

The *Principles and Standards* Number and Operations Standard has implications for the expectations of students at various levels. The expectations covered in this chapter include the following:

> *In grades preK–2, children should learn about number systems and use multiple models to develop understanding of place value and the base-ten number system. Children should develop fluency in adding, subtracting, multiplying, and dividing whole numbers.* (p. 392)
>
> *In grades 3–5, children should master the expectations of grades preK–2 and use the properties of operations such as the distributive property of multiplication over addition. Children should develop and use strategies for estimating the results of whole-number computations, and they should be able to judge if their estimates are reasonable. They should be able to select and use appropriate methods and tools for computing with whole numbers from among mental math, estimation, calculator, and paper and pencil.* (p. 392)

3-1 Numeration Systems

In the first section of this chapter, we introduce various number systems and compare them to the Hindu-Arabic system of numbers that we use today. By comparing the Hindu-Arabic system with ancient systems that used other bases, we may develop a clearer picture of how we do whole-number computation. The Hindu-Arabic system relies on ten digits—0, 1, 2, 3, 4, 5, 6, 7, 8, and 9—that represent the cardinal numbers of sets equivalent to those shown in Table 3-1.

Table 3-1

Digits for Cardinal Number of Set	Set
0	∅
1	$\{a\}$
2	$\{a, b\}$
3	$\{a, b, c\}$
4	$\{a, b, c, d\}$
5	$\{a, b, c, d, e\}$
6	$\{a, b, c, d, e, f\}$
7	$\{a, b, c, d, e, f, g\}$
8	$\{a, b, c, d, e, f, g, h\}$
9	$\{a, b, c, d, e, f, g, h, i\}$

numerals The written symbols in Table 3-1, such as 2 or 5, are **numerals**. Different cultures developed different numerals over the years to represent cardinal numbers. Table 3-2 shows other representations along with how they relate to the Hindu-Arabic digits 0 through 9 or 10.

Table 3-2

Babylonian		▼	▼▼	▼▼▼	▼▼▼▼	▼▼▼ ▼▼	▼▼▼ ▼▼▼	▼▼▼▼ ▼▼▼	▼▼▼▼ ▼▼▼▼	▼▼▼▼▼ ▼▼▼▼	<
Egyptian		I	II	III	IIII	III II	III III	IIII III	IIII IIII	III III III	∩
Mayan	𝋠	•	••	•••	••••	—	• —	•• —	••• —	•••• —	═
Greek		α	β	γ	δ	∊	ϕ	ζ	η	υ	ι
Roman		I	II	III	IV	V	VI	VII	VIII	IX	X
Hindu	०	१	२	३	४	५	६	७	८	९	
Arabic	٠	١	٢	٣	٤	٥	٦	٧	٨	٩	
Hindu-Arabic	0	1	2	3	4	5	6	7	8	9	10

As seen in Table 3-2, in the Hindu-Arabic system, using numerals to represent **numeration system** numbers greater than 9 requires a numeration system. A **numeration system** is a collection of properties and symbols agreed upon to represent numbers systematically. We will examine various numeration systems that have been used throughout history. Through the study of these systems, we can explore the evolution of the Hindu-Arabic system.

Hindu-Arabic Numeration System

The Hindu-Arabic numeration system that we use today was developed by the Hindus and transported to Europe by the Arabs—hence, the name *Hindu-Arabic.* The Hindu-Arabic system relies on the following properties:

1. All numerals are constructed from the 10 digits.
2. Place value is based on powers of 10, the number base of the system.

place value

face value

Because the Hindu-Arabic system is based on powers of 10, the system is a base-ten, or a decimal, system. **Place value** assigns a value to a digit depending on its placement in a numeral. To find the value of a digit in a whole number, we multiply the place value of the digit by its **face value**, where the face value is a digit. For example, in the numeral 5984, the 5 has place value "thousands," the 9 has place value "hundreds," the 8 has place value "tens," and the 4 has place value "units," as seen in Figure 3-1.

FIGURE 3-1

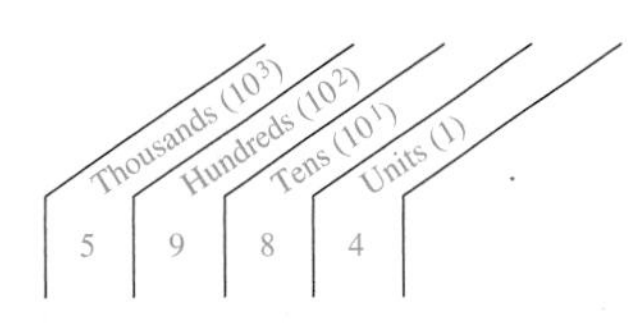

expanded form

factor

We could write 5984 in **expanded form** as $5 \cdot 10^3 + 9 \cdot 10^2 + 8 \cdot 10 + 4 \cdot 1$. In the expanded form of 5984, exponents have been used. For example, 1000, or $10 \cdot 10 \cdot 10$, is written as 10^3. In this case, 10 is a **factor** of the product. In general, we have the following definition.

Definition of a^n

If a is any number and n is any natural number, then

$$a^n = \underbrace{a \cdot a \cdot a \cdot \ldots \cdot a}_{n \text{ factors}}.$$

HISTORICAL NOTE

The invention of the Hindu-Arabic numeration system is considered one of the most important developments in mathematics. The system was first introduced in India, and then transmitted by the Arabs to North Africa and Spain and then to the rest of Europe. Historians trace the use of zero as a placeholder to the fourth century BCE (Before the Common Era). Arab mathematicians extended the decimal system to include fractions. The Italian mathematician Fibonacci, also known as Leonardo of Pisa (1170–1250), studied in Algeria and brought back with him the new numeration system, which he described and used in a book he published in 1202.◀

A set of base-ten blocks, shown in Figure 3-2 consists of *units, longs, flats,* and *blocks,* representing 1, 10, 100, and 1000, respectively. Such base-ten blocks, a subset of multibase blocks, can be used to teach place value.

FIGURE 3-2

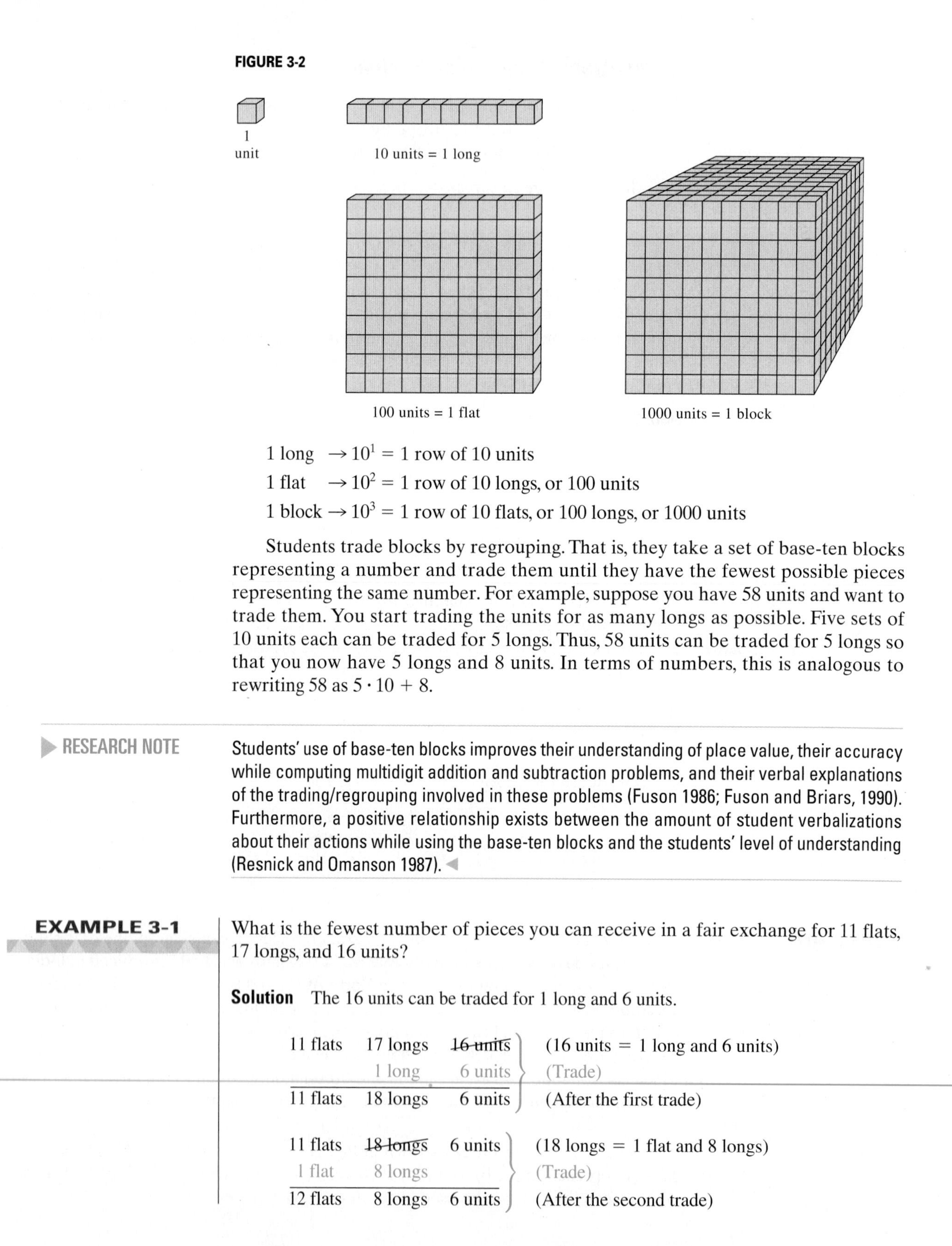

1 long $\rightarrow 10^1 = 1$ row of 10 units

1 flat $\rightarrow 10^2 = 1$ row of 10 longs, or 100 units

1 block $\rightarrow 10^3 = 1$ row of 10 flats, or 100 longs, or 1000 units

Students trade blocks by regrouping. That is, they take a set of base-ten blocks representing a number and trade them until they have the fewest possible pieces representing the same number. For example, suppose you have 58 units and want to trade them. You start trading the units for as many longs as possible. Five sets of 10 units each can be traded for 5 longs. Thus, 58 units can be traded for 5 longs so that you now have 5 longs and 8 units. In terms of numbers, this is analogous to rewriting 58 as $5 \cdot 10 + 8$.

RESEARCH NOTE

Students' use of base-ten blocks improves their understanding of place value, their accuracy while computing multidigit addition and subtraction problems, and their verbal explanations of the trading/regrouping involved in these problems (Fuson 1986; Fuson and Briars, 1990). Furthermore, a positive relationship exists between the amount of student verbalizations about their actions while using the base-ten blocks and the students' level of understanding (Resnick and Omanson 1987).

EXAMPLE 3-1

What is the fewest number of pieces you can receive in a fair exchange for 11 flats, 17 longs, and 16 units?

Solution The 16 units can be traded for 1 long and 6 units.

11 flats	17 longs	~~16 units~~	(16 units = 1 long and 6 units)
	1 long	6 units	(Trade)
11 flats	18 longs	6 units	(After the first trade)

11 flats	~~18 longs~~	6 units	(18 longs = 1 flat and 8 longs)
1 flat	8 longs		(Trade)
12 flats	8 longs	6 units	(After the second trade)

	~~12 flats~~	8 longs	6 units	(12 flats = 1 block and 2 flats)
1 block	2 flats			(Trade)
1 block	2 flats	8 longs	6 units	(After the third trade)

Therefore, the fewest number of pieces is $1 + 2 + 8 + 6 = 17$. This trading is analogous to rewriting $11 \cdot 100 + 17 \cdot 10 + 16$ as $1 \cdot 10^3 + 2 \cdot 10^2 + 8 \cdot 10 + 6$, which implies that there are 1286 units.

NOW TRY THIS 3-1

Use trading with base-ten blocks (shown in Figure 3-2) to write $3 \cdot 10^3 + 12 \cdot 10^2 + 11 \cdot 10 + 17$ in the Hindu-Arabic numeration system.

Next, we discuss other numeration systems. The study of such systems will give us a historical perspective on the development of numeration systems and will help us better understand our own Hindu-Arabic numeration system.

Tally Numeration System

tally numeration system

The **tally numeration system** used single strokes, or tally marks, to represent each object that was counted; for example, the first ten counting numbers are

|, ||, |||, ||||, |||||, ||||||, |||||||, ||||||||, |||||||||, ||||||||||

In a tally system there is a one-to-one correspondence between the marks and the items being counted. The system is simple but it requires many symbols, especially when numbers become greater. Also as numbers become greater, they are harder to read.

As we see in the "Wizard of Id" cartoon, the tally system can be improved by *grouping.* In the first frame, we see that the tallies are grouped into fives by placing a diagonal across four tallies to make a group of five. Grouping makes it easier to read the numeral. Later, as we see in the second frame, shortened numerals were made in another system (Roman).

THE WIZARD OF ID

Brant parker and Johnny hart

YOU @*☆!! COLLEGE GRADS THINK YOU'RE SO DANGED SMART!

©1992 by North America Syndicate, Inc. World rights reserved.

Egyptian Numeration System

The Egyptian numeration system, which dates back to about 3400 BCE, used tally marks. The first nine numerals in the Egyptian system in Table 3-2 show the use of tally marks. The Egyptians improved on the system based only on tally marks by developing a *grouping system* to represent certain sets of numbers. This makes the numbers easier to record. For example, the Egyptians used a heel bone symbol, ∩, to stand for a grouping of ten tally marks.

$$||||||||||\rightarrow \cap$$

Table 3-3 shows other numerals that the Egyptians used in their system.

Table 3-3

Egyptian Numeral	Description	Hindu-Arabic Equivalent
|	Vertical staff	1
∩	Heel bone	10
9	Scroll	100
𓆼	Lotus flower	1,000
𓂭	Pointing finger	10,000
𓆐	Polliwog or burbot	100,000
𓁨	Astonished man	1,000,000

additive property

In its simplest form, the Egyptian system involved an **additive property**; that is, the value of a number was the sum of the face values of the numerals. The Egyptians customarily wrote the numerals in decreasing order from left to right as in 𓆐999∩∩||. The number can be converted to base ten as shown here.

𓆐	represents	100,000	
999	represents	300	(100 + 100 + 100)
∩∩	represents	20	(10 + 10)
||	represents	2	(1 + 1)
𓆐999∩∩||	represents	100,322	

NOW TRY THIS 3-2

a. Use the Egyptian system to represent 1,312,322.
b. Use the Hindu-Arabic system to represent 𓆐𓆐𓆼𓆼𓆼∩∩∩||||.
c. What disadvantages do you see in the Egyptian system compared to the Hindu-Arabic system?

Babylonian Numeration System

The Babylonian numeration system was developed at about the same time as the Egyptian system. The symbols in Table 3-4 were made using a stylus either vertically or horizontally on clay tablets. The Babylonian system has been well preserved because clay tablets were used rather than papyrus used by the Egyptians.

Table 3-4

Babylonian Numeral	Hindu-Arabic Equivalent
▼	1
<	10

The Babylonian numerals 1 through 59 were similar to the Egyptian numerals, but the vertical staff and the heel bone were replaced by the symbols shown in Table 3-4. For example, << ▼▼ represented 22.

The Babylonian numeration system used a *place value system*. Numbers greater than 59 were represented by repeated groupings of 60, much as we use groupings of 10 today. For example ▼▼ << might represent $2 \cdot 60 + 20$, or 140. The space indicates that ▼▼ represents $2 \cdot 60$ rather than 2. Numerals immediately to the left of a second space have a value $60 \cdot 60$ times their face value, and so on.

<< ▼	represents	$20 \cdot 60 + 1$, or 1201
<▼ <▼ ▼	represents	$11 \cdot 60 \cdot 60 + 11 \cdot 60 + 1$, or $11 \cdot 60^2 + 11 \cdot 60 + 1$, or 40,261
▼ <▼ <▼ ▼	represents	$1 \cdot 60 \cdot 60 \cdot 60 + 11 \cdot 60 \cdot 60 + 11 \cdot 60 + 1$, or $1 \cdot 60^3 + 11 \cdot 60^2 + 11 \cdot 60 + 1$, or 256,261

The initial Babylonian system was inadequate by today's standards. For example, the symbol ▼▼ could have represented 2 or $2 \cdot 60$. Later, the Babylonians introduced the symbol ⧗ as a placeholder for missing position values. Using this symbol, < <<▼ represents $10 \cdot 60 + 21$ and < ⧗ <<▼ represents $10 \cdot 60^2 + 0 \cdot 60 + 21$. In this sense, ⧗ represented 0.

NOW TRY THIS 3-3

a. Use the Babylonian system to represent 12,321.

b. Use the Hindu-Arabic system to represent ▼▼ <▼ ▼.

c. What advantages does the Hindu-Arabic system have over the Babylonian system?

Mayan Numeration System

In the early development of numeration systems, people frequently used parts of their bodies to count. Fingers could be matched to objects to stand for one, two, three, four, or five objects. Two hands could then stand for a set of ten objects. In warmer climates where people went barefoot, people may have used their toes as well as their fingers for counting. The Mayans introduced an attribute that was not present in the Egyptian or early Babylonian systems, namely, a symbol for zero. The Mayan system used only three symbols, which Table 3-5 shows, and based their system on 20.

Table 3-5

Mayan Numeral	Hindu-Arabic Equivalent
•	1
—	5
	0

The symbols for the first 10 numerals in the Mayan system are shown in Table 3-2. Notice the groupings of five, where each horizontal bar represents a group of five. Thus the symbol for 19 was , or three 5s and four 1s. The symbol for 20 was , which represents one group of twenty plus zero 1s. In Figure 3-3(a), we have $2 \cdot 5 + 3 \cdot 1$, or thirteen groups of 20 plus $2 \cdot 5 + 1 \cdot 1$, or eleven 1s, for a total of 271. In Figure 3-3(b), we have $3 \cdot 5 + 1 \cdot 1$, or 16, groups of 20 and zero 1s, for a total of 320.

FIGURE 3-3

(a) $13 \cdot 20 + 11 \cdot 1 = 271$

(b) $16 \cdot 20 + 0 \cdot 1 = 320$

In a true base-twenty system, the place value of the symbols in the third position vertically from the bottom should be 20^2, or 400. However, the Mayans used $20 \cdot 18$, or 360, instead of 400. (The number 360 is an approximation of the length of a calendar year, which consisted of 18 months of 20 days each, plus 5 "unlucky" days.) Thus, instead of place values of 1, 20, 20^2, 20^3, 20^4, and so on, the Mayans used 1, 20, $20 \cdot 18$, $20^2 \cdot 18$, $20^3 \cdot 18$, and so on. For example, in Figure 3-4(a), we have $5 + 1$ (or 6) groups of 360, plus $5 + 5 + 2$ (or 12) groups of 20, plus $5 + 4$ (or 9) groups of 1, for a total of 2409. In Figure 3-4(b), we have $2 \cdot 5$ (or 10) groups of 360, plus 0 groups of 20, plus two 1s, for a total of 3602. Spacing is important in the Mayan system. For example, if two horizontal bars are placed close together, as in ═, the symbols represent $5 + 5 = 10$. If the bars are spaced apart, as in ═, then the value is $5 \cdot 20 + 5 \cdot 1 = 105$.

FIGURE 3-4

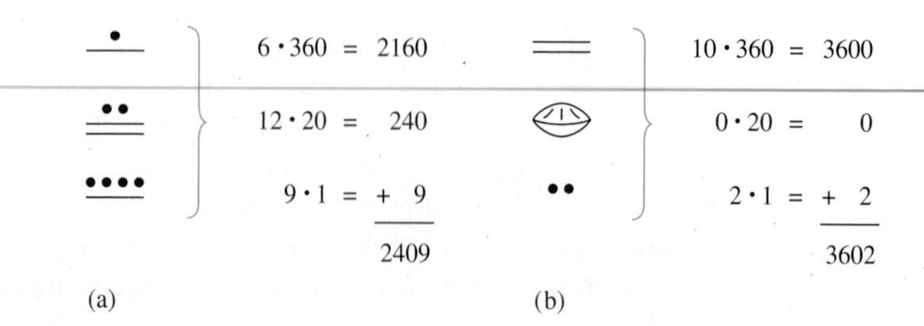

6 • 360 = 2160
12 • 20 = 240
9 • 1 = + 9
2409
(a)
10 • 360 = 3600
0 • 20 = 0
2 • 1 = + 2
3602
(b)

Roman Numeration System

The Roman numeration system was used in Europe in its early form from the third century BCE. It remains in use today, as seen on cornerstones, on the opening pages of books, and on the faces of some clocks. The Roman system uses only a few symbols, as shown in Table 3-6.

Table 3-6

Roman Numeral	Hindu-Arabic Equivalent
I	1
V	5
X	10
L	50
C	100
D	500
M	1000

Roman numerals can be combined by using an additive property. For example, MDCLXVI represents $1000 + 500 + 100 + 50 + 10 + 5 + 1 = 1666$, CCCXXVIII represents 328, and VI represents 6.

subtractive property

To avoid repeating a symbol more than three times, as in IIII, a **subtractive property** was introduced in the Middle Ages. For example, I is less than V, so if it is to the left of V, it is subtracted. Thus, IV has a value of $5 - 1$, or 4, and XC represents $100 - 10$, or 90. Some extensions of the subtractive property could lead to ambiguous results. For example, IXC could be 91 or 89. By custom, 91 is written XCI and 89 is written LXXXIX. In general, only one smaller number symbol can be to the left of a larger number symbol and the pair must be one of those listed in Table 3-7.

Table 3-7

Roman Numeral	Hindu-Arabic Equivalent
IV	$5 - 1$, or 4
IX	$10 - 1$, or 9
XL	$50 - 10$, or 40
XC	$100 - 10$, or 90
CD	$500 - 100$, or 400
CM	$1000 - 100$, or 900

multiplicative property

In the Middle Ages, a bar was placed over a Roman number to multiply it by 1000. The use of bars is based on a **multiplicative property**. For example, $\overline{\text{V}}$ represents $5 \cdot 1000$, or 5000, and $\overline{\text{CDX}}$ represents $410 \cdot 1000$, or 410,000. To indicate even greater numbers, more bars appear. For example, $\overline{\overline{\text{V}}}$ represents $(5 \cdot 1000) \cdot 1000$, or 5,000,000; $\overline{\overline{\overline{\text{CXI}}}}$ represents $111 \cdot 1000^3$, or 111,000,000,000; and $\overline{\text{CX}}\text{I}$ represents $110 \cdot 1000 + 1$, or 110,001.

Several properies might be used to represent some numbers, for example:

$$\overline{\text{D}}\text{CLIX} = \underbrace{\underbrace{(500 \cdot 1000)}_{\text{Multiplicative}} + \underbrace{(100 + 50)}_{\text{Additive}} + \underbrace{(10 - 1)}_{\text{Subtractive}}}_{\text{Additive}} = 500{,}159$$

Other Number Base Systems

To better understand our base-ten system and to investigate some of the problems that students might have when learning the Hindu-Arabic system, we will investigate similar systems but with different number bases.

Base Five

The Luo peoples of Kenya used a *quinary,* or base-five, system. A system of this type can be modeled by counting with only one hand. The digits available for counting are 0, 1, 2, 3, and 4. In the "one-hand system," or base-five system, you count 1, 2, 3, 4, 10, where 10 represents one hand and no fingers. Counting in base five proceeds as shown in Figure 3-5. We write the small "five" below the numeral as a reminder that the number is written in base five. If no base is written, a number is assumed to be in base ten.

FIGURE 3-5

One-Hand System	Base-Five Symbol	Base-Five Blocks
0 fingers	0_{five}	
1 finger	1_{five}	
2 fingers	2_{five}	
3 fingers	3_{five}	
4 fingers	4_{five}	
1 hand and 0 fingers	10_{five}	
1 hand and 1 finger	11_{five}	
1 hand and 2 fingers	12_{five}	
1 hand and 3 fingers	13_{five}	
1 hand and 4 fingers	14_{five}	
2 hands and 0 fingers	20_{five}	
2 hands and 1 finger	21_{five}	

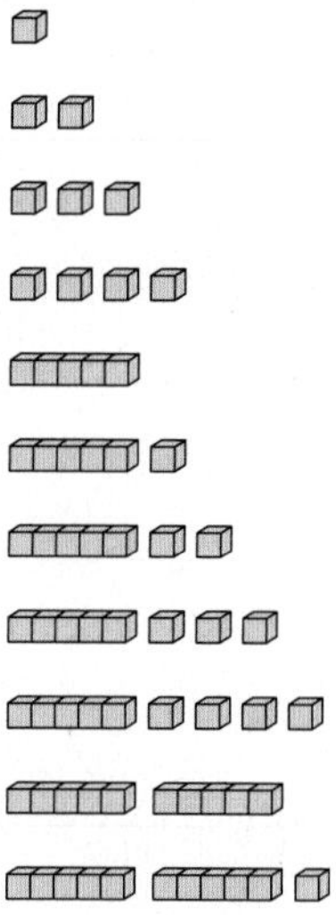

Counting in base five is similar to counting in base ten. Because we have only five digits (0, 1, 2, 3, and 4), 4_{five} plays the role of 9 in base ten. Figure 3-6 shows how we can find the number that comes after 34_{five} by using base-five blocks.

FIGURE 3-6

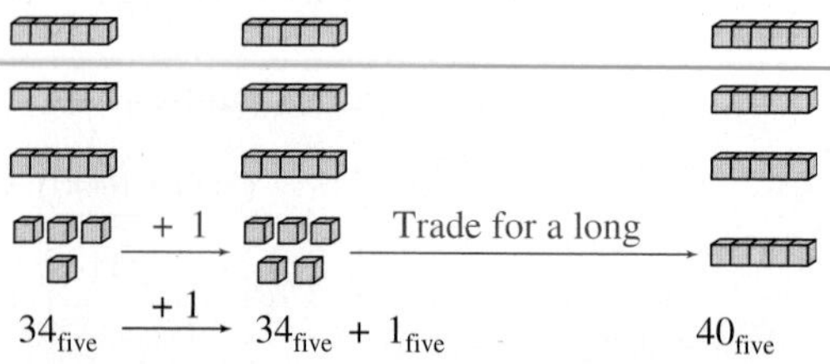

+ 1

Trade for a long

34_{five} $\xrightarrow{+1}$ $34_{five} + 1_{five}$ 40_{five}

What number follows 44_{five}? There are no more two-digit numbers in the system after 44_{five}. In base ten, the same situation occurs at 99. We use 100 to represent ten 10s, or one 100. In the base-five system, we need a symbol to represent five 5s. To continue the analogy with base ten, we use 100_{five} to represent one group of five 5s, zero groups of five, and zero units. To distinguish from "one hundred" in base ten, the name for 100_{five}, is read "one-zero-zero base five." The number 100 means $1 \cdot 10^2 + 0 \cdot 10^1 + 0$, whereas the number 100_{five} means $(1 \cdot 10^2 + 0 \cdot 10^1 + 0)_{\text{five}}$, or $(1 \cdot 5^2 + 0 \cdot 5^1 + 0)_{\text{five}}$, or 25.

Examples of base-five numerals along with their base-five block representations and conversions to base ten are given in Figure 3-7. Multibase blocks will be used throughout the text to illustrate various concepts.

FIGURE 3-7

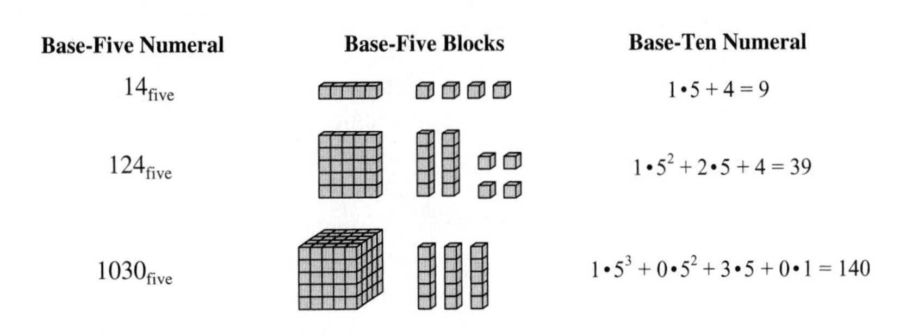

Base-Five Numeral
Base-Five Blocks
Base-Ten Numeral
14_{five}
$1 \cdot 5 + 4 = 9$
124_{five}
$1 \cdot 5^2 + 2 \cdot 5 + 4 = 39$
1030_{five}
$1 \cdot 5^3 + 0 \cdot 5^2 + 3 \cdot 5 + 0 \cdot 1 = 140$

EXAMPLE 3-2

Convert 11244_{five} to base ten.

Solution

$$\begin{aligned} 11244_{\text{five}} &= 1 \cdot 5^4 + 1 \cdot 5^3 + 2 \cdot 5^2 + 4 \cdot 5^1 + 4 \cdot 1 \\ &= 1 \cdot 625 + 1 \cdot 125 + 2 \cdot 25 + 4 \cdot 5 + 4 \cdot 1 \\ &= 625 + 125 + 50 + 20 + 4 \\ &= 824 \end{aligned}$$

Example 3-2 suggests a method for changing a base-ten number to a base-five number using powers of 5. To convert 824 to base five, we divide by successive powers of 5. A shorthand method for illustrating this conversion is the following:

		Divisor	Dividend	Quotient	
$5^4 = 625$	$\rightarrow$	625	824	1	How many groups of 625 in 824?
			-625		
$5^3 = 125$	$\rightarrow$	125	199	1	How many groups of 125 in 199?
			-125		
$5^2 = 25$	$\rightarrow$	25	74	2	How many groups of 25 in 74?
			-50		
$5^1 = 5$	$\rightarrow$	5	24	4	How many groups of 5 in 24?
			-20		
$5^0 = 1$	$\rightarrow$	1	4	4	How many 1s in 4?
			-4		
			0		

Thus, $824 = 11244_{\text{five}}$.

NOW TRY THIS 3-4

A different method of converting 824 to base five is shown using successive divisions by 5. The quotient in each case is placed below the dividend and the remainder is placed on the right, on the same line with the quotient. The answer is read from bottom to top, that is, as 11244_{five}.

5	824	
5	164	4
5	32	4
5	6	2
	1	1

a. Why does this method work?

b. Use this method to convert 728 to base five.

Calculators with the integer division feature—[INT÷] on a Texas Instrument calculator or [÷R] on a Casio—can be used to change base-ten numbers to different number bases. For example, to convert 8 to base five, we enter [8] [INT÷] [5] [=] and obtain $\underset{Q}{1}$ $\underset{R}{3}$. This implies that $8 = 13_{\text{five}}$. Will this technique work to convert 34 to base five? Why or why not?

Base Two

Historians tell of early tribes that used base two. Some aboriginal tribes still count "one, two, two and one, two twos, two twos and one, ..." Because base two has only two digits, it is called the **binary system**. Base two is especially important because of its use in computers. One of the two digits is represented by the presence of an electrical signal and the other by the absence of an electrical signal. Although base two works well for some purposes, it is inefficient for everyday use because multidigit numbers are reached very rapidly in counting in this system, as seen in the following cartoon where Peter is reading Jason's counting to five.

binary system

FOXTROT © 1996 Bill Amend. Reprinted with permission of UNIVERSAL PRESS SYNDICATE. All rights reserved.

NOW TRY THIS 3-5

In the "Foxtrot" cartoon, Jason has listed the first five counting numbers in base two. Continue the list until you have the first twenty counting numbers.

Conversions from base two to base ten, and vice versa, can be accomplished in a manner similar to that used for base-five conversions.

EXAMPLE 3-3

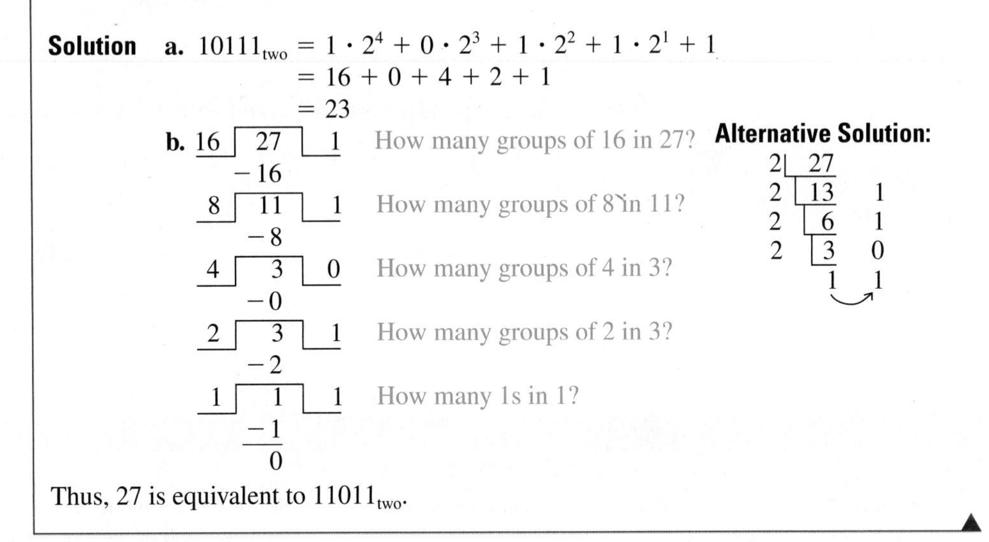

a. Convert 10111_{two} to base ten.
b. Convert 27 to base two.

Solution **a.** $10111_{two} = 1 \cdot 2^4 + 0 \cdot 2^3 + 1 \cdot 2^2 + 1 \cdot 2^1 + 1$
$= 16 + 0 + 4 + 2 + 1$
$= 23$

b.

Divisor	Dividend	Quotient	
16	27 − 16	1	How many groups of 16 in 27?
8	11 − 8	1	How many groups of 8 in 11?
4	3 − 0	0	How many groups of 4 in 3?
2	3 − 2	1	How many groups of 2 in 3?
1	1 − 1 = 0	1	How many 1s in 1?

Alternative Solution:

2	27	
2	13	1
2	6	1
2	3	0
	1	1

Thus, 27 is equivalent to 11011_{two}.

Base Twelve

Another commonly used number-base system is the base-twelve, or the duodecimal ("dozens"), system. Eggs are bought by the dozen, and pencils are bought by the *gross* (a dozen dozen). In base twelve, there are 12 digits, just as there are 10 digits in base ten, 5 digits in base five, and 2 digits in base two. In base twelve, new symbols are needed to represent the following groups of x's:

10 x's and 11 x's

xxxxxxxxxx *xxxxxxxxxxx*

The new symbols chosen are T and E, respectively, so that the base-twelve digits are 0, 1, 2, 3, 4, 5, 6, 7, 8, 9, T, and E. Thus, in base twelve we count "1, 2, 3, 4, 5, 6, 7, 8, 9, T, E, 10, 11, 12, . . . , 17, 18, 19, 1T, 1E, 20, 21, 22, . . . , 28, 29, 2T, 2E, 30, ..."

EXAMPLE 3-4

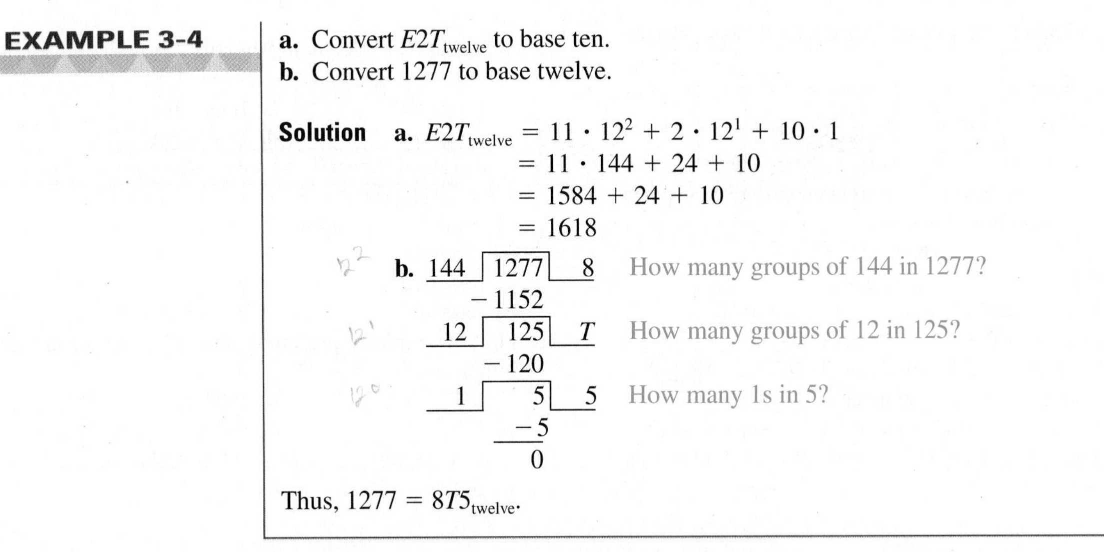

a. Convert $E2T_{twelve}$ to base ten.
b. Convert 1277 to base twelve.

Solution **a.** $E2T_{twelve} = 11 \cdot 12^2 + 2 \cdot 12^1 + 10 \cdot 1$
$= 11 \cdot 144 + 24 + 10$
$= 1584 + 24 + 10$
$= 1618$

b.

Divisor	Dividend	Quotient	
144	1277 − 1152	8	How many groups of 144 in 1277?
12	125 − 120	T	How many groups of 12 in 125?
1	5 − 5 = 0	5	How many 1s in 5?

Thus, $1277 = 8T5_{twelve}$.

EXAMPLE 3-5

Rob used base twelve to write the following:

$$g36_{\text{twelve}} = 1050_{\text{ten}}$$

What is the value of g?

Solution Using expanded form, we could write the following equations:

$$\begin{aligned} g \cdot 12^2 + 3 \cdot 12 + 6 \cdot 1 &= 1050 \\ 144g + 36 + 6 &= 1050 \\ 144g + 42 &= 1050 \\ 144g &= 1008 \\ g &= 7 \end{aligned}$$

ONGOING ASSESSMENT 3-1

1. For each of the following, tell which numeral represents the greater number and why:
   a. $\overline{M}$CDXXIV and $\overline{\overline{M}}$CDXXIV
   b. 4632 and 46,032
   c. <▼▼ and < ▼▼
   d. 999∩∩|| and 𓆼∩|
   e. Mayan numerals and Mayan numerals
2. For each of the following, name both the succeeding and preceding numbers (one more and one less):
   a. MCMXLIX **b.** $\overline{\text{MI}}$ c. CMXCIX
   d. << <▼ e. 𓆼99 **f.** Mayan numeral
3. If the cornerstone represents when a building was built and it reads MCMXXII, when was this building built?
4. Write each of the following in Roman symbols:
   a. 121 **b.** 42
   c. 89 **d.** 5282
5. Complete the following table, which compares symbols for numbers in different numeration systems:

	Hindu-Arabic	Babylonian	Egyptian	Roman	Mayan
a.	72				
b.		< ▼▼			
c.			𓆼99∩∩|||		

6. For each of the following decimal numerals, give the place value of the underlined numeral:
   a. 827,$\underline{3}$67 **b.** 8,421,0$\underline{0}$0
   c. 9$\underline{7}$,998 **d.** $\underline{8}$10,485
7. Rewrite each of the following as a base-ten numeral:
   a. $3 \cdot 10^6 + 4 \cdot 10^3 + 5$ **b.** $2 \cdot 10^4 + 1$
   c. $3 \cdot 10^3 + 5 \cdot 10^2 + 6 \cdot 10$ **d.** $9 \cdot 10^6 + 9 \cdot 10 + 9$
8. A certain three-digit whole number has the following properties: The hundreds digit is greater than 7; the tens digit is an odd number; and the sum of the digits is 10. What could the number be?
9. Study the following counting frame. In the frame, the value of each dot is represented by the number in the box below the dot. For example, the following figure represents the number 154:

••	•••	••
64	8	1

What numbers are represented in the frames in (a) and (b)?

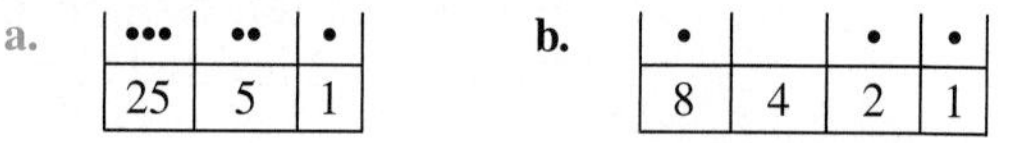

10. Write the base-four numeral for the base-four blocks shown.

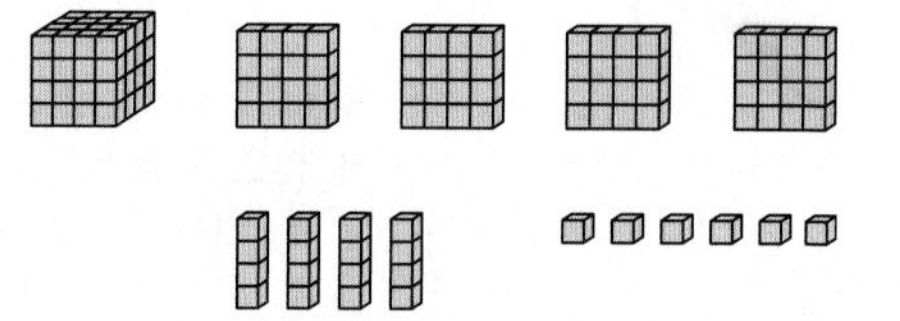

11. Write the first 15 counting numbers for each of the following bases:
    a. Base two **b.** Base three
    c. Base four **d.** Base eight
12. How many different digits are needed for base twenty?
13. Write 2032_{four} in expanded base-four notation.
14. Determine the greatest three-digit number in each of the following bases:
    a. Base two **b.** Base six
    c. Base ten **d.** Base twelve
15. Find the numbers preceding and succeeding each of the following:
    a. $EE0_{\text{twelve}}$ **b.** 100000_{two} c. 555_{six}
    d. 100_{seven} e. 1000_{five} **f.** 110_{two}
16. What, if anything, is wrong with the following numerals?
    a. 204_{four} **b.** 607_{five} c. $T12_{\text{three}}$

17. The smallest number of base-four blocks needed to represent 214 is ____ blocks ____ flats ____ longs ____ units.
18. Draw multibase blocks to represent
    a. 231_{five} **b.** 1101_{two}
19. An introduction to base five is especially suitable for early learning in elementary school, as children can think of making change using quarters, nickels, and pennies. Use only these coins to answer the following:
    a. What is the fewest number of quarters, nickels, and pennies you can receive in a fair exchange for two quarters, nine nickels, and eight pennies?
    b. How could you use the approach in (a) to write 73 in base five?
20. Without converting to base ten, tell which number is greater and why for each of the following:
    a. $3TE9E_{twelve}$ or $3E000_{twelve}$
    b. 110111_{two} or 101111_{two}
    c. 40444_{five} or 41011_{five}
21. Notice the following pattern and answer the questions that follow:

$$111_{two} + 1 = 1000_{two} = 2^3$$
$$444_{five} + 1 = 1000_{five} = 5^3$$
$$EEE_{twelve} + 1 = 1000_{twelve} = 12^3$$

    a. Explain why these statements are true without converting the number to base ten.
    b. Based on the pattern, write a similar statement in base eight.
    c. Make a conjecture based on the statements.
    d. Make similar statements for four-digit numbers in different bases.
    e. Explain why $1 + 2 + 2^2 + \ldots + 2^{10} = 2^{11} - 1$.
22. Without changing each number to base ten, find how much greater 344411_{five} is than 334411_{five}.
23. Recall that with base ten blocks, 1 long = 10 units, 1 flat = 10 longs, and 1 block = 10 flats (see Figure 3-2). In a set of multibase pieces make all possible exchanges to obtain the smallest number of pieces and write the corresponding numeral in the given base.
    a. Ten longs in base five b. Ten longs in base two
    c. Ten flats in base ten **d.** Twenty flats in base twelve
24. Convert each of the following base-ten numbers to numbers in the indicated bases:
    a. 432 to base five **b.** 1963 to base twelve
    c. 404 to base four **d.** 37 to base two
25. Change 42_{eight} to base two without first changing to base ten.
26. Write each of the following numbers in base ten:
    a. 432_{five} **b.** 101101_{two} c. $92E_{twelve}$
    d. $T0E_{twelve}$ e. 111_{twelve} **f.** 346_{seven}
27. You are asked to distribute \$900 in prize money. The dollar amounts for the prizes are \$625, \$125, \$25, \$5, and \$1. How should this \$900 be distributed in order to give the fewest number of prizes?
28. Convert each of the following:
    a. 58 days to weeks and days
    b. 54 months to years and months
    c. 29 hours to days and hours
    d. 68 inches to feet and inches
29. A bookstore ordered 11 gross, 6 dozen, and 6 pencils. Express the number of pencils in base twelve and in base ten.
30. For each of the following, find b if possible. If not possible, tell why.
    a. $b2_{seven} = 44_{ten}$
    b. $5b2_{twelve} = 734_{ten}$
    c. $23_{ten} = 25_b$
    d. $b2_{seven} = 2b_{ten}$
31. Write $12^5 + 25 \cdot 12^4 + 23$ in base-twelve notation without multiplying out 12^5 and 12^4.
32. Jillian is reading a book. Her sister Abby asks her what page she is on. Jillian, who likes riddles, says that the sum of the two page numbers that are open is 243 and that she is about to turn a page. What page number is Jillian on? Explain your reasoning.
33. Children learn the meaning of ten, hundred, thousand, and later a million. Some also learn that the name for 1000 million is a billion and 1000 billion is referred to as one trillion. Write the following as Hindu-Arabic numerals (base-ten notation).
    a. Six million six hundred thousand forty-five
    b. Nine billion four million five thousand six hundred forty-five
    c. Three trillion nine billion four hundred thousand sixty
    d. Hundred trillion
34. The Chinese abacus, depicted as follows, shows the number 5857. (*Hint:* The beads above the bar represent 5s, 50s, 500s, and 5000s.)

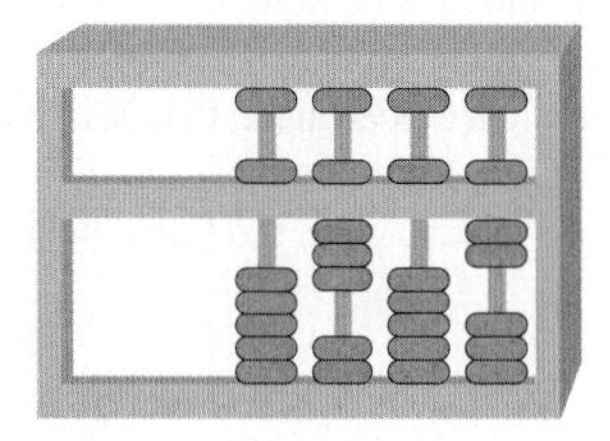

Discuss how the number 5857 is depicted and show how the number 4869 could be depicted.
35. On a calculator, using only the keys [1], [2], [3], [4], [5], [6], [7], [8], and [9], fill the calculator's display to show each of the following:
    a. The greatest number possible if each key may be used only once
    b. The least number possible if each key may be used only once

c. The greatest number possible if a key may be used more than once
d. The least number possible if a key may be used more than once

36. In a game called WIPEOUT, we are to "wipe out" digits from a calculator's display without changing any of the other digits. "Wipeout" in this case means to replace the chosen digit(s) with a 0. For example, if the initial number is 54,321 and we are to wipe out the 4, we could subtract 4000 to obtain 50,321. Complete the following two problems and then try other numbers or challenge another person to wipe out a number from the number you have placed on the screen:
a. Wipe out the 2s from 32,420.
b. Wipe out the 5 from 67,357.

Communication

37. Ben claims that zero is the same as nothing. Explain how you as a teacher would respond to Ben's statement.

38. What are the major drawbacks to each of the following systems?
a. Egyptian **b.** Babylonian **c.** Roman

39. a. Why are large numbers in the United States written with commas separating groups of three digits?
b. Find examples from other countries that do not use commas to separate groups of three digits.

40. Marcy bets that if you do a series of mathematical computations and activities, she can guess your color. First you must find your special number by completing the following:
Take the number of your birth month.
Add 24.
Add the difference you obtain when you subtract the number of your birth month from 12.
Divide by 3.
Add 13.
The result is your special number.
To each letter in the alphabet, assign a number that is the letter's order when alphabetical, that is, $a = 1, b = 2, c = 3, d = 4$, and so on. Find the letter that corresponds to your special number. Next write the name of a color that starts with this letter. What color will Marcy predict is your color? Explain why this works.

Open-Ended

41. An inspector of weights and measures uses a special set of weights to check the accuracy of scales. Various weights are placed on a scale to check accuracy of any amount from 1 oz through 15 oz. What is the least number of weights the inspector needs? What weights are needed to check the accuracy of scales from 1 oz through 15 oz? From 1 oz through 31 oz?

Cooperative Learning

42. a. Create a numeration system with unique symbols and write a paragraph explaining the properties of the system.
b. Complete the following table using the system:

Hindu-Arabic Numeral	Your System Numeral
1	
5	
10	
50	
100	
5,000	
10,000	
115,280	

Third International Mathematics and Science Study (TIMSS) Questions

Which digit is in the hundreds place in 2345?
a. 2 **b.** 3 **c.** 4 **d.** 5

TIMSS, 2003, Grade 4

Which of these is a name for 9740?
a. Nine thousand seventy-four
b. Nine thousand seven hundred forty
c. Nine thousand seventy-four hundred
d. Nine hundred seventy-four thousand

TIMSS, 2003, Grade 4

BRAIN TEASER

There are three nickels and three dimes concealed in three boxes. Two coins are placed in each of the boxes, which are labeled 10¢, 15¢, and 20¢. The coins are placed in such a way that no box contains the amount of money shown on its label; for example, the box labeled 10¢ does not really have a total of 10¢ in it. What is the minimum number of coins that you would have to remove from a box, and from which box or boxes, to determine which coins are in which boxes?

3-2 Algorithms for Whole-Number Addition and Subtraction

According to the *Principles and Standards*,

> *As children in prekindergarten through grade 2 develop an understanding of whole numbers and the operations of addition and subtraction, instructional attention should focus on strategies for computing with whole numbers so that students develop flexibility and computational fluency. Students will generate a range of interesting and useful strategies for solving computational problems, which should be shared and discussed. By the end of grade 2, students should know the basic addition and subtraction combinations, should be fluent in adding two-digit numbers, and should have methods for subtracting two-digit numbers. At the grades 3–5 level, as students develop the basic number combinations for multiplication and division, they should also develop reliable algorithms to solve arithmetic problems efficiently and accurately. These methods should be applied to larger numbers and practiced for fluency.*
>
> *Researchers and experienced teachers alike have found that when children in the elementary grades are encouraged to develop, record, explain, and critique one another's strategies for solving computational problems, a number of important kinds of learning can occur (see, e.g., Hiebert [1999]; Kamii, Lewis, and Livingston [1993]; Hiebert et al. [1997]). The efficiency of various strategies can be discussed. So can generalizability: Will this work for any numbers or only the two involved here? And experience suggests that in classes focused on the development and discussion of strategies, various "standard" algorithms either arise naturally or can be introduced by the teacher as appropriate. The point is that students must become fluent in arithmetic computation—they must have efficient and accurate methods that are supported by an understanding of numbers and operations. "Standard" algorithms for arithmetic computation are one means of achieving this fluency.* (p. 35)

The previous chapter introduced the operations of addition and subtraction of whole numbers and now, as pointed out in the *Standards*, it is time to focus on *computational fluency*—having and using efficient and accurate methods for computing. The *Standards* suggest that "standard algorithms" are one means to achieve this fluency. An **algorithm** (named for the ninth-century Arabian mathematician Mohammed al Khowarizmi) is a systematic procedure used to accomplish an operation. This section focuses on developing and understanding algorithms involving addition and subtraction. Alternative algorithms as well as standard algorithms will be developed.

algorithm

Addition Algorithms

In teaching mathematics to young children, it is important that we support them in the transition from concrete to abstract thinking by using techniques that parallel their developmental processes. In order for children to understand the use of paper-and-pencil algorithms, they should explore addition by first using manipulatives. If children can touch and move around items such as chips, bean sticks, and an abacus or use base-ten blocks, they can be led (and often will proceed naturally on their own) to the creation of algorithms for addition.

In what follows, we use base-ten blocks to illustrate the development of an algorithm for whole-number addition. Suppose we want to add $14 + 23$. We start with a concrete model in Figure 3-8(a), move to the expanded algorithm in Figure 3-8(b), and then to the standard algorithm in Figure 3-8(c).

FIGURE 3-8

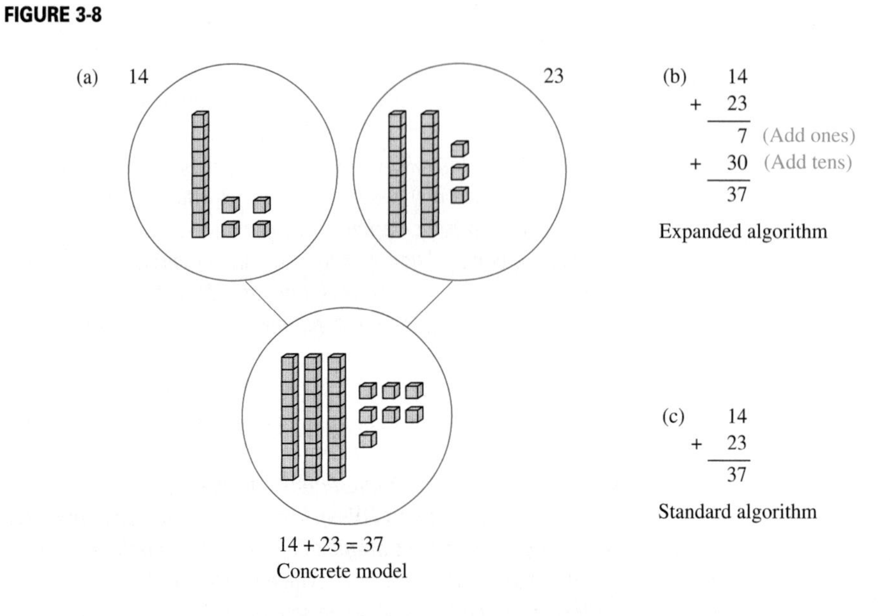

A more formal justification for this addition not usually presented at the elementary level is the following:

$14 + 23 = (1 \cdot 10 + 4) + (2 \cdot 10 + 3)$	Expanded form
$= (1 \cdot 10 + 2 \cdot 10) + (4 + 3)$	Commutative and associative properties of addition
$= (1 + 2) \cdot 10 + (4 + 3)$	Distributive property of multiplication over addition
$= 3 \cdot 10 + 7$	Single-digit addition facts
$= 37$	Place value

We see on the student page an example of using base-ten blocks with regrouping. Once children have mastered such use of concrete models with regrouping, they should be ready to use the expanded and standard algorithms. Figure 3-9 shows the computation 37 + 28 using both algorithms. In Figure 3-9(b), you will notice that when there were more than 10 ones, we regrouped 10 ones as a ten and then added the tens. Notice that the words *regroup* or *trade* are now commonly used in the elementary classroom to describe what we used to call *carrying*.

FIGURE 3-9

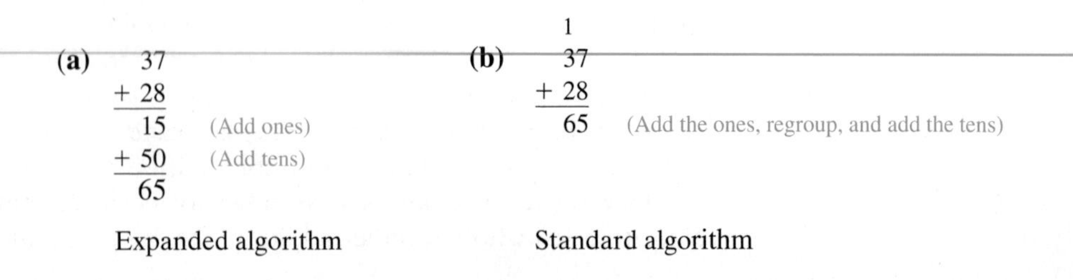

(a) 37
+ 28
15 (Add ones)
+ 50 (Add tens)
65
Expanded algorithm
(b) 1
37
+ 28
65 (Add the ones, regroup, and add the tens)
Standard algorithm

Students often develop algorithms on their own and learning can occur by investigating how and if various algorithms work. Addition of whole numbers using blocks has a natural carryover to the expanded form and trading used earlier. For example, consider the following addition:

$$\begin{array}{r} 376 \\ 459 \\ +\ 8716 \\ \hline \end{array} \quad \text{or} \quad \begin{array}{rrrr} & 3 \cdot 10^2 + & 7 \cdot 10 + & 6 \\ & 4 \cdot 10^2 + & 5 \cdot 10 + & 9 \\ 8 \cdot 10^3 + & 7 \cdot 10^2 + & 1 \cdot 10 + & 6 \\ \hline 8 \cdot 10^3 + & 14 \cdot 10^2 + & 13 \cdot 10 + & 21 \end{array}$$

To complete the addition, "fair" trading is used. However, consider an analogous algebra problem of adding polynomials:

$$(3x^2 + 7x + 6) + (4x^2 + 5x + 9) + (8x^3 + 7x^2 + 1 \cdot x + 6) \quad \text{or} \quad \begin{array}{rrrr} & 3x^2 + & 7x + & 6 \\ & 4x^2 + & 5x + & 9 \\ +\ 8x^3 + & 7x^2 + & 1 \cdot x + & 6 \\ \hline 8x^3 + & 14x^2 + & 13x + & 21 \end{array}$$

Note that if $x = 10$, the addition is the same as given earlier. Also note that knowledge of place value in addition problems aids in algebraic thinking. Next we explore several algorithms that have been used throughout history.

Left-to-Right Algorithm for Addition

Since children learn to read from left to right, it might seem natural that they may try to add from left to right. When working with base-ten blocks, many children in fact do combine the larger pieces first and then move to combining the smaller pieces. A left-to-right algorithm is as follows:

$$\begin{array}{r} 568 \\ +\ 757 \\ \hline 12 \\ 11 \\ 15 \\ \hline 1325 \end{array} \quad \longrightarrow \quad \begin{array}{r} 568 \\ +\ 757 \\ \hline 1\not{2}\not{1}5 \\ 32 \end{array} \quad \rightarrow \quad 1325$$

Explain why this technique works and try it with $9076 + 4689$.

Lattice Algorithm for Addition

We introduce this algorithm by working through an addition involving two four-digit numbers. For example,

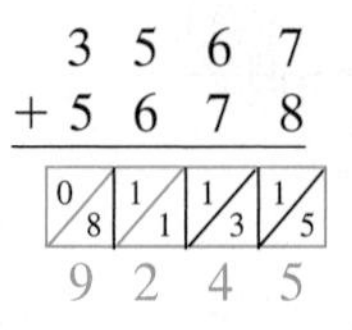

To use this algorithm, add the single-digit numbers on top to the single digit numbers on the bottom, from right to left, and record the results in a lattice. Then add the sums along the diagonals, as shown. You will see that this is very similar to the expanded algorithm introduced earlier. Try this technique with $4578 + 2691$.

SAMPLE SCHOOL BOOK PAGE:

ADDING TWO-DIGIT NUMBERS

Lesson 3-1

Key Idea
You can break apart numbers, using place value, to add.

Vocabulary
• regroup

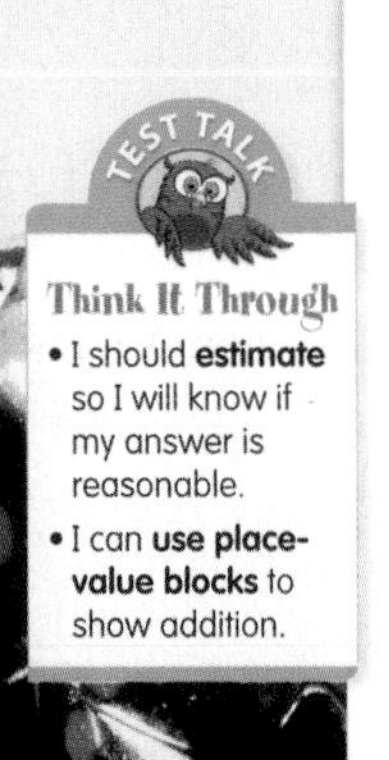

TEST TALK

Think It Through
• I should **estimate** so I will know if my answer is reasonable.
• I can **use place-value blocks** to show addition.

Adding Two-Digit Numbers

WARM UP
Use mental math.
1. 48 + 20 2. 63 + 11
3. 71 + 8 4. 53 + 5

LEARN

How do you add two-digit numbers?

Example

Cal counted 46 ladybugs on a log and 78 more on some bushes. How many ladybugs did he count all together?

Find 46 + 78.

Estimate: 46 rounds to 50. 78 rounds to 80.
50 + 80 = 130, so the answer should be about 130.

What You **Think**		What You **Write**
Kim's Way • Add the ones. 6 + 8 = **14** ones • Add the tens. **4** tens + **7** tens = **11** tens = **110** • Find the sum.	11 tens 14 ones	46 + 78 14 110 124
Henry's Way • Add the ones. 6 + 8 = **14** ones • **Regroup** 14 ones into **1** ten **4** ones. • Add the tens. **1** ten + **4** tens + 7 tens = 12 tens • Find the sum.	14 ones = **1** ten **4** ones	1 46 + 78 124

Cal counted 124 ladybugs all together.

Talk About It

1. Why did Henry write a small 1 above the 4 in the tens place?
2. Why should you estimate when adding two-digit numbers?

126

Source: *Scott Foresman–Addison Wesley Mathematics Grade 3, 2005* (p. 126).

Scratch Algorithm for Addition

The scratch algorithm for addition is often referred to as a low-stress algorithm because it allows students to perform complicated additions by doing a series of additions that involve only two single digits. An example follows:

1. $\begin{array}{r} 87 \\ 6\not{5}_2 \\ +\ 49 \\ \hline \end{array}$ Add the numbers in the units place starting at the top. When the sum is 10 or more, record this sum by scratching a line through the last digit added and writing the number of units next to the scratched digit. For example, since $7 + 5 = 12$, the "scratch" represents 10 and the 2 represents the units.

2. $\begin{array}{r} 87 \\ 6\not{5}_2 \\ +\ 4\not{9}_1 \\ \hline \end{array}$ Continue adding the units, including any new digits written down. When the addition again results in a sum of 10 or more, as with $2 + 9 = 11$, repeat the process described in (1).

3. $\begin{array}{r} {}^2 87 \\ 6\not{5}_2 \\ +\ 4\not{9}_1 \\ \hline 1 \end{array}$ When the first column of additions is completed, write the number of units, 1, below the addition line. Count the number of scratches, 2, and add this number to the second column.

4. $\begin{array}{r} {}^2\not{8}_0 7 \\ 6\ \not{5}_2 \\ \not{4}_0\not{9}_1 \\ \hline 2\,0\,1 \end{array}$ Repeat the procedure for each successive column.

Subtraction Algorithms

As with addition, we can use base-ten blocks to provide a concrete model for subtraction. Consider how the base-ten blocks are used to perform the subtraction $243 - 61$: First we represent 243 with 2 flats, 4 longs, and 3 units, as shown in Figure 3-10.

FIGURE 3-10

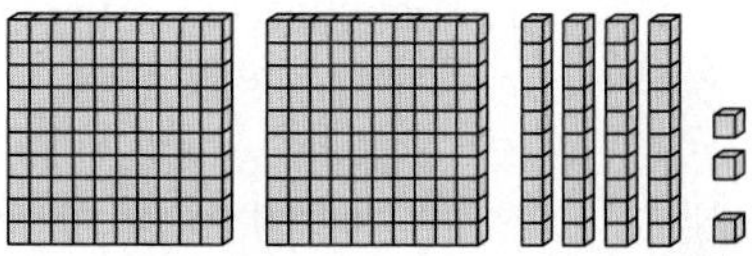

To subtract 61 from 243, we try to remove 6 longs and 1 unit from the blocks in Figure 3-10. We can remove 1 unit, as in Figure 3-11.

FIGURE 3-11

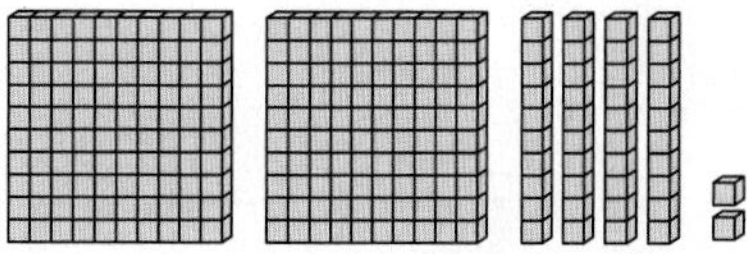

SAMPLE SCHOOL BOOK PAGE:

MODELS FOR SUBTRACTING THREE-DIGIT NUMBERS

Lesson 3-8

Key Idea
You can use blocks to show regrouping for subtraction.

Materials
- place-value blocks
or tools

Models for Subtracting Three-Digit Numbers

WARM UP
1. 96 − 42 2. 77 − 39
3. 43 − 8 4. 50 − 27

LEARN

Activity

How can you subtract with place-value blocks?

Find 255 − 163.

	What You Show	What You Write
a. Show 255 with place-value blocks.		255 − 163
b. Subtract the ones. Regroup if needed. 5 > 3. No regrouping is needed.	5 ones − 3 ones = 2 ones	255 − 163 = 2
c. Subtract the tens. Regroup if needed. 5 tens < 6 tens. So, regroup 1 hundred for 10 tens.	15 tens − 6 tens = 9 tens	1 15 ~~2~~~~5~~5 − 163 = 92
d. Subtract the hundreds.	1 hundred − 1 hundred = 0 hundreds	1 15 ~~2~~~~5~~5 − 163 = 92

e. Find the value of the remaining blocks in Step d: 9 tens 2 ones = 92, so 255 − 163 = 92.

f. In Step b, did you have to regroup to subtract the ones? Explain.

g. In Step c, did you have to regroup to subtract the tens? Explain.

h. Use place-value blocks to subtract.
243 − 72   145 − 126   223 − 156

TEST TALK
Think It Through
I can **use objects** to show a subtraction problem with regrouping.

150

Source: *Scott Foresman–Addison Wesley Mathematics Grade 3, 2005* (p. 150).

To remove 6 longs from Figure 3-11, we have to trade 1 flat for 10 longs, as shown in Figure 3-12.

FIGURE 3-12

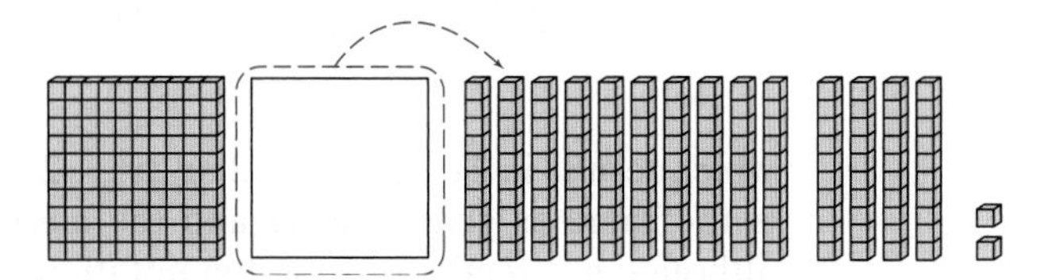

Now we can remove, or "take away," 6 longs, leaving 1 flat, 8 longs, and 2 units, or 182, as shown in Figure 3-13.

FIGURE 3-13

After students have worked with base-ten blocks for subtraction, they are ready to proceed to the standard algorithm as shown on the student page on page 174.

NOW TRY THIS 3-6

a. Use base-ten blocks and addition to check that $243 - 61 = 182$.

b. Do problems (f), (g), and (h) on the Student Page.

Equal-Addends Algorithm

The equal-addends algorithm for subtraction is based on the fact that the difference between two numbers does not change if we add the same amount to both numbers. For example, $93 - 27 = (93 + 3) - (27 + 3)$. Thus, the difference can be computed as $96 - 30 = 66$. Using this approach, the subtraction on the student page could be performed as follows.

$$\begin{array}{r} 255 \\ -\ 163 \\ \hline \end{array} \rightarrow \begin{array}{r} 255 + 7 \\ -\ (163 + 7) \\ \hline \end{array} \rightarrow \begin{array}{r} 262 \\ -\ 170 \\ \hline \end{array} \rightarrow \begin{array}{r} 262 + 30 \\ -\ (170 + 30) \\ \hline \end{array} \rightarrow \begin{array}{r} 292 \\ -\ 200 \\ \hline 92 \end{array}$$

Subtraction of whole numbers using blocks has a natural carryover to the expanded form and trading. For example, consider the following subtraction problem done earlier with blocks:

$$\begin{array}{r} 243 \\ -\ 61 \\ \hline \end{array} \text{ or } \begin{array}{r} 2 \cdot 10^2 + 4 \cdot 10 + 3 \\ -\ (6 \cdot 10 + 1) \\ \hline \end{array} \text{ or } \begin{array}{r} 1 \cdot 10^2 + 14 \cdot 10 + 3 \\ -\ (6 \cdot 10 + 1) \\ \hline 1 \cdot 10^2 + (14 - 6) \cdot 10 + (3 - 1) \end{array}$$

BRAIN TEASER

The number on a license plate consists of five digits. When the license plate is looked at upside down, you can still read it, but the value of the upside-down number is 78,633 greater than the real license number. What is the license number?

ONGOING ASSESSMENT 3-2

1. Find the missing digits in each of the following:

a.
```
   _ _ 1
 + 4 2 _
 _ 4 0 2
```
b.
```
   _ 0 2 5
   1 1 _ 6
 + 3 1 4 8
   6 _ 6 _
```
c.
```
   3 _ _
 - 1 5 9
   _ 2 4
```
d.
```
   1 _ _ _ 6
 -   8 3 0 9
     4 9 8 7
```

2. Make appropriate drawings like the ones in Figure 3-8 and Figure 3-10 that show the use of base-ten blocks to compute each of the following: **(a)** $29 + 37$ **(b)** $253 - 79$

3. Place the digits 7, 6, 8, 3, 5, and 2 in the boxes to obtain
a. The greatest sum **b.** The least sum

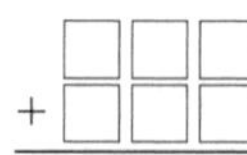

4. Place the digits 7, 6, 8, 3, 5, and 2 in the boxes to obtain
a. The greatest difference **b.** The least difference

5. Tom's diet allows only 1500 calories per day. For breakfast, Tom had skim milk (90 calories), a waffle with no syrup (120 calories), and a banana (119 calories). For lunch, he had $\frac{1}{2}$ cup of salad (185 calories) with mayonnaise (110 calories) and tea (0 calories). Then he had pecan pie (570 calories). Can he have dinner consisting of fish (250 calories), a $\frac{1}{2}$ cup of salad with no mayonnaise, and tea?

6. Wally kept track of last week's money transactions. His salary was \$150 plus \$54 in overtime and \$260 in tips. His transportation expenses were \$22, his food expenses were \$60, his laundry costs were \$15, his entertainment expenditures were \$58, and his rent was \$185. After expenses, did he have any money left? If so, how much?

7. In the following problem, the sum is correct but the order of the digits in each addend has been scrambled. Correct the addends to obtain the correct sum.

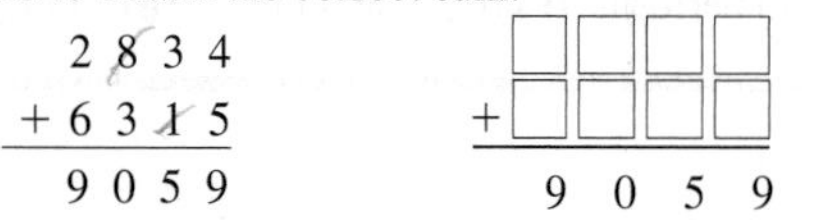

8. Use the equal-addends approach to compute each of the following:

a.
```
   93
 - 37
```
b.
```
  321
 - 38
```

9. Janet worked her addition problems by placing the partial sums as shown here:

```
   569
 + 645
    14
   10
  11
  1214
```

a. Use this method to work the following:

(i)
```
   687
 + 549
```
(ii)
```
   359
 + 673
```

b. Explain why this algorithm works.

10. Analyze the following computations. Explain what is wrong in each case.

a.
```
   135
 +  47
   172
```
b.
```
    87
 +  25
  1012
```
c.
```
   57
 - 38
   21
```
d.
```
   56
 - 18
   48
```

11. George is cooking an elaborate meal for Thanksgiving. He can cook only one thing at a time in his microwave oven. His turkey takes 75 min; the pumpkin pie takes 18 min; rolls take 45 sec; and a cup of coffee takes 30 sec to heat. How much time does he need to cook the meal?

12. Give reasons for each of the following steps:

$$\begin{aligned} 16 + 31 &= (1 \cdot 10 + 6) + (3 \cdot 10 + 1) \\ &= (1 \cdot 10 + 3 \cdot 10) + (6 + 1) \\ &= (1 + 3) \cdot 10 + (6 + 1) \\ &= 4 \cdot 10 + 7 \\ &= 47 \end{aligned}$$

13. In each of the following justify the standard addition algorithm using expanded form of the numbers, commutative and associative properties of addition, and the distributive property of multiplication over addition:
a. $68 + 23$
b. $174 + 285$
c. $2458 + 793$

14. Use the *lattice algorithm* to perform each of the following:
   a. 4358 + 3864 b. 4923 + 9897
   c. 2345 + 8888
15. Discuss the merit of the following algorithm for addition where we first add the ones, then the tens, then the hundreds, and then the total:

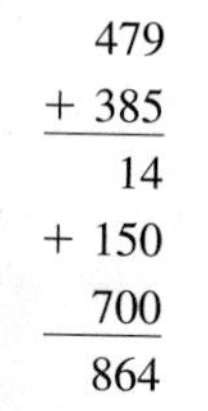

$$\begin{array}{r} 479 \\ +\ 385 \\ \hline 14 \\ +\ 150 \\ \hline 700 \\ \hline 864 \end{array}$$

16. The following example uses a regrouping approach to subtraction. Discuss the merit of this approach in teaching subtraction

$$\left.\begin{array}{r} 843 \\ -\ 568 \end{array}\right\} \rightarrow \left.\begin{array}{r} 800 + 40 + 3 \\ -\ (500 + 60 + 8) \end{array}\right\} \rightarrow$$

$$\left.\begin{array}{r} 800 + 30 + 13 \\ -\ (500 + 60 + 8) \end{array}\right\} \rightarrow \begin{array}{r} 700 + 130 + 13 \\ -\ (500 + \ \ 60 + \ \ 8) \\ \hline 200 + \ \ 70 + 5 = 275 \end{array}$$

17. Perform each of the following operations using the bases shown:
   a. $43_{five} + 23_{five}$ **b.** $43_{five} - 23_{five}$
   c. $432_{five} + 23_{five}$ **d.** $42_{five} - 23_{five}$
   e. $110_{two} + 11_{two}$ **f.** $10001_{two} - 111_{two}$
18. Construct an addition table for base eight.
19. Perform each of the following operations:

   a. $\begin{array}{r} \text{3 hr 36 min 58 sec} \\ +\ \text{5 hr 56 min 27 sec} \\ \hline \end{array}$ b. $\begin{array}{r} \text{5 hr 36 min 38 sec} \\ -\ \text{3 hr 56 min 58 sec} \\ \hline \end{array}$

20. Perform each of the following operations (2 c = 1 pt, 2 pt = 1 qt, 4 qt = 1 gal):

   a. $\begin{array}{r} \text{1 qt 1 pt 1 c} \\ +\ \ \text{1 pt 1 c} \\ \hline \end{array}$ **b.** $\begin{array}{r} \text{1 qt 1 c} \\ -\ \text{1 pt 1 c} \\ \hline \end{array}$

   c. $\begin{array}{r} \text{1 gal 3 qt 1 c} \\ -\ \ \text{4 qt 2 c} \\ \hline \end{array}$

21. Andrew's calculator was not functioning properly, When he pressed [8] [+] [6] [=], the numeral 20 appeared on the display. When he pressed [5] [+] [4] [=], 13 was displayed. When he pressed, [1] [5] [−] [3] [=], 9 was displayed. What do you think Andrew's calculator was doing?
22. The following is a supermagic square taken from an engraving called *Melancholia* by Dürer that includes the year (1514) it was constructed:

16	3	2	13
5	10	11	8
9	6	7	12
4	15	14	1

   a. Find the sum of each row, the sum of each column, and the sum of each diagonal.
   b. Find the sum of the four numbers in the center.
   c. Find the sum of the four numbers in each corner.
   d. Add 11 to each number in the square. Is the square still a magic square? Explain your answer.
   e. Subtract 11 from each number in the square. Is the square still a magic square?
23. Use scratch addition to perform the following:

   a. $\begin{array}{r} 432 \\ 976 \\ +\ 1418 \\ \hline \end{array}$ b. $\begin{array}{r} 32_{five} \\ 13_{five} \\ 22_{five} \\ 43_{five} \\ 23_{five} \\ +\ 12_{five} \\ \hline \end{array}$

24. Perform each of the following operations:

   a. $\begin{array}{r} \text{4 gross 4 doz 6 ones} \\ -\ \ \ \text{5 doz 9 ones} \\ \hline \end{array}$

   b. $\begin{array}{r} \text{2 gross 9 doz 7 ones} \\ +\ \text{3 gross 5 doz 9 ones} \\ \hline \end{array}$

25. Determine what is wrong with the following:

$$\begin{array}{r} 22_{five} \\ +\ 33_{five} \\ \hline 55_{five} \end{array}$$

26. Fill in the missing numbers in each of the following:

   a. $\begin{array}{r} 2\ \_\ \_{}_{five} \\ -\ \ 2\ 2_{five} \\ \hline \_\ 0\ 3_{five} \end{array}$ **b.** $\begin{array}{r} 2\ 0\ 0\ 1\ 0_{three} \\ -\ \ 2\ \_\ 2\ \_{}_{three} \\ \hline 1\ \_\ 2\ \_\ 1_{five} \end{array}$

27. Find the numeral to put in the blank to make each equation true. Do not convert to base ten.
   a. $3423_{five} -$ ________ $= 2132_{five}$
   b. $11011_{two} +$ ________ $= 100{,}000_{two}$
   c. $TEE_{twelve} -$ ________ $= 1$
   d. $1000_{five} +$ ________ $= 10{,}000_{five}$
28. The Hawks played the Elks in a basketball game. Based on the following information, complete the scoreboard showing the number of points scored by each team during each quarter and the final score of the game.

Teams	Quarters 1	2	3	4	Final Score
Hawks					
Elks					

   a. The Hawks scored 15 points in the first quarter.
   b. The Hawks were behind by 5 points at the end of the first quarter.
   c. The Elks scored 5 more points in the second quarter than they did in the first quarter.

d. The Hawks scored 7 more points than the Elks in the second quarter.

e. The Elks outscored the Hawks by 6 points in the fourth quarter.

f. The Hawks scored a total of 120 points in the game.

g. The Hawks scored twice as many points in the third quarter as the Elks did in the first quarter.

h. The Elks scored as many points in the third quarter as the Hawks did in the first two quarters combined.

29. A palindrome is any number that reads the same backward as forward, for example, 121 and 2332. Try the following. Begin with any number. Is it a palindrome? If not, reverse the digits and add this reversed number to the original number. Is the result a palindrome? If not, repeat the procedure until a palindrome is obtained. For example, start with 78. Because 78 is not a palindrome, we add: $78 + 87 = 165$. Because 165 is not a palindrome, we add: $165 + 561 = 726$. Again, 726 is not a palindrome, so we add $726 + 627$ to obtain 1353. Finally, $1353 + 3531$ yields 4884, which is a palindrome.

 a. Try this method with the following numbers:
 (i) 93 (ii) 588 (iii) 2003

 b. Find a number for which the procedure described takes more than five steps to form a palindrome.

30. a. Place the numbers 24 through 32 in the following circles so that the sums are the same in each direction:

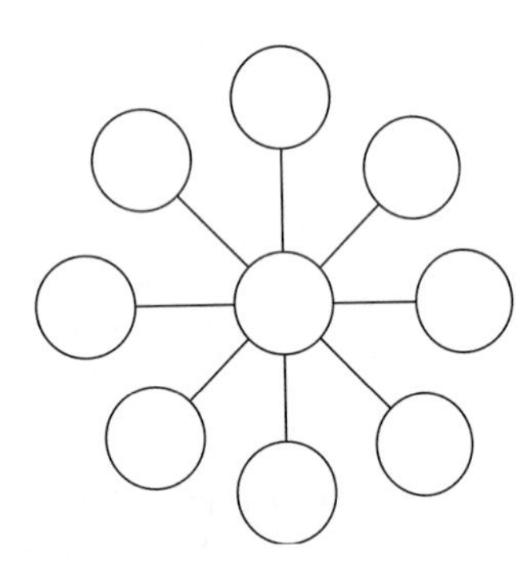

 b. How many different numbers can be placed in the middle to obtain a solution?

Communication

31. Tira, a fourth grader, performs addition by adding and subtracting the same number. She added as follows:

$$\begin{array}{r} 39 \\ +\,84 \\ \hline \end{array} \rightarrow \begin{array}{r} 39 + 1 \\ +\,84 - 1 \\ \hline \end{array} \rightarrow \begin{array}{r} 40 \\ +\,83 \\ \hline 123 \end{array}$$

How would you respond if you were her teacher?

32. Cathy found her own algorithm for subtraction. She subtracted as follows:

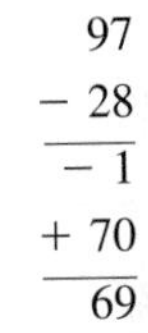

How would you respond if you were her teacher?

33. Discuss why the words *regroup* and *trade* are used rather than *carry* and *borrow* for whole-number addition and subtraction algorithms.

34. Discuss whether children should be encouraged to develop and use their own algorithms for whole-number addition and subtraction or whether they should be taught only one algorithm per operation and all students should use only the one algorithm.

35. Explain why the scratch addition algorithm works.

36. The *equal-addends* algorithm was introduced in this section. The following shows how this algorithm works for $1464 - 687$.

$$\begin{array}{r} 1\,4\,6\,{}^{1}4 \\ -\quad 6\,{}^{9}8\,7 \\ \hline 7 \end{array}$$
(Add 10 to the 4 ones to get 14 ones.)
(Add 1 ten to the 8 tens to get 9 tens.)
(Subtract the ones.)

Now we move to the next column.

$$\begin{array}{r} 1\,4\,{}^{1}6\,{}^{1}4 \\ -\ {}^{7}6\,{}^{9}8\,7 \\ \hline 7\,7\,7 \end{array}$$
(Add 10 tens to the 6 tens to get 16 tens.)
(Add 1 hundred to the 6 hundreds to get 7 hundreds.)
(Subract the 9 tens from the 16 tens and then the 7 hundreds from the 14 hundreds.)

 a. Try the technique on three more subtractions.

 b. Explain why the equal-addends algorithm works.

37. A child is asked to compute $7 + 2 + 3 + 8 + 11$ and writes $7 + 2 = 9 + 3 = 12 + 8 = 20 + 11 = 31$. Noticing that the answer is correct, if you were the teacher how would you react?

Open-Ended

38. Search for or develop an algorithm for whole-number addition or subtraction and write a description of your algorithm so that others can understand and use it.

Cooperative Learning

39. Investigate our money system and decide whether the system is a base-ten system.

Review Problems

40. Investigate the measuring of lengths in the metric system. Develop a plan for using place value with lengths to convert among different metric units.

41. Write 5280 in expanded form.

42. What is the value of MCDX in Hindu-Arabic numerals?

43. Convert each of the following to base ten:
 a. EOT_{twelve} b. 1011_{two} c. 43_{five}

44. If 1 mo is approximately 4 wk and 1 yr is approximately 365 days, or 52 wk, answer the following:
    a. Lewis and Clark spent approximately 2 yr, 4 mo, and 9 days exploring the territory in the Northwest. What is this time in weeks?
    b. It took Magellan 1126 days to circle the world. How many years is this?
    c. How many seconds old are you?
    d. Approximately how many times does your heart beat in 1 yr?

NAEP Questions (National Assessment of Educational Progress)

1. Add:

$$\begin{array}{r} 238 \\ +\ 462 \\ \hline \end{array}$$

a. 600 **b.** 690
c. 700 **d.** 790

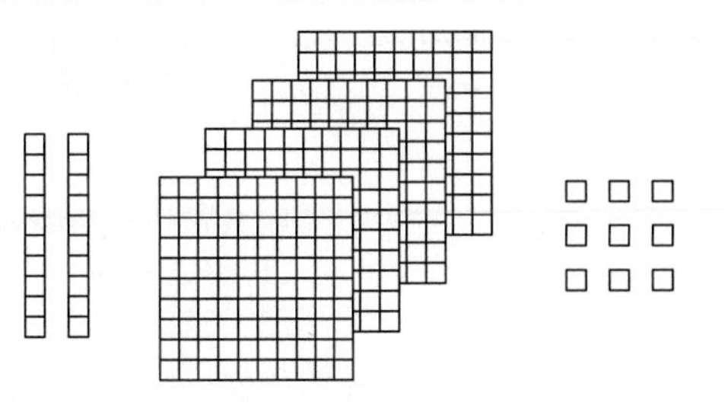

Each small square (□) above is equal to 1. There are 10 small squares in each strip. There are 100 small squares in each large square. What number is shown?

a. 4,029 **b.** 492
c. 429 **d.** 249

LABORATORY ACTIVITY

1. One type of Japanese abacus, *soroban*, is shown in Figure 3-16(a). In this abacus, a bar separates two sets of bead counters. Each counter above the bar represents five times the counter below the bar. Numbers are illustrated by moving the counter toward the bar. The number 7632 is pictured. Practice demonstrating and adding numbers on this abacus.

FIGURE 3-16

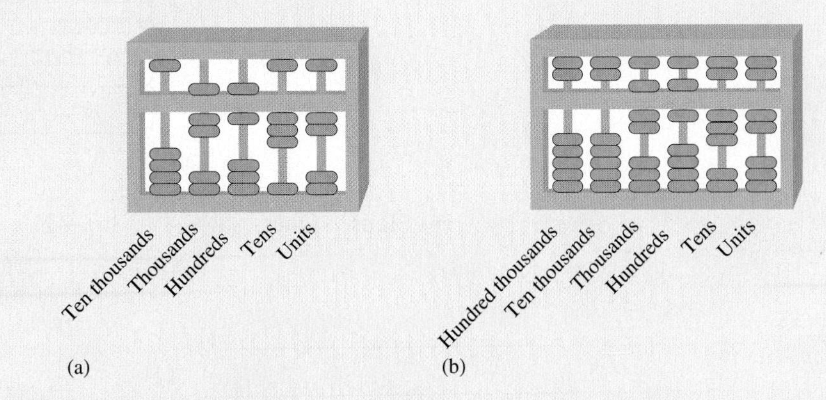

2. The Chinese abacus, *suan pan* (see Figure 3-16(b)), is still in use today. This abacus is similar to the Japanese abacus but has two counters above the bar. The number 7632 is also pictured on it. Practice demonstrating and adding numbers on this abacus. Compare the ease of using the two versions.

3-3 Algorithms for Whole-Number Multiplication and Division

In the *Principles and Standards*, we find the following regarding multiplication and division of whole numbers:

> *In grades 3–5, students should focus on the meanings of, and relationship between, multiplication and division. It is important that students understand what each number in a multiplication or division expression represents. For example, in multiplication, unlike addition, the factors in the problem can refer to different units. If students are solving the problem 29 × 4 to find out how many legs are on 29 cats, 29 is the number of cats (or number of groups), 4 is the number of legs on each cat (or number of items in each group), and 116 is the total number of legs on all the cats. Modeling multiplication problems with pictures, diagrams, or concrete materials helps students learn what the factors and their product represent in various contexts.* (p. 151)

In this section, multiplication and division algorithms will be introduced using various models.

Multiplication Algorithms

To develop algorithms for multiplying multidigit whole numbers, we use the strategy of *examining simpler computations first.* Consider $4 \cdot 12$. This computation could be pictured as in Figure 3-17(a) with 4 rows of 12 blocks, or 48 blocks. The blocks in Figure 3-17(a) can also be partitioned to show that $4 \cdot 12 = 4 \cdot (10 + 2) = 4 \cdot 10 + 4 \cdot 2$. The numbers $4 \cdot 10$ and $4 \cdot 2$ are *partial products.*

FIGURE 3-17

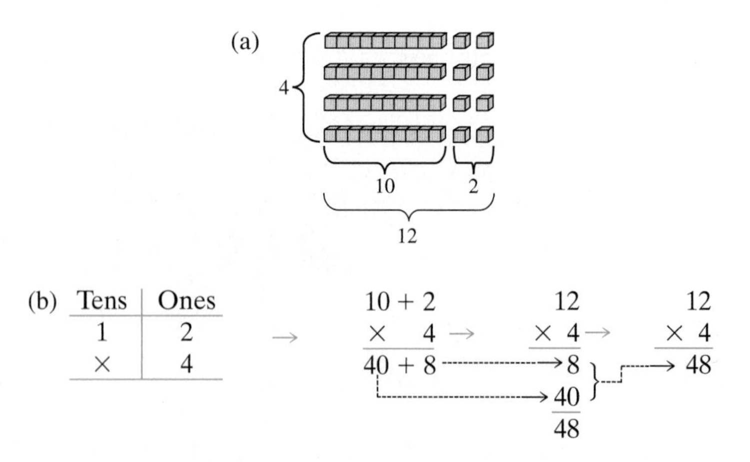

Figure 3-17(a) illustrates the distributive property of multiplication over addition on the set of whole numbers. The process leading to an algorithm for multiplying $4 \cdot 12$ is seen in Figure 3-17(b). Notice the similarity between the multiplication in Figure 3-17 and the following algebra multiplication:

$$\begin{aligned} 4(x + 2) &= 4x + 4 \cdot 2 \\ &= 4x + 8 \end{aligned}$$

Similarly, notice the analogy between the product

$$23 \cdot 14 = (2 \cdot 10 + 3)(1 \cdot 10 + 4) \quad \text{and} \quad (2x + 3)(1 \cdot x + 4).$$

The analogy is carried in the way we compute each product:

$$\begin{array}{r} 2 \cdot 10 + 3 \\ \times\,(1 \cdot 10 + 4) \\ \hline 12 \\ 8 \cdot 10 \\ 3 \cdot 10 \\ 2 \cdot 10^2 \\ \hline 2 \cdot 10^2 + 11 \cdot 10 + 12 \end{array} \qquad \begin{array}{r} 2x + 3 \\ \times\,(1 \cdot x + 4) \\ \hline 8x + 12 \\ 2x^2 + \ 3x \quad\ \ \\ \hline 2x^2 + 11x + 12 \end{array}$$

Multiplication of a three- or more digit numeral by a one-digit factor will be explored after we discuss multiplication by a power of 10.

Multiplication by 10^n

Next we consider multiplication by powers of 10. First consider what happens when we multiply a given number by 10, such as $23 \cdot 10$. If we start out with the base-ten block representation of 23, we have 2 longs and 3 units. To multiply by 10, we must replace each piece with a base-ten piece that represents the next higher power of 10. This is shown in Figure 3-18. Notice that the 3 units in 23 when multiplied by 10 become 3 longs or 3 tens. Therefore, after multiplication by 10 there are no units and hence we have 0 in the units place. In general, if we multiply any number by 10, we annex a 0 at the end of the number.

FIGURE 3-18

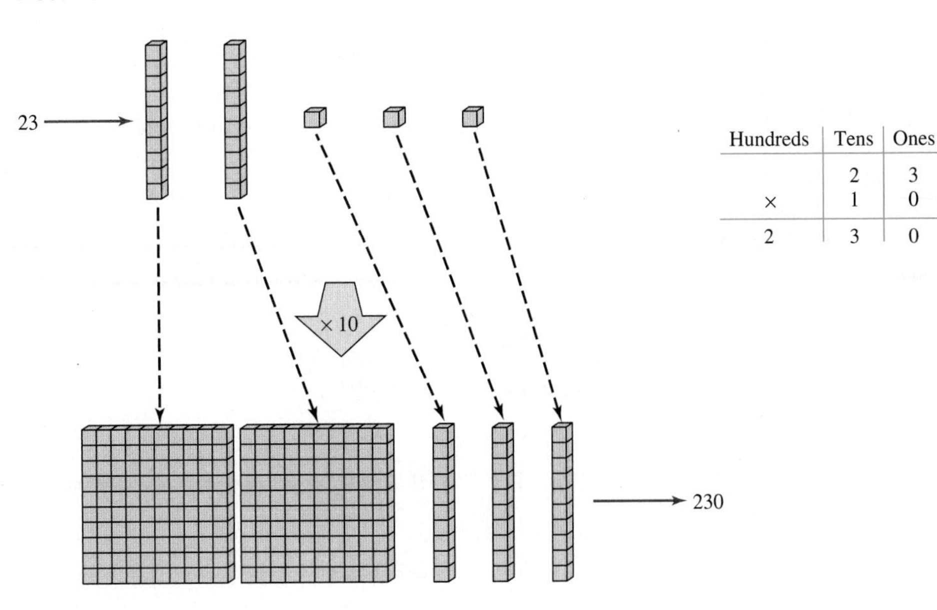

Hundreds	Tens	Ones
	2	3
×	1	0
2	3	0

The approach in Figure 3-18 can be explained by using the distributive property of multiplication over addition.

$$\begin{aligned} 23 \cdot 10 &= (2 \cdot 10 + 3)\,10 \\ &= (2 \cdot 10)10 + 3 \cdot 10 \\ &= 2(10 \cdot 10) + 3 \cdot 10 \\ &= 2 \cdot 10^2 + 3 \cdot 10 \\ &= 2 \cdot 10^2 + 3 \cdot 10 + 0 \cdot 1 \\ &= 230 \end{aligned}$$

To compute products such as $3 \cdot 200$, we proceed as follows:

$$\begin{aligned} 3 \cdot 200 &= 3(2 \cdot 10^2) \\ &= (3 \cdot 2)10^2 \\ &= 6 \cdot 10^2 \\ &= 6 \cdot 10^2 + 0 \cdot 10^1 + 0 \cdot 1 \\ &= 600 \end{aligned}$$

We see that multiplying 6 by 10^2 results in annexing two zeros to 6. This idea can be generalized to the statement that *multiplication of any natural number by* 10^n, *where* n *is a natural number, results in annexing* n *zeros to the number.*

REMARK The annexation of n zeros when multiplying by 10^n can also be explained as follows. We first multiply by 10, resulting in annexation of one zero (like in $23 \cdot 10 = 230$). When we multiply by another 10, another zero is annexed (as in $230 \cdot 10 = 2300$). Since we multiply n times by 10, n zeros are annexed.

RESEARCH NOTE
The ability to multiply and divide by powers of 10 is "fundamental" to the development and use of estimation skills (Rubenstein 1985).

When multiplying powers of 10, an extension of the definition of exponents is used. For example, $10^2 \cdot 10^1 = (10 \cdot 10)10 = 10^3$, or 10^{2+1}. In general, where a is a natural number and m and n are whole numbers, $a^m \cdot a^n$ is given by the following:

$$\begin{aligned} a^m \cdot a^n &= \underbrace{(a \cdot a \cdot a \cdot \ldots \cdot a)}_{m \text{ factors}} \cdot \underbrace{(a \cdot a \cdot a \cdot \ldots \cdot a)}_{n \text{ factors}} \\ &= \underbrace{a \cdot a \cdot a \cdot \ldots \cdot a}_{m + n \text{ factors}} = a^{m+n} \end{aligned}$$

Consequently, $a^m \cdot a^n = a^{m+n}$.

NOW TRY THIS 3-10

Does $a^n + a^m$ ever equal a^{m+n}? If so, when?

Multiplication by a power of 10 is helpful in calculating the product of a one-digit numeral and a three- or more digit numeral. In the following example we assume the previously developed algorithm for multiplying a one-digit numeral times a two-digit numeral.

$$\begin{aligned} 4 \cdot 367 &= 4(3 \cdot 10^2 + 6 \cdot 10 + 7) \\ &= (4 \cdot 3) \cdot 10^2 + (4 \cdot 6) \cdot 10 + 4 \cdot 7 \\ &= 1200 + 240 + 28 \\ &= 1468 \end{aligned}$$

NOW TRY THIS 3-11

Use expanded addition and an approach similar to the preceding to calculate $7 \cdot 4589$.

The cartoon shows that the application of multiplication to real-life situations can be tricky for some children.

Hi And Lois April 14, 2005

Multiplication with Two-Digit Factors

Consider $14 \cdot 23$. We first model this computation using base-ten blocks, as shown in Figure 3-19(a), and then showing all the partial products and adding, as shown in Figure 3-19(b). Another approach is to write 14 as $10 + 4$ and use the distributive property of multiplication over addition, as follows:

$$\begin{aligned} 14 \cdot 23 &= (10 + 4) \cdot 23 \\ &= 10 \cdot 23 + 4 \cdot 23 \\ &= 230 + 92 \\ &= 322 \end{aligned}$$

$5 per person per night. Meals for the trip will cost $28 per person. Has the band raised enough money yet? If not, how many more cars do they have to wash?

19. The following figure shows four function machines. The output from one machine becomes the input for the one below it. Complete the accompanying chart.

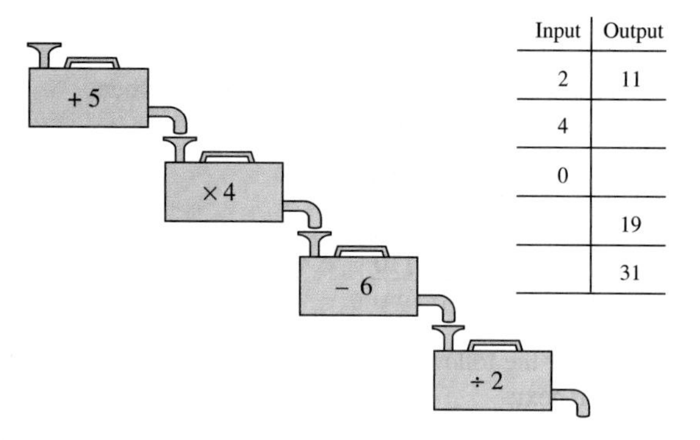

Input	Output
2	11
4	
0	
	19
	31

20. Choose three different digits.
   a. Form six different two-digit numbers from the numbers you chose. Each number can be used only once.
   b. Add the six numbers.
   c. Add the three digits you chose.
   d. Divide the answer in (b) by the answer in (c).
   e. Repeat (a) through (d) with three different numbers.
   f. Is the final result always the same? Why?

21. Consider the following multiplications. Notice that when the digits in the factors are reversed, the products are the same.

$$\begin{array}{r} 36 \\ \times\ 42 \\ \hline 1512 \end{array} \qquad \begin{array}{r} 63 \\ \times\ 24 \\ \hline 1512 \end{array}$$

   a. Find other multiplications where this procedure works.
   b. Find a pattern for the numbers that work in this way.

22. Molly read 160 pages in her book in 4 hr. Her sister Karly took 4 hr to read 100 pages in the same book. If the book is 200 pages long and if the two girls continued to read at these rates, how much longer would it take Karly to read the book than Molly?

23. Xuan saved $5340 in 3 years. If he saved $95 a month in the first year and a fixed amount per month for the next 2 years, how much did he save a month during the last 2 years?

24. Dan has 4520 pennies in three boxes. He says that there are 3 times as many pennies in the first box as in the third and twice as many in the second box as in the first. How much does he have in each box?

25. Gina buys apples from an orchard and then sells them at a country fair in bags of 3 for $1 a bag. She bought 50 boxes of apples, 36 apples in a box, and paid $452. If she sold all but 18 apples, what was her total profit?

26. A group of fourth-grade children had to cut four pieces of ribbon each 4 ft long from a ribbon of 44 yd. What is the length of the remaining ribbon?

27. How would you explain to children how to multiply $345 \cdot 678$, assuming that they know and understand multiplication by a single digit and multiplication by a power of 10?

28. Discuss possible error patterns in each of the following:

$$\textbf{a.}\ \begin{array}{r} 34 \\ \times\ 8 \\ \hline 2432 \end{array} \qquad \textbf{b.}\ \begin{array}{r} 35 \\ \times 26 \\ \hline 90 \end{array} \qquad \textbf{c.}\ \begin{array}{r} 34 \\ \times\ 6 \\ \hline 114 \end{array} \qquad \textbf{d.}\ \begin{array}{r} 5\ 3 \\ 5\overline{)2515} \\ -\ 25 \\ \hline 15 \\ -\ 15 \\ \hline 0 \end{array}$$

29. a. Give reasons for each of the following steps:

$$\begin{aligned} 56 \cdot 10 &= (5 \cdot 10 + 6) \cdot 10 \\ &= (5 \cdot 10) \cdot 10 + 6 \cdot 10 \\ &= 5 \cdot (10 \cdot 10) + 6 \cdot 10 \\ &= 5 \cdot 10^2 + 6 \cdot 10 \\ &= 5 \cdot 10^2 + 6 \cdot 10 + 0 \cdot 1 \\ &= 560 \end{aligned}$$

   b. Give reasons for each step in computing $34 \cdot 10^2$.

30. To transport the complete student body of 1672 students to a talk given by the governor, the school plans to rent buses that can hold 29 students each. How many buses are needed? Will all the buses be full?

31. **a.** If $a = 131 \cdot 4789 + 200$, without multiplying mentally find the quotient and remainder when a is divided by 131.
   b. Write the answer for part (a) in the form $a = bq + r$, where $0 \leq r < 131$.

32. If a leaves remainder 4 upon division by 17 and b leaves remainder 5 upon division by 17, without assuming particular values for a and b find the remainder when $a \cdot b$ is divided by 17. Explain your reasoning.

33. What is the remainder when x is divided by 13 if:
   a. $x + 9$ leaves remainder 0 upon division by 13
   b. $x - 9$ leaves remainder 0 upon division by 13

34. a. If x leaves remainder 0 upon division by 10, what can you tell about the last digit of x? Why?
   b. If x leaves remainder 0 upon division by 5, what can you tell about the last digit of x? Why?
   c. If x leaves remainder 13 upon division by 100, what are the last two digits of x? Why?

35. If $x + y$ leaves remainder 0 upon division by 10, what can you conclude about the last digit of x and the last digit of y? Why?

36. a. Find all the whole numbers that leave remainder 1 upon division by 4.
   b. Write the numbers from part (a) in a sequence starting from the smallest.
   c. What kind of sequence is the one in part (b)?

37. What happens when you multiply any two-digit number by 101? Explain why this happens.

38. Perform each of these operations using the bases shown:
   a. $32_{five} \cdot 4_{five}$
   b. $32_{five} \div 4_{five}$
   c. $43_{five} \cdot 23_{five}$
   d. $143_{five} \div 3_{five}$
   e. $10010_{two} \div 11_{two}$
   f. $10110_{two} \cdot 101_{two}$

39. For what possible bases are each of the following computations correct?

a.
```
  213
+ 308
  522
```
b.
```
  322
− 233
   23
```
c.
```
   213
 × 32
   430
 1043
 11300
```
d.
```
     101
 11)1111
   − 11
      11
    − 11
       0
```

40. a. Use lattice multiplication to compute $323_{five} \cdot 42_{five}$.

b. Find the smallest values of a and b so that $32_a = 23_b$.

41. Place the digits 7, 6, 8, and 3 in the boxes to obtain

```
  □□□
×   □
```

a. the greatest product. **b.** the least product.

42. Place the digits 7, 6, 8, 3, and 2 in the boxes to obtain

```
  □□□
×  □□
```

a. the greatest product. **b.** the least product.

43. Find the products of the following and describe the pattern that emerges:

a. 1×1, 11×11, 111×111, 1111×1111

b. 99×99, 999×999, 9999×9999

c. Test the patterns discovered for 30 products using a spreadsheet. If the patterns do not continue as expected, determine when the patterns stop.

Communication

44. Do you think it is valuable for students to see more than one method of doing computation problems? Why or why not?

45. Choose what you consider the "best" algorithm studied in this section. Explain the reasoning behind your choice.

46. Tom claims that long division should receive reduced attention in elementary classrooms. Do you agree or disagree? Defend your answer.

47. Pick a number. Double it. Multiply the result by 3. Add 24. Divide by 6. Subtract your original number. Is the result always the same? Write a convincing argument for your answer.

48. Prove that all numbers of the form *abba* (*a* and *b* are digits in base ten) leave remainder 0 upon division by 11. Is the same true for all the numbers of the form *abccba*? Why or why not?

Open-Ended

49. If a student presented a new "algorithm" for computing with whole numbers, describe the process you would recommend to the student to determine if the algorithm would always work.

Cooperative Learning

50. The traditional sequence for teaching operations in the elementary school is first addition, then subtraction, followed by multiplication, and finally division. Some educators advocate teaching addition followed by multiplication, then subtraction followed by division. Within your group prepare arguments for teaching the operations in either order listed.

Review Problems

51. Write the number succeeding 673 in Egyptian numerals.

52. Write $3 \cdot 10^5 + 2 \cdot 10^2 + 6 \cdot 10$ as a Hindu-Arabic numeral.

53. Illustrate the identity property of addition for whole numbers.

54. Rename each of the following using the distributive property of multiplication over addition:

a. $ax + bx + 2x$

b. $3(a + b) + x(a + b)$

55. At the beginning of a trip, the odometer registered 52,281. At the end of the trip, the odometer registered 59,260. How many miles were traveled on this trip?

Third International Mathematics and Science Study (TIMSS) Questions

Each student needs 8 notebooks for school. How many notebooks are needed for 115 students?

Use the tiles 1, 4, and 5. Write the numbers on the tiles in the boxes below to make the largest answer when you multiply.

```
  □□
×  □
```

$37 \times ■ = 703$.

What is the value of $37 \times ■ + 6$?

TIMSS, 2003, Grade 4

BRAIN TEASER

For each of the following, replace the letters with digits in such a way that the computation is correct. Each letter may represent only one digit.

a.
```
 LYNDON
    × B
JOHNSON
```
b.
```
   MA
   MA
 + MA
  EEL
```

LABORATORY ACTIVITY

1. Messages can be coded on paper tape in base two. A hole in the tape represents 1, whereas a space represents 0. The value of each hole depends on its position; from left to right, 16, 8, 4, 2, 1 (all powers of 2). Letters of the alphabet may be coded in base two according to their position in the alphabet. For example, G is the seventh letter. Since $7 = 1 \cdot 4 + 1 \cdot 2 + 1$, the holes appear as they do in Figure 3-25:

FIGURE 3-25

16 8 4 2 1

   a. Decode the message in Figure 3-26.

FIGURE 3-26

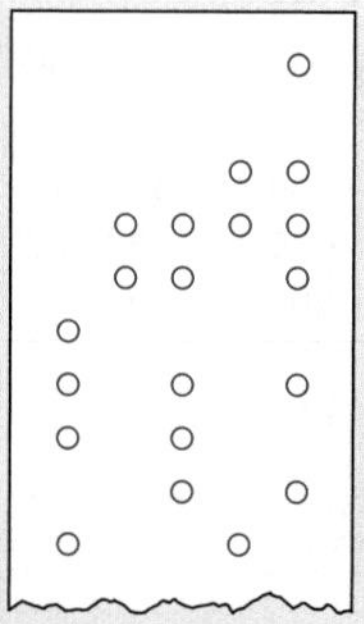

   b. Write your name on a tape using base two.

2. The number game in Figure 3-27 uses base-two arithmetic:

FIGURE 3-27

Card E		Card D		Card C		Card B		Card A	
16	24	8	24	4	20	2	18	1	17
17	25	9	25	5	21	3	19	3	19
18	26	10	26	6	22	6	22	5	21
19	27	11	27	7	23	7	23	7	23
20	28	12	28	12	28	10	26	9	25
21	29	13	29	13	29	11	27	11	27
22	30	14	30	14	30	14	30	13	29
23	31	15	31	15	31	15	31	15	31

   a. Suppose a person's age appears on cards E, C, and B, and the person is 22. Can you discover how this works and why?
   b. Design card F so that the numbers 1 through 63 can be used in the game. Note that cards A through E must also be changed.

3-4 Mental Mathematics and Estimation for Whole-Number Operations

In the *Principles and Standards* we find the following:

> *Part of being able to compute fluently means making smart choices about which tools to use and when. Students should have experiences that help them learn to choose among mental computation, paper-and-pencil strategies, estimation, and calculator use. The particular context, the question, and the numbers involved all play roles in those choices. Do the numbers allow a mental strategy? Does the context call for an estimate? Does the problem require repeated and tedious computations? Students should evaluate problem situations to determine whether an estimate or an exact answer is needed, using their number sense to advantage, and be able to give a rationale for their decision.* (p. 36)

RESEARCH NOTE

Good estimators tend to have strong self-concepts relative to mathematics, attribute their success in estimation to their ability rather than mere effort, and believe that estimation is an important tool. In contrast, poor estimators tend to have a weak self-concept relative to mathematics, attribute the success of others to effort, and believe that estimation is neither important nor useful (J. Sowder 1989).

The inability to use estimation skills is a direct consequence of student focus on mechanical manipulations of numbers, ignoring operational meaning, number sense, or concept of quantity/magnitude (Reys 1984).

mental mathematics

computational estimation

In the previous sections in this chapter we focused mainly on paper-and-pencil strategies. Next we focus on two other tools, namely, mental mathematics and computational estimation. **Mental mathematics** is the process of producing an answer to a computation without using computational aids. **Computational estimation** is the process of forming an approximate answer to a numerical problem. Facility with estimation strategies helps to determine whether or not an answer is reasonable. In the "Calvin and Hobbes" cartoon we see that Calvin does not have the number sense to tell whether his result was reasonable.

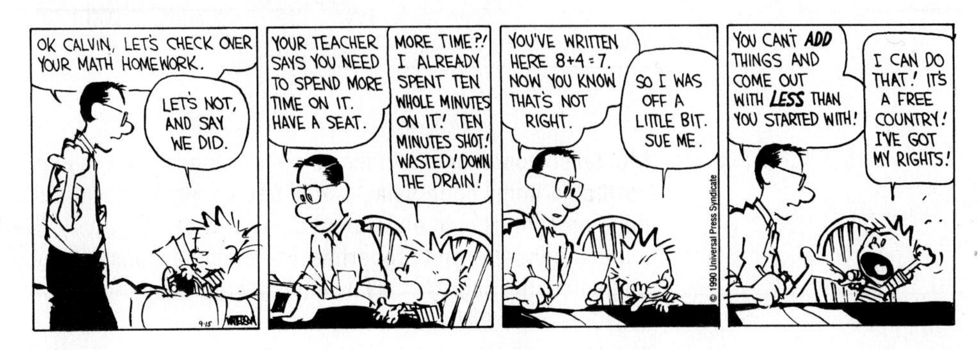

Proficiency in mental mathematics can help in your everyday estimation skills. It is essential that you have these skills even in a time when calculators are readily available. You must be able to judge the reasonableness of answers obtained on a calculator. Mental mathematics makes use of a variety of strategies and properties. Next we consider several of the most common strategies for performing operations mentally on whole numbers.

Mental Mathematics: Addition

1. *Adding from the left*

a. $\begin{array}{r} 67 \\ +\ 36 \\ \hline \end{array}$

$60 + 30 = 90$ (Add the tens.)
$7 + 6 = 13$ (Add the units.)
$90 + 13 = 103$ (Add the two sums.)

b. $\begin{array}{r} 36 \\ +\ 36 \\ \hline \end{array}$

$30 + 30 = 60$ (Double 30.)
$6 + 6 = 12$ (Double 6.)
$60 + 12 = 72$ (Add the doubles.)

2. *Breaking up and bridging*

$\begin{array}{r} 67 \\ +\ 36 \\ \hline \end{array}$

$67 + 30 = 97$ (Add the first number to the tens in the second number.)
$97 + 6 = 103$ (Add this sum to the units in the second number.)

3. *Trading off*

a. $\begin{array}{r} 67 \\ +\ 36 \\ \hline \end{array}$

$67 + 3 = 70$ (Add 3 to make a multiple of 10.)
$36 - 3 = 33$ (Subtract 3 to compensate for the 3 that was added.)
$70 + 33 = 103$ (Add the two numbers.)

b. $\begin{array}{r} 67 \\ +\ 29 \\ \hline \end{array}$

$67 + 30 = 97$ (Add 30 (next multiple of 10 greater then 29).)
$97 - 1 = 96$ (Subtract 1 to compensate for the extra 1 that was added.)

4. *Using compatible numbers*
Compatible numbers are numbers whose sums are easy to calculate mentally.

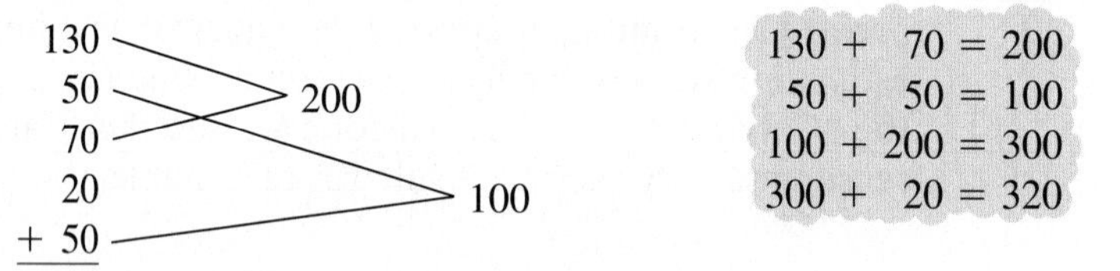

5. *Making compatible numbers*

$\begin{array}{r} 25 \\ +\ 79 \\ \hline \end{array}$

$25 + 75 = 100$ (25 + 75 adds to 100.)
$100 + 4 = 104$ (Add 4 more units.)

▶ **RESEARCH NOTE** Mental computation becomes efficient when it involves algorithms different from the standard algorithms done using pencil and paper. Also, mental computational strategies are quite personal, being dependent on a student's creativity, flexibility, and understanding of number concepts and properties. For example, consider the skills and thinking involved in computing the sum $74 + 29$ by mentally representing the problem as $70 + (29 + 1) + 3 = 103$ (J. Sowder 1988). ◀

Mental Mathematics: Subtraction

1. *Breaking up and bridging*

$$\begin{array}{r} 67 \\ -\ 36 \\ \hline \end{array} \quad \begin{array}{l} 67 - 30 = 37 \\ 37 - 6 = 31 \end{array}$$

(Subtract the tens in the second number from the first number.)
(Subtract the units in the second number from the difference.)

2. *Trading off*

$$\begin{array}{r} 71 \\ -\ 39 \\ \hline \end{array} \quad \begin{array}{l} (71 + 1) = 72; (39 + 1) = 40 \\ (72 - 40) = 32 \end{array}$$

(Add 1 to both numbers. Perform the subtraction, which is easier than the original problem.)

Notice that adding 1 to both numbers does not change the answer. Why?

3. *Drop the zeros*

$$\begin{array}{r} 8700 \\ -\ 500 \\ \hline \end{array} \quad \begin{array}{l} 87 - 5 = 82 \\ 82 \rightarrow 8200 \end{array}$$

(Notice that there are two zeros in each number. Drop these zeros and perform the computation. Then replace the two zeros to obtain proper place value.)

Another mental-mathematics technique for subtraction is called "adding up." This method is based on the *missing addend* approach and is sometimes referred to as the "cashier's algorithm." An example of the cashier's algorithm follows.

EXAMPLE 3-8

Noah owed \$11 for his groceries. He used a \$50 check to pay the bill. While handing Noah the change, the cashier said, "\$11, \$12, \$13, \$14, \$15, \$20, \$30, \$50." How much change did Noah receive?

Solution Table 3-10 shows what the cashier said and how much money Noah received each time. Since \$11 plus \$1 is \$12, Noah must have received \$1 when the cashier said \$12. The same reasoning follows for \$13, \$14, and so on. Thus, the total amount of change that Noah received is given by \$1 + \$1 + \$1 + \$1 + \$5 + \$10 + \$20 = \$39. In other words, \$50 − \$11 = \$39 because \$39 + \$11 = \$50.

Table 3-10

What the Cashier Said	\$11	\$12	\$13	\$14	\$15	\$20	\$30	\$50
Amount of Money Noah Received Each Time	0	\$1	\$1	\$1	\$1	\$5	\$10	\$20

NOW TRY THIS 3-13

Perform each of the following computations mentally and explain what technique you used to find the answer.

a. $40 + 160 + 29 + 31$
b. $3679 - 474$
c. $75 + 28$
d. $2500 - 700$

Mental Mathematics: Multiplication

As with addition and subtraction, mental mathematics is useful for multiplication. For example, consider 8×26. Students may think of this computation in a variety of ways, as shown here.

$26 = 20 + 6$	$26 = 25 + 1$	$26 = 30 - 4$
8×20 is 160 and 8×6 is 48, so 8×26 is $160 + 48$, or 208.	8×25 is 200, then 8×1 is 8 more, so 8×26 is $200 + 8$, or 208.	8×30 is 240, then take off $8 \times 4 = 32$, so 8×26 is $240 - 32 = 208$.

Next we consider several of the most common strategies for performing mental mathematics using multiplication.

1. *Front-end multiplying*

$$\begin{array}{r} 64 \\ \times 5 \\ \hline \end{array}$$

$60 \times 5 = 300$ (Multiply the number of tens in the first number by 5.)
$4 \times 5 = 20$ (Multiply the number of units in the first number by 5.)
$300 + 20 = 320$ (Add the two products.)

2. *Using compatible numbers*

$2 \times 9 \times 5 \times 20 \times 5$ Rearrange as $9 \times (2 \times 5) \times (20 \times 5) = 9 \times 10 \times 100 = 9000$.

3. *Thinking money*

a. $\begin{array}{r} 64 \\ \times 5 \\ \hline \end{array}$ Think of the product as 64 nickels, which can be thought of as 32 dimes, which is $32 \times 10 = 320$ cents.

b. $\begin{array}{r} 64 \\ \times 50 \\ \hline \end{array}$ Think of the product as 64 half-dollars, which is 32 dollars, or 3200 cents.

c. $\begin{array}{r} 64 \\ \times 25 \\ \hline \end{array}$ Think of the product as 64 quarters, which is 32 half-dollars, or 16 dollars. Thus we have 1600 cents.

Mental Mathematics: Division

1. *Breaking up the dividend*

$7\overline{)4256}$ $7\overline{)42|56}$ (Break up the dividend into parts.)

$$\begin{array}{r} 600 + 8 \\ 7\overline{)4200 + 56} \end{array}$$ (Divide both parts by 7.)

$600 + 8 = 608$ (Add the answers together.)

2. *Using compatible numbers*

a. $3\overline{)105}$ $105 = 90 + 15$ (Look for numbers that you recognize as divisible by 3 and having a sum of 105.)

$$\begin{array}{r} 30 + 5 \\ 3\overline{)90 + 15} \end{array} = 35$$ (Divide both parts and add the answers.)

b. $8\overline{)232}$ $232 = 240 - 8$ (Look for numbers that are easily divisible by 8 and whose difference is 232.)

$\dfrac{30 - 1}{8\overline{)240 - 8}} = 29$ (Divide both parts and take the difference.)

NOW TRY THIS 3-14

Perform each of the following computations mentally and explain what technique you used to find the answer:

a. $25 \cdot 32 \cdot 4$ **b.** $123 \cdot 3$
c. $25 \cdot 35$ **d.** $5075 \div 25$

Computational Estimation

Computational estimation may help determine whether an answer is reasonable or not. This is especially useful when the computation is done on a calculator. Some of the common estimation strategies for addition are given next.

1. *Front-end*
Front-end estimation begins by focusing on the lead, or front, digits of the addition. These front, or lead, digits are added and assigned an appropriate place value. At this point we may have an underestimate that needs to be adjusted. The adjustment is made by focusing on the next group of digits. The following example shows how front-end estimation works:

4 + 3 + 5
12 hundred

423 — 20
338 > 100 > 120
+ 561

Steps: **(1.) Add front-end digits** $4 + 3 + 5 = 12$.
(2.) Place value $= 1200$.
(3.) Adjust $61 + 38 \approx 100$ and $20 + 100$ is 120.
(4.) Adjusted estimate is $1200 + 120 = 1320$.

The student page shows how front-end estimation can be used to check whether computations are reasonable. Answer the *Talk About It* question on the student page.

2. *Grouping to nice numbers*
The strategy used to obtain the adjustment in the preceding example is the *grouping to nice numbers* strategy, which means that numbers that "nicely" fit together are grouped. Another example is given here.

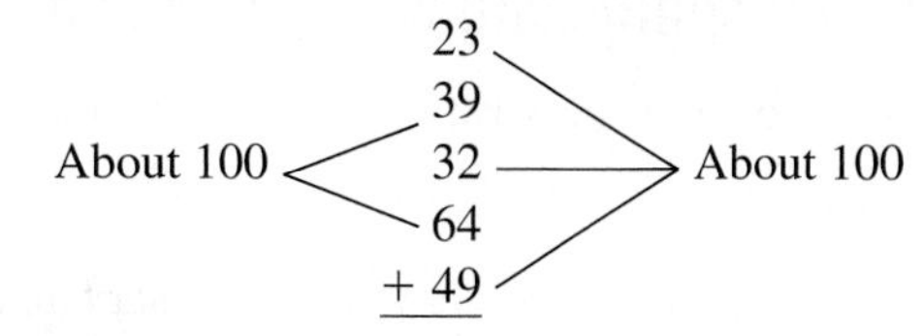

Therefore, the sum is about $100 + 100$, or 200.

3. *Clustering*
Clustering is used when a group of numbers cluster around a common value. This strategy is limited to certain kinds of computations. In the next example, the numbers seem to cluster around 6000.

```
  6200     (1.) Estimate the "average"—about 6000
  5842     (2.) Multiply the "average by the number of values to obtain
  6512          5 · 6000 = 30,000.
  5521
+ 6319
```

4. *Rounding*
Rounding is a way of cleaning up numbers so that they are easier to handle. Rounding enables us to find approximate answers to calculations, as follows:

```
  4724      5000     Round 4724 to 5000.
+ 3192    + 3000     Round 3192 to 3000.
            8000     Add the rounded numbers.

  1267      1300     Round 1267 to 1300.
-  510    -  500     Round 510 to 500.
             800     Subtract the rounded numbers.
```

Performing estimations requires a knowledge of place value and rounding techniques. We illustrate a rounding procedure that can be generalized to all rounding situations. For example, suppose we wish to round 4724 to the nearest thousand. We may proceed in four steps (see also Figure 3-28).

a. Determine between which two consecutive thousands the number lies.
b. Determine the midpoint between the thousands.
c. Determine which thousand the number is closer to by observing whether it is greater than or less than the midpoint. (Not all texts use the same rule for rounding when a number falls at a midpoint.)
d. If the number to be rounded is greater than or equal to the midpoint, round the given number to the greater thousand; otherwise, round to the lesser thousand. In this case, we round 4724 to 5000.

FIGURE 3-28

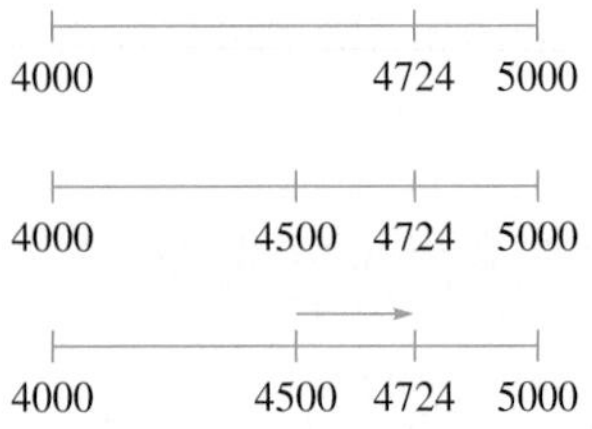

4000 4724 5000

4000 4500 4724 5000

4000 4500 4724 5000

5. *Using the range*
It is often useful to know into what *range* an answer falls. The range is determined by finding a low estimate and a high estimate and reporting that the answer falls in this interval. An example follows:

Problem	Low Estimate	High Estimate
378	300	400
+ 524	+ 500	+ 600
	800	1000

Thus, a range for this problem is from 800 to 1000.

Estimation: Multiplication and Division

Examples of estimation strategies for multiplication and division are given next.

1. *Front-end*

```
 524    500 × 8 = 4000        (Start multiplying at the front to obtain a
× 8                           first estimate.)
        20 × 8 = 160          (Multiply the next important digit 8 times.)
        4000 + 160 = 4160     (Adjust the first estimate by adding
                              the two numbers.)
```

SAMPLE SCHOOL BOOK PAGE:

ESTIMATING SUMS AND DIFFERENCES

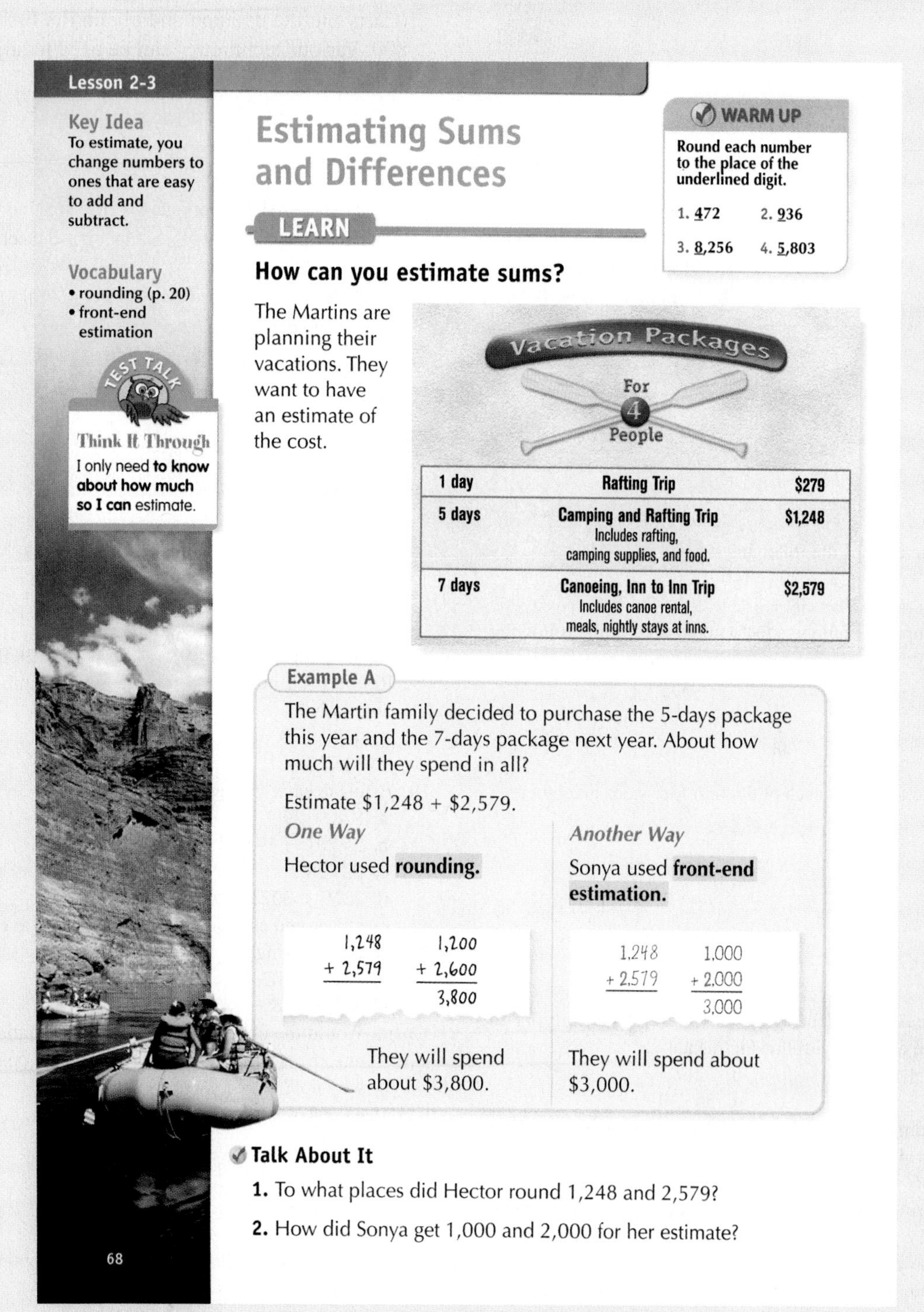
Lesson 2-3

Key Idea
To estimate, you change numbers to ones that are easy to add and subtract.

Vocabulary
• rounding (p. 20)
• front-end estimation

TEST TALK

Think It Through
I only need to know about how much so I can estimate.

Estimating Sums and Differences

WARM UP
Round each number to the place of the underlined digit.
1. 472 2. 936
3. 8,256 4. 5,803

LEARN

How can you estimate sums?

The Martins are planning their vacations. They want to have an estimate of the cost.

Vacation Packages
For 4 People

1 day	Rafting Trip	$279
5 days	Camping and Rafting Trip Includes rafting, camping supplies, and food.	$1,248
7 days	Canoeing, Inn to Inn Trip Includes canoe rental, meals, nightly stays at inns.	$2,579

Example A

The Martin family decided to purchase the 5-days package this year and the 7-days package next year. About how much will they spend in all?

Estimate $1,248 + $2,579.

One Way
Hector used **rounding.**

1,248 + 2,579 → 1,200 + 2,600 = 3,800

They will spend about $3,800.

Another Way
Sonya used **front-end estimation.**

1,248 + 2,579 → 1,000 + 2,000 = 3,000

They will spend about $3,000.

Talk About It

1. To what places did Hector round 1,248 and 2,579?
2. How did Sonya get 1,000 and 2,000 for her estimate?

68

Source: Scott Foresman–Addison Wesley Math, Grade 4, 2005 (p. 68)

2. *Compatible numbers*

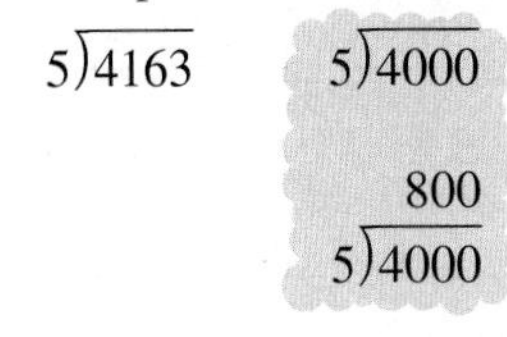

(Change 4163 to a number close to it that you know is divisible by 5.)

(Carry out the division and obtain the first estimate of 800. Various techniques can be used to adjust the first estimate.)

NOW TRY THIS 3-15

Estimate each of the following mentally and explain what technique you used to find the answer:

a. A sold-out concert was held in a theater with a capacity of 4525 people. Tickets were sold for $9 each. How much money was collected?

b. Fliers are to be delivered to 3625 houses and there are 42 people who will be doing the distribution. If distributed equally, how many houses will each person visit?

ONGOING ASSESSMENT 3-4

1. Compute each of the following mentally:
- **a.** $180 + 97 - 23 + 20 - 140 + 26$
- **b.** $87 - 42 + 70 - 38 + 43$

2. Use compatible numbers to compute each of the following mentally.
- **a.** $2 \cdot 9 \cdot 5 \cdot 6$
- **b.** $8 \cdot 25 \cdot 7 \cdot 4$
- **c.** $5 \cdot 11 \cdot 3 \cdot 20$
- **d.** $82 + 37 + 18 + 13$

3. Supply reasons for each of the first four steps given here.

$$\begin{aligned}(525 + 37) + 75 &= 525 + (37 + 75)\\ &= 525 + (75 + 37)\\ &= (525 + 75) + 37\\ &= 600 + 37\\ &= 637\end{aligned}$$

4. Use breaking and bridging or front-end multiplying to compute each of the following mentally:
- a. $567 + 38$
- **b.** $321 \cdot 3$
- **c.** $997 - 32$
- **d.** $56 \cdot 30$

5. Use trading off to compute each of the following mentally:
- a. $85 - 49$
- **b.** $87 + 33$
- c. $19 \cdot 6$
- **d.** $58 + 39$

6. A car trip took 8 hr of driving at an average of 62 mph. Mentally compute the total number of miles traveled. Describe your method.

7. Compute each of the following using the cashier's algorithm:
- **a.** $53 - 28$
- **b.** $63 - 47$

8. Compute each of the following mentally. In each case, briefly explain your method.
- a. $86 + 37$
- b. $97 + 54$
- c. $230 + 60 + 70 + 44 + 40 + 6$
- d. $81 - 46$
- **e.** $98 - 19$
- **f.** $9700 - 600$

9. Round each number to the place value indicated by the digit in bold.
- a. 5**2**80
- **b.** **1**15,234
- c. 1**1**5,234
- **d.** 2,3**2**5

10. Estimate each answer by rounding.
- **a.** $878 \div 29$
- **b.** $25{,}201 - 19{,}987$
- **c.** $32 \cdot 28$
- **d.** $2215 + 3023 + 5967 + 975$

11. Use front-end estimation with adjustment to estimate each of the following:
- a. $2215 + 3023 + 5987 + 975$
- **b.** $234 + 478 + 987 + 319 + 469$

12. a. Would the clustering strategy of estimation be a good one to use in each of the following cases? Why or why not?

(i)
```
   474
  1467
    64
+ 2445
```

(ii)
```
  483
  475
  530
  503
+ 528
```

b. Estimate each part of (a) using the following strategies:
- (i) Front-end
- (ii) Grouping to nice numbers
- (iii) Rounding

13. Use the range strategy to estimate each of the following. Explain how you arrived at your estimates.
   a. $22 \cdot 38$ **b.** $145 + 678$ c. $278 + 36$
14. Tom estimated $31 \cdot 179$ in the three ways shown.
   (i) $30 \cdot 200 = 6000$
   (ii) $30 \cdot 180 = 5400$
   (iii) $31 \cdot 200 = 6200$
   Without finding the actual product, which estimate do you think is closer to the actual product? Why?
15. About 3540 calories must be burned to lose 1 lb of body weight. Estimate how many calories must be burned to lose 6 lb.
16. Suppose you had a balance of \$3287 in your checking account and you wrote checks for \$85, \$297, \$403, and \$523. Estimate your balance and tell what you did and whether you think your estimate is too high or too low.
17. A theater has 38 rows with 23 seats in each row. Estimate the number of seats in the theater and tell how you arrived at your estimate.
18. Without computing, tell which of the following have the same answer. Describe your reasoning.
   a. $44 \cdot 22$ and $22 \cdot 11$
   b. $22 \cdot 32$ and $11 \cdot 64$
   c. $13 \cdot 33$ and $39 \cdot 11$
19. The following is a list of the areas in square miles of Europe's largest countries. Mentally use this information to decide if each of the given statements is true.

France	211,207
Spain	194,896
Sweden	173,731
Finland	130,119
Norway	125,181

   a. Sweden is less than 40,000 mi^2 larger than Finland.
   b. France is more than twice the size of Norway.
   c. France is more than 100,000 mi^2 larger than Norway.
   d. Spain is about 21,000 mi^2 larger than Sweden
20. The attendance at a World's Fair for one week follows:

Monday	72,250
Tuesday	63,891
Wednesday	67,490
Thursday	73,180
Friday	74,918
Saturday	68,480

   Estimate the week's attendance and tell what strategy you used and why you used it.
21. In each of the following, answer the question using estimation methods if possible. If estimation is not appropriate, explain why not.
   a. Josh has \$380 in his checking account. He wants to write checks for \$39, \$28, \$59, and \$250. Will he have enough money in his account to cover these checks?
   b. Gila deposited two checks into her account, one for \$981 and the other for \$1140. Does she have enough money in her account to cover a check for \$2000?
   c. Alberto and Juan are running for city council. They receive votes from two districts. Alberto receives 3473 votes from one district and 5615 votes from the other district. Juan receives 3463 votes from the first district and 5616 from the second. Who gets elected?
   d. Two rectangular parcels have dimensions 101 ft by 120 ft and 103 ft by 129 ft. Which parcel has greater area? (Recall that the area of a rectangle is length times width.)
22. In each of the following, determine if the estimate given in parentheses is high (higher than the actual answer) or low (lower than the actual answer). Justify your answers without computing the exact values.
   a. $299 \cdot 300$ (90,000)
   b. $6001 \div 299$ (20)
   c. $6000 \div 299$ (20)
   d. $10{,}000 \div 999$ (10)
   e. $999 \div 99$ (10)
   f. $1999 \div 201$ (10)
23. Use your calculator to calculate 25^2, 35^2, 45^2, and 55^2 and then see if you can find a pattern that will let you find 65^2 and 75^2 mentally.
24. Use your calculator to multiply several two-digit numbers times 99. Then see if you can find a pattern that will let you find the product of any two-digit number and 99 mentally.

Communication

25. What is the difference between mental mathematics and computational estimation?
26. Is the front-end estimate for addition before adjustment always less than the exact sum? Explain why or why not.
27. In the new textbooks, there is an emphasis on mental mathematics and estimation. Do you think these topics are important for today's students? Why?
28. Suppose x and y are positive (greater than 0) whole numbers. If x is greater than y and you estimate $x - y$ by rounding x up and y down, will your estimate always be too high or too low or could it be either? Explain.

Open-Ended

29. Give several examples from real-world situations where an estimate, rather than an exact answer, is sufficient.
30. Give a numerical example of when front-end estimation and rounding can produce the same estimate. Give an example of when they can produce a different estimate.

CHAPTER REVIEW

1. For each of the following base-ten numbers, tell the place value for each of the circled digits:
   a. 4(3)2 b. (3)432 c. 19(3)24
2. Convert each of the following to base ten:
   a. $\overline{CD}XLIV$ b. 432_{five} c. $ET0_{twelve}$
   d. 1011_{two} e. 4136_{seven}
3. Convert each of the following numbers to numbers in the indicated system:
   a. 999 to Roman
   b. 86 to Egyptian
   c. 123 to Mayan
   d. 346_{ten} to base five
   e. 27_{ten} to base two
4. Simplify each of the following, if possible. Write your answers in exponential form, a^b.
   a. $3^4 \cdot 3^7 \cdot 3^6$ b. $2^{10} \cdot 2^{11}$ c. $3^4 + 2 \cdot 3^4$
   d. $2^{80} + 3 \cdot 2^{80}$ e. $2^{100} + 2^{100}$
5. Write the base-three numeral for the base-three blocks shown.

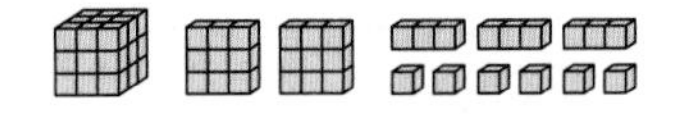

6. The smallest number of base-three blocks needed to represent 51 is _____ blocks _____ flats _____ longs _____ units.
7. Draw multibase blocks to represent
   a. 123_{four} b. 24_{five}
8. a. The first digit from the left (the lead digit) of a base-ten numeral is 4 followed by 10 zeros. What is the place value of 4?
   b. A number in base five has 10 digits. What is the place value of the second digit from the left?
   c. A number in base two has lead digit 1 followed by 30 zeros and units digit 1. What is the place value of the lead digit?
9. Write the following base-ten numerals in the indicated base without performing any multiplications:
   a. $10^{10} + 23$ in base ten b. $2^{10} + 1$ in base two
   c. $5^{10} + 1$ in base five d. $10^{10} - 1$ in base ten
   e. $2^{10} - 1$ in base two f. $12^5 - 1$ in base twelve
10. Use both the scratch and the traditional algorithms to perform the following:

$$\begin{array}{r} 316 \\ 712 \\ +\ 91 \\ \hline \end{array}$$

11. Use both the traditional and the lattice multiplication algorithms to perform the following:

$$\begin{array}{r} 613 \\ \times\ 98 \\ \hline \end{array}$$

12. Use both the repeated-subtraction and the conventional algorithms to perform the following:
   a. $912\overline{)4803}$ b. $11\overline{)1011}$
   c. $23_{five}\overline{)3312_{five}}$ d. $11_{two}\overline{)1011_{two}}$
13. Use the division algorithm to check your answers in problem 12.
14. a. Suppose a leaves remainder 0 upon division by b $(b \neq 0)$ and a as well as b leave remainder 0 upon division by c $(c \neq 0)$. Explain why

$$a \div b = (a \div c) \div (b \div c).$$

   b. Use the property in part (a) to calculate
      (i) $34000 \div 17000$
      (ii) $(176 \cdot 97) \div (44 \cdot 97)$
      (iii) $(a \cdot d) \div (b \cdot d)$ (You may assume that when a is divided by b the remainder is 0.)
   c. Suppose that when a is divided by b the remainder is r and $r \neq 0$. What is the remainder when $a \cdot d$ is divided by $b \cdot d$ $(d \neq 0)$? Justify your answer.
15. In some calculations a combination of mental math and a calculator is most appropriate. For example, because

$$200 \cdot 97 \cdot 146 \cdot 5 = 97 \cdot 146 \cdot (200 \cdot 5) = 97 \cdot 146 \cdot 1000,$$

   we can calculate $97 \cdot 146$ on a calculator and then mentally multiply by 1000. Show how to calculate each of the following using a combination of mental math and a calculator:
   a. $19 \cdot 5 \cdot 194 \cdot 2$ b. $379 \cdot 4 \cdot 193 \cdot 25$
   c. $8 \cdot 481 \cdot 73 \cdot 125$ d. $374 \cdot 200 \cdot 893 \cdot 50$
16. You had a balance in your checking account of $720 before writing checks for $162, $158, and $33 and making a deposit of $28. What is your new balance?
17. Jim was paid $320 a month for 6 mo and $410 a month for 6 mo. What were his total earnings for the year?
18. A soft-drink manufacturer produces 15,600 cans of his product each hour. Cans are packed 24 to a case. How many cases could be filled with the cans produced in 4 hr?
19. A limited partnership of 120 investors sold a piece of land for $461,040. If divided equally, how much did each investor receive?
20. How many 12-oz cans of juice would it take to give 60 people one 8-oz serving each?
21. I am thinking of a whole number. If I divide it by 13, then multiply the answer by 12, then subtract 20, and then add 89, I end up with 93. What was my original number?
22. Apples normally sell for 32¢ each. They go on sale for 3 for 69¢. How much money is saved if you purchase 2 doz apples while they are on sale?

23. A ski resort offers a weekend ski package for \$80 per person or \$6000 for a group of 80 people. Which would be the cheaper option for a group of 80?
24. The owner of a bicycle shop reported his inventory of bicycles and tricycles in an unusual way. He said he counted 126 wheels and 108 pedals. How many bikes and how many trikes did he have?
25. Josi has a job in which she works 30 hr/wk and gets paid \$15/hr. If she works more than 30 hr in a week, she receives \$18/hr for each hour over 30 hr. If she worked 38 hr at her job this week, how much did she earn?
26. In a television game show, there are five questions to answer. Each question is worth twice as much as the previous question. If the last question was worth \$6400, what was the first question worth?
27. Write an example of a base other than ten used in a real-life situation. How is it used?
28. Describe the important characteristics of each of the following systems:
    a. Egyptian
    b. Babylonian
    c. Roman
    d. Hindu-Arabic
29. Write 128 in each of the following bases:
    a. five
    b. two
    c. twelve
30. Perform each of the following computations:
    a. $\begin{array}{r} 123_{\text{five}} \\ +34_{\text{five}} \\ \hline \end{array}$
    b. $\begin{array}{r} 1010_{\text{two}} \\ -101_{\text{two}} \\ \hline \end{array}$
    c. $\begin{array}{r} 23_{\text{five}} \\ \times 34_{\text{five}} \\ \hline \end{array}$
    d. $\begin{array}{r} 1001_{\text{two}} \\ \times 101_{\text{two}} \\ \hline \end{array}$
31. Write each of the following in the indicated bases without multiplying out the various powers:
    a. $4 \cdot 5^6 + 11 \cdot 5^4 + 9$ in base five
    b. $2^{10} + 2^3$ in base two
    c. $11 \cdot 12^5 + 10 \cdot 12^3 + 20$ in base twelve
    d. $9 \cdot 8^5 + 8$ in base eight
32. Tell how to use compatible numbers mentally to perform each of the following:
    a. $26 + 37 + 24 - 7$
    b. $4 \cdot 7 \cdot 9 \cdot 25$
33. Compute each of the following mentally. Name the strategy you used to perform your mental math.
    a. $63 \cdot 7$
    b. $85 - 49$
    c. $(18 \cdot 5) \cdot 2$
    d. $2436 \div 6$
34. Estimate the following addition using (a) front-end estimation with adjustment and (b) rounding. Compare your answers and tell which you think is closer to the exact answer.

$$\begin{array}{r} 543 \\ 398 \\ 255 \\ 408 \\ +\ 998 \\ \hline \end{array}$$

35. Using clustering, estimate the sum $2345 + 2854 + 2234 + 2203$.
36. Explain how the standard division algorithm works for the following division:

$$\begin{array}{r} 23 \\ 14\overline{)322} \\ -\ 28 \\ \hline 42 \\ -\ 42 \\ \hline 0 \end{array}$$

37. In some cases the Distributive Property of Multiplication over Addition or Distributive Property of Multiplication over Subtraction can be used to obtain an answer quickly. Use one of the distributive properties to calculate each of the following in as simple a way as possible:
    a. $999 \cdot 47 + 47$
    b. $43 \cdot 59 + 41 \cdot 43$
    c. $1003 \cdot 79 - 3 \cdot 79$
    d. $1001 \cdot 113 - 113$
    e. $101 \cdot 35$
    f. $98 \cdot 35$
38. A square of a whole number is a *perfect square*; 49 is a perfect square because $49 = 7^2$. The sequence $1^2, 2^2, 3^2, 4^2, \ldots$ is a sequence of consecutive perfect squares. Janis claims that once she knows the value of a perfect square, she can easily find the value of the next perfect square. For example $25^2 = 625$, and to find the next perfect square she adds twice the number that was squared plus 1; that is, she adds $2 \cdot 25 + 1$, or 51 to 625, and gets $625 + 51$, or 676, for 26^2.
    a. Compute 27^2 using Janis's method.
    b. Why does the method work?
    c. Given a perfect square, describe a similar method to find the previous perfect square. Explain why your method works.
39. Recall that addition problems like $3478 + 521$ can be written and computed using expanded notation as shown here and answer the questions that follow.

$$\begin{array}{lr} & 3 \cdot 10^3 + 4 \cdot 10^2 + 7 \cdot 10 + 8 \\ + & 5 \cdot 10^2 + 2 \cdot 10 + 1 \\ \hline & 3 \cdot 10^3 + 9 \cdot 10^2 + 9 \cdot 10 + 9 \end{array}$$

    a. Write a corresponding addition algebra problem (use x for 10) and find the answer.
    b. Write a subtraction problem and the corresponding algebra problem and find the answer.
    c. Write a multiplication problem and the corresponding algebra problem and compute the answer.

SELECTED BIBLIOGRAPHY

Bass, H. "Computational Fluency, Algorithms, and Mathematical Proficiency: One Mathematician's Perspective." *Teaching Children Mathematics* 9 (February 2003): 322–329.

Berg, R. "Multiplication from Lilavati to the Summa." *Mathematics Teaching in the Middle School* 7 (December 2001): 226–231.

Bresser, R. "Helping English Language Learners Develop Computational Fluency." *Teaching Children Mathematics* 9 (February 2003): 294–299.

Broadent, F. "Lattice Multiplication and Division." *Arithmetic Teacher* 34 (January 1987): 28–31.

Brownell, W. "From NCTM's Archives: Meaning and Skill—Maintaining the Balance." *Teaching Children Mathematics* 9 (February 2003): 310–316.

Burns, M. "Introducing Division Through Problem-Solving Experiences." *Arithmetic Teacher* 38 (April 1991): 14–18.

Curcio, F., and S. Schwartz. "There Are No Algorithms for Teaching Algorithms." *Teaching Children Mathematics* 5 (September 1998): 26–30.

Ebdon, S. M. Coakley, and D. Legnard. "Mathematical Mind Journeys: Awakening Minds to Computational Fluency." *Teaching Children Mathematics* 9 (April 2003): 486–493.

Flowers, J., K. Kline, and R. Rubenstein. "Developing Teachers' Computational Fluency: Examples in Subtraction." *Teaching Children Mathematics* 9 (February 2003): 330–346.

Fuson, K. "Toward Computational Fluency." *Teaching Children Mathematics* 9 (February 2003): 300–309.

Gluck, D. "Helping Students Understand Place Value." *Arithmetic Teacher* 38 (March 1991): 10–13.

Graeber, A. "Misconceptions about Multiplication and Division." *Arithmetic Teacher* 40 (March 1993): 408–411.

Huinker, D. "Multiplication and Division Word Problems: Improving Students' Understanding." *Arithmetic Teacher* 37 (October 1989): 8–12.

Huinker, D., J. Freckman, and M. Steinmeyer. "Subtraction Strategies from Children's Thinking: Moving Toward Fluency with Greater Numbers." *Teaching Children Mathematics* 9 (February 2003): 347–353.

Kami, C., B. Lewis, and S. Livingston. "Primary Arithmetic: Children Inventing Their Own Procedures." *Arithmetic Teacher* 41 (December 1993): 200–203.

Kouba, V., and K. Franklin. "Multiplication and Division: Sense Making and Meaning." *Teaching Children Mathematics* 1 (May 1995): 574–577.

Lampert, M. "Teaching and Learning Long Division for Understanding in School." In *Analysis of Arithmetic for Mathematics Teaching*, edited by G. Leinhardt, R. Putnam, and R. Hattrup. Hillsdale, NJ: LEA, 1992.

Moldavan, C. "Culture in the Curriculum: Enriching Numeration and Number Operations." *Teaching Children Mathematics* 8 (December 2001): 238–243.

Nagel, N., and C. Swingen. "Students' Explanations of Place Value in Addition and Subtraction." *Teaching Children Mathematics* 5 (November 1998): 164–170.

National Council of Teachers of Mathematics. www.figurethis.org, 2002.

Postlewait, K., M. Adams, and J. Shih. "Promoting Meaningful Mastery of Addition and Subtraction." *Teaching Children Mathematics* 9 (February 2003): 354–357.

Randolph, T., and H. Sherman. "Alternative Algorithms: Increasing Options, Reducing Errors." *Teaching Children Mathematics* 7 (April 2001): 480–484.

Resnick, L., and S. Omanson. "Learning to Understand Arithmetic." *Advances in Instructional Psychology* (Vol. 3), edited by R. Glaser, Hillsdale, NJ: LEA, 1987.

Reyes, L. "Attitudes and Mathematics." In *Selected Issues in Mathematics Education*, edited by M. Lindquist. Reston, VA: NCTM, 1980.

Reys, R. "Mental Computation and Estimation: Past, Present, and Future." *Elementary School Journal* 84 (1984): 547–557.

Reys, R. "Computation Versus Number Sense." *Mathematics Teaching in the Middle School* 4 (October 1998): 110–112.

Reys, B., and R. Reys. "Computation in the Elementary Curriculum: Shifting the Emphasis." *Teaching Children Mathematics* 5 (December 1998): 236–241.

Ross, S. "Place Value: Problem Solving and Written Assessment." *Teaching Children Mathematics* 8 (March 2002): 419–423.

Russell, S. "Developing Computational Fluency with Whole Numbers." *Teaching Children Mathematics* 7 (November 2000): 154–158.

Scharton, S. "I Did it My Way." *Teaching Children Mathematics* 10 (January 2004): 278–283.

Silver, E., L. Shapiro, and A. Deutsch. "Sense Making and the Solution of Division Problems Involving Remainders: An Examination of Middle School Students' Solution Processes and Their Interpretations of Solutions." *Journal for Research in Mathematics Education* 24 (March 1993): 117–135.

Simonsen, L., and A. Teppo. "Using Alternative Algorithms with Preservice Teachers." *Teaching Children Mathematics* 5 (May 1999): 516–519.

Sowder, J. "Affective Factors and Computational Estimation Abilities." In *Affect and Problem Solving: A New Perspective*, edited by D. McLeod and V. Adams. New York: Springer-Verlag, 1989.

———."Mental Computation and Number Sense." *Arithmetic Teacher* 37 (March 1990): 18–20.

Tepper, J. "Basing a Career on Base Two." *Mathematics Teaching in the Middle School* 6 (October 2000): 116–119.

Whitenack, J., N. Knipping, S. Novinger, and G. Underwood. "Second Graders Circumvent Addition and Subtraction Difficulties." *Teaching Children Mathematics* 8 (December 2001): 228–233.

Wickett, M. "Discussion as a Vehicle for Demonstrating Computational Fluency in Multiplication." *Teaching Children Mathematics* 9 (February 2003): 318–321.

Zaslavsky, C. "Developing Number Sense: What Can Other Cultures Tell Us?" *Teaching Children Mathematics* 7 (February 2001): 312–319.

Integers and Number Theory

PRELIMINARY PROBLEM

Jillian loves math tricks. She asked her friend to choose any prime number greater than 3, square it, then subtract 4, then divide the new result by 12 and record the remainder. She then told her friend that the remainder was 9. How could she be sure that the remainder was 9 without knowing which prime her friend had picked?

The *Principles and Standards* expectations for students in Grades 3–5 include the following:

- *explore numbers less than 0 by extending the number line and through familiar applications;*
- *describe classes of numbers according to characteristics such as the nature of their factors.* (p. 148)

The expectations for students in Grades 6–8 include:

- *use factors, multiples, prime factorization, and relatively prime numbers to solve problems;*
- *develop meaning for integers and represent and compare quantities with them.* (p. 214)

In addition, the *Principles and Standards* point out that in Grades 6–8:

Students can also work with whole numbers in their study of number theory. Tasks, such as the following, involving factors, multiples, prime numbers, and divisibility can afford opportunities for problem solving and reasoning.

1. *Explain why the sum of the digits of any multiple of 3 is itself divisible by 3.*
2. *A number of the form abcabc always has several prime-number factors. Which prime numbers are always factors of a number of this form? Why?*

Middle-grades students should also work with integers. In lower grades, students may have connected negative integers in appropriate ways to informal knowledge derived from everyday experiences, such as below-zero winter temperatures or lost yards on football plays. In the middle grades, students should extend these initial understandings of integers. Positive and negative integers should be seen as useful for noting relative changes or values. Students can also appreciate the utility of negative integers when they work with equations whose solution requires them, such as $2x + 7 = 1$. (pp. 217–18)

In this chapter, we develop the necessary understanding that teachers need to help students reach these expectations. We start with the system of integers and develop an understanding of number theory.

Negative numbers are useful in everyday life. For example, Mount Everest is 29,028 ft above sea level, and the Dead Sea is 1293 ft below sea level. We may symbolize these elevations as 29,028 and $^{-}1293$.

In mathematics, the need for integers arises because subtractions cannot always be performed in only the set of whole numbers. To compute $4 - 6$ using the definition of subtraction for whole numbers, we must find a whole number n such that $6 + n = 4$. Because there is no such whole number n, the subtraction cannot be completed in the set of whole numbers. To perform the computation, we must invent a new number. This new number is a *negative integer*. If we attempt to calculate $4 - 6$ on a number line, then we must draw intervals to the left of 0. In Figure 4-1, $4 - 6$ is pictured as an arrow that starts at 0 and ends 2 units to the left of 0. The new number that corresponds to a point 2 units to the left of 0 is *negative two*, symbolized by $^{-}2$.

FIGURE 4-1

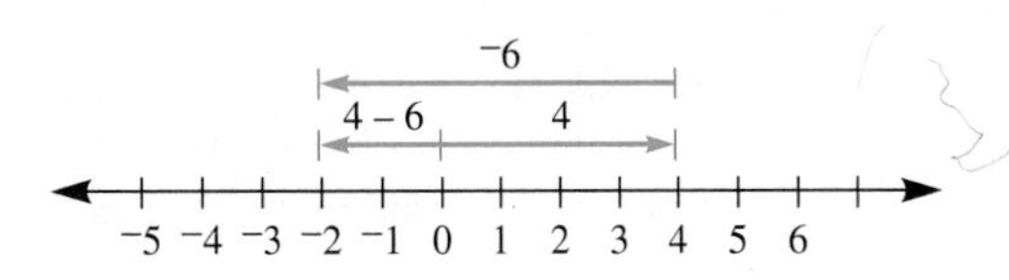

negative integers
positive integers
integers

Other numbers to the left of 0 are created similarly. The new set of numbers $\{^{-}1, ^{-}2, ^{-}3, ^{-}4, \ldots\}$ is the set of **negative integers**. The set $\{1, 2, 3, 4, \ldots\}$ is the set of **positive integers**. The integer 0 is neither positive nor negative. The union of the set of negative integers, the set of positive integers, and 0 is the set of **integers**. The set of integers is denoted I:

$$I = \{\ldots ^{-}4, ^{-}3, ^{-}2, ^{-}1, 0, 1, 2, 3, 4, \ldots\}$$

The mathematical treatment of questions concerning the integers is the subject of elementary number theory. As a field of study, number theory started to flourish in the seventeenth century with the work of Pierre de Fermat (1605–1665). Topics in number theory that occur in the elementary school curriculum include factors, multiples, divisibility tests, prime numbers, prime factorizations, greatest common divisors, and least common multiples. The topic of congruences, introduced by Carl Gauss (1777–1855), is also incorporated into the elementary curriculum through clock arithmetic and modular arithmetic. This topic of congruences gives students a look at a mathematical system.

▶ **HISTORICAL NOTE**

The Hindu mathematician Bhramagupta (ca. 598–665 CE) provided the first systematic treatment of negative numbers and of zero. The European mathematicians, not being aware of Bhramagupta's work for almost 1000 years after his death, shunned negative numbers. The Italian mathematician Gerolamo Cardano (1501–1576) was the first among European mathematicians to consider negative solutions to certain equations, but, still being uncomfortable with the concept of negative numbers, he called them "fictitious" numbers. ◀

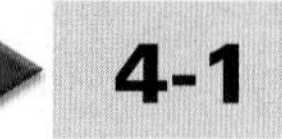

4-1 Integers and the Operations of Addition and Subtraction

Representations of Integers

It is unfortunate that we use the symbol "−" both to indicate a subtraction and a negative sign. To reduce confusion between the uses of this symbol in this text, a raised "$^{-}$" sign is used for negative numbers, as in $^{-}2$, in contrast to the lower sign for subtraction. To emphasize that an integer is positive, sometimes a raised plus sign is used, as in $^{+}3$. In this text, we use the plus sign for addition only and write $^{+}3$ simply as 3.

opposites

The negative integers are **opposites** of the positive integers. For example, the opposite of 5 is $^{-}5$. Similarly, the positive integers are the opposites of the negative integers. Because the opposite of 4 is denoted $^{-}4$, the opposite of $^{-}4$ can be denoted $^{-}(^{-}4)$, which equals 4. The opposite of 0 is 0. In the set of integers I, every element has an opposite that is also in I.

REMARK Using addition of integers, we shall soon see that when an opposite of an integer is added to the integer the sum is 0. In fact, if addition is introduced first, then ^{-}a can be defined as the solution of $x + a = 0$.

EXAMPLE 4-1 For each of the following, find the opposite of x:

a. $x = 3$ **b.** $x = {}^{-}5$ **c.** $x = 0$

Solution **a.** ${}^{-}x = {}^{-}3$ **b.** ${}^{-}x = {}^{-}({}^{-}5) = 5$ **c.** ${}^{-}x = {}^{-}0 = 0$ ▲

The value of ${}^{-}x$ in Example 4-1(b) is 5. Note that ${}^{-}x$ is the opposite of x and might *not represent a negative number*. In other words, x is a variable that can be replaced by some number either positive, zero, or negative. *Note:* ${}^{-}x$ is read "the opposite of x" not "minus x" or "negative x."

▶ **HISTORICAL NOTE**

The dash has not always been used for both the subtraction operation and the negative sign. Other notations were developed but never adopted universally. One such notation was used by Mohammed al-Khowârizmî (ca. 825), who indicated a negative number by placing a small circle over it. For example, ${}^{-}4$ was recorded as $\mathring{4}$. The Hindus denoted a negative number by enclosing it in a circle; for example, ${}^{-}4$ was recorded as ④. The symbols + and − first appeared in print in European mathematics in the late fifteenth century, at which time the symbols referred not to addition or subtraction nor positive or negative numbers, but to surpluses and deficits in business problems. ◀

Integer Addition

There are many ways to introduce operations on integers. Before formally defining the operations, we consider a more informal approach.

Chip Model for Addition

FIGURE 4-2

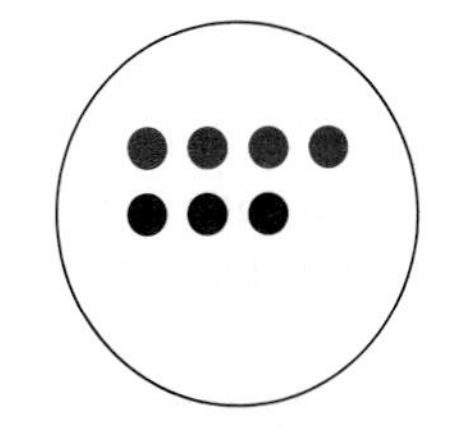

In the chip model, positive integers are represented by black chips and negative integers by red chips. One red chip neutralizes one black chip. Hence, the integer ${}^{-}1$ can be represented by 1 red chip, or 2 red and 1 black, or 3 red and 2 black, and so on. Similarly, every integer can be represented in many ways using chips. Figure 4-2 shows a chip model for the addition ${}^{-}4 + 3$. We put four red chips together with 3 black chips. Because 3 red chips neutralize 3 black ones, Figure 4-2 represents the equivalent of 1 red chip, or ${}^{-}1$.

Charged-Field Model for Addition

FIGURE 4-3

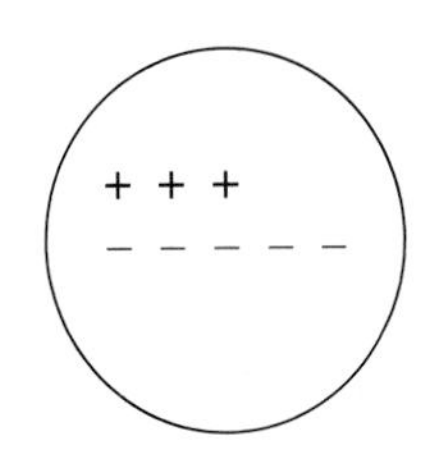

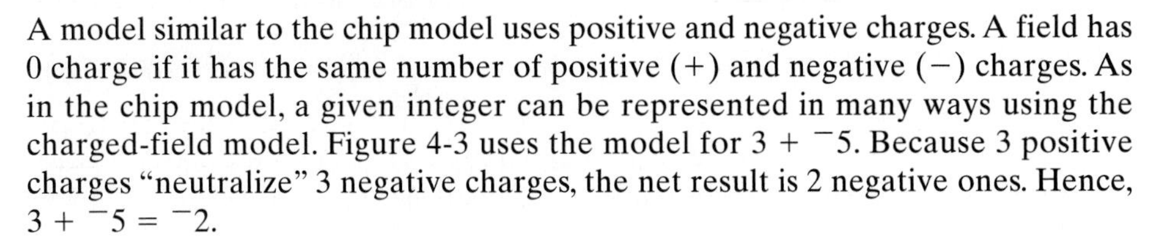
A model similar to the chip model uses positive and negative charges. A field has 0 charge if it has the same number of positive (+) and negative (−) charges. As in the chip model, a given integer can be represented in many ways using the charged-field model. Figure 4-3 uses the model for $3 + {}^{-}5$. Because 3 positive charges "neutralize" 3 negative charges, the net result is 2 negative ones. Hence, $3 + {}^{-}5 = {}^{-}2$.

Pattern Model for Addition

Addition of whole numbers was discussed in Chapter 2. Addition of integers can also be motivated by using patterns of addition of whole numbers. Notice that in the left column, the first four facts are known from whole-number addition. Also notice that the 4 stays fixed and as the numbers added to 4 decrease by 1, the sum decreases by 1. Following this pattern, $4 + {}^{-}1 = 3$ and we can complete the remainder of the first column. Similar reasoning can be used to complete the computations in the right column, where ${}^{-}2$ stays fixed and the other numbers decrease by 1 each time.

$$\begin{array}{rr} 4 + 3 = 7 & {}^{-}2 + 4 = 2 \\ 4 + 2 = 6 & {}^{-}2 + 3 = 1 \\ 4 + 1 = 5 & {}^{-}2 + 2 = 0 \\ 4 + 0 = 4 & {}^{-}2 + 1 = {}^{-}1 \\ 4 + {}^{-}1 = 3 & {}^{-}2 + 0 = {}^{-}2 \\ 4 + {}^{-}2 = 2 & {}^{-}2 + {}^{-}1 = {}^{-}3 \\ 4 + {}^{-}3 = 1 & {}^{-}2 + {}^{-}2 = {}^{-}4 \\ 4 + {}^{-}4 = 0 & {}^{-}2 + {}^{-}3 = {}^{-}5 \\ 4 + {}^{-}5 = {}^{-}1 & {}^{-}2 + {}^{-}4 = {}^{-}6 \\ 4 + {}^{-}6 = {}^{-}2 & {}^{-}2 + {}^{-}5 = {}^{-}7 \end{array}$$

TECHNOLOGY CORNER

On a spreadsheet, in column A enter 4 and fill down 20 rows. (For help on spreadsheets see Appendix I.) In column B, enter 3 as the first entry and then write a formula to add ${}^{-}1$ to 3 for the second entry, add ${}^{-}1$ to the second entry to get the third entry, and fill down continuing the pattern. In column C, find the sum of the respective entries in columns A and B. What patterns do you observe? Repeat the problem by changing the entries in column A to ${}^{-}4$ and repeating the process.

Number-Line Model

Another model for addition of integers involves a number line used with a moving object. For example, the car in Figure 4-4(a) starts at 0, facing in a positive direction

FIGURE 4-4

(a) $5 + 3$ is seen as moving the car forward 5 units and then 3 more units forward for a net move of 8 units to the right from 0. Thus $5 + 3 = 8$.

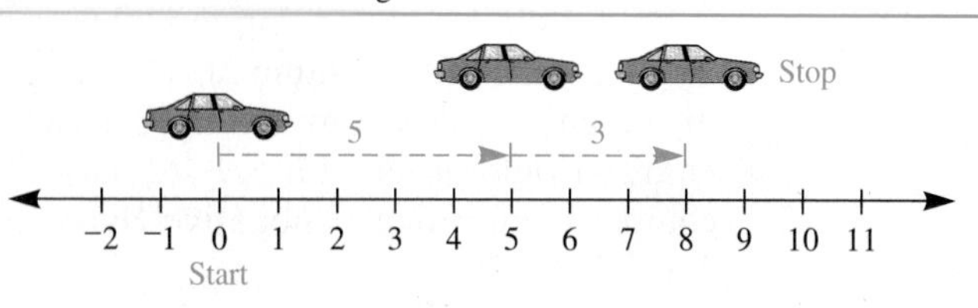

(b) $^-5 + 3$ is seen as moving the car 5 units in reverse and then moving it forward 3 units for a net move of 2 units to the left from 0. Thus, $^-5 + 3 = {}^-2$.

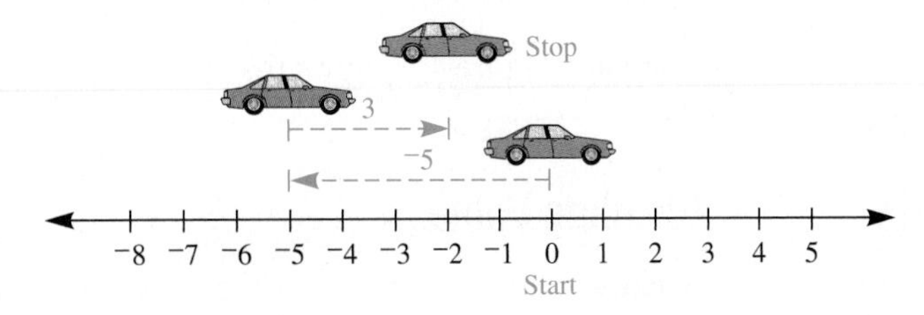

(c) $3 + {}^-5$ is seen as moving the car 3 units forward and then 5 units in reverse for a net move of 2 units to the left from 0. Thus, $3 + {}^-5 = {}^-2$.

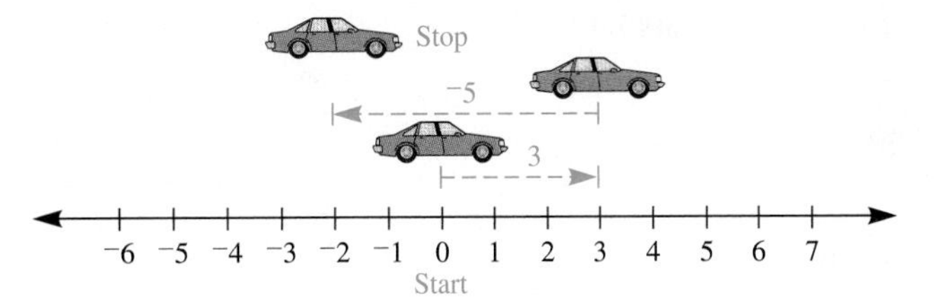

(d) $^-3 + {}^-5$ is seen as moving the car 3 units in reverse and then 5 more units in reverse for a net move of 8 units to the left from 0. Thus, $^-3 + {}^-5 = {}^-8$.

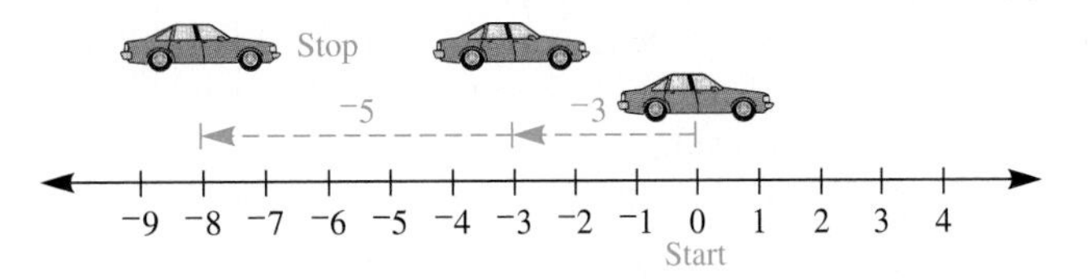

(to the right). To represent a positive integer, the car moves forward, and to represent a negative integer, it moves in reverse. For example, Figure 4-4(a) through (d) illustrates four different additions. Without the car, $^-3 + {}^-5$ can be pictured as in Figure 4-5.

FIGURE 4-5

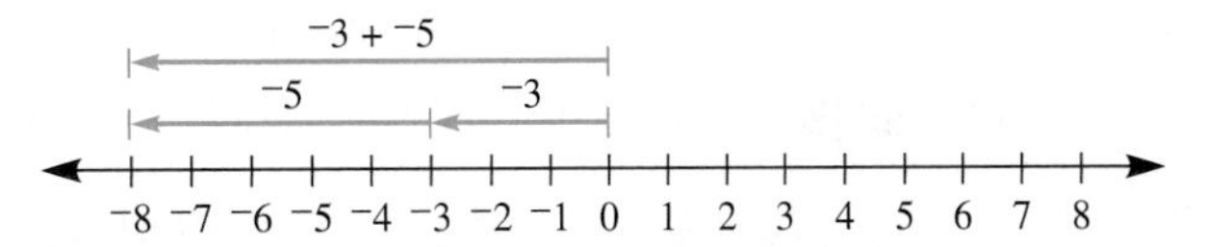

Figure 4-6 similarly depicts integer addition of $3 + {}^-5$.

FIGURE 4-6

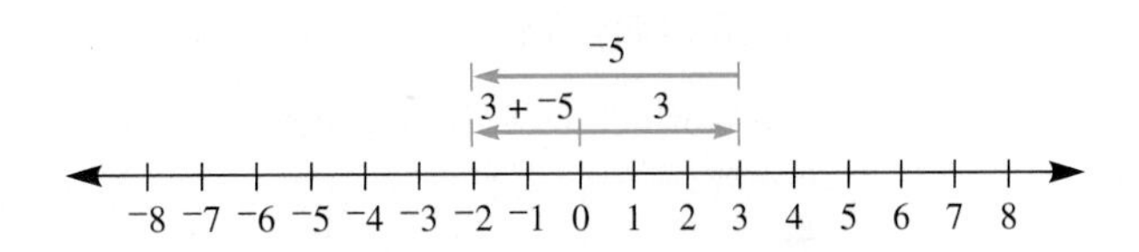

Example 4-2 involves a thermometer with a scale in the form of a vertical number line.

EXAMPLE 4-2 The temperature was $^-4$°C. In an hour, it rose 10°C. What is the new temperature?

FIGURE 4-7

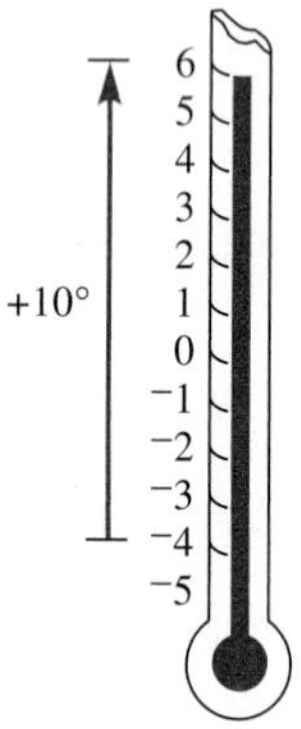

Solution Figure 4-7 shows that the new temperature is 6°C and that $^-4 + 10 = 6$.

Absolute Value

Because 4 and $^-4$ are opposites, they are on opposite sides of 0 on the number line and are the same distance (4 units) from 0, as shown in Figure 4-8.

FIGURE 4-8

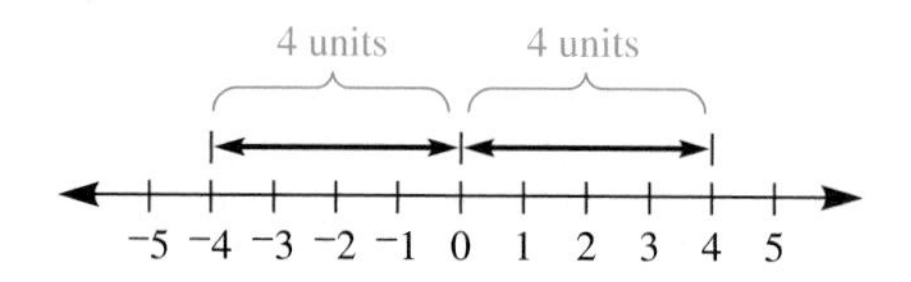

Distance is always a positive number or zero. The distance between the point corresponding to an integer and 0 is the **absolute value** of the integer. Thus, the absolute value of both 4 and $^-4$ is 4, written $|4| = 4$ and $|^-4| = 4$, respectively. Notice that if $x \geq 0$, then $|x| = x$, and if $x < 0$, then ^-x is positive. Therefore, we have the following:

Definition of Absolute Value

$$|x| = x \quad \text{if } x \geq 0$$
$$|x| = {}^-x \quad \text{if } x < 0$$

EXAMPLE 4-3 Evaluate each of the following:

a. $|20|$ **b.** $|^-5|$ **c.** $|0|$ **d.** $^-|^-3|$ **e.** $|2 - 5|$

Solution **a.** $|20| = 20$ **b.** $|^-5| = 5$ **c.** $|0| = 0$ **d.** $^-|^-3| = {}^-3$
e. $|2 + {}^-5| = |^-3| = 3$

NOW TRY THIS 4-1

Write each of the following in simplest form without the absolute value notation in the final answer:

a. $|x| + x$ if $x \leq 0$
b. $^-|x| + x$ if $x \leq 0$
c. $^-|x| + x$ if $x \geq 0$

REMARK It is possible to describe addition of integers as the process of finding the difference or the sum of the absolute values of the integers and attaching an appropriate sign.

Properties of Integer Addition

Integer addition has all the properties of whole-number addition. These properties are summarized next.

Properties

Given integers a, b, and c:

Closure property of addition of integers $a + b$ is a unique integer.

Commutative property of addition of integers $a + b = b + a$.

Associative property of addition of integers $(a + b) + c = a + (b + c)$.

Identity element of addition of integers 0 is the unique integer such that, for all integers a, $0 + a = a = a + 0$.

Notice the name *identity element* in the properties listed. Zero is the identity element of addition because when it is added to any integer it does not change the result; it leaves the integer unchanged.

additive inverse

We have seen that every integer has an opposite. This opposite is the **additive inverse** of the integer. The fact that each integer has a unique (one and only one) additive inverse is recorded as follows.

Uniqueness Property of the Additive Inverse

For every integer a, there exists a unique integer ^{-}a, the additive inverse of a, such that $a + {}^{-}a = 0 = {}^{-}a + a$.

REMARK Notice that by definition the additive inverse, ^{-}a, is the solution of the equation $x + a = 0$. The fact that the additive inverse is unique is equivalent to saying that the preceding equation has only one solution. In fact, for any integers a and b the equation $x + a = b$ has a unique solution.

The uniqueness of additive inverses can be used to justify other properties. For example, the opposite, or the additive inverse, of ^{-}a can be written $^{-}(^{-}a)$. However, because $a + {}^{-}a = 0$, the additive inverse of ^{-}a is also a. Because the additive inverse of ^{-}a must be unique, we have $^{-}(^{-}a) = a$. Other properties of addition of integers can be investigated by considering previously developed notions. For example, we saw that $^-2 + {}^-4 = {}^-6$, and we know that $^-6$ is the additive inverse of 6, or $2 + 4$. This leads us to the following:

$$^-2 + {}^-4 = {}^-(2 + 4)$$

This relationship is true in general and is stated along with its proof.

Properties of the Additive Inverse

For any integers, a and b:

1. $^-(^-a) = a$
2. $^-a + {}^-b = {}^-(a + b)$

We prove the second property of the additive inverse as follows: By definition $^-(a + b)$ is the additive inverse of $a + b$, that is, $(a + b) + {}^-(a + b) = 0$. If we could show that $^-a + {}^-b$ is also the additive inverse of $a + b$, the uniqueness of the additive inverse implies that $^-(a + b)$ and $^-a + {}^-b$ are equal. To show that $^-a + {}^-b$ is also the additive inverse of $a + b$, we need only to show that $(a + b) + (^-a + {}^-b) = 0$. This can be shown using the associative and commutative properties of integer addition and the definition of the additive inverse as follows:

$$\begin{aligned}(^-a + {}^-b) + (a + b) &= (^-a + a) + (^-b + b)\\ &= 0 + 0\\ &= 0\end{aligned}$$

Hence, $^-(a + b) = {}^-a + {}^-b$.

REMARK Notice that in the proof some of the steps involving the commutative and associative properties of integer addition were omitted. A more detailed proof showing all the steps follows.

$$\begin{aligned}(^-a + {}^-b) + (a + b) &= (^-a + {}^-b) + (b + a)\\ &= [(^-a + {}^-b) + b] + a\\ &= [^-a + (^-b + b)] + a\\ &= (^-a + 0) + a\\ &= {}^-a + a\\ &= 0\end{aligned}$$

EXAMPLE 4-4 Find the additive inverse of each of the following:

a. $^-(3 + x)$ **b.** $(a + {}^-4)$ **c.** $^-3 + (^-x)$

Solution **a.** $3 + x$

b. $^-(a + {}^-4)$, which by the preceding listed property (2), applied from right to left, can be written $^-a + {}^-(^-4)$, or $^-a + 4$.

c. $^-[^-3 + (^-x)]$, which can be written $^-(^-3) + {}^-(^-x)$, or $3 + x$.

FIGURE 4-9

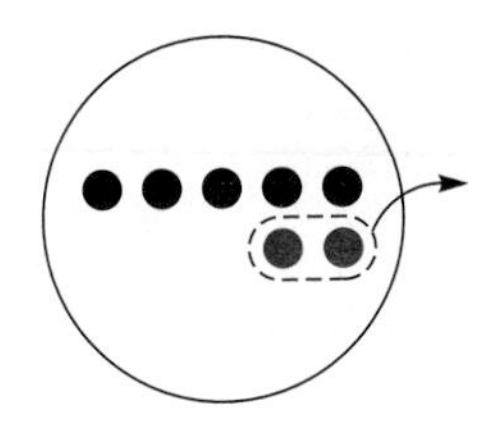

Integer Subtraction

As with integer addition, we explore several models for integer subtraction.

Chip Model for Subtraction

To find $3 - {}^{-}2$, we want to subtract ${}^{-}2$ (or remove 2 red chips) from 3 black chips. We need to represent 3 so that at least 2 red chips are present. In Figure 4-9, 3 is represented using 2 red and 5 black chips. When the 2 red chips are removed, 5 black ones are left and, hence, $3 - {}^{-}2 = 5$.

Charged-Field Model for Subtraction

Integer subtraction can be modeled with a charged field. For example, consider ${}^{-}3 - {}^{-}5$. To subtract ${}^{-}5$ from ${}^{-}3$, we first represent ${}^{-}3$ so that at least 5 negative charges are present. An example is shown in Figure 4-10(a). To subtract ${}^{-}5$, remove the 5 negative charges, leaving 2 positive charges, as in Figure 4-10(b). Hence, ${}^{-}3 - {}^{-}5 = 2$.

FIGURE 4-10

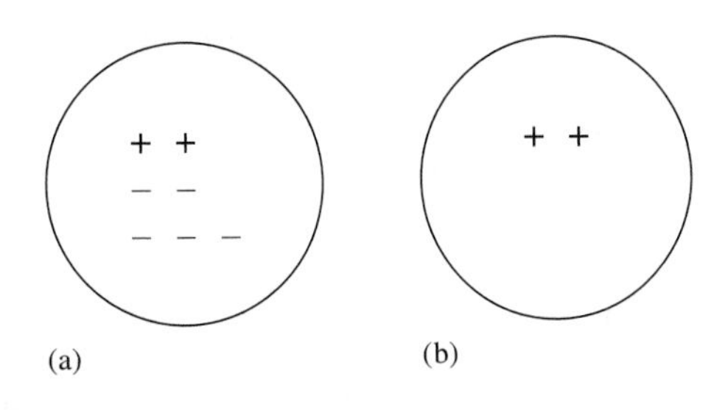

▶ RESEARCH NOTE
It is important to introduce children to negative numbers using manipulatives (Thompson 1988). ◀

Patterns Model for Subtraction

We can find the difference of two integers by considering the following patterns, where we start with subtractions that we already know how to do. Both the following pattern on the left and the pattern on the right start with $3 - 2 = 1$.

$3 - 2 = 1$	$3 - 2 = 1$
$3 - 3 = 0$	$3 - 1 = 2$
$3 - 4 = ?$	$3 - 0 = 3$
$3 - 5 = ?$	$3 - {}^{-}1 = ?$

In the pattern on the left, the difference decreases by 1. If we continue the pattern, we have $3 - 4 = {}^{-}1$ and $3 - 5 = {}^{-}2$. In the pattern on the right, the difference increases by 1. If we continue the pattern, we have $3 - {}^{-}1 = 4$ and $3 - {}^{-}2 = 5$.

Number-Line Model for Subtraction

The number-line model used for integer addition can also be used to model integer subtraction. While addition is modeled by maintaining the same direction and moving forward or backwards depending on whether a positive or negative integer is added, subtraction is modeled by turning around. In Figure 4-11, the car starts at 0 and is pointed in a positive direction (to the right). In this model, the operation of subtraction corresponds to facing the car in a negative direction. We subtract a positive integer by moving the car forward and a negative integer by moving the car in reverse. Figure 4-11 shows some examples.

FIGURE 4-11

(a) $5 - 3$ first tells you to move the car forward 5 units. The subtraction sign tells you to face the car in the opposite direction. Finally, move forward 3 units for a net move of 2 units to the right from 0. Thus, $5 - 3 = 2$.

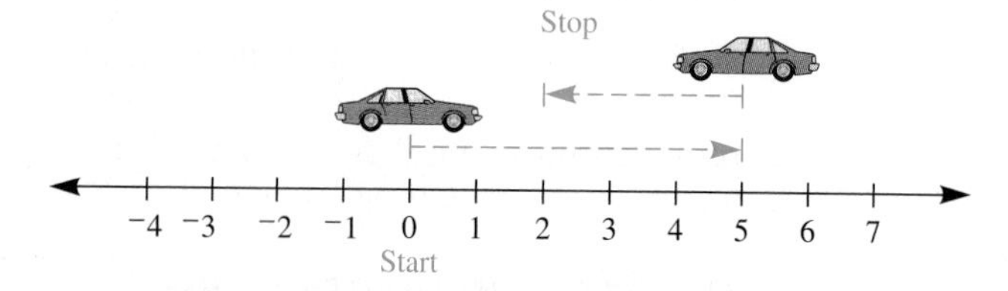

(b) $5 - {}^{-}3$ first tells you to move the car forward 5 units. Then face it in the opposite direction and move it in reverse 3 units. The net move is 8 units to the right of 0. Thus, $5 - {}^{-}3 = 8$.

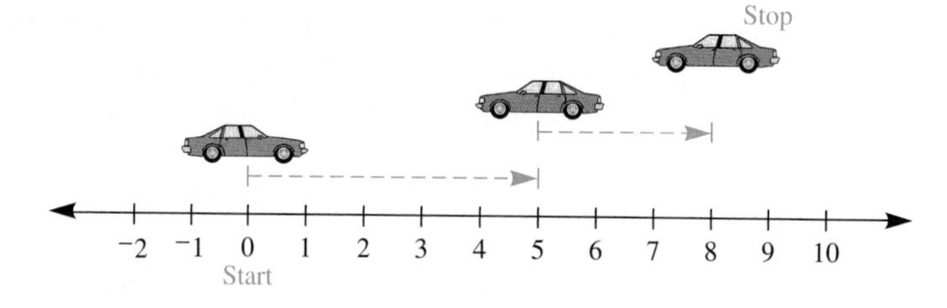

(c) In ${}^{-}5 - {}^{-}3$, move the car 5 units in reverse. Then face it in the opposite direction and move it in reverse 3 units. The net move is 2 units to the left of 0. Thus, ${}^{-}5 - {}^{-}3 = {}^{-}2$.

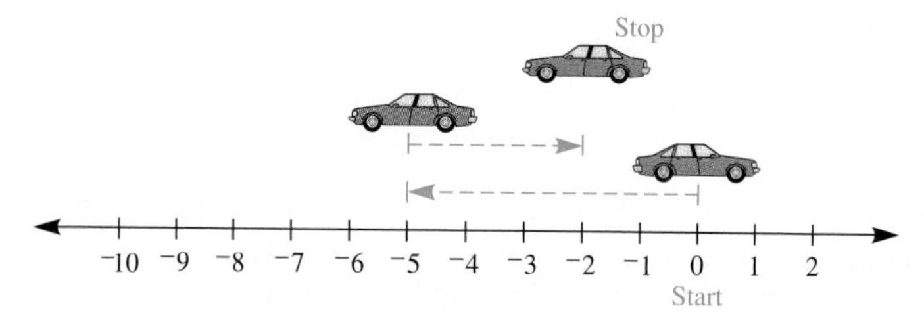

NOW TRY THIS 4-2

Suppose a mail carrier brings you three letters, one with a check for \$25 and the other two with bills for \$15 and \$20, respectively. You record this as $25 + {}^{-}15 + {}^{-}20$, or ${}^{-}10$; that is, you are \$10 poorer. Suppose that the next day you find out that the bill for \$20 was actually intended for someone else and therefore you give it back to the delivery person. You record your new balance as

$$-10 - {}^{-}20$$

or as

$$25 + {}^{-}15 + {}^{-}20 - {}^{-}20$$

which equals $25 + {}^{-}15$, or 10.

For each of the following, make up a mail delivery story and explain how your story can help to find the answer.

a. $23 + {}^{-}13 + {}^{-}12$

b. $18 - {}^{-}37$

Subtraction as the Inverse of Addition

Subtraction of integers, like subtraction of whole numbers, can be defined in terms of addition. Recall that $5 - 3$ can be computed by finding a whole number n as follows:

$$5 - 3 = n \quad \text{if and only if} \quad 5 = 3 + n$$

Because $3 + 2 = 5$, then $n = 2$.

Similarly, we compute $3 - 5$ as follows:

$$3 - 5 = n \quad \text{if and only if} \quad 3 = 5 + n$$

Because $5 + {}^{-}2 = 3$, then $n = {}^{-}2$. In general, for integers a and b, we have the following definition of *subtraction*.

Definition of Subtraction

For integers a and b, $a - b$ is the unique integer n such that $a = b + n$.

REMARK Addition "undoes" subtraction; that is, $(a - b) + b = a$. Also, subtraction "undoes" addition; that is, $(a + b) - b = a$.

EXAMPLE 4-5

Use the definition of subtraction to compute the following:

a. $3 - 10$ **b.** ${}^{-}2 - 10$

Solution **a.** Let $3 - 10 = n$. Then $10 + n = 3$, so $n = {}^{-}7$. Therefore, $3 - 10 = {}^{-}7$.

b. Let ${}^{-}2 - 10 = n$. Then $10 + n = {}^{-}2$, so $n = {}^{-}12$. Therefore, ${}^{-}2 - 10 = {}^{-}12$.

From our previous work with addition of integers, we know that $3 - 5 = {}^{-}2$ and $3 + {}^{-}5 = {}^{-}2$. Hence, $3 - 5 = 3 + {}^{-}5$. In general, the following is true.

Property

For all integers a and b, $a - b = a + ({}^{-}b)$.

The preceding property can be justified using the fact that the equation $b + x = a$ has a unique solution for x. From the definition of subtraction, the solution of the equation is $a - b$. To show that $a - b = a + {}^{-}b$, we only need to show that $a + {}^{-}b$ is also a solution. For that purpose, we substitute $a + {}^{-}b$ for x and check if $b + (a + {}^{-}b) = a$:

$$\begin{aligned} b + (a + {}^{-}b) &= b + ({}^{-}b + a) \\ &= (b + {}^{-}b) + a \\ &= 0 + a \\ &= a \end{aligned}$$

Consequently, $a - b = a + {}^{-}b$.

REMARK Sometimes the preceding property is used as the definition of subtraction.

NOW TRY THIS 4-3

a. Is the set of integers closed under subtraction? Why?

b. Do the commutative, associative, or identity properties hold for subtraction of integers? Why or why not?

Many early calculators had a change-of-sign key, either [CHS] or [+/−]. Later generations of calculators use [(−)], a key that allows computation with integers. For example, to compute $8 - (^{-}3)$, we would press [8] [−] [3] [+/−] [=]. Investigate what happens if you press [8] [−] [−] [3] [=].

EXAMPLE 4-6

Using the fact that $a - b = a + (^{-}b)$, compute each of the following:

a. $2 - 8$ **b.** $2 - (^{-}8)$ **c.** $^{-}12 - (^{-}5)$ **d.** $^{-}12 - 5$

Solution
a. $2 - 8 = 2 + {^{-}8} = {^{-}6}$
b. $2 - (^{-}8) = 2 + {^{-}(^{-}8)} = 2 + 8 = 10$
c. $^{-}12 - (^{-}5) = {^{-}12} + {^{-}(^{-}5)} = {^{-}12} + 5 = {^{-}7}$
d. $^{-}12 - 5 = {^{-}12} + {^{-}5} = {^{-}17}$

EXAMPLE 4-7

Use the fact that $a - b = a + (^{-}b)$ and the properties of additive inverse to write expressions equal to each of the following without parentheses in the final answer.

a. $^{-}(b - c)$ **b.** $a - (b + c)$

Solution
a. $^{-}(b - c) = {^{-}(b + {^{-}c})} = {^{-}b} + {^{-}(^{-}c)} = {^{-}b} + c$
b. $a - (b + c) = a + {^{-}(b + c)} = a + (^{-}b + {^{-}c}) = (a + {^{-}b}) + {^{-}c} = a + {^{-}b} + {^{-}c}$

REMARK It is possible to simplify the answers in parts (a) and (b) further, as follows: $^{-}b + c = c + {^{-}b} = c - b$, and $a + {^{-}b} + {^{-}c} = (a - b) - c$.

EXAMPLE 4-8

Simplify each of the following:

a. $2 - (5 - x)$ **b.** $5 - (x - 3)$ **c.** $^{-}(x - y) - y$

Solution
a.
$$\begin{aligned} 2 - (5 - x) &= 2 + {^{-}(5 + {^{-}x})} \\ &= 2 + {^{-}5} + {^{-}(^{-}x)} \\ &= 2 + {^{-}5} + x \\ &= {^{-}3} + x \end{aligned}$$

b. $$\begin{aligned} 5 - (x - 3) &= 5 + {}^{-}(x + {}^{-}3) \\ &= 5 + {}^{-}x + {}^{-}({}^{-}3) \\ &= 5 + {}^{-}x + 3 \\ &= 8 + {}^{-}x \\ &= 8 - x \end{aligned}$$

c. $$\begin{aligned} {}^{-}(x - y) - y &= {}^{-}(x + {}^{-}y) + {}^{-}y \\ &= [{}^{-}x + {}^{-}({}^{-}y)] + {}^{-}y \\ &= ({}^{-}x + y) + {}^{-}y \\ &= {}^{-}x + (y + {}^{-}y) \\ &= {}^{-}x + 0 \\ &= {}^{-}x \end{aligned}$$

Order of Operations

Subtraction in the set of integers is neither commutative nor associative, as illustrated in these counterexamples:

$$5 - 3 \neq 3 - 5 \quad \text{because} \quad 2 \neq {}^{-}2$$
$$(3 - 15) - 8 \neq 3 - (15 - 8) \quad \text{because} \quad {}^{-}20 \neq {}^{-}4$$

Remember that computations within parentheses must be completed before other computations.

An expression such as $3 - 15 - 8$ is ambiguous unless we know in which order to perform the subtractions. Mathematicians agree that $3 - 15 - 8$ means $(3 - 15) - 8$; that is, the subtractions in $3 - 15 - 8$ are performed in order from left to right. Similarly, $3 - 4 + 5$ means $(3 - 4) + 5$ and not $3 - (4 + 5)$. Thus, $(a - b) - c$ may be written without parentheses as $a - b - c$.

EXAMPLE 4-9

Compute each of the following:

a. $2 - 5 - 5$
b. $3 - 7 + 3$
c. $3 - (7 - 3)$

Solution **a.** $2 - 5 - 5 = {}^{-}3 - 5 = {}^{-}8$
b. $3 - 7 + 3 = {}^{-}4 + 3 = {}^{-}1$
c. $3 - (7 - 3) = 3 - 4 = {}^{-}1$

TECHNOLOGY CORNER

a. On a graphing calculator, graph the function with equation $y = x - {}^{-}4$.
b. Using the graph in (a), describe what happens as x takes on values that are less than ${}^{-}4$, equal to ${}^{-}4$, and greater than ${}^{-}4$.

ONGOING ASSESSMENT 4-1

1. Find the additive inverse of each of the following integers. Write your answer in the simplest possible form.
   a. 2 **b.** $^-5$ c. m
   d. 0 e. $-m$ **f.** $a + b$
2. Simplify each of the following:
   a. $^-(^-2)$ **b.** $^-(^-m)$ **c.** $^-0$
3. Evaluate each of the following:
   a. $|^-5|$ **b.** $|10|$ c. $^-|^-5|$ **d.** $^-|5|$
4. Demonstrate each of the following additions using the charged-field or chip model:
   a. $5 + {}^-3$ **b.** $^-2 + 3$
   c. $^-3 + 2$ **d.** $^-3 + {}^-2$
5. Compute each of the following using $a - b = a + {}^-b$.
   a. $3 - {}^-2$ **b.** $^-3 - 2$ c. $^-3 - {}^-2$
6. Answer each part of problem 5 using the definition of subtraction in terms of addition
7. Demonstrate each of the additions in problem 4 using a number-line model.
8. Write an addition fact that corresponds to each of the following sentences and then answer the question:
   a. A certain stock dropped 17 points and the following day gained 10 points. What was the net change in the stock's worth?
   b. The temperature was $^-10$°C and then it rose by 8°C. What is the new temperature?
   c. The plane was at 5000 ft and dropped 100 ft. What is the new altitude of the plane?
   d. A visitor in a Las Vegas casino lost \$200, won \$100, and then lost \$50. What is the change in the gambler's net worth?
   e. In four downs, the football team lost 2 yd, gained 7 yd, gained 0 yd, and lost 8 yd. What is the total gain or loss?
9. On January 1, Jane's bank balance was \$300. During the month, she wrote checks for \$45, \$55, \$165, \$35, and \$100 and made deposits of \$75, \$25, and \$400.
   a. If a check is represented by a negative integer and a deposit by a positive integer, express Jane's transactions as a sum of positive and negative integers.
   b. What was the balance in Jane's account at the end of the month?
10. Use a number-line model to find the following:
    a. $^-4 - {}^-1$ **b.** $^-4 - {}^-3$
11. Use patterns to show the following:
    a. $^-4 - {}^-1 = {}^-3$ **b.** $^-2 - 1 = {}^-3$
12. Perform each of the following:
    a. $^-2 + (3 - 10)$ **b.** $[8 - (^-5)] - 10$
    c. $(^-2 - 7) + 10$ **d.** $^-2 - (7 + 10)$
    e. $8 - 11 - 10$ **f.** $^-2 - 7 + 3$
13. In each of the following, write a subtraction problem that corresponds to the question and an addition problem that corresponds to the question and then answer the questions:
    a. The temperature is 55°F and is supposed to drop 60°F by midnight. What is the expected midnight temperature?
    b. Moses has overdraft privileges at his bank. If he had \$200 in his checking account and he wrote a \$220 check, what is his balance?
14. Answer each of the following:
    a. In a game of Triominoes, Jack's scores in five successive turns are 17, $^-8$, $^-9$, 14, and 45. What is his total at the end of five turns?
    b. The largest bubble chamber in the world is 15 ft in diameter and contains 7259 gal of liquid hydrogen at a temperature of $^-247$°C. If the temperature is dropped by 11°C per hour for 2 consecutive hours, what is the new temperature?
    c. The greatest recorded temperature ranges in the world are around the "cold pole" in Siberia. Temperatures in Verkhoyansk have varied from $^-94$°F to 98°F. What is the difference between the high and low temperatures in Verkhoyansk?
15. Motor oils protect car engines over a range of temperatures. These oils have names like 10W–40 or 5W–30. The following graph shows the temperatures, in degrees Fahrenheit, at which the engine is protected by a particular oil. Using the graph, find which oils can be used for the following temperatures:
    a. Between $^-5$° and 90° **b.** Below $^-20$°
    c. Between $^-10$° and 50° **d.** From $^-20$° to over 100°
    e. From $^-8$° to 90°

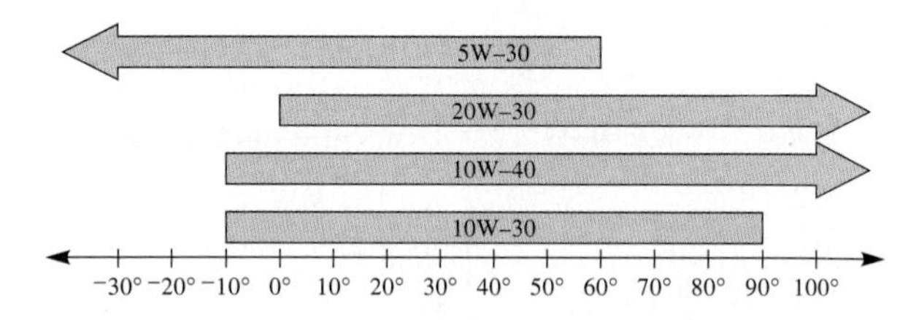

16. Apply the properties discussed in this section to simplify each of the following as much as possible. Show all work.
    a. $3 - (2 - 4x)$ **b.** $x - (^-x - y)$
    c. $4x - 2 - 3x$ **d.** $4x - (2 - 3x)$
17. For which integers a, b, c does $a - b - c = a - (b - c)$? Justify your answer.
18. Let W stand for the set of whole numbers, I the set of integers, I^+ the set of positive integers, and I^- the set of negative integers. Find each of the following:
    a. $W \cup I$ **b.** $W \cap I$ c. $I^+ \cup I^-$
    d. $I^+ \cap I^-$ e. $W - I$ **f.** $I - W$
    g. $W - I^+$ **h.** $W - I^-$ i. $I \cap I$

BRAIN TEASER

4-2

19. a. Use properties of addition to prove that for all integers x and y; $^{-}x - y = {}^{-}y - x$.
    b. Does part (a) imply that subtraction is commutative? Explain.
20. Complete the magic square using the following integers: $^{-}13, {}^{-}10, {}^{-}7, {}^{-}4, 2, 5, 8, 11$.

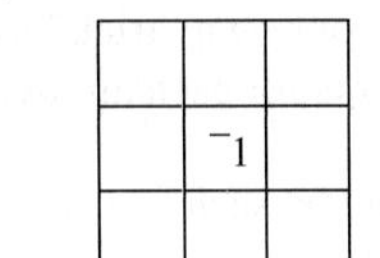

21. Let $f(x) = {}^{-}x - 1$ with domain I. Find the following:
    a. $f(^{-}1)$ b. $f(100)$
    c. $f(^{-}2)$ d. $f(^{-}a)$ in terms of a
    e. For which values of x will the output be 3?
22. Find all integers x, if there are any, such that the following are true:
    a. ^{-}x is positive.
    b. ^{-}x is negative.
    c. $^{-}x - 1$ is positive.
    d. $|x| = 2$.
    e. $^{-}|x| = 2$.
    f. $^{-}|x|$ is negative.
    g. $^{-}|x|$ is positive.
23. Let $f(x) = |1 - x|$ with domain I. Find the following:
    a. $f(10)$ b. $f(^{-}1)$
    c. All the inputs for which the output is 1
    d. The range
24. a. For each of the following functions find $f(f(x))$:
       i. $f(x) = x$
       ii. $f(x) = {}^{-}x$
       iii. $f(x) = {}^{-}x + 2$
    b. Interpret your answers in part (a) using the function machine model.
    ★c. Find other functions for which $f(f(x)) = x$. Justify your answer.
25. In each of the following, find all integers x satisfying the given equation:
    a. $|x - 6| = 6$ b. $|x| + 2 = 10$ c. $|^{-}x| = |x|$
26. By the definition of absolute value, the function $f(x) = |x|$ can be written as follows:

$$f(x) = \begin{cases} x, & \text{if } x \geq 0 \\ {}^{-}x, & \text{if } x < 0 \end{cases}$$

Write the function $f(x) = |x - 6|$ in a similar way without absolute value.

27. Determine how many integers there are between the following given integers (not including the given integers):
    a. 10 and 100 b. $^{-}30$ and $^{-}10$
    c. $^{-}10$ and 10 d. x and y (if $x < y$)
28. Suppose $a = 6$, $b = 5$, $c = 4$, and $d = {}^{-}3$. Insert parentheses in the expression $a - b - c - d$ to obtain the greatest possible and the least possible values. What are these values?
29. An arithmetic sequence may have a positive or negative difference. In each of the following arithmetic sequences, find the difference and write the next two terms:
    a. $0, {}^{-}3, {}^{-}6, {}^{-}9$
    b. $7, 3, {}^{-}1, {}^{-}5$
    c. $x + y, x, x - y$
    d. $1 - 3x, 1 - x, 1 + x$
30. Find the sums of the following arithmetic sequences:
    a. $^{-}20 + {}^{-}19 + {}^{-}18 + \ldots + 18 + 19 + 20$
    b. $100 + 99 + 98 + \ldots + {}^{-}50$
    c. $100 + 98 + 96 + \ldots + {}^{-}6$
31. In an arithmetic sequence, the eighth term minus the first term equals 21. The sum of the first and the eighth term is $^{-}5$. Find the fifth term of the sequence.
32. Classify each of the following as true or false. If false, give a counterexample.
    a. $|^{-}x| = |x|$
    b. $|x - y| = |y - x|$
    c. $|^{-}x + {}^{-}y| = |x + y|$
    d. $|x^2| = x^2$
    e. $|x^3| = x^3$
    f. $|x^3| = x^2|x|$
33. Solve the following equations:
    a. $x + 7 = 3$ b. $-10 + x = -7$
    c. $-x = 5$ d. $-x + 5 = 7$
    e. $1 - x = -13$ f. $-x - 8 = -9$
34. Assume that gear A has 56 teeth and gear B has 14 teeth. Suppose that the number of counterclockwise rotations is designated by a positive number and the number of clockwise rotations by a negative number. If gear A rotates 7 times per minute, how many times per minute does gear B rotate? Explain your reasoning.

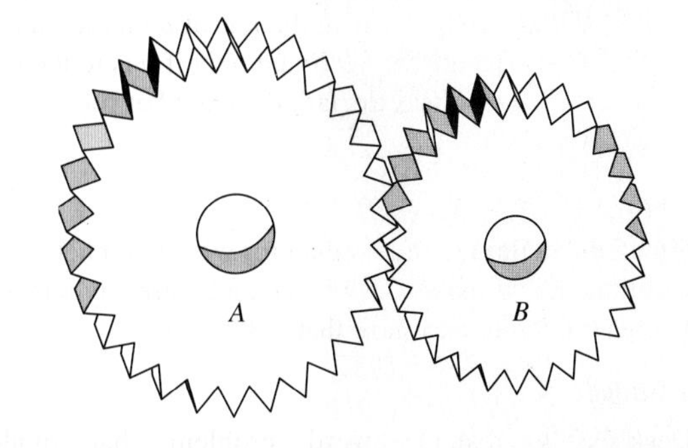

35. Complete each of the following integer arithmetic problems on your calculator, making use of the change-of-sign key. For example, on some calculators to find $^{-}5 + {}^{-}4$, press [5] [+/-] [+] [4] [+/-] [=].
    a. $^{-}12 + {}^{-}6$ b. $^{-}12 + 6$ c. $27 + {}^{-}5$
    d. $^{-}12 - 6$ e. $16 - {}^{-}7$
36. Estimate each of the following and then use a calculator to find the actual answer:
    a. $343 + {}^{-}42 - 402$ b. $^{-}1992 + 3005 - 497$
    c. $992 - {}^{-}10003 - 101$ d. $^{-}301 - {}^{-}1303 + 4993$

Communication

37. A turnpike driver had car trou
driven 12 mi from milepost 68
Assuming he is confused and di
his cellular phone for help, how
sible location? Explain.

38. Dolores claims that the best way
$a + (^-b)$, for all integers a and
add b to each expression you get
 a. Explain why you think Dolo
 b. Do you agree with Dolore
 "best way"? If not, what is a

39. Addition of integers with like si
absolute values as follows:
To add integers with like sig
of the integers. The sum has the
Describe in a similar way how
signs.

40. Explain why $b - a$ and $a - b$ a
other.

41. a. The absolute value of an int
 this contradict the fact that t
 be equal to ^-x? Explain wh
 b. Explain how to write the ac
 using the least number of sy

42. If an integer a is pictured on th
tance from the point on the r
the integer to the origin is $|a|$.
following.
 a. Explain why $|a - b|$ is the
 that represent the integers a
 b. One way to define "less tha
 $a < b$ if and only if a is to
 line. Consequently $b > a$ i
 of a. Use these ideas to ma
 gers x such that:
 (i) $|x| < 5$ (ii) $|x| < 1$
 (iii) $|x| \geq 5$ (iv) $|x| > -$

43. Recall the definition of less th
addition and define $a < b$ wh
Use your definition to show th

Open-Ended

44. Describe a realistic wor
$^-50 + (^-85) - (^-30)$.

45. In a library some floors are be
are above ground-level. If the
nated the zero floor, design a
number the floors and then de
the elevator to model addition

46. a. I am choosing an integer.
 integer, take the opposite c
 the opposite of the new re
 the original number?

29. In each of the following, find the next two terms. Assume the sequence is arithmetic or geometric, and find its difference or ratio and the nth term.
 a. $^-10, {}^-7, {}^-4, {}^-1, 2, 5, \_, \_$
 b. $10, 7, 4, 1, {}^-2, {}^-5, \_, \_$
 c. $^-2, {}^-4, {}^-8, {}^-16, {}^-32, {}^-64, \_, \_$
 d. $^-2, 4, {}^-8, 16, {}^-32, 64, \_, \_$
 e. $2, {}^-2^2, 2^3, {}^-2^4, 2^5, {}^-2^6, \_, \_$

30. Find the sum of the first 100 terms in parts (a) and (b) of problem 29.

31. Find the first five terms of the sequences whose nth term is
 a. $n^2 - 10$ b. $^-5n + 3$ c. $(^-2)^n - 1$
 d. $(^-2)^n + 2^n$ e. $n^2(^-1)^n$ f. $|10 - n^2|$

32. Find the first two terms of an arithmetic sequence in which the fourth term is $^-8$ and the 101st term is $^-493$.

33. In the geometric sequence $1, -2, 4, -8, \ldots$, determine if there is a term equal to the following numbers:
 a. 512 b. 1024

34. Tira noticed that every 30 sec, the temperature of a chemical reaction in her lab was decreasing by the same number of degrees. Initially, the temperature was 28°C and 5 min later, $^-12$°C. In a second experiment, Tira noticed that the temperature of the chemical reaction was initially $^-57$°C and was decreasing by 3°C every minute. If she started the two experiments at the same time, when were the temperatures of the reactions the same? What was that temperature?

35. If x and y are integers, classify each of the following as always true, sometimes true, or never true. Justify your answers.
 a. $xy = {}^-|x|\,|y|$ b. $(^-x)^3 = {}^-x^3$
 c. $^-x^2 = x^2$ d. $|x| > {}^-1$
 e. If $x > y$, then $x^2 > y^2$.
 f. If $x < y$ then $a - x < a - y$, for all integers a.

36. Jon has two checking accounts. In the first one, he is \$120 overdrawn, and in the second, his balance is \$300. If he deposits \$40 every day in the first account but withdraws \$20 daily from the second account, after how many days will the balance in each account be the same? Explain your solution.

37. If $x^2 < y^2$, what can you conclude about x and y?

Communication

38. Can $(^-x - y)(x + y)$ be multiplied by using the difference-of-squares formula? Explain why or why not.

39. Kahlil said that using the equation $(a + b)^2 = a^2 + 2ab + b^2$, he can find a similar equation for $(a - b)^2$. Examine his argument. If it is correct, supply any missing steps or justifications; if it is incorrect, point out why.

$$\begin{aligned}(a - b)^2 &= [a + (^-b)]^2\\ &= a^2 + 2a(^-b) + (^-b)^2\\ &= a^2 - 2ab + b^2\end{aligned}$$

40. a. Use the distributive property of multiplication over addition to show that $(^-1)a + a = 0$. (*Hint:* Write $a = 1 \cdot a$.)
 b. Use part (a) to show that $(^-1)a = {}^-a$.

41. Seventh-grader Nancy gave the following argument to show that $(^-a)b = {}^-(ab)$, for all integers a and b: *I know that* $(^-1)a = {}^-a$*; hence:*

$$\begin{aligned}(^-a)b &= [(^-1)a]b\\ &= (^-1)(ab)\\ &= {}^-(ab).\end{aligned}$$

If the argument is valid, complete its details; if it is not valid, explain why not.

42. Hosni gave the following argument that $^-(a + b) = {}^-a + {}^-b$, for all integers a and b. If the argument is correct, supply the missing reasons. If it is incorrect, explain why not.

$$\begin{aligned}-(a + b) &= (-1)(a + b)\\ &= (-1)a + (-1)b\\ &= -a + -b\end{aligned}$$

43. The Swiss mathematician Leonhard Euler (1707–1783) argued that $(^-1)(^-1) = 1$ as follows: "The result must be either $^-1$ or 1. If it is $^-1$, then $(^-1)(^-1) = {}^-1$. Because $^-1 = (^-1) \cdot 1$, we have $(^-1)(^-1) = (^-1) \cdot 1$. Now dividing both sides of the last equation by $^-1$ we get $^-1 = 1$, which of course cannot be true. Hence $(^-1)(^-1)$ must be equal to 1."
 a. What is your reaction to this argument? Is it logical? Why or why not?
 b. Can Euler's approach be used to justify other properties of integers? Explain.

44. The following graph shows the development of mathematics in different cultures. Explain the use of positive and negative numbers in the graph.

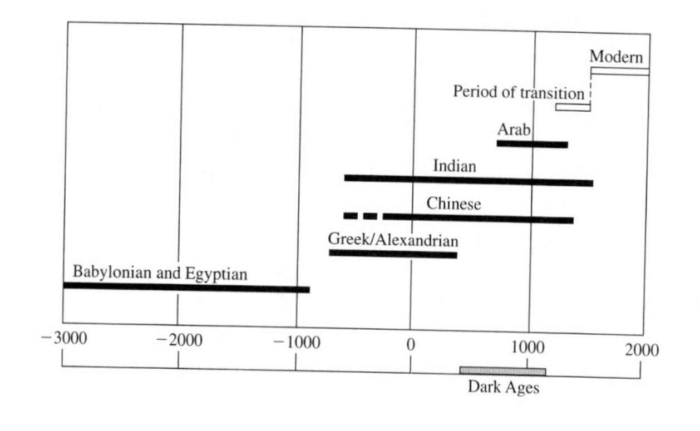

45. Each of the following sequences is an arithmetic sequence. For which values of n will the nth term of the second sequence exceed the nth term of the first sequence? Explain your solution and thought process.
 i. $1024, 1022, 1020, \ldots$
 ii. $^-2000, {}^-1996, {}^-1982, \ldots$

46. If $5x + 3 < {}^{-}20$, answer the following:
    a. Find the greatest integer x for which the inequality is true. Explain your thinking.
    b. Is there a least integer x for which the inequality is true? Explain why or why not.
47. Jill asks each of her classmates to choose a number, then multiply the number by $^{-}3$, add 2 to the product, multiply the result by $^{-}2$, and then subtract 14. Finally, each student is asked to divide the result by 6 and record the answer. When Jill gets an answer from a classmate, she just adds 3 to it in her head and announces the number each classmate originally chose. How did Jill know to add 3 to each answer?

Open-Ended

48. Make up a problem similar to problem 47 but with all numbers different and solve it.
49. On a national mathematics competition, scoring is accomplished using the formula 4 times the number done correctly minus the number done incorrectly. In this scheme, problems left blank are considered neither correct nor incorrect. Devise a scenario that would allow a student to have a negative score.
50. Select a current middle-school text that introduces multiplication and division of integers and discuss any models that were used and how effective you think they would be with a group of students.

Cooperative Learning

51. Devise a scheme for determining a grade-point average for a college student that allows negative quality points for a failing grade.
    a. Use your scheme to determine possible grades for students with positive, zero, and negative grade-point averages.
    b. Compare your scheme with that of another class group and write a rationale for the best scheme.
52. a. How would you introduce multiplication of integers in a middle-school class and how would you explain that a product of two negative numbers is positive? Write a rationale for your approach.
    b. Present your answers and compare them to those of another class group and together decide about the most appropriate way to introduce the concepts. Write a rationale for your approach.
53. Discuss in your group what each person's favorite approach is to justify the fact that $(^{-}1)(^{-}1) = 1$ and why it is the favorite.

Review Problems

54. Illustrate $^{-}8 + {}^{-}5$ on a number line.
55. Find the additive inverse of each of the following:
    a. $^{-}5$ b. 7 c. 0
56. Compute each of the following:
    a. $|{}^{-}14|$ b. $|{}^{-}14| + 7$
    c. $8 - |{}^{-}12|$ d. $|11| + |{}^{-}11|$
57. In the 1400s, European merchants used positive and negative numbers to label barrels of flour. For example, a barrel labeled +3 meant the barrel was 3 lb overweight, whereas a barrel labeled $^{-}5$ meant the barrel was 5 lb underweight. If the following numbers were found on 100-lb barrels, what was the total weight of the barrels?

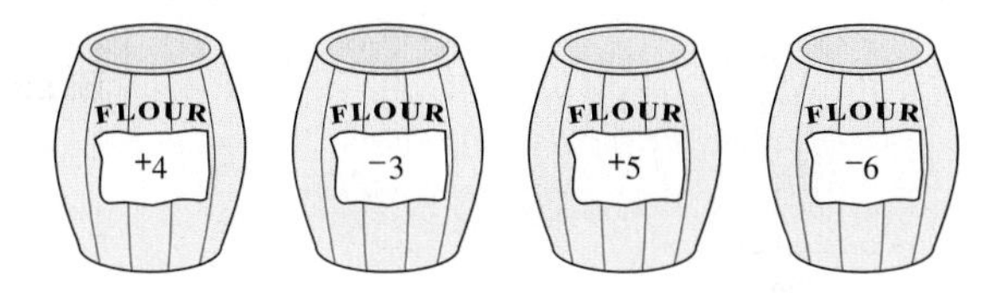

58. Write the function $f(x) = (x + |x|) \div 2$ without absolute value. (Distinguish two cases: $x \geq 0$ and $x < 0$.)
59. Solve for x if possible:
    a. $|x| = 3$ b. $|x| + 1 = 0$
    c. $|x| = x$ d. $|x| = {}^{-}x$

Third International Mathematics and Science Study (TIMSS) Questions

If n is a negative integer, which of these is the largest number?
a. $3 + n$
b. $3 \times n$
c. $3 - n$
d. $3 \div n$

If $x = {}^{-}3$, what is the value of $^{-}3x$?
a. -9
b. -6
c. -1
d. 1
e. 9

The numbers in the sequence 7, 11, 15, 19, 23, . . . increase by four. The numbers in the sequence 1, 10, 19, 28, 37, . . . increase by nine. The number 19 is in both sequences. If the two sequences are continued, what is the next number that is in BOTH the first and the second sequences?

Answer: ____________

TIMSS, Grade 8, 2003

BRAIN TEASER

If $a, \ldots, z$ are integers, find the product:

$$(x - a)(x - b)(x - c) \ldots (x - z)$$

4-3 Divisibility

The concepts of *even* and *odd* integers are commonly used. For example, during water shortages in the summer in some parts of the country, houses with even-number addresses can water on even-numbered days of the month and houses with odd-number addresses can water on odd-numbered days. An even integer is an integer that has 0 remainder when divided by 2. We say that it is divisible by 2. An odd integer is an integer that is not divisible by 2. The fact that 12 is divisible by 2 can be stated in the following equivalent statements in the left column:

Example	**General Statement**
12 is divisible by 2.	a is divisible by b.
2 is a divisor of 12.	b is a divisor of a.
12 is a multiple of 2.	a is a multiple of b.
2 is a factor of 12.	b is a factor of a.
2 divides 12.	b divides a.

divides

The statement that "2 divides 12" is written with a vertical segment, as in $2|12$, where the vertical segment means **divides**. Likewise, "b divides a" can be written $b|a$. Each statement in the previous right column can be written $b|a$. We write $5 \nmid 12$ to symbolize that 5 does not divide 12 or that 12 is not divisible by 5. The notation $5 \nmid 12$ also implies that 12 is not a multiple of 5 and 5 is not a factor of 12.

In general, if a is a nonnegative integer and b is a positive integer, we say that a is divisible by b or equivalently that b divides a if and only if the remainder when a is divided by b is 0. Using the division algorithm, this means that there is a unique q (quotient) such that $a = bq$. We extend this concept to divisibility for all integers in the following definition.

Definition of "Divides"

If a and b are any integers, then b divides a, written $b|a$, if, and only if, there is a unique integer q such that $a = bq$.

factor ◆ divisor ◆ multiple

If $b|a$, then b is a **factor**, or a **divisor**, of a, and a is a **multiple** of b.

Do not confuse $b|a$ with b/a, which is interpreted as $b \div a$. The former, a relation, is either true or false. The latter, an operation, has a numerical value if $a \neq 0$. Note that if b/a is an integer, then $a|b$. Also note that for positive integers $a \nmid b$ is equivalent to saying that the remainder when b is divided by a is not 0.

▶ **HISTORICAL NOTE**

Pierre de Fermat (1601–1665) was a lawyer and a magistrate who served in the provincial parliament in Toulouse, France. He devoted his leisure time to mathematics—a subject in which he had no formal training. After his death, his son decided to publish a new edition of Diophantus's *Arithmetica* with Fermat's notes. One of the notes in the margin of Fermat's copy asserted that the equation $x^n + y^n = z^n$ has no positive-integer solutions if n is an integer greater than 2 and commented, "I have found an admirable proof of this, but the margin is too narrow to contain it." Many great mathematicians spent years trying to prove Fermat's assertion, now called "Fermat's last theorem." In 1995, Andrew Wiles, a Princeton University mathematician, proved Fermat's last theorem. ◀

EXAMPLE 4-17

Classify each of the following as true or false. Explain your answers.

a. $^{-}3|12$ **b.** $0|2$ **c.** 0 is even. **d.** $8\nmid 2$ **e.** For all integers, a, $1|a$.
f. For all integers a, $^{-}1|a$. **g.** $3|6n$ for all integers n.
h. $(a-b)|(a^2-b^2)$ if a and b are integers and $a \neq b$. **i.** $0|0$

Solution
a. $^{-}3|12$ is true because $12 = {^{-}4}({^{-}3})$.
b. $0|2$ is false because there is no integer c such that $2 = c \cdot 0$.
c. $2|0$ is true because $0 = 0 \cdot 2$; therefore, 0 is even.
d. $8\nmid 2$ is true because there is no integer c such that $2 = c \cdot 8$.
e. $1|a$ is true for all integers a because $a = a \cdot 1$.
f. $^{-}1|a$ is true for all integers a because $a = ({^{-}a})({^{-}1})$.
g. $3|6n$ is true. Because $6n = 3 \cdot 2n$, $6n$ is a multiple of 3 and hence $3|6n$.
h. $(a-b)|(a^2-b^2)$ is true because $a^2 - b^2 = (a-b)(a+b)$ and $a \neq b$.
i. According to our definition, $0|0$ is false because $0 = 0 \cdot q$ for all integers q, so q is not unique.

▲

In Example 4-17(g), because $3|6$ we were able to show that $3|6n$ for all integers n. This can be generalized. If instead of $3|6$ we have $3|a$, then we can conclude that 3 divides any multiple of a. Even more generally, if $d|a$ then d divides any multiple of a. We state this fact in the following theorem.

▶ Theorem 4-1

For any integers a and d, if $d|a$ and n is any integer, then $d|na$.

The theorem can be stated in an equivalent form:

If d is a factor of a (that is, a equals some integer times d), then d is a factor of any multiple of a.

We can deduce other notions of divisibility from everyday models. Consider two packages of chewing gum each having five pieces, as in Figure 4-16. We can divide each package of gum evenly among five students. In addition, if we opened both packages and put all of the pieces in a bag, we could still divide the pieces of gum evenly among the five students. To generalize this notion, if we buy

FIGURE 4-16

gum in larger packages with a pieces in one package and b pieces in a second package with both a and b divisible by 5, we can record the preceding discussion as follows:

$$\text{If } 5 \mid a \text{ and } 5 \mid b, \text{ then } 5 \mid (a + b).$$

If the number, a, of pieces of gum in one package is divisible by 5, but the number, b, of pieces in the other package is not, then the total, $a + b$, cannot be divided evenly among the five students. This can be recorded as follows:

$$\text{If } 5 \mid a \text{ and } 5 \nmid b, \text{ then } 5 \nmid (a + b).$$

NOW TRY THIS 4-7

What, if anything, can you conclude if $5 \nmid a$ and $5 \nmid b$?

Since subtraction can be defined in terms of addition, results similar to addition hold for subtraction. These ideas are listed in Theorem 4-2.

Theorem 4-2

For any integers a, b, and d, the following holds:

a. If $d \mid a$ and $d \mid b$, then $d \mid (a + b)$.
b. If $d \mid a$ and $d \nmid b$, then $d \nmid (a + b)$.
c. If $d \mid a$ and $d \mid b$, then $d \mid (a - b)$.
d. If $d \mid a$ and $d \nmid b$, then $d \nmid (a - b)$.

The proofs of most theorems in this section are left as exercises, but the proof of Theorem 4-2(a) is given as an illustration.

Proof Theorem 4-2(a) is equivalent to the following:

If a is a multiple of d and b is a multiple of d, then $a + b$ is a multiple of d.

Notice that "a is a multiple of d" means $a = m \cdot d$, for some integer m. Similarly "b is a multiple of d" means $b = n \cdot d$, for some integer n. To show that $a + b$ is a multiple of d, we add these equations as follows:

$$a + b = md + nd$$

Is $md + nd$ a multiple of d? Notice that $md + nd = (m + n)d$, so $a + b = (m + n)d$. By the closure property of integer addition, $m + n$ is an integer. Also, $a + b$ is a multiple of d and therefore $d \mid (a + b)$.

EXAMPLE 4-18

Classify each of the following as true or false, where x, y, and z are integers. If a statement is true, prove it. If a statement is false, provide a counterexample.

a. If $3 \mid x$ and $3 \mid y$, then $3 \mid xy$.
b. If $3 \mid (x + y)$, then $3 \mid x$ and $3 \mid y$.
c. If $9 \nmid a$, then $3 \nmid a$.

Solution **a.** True. By Theorem 4-1, if $3|x$, then, for any integer y, $3|yx$ or $3|xy$.
b. False; for example, $3|(7 + 2)$, but $3 \nmid 7$ and $3 \nmid 2$.
c. False; for example, $9 \nmid 21$, but $3|21$.

NOW TRY THIS 4-8

If $3|x$, is it true that $3|xy$ regardless of whether $3|y$ or $3 \nmid y$? Why?

EXAMPLE 4-19

FIGURE 4-17

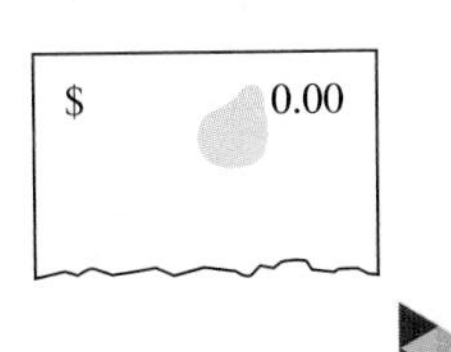

Five students found a padlocked money box that had a deposit slip attached to it. The deposit slip was water-spotted, so the currency total appeared as shown in Figure 4-17. One student remarked that if the money listed on the deposit slip was in the box, it could easily be divided equally among the five students without using coins. How did the student know this?

Solution Because the units digit of the amount of the currency is 0, the solution to the problem is to determine whether all natural numbers whose units digit is 0 are divisible by 5. To solve this problem, *look for a pattern.* Natural numbers whose units digit is 0 form a pattern, that is, 10, 20, 30, 40, 50, ... These numbers are multiples of 10. We are to determine whether 5 divides all multiples of 10. Since $5|10$, by Theorem 4-1, 5 divides any multiple of 10. Hence, 5 divides the amount of money in the box, and the student is correct.

Divisibility Rules

As shown in Example 4-19, sometimes it is handy to know if one number is divisible by another just by looking at it or by performing a simple test. We discovered that if a number ends in 0, then the number is divisible by 5. The same argument can be used to show that if a number ends in 5, it is divisible by 5. This is an example of a divisibility rule. Moreover, if the last digit of a number is neither 0 nor 5, then the number is not divisible by 5.

Elementary texts frequently state divisibility rules. However, such rules have limited use except for mental arithmetic. It is possible to determine whether 1734 is divisible by 17, either by using pencil and paper or a calculator. To check divisibility and avoid decimals, we can use a calculator with an integer division button, [INT÷]. On such a calculator, integer division can be performed using the following sequence of buttons:

[1] [7] [3] [4] [INT÷] [1] [7] [=]

to obtain the display $\underbrace{102}_{Q}$ $\underbrace{0}_{R}$.

This implies $1734/17 = 102$ with a remainder of 0, which, in turn, implies $17|1734$.

We could have determined this same result mentally by considering the following:

$$1734 = 1700 + 34$$

Because $17|1700$ and $17|34$, by Theorem 4-2(a), we have $17|(1700 + 34)$, or $17|1734$. Similarly, we could determine mentally that $17 \nmid 1735$.

REMARK Notice that $17|1734$ implies $17|(^-1)1734$, that is, $17|^-1734$. In general, $d|^-a$ if and only if $d|a$.

Divisibility Tests for 2, 5, and 10

To determine mentally whether a given integer n is divisible by another integer d, we think of n as the sum or difference of two integers where d divides at least one of these numbers. We try to choose numbers such that one of them is close to n and divisible by d and the other number is relatively small. As an example, consider the divisibility of 358 by 2:

$$\begin{aligned} 358 &= 350 + 8 \\ &= 35(10) + 8 \end{aligned}$$

We know that $2 \mid 10$, so that $2 \mid 35(10)$. We also know that $2 \mid 8$, which tells us that $2 \mid (35(10) + 8)$. Because $2 \mid 10$, 2 divides any multiple of 10, so to determine the divisibility of any integer by 2, we consider only whether the units digit is divisible by 2. If it is, then by Theorem 4-2(a) the number is divisible by 2. If not, then by Theorem 4-2(b) the number is not divisible by 2.

We can develop a similar test for divisibility by 10. In general, we have the following divisibility rules.

Divisibility Test for 2

An integer is divisible by 2 if and only if its units digit is divisible by 2.

Divisibility Test for 5

An integer is divisible by 5 if and only if its units digit is divisible by 5, that is if and only if the units digit is 0 or 5.

Divisibility Test for 10

An integer is divisible by 10 if and only if its units digit is divisible by 10, that is if and only if the units digit is 0.

Divisibility Tests for 4 and 8

When we consider divisibility rules for 4 and 8, we see that $4 \nmid 10$ and $8 \nmid 10$, so it is not a matter of checking the units digit for divisibility by 4 and 8. However, 4 (which is 2^2) divides 10^2, and 8 (which is 2^3) divides 10^3.

We first develop a divisibility rule for 4. Consider any four-digit number n such that $n = a \cdot 10^3 + b \cdot 10^2 + c \cdot 10 + d$. Our *subgoal* is to *write the given number as a sum of two numbers*, one of which is as great as possible and divisible by 4. We know that $4 \mid 10^2$ because $10^2 = 4 \cdot 25$ and, consequently, $4 \mid 10^3$. Because $4 \mid 10^2$, then $4 \mid b \cdot 10^2$; and because $4 \mid 10^3$, then $4 \mid a \cdot 10^3$. Finally, $4 \mid a \cdot 10^3$ and $4 \mid b \cdot 10^2$ imply $4 \mid (a \cdot 10^3 + b \cdot 10^2)$. Now the divisibility of $a \cdot 10^3 + b \cdot 10^2 + c \cdot 10 + d$ by 4 depends on the divisibility of $(c \cdot 10 + d)$ by 4. Notice that $c \cdot 10 + d$ is the number represented by the last two digits in the given number n. We summarize this in the following test.

Divisibility Test for 4

An integer is divisible by 4 if and only if the last two digits of the integer represent a number divisible by 4.

To investigate divisibility by 8, we note that the least positive power of 10 divisible by 8 is 10^3 since $10^3 = 8 \cdot 125$. Consequently, all integral powers of 10 greater than 10^3 also are divisible by 8. Hence, the following is a divisibility test for 8:

Divisibility Test for 8

An integer is divisible by 8 if and only if the last three digits of the integer represent a number divisible by 8.

EXAMPLE 4-20

a. Determine whether 97,128 is divisible by 2, 4, and 8.
b. Determine whether 83,026 is divisible by 2, 4, and 8.

Solution **a.** $2 \mid 97{,}128$ because $2 \mid 8$.
$4 \mid 97{,}128$ because $4 \mid 28$.
$8 \mid 97{,}128$ because $8 \mid 128$.

b. $2 \mid 83{,}026$ because $2 \mid 6$.
$4 \nmid 83{,}026$ because $4 \nmid 26$.
$8 \nmid 83{,}026$ because $8 \nmid 026$.

REMARK In Example 4-20(a), it would have been sufficient to check that the given number is divisible by 8 because if $8 \mid a$, then $2 \mid a$ and $4 \mid a$. Why? This relationship is shown in Figure 4-18.

FIGURE 4-18

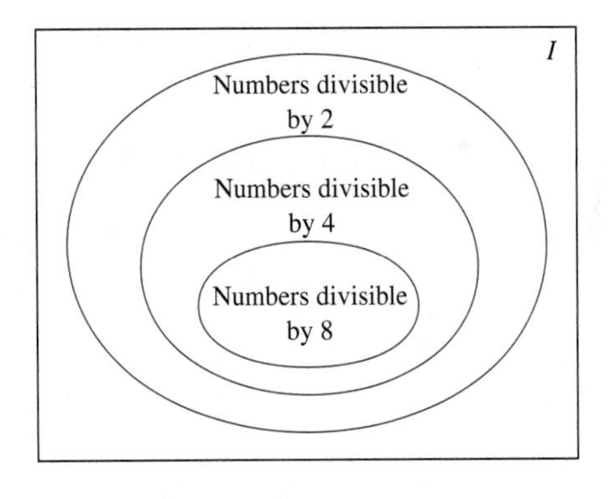

Notice that if $8 \nmid a$, we cannot conclude from this that $4 \nmid a$ or $2 \nmid a$. Why?

Divisibility Tests for 3 and 9

Next, we consider a divisibility test for 3. No power of 10 is divisible by 3, but the numbers 9, and 99, and 999, and others of this type are close to powers of 10 and are divisible by 3. For example, to determine whether 5721 is divisible by 3, we rewrite the number using 999, 99, and 9, as follows:

$$\begin{aligned} 5721 &= 5 \cdot 10^3 + 7 \cdot 10^2 + 2 \cdot 10 + 1 \\ &= 5(999 + 1) + 7(99 + 1) + 2(9 + 1) + 1 \\ &= 5 \cdot 999 + 5 \cdot 1 + 7 \cdot 99 + 7 \cdot 1 + 2 \cdot 9 + 2 + 1 \\ &= (5 \cdot 999 + 7 \cdot 99 + 2 \cdot 9) + (5 + 7 + 2 + 1) \end{aligned}$$

The sum in the first set of parentheses in the last line is divisible by 3, so the divisibility of 5721 by 3 depends on the sum in the second set of parentheses. In this case, $5 + 7 + 2 + 1 = 15$ and $3 | 15$, so $3 | 5721$. Hence, to test 5721 for divisibility by 3, we test $5 + 7 + 2 + 1$ for divisibility by 3. Notice that $5 + 7 + 2 + 1$ is the sum of the digits of 5721. The example suggests the following test for divisibility by 3:

Divisibility Test for 3

An integer is divisible by 3 if and only if the sum of its digits is divisible by 3.

We can use an argument similar to the one used to demonstrate that $3 | 5721$ to prove the test for divisibility by 3 on any integer and in particular for any four-digit number $n = a10^3 + b10^2 + c10 + d$. Even though $a10^3 + b10^2 + c10 + d$ is not necessarily divisible by 3, the number $a999 + b99 + c9$ is close to n and *is* divisible by 3. We have the following:

$$\begin{aligned} a10^3 + b10^2 + c10 + d &= a1000 + b100 + c10 + d \\ &= a(999 + 1) + b(99 + 1) + c(9 + 1) + d \\ &= (a999 + b99 + c9) + (a1 + b1 + c1 + d) \\ &= (a999 + b99 + c9) + (a + b + c + d) \end{aligned}$$

Because $3 | 9$, $3 | 99$, and $3 | 999$, it follows that $3 | (a999 + b99 + c9)$. If $3 | (a + b + c + d)$, then $3 | [(a999 + b99 + c9) + (a + b + c + d)]$; that is, $3 | n$. If, on the other hand, $3 \nmid (a + b + c + d)$, it follows from Theorem 4-2(b) that $3 \nmid n$.

Since $9 | 9$, $9 | 99$, $9 | 999$, and so on, a test similar to that for divisibility by 3 applies to divisibility by 9. Why?

Divisibility Test for 9

An integer is divisible by 9 if and only if the sum of the digits of the integer is divisible by 9.

EXAMPLE 4-21

Use divisibility tests to determine whether each of the following numbers is divisible by 3 and divisible by 9:

a. 1002 **b.** 14,238

Solution **a.** Because $1 + 0 + 0 + 2 = 3$ and $3 | 3$, it follows that $3 | 1002$. Because $9 \nmid 3$, it follows that $9 \nmid 1002$.

b. Because $1 + 4 + 2 + 3 + 8 = 18$ and $3 | 18$, it follows that $3 | 14{,}238$. Because $9 | 18$, it follows that $9 | 14{,}238$.

▲

EXAMPLE 4-22

The store manager has an invoice for 72 four-function calculators. The first and last digits on the receipt are illegible. The manager can read

$■67.9■

What are the missing digits, and what is the cost of each calculator?

Solution Let the missing digits be x and y so that the number is $x67.9y$ dollars, or $x679y$ cents. Because there were 72 calculators sold, the number on the invoice must be divisible by 72. Because the number is divisible by 72 and $72 = 8 \cdot 9$, it must be divisible by 8 and 9, which are factors of 72. For the number on the invoice to be divisible by 8, the three-digit number $79y$ must be divisible by 8. Because $79y$ must be divisible by 8, it is an even number. Therefore, $79y$ must be either 790, 792, 794, 796, or 798. Only the number 792 is divisible by 8, so we know the last digit, y, on the invoice must be 2.

Because the number on the invoice must be divisible by 9, we know that 9 must divide $x + 6 + 7 + 9 + 2$, or $(x + 24)$. Since 3 is the only single digit that will make $(x + 24)$ divisible by 9, then x must be 3. Therefore, the number on the invoice must be \$367.92. The calculators must cost \$367.92/72, or \$5.11, each.

▲

Divisibility Tests for 11 and 6

The divisibility test for 7 is usually harder to use than actually performing the division, so we omit the test. We state the divisibility test for 11 but omit the proof. Interested readers might try to find a proof.

Divisibility Test for 11

An integer is divisible by 11 if and only if the sum of the digits in the places that are even powers of 10 minus the sum of the digits in the places that are odd powers of 10 is divisible by 11.

For example, to test whether 8,471,986 is divisible by 11, we check whether 11 divides the difference $(6 + 9 + 7 + 8) - (8 + 1 + 4)$, or 17. Because $11 \nmid 17$, it follows from the divisibility test for 11 that $11 \nmid 8{,}471{,}986$. A number like 2772 is divisible by 11 because $(2 + 7) - (7 + 2) = 9 - 9 = 0$ and 0 is divisible by 11.

The divisibility test for 6 is related to the divisibility tests for 2 and 3. In Section 4-4, we will ask you to show that if $2|n$ and $3|n$, then $(2 \cdot 3)|n$, and in general: if a and c have no factors in common, then if $a|b$ and $c|b$ we can conclude that $ac|b$. Consequently, the following divisibility test is true.

Divisibility Test for 6

An integer is divisible by 6 if and only if the integer is divisible by both 2 and 3.

Divisibility tests for other numbers are explored in the Ongoing Assessment 4-3.

The number 57,729,364,583 has too many digits for most calculator displays. Determine whether it is divisible by each of the following:

a. 2 **b.** 3 **c.** 5 **d.** 6 **e.** 8 **f.** 9 **g.** 10 **h.** 11

Solution
a. No, the last digit, 3, is not divisible by 2.
b. No, the sum of the digits is 59, which is not divisible by 3.
c. No, the last digit is neither 0 nor 5.

e. No, because the number formed by the last three digits, 583, is not divisible by 8.
f. No, because the sum of the digits is 59, which is not divisible by 9.
g. No, because the units digit is not 0.
h. Yes, because $(3 + 5 + 6 + 9 + 7 + 5) - (8 + 4 + 3 + 2 + 7) = 35 - 24 = 11$ and 11 is divisible by 11.

▲

NOW TRY THIS 4-9

Fill in the following blanks so that the number is divisible by 9. List all possibilities.

12,506,5_ _.

PROBLEM SOLVING **A Mistake in the Inventory**

A class from Washington School visited a neighborhood cannery warehouse. The warehouse manager told the class that there were 11,368 cans of juice in the inventory and that the cans were packed in boxes of 6 or 24, depending on the size of the can. One of the students, Sam, thought for a moment and announced that there was a mistake in the inventory. Is Sam's statement correct? Why or why not?

Understanding the Problem The problem is to determine if the manager's inventory of 11,368 cans was correct. To solve the problem, we must assume there are no partial boxes of cans; that is, a box must contain exactly 6 or exactly 24 cans of juice.

Devising a Plan We know that the boxes contain either 6 cans or 24 cans, but we do not know how many boxes of each type there are. One strategy for solving this problem is to *find an equation* that involves the total number of cans in all the boxes.

The total number of cans, 11,368, equals the number of cans in all the 6-can boxes plus the number of cans in all the 24-can boxes. If there are n boxes containing 6 cans each, there are $6n$ cans altogether in those boxes. Similarly, if there are m boxes with 24 cans each, those boxes contain a total of $24m$ cans. Because the total was reported to be 11,368 cans, we have the equation $6n + 24m = 11{,}368$. Sam claimed that $6n + 24m \neq 11{,}368$.

One way to show that $6n + 24m \neq 11{,}368$ is to show that $6n + 24m$ and 11,368 do not have the same divisors. Both $6n$ and $24m$ are divisible by 6. This implies that $6n + 24m$ must be divisible by 6. If 11,368 is not divisible by 6, then Sam is correct.

Carrying Out the Plan The divisibility test for 6 states that a number is divisible by 6 if and only if the number is divisible by both 2 and 3. Because 11,368 is an even number, it is divisible by 2. Is it divisible by 3?

The divisibility test for 3 states that a number is divisible by 3 if and only if the sum of the digits in the number is divisible by 3. We see that $1 + 1 + 3 + 6 + 8 = 19$, which is not divisible by 3, so 11,368 is not divisible by 3. Hence, Sam is correct.

Looking Back Suppose 11,368 had been divisible by 6. Would that have implied that the manager was correct? The answer is no; it would have implied only that we would have to change our approach to the problem.

As a further Looking Back activity, suppose that, given different data, the manager is correct. Can we determine values for m and n? In fact, this can be done. If a computer is available, a program can be written to determine all possible natural-number values of m and n.

RESEARCH NOTE

Teachers' emphasis on specific problem solving heuristics (for example, drawing a diagram, constructing a chart, working backwards) as an integral part of instruction does significantly impact their students' problem-solving performance. Students who received such instruction made more effective use of these problem-solving behaviors in new situations when compared to students not receiving such instruction (Vos 1976; Suydam 1987).

HISTORICAL NOTE

A modern mathematician who worked in the area of number theory was American Julia Robinson (1919–1985). Robinson's work with the Russian mathematician Yuri Matijasevic on Diophantine equations led directly to the solution of the 10th of the famous set of 23 problems the German mathematician David Hilbert posed. Robinson was the first woman mathematician to be elected to the National Academy of Sciences and the first woman president of the American Mathematical Society. She died of leukemia at the age of 65.

BRAIN TEASER

The following is an argument to show that an ant weighs as much as an elephant. What is wrong?

Let e be the weight of the elephant and a the weight of the ant. Let $e - a = d$. Consequently, $e = a + d$. Multiply each side of $e = a + d$ by $e - a$. Then simplify.

$$\begin{aligned} e(e - a) &= (a + d)(e - a) \\ e^2 - ea &= ae + de - a^2 - da \\ e^2 - ea - de &= ae - a^2 - da \\ e(e - a - d) &= a(e - a - d) \\ e &= a \end{aligned}$$

Thus, the weight of the elephant equals the weight of the ant.

ONGOING ASSESSMENT 4-3

1. Classify each of the following as true or false. If false, tell why.
   a. 6 is a factor of 30.
   b. 6 is a divisor of 30.
   c. $6|30$.
   d. 30 is divisible by 6.
   e. 30 is a multiple of 6.
   f. 6 is a multiple of 30.
2. Using divisibility tests, answer each of the following:
   a. There are 1379 children signed up to play in a baseball league. If exactly 9 players are to be placed on each team, will any team be short of players?
   b. A forester has 43,682 seedlings to be planted. Can these be planted in an equal number of rows with 11 seedlings in each row?
   c. There are 261 students to be assigned to 9 teachers so that each teacher has the same number of students. Is this possible?
   d. Six friends win with a lottery ticket. The payoff is $242,800. Can the money be divided evenly?
   e. Jack owes $7812 on a new car. Can this amount be paid in 12 equal monthly installments?
3. Without using a calculator, test each of the following numbers for divisibility by 2, 3, 4, 5, 6, 8, 9, 10, and 11:
   a. 746,988
   b. 81,342
   c. 15,810
   d. 4,201,012
   e. 1001
   f. 10,001
4. Determine each of the following without actually performing the division. Explain how you did it in each case.
   a. Is 34,015 divisible by 17?
   b. Is 34,051 divisible by 17?
   c. Is 19,031 divisible by 19?
   d. Is $2 \cdot 3 \cdot 5 \cdot 7$ divisible by 5?
   e. Is $(2 \cdot 3 \cdot 5 \cdot 7) + 1$ divisible by 5?

5. Justify each of the following using some of the theorems in this section.
   a. $a^3|a^4$, if $a \neq 0$.
   b. $a^4|a^{10}$, if $a \neq 0$.
   c. $a^n|a^m$, if $0 \leq n \leq m$, if $a \neq 0$.
   d. If $b|a$ and $c \neq 0$, then $bc|ac$.
6. Justify each of the given statements, assuming that *a*, *b*, and *c* are integers. If a statement cannot be justified by one of the theorems in this section, answer "none."
   a. $4|20$ implies $4|113 \cdot 20$.
   b. $4|100$ and $4 \nmid 13$ imply $4 \nmid (100 + 13)$.
   c. $4|100$ and $4 \nmid 13$ imply $4 \nmid 1300$.
   d. $3|(a + b)$ and $3 \nmid c$ imply $3 \nmid (a + b + c)$.
   e. $3|a$ implies $3|a^2$.
7. Classify each of the following as always true, sometimes true, or never true. Justify your answers.
   a. If $b|a$, then $b + c|a + c$.
   b. If $b|a$, then $b^2|a^3$.
   c. If $b|a$, then $b|(^-a)$ and $(-b)|(-a)$.
8. Use the definition of "divides" or Theorem 4-2 to justify each of the following:
   a. $7|210$
   b. $19|(1900 + 38)$
   c. $6|2^3 \cdot 3^2 \cdot 17^4$
   d. $7 \nmid (4200 + 22)$
   e. $26|13^4 \cdot 100$
   f. $13 \nmid (2^4 \cdot 5^3 \cdot 26 + 1)$
   g. $2^4 \nmid (2 \cdot 4 \cdot 6 \cdot 8 \cdot 17^{10} + 1)$
   h. $2^4|(10^4 + 6^4)$
9. Classify each of the following as true or false:
   a. If every digit of a number is divisible by 3, the number itself is divisible by 3.
   b. If a number is divisible by 3, then every digit of the number is divisible by 3.
   c. A number is divisible by 3 if and only if every digit of the number is divisible by 3.
   d. If a number is divisible by 6, then it is divisible by 2 and by 3.
   e. If a number is divisible by 2 and 3, then it is divisible by 6.
   f. If a number is divisible by 2 and 4, then it is divisible by 8.
   g. If a number is divisible by 8, then it is divisible by 2 and 4.
10. Classify each of the statements in problem 9 as sometimes, always, or never true.
11. Devise a test for divisibility by each of the following numbers:
    a. 16
    b. 25
12. When the two missing digits in the following number are replaced, the number is divisible by 99. What is the number?

    85__ __1

13. Fill each of the following blanks with the greatest digit that makes the statement true:
    a. 3|74_
    b. 9|83_45
    c. 11|6_55
14. Place a digit in the square, if possible, so that the number

    527,4□2

    is divisible by
    a. 2 b. 3 c. 4 d. 9 e. 11
15. Without multiplying or using a calculator, classify each of the following as true or false. Justify your answers.
    a. $7|280021$
    b. $19 \nmid 3,800,018$
    c. $23|46^{10}$
    d. $23 \nmid 460,046$
16. The bookstore marked some notepads down from \$2.00 but still kept the price over \$1.03. It sold all of them. The total amount of money from the sale of the pads was \$31.45. How many notepads were sold?
17. A group of people ordered No-Cal candy bars. The bill was \$2.09. If the original price of each was 12¢ but the price has been inflated, how much does each cost?
18. Leap years occur in years that are divisible by 4. However, if the year ends in two zeros, in order for the year to be a leap year, it must be divisible by 400. Determine which of the following are leap years:
    a. 1776 b. 1986 c. 2000 d. 2100
19. In a football game, a touchdown with an extra point is worth 7 points and a field goal is worth 3 points. Suppose that in a game the only scoring done by teams are touchdowns with extra points and field goals.
    a. Which of the scores 1 to 25 are impossible for a team to score?
    b. List all possible ways for a team to score 40 points.
    c. A team scored 57 points with 6 touchdowns and 6 extra points. How many field goals did the team score?
20. Complete the following table where *n* is the given integer.

	n	Remainder When *n* is Divided by 9	Sum of the Digits of *n*	Remainder When the Sum of the Digits of *n* Is Divided by 9
a.	31			
b.	143			
c.	345			
d.	2987			
e.	7652			

    f. Make a conjecture about the remainder and the sum of the digits in an integer when it is divided by 9.
21. A test for checking computations is called *casting out nines*. Consider the sum $193 + 24 + 786 = 1003$. The remainders when 193, 24, and 786 are divided by 9 are 4, 6, and 3, respectively. The sum of the remainders, 13, has a remainder of 4 when divided by 9, as does 1003.

Checking the remainders in this manner provides a quasi-check for the computation. Find the following sums and use casting out nines to check your sums:

a. 12,343 + 4546 + 56
b. 987 + 456 + 8765
c. 10,034 + 3004 + 400 + 20
d. Will this check always work for addition? Give an example to illustrate your answer.
e. Try the check on the subtraction 1003 − 46.
f. Try the check on the multiplication $345 \cdot 56$.
g. Would it make sense to try the check on division? Why or why not?

22. **a.** If 21 divides n, what other natural numbers divide n? Why?
b. If 16 divides n, what other natural numbers divide n? Why?

23. A palindrome is a number that reads the same forward as backward.
a. Check the following four-digit palindromes for divisibility by 11:
i. 4554 ii. 9339 iii. 2002 iv. 2222
b. Are all four-digit palindromes divisible by 11? Why or why not?
c. Are all five-digit palindromes divisible by 11? Why or why not?
d. Are all six-digit palindromes divisible by 11? Why or why not?

24. The numbers 5872 and 2785 are a palindromic pair of numbers because reversing the order of the digits of one number gives the other number. Explain why in a palindromic pair, if one number is divisible by 3, then so is the other.

25. The numbers x and y are divisible by 5.
a. Is the sum of x and y divisible by 5? Why?
b. Is the difference of x and y divisible by 5? Why?
c. Is the product of x and y divisible by 5? Why?
d. Is the quotient of x and y divisible by 5? Why?

26. Using only divisibility tests, explain whether 6,868,395 is divisible by 15.

27. Classify each of the following as true or false, assuming that a, b, c, and d are integers and $d \neq 0$. If a statement is false, give a counterexample.
a. If $d|(a + b)$, then $d|a$ and $d|b$.
b. If $d|(a + b)$, then $d|a$ or $d|b$.
c. If $d|ab$, then $d|a$ or $d|b$.
d. If $ab|c$, $a \neq 0$, and $b \neq 0$, then $a|c$ and $b|c$.
e. If $a|b$ and $b|a$, then $a = b$.
f. If $d|a$ and $d|b$, then $d|(ax + by)$ for any integers x and y.
g. If $d \nmid a$ and $d \nmid b$, then $d \nmid (a + b)$.
h. If $d|a^2$, then $d|a$.
i. If $d \nmid a$, then $d \nmid a^2$.

★**28.** Prove Theorem 4-2(b).

29. Prove the test for divisibility by 9 for any five-digit number.

30. **a.** Choose a two-digit number such that the number in the tens place is 1 greater than the number in the units place. Reverse the digits in your number, and subtract this number from your original number, for example, 87 − 78 = 9. Make a conjecture concerning the results of performing these kinds of operations.
b. Choose any two-digit number such that the number in the tens place is 2 greater than the number in the units place. Reverse the digits in your number, and subtract this number from your original number; for example, 31 − 13 = 18. Make a conjecture concerning the results of performing these kinds of operations.
★**c.** Prove that for any two-digit number, if the digits are reversed and the numbers subtracted, the difference is a multiple of 9.
d. Investigate what happens whenever two-digit numbers with equal digit sums are subtracted; for example, 62 − 35 = 27.

Communication

31. A customer wants to mail a package. The postal clerk determines the cost of the package to be \$18.95, but only 6¢ and 9¢ stamps are available. Can the available stamps be used for the exact amount of postage for the package? Why or why not?

32. **a.** Jim uses his calculator to see if a number n having eight or fewer digits is divisible by a number d. He finds that $n \div d$ has a display of 32. Does $d|n$? Why?
b. If $n \div d$ gives a display of 16.8, does $d|n$? Why?

33. Is the area of each of the following rectangles divisible by 4? Explain why or why not.

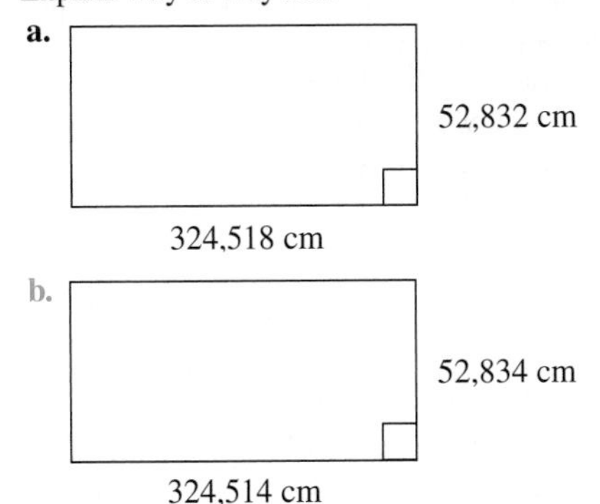

34. Can you find three consecutive natural numbers none of which is divisible by 3? Explain your answer.

35. Answer each of the following and justify your answers.
a. If a number is not divisible by 5, can it be divisible by 10?
b. If a number is not divisible by 10, can it be divisible by 5?

36. A number in which each digit except 0 appears exactly 3 times is divisible by 3. For example, 777,555,222 and 414,143,313 are divisible by 3. Explain why this statement is true.

37. Enter any three-digit number on the calculator; for example, enter 243. Repeat it: 243,243. Divide by 7. Divide by 11. Divide by 13. What is the answer? Try it again with any other three-digit number. Will this always work? Why?

38. Alexa claims that she can justify the divisibility test by 11. She says: *I noticed that each even power of 10 can be written as a multiple of 11 plus 1 and every odd power of 10 can be written as a multiple of 11 minus 1. In fact:*

$$\begin{aligned}10 &= 11 - 1\\ 10^2 &= 99 + 1 = 9 \cdot 11 + 1\\ 10^3 &= 10 \cdot 10^2 = 10(9 \cdot 11 + 1) = 90 \cdot 11 + 10\\ &= 90 \cdot 11 + 11 - 1 = 91 \cdot 11 - 1\\ 10^4 &= 10^2 \cdot 10^2 = 100(9 \cdot 11 + 1)\\ &= 900 \cdot 11 + 9 \cdot 11 + 1 = 909 \cdot 11 + 1\end{aligned}$$

and so on.

Now I look at a four-digit number abcd and proceed as in the divisibility by 3. I collect the parts that are divisible by 11 regardless of what the digits are and put together the rest, which is

$$d - c + b - a$$

Complete the details of Alexa's argument and justify the test for divisibility by 11.

39. Take a number written in base ten with three or more digits and subtract from it the indicated expression. By what numbers can you be sure that the difference is divisible? Justify your answers.
   a. The unit digit
   b. The number formed by the last two digits (that is, the 10 digit followed by the unit digit).
   c. The sum of the digits
   d. Answer the preceding questions for a three- or more digit number written in base five.

40. a. In what bases will divisibility by 2 depend only on the last digit? Justify your answer.
   b. In what bases will divisibility by 2 depend only on the sum of the digits being even or odd? Justify your answer.

Open-Ended

41. A breakfast-food company had a contest in which numbers were placed in breakfast-food boxes. A prize of $1000 was awarded to anyone who could collect numbers whose sum was 100. The company had thousands of cards made with the following numbers on them:

3 12 15 18 27 33 45 51 66 75 84 90

   a. If the company did not make any more cards, is there a winning combination?
   b. If the company is going to add one more number to the list and it wants to make sure the contest has at most 1000 winners, suggest a strategy for it to use.

42. How would you use concrete materials to explain to young children the following:
   a. A number being even or odd
   b. A number being divisible by 3 or not being divisible by 3
   c. That if $4|a$, then $2|a$

Cooperative Learning

43. In your group, discuss the value of teaching various divisibility tests in middle school. If a teacher decides to discuss the various tests, how should they be introduced?

Review Problems

44. Find all integers x (if possible) that make each of the following true:
   a. $3(^-x) = 6$
   b. $(^-2)|x| = 6$
   c. $(^-x) \div 0 = {}^-1$
   d. $^-(x - 1) = 1 - x$
   e. $^-|^-x| = 5$
   f. $^-x < 0$

45. Simplify each of the following:
   a. $3x - (1 - 2x)$
   b. $(^-2x)^2 - 3x^2$
   c. $y - x - 2(y - x)$
   d. $(x - 1)^2 - x^2 + 2x$

46. Consider the function $y = f(x) = {}^-2x - 3$ whose domain is the set of integers and answer the following questions:
   a. Find $f(^-5)$.
   b. For what value of x will the value of y be 17?
   c. Is 2 a possible output? Explain your answer.
   d. Find all possible outputs.

47. Jerry decides to start a juice bar. He figures out that his start-up cost will be $1800 (to buy the equipment and pay for permits.) If he charges $2 for each cup of juice, answer the following:
   a. What will his profit (or loss) be after he sells 700 cups of juice?
   b. If he sells x cups of juice and his profit is $P(x)$, write an equation that will show $P(x)$ as a function of x (in terms of x).
   c. What is Jerry's *break-even point;* that is, how many cups of juice must he sell before he makes a profit?

BRAIN TEASER

Dee finds that she has an extraordinary Social Security number. Its nine digits contain all the numbers from 1 through 9. They also form a number with the following characteristics: When read from left to right, its first two digits form a number divisible by 2, its first three digits form a number divisible by 3, its first four digits form a number divisible by 4, and so on, until the complete number is divisible by 9. What is Dee's Social Security number?

4-4 Prime and Composite Numbers

One method used in elementary schools to determine the positive divisors of a number is to use squares of paper and to represent the number as a rectangle. Such a rectangle resembles a candy bar formed with small squares. The dimensions of the rectangle are divisors of the number. For example, Figure 4-19 shows rectangles to represent 12.

FIGURE 4-19

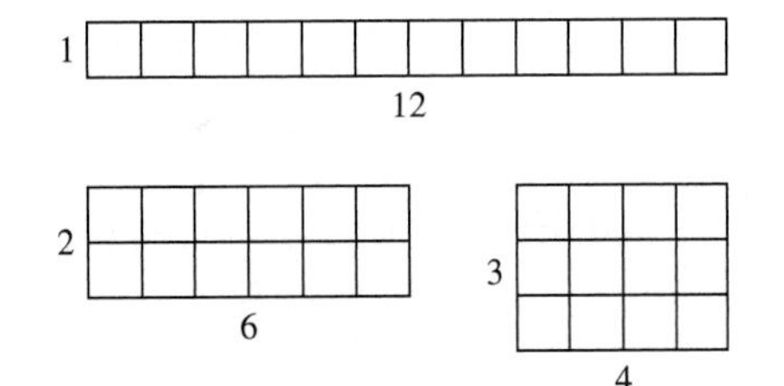

FIGURE 4-20

As the figure shows, the number 12 has six positive divisors: 1, 2, 3, 4, 6, and 12. If rectangles were used to find the divisors of 7, then we would find only a 1×7 rectangle, as Figure 4-20 shows. Thus, 7 has exactly two divisors: 1 and 7.

To illustrate further the number of divisors of a number, we construct Table 4-1. Below each number listed across the top, we identify numbers less than or equal to 37 that have that number of positive divisors. For example, 12 is in the 6 column because it has six positive divisors, and 7 is in the 2 column because it has only two positive divisors.

Table 4-1 Number of Factors

1	2	3	4	5	6	7	8	9
1	2	4	6	16	12		24	36
	3	9	8		18		30	
	5	25	10		20			
	7		14		28			
	11		15		32			
	13		21					
	17		22					
	19		26					
	23		27					
	29		33					
	31		34					
	37		35					

NOW TRY THIS 4-10

a. What patterns do you see forming in Table 4-1?
b. Will there be other entries in the 1 column? Why?
c. What are the next three numbers in the 3 column?
d. Find an entry for the 7 column.
e. What kinds of numbers have an odd number of factors? Why?

The numbers in the 2 column in Table 4-1 are of particular importance. Notice that they have exactly two positive divisors, namely, 1 and themselves. Any positive integer with exactly two distinct, positive divisors is a *prime number*, or a **prime**. Any integer greater than 1 that has a positive factor other than 1 and itself is a *composite number*, or a **composite**. For example, 4, 6, and 16 are composites because they have positive factors other than 1 and themselves. The number 1 has only one positive factor, so it is neither prime nor composite. From the 2 column in Table 4-1, we see that the first 12 primes are 2, 3, 5, 7, 11, 13, 17, 19, 23, 29, 31, and 37. Other patterns in the table are explored in the problem set.

prime

composite

EXAMPLE 4-24

Show that the following numbers are composite:

a. 1564 **b.** 2781 **c.** 1001 **d.** $3 \cdot 5 \cdot 7 \cdot 11 \cdot 13 + 1$

Solution

a. Since $2|4$, 1564 is divisible by 2 and is composite.

b. Since $3|(2 + 7 + 8 + 1)$, 2781 is divisible by 3 and is composite.

c. Since $11|[(1 + 0) - (0 + 1)]$, 1001 is divisible by 11 and is composite.

d. Because a product of odd numbers is odd (why?), $3 \cdot 5 \cdot 7 \cdot 11 \cdot 13$ is odd. If we add 1 to an odd number, the sum is even. An even number (other than 2) has a factor of 2 and is therefore composite.

Prime Factorization

Composite numbers can be expressed as products of two or more whole numbers greater than 1. For example, $18 = 2 \cdot 9$, $18 = 3 \cdot 6$, or $18 = 2 \cdot 3 \cdot 3$. Each expression of 18 as a product of factors is a **factorization**.

factorization

prime factorization

A factorization containing only prime numbers is a **prime factorization**. To find a prime factorization of a given composite number, first rewrite the number as a product of two smaller numbers greater than 1. Continue the process, factoring the lesser numbers until all factors are primes. For example, consider 260:

$$260 = 26 \cdot 10 = 2 \cdot 13 \cdot 2 \cdot 5 = 2 \cdot 2 \cdot 5 \cdot 13 = 2^2 \cdot 5 \cdot 13$$

factor tree

The procedure for finding a prime factorization of a number can be organized using a **factor tree**, as Figure 4-21(a) demonstrates. The last branches of the tree display the prime factors of 260. A second way to factor 260 is shown in Figure 4-21(b). The two trees produce the same prime factorization, except for the order in which the primes appear in the products.

The *Fundamental Theorem of Arithmetic*, or the *Unique Factorization Theorem*, states that in general, if order is disregarded, the prime factorization of a number is unique.

Theorem 4-3

Fundamental Theorem of Arithmetic Each composite number can be written as a product of primes in one and only one way except for the order of the prime factors in the product.

FIGURE 4-21

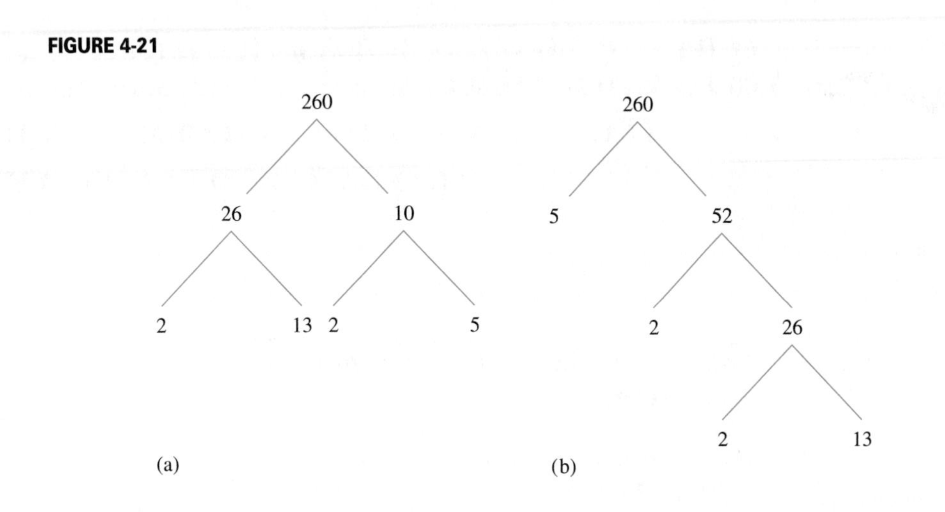

The Fundamental Theorem of Arithmetic assures us that once we find a prime factorization of a number, a different prime factorization of the same number cannot be found. For example, consider 260. We start with the least prime, 2, and see whether it divides 260. If not, we try the next greater prime and check for divisibility by this prime. Once we find a prime that divides the number in question, we must find the quotient of the number divided by the prime. This step in the prime factorization of 260 is shown in Figure 4-22(a). Next we check whether the prime divides the quotient. If so, we repeat the process; if not, we try the next greater prime, 3, and check to see if it divides the quotient. We see that 260 divided by 2 yields 130, as shown in Figure 4-22(b). We continue the procedure, using greater primes, until a quotient of 1 is reached. The original number is the product of all the prime divisors used. The complete procedure for 260 is shown in Figure 4-22(c). An alternative form is shown in Figure 4-22(d).

FIGURE 4-22

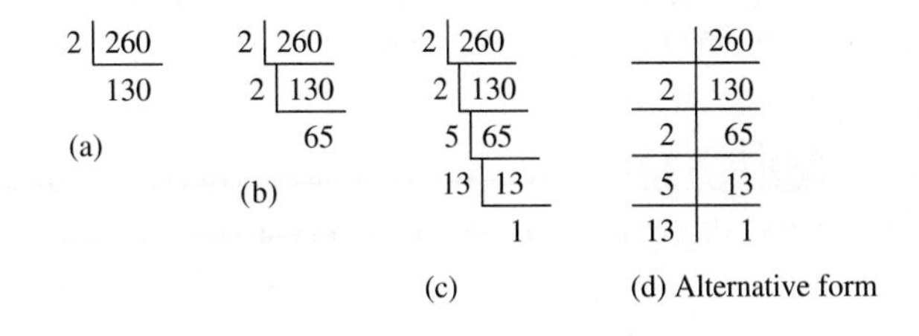

The primes in the prime factorization of a number are typically listed in increasing order from left to right, and if a prime appears in a product more than once, exponential notation is used. Thus, the factorization of 260 is written as $2^2 \cdot 5 \cdot 13$.

Prime factorization is demonstrated in the partial student page on page 262.

SAMPLE SCHOOL BOOK PAGE:

HOW CAN YOU WRITE A NUMBER AS A PRODUCT OF PRIME FACTORS?

How can you write a number as a product of prime factors?

Every composite number can be written as a product of prime factors. This product is called the **prime factorization** of a number. You can use a "factor tree" to find a prime factorization.

Example C

Write 18 as a product of prime factors.

First write 18 as a product of two factors. This can be done in more than one way. Then write each factor that is not prime as a product. Continue until all branches end in prime numbers.

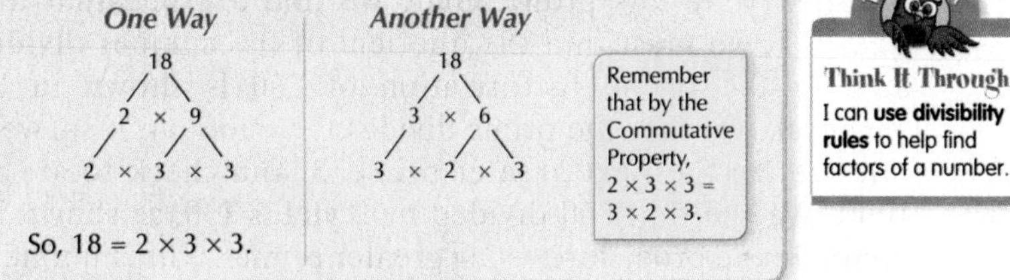

So, 18 = 2 × 3 × 3.

TEST TALK — Think It Through
I can **use divisibility rules** to help find factors of a number.

Talk About It

2. In the first factor tree for 18, why was the 2 written in the last row, but the 9 was not?

3. Do the first two factors you select for a number in a factor tree, such as 2 and 9, change the final result? Explain.

4. If the last row in a factor tree is 2 × 3 × 10, how do you know the factor tree is not complete?

Take It to the NET
More Examples
www.scottforesman.com

CHECK

For another example, see Set 3-11 on p. 194.

Write whether each number is prime or composite.

1. 13 **2.** 54 **3.** 23 **4.** 675 **5.** 41

Use factor trees to find the prime factorization of each number.

6. 15 **7.** 22 **8.** 28 **9.** 36 **10.** 40

11. Number Sense It is not always easy to tell if a number is prime. But some numbers are easily seen to be composite. How do you immediately know that 12,345,755 is composite?

Section B Lesson 3-11 165

Source: Scott Foresman–Addison Wesley Mathematics 2005, Grade 5 (p. 165).

NOW TRY THIS 4-11

Colored rods are used in the elementary-school classroom to teach many concepts. The rods vary in length from 1 cm to 10 cm. Various lengths have colors associated with them. For example, the 5 rod is yellow. Rods are shown in Figure 4-23 with their appropriate colors. A row with all the same color rods is called a *one-color train*.

a. What rods can be used to form a one-color train for 18?

b. What one-color trains are possible for 24?

c. How many one-color trains of two or more rods are possible for each prime number?

d. If a number can be represented by an all-red train, an all-green train, and an all-yellow train, what is the least number of factors it must have? What are they?

FIGURE 4-23

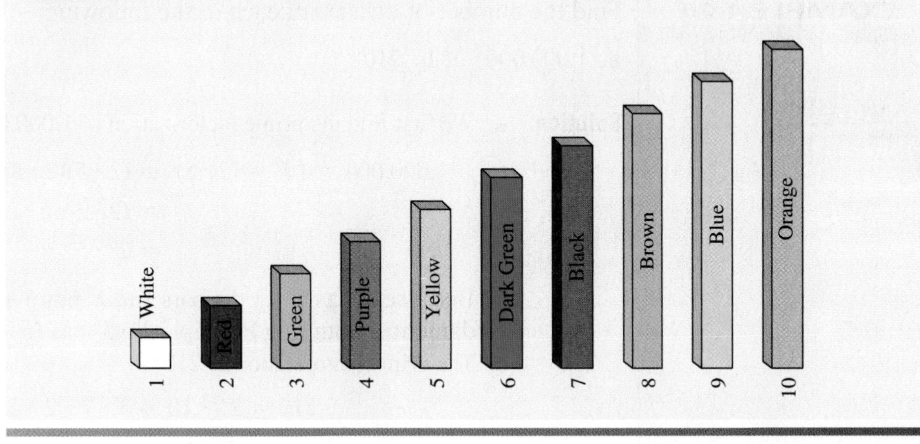

Number of Divisors

How many positive divisors does 24 have? Note that the question asks for the number of divisors, not just prime divisors. To aid in the listing, we group divisors as follows:

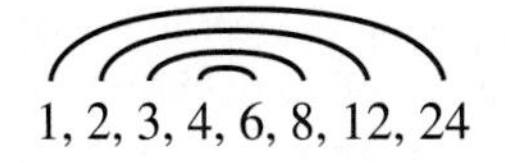

The divisors of 24 occur in pairs, where the product of the divisors in each pair is 24. If 3 is a divisor of 24, then 24/3, or 8, is also a divisor of 24. In general, if a natural number k is a divisor of 24, then $24/k$ is also a divisor of 24.

Another way to think of the number of divisors of 24 is to consider the prime factorization $24 = 2^3 \cdot 3$. The divisors of 2^3 are $2^0, 2^1, 2^2$, and 2^3. The divisors of 3 are 3^0 and 3^1. We know that 2^3 has $(3 + 1)$, or 4, divisors and 3^1 has $(1 + 1)$, or 2, divisors. Because each divisor of 24 is the product of a divisor of 2^3 and a divisor of 3^1, then we use the Fundamental Counting Principle (discussed in Section 2-1) to conclude that 24 has $4 \cdot 2$, or 8, divisors. This is summarized in Table 4-2.

This discussion can be generalized as follows: If p is any prime and n is any natural number, then the divisors of p^n are $p^0, p^1, p^2, p^3, \ldots, p^n$. Therefore, there are $(n + 1)$ divisors of p^n. Now, using the Fundamental Counting Principle, we can find the number of divisors of any number whose prime factorization is known.

Table 4-2

Divisors of 2^3	$2^0 = 1$	$2^1 = 2$	$2^2 = 4$	$2^3 = 8$
Divisors of 3^1	$3^0 = 1$	$3^1 = 3$		
Divisors of 3^1 · Divisors of 2^3 (Divisors of 24)	$3^0 \cdot 2^0 = 1$ $3^1 \cdot 2^0 = 3$	$3^0 \cdot 2^1 = 2$ $3^1 \cdot 2^1 = 6$	$3^0 \cdot 2^2 = 4$ $3^1 \cdot 2^2 = 12$	$3^0 \cdot 2^3 = 8$ $3^1 \cdot 2^3 = 24$

For example, if p and q are different primes, then we use the Fundamental Counting Principle to find that $p^n q^m$ will have $(n + 1)(m + 1)$ positive divisors.

EXAMPLE 4-25

Find the number of divisors of each of the following:

a. 1,000,000 **b.** 210^{10}

Solution **a.** We first find the prime factorization of 1,000,000.

$$\begin{aligned} 1{,}000{,}000 = 10^6 = (2 \cdot 5)^6 &= (2 \cdot 5)(2 \cdot 5)(2 \cdot 5)(2 \cdot 5)(2 \cdot 5)(2 \cdot 5) \\ &= (2 \cdot 2 \cdot 2 \cdot 2 \cdot 2 \cdot 2)\,(5 \cdot 5 \cdot 5 \cdot 5 \cdot 5 \cdot 5) \\ &= 2^6 \cdot 5^6 \end{aligned}$$

Because 2^6 has $6 + 1$ divisors and 5^6 has $6 + 1$ divisors, then by the Fundamental Counting Principle $2^6 \cdot 5^6$ has $(6 + 1)(6 + 1)$, or 49, divisors.

b. The prime factorization of 210 is

$$210 = 21 \cdot 10 = 3 \cdot 7 \cdot 2 \cdot 5 = 2 \cdot 3 \cdot 5 \cdot 7,$$
$$210^{10} = (2 \cdot 3 \cdot 5 \cdot 7)^{10} = 2^{10} \cdot 3^{10} \cdot 5^{10} \cdot 7^{10}.$$

By the Fundamental Counting Principle, the number of divisors of 210^{10} is $(10 + 1)(10 + 1)(10 + 1)(10 + 1)$, 11^4, 14,641.

NOW TRY THIS 4-12

To determine whether it is necessary to divide 97 by 2, 3, 4, 5, 6, . . . , 96 to check if it is prime, answer the following (justify your answers):

a. If 2 is not a divisor of 97, could any multiple of 2 be a divisor of 97?

b. If 3 is not a divisor of 97, what other numbers could not be divisors of 97?

c. If 5 is not a divisor of 97, what other numbers could not be divisors of 97?

d. If 7 is not a divisor of 97, what other numbers could not be divisors of 97?

e. Conjecture what numbers we have to check for divisibility in order to determine if 97 is prime.

Determining if a Number Is Prime

As depicted in the following cartoon by Sidney Harris, prime numbers have fascinated people of various backgrounds. In Now Try This 4-12, you might have found that to determine if a number is prime, you must check only divisibility by prime numbers less than the given number. Why? However, do we need to check all the primes less than the number? Suppose we want to check if 97 is prime and we find that 2, 3, 5, and 7 do not divide 97. Could a greater prime divide 97? If p is a prime greater than 7,

then $p \geq 11$. If $p|97$, then $97/p$ also divides 97. However, because $p \geq 11$ then $97/p$ must be less than 10 and hence cannot divide 97. Why? So we see that there is no need to check for divisibility by numbers other than 2, 3, 5, and 7. These ideas are generalized in the following theorems.

Theorem 4-4

If d is a divisor of n, then $\frac{n}{d}$ is also a divisor of n.

Suppose that p is the *least* divisor of n (greater than 1). Such a divisor must be prime (why?). Then by Theorem 4–4, n/p is also a divisor of n, and because p is the least divisor of n, then $p \leq n/p$. If $p \leq n/p$, then $p^2 \leq n$. This idea is summarized in the following theorem.

Theorem 4-5

If n is composite, then n has a prime factor p such that $p^2 \leq n$.

Theorem 4–5 can be used to help determine whether a given number is prime or composite. For example, consider the number 109. If 109 is composite, it must have a prime divisor p such that $p^2 \leq 109$. The primes whose squares do not exceed 109 are 2, 3, 5, and 7. Mentally, we can see that $2 \nmid 109$, $3 \nmid 109$, $5 \nmid 109$, and $7 \nmid 109$. Hence, 109 is prime. The argument used leads to the following theorem.

Theorem 4-6

If n is an integer greater than 1 and not divisible by any prime p, such that $p^2 \leq n$, then n is prime.

REMARK Because $p^2 \leq n$ implies that $p \leq \sqrt{n}$, Theorem 4-6 says that to determine if a number n is prime, it is enough to check if any prime less than or equal to $\sqrt{n}$ is a divisor of n.

EXAMPLE 4-26

a. Is 397 composite or prime? **b.** Is 91 composite or prime?

Solution **a.** The possible primes p such that $p^2 \leq 397$ are 2, 3, 5, 7, 11, 13, 17, and 19. Because $2 \nmid 397$, $3 \nmid 397$, $5 \nmid 397$, $7 \nmid 397$, $11 \nmid 397$, $13 \nmid 397$, $17 \nmid 397$, and $19 \nmid 397$, the number 397 is prime.

b. The possible primes p such that $p^2 \leq 91$ are 2, 3, 5, and 7. Because 91 is divisible by 7, it is composite.

More About Primes

One way to find all the primes less than a given number is to use the *Sieve of Eratosthenes*, named after the Greek mathematician Eratosthenes (ca. 276–194 BCE). If all the natural numbers greater than 1 are considered (or placed in the sieve), the numbers that are not prime are methodically crossed out (or drop through the holes of the sieve). The remaining numbers are prime. The following partial student page illustrates this process. Before reading on, answer the questions on the student page.

Notice that we can stop crossing out multiples of primes after all the multiples of 5 and all the multiples of 7 have been crossed out because 7 is the greatest prime whose square, 49, is less than 100. All the numbers remaining in the list and not crossed out are prime.

HISTORICAL NOTE

Eratosthenes (276–194 BCE), a Greek scholar, was born in Cyrene, then a Greek colony in North Africa now in Libya, but spent most of his life in Alexandria as the chief librarian at the museum there. He wrote treatises on a variety of topics, including geography, philosophy, astronomy, poetry, and mathematics. In his work *Geographica*, he gave arguments for the spherical shape of the Earth. Today, Eratosthenes is best known for his "sieve"—a systematic procedure for isolating the prime numbers—and for a simple method for calculating the circumference of the Earth.

SAMPLE SCHOOL BOOK PAGE: ENRICHMENT

SIEVE OF ERATOSTHENES

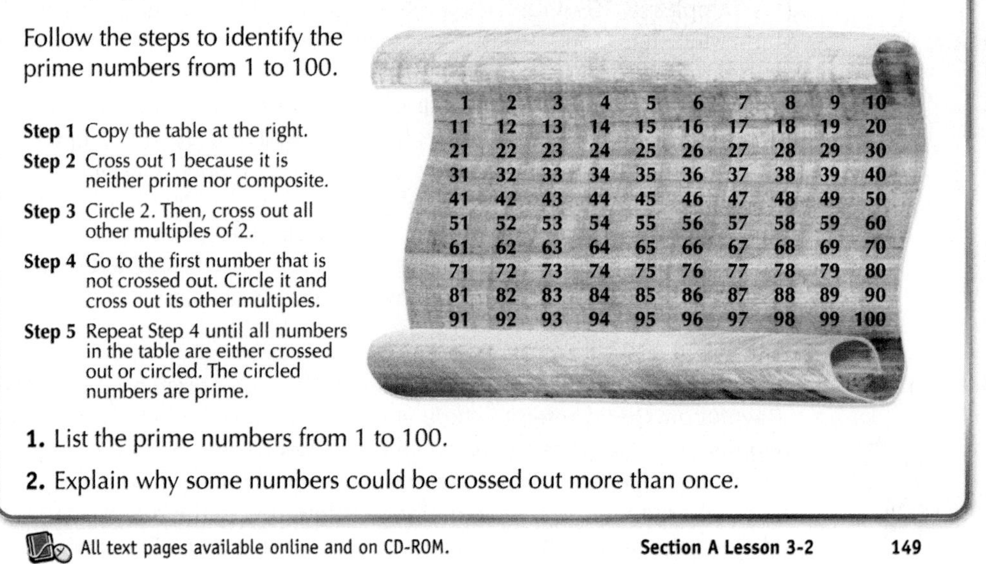

Enrichment

Sieve of Eratosthenes

About 230 B.C., the Greek mathematician Eratosthenes developed a method for identifying prime numbers. The method is called the **Sieve of Eratosthenes.**

Follow the steps to identify the prime numbers from 1 to 100.

Step 1 Copy the table at the right.

Step 2 Cross out 1 because it is neither prime nor composite.

Step 3 Circle 2. Then, cross out all other multiples of 2.

Step 4 Go to the first number that is not crossed out. Circle it and cross out its other multiples.

Step 5 Repeat Step 4 until all numbers in the table are either crossed out or circled. The circled numbers are prime.

1	2	3	4	5	6	7	8	9	10
11	12	13	14	15	16	17	18	19	20
21	22	23	24	25	26	27	28	29	30
31	32	33	34	35	36	37	38	39	40
41	42	43	44	45	46	47	48	49	50
51	52	53	54	55	56	57	58	59	60
61	62	63	64	65	66	67	68	69	70
71	72	73	74	75	76	77	78	79	80
81	82	83	84	85	86	87	88	89	90
91	92	93	94	95	96	97	98	99	100

1. List the prime numbers from 1 to 100.
2. Explain why some numbers could be crossed out more than once.

All text pages available online and on CD-ROM. Section A Lesson 3-2 149

Source: Scott Foresman–Addison Wesley Mathematics 2005, Grade 6 (p. 149)

Primes have been treasured since antiquity, as the following cartoon depicts.

There are infinitely many whole numbers, infinitely many odd whole numbers, and infinitely many even whole numbers. Are there infinitely many primes? Because prime numbers do not appear in any known pattern, the answer to this question is not obvious. Euclid was the first to prove that there are infinitely many primes.

Mathematicians have long looked for a formula that produces only primes, but no one has found it. One result was the expression $n^2 - n + 41$, where n is a whole number. Substituting $0, 1, 2, 3, \ldots, 40$ for n in the expression always results in a prime number. However, substituting 41 for n gives $41^2 - 41 + 41$, or 41^2, a composite number.

In 1998, Roland Clarkson, a 19-year-old student at California State University, showed that $2^{3021377} - 1$ is prime. The number has 909,526 digits. The full decimal expansion of the number would fill several hundred pages. Since then at least two more large primes have been discovered: $2^{24036583} - 1$ and $2^{25964951} - 1$. The latter was found in 2005 by Dr. Martin Nowak, a German eye surgeon and a mathematical hobbyist. These are examples of *Mersenne primes*. A Mersenne prime, named after the French monk Marin Mersenne (1588–1648), is a prime of the form $2^n - 1$, where n is prime.

Searching for large primes has led to advances in *distributed computing*, that is, using the Internet to utilize the unused computing power of great numbers of computers. Searching for Mersenne primes has been used as a test for computer hardware.

Another type of interesting prime is a *Sophie Germain prime*, which is an odd prime p for which $2p + 1$ is also a prime. Notice that $p = 3$ is a Sophie Germain prime, since $2 \cdot 3 + 1$, or 7, is also a prime. Check that 5, 11, and 23 are also such primes. The primes were named after the French mathematician Sophie Germain. In 2005, the greatest Sophie Germain prime discovered had 36,523 digits.

▶ **HISTORICAL NOTE**

Sophie Germain (1776–1831) was born in Paris and grew up during the French Revolution. She wanted to study at the prestigious École Polytechnique but women were not allowed as students. Consequently, she studied from lecture notes and from Gauss's monograph on number theory. In addition to her work in number theory, she made major contributions to the mathematical theory of elasticity, for which she was awarded the prize of the French Academy of Sciences. Germain's work was highly regarded by Gauss, who recommended her for an honorary degree from the University of Göttingen. She died before the degree could be awarded. ◀

PROBLEM SOLVING **How Many Bears?**

A large toy store carries one kind of stuffed bear. On Monday the store sold a certain number of the stuffed bears for a total of \$1843 and on Tuesday, without changing the price, the store sold a certain number of the stuffed bears for a total of \$1957. How many toy bears were sold each day if the price of each bear is a whole number and greater than \$1?

Understanding the Problem One day a store sold a number of stuffed bears for \$1843 and on the next day a number of them for a total of \$1957. We need to find the number of bears sold on each day.

Devising a Plan If x bears were sold the first day and y bears the second day, and if the price of each bear was c dollars, we would have $cx = 1843$ and $cy = 1957$. Thus, 1843 and 1957 should have a common factor—the price c. We could factor each number and find the possible factors. If the problem is to have a unique solution, the two numbers should have only one common factor other than 1. Any common factor of 1957 and 1843 will also be a factor of $1957 - 1843 = 114$ and the factors of 114 are easier to find.

Carrying Out the Plan We have $114 = 2 \cdot 57 = 2 \cdot 3 \cdot 19$. Thus, if 1957 and 1843 have a common prime factor, it must be 2, 3, or 19. But neither 2 nor 3 divides the number, hence the only possible common factor is 19. We divide each number by 19 and find

$$1843 = 19 \cdot 97.$$
$$1957 = 19 \cdot 103.$$

Notice that neither 97 nor 103 is divisible by 2, 3, 5, or 7. Hence 97 and 103 are primes (why?) and therefore the only common factor (greater than 1) of 1843 and 1957 is 19. Consequently, the price of each bear was \$19. The first day 97 bears were sold and the next day 103 bears were sold.

Looking Back Notice that the problem had a unique solution because the only common factor (greater than 1) of the two numbers was 19. We could create similar problems by having the price of the item be a prime number and the number of items sold each day also be prime numbers. For example, the total sale on the first day could have been $23 \cdot 101$ or \$2323 and on the second day $23 \cdot 107$ or \$2461 (notice that 23, 101, and 107 are prime numbers).

To find a common factor of 1957 and 1843, we found all the common factors of $1957 - 1843 = 114 = 2 \cdot 3 \cdot 19$ and checked which of the factors of the difference was a common factor of the original numbers. We have used Theorem 4-2: property: If $d|a$ and $d|b$, then $d|(a - b)$. This theorem assures us that every common factor of a and b will also be a factor of $a - b$. ▲▲

▶ **HISTORICAL NOTE**

During World War II, Alan Turing (1912–1954), Peter Hilton, and other British analysts helped crack the codes developed on the German Enigma cipher machine. In the 1970s, determining large prime numbers became extremely useful in coding and decoding secret messages. In all coding and decoding, the letters of an alphabet correspond in some way to nonnegative integers. A "safe" coding system, in which messages are unintelligible to everyone except the intended receiver, was devised by three Massachusetts Institute of Technology scientists (Ronald Rivest, Adi Shamir, and Leonard Adleman) and is referred to as the RSA (their initials) system. The secret deciphering key consists of two large prime numbers chosen by the user. The enciphering key is the product of these two primes. Because it is extremely difficult and time-consuming to factor large numbers, it was practically impossible to recover the deciphering key from a known enciphering key. In 1982, new methods for factoring large numbers were invented, which resulted in the use of even greater primes to prevent the breaking of decoding keys. ◀

ONGOING ASSESSMENT 4-4

1. Find the least positive number that is divisible by three different primes.
2. Determine which of the following numbers are primes:
   a. 109 **b.** 119
   c. 33 **d.** 101
   e. 463 **f.** 97
   g. $2 \cdot 3 \cdot 5 \cdot 7 + 1$ **h.** $2 \cdot 3 \cdot 5 \cdot 7 - 1$
3. Use a factor tree to find the prime factorization for each of the following:
   a. 504 b. 2475 **c.** 11,250

4. a. Fill in the missing numbers in the following factor tree:

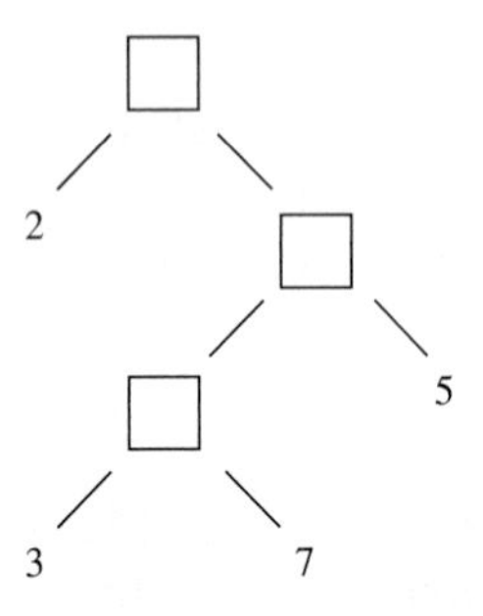

b. How could you find the top number without finding the other two numbers?

5. What is the greatest prime you must consider to test whether 5669 is prime?

6. Find the prime factorizations of the following:
   a. $1 \cdot 2 \cdot 3 \cdot 4 \cdot 5 \cdot 6 \cdot 7 \cdot 8 \cdot 9 \cdot 10$
   b. $10^2 \cdot 26 \cdot 49^{10}$
   c. 101
   d. 1001
   e. 10011001
   f. 1001^2
   g. 999^{10}
   h. $111^{10} - 111^9$

7. a. When the U.S. flag had 48 stars, the stars formed a 6×8 rectangular array. In what other rectangular arrays could they have been arranged?
   b. How many rectangular arrays of stars could there be if there were only 47 states?

8. If the Spanish Armada had consisted of 177 galleons, could it have sailed in an equal number of small flotillas? If so, how many ships would have been in each?

9. Suppose the 435 members of the House of Representatives are placed on committees consisting of more than 2 members but fewer than 30 members. Each committee is to have an equal number of members and each member is to be on only one committee.
   a. What size committees are possible?
   b. How many committees are there of each size?

10. Mr. Arboreta wants to plant fruit trees in a rectangular array. For each of the following numbers of trees, find all possible numbers of rows if each row is to have the same number of trees:
    a. 36 b. 28 c. 17 d. 144

11. Find the least number divisible by each natural number less than or equal to 12.

12. Some of the divisors of a locker number are 2, 5, and 9. If there are exactly nine additional divisors, what is the locker number?

13. a. Use the Fundamental Theorem of Arithmetic to justify that if $2|n$ and $3|n$, then $6|n$.
    b. Is it always true that if $a|n$ and $b|n$, then $ab|n^2$? Either prove the statement or give a counterexample.

14. Find the greatest four-digit number that has exactly three factors.

15. Extend the Sieve of Eratosthenes to find all primes less than 200.

16. The prime numbers 11 and 13 are called *twin primes* because they differ by 2. (The existence of infinitely many twin primes has not been proved.) Find all the twin primes less than 200.

17. Show that if 1 were considered a prime, every number would have more than one prime factorization.

18. If $42|n$, what other positive integers divide n?

19. If 1000 is a factor of n, what other positive integers divide n? How many such integers are there?

20. Is it possible to find positive integers x, y, and z such that $2^x \cdot 3^y = 5^z$? Why or why not?

21. It is not known whether there are infinitely many primes in the infinite sequence consisting only of ones: 1, 11, 111, 1111, ... Find infinitely many composite numbers in the sequence.

22. a. Show that there are infinitely many composite numbers in the arithmetic sequence 1, 5, 9, 13, 17, ...
    ★b. Does every arithmetic sequence consisting of integers with difference greater than 0 have infinitely many composite numbers? Justify your answer.

23. If $2N = 2^6 \cdot 3^5 \cdot 5^4 \cdot 7^3 \cdot 11^7$, explain why $2 \cdot 3 \cdot 5 \cdot 7 \cdot 11$ is a factor of N.

24. Is $3^2 \cdot 2^4$ a factor of $3^4 \cdot 2^7$? Explain why or why not.

25. Explain why each of the following numbers is composite:
    a. $3 \cdot 5 \cdot 7 \cdot 11 \cdot 13$
    b. $(3 \cdot 4 \cdot 5 \cdot 6 \cdot 7 \cdot 8) + 2$
    c. $(3 \cdot 5 \cdot 7 \cdot 11 \cdot 13) + 5$
    d. $10! + 7$ (*Note:* $10! = 1 \cdot 2 \cdot 3 \cdot 4 \cdot 5 \cdot 6 \cdot 7 \cdot 8 \cdot 9 \cdot 10$.)
    e. $10! + k$, where $k = 2, 3, 4, 5, 6, 7, 8, 9$, or 10

26. Explain why $2^3 \cdot 3^2 \cdot 25^3$ is not a prime factorization and find the prime factorization of the number.

27. A prime such as 7331 is a *superprime* because any integers obtained by deleting digits from the right of 7331 are prime; for example, 733, 73, and 7.
    a. For a prime to be a superprime, what digits cannot appear in the number?
    b. Of the digits that can appear in a superprime, what digit cannot be the leftmost digit of a superprime?
    c. Find all of the two-digit superprimes.
    d. Find a three-digit superprime.

28. Is the following always true? (Justify your answer) If $m|ab$, then $m|a$ or $m|b$.

29. Find the prime factorizations of each of the following:
    a. $36^{10} \cdot 49^{20} \cdot 6^{15}$
    b. $100^{60} \cdot 300^{40}$
    c. $2 \cdot 3^4 \cdot 5^{110} \cdot 7 + 4 \cdot 3^4 \cdot 5^{110}$
    d. $2 \cdot 3 \cdot 5 \cdot 7 \cdot 11 + 1$

★30. Use the Fundamental Theorem of Arithmetic to justify the following statement for whole numbers a and b greater than 1. If p is prime and $p|ab$, then $p|a$ or $p|b$.

Communication

31. Explain why the product of any three consecutive integers is divisible by 6.

32. Explain why the product of any four consecutive integers is divisible by 24.

33. In order to test for divisibility by 12, one student checked to determine divisibility by 3 and 4; another checked for divisibility by 2 and 6. Are both students using a correct approach to divisibility by 12? Why or why not?

34. In the Sieve of Eratosthenes for numbers less than 100, explain why, after we cross out all the multiples of 2, 3, 5, and 7, the remaining numbers are primes.

35. Let $M = 2 \cdot 3 \cdot 5 \cdot 7 + 11 \cdot 13 \cdot 17 \cdot 19$. Without multiplying, show that none of the primes less than or equal to 19 divides M.

★36. A woman with a basket of eggs finds that if she removes the eggs from the basket 3 or 5 at a time, there is always 1 egg left. However, if she removes the eggs 7 at a time, there are no eggs left. If the basket holds up to 100 eggs, how many eggs does she have? Explain your reasoning.

37. Explain why, when a number is composite, its least positive divisor, other than 1, must be prime.

38. a. An eighth grader at the Roosevelt Middle School claims that because there are as many even numbers as odd numbers between 1 and 1000, there must be as many numbers that have an even number of divisors as numbers that have an odd number of divisors between 1 and 1000. Is the student correct? Why or why not?

b. How many numbers between 1 and 1000 have an odd number of divisors?

★39. Euclid proved that given any finite list of primes, there exists a prime not in the list. Read the following argument and answer the questions that follow.

Let 2, 3, 5, 7, . . . , p be a list of all the primes less than or equal to a certain prime p. We will show that there exists a prime not on the list. Consider the product

$$2 \cdot 3 \cdot 5 \cdot 7 \cdot \ldots \cdot p$$

Notice that every prime in our list divides that product. However, if we add 1 to the product, that is, form the number $N = (2 \cdot 3 \cdot 5 \cdot 7 \cdot \ldots \cdot p) + 1$, then none of the primes in the list will divide N. Notice that whether N is prime or composite, some prime q must divide N. Because no prime in our list divides N, q is not one of the primes in our list. Consequently $q > p$. We have shown that there exists a prime greater than p.

a. Explain why no prime in the list will divide N.

b. Explain why some prime must divide N.

c. Someone discovered a prime that has 65,050 digits. How does the preceding argument assure us that there exists an even larger prime?

d. Does the argument show that there are infinitely many primes? Why or why not?

e. Let $M = 2 \cdot 3 \cdot 5 \cdot 7 \cdot 11 \cdot 13 \cdot 17 \cdot 19 + 1$. Without multiplying, explain why some prime greater than 19 will divide M.

40. Jose says that he can quickly create primes less than 120 using products or sums involving 1, 2, 3, 5, or 7. He claims that each of the following are prime numbers:

$$\begin{array}{ll} 2 \cdot 3 \cdot 5 - 1 = 29 & 2 \cdot 3 \cdot 7 + 5 = 47 \\ 2 \cdot 3 \cdot 5 + 7 = 37 & 2 \cdot 5 \cdot 7 + 3 = 73 \\ 2 \cdot 5 + 7 \cdot 3 = 31 & 3^2 \cdot 2 \cdot 5 + 7 = 97 \\ 2 \cdot 7 + 5 \cdot 3 = 29 & 2^2 \cdot 3 \cdot 7 + 5 = 89 \\ 2 \cdot 3 + 5 \cdot 7 = 41 & 2^2 \cdot 3^2 + 5 \cdot 7 = 71 \end{array}$$

a. Explain how Jose can be sure that each of the numbers is a prime.

b. Create three more primes less than 120 using Jose's approach and explain how you can be immediately sure that they are prime numbers.

c. How could you use an approach similar to Jose's to create primes less than 289?

Open-Ended

41. a. In which of the following intervals do you think there are the most primes? Why? Check to see if you were correct.

i. 0–99 **ii.** 100–199
iii. 200–299 **iv.** 300–399

b. What is the longest string of consecutive composite numbers in the intervals?

c. How many twin primes (see problem 16) are there in each interval?

d. What patterns, if any, do you see for any of the preceding questions? Predict what might happen in other intervals.

42. One of the recently discovered primes has 65,050 digits. If you wrote this number out, how long a sheet of paper would you need?

43. A number is a *perfect number* if the sum of its factors (other than the number itself) is equal to the number. For example, 6 is a perfect number because its factors sum to 6, that is, $1 + 2 + 3 = 6$. An *abundant number* has factors whose sum is greater than the number itself. A *deficient number* is a number with factors whose sum is less than the number itself.

a. Classify each of the following numbers as perfect, abundant, or deficient:

i. 12 ii. 28 iii. 35

b. Find at least one more number that falls in each class.

Cooperative Learning

44. A class of 23 students was using square tiles to build rectangular shapes. Each student had more than 1 tile and each had a different number of tiles. Each student was able to build only one shape of rectangle. All tiles had to be used to build a rectangle and the rectangle could not have holes. For example, a 2 by 6 rectangle uses 12 tiles and is considered the same as a 6 by 2 rectangle but is different from a 3 by 4 rectangle. The class did the activity using the least number of tiles. How many tiles did the class use? Divide the work among the members of your group to explore the various rectangles that could be made.

Review Problems

45. Classify the following as true or false:
    a. 11 is a factor of 189.
    b. 1001 is a multiple of 13.
    c. $7|1001$ and $7\nmid 12$ imply $7\nmid(1001 - 12)$.
    d. If a number is divisible by both 7 and 11, then its prime factorization contains 7 and 11.
46. Test each of the following for divisibility by 2, 3, 4, 5, 6, 7, 8, 9, 10, and 11:
    a. 438,162
    b. 2,345,678,910
47. Prove that if a number is divisible by 12, then it is divisible by 3.
48. Could $3376 be divided exactly among either seven or eight people?

LABORATORY ACTIVITY

In Figure 4-24, a spiral starts with 41 at its center and continues in a counterclockwise direction. Primes are written in and squares that represent composites are shaded. Continue the spiral until you reach the prime 439. Check the primes along the diagonal. Can you find each of the primes from the formula $n^2 + n + 41$ by substituting appropriate values for n?

FIGURE 4-24

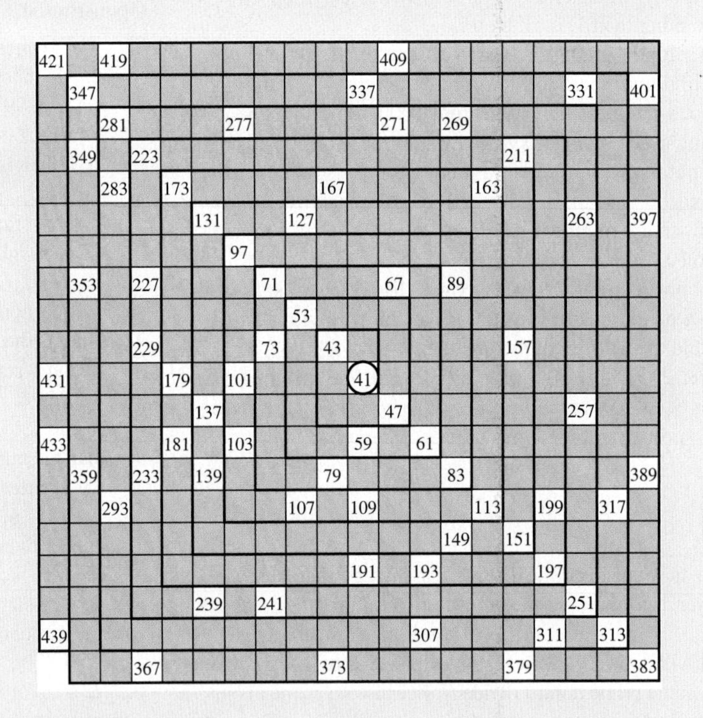

BRAIN TEASER

One Saturday Jody cut short her visit with her friend Natasha to take three other friends to a movie. "How old are they?" asked Natasha. "The product of their ages is 2450 and the sum is exactly twice your age," replied Jody. Natasha thought for a moment and said: "I need more information." To that Jody replied, "I should have mentioned that I am at least one year younger than the oldest of my three friends." With this information Natasha immediately found the ages of the friends. How did Natasha figure the ages of the friends and what were their ages?

4-5 Greatest Common Divisor and Least Common Multiple

Consider the following situation:

Two bands are to be combined to march in a parade. A 24-member band will march behind a 30-member band. The combined bands must have the same number of columns and the same number of members in each column. What is the greatest number of columns in which they can march?

The bands could each march in two columns, and we would have the same number of columns, but this does not satisfy the condition of having the greatest number of columns. The number of columns must divide both 24 and 30. Why? Numbers that divide both 24 and 30 are 1, 2, 3, and 6. The greatest of these numbers is 6, so the bands should each march in columns of 6. The first band would have 6 columns with 4 members in each column, and the second band would have 6 columns with 5 members in each column.

In this problem, we have found the greatest number that divides both 24 and 30, that is, the **greatest common divisor (GCD)**.

greatest common divisor

RESEARCH NOTE
Possibly because students often confuse factors and multiples, the greatest common factor and the least common multiple are difficult topics for students to grasp (Greavis and Greaver 1992).

Definition

The **greatest common divisor (GCD)** of two integers a and b is the greatest integer that divides both a and b.

We can find the GCD in many ways. We show several ways next.

Colored Rods Method

We can build a model of two or more integers with colored rods to determine the GCD of two positive integers. For example, consider finding the GCD of 6 and 8 using the 6 rod and the 8 rod, as in Figure 4-25.

FIGURE 4-25

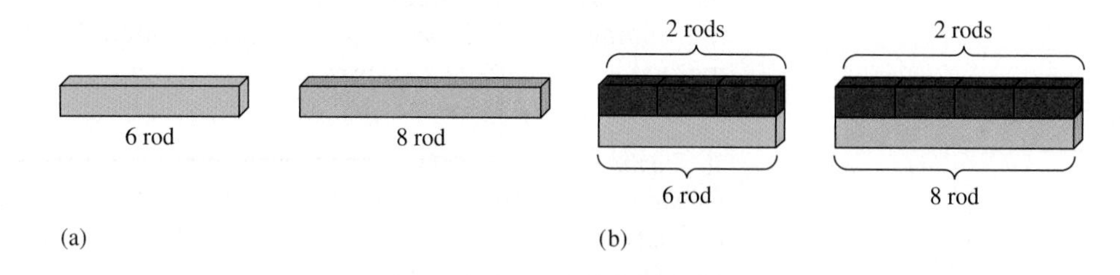

To find the GCD of 6 and 8, we must find the longest rod such that we can use multiples of that rod to build both the 6 rod and the 8 rod. The 1 rods and the 2 rods can be used to build both the 6 and 8 rods, as shown in Figure 4-25(b); the 3 rods can be used to build the 6 rod but not the 8 rod; the 4 rods can be used to build the 8 rod but not the 6 rod; the 5 rods can be used to build neither; and the 6 rods cannot be used to build the 8 rod. Therefore GCD(6, 8) = 2.

NOW TRY THIS 4-13

Explain how you could use colored rods to solve the marching bands' problem, stated at the beginning of this section.

The Intersection-of-Sets Method

In the *intersection-of-sets* method, we list all members of the set of positive divisors of the two integers, then find the set of all *common divisors*, and, finally, pick the *greatest* element in that set. For example, to find the GCD of 20 and 32, denote the sets of divisors of 20 and 32 by D_{20} and D_{32}, respectively.

$$D_{20} = \{1, 2, 4, 5, 10, 20\}$$
$$D_{32} = \{1, 2, 4, 8, 16, 32\}$$

The set of all common positive divisors of 20 and 32 is

$$D_{20} \cap D_{32} = \{1, 2, 4\}.$$

Because the greatest number in the set of common positive divisors is 4, the GCD of 20 and 32 is 4, written $\text{GCD}(20, 32) = 4$.

The Prime Factorization Method

The intersection-of-sets method is rather time-consuming and tedious if the numbers have many divisors. Another, more efficient, method is the prime factorization method. To find GCD(180, 168), first notice that

$$180 = 2 \cdot 2 \cdot 3 \cdot 3 \cdot 5$$

and

$$\updownarrow \quad \updownarrow \quad \searrow$$

$$168 = 2 \cdot 2 \cdot 2 \cdot 3 \cdot 7.$$

We see that 180 and 168 have two factors of 2 and one of 3 in common. These common primes divide both 180 and 168. In fact, the only numbers other than 1 that divide both 180 and 168 must have no more than two 2s and one 3 and no other prime factors in their prime factorizations. The possible common divisors are $1, 2, 2^2, 3, 2 \cdot 3$, and $2^2 \cdot 3$. Hence, the greatest common divisor of 180 and 168 is $2^2 \cdot 3$. The procedure for finding the GCD of two or more numbers by using the prime factorization method is summarized as follows:

To find the GCD of two or more positive integers, first find the prime factorizations of the given numbers and then identify each common prime factor of the given numbers. The GCD is the product of the common factors, each raised to the lowest power of that prime that occurs in any of the prime factorizations.

If we apply the prime factorization technique to finding GCD(4, 9), we see that 4 and 9 have no common prime factors. But that does not mean there is no GCD. We still have 1 as a common divisor, so $\text{GCD}(4, 9) = 1$. Numbers, such as 4 and 9, whose GCD is 1 are **relatively prime**. Both the intersection-of-sets method and the prime factorization method are found on the student page.

relatively prime

SAMPLE SCHOOL BOOK PAGE:

GREATEST COMMON FACTOR

Lesson 3-3

Key Idea
There are different ways to find the factors that are common to two or more numbers.

Vocabulary
- common factor
- greatest common factor (GCF)
- prime factorization (p. 147)

TEST TALK

Think It Through
- I can **use factors** to identify equal groups for sharing.
- I can **make an organized list** to find the common factors and GCF.

Greatest Common Factor

WARM UP

List the factors of each number.

1. 12
2. 31
3. 54
4. 100

LEARN

How can you use factors?

Janelle is making snack packs for a group hike. Each pack should have the same number of bags of trail mix and the same number of bottles of water. What is the greatest number of snack packs that she can make with no refreshments left over?

Hike Refresments
60 bags of trail mix
90 bottles of water

To solve this problem, you need to find the numbers that are factors of both 60 and 90. These are the **common factors** of 60 and 90. The **greatest common factor (GCF)** is the *greatest* number that is a factor of both 60 and 90.

Example

Find the GCF of 60 and 90.

One Way

List the factors of each number.

60: 1, 2, 3, 4, 5, 6, 10, 12, 15, 20, 30, 60
90: 1, 2, 3, 5, 6, 9, 10, 15, 18, 30, 45, 90

Circle pairs of common factors. Select the greatest one.

60: 1, 2, 3, 4, 5, 6, 10, 12, 15, 20, 30, 60
90: 1, 2, 3, 5, 6, 9, 10, 15, 18, 30, 45, 90

The GCF is 30.

Another Way

Use prime factorization.

$60 = 2 \times 2 \times 3 \times 5$
$90 = 2 \times 3 \times 3 \times 5$

Find the product of the common prime factors. If there are no common prime factors, the GCF is 1.

$60 = 2 \times 2 \times 3 \times 5$
$90 = 2 \times 3 \times 3 \times 5$ $\quad 2 \times 3 \times 5 = 30$

The GCF is 30.

The greatest number of snack packs she can make is 30.

Talk About It

1. In the second method, why is 2 used only once as a factor of the GCF?
2. In each of the 30 snack packs, how many bags of trail mix and how many bottles of water will there be?
3. **Reasoning** Find the GCF of 48 and 120. Which method did you use? Why?

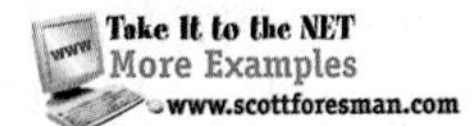

150

Source: *Scott Foresman–Addison Wesley Mathematics 2005 Grade 6* (p. 150).

(GCF in the student page stands for *Greatest Common Factor*, which is the same as GCD.)

EXAMPLE 4-27

Find each of the following:

a. GCD(108, 72)
b. GCD(0, 13)
c. GCD(x, y) if $x = 2^3 \cdot 7^2 \cdot 11 \cdot 13$ and $y = 2 \cdot 7^3 \cdot 13 \cdot 17$
d. GCD(x, y, z) if $z = 2^2 \cdot 7$, using x and y from (c)
e. GCD(x, y), where $x = 5^4 \cdot 13^{10}$ and $y = 3^{10} \cdot 11^{20}$

Solution **a.** Since $108 = 2^2 \cdot 3^3$ and $72 = 2^3 \cdot 3^2$, it follows that GCD(108, 72) $= 2^2 \cdot 3^2 = 36$.
b. Because $13|0$ and $13|13$, it follows that GCD(0, 13) = 13.
c. GCD(x, y) $= 2 \cdot 7^2 \cdot 13 = 1274$.
d. Because $x = 2^3 \cdot 7^2 \cdot 11 \cdot 13$, $y = 2 \cdot 7^3 \cdot 13 \cdot 17$, and $z = 2^2 \cdot 7$, then GCD(x, y, z) $= 2 \cdot 7 = 14$. Notice that GCD(x, y, z) can also be obtained by finding the GCD of z and 1274, the answer from (c).
e. Because x and y have no common prime factors, GCD(x, y) = 1. ▲

Calculator Method

Calculators with a [Simp] key can be used to find the GCD of two numbers. For example, to find the GCD(120, 180), use the following sequence of buttons to start: First, press [1] [2] [0] [/] [1] [8] [0] [Simp] [=] to obtain the display [N/D → n/d 60/90]. By pressing the [x⇔y] button, we see [2] on the display as a common divisor of 120 and 180. By pressing the [x⇔y] button again and pressing [Simp] [=] [x⇔y], we see 2 again as a factor. The process is repeated to reveal 3 and 5 as other common factors. The GCD(120, 180) is the product of the common prime factors $2 \cdot 2 \cdot 3 \cdot 5$, or 60.

Euclidean Algorithm Method

Large numbers may be hard to factor. For these numbers, another method is more efficient than factorization for finding the GCD. For example, suppose we want to find GCD(676, 221). If we could find two smaller numbers whose GCD is the same as GCD(676, 221), our task would be easier. From Theorem 4-2(c), every divisor of 676 and 221 is also a divisor of $676 - 221$ and 221. Conversely, every divisor of $676 - 221$ and 221 is also a divisor of 676 and 221. Thus, the set of all the common divisors of 676 and 221 is the same as the set of all common divisors of $676 - 221$ and 221. Consequently, GCD(676, 221) = GCD($676 - 221$, 221). This process can be continued to subtract three 221s from 676 so that GCD(676, 221) = GCD($676 - 3 \cdot 221$, 221) = GCD(13, 221). To determine how many 221s can be subtracted from 676, we could have divided as follows:

$$\begin{array}{r} 3 \\ 221\overline{)676} \\ \underline{663} \\ 13 \end{array}$$

Continuing, we see that GCD(13, 221) = GCD(0, 13) from the following division:

$$\begin{array}{r} 17 \\ 13\overline{)221} \\ \underline{13} \\ 91 \\ \underline{91} \\ 0 \end{array}$$

Because GCD(0, 13) = 13, the GCD(676, 221) = 13. Based on this illustration, we make the generalization outlined in the following theorem.

Theorem 4-7

If a and b are any whole numbers greater than 0 and $a \geq b$, then GCD(a, b) = GCD(r, b), where r is the remainder when a is divided by b.

REMARK Because GCD(x, y) = GCD(y, x) for all integers x and y not both 0, Theorem 4–7 can be written

$$\text{GCD}(a, b) = \text{GCD}(b, r).$$

Euclidean algorithm

Finding the GCD of two numbers by repeatedly using Theorem 4-7 until the remainder 0 is reached is referred to as the **Euclidean algorithm**.

EXAMPLE 4-28

Use the Euclidean algorithm to find GCD(10764, 2300).

Solution

$$\begin{array}{r} 4 \\ 2300\overline{)10764} \\ \underline{9200} \\ 1564 \end{array}$$ Thus, GCD(10764, 2300) = GCD(2300, 1564).

$$\begin{array}{r} 1 \\ 1564\overline{)2300} \\ \underline{1564} \\ 736 \end{array}$$ Thus, GCD(2300, 1564) = GCD(1564, 736).

$$\begin{array}{r} 2 \\ 736\overline{)1564} \\ \underline{1472} \\ 92 \end{array}$$ Thus, GCD(1564, 736) = GCD(736, 92).

$$\begin{array}{r} 8 \\ 92\overline{)736} \\ \underline{736} \\ 0 \end{array}$$ Thus, GCD(736, 92) = GCD(92, 0).

Because GCD(92, 0) = 92, it follows that GCD(10764, 2300) = 92. ▲

REMARK The procedure for finding the GCD by using the Euclidean algorithm can be stopped at any step at which the GCD is obvious.

A calculator with the integer division feature can also be used to perform the Euclidean algorithm. This feature yields the quotient and the remainder when doing a division. For example, if the integer division key looks like INT ÷, then to find GCD(10, 764, 2300) we proceed as follows:

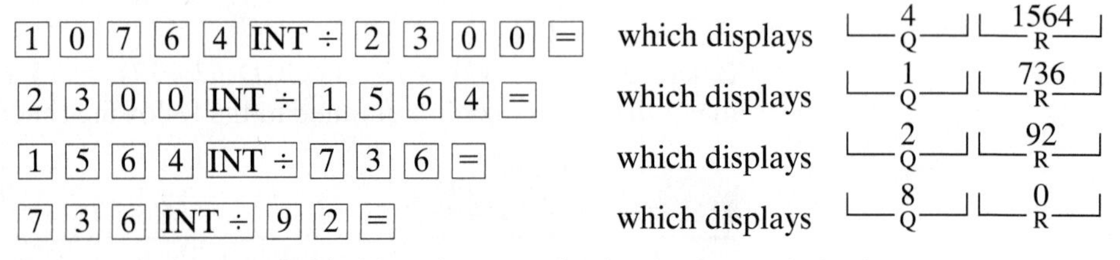

The last number we divided by when we obtained a 0 remainder is 92, so

$$\text{GCD}(10{,}764,\ 2300) = 92.$$

Sometimes shortcuts can be used to find the GCD of two or more numbers, as in the following example.

EXAMPLE 4-29

Find each of the following:

a. GCD(134791, 6341, 6339)
b. The GCD of any two consecutive integers

Solution **a.** Any common divisor of three numbers is also a common divisor of any two of them (why?). Consequently, the GCD of three numbers cannot be greater than the GCD of any two of the numbers. The numbers 6341 and 6339 are close to each other and therefore it is easy to find their GCD:

$$\begin{aligned}\text{GCD}(6341, 6339) &= \text{GCD}(6341 - 6339, 6339)\\ &= \text{GCD}(2, 6339)\\ &= 1.\end{aligned}$$

Because GCD(134791, 6341, 6339) cannot be greater than 1, it follows that it must equal 1.

b. Notice that GCD(4, 5) = 1, GCD(5, 6) = 1, GCD(6, 7) = 1, and GCD(99, 100) = 1. It seems that the GCD of any two consecutive integers is 1. To justify this conjecture, we need to show that for all integers n, $\text{GCD}(n, n+1) = 1$. We have

$$\begin{aligned}\text{GCD}(n, n+1) = \text{GCD}(n+1, n) &= \text{GCD}(n+1-n, n)\\ &= \text{GCD}(1, n)\\ &= 1.\end{aligned}$$

Least Common Multiple

Hot dogs are usually sold 10 to a package, while hot dog buns are usually sold 8 to a package. This mismatch causes troubles when one is trying to match hot dogs and buns. What is the least number of packages of each you could order so that there is an equal number of hot dogs and buns? The numbers of hot dogs that we could have are just the multiples of 10, that is, 10, 20, 30, 40, 50, ... Likewise, the possible numbers of buns are 8, 16, 24, 32, 40, 48, ... We can see that the number of hot dogs matches the number of buns whenever 10 and 8 have multiples in common. This occurs at 40, 80, 120, ... In this problem, we are interested in the least of these multiples, 40. Therefore, we could obtain the same number of hot dogs and buns in the least amount by buying four packages of hot dogs and five packages of buns. The answer 40 is the **least common multiple (LCM)** of 8 and 10.

least common multiple

Definition

Suppose that a and b are positive integers. Then the least common multiple (LCM) of a and b is the least positive integer that is simultaneously a multiple of a and a multiple of b.

As with GCDs, there are several methods for finding least common multiples.

Colored Rods Method

We can use colored rods to determine the LCM of two numbers. For example, consider the 3 rod and the 4 rod in Figure 4-26(a). We build trains of 3 rods and 4 rods until they are the same length, as shown in Figure 4-26(b). The LCM is the common length of the train.

FIGURE 4-26

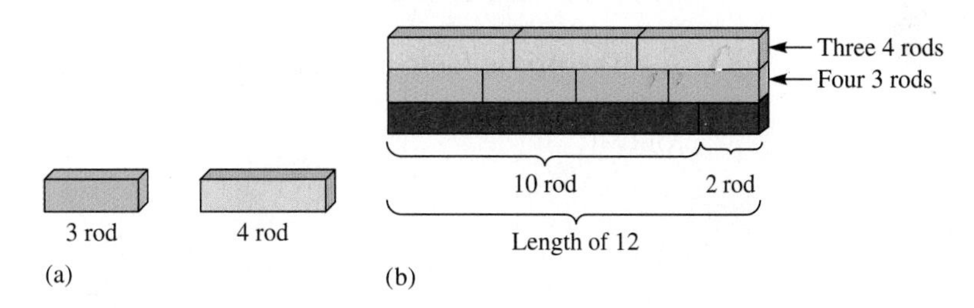

The Intersection-of-Sets Method

In the *intersection-of-sets* method, we first find the set of all positive *multiples* of both the first and second numbers, then find the set of all *common multiples* of both numbers, and finally pick the *least* element in that set. For example, to find the LCM of 8 and 12, denote the sets of positive multiples of 8 and 12 by M_8 and M_{12}, respectively.

$$M_8 = \{8, 16, 24, 32, 40, 48, 56, 64, 72, \ldots\}$$
$$M_{12} = \{12, 24, 36, 48, 60, 72, 84, 96, 108, \ldots\}$$

The set of common multiples is

$$M_8 \cap M_{12} = \{24, 48, 72, \ldots\}.$$

Because the least number in $M_8 \cap M_{12}$ is 24, the LCM of 8 and 12 is 24, written $\text{LCM}(8, 12) = 24$.

The Prime Factorization Method

The intersection-of-sets method for finding the LCM is often lengthy, especially when it is used to find the LCM of three or more natural numbers. Another, more efficient, method for finding the LCM of several numbers is the *prime factorization method*. For example, to find LCM(40, 12), first find the prime factorizations of 40 and 12, namely, $2^3 \cdot 5$ and $2^2 \cdot 3$, respectively.

If $m = \text{LCM}(40, 12)$, then m is a multiple of 40 and must contain both 2^3 and 5 as factors. Also, m is a multiple of 12 and must contain 2^2 and 3 as factors. Since 2^3 is a multiple of 2^2, then $m = 2^3 \cdot 5 \cdot 3 = 120$. In general, we have the following:

To find the LCM of two natural numbers, first find the prime factorization of each number. Then take each of the primes that are factors of either of the given numbers. The LCM is the product of these primes, each raised to the greatest power of the prime that occurs in either of the prime factorizations.

EXAMPLE 4-30

Find the LCM of 2520 and 10,530.

Solution

$$2520 = 2^3 \cdot 3^2 \cdot 5 \cdot 7.$$
$$10{,}530 = 2 \cdot 3^4 \cdot 5 \cdot 13.$$
$$\text{LCM}(2520, 10530) = 2^3 \cdot 3^4 \cdot 5 \cdot 7 \cdot 13 = 294{,}840$$

The prime factorization method can also be used to find the LCM of more than two numbers. For example, to find LCM(12, 108, 120), we can proceed as follows:

$$12 = 2^2 \cdot 3$$
$$108 = 2^2 \cdot 3^3$$
$$120 = 2^3 \cdot 3 \cdot 5$$

Then, $\text{LCM}(12, 108, 120) = 2^3 \cdot 3^3 \cdot 5 = 1080$.

To see the connection between the GCD and LCM, consider the GCD and LCM of 6 and 9. Because $6 = 2 \cdot 3$ and $9 = 3^2$, it follows that $\text{GCD}(6, 9) = 3$ and $\text{LCM}(6, 9) = 18$. Notice that $\text{GCD}(6, 9) \cdot \text{LCM}(6, 9) = 3 \cdot 18 = 54$, and 54 is the product of the original numbers 6 and 9. In general, for any two natural numbers a and b, the connection between their GCD and LCM is given by Theorem 4–8.

Theorem 4–8

For any two natural numbers a and b,

$$\text{GCD}(a, b) \cdot \text{LCM}(a, b) = ab.$$

Theorem 4-8 can be justified in several ways. Here is a specific example that suggests how the theorem might be proved.

Suppose $a = 5^{13} \cdot 7^{20} \cdot 11^4$ and

$b = 5^{10} \cdot 7^{25} \cdot 11^6 \cdot 13.$

Then, $\text{LCM}(a, b) = 5^{13} \cdot 7^{25} \cdot 11^6 \cdot 13$ and

$\text{GCD}(a, b) = 5^{10} \cdot 7^{20} \cdot 11^4.$

Now we have

$$\text{LCM}(a, b) \cdot \text{GCD}(a, b) = 5^{13+10} \cdot 7^{25+20} \cdot 11^{6+4} \cdot 13 \quad \text{and}$$
$$ab = 5^{13+10} \cdot 7^{20+25} \cdot 11^{4+6} \cdot 13$$

For the preceding values of a and b, Theorem 4-8 is true. Notice, however, that in the product $\text{LCM}(a, b) \cdot \text{GCD}(a, b)$ we have all the powers of the primes appearing in a or in b, because for the LCM we take the greater of the powers of the common primes and for GCD the lesser. Also in ab we have all the powers. Hence, Theorem 4-8 is true in general.

The Euclidean Algorithm Method

Theorem 4-8 is useful for finding the LCM of two numbers a and b when their prime factorizations are not easy to find. $\text{GCD}(a, b)$ can be found by the Euclidean algorithm, the product ab can be found by simple multiplication, and $\text{LCM}(a, b)$ can be found by division.

EXAMPLE 4-31

Find LCM(731, 952).

Solution By the Euclidean algorithm, GCD(731, 952) = 17. By Theorem 4-8,

$$17 \cdot \text{LCM}(731, 952) = 731 \cdot 952.$$

Consequently,

$$\text{LCM}(731, 952) = \frac{731 \cdot 952}{17} = 40{,}936.$$

▲

The Division-by-Primes Method

Another procedure for finding the LCM of several natural numbers involves *division by primes*. For example, to find LCM(12, 75, 120), we start with the least prime that divides at least one of the given numbers and divide as follows:

$$\begin{array}{r|l} 2 & 12,\ 75,\ 120 \\ \hline & 6,\ 75,\ 60 \end{array}$$

Because 2 does not divide 75, simply bring down the 75. To obtain the LCM using this procedure, continue the division process until the row of answers consists of relatively prime numbers.

$$\begin{array}{r|rrr} 2 & 12, & 75, & 120 \\ 2 & 6, & 75, & 60 \\ 2 & 3, & 75, & 30 \\ 3 & 3, & 75, & 15 \\ 5 & 1, & 25, & 5 \\ & 1, & 5, & 1 \end{array}$$

Thus, $\text{LCM}(12, 75, 120) = 2 \cdot 2 \cdot 2 \cdot 3 \cdot 5 \cdot 1 \cdot 5 \cdot 1 = 2^3 \cdot 3 \cdot 5^2 = 600$.

ONGOING ASSESSMENT 4-5

1. Find the GCD and the LCM for each of the following using the intersection-of-sets method:
   a. 18 and 10
   b. 24 and 36
   c. 8, 24, and 52
   d. 7 and 9
2. Find the GCD and the LCM for each of the following using the prime factorization method:
   a. 132 and 504 **b.** 65 and 1690
   c. 900, 96, and 630 **d.** 108 and 360
   e. 11 and 19
3. Find the GCD for each of the following using the Euclidean algorithm:
   a. 220 and 2924
   b. 14,595 and 10,856
   c. 122,368 and 123,152
4. Find the LCM for each of the following using any method:
   a. 24 and 36 **b.** 72 and 90 and 96
   c. 90 and 105 and 315 **d.** 9^{100} and 25^{100}
5. Find the LCM for each of the following pairs of numbers using Theorem 4–8 and the answers from problem 3:
   a. 220 and 2924
   b. 14,595 and 10,856
   c. 122,368 and 123,152
6. Use colored rods to find the GCD and the LCM of 6 and 10.
7. In Quinn's dormitory room, there are three snooze-alarm clocks, each of which is set at a different time. Clock A goes off every 15 min, clock B goes off every 40 min, and clock C goes off every 60 min. If all three clocks go off at 6:00 A.M., answer the following:
   a. How long will it be before the clocks go off together again after 6:00 A.M.?
   b. Would the answer to (a) be different if clock B went off every 15 min and clock A went off every 40 min?
8. Midas has 120 gold coins and 144 silver coins. He wants to place his gold coins and his silver coins in stacks so that there are the same number of coins in each stack. What is the greatest number of coins that he can place in each stack?
9. Bill and Sue both work at night. Bill has every sixth night off and Sue has every eighth night off. If they are both off tonight, how many nights will it be before they are both off again?
10. By selling cookies at 24¢ each, José made enough money to buy several cans of pop costing 45¢ per can. If he had no money left over after buying the pop, what is the least number of cookies he could have sold?
11. Bijous I and II start their movies at 7:00 P.M. The movie at Bijou I takes 75 min, while the movie at Bijou II takes 90 min. If the shows run continuously, when will they start at the same time again?
12. Two bike riders ride around in a circular path. The first rider completes one round in 12 min and the second rider completes it in 18 min. If they both start at the same place and the same time and go in the same direction, after how many minutes will they meet again at the starting place?
13. Three motorcyclists ride around a circular race course starting at the same place and the same time. The first passes the starting point every 12 min, the second every 18 min and the third every 16 min. After how many minutes will all three pass the starting point again at the same time? Explain your reasoning.
14. A rectangular field with dimensions 75 ft by 625 ft is to be divided into same-size square plots. If the sides of the squares need to be whole numbers of feet long, what are:
    a. The largest squares possible and how many such squares will fit in the field?
    b. What are the smallest squares possible?
    c. What other size squares are possible?
15. The principal of Valley Elementary School wants to divide each of the three fourth-grade classes into small same size groups with at least 2 students in each. If the classes have 18, 24, and 36 students, respectively, what size groups are possible?
16. Assume a and b are natural numbers and answer the following:
    a. If $\text{GCD}(a, b) = 1$, find $\text{LCM}(a, b)$.
    b. Find $\text{GCD}(a, a)$ and $\text{LCM}(a, a)$.

c. Find GCD(a^2, a) and LCM(a^2, a).
d. If $a|b$, find GCD(a, b) and LCM(a, b).
e. If a and b are two primes, find GCD(a, b) and LCM(a, b).
f. What is the relationship between a and b if GCD(a, b) $= a$?
g. What is the relationship between a and b if LCM(a, b) $= a$?

17. Classify each of the following as true or false:
a. If GCD(a, b) $= 1$, then a and b cannot both be even.
b. If GCD(a, b) $= 2$, then both a and b are even.
c. If a and b are even, then GCD(a, b) $= 2$.
d. For all natural numbers a and b, LCM(a, b)|GCD(a, b).
e. For all natural numbers a and b, LCM(a, b)|ab.
f. GCD(a, b) $\leq a$.
g. LCM(a, b) $\geq a$.

18. To find GCD(24, 20, 12), it is possible to find GCD(24, 20), which is 4, and then find GCD(4, 12), which is 4. Use this approach and the Euclidean algorithm to find
a. GCD(120, 75, 105)
b. GCD(34578, 4618, 4619)

19. a. Show that 97,219,988,751 and 4 are relatively prime.
b. Show that 181,345,913 and 11 are relatively prime.

20. The radio station gave away a discount coupon for every twelfth and 13th caller. Every twentieth caller received free concert tickets. Which caller was first to get both a coupon and a concert ticket?

21. Jackie spent the same amount of money on DVDs that she did on compact discs. If DVDs cost \$12 and CDs \$16, what is the least amount she could have spent on each?

22. Larry and Mary bought a special 360-day joint membership to a tennis club. Larry will use the club every other day, and Mary will use the club every third day. They both use the club on the first day. How many days will neither person use the club in the 360 days?

23. At the Party Store, paper plates come in packages of 30, paper cups in packages of 15, and napkins in packages of 20. What is the least number of plates, cups, and napkins that can be purchased so that there is an equal number of each?

24. Determine how many complete revolutions gear 2 in the following must make before the arrows are lined up again.

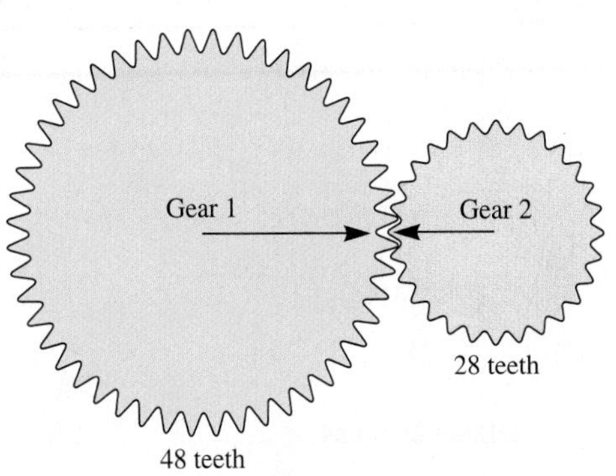

25. Determine how many complete revolutions each gear in the following must make before the arrows are lined up again:

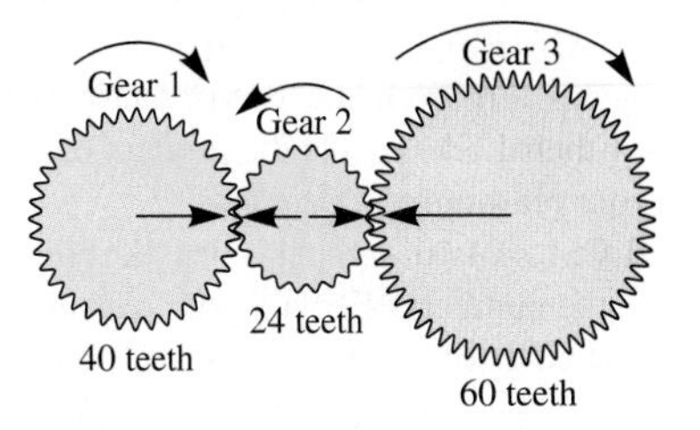

26. Venn diagrams can be used to show factors of two or more numbers. Draw Venn diagrams to show the common prime factors for each of the following sets of three numbers:
a. 10, 15, 60
b. 8, 16, 24

27. What are the factors of 4^{10}?

28. Find all natural numbers x such that GCD(25, x) $= 1$ and $1 \leq x \leq 25$.

29. In algebra it is often necessary to factor an expression as much as possible. For example, $a^3b^2 + a^2b^3 = a^2b^2(a + b)$ and no further factoring is possible without knowing the values of a and b. Factor each of the following as much as possible:
a. $12x^4y^3 + 18x^3y^4$
b. $12x^3y^2z^2 + 18x^2y^4z^3 + 24x^4y^3z^4$
c. $(6x - 6y)^2 + (3x - 3y)^2 + 9(x - y)^2$
d. $6(x^2 - y^2) - 3(x - y) + 9(y - x)$
e. $6(x^2 - y^2) + 12(x^2 - y^2) + 18(y^2 - x^2)$

30. Label the following statements as "always true," "sometimes true," or "never true." Justify your answers.
a. GCD(a, b) = GCD($|a|$, b) = GCD($|a|$, $|b|$)
b. GCD($-a$, b) = GCD(a, $-b$) = GCD($-a$, $-b$)
c. If GCD(a, b) $= 1$, then GCD(a^2, b^2) = GCD(a, b^3).
d. If LCM(a, b, c) $= abc$, then GCD(a, b, c) $= 1$.

31. Find the GCD and the LCM of each of the following. (Do not compute the products.)
a. 10!, 11!
b. 10!, 10! + 1
c. pqr, qrs (where p, q, r, s are prime numbers)
d. 2^{10}, 2^8

Communication

32. Can two whole numbers have a greatest common multiple? Explain your answer.

33. Describe to a sixth-grade student the difference between a divisor and a multiple.

34. Is it true that GCD(a, b, c) $\cdot$ LCM(a, b, c) $= abc$? Explain your answer.

35. A rectangular plot of land is 558 m by 1212 m. A surveyor needs to divide the plot into the largest possible square plots of the same size, being a whole number of meters long. What is the size of each square and how many square plots can be created? Explain your reasoning.

36. Suppose that GCD(a, b, c) = 1. Is it necessarily true that GCD(a, b) = GCD(b, c) = 1? Explain your reasoning.
37. Suppose GCD(a, b) = GCD(b, c) = 2. Does that always imply that GCD(a, b, c) = 2? Justify your answer.
38. How can you tell from the prime factorization of two numbers if their LCM equals the product of the numbers? Explain your reasoning.
39. Can the LCM of two numbers ever be greater than the product of the numbers? Explain your reasoning.
40. Jill finds the GCD $(2^{20} + 1, 2^{18} - 1)$ as follows:

$$\begin{aligned} \text{GCD}(2^{20} + 1, 2^{18} - 1) &= \text{GCD}(2^{20} + 1 - 2^2(2^{18} - 1), 2^{18} - 1) \\ &= \text{GCD}(5, 2^{18} - 1) \\ &= 1 \end{aligned}$$

a. Justify Jill's approach.
b. Show that GCD$(5, 2^{18} - 1) = 1$ without computing $2^{18} - 1$

41. Let GCD(m, n) = g and LCM(m, n) = l. Jackie conjectures that GCD($m + n$, l) = g for all integers m and n. Check Jackie's conjecture for three different pairs of integers.

Open-Ended

42. Examine three elementary-school textbooks and report on how the introduction of the topics of GCD and LCM differ in the different textbooks.
43. Make up a word problem that can be solved by finding the GCD and another that can be solved by finding the LCM. Solve your problems and explain why you are sure that your approach is correct.
44. Find three pairs of numbers for which the LCM of the numbers in a pair is smaller than the product of the two numbers.
45. Describe infinitely many pairs of numbers whose GCD is
a. 2 **b.** 6 **c.** 91

Cooperative Learning

46. **a.** In your group, discuss whether the Euclidean algorithm for finding the GCD of two numbers should be introduced in middle school (to all students? to some?) Why or why not?
b. If you decide that it should be introduced in middle school, discuss how it should be introduced. Report your group's decision to the class.

Review Problems

47. Find two whole numbers x and y such that

$$x \cdot y = 1{,}000{,}000$$

and neither x nor y contains any zeros as digits.

48. Fill each blank space with a single digit that makes the corresponding statement true. Find all possible answers.
a. 3|83_51
b. 11|8_691
c. 23|103_6
49. Is 3111 a prime? Prove your answer.
50. Find a number that has exactly six prime factors.
51. Produce the least positive number that is divisible by 2, 3, 4, 5, 6, 7, 8, 9, 10, and 11.
52. What is the greatest prime that must be used to determine if 2089 is prime?

NAEP Questions

The least common multiple of 8, 12, and a third number is 120. Which of the following could be the third number?

a. 15 **b.** 16
c. 24 **d.** 32
e. 48

TECHNOLOGY CORNER

1. Write a spreadsheet to generate the first 50 multiples of 3 and the first 50 multiples of 4. Describe the intersection of the two sets.
2. Use a spreadsheet to find the factors of 2486. How far down do you need to copy the formula to be sure you have found all the divisors?

	A	B
1	1	= 2486/A1
2	2	
3	3	

3. Make a spreadsheet with four columns:
Column A—the multiples of 6
Column B—the multiples of 9
Column C—the multiples of 12
Column D—the multiples of 15

a. What is the least number that appears in all four columns?
b. Explain how to find this number without using a spreadsheet.

For any $n \times m$ rectangle such that $\text{GCD}(n, m) = 1$, find a rule for determining the number of unit squares (1×1) that a diagonal passes through. For example, in the drawings in Figure 4-27 the diagonal passes through 8 and 6 unit squares, respectively.

FIGURE 4-27

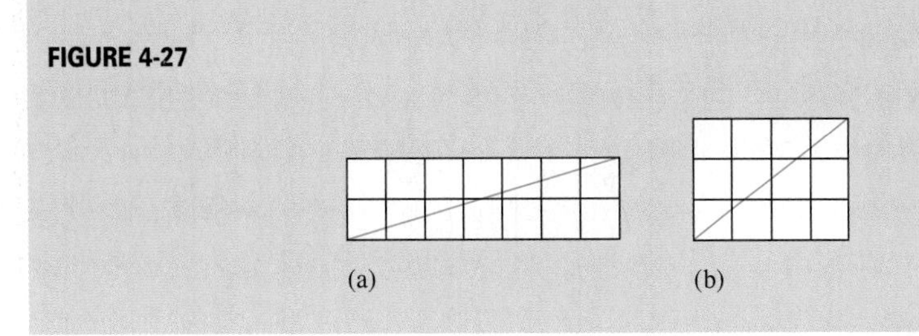

4-6 *Clock and Modular Arithmetic

In this section, we investigate clock arithmetic. Consider the following:

a. A doctor's prescription says to take a pill every 8 hr. If you take the first pill at 7:00 A.M., when should you take the next two pills?

b. Suppose you are following a bean soup recipe that calls for letting the beans soak for 12 hr. If you begin soaking them at 8:00 P.M., when should you take them out?

c. The odometer on a car gives the total miles traveled up to 99,999 miles and then starts counting from 0. If the odometer shows 99,124 miles, what will it show after a trip of 2,116 miles?

Some of these situations involve the ability to solve arithmetic problems using clocks. Most people can solve these problems without thinking much about what they are doing. It is possible to use the clock in Figure 4-28 to determine that 8 hr after 7:00 A.M. is 3:00 P.M. and 8 hr after that is 11:00 P.M. Also, 12 hr after 8:00 P.M. is 8:00 A.M. We could record these additions on the clock as

$$7 \oplus 8 = 3, \qquad 3 \oplus 8 = 11, \qquad 8 \oplus 12 = 8$$

where $\oplus$ indicates addition on a 12-hr clock.

FIGURE 4-28

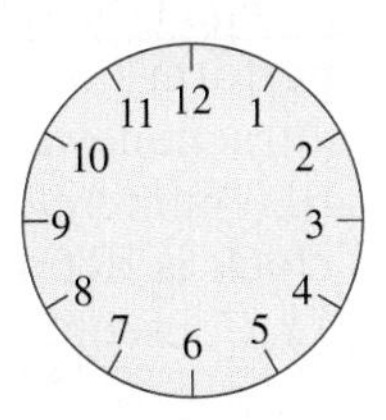

You probably noticed the special role of 12 when you found that $8 \oplus 12 = 8$. In the 12-hr clock arithmetic, 12 acts like a 0 if you were adding in the set of whole numbers. An addition table for the finite system based on the clock is shown in Table 4-3.

Table 4-3

+	12	1	2	3	4	5	6	7	8	9	10	11
12	12	1	2	3	4	5	6	7	8	9	10	11
1	1	2	3	4	5	6	7	8	9	10	11	12
2	2	3	4	5	6	7	8	9	10	11	12	1
3	3	4	5	6	7	8	9	10	11	12	1	2
4	4	5	6	7	8	9	10	11	12	1	2	3
5	5	6	7	8	9	10	11	12	1	2	3	4
6	6	7	8	9	10	11	12	1	2	3	4	5
7	7	8	9	10	11	12	1	2	3	4	5	6
8	8	9	10	11	12	1	2	3	4	5	6	7
9	9	10	11	12	1	2	3	4	5	6	7	8
10	10	11	12	1	2	3	4	5	6	7	8	9
11	11	12	1	2	3	4	5	6	7	8	9	10

NOW TRY THIS 4-13

Examine Table 4-3 to determine if the following properties hold for $\oplus$ on the set of numbers in the table:

a. Commutative property of addition
b. Identity property of addition
c. Inverse property of addition

When we allow numbers other than those on the 12-hr clock to be added, such as $8 \oplus 24 = 8$, we find that numbers such as 24, 36, 48, . . . act like 12 (or 0). Likewise, the numbers 13, 25, 37, . . . act like the number 1. Similarly, we can generate classes of numbers that act like each of the numbers on the 12-hr clock. The members of any one class differ by multiples of 12. Consequently, to perform additions on a 12-hr clock we perform regular addition, divide by 12, and record the remainder as the answer. For example, we can find $11 \oplus 8$ and $8 \oplus 12$ as follows:

> $11 + 8 = 19$. Next divide $19 \div 12$. The quotient is 1 with a remainder of 7, which is the answer.
>
> $8 + 12 = 20$. Next divide $20 \div 12$. The quotient is 1 with a remainder of 8, which is the answer.

Whenever the sum of digits on a 12-hr clock exceeds 12, add the numbers normally and then obtain the remainder when the sum is divided by 12.

To perform other operations on the clock, such as $2 \ominus 9$, where $\ominus$ denotes clock subtraction, we could interpret it as the time 9 hr before 2 o'clock. Counting backward (counterclockwise) 9 units from 2 reveals that $2 \ominus 9 = 5$. If subtraction on the clock is defined in terms of addition, we have $2 \ominus 9 = x$ if and only if $2 = 9 \oplus x$. Consequently, $x = 5$.

EXAMPLE 4-32

Perform each of the following computations on a 12-hr clock:

a. $8 \oplus 8$ **b.** $4 \ominus 12$ **c.** $4 \ominus 4$ **d.** $4 \ominus 8$

Solution **a.** $(8 + 8) \div 12$ has remainder 4. Hence, $8 \oplus 8 = 4$.

b. $4 \ominus 12 = 4$, since by counting forward or backward 12 hr, you arrive at the original position.

c. $4 \ominus 4 = 12$. This should be clear from looking at the clock, but it can also be found by using the definition of subtraction in terms of addition.

d. $4 \ominus 8 = 8$ because $8 \oplus 8 = 4$. ▲

Clock multiplication can be defined using repeated addition, as with whole numbers. For example, $2 \otimes 8 = 8 \oplus 8 = 4$, where $\otimes$ denotes clock multiplication. Similarly, $3 \otimes 5 = (5 \oplus 5) \oplus 5 = 10 \oplus 5 = 3$.

Clock division can be defined in terms of multiplication. For example, 8 ⨸ 5 = x, where ⨸ denotes clock division, if and only if $8 = 5 \otimes x$ for a unique x in the set $\{1, 2, 3, \ldots, 12\}$. Because $5 \otimes 4 = 8$, then 8 ⨸ 5 = 4.

EXAMPLE 4-33

Perform the following operations on a 12-hr clock, if possible:

a. $3 \otimes 11$ **b.** 2 ⨸ 7 **c.** 3 ⨸ 2 **d.** 5 ⨸ 12

Solution **a.** $3 \otimes 11 = (11 \oplus 11) \oplus 11 = 10 \oplus 11 = 9$.

b. 2 ⨸ 7 = x if and only if $2 = 7 \otimes x$. Consequently, $x = 2$.

c. 3 ⨸ 2 = x if and only if $3 = 2 \otimes x$. Multiplying each of the numbers 1, 2, 3, 4, . . . , 12 by 2 shows that none of the multiplications yields 3. Thus, the equation $3 = 2 \otimes x$ has no solution, and consequently, 3 ⨸ 2 is undefined.

d. 5 ⨸ 12 = x if and only if $5 = 12 \otimes x$. However, $12 \otimes x = 12$ for every x in the set $\{1, 3, 4, \ldots, 12\}$. Thus, $5 = 12 \otimes x$ has no solution on the clock, and therefore 5 ⨸ 12 is undefined. ▲

FIGURE 4-29

Table 4-4

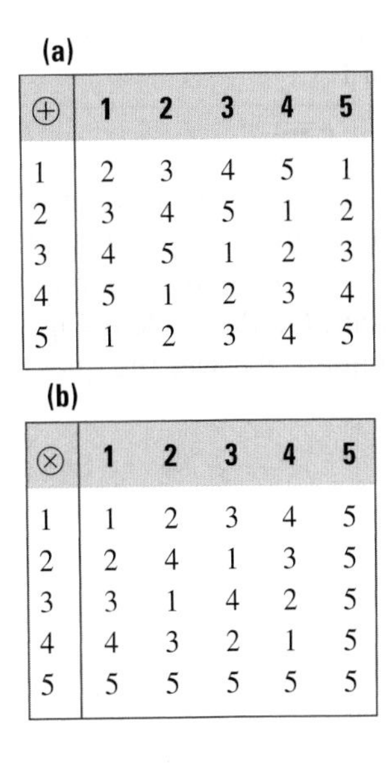

(a)

$\oplus$	1	2	3	4	5
1	2	3	4	5	1
2	3	4	5	1	2
3	4	5	1	2	3
4	5	1	2	3	4
5	1	2	3	4	5

(b)

$\otimes$	1	2	3	4	5
1	1	2	3	4	5
2	2	4	1	3	5
3	3	1	4	2	5
4	4	3	2	1	5
5	5	5	5	5	5

Adding or subtracting 12 on a 12-hr clock gives the same result. Thus 12 behaves as 0 does in a base-ten addition or subtraction and is the additive identity for addition on the 12-hr clock. Similarly, on a 5-hr clock 5 behaves as 0 does.

Addition, subtraction, and multiplication on a 12-hr clock can be performed for any two numbers but, as shown in Example 4-33(d), not all divisions can be performed. Division by 12, the additive identity, on a 12-hr clock either can never be performed or is not meaningful, since it does not yield a unique answer. However, there are clocks on which all divisions can be performed, except by the corresponding additive identities. One such clock is a 5-hr clock, shown in Figure 4-29.

On this clock, $3 \oplus 4 = 2$, $2 \ominus 3 = 4$, $2 \otimes 4 = 3$, and 3 ⨸ 4 = 2. Since adding 5 to any number yields the original number, 5 is the additive identity for this 5-hr clock, as seen in Table 4-4(a). Consequently, you might suspect that division by 5 is not possible on a 5-hour clock. To determine which divisions are possible, consider Table 4-4(b), a multiplication table for 5-hr clock arithmetic. To find 1 ⨸ 2, we write 1 ⨸ 2 = x, which is equivalent to $1 = 2 \otimes x$. The second row of Table 4-4(b) shows that $2 \otimes 1 = 2$, $2 \otimes 2 = 4$, $2 \otimes 3 = 1$, $2 \otimes 4 = 3$, and $2 \otimes 5 = 5$. The solution of

QUESTIONS FROM THE CLASSROOM

1. A fourth-grade student devised the following subtraction algorithm for subtracting 84 − 27.
4 minus 7 equals negative 3.

$$\begin{array}{r} 84 \\ -27 \\ \hline {}^{-}3 \end{array}$$

80 minus twenty equals 60.

$$\begin{array}{r} 84 \\ -27 \\ \hline {}^{-}3 \\ 60 \end{array}$$

60 plus negative 3 equals 57.

$$\begin{array}{r} 84 \\ -27 \\ \hline {}^{-}3 \\ +60 \\ \hline 57 \end{array}$$

Thus the answer is 57. What is your response as a teacher?
2. A seventh-grade student does not believe that $^{-}5 < {}^{-}2$. The student argues that a debt of \$5 is greater than a debt of \$2. How do you respond?
3. An eighth-grade student claims she can prove that subtraction of integers is commutative. She points out that if a and b are integers, then $a - b = a + {}^{-}b$. Since addition is commutative, so is subtraction. What is your response?
4. A student computes $^{-}8 - 2({}^{-}3)$ by writing $^{-}10({}^{-}3) = 30$. How would you help this student?
5. A student says that his father showed him a very simple method for dealing with expressions like $^{-}(a - b + 1)$ and $x - (2x - 3)$. The rule is, if there is a negative sign before the parentheses, change the signs of the expressions inside the parentheses. Thus, $^{-}(a - b + 1) = {}^{-}a + b - 1$ and $x - (2x - 3) = x - 2x + 3$. What is your response?
6. A student had the following picture of an integer and its opposite. Other students in the class objected, saying that ^{-}a should be to the left of 0. How do you respond?

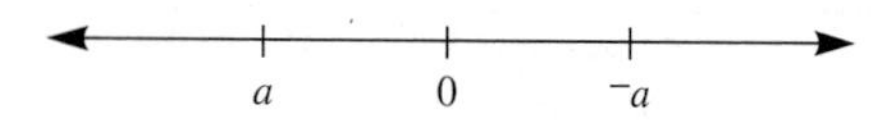

7. A student found that addition of integers can be performed by finding the sum or the difference of the absolute values of these integers and then attaching the "−" sign if necessary. She would like to know if this is always true. How do you respond?
8. A student claims that $a|a$ and $a|a$ implies $a|(a - a)$, and hence, $a|0$. Is the student correct?
9. A student writes, "If $d \nmid a$ and $d \nmid b$, then $d \nmid (a + b)$." How do you respond?
10. Your seventh-grade class has just completed a unit on divisibility rules. One of the better students asks why divisibility by numbers other than 3 and 9 cannot be tested by dividing the sum of the digits by the tested number. How should you respond?
11. A student says that a number with an even number of digits is divisible by 7 if and only if each of the numbers formed by pairing the digits into groups of two is divisible by 7. For example, 49,562,107 is divisible by 7, since each of the numbers 49, 56, 21, and 07 is divisible by 7. Is this true?
12. A sixth-grade student argues that there are infinitely many primes because "there is no end to numbers." How do you respond?
13. A student claims that a number is divisible by 21 if and only if it is divisible by 3 and by 7, and, in general, a number is divisible by $a \cdot b$ if, and only if it is divisible by a and by b. What is your response?
14. A student says that for any two integers a and b, GCD(a, b) divides LCM(a, b) and, hence, GCD$(a, b) <$ LCM(a, b). Is the student correct? Why or why not?
15. A student asks about the relation between least common multiple and least common denominator. How do you respond?
16. A student found that all three-digit numbers of the form aba, where $a + b$ is a multiple of 7, are divisible by 7. She would like to know why this is so. How do you respond?
17. A student wants to know how many integers between 1 and 10,000 are either multiples of 3 or multiples of 5. She wonders if it is correct to find the number of those integers that are multiples of 3 and add the number of those that are multiples of 5. How do you respond?
18. A student claims that every prime greater than 3 is a term in the arithmetic sequence whose nth term is $6n + 1$ or in the arithmetic sequence whose nth term is $6n - 1$. Is this true? If so why?
19. Dolores claims that she has a shortcut for finding the GCD using the Euclidean algorithm. She says when the remainder is large she uses a negative "remainder." For example, to find GCD(2132, 534), she divides 2132 by 534 and gets $2132 = 3 \cdot 534 + 530$, which gives remainder 530. In such case she writes $2132 = 4 \cdot 534 - 4$ and claims that

$$\begin{aligned} \text{GCD}(2132,534) &= \text{GCD}(-4, 534) \\ &= \text{GCD}(4, 534) \\ &= 2 \ \ (\text{because } 4 \nmid 534). \end{aligned}$$

Is the approach correct and if so, why?
20. Gwen claims that she found a proof that all integers greater than 2 are composite. Examine her "proof" and tell what is wrong with it.
Any prime number p greater than 2 is odd, hence $p + 1$ and $p - 1$ are even numbers. Thus, $\frac{p + 1}{2}$ and $\frac{p - 1}{2}$ are integers. You can check that

$p = \left(\frac{p+1}{2}\right)^2 - \left(\frac{p-1}{2}\right)^2$. If we let $\frac{p+1}{2} = m$ and $\frac{p-1}{2} = n$, then

$$p = m^2 - n^2 = (m - n)(m + n)$$

Since p is a product of two integers, $(m - n)$ and $(m + n)$, it is composite.

CHAPTER OUTLINE

I. Basic concepts of integers

A. The set of **integers**, I, is $\{\ldots, {}^-3, {}^-2, {}^-1, 0, 1, 2, 3, \ldots\}$.

B. The distance from any integer to 0 is the **absolute value** of the integer. The absolute value of an integer x is denoted $|x|$. If $x \geq 0$, then $|x| = x$ and if $x < 0$, then $|x| = {}^-x$.

C. Operations with integers

1. **Addition:** For any integers a and b,
$${}^-a + {}^-b = {}^-(a + b)$$
2. **Subtraction**
   - **a.** If a and b are any integers, then $a - b = n$ if and only if $a = b + n$.
   - **b.** For all integers a and b, $a - b = a + {}^-b$.
3. **Multiplication:** For any integers a and b,
   - **a.** $({}^-a) \cdot ({}^-b) = ab$.
   - **b.** $({}^-a) \cdot b = b \cdot ({}^-a) = {}^-(ab)$.
4. **Division:** If a and b are any integers with $b \neq 0$, then $a \div b$ is the unique integer c, if it exists, such that $a = bc$.
5. **Order of operations:** When addition, subtraction, multiplication, and division appear without parentheses, multiplications and divisions are done first in the order of their appearance from left to right and then additions and subtractions are done in the order of their appearance from left to right. Any arithmetic in parentheses is done first.

II. The system of integers

A. The set of integers, $I = \{\ldots, {}^-3, {}^-2, {}^-1, 0, 1, 2, 3, \ldots\}$, along with the operations of addition and multiplication, satisfy the following properties:

Property	+	×
Closure	Yes	Yes
Commutative	Yes	Yes
Associative	Yes	Yes
Identity	Yes, 0	Yes, 1
Inverse	Yes	No
Distributive Property of Multiplication over Addition		

B. **Zero multiplication property of integers** For any integer a, $a \cdot 0 = 0 = 0 \cdot a$.

C. **Addition property of equality:** For any integers a, b, and c, if $a = b$, then $a + c = b + c$.

D. **Multiplication property of equality:** For any integers a, b, and c, if $a = b$, then $ac = bc$.

E. **Substitution property:** Any number may be substituted for its equal.

F. **Cancellation properties of equality**

1. For any integers a, b, and c, if $a + c = b + c$, then $a = b$.
2. For any integers a, b, and c, if $c \neq 0$ and $ac = bc$, then $a = b$.

G. For all integers a, b, and c,

1. ${}^-({}^-a) = a$.
2. $a - (b - c) = a - b + c$
3. $(a + b)(a - b) = a^2 - b^2$ **(difference-of-squares formula)**.

III. Divisibility

A. If a and b are any integers, then b **divides** a, denoted $b|a$, if and only if there is a unique integer c such that $a = cb$.

B. The following are basic divisibility theorems for integers a, b, and d:

1. If $d|a$ and k is any integer, then $d|ka$.
2. If $d|a$ and $d|b$, then $d|(a + b)$ and $d|(a - b)$.
3. If $d|a$ and $d \nmid b$, then $d \nmid (a + b)$ and $d \nmid (a - b)$.

C. Divisibility tests

1. An integer is divisible by 2, 5, or 10 if and only if its units digit is divisible by 2, 5, or 10, respectively.
2. An integer is divisible by 4 if and only if the last two digits of the integer represent a number divisible by 4.
3. An integer is divisible by 8 if and only if the last three digits of the integer represent a number divisible by 8.
4. An integer is divisible by 3 or by 9 if and only if the sum of its digits is divisible by 3 or 9, respectively.
5. An integer is divisible by 11 if and only if the sum of the digits in the places that are even powers of 10 minus the sum of the digits in the places that are odd powers of 10 is divisible by 11.
6. An integer is divisible by 6 if and only if the integer is divisible by both 2 and 3.

IV. Prime and composite numbers

A. Positive integers that have exactly two positive divisors are **primes**. Integers greater than 1 and not primes are **composites**.

B. **Fundamental Theorem of Arithmetic:** Every composite number has one and only one prime factorization, aside from variation in the order of the prime factors.

2. Rewrite both fractions with the same least common denominator. Since LCM(42, 35) = 210, then

$$\frac{12}{42} = \frac{60}{210} \text{ and } \frac{10}{35} = \frac{60}{210}.$$

Thus,

$$\frac{12}{42} = \frac{10}{35}.$$

3. Rewrite both fractions with a common denominator (not necessarily the least). A common multiple of 42 and 35 may be found by finding the product $42 \cdot 35$, or 1470. Now,

$$\frac{12}{42} = \frac{420}{1470} \text{ and } \frac{10}{35} = \frac{420}{1470}.$$

Hence,

$$\frac{12}{42} = \frac{10}{35}.$$

The third method suggests a general algorithm for determining if two fractions $\frac{a}{b}$ and $\frac{c}{d}$ are equal. Rewrite both fractions with common denominator bd. That is,

$$\frac{a}{b} = \frac{ad}{bd} \text{ and } \frac{c}{d} = \frac{bc}{bd}.$$

Because the denominators are the same, $\frac{ad}{bd} = \frac{bc}{bd}$ if and only if $ad = bc$. For example, $\frac{24}{36} = \frac{6}{9}$ because $24 \cdot 9 = 216 = 36 \cdot 6$. In general, the following property results.

Property

Two fractions $\frac{a}{b}$ and $\frac{c}{d}$ are equal if and only if $ad = bc$.

Using a calculator, we can determine if two fractions are equal by using the preceding property. Since both [2] [×] [2] [1] [9] [6] [=] and [4] [×] [1] [0] [9] [8] [=] yield a display of 4392, we see that $\frac{2}{4} = \frac{1098}{2196}$.

Ordering Rational Numbers

As discussed in the *Principles and Standards* (p. 392), students in lower grades should experience comparing fractions between 0 and 1 in relation to such benchmarks as 0, $\frac{1}{2}$, $\frac{3}{4}$, and 1. In the middle grades, the comparison of fractions becomes more difficult. We first consider the comparison of fractions with like denominators.

Children know that $\frac{7}{8} > \frac{5}{8}$ because if a pizza is divided into 8 parts, then 7 parts of a pizza is more than 5 parts. Similarly, $\frac{3}{7} < \frac{4}{7}$. Thus, given two fractions with common positive denominators and nonnegative numerators, the one with the greater numerator is the greater fraction. This can be stated as follows.

Theorem 5–1

If a, b, and c are integers and $b > 0$, then $\frac{a}{b} > \frac{c}{b}$ if and only if $a > c$.

NOW TRY THIS 5-5

Determine if Theorem 5–1 is true if $b < 0$.

Comparing fractions with unlike denominators is more difficult. For example, students may incorrectly reason that $\frac{1}{8} > \frac{1}{7}$ because 8 is greater than 7. In other cases, they might falsely believe that $\frac{6}{7}$ is equal to $\frac{7}{8}$ because in both cases the difference between the numerator and the denominator is 1. Comparing fractions with unlike denominators may be aided by using fraction strips to compare the fractions visually. For example, consider the fractions $\frac{4}{5}$ and $\frac{11}{12}$ shown in Figure 5-5.

FIGURE 5-5

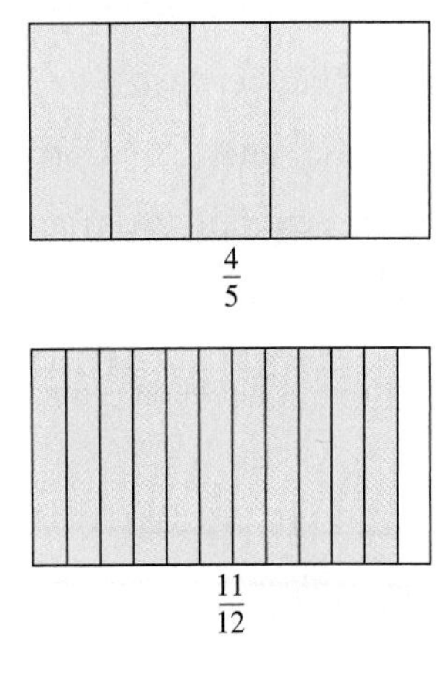

From Figure 5-5, students see that each fraction is one piece less than the same-size whole unit. However, they see that the missing piece for $\frac{11}{12}$ is smaller than the missing piece for $\frac{4}{5}$, so $\frac{11}{12}$ must be greater than $\frac{4}{5}$.

Comparing fractions with unlike denominators can be accomplished by rewriting the fractions with the same denominator. Then we can compare them using a previously learned technique. Using the common denominator bd, we can write the fractions $\frac{a}{b}$ and $\frac{c}{d}$ as $\frac{ad}{bd}$ and $\frac{bc}{bd}$. Because $b > 0$ and $d > 0$, then $bd > 0$ and we can apply Theorem 5–1 as follows:

$$\frac{a}{b} > \frac{c}{d} \text{ if and only if } \frac{ad}{bd} > \frac{bc}{bd} \quad \text{and} \quad \frac{ad}{bd} > \frac{bc}{bd} \text{ if and only if } ad > bc$$

Therefore, we have the following theorem.

Theorem 5–2

If a, b, c, and d are integers and $b > 0$, $d > 0$, $\frac{a}{b} > \frac{c}{d}$ if and only if $ad > bc$.

NOW TRY THIS 5-6

Order the fractions $\frac{3}{4}$, $\frac{9}{16}$, $\frac{5}{8}$, and $\frac{2}{3}$ from least to greatest.

RESEARCH NOTE Students taught the common denominator method for comparing two fractions tend to ignore it and focus on rules associated with ordering whole numbers. Students who correctly compare numerators if the denominators are equal often compare denominators if the numerators are equal (Behr et al. 1984).

As suggested in the Research Note above, if students compare denominators of fractions if the numerators are the same, then this can be a good strategy if handled properly. For example, consider $\frac{3}{4}$ and $\frac{3}{10}$. If the whole is the same for both fractions, this means that we have three $\frac{1}{4}$s and three $\frac{1}{10}$s. Because $\frac{1}{4}$ is greater than $\frac{1}{10}$, then three of the larger parts is greater than three of the smaller parts, so $\frac{3}{4} > \frac{3}{10}$.

Denseness of Rational Numbers

The set of rational numbers has a property that the set of whole numbers and the set of integers do not have. Consider $\frac{1}{2}$ and $\frac{2}{3}$. To find a rational number between $\frac{1}{2}$ and $\frac{2}{3}$, we first rewrite the fractions with a common denominator, as $\frac{3}{6}$ and $\frac{4}{6}$. Because there is no whole number between the numerators 3 and 4, we next find two fractions equal, respectively, to $\frac{1}{2}$ and $\frac{2}{3}$ with greater denominators. For example, $\frac{1}{2} = \frac{6}{12}$ and $\frac{2}{3} = \frac{8}{12}$, and $\frac{7}{12}$ is between the two fractions $\frac{6}{12}$ and $\frac{8}{12}$. So $\frac{7}{12}$ is between $\frac{1}{2}$ and $\frac{2}{3}$. This property is generalized as follows.

Denseness Property for Rational Numbers

Given rational numbers $\frac{a}{b}$ and $\frac{c}{d}$, there is another rational number between these two numbers.

NOW TRY THIS 5-7

Explain why there are infinitely many rational numbers between any two rational numbers.

EXAMPLE 5-3

Find two fractions between $\frac{7}{18}$ and $\frac{1}{2}$.

Solution Because $\frac{1}{2} = \frac{1 \cdot 9}{2 \cdot 9} = \frac{9}{18}$, we see that $\frac{8}{18}$, or $\frac{4}{9}$, is between $\frac{7}{18}$ and $\frac{9}{18}$. To find another fraction between the given fractions, we find two fractions equal to $\frac{7}{18}$ and $\frac{9}{18}$, respectively, but with greater denominators; for example, $\frac{7}{18} = \frac{14}{36}$ and $\frac{9}{18} = \frac{18}{36}$.

We now see that $\frac{15}{36}, \frac{16}{36}$, and $\frac{17}{36}$ are all between $\frac{14}{36}$ and $\frac{18}{36}$ and thus between $\frac{7}{18}$ and $\frac{1}{2}$.

Some students incorrectly add $\frac{a}{b} + \frac{c}{d}$ as $\frac{a + c}{b + d}$. Although the technique does not produce the correct sum, as will be seen in Section 5-2, it provides a way to find a number between any two rational numbers. In Example 5-3, to find a number between $\frac{7}{18}$ and $\frac{1}{2}$ we could add the numerators and add the denominators to produce $\frac{7 + 1}{18 + 2} = \frac{8}{20}$. We see that $\frac{7}{18} < \frac{8}{20}$ because $140 < 144$. Also, $\frac{8}{20} < \frac{1}{2}$ because $16 < 20$. We state the general property in the following theorem, whose proof is explored in Now Try This 5-8.

Theorem 5-3

Let $\frac{a}{b}$ and $\frac{c}{d}$ be any rational numbers with positive denominators, where $\frac{a}{b} < \frac{c}{d}$. Then, $\frac{a}{b} < \frac{a + c}{b + d} < \frac{c}{d}$.

NOW TRY THIS 5-8

Prove Theorem 5–3; that is, if $\frac{a}{b}$ and $\frac{c}{d}$ are any rational numbers, where $\frac{a}{b} < \frac{c}{d}$, then $\frac{a}{b} < \frac{a + c}{b + d} < \frac{c}{d}$. (*Hint:* If $\frac{a}{b} < \frac{c}{d}$, then $ad < bc$. Now add ab to both sides and use this to show that $\frac{a}{b} < \frac{a + c}{b + d}$. Then finish the proof in a similar way.)

LABORATORY ACTIVITY

Obtain tangram pieces or build them and cut them out as shown in Figure 5-6. Answer each of the following.

a. If the area of the entire square is 1 square unit, find the area of each tangram piece.
b. If the area of piece a is 1 square unit, find the area of each tangram piece.

FIGURE 5-6

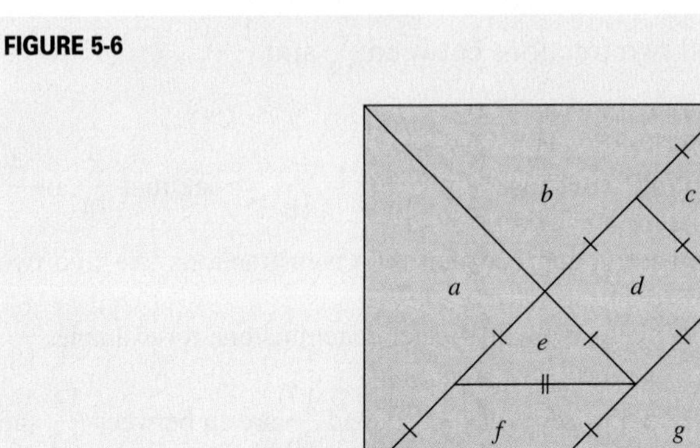

ONGOING ASSESSMENT 5-1

1. Write a sentence that illustrates the use of $\frac{7}{8}$ in each of the following ways:
   a. As a division problem
   b. As part of a whole
   c. As a ratio
2. For each of the following, write a fraction to represent the shaded portion:

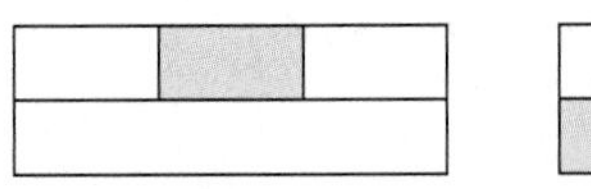
(a)

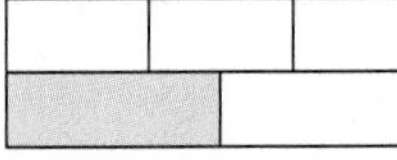
(b)

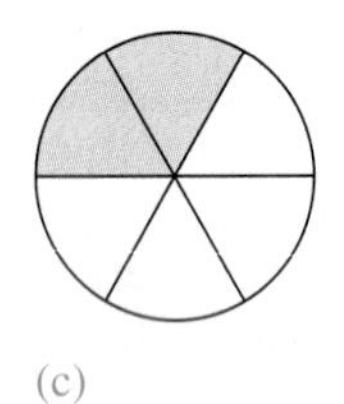
(c)

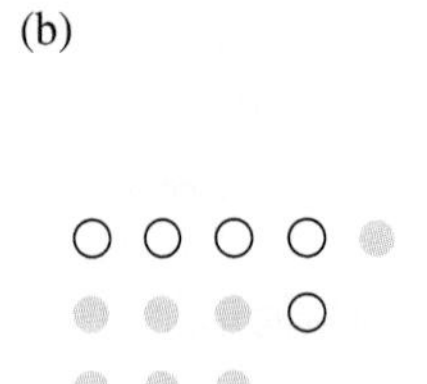
(d)

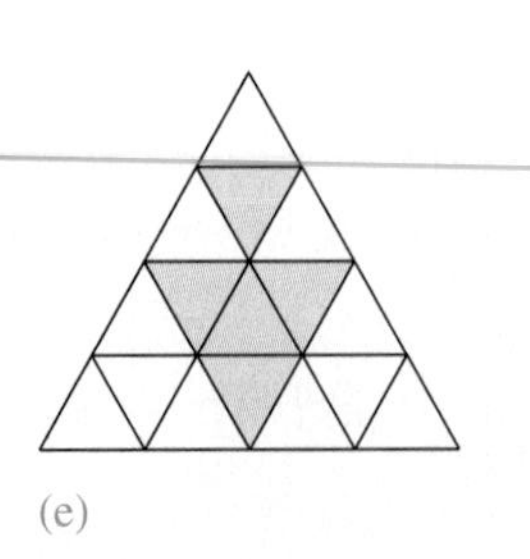
(e)

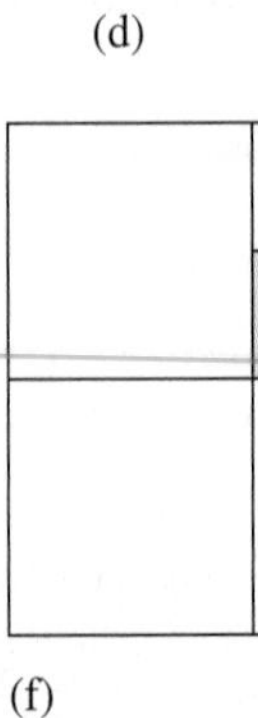
(f)

3. For each of the following four squares, write a fraction to represent the shaded portion. What property of fractions does the diagram illustrate?

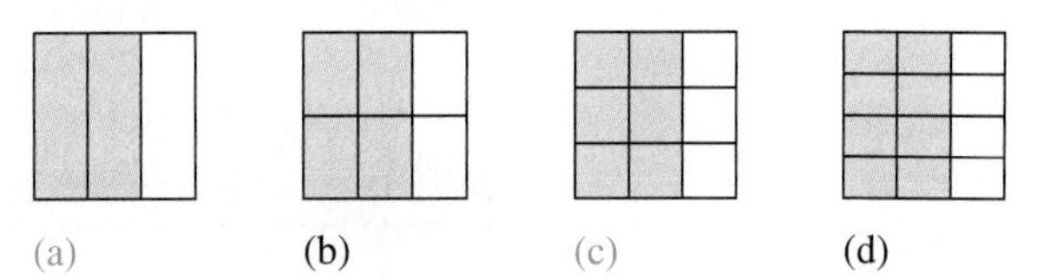
(a) (b) (c) (d)

4. Complete each of the following figures so that it shows $\frac{3}{5}$:

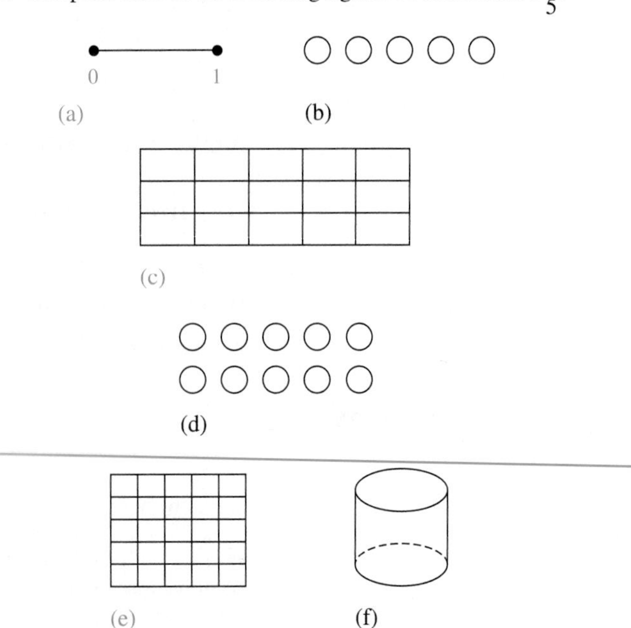

5. Could the shaded portions in the following figures represent the indicated fractions? Tell why.

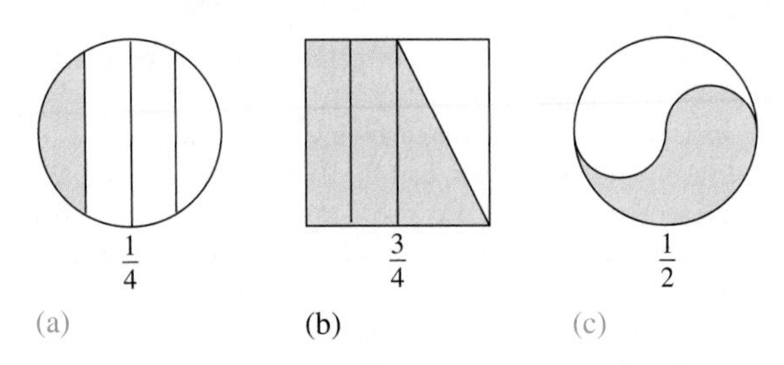

(a) $\frac{1}{4}$ (b) $\frac{3}{4}$ (c) $\frac{1}{2}$

6. If each of the following models represents the given fraction, draw a model that represents the *whole*.

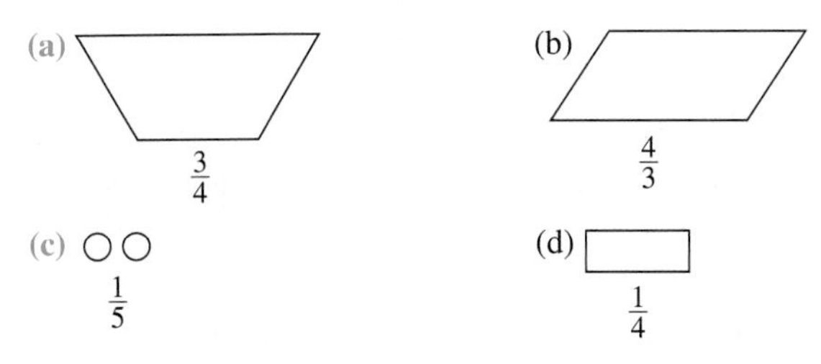

(a) $\frac{3}{4}$ (b) $\frac{4}{3}$ (c) $\frac{1}{5}$ (d) $\frac{1}{4}$

7. In each case, subdivide the *whole* shown on the right to show the equivalent fraction.

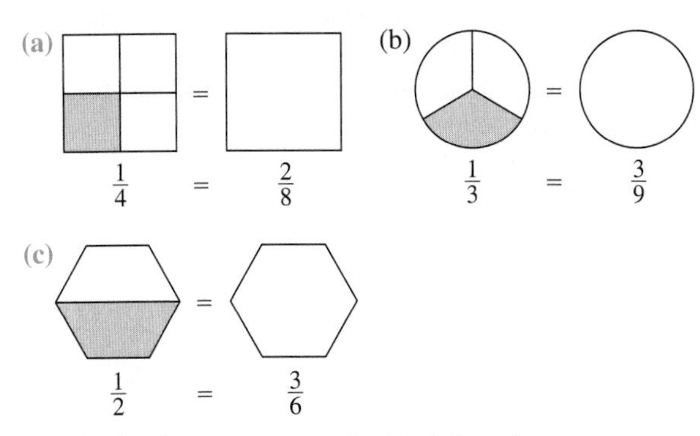

(a) $\frac{1}{4} = \frac{2}{8}$ (b) $\frac{1}{3} = \frac{3}{9}$ (c) $\frac{1}{2} = \frac{3}{6}$

8. Write a fraction to represent the shaded portion.

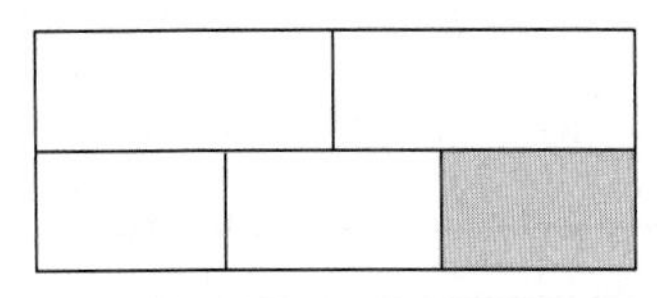

9. Referring to the figure, represent each of the following as a fraction:

a. The dots inside the circle as a part of all the dots
b. The dots inside the rectangle as a part of all the dots
c. The dots in the intersection of the rectangle and the circle as a part of all the dots
d. The dots outside the circle but inside the rectangle as part of all the dots

10. For each of the following, write three fractions equal to the given fraction:

a. $\frac{2}{9}$ **b.** $\frac{^-2}{5}$ c. $\frac{0}{3}$ **d.** $\frac{a}{2}$

11. Find the simplest form for each of the following fractions:

a. $\frac{156}{93}$ **b.** $\frac{27}{45}$ c. $\frac{^-65}{91}$
d. $\frac{0}{68}$ e. $\frac{84^2}{91^2}$ **f.** $\frac{662}{703}$

12. Mr. Gonzales and Ms. Price gave the same test to their fifth-grade classes. In Mr. Gonzales's class, 20 out of 25 students passed the test, and in Ms. Price's class, 24 out of 30 students passed the test. One of Ms. Price's students heard about the results of the tests and claimed that the classes did equally well. Is the student right? Explain.

13. For each of the following, choose the expression in parentheses that equals or describes best the given fraction:

a. $\frac{0}{0}$ (1, undefined, 0) **b.** $\frac{5}{0}$ (undefined, 5, 0)
c. $\frac{0}{5}$ (undefined, 5, 0)
d. $\frac{2+a}{a}$ (2, 3, cannot be simplified)
e. $\frac{15+x}{3x}$ $\left(\frac{5+x}{x}, 5, \text{cannot be simplified}\right)$
f. $\frac{2^6+2^5}{2^4+2^7}$ $\left(1, \frac{2}{3}, \text{cannot be simplified}\right)$
g. $\frac{2^{100}+2^{98}}{2^{100}-2^{98}}$ $\left(2^{196}, \frac{5}{3}, \text{too large to simply}\right)$

14. Find the simplest form for each of the following:

a. $\frac{a^2-b^2}{3a+3b}$ **b.** $\frac{14x^2y}{63xy^2}$ c. $\frac{a^2+ab}{a+b}$ **d.** $\frac{a}{3a+ab}$

15. Determine if the following pairs are equal:

a. $\frac{3}{8}$ and $\frac{375}{1000}$ **b.** $\frac{18}{54}$ and $\frac{23}{69}$
c. $\frac{6}{10}$ and $\frac{600}{1000}$ **d.** $\frac{17}{27}$ and $\frac{25}{45}$

16. Determine if the following pairs are equal by changing both to the same denominator:

a. $\frac{10}{16}$ and $\frac{12}{18}$ **b.** $\frac{3}{12}$ and $\frac{41}{154}$
c. $\frac{3}{^-12}$ and $\frac{^-36}{144}$ **d.** $\frac{^-21}{86}$ and $\frac{^-51}{215}$

17. A board is needed that is exactly $\frac{11}{32}$ in. wide to fill a hole. Can a board that is $\frac{3}{8}$ in. be shaved down to fit the hole? If so, how much must be shaved from the board?

18. Draw an area model to show that $\frac{3}{4} = \frac{6}{8}$.

19. If a fraction is equal to $\frac{3}{4}$ and the sum of the numerator and denominator is 84, what is the fraction?

20. The following two parking meters are next to each other with the times left as shown. Which meter has more time left on it? How much more?

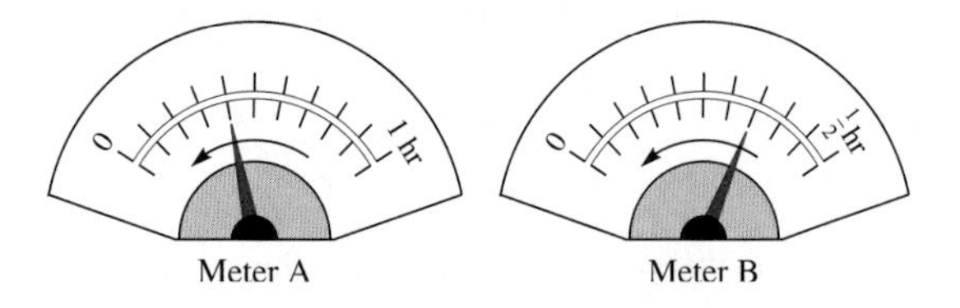

21. Mr. Gomez filled his car's 16-gal gas tank. He took a short trip and used 6 gal of gas. Draw an arrow in the following figure to show what his gas gauge looked like after the trip:

22. Read each measurement as shown on the following ruler:

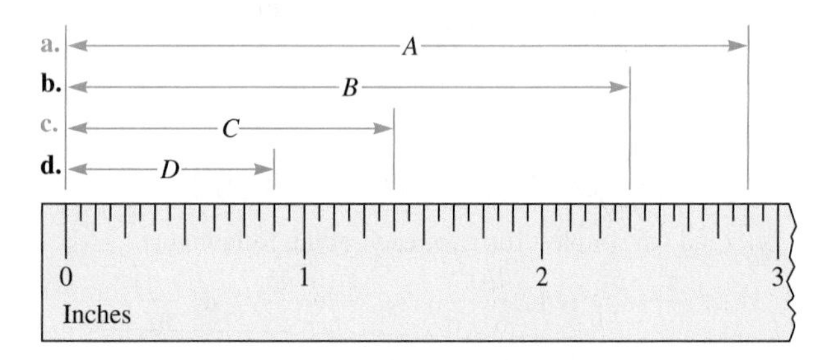

23. Determine if each of the following is correct. Explain.

a. $\dfrac{ab + c}{b} = a + c$

b. $\dfrac{a + b}{a + c} = \dfrac{b}{c}$

c. $\dfrac{ab + ac}{ac} = \dfrac{b + c}{c}$

24. Solve for x in each of the following:

a. $\frac{2}{3} = \frac{x}{16}$ **b.** $\frac{3}{4} = \frac{^-27}{x}$ c. $\frac{3}{x} = \frac{3x}{x^2}$

25. a. If $\frac{a}{c} = \frac{b}{c}$, what must be true?

b. If $\frac{a}{b} = \frac{a}{c}$, what must be true?

26. Let W be the set of whole numbers, N be the set of natural numbers, I be the set of integers, and Q the set of rational numbers.

a. Write sentences to describe the relationships among N, W, I, and Q.

b. Draw a Venn diagram showing the relationships.

27. In Amy's algebra class, 6 of the 31 students received A's on a test. The same test was given to Bren's class and 5 of the 23 students received A's. Which class had the higher rate of A's?

28. For each of the following pairs of fractions, replace the comma with the correct symbol (<, =, >) to make a true statement:

a. $\frac{7}{8}, \frac{5}{6}$ **b.** $2\frac{4}{5}, 2\frac{3}{6}$ c. $\frac{^-7}{8}, \frac{^-4}{5}$

d. $\frac{1}{^-7}, \frac{1}{^-8}$ e. $\frac{2}{5}, \frac{4}{10}$ **f.** $\frac{0}{7}, \frac{0}{17}$

29. Arrange each of the following in decreasing order:

a. $\frac{11}{22}, \frac{11}{16}, \frac{11}{13}$ **b.** $\frac{^-1}{5}, \frac{^-19}{36}, \frac{^-17}{30}$

30. If $\frac{a}{b} < 1$ and $\frac{c}{d} > 0$, compare the size of $\frac{c}{d}$ with $\frac{ac}{bd}$.

31. Show that the sequence $\frac{1}{2}, \frac{2}{3}, \frac{3}{4}, \frac{4}{5}, \frac{5}{6}, \frac{6}{7}, \ldots$ is an increasing sequence; that is, show that each term in the sequence is greater than the preceding one.

32. For each of the following, find two rational numbers between the given fractions:

a. $\frac{3}{7}$ and $\frac{4}{7}$ **b.** $\frac{^-7}{9}$ and $\frac{^-8}{9}$

c. $\frac{5}{6}$ and $\frac{83}{100}$ **d.** $\frac{^-1}{3}$ and $\frac{3}{4}$

33. Consider the following number grid. The circled numbers form a rhombus (that is, all sides are the same length).

1	2	3	4	5	6	7	(8)	9	10
11	12	13	14	15	16	17	18	19	(20)
21	22	23	24	25	26	(27)	28	29	30
31	32	33	34	35	36	37	38	(39)	40
41	42	43	44	45	46	47	48	49	50

a. If A is the sum of the four circled numbers and B is the sum of the four interior numbers, find $\frac{A}{B}$.

b. Form a rhombus by circling the numbers 6, 18, 25, and 37. Compute A and B as in (a) and then find $\frac{A}{B}$.

c. How do the answers in (a) and (b) compare? Why does this happen?

34. A scale on a map is 12 mi to the inch. What is the airline mileage between two cities that are 38 in. apart on the map?

35. a. Six oz is what part of a pound? A ton?
    b. A dime is what fraction of a dollar?
    c. 15 min is what fraction of an hour?
    d. 8 hr is what fraction of a day?

Communication

36. Jane has a recipe that calls for 3 c of flour. She wants to make $\frac{3}{4}$ of the recipe. Instead of determining directly how many cups are needed for the new recipe, she fills $\frac{3}{4}$ of a cup 3 times. Explain why Jane's method works.

37. Ann claims that she can't show $\frac{3}{4}$ of the following faces because some are big and some are small. What do you tell her?

38. A student claims that $\frac{2}{3} = \frac{6}{7}$ because if you add 4 to both the top and the bottom, the fractions must be equal. How do you respond?

39. If $\frac{1}{3}$ of each of two classes is female, explain whether each class must contain the same number of females.

40. Consider the set of all fractions equal to $\frac{1}{2}$. If you take any 10 of those fractions, add their numerators to obtain the numerator of a new fraction and add their denominators to obtain the denominator of a new fraction, how does the new fraction relate to $\frac{1}{2}$? Generalize what you found and explain.

41. Should fractions always be reduced to their simplest form? Why or why not?

42. How would you respond to each of the following students?
    a. Iris claims that if we have two positive rational numbers, the one with the greater numerator is the greater.
    b. Shirley claims that if we have two positive rational numbers, the one with the greater denominator is the lesser.

43. If we take the set of fractions equivalent to $\frac{1}{3}$ and graph them as points on a coordinate system so that the numerator becomes the x-coordinate and the denominator becomes the y-coordinate for that point, explain what type of graph we would get.

44. Write an explanation of how to convert inches to yards and vice versa.

Open-Ended

45. List three types of measure that require rational numbers as the appropriate number of units in the measurements.

46. Some people have argued that the system of integers is more understandable than the system of positive rational numbers. If you could decide which should be taught first in school, which would you choose and why?

Cooperative Learning

47. Assume the tallest person in your group is 1 unit tall and do the following:
    a. Find rational numbers to approximately represent other members of the group.
    b. Make a number line using the rational numbers for each person ordered according to height.

Third International Mathematics and Science Study (TIMSS) Questions

Which shows $\frac{2}{3}$ of the square shaded?

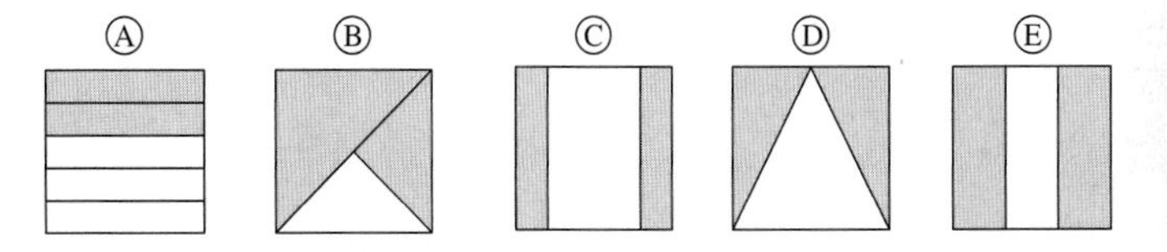

TIMSS 2003, Grade 4

In the figure, how many MORE small squares need to be shaded so that $\frac{4}{5}$ of the small squares are shaded?

a. 5 **b.** 4 **c.** 3 **d.** 2 **e.** 1

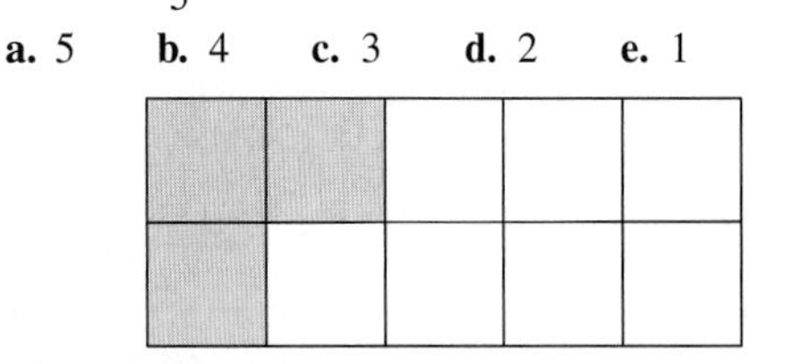

TIMSS 2003, Grade 8

BRAIN TEASER

In an old Sam Loyd puzzle, a watch is described as having stopped when the minute and hour hands formed a straight line, and the second hand was not on 12. At what times can this happen?

5-2 Addition and Subtraction of Rational Numbers

Addition and subtraction of rational numbers is very much like addition and subtraction of whole numbers and integers. We first demonstrate the addition of two rational numbers with like denominators, $\frac{2}{5} + \frac{1}{5}$, using an area model in Figure 5-7(a) and a number-line model in Figure 5-7(b).

FIGURE 5-7

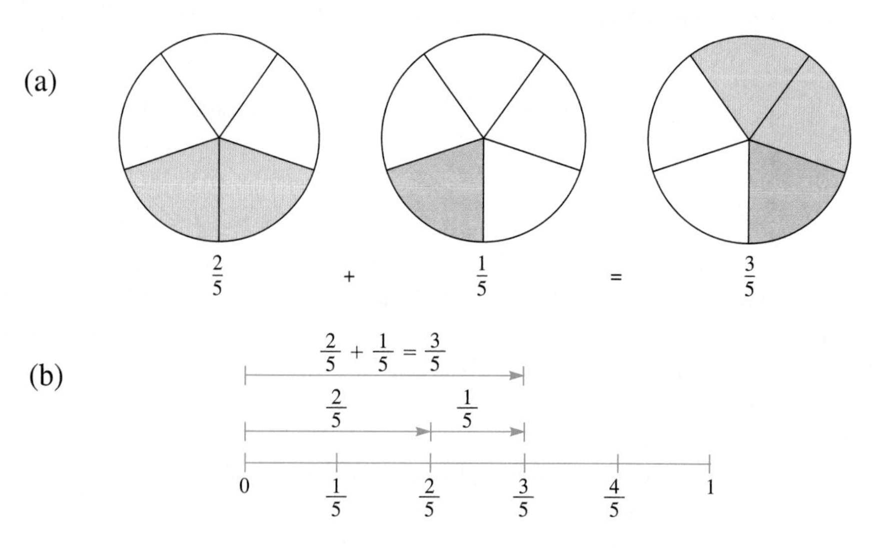

Why does the area model in Figure 5-7(a) make sense? Suppose that someone gives you $\frac{2}{5}$ of a pie to start with and then gives you another $\frac{1}{5}$ of the pie. In Figure 5-7(a), $\frac{2}{5}$ is represented by 2 pieces when the pie is cut into 5 equal-size pieces and $\frac{1}{5}$ is represented by 1 piece of the 5 equal-size pieces. So all together you have $2 + 1 = 3$ pieces of the 5 equal-size pieces, or $\frac{3}{5}$ of the total (*whole*) pie. The number-line model in Figure 5-7(b) works the same as the number-line model for whole numbers.

The ideas illustrated in Figure 5-7 can be applied to the sum of two rational numbers with like denominators and are summarized in the following definition.

Definition of Addition of Rational Numbers with Like Denominators

If $\frac{a}{b}$ and $\frac{c}{b}$ are rational numbers, then $\frac{a}{b} + \frac{c}{b} = \frac{a + c}{b}$.

RESEARCH NOTE

Students learning computational algorithms involving fractions have difficulty connecting their models involving manipulatives with symbolic procedures. Often, students' personal competence with a rote procedure "outstrips" their conceptual understanding of fractions; the unfortunate result is that students cannot monitor their work and can check their work only by repeating the rote procedure. They are unable to judge the reasonableness of the answer (Wearne and Hiebert 1988).

Next we consider the addition of two rational numbers with unlike denominators, using Polya's four-step, problem-solving process.

PROBLEM SOLVING **Adding Rational Numbers Problem**

Determine how to add the rational numbers $\frac{2}{3}$ and $\frac{1}{4}$.

Understanding the Problem We can model $\frac{2}{3}$ and $\frac{1}{4}$ as parts of a whole, as seen in Figure 5-8, but we need a way to combine the two drawings to find the sum.

FIGURE 5-8

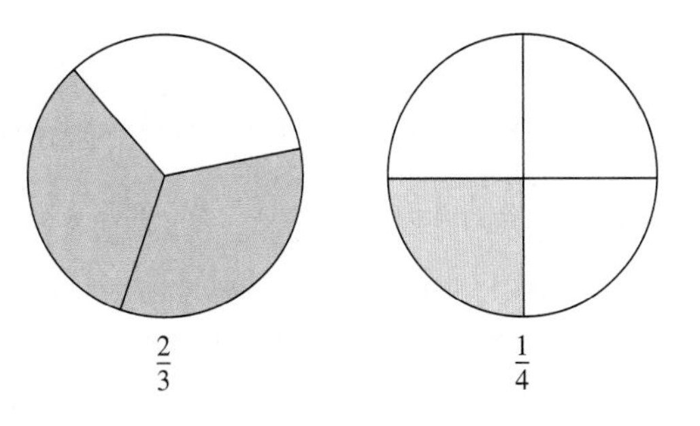

Devising a Plan We use the strategy of *solving a related problem* and consider adding rational numbers with the same denominators. We can find the sum by writing each fraction with a common denominator and then completing the computation.

Carrying Out the Plan From earlier work in the chapter, we know that $\frac{2}{3}$ is equal to infinitely many representations of the rational numbers $\frac{2}{3}$, including $\frac{4}{6}, \frac{6}{9}, \frac{8}{12}$, and so on. Also $\frac{1}{4}$ is equal to infinitely many rational numbers, including $\frac{2}{8}, \frac{3}{12}, \frac{4}{16}$, and so on. By comparing the two sets of rational numbers, we see that $\frac{8}{12}$ and $\frac{3}{12}$ have the same denominator. One is 8 parts of 12 equal parts, while the other is 3 parts of 12 equal parts. Consequently, the sum is $\frac{2}{3} + \frac{1}{4} = \frac{8}{12} + \frac{3}{12} = \frac{11}{12}$. Figure 5-9 illustrates the addition.

FIGURE 5-9

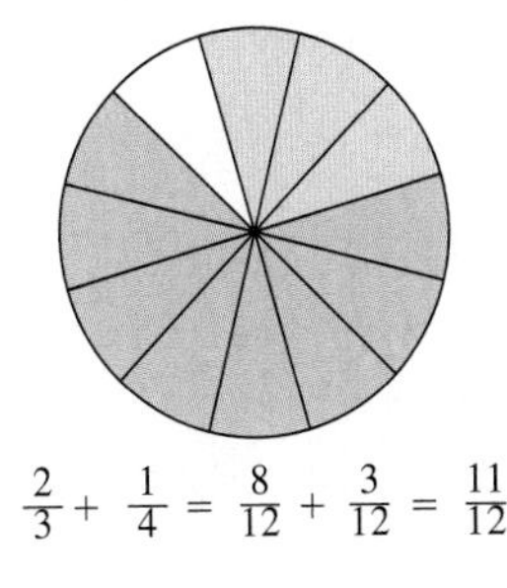

$\frac{2}{3} + \frac{1}{4} = \frac{8}{12} + \frac{3}{12} = \frac{11}{12}$

Looking Back To add two rational numbers of unlike denominators, we considered equal rational numbers with like denominators. The common denominator for $\frac{2}{3}$ and $\frac{1}{4}$ is 12. This is also the least common denominator, or the LCM(3, 4). To add two fractions with unequal denominators such as $\frac{5}{12}$ and $\frac{7}{18}$, we could find equal fractions with the denominator as the LCM(12, 18), or 36. However, any common denominator will work as well, for example, 72 or even $12 \cdot 18$.

By considering the sum $\frac{2}{3} + \frac{1}{4} = \frac{2 \cdot 4}{3 \cdot 4} + \frac{1 \cdot 3}{4 \cdot 3} = \frac{8}{12} + \frac{3}{12} = \frac{11}{12}$, we can generalize to the sum of two rational numbers with unlike denominators, as in the following property.

Property

If $\frac{a}{b}$ and $\frac{c}{d}$ are any two rational numbers, then $\frac{a}{b} + \frac{c}{d} = \frac{ad + bc}{bd}$.

NOW TRY THIS 5-9

Find the sum of $\frac{a}{b} + \frac{c}{b}$ using the preceding property for unlike denominators. Show the result is equivalent to that obtained by using the definition of addition involving like denominators.

EXAMPLE 5-4

Find each of the following sums:

a. $\frac{1}{15} + \frac{4}{21}$

b. $\frac{2}{^{-}3} + \frac{1}{5}$

c. $\left(\frac{3}{4} + \frac{1}{5}\right) + \frac{1}{6}$

d. $\frac{3}{x} + \frac{4}{y}$

Solution **a.** Because LCM(15, 21) $= 3 \cdot 5 \cdot 7$, then $\frac{2}{15} + \frac{4}{21} = \frac{2 \cdot 7}{15 \cdot 7} + \frac{4 \cdot 5}{21 \cdot 5} = \frac{14}{105} + \frac{20}{105} = \frac{34}{105}$

b. $\dfrac{2}{^-3} + \dfrac{1}{5} = \dfrac{(2)(5) + (^-3)(1)}{(^-3)(5)} = \dfrac{10 + {^-3}}{^-15} = \dfrac{7}{^-15} = \dfrac{7(^-1)}{^-15(^-1)} = \dfrac{^-7}{15}$

c. $\dfrac{3}{4} + \dfrac{1}{5} = \dfrac{3 \cdot 5 + 4 \cdot 1}{4 \cdot 5} = \dfrac{19}{20}$; hence, $\left(\dfrac{3}{4} + \dfrac{1}{5}\right) + \dfrac{1}{6} = \dfrac{19}{20} + \dfrac{1}{6} = \dfrac{19 \cdot 6 + 20 \cdot 1}{20 \cdot 6} = \dfrac{134}{120}$, or $\dfrac{67}{60}$.

d. $\dfrac{3}{x} + \dfrac{4}{y} = \dfrac{3y}{xy} + \dfrac{4x}{xy} = \dfrac{3y + 4x}{xy}$

Mixed Numbers

mixed numbers

In everyday life, we often use **mixed numbers**, that is, numbers that are made up of an integer and a fractional part of an integer. For example, Figure 5-10 shows that the nail is $2\frac{3}{4}$ in. long. The mixed number $2\frac{3}{4}$ means $2 + \frac{3}{4}$. It is sometimes inferred that $2\frac{3}{4}$ means 2 times $\frac{3}{4}$, since xy means $x \cdot y$, but this is not correct. Also, the number $^-4\frac{3}{4}$ means $-\left(4\frac{3}{4}\right)$, or $^-4 - \frac{3}{4}$, not $^-4 + \frac{3}{4}$.

FIGURE 5-10

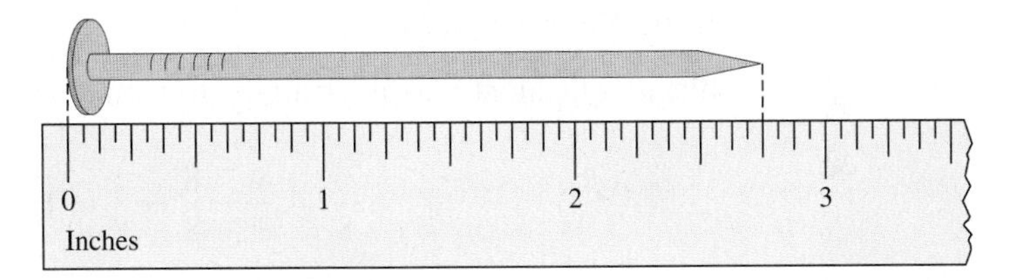

In a National Assessment of Educational Progress (NAEP) test, students were given the following problem:

$5\frac{1}{4}$ is the same as:

(a) $5 + \frac{1}{4}$ (b) $5 - \frac{1}{4}$ (c) $5 \times \frac{1}{4}$ (d) $5 \div \frac{1}{4}$

Only 47% of the seventh graders chose the correct response, (a), and only 44% of the eleventh graders chose the correct response.

A mixed number is a rational number, and therefore it can always be written in the form $\frac{a}{b}$. For example,

$$2\frac{3}{4} = 2 + \frac{3}{4} = \frac{2}{1} + \frac{3}{4} = \frac{2 \cdot 4 + 1 \cdot 3}{1 \cdot 4} = \frac{8 + 3}{4} = \frac{11}{4}.$$

EXAMPLE 5-5

Change each of the following mixed numbers to the form $\frac{a}{b}$, where a and b are integers:

a. $4\frac{1}{3}$ **b.** $^{-}3\frac{2}{5}$

Solution **a.** $4\frac{1}{3} = 4 + \frac{1}{3} = \frac{4}{1} + \frac{1}{3} = \frac{4 \cdot 3 + 1 \cdot 1}{1 \cdot 3} = \frac{12 + 1}{3} = \frac{13}{3}$

b. $^{-}3\frac{2}{5} = ^{-}\left(3 + \frac{2}{5}\right) = ^{-}\left(\frac{3}{1} + \frac{2}{5}\right) = ^{-}\left(\frac{3 \cdot 5 + 1 \cdot 2}{1 \cdot 5}\right) = \frac{^{-}17}{5}$

NOW TRY THIS 5-10

Use the ideas in Example 5-5 to convert a general mixed number $A\frac{b}{c}$ $\left(\text{notation for } A + \frac{b}{c}\right)$ to an improper fraction.

Now that we know how to change a mixed number to an improper fraction, we consider the following student page, which explores the changing of an improper fraction to a mixed number. The page in the student book that preceded this student page discussed the "universal language" of trade and how the Spanish dollar was the main money used for trade in colonial America. It talked about how the dollar coin was often cut into 2, 4, or 8 equal pieces to make change (8 bits = 1 dollar). Many of us have probably referred to a quarter as "two bits" but never understood where this term came from. The example on the student page of converting 21 bits for whole dollars and bits is equivalent to converting $\frac{21}{8}$ to a mixed number.

$$\frac{21}{8} = \frac{8}{8} + \frac{8}{8} + \frac{5}{8} = 1 + 1 + \frac{5}{8} = 2\frac{5}{8}$$

Next try problem 5 on the student page on your own.

EXAMPLE 5-6

Change $\frac{29}{5}$ to a mixed number.

Solution $\frac{29}{5} = \frac{5 \cdot 5 + 4}{5} = \frac{5 \cdot 5}{5} + \frac{4}{5} = 5 + \frac{4}{5} = 5\frac{4}{5}$

REMARK In elementary schools, problems like Example 5-6 are usually computed using division, as follows:

$$\begin{array}{r} 5 \\ 5\overline{)29} \\ \underline{25} \\ 4 \end{array}$$

Hence, $\frac{29}{5} = 5 + \frac{4}{5} = 5\frac{4}{5}$.

SAMPLE SCHOOL BOOK PAGE:

FRACTIONS AND MIXED NUMBERS

GOAL

LEARN HOW TO...
- write fractions as mixed numbers
- write mixed numbers as fractions

AS YOU...
- exchange colonial money

KEY TERM
- mixed number

Exploration 1

Renaming Fractions and Mixed Numbers

If you had lived in colonial America, a knowledge of fractions and *mixed numbers* would have helped you when you made purchases with Spanish dollars.

4 Suppose a colonial merchant had 11 bits.

a. If the merchant combined bits to make Spanish dollars, how many Spanish dollars and how many leftover bits did the merchant have?

b. What fraction represents the leftover bits?

▶ **Fractions as Mixed Numbers** **You can use the same thinking as in Question 4 to write fractions as *mixed numbers*. A mixed number is the sum of a nonzero whole number and a fraction between 0 and 1.**

EXAMPLE

Suppose you have 21 bits and you want to exchange the bits for whole Spanish dollars and bits.

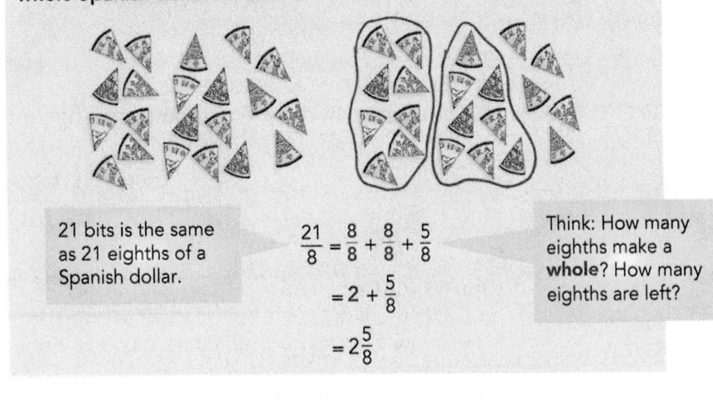

21 bits is the same as 21 eighths of a Spanish dollar.

$$\frac{21}{8} = \frac{8}{8} + \frac{8}{8} + \frac{5}{8}$$
$$= 2 + \frac{5}{8}$$
$$= 2\frac{5}{8}$$

Think: How many eighths make a **whole**? How many eighths are left?

5 What mixed number is equal to $\frac{15}{4}$? If necessary, use the coins or make a sketch.

Source: Middle Grade Mathematics, Book 2, McDougal Littell Publishing Co., 2005 (p. 178).

We can use scientific/fraction calculators to change improper fractions to mixed numbers. For example, if we enter 2 9 / 5 and press Ab/c, then 5 ⊔ 4/5 appears, which means $5\frac{4}{5}$.

We can also use scientific/fraction calculators to add mixed numbers. For example, to add $2\frac{4}{5} + 3\frac{5}{6}$, we enter 2 Unit 4 / 5 + 3 Unit 5 / 6 =, and the display reads 5 ⊔ 49/30. We then press Ab/c to obtain 6 ⊔ 19/30, which means $6\frac{19}{30}$.

Because mixed numbers are rational numbers, the methods of adding rationals can be extended to include mixed numbers. The following student page shows a method for computing sums of mixed numbers. Work the three problems on the bottom of the student page.

Properties of Addition for Rational Numbers

Rational numbers have the following properties for addition: *closure, commutative, associative, additive identity*, and *additive inverse*. To emphasize the additive inverse property of rational numbers, we state it explicitly, as follows.

Property

Additive inverse property of rational numbers For any rational number $\frac{a}{b}$, there exists a unique rational number $-\frac{a}{b}$, the additive inverse of $\frac{a}{b}$, such that

$$\frac{a}{b} + \left(-\frac{a}{b}\right) = 0 = \left(-\frac{a}{b}\right) + \frac{a}{b}.$$

Another form of $-\frac{a}{b}$ can be found by considering the sum $\frac{a}{b} + \frac{^-a}{b}$. Because

$$\frac{a}{b} + \frac{^-a}{b} = \frac{a + {^-a}}{b} = \frac{0}{b} = 0,$$

it follows that $-\frac{a}{b}$ and $\frac{^-a}{b}$ are both additive inverses of $\frac{a}{b}$, so $-\frac{a}{b} = \frac{^-a}{b}$.

EXAMPLE 5-7 Find the additive inverses for each of the following:

a. $\frac{3}{5}$ **b.** $\frac{^-5}{11}$ **c.** $4\frac{1}{2}$

Solution **a.** $-\frac{3}{5}$ or $\frac{^-3}{5}$ **b.** $-\left(\frac{^-5}{11}\right) = \frac{^-(^-5)}{11} = \frac{5}{11}$ **c.** $-4\frac{1}{2}$, or $\frac{^-9}{2}$

SAMPLE SCHOOL BOOK PAGE:

ADDING MIXED NUMBERS

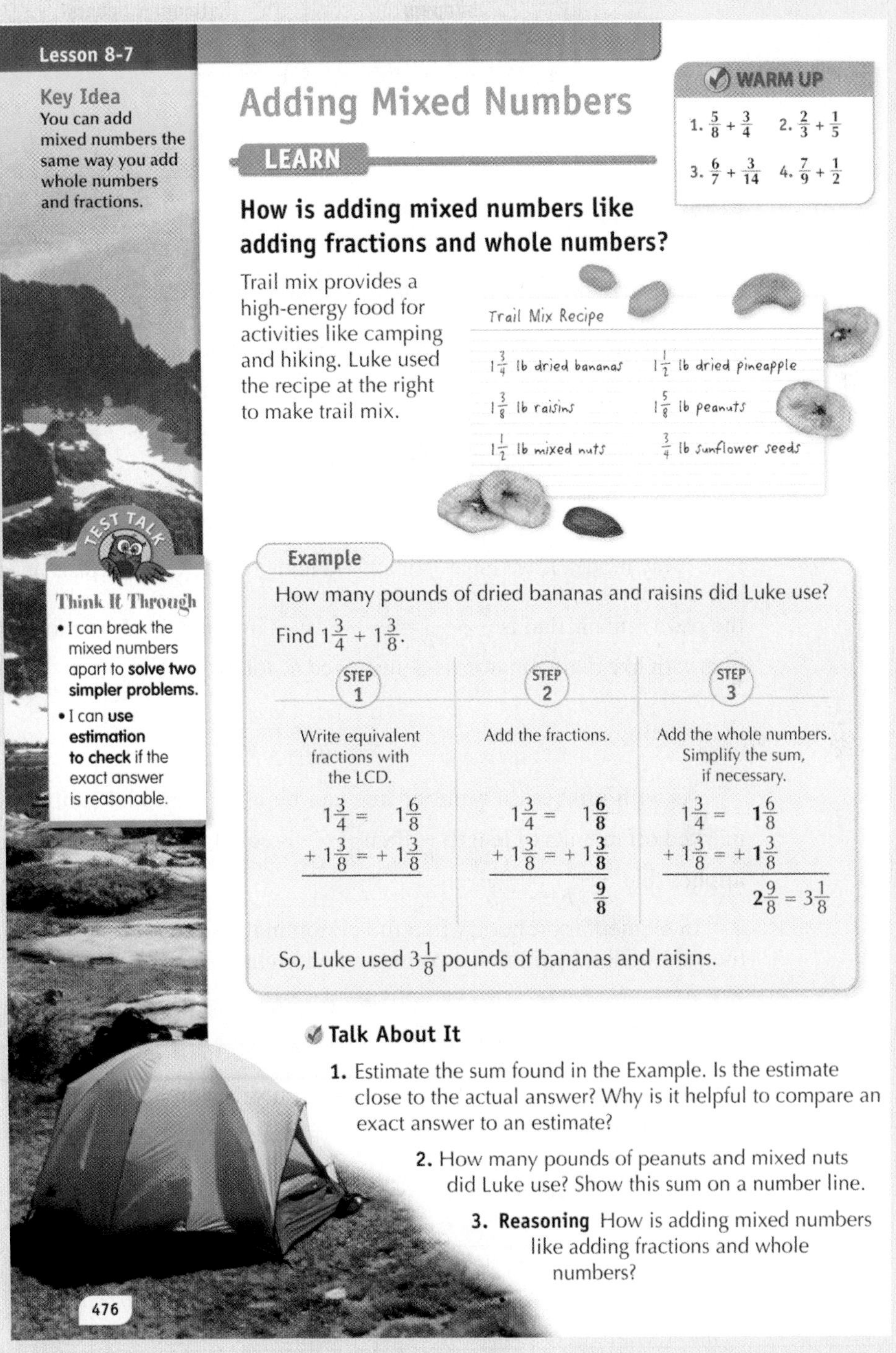

Lesson 8-7

Key Idea
You can add mixed numbers the same way you add whole numbers and fractions.

Adding Mixed Numbers

WARM UP

1. $\frac{5}{8}+\frac{3}{4}$ 2. $\frac{2}{3}+\frac{1}{5}$
3. $\frac{6}{7}+\frac{3}{14}$ 4. $\frac{7}{9}+\frac{1}{2}$

LEARN

How is adding mixed numbers like adding fractions and whole numbers?

Trail mix provides a high-energy food for activities like camping and hiking. Luke used the recipe at the right to make trail mix.

Trail Mix Recipe

$1\frac{3}{4}$ lb dried bananas	$1\frac{1}{2}$ lb dried pineapple
$1\frac{3}{8}$ lb raisins	$1\frac{5}{8}$ lb peanuts
$1\frac{1}{2}$ lb mixed nuts	$\frac{3}{4}$ lb sunflower seeds

TEST TALK

Think It Through

- I can break the mixed numbers apart to **solve two simpler problems.**
- I can **use estimation to check** if the exact answer is reasonable.

Example

How many pounds of dried bananas and raisins did Luke use?

Find $1\frac{3}{4}+1\frac{3}{8}$.

STEP 1	STEP 2	STEP 3
Write equivalent fractions with the LCD.	Add the fractions.	Add the whole numbers. Simplify the sum, if necessary.
$1\frac{3}{4} = 1\frac{6}{8}$ $+1\frac{3}{8} = +1\frac{3}{8}$	$1\frac{3}{4} = 1\frac{6}{8}$ $+1\frac{3}{8} = +1\frac{3}{8}$ $\frac{9}{8}$	$1\frac{3}{4} = 1\frac{6}{8}$ $+1\frac{3}{8} = +1\frac{3}{8}$ $2\frac{9}{8} = 3\frac{1}{8}$

So, Luke used $3\frac{1}{8}$ pounds of bananas and raisins.

Talk About It

1. Estimate the sum found in the Example. Is the estimate close to the actual answer? Why is it helpful to compare an exact answer to an estimate?
2. How many pounds of peanuts and mixed nuts did Luke use? Show this sum on a number line.
3. **Reasoning** How is adding mixed numbers like adding fractions and whole numbers?

476

Source: Scott Foresman–Addison Wesley Mathematics, Grade 5, 2005 (p. 476).

Properties of the additive inverse for rational numbers are analogous to those of the additive inverse for integers, as shown in Table 5-2. As with the set of integers, the set of rational numbers also has the addition property of equality.

Table 5-2

Integers	Rational Numbers
1. $^{-}(^{-}a) = a$	1. $-\left(-\frac{a}{b}\right) = \frac{a}{b}$
2. $^{-}(a + b) = {}^{-}a + {}^{-}b$	2. $-\left(\frac{a}{b} + \frac{c}{d}\right) = \frac{^{-}a}{b} + \frac{^{-}c}{d}$

Property

Addition property of equality If $\frac{a}{b}$ and $\frac{c}{d}$ are any rational numbers such that $\frac{a}{b} = \frac{c}{d}$, and if $\frac{e}{f}$ is any rational number, then $\frac{a}{b} + \frac{e}{f} = \frac{c}{d} + \frac{e}{f}$.

Subtraction of Rational Numbers

In elementary school, subtraction of rational numbers is usually introduced by using a take-away model. If we have $\frac{6}{7}$ of a pizza and $\frac{2}{7}$ of the original pizza is taken away, $\frac{4}{7}$ of the pizza remain: that is, $\frac{6}{7} - \frac{2}{7} = \frac{6-2}{7} = \frac{4}{7}$. In general, subtraction of rational numbers with like denominators is determined as follows:

$$\frac{a}{b} - \frac{c}{b} = \frac{a-c}{b}.$$

As with integers, a number line can be used to model subtraction. If a line is marked off in units of length $\frac{1}{b}$, then $\frac{a}{b} - \frac{c}{b}$ is equal to $(a - c)$ units of length $\frac{1}{b}$, which implies that $\frac{a}{b} - \frac{c}{b} = \frac{a-c}{b}$.

In elementary school, when the denominators are not the same we can perform the subtraction by finding a common denominator. For example,

$$\frac{3}{4} - \frac{2}{3} = \frac{3 \cdot 3}{4 \cdot 3} - \frac{2 \cdot 4}{3 \cdot 4} = \frac{9}{12} - \frac{8}{12} = \frac{9-8}{12} = \frac{1}{12}.$$

Subtraction of rational numbers, like subtraction of integers, can be defined in terms of addition as follows.

Definition of Subtraction of Rational Numbers in Terms of Addition

If $\frac{a}{b}$ and $\frac{c}{d}$ are any rational numbers, then $\frac{a}{b} - \frac{c}{d}$ is the unique rational number $\frac{e}{f}$ such that $\frac{a}{b} = \frac{c}{d} + \frac{e}{f}$.

As with integers, we can see that subtraction of rational numbers can be performed by adding the additive inverses. The following theorem states this.

Theorem 5–4

If $\frac{a}{b}$ and $\frac{c}{d}$ are any rational numbers, then $\frac{a}{b} - \frac{c}{d} = \frac{a}{b} + \frac{^-c}{d}$.

Now, using the definition of addition of rational numbers, we obtain the following:

$$\begin{aligned}\frac{a}{b} - \frac{c}{d} &= \frac{a}{b} + \frac{^-c}{d}\\ &= \frac{ad + b(^-c)}{bd}\\ &= \frac{ad + {}^-(bc)}{bd}\\ &= \frac{ad - bc}{bd}.\end{aligned}$$

We summarize this result in the following.

Property

If $\frac{a}{b}$ and $\frac{c}{d}$ are any rational numbers, then $\frac{a}{b} - \frac{c}{d} = \frac{ad - bc}{bd}$.

EXAMPLE 5-8

Find each difference in the following:

a. $\frac{5}{8} - \frac{1}{4}$ **b.** $5\frac{1}{3} - 2\frac{3}{4}$

Solution **a.** One approach is to find the LCM for the fractions. Because LCM(8, 4) = 8, we have

$$\frac{5}{8} - \frac{1}{4} = \frac{5}{8} - \frac{2}{8} = \frac{3}{8}.$$

An alternative approach is as follows:

$$\frac{5}{8} - \frac{1}{4} = \frac{5 \cdot 4 - 8 \cdot 1}{8 \cdot 4} = \frac{20 - 8}{32} = \frac{12}{32}, \text{ or } \frac{3}{8}.$$

b. Two methods of solution are given:

$$\begin{aligned}5\frac{1}{3} &= 5\frac{4}{12} = 4 + 1\frac{4}{12} = 4\frac{16}{12}\\ -2\frac{3}{4} &= -2\frac{9}{12} = -2\frac{9}{12} = -2\frac{9}{12}\\ &\qquad\qquad\qquad\qquad\quad 2\frac{7}{12}\end{aligned}$$

$$\begin{aligned}5\frac{1}{3} - 2\frac{3}{4} &= \frac{16}{3} - \frac{11}{4}\\ &= \frac{16 \cdot 4 - 3 \cdot 11}{3 \cdot 4}\\ &= \frac{64 - 33}{12}\\ &= \frac{31}{12}, \text{ or } 2\frac{7}{12}\end{aligned}$$

Estimation with Rational Numbers

Estimation helps us make practical decisions in our everyday lives. For example, suppose we need to double a recipe that calls for $\frac{3}{8}$ of a cup of flour. Will we need more or less than a cup of flour? Many of the estimation and mental math techniques that we learned to use with whole numbers also work with rational numbers.

Estimation plays an important role in judging the reasonableness of computations. Students do not necessarily have this skill. For example, when asked to estimate $\frac{12}{13} + \frac{7}{8}$, only 24% of 13-year-old students on a national assessment said the answer was close to 2. Most said it was close to 1, 19, or 21. These incorrect estimates suggest common computational errors in adding fractions and a lack of understanding of the operation being carried out. These incorrect estimates also suggest a lack of number sense.

NOW TRY THIS 5-11

A student added $\frac{3}{4} + \frac{1}{2}$ and obtained $\frac{4}{6}$. How would you use estimation to show this student that his answer could not be correct?

Sometimes to obtain an estimate it is desirable to round fractions to a *convenient* fraction, such as $\frac{1}{2}, \frac{1}{3}, \frac{1}{4}, \frac{1}{5}, \frac{2}{3}, \frac{3}{4}$, or 1. For example, if a student had 59 correct answers out of 80 questions, the student answered $\frac{59}{80}$ of the questions correctly, which is approximately $\frac{60}{80}$, or $\frac{3}{4}$. Intuitively, we know $\frac{60}{80}$ is greater than $\frac{59}{80}$. On a number line, the greater fraction is to the right of the lesser. The estimate $\frac{3}{4}$ for $\frac{59}{80}$ is a high estimate. In a similar way, we can estimate $\frac{31}{90}$ by $\frac{30}{90}$, or $\frac{1}{3}$. In this case, the estimate of $\frac{1}{3}$ is a low estimate.

EXAMPLE 5-9

A sixth-grade class is collecting cans to take to the recycling center. Becky's group brought the following amounts (in pounds). About how many pounds does her group have all together?

$$1\frac{1}{8}, 3\frac{4}{10}, 5\frac{7}{8}, \frac{6}{10}$$

Solution We can estimate the amount by using front-end estimation and then adjusting by using 0, $\frac{1}{2}$, and 1 as reference points. The front-end estimate is $(1 + 3 + 5)$, or 9. The adjustment is $\left(0 + \frac{1}{2} + 1 + \frac{1}{2}\right)$, or 2. An adjusted estimate would be 11 lb.

EXAMPLE 5-10

Estimate each of the following:

a. $\frac{27}{13} + \frac{10}{9}$

b. $3\frac{9}{10} + 2\frac{7}{8} + \frac{11}{12}$

Solution **a.** Because $\frac{27}{13}$ is more than 2 and $\frac{10}{9}$ is more than 1, an estimate is a number close to but more than 3.

b. We first add the whole-number parts to obtain 3 + 2, or 5. Because each of the fractions, $\frac{9}{10}, \frac{7}{8}$, and $\frac{11}{12}$, is close to but less than 1, their sum is less than 3. The approximate answer is a number close to but less than 8.

ONGOING ASSESSMENT 5-2

1. Compute each of the following using any method:
   a. $\frac{1}{2} + \frac{2}{3}$
   b. $\frac{4}{12} - \frac{2}{3}$
   c. $\frac{5}{x} + \frac{^-3}{y}$
   d. $\frac{^-3}{2x^2y} + \frac{5}{2xy^2} + \frac{7}{x^2}$
   e. $\frac{5}{6} + 2\frac{1}{8}$
   f. $^-4\frac{1}{2} - 3\frac{1}{6}$
2. Change each of the following fractions to mixed numbers:
   a. $\frac{56}{3}$
   b. $\frac{14}{5}$
   c. $-\frac{293}{100}$
   d. $-\frac{47}{8}$
3. Change each of the following mixed numbers to fractions in the form $\frac{a}{b}$, where a and b are integers:
   a. $6\frac{3}{4}$
   b. $7\frac{1}{2}$
   c. $^-3\frac{5}{8}$
   d. $^-4\frac{2}{3}$
4. Place the numbers 2, 5, 6, and 8 in the following boxes to make the equation true:

$$\frac{\square}{\square} + \frac{\square}{\square} = \frac{23}{24}$$

5. Approximate each of the following situations with a convenient fraction. Explain your reasoning. Tell whether your estimate is high or low.
   a. Giorgio had 15 base hits out of 46 times at bat.
   b. Ruth made 7 goals out of 41 shots.
   c. Laura answered 62 problems correctly out of 80.
   d. Jonathan made 9 baskets out of 19.
6. Use the information in the table to answer each of the following questions:

Team	Games Played	Games Won
Ducks	22	10
Beavers	19	10
Tigers	28	9
Bears	23	8
Lions	27	7
Wildcats	25	6
Badgers	21	5

   a. Which team won just over $\frac{1}{2}$ of its games?
   b. Which team won just under $\frac{1}{2}$ of its games?
   c. Which team won just over $\frac{1}{3}$ of its games?
   d. Which team won just under $\frac{1}{3}$ of its games?
   e. Which team won just over $\frac{1}{4}$ of its games?
   f. Which team won just under $\frac{1}{4}$ of its games?
7. Sort the following fraction cards into the ovals by estimating in which oval the fraction belongs:

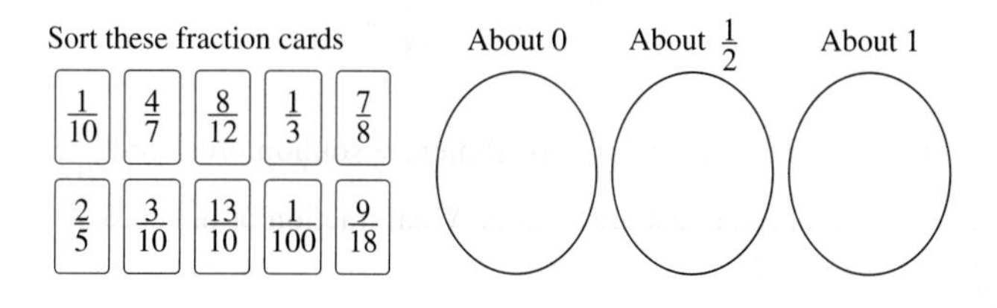

Property

For any nonzero rational number a and any integers m and n, $a^m \cdot a^n = a^{m+n}$.

Other properties of exponents can be developed by using the properties of rational numbers. For example,

$$\frac{2^5}{2^3} = \frac{2^3 \cdot 2^2}{2^3} = 2^2 = 2^{5-3} \qquad \frac{2^5}{2^8} = \frac{2^5}{2^5 \cdot 2^3} = \frac{1}{2^3} = 2^{-3} = 2^{5-8}.$$

With integer exponents, the following property holds.

Property

For any rational number a such that $a \neq 0$ and for any integers m and n, $\frac{a^m}{a^n} = a^{m-n}$.

At times, we may encounter an expression like $(2^4)^3$. This expression can be written as a single power of 2 as follows:

$$(2^4)^3 = 2^4 \cdot 2^4 \cdot 2^4 = 2^{4+4+4} = 2^{3 \cdot 4} = 2^{12}.$$

In general, if a is a nonzero rational number and m and n are positive integers, then

$$(a^m)^n = \underbrace{a^m \cdot a^m \cdot a^m \cdot \ldots \cdot a^m}_{n \text{ factors}} = a^{\overbrace{m+m+\ldots+m}^{n \text{ terms}}} = a^{nm} = a^{mn}$$

Does this property hold for negative-integer exponents? For example, does $(2^3)^{-4} = 2^{(3)(-4)} = 2^{-12}$? The answer is yes because $(2^3)^{-4} = \frac{1}{(2^3)^4} = \frac{1}{2^{12}} = 2^{-12}$. Also, $(2^{-3})^4 = \left(\frac{1}{2^3}\right)^4 = \frac{1}{2^3} \cdot \frac{1}{2^3} \cdot \frac{1}{2^3} \cdot \frac{1}{2^3} = \frac{1^4}{(2^3)^4} = \frac{1}{2^{12}} = 2^{-12}$.

Property

For any rational number $a \neq 0$ and any integers m and n,

$$(a^m)^n = a^{mn}.$$

Using the definitions and properties developed, we can derive additional properties. Notice, for example, that

$$\left(\frac{2}{3}\right)^4 = \frac{2}{3} \cdot \frac{2}{3} \cdot \frac{2}{3} \cdot \frac{2}{3} = \frac{2 \cdot 2 \cdot 2 \cdot 2}{3 \cdot 3 \cdot 3 \cdot 3} = \frac{2^4}{3^4}.$$

This property can be generalized as follows.

Property

For any nonzero rational number $\frac{a}{b}$ and any integer m,

$$\left(\frac{a}{b}\right)^m = \frac{a^m}{b^m}.$$

From the definition of negative exponents, the preceding property, and division of fractions, we have

$$\left(\frac{a}{b}\right)^{-m} = \frac{1}{\left(\frac{a}{b}\right)^m} = \frac{1}{\frac{a^m}{b^m}} = \frac{b^m}{a^m} = \left(\frac{b}{a}\right)^m.$$

Property

For any nonzero rational number $\frac{a}{b}$ and any integer m, $\left(\frac{a}{b}\right)^{-m} = \left(\frac{b}{a}\right)^m$.

A property similar to this holds for multiplication. For example,

$$(2 \cdot 3)^{-3} = \frac{1}{(2 \cdot 3)^3} = \frac{1}{2^3 \cdot 3^3} = \left(\frac{1}{2^3}\right) \cdot \left(\frac{1}{3^3}\right) = 2^{-3} \cdot 3^{-3},$$

and in general, it is true that $(a \cdot b)^m = a^m \cdot b^m$ if a and b are rational numbers and m is an integer.

The definitions and properties of exponents are summarized in the following list. For any rational numbers a and b and integers m and n (as long as 0^0 does not appear), we have the following.

Properties of Exponents

1. $a^m = \underbrace{a \cdot a \cdot a \cdot \ldots \cdot a}_{m \text{ factors}}$, where m is a positive integer
2. $a^0 = 1$, where $a \neq 0$
3. $a^{-m} = \frac{1}{a^m}$, where $a \neq 0$
4. $a^m \cdot a^n = a^{m+n}$
5. $\frac{a^m}{a^n} = a^{m-n}$, where $a \neq 0$
6. $(a^m)^n = a^{nm}$
7. $\left(\frac{a}{b}\right)^m = \frac{a^m}{b^m}$, where $b \neq 0$
8. $(ab)^m = a^m \cdot b^m$
9. $\left(\frac{a}{b}\right)^{-m} = \left(\frac{b}{a}\right)^m$, where $a \neq 0, b \neq 0$

Observe that all the properties of exponents refer to powers with either the same base or the same exponent. To evaluate expressions using exponents where different bases or powers are used, perform all the computations or rewrite the expressions in either the same base or the same exponent if possible. For example, $\frac{27^4}{81^3}$ can be rewritten $\frac{27^4}{81^3} = \frac{(3^3)^4}{(3^4)^3} = \frac{3^{12}}{3^{12}} = 1$.

EXAMPLE 5-18

Write each of the following in simplest form using positive exponents in the final answer:

a. $16^2 \cdot 8^{-3}$ **b.** $20^2 \div 2^4$ **c.** $(10^{-1} + 5 \cdot 10^{-2} + 3 \cdot 10^{-3}) \cdot 10^3$ **d.** $(x^3y^{-2})^{-4}$

Solution **a.** $16^2 \cdot 8^{-3} = (2^4)^2 \cdot (2^3)^{-3} = 2^8 \cdot 2^{-9} = 2^{8+-9} = 2^{-1} = \frac{1}{2}$

b. $\frac{20^2}{2^4} = \frac{(2^2 \cdot 5)^2}{2^4} = \frac{2^4 \cdot 5^2}{2^4} = 5^2$

c. $(10^{-1} + 5 \cdot 10^{-2} + 3 \cdot 10^{-3}) \cdot 10^3 = 10^{-1} \cdot 10^3 + 5 \cdot 10^{-2} \cdot 10^3 + 3 \cdot 10^{-3} \cdot 10^3$
$= 10^{-1+3} + 5 \cdot 10^{-2+3} + 3 \cdot 10^{-3+3}$
$= 10^2 + 5 \cdot 10^1 + 3 \cdot 10^0$
$= 153$

d. $(x^3y^{-2})^{-4} = x^{-12}y^8 = \frac{1}{x^{12}} \cdot y^8 = \frac{y^8}{x^{12}}$

BRAIN TEASER

A castle in the faraway land of Aluossim was surrounded by four moats. One day, the castle was attacked and captured by a fierce tribe from the north. Guards were stationed at each bridge. Prince Juan was allowed to take a number of bags of gold from the castle as he went into exile. However, the guard at the first bridge demanded half the bags of gold plus one more bag. Juan met this demand and proceeded to the next bridge. The guards at the second, third, and fourth bridges made identical demands, all of which the prince met. When Juan finally crossed all the bridges, a single bag of gold was left. With how many bags did Juan start?

ONGOING ASSESSMENT 5-3

1. In the following figures, a unit rectangle is used to illustrate the product of two fractions. Name the fractions and their products.

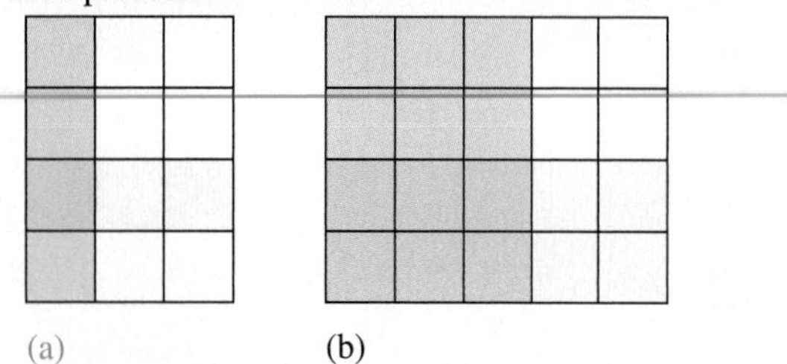

(a) (b)

2. Use a rectangular region to illustrate each of the following products:
   a. $\frac{1}{3} \cdot \frac{3}{4}$ **b.** $\frac{2}{3} \cdot \frac{1}{5}$ c. $\frac{2}{5} \cdot \frac{1}{3}$
3. Find each of the following products. Write your answers in simplest form.
   a. $\frac{49}{65} \cdot \frac{26}{98}$ **b.** $\frac{a}{b} \cdot \frac{b^2}{a^2}$ c. $\frac{xy}{z} \cdot \frac{z^2a}{x^3y^2}$
   d. $2\frac{1}{3} \cdot 3\frac{3}{4}$ e. $\frac{22}{7} \cdot 4\frac{2}{3}$ **f.** $\frac{-5}{2} \cdot 2\frac{1}{2}$

4. Use the distributive property to find each product.
 a. $4\frac{1}{2} \cdot 2\frac{1}{3}$ $\left[\textit{Hint: } \left(4 + \frac{1}{2}\right) \cdot \left(2 + \frac{1}{3}\right).\right]$
 b. $3\frac{1}{3} \cdot 2\frac{1}{2}$ c. $248\frac{2}{5} \cdot 100\frac{1}{8}$
5. Find the multiplicative inverse of each of the following:
 a. $\frac{^-1}{3}$ **b.** $3\frac{1}{3}$
 c. $\frac{x}{y}$, if $x \neq 0$ and $y \neq 0$ **d.** $^-7$
6. Solve for x in each of the following:
 a. $\frac{2}{3}x = \frac{7}{6}$ **b.** $\frac{3}{4} \div x = \frac{1}{2}$
 c. $\frac{5}{6} + \frac{2}{3}x = \frac{3}{4}$ **d.** $\frac{2x}{3} - \frac{1}{4} = \frac{x}{6} + \frac{1}{2}$
7. Find a fraction such that if you add the denominator to the numerator and place it over the original denominator, the new fraction has triple the value of the original fraction.
8. Show that the following properties do not hold for the division of rational numbers:
 a. Commutative **b.** Associative
 c. Identity **d.** Inverse
9. Compute the following mentally. Find the exact answers.
 a. $3\frac{1}{4} \cdot 8$ **b.** $7\frac{1}{4} \cdot 4$ c. $9\frac{1}{5} \cdot 10$ **d.** $8 \cdot 2\frac{1}{4}$
 e. $3 \div \frac{1}{2}$ **f.** $3\frac{1}{2} \div \frac{1}{2}$ g. $3 \div \frac{1}{3}$ **h.** $4\frac{1}{2} \div 2$
10. Choose the number from among the numbers in parentheses that best approximates each of the following:
 a. $3\frac{11}{12} \cdot 5\frac{3}{100}$ (8, 20, 15, 16)
 b. $2\frac{1}{10} \cdot 7\frac{7}{8}$ (16, 14, 4, 3)
 c. $20\frac{2}{3} \div 9\frac{7}{8}$ $\left(2, 180, \frac{1}{2}, 10\right)$
 d. $\frac{1}{101} \div \frac{1}{103}$ $\left(0, 1, \frac{1}{2}, \frac{1}{4}\right)$
11. Estimate the following:
 a. $5\frac{4}{5} \cdot 3\frac{1}{10}$ **b.** $4\frac{10}{11} \cdot 5\frac{1}{8}$
 c. $\dfrac{20\frac{8}{9}}{3\frac{1}{12}}$ **d.** $\dfrac{12\frac{1}{3}}{1\frac{7}{8}}$
12. Without actually doing the computations, choose the phrase in parentheses that correctly describes each:
 a. $\frac{13}{14} \cdot \frac{17}{19}$ (greater than 1, less than 1)
 b. $3\frac{2}{7} \div 5\frac{1}{9}$ (greater than 1, less than 1)
 c. $4\frac{1}{3} \div 2\frac{3}{100}$ (greater than 2, less than 2)
 d. $16 \div 4\frac{3}{18}$ (greater than 4, less than 4)
 e. $16 \div 3\frac{8}{9}$ (greater than 4, less than 4)
13. A sewing project requires $6\frac{1}{8}$ yd of material that sells for 62¢ per yard and $3\frac{1}{4}$ yd that sells for 81¢ per yard. Choose from the following the best estimate for the cost of the project:
 a. Between \$2 and \$4 b. Between \$4 and \$6
 c. Between \$6 and \$8 d. Between \$8 and \$10
14. When you multiply a certain number by 3 and then subtract $\frac{7}{18}$, you get the same result as when you multiply the number by 2 and add $\frac{5}{12}$. What is the number?
15. Five-eighths of the students at Salem State College live in dormitories. If 6000 students at the college live in dormitories, how many students are there in the college?
16. Di Paloma University had a faculty reduction and lost $\frac{1}{5}$ of its faculty. If 320 faculty members were left after the reduction, how many members were there originally?
17. Alberto owns $\frac{5}{9}$ of the stock in the N.W. Tofu Company. His sister Renatta owns half as much stock as Alberto. What part of the stock is owned by neither Alberto nor Renatta?
18. A person has $29\frac{1}{2}$ yd of material available to make doll uniforms. Each uniform requires $\frac{3}{4}$ yd of material.
 a. How many uniforms can be made?
 b. How much material will be left over?
19. A suit is on sale for \$180. What was the original price of the suit if the discount was $\frac{1}{4}$ of the original price?
20. Every employee's salary at the Sunrise Software Company increases each year by $\frac{1}{10}$ of that person's salary the previous year.
 a. If Martha's present annual salary is \$100,000, what will her salary be in 2 yr?
 b. If Aaron's present salary is \$99,000, what was his salary 1 yr ago?
 c. If Juanita's present salary is \$363,000, what was her salary 2 yr ago?
21. Jasmine is reading a book. She has finished $\frac{3}{4}$ of the book and has 82 pages left to read. How many pages has she read?

22. John took all his money out of his bank savings account. He spent \$50 on a radio and $\frac{3}{5}$ of what remained on presents. Half of what was left he put back in his checking account, and the remaining \$35 he donated to charity. How much money did John originally have in his savings account?

23. Peter, Paul, and Mary start at the same time walking around a circular track in the same direction. Peter takes $\frac{1}{2}$ hr to walk around the track. Paul takes $\frac{5}{12}$ hr, and Mary takes $\frac{1}{3}$ hr.
   a. How many minutes does it take each person to walk around the track?
   b. How many times will each person go around the track before all three meet again at the starting line?

24. The formula for converting degrees Celsius (C) to degrees Fahrenheit (F) is $F = \left(\frac{9}{5}\right) \cdot C + 32$.
   a. If Samantha reads that the temperature is 32°C in Spain, what is the Fahrenheit temperature?
   b. If the temperature dropped to $^{-}40$°F in West Yellowstone, what is the temperature in degrees Celsius?

25. Al gives $\frac{1}{2}$ of his marbles to Bev. Bev gives $\frac{1}{2}$ of these to Carl. Carl gives $\frac{1}{2}$ of these to Dani. If Dani has four marbles, how many did Al have originally?

26. The normal brain weight for an African bull elephant is 9 1/4 lb. Approximately how much would be the weight of the brains of 13 of these elephants?

27. Write each of the following in simplest form using positive exponents in the final answer:
   a. $3^{-7} \cdot 3^{-6}$ **b.** $3^7 \cdot 3^6$
   c. $5^{15} \div 5^4$ **d.** $5^{15} \div 5^{-4}$
   e. $(^{-}5)^{-2}$ **f.** $\frac{a^2}{a^{-3}}$, where $a \neq 0$
   g. $\frac{a}{a^{-1}}$, where $a \neq 0$ **h.** $\frac{a^{-3}}{a^{-2}}$, where $a \neq 0$

28. Write each of the following in simplest form using positive exponents in the final answer:
   a. $\left(\frac{1}{2}\right)^3 \cdot \left(\frac{1}{2}\right)^7$ **b.** $\left(\frac{1}{2}\right)^9 \div \left(\frac{1}{2}\right)^6$
   c. $\left(\frac{2}{3}\right)^5 \cdot \left(\frac{4}{9}\right)^2$ **d.** $\left(\frac{3}{5}\right)^7 \div \left(\frac{3}{5}\right)^7$
   e. $\left(\frac{3}{5}\right)^7 \div \left(\frac{5}{3}\right)^4$ **f.** $\left[\left(\frac{5}{6}\right)^7\right]^3$

29. If a and b are rational numbers, with $a \neq 0$ and $b \neq 0$, and if m and n are integers, which of the following are true and which are false? Justify your answers.
   a. $a^m \cdot b^n = (ab)^{m+n}$ **b.** $a^m \cdot b^n = (ab)^{mn}$
   c. $a^m \cdot b^m = (ab)^{2m}$ **d.** $a^0 = 0$
   e. $(a+b)^m = a^m + b^m$ **f.** $(a+b)^{-m} = \frac{1}{a^m} + \frac{1}{b^m}$
   g. $a^{mn} = a^m \cdot a^n$ **h.** $\left(\frac{a}{b}\right)^{-1} = \frac{b}{a}$

30. Solve for the integer n in each of the following:
   a. $2^n = 32$ **b.** $n^2 = 36$
   c. $2^n \cdot 2^7 = 2^5$ **d.** $2^n \cdot 2^7 = 8$
   e. $(2+n)^2 = 2^2 + n^2$ **f.** $3^n = 27^5$

31. A human being has approximately 25 trillion $(25 \cdot 10^{12})$ red blood cells, and each has an average radius of $4 \cdot 10^{-3}$ mm (millimeters).
   a. If these cells were placed end to end in a line, how long would the line be in millimeters?
   b. If 1 km is 10^6 mm, how long would the line be in kilometers?

32. Solve each of the following inequalities for x, where x is an integer:
   a. $3^x \leq 81$ **b.** $4^x < 8$ c. $3^{2x} > 27$ **d.** $2^x > 1$

33. Determine which fraction in each of the following pairs is greater:
   a. $\left(\frac{1}{2}\right)^3$ or $\left(\frac{1}{2}\right)^4$ **b.** $\left(\frac{3}{4}\right)^{10}$ or $\left(\frac{3}{4}\right)^8$
   c. $\left(\frac{4}{3}\right)^{10}$ or $\left(\frac{4}{3}\right)^8$ **d.** $\left(\frac{3}{4}\right)^{10}$ or $\left(\frac{4}{5}\right)^{10}$
   e. $\left(\frac{4}{3}\right)^{10}$ or $\left(\frac{5}{4}\right)^{10}$ **f.** $\left(\frac{3}{4}\right)^{100}$ or $\left(\frac{3}{4} \cdot \frac{9}{10}\right)^{100}$

34. Suppose the amount of bacteria in a certain culture is given as a function of time by $Q(t) = 10^{10}\left(\frac{6}{5}\right)^t$, where t is the time in seconds and $Q(t)$ is the amount of bacteria after t sec. Find the following:
   a. The initial number of bacteria (that is, the number of bacteria at $t = 0$)
   b. The number of bacteria after 2 sec

35. Let $S = \frac{1}{2} + \frac{1}{2^2} + \frac{1}{2^3} + \ldots + \frac{1}{2^{64}}$.
   a. Use the distributive property of multiplication over addition to find an expression for $2S$.
   b. Show that $2S - S = S = 1 - \left(\frac{1}{2}\right)^{64}$.
   c. Find a simple expression for the sum

$$\frac{1}{2} + \frac{1}{2^2} + \frac{1}{2^3} + \ldots = \frac{1}{2^n}.$$

36. In an arithmetic sequence, the first term is 1 and the hundredth term is 2. Find the following:
   a. The 50th term
   b. The sum of the first 50 terms

37. If $f(n) = \frac{3}{4} \cdot 2^n$, find the following:
   a. $f(0)$ **b.** $f(5)$ c. $f(^{-}5)$
   d. The greatest integer value of n for which $f(n) < \frac{3}{1400}$

38. If the nth term of a sequence is given by $a_n = 3 \cdot 2^{-n}$, answer the following:
    a. Find the first five terms.
    b. Show that the first five terms are in a geometric sequence.
    c. Find the first term that is less than $\frac{3}{1000}$.
39. In the following, determine which number is greater:
    a. 32^{50} or 4^{100} **b.** $(^{-}27)^{^{-}15}$ or $(^{-}3)^{^{-}75}$
40. There is a simple method for squaring any number that consists of a whole number and $\frac{1}{2}$. For example $\left(3\frac{1}{2}\right)^2 = 3 \cdot 4 + \left(\frac{1}{2}\right)^2 = 12\frac{1}{4}$; $\left(4\frac{1}{2}\right)^2 = 4 \cdot 5 + \left(\frac{1}{2}\right)^2 = 20\frac{1}{4}$; $\left(5\frac{1}{2}\right)^2 = 5 \cdot 6 + \left(\frac{1}{2}\right)^2 = 30\frac{1}{4}$.
    a. Write a statement for $\left(n + \frac{1}{2}\right)^2$ that generalizes these examples, where n is a whole number.
    ★**b.** Justify this procedure.
41. Consider these products:

    First product: $\left(1 + \frac{1}{1}\right)\left(1 + \frac{1}{2}\right)$

    Second product: $\left(1 + \frac{1}{1}\right)\left(1 + \frac{1}{2}\right)\left(1 + \frac{1}{3}\right)$

    Third product: $\left(1 + \frac{1}{1}\right)\left(1 + \frac{1}{2}\right)\left(1 + \frac{1}{3}\right)\left(1 + \frac{1}{4}\right)$

    a. Calculate the value of each product. Based on the pattern in your answers, guess the value of the fourth product. Then check to determine if your guess is correct.
    b. Guess the value of the 100th product.
    c. Find as simple an expression as possible for the nth product.
42. Show that the arithmetic mean of two rational numbers is between the two numbers; that is, for $0 < \frac{a}{b} < \frac{c}{d}$, prove that $0 < \frac{a}{b} < \frac{1}{2}\left(\frac{a}{b} + \frac{c}{d}\right) < \frac{c}{d}$.

Communication

43. Amy says that dividing a number by $\frac{1}{2}$ is the same as taking half of a number. How do you respond?
44. Suppose you divide a natural number, n, by a positive rational number less than 1. Will the answer always be less than n, sometimes less than n, or never less than n? Why?
45. If the fractions represented by points C and D on the following number line are multiplied, what point best represents the product? Explain why.

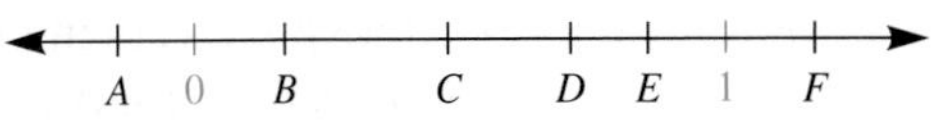

46. If the product of two numbers is 1 and one of the numbers is greater than 1, what do you know about the other number? Explain your answer.
47. Bente says to do the problem $12\frac{1}{4} \div 3\frac{3}{4}$ you just find $12 \div 3 = 4$ and $\frac{1}{4} \div \frac{3}{4} = \frac{1}{3}$ to get $4\frac{1}{3}$. How do you respond?
48. Carl says that every rational number has a multiplicative inverse. How do you respond?
49. Dani says that if we have $\frac{3}{4} \cdot \frac{2}{5}$, we could just do $\frac{3}{5} \cdot \frac{2}{4} = \frac{3}{5} \cdot \frac{1}{2} = \frac{3}{10}$. Is she correct? Explain why.

Open-Ended

50. In the book *Knowing and Teaching Elementary Mathematics*, Liping Ma presents the following scenario.

    Imagine that you are teaching division with fractions. To make this situation meaningful for kids, something that many teachers try to do is relate mathematics to other things. Sometimes they try to come up with real-world situations or story problems to show the application of some particular piece of content. What would you say would be a good story or model for $1\frac{3}{4} \div \frac{1}{2}$*?* (p. 55)

    a. How would you respond to her question?
    b. If possible, obtain a copy of Ma's book and read how U.S. teachers responded to this task compared to Chinese teachers. Report your findings.
51. Would you use the problem in the following cartoon in your class? Why or why not? Solve the problem.

Cooperative Learning

52. Choose a brick building on your campus. Measure the height of one brick and the thickness of mortar between bricks. Estimate the height of the building and then calculate the height of the building. Were rational numbers used in your computations?

Review Problems

53. Perform each of the following computations. Leave your answers in simplest form.
    a. $\frac{^-3}{16} + \frac{7}{4}$
    b. $\frac{1}{6} + \frac{^-4}{9} + \frac{5}{3}$
    c. $\frac{^-5}{2^3 \cdot 3^2} - \frac{^-5}{2 \cdot 3^3}$
    d. $3\frac{4}{5} + 4\frac{5}{6}$
    e. $5\frac{1}{6} - 3\frac{5}{8}$
    f. $-4\frac{1}{3} - 5\frac{5}{12}$
54. Each student at Sussex Elementary School takes one foreign language. Two-thirds of the students take Spanish, $\frac{1}{9}$ take French, $\frac{1}{18}$ take German, and the rest take some other foreign language. If there are 720 students in the school, how many do not take Spanish, French, or German?

Third International Mathematics and Science Study (TIMSS) Questions

There are 600 balls in a box, and $\frac{1}{3}$ of the balls are red. How many red balls are in the box?

Answer: ___________ red balls

TIMSS 2003, Grade 4

A scoop holds $\frac{1}{5}$ kg of flour. How many scoops of flour are needed to fill a bag with 6 kg of flour?

Answer: ___________

TIMSS 2003, Grade 8

BRAIN TEASER

A woman's will decreed that her cats be shared among her three daughters as follows: $\frac{1}{2}$ of the cats to the eldest daughter, $\frac{1}{3}$ of the cats to the middle daughter, and $\frac{1}{9}$ of the cats to the youngest daughter. Since the woman had 17 cats, the daughters decided that they could not carry out their mother's wishes. The judge who held the will agreed to lend the daughters a cat so that they could share the cats as their mother wished. Now, $\frac{1}{2}$ of 18 is 9; $\frac{1}{3}$ of 18 is 6; and $\frac{1}{9}$ of 18 is 2. Since $9 + 6 + 2 = 17$, the daughters were able to divide the 17 cats and return the borrowed cat. They obviously did not need the extra cat to carry out their mother's bequest, but they could not divide 17 into halves, thirds, and ninths. Has the woman's will really been followed?

5-4 Proportional Reasoning

In the *Principles and Standards* we find the following:

> *Working with proportions is a major focus proposed in these Standards for the middle grades. Students should become proficient in creating ratios to make comparisons in situations that involve pairs of numbers, as in the following problem:*
>
> *If three packages of cocoa make fifteen cups of hot chocolate, how many packages are needed to make sixty cups?* (p. 34)

As seen in the quote from the *Standards*, ratios and proportions are a very important part of the middle-grades curriculum. Ratios are encountered in everyday life.

For example, there may be a 2-to-3 ratio of Democrats to Republicans on a certain legislative committee, a friend may be given a speeding ticket for driving 69 miles per hour, or eggs may cost 98¢ a dozen. Each of these illustrates a **ratio**. Ratios are written $\frac{a}{b}$ or $a:b$ and are usually used to compare quantities.

ratio

A ratio of $1:3$ for boys to girls in a class means that the number of boys is $\frac{1}{3}$ that of girls; that is, there is 1 boy for every 3 girls. Notice we could also say that the ratio of girls to boys is $3:1$, or that there are 3 times as many girls as boys. Some ratios give **part-to-part** comparisons such as the ratio of the number of boys to girls or the number of students to one teacher. For example, a school might say that the average ratio of students to teachers cannot exceed $24:1$. Ratios can also represent **part-to-whole** or **whole-to-part** comparisons. For example, if the ratio of boys to girls in a class is $1:3$, then the ratio of boys (part) to children (whole) is $1:4$. We could also say that the ratio of children (whole) to boys (part) is $4:1$.

part-to-part comparison

part-to-whole comparison ◆

whole-to-part comparison

Notice that the ratio of $1:3$ for boys to girls in a class does not tell us how many boys and how many girls there are in the class. It only tells us the relative size of the groups. There could be 2 boys and 6 girls or 3 boys and 9 girls or 4 boys and 12 girls or some other number that gives a fraction equivalent to $\frac{1}{3}$.

EXAMPLE 5-19

There were 7 males and 12 females in the Dew Drop Inn on Monday evening. In the Game Room next door were 14 males and 24 females.

a. Express the number of males to females at the Inn as a ratio (part-to-part).
b. Express the number of males to females at the Game Room as a ratio (part-to-part).
c. Express the number of males in the Game Room to the number of people in the Game Room as a ratio (part-to-whole).

Solution **a.** The ratio is $\frac{7}{12}$. **b.** The ratio is $\frac{14}{24}$ or $\frac{7}{12}$. **c.** The ratio is $\frac{14}{38}$ or $\frac{7}{19}$. ▲

Proportions

In Example 5-19, the ratios $\frac{7}{12}$ and $\frac{14}{24}$ are equal and proportional to each other. In general, two ratios are **proportional** if and only if the fractions representing them are equal. Two equal ratios form a **proportion**.

proportional

proportion

For example, $\frac{2}{3}$ and $\frac{8}{12}$ form a proportion because $\frac{2}{3} = \frac{8}{12}$, since $2 \cdot 12 = 8 \cdot 3$. Also $\frac{3}{4} \neq \frac{4}{5}$ because $3 \cdot 5 \neq 4 \cdot 4$. In general, we have the following property, that follows directly from the property of *equal fractions* in Section 5.1.

Property of Proportions

If a, b, c, and d are all real numbers and $b \neq 0$ and $d \neq 0$, then the proportion

$$\frac{a}{b} = \frac{c}{d} \quad \text{is true if and only if} \quad ad = bc.$$

NOW TRY THIS 5-14

Justify the property of proportions by multiplying each side of the proportion by bd.

RESEARCH NOTE The cross product algorithm for evaluating a proportion (using equality of fractions) is (1) an extremely efficient algorithm but rote and without meaning, (2) usually misunderstood, (3) rarely generated by students independently, and (4) often used as a "means of avoiding proportional reasoning rather than facilitating it" (Cramer and Post, 1993; Post et al. 1988; Hart 1984; Lesh et al. 1988).

Students in the lower grades typically experience problems that are additive. The relationship between two ratios in a proportion is *multiplicative*, not *additive*. To see the difference, consider the following two problems.

Allie and Bente type at the same speed. Allie started typing first. When Allie had typed 8 pages. Bente had typed 4 pages. When Bente has typed 10 pages, how many has Allie typed?

This is an example of the *additive* relationship. Students should reason that since the two people type at the same speed, when Bente has typed an additional 6 pages, Allie should have also typed an additional 6 pages, so she should have typed $8 + 6$, or 14, pages.

Next consider the following problem.

Carl can type 8 pages for every 4 pages that Dan can type. If Dan has typed 12 pages, how many pages has Carl typed?

If students try an *additive* approach, they will conclude that since Dan has typed an additional 4 pages, then Carl should have typed an additional 4 pages. However, the correct reasoning is that since Carl types twice as fast as Dan he will type twice as many pages as Dan. Therefore, when Dan has typed 12 pages, Carl has typed 24 pages. The relationship between the ratios is *multiplicative*. Another way to solve this problem is to set up the proportion $\frac{8}{4} = \frac{x}{12}$, where x is the number of pages that Carl will type, and solve for x.

In the problem introduced in the quote from the *Principles and Standards*, one term in the proportion is missing:

$$\frac{3}{15} = \frac{x}{60}.$$

One way to solve the equation is to multiply both sides by 60, as follows:

$$\begin{aligned} \frac{3}{15} \cdot 60 &= \frac{x}{60} \cdot 60 \\ 3 \cdot 4 &= x \\ 12 &= x. \end{aligned}$$

Therefore, 12 packages of cocoa are needed to make 60 cups of hot chocolate.

Another method of solution, called the "cross-multiplication method" or "cross-product method," is given next.

This equation is a proportion if and only if

$$3 \cdot 60 = 15 \cdot x$$
$$180 = 15 \cdot x$$
$$12 = x.$$

EXAMPLE 5-20

If there are 3 cars for every 8 students at a high school, how many cars are there for 1200 students?

Solution We use the strategy of *setting up a table*, as shown in Table 5-3.

Table 5-3

Number of cars	3	x
Number of students	8	1200

The ratio of cars to students should always be the same.

$$\begin{array}{l} \text{Cars} \to \\ \text{Students} \to \end{array} \frac{3}{8} = \frac{x}{1200}$$
$$3 \cdot 1200 = 8 \cdot x$$
$$3600 = 8x$$
$$450 = x$$

Thus, there are 450 cars.

Next consider the student page. It shows the cost to rent a car from two different companies. The ratios for the What-A-Deal company do not form a proportion because $\frac{1}{20} \neq \frac{2}{35}$. The Value Vehicle ratios do form a proportion because $\frac{1}{20} = \frac{2}{40} = \frac{3}{60} = \frac{4}{80}$. To check for proportions, the student page uses *cross products*. Read through the example to see how attention is paid to the units in the problem. Work the *Think About It* on the bottom of the student page.

On the student page, we find "We say these quantities **vary proportionally**." What does this mean? Consider Table 5-4.

Table 5-4

Days (d)	1	2	3	4
Cost (c)	20	40	60	80

The ratios $\frac{c}{d}$ are all equal, that is, $\frac{20}{1} = \frac{40}{2} = \frac{60}{3} = \frac{80}{4}$. Thus, each pair of ratios forms a proportion. In this case, $\frac{c}{d} = \frac{20}{1}$, or 20, for all values of c and d. This is also expressed by saying that *c is proportional to d* or *c varies proportionally to d* or *c*

SAMPLE SCHOOL BOOK PAGE:

UNDERSTANDING PROPORTIONS

Lesson 6-5

Key Idea
A proportion is a statement that two ratios are equal.

Vocabulary
- proportion
- cross products

Understanding Proportions

WARM UP

Write = or ≠.

1. $\frac{3}{4}$ ▪ $\frac{9}{12}$ 2. $\frac{8}{16}$ ▪ $\frac{4}{8}$
3. $\frac{20}{12}$ ▪ $\frac{5}{3}$ 4. $\frac{3}{1}$ ▪ $\frac{1}{3}$
5. $\frac{24}{6}$ ▪ $\frac{4}{1}$ 6. $\frac{12}{15}$ ▪ $\frac{3}{5}$

LEARN

What is a proportion?

A **proportion** states that two ratios are equal. You can use the first two Value Vehicle ratios to write a proportion: $\frac{1 \text{ day}}{\$20} = \frac{2 \text{ days}}{\$40}$.

Value Vehicle

Days	Cost
1	\$20
2	\$40
3	\$60
4	\$80

In a proportion, the units must be the same across the top and bottom, or down the left and right sides. In this case, days are across the top and dollars across the bottom.

For Value Vehicle, the ratios of days to the cost of renting a car are all equal. We say that these quantities **vary proportionally.** The What-A-Deal Wheels ratios are NOT all equal, so these quantities do NOT vary proportionally.

What-A-Deal Wheels

Days	Cost
1	\$20
2	\$35
3	\$48
4	\$59

In the proportion $\frac{1}{20} = \frac{2}{40}$, 1×40 and 20×2 are cross products. The **cross products** of the terms in a proportion are equal.

Example

Decide if the ratios $\frac{3 \text{ ft}}{8 \text{ sec}}$ and $\frac{9 \text{ ft}}{24 \text{ sec}}$ form a proportion.

What You Do

Look at the units.

$\frac{3 \text{ ft}}{8 \text{ sec}} \stackrel{?}{=} \frac{9 \text{ ft}}{24 \text{ sec}}$ The units are the same across the top and bottom.

Look at the cross products.

$3 \times 24 \stackrel{?}{=} 8 \times 9$ ← The cross products are equal.

$72 = 72$

Why It Works

Multiply both sides of the proportion by 24×8 and simplify.

$\frac{24 \times \cancel{8} \times 3}{\cancel{8}} = \frac{9 \times \cancel{24} \times 8}{\cancel{24}}$

$24 \times 3 = 9 \times 8$ ← cross products

$72 = 72$

Since the units are the same and the cross products are equal, the ratios form a proportion.

Talk About It

1. Do the ratios $\frac{4 \text{ ft}}{6 \text{ ft}}$ and $\frac{12 \text{ sec}}{18 \text{ sec}}$ form a proportion? Why or why not?

316

Source: Scott Foresman–Addison Wesley Mathematics, Grade 6, 2005 (p. 316).

varies directly to d. In this case, $c = 20d$ for every c and d. The number 20 is the **constant of proportionality**. For example, we can say that *gas used by a car is proportional to the miles traveled* or *lottery profits vary directly with the number of tickets sold.*

Definition

If the variables x and y are related in the equality $y = kx$, $\left(k = \frac{y}{x}\right)$, then **$y$ is said to be proportional to x** and k is the **constant of proportionality** between y and x.

It is important to remember that in the ratio $a \div b$, a and b do not have to be integers. For example, if in Eugene, Oregon, $\frac{7}{10}$ of the population exercises regularly, then $\frac{3}{10}$ of the population do not exercise regularly, and the ratio of those who do to those who do not is $\frac{7}{10} : \frac{3}{10}$. This ratio can be written $7:3$.

It is also important to notice units of measure when we work with proportions. For example, if a turtle travels 5 in. every 10 sec, how many feet does it travel in 50 sec? If units of measure are ignored, we might set up the following proportion:

$$\frac{5}{10} = \frac{x}{50}.$$

This statement is incorrect. A correct statement must involve the same units in each ratio. We may write the following:

$$\frac{5 \text{ in.}}{10 \text{ sec}} = \frac{x \text{ in.}}{50 \text{ sec}}.$$

This implies that $x = 25$. Consequently, since 12 in. = 1 ft, the turtle travels $\frac{25}{12}$ ft, or $2\frac{1}{12}$ ft, or 2 ft 1 in.

The *Principles and Standards* point out the following concerning the cross-multiplication method of solving proportions.

> *Instruction in solving proportions should include methods that have a strong intuitive basis. The so-called cross-multiplication method can be developed meaningfully if it arises naturally in students' work, but it can also have unfortunate side effects when students do not adequately understand when the method is appropriate to use. Other approaches to solving proportions are often more intuitive and also quite powerful. For example, when trying to decide which is the better buy—12 tickets for 15.00 or 20 tickets for \$23.00—students might choose to use a scaling strategy (finding the cost for a common number of tickets) or a unit-rate strategy (finding the cost for one ticket).* (p. 221)

scaling strategy

The **scaling strategy** for solving the problem would involve finding the cost for a common number of tickets. Because LCM(12, 20) = 60, we could choose to find the cost of 60 tickets under each plan.

In the first plan, since 12 tickets cost \$15, then 60 tickets cost \$75.

In the second plan, since 20 tickets cost \$23, then 60 tickets cost \$69.

which also implies that the ratio of person time to dog time is

$$\frac{1 \text{ day to a person}}{7 \text{ days to a dog}}.$$

If a person eats 3 meals per day and the author of the cartoon assumes that the ratio of person time to dog time carries over to meals, then we have the following:

$$\frac{3 \text{ meals for a person}}{x \text{ meals for a dog}} = \frac{1 \text{ day for a person}}{7 \text{ days for a dog}}.$$

Solving for x, we have $x = 21$ meals for a dog. Thus, the ratio in the cartoon's last frame follows from the ratio in the first frames.

Looking Back Each statement in the cartoon appears to be consistent with the other statements. Either frame 1 or frame 2 of the cartoon is not needed to analyze the problem.

Scale Drawings

scale

Ratio and proportions are used in scale drawings. For example, if the scale is 1 : 300, then the length of 1 cm in such a drawing represents 300 cm, or 3 m in true size. The **scale** is the ratio of the size of the drawing to the size of the object. The following example shows the use of scale drawings.

EXAMPLE 5-22

The floor plan of the main floor of a house in Figure 5-16 is drawn in the scale of 1 : 300. Find the dimensions in meters of the living room.

FIGURE 5-16

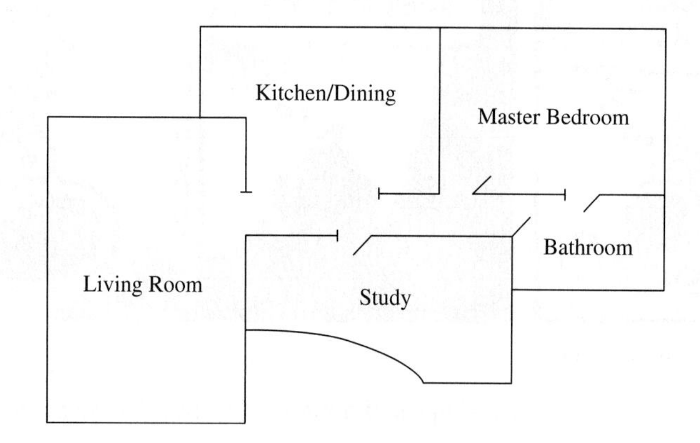

Solution In Figure 5-16, the dimensions of the living room measured with a centimeter ruler are approximately 3.7 cm by 2.5 cm. Because the scale is 1 : 300, 1 cm in the drawing represents 300 cm, or 3 m in true size. Hence, 3.7 cm represents $3.7 \cdot 3$, or 11.1 m, and 2.5 cm represents $2.5 \cdot 3$, or 7.5 m. Hence, the dimensions of the living room are approximately 11.1 m by 7.5 m.

ONGOING ASSESSMENT 5-4

1. Answer the following regarding the English alphabet:
   a. Determine the ratio of vowels to consonants.
   b. What is the ratio of consonants to vowels?
   c. What is the ratio of consonants to letters in the English alphabet?
   d. Write a word that has a ratio of $2:3$ of vowels to consonants.
2. Solve for x in each of the following proportions:
   a. $\frac{12}{x} = \frac{18}{45}$ **b.** $\frac{x}{7} = \frac{-10}{21}$
   c. $\frac{5}{7} = \frac{3x}{98}$ **d.** $3\frac{1}{2}$ is to 5 as x is to 15.
3. There are approximately 2 lb of muscle for every 5 lb of body weight. For a 90-lb person, how much of the weight is muscle?
4. There are 5 adult drivers to each teenage driver in Aluossim. If there are 12,345 adult drivers in Aluossim, how many teenage drivers are there?
5. If 4 grapefruits sell for 79¢, how much do 6 grapefruits cost?
6. On a map, $\frac{1}{3}$ in. represents 5 mi. If New York and Aluossim are 18 in. apart on the map, what is the actual distance between them?
7. David read 40 pages of a book in 50 min. How many pages should he be able to read in 80 min if he reads at a constant rate?
8. A candle is 30 in. long. After burning for 12 min, the candle is 25 in. long. How long will it take for the whole candle to burn at the same rate?
9. Two numbers are in the ratio $3:4$. Find the numbers if
   a. Their sum is 98.
   b. Their product is 768.
10. A rectangular yard has a width-to-length ratio of $5:9$. If the distance around the yard is 2800 ft, what are the dimensions of the yard?
11. Gary, Bill, and Carmella invested in a corporation in the ratio of $2:4:5$, respectively. If they divide the profit of \$82,000 proportionally to their investment, how much will each receive?
12. Sheila and Dora worked $3\frac{1}{2}$ hr and $4\frac{1}{2}$ hr, respectively, on a programming project. They were paid \$176 for the project. How much did each earn?
13. Vonna scored 75 goals in her soccer kicking practice. If her success-to-failure rate is $5:4$, how many times did she attempt a goal?
14. Express each of the following as a ratio whose terms are whole numbers:
   a. $\frac{1}{6}:1$ **b.** $\frac{1}{3}:\frac{1}{3}$ c. $\frac{1}{6}:\frac{2}{7}$
15. Write three other proportions that follow from the following proportion:

$$\frac{12¢}{36 \text{ oz}} = \frac{16¢}{48 \text{ oz}}$$

16. The rise and span for a house roof are identified as shown on the drawing. The pitch of a roof is the ratio of the rise to the half-span.
   a. If the rise is 10 ft and the span is 28 ft, what is the pitch?
   b. If the span is 16 ft and the pitch is $\frac{3}{4}$, what is the rise?

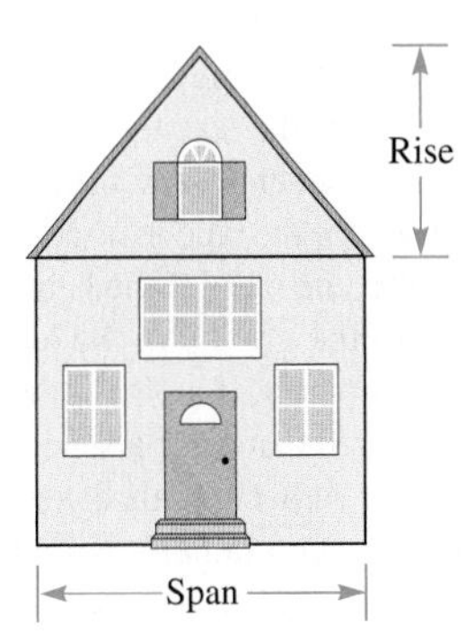

17. A grasshopper can jump 20 times its length. If jumping ability in humans were proportional to a grasshopper's, how far could a 6-ft-tall person jump?
18. Jim found out that after working for 9 months he had earned 6 days of vacation time. How many days per year does he earn at this rate?
19. Gear ratios are used in industry. A gear ratio is the comparison of the number of teeth on two gears. When two gears are meshed, the revolutions per minute (rpm) are inversely proportional to the number of teeth; that is,

$$\frac{\text{rpm of large gear}}{\text{rpm of small gear}} = \frac{\text{number of teeth on small gear}}{\text{number of teeth on large gear}}$$

   a. The rpm ratio of the large gear to the small gear is 4:6. If the small gear has 18 teeth, how many teeth does the large gear have?
   b. The large gear revolves at 200 rpm and has 60 teeth. How many teeth are there on the small gear that has an rpm of 600?
20. A Boeing 747 jet is approximately 230 ft long and has a wingspan of 195 ft. If a scale model of the plane is about 40 cm long, what is the model's wingspan?
21. Jennifer weighs 160 lb on Earth and 416 lb on Jupiter. Find Amy's weight on Jupiter if she weighs 120 lb on Earth.
22. a. If the ratio of boys to girls in a class is $2:3$, what is the ratio of boys to all the students in the class? Why?
   b. If the ratio of boys to girls in a class is $m:n$, what is the ratio of boys to all the students in the class?
   c. If $\frac{3}{5}$ of the class are girls, what is the ratio of girls to boys?

23. A recipe calls for 1 tsp of mustard seeds, 3 c of tomato sauce, $1\frac{1}{2}$ c of chopped scallions, and $3\frac{1}{4}$ c of beans. If one ingredient is altered as specified, how must the other ingredients be changed to keep the proportions the same? Explain your reasoning.
   a. 2 c of tomato sauce
   b. 1 c of chopped scallions
   c. $1\frac{3}{4}$ c of beans

24. The electrical resistance of a wire, measured in ohms (Ω), is equal to a constant times the length of the wire. If the electrical resistance of a 5-ft wire is 4.2 Ω, what is the resistance of 18 ft of the same wire?

25. At Rattlesnake School the teacher-student ratio is 1:30. If the school has 1200 students, how many additional teachers must be hired to reduce the ratio to 1:20?

26. At a particular time, the ratio of the height of an object that is perpendicular to the ground to the length of its shadow is the same for all objects. If a 30-ft tree casts a shadow of 12 ft, how tall is a tree that casts a shadow of 14 ft?

27. In a photograph of a father and his daughter, the daughter's height is 2.3 cm and the father's height is 5.8 cm. If the father is actually 188 cm tall, how tall is the daughter?

28. The amount of gold jewelry and other products is measured in karats (K), where 24K represents pure gold. The mark 14K on a chain indicates that the ratio between the mass of the gold in the chain and the mass of the chain is 14:24. If a gold ring is marked 18K and it weighs 0.4 oz, what is the value of the gold in the ring if pure gold is valued at $300 per oz?

29. If Amber is paid $8.00 per hour for typing, the table shows how much she earns.

Hours (h)	1	2	3	4	5
Wages (w)	8	16	24	32	40

   a. How much did Amber make for a 40-hr work week?
   b. What is the constant of proportionality?

30. The following table shows several possible widths W and corresponding lengths L of a rectangle whose area is 10 ft^2.

Width (W) (Feet)	Length (L) (Feet)	Area (Square Feet)
0.5	20	$0.5 \cdot 20 = 10$
1	10	$1 \cdot 10 = 10$
2	5	$2 \cdot 5 = 10$
2.5	4	$2.5 \cdot 5 = 10$
4	2.5	$4 \cdot 2.5 = 10$
5	2	$5 \cdot 2 = 10$
10	1	$10 \cdot 1 = 10$
20	0.5	$20 \cdot 0.5 = 10$

L
Area = 10 ft^2 W

   a. Use the values in the table and some additional values to graph the length L on the vertical axis versus the width W on the horizontal axis.
   b. What is the algebraic relationship between L and W?
   c. Write W as a function of L; that is, express W in terms of L.
   d. Write L as a function of W; that is, express L in terms of W.

31. **a.** In Room A of the University Center are 1 man and 2 women; in Room B are 2 men and 4 women; and in Room C are 5 men and 10 women. If all the people in Rooms B and C go to Room A, what will be the ratio of men to women in Room A?
   b. Prove the following generalization of the proportions used in (a):

$$\text{If } \frac{a}{b} = \frac{c}{d} = \frac{e}{f}, \text{ then } \frac{a}{b} = \frac{c}{d} = \frac{e}{f} = \frac{a+c+e}{b+d+f}.$$

★ 32. Prove that if $\frac{a}{b} = \frac{c}{d}$, then the following are true:
   a. $\frac{a+b}{b} = \frac{c+d}{d}$ $\left(\textit{Hint: } \frac{a}{b} + 1 = \frac{c}{d} + 1\right)$
   b. $\frac{a}{a+b} = \frac{c}{c+d}$
   c. $\frac{a-b}{a+b} = \frac{c-d}{c+d}$

Communication

33. Iris has found some dinosaur bones and a fossil footprint. The length of the footprint is 40 cm, the length of the thigh bone is 100 cm, and the length of the body is 700 cm.
   a. What is the ratio of the footprint's length to the dinosaur's length?
   b. Iris found a new track that she believes was made by the same species of dinosaur. If the footprint was 30 cm long and if the same ratio of foot length to body length holds, how long is the dinosaur?
   c. In the same area, Iris also found a 50-cm thigh bone. Do you think this thigh bone belonged to the same dinosaur that made the 30-cm footprint that Iris found? Why or why not?

34. Suppose a 10-in. pizza costs $4. For you to find the price x of a 14-in. pizza, is it correct to set up the proportion $\frac{x}{4} = \frac{14}{10}$? Why or why not?

35. Amy's friend told her the ratio of girls to boys in her new class is 5:6. Amy was very surprised to think her class would only have 11 students. What do you tell her?

36. In a condo complex $\frac{2}{3}$ of the men were married to $\frac{3}{4}$ of the women. What is the ratio of married people to the total adult population of the condo complex? Explain how you can obtain this ratio without knowing the actual number of men or women.

17. Steve claims that the shac
since there are 5 circles sha
do you respond?

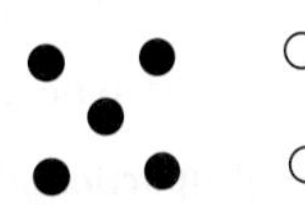

18. Daryl says that each piece
the pie. How do you respo

I. Fractions and rational nu
A. Numbers of the form
$b \neq 0$, are **rational n**
is the denominator.
B. A rational number ca
1. A division probl
tion problem
2. A partition, or pa
3. A ratio
4. A probability
C. Fundamental Law
and any number $c \neq$
D. Two fractions $\frac{a}{b}$ and
if $ad = bc$.
E. If GCD$(a, b) = 1$, th
F. If $0 \leq |a| < |b|$, then
If $|a| \geq |b| > 0$, $\frac{a}{b}$ is a
G. $\frac{a}{b} > \frac{c}{d}$ if and only if
H. If $\frac{a}{b} < \frac{c}{d}$, then $\frac{a}{b} <$
II. Operations on rational n
A. $\frac{a}{b} + \frac{c}{b} = \frac{a+c}{b}$
B. $\frac{a}{b} + \frac{c}{d} = \frac{ad+bc}{bd}$
C. $\frac{a}{b} - \frac{c}{d} = \frac{ad-bc}{bd}$
D. $\frac{a}{b} \cdot \frac{c}{d} = \frac{ac}{bd}$
E. $\frac{a}{b} \div \frac{c}{d} = \frac{a}{b} \cdot \frac{d}{c} =$

37. Write a paragraph in which you use the terms *ratio* and *proportion* correctly.

Open-Ended

38. List three real-world situations that involve ratio and proportion.

39. Find examples of ratios in a newspaper.

40. Boyle's Law states that at a given temperature, the product of the volume V of a gas and the pressure P is a constant c as follows:

$$PV = c$$

a. If at a given temperature, a pressure of 48 lb/in^2 compresses a certain gas to a volume of 960 in^3, what pressure would be necessary to compress the gas to a volume of 800 in^3 at the same temperature?

b. Find three other real-world situations in which the variables are related mathematically like the variables in Boyle's Law. In each case, describe how the variables are related using ratio and proportion.

41. Research the Golden Ratio that the Greeks used in the design of the Parthenon. Write a report on this ratio and include a drawing of a golden rectangle.

Cooperative Learning

42. Look at various elementary science and mathematics books and determine both if and how *dimensional analysis* is treated in those texts. How is dimensional analysis related to ratio and proportion?

43. In *Gulliver's Travels* by Jonathan Swift we find the following:

The seamstresses took my measure as I lay on the ground, one standing at my neck and another at mid-leg, with a strong cord extended, that each held by the end, while the third measured the length of the cord with a rule of an inch long. Then they measured my right thumb and desired no more; for by a mathematical computation, that twice around the thumb is once around the wrist, and so on to the neck and the waist; and with the help of my old shirt, which I displayed on the ground before them for a pattern, they fitted me exactly.

a. Explore the measurements of those in your group to see if you believe the ratios mentioned for Gulliver.

b. Suppose the distance around a person's thumb is 9 cm. What is the distance around the person's neck?

c. What ratio could be used to compare a person's height to his or her double arm span? Does this ratio have anything to do with da Vinci's famous painting *Vitruvian Man*?

d. Do you think there is a ratio between foot length and height? If so, what might it be?

e. Estimate other body ratios and then see how close you are to actual measurements.

Review Problems

44. Find all values of x that satisfy the following equations or inequalities:

a. $\frac{x}{3} = \frac{3x}{4}$

b. $3^x = 243$

c. $\frac{1}{3}x + \frac{3}{4}x = 49$

d. $\frac{1}{3}x \leq 49 - \frac{3}{4}x$

45. Do all rational numbers have a multiplicative inverse?

46. Find three rational numbers between 2 and 3 so that the numbers, including 2 and 3, form an arithmetic sequence.

47. If a geometric sequence has $\frac{4}{9}$ as its seventh term and the ratio of the sequence is $\frac{1}{2}$, what is the first term?

48. Determine what, if anything, is wrong with each of the following:

a. $2 = \frac{6}{3} = \frac{3+3}{3} = \frac{3}{3} + 3 = 1 + 3 = 4$

b. $1 = \frac{4}{2+2} = \frac{4}{2} + \frac{4}{2} = 2 + 2 = 4$

c. $\frac{ab+c}{a} = \frac{\cancel{a}b+c}{\cancel{a}} = b + c$

d. $\frac{a^2-b^2}{a-b} = \frac{a \cdot \cancel{a} - b \cdot \cancel{b}}{\cancel{a} - \cancel{b}} = a - b$

e. $\frac{a+c}{b+c} = \frac{a+\cancel{c}}{b+\cancel{c}} = \frac{a}{b}$

Third International Mathematics and Science Study (TIMSS) Questions

For every soft drink bottle that Fred collected, Maria collected 3. Fred collected a total of 9 soft drink bottles. How many did Maria collect?

a. 3 **b.** 12 **c.** 13 **d.** 27

TIMSS 2003, Grade 4

Three brothers, Bob, Dan, and Mark, receive a gift of 45 000 zeds from their father. The money is shared between the brothers in proportion to the number of children each one has. Bob has 2 children, Dan has 3 children, and Mark has 4 children.

How many zeds does Mark get?

a. 5000
b. 10,000
c. 15,000
d. 20,000

TIMSS 2003, Grade 8

SAMPLE SCHOOL BOOK PAGE: DECIMAL PLACE VALUE

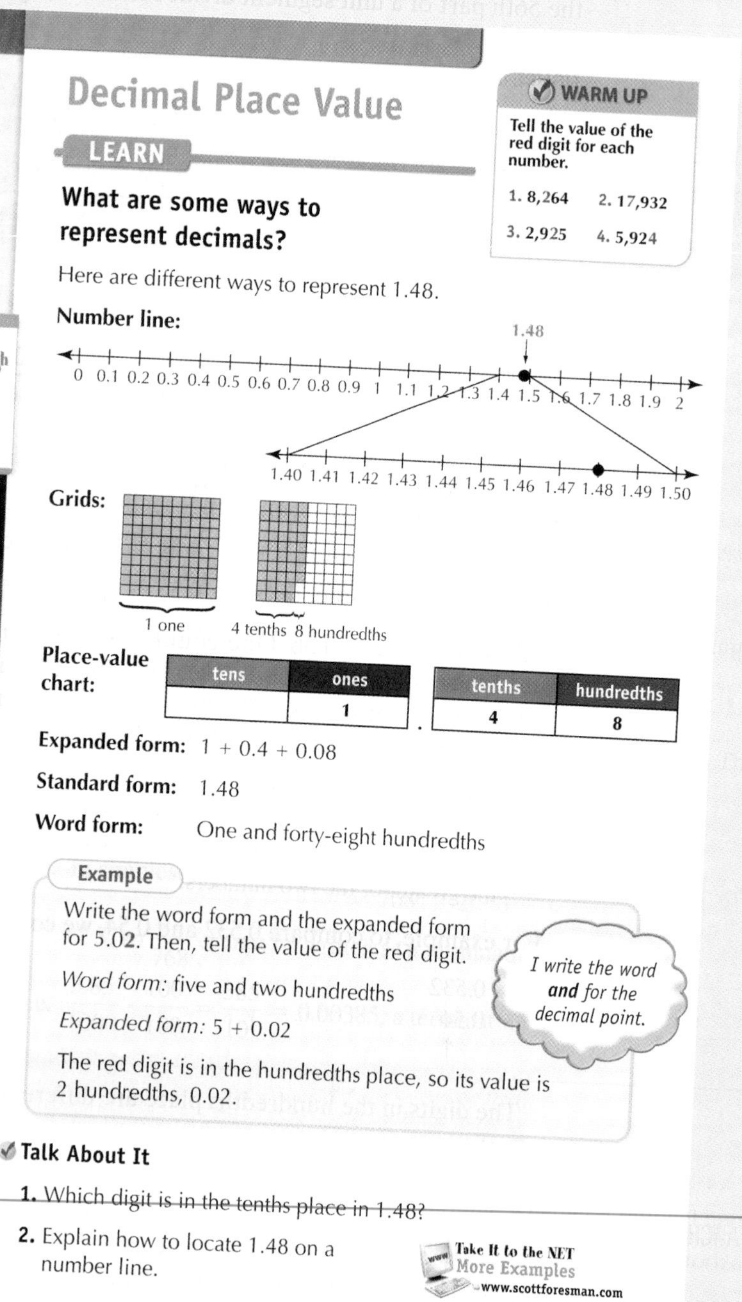

Lesson 11-2

Key Idea
There are many ways to represent decimal numbers.

Decimal Place Value

WARM UP

Tell the value of the red digit for each number.

1. 8,264 2. 17,932
3. 2,925 4. 5,924

LEARN

What are some ways to represent decimals?

Here are different ways to represent 1.48.

Number line:

Grids:

Place-value chart:

tens	ones	.	tenths	hundredths
	1	.	4	8

Expanded form: 1 + 0.4 + 0.08

Standard form: 1.48

Word form: One and forty-eight hundredths

Think It Through
I can **use objects, draw pictures,** or **make a chart** to represent 1.48.

Example

Write the word form and the expanded form for 5.02. Then, tell the value of the red digit.

Word form: five and two hundredths

Expanded form: 5 + 0.02

The red digit is in the hundredths place, so its value is 2 hundredths, 0.02.

I write the word ***and*** *for the decimal point.*

Talk About It

1. Which digit is in the tenths place in 1.48?
2. Explain how to locate 1.48 on a number line.

Take It to the NET
More Examples
www.scottforesman.com

628

Source: Scott Foresman–Addison Wesley Mathematics, Grade 4, 2005 (p. 628).

ONGOING ASSESSMENT 6-1

1. Write each of the following as a sum, using powers of 10 to show place value:
   a. 0.023 **b.** 206.06
   c. 312.0103 **d.** 0.000132
2. Rewrite each of the following as decimals:
   a. $4 \cdot 10^3 + 3 \cdot 10^2 + 5 \cdot 10 + 6 + 7 \cdot 10^{-1} + 8 \cdot 10^{-2}$
   b. $4 \cdot 10^3 + 6 \cdot 10^{-1} + 8 \cdot 10^{-3}$
   c. $4 \cdot 10^4 + 3 \cdot 10^{-2}$
   d. $2 \cdot 10^{-1} + 4 \cdot 10^{-4} + 7 \cdot 10^{-7}$
3. Write each of the following as numerals:
   a. Five hundred thirty-six and seventy-six ten thousandths
   b. Three and eight thousandths
   c. Four hundred thirty-six millionths
   d. Five million and two tenths
4. Write each of the following in words:
   a. 0.34 **b.** 20.34
   c. 2.034 **d.** 0.000034
5. Write each of the following terminating decimals in common fraction notation:
   a. 0.436 **b.** 25.16 c. $^-316.027$
   d. 28.1902 e. $^-4.3$ **f.** $^-62.01$
6. Mentally determine which of the following represent terminating decimals:
   a. $\frac{4}{5}$ **b.** $\frac{61}{2^2 \cdot 5}$ c. $\frac{3}{6}$ **d.** $\frac{1}{2^5}$ e. $\frac{36}{5^5}$
   f. $\frac{133}{625}$ g. $\frac{1}{3}$ **h.** $\frac{2}{25}$ i. $\frac{1}{13}$ **j.** $\frac{26}{65}$
7. Where possible, write each of the numbers in problem 6 as a terminating decimal.
8. a. Seven minutes is part of an hour. If 7 min were to be expressed as a decimal part of an hour, explain whether or not it would be a terminating decimal.
   b. What whole number of minutes (less than 60) could be expressed as terminating decimal parts of an hour?
9. A unit segment has length of 1. If a unit segment had one endpoint at 3, the other at 4 on a number line, and we considered the $\frac{56}{100}$ part of that unit segment as in Figure 6.3, what decimal would the rightmost endpoint of the $\frac{56}{100}$ segment represent?
10. Given the U.S. monetary system, what decimal reason can you think of for having coins only for a penny, nickel, dime, quarter, and half-dollar as coins less than $1.00?
11. In each of the following, order the decimals from greatest to least:
    a. 13.4919, 13.492, 13.49183, 13.49199
    b. $^-1.453$, $^-1.45$, $^-1.4053$, $^-1.493$
12. Write the numbers in each of the following sentences as decimals in symbols:
    a. A mite has body length about fourteen thousandths of an inch.
    b. The Earth goes around the Sun once every three hundred sixty-five and twenty-four hundredths days.
13. Use a grid with 100 squares and represent 0.32. Explain your representation.
14. If the decimals 0.804, 0.84, and 0.8399 are arranged on a typical number line, which is furthest to the right?
15. Write a decimal number that has a ten-thousandths place and is between 8.34 and 8.341.
16. a. Show that between any two terminating decimals, there is another terminating decimal.
    b. Argue that part (a) can be used to show that there are infinitely many terminating decimals between any two terminating decimals.
17. a. Describe the decimal 0.613 using base-ten blocks.
    b. Explain whether you can describe a decimal such as 0.6134 using base-ten blocks.
18. Explain the mathematical meaning of a sign on a copy machine reading ".05¢ a copy."
19. Describe in words the location of 0.056 on a number line.
20. Given any reduced rational number $\frac{a}{b}$ with $0 < a < b$, where b is of the form $2^m \cdot 5^n$ (m and n are whole numbers), determine a relationship between m and/or n and the number of digits in the terminating decimal. Justify your answer.
21. Which of the following numbers is the greatest: $100{,}000^3$, 1000^5, $100{,}000^2$? Justify your answer.
22. A baseball player's batting average was reported as "three-twenty-two." A batting average is essentially determined when you divide the number of times that the player has hits by the number of times at bat. Explain why the reported batting average is not mathematically correct.
23. If "decimals" in other number bases work the same as in base ten, explain the meaning of the following:
    a. 3.145_{six} **b.** 0.00334_{seven}
24. The five top swimmers in an event had the following times:

Emily	64.54 sec	Kathy	64.02 sec
Molly	64.46 sec	Rhonda	63.54 sec
Martha	63.59 sec		

List them in the order they placed.

Communication

25. Using Simon Stevin's notation, how would you write the following:
    a. 0.3256 **b.** 0.0032
26. If 1 mL is 0.001 L, how should 18 mL be expressed as a terminating decimal number of liters?
27. Explain whether 1 day can be expressed as a terminating decimal part of a year.
28. Using the number-line model to depict terminating decimals, explain whether you think that there is a greatest terminating decimal less than 1.

29. Look up the meaning of *tithe*. How is that related to decimals?

30. Explain how you would use base-ten blocks to represent two and three hundred forty-five thousandths.

31. Explain why in Theorem 6-1 the rational number must be in simplest form before examining the denominator.

32. In your own words, explain what a decimal point is and what it does.

Open-Ended

33. Determine how decimal notation is symbolized in different countries.

34. Examine three elementary school textbooks and report how the introductions of the topics of exponents and decimals differ, if they do. Report the grade level that decimals are introduced.

Cooperative Learning

35. Based on a system similar to that of base ten, determine how you might introduce a "decimal" notation in base five.

36. Simon Stevin is credited with propagating the decimal system. In small groups, research the history of decimals to find contributions of the Arabs, the Chinese, and Renaissance mathematicians. Explain whether or not you believe Stevin "invented" the decimal system.

BRAIN TEASER

A U.S. Post Office advertised stamps for \$.37¢ each. Was this "true" advertising?

6-2 Operations on Decimals

To develop an algorithm for addition of terminating decimals, consider the sum 2.16 + 1.73. In elementary school, base-ten blocks are recommended to demonstrate such an addition problem. Figure 6-4 shows how the addition can be performed.

FIGURE 6-4

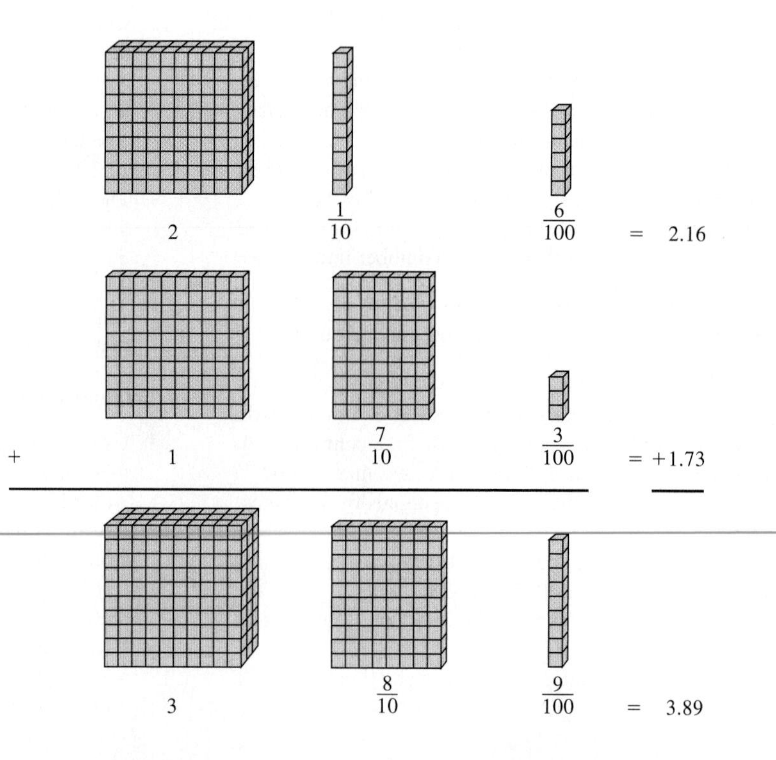

The computation in Figure 6-4 can be explained by *changing it to a problem we already know how to solve*, that is, to a sum involving fractions. We then use the commutative and associative properties of addition to aid in the computation, as follows:

$$\begin{aligned} 2.16 + 1.73 &= \left(2 + \frac{1}{10} + \frac{6}{100}\right) + \left(1 + \frac{7}{10} + \frac{3}{100}\right) \\ &= (2 + 1) + \left(\frac{1}{10} + \frac{7}{10}\right) + \left(\frac{6}{100} + \frac{3}{100}\right) \\ &= 3 + \frac{8}{10} + \frac{9}{100} \\ &= 3.89 \end{aligned}$$

In the computation of Figure 6-4, we see that we are adding units to units, tenths to tenths, and hundredths to hundredths. This can be accomplished most efficiently by keeping the numbers in their decimal forms, lining up the decimal points, and adding as if the numbers were whole numbers. This technique works for both addition and subtraction, as demonstrated on the student page. Note also the use of a calculator on the decimal addition, Example C.

Multiplying Decimals

Just as we explained an algorithm for adding and subtracting terminating decimals by representing them as fractions, we develop and explain an algorithm for multiplication of decimals. Consider the product $4.62 \cdot 2.4$:

$$(4.62)(2.4) = \frac{462}{100} \cdot \frac{24}{10} = \frac{462}{10^2} \cdot \frac{24}{10^1} = \frac{462 \cdot 24}{10^2 \cdot 10^1} = \frac{11{,}088}{10^3} = 11.088$$

The answer to this computation was obtained by multiplying the whole numbers 462 and 24 and then dividing the result by 10^3.

The algorithm for multiplying decimals can be stated as follows:

If there are n digits to the right of the decimal point in one number and m digits to the right of the decimal point in a second number, multiply the two numbers, ignoring the decimals, and then place the decimal point so that there are n + m digits to the right of the decimal point in the product.

REMARK There are $n + m$ digits to the right of the decimal point in the product because $10^n \cdot 10^m = 10^{n+m}$.

SAMPLE SCHOOL BOOK PAGE:

ADDING AND SUBTRACTING WHOLE NUMBERS AND DECIMALS

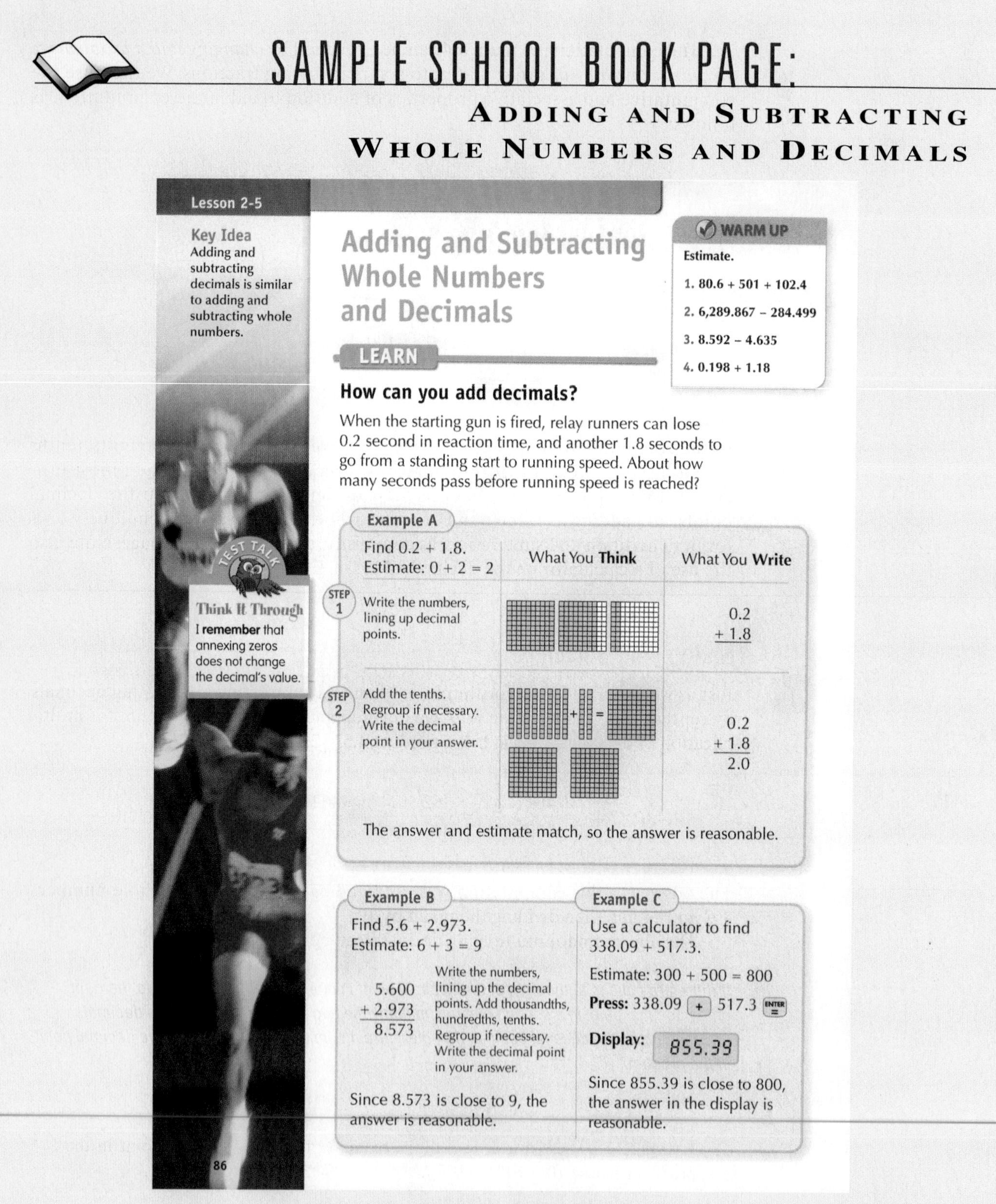

Lesson 2-5

Key Idea
Adding and subtracting decimals is similar to adding and subtracting whole numbers.

Adding and Subtracting Whole Numbers and Decimals

WARM UP
Estimate.
1. 80.6 + 501 + 102.4
2. 6,289.867 − 284.499
3. 8.592 − 4.635
4. 0.198 + 1.18

LEARN

How can you add decimals?

When the starting gun is fired, relay runners can lose 0.2 second in reaction time, and another 1.8 seconds to go from a standing start to running speed. About how many seconds pass before running speed is reached?

TEST TALK
Think It Through
I **remember** that annexing zeros does not change the decimal's value.

Example A

Find 0.2 + 1.8. Estimate: 0 + 2 = 2	What You **Think**	What You **Write**
STEP 1 Write the numbers, lining up decimal points.		0.2 + 1.8
STEP 2 Add the tenths. Regroup if necessary. Write the decimal point in your answer.		0.2 + 1.8 = 2.0

The answer and estimate match, so the answer is reasonable.

Example B

Find 5.6 + 2.973.
Estimate: 6 + 3 = 9

5.600
+ 2.973
8.573

Write the numbers, lining up the decimal points. Add thousandths, hundredths, tenths. Regroup if necessary. Write the decimal point in your answer.

Since 8.573 is close to 9, the answer is reasonable.

Example C

Use a calculator to find 338.09 + 517.3.

Estimate: 300 + 500 = 800

Press: 338.09 [+] 517.3 [ENTER =]

Display: 855.39

Since 855.39 is close to 800, the answer in the display is reasonable.

86

Source: Scott Foresman–Addison Wesley Mathematics, Grade 6, 2005 (p. 86).

EXAMPLE 6-4

Compute each of the following:

a. (6.2)(1.43) **b.** (0.02)(0.013) **c.** (1000)(3.6)

Solution **a.**

```
  1.4 3     (2 digits after the decimal point)
 × 6.2      (1 digit after the decimal point)
  2 8 6
 8 5 8
 8.8 6 6    (2 + 1, or 3 digits after the decimal point)
```

b.

```
     0.0 1 3
 ×     0.0 2
 0.0 0 0 2 6
```

c.

```
        3.6
 ×  1 0 0 0
   3 6 0 0.0
```

NOW TRY THIS 6-2

Example 6-4(c) suggests that multiplication by 1000, or 10^3, results in moving the decimal point in the product three places to the right. **(a)** Explain why this is true using expanded notation and the distributive property of multiplication over addition. **(b)** In general, how does multiplication by 10^n, where n is a positive integer, affect the product? Why?

Scientific Notation

Many calculators display the decimals for the fractions $\frac{3}{45{,}689}$ or $\frac{5}{76{,}146}$ as [6.5661319 −05] and [6.566333 −05], respectively. The displays are in *scientific notation*. The first display is a notation for $6.5661319 \cdot 10^{-5}$ and the second for $6.566333 \cdot 10^{-5}$.

Scientists use scientific notation to handle either very small or very large numbers. For example, "the Sun is 93,000,000 mi from Earth" is expressed as "the Sun is $9.3 \cdot 10^7$ mi from Earth." A micron, a metric unit of measure that is 0.000001 m, is written $1 \cdot 10^{-6}$m.

Definition of Scientific Notation

In **scientific notation**, a positive number is written as the product of a number greater than or equal to 1 and less than 10 and an integer power of 10.

The following numbers are in scientific notation:

$$8.3 \cdot 10^8, \quad 1.2 \cdot 10^{10}, \quad \text{and} \quad 7.84 \cdot 10^{-6}$$

The numbers $0.43 \cdot 10^9$ and $12.3 \cdot 10^{-6}$ are not in scientific notation because 0.43 and 12.3 are not greater than or equal to 1 and less than 10. To write a number like 934.5 in scientific notation, we divide by 10^2 to get 9.345 and then multiply by 10^2 to retain the value of the original number:

$$934.5 = \left(\frac{934.5}{10^2}\right)10^2 = 9.345 \cdot 10^2$$

378 CHAPTER 6

8. If each of the following sequences is either arithmetic or geometric, continue the following decimal patterns:
   a. 0.9, 1.8, 2.7, 3.6, 4.5, ____, ____, ____
   b. 0.3, 0.5, 0.7, 0.9, 1.1, ____, ____, ____
   c. 1, 0.5, 0.25, 0.125, ____, ____, ____
   d. 0.2, 1.5, 2.8, 4.1, 5.4, ____, ____, ____
9. If the first term of a finite geometric sequence is 0.9 and its ratio is 0.2, what is the sum of the first five terms?
10. Interpret the decimal 0.3333333 as a sum of a finite geometric sequence whose first term is 0.3. (*Hint:* Write 0.3333333 as the sum of fractions whose denominators are powers of 10.)

EXAMPLE 6-5

11. In a finite geometric sequence, the first term is 0.2 and the sixth term is 0.000486. What are the second through fifth terms?
12. Estimate the placement of each of the following on the given number line by placing the letter for each computation in the appropriate box.
   a. $0.3 \div 0.31$ b. $0.3 \cdot 0.31$
   c. $0.3 + 0.31$ d. $0.3 - 0.31$

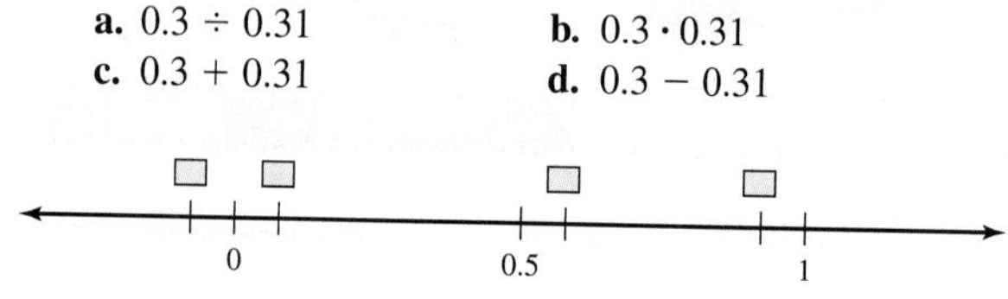

EXAMPLE 6-6

13. A bank statement from a local bank shows that a checking account has a balance of \$83.62. The balance recorded in the checkbook shows only \$21.69. After checking the canceled checks against the record of these checks, the customer finds that the bank has not yet recorded six checks in the amounts of \$3.21, \$14.56, \$12.44, \$6.98, \$9.51, and \$7.49. Is the bank record correct? (Assume the person's checkbook records *are* correct.)
14. Convert each of the following to standard numerals:
   a. $3.2 \cdot 10^{-9}$ b. $3.2 \cdot 10^{9}$
   c. $4.2 \cdot 10^{-1}$ d. $6.2 \cdot 10^{5}$
15. Write the numerals in each of the following sentences in scientific notation:
   a. The diameter of Earth is about 12,700,000 m.
   b. The distance from Pluto to the Sun is about 4,486,000,000 km.
   c. Each year, about 50,000,000 cans are discarded in the United States.
16. Write the numerals in each of the following sentences in standard form:
   a. A computer requires $4.4 \cdot 10^{-6}$ sec to do an addition problem.
   b. There are about $1.99 \cdot 10^{4}$ km of coastline in the United States.
   c. Earth has existed for approximately $3 \cdot 10^{9}$ yr.
17. Write the results of each of the following in scientific notation:
   a. $(8 \cdot 10^{12}) \cdot (6 \cdot 10^{15})$
   b. $(16 \cdot 10^{12}) \div (4 \cdot 10^{5})$
   c. $(5 \cdot 10^{8}) \cdot (6 \cdot 10^{9}) \div (15 \cdot 10^{15})$
18. Round each of the following numbers as specified:
   a. 203.651 to the nearest hundred
   b. 203.651 to the nearest ten
   c. 203.651 to the nearest unit
   d. 203.651 to the nearest tenth
   e. 203.651 to the nearest hundredth
19. Jane's car travels 224 mi on 12 gal of gas. Rounded to the nearest mile, how many miles to the gallon does her car get?
20. Audrey wants to buy some camera equipment to take pictures on her daughter's birthday. To estimate the total cost, she rounds each price to the nearest dollar and adds the rounded prices. What is her estimate for the items listed?

Camera	\$54.56
Film	\$4.50
Case	\$17.85

21. Estimate the sum or difference in each of the following by using (i) rounding and (ii) front-end estimation. Then perform the computations to see how close your estimates are to the actual answers.

a.	b.	c.	d.
65.84	89.47	5.85	223.75
24.29	− 32.16	6.13	− 87.60
12.18		9.10	
+ 19.75		+ 4.32	

22. Mary Kim invested \$964 in 18 shares of stock. A month later, she sold the 18 shares at \$61.48 per share. She also invested in 350 shares of stock for a total of \$27,422.50. She sold this stock for \$85.35 a share and paid \$495 in total commissions. What was Mary Kim's profit or loss on the transactions to the nearest dollar?
23. Find the least and the greatest possible products for each expression using the digits 1 through 9. Each digit may be used only once in each part.
   a. □.□ × □ b. □.□ × □.□
24. Some digits in the following number are covered by squares:

If each of the digits 1 through 9 is used exactly once in the number, determine the number in each of these cases:
   a. The number is the greatest possible.
   b. The number is the least possible.
25. Iris worked a 40-hr week at \$6.25/hr. Mentally compute her salary for the week and explain how you did it.
26. Mentally compute the number to fill in the blank in each of the following:
   a. $8.4 \cdot 6 = 4.2 \cdot$ ____
   b. $10.2 \div 0.3 = 20.4 \div$ ____
   c. $a \cdot b = (a/2) \cdot$ ____
   d. $a \div b = 2a \div$ ____
27. Which of the following result in equal quotients?
   a. $7 \div 0.25$ b. $70 \div 2.5$
   c. $0.7 \div 0.25$ d. $700 \div 25$

28. Use estimation to place the decimal point in the correct position in each of the following products:
    a. $534 \cdot 0.34 = 18156$
    b. $5.07 \cdot 29.3 = 148551$
29. Is subtracting 0.3 the same as
    a. subtracting 0.30?
    b. subtracting 0.03? Explain.
30. Use estimation to choose a decimal to multiply by 9 in order to get within 1 of 93. Explain how you made your choice and check your estimate.
31. a. Fill in the parentheses in each of the following to write a true equation:

$$1 \cdot 2 + 0.25 = (\ \ )^2$$

$$2 \cdot 3 + 0.25 = (\ \ )^2$$

Conjecture what the next two equations in this pattern will be.
    b. Do the computations to determine if your next two equations are correct.
    c. Generalize your answer in part (a) by filling in an appropriate expression in the equation

$$n\,(n + 1) + 0.25 = (\ \ )^2$$

where n is the number of the terms in this sequence of equations.

Communication

32. Give an example of a balanced checkbook where entries could be incorrect.
33. How is multiplication of decimals like multiplication of whole numbers? How is it different?
34. Why are estimation skills important in dividing decimals?
35. In the text, multiplication and division were done using both fractional and decimal forms. Discuss the advantages and disadvantages of each.
36. Explain why subtraction of terminating decimals can be accomplished by lining up the decimal points, subtracting as if the numbers were whole numbers, and then placing the decimal point in the difference.

Open-Ended

37. Find several examples of the use of decimals in the newspaper. Tell whether you think the numbers are exact or estimates. Also tell why you think decimals were used instead of fractions.
38. How could a calculator be used to develop or reinforce the understanding of multiplication of decimals?
39. How might certain calculators impede the learning of multiplication of decimals if it displays the simplest form of a terminating decimal, for example, 0.3 instead of 0.30?
40. Wearne and Hiebert (1988) reported that students who connect physical representations of decimals with decimal notation are more likely to create their own procedures for converting a fraction to its decimal notation. Do you think that this research supports or does not support the conversion of decimals to fractions to explain computational procedures?

Cooperative Learning

41. In your group, decide on all the prerequisite skills that students need before learning to perform arithmetic operations on decimals.
42. You will need a calculator and a partner to play the game described on page 388 on the student page from McDougal Littell *Middle Grades MATH Thematics, Book 1*, 2005.

Review Problems

43. Write 14.0479 in expanded form.
44. Without dividing, determine which of the following represent terminating decimals:
    a. $\frac{24}{36}$
    b. $\frac{35}{56}$
45. If the denominator of a fraction is 26, is it possible that the fraction could be written as a terminating decimal? Why or why not?
46. $\frac{35}{56}$ can be written as a decimal that terminates. Explain why.

Third International Mathematics and Science Study (TIMSS) Questions

A rubber ball rebounds to half the height it drops. If the ball is dropped from a rooftop 18 m above the ground, what is the total distance traveled by the time it hits the ground the third time?
a. 31.5 m
b. 40.5 m
c. 45 m
d. 63 m

In a discus-throwing competition, the winning throw was 61.60 m. The second-place throw was 59.72 m. How much longer was the winning throw than the second-place throw?
a. 1.18 m
b. 1.88 m
c. 1.98 m
d. 2.18 m

TIMSS, Grade 8, 1994

SAMPLE SCHOOL BOOK PAGE:

Estimating Decimal Products

LEARN HOW TO...
- improve your estimating skills

AS YOU...
- play *Target Number Plus or Minus 1*

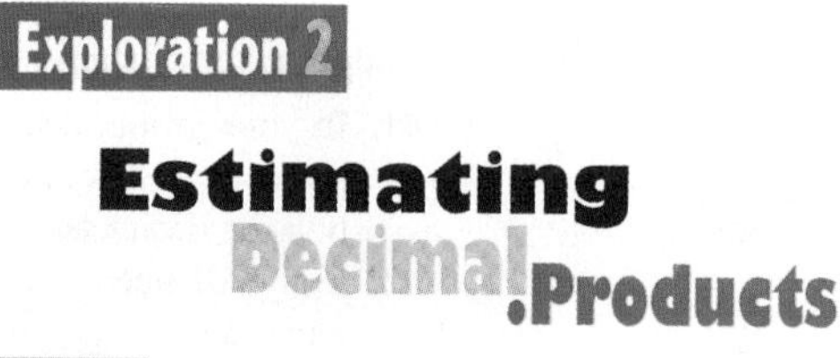

SET UP *Work with a partner. You will need one calculator.*

In a new game, *Target Number Plus or Minus 1*, the goal is to find a product that is within 1 of the target number. To play the game, follow the flowchart below. You must use estimation and mental math to decide what numbers to enter in the calculator.

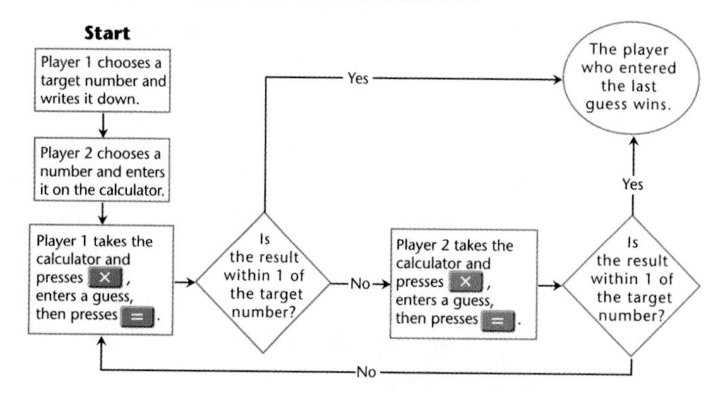

Sample Game

Player 1 chose 196 for the target number.
Player 2 chose the number 7.

Player	Keys pressed	Display
2	7	7.
1	× 2 5 =	175.
2	× 1 . 1 =	192.5
1	× 1 . 0 2 =	196.35

Since 196.35 is within 1 of 196, Player 1 wins!

276 **Module 4** Mind Games

Source: Math Thematics, Book 1, McDougal Littell, 2005 (p. 276).

6-3 Nonterminating Decimals

Earlier in the chapter, procedures for converting some rational numbers to decimals were developed. For example, $\frac{7}{8}$ can be written as a terminating decimal as follows:

$$\frac{7}{8} = \frac{7}{2^3} = \frac{7 \cdot 5^3}{2^3 \cdot 5^3} = \frac{875}{1000} = 0.875$$

The decimal for $\frac{7}{8}$ can also be found by division:

$$\begin{array}{r} 0.875 \\ 8\overline{)7.000} \\ \underline{6\,4} \\ 60 \\ \underline{56} \\ 40 \\ \underline{40} \\ 0 \end{array}$$

Repeating Decimals

If we use a calculator to find a decimal representation for $\frac{2}{11}$, the calculator may display 0.1818181. It seems that 18 repeats itself. To examine what digits, if any, the calculator did not display, consider the following division:

$$\begin{array}{r} 0.18 \\ 11\overline{)2.00} \\ \underline{1\,1} \\ 90 \\ \underline{88} \\ 2 \end{array}$$

At this point, if the division is continued, the division pattern would repeat, since the remainder 2 divided by 11 repeats the division. Thus the quotient is 0.181818 .... A decimal of this type is a **repeating decimal**, and the repeating block of digits is the **repetend**. The repeating decimal is written $0.\overline{18}$, where the bar indicates that the block of digits underneath is repeated continuously.

repeating decimal

repetend

EXAMPLE 6-10

Convert the following to decimals:

a. $\frac{1}{7}$ **b.** $\frac{2}{13}$

Solution We use a calculator to divide, and it seems that the division pattern repeats.

a. $\frac{1}{7} = 0.\overline{142857}$ **b.** $\frac{2}{13} = 0.\overline{153846}$ ▲

To see why in Example 6-10 the division pattern repeats as predicted, consider the following divisions:

```
a.    0.142857         b.     0.153846
    7)1.000000            13)2.000000
      7                      1 3
      30                       70
      28                       65
       20                       50
       14                       39
        60                      110
        56                      104
         40                       60
         35                       52
          50                       80
          49                       78
           1                        2
```

In $\frac{1}{7}$, the remainders obtained in the division are 3, 2, 6, 4, 5, and 1. These are all the possible nonzero remainders that can be obtained when dividing by 7. If we had obtained a remainder of 0, the decimal would terminate. Consequently, the seventh division cannot produce a new remainder. Whenever a remainder recurs, the process repeats itself. Using similar reasoning, we could predict that the repetend for $\frac{2}{13}$ could not have more than 12 digits, because there are only 12 possible nonzero remainders. However, one of the remainders could repeat sooner than that, which was actually the case in part (b). In general, if $\frac{a}{b}$ is any rational number in simplest form with $b \neq 0$ and $b > a$, and it does not represent a terminating decimal, the repetend has at most $b - 1$ digits. Therefore, *a rational number may always be represented either as a terminating decimal or as a repeating decimal.*

NOW TRY THIS 6-6

a. Write $\frac{1}{9}$ as a decimal.

b. Based on your answer in part (a), mentally compute the decimal representation for each of the following.

(i) $\frac{2}{9}$ (ii) $\frac{3}{9}$

(iii) $\frac{5}{9}$ (iv) $\frac{8}{9}$

EXAMPLE 6-11

Use a calculator to convert $\frac{1}{17}$ to a repeating decimal.

Solution In using a calculator, if we press [1] [÷] [1] [7] [=], we obtain the following, shown as part of a division problem:

$$17\overline{)1.}\quad 0.0588235$$

Without knowing whether the calculator has an internal round-off feature and with the calculator's having an eight-digit display, we find the greatest number of digits to be trusted in the quotient is six following the decimal point. (Why?) If we use those six places and multiply 0.058823 times 17, we may continue the operation as follows:

We then obtain 0.999991, which we may place in the preceding division:

$$\begin{array}{r} 0.058823 \\ 17\overline{)1.000000} \\ \underline{999991} \\ 9 \end{array}$$

Next, we divide 9 by 17 to obtain 0.5294118. Again ignoring the rightmost digit, we continue as before, completing the division as follows, where the repeating pattern is apparent:

$$\begin{array}{l} \quad\;\; 0.05882352941176470588235 \\ 17\overline{)1.0000000000000000000000} \\ \quad\;\; \underline{999991} \\ \qquad\qquad 9000000 \\ \qquad\qquad \underline{8999987} \\ \qquad\qquad\qquad 13000000 \\ \qquad\qquad\qquad \underline{12999985} \\ \qquad\qquad\qquad\qquad\;\; 15 \end{array}$$

Thus, $\dfrac{1}{17} = 0.\overline{0588235294117647}$, and the repetend is 16 digits long.

Writing a Repeating Decimal in the Form $\frac{a}{b}$ Where $a, b \in I, b \neq 0$

We have already considered how to write terminating decimals in the form $\frac{a}{b}$, where a, b are integers and $b \neq 0$. For example,

$$0.55 = \frac{55}{10^2} = \frac{55}{100}$$

To write $0.\overline{5}$ in a similar way, we see that because the repeating decimal has infinitely many digits, there is no single power of 10 that can be placed in the denominator. To overcome this difficulty, we must somehow eliminate the infinitely repeating part of the decimal. If we let $n = 0.\overline{5}$, then our *subgoal* is to write an equation for n without a repeating decimal. It can be shown that $10(0.555\ldots) = 5.555\ldots = 5.\overline{5}$. Hence, $10n = 5.\overline{5}$. Using this information, we subtract the corresponding sides of the equations to obtain an equation whose solution can be written without a repeating decimal.

$$\begin{aligned} 10n &= 5.\overline{5} \\ -\ \ n &= 0.\overline{5} \\ \hline 9n &= 5 \\ n &= \frac{5}{9} \end{aligned}$$

so we choose numbers closer to 1.4 in order to find the next approximation. We find the following:

$$(1.42)^2 = 2.0164$$
$$(1.41)^2 = 1.9981$$

Thus, $1.41 < \sqrt{2} < 1.42$. We can continue this process until we obtain the desired approximation. Note that if your calculator has a square-root key, we can obtain an approximation directly.

NOW TRY THIS 6-10

One algorithm for calculating square roots is sometimes attributed to Archimedes. The algorithm makes use of making closer and closer estimates to the square root. To find the square root of a positive number, n, first make a guess. Call the first guess *Guess1*. Now compute as follows:

Step 1: Divide n by *Guess1*.
Step 2: Now add *Guess1* to the quotient obtained in Step 1.
Step 3: Divide the sum in Step 2 by 2. The quotient becomes *Guess2*.

Repeat the steps using *Guess2* to obtain successive guesses or until the desired accuracy is achieved.

a. Use the method described to find the square root of 13.
b. Write the steps for the algorithm in a recursive formula.

The System of Real Numbers

real numbers

The set of **real numbers** R, is the union of the set of rational numbers and the set of irrational numbers. Real numbers represented as decimals can be terminating, repeating, or nonterminating and nonrepeating.

Every integer is a rational number as well as a real number. Every rational number is a real number, but not every real number is rational, as has been shown with $\sqrt{2}$. The relationships among sets of numbers are summarized in the tree in Figure 6-9.

FIGURE 6-9

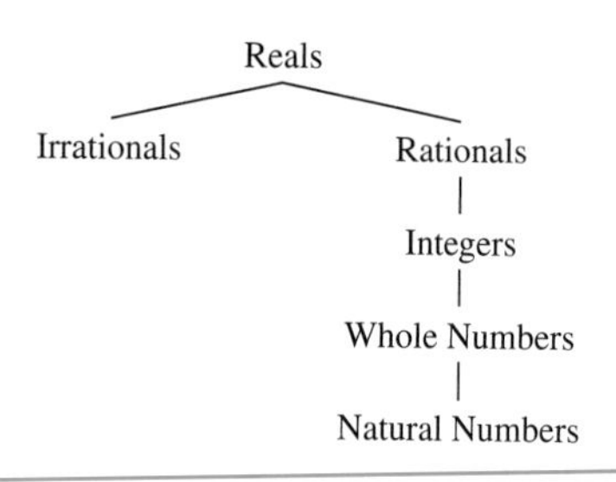

The concept of fractions can now be extended to include all numbers of the form $\frac{a}{b}$, where a and b are real numbers with $b \neq 0$, such as $\frac{\sqrt{3}}{5}$. Addition, subtraction, multiplication, and division are defined on the set of real numbers in such a way that all the properties of these operations on rationals still hold. The properties are summarized next.

Properties

Closure properties For real numbers a and b, $a + b$ and ab are unique real numbers.
Commutative properties For real numbers a and b, $a + b = b + a$ and $ab = ba$.
Associative properties For real numbers a, b, and c, $a + (b + c) = (a + b) + c$ and $a(bc) = (ab)c$.
Identity properties The number 0 is the unique additive identity and 1 is the unique multiplicative identity such that, for any real number a, $0 + a = a = a + 0$ and $1 \cdot a = a = a \cdot 1$.
Inverse properties (1) For every real number a, ^{-}a is its unique additive inverse; that is, $a + {}^{-}a = 0 = {}^{-}a + a$. (2) For every nonzero real number a, $\frac{1}{a}$ is its unique multiplicative inverse; that is, $a\left(\frac{1}{a}\right) = 1 = \left(\frac{1}{a}\right)a$.
Distributive property of multiplication over addition For real numbers a, b, and c, $a(b + c) = ab + ac$.
Denseness property For real numbers a and b, there exists a real number c such that $a < c < b$.

Radicals and Rational Exponents

Scientific calculators have a $\boxed{y^x}$ key with which we can find the values of expressions like $3.41^{2/3}$ and $4^{1/2}$. What does $4^{1/2}$ mean? By extending the properties of exponents previously developed for integer exponents, we have $4^{1/2} \cdot 4^{1/2} = 4^{1/2+1/2} = 4^1$. This implies that $(4^{1/2})^2 = 4$, or $4^{1/2}$, is a square root of 4. The number $4^{1/2}$ is assumed to be the principal square root of 4, that is, $4^{1/2} = \sqrt{4}$. *In general, if x is a nonnegative real number, then $x^{1/2} = \sqrt{x}$.* Similarly, $(x^{1/3})^3 = x^{(1/3)3} = x^1$, and $x^{1/3} = \sqrt[3]{x}$. This discussion leads to the following:

1. $x^{1/n} = \sqrt[n]{x}$, where $\sqrt[n]{x}$ is meaningful.
2. $(x^m)^{1/n} = \sqrt[n]{x^m}$ if gcd $(m, n) = 1$.
3. $x^{m/n} = \sqrt[n]{x^m}$, if $\frac{m}{n}$ is in simplest form.

More Properties of Exponents

The properties of integer exponents also hold for rational exponents. These properties are equivalent to the corresponding properties of radicals if the expressions involving radicals are meaningful.

Let r and s be any rational numbers, x and y be any real numbers, and n be any nonzero integer.

a. $x^{-r} = \frac{1}{x^r}$.
b. $(xy)^r = x^r y^r$ implies $(xy)^{1/n} = x^{1/n}y^{1/n}$ and $\sqrt[n]{xy} = \sqrt[n]{x}\,\sqrt[n]{y}$.
c. $\left(\frac{x}{y}\right)^r = \frac{x^r}{y^r}$ implies $\left(\frac{x}{y}\right)^{1/n} = \frac{x^{1/n}}{y^{1/n}}$ and $\sqrt[n]{\frac{x}{y}} = \frac{\sqrt[n]{x}}{\sqrt[n]{y}}$.
d. $(x^r)^s = x^{rs}$ implies $(x^{1/n})^s = x^{s/n}$ and hence, $(\sqrt[n]{x})^s = \sqrt[n]{x^s}$.

REMARK The preceding properties can be used to write equivalent expressions for the roots of many numbers. For example, $\sqrt{96} = \sqrt{16 \cdot 6} = \sqrt{16} \cdot \sqrt{6} = 4 \cdot \sqrt{6}$. Similarly, $\sqrt[3]{54} = \sqrt[3]{27 \cdot 2} = \sqrt[3]{27} \cdot \sqrt[3]{2} = 3 \cdot \sqrt[3]{2}$.

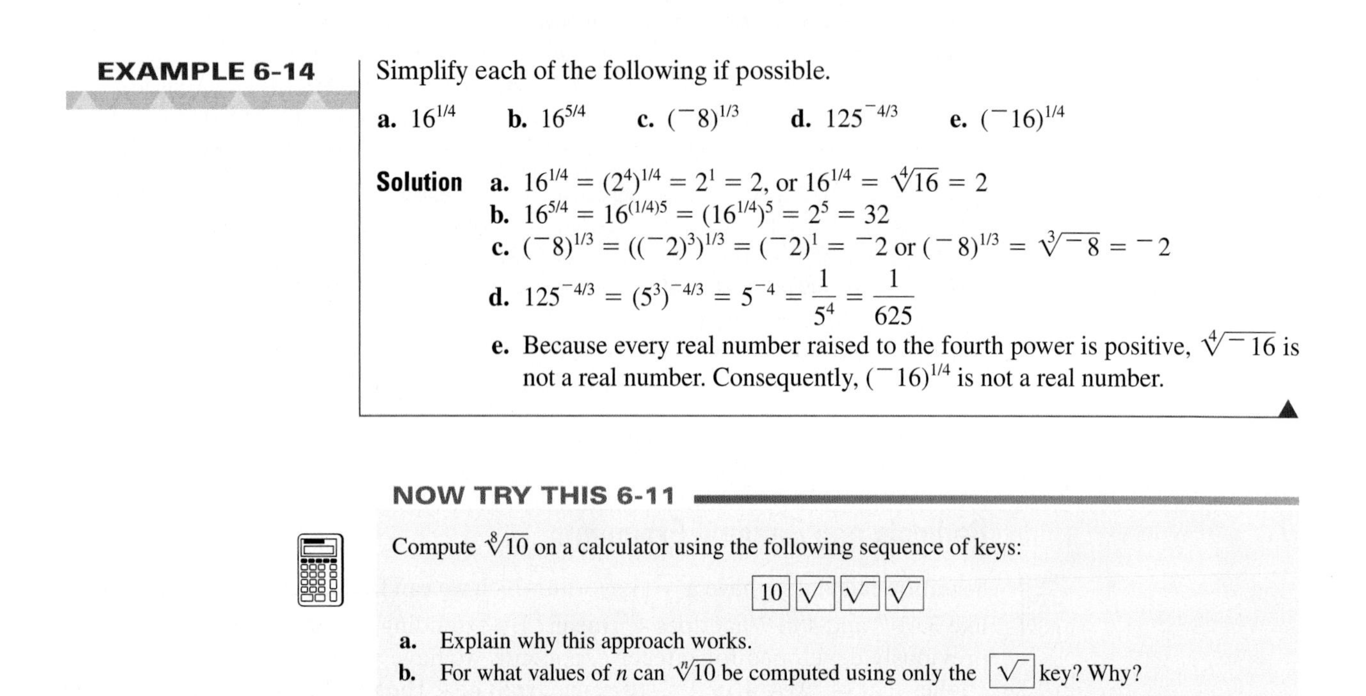

EXAMPLE 6-14 Simplify each of the following if possible.

a. $16^{1/4}$ **b.** $16^{5/4}$ **c.** $(^-8)^{1/3}$ **d.** $125^{^-4/3}$ **e.** $(^-16)^{1/4}$

Solution
a. $16^{1/4} = (2^4)^{1/4} = 2^1 = 2$, or $16^{1/4} = \sqrt[4]{16} = 2$
b. $16^{5/4} = 16^{(1/4)5} = (16^{1/4})^5 = 2^5 = 32$
c. $(^-8)^{1/3} = ((^-2)^3)^{1/3} = (^-2)^1 = {}^-2$ or $(^-8)^{1/3} = \sqrt[3]{^-8} = {}^-2$
d. $125^{^-4/3} = (5^3)^{^-4/3} = 5^{^-4} = \dfrac{1}{5^4} = \dfrac{1}{625}$
e. Because every real number raised to the fourth power is positive, $\sqrt[4]{^-16}$ is not a real number. Consequently, $(^-16)^{1/4}$ is not a real number. ▲

NOW TRY THIS 6-11

Compute $\sqrt[8]{10}$ on a calculator using the following sequence of keys:

[10] [√] [√] [√]

a. Explain why this approach works.
b. For what values of n can $\sqrt[n]{10}$ be computed using only the [√] key? Why?

ONGOING ASSESSMENT 6-4

1. Write an irrational number whose digits are 2s and 3s.
2. Use the Pythagorean Theorem to find x.

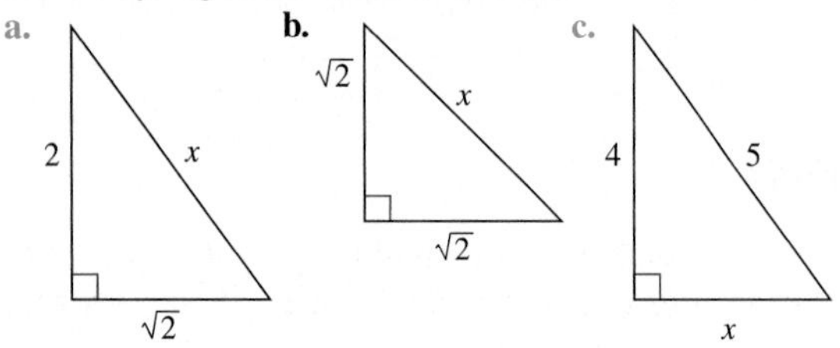

3. Arrange the following real numbers in order from greatest to least:
$0.9,\ 0.\overline{9},\ 0.\overline{98},\ 0.9\overline{88},\ 0.99\overline{8},\ 0.\overline{898},\ \sqrt{0.98}$
4. Determine which of the following represent irrational numbers:
a. $\sqrt{51}$ b. $\sqrt{64}$ c. $\sqrt{324}$
d. $\sqrt{325}$ e. $2 + 3\sqrt{2}$ f. $\sqrt{2} \div 5$
5. If possible, find the square roots for each of the following without using a calculator:
a. 225 b. 169 c. $^-81$ d. 625
6. Find the approximate square roots for each of the following, rounded to hundredths, by using the squeezing method:
a. 7 b. 0.0120 c. 20.3 d. 1.64
7. Classify each of the following as true or false. If false, give a counterexample.
a. The sum of any rational number and any irrational number is a rational number.
b. The sum of any two irrational numbers is an irrational number.
c. The product of any two irrational numbers is an irrational number.
d. The difference of any two irrational numbers is an irrational number.
8. Find three irrational numbers between 1 and 3.
9. Find an irrational number between $0.\overline{53}$ and $0.\overline{54}$.
10. Between the terminating decimals 0.5 and 0.6, find an irrational number.
11. Between the repeating decimals $0.\overline{5}$ and $0.\overline{6}$, find an irrational number.

12. If one adds a rational number and an irrational number, what type of number results? Prove your answer.

13. Based on your answer in problem 12, argue that there are infinitely many irrational numbers.

14. If R is the set of real numbers, Q is the set of rational numbers, I is the set of integers, W is the set of whole numbers, and S is the set of irrational numbers, find each of the following:

a. $Q \cup S$ **b.** $Q \cap S$ c. $Q \cap R$
d. $S \cap W$ e. $W \cup R$ **f.** $Q \cup R$

15. If the following letters correspond to the sets listed in problem 14, complete the following table by placing checkmarks in the appropriate columns. (N is the set of natural numbers.)

	N	I	Q	R	S
a. 6.7			✔	✔	
b. 5					
c. $\sqrt{2}$					
d. $^{-}5$					
e. $3\frac{1}{7}$					

16. If the following letters correspond to the sets listed in problem 14, put a checkmark under each set of numbers for which a solution to the problem exists. (N is the set of natural numbers).

	N	I	Q	R	S
a. $x^2 + 1 = 5$					
b. $2x - 1 = 32$					
c. $x^2 = 3$					
d. $x^2 = 4$					
e. $\sqrt{x} = {}^{-}1$					
f. $\frac{3}{4}x = 4$					

17. Determine for what real values of x, if any, each of the following statements is true:

a. $\sqrt{x} = 8$ **b.** $\sqrt{x} = {}^{-}8$ c. $\sqrt{{}^{-}x} = 8$
d. $\sqrt{{}^{-}x} = {}^{-}8$ e. $\sqrt{x} > 0$ **f.** $\sqrt{x} < 0$

18. A diagonal brace is placed in a 4 ft × 5 ft rectangular gate. What is the length of the brace to the nearest tenth of a foot? (*Hint:* Use the Pythagorean Theorem.)

19. Write each of the following square roots in the form $a\sqrt{b}$, where a and b are integers and b has the least value possible:

a. $\sqrt{180}$ **b.** $\sqrt{363}$ c. $\sqrt{252}$

20. Write each of the following in the simplest form $a\sqrt[n]{b}$, where a and b are integers, $b > 0$, and b has the least value possible:

a. $\sqrt[3]{{}^{-}54}$ **b.** $\sqrt[5]{96}$ c. $\sqrt[3]{250}$ **d.** $\sqrt[5]{{}^{-}243}$

21. In each of the following geometric sequences, find the missing terms:

a. 5, _, _, 10 **b.** 2, _, _, _, 1

22. The following exponential function approximates the number of bacteria after t hr: $E(t) = 2^{10} \cdot 16^t$.

a. What is the initial number of bacteria, that is, the number when $t = 0$?
b. After $\frac{1}{4}$ hr, how many bacteria are there?
c. After $\frac{1}{2}$ hr, how many bacteria are there?

23. Solve for x in the following, where x is a rational number:

a. $3^x = 81$ **b.** $4^x = 8$
c. $128^{-x} = 16$ **d.** $\left(\frac{4}{9}\right)^{3x} = \frac{32}{243}$

24. Classify each of the following numbers as rational or irrational:

a. $\sqrt{2} - \frac{2}{\sqrt{2}}$ **b.** $(\sqrt{2})^{-4}$
c. $\frac{1}{1 + \sqrt{2}}$ **d.** $\frac{4}{\sqrt{2}} - \sqrt{2}$

25. a. Is it ever true that $\sqrt{a^2} = {}^{-}a$? Explain your answer.
b. For which values of x, if any, is
(i) $x^2 \geq 2$? **(ii)** $x^2 \geq {}^{-}2$?

26. Figure 6-7 showed a way to construct the square root of any positive integer. Figure 6-6 showed how the square root of a positive integer could be placed on a number line. Using these methods, describe where $\sqrt{237}$ would be on a number line.

27. If $a > 2$, describe where $\frac{1}{\sqrt{a}}$ would be located on a number line.

28. Describe how you would decide how $\frac{2}{\sqrt{3}}$ compares to $\frac{3}{\sqrt{5}}$.

★29. Use the steps for the recursive formula found in Now Try This 6-10 to show that the algorithm could produce the square root of a positive real number.

Communication

30. A mathematician once described the set of rational numbers as the stars and the set of irrational numbers as the black in the night sky. What do you think the mathematician meant?

31. A student argues that the only solution to $x^2 = 4$ is 2 because $\sqrt{4}$ is equal to $^{+}2$ and not $^{-}2$. How do you respond?

32. Jim asked, if $\sqrt{2}$ can be written $\frac{\sqrt{2}}{1}$, why isn't it rational? How would you answer him?

33. Find the value of $\sqrt{3}$ on a calculator. Explain why this can't be the exact value of $\sqrt{3}$.

34. Is it true that $\sqrt{a + b} = \sqrt{a} + \sqrt{b}$? Explain.

35. Pi (π) is an irrational number. Could $\pi = \frac{22}{7}$? Why or why not?

36. Without using a calculator or doing any computation, determine if $\sqrt{13} = 3.60\overline{5}$. Explain why or why not.

37. Is $\sqrt{x^2 + y^2} = x + y$ for all values of x and y? Explain your reasoning.

38. Without using a calculator, arrange the following in increasing order. Explain your reasoning.
$(4/25)^{-1/3}$, $(25/4)^{1/3}$, $(4/25)^{-1/4}$

Open-Ended

39. The sequence 1, 1.01, 1.001, 1.0001, . . . is an infinite sequence of rational numbers.
   a. Write several other infinite sequences of rational numbers.
   b. Write an infinite sequence of irrational numbers.

40. a. Place five irrational numbers between $\frac{1}{2}$ and $\frac{3}{4}$.
   b. Write an infinite sequence of irrational numbers all of whose terms are between $\frac{1}{2}$ and $\frac{3}{4}$.

Cooperative Learning

41. Let each member of a group choose a number between 0 and 1 on a calculator and check what happens when the $\boxed{x^2}$ key is pressed in succession until it is clear that there is no reason to go on.
   a. Compare your answers and write a conjecture based on what you observe.
   b. Use other keys on the calculator in a similar way. Describe the process and state a corresponding conjecture.
   c. Why do you get the result you do?

42. A calculator displays the following: $(3.7)^{2.4} = 23.103838$. In your group, discuss the meaning of the expression $(3.7)^{2.4}$ in view of what you know about exponents. Compare your findings with those of other groups.

43. George Cantor compared the number of elements in the set of rational and the set of irrational numbers. What did he discover?

Review Problems

44. a. Human bones make up 0.18 of a person's body weight. How much do the bones of a 120-lb person weigh?
   b. Muscles make up about 0.4 of a person's body weight. How much do the muscles of a 120-lb person weigh?

45. Write each of the following decimals in the form $\frac{a}{b}$, where $a, b \in I, b \neq 0$:
   a. 16.72 b. 0.003 c. $^-5.07$ d. 0.123

46. Write a repeating decimal equal to each of the following without using more than one zero:
   a. 5 b. 5.1 c. $\frac{1}{2}$

47. Write 0.00024 as a fraction in simplest form.

48. Write $0.\overline{24}$ as a fraction in simplest form.

49. Write each of the following as a standard numeral:
   a. $2.08 \cdot 10^5$ b. $3.8 \cdot 10^{-4}$

BRAIN TEASER

Many schools celebrate Pi Day as suggested in the cartoon. When do you think it is celebrated? Why?

FOXTROT © Bill Amend. Reprinted with permission of UNIVERSAL PRESS SYNDICATE. All rights reserved.

6-5 Percents

Percents are very useful in conveying information. People hear that there is a 60 percent chance of rain or that their savings accounts are drawing 6 percent interest. The word **percent** comes from the Latin phrase *per centum*, which means *per hundred*. For example, a bank that pays 6 percent simple interest on a savings account pays \$6 for each \$100 in the account for 1 yr; that is, it pays $\frac{6}{100}$ of whatever amount is in the account for 1 yr. The symbol, %, indicates percent. For example, we write 6% for $\frac{6}{100}$.

percent

In general, we have the following definition.

Definition of Percent

$$n\% = \frac{n}{100}$$

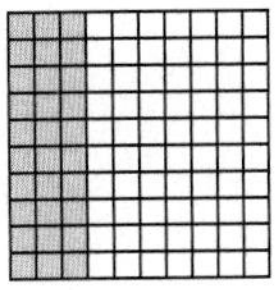

Thus, $n\%$ of a quantity is $\frac{n}{100}$ of the quantity. Therefore, 1% is one-hundredth of a whole and 100% represents the entire quantity, whereas 200% represents $\frac{200}{100}$, or 2 times, the given quantity. Percents can be illustrated by using a hundreds grid. For example, what percent of the grid is shaded in Figure 6-10? Because 30 out of the 100, or $\frac{30}{100}$, of the squares are shaded, we say that 30% of the grid is shaded (or similarly, 70% of the grid is not shaded).

FIGURE 6-10

Because $n\% = \frac{n}{100}$, to convert a number to a percent we write it as a fraction with denominator 100; the numerator gives the amount of the percent. For example, $\frac{3}{4} = \frac{3 \cdot 25}{4 \cdot 25} = \frac{75}{100}$. Hence, $\frac{3}{4} = 75\%$.

EXAMPLE 6-15

Write each of the following as a percent:

a. 0.03 **b.** $0.\overline{3}$ **c.** 1.2 **d.** 0.00042

e. 1 **f.** $\frac{3}{5}$ **g.** $\frac{2}{3}$ **h.** $2\frac{1}{7}$

Solution **a.** $0.03 = 100\left(\frac{0.03}{100}\right) = \frac{3}{100} = 3\%$

b. $0.\overline{3} = 100\left(\frac{0.\overline{3}}{100}\right) = \frac{33.\overline{3}}{100} = 33.\overline{3}\%$

c. $1.2 = 100\left(\frac{1.2}{100}\right) = \frac{120}{100} = 120\%$

d. $0.00042 = 100\left(\frac{0.00042}{100}\right) = \frac{0.042}{100} = 0.042\%$

e. $1 = 100\left(\frac{1}{100}\right) = \frac{100}{100} = 100\%$

f. $\frac{3}{5} = 100\left[\frac{\left(\frac{3}{5}\right)}{100}\right] = \frac{60}{100} = 60\%$

g. $\frac{2}{3} = 100\left[\frac{\left(\frac{2}{3}\right)}{100}\right] = \frac{\left(\frac{200}{3}\right)}{100} = \frac{66.\overline{6}}{100} = 66.\overline{6}\%$

h. $2\frac{1}{7} = 100\left[\frac{\left(2\frac{1}{7}\right)}{100}\right] = \frac{\left(\frac{1500}{7}\right)}{100} = \frac{1500}{7}\%$, or $214\frac{2}{7}\%$

A number can also be converted to a percent by using a *proportion*. For example, to write $\frac{3}{5}$ as a percent, find the value of n in the following proportion:

$$\frac{3}{5} = \frac{n}{100}$$

$$\left(\frac{3}{5}\right) \cdot 100 = n$$

$$n = 60$$

Therefore,
$$\frac{3}{5} = 60\%$$

Still another way to convert a number to a percent is to recall that $1 = 100\%$. Thus, for example, $\frac{3}{4} = \frac{3}{4}$ of $1 = \frac{3}{4} \cdot 1 = \frac{3}{4} \cdot 100\% = 75\%$.

REMARK The % symbol is crucial in identifying the meaning of a number. For example, $\frac{1}{2}$ and $\frac{1}{2}\%$ are different numbers: $\frac{1}{2} = 50\%$, which is not equal to $\frac{1}{2}\%$. Similarly, 0.01 is different from 0.01%, which is 0.0001.

In the following student page on finding percents, an algorithmic method for converting a proportion to a percentage is seen in the example. Complete the example.

In our computations, it is sometimes useful to convert percents to decimals. This can be done by writing the percent as a fraction in the form $\frac{n}{100}$ and then converting the fraction to a decimal.

SAMPLE SCHOOL BOOK PAGE:

GOAL

LEARN HOW TO...
- find percents

AS YOU...
- find Audience Approval ratings

Exploration 2

Finding Percents

You will need the frequency table from the Setting the Stage *on page 344.*

11 **Try This as a Class** Suppose a class gave two movies $3\frac{1}{2}$ or 4 stars as shown in the ratios below.

a. Estimate the Audience Approval rating for each movie. Explain the reasoning you used.

b. Using estimation, can you determine which movie had a higher Audience Approval rating? Explain.

▶ **A percent bar model can help you see how to set up a proportion to find the exact percent equivalent for a ratio.**

EXAMPLE

Set up a proportion to find the Audience Approval rating for Movie A, represented by the ratio $\frac{19}{30}$.

SAMPLE RESPONSE

Rotate the bar 90° so the parts are over the wholes.

The shaded part represents the students who liked the movie.

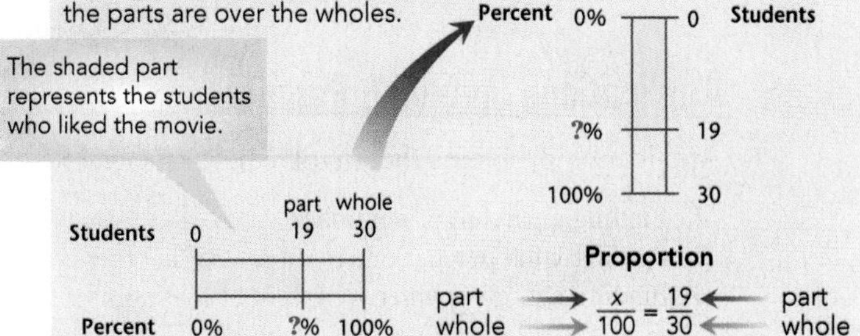

Source: Math Thematics, Book 2, McDougal Littell, 2005 (p. 348).

EXAMPLE 6-16

Write each of the following percents as a decimal:

a. 5% **b.** 6.3% **c.** 100% **d.** 250% **e.** $\frac{1}{3}\%$ **f.** $33\frac{1}{3}\%$

Solution **a.** $5\% = \frac{5}{100} = 0.05$ **b.** $6.3\% = \frac{6.3}{100} = 0.063$

c. $100\% = \frac{100}{100} = 1$ **d.** $250\% = \frac{250}{100} = 2.50$

e. $\frac{1}{3}\% = \frac{\frac{1}{3}}{100} = \frac{0.\overline{3}}{100} = 0.00\overline{3}$ **f.** $33\frac{1}{3}\% = \frac{33\frac{1}{3}}{100} = \frac{33.\overline{3}}{100} = 0.\overline{3}$

Another approach to writing a percent as a decimal is first to convert 1% to a decimal. Because $1\% = \frac{1}{100} = 0.01$, we can conclude that $5\% = 5 \cdot 0.01 = 0.05$ and that $6.3\% = 6.3 \cdot 0.01 = 0.063$.

NOW TRY THIS 6-12

a. Investigate how your calculator handles percents and tell what the calculator does when the % key is pushed.

b. Use your calculator to change $\frac{1}{3}$ to a percent.

In the *Principles and Standards* we find the following under the Representation Standard for Grades 6–8.

Middle-grades students who are taught with this Standard in mind will learn to recognize, compare, and use an array of representational forms for fractions, decimals, percents and integers. They also will learn to use representational forms such as exponential and scientific notation when working with large and small numbers and to use a variety of graphical tools to represent and analyze data sets. (p. 280)

Applications Involving Percent

Application problems that involve percents usually take one of the following forms:

1. Finding a percent of a number
2. Finding what percent one number is of another
3. Finding a number when a percent of that number is known

Before we consider examples illustrating these forms, recall what it means to find a fraction "of" a number. For example, $\frac{2}{3}$ of 70 means $\frac{2}{3} \cdot 70$. Similarly, to find 40% of 70, we have $\frac{40}{100}$ of 70, which means $\frac{40}{100} \cdot 70$, or $0.40 \cdot 70 = 28$.

A different way to think about 40% of 70 is to consider that 70 represents 100 parts (or the whole) and 40% only requires 40 of those 100 parts. For example, if

$$\begin{aligned} 100 \text{ parts} &= 70 \\ 1 \text{ part} &= \left(\frac{70}{100}\right), \text{ or } 0.7 \\ 40 \text{ parts} &= 40(0.7), \text{ or } 28.0 \end{aligned}$$

Thus, 40% of 70 = 28.0.

In Figure 6-11, consider the percent bar that represents 100% of the whole with 40% of the whole shaded. Note that 100% of the bar represents 70.

FIGURE 6-11

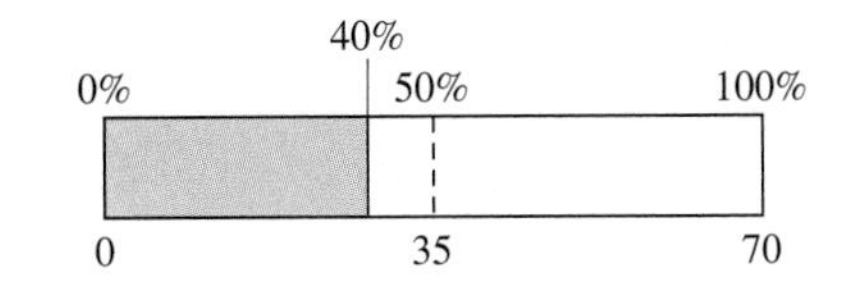

Also, half of the percent bar (50% denoted by the dotted segment) represents half of 70, or 35. Thus, we know that 40% of the bar (or 40% of 70) is less than 35. In fact, if the top of the bar is thought of as being marked off in 1% intervals, there are 100 intervals marking whole numbers of percentages. If at the same time, the bottom of the bar is considered to be marked in intervals of 1, there would be only 70 intervals marked at the bottom. Where would you expect the two sets of intervals to align?

Table 6-3

0%	0
10%	
20%	
30%	
40%	?
50%	35
60%	
70%	
80%	
90%	
100%	70

Suppose that we know that as in Table 6-3, 0% corresponds to 0; 50% corresponds to 35; and 100% corresponds to 70. What percentages correspond to 10%, 20%, 30%, and so on? Now if there are 100 intervals marking percentages compared to only 70 intervals marking the corresponding length, there must be a ratio of $\frac{100}{70}$ or $\frac{10}{7}$. Thus, 10% should compare to 7; 20% to $2 \cdot 7$, or 14; and so on. Hence, 40% corresponds to $4 \cdot 7$, or 28.

EXAMPLE 6-17

A house that sells for \$92,000 requires a 20% down payment. What is the amount of the down payment?

Solution The down payment is 20% of \$92,000, or $0.20 \cdot \$92{,}000 = \$18{,}400$. Hence, the amount of the down payment is \$18,400.

EXAMPLE 6-18

If Alberto has 45 correct answers on an 80-question test, what percent of his answers are correct?

Solution Alberto has $\frac{45}{80}$ of the answers correct. To find the percent of correct answers, we need to convert $\frac{45}{80}$ to a percent. We can do this by multiplying the fraction by 100 and attaching the % symbol as follows:

$$\begin{aligned}\frac{45}{80} &= 100 \cdot \frac{45}{80}\% \\ &= 56.25\%\end{aligned}$$

Thus 56.25% of the answers are correct.

An alternative solution uses proportion. Let n be the percent of correct answers and proceed as follows:

$$\frac{45}{80} = \frac{n}{100}$$

$$\frac{45}{80} \cdot 100 = n$$

$$n = \frac{4500}{80} = 56.25$$

EXAMPLE 6-19

Forty-two percent of the parents of the schoolchildren in the Paxson School District are employed at Di Paloma University. If the number of parents employed by the university is 168, how many parents are in the school district?

Solution Let n be the number of parents in the school district. Then 42% of n is 168. We *translate this information into an equation* and solve for n.

$$\begin{aligned}42\% \text{ of } n &= 168 \\ \frac{42}{100} \cdot n &= 168 \\ 0.42 \cdot n &= 168 \\ n &= \frac{168}{0.42} = 400\end{aligned}$$

There are 400 parents in the school district.

Example 6–19 can be solved using a proportion. Forty-two percent, or $\frac{42}{100}$, of the parents are employed at the university. If n is the total number of parents, then $\frac{168}{n}$ also represents the fraction of parents employed there. Thus,

$$\frac{42}{100} = \frac{168}{n}$$
$$42n = 100 \cdot 168$$
$$n = \frac{16{,}800}{42} = 400$$

We can also solve the problem as follows:

$$42\% \text{ of } n \text{ is } 168$$
$$1\% \text{ of } n \text{ is } \frac{168}{42}$$
$$100\% \text{ of } n \text{ is } 100\left(\frac{168}{42}\right)$$

Therefore,

$$n = 100\left(\frac{168}{42}\right), \text{ or } 400.$$

EXAMPLE 6-20

Kelly bought a bicycle and a year later sold it for 20% less than what she paid for it. If she sold the bike for $144, what did she pay for it?

Solution We are looking for the original price, P, that Kelly paid for the bike. We know that she sold the bike for $144 and that this included a 20% loss. Thus, we can *write the following equation*:

$$\$144 = P - \text{Kelly's loss}$$

Because Kelly's loss is 20% of P, we proceed as follows:

$$\$144 = P - 20\% \cdot P$$
$$\$144 = P - 0.20 \cdot P$$
$$\$144 = (1 - 0.20)P$$
$$\$144 = 0.80P$$
$$\frac{\$144}{0.80} = P$$
$$\$180 = P$$

Thus, she paid $180 for the bike.

EXAMPLE 6-21

Westerner's Clothing Store advertised a suit for 10% off, for a savings of $15. Later, the manager marked the suit at 30% off the original price. What is the amount of the current discount?

Solution A 10% discount amounts to a $15 savings. We could find the amount of the current discount if we knew the original price. Thus finding the original price becomes our *subgoal*. Because 10% of P is $15, we have the following:

$$10\% \cdot P = \$15$$
$$0.10 \cdot P = \$15$$
$$P = \$150$$

To find the current discount, we calculate 30% of \$150. Because $0.30 \cdot \$150 = \$45$, the amount of the 30% discount is \$45.

In the *Looking Back* stage of problem solving, we check the answer and look for other ways to solve the problem. A different approach leads to a more efficient solution and confirms the answer. If 10% of the price is \$15, then 30% of the price is 3 times \$15, or \$45.

RESEARCH NOTE In a study of Grades 4, 6, and 8 Japanese students, a low percentage of students was able to correctly answer "What is 100% of 48?" Further research was called for to determine if this low score was an anomaly or if it was representative of Japanese students' knowledge of percent. None of the students made any comments to suggest any conceptual links, connections, or similarities between the fraction and decimal computations (Reys et al. 1995).

NOW TRY THIS 6-13

In the "Hi and Lois" cartoon, Dot is babysitting Trixie and presents Lois a bill. What percentage of the total bill is her tip?

Mental Math with Percents

Mental math may be helpful when working with percents. Two techniques follow:

1. *Using fraction equivalents*
   Knowing fraction equivalents for some percents can make some computations easier. Table 6-4 gives several fraction equivalents.

Table 6-4

Percent	25%	50%	75%	$33\frac{1}{3}\%$	$66\frac{2}{3}\%$	10%	1%
Fraction Equivalent	$\frac{1}{4}$	$\frac{1}{2}$	$\frac{3}{4}$	$\frac{1}{3}$	$\frac{2}{3}$	$\frac{1}{10}$	$\frac{1}{100}$

These equivalents can be used in such computations as the following:

$$50\% \text{ of } \$80 = \left(\frac{1}{2}\right)80 = \$40$$

$$66\frac{2}{3}\% \text{ of } 90 = \left(\frac{2}{3}\right)90 = 60$$

2. *Using a known percent*
Frequently, we may not know a percent of something, but we know a close percent of it. For example, to find 55% of 62, we might do the following:

$$50\% \text{ of } 62 = \left(\frac{1}{2}\right)(62) = 31$$

$$5\% \text{ of } 62 = \left(\frac{1}{2}\right)(10\%)(62) = \left(\frac{1}{2}\right)(6.2) = 3.1$$

Adding, we see that 55% of 62 is $31 + 3.1 = 34.1$.

Estimations with Percents

Estimations with percents can be used to determine whether answers are reasonable. Following are two examples:

1. To estimate 27% of 598, note that 27% of 598 is a little more than 25% of 598, but 25% of 598 is approximately the same as 25% of 600, or $\frac{1}{4}$ of 600, or 150. Here, we have adjusted 27% downward and 598 upward, so 150 should be a reasonable estimate. A better estimate might be obtained by estimating 30% of 600 and then subtracting 3% of 600 to obtain 27% of 600, giving $180 - 18$, or 162.
2. To estimate 148% of 500, note that 148% of 500 should be slightly less than 150% of 500. 150% of 500 is $1.5(500) = 750$. Thus, 148% of 500 should be a little less than 750.

EXAMPLE 6-22

Laura wants to buy a blouse originally priced at \$26.50 but now on sale at 40% off. She has \$17 in her wallet and wonders if she has enough cash. How can she mentally find out?

Solution It is easier to find 40% of \$25 (versus \$26.50) mentally. One way is to find 10% of \$25, which is \$2.50. Now, 40% is 4 times that much, that is, $4 \cdot \$2.50$, or \$10. Thus, Laura estimates that the blouse will cost $\$26.50 - \$10$, or \$16.50. Since the actual discount is greater than \$10 (40% of 26.50 is greater than 40% of 25), Laura will have to pay less than \$16.50 for the blouse and, hence, she has enough cash. ▲

Sometimes it may not be clear, which operations to perform with percent. The following example investigates this.

EXAMPLE 6-23

Which of the following statements are true and which are false? Explain your answers.

a. Leonardo got a 10% raise at the end of his first year on the job and a 10% raise after another year. His total raise was 20% of his original salary.
b. Jung and Dina paid 45% of their first department store bill of \$620 and 48% of the second department store bill of \$380. They paid 45% + 48% = 93% of the total bill of \$1000.
c. Bill spent 25% of his salary on food and 40% on housing. Bill spent 25% + 40% = 65% of his salary on food and housing.
d. In Bordertown, 65% of the adult population works in town, 25% works across the border, and 15% is unemployed.
e. In Clean City, the fine for various polluting activities is a certain percentage of one's monthly income. The fine for smoking in public places is 40%, for driving a polluting car is 50%, and for littering is 30%. Mr. Schmutz committed all three polluting crimes in one day and paid a fine of 120% of his monthly salary.

Solution **a.** In applications, percent has meaning only when it represents part of a quantity. For example, 10% of a quantity plus another 10% of the same quantity is 20% of that quantity. In Leonardo's case, the first 10% raise was calculated based on his original salary and the second 10% raise was calculated on his new salary. Consequently, the percentages cannot be added, and the statement is false.

b. The answer does not make sense. Jung and Dina paid less than $\frac{1}{2}$ of each bill, so they could not have paid 93% (almost all) of the total. In fact, $\frac{1}{2}$ of one bill plus $\frac{1}{2}$ of the other bill is not the full amount of the total bill, because the bills are different.

c. Because the percentages are of the same quantity, the statement is true.

d. Because the percentages are of the same quantity, that is, the number of adults, we can add them: 65% + 25% + 15% = 105%. But 105% of the population accounts for more (5% more) than the town's population, which is impossible. Hence, the statement is false.

e. Again, the percentages are of the same quantity, that is, the individual's monthly income. Hence, we can add them: 120% of one's monthly income is a stiff fine, but possible.

RESEARCH NOTE Formal instruction in the application of percent tends to make the students' concepts of percent less intuitive and more rule-driven, actually narrowing rather than expanding the strategies and computational methods students use in working with percents (Lembke et al. 1994).

ONGOING ASSESSMENT 6-5

1. Express each of the following as percents:
   a. 7.89 **b.** 0.032 c. 193.1 **d.** 0.2
   e. $\frac{5}{6}$ **f.** $\frac{3}{20}$ g. $\frac{1}{8}$ **h.** $\frac{3}{8}$
   i. $\frac{5}{8}$ **j.** $\frac{1}{6}$ k. $\frac{4}{5}$ **l.** $\frac{1}{40}$

2. Convert each of the following percents to decimals:
   a. 16% **b.** $4\frac{1}{2}\%$ c. $\frac{1}{5}\%$ **d.** $\frac{2}{7}\%$
   e. $13\frac{2}{3}\%$ **f.** 125% g. $\frac{1}{3}\%$ **h.** $\frac{1}{4}\%$

3. Fill in the following blanks to find other expressions for 4%:
   a. _______ for every 100
   b. _______ for every 50
   c. 1 for every _______
   d. 8 for every _______
   e. 0.5 for every _______
4. Different calculators compute percents in various ways. To investigate this, consider $5 \cdot 6\%$.
   a. If the following sequence of keys is pressed, is the correct answer of 0.3 displayed on your calculator?

   [5] [×] [6] [%] [=]

   b. Press [6] [%] [×] [5] [=]. Is the answer 0.3?
5. Answer each of the following:
   a. What is 6% of 34?
   b. 17 is what percent of 34?
   c. 18 is 30% of what number?
   d. What is 7% of 49?
   e. 61.5 is what percent of 20.5?
   f. 16 is 40% of what number?
6. a. Write a fraction representing 5% of x.
   b. If 10% of an amount is a, what is the amount in terms of a?
7. Marc had 84 boxes of candy to sell. He sold 75% of the boxes. How many did he sell?
8. Gail made $16,000 last year and received a 6% raise. How much does she make now?
9. Gail received a 7% raise last year. If her salary is now $27,285, what was her salary last year?
10. Joe sold 180 newspapers out of 200. Bill sold 85% of his 260 newspapers. Ron sold 212 newspapers, 80% of those he had.
   a. Who sold the most newspapers? How many?
   b. Who sold the greatest percentage of his newspapers? What percent?
   c. Who started with the greatest number of newspapers? How many?
11. If a dress that normally sells for $35 is on sale for $28, what is the "percent off"? (This could be called a *percent of decrease*, or a *discount*.)
12. A used car originally cost $1700. One year later, it was worth $1400. What is the percentage of depreciation?
13. On a certain day in Glacier Park, 728 eagles were counted. Five years later, 594 were counted. What was the percentage of decrease in the number of eagles counted?
14. Mort bought his house in 2000 for $59,000. It was recently appraised at $95,000. What is the *percent of increase* in value?
15. Xuan weighed 9 lb when he was born. At 6 mo, he weighed 18 lb. What was the percent of increase in Xuan's weight?
16. Sally bought a dress marked 20% off. If the regular price was $28.00, what was the sale price?
17. What is the sale price of a softball if the regular price is $6.80 and there is a 25% discount?
18. If a $\frac{1}{4}$-c serving of Crunchies breakfast food has 0.5% of the minimum daily requirement of vitamin C, how many cups would you have to eat to obtain the minimum daily requirement of vitamin C?
19. An airline ticket costs $320 without the tax. If the tax rate is 5%, what is the total bill for the airline ticket?
20. Bill got 52 correct answers on an 80-question test. What percent of the questions did he answer incorrectly?
21. A real-estate broker receives 4% of an $80,000 sale. How much does the broker receive?
22. A survey reported that $66\frac{2}{3}\%$ of 1800 employees favored a new insurance program. How many employees favored the new program?
23. Which represents the greater percent: $\frac{325}{500}$ or $\frac{600}{1000}$? How can you tell?
24. **a.** How can an estimate of 10% of a number help you estimate 35% of the number?
   b. Mentally compute 35% of $8.00.
25. If 30 is 150% of a number, is the number greater than or less than 30? Why?
26. What is 40% of 50% of a number?
27. If you add 20% of a number to the number itself, what percent of the result would you have to subtract to get the original number back?
28. An advertisement reads that if you buy 10 items, you get 20% off your total purchase price. You need 8 items that cost $9.50 each.
   a. How much would 8 items cost? 10 items?
   b. Is it more economical to buy 8 items or 10 items?
29. Soda is advertised at 45¢ a can or $2.40 a six-pack. If 6 cans are to be purchased, what percent is saved by purchasing the six-pack?
30. John paid $330 for a new mountain bicycle to sell in his shop. He wants to price it so that he can offer a 10% discount and still make 20% of the price he paid for it. At what price should the bike be marked?
31. The price of a suit that sold for $200 was reduced by 25%. By what percent must the price of the suit be increased to bring the price back to $200?
32. The car Elsie bought 1 yr ago has depreciated by $1116.88, which is 12.13% of the price she paid for it. How much did she pay for the car, to the nearest cent?
33. Solve each of the following using mental mathematics:
   a. 15% of $22 **b.** 20% of $120
   c. 5% of $38 **d.** 25% of $98
34. If we build a 10×10 model with blocks, as shown in the following figure, and paint the entire model, what percent of the cubes will have each of the following?
   a. Four faces painted

b. Three faces painted
c. Two faces painted

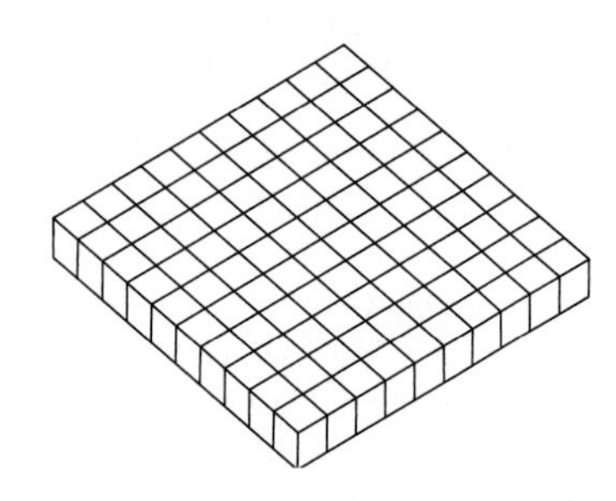

35. For people to be safe but still achieve a cardiovascular training effect, they should monitor their heart rates while exercising. The maximum heart rate can be approximated by subtracting your age from 220. You can achieve a safe training effect if you maintain your heart rate between 60% and 80% of that number for at least 20 min three times a week.
a. Determine the range for your age.
b. At the top of a long hill, Jeannie slows her bike and takes her pulse. She counts 41 beats in 15 sec.
(i) Express in decimal form the amount of time in seconds between successive beats.
(ii) Express the amount in terms of minutes.

36. A crew consists of 1 apprentice, 1 journeyman, and 1 master carpenter. The crew receives a check for $4200 for a job they just finished. A journeyman makes 200% of what an apprentice makes, and a master makes 150% of what a journeyman makes. How much does each person in the crew earn?

37. a. In an incoming freshman class of 500 students, only 20 claimed to be math majors. What percent of the freshman class is this?
b. When the survey was repeated the next year, 5% of nonmath majors had decided to switch and become math majors.
(i) How many math majors are there now?
(ii) What percent of the former freshman class do they represent?

38. Ms. Price has received a 10% raise in salary in each of the last 2 yr. If her annual salary this year is $100,000, what was her salary 2 yr ago, rounded to the nearest penny?

39. *USA Today* (2005) reported that the U.S. Congress was sent a $2.57 trillion budget for fiscal year 2006. It further reported that one would have to purchase a $100 item every second for 815 years to spend that much money.
a. Decide whether or not you believe this report and defend your answer.
b. What percentage of the money would be spent each year at the rate described?

40. The U.S. Census Bureau's *The American Indian and Alaska Native Population: 2000* reports the following populations for the largest American Indian tribes:

Cherokee	729,533
Navajo	298,197
Latin	180,940
Choctaw	158,774
Sioux	153,360
Chippewa	149,669
Apache	96,833

a. Round each population to the nearest 100,000.
b. Use the rounded populations in part (a) to write each separate population as a decimal part of the whole population.
c. Convert the decimals in part (b) to percentages.
d. Display the percentages found in part (c) on a percent bar representing all of the given tribes.

41. If you wanted to spend 25% of your monthly salary on entertainment and 56% of the salary on rent, could those amounts be $500 and $950? Why or why not?

42. If 70% of the girls in a class wanted to have a prom and 40% of the boys wanted a prom, is it possible that only 50% of the students in the class wanted a prom? Explain your answer.

43. If 70% of the girls in a class wanted to have a prom and 60% of the boys wanted a prom, is it possible that only 50% of the students in the class wanted a prom? Explain your answer.

44. An organization has 100,000 members. A bylaw change can be made at the annual business meeting held once each year, and a bylaw change must be approved by the majority of those attending the meeting. The chair of the meeting cannot vote unless there is a tied vote but does count as an attendee at the meeting.
a. With these rules, what is the minimum number required at the meeting to make a bylaw change?
b. Based on your answer to part (a), what percentage of the membership can change the bylaws of the organization?

45. A tip in a restaurant has been typically figured at 15% of the total bill.
a. If the bill is $30, what would be the typical tip?
b. If the patron receiving the bill gave a tip that was half the bill, what is the percentage of the tip?
c. If the patron receiving the bill gave a tip that was equal to the bill, what is the percentage of the tip?

Communication

46. Draw a bar representing the percentages of A's, B's, C's, D's, and F's expected in this class. How would you defend to a collegiate administrator the percentages of grades given?

47. Is 40% of 30 equal to 30% of 40? Explain why or why not.

48. Why is it possible to have an increase of 150% in a price but not a 150% decrease in price?

49. A student asks if 90% means 90 out of 100, how can she possibly score 90% on a test that has only 20 questions. How do you answer her?

50. Two equal amounts of money were invested in two different stocks. The value of the first stock increased by 15% the first year and then decreased by 15% the second year.

The second stock decreased by 15% the first year and increased by 15% the second year. Was one investment better than the other? Explain your reasoning.

Open-Ended

51. Write and solve a word problem whose solution involves the following. If one of these tasks is impossible, explain why.

a. Addition of percent
b. Subtraction of percent
c. Multiplication of percent
d. Division of percent
e. A percent whose decimal representation is raised to the second power
f. A percent greater than 100

52. Look at newspapers and magazines for information given in percents.

a. Based on your findings, write a problem that involves social science as well as mathematics.
b. Write a clear solution to your problem in (a).

53. Write a percentage problem whose answer is the solution of the following equations:

a. $\frac{37}{100} = \frac{115}{x}$ **b.** $\frac{p}{100} = \frac{a}{x}$

Cooperative Learning

54. Find the percentage of students in your class that engage in each of the following activities:

a. Studying and doing homework

Number of Hours per Week (h)	Percent
$h < 1$	
$1 \le h < 3$	
$3 \le h < 5$	
$5 \le h < 10$	
$h \ge 10$	
Total	

b. Watching TV

Number of Hours per Week (h)	Percent
$h < 1$	
$1 \le h < 5$	
$5 \le h < 10$	
$h \ge 10$	
Total	

c. Did your totals add up to 100%? Why or why not?

Review Problems

55. a. Find $0.8 \div 0.32$ by converting these numbers to rational numbers in the form $\frac{a}{b}$, where a and b are integers and $b \neq 0$, dividing the rational numbers, and then changing the quotient back to decimal form.

b. Find $0.8 \div 0.32$ using decimal division and compare your answer with the answer from part (a).

56. Change $7.27\overline{1}$ to a rational number in the form $\frac{a}{b}$, where a and b are integers ($b \neq 0$).

57. Find a repeating decimal between 0.2 and $0.\overline{2}$.

58. Order the following from least to greatest:

$2.5,\quad 5/3,\quad 2.0\overline{5},\quad 2.\overline{15},\quad 7/3$

59. Show that $0.\overline{9} = 1$ in three ways.

Third International Mathematics and Science Study (TIMSS) Questions

Experts say that 25% of all serious bicycle accidents involve head injuries and that, of all head injuries, 80% are fatal. What percentage of all serious bicycle accidents involve fatal head injuries?

a. 16%
b. 20%
c. 55%
d. 105%

Last year there were 1172 students at Beaton High School. This year there are 15 percent more students than last year. Approximately how many students are at Beaton High School this year?

a. 1800
b. 1600
c. 1500
d. 1400
e. 1200

In a school election with three candidates, Joe received 120 votes, Mary received 50 votes, and George received 30 votes. What percentage of the total number of votes did Joe receive?

a. 60%
b. $66\frac{2}{3}\%$
c. 80%
d. 120%

TIMSS, Grade 8, 1995

BRAIN TEASER

The crust of a certain pumpkin pie is 25% of the pie. By what percent should the amount of crust be reduced in order to make it constitute 20% of the pie?

TECHNOLOGY CORNER

We can use spreadsheets to solve mixture problems. For example, consider the problem of finding out how many liters of water must be added to 5 L of pure lemon juice to change its concentration from 100% to less than 30% lemon juice.

Six lemonade mixtures were prepared starting from 5 L of pure lemon juice and adding water in 2-L increments. At each step, the percent of lemon juice in the mixture was calculated. The results of the process are summarized in the spreadsheet in Figure 6-12. The formulas used to obtain the results in a particular column are given in row 12.

a. Explain how this spreadsheet can be used to help students solve the problem.

b. Explain the formulas in row 12.

	A	B	C	D
1	Liters of	Liters of	Total Liters	% Lemon Juice
2	Lemon Juice	Water Added	in Mixture	in Mixture
3	(L)	(L)	(L)	
4	5	0	5	100.00
5	5	2	7	71.43
6	5	4	9	55.56
7	5	6	11	45.45
8	5	8	13	38.46
9	5	10	15	33.33
10	5	12	17	29.41
11				
12	5	x	$5 + x$	$5/(5 + x)*100$
13		(where x is a		
14		multiple of 2)		
15				
16				
17				

Sheet1 Sheet2

RESEARCH NOTE At grade 8, students who used calculators frequently both during regular class and on tests, as reported by both teachers and students, had higher scale scores than those who used calculators little if at all (Sowder et al. 2004). NAEP data does not provide the type of data necessary to argue either for or against calculators at grade four (Kloosterman, Warfield et al. 2004).

6-6 *Computing Interest

interest When a bank advertises a $5\frac{1}{2}\%$ interest rate on a savings account, the **interest** is the amount of money the bank will pay for using that money. The original amount **principal** deposited or borrowed is the **principal**. The percent used to determine the interest is **interest rate** the **interest rate**. Interest rates are given for specific periods of time, such as years, **simple interest** months, or days. Interest computed on the original principal is **simple interest**. For example, suppose we borrow \$5000 from a company at a simple interest rate of 9% for 1 yr. The interest we owe on the loan for 1 yr is 9% of \$5000, or $\$5000 \cdot 0.09$. In general, if a principal, P, is invested at an annual interest rate of r, then the simple interest after 1 yr is $Pr \cdot 1$; after t years, it is Prt. Thus, if I represents simple interest, we have

$$I = Prt$$

The amount needed to pay off a \$5000 loan at 9% simple interest is the \$5000 borrowed plus the interest on the \$5000, that is, $5000 + 5000 \cdot 0.09$, or \$5450. In general, **amount/balance** *an* **amount** (*or* **balance**) *A is equal to the principal P plus the interest I*; that is,

$$A = P + I = P + Prt = P(1 + rt)$$

EXAMPLE 6-24

Vera opened a savings account that pays simple interest at the rate of $5\frac{1}{4}\%$ per year. If she deposits \$2000 and makes no other deposits, find the interest and the final amount for the following time periods:

a. 1 year **b.** 90 days

Solution **a.** To find the interest for 1 yr, we proceed as follows:

$$I = \$2000 \cdot 5\frac{1}{4}\% \cdot 1 = \$2000 \cdot 0.0525 \cdot 1 = \$105$$

Her final amount at the end of 1 yr is

$$\$2000 + \$105 = \$2105$$

b. When the interest rate is annual and the interest period is given in days, we represent the time as a fractional part of a year by dividing the number of days by 365. Thus

$$I = \$2000 \cdot 5\frac{1}{4}\% \cdot \frac{90}{365}$$

$$= \$2000 \cdot 0.0525 \cdot \frac{90}{365} = \$25.89$$

Hence,

$$A \doteq \$2000 + \$25.89$$

$$A \doteq \$2025.89$$

Thus, Vera's amount after 90 days is approximately \$2025.89.

EXAMPLE 6-25

Find the annual interest rate if a principal of \$10,000 increased to \$10,900 at the end of 1 yr.

Solution Let the annual interest be $x\%$. We know that $x\%$ of \$10,000 is the increase. Because the increase is \$10,900 − \$10,000 = \$900, we use the strategy of *writing an equation* for x as follows:

$$x\% \text{ of } 10{,}000 = 900$$

$$\frac{x}{100} \cdot 10{,}000 = 900$$

$$x = 9$$

Thus, the interest is 9%. We can also solve this problem mentally by asking, "What percent of 10,000 is 900?" Because 1% of 10,000 is 100, to obtain 900, we take 9% of 10,000.

Compound Interest

In business transactions, interest is sometimes calculated daily (365 times a year). In the case of savings, the earned interest is added daily to the principal, and each day the interest is earned on a different amount; that is, it is earned on the previous interest as well as the principal. When interest is computed in this way, it is called **compound interest**. Compounding usually is done annually (once a year), semiannually (twice a year), quarterly (4 times a year), or monthly (12 times a year). However, even when the interest is compounded, it is given as an annual rate. For example, if the annual rate is 6% compounded monthly, the interest per month is $\frac{6}{12}\%$, or 0.5%. If it is compounded daily, the interest per day is $\frac{6}{365}\%$. In general, *the interest rate per period is the annual interest rate divided by the number of periods in a year.*

compound interest

We can use a spreadsheet to compare various compound interest rates. Work through the student page on page 423 and answer the questions in the *TRY IT* and the *ON YOUR OWN* sections.

EXAMPLE 6-26

If you invest \$100 at 8% compounded quarterly, how much will you have in the account after 1 yr?

Solution The quarterly interest rate is $\frac{1}{4} \cdot 8\%$, or 2%. It seems that we would have to calculate the interest 4 times. But we can also reason as follows. If at the beginning of any of the four periods there are x dollars in the account, at the end of that period there will be

$$\begin{aligned} x + 2\% \text{ of } x &= x + 0.02x \\ &= x(1 + 0.02) \\ &= x \cdot 1.02 \text{ dollars.} \end{aligned}$$

Hence, to find the amount at the end of any period, we need only multiply the amount at the beginning of the period by 1.02. From Table 6-4, we see that the amount at the end of the fourth period is $\$100 \cdot 1.02^4$. On a scientific calculator, we can find the amount using

COMPOUND INTEREST RATES

Using a Spreadsheet • Compound Interest

Problem: Which investment strategy will cause your $100 investment to increase the most in 4 years: if you earn 4% interest compounded annually or if you earn 3.75% interest compounded monthly?

A spreadsheet can help you find the answer to this problem.

1 **Enter the following information in your spreadsheet as shown.**

	A	B	C	D	E	F
1	Year	Amount	Annual Rate	Month	Amount	Monthly Rate
2	0	100		0	100	

2 **Enter the following formulas.**
In cell C2, enter =.04.
In cell F2, enter =.0375/12.
In cell A3, enter =A2+1.
In cell B3, enter =B2+B2*C$2.
In cell D3, enter =D2+1.
In cell E3, enter =E2+E2*F$2.

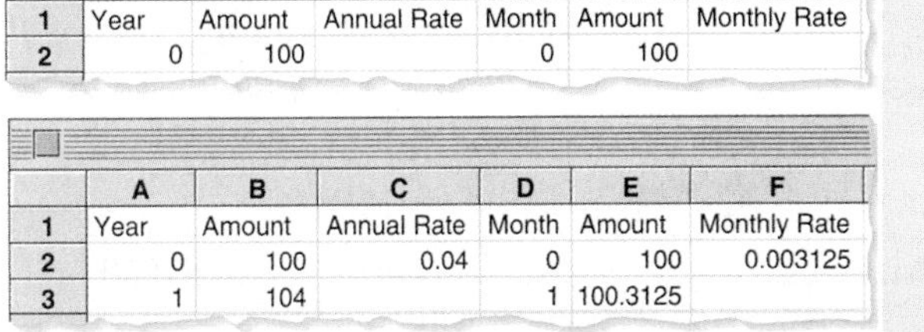

	A	B	C	D	E	F
1	Year	Amount	Annual Rate	Month	Amount	Monthly Rate
2	0	100	0.04	0	100	0.003125
3	1	104		1	100.3125	

3 **Select cells A3 to F50 and use the Fill Down command.**

	A	B	C	D	E	F
1	Year	Amount	Annual Rate	Month	Amount	Monthly Rate
2	0	100	0.04	0	100	0.003125
3	1	104		1	100.3125	
4	2	108.16		2	100.6259	
5	3	112.4864		3	100.9404	
6	4	116.9858		4	101.2558	
50	48	657.0528		48	116.1562	

Solution: $100 at 4% interest compounded annually is $116.99; $100 at 3.75% interest compounded monthly is only $116.16.

TRY IT

Which investment strategy will cause your $100 investment to increase the most in 4 years: if you earn 5% interest compounded annually or if you earn 5% interest compounded quarterly (4 times a year)?

ON YOUR OWN

- Which is easier, computing compound interest by calculator or by using a spreadsheet? Explain.
- Why must you divide the interest rate by the number of compounding periods?
- Why do you have to enter formulas by using the "=" symbol?

312

Source: Scott Foresman–Addison Wesley Middle School Math, Course 3, 2002 (p. 312).

[1] [0] [0] [×] [1] [.] [0] [2] [y^x] [4] [=]. The calculator displays 108.24322. Thus, the amount at the end of 1 yr is approximately $108.24.

Table 6-4

Period	Initial Amount	Final Amount
1	100	$100 \cdot 1.02$
2	$100 \cdot 1.02$	$(100 \cdot 1.02) \cdot 1.02$ or $100 \cdot 1.02^2$
3	$100 \cdot 1.02^2$	$(100 \cdot 1.02^2) \cdot 1.02$ or $100 \cdot 1.02^3$
4	$100 \cdot 1.02^3$	$(100 \cdot 1.02^3) \cdot 1.02$ or $100 \cdot 1.02^4$

In Example 6-26, finding the final amount at the end of the nth period amounts to finding the nth term of a geometric sequence whose first term is $100 \cdot 1.02$ (amount at the end of the first period) and whose ratio is 1.02. Thus, the amount at the end of the nth period is given by $(100 \cdot 1.02) \cdot (1.02)^{n-1} = 100 \cdot 1.02^n$. We can generalize this discussion. If the principal is P and the interest rate per period is r, then the amount A after n periods is $P(1 + r) \cdot (1 + r)^{n-1}$ or $P(1 + r)^n$. Therefore, we have a *formula for computing the amount at the end of the nth period, namely* $A = P(1 + r)^n$.

EXAMPLE 6-27

Suppose you deposit $1000 in a savings account that pays 6% interest compounded quarterly.

a. What is the balance at the end of 1 yr?
b. What is the *effective annual yield* on this investment; that is, what is the rate that would have been paid if the amount had been invested using simple interest?

Solution **a.** An annual interest rate of 6% earns $\frac{1}{4}$ of 6%, or an interest rate of $\frac{0.06}{4}$, in 1 quarter. Because there are 4 periods, we have the following:

$$A = 1000\left(1 + \frac{0.06}{4}\right)^4 \doteq \$1061.36$$

The balance at the end of 1 yr is approximately $1061.36.

b. Because the interest earned is $1061.36 − $1000.00 = $61.36, the effective annual yield can be computed by using the simple interest formula, $I = Prt$.

$$61.36 = 1000 \cdot r \cdot 1$$
$$\frac{61.36}{1000} = r$$
$$0.06136 = r$$
$$6.136\% = r$$

The effective annual yield is 6.136%.

EXAMPLE 6-28

To save for their child's college education, a couple deposits \$3000 into an account that pays 7% annual interest compounded daily. Find the amount in this account after 8 yr.

Solution The principal in the problem is \$3000, the daily rate i is 0.07/365, and the number of compounding periods is $8 \cdot 365$, or 2920. Thus we have

$$A = \$3000\left(1 + \frac{0.07}{365}\right)^{2920} = \$5251.74$$

Thus, the amount in the account is approximately \$5251.74.

ONGOING ASSESSMENT 6-6

You will need a calculator to do most of the following problems.

1. Complete the following compound interest chart.

Compounding Period	Principal	Annual Rate	Length of Time (Years)	Interest Rate per Period	Number of Periods	Amount of Interest Paid
a. Semiannual	\$1000	6%	2			
b. Quarterly	\$1000	8%	3			
c. Monthly	\$1000	10%	5			
d. Daily	\$1000	12%	4			

2. Ms. Jackson borrowed \$42,000 at 8.75% annual simple interest to buy her house. If she won the Irish Sweepstakes exactly 1 yr later and was able to repay the loan without penalty, how much interest would she owe?
3. Carolyn went on a shopping spree with her Bankamount card and made purchases totaling \$125. If the interest rate is 1.5% per month on the unpaid balance and she does not pay this debt for 1 yr, how much interest will she owe at the end of the year?
4. A man collected \$28,500 on a loan of \$25,000 he made 4 yr ago. If he charged simple interest, what was the rate he charged?
5. Burger Queen will need \$50,000 in 5 yr for a new addition. To meet this goal, the company deposits money in an account today that pays 3% annual interest compounded quarterly. Find the amount that should be invested to total \$50,000 in 5 yr.
6. A company is expanding its line to include more products. To do so, it borrows \$320,000 at 13.5% annual simple interest for a period of 18 mo. How much interest must the company pay?
7. To save for their retirement, a couple deposits \$4000 in an account that pays 5.9% interest compounded quarterly. What will be the value of their investment after 20 yr?
8. A car company is offering car loans at a simple interest rate of 4.7%. Find the interest charged to a customer who finances a car loan of \$7200 for 3 yr.
9. Johnny and Carolyn have three savings plans, which accumulated the following amounts of interest for 1 yr:
   a. A passbook savings account that accumulated \$53.90 on a principal of \$980
   b. A certificate of deposit that accumulated \$55.20 on a principal of \$600
   c. A money market certificate that accumulated \$158.40 on a principal of \$1200

   Which of these accounts paid the best interest rate for the year?
10. A hamburger costs \$1.35 and the price continues to rise at a rate of 11% a year for the next 6 yr. What will the price of a hamburger be at the end of 6 yr?
11. If college tuition is \$10,000 this year, what will it be 10 yr from now, assuming a constant inflation rate of 9% a year?

12. Sara invested money at a bank that paid 3.5% compounded quarterly. If she had \$4650 at the end of 4 yr, what was her initial investment?
13. Adrien and Jarrell deposit \$300 on January 1 in a holiday savings account that pays 1.1% per month interest and they withdraw the money on December 1 of the same year. What is the effective annual yield?
14. The number of trees in a rain forest decreases each month by 0.5%. If the forest has approximately $2.34 \cdot 10^9$ trees, how many trees will be left after 20 yr?
15. An amount of \$3000 was deposited in a bank at a rate of 2% compounded quarterly for 3 yr. The rate then increased to 3% and was compounded quarterly for the next 3 yr. If no money was withdrawn, what was the balance at the end of this time?
16. A money market fund pays 14% annual interest compounded daily. What is the value of \$10,000 invested in this fund after 15 yr?
17. The New Age Savings Bank advertises 4% interest rates compounded daily, while the Pay More Bank pays 5.2% interest compounded annually. Which bank offers a better rate for a customer who plans to leave her money in for exactly 1 yr?
18. A car is purchased for \$15,000. If each year the car depreciates by 10% of its value the preceding year, what will its value be at the end of 3 yr?
19. If a fixed simple interest rate of 4% is paid on a savings account for a period of n years and the interest remains in the account, the amount every year would represent what type of sequence?
20. If a fixed simple interest rate is paid on a savings account for a period of years and enough money is withdrawn from the account that allows the principal to remain fixed but no other money is withdrawn, what type of sequence does the money earned every year represent?
21. If a publishing company signed an agreement to allow a textbook (originally published with 500 pages) to expand over several editions to 1000 pages and the book was growing at approximately 10% in the number of pages over each edition, how many editions could be published before it reached the contractual limit?
22. According to some reports, the number of students taking mathematics each year after they reach the stage that they are not required to take it drops by half. In an urban school system, only 2 yr of mathematics is required for graduation. If the system allows the second year to be completed by the ninth grade, when students take a high school course as an eighth grader, how many students of the original 200,000 in the system at eighth grade could be expected to take math their senior year?

Communication

23. According to a newspaper report, trees on a certain land area are being cut at a rate of 12% per year. The lumber company involved claimed to replant 2000 trees every year in this area. Discuss the future tree production of this land area if this plan continues.
24. Because of a recession, the value of a new house depreciated 10% each year for 3 yr in a row. Then, for the next 3 yr, the value of the house increased 10% each year. Did the value of the house increase or decrease after 6 yr? Explain.
25. Determine the number of years (to the nearest tenth) it would take for any amount of money to double if it were deposited at a 10% interest rate compounded annually. Explain your reasoning.
26. A car parts manufacturer advertised three devices that could be installed in a car to save gas. The first could save 15% on fuel, the second 35%, and the third 50%. Explain whether you could conclude that when all three devices were installed, the savings on fuel would be 100%?
27. Each year a car's value depreciated 20% from the previous year. Mike claims that after 5 yr the car would depreciate 100% and would not be worth anything. Is Mike correct? Explain why or why not. If not, find the actual percent the car would depreciate after 5 yr.

Open-Ended

28. The effect of depreciation can be computed using a formula similar to the formula for compound interest.
    a. Assume depreciation is the same each month. Write a problem involving depreciation and solve it.
    b. Develop a general formula for depreciation defining what each variable in the formula stands for.
29. Find four large cities around the world and an approximate percentage rate of population growth for the countries in which the cities are located. Estimate the population in each of the four cities in 25 yr.
30. State different situations that do not involve money in which a formula like the one for compound interest is used. In each case, state a related problem and write its solution.

Cooperative Learning

31. The federal Truth in Lending Act, passed in 1969, requires lending institutions to quote an annual percentage rate (APR) that helps consumers compare the true cost of loans regardless of how each lending institution computes the interest and adds on costs.
    a. Call different banks and ask for their APR on some loans and the meaning of APR.
    b. Based on your findings in (a), write a clear definition of APR.
    c. Use the information given by your credit card (you may need to call the bank) and compute the APR on cash advances. Is your answer the same as that given by the bank? Compare the APR for different credit cards.

BRAIN TEASER

In the cartoon, which of the people talking in the first two cells is correct? Why?

HINT FOR SOLVING THE PRELIMINARY PROBLEM

Hint: To consider whether or not the server's statement is correct, one needs to know that the tip is considered to be a percentage of the bill. If the bill were \$10.00 and the tip were to be 15%, then it would be 0.15 · \$10.00, or \$1.50. If the bill were doubled as described, the tip is as much as the original bill. What percentage of the original bill is that?

QUESTIONS FROM THE CLASSROOM

1. A student claims that 0.36 is greater than 0.9 because 36 is greater than 9. How do you respond?

2. A student reports that $^-438{,}340{,}000$ cannot be written in scientific notation. How do you respond?

3. A student multiplies (6.5)(8.5) to obtain the following:

$$\begin{array}{r} 8.5 \\ \times 6.5 \\ \hline 4\,2\,5 \\ 5\,1\,0 \\ \hline 5\,5.2\,5 \end{array}$$

However, when the student multiplies $8\frac{1}{2} \cdot 6\frac{1}{2}$, she obtains the following:

$$\begin{array}{rl} 8\frac{1}{2} & \\ \times 6\frac{1}{2} & \\ \hline 4\frac{1}{4} & \left(\frac{1}{2} \cdot 8\frac{1}{2}\right) \\ 48 & (6 \cdot 8) \\ \hline 52\frac{1}{4} & \end{array}$$

How is this possible?

4. A student says that $3\frac{1}{4}\% = 0.03 + 0.25 = 0.28$. Is this correct? Why?

5. A student tries to calculate $0.999^{10{,}000}$ on a calculator and finds the answer to be $4.5173346 \cdot 10^{-5}$. The student wonders how it could be that a number like 0.999, so close to 1, when raised to some power could result in a number close to 0. How do you respond·

6. Explain how you would respond to the following:

a. A student claims that $\frac{9443}{9444}$ and $\frac{9444}{9445}$ are equal because both display 0.9998941 on his scientific calculator when the divisions are performed.

b. Another student claims that the fractions are not equal and wants to know if there is any way the same calculator can determine which is greater.

7. Why is $\sqrt{25} \neq {}^-5$?

8. A student claims that $\sqrt{(^-5)^2} = {}^-5$ because $\sqrt{a^2} = a$. Is this correct?

9. Another student says that $\sqrt{(^-5)^2} = [(^-5)^2]^{1/2} = (^-5)^{2/2} = (^-5)^1 = {}^-5$. Is this correct?

10. A student claims that the equation $\sqrt{^-x} = 3$ has no solution, since the square root of a negative number does not exist. Why is this argument wrong?

11. A student reports that it is impossible to mark a product up 150% because 100% of something is all there is. What is your response?

12. A student argues that a $p\%$ increase in salary followed by a $q\%$ decrease is equivalent to a $q\%$ decrease followed by a $p\%$ increase because of the commutative property of multiplication. How do you respond?

13. A student argues that $0.01\% = 0.01$ because in 0.01%, the percent is already written as a decimal. How do you respond?

14. A student claims that if the value of an item increases by 100% each year from its value the previous year and if the original price is d dollars, then the value after n years will be $d \cdot 2^n$ dollars. Is the student correct? Why or why not?

15. A student argues that in the real world, irrational numbers are never used. How do you respond?

16. A student argued that there were infinitely many irrational numbers between 0 and 1 because $\frac{\pi}{n}$, where $n \geq 4$ and $n \in N$, is in the interval. Why is the student correct?

17. A student argued that 200% of a quantity could not exist because 100% of a quantity is all that there is. His example used his class as the quantity and he claimed that 200% of his class could not exist. How would you respond?

18. Brown's research in 1981 argued that students have significantly more difficulty with computations in word problems if there are decimals in the problems. What is your reaction to that research?

19. A student argues that fractions should no longer be taught once students learn how to work with decimal numbers. How do you respond?

CHAPTER REVIEW

1. a. On the number line, find the decimals that correspond to points A, B, and C.
   b. Indicate by D the point that corresponds to 0.09 and by E the point that corresponds to 0.15.

   0 0.1 0.2

   A B C

2. Write each of the following as a rational number in the form a/b, where a and b are integers and $b \neq 0$:
   a. 32.012 **b.** 0.00103

3. Give a test to determine if a fraction can be written as a terminating decimal without one's actually performing the division. Explain why this test is valid.

4. A board is 442.4 cm long. How many shelves can be cut from it if each shelf is 55.3 cm long? (Disregard the width of the cuts.)

5. Write each of the following as a decimal:
   a. $\frac{4}{7}$ **b.** $\frac{1}{8}$ c. $\frac{2}{3}$ **d.** $\frac{5}{8}$

6. Write each of the following as a fraction in simplest form:
   a. 0.28 **b.** $^-5.07$ c. $0.\overline{3}$ **d.** $2.0\overline{8}$

7. Round each of the following numbers as specified:
   a. 307.625 to the nearest hundredth
   b. 307.625 to the nearest tenth
   c. 307.625 to the nearest unit
   d. 307.625 to the nearest hundred

8. Rewrite each of the following in scientific notation:
   a. 426,000 **b.** $324 \cdot 10^{^-6}$
   c. 0.00000237 **d.** $^-0.325$

9. Order each of the following decimals from greatest to least:
   $1.4\overline{519}$, $1.451\overline{9}$, 1.4519, $1.45\overline{19}$, $^-0.134$, $^-0.13\overline{401}$, $0.13\overline{401}$

10. Each of the following is a geometric sequence. Find the missing terms.
   a. 5, _, 10 **b.** 1, _, _, _, 1/4

11. Write each of the following in scientific notation without using a calculator:
   a. 1783411.56 **b.** $\frac{347}{10^8}$
   c. $49.3 \cdot 10^8$ **d.** $29.4 \cdot \frac{10^{12}}{10^{^-4}}$
   e. $0.47 \cdot 1000^{12}$ **f.** $\frac{3}{5^9}$

12. **a.** Find five decimals between 0.1 and 0.11 and order them from greatest to least.
   b. Find four decimals between 0 and 0.1 listed from least to greatest so that each decimal starting from the second is twice as large as the preceding one.
   c. Find four decimals between 0.1 and 0.2 and list them in increasing order so that the first one is halfway between 0.1 and 0.2, the second halfway between the first and 0.2, the third halfway between the second and 0.2, and similarly for the fourth one.

13. Classify each of the following as rational or irrational (assume the patterns shown continue):
   a. 2.191199119991199991 19 . . .
   b. $\frac{1}{\sqrt{2}}$ c. $\frac{4}{9}$
   d. 0.0011001100110011 . . .
   e. 0.001100011000011 . . .

14. Write each of the following in the form $a\sqrt{b}$ or $a\sqrt[3]{b}$, where a and b are integers and b has the least value possible:
   a. $\sqrt{242}$ **b.** $\sqrt{288}$
   e. $\sqrt{360}$ **d.** $\sqrt[3]{162}$

15. Answer each of the following and explain your answers:
   a. Is the set of irrational numbers closed under addition?
   b. Is the set of irrational numbers closed under subtraction?
   c. Is the set of irrational numbers closed under multiplication?
   d. Is the set of irrational numbers closed under division?

16. Find an approximation for $\sqrt{23}$ correct to three decimal places without using the y^x or the $\sqrt{\ }$ keys.

17. Answer each of the following:
   a. 6 is what percent of 24?
   b. What is 320% of 60?
   c. 17 is 30% of what number?
   d. 0.2 is what percent of 1?

18. Change each of the following to percents:
   a. $\frac{1}{8}$ **b.** $\frac{3}{40}$ c. 6.27
   d. 0.0123 e. $\frac{3}{2}$

19. Change each of the following percents to decimals:
   a. 60% **b.** $\frac{2}{3}\%$ c. 100%

20. Sandy received a dividend that equals 11% of the value of her investment. If her dividend was $1020.80, how much was her investment?

21. Five computers in a shipment of 150 were found to be defective. What percent of the computers were defective?

22. On a mathematics examination, a student missed 8 of 70 questions. What percent of the questions, rounded to the nearest tenth, did the student do correctly?

23. A microcomputer system costs $3450 at present. This is 60% of the cost 4 yr ago. What was the cost of the system 4 yr ago? Explain your reasoning.

24. If, on a purchase of one new suit, you are offered successive discounts of 5%, 10%, or 20% in any order you wish, what order should you choose?

25. Jane bought a bicycle and sold it for 30% more than she paid for it. She sold it for $104. How much did she pay for it?

26. In the southern part of the United States, "sweet tea" is commonly seen on menus. If "sweet tea" is made by adding 1 c of sugar for each 4 c of regular tea, what percentage of the final mixture of a 1-gal jar of sweet tea (16 c) is sugar?

27. A school district had financial difficulties and discussed reducing the beginning teacher salary from $28,000 in 2003–4 by 5% for 2004–5. The school system had a strike in 2004–5 and the salary in 2005–6 was 5% higher than in 2004–5.
   a. Determine the 2004–5 beginning teacher salary.
   b. What was the percent of increase in salary for beginning teachers from 2003–4 to 2005–6?

28. The student bookstore had a textbook for sale at $89.95. A student found the book on eBay for $62.00. If the student bought the book on eBay, what percentage of the cost of the bookstore book did she save?

29. When a store had a 60% off sale, Dori had a coupon for an additional 40% off any item, and thought she should be able to obtain the dress that she wanted for free. If you were the store manager, how would you explain the mathematics of the situation to her?

30. Explain whether or not you could have each of the following as mathematically meaningful percentages:
   a. $\pi\%$
   b. $\sqrt{2}\%$
   c. $0.3\overline{4}\%$
   d. $(1 + 0.\overline{3})\%$

31. What would be the third term of a geometric sequence whose first term is π and whose ratio is $\frac{1}{\pi}$?

32. Explain whether or not you would ever expect a price in a store to be an irrational number.

33. Are the irrational numbers closed under addition? Prove your answer.

★ 34. A company was offered a $30,000 loan at a 12.5% annual interest rate for 4 yr. Find the simple interest due on the loan at the end of 4 yr.

★ 35. A money market fund pays 14% annual interest compounded quarterly. What is the value of a $10,000 investment after 3 yr?

CHAPTER OUTLINE

I. Decimals
   A. Every rational number can be represented as a terminating or repeating decimal.
   B. A rational number $\frac{a}{b}$, in simplest form whose denominator is of the form $2^m \cdot 5^n$, where m and n are whole numbers, can be expressed as a **terminating decimal**.
   C. A **repeating decimal** is a decimal with a block of digits, called the **repetend,** that repeat infinitely many times.
   D. A number is in **scientific notation** if it is written as the product of a number n that is greater than or equal to 1 and less than 10 and an integer power of 10.
   E. An **infinite geometric sequence** with first term a and ratio r, where $0 < r < 1$, has sum $S = \frac{a}{1-r}$.

II. Real numbers
   A. An **irrational number** is represented by a nonterminating, nonrepeating decimal.
   B. The set of **real numbers** is the set of all decimals, namely, the union of the set of rational numbers and the set of irrational numbers.
   C. If a is any whole number, then the **principal square root** of a, denoted by $\sqrt{a}$, is the nonnegative number b such that $b \cdot b = b^2 = a$.
   D. Square roots and nth roots can be found by using the **squeezing method** or the Archimedean method.
   E. Computation methods with terminating and repeating decimals can be verified by converting the decimals to rational numbers in the form $\frac{a}{b}$, where a and b are integers and $b \neq 0$.

F. Estimations for decimal arithmetic can be accomplished following the same general rules as for other sets of numbers.

III. Radicals and rational exponents

A. $\sqrt[n]{x}$ or $x^{1/n}$, is the ***n*****th root** of x and n is the **index**.

B. The following properties hold for all radicals if the expressions involving radicals are meaningful:

a. $\sqrt[n]{xy} = \sqrt[n]{x} \cdot \sqrt[n]{y}$

b. $\sqrt[n]{\frac{x}{y}} = \frac{\sqrt[n]{x}}{\sqrt[n]{y}}$

c. $x^{m/n} = (\sqrt[n]{x})^m = \sqrt[n]{x^m}$, if m/n is in simplest form.

IV. Percent and interest

A. Percent means *per hundred*. Percent is written using the % symbol: $x\% = \frac{x}{100}$.

*__B. Simple interest__ is computed using the formula $I = Prt$, where I is the interest, P is the principal, r is the annual interest rate, and t is the time in years.

*__C.__ When **compound interest** is involved, we use the formula $A = P(1 + i)^n$, where A is the balance, P is the principal, i is the interest rate per period, and n is the number of periods.

SELECTED BIBLIOGRAPHY

Behr, M., I. Wachsmuth, T. Post, and R. Lesh. "Order and Equivalence of Rational Numbers: A Clinical Teaching Experiment." *Journal for Research in Mathematics Education* 15 (July 1984): 323–341.

Beigie, D. "Investigating Limits in Number Patterns." *Mathematics Teaching in the Middle School* 7 (April 2002): 438–443.

Bennett, A., and T. Nelson. "A Conceptual Model for Solving Percent Problems." *Mathematics Teaching in the Middle School* 1 (April 1994): 20–25.

Brown, M. "Place Value and Decimals." In *Children's Understanding of Mathematics*. London: John Murray, 1981, pp. 11–16.

Drum, R., and W. Petty, Jr. "2 Is Not the Same as 2.0!" *Mathematics Teaching in the Middle School* 6 (September 2000): 34–38.

Glasgow, R., G. Ragan, W. Fields, R. Reys, and D. Wasman. "The Decimal Dilemma." *Teaching Children Mathematics* 7 (October 2000): 89–93.

Hiebert, J. "Mathematical, Cognitive, and Instructional Analyses of Decimal Fractions." In *Analysis of Artithmetic for Mathematics Teaching*, edited by G. Leinhardt, R. Putman, and R. Hattrup. Hillsdale, N.J.: LEA, 1992.

Lembke, L., and B. Reys. "The Development of, and Interaction between, Intuitive and School-taught Ideas about Percent." *Journal of Research in Mathematics Education* 25 (May 1994): 237–259.

Martine, S., and J. Bay-Williams. "Investigating Students' Conceptual Understanding of Decimal Fractions Using Multiple Representations." *Mathematics Teaching in the Middle School* 8 (January 2003): 244–247.

Moss, J., and B. Caswell. "Building Percent Dolls: Connecting Linear Measurement to Learning Ratio and Proportion." *Mathematics Teaching in the Middle School* 10 (September 2004): 68–74.

Neuschwander, C. *Sir Cumference and the Dragon of Pi: A Math Adventure*. Watertown, MA: Cambridge Publishing, Inc., 1999.

Oppenheimer, L., and R. Hunting. "Relating Fractions and Decimals: Listening to Students Talk." *Mathematics Teaching in the Middle School* 4 (February 1999): 318–321.

Reys, B., and F. Arbaugh. "Clearing Up the Confusion over Calculator Use in Grades K–5." *Teaching Children Mathematics* 8 (October 2001): 90–94.

Reys, R., B. Reys, N. Nohda, and H. Emori. "Mental Computation Performance and Strategy Use of Japanese Students in Grades 2, 4, 6, and 8." In *Journal for Research in Mathematics Education* 26 (July 1995): 304–326.

Sowder, J., D. Wearne, W. Martin, and M. Strutchens. "What Do 8th-Grade Students Know about Mathematics?" In *Results and Interpretations of the 1990–2000 Mathematics Assessments of the National Assessment of Educational Progress*, edited by P. Kloosterman and F. Lester. Reston, VA.: NCTM, 2004, pp. 105–144.

Sweeney, E., and R. Quinn. "Concentration: Connecting Fractions, Decimals, & Percents." *Mathematics Teaching in the Middle School* 5 (January 2000): 324–328.

Thompson, A., and S. Sproule. "Deciding When to Use Calculators." *Mathematics Teaching in the Middle School* 6 (October 2000): 12–129.

Thompson, C., and V. Walker. "Connecting Decimals and Other Mathematical Content." *Teaching Children Mathematics* 2 (April 1996): 496–502.

Wearne, D., and J. Hiebert. "A Cognitive Approach to Meaningful Mathematics Instruction: Testing a Local Theory Using Decimal Numbers." *Journal for Research in Mathematics Education* 19 (November 1988): 371–384.

Weibe, J. "Manipulating Percentages." *Mathematics Teacher* 79 (January 1986): 21, 23–26.

Williams, S., and J. Copley. "Using Calculators to Discover Patterns in Dividing Decimals." *Mathematics Teaching in the Middle School* 1 (April 1994): 72–75.

Yang, D., and R. Reys. "One Fraction Problem, Many Solution Paths." *Mathematics Teaching in the Middle School* 7 (November 2001): 164–166.

Probability

PRELIMINARY PROBLEM

Al tosses one quarter and at the same time Betty tosses two quarters. What is the probability that Betty gets the same number of heads as Al?

Probability, with its roots in gambling, is used in such areas as predicting sales, planning political campaigns, determining insurance premiums, making investment decisions, and testing experimental drugs. Some examples of uses of probability in everyday conversations include the following:

What is the probability that the Red Sox will win the World Series?

There is no chance you will get a raise.

There is a 50% chance of rain today.

The *Principles and Standards* call for an increased emphasis on data analysis and probability. In the *Standards*, we find the following:

Although the computation of probabilities can appear to be simple work with fractions, students must grapple with many conceptual challenges in order to understand probability. Misconceptions about probability have been held not only by many students but also by many adults (Konold 1989). To correct misconceptions, it is useful for students to make predictions and then compare the predictions with actual outcomes. (p. 254)

Probability Expectations

The *Principles and Standards* Data Analysis and Probability Standard describes its expectations for students with regard to the study of probability. The expectations covered in this chapter include the following:

In grades K–2, children should discuss events related to their experience as likely or unlikely. (p. 400)

In grades 3–5, children should be able to describe events as likely or unlikely and discuss degrees of likelihood using words such as certain, equally likely, *and* impossible. *They should be able to predict the probability of simple experiments and test the predictions. They should understand that the measure of likelihood of an event can be represented by a number from 0 to 1.* (p. 400)

In grades 6–8, children should understand complementary and mutually exclusive events. They should be able to make and test conjectures about the results of experiments and simulations. They should be able to compute probabilities of compound events using methods such as organized lists, tree diagrams, and area models. (p. 401)

In this chapter, we introduce all of the topics in the preceding expectations along with several others. We use tree diagrams and geometric probabilities (area models) to solve problems and to analyze games that involve spinners, cards, and dice. We introduce counting techniques and discuss the role of simulations in probability.

▶ **HISTORICAL NOTE**

The first time probability appears to have been mentioned in the Western history of mathematics is in a 1477 commentary on Dante's *Divine Comedy*, but most historians think that it originated in an unfinished dice game. The French mathematician Blaise Pascal (1623–1665) received a letter from his friend Chevalier de Méré, a professional gambler, who asked how to divide the stakes if two players start, but fail to complete, a game consisting of five matches in which the winner is the one who wins three out of five matches. The players decided to divide the stakes according to their chances of winning the game. Pascal shared the problem with Pierre de Fermat (1601–1665) and together they solved the problem, which prompted the development of probability.

Since the work of the French mathematician Pierre Simon de Laplace (1749–1827), probability theory has become a major mathematical tool in science. In 1905 Albert Einstein (1879–1955) used the theory of probability in his work in physics. ◀

7-1 How Probabilities Are Determined

Probabilities are ratios, expressed as fractions, decimals, or percents, determined by considering results or outcomes of experiments. An **experiment** is an activity whose results can be observed and recorded. Each of the possible results of an experiment is an **outcome**. If we toss a coin that cannot land on its edge, there are two distinct possible outcomes: heads (H) and tails (T).

experiment

outcome

sample space

A set of all possible outcomes for an experiment is a **sample space**. The outcomes in the sample space cannot overlap. In a single coin toss, the sample space S is given by $S = \{H, T\}$. The sample space can be modeled by a tree diagram, as shown in Figure 7-1. Each outcome of the experiment is designated by a separate branch in the tree diagram. The sample space S for rolling the standard die in Figure 7-2(a) is $S = \{1, 2, 3, 4, 5, 6\}$. Figure 7-2(b) gives a tree diagram for the sample space.

FIGURE 7-1

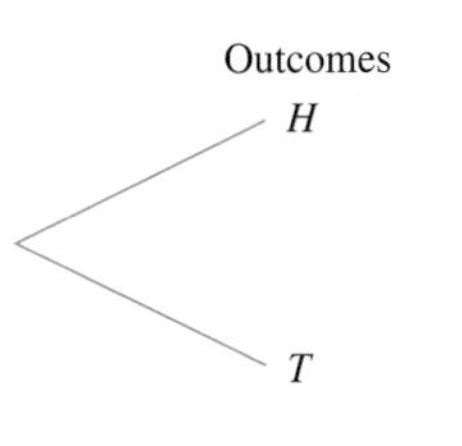

FIGURE 7-2

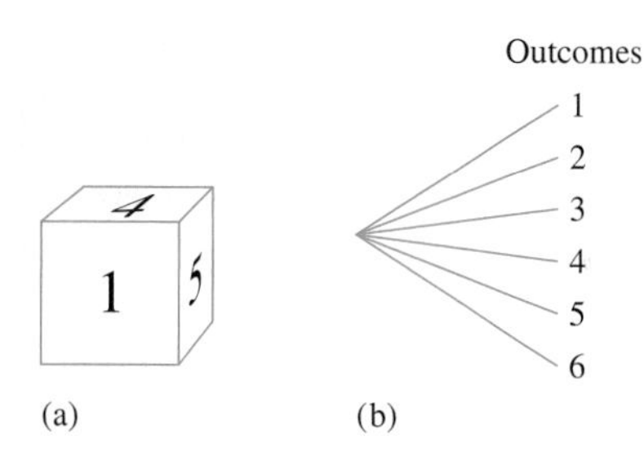

event

Any subset of a sample space is an **event**. For example, the set of all even-numbered rolls $\{2, 4, 6\}$ is a subset of all possible rolls of a die $\{1, 2, 3, 4, 5, 6\}$ and is an event.

EXAMPLE 7-1

Suppose an experiment consists of drawing 1 slip of paper from a jar containing 12 slips of paper, each with a different month of the year written on it. Find each of the following:

a. The sample space S for the experiment
b. The event A consisting of outcomes having a month beginning with J
c. The event B consisting of outcomes having the name of a month that has exactly 4 letters
d. The event C consisting of outcomes having a month that begins with M or N

Solution **a.** $S =$ {January, February, March, April, May, June, July, August, September, October, November, December}
b. $A =$ {January, June, July}
c. $B =$ {June, July}
d. $C =$ {March, May, November}

▲

EXAMPLE 7-2

Suppose we toss a fair coin 3 times and record the results. Find each of the following:

a. The sample space for this experiment
b. The event A of tossing two heads and one tail
c. The event B of tossing no tails
d. The event C of tossing a head on the last toss

Solution **a.** $S = \{HHH, HHT, HTH, THH, TTT, TTH, THT, HTT\}$
b. $A = \{HHT, HTH, THH\}$
c. $B = \{HHH\}$
d. $C = \{HHH, HTH, THH, TTH\}$

Determining Probabilities

Around 1900, the English statistician Karl Pearson tossed a coin 24,000 times and recorded 12,012 heads. During World War II, John Kerrich, a Dane and a prisoner of war, tossed a coin 10,000 times. A subset of his results is in Table 7-1. The *relative frequency* column on the right is obtained by dividing the number of heads by the number of tosses of the coin.

Table 7-1

Number of Tosses	Number of Heads	Relative Frequency (rounded)
10	4	0.400
50	25	0.500
100	44	0.440
500	255	0.510
1,000	502	0.502
5,000	2,533	0.507
8,000	4,034	0.504
10,000	5,067	0.507

As the number of Kerrich's tosses increased, he obtained heads close to half the time. The relative frequency for Pearson's 24,000 tosses gives a similar result of 12,012/24,000, or approximately $\frac{1}{2}$. Kerrich was using the relative frequency interpretation of probability. In this interpretation, a probability is the long-run proportion of times that an event will occur given many repetitions under identical circumstances.

experimentally ◆ empirically

When a probability is determined by observing outcomes of experiments, it is said to be determined **experimentally**, or **empirically**. The exact number of heads that occurs when a fair coin is tossed a few times cannot be predicted accurately. A *fair coin* is a coin that is just as likely to land "heads" as it is to land "tails." Probabilities only suggest what will happen in the "long run." This concept is called *The Law of Large Numbers* or is sometimes called *Bernoulli's Theorem*.

Law of Large Numbers (Bernoulli's Theorem)

If an experiment is repeated a large number of times, the *experimental* or *empirical* probability of a particular outcome approaches a fixed number as the number of repetitions increases.

HISTORICAL NOTE The Bernoulli family, a famous family in mathematics in Europe in the late seventeenth and early eighteenth centuries, produced a number of prominent mathematicians. Of Belgian origin, the family fled Amsterdam for Basel, Switzerland, to escape religious persecution and continued in the spice business. The eldest son, Jakob (1654–1705), was the first of several family members who, against the wishes of their parents, pursued the study of mathematics. After taking his degree in theology, Jakob traveled for 7 years studying mathematics. He returned to Basel in 1683, and in 1687 became the chair of mathematics at the University of Basel. His greatest and most original work, *Ars Conjectandi* (*The Art of Conjecturing*), laid the foundation for the modern theory of probability.

When a fair coin is tossed many times and the fraction (or proportion) of heads is near $\frac{1}{2}$, we say that the probability of heads occurring is $\frac{1}{2}$ and write $P(\{H\}) = \frac{1}{2}$, or we shorten the symbolism to $P(H) = \frac{1}{2}$.

In this text, by "probability" we mean *theoretical probability*. We assign **theoretical probabilities** to the outcomes under ideal conditions. For example, we could argue that since an ideal coin is symmetric and has two sides, then each side should appear about the same number of times if the coin is tossed many times. Again we would conclude that

theoretical probabilities

$$P(H) = P(T) = \frac{1}{2}.$$

When one outcome is just as likely as another, as in coin tossing, the outcomes are **equally likely**. If an experiment is repeated many times, the experimental probability of the event's occurring should be approximately equal to the theoretical probability of the event's occurring.

equally likely

RESEARCH NOTE Students tend to categorize events as equally likely because of their mere listing in the sample space. For example, on a single roll of a die a student claims the sample space consists of either a prime number or a composite number and says that the probability of rolling a prime number is the same as the probability of rolling a composite number (Lecoutre 1992).

A *fair* die is a die that is just as likely to land showing any of the numerals 1 through 6. Its sample space S is given by $S = \{1, 2, 3, 4, 5, 6\}$, and $P(1) = P(2) = P(3) = P(4) = P(5) = P(6) = \frac{1}{6}$. The probability of rolling an even number, that is, the probability of the event $E = \{2, 4, 6\}$, is $\frac{3}{6}$, or $\frac{1}{2}$. For a sample space with equally likely outcomes, the probability of an event A can be defined as follows.

Definition of Probability of an Event with Equally Likely Outcomes

For an experiment with sample space S with equally likely outcomes, the **probability of an event A** is given by

$$P(A) = \frac{\text{Number of elements of } A}{\text{Number of elements of } S} = \frac{n(A)}{n(S)}.$$

Note that this definition applies only to a sample space that has equally likely outcomes. If each possible outcome of the sample space is equally likely, the sample space is a **uniform sample space**. From the preceding definition it follows that if a uniform sample space has n outcomes, the probability of each outcome is $\frac{1}{n}$. Applying the preceding definition to a space with outcomes that are not equally likely (nonuniform) leads to incorrect conclusions. For example, the sample space for spinning the spinner in Figure 7-3 is given by $S = \{\text{Red, Blue}\}$, but the outcome Blue is more likely to occur than is the outcome Red, so $P(\text{Red})$ is not equal to $\frac{1}{2}$ but $\frac{1}{4}$ (why?). If the spinner were spun 100 times, we could reasonably expect that about $\frac{1}{4}$ of 100, or 25, of the outcomes would be Red, whereas about $\frac{3}{4}$ of 100, or 75, of the outcomes would be Blue.

uniform sample space

FIGURE 7-3

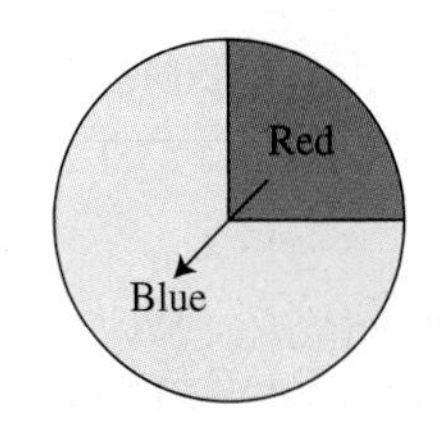

Notice that $P(\text{Red}) = \frac{Area(\text{Red})}{Area(\text{Circle})}$ because $Area(\text{Red}) = \frac{1}{4} Area(\text{Circle})$. Later in this chapter, we investigate more examples of computing probabilities involving areas.

NOW TRY THIS 7-1

a. In an experiment of tossing a fair coin once, what is the sum of the probabilities of all the distinct outcomes in the sample space?

b. In an experiment of tossing a fair die once, what is the sum of the probabilities of all the distinct outcomes in the sample space?

c. Does the sum of the probabilities of all the distinct outcomes of any sample space always result in the same number? Why?

EXAMPLE 7-3

at random

Let $S = \{1, 2, 3, 4, 5, \ldots, 25\}$. If a number is chosen **at random**, that is, with the same chance of being drawn as all other numbers in the set, calculate each of the following probabilities:

Mutually Exclusive Events

Consider one spin of the wheel in Figure 7-4. For this expe
2, 3, 4, 5, 6, 7, 8, 9}. If $A = \{0, 1, 2, 3, 4\}$ and $B = \{5, 7\}$, then A

mutually exclusive

are **mutually exclusive** events. If event A occurs, then ever
have the following definition.

Definition of Mutually Exclusive

Events A and B are **mutually exclusive** if they have no elen
$A \cap B = \varnothing$.

Each outcome in the space $S = \{0, 1, 2, 3, 4, 5, 6, 7, 8, 9\}$ is
bility $\frac{1}{10}$. Now with mutually exclusive sets A and B give
bility of A or B as $P(A \cup B)$, we have the following:

$$P(A \cup B) = \frac{n(A \cup B)}{n(S)} = \frac{7}{10} = \frac{5+2}{10} =$$

$$= \frac{n(A)}{n(S)} + \frac{n(B)}{n(S)} = P(A) + P(B)$$

The result developed in this example is true for all mut
general, we have the following property.

Property

If events A and B are mutually exclusive, then $P(A \text{ or } B) = P(A \cup$

For a sample space with equally likely outcomes, this pro
from the fact that if $A \cap B = \varnothing$, then $n(A \cup B) = n(A) + n$
bility of the union of events such that any two are mutua
the probabilities of those events.

Complementary Events

If the weather forecaster tells us that the probability of rai

complements

ability that it will not rain? These two events—rain and n
of each other. Therefore, if the probability of rain is 25%,
not rain is $100\% - 25\% = 75\%$, or $1 - \frac{1}{4} = \frac{3}{4}$. Notice tha
The two events rain and no rain are mutually exclusive b
other cannot. Two mutually exclusive events whose unic

complementary events

complementary events. If A is an event, the compleme
an event. For example, consider the event $A = \{2, 4\}$ of

a. The event A that an even number is drawn
b. The event B that a number less than 10 and greater than 20 is drawn
c. The event C that a number less than 26 is drawn
d. The event D that a prime number is drawn
e. The event E that a number both even and prime is drawn

Solution Each of the 25 numbers in set S has an equal chance of being drawn.

a. $A = \{2, 4, 6, 8, 10, 12, 14, 16, 18, 20, 22, 24\}$, so $n(A) = 12$. Thus,

$$P(A) = \frac{n(A)}{n(S)} = \frac{12}{25}.$$

b. $B = \varnothing$, so $n(B) = 0$. Thus, $P(B) = \frac{0}{25} = 0$.

c. $C = S$ and $n(C) = 25$. Thus, $P(C) = \frac{25}{25} = 1$.

d. $D = \{2, 3, 5, 7, 11, 13, 17, 19, 23\}$, so $n(D) = 9$. Thus,

$$P(D) = \frac{n(D)}{n(S)} = \frac{9}{25}.$$

e. $E = \{2\}$, so $n(E) = 1$. Thus, $P(E) = \frac{1}{25}$.

impossible event

In Example 7-3(b), event B is the empty set. An event such as B that has no outcomes is an **impossible event** *and has probability* 0. If the word *and* were replaced by *or* in Example 7-3(b), then event B would no longer be the empty set. In Example 7-3(c), event C consists of drawing a number less than 26 on a single draw. Because every number in S is less than 26, $P(C) = \frac{25}{25} = 1$. An event that has probability 1 is a **certain event**.

certain event

Any event A is a subset of a sample space S, $\varnothing \subseteq A \subseteq S$. Hence, $0 = n(\varnothing) \le n(A) \le n(S)$ and $0 = \frac{0}{n(S)} \le \frac{n(A)}{n(S)} \le \frac{n(S)}{n(S)} = 1$. In general, we have the following property.

Property

If A is any event and S is the sample space, then $0 \le P(A) \le 1$.

FIGURE 7-4

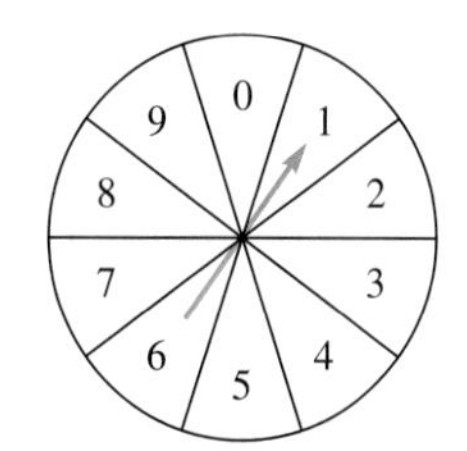

Consider one spin of the wheel shown in Figure 7-4. For this experiment, $S = \{0, 1, 2, 3, 4, 5, 6, 7, 8, 9\}$. If A is the event of spinning a number in the set $\{0, 1, 2, 3, 4\}$ and B is the event of spinning a number in the set $\{5, 7\}$, then using the definition of probability for equally likely events, $P(A) = \frac{n(A)}{n(S)}$ =

$= \frac{2}{10}$. The probability of an event can be found by adding the p
events representing the various outcomes in the set. For exampl
be represented as the union of two disjoint events, that is, spinni
Then $P(B)$ can be found by adding the probabilities of each eve

$$P(B) = P(5) + P(7) = \frac{1}{10} + \frac{1}{10} = \frac{2}{10}$$

Likewise,

$$P(A) = \frac{1}{10} + \frac{1}{10} + \frac{1}{10} + \frac{1}{10} + \frac{1}{10} = \frac{5}{10}$$

These are special cases of the following property, which holds

Property

The probability of an event is equal to the sum of the probabilities o
making up the event.

EXAMPLE 7-4 If we draw a card at random from an ordinary deck of pla
probability that

a. The card is an ace?
b. The card is an ace or a queen?

Solution **a.** There are 52 cards in a deck, 4 of which are aces.
ace, then $A = \{$♠, ♣, ♦, ♥$\}$. We use the de
equally likely outcomes to compute the following:

$$P(A) = \frac{n(A)}{n(S)} = \frac{4}{52}$$

An alternative approach is to find the sum of the p
each of the outcomes in the event, where the pro
single ace from the deck is $\frac{1}{52}$:

$$P(A) = \frac{1}{52} + \frac{1}{52} + \frac{1}{52} + \frac{1}{52}$$

b. The event E of getting an ace or a queen consis
4 queens. Hence,

$$P(E) = \frac{n(E)}{n(S)} = \frac{8}{52}.$$

Independent Events

In Figure 7-13, the fact that the first ball was not replaced affects the probability of the color of the second ball drawn. When the outcome of one event has no influence on the outcome of a second event, the events are **independent**. For example, if two coins are flipped and event E_1 is obtaining a head on the first coin and E_2 is obtaining a tail on the second coin, then E_1 and E_2 are independent events because one event has no influence on the second. Notice that $P(E_1) = \frac{1}{2}$, $P(E_2) = \frac{1}{2}$, and $P(E_1 \cap E_2) = \frac{1}{4}$. So in this case, $P(E_1 \cap E_2) = P(E_1) \cdot P(E_2)$.

independent

Next, consider two boxes: box 1 contains two white and two black balls, and box 2 contains two white balls and three black balls. Let B_1 be the event of drawing a black ball from box 1 and B_2 the event of drawing a black ball from box 2. Notice that the events are independent. Suppose a ball is drawn from each box and we are interested in the probability that each ball is black; that is, $P(B_1 \cap B_2)$. We know that $P(B_1) = \frac{2}{4}$, or $\frac{1}{2}$, and $P(B_2) = \frac{3}{5}$.

Is $P(B_1 \cap B_2) = P(B_1) \cdot P(B_2)$ in this case as well? We can answer this question by computing $P(B_1 \cap B_2)$ using a familiar approach. If we consider all the black balls to be different and all the white balls to be different, there are $4 \cdot 5$ different pairs of balls. (Why?) Among these pairs, we are interested in pairs consisting only of black balls. There are $2 \cdot 3$ such pairs. Why? Hence,

$$P(B_1 \cap B_2) = \frac{2 \cdot 3}{4 \cdot 5} = \frac{2}{4} \cdot \frac{3}{5} = \frac{1}{2} \cdot \frac{3}{5}.$$

Thus we see that for the independent events B_1 and B_2, we have $P(B_1 \cap B_2) = P(B_1) \cdot P(B_2)$. The property can be generalized as follows.

Property

For any independent events E_1 and E_2,

$$P(E_1 \cap E_2) = P(E_1) \cdot P(E_2).$$

RESEARCH NOTE Students often will assign a higher probability to the **conjunction of two events** than to either of the two events individually. This conjunction fallacy occurs even if students have had a probability course. For example, students rate the probability of "being 55 **and** having a heart attack" as more likely than the probability of either "being 55" or "having a heart attack." An explanation for the error is that students may confuse the conjunction form (e.g., "being 55 and having a heart attack") with the conditional form (e.g., "had a heart attack given that they are over 55") (Kahneman and Tversky 1983).

The student page gives examples of problems that involve independent and dependent events. Answer question 1 at the bottom of the student page.

SAMPLE SCHOOL BOOK PAGE:

INDEPENDENT EVENTS

Lesson 11-15

Key Idea
To find the probability that two events will occur, you need to know whether the occurrence of one event has an effect on the occurrence of the other.

Vocabulary
- independent events
- dependent events

Think It Through
I need to **remember vocabulary terms.** Mutually exclusive events cannot happen at the same time. Independent and dependent events can happen at the same time.

Multiplying Probabilities

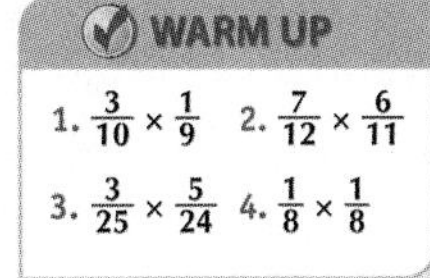

WARM UP
1. $\frac{3}{10} \times \frac{1}{9}$ 2. $\frac{7}{12} \times \frac{6}{11}$
3. $\frac{3}{25} \times \frac{5}{24}$ 4. $\frac{1}{8} \times \frac{1}{8}$

LEARN

How do you find the probability that two events will occur?

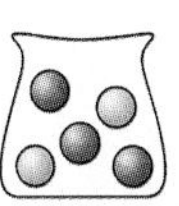

Roxanne draws one marble from this bag without looking. Then she draws a second marble.

Example A

What is the probability that Roxanne will get a red marble each time if she replaces the first marble before she draws the second one?

1st draw

$P(\text{red}) = \frac{3}{5}$

Since she replaces the first marble before the second draw, the outcome of the first draw has no effect on the outcome of the second draw. The two draws are **independent events.**

2nd draw

To find $P(\text{red, red})$, find the probability of each event and multiply.

$P(\text{red, red}) = P(\text{red on 1st draw}) \times P(\text{red on 2nd draw})$

$= \frac{3}{5} \times \frac{3}{5} = \frac{9}{25} = 36\%$

$P(\text{red}) = \frac{3}{5}$

Example B

What is the probability that Roxanne will get a red marble each time if she does not replace the first marble before she draws second one?

1st draw

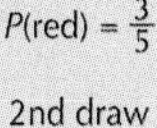

Since the outcome of the first draw affects the outcome of the second draw, the two draws are **dependent events.**

$P(\text{red}) = \frac{3}{5}$

2nd draw

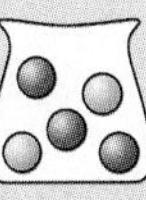

To find $P(\text{red, red})$, find the probability of each event and multiply.

$P(\text{red, red}) = P(\text{red on 1st draw}) \times P(\text{red on 2nd draw})$

$= \frac{3}{5} \times \frac{1}{2} = \frac{3}{10} = 30\%$

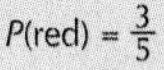

Talk About It

1. In Example B, why is the probability of the first draw different from the second draw?

Take It to the NET
More Examples
www.scottforesman.com

672

Source: Scott Foresman–Addison Wesley Mathematics, Grade 6, 2005 (p. 672).

NOW TRY THIS 7-3

In the following cartoon, assume that the events are independent and that there is a 30% chance of rain tonight and a 60% chance of rain tomorrow.

a. What is the probability that it will rain both times?
b. What is the probability that it will not rain either of the times?
c. What is the probability that it will rain exactly one of the times?
d. What is the probability that it will rain at least one of the times?
e. In real life, do you think that the events of rain tonight and rain tomorrow are independent events? Why or why not?

EXAMPLE 7-8

Figure 7-14 shows a box with 11 letters. Some letters are repeated. Suppose 4 letters are drawn at random from the box one-by-one without replacement. What is the probability of the outcome *BABY*, with the letters chosen in exactly the order given?

FIGURE 7-14

P R O B A B I L I T Y

Solution We do not need the entire tree diagram to find this probability because we are interested in only the branch leading to the outcome *BABY*. The portion needed is shown in Figure 7-15.

FIGURE 7-15

$$\xrightarrow{\frac{2}{11}} B \xrightarrow{\frac{1}{10}} A \xrightarrow{\frac{1}{9}} B \xrightarrow{\frac{1}{8}} Y \qquad \text{Probability of Outcome: } \frac{2}{11}\cdot\frac{1}{10}\cdot\frac{1}{9}\cdot\frac{1}{8} = \frac{2}{7920}$$

The probability of the first B is $\frac{2}{11}$ because there are 2 B's out of 11 letters. The probability of the second B is $\frac{1}{9}$ because there are 9 letters left after 1 B and 1 A have been chosen. Then, $P(BABY)$ is $\frac{2}{7920}$, as shown.

In Example 7-8, suppose 4 letters are drawn one-by-one from the box and the letters are replaced after each drawing. In this case, the branch needed to find $P(BABY)$ in the order drawn is pictured in Figure 7-16. Then, $P(BABY) = \left(\frac{2}{11}\right) \cdot \left(\frac{1}{11}\right) \cdot \left(\frac{2}{11}\right) \cdot \left(\frac{1}{11}\right)$, or $\frac{4}{14{,}641}$.

FIGURE 7-16

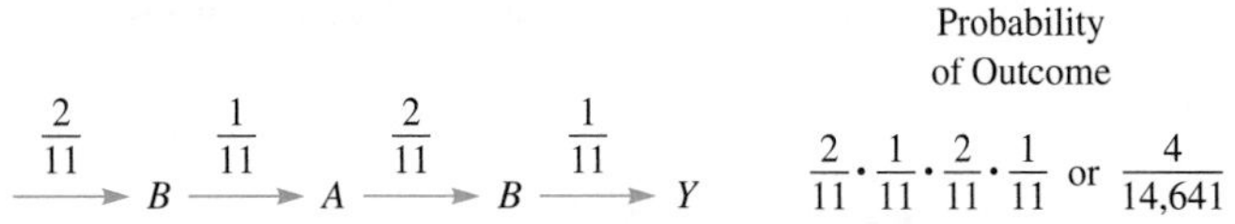

Probability of Outcome

$$\xrightarrow{\frac{2}{11}} B \xrightarrow{\frac{1}{11}} A \xrightarrow{\frac{2}{11}} B \xrightarrow{\frac{1}{11}} Y \qquad \frac{2}{11} \cdot \frac{1}{11} \cdot \frac{2}{11} \cdot \frac{1}{11} \text{ or } \frac{4}{14{,}641}$$

EXAMPLE 7-9

Consider the three boxes in Figure 7-17. A letter is drawn from box 1 and placed in box 2. Then, a letter is drawn from box 2 and placed in box 3. Finally, a letter is drawn from box 3. What is the probability that the letter drawn from box 3 is B?

FIGURE 7-17

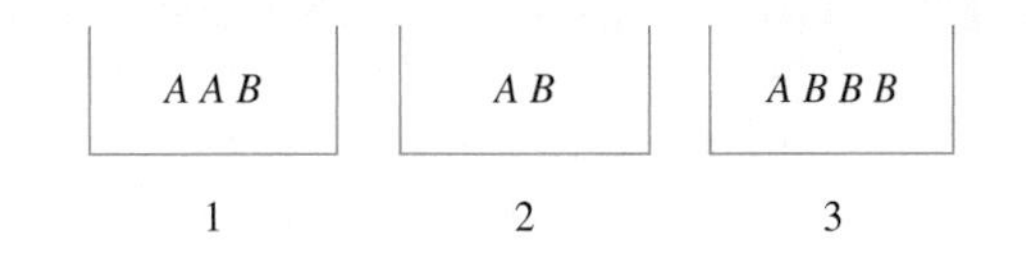

Solution A tree diagram for this experiment is given in Figure 7-18. Notice that the denominators in the second stage are 3 rather than 2 because in this stage, there are now three letters in box 2. The denominators in the third stage are 5 because in this stage, there are five letters in box 3. To find the probability that a B is drawn from box 3, add the probabilities for the outcomes AAB, ABB, BAB, and BBB that make up this event.

FIGURE 7-18

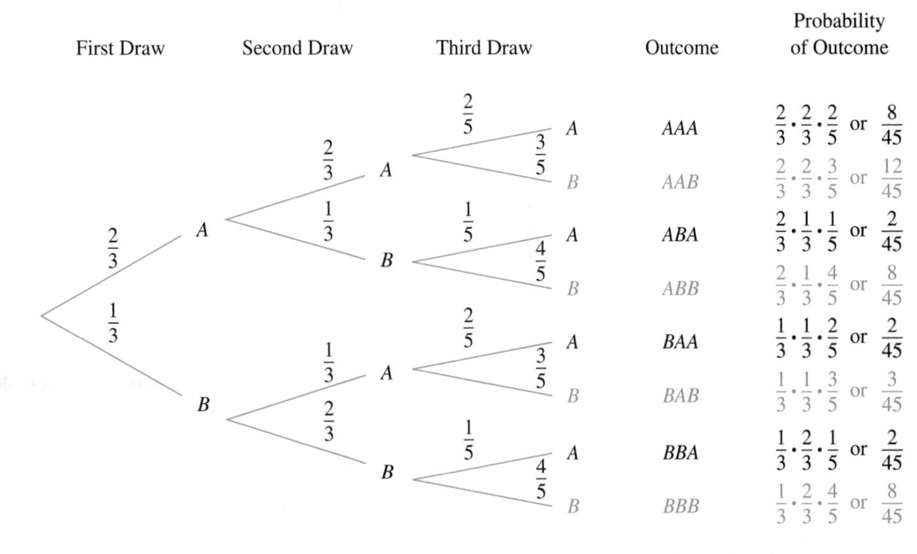

Thus, the probability of obtaining a B on the draw from box 3 in this experiment is

$$\frac{12}{45} + \frac{8}{45} + \frac{3}{45} + \frac{8}{45} = \frac{31}{45}.$$

NOW TRY THIS 7-4

Suppose that in Example 7–9 it is known that the letter A was drawn on the first draw. What is the probability that

a. The last letter drawn is a B?
b. The last letter drawn is an A?
c. The last two letters drawn will match, that is, 2 A's or 2 B's?

Modeling Games

We can use models to analyze games that involve probability. Consider the following game, which Arthur and Gwen play: There are two colored marbles and one white marble in a box. Gwen mixes the marbles, and Arthur draws two marbles at random without replacement. If the two marbles match, Arthur wins; otherwise, Gwen wins. Does each player have an equal chance of winning? We *develop a model* to analyze the game. One possible model is a tree diagram, as shown in Figure 7-19(a). Because the outcome ○ ○ can't happen, the tree diagram could also be shortened as in Figure 7-19(b).

FIGURE 7-19

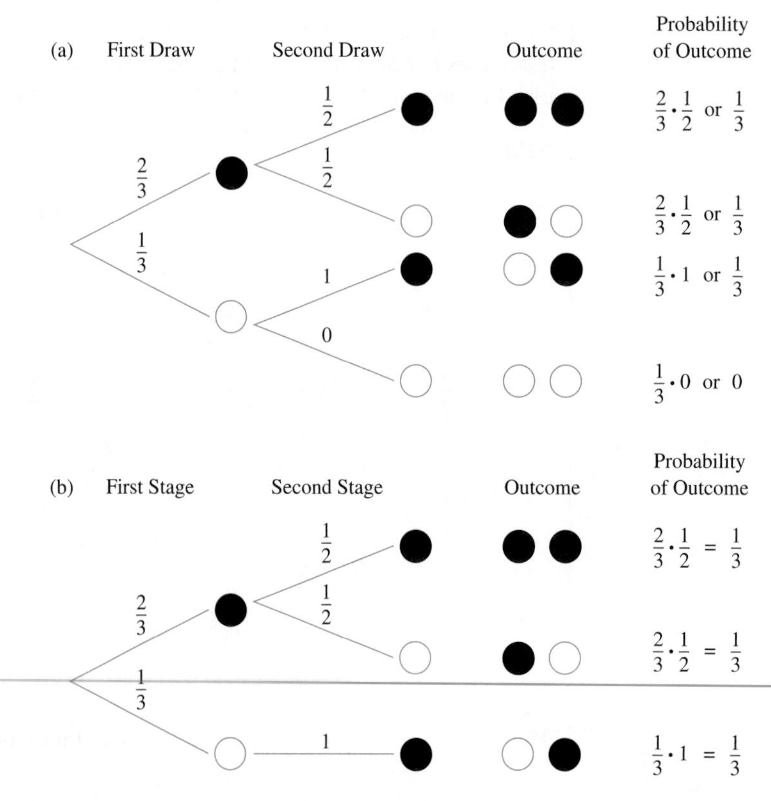

The probability that the marbles are the same color is $\frac{1}{3} + 0$, or $\frac{1}{3}$, and the probability that they are not the same color is $\frac{1}{3} + \frac{1}{3}$, or $\frac{2}{3}$. Because $\frac{1}{3} \neq \frac{2}{3}$, the players do not have the same chance of winning.

FIGURE 7-20

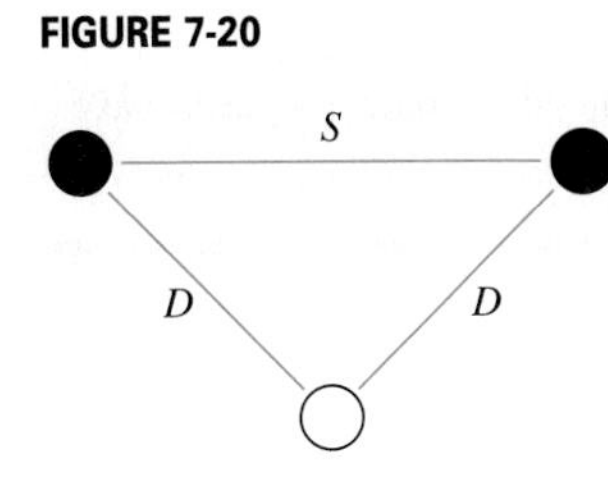

An alternative model for analyzing this game is given in Figure 7-20, where the colored and white marbles are shown along with the possible ways of drawing two marbles. Each line segment in the diagram represents one pair of marbles that could be drawn. *S* indicates that the marbles in the pair are the same color, and *D* indicates that the marbles are different colors. Because there are two *D*'s in Figure 7-20, we see that the probability of drawing two different-colored marbles is $\frac{2}{3}$. Likewise, the probability of drawing two marbles of the same color is $\frac{1}{3}$. Because $\frac{2}{3} \neq \frac{1}{3}$, the players do not have an equal chance of winning.

Will adding another white marble give each player an equal chance of winning? With two white and two colored marbles, we have the model in Figure 7-21. Therefore, $P(D) = \frac{4}{6}$, or $\frac{2}{3}$, and $P(S) = \frac{2}{6}$, or $\frac{1}{3}$. We see that adding a white marble does not change the probabilities.

FIGURE 7-21

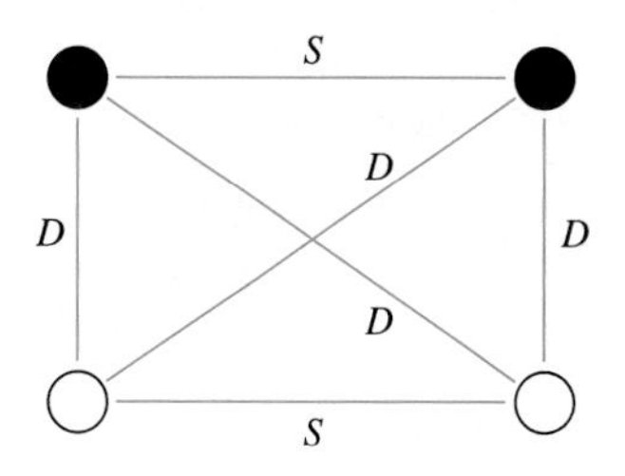

Next, consider a game with the same rules but using three colored marbles and one white marble. Figure 7-22 shows a model for this situation.

FIGURE 7-22

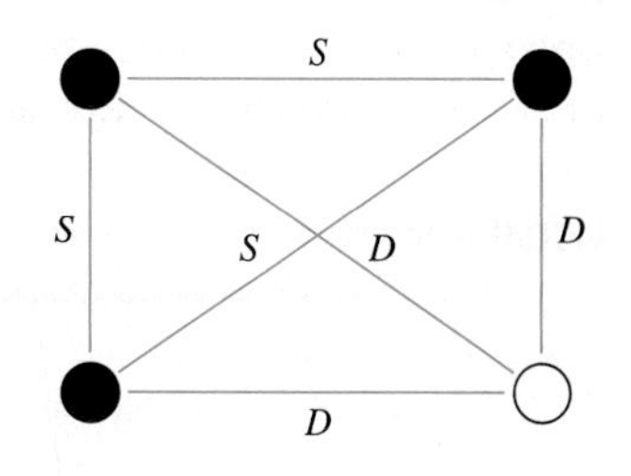

Thus, the probability of drawing two marbles of the same color is $\frac{3}{6}$, and the probability of drawing two marbles of different colors is $\frac{3}{6}$. Finally, we have a game in which each player has an equal chance of winning.

NOW TRY THIS 7-5

Referring to the games described on the previous page, answer the following:

a. Does each player have an equal chance of winning if only one white marble and one colored marble are used and the marble is replaced after the first draw?

b. Find games that involve different numbers of marbles in which each player has an equal chance of winning.

c. Find a pattern for the numbers of colored and white marbles that allow each player to have an equal chance of winning.

PROBLEM SOLVING — A String-Tying Game

In a party game, a child is handed six strings, as shown in Figure 7-23(a). Another child ties the top ends two at a time, forming three separate knots, and the bottom ends, forming three separate knots, as in Figure 7-23(b). If the strings form one closed ring, as in Figure 7-23(c), the child tying the knots wins a prize. What is the probability that the child wins a prize on the first try? Guess the probability before working the problem.

FIGURE 7-23

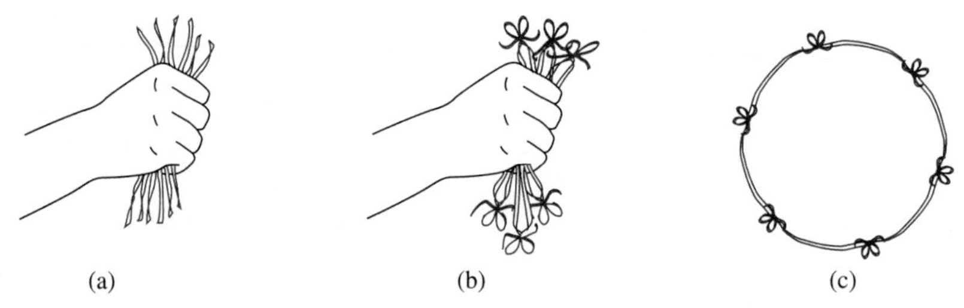

(a) (b) (c)

Understanding the Problem The problem is to determine the probability that one closed ring will be formed. One closed ring means that all six pieces are joined end-to-end to form one, and only one, ring, as shown in Figure 7-23(c).

Devising a Plan Figure 7-24(a) shows what happens when the ends of the strings of one set are tied in pairs at the top. Notice that no matter in what order those ends are tied, the result appears as in Figure 7-24(a).

FIGURE 7-24

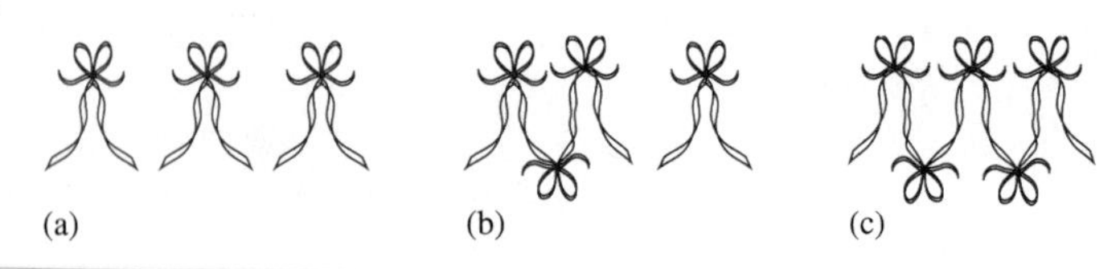

(a) (b) (c)

Then, the other ends are tied in a three-stage experiment. If we pick any string in the first stage, then there are five choices for its mate. Four of these choices are favorable choices for forming a ring. Why? Thus, the probability of forming a favorable first tie is $\frac{4}{5}$. Figure 7-24(b) shows a favorable tie at the first stage.

For any one of the remaining four strings, there are three choices for its mate. Two of these choices are favorable ones. Why? Thus, the probability of forming a favorable second tie is $\frac{2}{3}$. Figure 7-24(c) shows a favorable tie at the second stage.

Now, two ends remain. Since nothing can go wrong at the third stage, the probability of making a favorable tie is 1. If we use the probabilities completed at each stage and a single branch of a tree diagram, we can calculate the probability of performing three successful ties in a row and hence the probability of forming one closed ring.

Carrying Out the Plan If we let S represent a successful tie at each stage, then the branch of the tree with which we are concerned is the one shown in Figure 7-25.

FIGURE 7-25

First Tie	Second Tie	Third Tie
$\xrightarrow{\frac{4}{5}} S$	$\xrightarrow{\frac{2}{3}} S$	$\xrightarrow{\frac{1}{1}} S$

Thus, the probability of forming one ring is $P(\text{ring}) = \frac{4}{5} \cdot \frac{2}{3} \cdot \frac{1}{1} = \frac{8}{15} = 0.5\overline{3}$.

Looking Back The probability that a child will form a ring on the first try is $\frac{8}{15}$. A class might simulate this problem several times with strings to see how the fraction of successes compares with the theoretical probability of $\frac{8}{15}$.

Related problems that could be posed for solution include the following:

1. If a child fails to get a ring 10 times in a row, the child may not play again. What is the probability of such a streak of bad luck?
2. If the number of strings is reduced to three and the rule is that an upper end must be tied to a lower end, what is the probability of a single ring?
3. If the number of strings is three, but an upper end can be tied to either an upper or a lower end, what is the probability of a single ring?
4. What is the probability of forming three rings in the original problem?
5. What is the probability of forming two rings in the original problem?

Geometric Probability (Area Models)

FIGURE 7-26

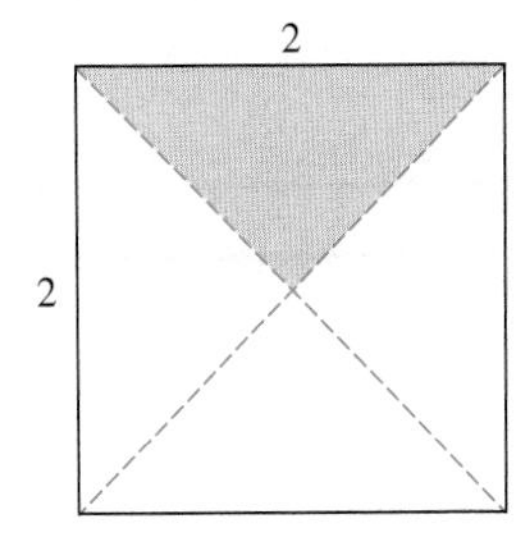

A probability model that uses geometric shapes is an *area model*. When area models are used to determine probabilities geometrically, outcomes are associated with points chosen at random in a geometric region that represents the sample space. For example, suppose we throw darts at a square target 2 units long on a side and divided into four congruent triangles, as shown in Figure 7-26. If the dart must hit the target somewhere and if all spots can be hit with equal probability, what is the probability that the dart will land in the shaded region? The entire target, which has an area of 4 square units, represents the sample space. The shaded area is the event of a successful toss. The area of the shaded part is $\frac{1}{4}$ of the sample space. Thus, the probability of the dart's landing in the shaded region is the ratio of the area of the event to the area of the sample space, or $\frac{1}{4}$.

PROBLEM SOLVING **A Quiz-Show Game**

On a quiz show, a contestant stands at the entrance to a maze that opens into two rooms, labeled A and B in Figure 7-27. The master of ceremonies' assistant is to place a new car in one room and a donkey in the other. The contestant must walk through the maze into one of the rooms and will win whatever is in that room. If the contestant makes each decision in the maze at random, in which room should the assistant place the car to give the contestant the best chance to win?

FIGURE 7-27

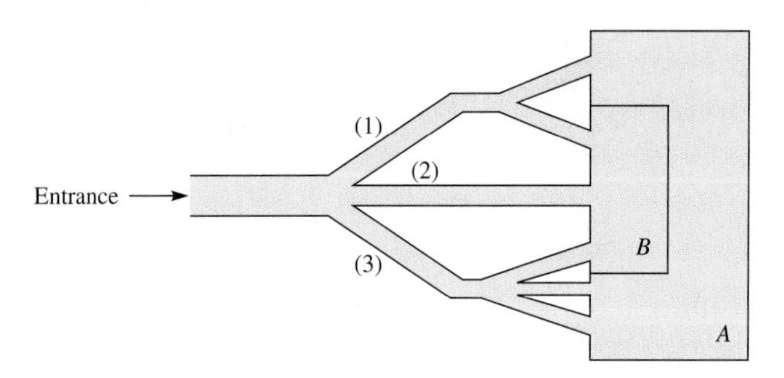

Understanding the Problem The contestant must first choose one of the paths marked 1, 2, or 3 and then choose other paths as she proceeds through the maze. To determine the room the contestant is most likely to choose, the assistant must be able to determine the probability of the contestant's reaching each room.

Devising a Plan One way to determine where the car should be placed is to *model the choices with a tree diagram* and to compute the probabilities along the branches of the tree.

Carrying Out the Plan A tree diagram for the maze is shown in Figure 7-28, along with the possible outcomes and the probabilities of each branch.

FIGURE 7-28

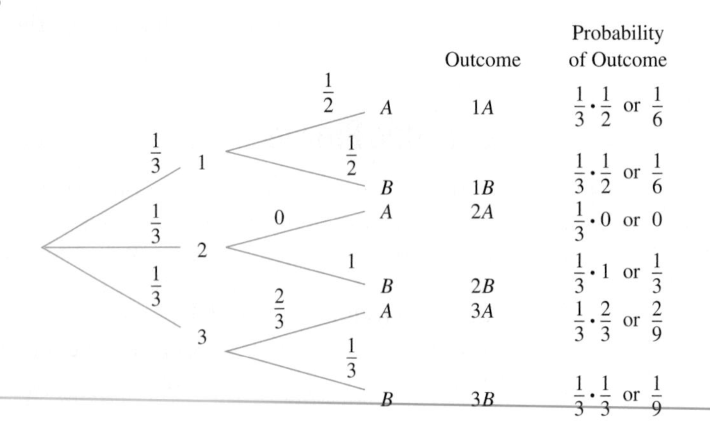

The probability that room A is chosen is $\frac{1}{6} + 0 + \frac{2}{9} = \frac{7}{18}$. Hence the probability that room B is chosen is $1 - \frac{7}{18} = \frac{11}{18}$. Thus room B has the greater probability of being chosen. This is where the car should be placed for the contestant to have the best chance of winning it.

Looking Back An alternative model for this problem and for many probability problems is an area model. The rectangle in Figure 7-29(a) represents the first three choices that the contestant can make. Because each choice is equally likely, each is represented by an equal area. If the contestant chooses the upper path, then rooms *A* and *B* have an equal chance of being chosen. If she chooses the middle path, then only room *B* can be entered. If she chooses the lower path, then room *A* is entered $\frac{2}{3}$ of the time. This can be expressed in terms of the area model shown in Figure 7-29(b). Dividing the rectangle into pieces of equal area, we obtain the model in Figure 7-29(c), in which the area representing room *B* is shaded. Because the area representing room *B* is greater than the area representing room *A*, room *B* has the greater probability of being chosen. Figure 7-29(c) can enable us to find the probability of choosing room *B*. Because the shaded area consists of 11 rectangles out of a total of 18 rectangles, the probability of choosing room *B* is $\frac{11}{18}$. We can vary the problem by changing the maze or by changing the locations of the rooms.

FIGURE 7-29

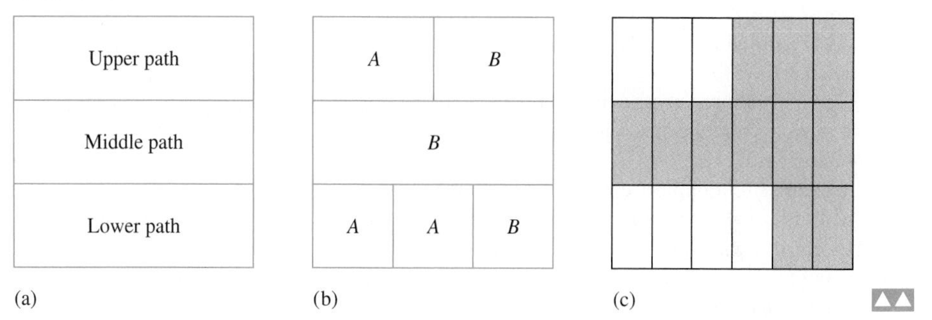

ONGOING ASSESSMENT 7-2

1. a. A box contains three white balls and two red balls. A ball is drawn at random from the box and not replaced. Then a second ball is drawn from the box. Draw a tree diagram for this experiment and find the probability that the two balls are of different colors.

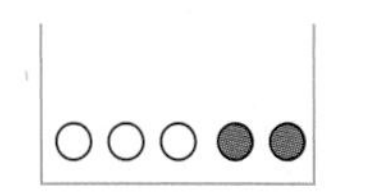

   b. Suppose that a ball is drawn at random from the box in part (a), its color is recorded, and then it is put back in the box. Draw a tree diagram for this experiment and find the probability that the two balls are of different colors.

2. Suppose an experiment consists of spinning *X* and then spinning *Y*, as follows:

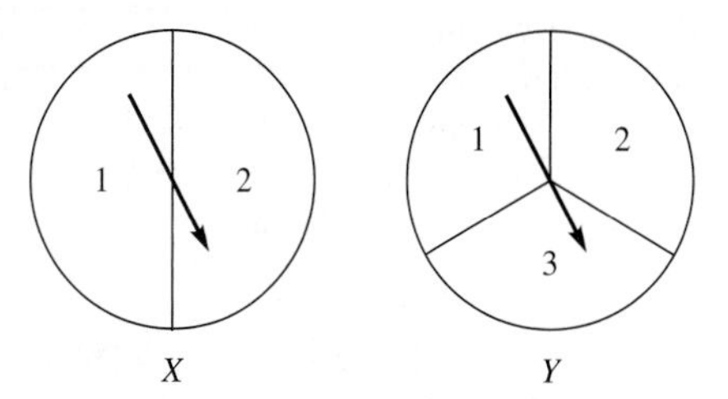

Find the following:

   a. The sample space *S* for the experiment
   b. The event *A* consisting of outcomes from spinning an even number followed by an even number

c. The event B consisting of outcomes from spinning at least one 2

d. The event C consisting of outcomes from spinning exactly one 2

3. A box contains six letters, shown as follows. What is the probability of the outcome *DAN* in that order if three letters are drawn one-by-one

a. With replacement?

b. Without replacement?

| R A N D O M |

4. Following are three boxes containing letters:

M A T H	A N D	H I S T O R Y
1	2	3

a. From box 1, three letters are drawn one-by-one without replacement and recorded in order. What is the probability that the outcome is *HAT*?

b. From box 1, three letters are drawn one-by-one with replacement and recorded in order. What is the probability that the outcome is *HAT*?

c. One letter is drawn at random from box 1, then another from box 2, and then another from box 3, with the results recorded in order. What is the probability that the outcome is *HAT*?

d. If a box is chosen at random and then a letter is drawn at random from the box, what is the probability that the outcome is *A*?

5. An executive committee consisted of 10 members: 4 women and 6 men. Three members were selected at random to be sent to a meeting in Hawaii. A blindfolded woman drew 3 of the 10 names from a hat. All 3 names drawn were women's. What was the probability of such luck?

6. Two boxes with letters follow. You are to choose a box and draw three letters at random, one-by-one, without replacement. If the outcome is SOS, you win a prize.

S O S	S O S S O S
1	2

a. Which box should you choose?

b. Which box would you choose if the letters are to be drawn with replacement?

7. Following are three boxes containing balls. Draw a ball from box 1 and place it in box 2. Then draw a ball from box 2 and place it in box 3. Finally, draw a ball from box 3.

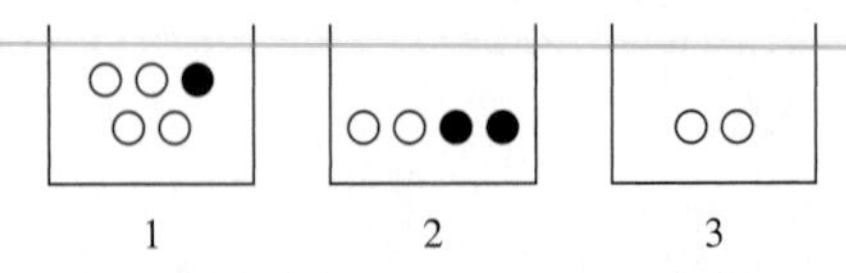

a. What is the probability that the last ball, drawn from box 3, is white?

b. What is the probability that the last ball drawn is colored?

8. An assembly line has two inspectors. The probability that the first inspector will miss a defective item is 0.05. If the defective item passes the first inspector, the probability that the second inspector will miss it is 0.01. What is the probability that a defective item will pass by both inspectors?

9. Following are two boxes containing colored and white balls. A ball is drawn at random from box 1. Then a ball is drawn at random from box 2, and the colors of balls from both boxes are recorded in order.

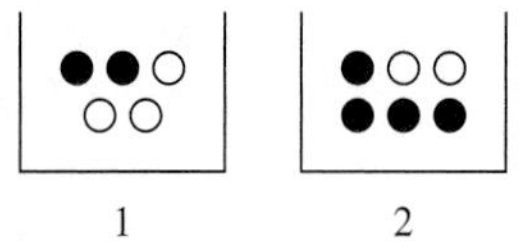

Find each of the following:

a. The probability of two white balls

b. The probability of at least one colored ball

c. The probability of at most one colored ball

d. The probability of ● ○ or ○ ●

10. A penny, a nickel, a dime, and a quarter are tossed. What is the probability of obtaining at least three heads?

11. Assume the probability is $\frac{1}{2}$ that a child born is a boy. What is the probability that if a family is going to have four children, they will all be boys?

12. Brittany is going to ascend a four-step staircase. At any time, she is just as likely to stride up one step or two steps. Find the probability that she will ascend the four steps in

a. Two strides   b. Three strides   c. Four strides

13. A box contains five slips of paper. Each slip has one of the numbers 4, 6, 7, 8, or 9 written on it. There are two players for the game. The first player reaches into the box and draws two slips and adds the two numbers. If the sum is even, the player wins. If the sum is odd, the player loses.

a. What is the probability that the first player wins?

b. Does the probability change if the two numbers are multiplied? Explain.

14. Suppose we spin the following spinner with the first spin giving the numerator and the second spin giving the denominator of a fraction. What is the probability that the fraction will be greater than $1\frac{1}{2}$?

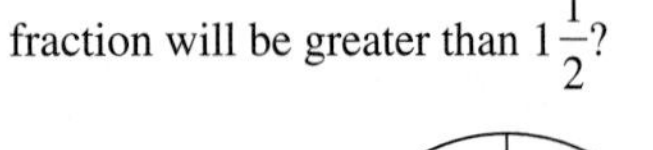

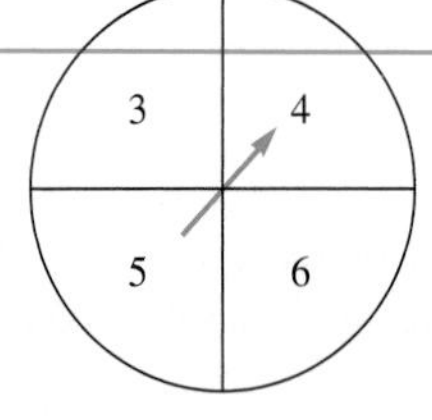

15. The following shows the numbers of symbols on each of the three dials of a standard slot machine:

Symbol	Dial 1	Dial 2	Dial 3
Bar	1	3	1
Bell	1	3	3
Plum	5	1	5
Orange	3	6	7
Cherry	7	7	0
Lemon	3	0	4
Total	20	20	20

Find the probability for each of the following:

a. Three plums **b.** Three oranges
c. Three lemons **d.** No plums

16. You play a game in which you first choose one of the two spinners shown. You then spin your spinner and a second person spins the other spinner. The one with the greater number wins.

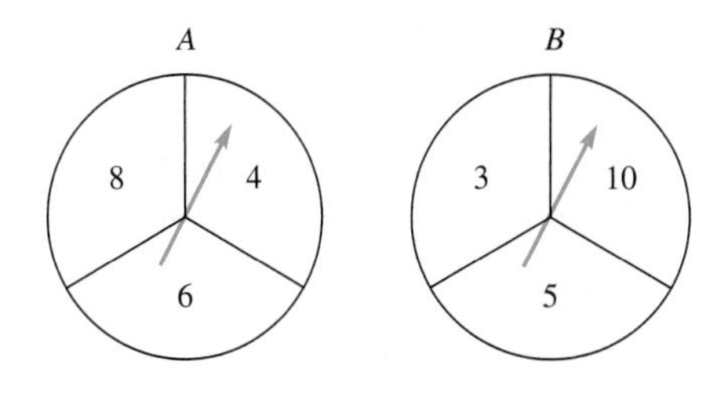

a. Which spinner should you choose? Why?

b. Notice that the sum of the numbers on each spinner is 18. Design two spinners with unequal sums so that choosing the spinner with the least sum will give the player a greater probability of winning.

17. An experiment consists of spinning the spinner shown and then flipping a coin with sides numbered 1 and 2.

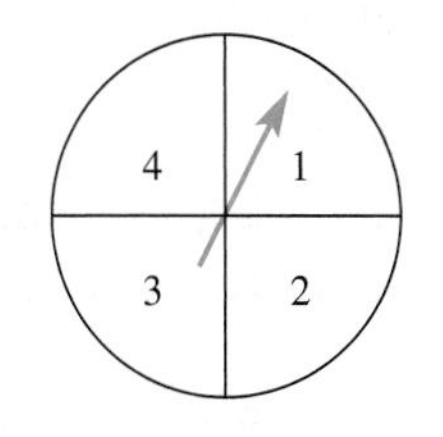

What is the probability that

a. The number on the spinner will be greater than the number on the coin?

b. The outcome will consist of two consecutive integers in any order?

18. If a person takes a five-question true-false test, what is the probability that the score is 100% correct if the person guesses on every question?

19. Rattlesnake and Paxson Colleges play four games against each other in a chess tournament. Rob Fisher, the chess whiz from Paxson, withdrew from the tournament, so the probabilities that Rattlesnake and Paxson will win each game are $\frac{2}{3}$ and $\frac{1}{3}$, respectively. Determine the following probabilities:

a. Paxson loses all four games.

b. The match is a draw with each school winning two games.

20. The combinations on the lockers at the high school consist of three numbers, each ranging from 0 to 39. If a combination is chosen at random, what is the probability that the first two numbers are multiples of 9 and the third number is a multiple of 4?

21. The following box contains the 11 letters shown. The letters are drawn one-by-one without replacement, and the results are recorded in order. Find the probability of the outcome MISSISSIPPI.

22. Consider the following dartboard: (Assume that all quadrilaterals are squares and that the x's represent equal measures.)

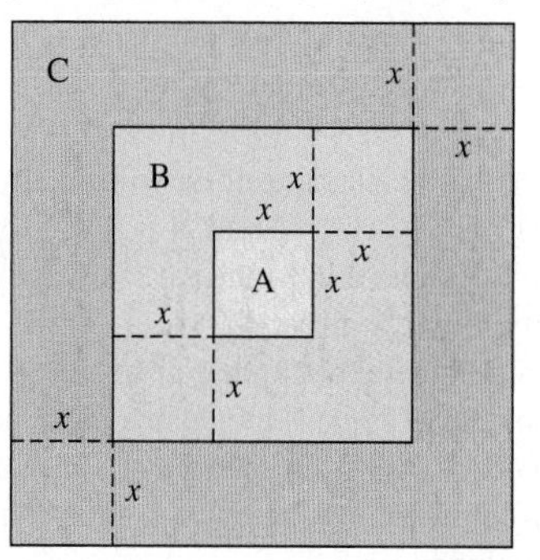

If a dart may hit any point on the board with equal probability, what is the probability that it will land in

a. Section *A*? **b.** Section *B*? c. Section *C*?

23. In the following square dartboard, suppose a dart is equally likely to land in any region of the board.

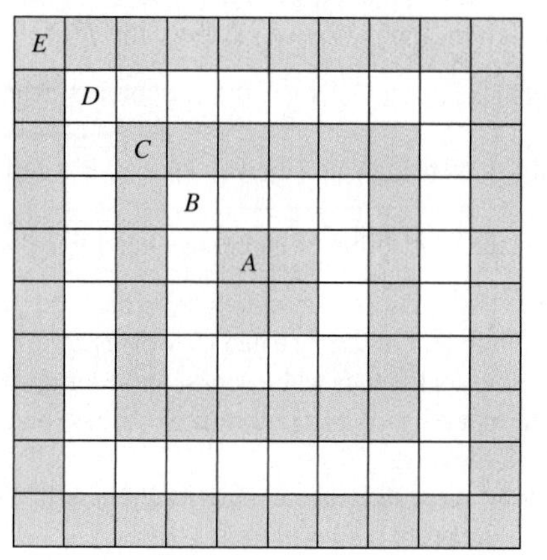

Points are given as follows:

Region	Points
A	10
B	8
C	6
D	4
E	2

a. What is the total area of the board?

b. What is the probability of a dart's landing in each region of the board?

c. If two darts are tossed, what is the probability of scoring 20 points?

d. What is the probability that the dart will land in neither *D* nor *E*?

24. The land area of Earth is approximately 57,500,000 mi^2. The water area of Earth is approximately 139,600,000 mi^2. If a meteor lands at random on the planet, what is the probability, to the nearest tenth, that it will hit water?

25. An electric clock is stopped by a power failure. What is the probability that the second hand is stopped between the 3 and the 4?

26. A husband and wife discover that there is a 10% probability of their passing on a hereditary disease to one of their children. If they plan to have three children, what is the probability that at least one child will inherit the disease?

27. Let $A = \{x | {}^-1 < x < 1\}$ and $B = \{x | {}^-3 < x < 2\}$. If a real number is picked at random from set B, what is the probability it will be in set A?

28. At a certain hospital, 40 patients have lung cancer, 30 patients smoke, and 25 have lung cancer and smoke. Suppose the hospital contains 200 patients. If a patient chosen at random is known to smoke, what is the probability that the patient has lung cancer?

29. There are 40 employees in a certain firm. We know that 28 of these employees are males, 2 of these males are secretaries, and 10 secretaries are employed by the firm. What is the probability that an employee chosen at random is a secretary, given that the person is a male?

30. In a certain population of caribou, the probability of an animal's being sickly is $\frac{1}{20}$. If a caribou is sickly, the probability of its being eaten by wolves is $\frac{1}{3}$. If a caribou is not sickly, the probability of its being eaten by wolves is $\frac{1}{150}$. If a caribou is chosen at random from the herd, what is the probability that it will be eaten by wolves?

31. Four blue socks, four white socks, and four gray socks are mixed in a drawer. You pull out two socks, one at a time, without looking.

a. Draw a tree diagram along with the possible outcomes and the probabilities of each branch.

b. What is the probability of getting a pair of socks of the same color?

c. What is the probability of getting two gray socks?

d. Suppose that, instead of pulling out two socks, you pull out four socks. What is now the probability of getting two socks of the same color?

32. Solve the Quiz-Show Game in this section by replacing Figure 7-27 with the following maze:

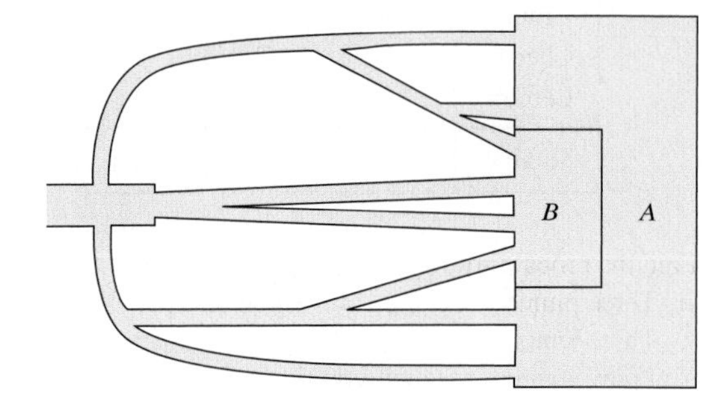

33. When you toss a quarter 4 times, what is the probability that you get at least

a. As many heads as tails?

b. As many tails as heads?

34. A manufacturer found that among 500 randomly selected smoke detectors only 450 worked properly. Based on this information, how many smoke detectors would you have to install to be sure that the probability that at least one of them will work will be greater than 99.99%?

★ 35. Carolyn will win a large prize if she wins two tennis games in a row out of three games. She is to play alternately against Billie and Bobby. She may choose to play Billie-Bobby-Billie or Bobby-Billie-Bobby. She wins against Billie 50% of the time and against Bobby 80% of the time. Which alternative should she choose, and why?

★ **36.** Jane has two tennis serves, a hard serve and a soft serve. Her hard serve is in (a good serve) 50% of the time, and her soft serve is in (good) 75% of the time. If her hard serve is in, she wins 75% of her points. If her soft serve is in, she wins 50% of her points. Since she is allowed to reserve one time if her first serve is out, what should her serving strategy be? That is, should she serve hard followed by soft; both hard; soft followed by hard; or both soft?

Communication

37. Jim rolled a fair die 5 times and obtained a 3 every time. He concluded that on the next roll, a 3 is more likely to occur than the other numbers. Explain whether this is true.

38. A witness to a crime observed that the criminal had blond hair and blue eyes and drove a red car. When the police look for a suspect, is the probability greater that they will find someone with blond hair and blue eyes or that they will find someone with blond hair and blue eyes who drives a red car? Explain your answer.

39. You are given three white balls, one red ball, and two identical boxes. You are asked to distribute the balls in the boxes in any way you like. You then are asked to select a box (after the boxes have been shuffled) and to pick a ball at random from that box. If the ball is red, you win a prize. How should you distribute the balls in the boxes to maximize your chances of winning? Justify your reasoning.

Open-Ended

40. Make up a game in which the players have an equal chance of winning and that involves rolling two regular dice.

41. How can the faces of two cubes be numbered so that when they are rolled, the resulting sum is a number 1 to 12 inclusive and each sum has the same probability?

42. Use graph paper to design a dartboard such that the probability of hitting a certain part of the board is $\frac{3}{5}$. Explain your reasoning.

Cooperative Learning

43. Use two spinners divided into four equal areas with the areas numbered 1–4. A player spins both spinners and computes the product of the two numbers. If the product is 1, 2, 3, or 4, player A wins and places a marker on the appropriate square on the game board shown. In the same way, player B wins if the product is 6, 8, 9, 12, or 16 and places a marker in the appropriate square. The game ends when one player crosses the finish line. Each player receives 1 point for each marker, and the game winner is the person with the most points when the game ends.

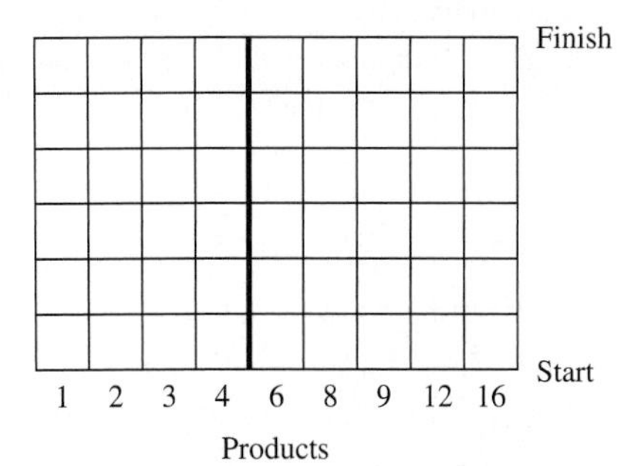

a. Do you think the game looks fair? Why or why not? If not, who do you think has the best chance of winning?

b. Play the game to see if it seems fair. Do you think it is fair based on playing it? Why?

c. Complete the following table to determine possible products.

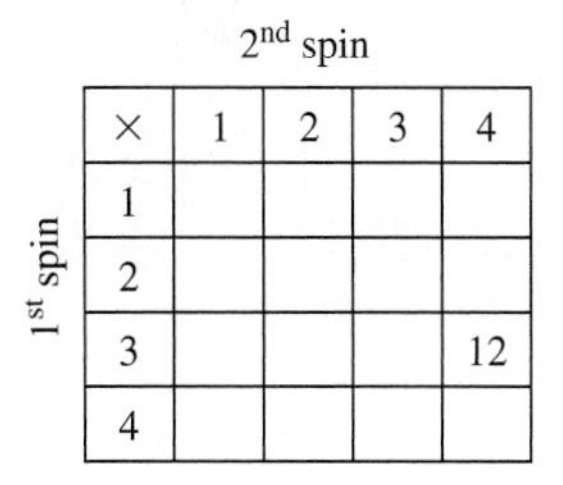
2nd spin (columns); 1st spin (rows)

×	1	2	3	4
1				
2				
3				12
4				

d. Based on the table in part (c), is this a fair game? Explain why.

e. Replace one of the spinners in the preceding game with one divided into six equal areas with the numbers 1, 2, 3, 4, 5, and 6 in the regions. Design a similar game and decide what products each player should use so that this is a fair game. Explain how your game is fair.

44. Play the following game in pairs. One player chooses one of four equally likely outcomes from the sample space {*HH, HT, TH, TT*}, obtained by tossing a fair coin twice. The other player then chooses one of the other outcomes. A coin is flipped until either player's choice appears. For example, the first player chooses *TT* and the second player chooses *HT*. If the first two flips yield *TH*, then no one wins and the game continues. If, after five flips, the string *THHHT* appears, the second player is the winner because the sequence *HT* finally appeared. Play the game 10 times. Does each player appear to have the same chance of winning? Analyze the game for the case in which the first player chooses *TT* and the second *HT*, and explain whether the game is fair. (*Hint*: Find the probability that the first player wins the game by showing that the first player will win if and only if "tails" appears on the first and on the second flip.)

45. Consider the three spinners *A*, *B*, and *C* shown in the following figure:

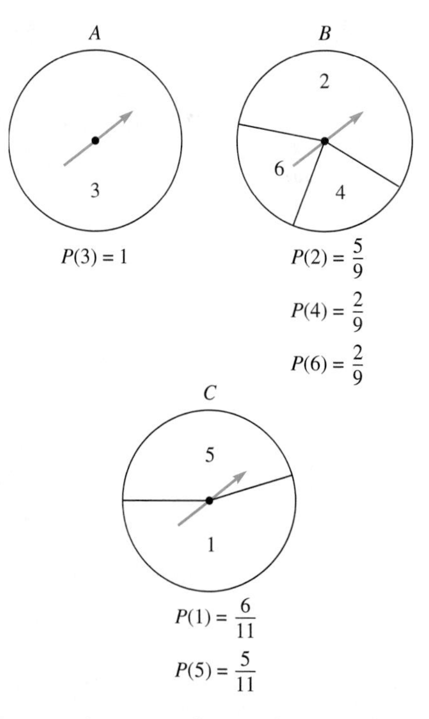

a. Suppose there are only two players and that the first player chooses a spinner, then the second player chooses a different spinner, and each person spins his or her spinner, with the highest number winning. Play the game several times to get a feeling for it. Determine if each player appears to have the same chances of winning the game. If not, which spinner should you choose in order to win?

b. This time play the same game with three players. If each player must choose a different spinner, is the winning strategy the same as it was in (a)? Why or why not?

Review Problems

46. Match the following phrase to the probability that describes it:

a. A certain event	**(i)** $\frac{1}{1000}$
b. An impossible event	**(ii)** $\frac{999}{1000}$
c. A very likely event	**(iii)** 0
d. An unlikely event	**(iv)** $\frac{1}{2}$
e. A 50% chance	**(v)** 1

47. A date in the month of April is chosen at random. Find the probability of the date's being each of the following:
 a. April 7
 b. April 31
 c. Before April 20

Third International Mathematics and Science Study (TIMSS) Questions

The figure below shows a spinner with 24 sectors. When someone spins the arrow, it is equally likely to stop on any sector.

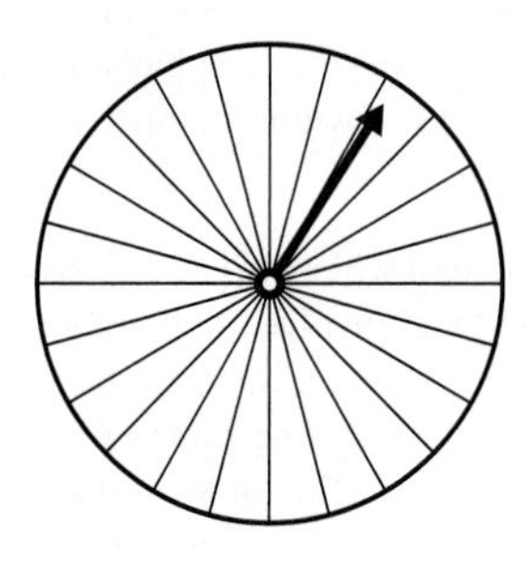

$\frac{1}{8}$ of the sectors are blue, $\frac{1}{24}$ are purple, $\frac{1}{2}$ are orange, and $\frac{1}{3}$ are red. If a person spins the arrow, on which color sector is the spinner LEAST likely to stop?

a. blue
b. purple
c. orange
d. red

TIMSS, Grade 8, 2003

BRAIN TEASER

Bradley Efron, a Stanford University statistician, designed a set of nonstandard dice whose faces are numbered as shown in Figure 7-30. The dice are to be used in a game in which each player chooses a die and then rolls it. Whoever rolls the greatest number is the winner. What strategy should you use so that you have the best chance of winning this game?

FIGURE 7-30

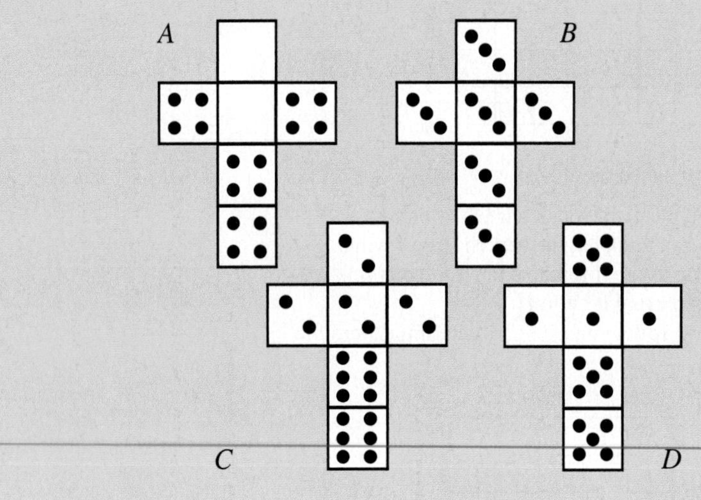

7-3 Using Simulations in Probability

simulation

Students can often use simulations to study phenomena too complex to analyze by other means. A **simulation** is a technique used to act out a problem by conducting experiments whose outcomes are analogous to the original problem. Using simulations, students can estimate rather than determine probabilities theoretically. Simulations can take various forms. For example, consider the simulation on the student page. In this case, the students use a simulation with a globe to find the probability that a meteorite will hit land. Follow the instructions on the student page and then answer the questions in Now Try This 7-6.

NOW TRY THIS 7-6

a. After doing the simulation on the student page, what percentage of Earth do you think is covered by land?

b. What is the probability that a meteorite will hit water?

c. It is estimated that approximately 500 meteorites hit Earth each year. How many would you expect to hit land each year?

In the *Principles and Standards*, we find the following:

If simulations are used, teachers need to help students understand what the simulation data represent and how they relate to the problem situation, such as flipping coins. (p. 254)

Suppose we want to simulate the results of flipping a coin 100 times. We could do this using random digits, as in Table 7-3. Random-digit tables are lists of digits selected at random, often by a computer or calculator. To simulate the coin toss, pick a number at random to start and then read across the table, letting an even digit represent heads and an odd digit represent tails. Continue this process for 100 digits. The simulated probability of heads is the ratio of the number of even digits found (heads) to 100. For example, if we choose the top two rows in Table 7-3 and use the first 100 numbers, we find that there are 44 even numbers. Because even numbers represent heads, the simulated probability of tossing a head is $P(H) = \frac{44}{100}$. Notice that we did not obtain the theoretical probability of $\frac{1}{2}$. However, if the number of random digits chosen was much greater, then the simulated probability should be closer to the theoretical probability.

Similarly, to simulate the probability of a couple's having two girls (*GG*) in an expected family, we could use the random-digit table with an even digit representing a girl and an odd digit representing a boy. Because there are two children, we need to consider pairs of digits. If we examine 100 pairs, then the simulated probability of *GG* will be the number of pairs of even digits divided by 100, the total number of pairs considered.

SAMPLE SCHOOL BOOK PAGE:

GEOMETRIC PROBABILITY

GOAL

LEARN HOW TO...
- find geometric probabilities
- use probability to predict

AS YOU...
- describe the chance of an object falling in a particular area

KEY TERM
- geometric probability

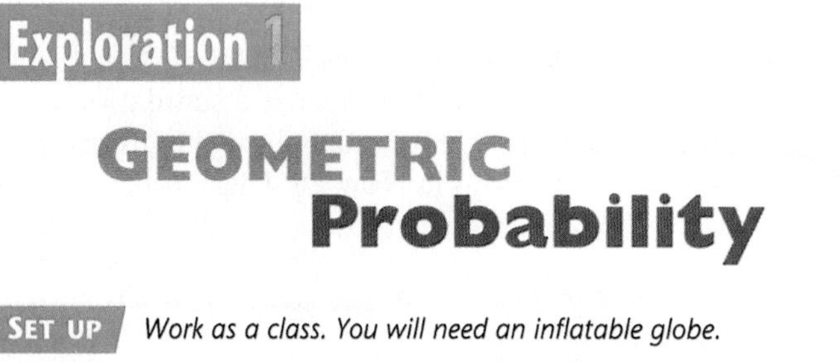

SET UP *Work as a class. You will need an inflatable globe.*

▶ **A meteorite has an equally likely chance of hitting anywhere on Earth. To find the probability that a meteorite will hit land you will conduct an experiment with a globe.**

2 **Try This as a Class** Follow the steps below to simulate a meteorite falling to Earth.

First Make a table like the one shown to record the results of your experiment.

Next Carefully toss an inflated globe from one person to another. Each time the globe is caught, make a tally mark to record whether the left index finger is touching land or touching water before tossing it again. Record the results of 30 tosses.

Then Complete your table by calculating the percent of the tosses that hit land and the percent that hit water.

	Tally	Number of tosses	Percent of tosses
Land			
Water			
Total		30	

FOR ◀ HELP with *finding probabilities,* see **MODULE 4, p. 242**

Source: Math Thematics, Book 1, McDougal Littell, 2005 (p. 566).

Table 7-3 Random Digits

36422	93239	76046	81114	77412	86557	19549	98473	15221	87856
78496	47197	37961	67568	14861	61077	85210	51264	49975	71785
95384	59596	05081	39968	80495	00192	94679	18307	16265	48888
37957	89199	10816	24260	52302	69592	55019	94127	71721	70673
31422	27529	95051	83157	96377	33723	52902	51302	86370	50452
07443	15346	40653	84238	24430	88834	77318	07486	33950	61598
41348	86255	92715	96656	49693	99286	83447	20215	16040	41085
12398	95111	45663	55020	57159	58010	43162	98878	73337	35571
77229	92095	44305	09285	73256	02968	31129	66588	48126	52700
61175	53014	60304	13976	96312	42442	96713	43940	92516	81421
16825	27482	97858	05642	88047	68960	52991	67703	29805	42701
84656	03089	05166	67571	25545	26603	40243	55482	38341	97782
03872	31767	23729	89523	73654	24626	78393	77172	41328	95633
40488	70426	04034	46618	55102	93408	10965	69744	80766	14889
98322	25528	43808	05935	78338	77881	90139	72375	50624	91385
13366	52764	02407	14202	74172	58770	65348	24115	44277	96735
86711	27764	86789	43800	87582	09298	17880	75507	35217	08352
53886	50358	62738	91783	71944	90221	79403	75139	09102	77826
99348	21186	42266	01531	44325	61042	13453	61917	90426	12437
49985	08787	59448	82680	52929	19077	98518	06251	58451	91140
49807	32863	69984	20102	09523	47827	08374	79849	19352	62726
46569	00365	23591	44317	55054	99835	20633	66215	46668	53587
09988	44203	43532	54538	16619	45444	11957	69184	98398	96508
32916	00567	82881	59753	54761	39404	90756	91760	18698	42852
93285	32297	27254	27198	99093	97821	46277	10439	30389	45372
03222	39951	12738	50303	25017	84207	52123	88637	19369	58289
87002	61789	96250	99337	14144	00027	43542	87030	14773	73087
68840	94259	01961	42552	91843	33855	00824	48733	81297	80411
88323	28828	64765	08244	53077	50897	91937	08871	91517	19668
55170	71062	64159	79364	53088	21536	39451	95649	65256	23950

NOW TRY THIS 7-7

a. Use the random-digit table to estimate the probability that in a family of three, there are two girls and one boy.

b. Determine the theoretical probability of the family's having two girls and one boy and compare the answer to the simulated probability in (a).

c. Should the answers in (a) and (b) always be exactly the same? Why? How can you make sure that the answers are approximately the same?

EXAMPLE 7-10

A baseball player, Reggie, has a batting average of 0.400; that is, his probability of getting a hit on any particular time at bat is 0.400. Estimate the probability that he will get at least one hit in his next three times at bat.

Solution We use a random-digit table to simulate this example. We choose a starting point and place the random digits in groups of three. Because Reggie's probability of getting a hit on any particular time at bat is 0.400, we could use the occurrence of four numbers from 0 through 9 to represent a hit. Suppose a hit is represented by the digits 0, 1, 2, and 3. At least one hit is obtained in three times at bat if, in any sequence of three digits, a 0, 1, 2, or 3 appears. Data for 50 trials are given next:

780	862	760	580	783	720	590	506	021	366
848	118	073	077	042	254	063	667	374	153
377	883	573	683	780	115	662	591	685	274
279	652	754	909	754	892	310	673	964	351
803	034	799	915	059	006	774	640	298	961

We see that a 0, 1, 2, or 3 appears in 42 out of the 50 trials. Thus, an estimate for the probability of at least one hit on Reggie's next three times at bat is $\frac{42}{50}$. Try to determine the theoretical probability for this experiment.

From a random sample, we may be able to deduce information about the population from which the sample was taken. To see how this can be done, consider Example 7–11.

EXAMPLE 7-11

To determine the number of fish in a certain pond, suppose we capture 300 fish, mark them, and throw them back into the pond. Suppose that the next day, 200 fish are caught and 20 of these are already marked. These 200 fish are then thrown back into the pond. Estimate how many fish are in the pond.

Solution Because 20 of the 200 fish are marked, we assume that $\frac{20}{200}$, or $\frac{1}{10}$, of the fish are marked. Thus, $\frac{1}{10}$ of the population is marked. If n represents the population, then $\frac{1}{10}n = 300$ and $n = 300 \cdot 10 = 3000$. Hence, an estimate for the fish population of the pond is 3000.

The following "Peanuts" cartoon suggests a simulation problem concerning chocolate chip cookies.

EXAMPLE 7-12

Suppose Lucy makes enough batter for exactly 100 chocolate chip cookies and mixes 100 chocolate chips into the batter. If the chips are distributed at random and Charlie Brown chooses a cookie at random from the 100 cookies, estimate the probability that it will contain exactly one chocolate chip.

Solution We can use a simulation to estimate the probability of choosing a cookie with exactly one chocolate chip. We construct a 10×10 grid, as shown in Figure 7-31(a), to represent the 100 cookies Lucy made. Each square (cookie) can be associated with some ordered pair, where the first component is for the horizontal scale and the second is for the vertical scale. For example, the squares (0, 2) and (5, 3) are pictured in Figure 7-31(a). Using the random-digit table, close your eyes and then take a pencil and point to one number to start. Look at the number and the number immediately following it. Consider these numbers an ordered pair and continue until you obtain 100 ordered pairs to represent the 100 cookies. For example, suppose we start at a 3 and the numbers following 3 are as follows:

39968 80495 00192...

Then the ordered pairs would be given as (3, 9), (9, 6), (8, 8), (0, 4), and so on. Use each pair of numbers as the coordinates for the square (cookie) and place a tally on the grid to represent each chip, as shown in Figure 7-31(b). We estimate the probability that a cookie has exactly one chip by counting the number of squares with exactly one tally and dividing by 100.

FIGURE 7-31

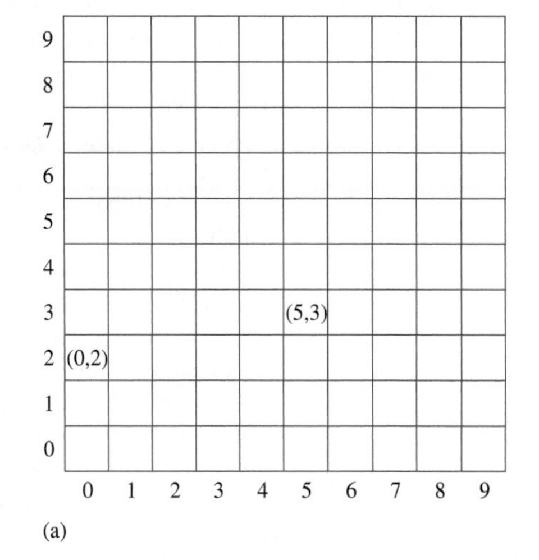

(a)

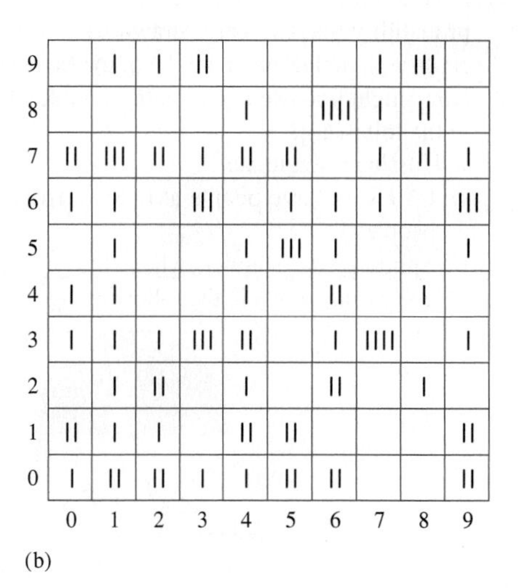

(b)

Table 7-4 shows the results of one simulation. Thus the estimate for the probability of Charlie Brown receiving a cookie with exactly one chip is $\frac{34}{100}$.

Table 7-4

Number of Chips	Number of Cookies
0	38
1	34
2	20
3	6
≥ 4	2

Try a simulation on your own and compare your results with the preceding ones and with the results given in Table 7-5, obtained by theoretical methods.

Table 7-5

Number of Chips	Number of Cookies
0	36.8
1	36.8
2	18.4
3	6.1
≥ 4	1.9

ONGOING ASSESSMENT 7-3

1. How could you use a deck of cards to simulate the birth of boys and girls?
2. How could you use a random-number table to estimate the probability that two cards drawn from a standard deck of cards with replacement will be of the same suit?
3. How might you use a random-digit table to simulate each of the following?
   a. Tossing a single die
   b. Choosing three people at random from a group of 20 people
   c. Spinning the spinner, where the probability of each color is as shown in the following figure:

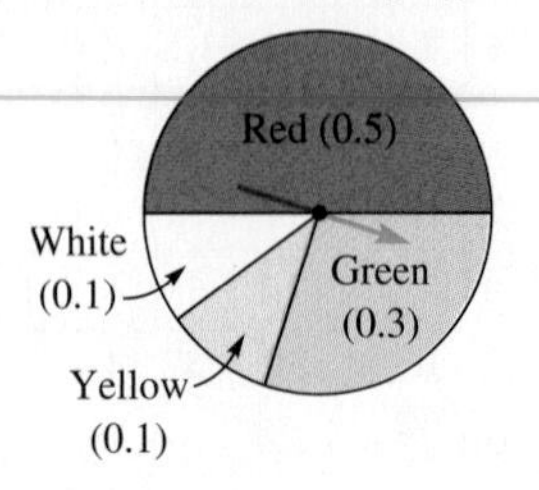

4. A school has 500 students. The principal is to pick 30 students at random from the school to go to the Rose Bowl. How can this be done by using a random-digit table?
5. In a certain city, the probability that it will rain on a certain day is 0.8 if it rained the day before. The probability that it will be dry on a certain day is 0.3 if it was dry the day before. It is now Sunday, and it is raining. Use the random-digit table to simulate the weather for the rest of the week.
6. It is reported that 15% of people who come into contact with a person infected with strep throat contract the disease. How might you use the random-digit table to simulate the probability that at least one child in a three-child family will catch the disease, given that each child has come into contact with the infected person?
7. Pick a block of two digits from the random-digit table. What is the probability that the block picked is less than 30?
8. An estimate of the fish population of a certain pond was found by catching 200 fish and marking and returning them to the pond. The next day, 300 fish were caught, of

which 50 had been marked the previous day. Estimate the fish population of the pond.

9. Suppose that in the World Series, the two teams are evenly matched. The two teams play until one team wins four games, and no ties are possible.
   a. What is the maximum number of games that could be played?
   b. Use simulation to approximate the probabilities that the series will end in (i) four games and in (ii) seven games.

10. Assume Carmen Smith, a basketball player, makes free throws with 80% probability of success and is placed in a one-and-one situation where she is given a second foul shot only if the first shot goes through the basket. Simulate the 25 attempts from the foul line in one-and-one situations to determine how many times we would expect Carmen to score 0 points, 1 point, and 2 points.

Communication

11. In an attempt to reduce the growth of its population, China instituted a policy limiting a family to one child. Rural Chinese suggested revising the policy to limit families to one son. Assuming the suggested policy is adopted and that any birth is as likely to produce a boy as a girl, explain how to use simulation to answer the following:
   a. What would be the average family size?
   b. What would be the ratio of newborn boys to newborn girls?

12. Consider a "walk" on the following grid starting out at the origin 0 and "walking" one unit (block) north, and at each intersection turning left with probability $\frac{1}{2}$, turning right with probability $\frac{1}{6}$, and moving straight with probability $\frac{1}{3}$. Explain how to simulate the "walk" using a regular six-sided die.

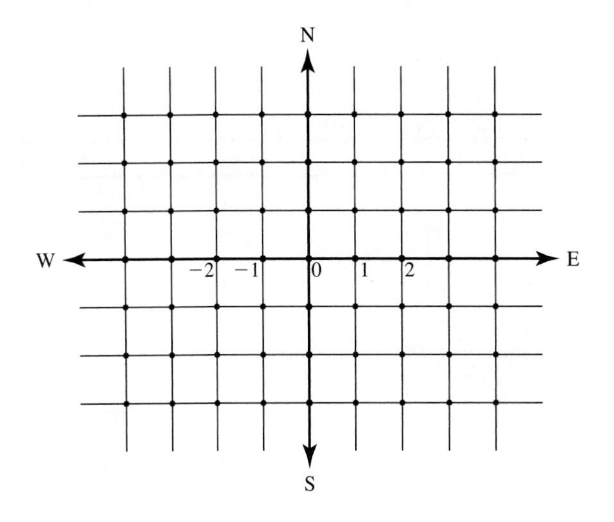

Open-Ended

13. What is the probability that in a group of five people chosen at random, at least two will have birthdays in the same month? Design a simulation for this problem and try your simulation 10 times.

14. The probability of the home team's winning a basketball game is 80%. Describe a simulation of the probability that the home team will win three home games in a row.

15. Montana duck hunters are all perfect shots. Ten Montana hunters are in a duck blind when 10 ducks fly over. All 10 hunters pick a duck at random to shoot at, and all 10 hunters fire at the same time. How many ducks could be expected to escape, on the average, if this experiment were repeated a large number of times? How could this problem be simulated?

Cooperative Learning

16. The sixth-grade class decided that the ideal number of children in a family is four: two boys and two girls.
   a. As a group, design a simulation to determine the probability of two boys and two girls in a family of four.
   b. Have each person in your group try the simulation 25 times and compare the probabilities.
   c. Combine the results of all the members of your group and use this information to find a simulated probability.
   d. Compute the theoretical probability of having two boys and two girls in a family of four and compare your answer to the simulated probability.

17. Have you ever wondered how you would score on a 10-item true-false test if you guessed at every answer?
   a. Simulate your score by tossing a coin 10 times, with heads representing true and tails representing false. Check your answers by using the following key and score yourself:

1	2	3	4	5	6	7	8	9	10
F	T	F	F	T	F	T	T	T	F

   b. Combine your results with those of others in your group to find the average (mean) number of correct answers for your group.
   c. How many items would you expect to get correct if the number of items were 30 instead of 10?
   d. What is the theoretical probability of your getting all the answers correct on a 10-item true-false test if all answers were chosen by flipping a coin?

18. a. Estimate how many cards you would expect to have to turn over on the average in an ordinary playing deck before an ace appeared.
   b. Have each person in your group repeat the experiment 10 times or simulate it for 10 trials. Then find the average number of aces.

c. Combine the results of all the members of your group and find the average number of cards that needs to be drawn. How does this average compare to your individual estimate?

★ **19.** A cereal company places a coupon bearing a number from 1 to 9 in each box of cereal. If the numbers are distributed at random in the boxes, estimate the number of boxes you would have to purchase in order to obtain all nine numbers. Explain how the random-digit table could be used to estimate the number of coupons. Each person in the group should simulate 10 trials, and the results in the group should be combined to find the estimate.

Review Problems

20. In a two-person game, four coins are tossed. If exactly two heads come up, you win. If anything else comes up, you lose. Does each player have an equal chance of winning the game? Explain why or why not.

21. A single card is drawn from an ordinary deck. What is the probability of obtaining each of the following?
   a. A club
   b. A queen and a spade
   c. Not a queen
   d. Not a heart
   e. A spade or a heart
   f. The 6 of diamonds
   g. A queen or a spade
   h. Either red or black

22. From a sack containing seven red marbles, eight blue marbles, and four white marbles, marbles are drawn at random for several experiments. Determine the probability of each of the following events:
   a. One marble drawn at random is either red or blue.
   b. The first draw is red and the second is blue, where one marble is drawn at random, its color is recorded, the marble is replaced, and another marble is drawn.
   c. The event in (b) where the first marble is not replaced.

TECHNOLOGY CORNER

RANDINT(or RANDI(

1. On some graphing calculators, if we choose the MATH menu and then select PRB, which stands for PROBABILITY, we find **RANDINT(or RANDI(**, the random integer feature. RANDINT(generates a random integer within a specified range. It requires two inputs that are the upper and lower boundaries for the integers. For example, RANDINT(1, 10) generates a random integer from 1 through 10.
   a. How could you use RANDINT(to simulate tossing a single die?
   b. How could you use RANDINT(to simulate the sum of the numbers when tossing two dice?

RAND

int

2. Some graphing calculators have a **RAND** function, a random-digit generator. RAND generates and returns a random number greater than 0 and less than 1. For example, RAND might produce the numbers .5956605, .049599836, or .876572691. To have RAND produce random numbers from 1 to 10 as in (1), we enter int (10 *RAND) + 1. The **int** (greatest integer) feature is found in the MATH menu under NUM. The feature int returns the greatest integer less than or equal to a number.
   a. How could you use RAND to simulate tossing a single die?
   b. How could you use RAND to simulate tossing two dice and taking the sum of the numbers?

3. Use one of these random features to simulate tossing two dice 30 times. Based on your simulation, what is the probability that a sum of 7 will occur?

7-4 Odds, Conditional Probability, and Expected Value

Computing Odds

odds in favor

People talk about the *odds in favor of* and the *odds against* a particular event's happening. When the **odds in favor** of the president's being reelected are 4 to 1, this refers to how likely the president is to win the election relative to how likely the president is to lose. The probability of the president's winning is 4 times the probability of losing. If W represents the event the president wins the election and L represents the event the president loses, then $P(W) = 4P(L)$ or as a proportion, we have

$$\frac{P(W)}{P(L)} = \frac{4}{1}, \text{ or } 4:1.$$

Because W and L are complements of each other, $L = \overline{W}$, we have

$$\frac{P(W)}{P(\overline{W})} = \frac{P(W)}{1-P(W)} = \frac{4}{1}, \text{ or } 4:1.$$

odds against

The **odds against** the president's winning are how likely the president is to lose relative to how likely the president is to win. Using the preceding information, we have

$$\frac{P(L)}{P(W)} = \frac{1}{4}, \text{ or } 1:4.$$

Because $L = \overline{W}$, we have

$$\frac{P(\overline{W})}{P(W)} = \frac{1-P(W)}{P(W)} = \frac{1}{4}, \text{ or } 1:4.$$

Formally, odds are defined as follows.

Definition of Odds

Let $P(A)$ be the probability that A occurs and $P(\overline{A})$ be the probability that A does not occur. Then the **odds in favor** of an event A are

$$\frac{P(A)}{P(\overline{A})}, \text{ or } \frac{P(A)}{1-P(A)},$$

and the **odds against** an event A are

$$\frac{P(\overline{A})}{P(A)}, \text{ or } \frac{1-P(A)}{P(A)}.$$

When odds are calculated for equally likely outcomes, we have the following:

Odds in favor of an event A: $\dfrac{P(A)}{P(\overline{A})} = \dfrac{n(A)}{n(S)} \div \dfrac{n(\overline{A})}{n(S)} = \dfrac{n(A)}{n(S)} \cdot \dfrac{n(S)}{n(\overline{A})} = \dfrac{n(A)}{n(\overline{A})}$

Odds against an event A: $\dfrac{n(\overline{A})}{n(A)}$

Thus, in the case of equally likely outcomes, we have

$$\text{Odds in favor} = \frac{\text{Number of favorable outcomes}}{\text{Number of unfavorable outcomes}}$$

$$\text{Odds against} = \frac{\text{Number of unfavorable outcomes}}{\text{Number of favorable outcomes}}.$$

When you roll a die, the number of favorable ways of rolling a 4 in one throw of a die is 1, and the number of unfavorable ways is 5. Thus the odds in favor of rolling a 4 are 1 to 5.

EXAMPLE 7-13

For each of the following, find the odds in favor of the event's occurring:

a. Rolling a number less than 5 on a die
b. Tossing heads on a fair coin
c. Drawing an ace from an ordinary 52-card deck
d. Drawing a heart from an ordinary 52-card deck

Solution **a.** The probability of rolling a number less than 5 is $\frac{4}{6}$; the probability of rolling a number not less than 5 is $\frac{2}{6}$. The odds in favor of rolling a number less than 5 are $\left(\frac{4}{6}\right) \div \left(\frac{2}{6}\right)$, or $4:2$, or $2:1$.

b. $P(H) = \frac{1}{2}$ and $P(\overline{H}) = \frac{1}{2}$. The odds in favor of getting heads are $\left(\frac{1}{2}\right) \div \left(\frac{1}{2}\right)$, or $1:1$.

c. The probability of drawing an ace is $\frac{4}{52}$, and the probability of not drawing an ace is $\frac{48}{52}$. The odds in favor of drawing an ace are $\left(\frac{4}{52}\right) \div \left(\frac{48}{52}\right)$, or $4:48$, or $1:12$.

d. The probability of drawing a heart is $\frac{13}{52}$, or $\frac{1}{4}$, and the probability of not drawing a heart is $\frac{39}{52}$, or $\frac{3}{4}$. The odds in favor of drawing a heart are $\left(\frac{13}{52}\right) \div \left(\frac{39}{52}\right) = \frac{13}{39}$, or $13:39$, or $1:3$.

NOW TRY THIS 7-8

In Example 7-13(a), there are four ways to roll a number less than 5 on a die (favorable outcomes) and two ways of not rolling a number less than 5 (unfavorable outcomes), so the odds in favor of rolling a number less than 5 are $4:2$, or $2:1$. Work the other three parts of Example 7-13 using this approach.

Given the probability of an event, it is possible to find the odds in favor of (or against) the event and vice versa. For example, if the odds in favor of an event A are $5:1$, then we have the following:

$$\frac{P(A)}{1 - P(A)} = \frac{5}{1}$$
$$P(A) = 5[1 - P(A)]$$
$$P(A) = 5 - 5P(A)$$
$$6P(A) = 5$$
$$P(A) = \frac{5}{6}$$

REMARK The probability $\frac{5}{6}$ is a ratio. The exact number of favorable outcomes and the exact total of all outcomes are not necessarily known.

EXAMPLE 7-14

In the following cartoon, find the probability of making totally black copies if the odds are 3 to 1 against making totally black copies:

Different people could have different interpretations of what "splitting fairly" means. Possibly, though, the best is to split the pot in proportion to the probabilities of each player's winning the game when play was halted. We must calculate the expected value for each player and split the pot accordingly.

Devising a Plan A third head would make Al the winner, whereas Betsy needs two more tails to win. A *tree diagram* that simulates the completion of the game allows us to find the probability of each player's winning the game. Once we find the probabilities, all we need do is multiply the probabilities by the amount of the pot, \$100, to determine each player's fair share.

Carrying Out the Plan The tree diagram in Figure 7-34 shows the possibilities for game winners if the game is completed. We can find the probabilities of each player's winning as follows:

FIGURE 7-34

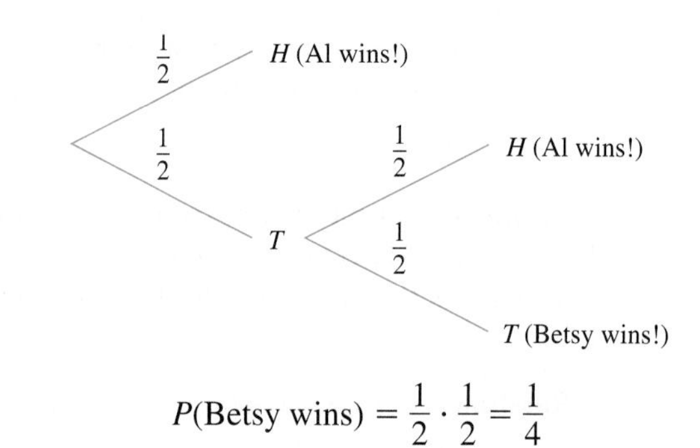

$$P(\text{Betsy wins}) = \frac{1}{2} \cdot \frac{1}{2} = \frac{1}{4}$$

$$P(\text{Al wins}) = 1 - \frac{1}{4} = \frac{3}{4}$$

Hence, the fair way to split the stakes is for Al to receive $\frac{3}{4}$ of \$100, or \$75, and Betsy should receive $\frac{1}{4}$ of \$100, or \$25.

Looking Back The problem could be made even more interesting by assuming that the coin is not fair so that the probability is not $\frac{1}{2}$ for each branch in the tree diagram. Other possibilities arise if the players have unequal amounts of money in the pot or if more tosses are required in order to win.

ONGOING ASSESSMENT 7-4

1. a. What are the odds in favor of drawing a face card from an ordinary deck of playing cards?
   b. What are the odds against drawing a face card?
2. On a single roll of a pair of dice, what are the odds against rolling a sum of 7?
3. If the probability of a boy's being born is $\frac{1}{2}$, and a family plans to have four children, what are the odds against having all boys?

4. Diane tossed a coin nine times and got 9 tails. Assume that Diane's coin is fair and answer each of the following questions:
   a. What is the probability of tossing a tail on the 10th toss?
   b. What is the probability of tossing 10 more tails in a row?
   c. What are the odds against tossing 10 more tails in a row?
5. If the odds against Deborah's winning first prize in a chess tournament are 3 to 5, what is the probability that she will win first prize?
6. What are the odds in favor of tossing at least two heads if a fair coin is tossed 3 times?
7. If the probability of rain for the day is 60%, what are the odds against its raining?
8. On an American roulette wheel, half of the slots numbered 1 through 36 are red and half are black. Two slots, numbered 0 and 00, are green. What are the odds against a red slot's coming up on any spin of the wheel?
9. On a tote board at a racetrack, the odds for Gameylegs are listed as 26 : 1. Tote boards list the odds that the horse will lose the race. If this is the case, what is the probability of Gameylegs's winning the race?
10. If a whole number less than 10 is chosen at random, what are the odds that the number is greater than 5?
11. If the probability that a randomly chosen household has a cat is 0.27, what are the odds against a chosen household having a cat?
12. If the probability of an event is $0.\overline{3}$, what are the odds against the event?
13. What are the odds in favor of randomly drawing the letter S from the letters in the word MISSISSIPPI?
14. A container has three white balls and two red balls. A first ball is drawn at random and not replaced. Then a second ball is drawn. Given the following conditions, what is the probability that the second ball was red?
   a. The first ball was white.
   b. The first ball was red.
15. The following spinner is spun. Given the conditions listed below, what is the probability that it lands on 2?

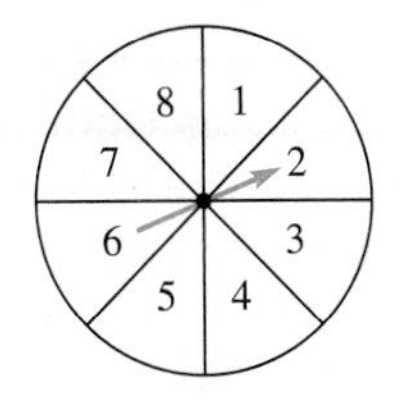

   a. It lands on an odd number.
   b. It lands on a number divisible by 3.
   c. It does not land on 5, 6, or 7.
   d. It lands on a number less than 4.

16. If $P(A) = \frac{2}{3}$, $P(B) = \frac{1}{2}$, and $P(B|A) = \frac{1}{3}$, find $P(A|B)$.

17. You play a game in which two dice are rolled. If a sum of 7 appears, you win $10; otherwise, you lose $2.00. If you intend to play this game for a long time, should you expect to make money, lose money, or come out about even? Explain.
18. On a roulette wheel are 36 slots numbered 1 through 36 and 2 slots numbered 0 and 00. You can bet on a single number. If the ball lands on your number, you receive 35 chips plus the chip you played.
   a. What is the probability that the ball will land on 17?
   b. What are the odds against the ball landing on 17?
   c. If each chip is worth $1, what is the expected payoff for a player who plays the number 17 for a long time?
19. Suppose five quarters, five dimes, five nickels, and ten pennies are in a box. One coin is selected at random. What is the expected value of this experiment?
20. If the odds in favor of Fast Leg's winning a horse race are 5 to 2 and the first prize is $14,000, what is the expected value of Fast Leg's winning?
21. Suppose a standard six-sided die is rolled and you receive $1 for every dot showing on the top of the die. What should the cost of playing the game be in order to make it a fair game?
22. Sweepstakes are required by law to display the odds of winning as well as the payoffs. That information is sufficient to calculate the expected value of a sweepstakes. Suppose that mailing the sweepstakes costs $0.37 and the odds in favor of winning the various prizes are as follows:

Odds	Prize	Quantity
1 to 20,000,000	$1,000,000	1
1 to 20,000,000	$100,000	10
1 to 1,000,000	$1,000	100

   a. What is the expected value of the sweepstakes for any individual?
   b. At what postage rate would the drawing be fair?
23. Suppose it costs $8 to roll a pair of dice. You get paid the sum of the numbers in dollars that appear on the dice.
   a. What is the expected value of the game?
   b. Is it a fair game?

Communication

24. Explain the difference between odds and probability.
25. A game involves tossing two coins. A player wins $1.00 if both tosses result in heads. What should you pay to play this game in order to make it a fair game? Explain your answer.

Open-Ended

26. Suppose you flip two fair coins. Design a fair game with a payoff based on the number of heads that are shown.

27. An insurance company sells a policy that pays $50,000 in case of accidental death. According to company figures, the rate of accidental death is 47 per 100,000 population. What annual premium should the company charge for this coverage? Explain how much profit the company will make under your plan, how you determined the amount of profit needed for the company, and how the annual premium was computed.
28. Write a game-type problem about odds and payoffs so that the odds in favor of an event are 2 : 3 and the game is a fair game.

Cooperative Learning

29. As a group, design a game that involves cards, dice, or spinners.
    a. Write the rules so that any person who wants to play can understand the game.
    b. Write a description explaining whether the game is fair and how you arrived at your conclusion.
    c. Calculate the odds of each player's winning.
    d. If betting is involved, discuss expected values.
    e. Exchange a game with another group and compare your analysis of their game with their analysis of your group's game.

Review Problems

30. Refer to the following spinners and write the sample space for each of the following experiments:

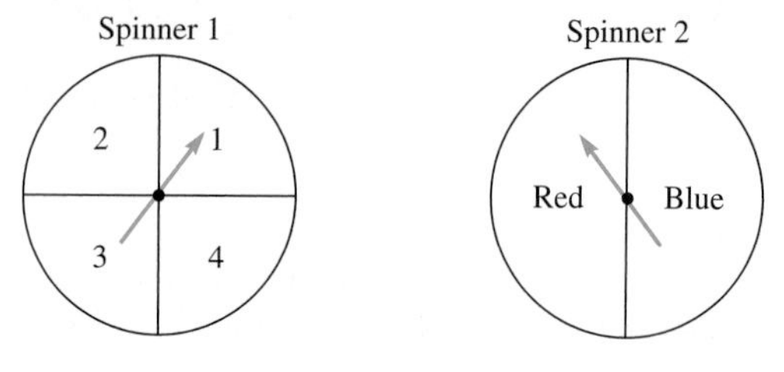

    a. Spin spinner 1 once.
    b. Spin spinner 2 once.
    c. Spin spinner 1 once and then spin spinner 2 once.
    d. Spin spinner 2 once and then roll a die.
    e. Spin spinner 1 twice.
    f. Spin spinner 2 twice.
31. Draw a spinner with two sections, red and blue, such that the probability of getting (Blue, Blue) on two spins is $\frac{25}{36}$.
32. Find the probability of getting two vowels when someone draws two letters from the English alphabet with replacement.

BRAIN TEASER

It is your first day of class; your class has 40 students. A friend who does not know any students in your class bets you that at least 2 of them share a birthday (month and day). What are your friend's chances of winning the bet?

7-5 Using Permutations and Combinations in Probability

Permutations of Unlike Objects

permutation

An arrangement of things in a definite order with no repetitions is a **permutation**. For example, RAT, RTA, ART, ATR, TRA, and TAR are all different arrangements of the three letters R, A, and T. Notice that order is important and there are no repetitions. Determining the number of possible arrangements of the three letters without making a list can be done using the Fundamental Counting Principle. (Recall the Fundamental Counting Principle states that if an event M can occur in m ways and, after M has occurred, event N can occur in n ways, then event M followed by event N can occur in $m \cdot n$ ways.) Because there are three ways to choose the first letter, two ways to choose the second letter, and one way to choose the third letter, there are $3 \cdot 2 \cdot 1$, or 6, ways to arrange the letters. It is common to record the number of permutations of three objects taken three at a time as ${}_3P_3$. Therefore, ${}_3P_3 = 6$.

Consider how many ways the owner of an ice cream parlor can display 10 flavors in a row along the front of the display case. The first position can be filled in 10 ways, the second position in 9 ways, the third position in 8 ways, and so on. By the Fundamental Counting Principle, there are $10 \cdot 9 \cdot 8 \cdot 7 \cdot 6 \cdot 5 \cdot 4 \cdot 3 \cdot 2 \cdot 1$, or 3,628,800, ways to display the flavors. If there were 16 flavors, there would be $16 \cdot 15 \cdot 14 \cdot 13 \cdot \ldots \cdot 3 \cdot 2 \cdot 1$ ways to arrange them. In general, *if there are n objects, then the number of possible ways to arrange the objects in a row is the product of all the natural numbers from n to 1, inclusive.* This expression, ***n* factorial**, is denoted ***n*!** as shown next.

***n* factorial (*n*!)**

$$n! = n \cdot (n - 1) \cdot (n - 2) \cdot \ldots \cdot 3 \cdot 2 \cdot 1$$

For example, $5! = 5 \cdot 4 \cdot 3 \cdot 2 \cdot 1$, $3! = 3 \cdot 2 \cdot 1$, and $1! = 1$. Using factorial notation is helpful in counting and probability problems.

Many calculators have a factorial key such as $\boxed{x!}$. To use this key, enter a whole number and then press the factorial key. For example, to compute 5!, press $\boxed{5}\ \boxed{x!}$ and 120 will appear on the display.

Consider the set of people in a small club, {Al, Betty, Carl, Dan}. For them to elect a president and a secretary, order is important and no repetitions are possible. How many ways are there to elect a committee of two in which one person is president and the other secretary? One way to answer the question is to agree that the choice "Al, Betty" denotes Al as president and Betty as secretary, while the choice "Betty, Al" indicates that Betty is president and Al is secretary. Thus, order is important and no repetitions are possible. Consequently, counting the number of possibilities is a permutation problem. Since there are four ways of choosing a president and then three ways of choosing a secretary, by the Fundamental Counting Principle, there are $4 \cdot 3$, or 12, ways of choosing a president and a secretary. Choosing two officers from a club of four is a permutation of four people chosen two at a time. The number of possible permutations of four objects taken two at a time, denoted ${}_4P_2$, may be counted using the Fundamental Counting Principle. Therefore, we have ${}_4P_2 = 4 \cdot 3$, or 12.

NOW TRY THIS 7-9

a. Write ${}_nP_2$, ${}_nP_3$, and ${}_nP_4$ in terms of n.

b. Based on your answers in part (a), write ${}_nP_r$ in terms of n and r.

c. In the club mentioned, how many ways are there to choose a president, vice president, and secretary?

The number of permutations can be written in terms of factorials. Consider the number of permutations of 20 objects chosen 3 at a time:

$$\begin{aligned} {}_{20}P_3 &= 20 \cdot 19 \cdot 18 \\ &= \frac{20 \cdot 19 \cdot 18 \cdot (17 \cdot \ldots \cdot 3 \cdot 2 \cdot 1)}{(17 \cdot \ldots \cdot 3 \cdot 2 \cdot 1)} \\ &= \frac{20!}{17!} \\ &= \frac{20!}{(20 - 3)!} \end{aligned}$$

This can be generalized as follows:

Permutation of Objects in a Set

If a set has n elements, then the number of permutations of n objects taken r at a time is given by

$$_nP_r = \frac{n!}{(n-r)!}.$$

$_nP_n$ is the number of permutations of n objects chosen n at a time—that is, the number of ways of rearranging n objects in a row. We have seen that this number is $n!$. If we use the formula for $_nP_r$ to compute $_nP_n$, we obtain

$$_nP_n = \frac{n!}{(n-n)!} = \frac{n!}{0!}.$$

Consequently, $n! = n!/0!$. To make this equation true, *we define 0! to be 1.*

Many calculators, especially graphing calculators, can calculate the number of permutations of n objects taken r at a time. This feature is usually denoted $\boxed{_nP_r}$. To use this key, enter the value of n, then press $\boxed{_nP_r}$, followed by the value of r. If you then press $\boxed{=}$ or $\boxed{\text{ENTER}}$, the number of permutations is displayed.

NOW TRY THIS 7-10

a. Try to use a factorial key $\boxed{x!}$ on your calculator to compute $\frac{100!}{98!}$. What happens? Why?

b. Without using a calculator, use the definition of factorials to compute the expression in part (a).

EXAMPLE 7-17

a. A baseball team has nine players. Find the number of ways the manager can arrange the batting order.

b. Find the number of ways of choosing three initials from the alphabet if none of the letters can be repeated.

Solution **a.** Because there are nine ways to choose the first batter, eight ways to choose the second batter, and so on, there are $9 \cdot 8 \cdot 7 \cdot \ldots \cdot 2 \cdot 1 = 9!$, or 362,880, ways of arranging the batting order. Using the formula for permutations, we have $_9P_9 = \frac{9!}{0!} = 362{,}880$.

b. There are 26 ways of choosing the first letter, 25 ways of choosing the second letter, and 24 ways of choosing the third letter. Hence, there are $26 \cdot 25 \cdot 24$, or 15,600, ways of choosing the three letters. Alternately, if we use the formula for permutations, we have

$$_{26}P_3 = \frac{26!}{23!} = \frac{26 \cdot 25 \cdot 24 \cdot (23 \cdot 22 \cdot 21 \cdot \ldots \cdot 1)}{23 \cdot 22 \cdot 21 \cdot \ldots \cdot 1}$$
$$= 26 \cdot 25 \cdot 24$$
$$= 15{,}600.$$

NOW TRY THIS 7-11

If the digits 1 through 9 inclusive are used with no repeats to form a four-digit code, what is the probability that a code number selected at random starts with 9?

Permutations Involving Like Objects

In the previous counting examples, each object to be counted was distinct. Suppose we wanted to rearrange the letters in the word *ZOO*. How many choices would we have? A tree diagram, as in Figure 7-35, suggests that there might be $3 \cdot 2 \cdot 1 = 3!$, or 6, possibilities. However, looking at the list of possibilities shows that *ZOO, OZO,* and *OOZ* each appear twice because the *O*'s are not different. We need to determine how to remove the duplication in arrangements such as this where some objects are the same. To eliminate the duplication, we divide the number of arrangements shown by the number of ways the two *O*'s can be rearranged, which is 2!. Consequently, there are $\frac{3!}{2!}$, or 3, ways of arranging the letters in *ZOO*. The arrangements are *ZOO, OZO,* and *OOZ*.

FIGURE 7-35

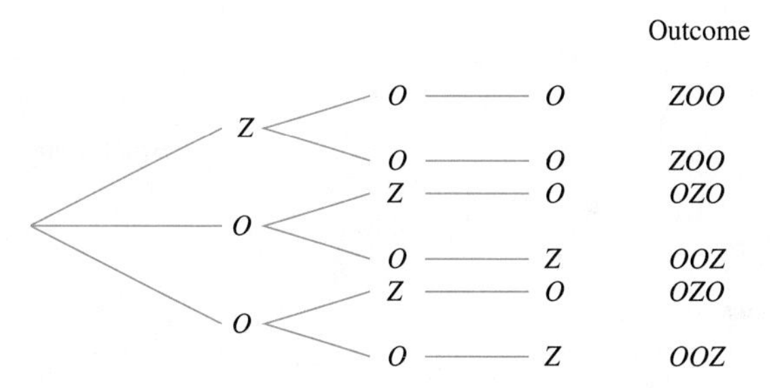

Permutations of Like Objects

If a set contains n elements, of which r_1 are of one kind, r_2 are of another kind, and so on through r_k, then the number of different arrangements of all n elements is equal to

$$\frac{n!}{r_1! \cdot r_2! \cdot r_3! \cdot \ldots \cdot r_k!}.$$

EXAMPLE 7-18

Find the number of rearrangements of the letters in each of the following words:

a. bubble
b. statistics

Solution **a.** There are 6 letters with *b* repeated 3 times. Hence, the number of arrangements is

$$\frac{6!}{3!} = 6 \cdot 5 \cdot 4 = 120.$$

b. There are 10 letters in the word *statistics*, with three *s*'s, three *t*'s, and two *i*'s in the word. Hence, the number of arrangements is

$$\frac{10!}{3! \cdot 3! \cdot 2!} = \frac{10 \cdot 9 \cdot 8 \cdot 7 \cdot 6 \cdot 5 \cdot 4 \cdot 3 \cdot 2 \cdot 1}{3 \cdot 2 \cdot 1 \cdot 3 \cdot 2 \cdot 1 \cdot 2 \cdot 1} = 50{,}400.$$

Combinations

Reconsider the club {Al, Betty, Carl, Dan}. Suppose a two-person committee is selected with no chair. In this case, order is not important, and an Al-Betty choice is the same as a Betty-Al choice. An arrangement of objects in which the order makes no difference is a **combination**. A comparison of the results of electing a president and a secretary for the club and the results of simply selecting a two-person committee are shown in Figure 7-36. Because each two permutations "shrink" into one combination, we see that the number of combinations is the number of permutations divided by 2, or

combination

$$\frac{4 \cdot 3}{2} = 6.$$

FIGURE 7-36

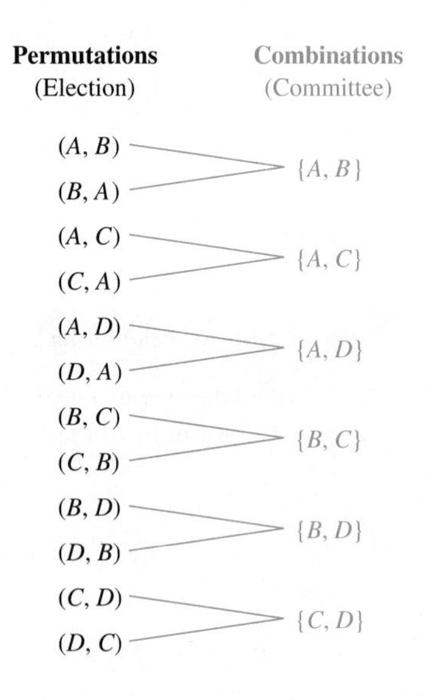

In how many ways can a committee of three people be selected from the club {Al, Betty, Carl, Dan}? To solve this problem, we proceed as we did earlier and find the number of ways to select three people from a group of four for three offices, say president, vice president, and secretary (a permutation problem) and then use this result to see how many combinations of people are possible for the committee. Figure 7-37 shows a partial list for both problems.

FIGURE 7-37

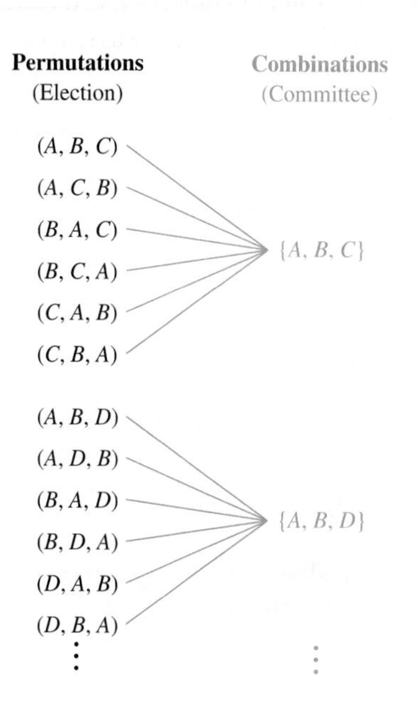

By the Fundamental Counting Principle, if order is important, the number of ways to choose three people from the list of four is $4 \cdot 3 \cdot 2$, or 24. However, with each triple chosen, there are 3!, or 6, ways to rearrange the triple, as seen in Figure 7-37. Therefore, there are 3! times as many permutations as combinations, or equivalently, each 3!, or 6, permutations "shrink" into one combination. Therefore, to find the number of combinations, we divide the number of permutations, 24, by 3!, or 6, to obtain 4. The four committees are $\{A, B, C\}$, $\{A, B, D\}$, $\{B, C, D\}$, and $\{A, C, D\}$.

In general, we use the following rule to count combinations.

Counting Combinations

To find the number of combinations possible in a counting problem, first use the Fundamental Counting Principle to find the number of permutations and then divide by the number of ways in which each choice can be arranged.

Symbolically, the number of combinations of n objects taken r at a time is denoted ${}_nC_r$. From the preceding discussion, we develop the following formula:

$$ {}_nC_r = \frac{{}_nP_r}{{}_rP_r} = \frac{\dfrac{n!}{(n-r)!}}{r!} = \frac{n!}{r!(n-r)!} $$

It is not necessary to memorize this formula; we can always find the number of combinations by using the reasoning developed in the committee example.

EXAMPLE 7-19

The Library of Science Book Club offers 3 free books from a list of 42. If you circle 3 choices from a list of 42 numbers on a postcard, how many possible choices are there?

Solution Order is not important, so this is a combination problem. By the Fundamental Counting Principle, there are $42 \cdot 41 \cdot 40$ ways to choose the 3 free books. Because each set of 3 circled numbers could be rearranged $3 \cdot 2 \cdot 1$ ways, there is an extra factor of 3! in the original $42 \cdot 41 \cdot 40$ ways. Therefore, the number of combinations possible for 3 books is

$$\frac{42 \cdot 41 \cdot 40}{3!} = 11{,}480.$$

If we use the formula for ${}_nC_r$, we have ${}_{42}C_3 = \frac{42!}{3!\,39!}$. Show this formula gives the same result as above.

EXAMPLE 7-20

At the beginning of the second quarter of a mathematics class for elementary school teachers, each of the class's 25 students shook hands with each of the other students exactly once. How many handshakes took place?

Solution Since the handshake between persons A and B is the same as that between persons B and A, this is a problem of choosing combinations of 25 people 2 at a time. There are

$${}_{25}C_2 = \frac{25 \cdot 24}{2!} = 300.$$

different handshakes.

EXAMPLE 7-21

Given a class of 12 girls and 10 boys, answer each of the following:

a. In how many ways can a committee of 5 consisting of 3 girls and 2 boys be chosen?
b. What is the probability that a committee of 5, chosen at random from the class, consists of 3 girls and 2 boys?
c. How many of the possible committees of 5 have no boys?
d. What is the probability that a committee of 5, chosen at random from the class, consists only of girls?

Solution **a.** Based on the information in the problem, we do not assign special functions to members on a committee and, hence, the order of the children on a committee does not matter. From 12 girls we can choose 3 girls in ${}_{12}C_3$ ways. Each of these choices can be paired with ${}_{10}C_2$ combinations of boys. By the Fundamental Counting Principle, the total number of committees is

$${}_{12}C_3 \cdot {}_{10}C_2 = \frac{12 \cdot 11 \cdot 10}{3!} \cdot \frac{10 \cdot 9}{2} = 9900.$$

b. The total number of committees of 5 is ${}_{22}C_5$ or 26,334. Using part (a), we find the probability that a committee of 5 will consist of 3 girls and 2 boys to be

$$\frac{{}_{12}C_3 \cdot {}_{10}C_2}{{}_{22}C_5} = \frac{9900}{26{,}334} \doteq 0.3759.$$

c. The number of ways to choose 0 boys and 5 girls from the 12 girls in the class is

$$_{10}C_0 \cdot {}_{12}C_5 = 1 \cdot {}_{12}C_5 = \frac{12 \cdot 11 \cdot 10 \cdot 9 \cdot 8}{5 \cdot 4 \cdot 3 \cdot 2 \cdot 1} = 792.$$

d. $\dfrac{_{12}C_5}{_{22}C_5} = \dfrac{792}{26334} \doteq 0.030075$

▲

PROBLEM SOLVING **A True-False Test Problem**

In the following "Peanuts" cartoon, suppose Peppermint Patty took a six-question true-false test. If she answered each question true or false at random, what is the probability that she answered 50% of the questions correctly?

Understanding the Problem A score of 50% indicates that Peppermint Patty answered $\frac{1}{2}$ of the six questions, or three questions, correctly. She answered the questions true or false at random, so the probability that she answered a given question correctly is $\frac{1}{2}$. We are asked to determine the probability that Patty answered exactly three of the questions correctly.

Devising a Plan We do not know which three questions Patty missed. She could have missed any three out of six on the test. Suppose she answered questions 2, 4, and 5 incorrectly. In this case, she would have answered questions 1, 3, and 6 correctly. We can compute the probability of this set of answers by *using a branch of a tree diagram*, as in Figure 7-38, where *C* represents a correct answer and *I* represents an incorrect answer.

FIGURE 7-38

Question:	1	2	3	4	5	6	Probability of Outcome
	$\xrightarrow{\frac{1}{2}} C$	$\xrightarrow{\frac{1}{2}} I$	$\xrightarrow{\frac{1}{2}} C$	$\xrightarrow{\frac{1}{2}} I$	$\xrightarrow{\frac{1}{2}} I$	$\xrightarrow{\frac{1}{2}} C$	$\left(\frac{1}{2}\right)^6$

Multiplying the probabilities along the branches, we obtain $\left(\frac{1}{2}\right)^6$ as the probability of answering questions 1 through 6 in the following way: *C I C I I C*. There are other ways to answer exactly three questions correctly: for example, *C C C I I I*. The probability of answering questions 1 through 6 in this way is also $\left(\frac{1}{2}\right)^6$. The number of ways to answer the questions is simply the number of ways of arranging three *C*'s and three *I*'s in a row, which is also the number of ways of choosing three correct questions out of six, that is, ${}_6C_3$. Because all these arrangements give Patty a score of 50%, the desired probability is the sum of the probabilities for each arrangement.

Carrying Out the Plan There are ${}_6C_3$, or 20, sets of answers similar to the one in Figure 7-38, with three correct and three incorrect answers. The product of the probabilities for each of these sets of answers is $\left(\frac{1}{2}\right)^6$, so the sum of the probabilities for all 20 sets is $20 \cdot \left(\frac{1}{2}\right)^6$, or approximately 0.3125. Thus, Peppermint Patty has a probability of 0.3125 of obtaining a score of exactly 50% on the test.

Looking Back It seems paradoxical to learn that the probability of obtaining a score of 50% on a six-question true-false test is not close to $\frac{1}{2}$. As an extension of the problem, suppose a passing score is a score of at least 70%. Now what is the probability that Peppermint Patty will pass? What is the probability of her obtaining a score of at least 50% on the test? If the test is a six-question multiple-choice test with five alternative answers for each question, what is the probability of obtaining a score of at least 50% by random guessing?

PROBLEM SOLVING **Matching Letters to Envelopes**

Stephen placed three letters in envelopes while he was having a telephone conversation. He addressed the envelopes and sealed them without checking if each letter was in the correct envelope. What is the probability that each of the letters was inserted correctly?

Understanding the Problem Stephen sealed three letters in addressed envelopes without checking to see if each was in the correct envelope. We are to determine the probability that each of the three letters was placed correctly. This probability could be found if we knew the sample space, or at least how many elements are in the sample space.

Devising a Plan To aid in solving the problem, we represent the respective letters as *a, b,* and *c* and the respective envelopes as *A, B,* and *C*. For example, a correctly placed letter *a* would be in envelope *A*. To construct the sample space, we use the strategy of *making a table*. The table should show all the possible permutations of letters in envelopes. Once the table is completed, we can determine the probability that each letter is placed correctly.

Carrying Out the Plan Table 7-6 is constructed by using the envelope labels *A, B,* and *C* as headings and listing all ways that letters *a, b,* and *c* could be placed in the envelopes. Case 1 is the only case out of six in which each of the envelopes is labeled correctly, so the probability that each envelope is labeled correctly is $\frac{1}{6}$.

Table 7-6

Letters	Addresses A	B	C
1	*a*	*b*	*c*
2	*a*	*c*	*b*
3	*b*	*a*	*c*
4	*b*	*c*	*a*
5	*c*	*a*	*b*
6	*c*	*b*	*a*

Looking Back Is the probability of having each letter placed incorrectly the same as the probability of having each letter placed correctly? A first guess might be that the probabilities are the same, but that is not true. Why?

We also could have used a counting argument to solve the problem. Given an envelope, there is only one correct letter to place in the envelope. Thus, there is one correct way to place the letters in the envelopes. By the Fundamental Counting Principle, there are $3 \cdot 2 \cdot 1$ ways of choosing the letters to place in the envelopes, so the probability of having the letters correctly placed is $\frac{1}{6}$. ▲▲

ONGOING ASSESSMENTS 7-5

1. The eighth-grade class at a grade school has 16 girls and 14 boys. How many different boy-girl dates can be arranged?
2. If a coin is tossed 5 times, in how many different ways can the sequence of heads and tails appear?
3. The telephone prefix for a university is 243. The prefix is followed by four digits. How many telephones are possible before a new prefix is needed?
4. Radio stations in the United States have call letters that begin with either *K* or *W*. Some have three letters; others have four letters. How many sets of three-letter call letters are possible? How many sets of four-letter call letters are possible?
5. Carlin's Pizza House offers 3 kinds of salad, 15 kinds of pizza, and 4 kinds of dessert. How many different three-course meals can be ordered?
6. Decide whether each of the following is true or false:
   a. $6! = 6 \cdot 5!$
   b. $3! + 3! = 6!$
   c. $\frac{6!}{3!} = 2!$
   d. $\frac{6!}{3} = 2!$
   e. $\frac{6!}{5!} = 6!$
   f. $\frac{6!}{4!2!} = 15$
   g. $n!(n + 1) = (n + 1)!$
7. In how many ways can the letters in the word *SCRAMBLE* be rearranged?
8. How many two-person committees can be formed from a group of six people?
9. Find the number of ways to rearrange the letters in the following words:
   a. *OHIO*
   b. *ALABAMA*
   c. *ILLINOIS*
   d. *MISSISSIPPI*
   e. *TENNESSEE*

10. In a car race, there are 6 Chevrolets, 4 Fords, and 2 Pontiacs. In how many ways can the 12 cars finish if we consider only the makes of the cars?

11. Assume a class has 30 members.
   a. In how many ways can a president, a vice president, and a secretary be selected?
   b. How many committees of three people can be chosen?

12. A basketball coach was criticized in the newspaper for not trying out every combination of players. If the team roster has 12 players, how many 5-player combinations are possible?

13. A five-volume numbered set of books is placed randomly on a shelf. What is the probability that the books will be numbered in the correct order from left to right?

14. Take 10 points in a plane, no 3 of them on a line. How many straight lines can be drawn if each line is drawn through a pair of points?

15. Sally has four red flags, three green flags, and two white flags. How many nine-flag signals can she run up a flagpole?

16. Find the number of shortest paths from point A to point B along the edges of the cubes in each of the following. (For example, in (a) one shortest path is A-C-D-B.)

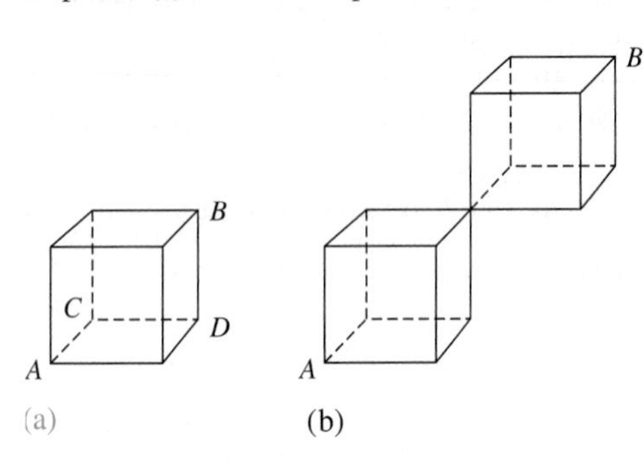

17. At a party, 28 handshakes took place. Each person shook hands exactly once with each of the others present. How many people were at the party?

18. A committee of three people is selected at random from a set consisting of seven Americans, five French people, and three English people.
   a. What is the probability that the committee consists of all Americans?
   b. What is the probability that the committee has no Americans?

19. How many different 5-card hands can be dealt from a standard deck of 52 playing cards?

20. License plates in a certain state have three letters followed by three digits. How many different plates are possible if no repetitions of letters or numbers are allowed?

21. In a certain lottery game, 54 numbers are randomly mixed and 6 are selected. A person must pick all 6 numbers to win. Order is not important. What is the probability of winning?

22. Social Security numbers are in the form ###-##-####, where each symbol represents a number 0 through 9. How many Social Security numbers are possible using this format?

23. The probability of a basketball player's making a free throw successfully at any time in a game is $\frac{2}{3}$. If the player attempts 10 free throws in a game, what is the probability that exactly 6 are made?

24. A fair die is rolled 8 times. What is the probability of getting
   a. 1 on each of the eight rolls?
   b. 6 exactly twice in the eight rolls?
   c. 6 at least once in the eight rolls?

25. Two fair dice are rolled 5 times and the sum of the numbers that come up is recorded. Find the probability of getting
   a. A sum of 7 on each of the five rolls
   b. A sum of 7 exactly twice in the five rolls

26. From a group of 10 boys and 12 girls, a committee of 4 students is chosen at random. What is the probability that
   a. All 4 members on the committee will be girls?
   b. There will be at least 1 girl on the committee?
   c. All 4 members of the committee will be boys?

27. From a group of 20 Britons, 21 Italians, and 4 Danes, a committee of 8 people is chosen at random. Find the probability that (express your answers using notation for combinations)
   a. The committee will consist of 2 Britons, 4 Italians, and 2 Danes.
   b. The committee will have no Britons.
   c. There will be at least one Briton on the committee.
   d. All members of the committee will be Britons.

28. What are the odds against a royal flush in poker, that is, a 10, jack, queen, king, and ace all of the same suit?

29. How many different 12-person juries can be selected from a pool of 24 people?

30. Stephen placed five letters in envelopes while he was watching television and addressed the envelopes without checking if each letter was in the correct envelope.
   a. What is the probability that all five letters were in the correct envelopes?
   b. What is the probability each of the five letters were addressed incorrectly?
   c. What is the probability that exactly four letters were addressed correctly?
   d. If n letters were put in n envelopes at random, what is the probability that exactly $n - 1$ of them were in the correct envelopes?

31. A company is setting up four-digit ID numbers for employees.
   a. How many four-digit numbers are there if numbers can start with 0 and numbers can be repeated?

b. How many four-digit numbers are there if numbers can start with 0 and all the digits must be different?

c. If you randomly assign a four-digit ID number that can start with 0 and numbers can be repeated, what is the probability that all the digits are even?

Communication

32. The terms *Fundamental Counting Principle, permutations,* and *combinations* are all used to work with counting problems. In your own words, explain how all these terms are related and how they are used.

33. a. A bicycle lock has three reels, each of which contains the numbers 0 through 9. To open the lock, you must enter the numbers in the correct order, such as 369 or 455, where one number is chosen from each reel. How many different possibilities are there for the numbers to open the lock? Explain how you arrived at your answer.

b. The lock in the cartoon is called a *combination* lock. Explain why this is probably not a good name for this lock for someone who has studied counting problems.

REAL LIFE ADVENTURES by Gary Wise and Lance Aldrich

The hacksaw industry would be in serious trouble without the combination-lock industry.

REAL LIFE ADVENTURES © 1995 GarLanco. Reprinted with permission of UNIVERSAL PRESS SYNDICATE. All rights reserved.

34. a. Ten people are to be seated on 10 chairs in a line. Among them is a family of 3 that does not want to be separated. How many different seating arrangements are possible? Explain how you arrived at your answer.

b. How many possible seating arrangements are there in part (a) in which the family members do not sit all together? Explain how you arrived at your answer.

35. In how many ways can five couples be seated in a row of 10 chairs if no couple is separated? Explain how you arrived at your answer.

Open-Ended

36. Suppose the Department of Motor Vehicles uses only six spaces and the numbers 0 through 9 to create its license plates. Numbers can be repeated.

a. How many license plates are possible?

b. Based on the 2000 census, determine whether there are any states in which the answer in (a) might provide enough license plates.

c. If you were in charge of making license plates for the state of California, describe the method you would use to ensure you would have enough license plates.

Cooperative Learning

37. The following triangular array of numbers is a part of **Pascal's triangle**:

	Row
1	(0)
1 1	(1)
1 2 1	(2)
1 3 3 1	(3)
1 4 6 4 1	(4)
1 5 10 10 5 1	(5)
1 6 15 20 15 6 1	(6)

a. In your group, decide how the triangle was constructed and complete the next two rows.

b. Describe at least three number patterns in Pascal's triangle.

c. Find the sum of the numbers in each row. Predict the sum of the numbers in row 10.

d. The entries in row 2 are just ${}_2C_0$, ${}_2C_1$, and ${}_2C_2$. Have different members of your group investigate whether a similar pattern holds for other rows in Pascal's triangle.

e. Describe how you could use combinations to find any entry in Pascal's triangle.

Review Problems

38. Two cards are drawn at random without replacement from a deck of 52 cards. What is the probability that

a. At least 1 card is an ace?

b. Exactly 1 card is red?

39. If two regular dice are tossed, what is the probability of tossing a sum greater than 10?

40. Two coins are tossed. You win \$5.00 if both coins are heads and \$3.00 if both coins are tails and lose \$4.00 if the coins do not match. What is the expected value of this game? Is this a fair game?

41. On a roulette wheel, the probability of winning when you pick a particular number is $\frac{1}{38}$. Suppose you pay \$1.00 to play the game, and if your number is picked, you win \$36. (a) Is this a fair game? (b) What would happen if you played this game a large number of times?

BRAIN TEASER

An airplane can complete its flight if at least $\frac{1}{2}$ of its engines are working. If the probability that an engine fails is 0.01 and all engine failures do not depend on each other, what is the probability of a successful flight if the plane has

a. Two engines? **b.** Four engines?

HINT FOR SOLVING THE PRELIMINARY PROBLEM

We must first determine when Al and Betty have the same number of heads. This happens at 0 heads and at 1 head. Next we must determine the probability of each of these events for each person. Then we must use these probabilities to compute the probability that they both have the same number of heads.

QUESTIONS FROM THE CLASSROOM

1. A student claims that if a fair coin is tossed and comes up heads 5 times in a row, then, according to the law of averages, the probability of tails on the next toss is greater than the probability of heads. What is your reply?

2. A student observes the following spinner and claims that the color red has the highest probability of appearing, since there are two red areas on the spinner. What is your reply?

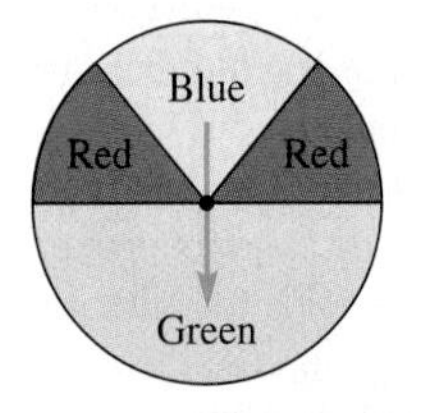

3. A student tosses a coin 3 times, and tails appears each time. The student concludes that the coin is not fair. What is your response?

4. An experiment consists of tossing a coin twice. The student reasons that there are three possible outcomes: two heads, one head and one tail, or two tails. Thus, $P(HH) = \frac{1}{3}$. What is your reply?

5. A student would like to know the difference between two events being independent and two events being mutually exclusive. How would you answer her?

6. In response to the question, "If a fair die is rolled twice, what is the probability of rolling a pair of 5s?" a student replies, "One-third, because $\frac{1}{6} + \frac{1}{6} = \frac{1}{3}$." How do you respond?

7. A student wonders why probabilities cannot be negative. What is your response?

8. A student claims that "if the probability of an event is $\frac{3}{5}$, then there are three ways the event can occur and only five elements in the sample space." How do you respond?

9. A student claims that if the odds in favor of winning a game are $a{:}b$, then out of every $a + b$ games she would win a games. Hence, the probability of winning the game is $\frac{a}{a+b}$. Is the student's reasoning correct? Why or why not?

10. A student does not understand the meaning of ${}_4P_0$. The student wants to know how we can consider permutations of four objects chosen zero at a time. How do you respond?

11. A student wants to know why, if we can define 0! as 1, we cannot define $\frac{1}{0}$ as 1. How do you respond?

12. A student is not sure when to add and when to multiply probabilities. How do you respond?

13. Alberto is to spin the spinners shown and compute the probability of two blacks. He looks at the spinners and says the answer is $\frac{1}{2}$ because $\frac{1}{2}$ of the areas of the circles are black. How do you respond?

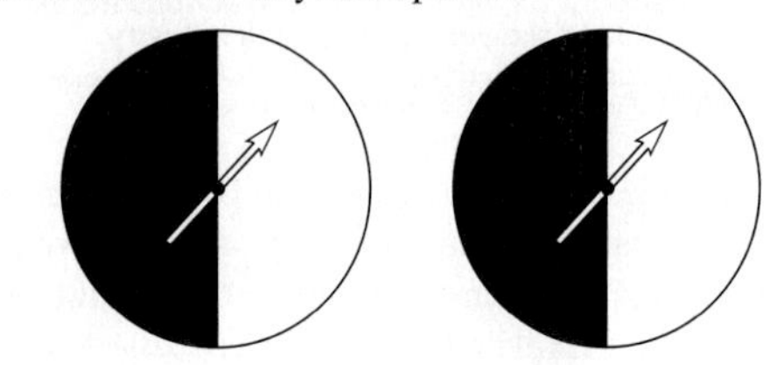

14. There are two bags each containing red balls and yellow balls. Bag A contains 1 red and 4 yellow balls. Bag B contains 3 red and 13 yellow balls. Alva says that you should always choose bag B if you are trying to draw a red ball because it has more red balls than bag A. How do you respond?

CHAPTER OUTLINE

I. Probability

A. Probabilities can be determined **experimentally (empirically)** or **theoretically**.

B. A **sample space** is the set of all possible **outcomes** of an **experiment**.

C. An **event** is a subset of a sample space.

D. Outcomes are **equally likely** if each outcome is as likely to occur as another.

E. If all outcomes of an experiment are *equally likely*, the **probability of an event** A from sample space S is given by

$$P(A) = \frac{n(A)}{n(S)}.$$

F. If each possible outcome of the sample space is equally likely the sample space is a **uniform sample space**.

G. Law of Large Numbers If an experiment is repeated a large number of times, the experimental (empirical) probability of a particular outcome approaches a fixed number as the number of repetitions increases

H. An **impossible event** is an event with a probability of 0. An impossible event can never occur.

I. A **certain event** is an event with a probability of 1. A certain event is sure to happen.

J. If A is an event and $A \subseteq S$ and S is the sample space, then $0 \leq P(A) \leq 1$.

K. Two events are **mutually exclusive** if and only if exactly one of the events can occur at any given time—that is, if and only if the events are disjoint.

L. The probability of the **complement of an event** is given by $P(\overline{A}) = 1 - P(A)$, where A is the event and $\overline{A}$ is its complement.

M. Multiplication rule for probabilities For all **multistage experiments**, the probability of the outcome along any path of a tree diagram is equal to the product of all the probabilities along the path.

N. If A and B are independent events, then $P(A \cup B) = P(A) + P(B) - P(A \cap B)$.

O. If events E_1 and E_2 are **independent**—that is, the outcome of one does not depend on the outcome of the other—then $P(E_1 \text{ and } E_2) = P(E_1) \cdot P(E_2)$.

P. Simulations can play an important part in probability. Fair coins, dice, spinners, and random-digit tables are useful in performing simulations.

II. Odds, expected value, and conditional probability

A. The **odds in favor** of an event A are given by

$$\frac{P(A)}{P(\overline{A})} = \frac{P(A)}{1 - P(A)}.$$

B. The **odds against** an event A are given by

$$\frac{P(\overline{A})}{P(A)} = \frac{1 - P(A)}{P(A)}.$$

C. If events A and B are events in sample space S and $P(A) \neq 0$, then the **conditional probability** that event B occurs given that event A occurs is

$$P(B|A) = \frac{P(A \cap B)}{P(A)}.$$

D. If, in an experiment, the possible outcomes are numbers $a_1, a_2, \ldots, a_n$, occurring with probabilities $p_1, p_2, \ldots, p_n$, respectively, then the **expected value (mathematical expectation)** E is defined as

$$E = a_1 \cdot p_1 + a_2 \cdot p_2 + a_3 \cdot p_3 + \ldots + a_n \cdot p_n.$$

E. A **fair game** is a game in which the expected net winnings or expected value is \$0.

III. Counting principles

A. Fundamental Counting Principle If an event M can occur in m ways and, after it has occurred, event N can occur in n ways, then event M followed by event N can occur in $m \cdot n$ ways.

B. Permutations are arrangements in which order is important. The number of permutations of r elements chosen from n elements is given by:

$${}_nP_r = \frac{n!}{(n-r)!}$$

C. The expression $\boldsymbol{n!}$, called $\boldsymbol{n}$ **factorial**, represents the product of all the natural numbers less than or equal to n. $0!$ is defined as 1.

D. Permutations of like objects If a set contains n elements, of which r_1 are of one kind, r_2 are of another kind, and so on through r_k, then the number of different arrangements of all n elements is equal to

$$\frac{n!}{r_1! \cdot r_2! \cdot r_3! \cdot \ldots \cdot r_k!}.$$

E. Combinations are arrangements in which order is *not* important. To find the number of combinations possible, first use the Fundamental Counting Principle to find the number of permutations and then divide by the number of ways in which each choice can be arranged:

$${}_nC_r = \frac{{}_nP_r}{{}_rP_r}$$

CHAPTER REVIEW

1. A coin is flipped 3 times and heads (H) or tails (T) are recorded.
   a. List all the elements in the sample space.
   b. List the elements in the event "at least two heads appear."
   c. Find the probability that the event in part (b) occurs.
2. Suppose the names of the days of the week are placed in a box and one name is drawn at random.
   a. List the sample space for this experiment.
   b. List the event consisting of outcomes that the day drawn starts with the letter T.
   c. What is the probability of drawing a day that starts with T?
3. If you have a jar of 1000 jelly beans and you know that $P(\text{Blue}) = \frac{4}{5}$ and $P(\text{Red}) = \frac{1}{8}$, list several things you can say about the beans in the jar.
4. In the 1960 presidential election, John F. Kennedy received 34,226,731 votes and Richard M. Nixon received 34,108,157. If a 1960 voter is chosen at random, answer the following:
   a. What is the probability that the person voted for Kennedy?
   b. What is the probability that the person voted for Nixon?
   c. What are the odds that a person chosen at random did not vote for Nixon?
5. A box contains three red balls, five black balls, and four white balls. Suppose one ball is drawn at random. Find the probability of each of the following events:
   a. A black ball is drawn.
   b. A black or a white ball is drawn.
   c. Neither a red nor a white ball is drawn.
   d. A red ball is not drawn.
   e. A black ball and a white ball are drawn.
   f. A black or white or red ball is drawn.
6. One card is selected at random from an ordinary set of 52 cards. Find the probability of each of the following events:
   a. A club is drawn.
   b. A spade and a 5 are drawn.
   c. A heart or a face card is drawn.
   d. A jack is not drawn.
7. A box contains five colored balls and four white balls. If three balls are drawn one-by-one, find the probability that they are all white if the draws are made as follows:
   a. With replacement
   b. Without replacement
8. Consider the following two boxes. If a letter is drawn from box 1 and placed into box 2 and then a letter is drawn from box 2, what is the probability that the letter is an L?

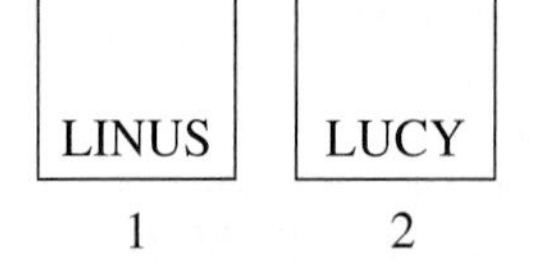

9. Use the following boxes for a two-stage experiment. First select a box at random and then select a letter at random from the box. What is the probability of drawing an A?

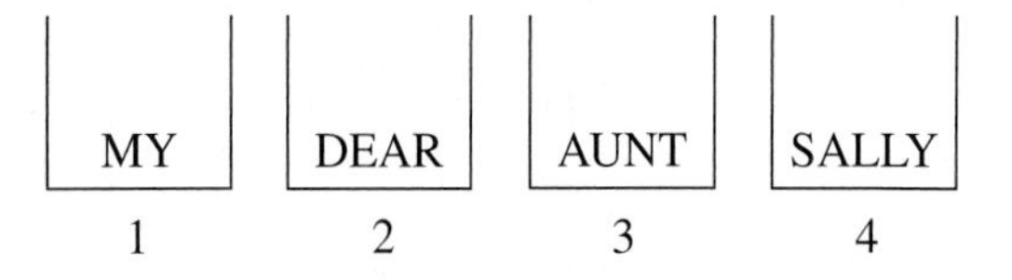

10. Consider the following boxes. Draw a ball from box 1 and put it into box 2. Then draw a ball from box 2 and put it into box 3. Finally, draw a ball from box 3. Construct a tree diagram for this experiment and calculate the probability that the last ball chosen is colored.

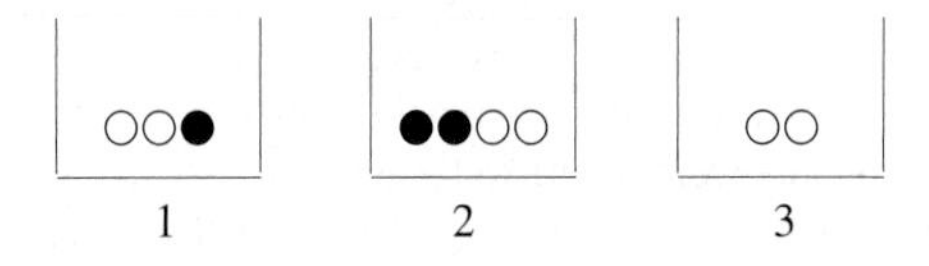

11. What are the odds in favor of drawing a jack when one card is drawn from an ordinary deck of playing cards?
12. A die is rolled once. What are the odds against rolling a prime number?
13. If the odds in favor of a particular event are 3 to 5, what is the probability that the event will occur?
14. A game consists of rolling two dice. Rolling double 1s pays \$7.20. Rolling double 6s pays \$3.60. Any other roll pays nothing. What is the expected value for this game?
15. A total of 3000 tickets have been sold for a drawing. If one ticket is drawn for a single prize of \$1000, what is a fair price for a ticket?
16. In a special raffle, a ticket costs \$2. You mark any four digits on a card (repetition and 0 are allowed). If you select the winning number, you win \$15,000. What is the expected value?
17. How many four-digit numbers can be formed if the first digit cannot be 0 and the last digit must be 2?
18. A club consists of 10 members. In how many different ways can a group of 3 people be selected to go on a European trip?

19. Find the number of ways that 4 flags can be displayed on a flagpole, one above the other, if 10 different flags are available.
20. Five women live together in an apartment. Two have blue eyes. If two of the women are chosen at random, what is the probability that they both have blue eyes?
21. Five horses (Applefarm, Bandy, Cash, Deadbeat, and Egglegs) run in a race.
    a. In how many ways can the first-, second-, and third-place horses be determined?
    b. Find the probability that Deadbeat finishes first and Bandy finishes second in the race.
    c. Find the probability that the first-, second-, and third-place horses are Deadbeat, Egglegs, and Cash, in that order.
22. Al and Ruby each roll an ordinary die once. What is the probability that the number of Ruby's roll is greater than the number of Al's roll?
23. Amy has a quiz on which she is to answer any three of the five questions. If she is equally well versed on all questions and chooses three questions at random, what is the probability that question 1 is not chosen?
24. How many batting lineups are there for the nine players of a baseball team if the center fielder must bat fourth and the pitcher last?
25. For all values of n and r, when $r \leq n$, does ${}_nC_r = {}_nC_{n-r}$? Explain.
26. On a certain street are three traffic lights. At any given time, the probability that a light is green is 0.3. What is the probability that a person will hit all three lights when they are green?
27. A three-stage rocket has the following probabilities for failure: The probability for failure at stage one is $\frac{1}{6}$; at stage two, $\frac{1}{8}$; and at stage three, $\frac{1}{10}$. What is the probability of a successful flight, given that the first stage was successful?
28. How could each of the following be simulated by using a random-digit table?
    a. Tossing a fair die
    b. Picking 3 months at random from the 12 months of the year
    c. Spinning the spinner shown

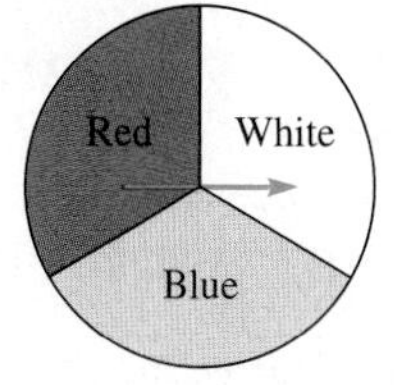

29. Otto says that if you toss three coins you get $3H$, $2H$, $1H$, or $0H$, so the probability of getting three heads is $\frac{1}{4}$. How do you respond?
30. If a dart is thrown at the following tangram dartboard and we assume the dart lands at random on the board, what is the probability of its landing in each of the following areas?
    a. Area A **b.** Area B c. Area C

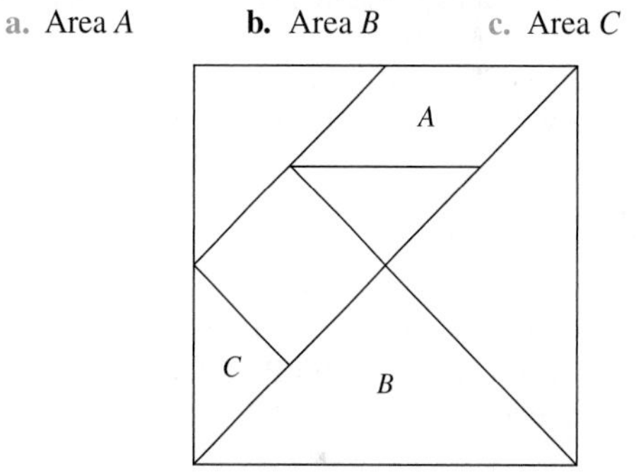

31. The points M, N, O, P, and Q in the following figure represent exits on a highway (the numbers represent miles). An accident occurs at random between points M and Q. What is the probability that it has occurred between N and O?

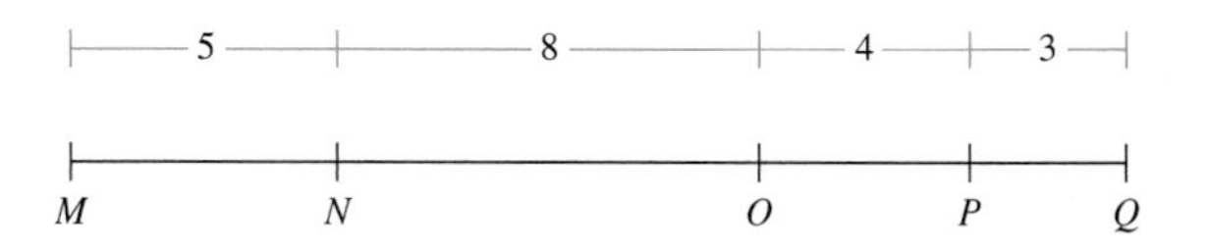

SELECTED BIBLIOGRAPHY

Aspinwall, L., and K. Shaw. "Enriching Students' Mathematical Intuition with Probability Games and Tree Diagrams." *Mathematics Teaching in the Middle School* 6 (December 2000): 214–220.

Bintz, W., and S. Moore. "Using Literature to Teach Factorials." *Mathematics Teaching in the Middle School* 8 (May 2003): 461–465.

Brahier, D. "Genetics as a Context for the Study of Probability." *Mathematics Teaching in the Middle School* 5 (December 1999): 214–221.

Coffey, D., and M. Richardson. "Rethinking Fair Games." *Mathematics Teaching in the Middle School* 10 (February 2005): 298–303.

Edwards, T., and S. Hensien. "Using Probability Experiments to Foster Discourse." *Teaching Children Mathematic* 6 (April 2000): 524–529.

Ewbank, W., and J. Ginther. "Probability on a Budget." *Mathematics Teaching in the Middle School* 7 (January 2002): 280–283.

Freda, A. "Roll the Dice; An Introduction to Probability." *Mathematics Teaching in the Middle School* 4 (October 1998): 85–89.

Hardy, M. "Burgers, Graphs, and Combinations." *Mathematics Teaching in the Middle School* 7 (October 2001): 72–76.

Hoiberg, K., J. Sharp, T. Hodgson, and J. Colbert. "Geometric Probability and the Areas of Leaves." *Mathematics Teaching in the Middle School* 10 (March 2005): 326–332.

Jardine, D. "Looking at Probability through a Historical Lens." *Mathematics Teaching in the Middle School* 6 (September 2000): 50–54.

Jones, G., C. Thornton, C. Langrall and J. Tarr. "Understanding Students' Probabilistic Reasoning." In *Developing Mathematical Reasoning in Grades K–12*, edited by L. Stiff. Reston, Va. NCTM, 1999.

Kader, G., and M. Perry. "Push-Penny: What Is Your Expected Score?" *Mathematics Teaching in the Middle School* 3 (February 1998): 370–377.

Kahneman, D., and A. Tversky, "Extensional versus Intuitive Reasoning: The Conjunction Fallacy in Probability Judgement." *Psychological Review*, 1983, 90 (4): 293–315.

Lappan, G., et al. "Area Models for Probability." *Mathematics Teacher* 80 (November 1987): 650–654.

Lawrence, A. "From *The Giver* to *The Twenty-One Balloons:* Explorations with Probability." *Mathematics Teaching in the Middle School* 4 (May 1999): 504–509.

Lecoutre, M. "Cognitive Models and Problem Spaces in 'Purely Random' Situations." *Educational Studies in Mathematics*, 1992, 23: 557–568.

Norton, R. "Determining Probabilities by Examining Underlying Structure." *Mathematics Teaching in the Middle School* 7 (October 2001): 78–82.

Quinn, R. "Using Attribute Blocks to Develop a Conceptual Understanding of Probability." *Mathematics Teaching in the Middle School* 6 (January 2001): 290–294.

Quinn, R. "Having Fun with Baseball Statistics." *Mathematics Teaching in the Middle School* 1 (May 1996): 780–785.

Shaughnessy, J., and T. Dick. "Monty's Dilemma: Should You Stick or Switch?" *Mathematics Teacher* 84 (April 1991): 252–256.

Tarr, J. "Providing Opportunities to Learn Probability Concepts." *Teaching Children Mathematic* 8 (April 2002): 482–487.

Thompson, D., and R. Austin. "Socrates and the Three Little Pigs: Connecting Patterns, Counting Trees, and Probability." *Mathematics Teaching in the Middle School* 5 (November 1999): 156–161.

Van Zoest, L., and R. Walker. "Racing to Understand Probability." *Mathematics Teaching in the Middle School* 3 (October 1997): 162–170.

Wiest, L., and R. Quinn. "Exploring Probability through an Evens-Odds Dice Game." *Mathematics Teaching in the Middle School* 4 (March 1999): 358–362.

Data Analysis/ Statistics: An Introduction

PRELIMINARY PROBLEM

Mr. Ramirez and Ms. Jonsey gave tests to their classes with the results seen in the table.

	Overall Mean	Mean for Females	Mean for Males	Percent Females
Ramirez	218	230	205	
Jonsey	221	224		88

What are the missing entries? For an overall mean, Ms. Jonsey's class is higher, but for females, the mean is higher in Mr. Ramirez's class. Is it possible that the mean for males is higher in Mr. Ramirez's class as well?

In *Principles and Standards*, we find the following:

> *Instructional programs from prekindergarten through grade 12 should enable all students to*
>
> - *formulate questions that can be addressed with data and collect, organize, and display relevant data to answer them;*
> - *select and use appropriate statistical methods to analyze data;*
> - *develop and evaluate inferences and predictions that are based on data . . .* (p. 48)

In addition, we find the following:

> *Students need to know about data analysis and related aspects of probability in order to reason statistically—skills necessary to becoming informed citizens and intelligent consumers.* (p. 48)

Data analysis, which usually refers to a more informal approach to statistics, is a relatively new term in mathematics. However, *statistics* once referred to numerical information about state or political territories. The word itself comes from the Latin *statisticus*, meaning "of the state." In today's world, much of statistics involves making sense of data.

In *A Curriculum Framework for PreK–12 Statistics Education* (March 2005), presented to the American Statistical Association (hereafter referred to as the *Curriculum Framework*), recommendations are made for statistics education that complement the *Principles and Standards for School Mathematics* (NCTM 2000). In the *Curriculum Framework*, specific data analysis recommendations are divided into three parts—A, B and C—with the more intuitive parts for early grades being in part A and the more advanced ideas being in part C. According to the *Curriculum Framework*, "Sound statistical reasoning skills take a long time to develop. . . . The surest way to reach the necessary skill level is to begin the educational process in the elementary grades and keep strengthening and expanding these skills throughout the middle and high school years" (p. 3). The combination of documents recommends a solid background for both elementary students and prospective teachers of elementary students.

In *Principles and Standards*, for the early grades, students explore the basic ideas of statistics by collecting data, organizing the data pictorially, and then interpreting information from their displays. The ideas of gathering, representing, and analyzing data are expanded in the later grades. In this chapter, we deal with categorical and numerical data, and then with representations of data, and also key statistical concepts including measures of central tendency and of variation. We also address uses and misuses of statistics.

▶ HISTORICAL NOTE

The seventeenth-century work of John Graunt (1620–1674) and the nineteenth-century work of Adolph Quetelet (1796–1874) involved making predictions on the collection of data. Graunt dealt with birth and death records, and Quetelet dealt with crime and mortality rates. Florence Nightingale (1820–1910) worked with mortality tables during the Crimean War to get British hospitals to improve care. Other notables who worked with data collection and analysis include Sir Francis Galton (1822–1911) and Gregor Mendel (1822–1884). In the twentieth century, work continued by Ronald Fisher (1890–1962) in genetics and Andrei Nikolaevich Kolmogorov (1903–1987), who also was chairman of the Commission for Mathematical Education under the Presidium of the Academy of Sciences of the former Soviet Union. ◀

8-1 Statistical Graphs of Categorical and Numerical Data

Visual illustrations are an important part of statistics. Such visual illustrations or graphs take many forms: pictographs, circle graphs, pie charts, dot plots, line plots, scatterplots, stem-and-leaf plots, box plots, frequency tables, histograms, bar graphs, and frequency polygons or line graphs. A *graph* is a picture that displays data. Graphs are used to try to tell a story. In the "Herman" cartoon, we see a graph being used to display some particular data to make a point to an audience. What message do you think the presenter is trying to get across? What labels might appear on the vertical and horizontal axes?

"That's the last time I go on vacation."

RESEARCH NOTE Describing data frequently involves reading information from graphical displays, tables, lists, and so on. The majority of students in elementary and middle-school grades can read data from these representations accurately (Beaton, Mullis, Martin et al. 1996; Bright and Friel 1998; Jones, Thornton, Langrall, et al. 1999; Jones, Thornton, Langrall, et al. 2000; Periera-Mendoza and Mellor 1991).

In *Principles and Standards*, we find the following:

A fundamental idea in prekindergarten through grade 2 is that data can be organized or ordered and that this "picture" of the data provides information about the phenomenon or question. In grades 3–5, students should develop skill in representing their data, often using bar graphs, tables, or line plots. . . . Students in grades 6–8 should begin to compare the effectiveness of various types of displays in organizing the data for further analysis or in presenting the data clearly to an audience. (p. 49)

In particular, we find the following by grade level.

In prekindergarten through grade 2 all students should

- *pose questions and gather data about themselves and their surroundings;*
- *sort and classify objects according to their attributes and organize data about the objects;*
- *represent data using concrete objects, pictures, and graphs;*
- *describe parts of the data and the set of data as a whole to determine what the data show.*

In grades 3–5 all students should

- *design investigations to address a question and consider how data-collection methods affect the nature of the data set;*
- *collect data using observations, surveys, and experiments;*
- *represent data using tables and graphs such as line plots, bar graphs, and line graphs;*
- *recognize the differences in representing categorical and numerical data;*
- *describe the shape and important features of a set of data and compare related data sets, with an emphasis on how the data are distributed. . . .*

In grades 6–8 all students should

- *formulate questions, design studies, and collect data about a characteristic shared by two populations or different characteristics within one population;*
- *select, create, and use appropriate graphical representations of data, including histograms, box plots, and scatterplots; . . .*
- *discuss and understand the correspondence between data sets and their graphical representations, especially histograms, stem-and-leaf plots, box plots, and scatterplots; . . .*
- *make conjectures about possible relationships between two characteristics of a sample on the basis of scatterplots of the data and approximate lines of fit; . . .* (p. 401)

Data: Categorical and Numerical

Statistical thinking begins in early grades with a need to know such things as "the most popular" pet, the favorite type of shoe, the most used color to paint, and so on. **Data** may be collected to find answers to such questions. Data collected may be either categorical or numerical depending on the questions being answered. For example, according to the *Curriculum Framework*, in the elementary grades, students may

be interested in the favorite type of music among students at a certain grade level. . . . The class might investigate the question: What type of music is most popular among students? *This question attempts to measure a characteristic in the population of school-grade children that will have the dance. The characteristic, favorite music type is a categorical variable—each child in that grade would be placed in a particular nonnumerical category based on his or her favorite music type. The resulting data are often called* Categorical Data (p. 22).

Categorical data are data that represent characteristics of objects or individuals in groups (or categories), such as black or white, inside or outside, male or female.

Numerical data are data collected on numerical variables. For example, in grade school, students may ask whether there is a difference in the distance that girls and boys can jump. The distance jumped is a numerical variable and the data collected is called numerical data.

Frequently, *different representations are needed for the different types of data, but in common usage, there are some representations that are used for both types of*

data

categorical data

numerical data

data. We will see different representations both for categorical and numerical data in the following sections.

▶ RESEARCH NOTE

Explorations with categorical and numerical data in instruction that uses technology with primary-aged children produce more focused and less idiosyncratic descriptions of the data (Jones, Thornton, Langrall, et al. 1999). ◀

Pictographs

pictograph

An elementary student might use a picture graph, or **pictograph**, to represent tallies of categories. For example, categorical data might be seen in the determination of the month in which the most newspapers were recycled. The month is the *category* and the data might be depicted in a pictograph. Pictographs are often seen in newspapers and magazines.

In a pictograph, a symbol or an icon is used to represent a quantity of items. A *legend* tells what the symbol represents. Pictographs are used frequently to show comparisons of outputs, as in Figure 8-1. A major disadvantage of pictographs is evident in Figure 8-1. The month of September contains a partial bundle of newspapers. It is impossible to tell from the graph the weight of that bundle with any accuracy.

FIGURE 8-1

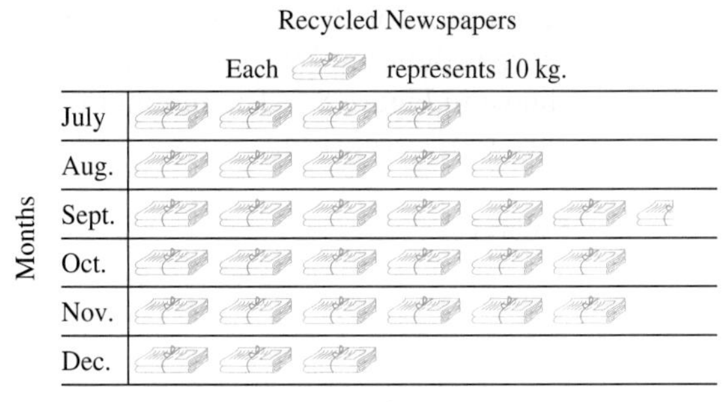

REMARK All graphs need titles; if needed, legends should be shown.

In Figure 8-2(a), the number of students in each teacher's fifth-grade class at Hillview is depicted in the tabular representation. Each teacher is a category and the frequency count table provides a method to summarize the categorical data. Figure 8-2(b) depicts the information in a pictograph.

Dot Plots

dot plot ◆ line plot

Next we examine a **dot plot**, sometimes called a **line plot**. Dot plots provide a quick, simple way of organizing numerical data. Typically, we use them when there is only one group of data with fewer than 50 values.

Choosing a Data Display

Choosing an appropriate data display is not always easy. Each type of graph is suitable for presenting certain kinds of data. In this section, you have seen pictographs, dot plots, stem-and-leaf plots, histograms, bar graphs, line graphs, scatterplots, and circle graphs (pie charts). Some uses follow:

Bar graph—Used to compare numbers of data items in grouped categories; order of data does not matter except for convenience.

Histogram—Used to compare numbers of data items grouped in numerical intervals; order matters in the data depicted.

Box plot—Used to show median, quartile, and extremes of data set (discussed later).

Stem-and-leaf plot—Used to show each value in a data set and to group values into intervals.

Scatterplot—Used to show the relationship between two sets of data.

Line graph—Used to show how data values change over time; normally used for continuous data.

Circle graph—Used to show the division of a whole into parts.

RESEARCH NOTE

Many elementary students have difficulty creating visual displays of data (Mullis, Martin, et al. 1997). Middle-school students have substantial gaps in abilities to construct graphs from given data (Berg and Phillips 1994).

NOW TRY THIS 8-3

a. Which graph in Figure 8-16(a) or (b) displays the data more effectively? Why?

FIGURE 8-16

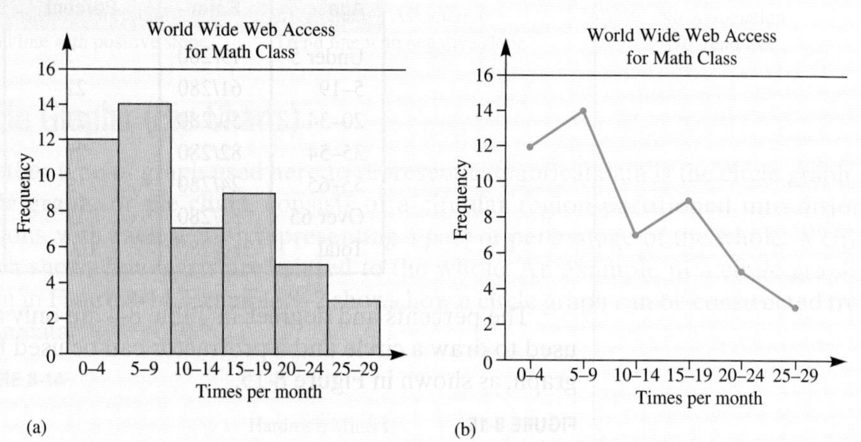

Note: Figure 8-16(b) could have been formed by connecting the midpoints of the tops of the bars in Figure 8-16(a).

b. Explain whether connecting the dots with line segments as in Figure 8-16(b) is meaningful.

c. To show each of the following, which graph is the best choice: line graph, bar graph, or circle graph? Why?

(i) The percentage of a college student's budget devoted to housing, clothing, food, tuition and books, taxes

(ii) Showing the change in the cost of living over the past 12 months

ONGOING ASSESSMENT 8-1

1. Make a pictograph to represent the categorical data in the following table. Use to represent 10 glasses of lemonade sold.

Glasses of Lemonade Sold

Day	Tally	Frequency
Monday	卌 卌 卌	15
Tuesday	卌 卌 卌 卌	20
Wednesday	卌 卌 卌 卌 卌 卌	30
Thursday	卌	5
Friday	卌 卌	10

2. The following pictograph shows the approximate number of people who speak the six most common languages on Earth.
   a. About how many people speak Spanish?
   b. About how many people speak English?
   c. About how many more people speak Mandarin than Arabic?

Number of People Speaking the Six Most Common Languages

Arabic
English
Hindi
Mandarin
Russian
Spanish

Each ● represents 100 million people.

3. Draw a double-bar graph showing that in a certain city, the *New York Times* is much more popular than the *Wall Street Journal*. Show the title, legend, and any other needed information so that the graph could be understood only by looking at it.
4. Following are the ages of the 30 students from Washington School who participated in the city track meet. Draw a dot plot to represent these data.

10	10	11	10	13	8	10	13	14	9
14	13	10	14	11	9	13	10	11	12
11	12	14	13	12	8	13	14	9	14

5. The Automobile Club of America considers that its holiday weekends start on Friday afternoon at 6:00 P.M. They last until Monday night at midnight.
   a. If 150 people died on one such weekend by Saturday at 6:00 P.M. and the death rate is constant, draw a graphical representation showing how many people might die on highways in the entire weekend.
   b. Discuss the reasonableness of your graph and the estimated total.
6. If a 3-in-long rectangular bar represents 100% of a population, divide the bar into segments to represent the budget of a family whose total monthly income is \$4500 and who spends the following:

Rent	\$1800
Food	\$1500
Transportation	\$200
Entertainment	\$400
Utilities	\$300
Savings	All that is left

7. The following stem-and-leaf plot gives the weight in pounds of all 15 students in the Algebra 1 class at East Junior High:

Weights of Students in East Junior High Algebra 1 Class

7	24
8	112578
9	2478
10	3
11	
12	35

10 | 3 represents 103 lb

   a. Write the weights of the 15 students.
   b. What is the weight of the lightest student in the class?
   c. What is the weight of the heaviest student in the class?
8. Draw a histogram based on the stem-and-leaf plot in problem 7.
9. Toss a coin 30 times.
   a. Construct a dot plot for the data.
   b. Draw a bar graph for the data.
10. The following figure shows a bar graph of the rainfall in centimeters during the last school year:

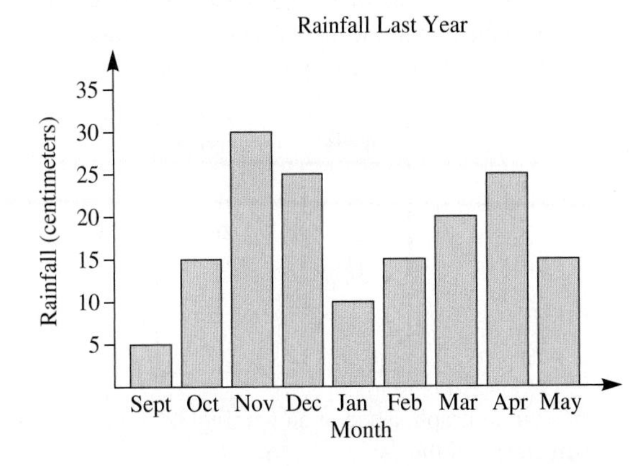

   a. Which month had the most rainfall, and how much did it have?
   b. How much total rain fell in October, December, and January?

11. HKM Company employs 40 people of the following ages:

34	58	21	63	48	52	24	52	37	23
23	34	45	46	23	26	21	18	41	27
23	45	32	63	20	19	21	23	54	62
41	32	26	41	25	18	23	34	29	26

a. Draw a stem-and-leaf plot for the data.
b. Are more employees in their 40s or in their 50s?
c. How many employees are less than 30 years old?
d. What percentage of the people are 50 years or older?

12. Given the following bar graph, estimate the length of the following rivers:
a. Mississippi **b.** Columbia

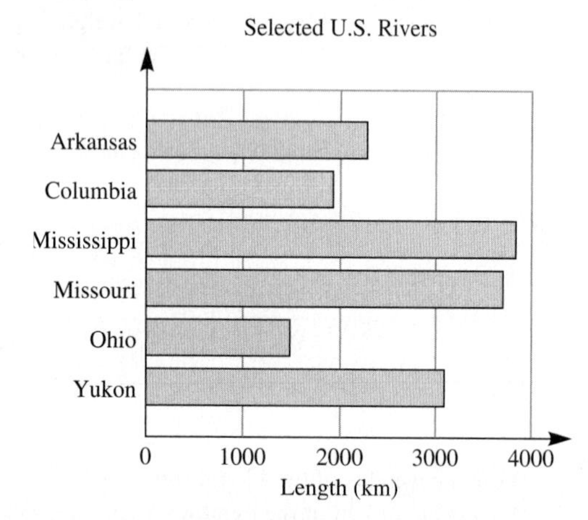

13. Five coins are tossed 64 times. A distribution for the number of heads obtained is shown in the following table. Draw a bar graph for the data.

Number of Heads	0	1	2	3	4	5
Frequency	2	10	20	20	10	2

14. The following table shows the grade distribution for the final examination in the mathematics course for elementary teachers. Draw a circle graph for the data.

Grade	Frequency
A	4
B	10
C	37
D	8
F	1

15. In a circle graph, a sector containing 32° represents what percentage of the data?

16. a. What type of graph might be used to depict the changing life expectancies for women in the United States over the past 200 years?
b. Justify your use of the graph listed in part (a).

17. A single bar graph is sometimes used like a circle graph. If the bar is 8 cm, how long is each piece?

Savings	Rent	Food	Auto Payment	Tuition
10%	30%	12%	27%	x%

18. In the following graph, a book club of 21 people chose their favorite book type, such as detective story, romance, and so on.
a. How many people are represented in each sector of the circle graph?
b. If *you* were describing the types of books that were chosen in the 5% sectors, what would they be and why?

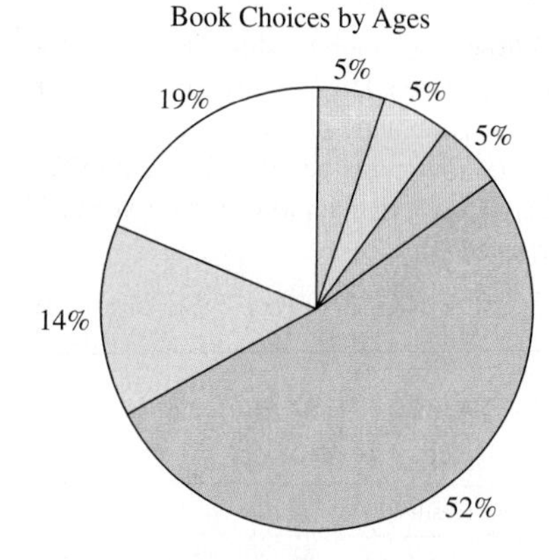

19. If the choices of book types by people in different age groups are depicted as in the bar graph given, explain whether or not you think that the graph is accurate and why your answer is what it is.

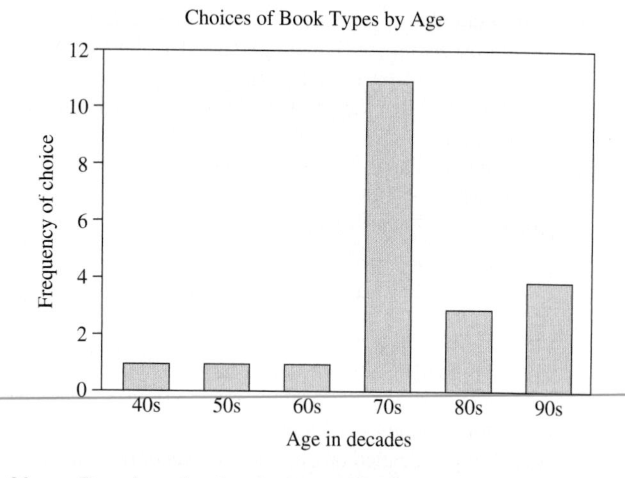

20. a. Based on the data in the graph of problem 19, in what age decade do people read the most?
b. Based on the data in the graph of problem 19, what is the mode? Is it a frequency of choice or is it an age in decades? Explain your reasoning.

21. The following are the amounts (to the nearest dollar) paid by 25 students for textbooks during the fall term:

35	42	37	60	50
42	50	16	58	39
33	39	23	53	51
48	41	49	62	40
45	37	62	30	23

   a. Draw an ordered stem-and-leaf plot to illustrate the data.
   b. Construct a grouped frequency table for the data, starting the first class at $15.00 with intervals of $5.00 each.
   c. Draw a bar graph of the data.

22. The following horizontal bar graph gives the top speeds of several animals:

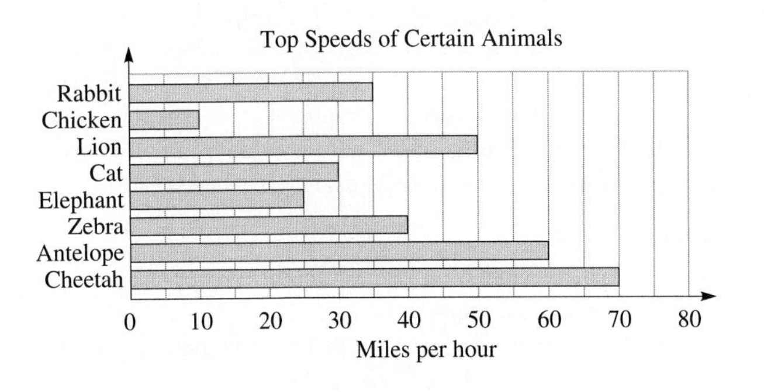

   a. Which is the slowest animal shown?
   b. How fast can a chicken run?
   c. Which animal can run twice as fast as a rabbit?
   d. Can a lion outrun a zebra?

23. The following bar graph shows the life expectancies for men and women:

Life Expectancy at Birth

Life expectancy: 50, 55, 60, 65, 70, 75, 80

Year of birth: 1925, 1935, 1945, 1955, 1965, 1975, 1985, 1995, 2005

Men Women

   a. Whose life expectancy has changed the most since 1925?
   b. In 1925, about how much longer was a woman expected to live than a man?
   c. In 2005, about how much longer was a woman expected to live than a man?

24. The following graph shows how the value of a car depreciates each year. This graph allows us to find the trade-in value of a car for each of 5 yr. The percents given in the graph are based on the selling price of the new car.

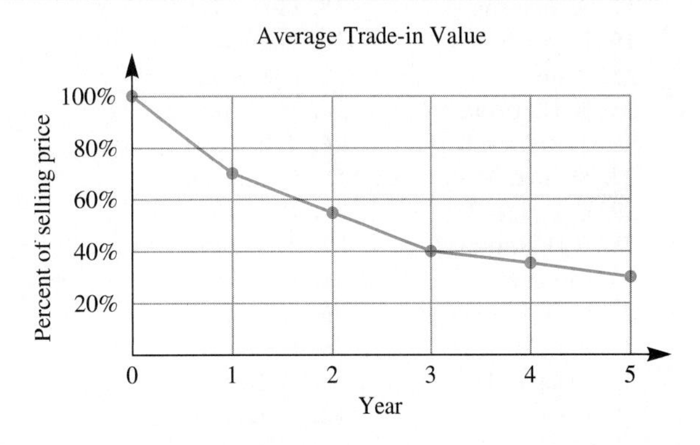

   a. What is the approximate trade-in value of a $12,000 car after 1 yr?
   b. How much has a $20,000 car depreciated after 5 yr?
   c. What is the approximate trade-in value of a $20,000 car after 4 yr?
   d. Dani wants to trade in her car before it loses half its value. When should she do this?

25. Use the circle graph to answer the following questions:

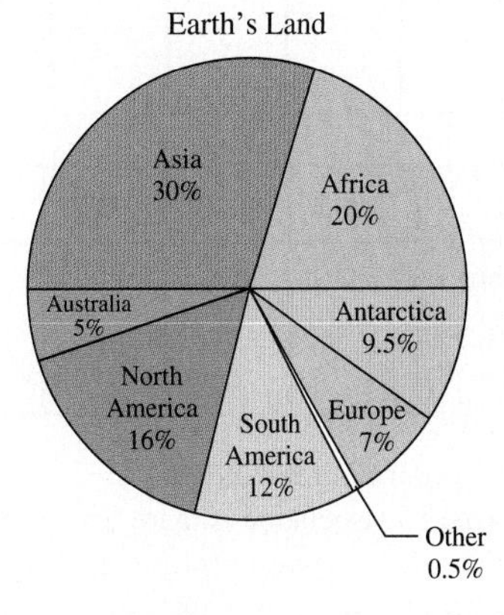

   a. Which is the largest continent?
   b. Which continent is about twice the size of Antarctica?
   c. How does Africa compare in size to Asia?
   d. Which two continents make up about half of Earth's surface?
   e. What is the ratio of the size of Australia to North America?
   f. If Europe has approximately 4.1 million mi^2 of land, what is the total area of the land on Earth?

26. A list of presidents, with the number of children for each, follows:

   1. Washington, 0 **2.** J. Adams, 5
   3. Jefferson, 6 **4.** Madison, 0
   5. Monroe, 2 **6.** J. Q. Adams, 4
   7. Jackson, 0 **8.** Van Buren, 4

9. W. H. Harrison, 10	**10.** Tyler, 14
11. Polk, 0	**12.** Taylor, 6
13. Fillmore, 2	**14.** Pierce, 3
15. Buchanan, 0	**16.** Lincoln, 4
17. A. Johnson, 5	**18.** Grant, 4
19. Hayes, 8	**20.** Garfield, 7
21. Arthur, 3	**22.** Cleveland, 5
23. B. Harrison, 3	**24.** McKinley, 2
25. T. Roosevelt, 6	**26.** Taft, 3
27. Wilson, 3	**28.** Harding, 0
29. Coolidge, 2	**30.** Hoover, 2
31. F. D. Roosevelt, 6	**32.** Truman, 1
33. Eisenhower, 2	**34.** Kennedy, 4
35. L. B. Johnson, 2	**36.** Nixon, 2
37. Ford, 4	**38.** Carter, 4
39. Reagan, 4	**40.** G. Bush, 5
41. Clinton, 1	**42.** G. W. Bush, 2

a. Construct a dot plot for these data.
b. Make a frequency table for these data.
c. What is the most frequent number of children?

27. Coach Lewis kept track of the basketball team's jumping records for a 10-year period, as follows:

Year	1996	1997	1998	1999	2000	2001
Record (nearest in.)	65	67	67	68	70	74

Year	2002	2003	2004	2005
Record (nearest in.)	77	78	80	81

a. Draw a scatterplot for the data.
b. What kind of association is there for these data?

28. Refer to the following scatterplot regarding movie attendance in a certain city:

Yearly Movie Attendance

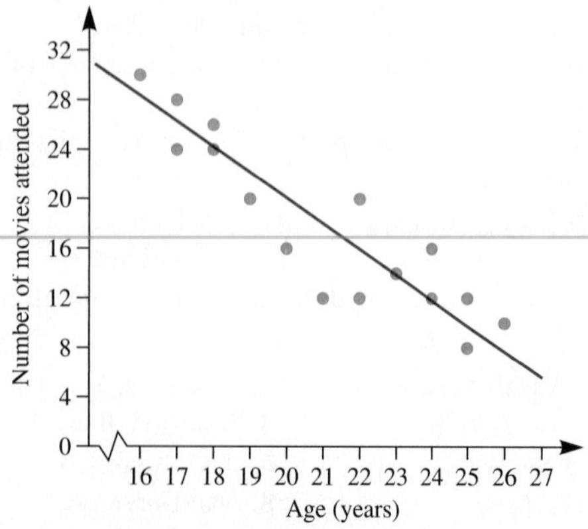

a. What type of association exists for these data?
b. About how many movies does an average 25-year-old attend?
c. From the data in the scatterplot, conjecture how old you think a person is who attends 16 movies a year.

29. The following table depicts the number of deaths in males in the United States from Acquired Immunodeficiency Syndrome (AIDS) by age in 1985 and in 2002:

Age (yr)	**1985**	**2002**
13–19	27	199
20–29	1501	3433
30–39	3588	12,101
40–49	1634	10,658
50–59	597	3959
60–69	157	1294

a. Choose and construct a graph to display the data.
b. Are any patterns of difference evident in the comparison of the two groups of data?

Communication

30. a. Discuss when a pictograph might be more appropriate than a circle graph.
b. Discuss when a circle graph might be more appropriate than a bar graph or a line graph.
c. Give an example of a set of data for which a stem-and-leaf plot would be more informative than a histogram.

31. Explain whether a circle graph would change if the amount of data in each category was doubled.

32. Explain why the sum of the percents in a circle graph should always be 100. How could it happen that the sum is close to 100%?

33. The federal budget for 1 yr is typically depicted with one type of visual representation. Which one is used and why?

34. Tell whether it is appropriate to use a bar graph for each of the following. If so, draw the appropriate graph.
a. U.S. population
b. Continents of the world

Year	**U.S. Population**
1920	105,710,620
1930	122,775,046
1940	131,669,275
1950	150,697,361
1960	179,323,175
1970	203,302,031
1980	226,542,203
1990	248,765,170
2000	281,421,906

Continent	**Area in Square Miles (mi^2)**
Africa	11,694,000
Antarctica	5,100,000
Asia	16,968,000
Australia	2,966,000
Europe	4,066,000
North America	9,363,000
South America	6,886,000

35. Discuss the trend in U.S. car sales from 1985 to 2003 based on the information in the given circle graphs.

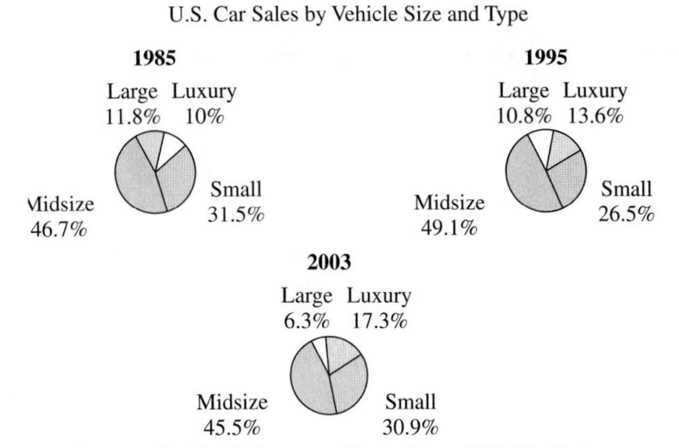

Data taken from *The World Almanac and Book of Facts 2005*. World Almanac Books, 2005.

36. The following graphs give the temperatures for a certain day. Which graph is more helpful for guessing the actual temperature at 10:00 A.M.? Why?

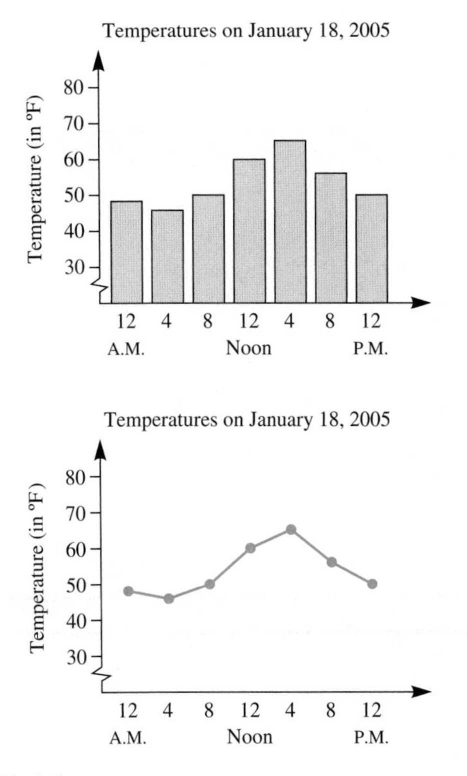

Open-Ended

37. If you measured the circumference of each person's head in your class, measured the width of the face at approximately the middle of the forehead, plotted the face width on a horizontal axis, and the circumference on the vertical axis,

a. Explain whether or not you would expect a positive association between the two variables.

b. If you expected a positive association in part (a), is there a mathematical basis for that expectation? If so, what is it?

38. Find six recent examples of different types of visual representations of data in your newspaper. Explain whether you think the representations are appropriate.

39. Choose a topic, describe how you would go about collecting data on the topic, and then explain how you would display your data in a graph. Tell why you chose the particular graph.

40. A graph similar to the following one was depicted on a package of cigarettes in Canada.

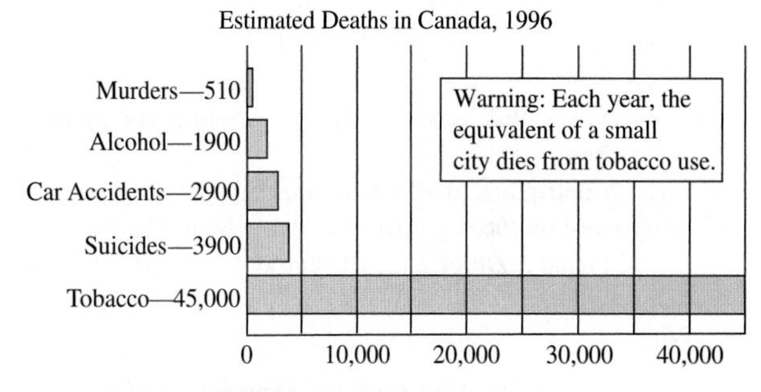

a. Make a circle graph of the data.

b. Explain which graph you think is more effective.

c. Explain whether you would like to see this labeling on cigarette packages in the United States.

41. The following problem from *Middle Grades Math Thematics, Book 2*, McDougal Littell, 2005 (page 332) depicts a way of finding a trend line (fitted line). Choose any scatterplot and answer questions 18a–c.

18 a. Place a piece of spaghetti on the scatter plot so that the data points are close to the spaghetti and there are about the same number of points on one side of the spaghetti as on the other side.

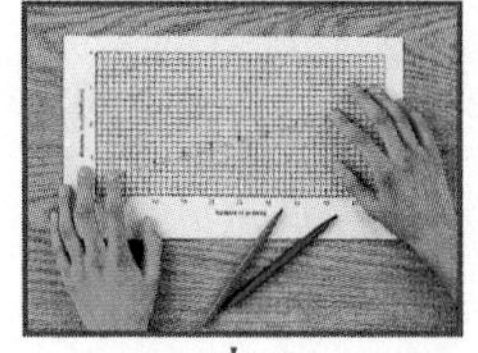

b. Draw a line segment along the piece of spaghetti. This is a **fitted line** for the scatter plot.

c. Discussion Do you think every person will draw the same line? Explain.

42. Study the following graph and draw some conclusions about the death rates for the selected causes:

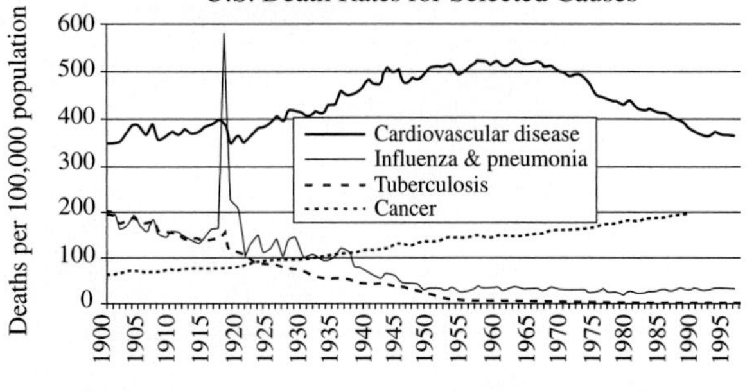

Data taken from *The World Almanac and Book of Facts 1999*, World Almanac Books, 1999.

Cooperative Learning

43. a. Perform the experiment in problem 37.
    b. If you were a seller of hats, what is the most common size head for which you would purchase hats?
44. Find data regarding the cost of running for the U.S. Senate. Use the data to argue that only wealthy people may run for the Senate.
45. Decide on and give a rationale for the type of graph that you would use to show the percentage of time that professors spend on teaching, service, and research.
46. a. Draw a circle graph depicting the percentage of pages devoted to each chapter in this book.
    b. Based on your circle graph, which chapter do you think is the most important in the book?
    c. Arrange a debate with your classmates to argue the merits of this type of use of data and conclusions reached.
47. Choose one page of this text. Find the word length of every word on the page. Draw a graph depicting this data. If you chose another page of the book at random, what would you expect the most common word length to be? Why?

National Assessment of Educational Progress (NAEP)

FINAL TEST SCORES	
Score	**Number of Students**
95	50
90	120
85	170
80	60
75	10

Use the information in the preceding table to complete the following bar graph.

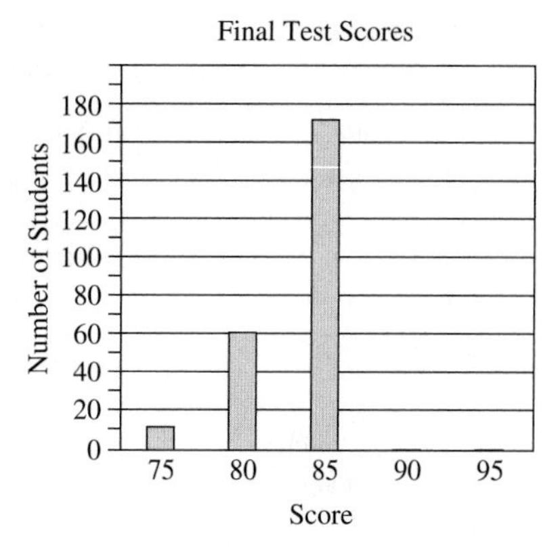

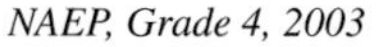

NAEP, Grade 4, 2003

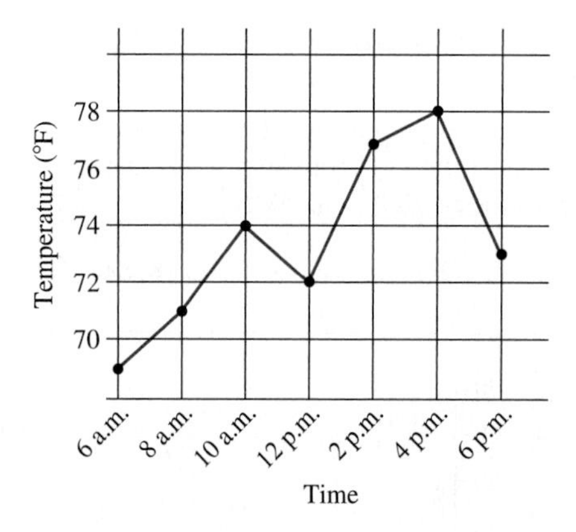

According to the graph above, the temperature at 10 A.M. is approximately how many degrees greater than the temperature at 8 A.M.?

a. 1 b. 1.5
c. 2 d. 2.5
e. 3

NAEP, Grade 12, 2003

TECHNOLOGY CORNER

The student page shows how scatterplots may be drawn with a spreadsheet. See how the technology you use allows you to find a trend line or fitted line.

SCATTER PLOTS

TECHNOLOGY Using Technology to Make Scatter Plots

You can use spreadsheet software or other graphing technology to make your scatter plot for Question 17 on page 332.

Enter the data for the graph into the spreadsheet.

Enter the data for the horizontal axis in column A.

File Edit Format Calculate Options View

NUMBER OF PENNIES VS. DISTANCE

B6 | × | ✓ | 1.5

	A	B	C	D	E
1	Number of pennies	Distance (cm)			
2	5	0.5			
3	10	0.8			
4	15	1.2			
5	20	1.4			
6	25	1.5			

Enter the corresponding data for the vertical axis in column B.

To make a graph, highlight the data you entered and choose the option that makes a chart. Then select the type of graph you want to make.

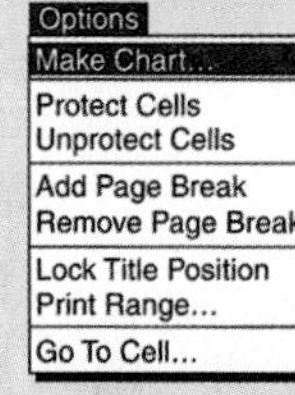

Chart Options

Modify: Axis, Series, Labels, General

Gallery: Bar, Line, Scatter, Pie, Stacked Bar, X–Y Line, X–Y Scatter, Pictogram

Experiment with the labels, grid lines, and scale until the graph appears the way you want it to. Be sure to include a title.

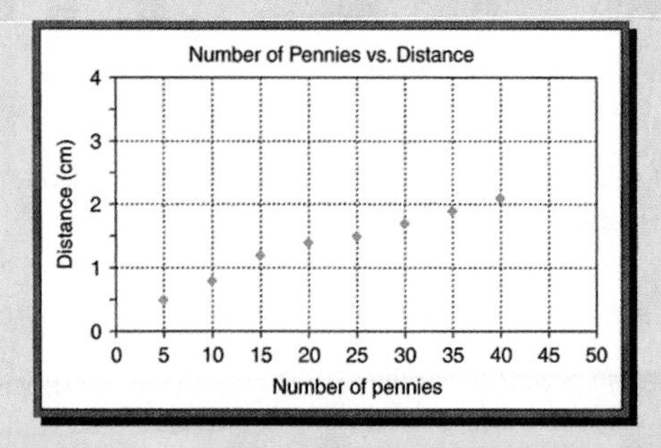

Step 4 You can print out the graph and draw a fitted line through your scatter plot as you did in Question 18 on page 332. Some graphing technology will also let you draw a fitted line or it will draw one for you.

Source: Math Thematics, Book 3, McDougal Littell, 2005 (p. 333).

BRAIN TEASER

In the "Fox Trot" cartoon, Peter describes one situation where a line-of-best fit might be used. Decide whether you think this is appropriate or not.

FOXTROT © Bill Amend. Reprinted with permission of UNIVERSAL PRESS SYNDICATE. All rights reserved.

8-2 Measures of Central Tendency and Variation

In *Principles and Standards*, we find the following by grade level.

In grades 3–5 all students should

. . .

- *use measures of center, focusing on the median, and understand what each does and does not indicate about the data sets;*
- *compare different representations of the same data and evaluate how well each representation shows important aspects of the data;. . .* (p. 400)

In grades 6–8 all students should

. . .

- *find, use, and interpret measures of center and spread, including mean and interquartile range;. . .* (p. 401)

The media present us with a variety of data and statistics. For example, we find in the *World Almanac* that the average person's lifetime includes 6 yr of eating, 4 yr of cleaning, 2 yr of trying to return telephone calls to people who never seem to be in, 6 mo waiting at stop lights, 1 yr looking for misplaced objects, and 8 mo opening junk mail. In the previous section, we examined data by looking at graphs to display the overall distribution of values. In this section, we describe specific aspects of data

▶ RESEARCH NOTE

measures of central tendency
measures of center

by using a few carefully chosen numbers. Two important aspects of data are its *center* and its *spread.* The mean and median are **measures of central tendency** or of **center** that describe where data are centered. Each of these measures is a single number that describes the data. However, each does it slightly differently. The *range, interquartile range, variance, mean absolute deviation*, and *standard deviation* introduced later in this section describe the spread of data and should be used with measures of central tendency.

A word that is often used in statistics is *average*. For example, suppose that, as the "Far Side" cartoon below suggests, the average number of children in a family is 1.5. What does this mean? How can the average number of children be 1.5?

"Bob and Ruth! Come on in. ... Have you met Russell and Bill, our 1.5 children?"

The Far Side® by Gary Larson © 1988 FarWorks, Inc. All Rights Reserved. Used with permission.

To explore more about averages, examine the following set of data for three teachers, each of whom claims that his or her class scored better *on the average* than the other two classes did:

Mr. Smith:	62, 94, 95, 98, 98
Mr. Jones:	62, 62, 98, 99, 100
Ms. Rivera:	40, 62, 85, 99, 99

All of these teachers are correct in their assertions because each has used a different number to characterize the scores in the class. In the following, we examine how each teacher can justify the claim.

Choosing the Most Appropriate Average

Although the *mean* is the number most commonly used "average" to describe a set of data, it may not always be the most appropriate choice.

EXAMPLE 8-4

Suppose a company employs 20 people. The president of the company earns $200,000, the vice president earns $75,000, and 18 employees earn $10,000 each. Is the mean the best number to choose to represent the "average" salary for the company?

Solution The mean salary for this company is

$$\frac{\$200{,}000 + \$75{,}000 + 18(\$10{,}000)}{20} = \frac{\$455{,}000}{20} = \$22{,}750.$$

In this case, the mean salary of $22,750 is not representative. Either the median or mode, both of which are $10,000, would describe the typical salary better.

▲

In Example 8-4, notice that *the mean is affected by extreme values*. In most cases, the *median* is not affected by extreme values. The median, however, can also be misleading, as shown in the following example.

EXAMPLE 8-5

Suppose nine students make the following scores on a test:

30, 35, 40, 40, 92, 92, 93, 98, 99

Is the median the best "average" to represent the set of scores?

Solution The median score is 92. From that score, one might infer that the individuals all scored very well, yet 92 is certainly not a typical score. In this case, the mean of approximately 69 might be more appropriate than the median. However, with the spread of the scores, neither is very appropriate for this distribution.

▲

The *mode*, too, can be misleading in describing a set of data with very few items that occur frequently, as shown in the following example.

EXAMPLE 8-6

Is the mode an appropriate "average" for the following test scores?

40, 42, 50, 62, 63, 65, 98, 98

Solution The mode of the set of scores is 98 because this score occurs most frequently. The score of 98 is not representative of the set of data because of the large spread of scores.

▲

The choice of which number to use to represent a particular set of data is not always easy. In the example involving the three teachers, each teacher chose the average that best suited his or her claim. The type of average should always be specified along with a measure of spread (*Curriculum Framework*, p. 29).

PROBLEM SOLVING **The Missing Grades**

Students of Dr. Van Horn were asked to keep track of their own grades. One day, Dr. Van Horn asked the students to report their grades. One student had lost the papers but claims to remember the grades on four of six assignments: 100, 82, 74, and 60. In addition, the student remembered that the mean of all six papers was 69, and the other two papers had identical grades. What were the grades on the other two homework papers?

Understanding the Problem The student had scores of 100, 82, 74, and 60 on four of six papers. The mean of all six papers was 69, and two identical scores were missing. The missing scores must be less than 60; otherwise, from observation, the mean could not be less than three of the four known scores and greater than the fourth one given.

Devising a Plan To find the missing grades, we use the strategy of *writing an equation* for x. The mean is obtained by finding the sum of the scores and then dividing by the number of scores, which is 6. So if we let x stand for each of the two missing grades, we have

$$69 = \frac{100 + 82 + 74 + 60 + x + x}{6}$$

Carrying Out the Plan We now solve the equation as follows:

$$69 = \frac{100 + 82 + 74 + 60 + x + x}{6}$$

$$69 = \frac{316 + 2x}{6}$$

$$49 = x$$

Since the solution to the equation is $x = 49$, each of the two missing scores was 49.

Looking Back The answer of 49 seems reasonable, since the mean of 69 is less than three of the four given scores. We can check this by computing the mean of the scores 100, 82, 74, 60, 49, 49 and showing that it is 69.

Measures of Spread

The mean and median provide limited information about a whole distribution of data. For example, if you sit in a sauna for 30 min and then in a refrigerated room for 30 min, an average temperature of your surroundings for that hour might sound comfortable. To tell how much the data are scattered, we develop measures of *spread* or *dispersion*. Perhaps the easiest way to measure spread is the **range**, the difference between the greatest and the least values in a data set. For example, the range in the set of data 1, 3, 7, 8, 10 is $10 - 1 = 9$. However, just because the range of two sets of data is the same, the data do not have to have the same dispersion. For example, the data set 1, 10, 10, 10, 10 also has a range of 9 and is spread quite differently from the first collection of data. For this reason we need other measures of spread besides the range.

range

Another measure of spread is the *interquartile range (IQR)*. The IQR is the range of the middle half of the data. Consider the following set of test scores:

20 25 40 50 50 60 70 75 80 80 90 100 100

The range for this set of scores is 100 − 20 = 80. The median score for this set of data is 70. We mark this location with a vertical bar between the 7 and the 0 as shown.

20 25 40 50 50 60 7 | 0 75 80 80 90 100 100

vertical bar

Next, we consider only the data values to the left of the **vertical bar** and draw another vertical bar where the median of those values is located:

20 25 40 | 50 50 60

lower quartile ◆ first quartile (Q_1)

upper quartile (Q_3)

The score of 45 = (40 + 50)/2 is the median of the scores less than the median of all scores and is the **lower quartile**. The lower quartile, or the **first quartile**, is denoted Q_1. One quarter, or 25%, of the scores lie at or below Q_1. Similarly, we can find the upper, or third, quartile (Q_3), which is (80 + 90)/2, or 85. The **upper quartile (Q_3)** is the median of the scores greater than the median of all scores. Three-quarters, or 75%, of the scores lie at or below Q_3. Thus we have divided the scores into four groups of three scores each:

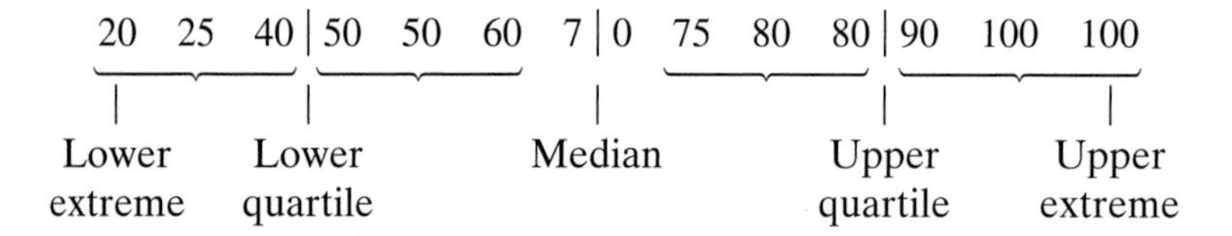

IQR

The *interquartile range* (**IQR**) is the difference between the upper quartile and the lower quartile. In this case, IQR = 85 − 45 = 40. The IQR is itself another useful measure of spread because it is less influenced by extreme values. The IQR contains the middle 50% of the values.

The interquartile range is the measure of spread most often reported with the median. This is done because the median marks the exact center of the data. However, as we saw earlier, the same median could be reported with two sets of data that have very different spreads. With the interquartile range reported along with the median, not only do we know the middle, we know how spread the middle 50% of the data are. If it is widespread, then we are aware of it and can assess and use the data accordingly. For example, in the data set just presented, the median of 70 should be reported with the interquartile range of 40. In Now Try This 8-7, we are asked to create a distorted set of data that has these two measures associated with it.

NOW TRY THIS 8-7

Create a set of data that ranges from 0 to 100 and is widely spread yet has median 70 and an interquartile range of 70.

Box Plots

box plot ◆ box-and-whisker plot

Reporting data displayed in a **box plot** (or a **box-and-whisker plot**) is a way to avoid issues such as found in Now Try This 8-7. A box plot is a way to display data visually and draw informal conclusions. Box plots show only certain data. These plots are visual representations of the *five-number summary* of the data. The five numbers are the median, the upper and lower quartiles (and, hence, the interquartile range information), and the least and greatest values in the distribution. The center, the spread, and the overall range are immediately evident by looking at the plot.

To construct a box plot, we need data along with its median, upper and lower quartiles, and extremes marked with vertical bars. To construct the box, we connect the vertical bars at the quartiles to form a box. We draw segments from each end of the box to the extreme values to form the whiskers. The box plot can be either vertical or horizontal. A vertical version of the box plot for the given data is shown in Figure 8-19.

FIGURE 8-19

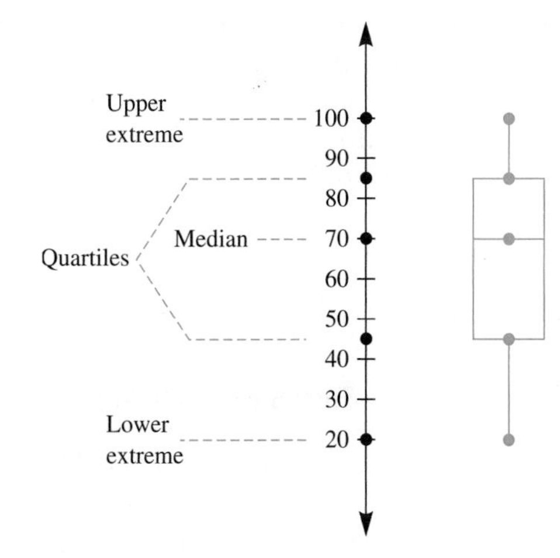

The box plot gives a fairly clear picture of the spread of the data. If we look at the graph in Figure 8-19, we can see that the median is 70, the maximum value is 100, the minimum value is 20, and the upper and lower quartiles are 45 and 85. The median is above the center of the box and so there are more scores above than below it.

EXAMPLE 8-7

What are the minimum and maximum values, the median, and the lower and upper quartiles of the box plot shown in Figure 8-20?

FIGURE 8-20

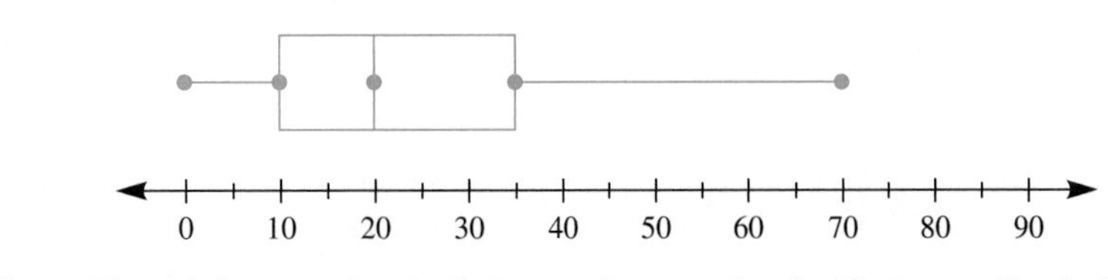

Solution The minimum value is 0, the maximum value is 70, the median is 20, the lower quartile is 10, and the upper quartile is 35. ▲

Outliers

An *outlier* is a value that is widely separated from the rest of a group of data. For example, in a set of scores such as

91 92 92 93 93 93 94

all data are grouped close together and no values are widely separated. However, in a set of scores such as

21 92 92 93 93 93 95 150

both 21 and 150 are widely separated from the rest of the data. These values are potential outliers. The upper and lower extreme values are not necessarily outliers. In data such as

75 90 91 92 92 93 93,

it is not easy to decide, so we develop a convention for determining outliers. *An* **outlier** *is any value that is more than 1.5 times the interquartile range above the upper quartile or below the lower quartile.* Statisticians sometimes use values different from 1.5 to determine outliers.

outlier

It is common practice to indicate outliers with asterisks. Whiskers are then drawn to the extreme points that are not *outliers.* To investigate how this works, consider Example 8-8.

EXAMPLE 8-8

Draw a box plot of the data in Table 8-6 and identify possible outliers.

Table 8-6 Final Medal Standings for Top 20 Countries—2000 Summer Olympics

Country	Medals	
United States	97	
Russia	88	
China	59	
Australia	58	
Germany	57	$Q_3 = 45.5$
France	38	
Italy	34	
Cuba	29	
Britain	28	
South Korea	28	Median (Q_2) = 27
Romania	26	
Netherlands	25	
Ukraine	23	
Japan	18	
Hungary	17	$Q_1 = 17$
Belarus	17	
Poland	14	
Canada	14	
Bulgaria	13	
Greece	13	

Data taken from *The World Almanac and Book of Facts 2002*, World Almanac Books, 2002.

Solution The extreme scores are 97 and 13, the median is 27, $Q_1 = 17$, and $Q_3 = 45.5$. The IQR is 45.5 − 17, or 28.5. Outliers are scores that are greater than 45.5 + 1.5(28.5), or 88.25, or less than 17 − 1.5(28.5), or $^{-}25.75$. Therefore, in this data 97 is the only outlier. A box plot is given in Figure 8-21. Notice that the whisker stops at the extreme point 13 on the lower end and at 88 on the upper end. The outlier is indicated with an asterisk.

FIGURE 8-21

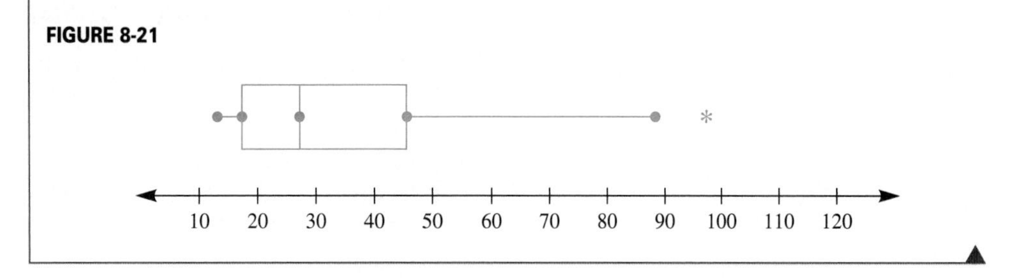

Comparing Sets of Data

Box plots are used primarily for large sets of data or for comparing several distributions. The stem-and-leaf plot is usually a much clearer display for a single distribution. Parallel box plots drawn using the same number line give us the easiest comparison of medians, extreme scores, and the quartiles for the sets of data. As an example, we construct parallel box plots comparing the data in Table 8-7.

Table 8-7 Mean Earnings of Males and Females, 1967–2003 (full-time workers age 15 and older; income in current dollars)

	Male	Female		Male	Female
2003	$ 41,483	$ 24,630	1984	19,438	9,584
2002	41,057	23,619	1983	18,109	8,780
2001	40,859	23,602	1982	17,381	8,195
2000	40,254	22,428	1981	16,515	7,440
1999	38,384	21,312	1980	15,340	6,772
1998	36,315	20,462	1979	14,311	6,026
1997	34,794	19,511	1978	13,113	5,599
1996	32,800	18,369	1977	12,063	5,291
1995	31,454	17,265	1976	11,165	4,875
1994	30,367	16,478	1975	10,429	4,513
1993	28,939	15,761	1974	9,861	4,161
1992	26,810	14,922	1973	9,289	3,799
1991	26,369	14,449	1972	8,635	3,577
1990	26,041	13,913	1971	7,892	3,333
1989	25,746	13,226	1970	7,537	3,138
1988	24,054	12,311	1969	7,202	2,945
1987	22,798	11,538	1968	6,626	2,732
1986	21,822	10,741	1967	6,054	2,483
1985	20,652	10,173			

Source: Data taken from U.S. Census Data, 2004.

Before constructing parallel box plots, we find the five important values for each group of data. These values are given in Table 8-8.

Table 8-8

Value	Female	Male
Maximum	$24,360.00	$41,483.00
Upper quartile	16,876.50	30,910.50
Median	10,173.00	20,652.50
Lower quartile	4,604.00	10,797.00
Minimum	2,483.00	6,054.00

In this example, the IQR for males is $20,117.50; the IQR for females is $12,272.50. Checking reveals there are no outliers.

Next we draw the horizontal scale and construct the box plots for the female and male population using the data in Table 8-8, as shown in Figure 8-22.

FIGURE 8-22

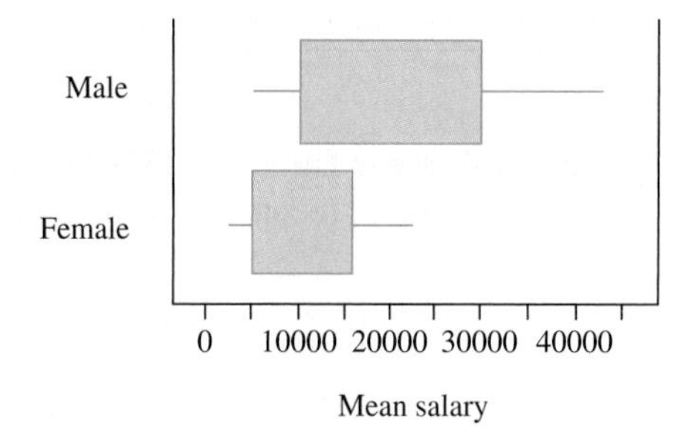

From the box plot in Figure 8-22, we can see that the mean salaries for males have been higher than those for females, since the extreme values, median, and quartiles for the males are greater than those for females. Also, more than 50% of the mean salaries for males are greater than those for the mean salaries of females over the time period.

Although we cannot spot clusters or gaps in box plots as we can with stem-and-leaf or line plots, we can more easily compare data from different sets. With box plots, we do not need to have sets of data that are approximately the same size, as we did for stem-and-leaf plots. To compare data from two or more sets using their box plots, we first study the boxes to see if they are located in approximately the same places. Next, we consider the lengths of the boxes to see if the variability of the data is about the same. We also check whether the median, the quartiles, and the extreme values in one set are greater than those in another set. If they are, the data in the first set are greater than those of the other set, no matter how we compare them. If they are not, we can continue to study the data for other similarities and differences.

NOW TRY THIS 8-8

Using only the data from 1983 through 1992 in Table 8-7, decide whether there are outliers. How do you think the two box plots would compare for this data with the comparison made in Figure 8-22?

Read carefully the box plots from the student text on page 538 and complete problem 26.

Variation: Mean Absolute Deviation, Variance, and Standard Deviation

Earlier in the chapter, it was noted that the measure of spread is needed when data are summarized with a single number, such as the mean or median. There are three measures of spread that are generally discussed and each has different uses. The most sophisticated is the standard deviation, as will be seen in the following paragraphs. To examine these measures of spread, we start again with a simple test-score example.

Suppose Professors Abel and Babel each taught a section of a graduate statistics course and each had six students. Both professors gave the same final exam. The results, along with the means for each group of scores, are given in Table 8-9, with stem-and-leaf plots in Figure 8-23(a) and (b), respectively. As the stem-and-leaf plots show, the sets of data are very different. The first is more spread out, or varies more, than the second. However, each set has 60 as the mean. Each median also equals 60. Although the mean and the median for these two groups are the same, the two distributions of scores are very different.

Table 8-9

Abel's Class Scores	Babel's Class Scores
100	70
80	70
70	60
50	60
50	60
10	40
$\bar{x} = \frac{360}{6} = 60$	$\bar{x} = \frac{360}{6} = 60$

FIGURE 8-23

Professor Abel's Class Scores

Stem	Leaf
1	0
2	
3	
4	
5	00
6	
7	0
8	0
9	
10	0

7 | 0 represents a score of 70

(a)

Professor Babel's Class Scores

Stem	Leaf
4	0
5	
6	000
7	00

7 | 0 represents a score of 70

(b)

SAMPLE SCHOOL BOOK PAGE:

BOX PLOTS

▶ **Box-and-whisker plots are useful for comparing two or more sets of data. The box-and-whisker plots below show how two stone-skipping rivals have performed in competition from 1992–1995.**

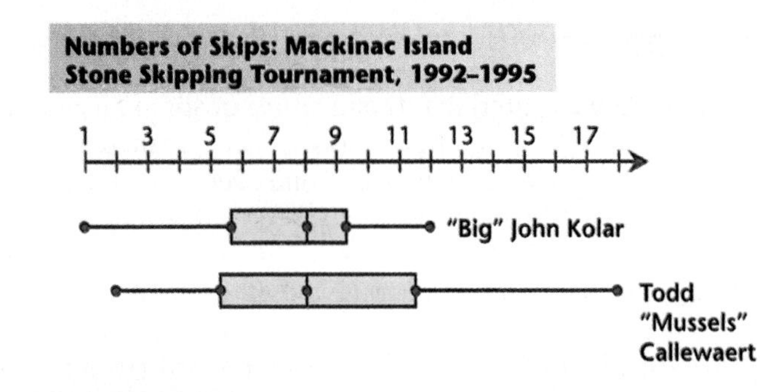

26 ✓ **CHECKPOINT** Use the box-and-whisker plots above.

a. Each year contestants make 6 throws. Each plot above shows 24 throws for the four years of competition. About how many data points are in each box? in each whisker?

b. Who had the most skips in 1992–1995 tournament play? Who had the fewest? How many skips was this for each?

c. Compare the median number of skips made by each stone skipper during the four years of tournament play.

d. The best 25% of Todd Callewaert's throws from 1992–1995 made 11.5 skips or more. How does this compare with John Kolar's throws?

e. Is it possible to tell who won most often in the years 1992–1995? Explain.

✓ **QUESTION 26**

...checks that you can interpret box-and-whisker plots.

Source: Math Thematics, Book 2, McDougal Littell, 2005 (p. 335).

As we have seen, there are many ways to measure the spread of data. The simplest way is to find the range. The range for Professor Abel's class is 100 − 10, or 90. The range for Professor Babel's class is 70 − 40, or 30. If we use the range as our measure of dispersion, we see that Abel's class is much more spread out than Babel's class. If we use the interquartile range, the IQR for Abel's class is 30, and for Babel's class, 10. Again, these measures of spread show more of a spread for Abel's class than for Babel's.

Mean Absolute Deviation

mean absolute deviation (MAD)

One of the most basic ways to measure the spread of data is to measure the distance that each data point is away from the mean. As noted in Chapter 4, the absolute value is one method used for finding distance. The **mean absolute deviation (MAD)** makes use of the absolute value to find the distance each data point is away from the mean; then the mean of those distances is found to give an "average distance from the mean" for each of the points. For example, consider the mean test scores for a class where we can find the mean absolute deviation of the scores from the mean by using the following steps:

- Measure the distance from the mean by simply subtracting the score minus the mean.
- Find the absolute value of the differences.
- Sum those absolute values.
- Find the mean by dividing the sum by the number of scores.

In Table 8-10, we give a sample set of data and compute the mean absolute deviation.

Table 8-10

Test Scores	$\lvert$Test Score − Mean$\rvert$
99	$\lvert 99 - 83.2 \rvert$, or 15.8
67	$\lvert 67 - 83.2 \rvert$, or 16.2
84	$\lvert 84 - 83.2 \rvert$, or 0.8
99	$\lvert 99 - 83.2 \rvert$, or 15.8
67	$\lvert 67 - 83.2 \rvert$, or 16.2
$\bar{x} = \frac{(99 + 67 + 84 + 99 + 67)}{5} = 83.2$	Absolute deviation = 64.8
	$\text{MAD} = \frac{64.8}{5}$, or 12.96

The MAD may be summarized as follows for the numbers $x_1, x_2, x_3, \ldots x_n$, *where* $\bar{x}$ *is the mean of the numbers*:

$$\text{MAD} = \frac{|x_1 - \bar{x}| + |x_2 - \bar{x}| + \ldots + |x_n - \bar{x}|}{n}$$

A visual picture of the mean absolute deviation for the given set of test scores is seen in Figure 8-24.

FIGURE 8-24

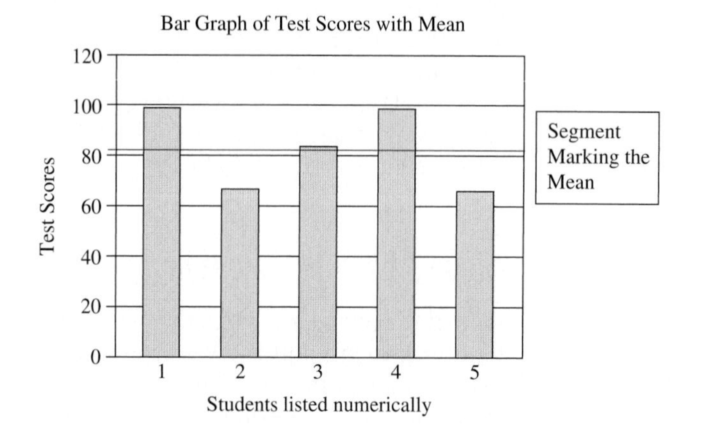

The mean absolute deviation is a rough measure of the "average" distance that a score is from the mean and hence gives an idea of how far away from the mean that the test scores are. The MAD is a relatively easy measure for students to find, is recommended for use in the *Curriculum Framework* (p. 43), and serves as a precursor to the standard deviation. One reason for developing other measures of deviation is because the MAD is not always a great measure of spread, as can be seen using the data for Abel from Table 8-9 and displayed in Table 8-11.

Table 8-11

Abel's Class Scores	|Abel's Scores − 60|	Babel's Class Scores	|Babel's Scores − 60|
100	40	70	10
80	20	70	10
70	10	60	0
50	10	60	0
50	10	60	0
10	50	40	20
Mean = 60	MAD = $\frac{140}{6}$, or approximately 23.3	Mean = 60	MAD = $\frac{40}{6}$, or approximately 6.7

Using the mean absolute deviation on the two sets of scores, it is easy to see that even though the means are 60, the spread is very different. For Abel's class, the scores are an average 23.3 points from the mean, while in Babel's class, the scores are an average of 6.7 points from the mean. Babel's test scores are "less spread out."

Two other commonly used measures of spread are the *variance* and the *standard deviation*. These measures are based on how far the scores are from the mean. To find out how far each value differs from the mean, we subtract each value in the data from the mean to obtain the deviation. Some of these deviations may be positive, and others may be negative. Because the mean is the balance point, the total of the deviations above the mean equals the total of the deviations below the mean. (The mean of the deviations is 0 because the sum of the deviations is 0.)

Squaring the deviations makes them all positive. The mean of the squared deviations is the **variance**. Because the variance involves squaring the deviations, it does not have the same units of measurement as the original observations. For example, lengths measured in feet have a variance measured in square feet. To obtain the same units as the original observations, we take the square root of the variance and obtain the **standard deviation**.

variance

standard deviation

The steps involved in calculating the variance, v, and standard deviation, s, of n numbers are as follows:

1. Find the mean of the numbers.
2. Subtract the mean from each number.
3. Square each difference found in step 2.
4. Find the sum of the squares in step 3.
5. Divide by n to obtain the variance, v.
6. Find the square root of v to obtain the standard deviation, s.

These six steps can be summarized for the numbers $x_1, x_2, x_3, \ldots, x_n$ as follows, where $\bar{x}$ is the mean of these numbers:

$$s = \sqrt{v} = \sqrt{\frac{(x_1 - \bar{x})^2 + (x_2 - \bar{x})^2 + (x_3 - \bar{x})^2 + \cdots + (x_n - \bar{x})^2}{n}}$$

REMARK In some textbooks, this formula involves division by $n - 1$ instead of by n. Division by $n - 1$ is more useful for advanced work in statistics. Also, statisticians prefer the standard deviation to the mean absolute deviation because of its useful properties.

The variances and standard deviations for the final exam data from the classes of Professors Abel and Babel are calculated by using Tables 8-12 and 8-13, respectively.

Table 8-12 Abel's Scores

x	$x - \bar{x}$	$(x - \bar{x})^2$
100	40	1600
80	20	400
70	10	100
50	$^-10$	100
50	$^-10$	100
10	$^-50$	2500
Totals: 360	0	4800

$$\bar{x} = \frac{360}{6} = 60$$

$$v = \frac{4800}{6} = 800$$

$$x = \sqrt{800} \doteq 28.3$$

Table 8-13 Babel's Scores

x	$x - \bar{x}$	$(x - \bar{x})^2$
70	10	100
70	10	100
60	0	0
60	0	0
60	0	0
40	$^-20$	400
Totals: 360	0	600

$$\bar{x} = \frac{360}{6} = 60$$

$$v = \frac{600}{6} = 100$$

$$s = \sqrt{100} = 10$$

Values far from the mean on either side will have larger positive squared deviations, whereas values close to the mean will have smaller positive squared deviations. Therefore, the standard deviation is a greater number when the values from a set of data are widely spread and a lesser number (close to 0) when the data values are close together.

EXAMPLE 8-9

Professor Abel gave two group exams. Exam A had grades of 0, 0, 0, 100, 100, 100, and exam B had grades of 50, 50, 50, 50, 50, 50. Find the following for each exam:

a. Mean
b. Range
c. Mean absolute deviation
d. Standard deviation
e. Median
f. Interquartile range

Solution **a.** The means for exams A and B are each 50.

b. The range for exams A and B are 100 and 0, respectively.

c. The mean absolute deviations for the two exams are as follows:

$$\text{MAD}_A = \frac{|0-50| + |0-50| + |0-50| + |100-50| + |100-50| + |100-50|}{6} = 50$$

$$\text{MAD}_B = \frac{|50-50| + |50-50| + |50-50| + |50-50| + |50-50| + |50-50|}{6} = 0$$

d. The standard deviations for exams A and B are as follows:

$$s_A = \sqrt{\frac{3(0-50)^2 + 3(100-50)^2}{6}} = 50$$

$$s_B = \sqrt{\frac{6(50-50)^2}{6}} = 0$$

e. The medians for the exams are each 50.

f. The interquartile range for exams A and B, respectively, are 100 and 0. ▲

In Example 8–9, exam A has mean 50, mean absolute deviation of 50, and standard deviation of 50. Exam A also had median 50 and interquartile range 100. Reporting the three measures together in each case demonstrates that the scores are widely spread away from the mean and median. Exam B has mean 50, mean absolute deviation 0, and standard deviation of 0. Its median is 50 with interquartile range 0. All of the descriptors for Exam B show that there is very little spread of data from the mean and median. (In this case there is none.) Example 8–9 was chosen with extreme scores to illustrate what can happen with measures of central tendency and measures of spread. Normally, one can tell more about the distribution of data points when the mean, mean absolute deviation, or standard deviation are reported together, as well as when the median and interquartile range are reported together.

Normal Distributions

To better understand how standard deviations are used as measures of spread, we next consider normal distributions. The graphs of normal distributions are the bell-shaped curves called *normal curves*. These curves often describe distributions such as IQ scores for the population of the United States.

normal curve

A **normal curve** is a smooth, bell-shaped curve that depicts frequency values distributed symmetrically about the mean. (Also, the mean, median, and mode all have the same value.) The normal curve is a theoretical distribution that extends infinitely in both directions. It gets closer and closer to the x-axis but never reaches it.

On a normal curve, about 68% of the values lie within 1 standard deviation of the mean, about 95% lie within 2 standard deviations, and about 99.8% are within 3 standard deviations. The percentages represent approximations of the total percent of area under the curve. The curve and the percentages are illustrated in Figure 8-25.

FIGURE 8-25

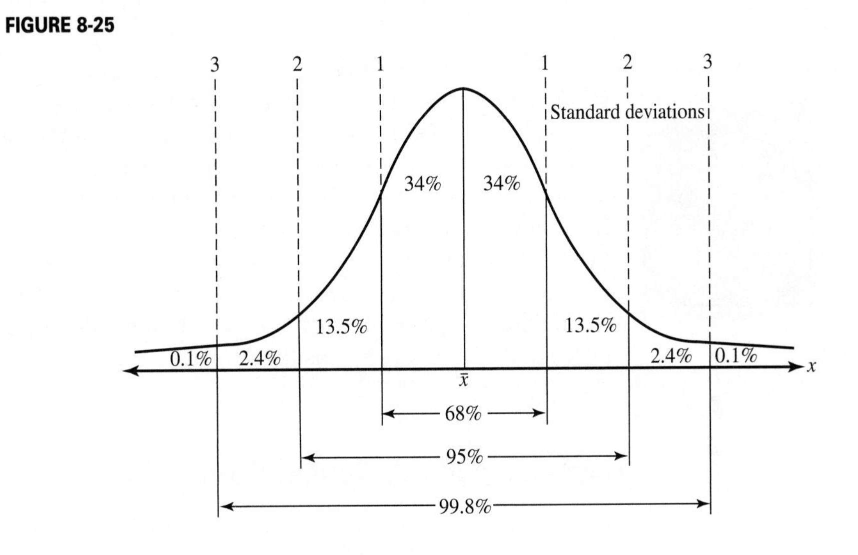

REMARK The normal curve, or normal-probability curve, has equation $y = \frac{1}{s\sqrt{2\pi}} e^{(-1/2)(x-m)^2/s^2}$, where $-\infty < x < \infty$, s is the standard deviation, and m is the mean; e is approximately 2.718281828 . . . .

Suppose the area under the curve represents the population of the United States. Psychologists claim that the mean IQ is 100 and the standard deviation is 15. They also claim that an IQ score of over 130 represents a superior score. Because 130 is equal to the mean plus 2 standard deviations, we see from Figure 8-25 that only 2.5% of the population fall into this category.

EXAMPLE 8-10

When a standardized test was scored, there was a mean of 500 and a standard deviation of 100. Suppose that 10,000 students took the test and their scores had a mound-shaped distribution, making it possible to use a normal curve to approximate the distribution.

a. How many scored between 400 and 600?
b. How many scored between 300 and 700?
c. How many scored between 200 and 800?

Solution

a. Since 1 standard deviation on either side of the mean is from 400 to 600, about 68% of the scores fall in this interval. Thus, 0.68(10,000), or 6800, students scored between 400 and 600.
b. About 95% of 10,000, or 9500, students scored between 300 and 700.
c. About 99.8% of 10,000, or 9980, students scored between 200 and 800.

REMARK About 0.2%, or 20, students' scores in Example 8–10 fall outside 3 standard deviations. About 10 of these students did very well on the test, and about 10 students did very poorly.

▶ **HISTORICAL NOTE**

Abraham De Moivre (1667–1754), a French Huguenot, was the first to develop and study the normal curve. He was one of the first to study actuarial information, in his book *Annuities upon Lives.* He also worked in trigonometry and complex numbers. There is an interesting fable about the death of De Moivre. It is reported that he noticed that each day he required one-quarter of an hour more sleep than he had on the previous day; when the arithmetic progression for sleep reached 24 hr, he died. De Moivre's work with the normal curve went essentially unnoticed. Later, the normal curve was developed independently by Pierre Laplace (1749–1827) and Karl Friedrich Gauss (1777–1855), who found so many applications for the normal curve that it is referred to as the *Gaussian curve.* ◀

Application of the Normal Curve

Suppose that a group of students asked their teacher to grade "on a curve." If the teacher gave a test to 200 students and the mean on the test was 71, with a standard deviation of 7, the graph in Figure 8-26 shows how the grades could be assigned. In Figure 8-26, the teacher has used the normal curve in grading. (The use of the normal curve presupposes that the teacher had a mound-shaped distribution of scores and also that the teacher arbitrarily decided to use the lines marking standard deviations to determine the boundaries of the A's, B's, C's, D's, and F's.) Thus, based on the normal curve in Figure 8-26, Table 8-14 shows the range of grades that the teacher might assign if the grades are rounded. Students who ask their teachers to grade on the curve may wish to reconsider if the normal curve is to be used.

FIGURE 8-26

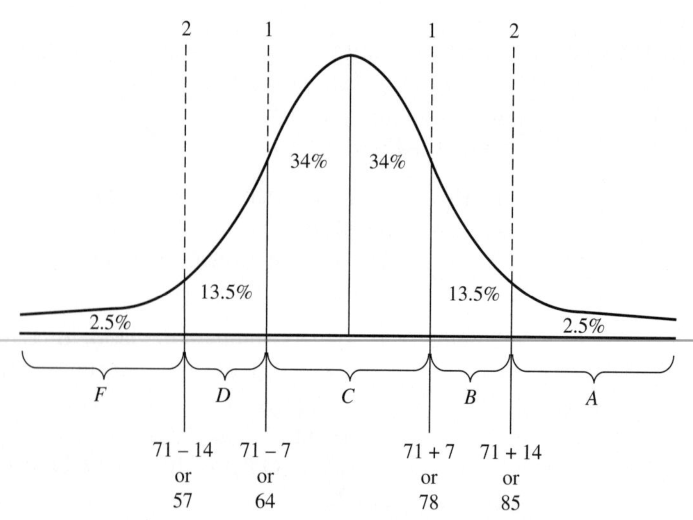

Table 8-14

Test Score	Grade	Number of People per Grade	Percentage Receiving Grade
85 and above	A	5	2.5%
78–84	B	27	13.5%
64–77	C	136	68%
57–63	D	27	13.5%
Below 57	F	5	2.5%

percentiles

When students take a standardized test such as the ACT or SAT, their scores are often reported in **percentiles.** A percentile shows a person's score relative to other scores. For example, if a student's score is at the 82nd percentile, this means that approximately 82% of those taking the test scored lower than the student and approximately 18% had higher scores.

BRAIN TEASER

At a birthday party, the honoree would not tell his age but agreed to give hints. He computed and announced that the mean age of his seven party guests was 21. When 29-year-old Jill arrived at the party, the honoree announced that the mean age of the eight people was now 22. Jack, another 29-year-old, arrived next. The honoree then added his age to the set of ages of the other nine people and announced that the mean was now 27. How old was the honoree?

ONGOING ASSESSMENT 8-2

1. Calculate the mean, the median, and the mode for each of the following data sets:
   a. 2, 8, 7, 8, 5, 8, 10, 5
   b. 10, 12, 12, 14, 20, 16, 12, 14, 11
   c. 18, 22, 22, 17, 30, 18, 12
   d. 82, 80, 63, 75, 92, 80, 92, 90, 80, 80
   e. 5, 5, 5, 5, 5, 10
2. Write an example of a data set with seven data points for which the mode is not a good descriptor of the center of the data.
3. a. If each of six students scored 80 on a test, find each of the following for the set of six scores:
   (i) Mean **(ii)** Median **(iii)** Mode
   b. Make up another set of six scores that are not all the same but in which the mean, median, and mode are all 80.
4. The mean score on a set of 20 tests is 75. What is the sum of the 20 test scores?
5. The tram at a ski area has a capacity of 50 people with a load limit of 7500 lb. What is the mean weight of the passengers if the tram is loaded to capacity?
6. The mean for a set of 28 scores is 80. Suppose two more students take the test and score 60 and 50. What is the new mean?
7. The names and ages for each person in a family of five follow:

Name	Dick	Jane	Kirk	Jean	Scott
Age	40	36	8	6	2

   a. What is the mean age?
   b. Find the mean of the ages 5 yr from now.
   c. Find the mean 10 yr from now.
   d. Describe the relationships among the means found in (a), (b), and (c).

8. Suppose you own a hat shop and decide to order hats in only *one* size for the coming season. To decide which size to order, you look at last year's sales figures, which are itemized according to size. Should you find the mean, median, or mode for the data? Why?

9. Selina claimed that in her class all of the scores on a test were either 100 or 50, so the mean must be 75.
   a. Explain with an example how this reasoning could be true.
   b. Explain with an example how this reasoning could be false.

10. Suppose in Selina's class there were three students who scored 100 and nine who scored 50. What is the mean of the scores in the class?

11. Suppose there were m students in a class who scored 100 and n students who scored 50. Write an algebraic expression for the mean in terms of m and n.

12. Suppose there were n students in a class with h that scored 100 while the rest scored 50. Write an algebraic expression for the mean for the class in terms of n and h.

13. a. Mr. Alberto wanted to count the score on a term paper as 60% of a final grade, homework as 25% of the grade, and the final exam as the remainder of the grade. If you were in the class, made 85 on the term paper, had a 78 average on homework, and scored 90 on the final exam, what number could Mr. Alberto use to determine your grade *if* he used his grading scheme?
   b. Write an algebraic expression to generalize the scoring procedure for Mr. Alberto's class to allow him to use any percentages he likes for the scoring scheme. Clearly describe what the variables represent.

14. A table showing Jon's fall-quarter grades follows. Find his grade point average for the term (A = 4, B = 3, C = 2, D = 1, F = 0).

Course	Credits	Grade
Math	5	B
English	3	A
Physics	5	C
German	3	D
Handball	1	A

15. If the mean weight of seven tackles on a team is 230 lb and the mean weight of the four backfield members is 190 lb, what is the mean weight of the 11-person team?

16. If 99 people had a mean income of $12,000, how much is the mean income increased by the addition of a single income of $200,000?

17. The following table gives the annual salaries of the 40 dancers of a certain troupe.
   a. Find the mean annual salary for the troupe.
   b. Find the median annual salary.
   c. Find the mode.

Salary	Number of Dancers
$ 18,000	2
22,000	4
26,000	4
35,000	3
38,000	12
44,000	8
50,000	4
80,000	2
150,000	1

18. Use the data in problem 17 to find the following:
   a. Range
   b. Mean absolute deviation
   c. Standard deviation
   d. Interquartile range

19. Use the data in problem 17 and the values in problem 18 to answer the following:
   a. Which values from problem 18 should be reported to best describe the data in problem 17?
   b. Explain why you made the choice you did in part (a).

20. Refer to the following chart. In a gymnastics competition, each competitor receives six scores. The highest and lowest scores are eliminated, and the official score is the mean of the four remaining scores.

Gymnast	Scores					
Balance Beam						
Meta	9.2	9.2	9.1	9.3	9.8	9.6
Lisa	9.3	9.1	9.4	9.6	9.9	9.4
Olga	9.4	9.5	9.6	9.6	9.9	9.6
Uneven Bars						
Meta	9.2	9.1	9.3	9.2	9.4	9.5
Lisa	10.0	9.8	9.9	9.7	9.9	9.8
Olga	9.4	9.6	9.5	9.4	9.4	9.4
Floor Exercises						
Meta	9.7	9.8	9.4	9.8	9.8	9.7
Lisa	10.0	9.9	9.8	10.0	9.7	10.0
Olga	9.4	9.3	9.6	9.4	9.5	9.4

   a. If the only events in the competition are the balance beam, the uneven bars, and the floor exercise, find the winner of each event.
   b. Find the overall winner of the competition if the overall winner is the person with the highest combined official scores.

21. Maria needed 8 gal of gas to fill her car's gas tank. The mileage odometer read 42,800 mi. When the odometer read 43,030, Maria filled the tank with 12 gal. At the end of the trip, she filled the tank with 18 gal and the odometer read 43,390 mi. How many miles per gallon did she get for the entire trip?

22. The youngest person in a company is 24 years old. The range of ages is 34 yr. How old is the oldest person in the company?

23. Choose the data set(s) that fit the descriptions given in each of the following:

a. The mean is 6.
The range is 6.
Set *A*: 3, 5, 7, 9
Set *B*: 2, 4, 6, 8
Set *C*: 2, 3, 4, 15

b. The mean is 11.
The median is 11.
The mode is 11.
Set *A*: 9, 10, 10, 11, 12, 12, 13
Set *B*: 11, 11, 11, 11, 11, 11, 11
Set *C*: 9, 11, 11, 11, 11, 12, 12

c. The mean is 3.
The median is 3.
It has no mode.
Set *A*: $0, 2\frac{1}{2}, 6\frac{1}{2}$
Set *B*: 3, 3, 3, 3
Set *C*: 1, 2, 4, 5

d. The box plot shown next.

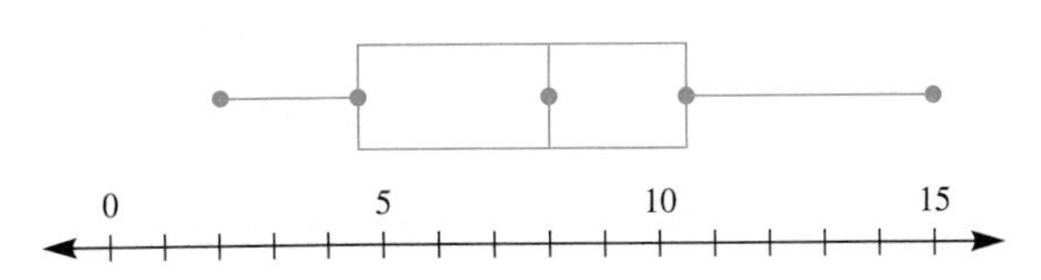

Set *A*: 2, 3, 4, 4, 6, 6, 7, 15
Set *B*: 2, 3, 6, 6, 8, 9, 12, 14, 15
Set *C*: 2, 4.5, 8, 13, 15
Set *D*: 2, 3, 6, 6, 8, 9, 10, 11, 15

24. To receive an A in a class, Willie needs at least a mean of 90 on five exams. Willie's grades on the first four exams were 84, 95, 86, and 94. What minimum score does he need on the fifth exam to receive an A in the class?

25. Ginny's median score on three tests was 90. Her mean score was 92 and her range was 6. What were her three test scores?

26. The mean of five numbers is 6. If one of the five numbers is removed, the mean becomes 7. What is the value of the number that was removed?

27. a. Find the mean and the median of the following arithmetic sequences:
(i) 1, 3, 5, 7, 9 (ii) 1, 3, 5, 7, 9, . . . , 199
(iii) 7, 10, 13, 16, . . . , 607

b. Based on your answers in (a), make a conjecture about the mean and the median of any arithmetic sequence.

28. Construct a box plot for the following gas mileages per gallon of various company cars:

22 18 14 28 30 12 38 22
30 39 20 18 14 16 10

29. Following are box plots comparing the ticket prices of two performing arts theaters:

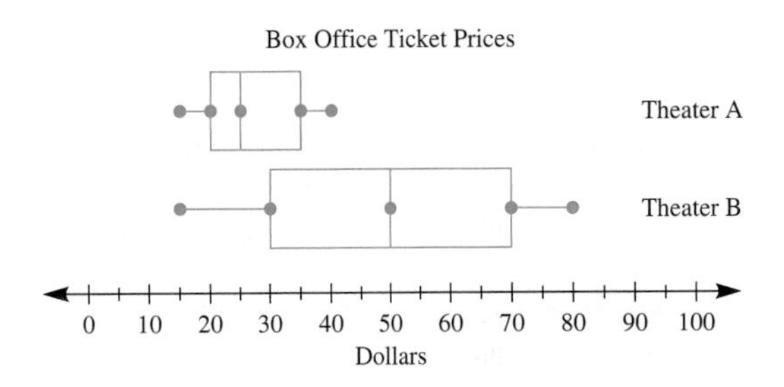

a. What is the median ticket price for each theater?
b. Which theater has the greatest range of prices?
c. What is the highest ticket price at either theater?
d. Make some statements comparing the ticket prices at the two theaters.

30. Construct a box plot for the following set of test scores. Indicate outliers, if any, with asterisks.

20 95 40 70 90 70 80 80 90 95

31. The following table shows the heights in feet of the tallest 10 buildings in Los Angeles and in Minneapolis:

Los Angeles	Minneapolis
858	950
750	775
735	668
699	579
625	561
620	447
578	440
571	416
534	403
516	366

a. Draw horizontal box plots to compare the data.
b. Are there any outliers in this data? If so, which values are they?
c. Based on your box plots from (a), make some comparisons of the heights of the buildings in the two cities.

32. The following shows box plots for the piano practice times in hours per week for Tom and Dick. Make some comparisons for their practice times.

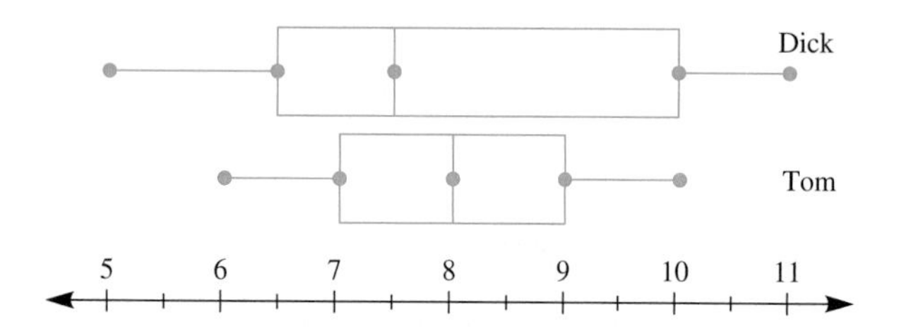

33. What is the standard deviation of the heights of seven trapeze artists if their heights are 175 cm, 182 cm, 190 cm, 180 cm, 192 cm, 172 cm, and 190 cm?

34. **a.** If all the numbers in a set are equal, what is the standard deviation?

b. If the standard deviation of a set of numbers is zero, must all the numbers in the set be equal?

35. In a Math 131 class at DiPaloma University, the grades on the first exam were as follows:

96	71	43	77	75	76	61
83	71	58	97	76	74	91
74	71	77	83	87	93	79

a. Find the mean. **b.** Find the median.
c. Find the mode. **d.** Find the IQR.
e. Find the variance of the scores.
f. Find the standard deviation of the scores.
g. Find the mean absolute deviation of the scores.

36. In a school system, teachers start at a salary of \$25,200 and have a top salary of \$51,800. The teachers' union is bargaining with the school district for next year's salary increment.

a. If every teacher is given a \$1000 raise, what happens to each of the following?
(i) Mean (ii) Median
(iii) Extremes (iv) Quartiles
(v) Standard deviation (vi) IQR

b. If every teacher received a 5% raise, what does this do to the following?
(i) Mean (ii) Standard deviation

37. The mean IQ score for 1500 students is 100, with a standard deviation of 15. Assuming the scores have a normal curve,

a. How many have an IQ between 85 and 115?
b. How many have an IQ between 70 and 130?
c. How many have an IQ over 145?

38. Sugar Plops boxes say they hold 16 oz. To make sure they do, the manufacturer fills the box to a mean weight of 16.1 oz, with a standard deviation of 0.05 oz. If the weights have a normal curve, what percentage of the boxes actually contains 16 oz or more?

39. For certain workers, the mean wage is \$5.00/hr, with a standard deviation of \$0.50. If a worker is chosen at random, what is the probability that the worker's wage is between \$4.50 and \$5.50? Assume a normal distribution of wages.

40. Assume a normal distribution and that the average phone call in a certain town lasted 4 min, with a standard deviation of 2 min. What percentage of the calls lasted fewer than 2 min?

41. In a normal distribution, how are the mean and the median related? Why?

42. According to psychologists, IQs are normally distributed, with a mean of 100 and a standard deviation of 15.

a. What percentage of the population has IQs between 100 and 130?

b. What percentage of the population has IQs lower than 85?

43. The weights of newborn babies are distributed normally, with a mean of approximately 105 oz and a standard deviation of 20 oz. If a newborn is selected at random, what is the probability that the baby weighs less than 125 oz?

44. On a certain exam, the mean is 72 and the standard deviation is 9. If a grade of A is given to any student who scores at least 2 standard deviations above the mean, what is the lowest score that a person could receive and still get an A?

45. Assume that the heights of American women are approximately normally distributed, with a mean of 65.5 in. and a standard deviation of 2.5 in. Within what range are the heights of 95% of American women?

46. A tire company tests a particular model of tire and found the tires to be normally distributed with respect to wear. The mean was 28,000 mi and the standard deviation was 2500 mi. If 2000 tires are tested, about how many are likely to wear out before 23,000 mi?

47. A standardized mathematics test was given to 10,000 students and the scores were normally distributed. The mean was 500 and the standard deviation was 60. If a student scored below 440 points, the student was considered deficient in mathematics. About how many students were rated deficient?

★ **48.** Show that the following formula for variance is equivalent to the one given in the text:

$$v = \frac{x_1^2 + x_2^2 + \ldots + x_n^2}{n} - \bar{x}^2$$

Communication

49. If you were considering ages and wanted one number to represent the age at which a person can get a driver's license, which "average" would you use and why?

50. A movie chain conducts a popcorn poll in which each person entering a theater and buying a box of popcorn is asked a yes-no question. Which "average" do you think is used to report the result and why?

51. When a government agency reports the rainfall for a state for a year, which "average" do you think it uses and why?

52. Carl had scores of 90, 95, 85, and 90 on his first four tests.

a. Find the median, mean, and mode.

b. Carl scored a 20 on his fifth exam. Which of the three averages would Carl want the instructor to use to compute his grade? Why?

c. Which measure is affected most by an extreme score?

53. The mean of the five numbers given is 50:

20 35 50 60 85

a. Add four numbers to the list so that the mean of the nine numbers is still 50.

b. Explain how you could choose the four numbers to add to the list so that the mean did not change.

c. How does the mean of the four numbers you added to the list compare to the original mean of 50? Why?

54. Sue drives 5 mi at 30 mph and then 5 mi at 50 mph. Is the mean speed for the trip 40 mph? Why or why not?

55. Explain why the mode could be a less-than-adequate measure of center for a data set.

56. If a custodian were to report the amount of tissue in a set of bathroom stalls, what type of measure would you want reported to convince you that there was an adequate supply until the stalls were checked again? Explain your answer.

57. If you were to make an argument that there were not enough women's bathrooms in a theater, what type of data would you use and why? Explain the types of measures of center and spread that you would report for your data.

58. The mean of 5, 7, 9 is 7. The mean of 67, 72, 77 is 72.

a. Find two more examples where the mean of an ordered set of data is the "middle" data point.

b. Suppose that a data set consists of $a_1, a_2, a_3, \ldots a_n$, an arithmetic sequence. Explain why the mean of this data set is $\frac{a_1 + a_n}{2}$.

Open-Ended

59. In 1991, students from countries around the world took the International Assessment of Educational Progress test. The following table gives the average mathematics scores for 15 countries, along with the number of days that students in each country spend in school each year.

(a) Make a conjecture about the relationship between the country's mathematics score and the number of days spent in school.

(b) Test your conjecture by using a box plot to compare students who spend 190 days or fewer in school with those who spend more than 190 days in school.

(c) Explain how your graph supports or disproves your conjecture.

Country	Days of School	Math Score
Canada	188	62
France	174	64
Hungary	177	68
Ireland	173	61
Israel	215	63
Italy	204	64
Jordan	191	40
Korea	222	73
Scotland	191	61
Slovenia	190	57
Soviet Union	198	70
Spain	188	55
Switzerland	207	71
Taiwan	222	73
United States	178	55

(d) Draw a scatterplot for the data and see if there seems to be a correlation between mathematics scores and days in school. Is this the same message your box plot gave you? Which plot would you use to support your conjecture?

60. Use the data in the following table to compare the number of people living in the United States from 1810 through 1900 and the number living in the United States from 1910 through 2000. Use any form of graphical representation to make the comparison and explain why you chose the representation that you did.

Year	Number (thousands)	Year	Number (thousands)
1810	7,239	1910	92,228
1820	9,638	1920	106,021
1830	12,866	1930	123,202
1840	17,068	1940	132,164
1850	23,191	1950	151,325
1860	31,443	1960	179,323
1870	38,558	1970	203,302
1880	50,189	1980	226,542
1890	62,979	1990	248,765
1900	76,212	2000	281,422

Data taken from *The World Almanac and Book of Facts 2002,* World Almanac Books, 2002.

Cooperative Learning

61. In small groups, determine a method of finding the number and types of graphs and statistical representations used in at least two newspapers in your campus library. Based on your findings, write a report defending which type(s) of representation should be emphasized in a journalistic statistics class.

Review Problems

62. Given the following double-bar graph, make some comparisons of the number of men and women in the labor force over the years.

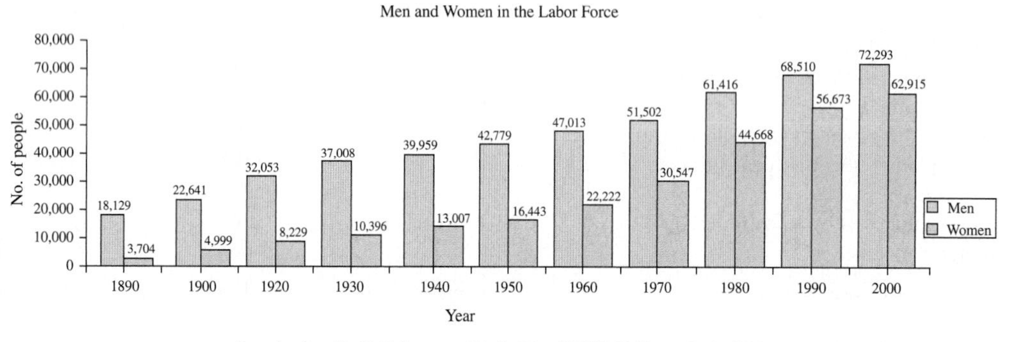

Data taken from *The World Almanac and Book of Facts 2002*, World Almanac Books, 2002.

63. Consider the following circle graph. What is the number of degrees in each sector of the graph?

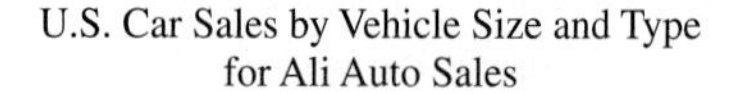

U.S. Car Sales by Vehicle Size and Type
for Ali Auto Sales

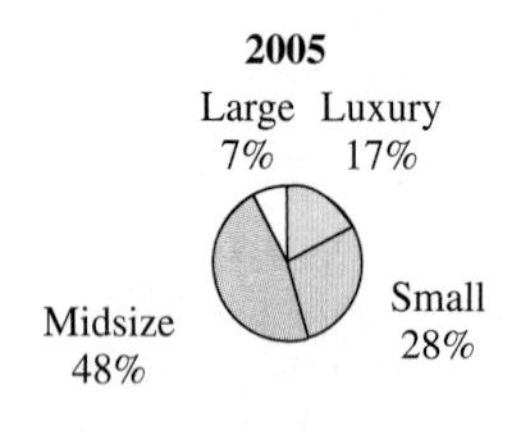

64. Given the bar graph shown, answer the following:
   a. Which mountain is the highest? Approximately how high is it?
   b. Which mountains are higher than 6000 m?

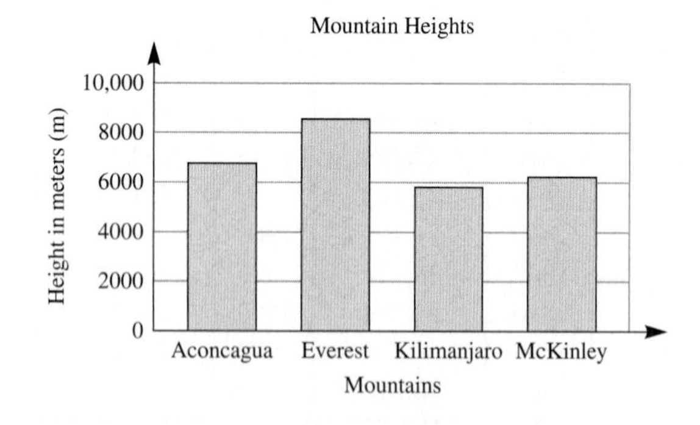

65. Following are raw test scores from a history test:

86 85 87 96 55
90 94 82 68 77
88 89 85 74 90
72 80 76 88 73
64 79 73 85 93

   a. Construct an ordered stem-and-leaf plot for the given data.
   b. Construct a grouped frequency table for these scores with intervals of 5, starting the first class at 55.
   c. Draw a histogram of the data.
   d. If a circle graph of the grouped data in (b) were drawn, how many degrees would be in the section representing the 85 through 89 interval?

National Assessment of Educational Progress (NAEP)

Score	Number of Students
90	1
80	3
70	4
60	0
50	3

The table above shows the scores of a group of 11 students on a history test. What is the average (mean) score of the group to the nearest whole number?

NAEP, 2003, Grade 8

Tetsu rides his bicycle x miles the first day, y miles the second day, and z miles the third day. Which of the following expressions represents the average number of miles per day that Tetsu travels?

a. $x + y + z$
b. xyz
c. $3(x + y + z)$
d. $3(xyz)$
e. $(x + y + z)/3$

NAEP, 2003, Grade 8

LABORATORY ACTIVITY

We can model the mean as a measure of central tendency using a strip of cardboard that is 1 in. wide and 1 ft long with holes 1 in. apart and $\frac{1}{8}$ in. from the edge, as in Figure 8-27.

FIGURE 8-27

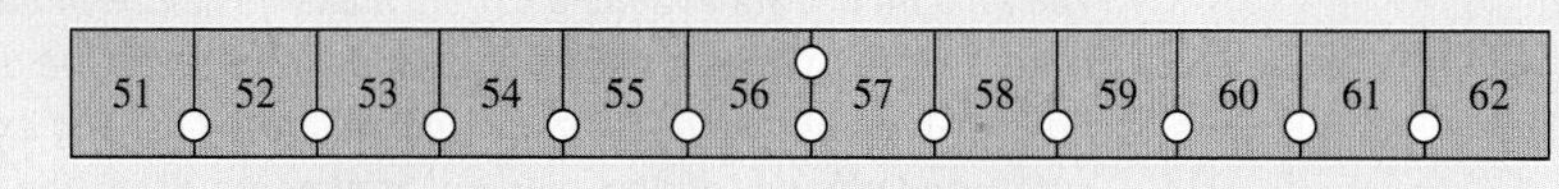

Use string and tape to suspend the strip from a desk with the string tied through a hole punched between 56 and 57. Then use paper clips of equal size to investigate means, as in Figure 8-28.

FIGURE 8-28

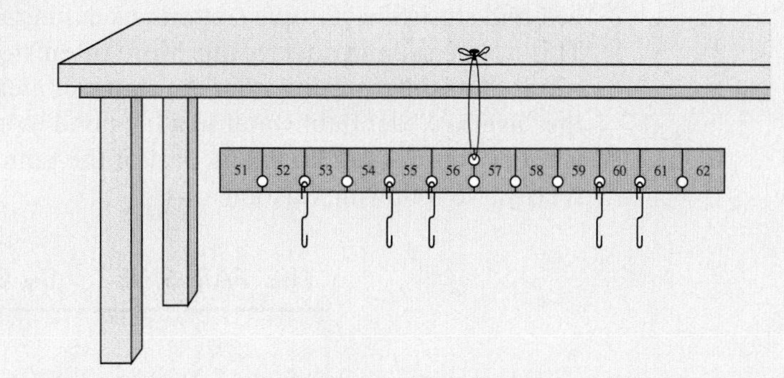

1. **a.** If paper clips are hung in the holes at 51, 53, and 60, where should an additional clip be hung in order to achieve a balance?
   b. If paper clips are hung at 51, 54, and 60, where should two additional paper clips be hung to achieve a balance?
   c. If paper clips are hung at 51, 53, 54, and 55, where should four additional clips be hung to achieve a balance?
2. Find the mean of the data of all the numbers in each part in (1) and compare this answer with the number in the center of the strip.
3. Would any of the means in (2) change if we hung an additional paper clip in the center of the strip?
4. Find the median and mode for 51, 53, 54, 56, 58, 58, and 59.
5. Hang paper clips in each hole in (4). If a number appears more than once, hang that number of paper clips in the hole. Is there a balance around the median? Is there a balance around the mode?
6. Under what conditions do you think there would be a balance around (a) the mean, (b) the median, and (c) the mode? Test your conjecture using the cardboard strip.

TECHNOLOGY CORNER

Use a graphing calculator or create a spreadsheet to find the mean absolute deviation, the variance, and standard deviation of the set of scores 32, 41, 47, 53, and 57. If you use a spreadsheet, make column A be the set of scores, column B the scores minus the mean, column C the absolute value of the differences, and column D the squares of the difference of the scores and the mean. Compute the mean absolute deviation, variance, and the standard deviation.

8-3 Abuses of Statistics

In *Principles and Standards*, we find the following:

> *The amount of data available to help make decisions in business, politics, research, and everyday life is staggering. Consumer surveys guide the development and marketing of products. Polls help determine political-campaign strategies, and experiments are used to evaluate the safety and efficacy of new medical treatments. Statistics are often misused to sway public opinion on issues or to misrepresent the quality and effectiveness of commercial products.* (p. 48)

As noted, statistics are frequently abused. Benjamin Disraeli (1804–1881), an English prime minister, once remarked, "There are three kinds of lies: lies, damned lies, and statistics." People sometimes deliberately use statistics to mislead others. This can be seen in advertising. More often, however, the misuse of statistics is the result of misinterpreting what the statistics mean. For example, if we were told that the "average" depth of water in a lily pond was 2 ft, most of us would presume that a heron could stand up in any part of the pond. This is not necessarily the case, as seen in the following cartoon.

Consider an advertisement reporting that of the people responding to a recent survey, 98% said that Buffepain is the most effective pain reliever of headaches and arthritis. To certify that the statistics are not being misused, the following information should have been reported:

1. The number of people surveyed
2. The number of people who responded

3. How the people who participated in the survey were chosen
4. The number and type of pain relievers tested

Without the information listed, the following situations are possible, all of which could cause the advertisement to be misleading:

1. Suppose 1,000,000 people nationwide were sent the survey, and only 50 responded. This would mean that there was only a 0.005% response, which would certainly cause us to mistrust the ad.
2. Of the 50 responding in (1), suppose 49 responses were affirmative. The 98% claim is true, but 999,950 people did not respond.
3. Suppose all the people who received the survey were chosen from a town in which the major industry was the manufacture of Buffepain. It is very doubtful that the survey would represent an unbiased sample.
4. Suppose only two "pain relievers" were tested: Buffepain, whose active ingredient is 100% aspirin, and a placebo containing only powdered sugar.

This is not to say that advertisements of this type are all misleading or dishonest, but simply that statistics are only as honest as their users.

In the Buffepain report, a primary issue deals with the survey conducted, the number of people involved in the survey, how they were chosen, and how results of the survey were interpreted. Surveys are common for gaining information from a population. However, there are classic examples in history of how surveys and survey information have been either misused or misinterpreted. A well-known example was seen in the predictions of the winner of the Harry S. Truman/Thomas E. Dewey 1948 U. S. presidential election. A leading pollster, Elmo Roper, was so confident of a loss by Truman that on September 9, 1948, approximately 2 months before the election, he announced that there would be no more polls on the election. Additionally, while Truman was still on the campaign trail, *Newsweek*, after polling 50 key political journalists, stated on October 11, 1948, approximately 1 month before the election, that Dewey would win.

In the *Newsweek* survey, a very select group was surveyed. Questions that should have been asked include (1) How was the group chosen? (2) Was the group representative of the voters in the United States? (3) Could the result of this survey appropriately have been generalized to all voters? All of these types of questions are important to surveyors. What is the population to which the results are being generalized? Is the entire population to be surveyed, or is the survey given only to a sample of the population? If only a sample will be used, how is the sample chosen? Is the sample of the population randomly chosen so that each person being surveyed has an equally likely chance of being chosen? How large is the population and how large a sample must be used so that the sample is representative of the population?

A version of the survey is sometimes seen in the early grades when students try to determine the favorite type of pet of the students in their school. A very simple version of this type of survey uses only the students in one class. If that is done, then the answer is known for that entire population. Note that the answer is not generalizable to the whole school. How might the sample be taken if students wanted to find the favorite pet of all the students in the school?

NOW TRY THIS 8-9

Based on the graph in Figure 8-29, write arguments to determine whether or not the following inferences are correct:

a. 82% of the parents surveyed say that there is not too much focus on preparing for tests.
b. Only 9% of the parents say that learning is thwarted.
c. 12% of the parents say that it is because test questions are too difficult and that expectations are unreasonable.
d. At least 89% of the parents believe that schools require too many tests.
e. The vast majority of parents say that teachers are not putting too much academic pressure on their children.

FIGURE 8-29

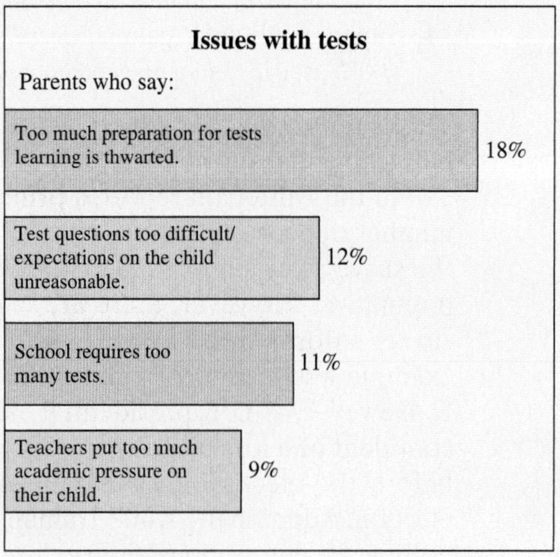

Issues with tests

Parents who say:

Too much preparation for tests learning is thwarted. 18%

Test questions too difficult/ expectations on the child unreasonable. 12%

School requires too many tests. 11%

Teachers put too much academic pressure on their child. 9%

Source: Data from *USA Today*. November 21, 2000.

What other information is needed to draw conclusions from the graph?

A different type of misuse of statistics involves graphs. Among the things to look for in a graph are the following. If they are not there, then the graph may be misleading.

1. Title
2. Labels on both axes of a line or bar chart and on all sections of a pie chart
3. Source of the data
4. Key to a pictograph
5. Uniform size of symbols in a pictograph
6. Scale: Does it start with zero? If not, is there a break shown?
7. Scale: Are the numbers equally spaced?

To see an example of a misleading use of graphs, consider how graphs can be used to distort data or exaggerate certain pieces of information. Graphs using a break in the vertical axis can be used to create different visual impressions, which are sometimes misleading. For example, consider the two graphs in Figure 8-30, which represent the number of girls trying out for basketball at each of three middle schools. As we can see, the graph in Figure 8-30(a) portrays a different picture from the one in Figure 8-30(b).

FIGURE 8-30

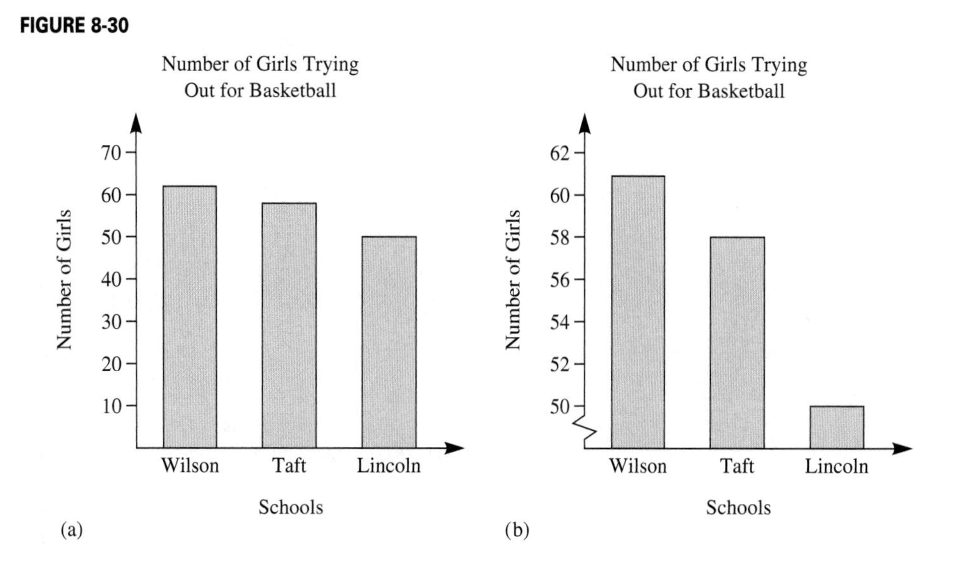

RESEARCH NOTE Elementary students have trouble interpreting data; 80% of the first and second graders in one study gave idiosyncratic or incomplete responses when analyzing data from a line graph or bar chart (Putt, Jones, et al. 1999). Almost all fourth and sixth graders could read bar graphs, but many fewer could use the graphs to predict (Pereira-Mendoza and Mellor 1991).

A line graph, histogram, or bar graph can be altered by changing the scale of the graph. For example, consider the data in Table 8-15 for the number of graduates from a community college for the years 2001 through 2005.

Table 8-15

Year	2001	2002	2003	2004	2005
Number of Graduates	140	180	200	210	160

The graphs in Figure 8-31(a) and (b) represent the same data, but different scales are used in each. The statistics presented are the same, but these graphs do not convey the same psychological message. In Figure 8-31(b), the spacing of the years on the horizontal axis of the graph is more spread out and that for the numbers on the vertical axis is more condensed than in Figure 8-31(a). Both of these changes minimize the variability of the data. A college administrator might use a graph like the one in Figure 8-31(b) to convince people that the college was not in serious enrollment trouble.

FIGURE 8-31

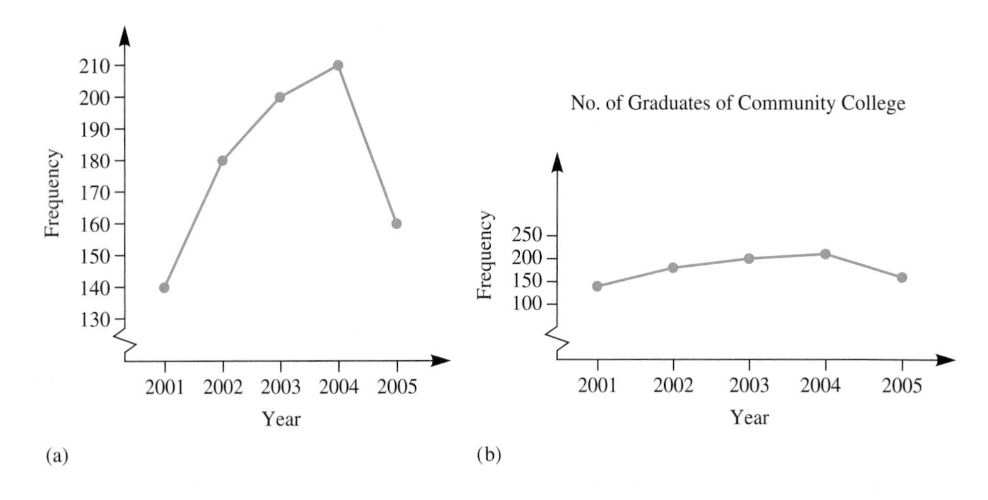

Another error that frequently occurs in all arenas is the use of continuous-curve graphs, as in Figure 8-31, to depict data that are discrete (a finite number of data values). In Figure 8-32, it may or may not make sense to discuss the enrollment at 2001.5, yet the way in which the graph is constructed leads us to believe that such a value exists. Other ways to distort graphs include omitting a scale, as in Figure 8-32(a). The scale is given in Figure 8-32(b).

FIGURE 8-32

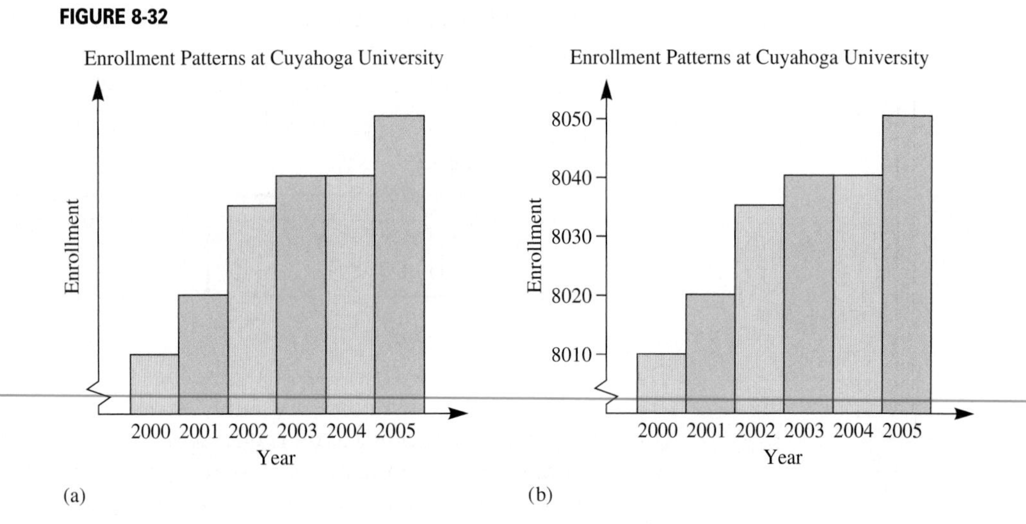

Other graphs can also be misleading. Suppose, for example, that the number of boxes of cereal sold by Sugar Plops last year was 2 million and the number of boxes of cereal sold by Korn Krisps was 8 million. The Korn Krisps executives prepared the graph in Figure 8-33 to demonstrate the data. The Sugar Plops people objected. Do you see why?

FIGURE 8-33

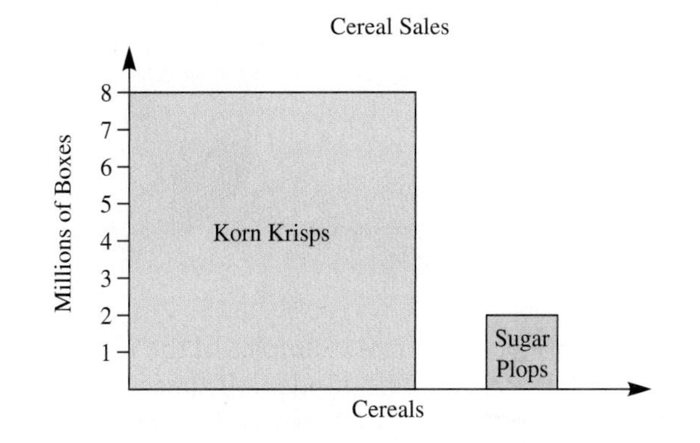

The graph in Figure 8-33 clearly distorts the data, since the figure for Korn Krisps is both 4 times as high and 4 times as wide as the bar for Sugar Plops. Thus, the area of the box face representing Korn Krisps is 16 times the comparable area representing Sugar Plops, rather than 4 times the area, as would be justified by the original data.

Figure 8-34 shows how the comparison of Sugar Plops and Korn Krisps cereals might look if the figures were made three-dimensional. The figure for Korn Krisps has a volume 64 times the volume of the Sugar Plops figure.

FIGURE 8-34

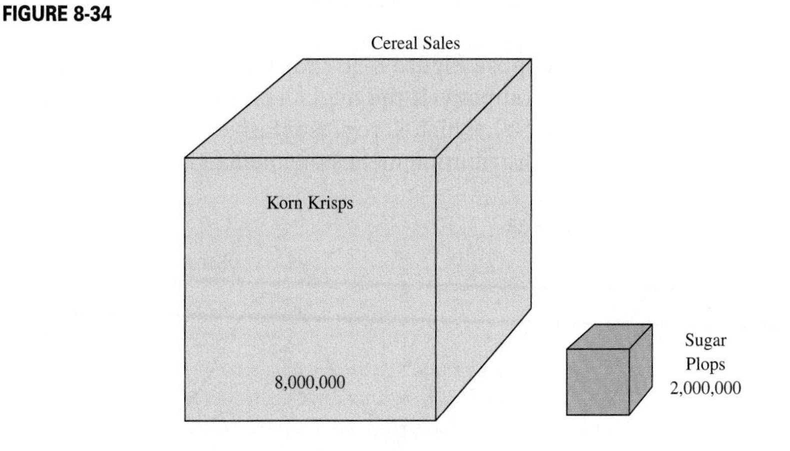

Circle graphs easily become distorted when attempts are made to depict them as three-dimensional. Many graphs of this type do not acknowledge either the variable thickness of the depiction or the distortion due to perspective. Observe that the 27% sector pictured in Figure 8-35 looks far greater than the 23% sector, although they should be very nearly the same size.

FIGURE 8-35

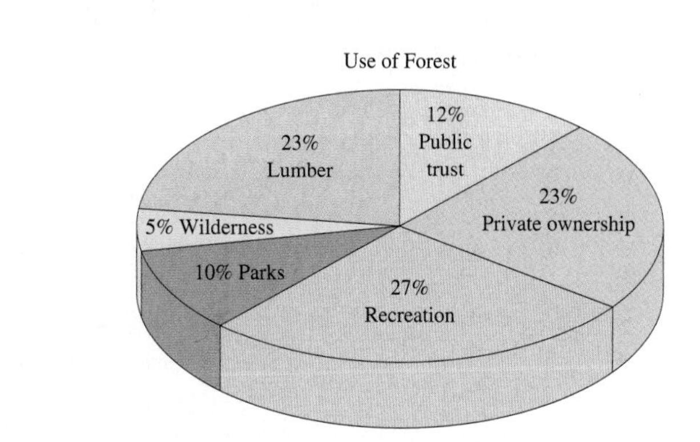

The final examples of the misuses of statistics involve misleading uses of mean, median, and mode. All these are "averages" and can be used to suit a person's purposes. As discussed in Section 8-2 in the example involving the teachers Smith, Jones, and Rivera, each teacher had reported that his or her class had done better than the other two. Each of the teachers used a different number to represent the test scores.

As another example, company administrators wishing to portray to prospective employees a rosy salary picture may find a mean salary of \$58,000 for line workers as well as upper management in the schedule of salaries. At the same time, a union that is bargaining for salaries may include part-time employees as well as line workers and will exclude management personnel in order to present a mean salary of \$29,000 at the bargaining table. The important thing to watch for when a mean is reported is disparate cases in the reference group. If the sample is small, then a few extremely high or low scores can have a great influence on the mean.

Suppose Figure 8-36 shows the salaries of both management and line workers of the company. If the median is being used as the average, then the median might be \$33,500, which is representative of neither major group of employees. The bimodal distribution means the median is nonrepresentative of the distribution.

FIGURE 8-36

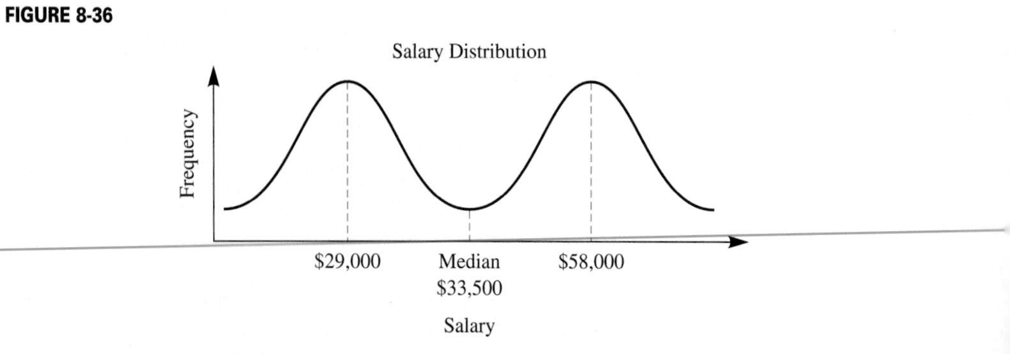

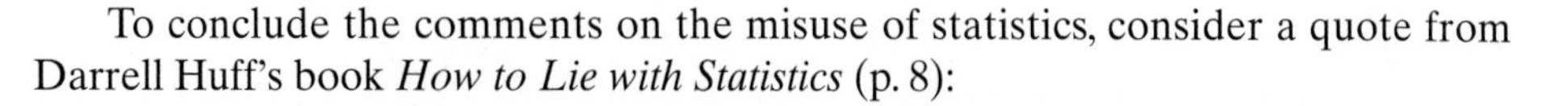

To conclude the comments on the misuse of statistics, consider a quote from Darrell Huff's book *How to Lie with Statistics* (p. 8):

So it is with much that you read and hear. Averages and relationships and trends and graphs are not always what they seem. There may be more in them than meets the eye, and there may be a great deal less.

The secret language of statistics, so appealing in a fact-minded culture, is employed to sensationalize, inflate, confuse, and oversimplify. Statistical methods and statistical terms are necessary in reporting the mass data of social and economic trends, business conditions, "opinion" polls, census. But without writers who use the words with honesty and understanding and readers who know what they mean, the result can be semantic "nonsense."

ONGOING ASSESSMENT 8-3

This entire set of assessment items is appropriate for communication and cooperative learning. Many items are open-ended and several lend themselves to further investigation.

1. Discuss whether the following claims could be misleading. Explain why and how.
   a. A car manufacturer claims its car is quieter than a glider.
   b. A motorcycle manufacturer claims that more than 95% of its cycles sold in the United States in the last 15 yr are still on the road.
   c. A company claims its fruit juice has 10% more fruit solids than is required by U.S. government standards. (The government requires 10% fruit solids.)
   d. A brand of bread claims to be 40% fresher.
   e. A used-car dealer claims that a car she is trying to sell will get up to 30 mpg.
   f. Sudso claims that its detergent will leave your clothes brighter.
   g. A sugarless gum company claims that 8 of every 10 dentists responding to the survey recommend sugarless gum.
   h. Most accidents occur in the home. Therefore, to be safer, you should stay out of your house as much as possible.
   i. More than 95% of the people who fly to a certain city do so on Airline A. Therefore, most people prefer Airline A to other airlines.
2. The city of Podunk advertised that its temperature was the ideal temperature in the country because its mean temperature was 25°C. What possible misconceptions could people draw from this advertisement?
3. Jenny averaged 70 on her quizzes during the first part of the quarter and 80 on her quizzes during the second part of the quarter. When she found out that her final average for the quarter was not 75, she went to argue with her teacher. Give a possible explanation for Jenny's misunderstanding.
4. Suppose the following circle graphs are used to illustrate the fact that the number of elementary teaching majors at teachers' colleges has doubled between 2000 and 2005, while the percent of male elementary teaching majors has stayed the same. What is misleading about the way the graphs are constructed?

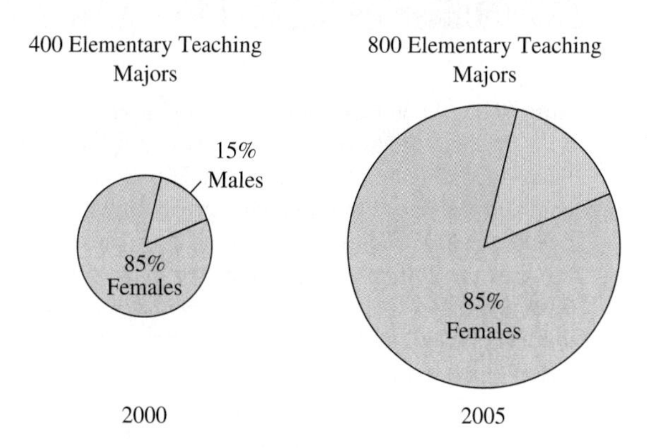

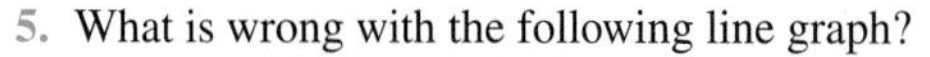
5. What is wrong with the following line graph?

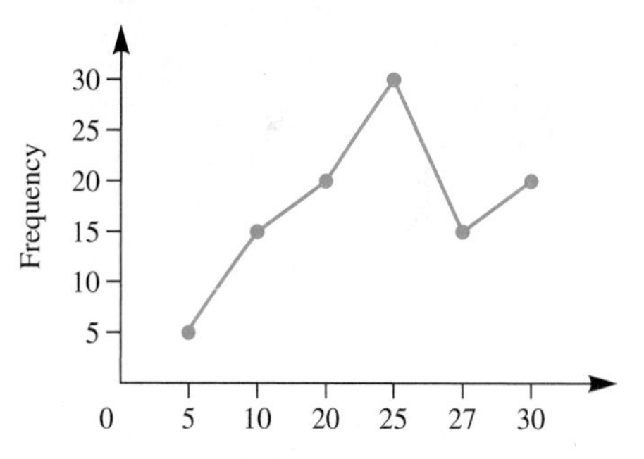

6. Can you draw any valid conclusions about a set of data in which the mean is less than the median?
7. A student read that 9 out of 10 pickup trucks sold in the last 10 yr are still on the road. She concluded that the average life of a pickup is around 10 yr. Is she correct?
8. General Cooster once asked a person by the side of a river if the river was too deep to ride his horse across. The person responded that the average depth was 2 ft. If General Cooster rode out across the river, what assumptions did he make on the basis of the person's information?
9. Which of the following pieces of information would not be helpful in deciding the type of automobile that is the most economical to drive?
   a. Range of insurance costs
   b. Modes of drivers' ages for specific vehicle types
   c. Mean miles per gallon
   d. Typical cost of repairs per year
   e. Cost of routine maintenance

10. In the Shoe cartoon shown, explain the intended meaning of the second panel.

11. Doug's Dog Food Company wanted to impress the public with the magnitude of the company's growth. Sales of Doug's Dog Food had doubled from 2003 to 2005, so the company displayed the following graph, in which the radius of the base and the height of the 2005 can are double those of the 2003 can. What does the graph really show with respect to the growth of the company? (*Hint:* The volume of a cylinder is given by $V = \pi r^2 h$, where r is the radius of the base and h is the height.)

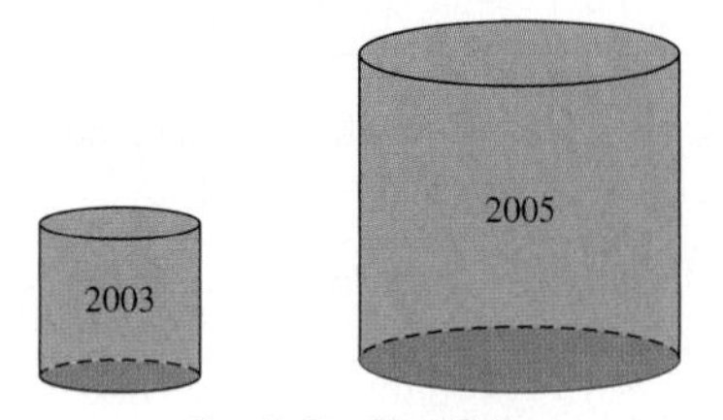

Doug's Dog Food Sales

12. Explain what is wrong with the following graph:

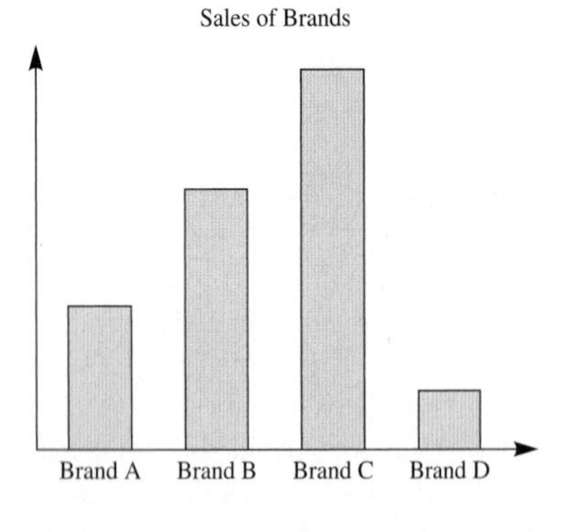

13. Refer to the following circle graph.

Drivers in Fatal Accidents

Men

Women

Ms. McNulty claims that on the basis of this information, we can conclude that men are worse drivers than women. Discuss whether you can reach that conclusion from the pictograph or you need more information. If more information is needed, what would you like to know?

14. The following graph was prepared to compare prices of washing machines at three stores:

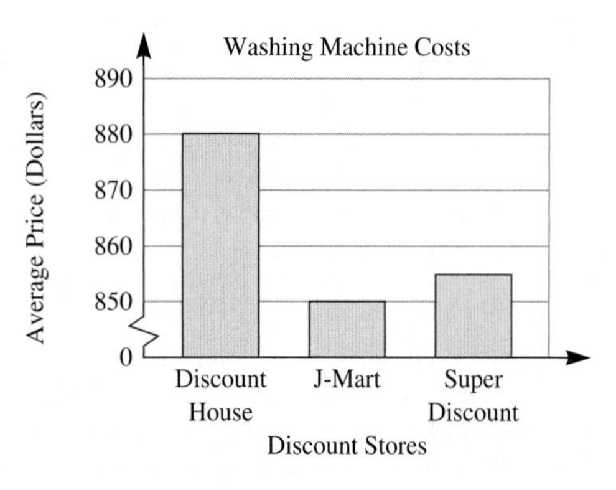

Which of the following statements is true? Explain why or why not.

a. Prices vary widely at the three stores.
b. The price at Discount House is 4 times as great as that at J-Mart.
c. The prices at J-Mart and Super Discount differ by less than $10.

15. The following table gives the number of accidents per year on a certain highway for a 5-yr period:

Year	2001	2002	2003	2004	2005
Number of Accidents	24	26	30	32	38

a. Draw a bar graph to convince people that the number of accidents is on the rise and that something should be done about it.

b. Draw a bar graph to show that the rate of accidents is almost constant, and that nothing needs to be done.

16. Write a list of scores for which the mean and median are not representative of the list.

17. The following graph depicts the mean center of population of the United States and shows how the center has shifted from 1790 to 2000. Based solely on this graph, could you conclude that the population of the West Coast has increased since 1790?

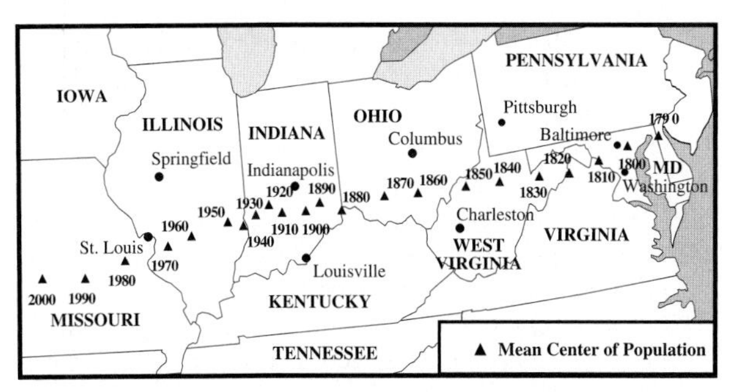

Source: U.S. Bureau of the Census, *2000 Census of Population and Housing, Population and Housing Unit Counts, United States* (2000 CPH-2-1).

18. The data in the graph of problem 17 roughly follows a line. If the mean center of the population continues to move westward in the next 200 years, where would you expect it to be at the end of that period? At the end of 400 years?

19. In a recent survey, teachers rated their mathematics textbooks as follows:

National Survey of Teachers of Mathematics

	Grades K–4 (percentage)	Grades 5–8 (percentage)
Very poor	1	2
Poor	3	5
Fair	18	16
Good	35	33
Very good	36	33
Excellent	8	10

At about the same time that the survey was being done, newspapers printed articles saying that national experts in mathematics have been very critical of mathematics textbooks at all levels.

a. What is your reaction to the survey and the reports?

b. Does the national survey agree with the newspaper articles? Explain your thinking as you react.

20. a. Use the following data to justify the amount of time that you expect to assign for weekly homework to classes in Grades K–4 and Grades 5–8.

National Survey of Teachers Concerning the Amount of Homework Assigned per Week

	K–4 (percentages)	5–8 (percentages)
0–30 min	48	8
31–60 min	27	21
61–90 min	13	26
91–120 min	8	24
2–3 hr	3	17
More than 3 hr	1	5

b. How might the survey data be misused to justify assigning at least 2 hr of homework per week?

21. If you knew that 43% and 57% of mathematics teachers of Grades K–4 and 5–8, respectively, considered themselves well prepared to use calculators/computers to demonstrate mathematical principles, and you assume each of the following roles, what should you do based on these data? In each case, how might you be misusing the data?

a. Teacher **b.** Principal
c. Parent

22. The No Child Left Behind Act (2000) signed into law in the United States demands that teachers be highly qualified. One aspect of that law said that teachers of middle-school mathematics should have a major in the subject they teach. In January 2003, a survey revealed that 61% of middle-school mathematics classes are led by teachers who lack even a minor in mathematics.

a. Using these data, what actions should a state superintendent take?

b. How might you be misusing the data to justify your actions?

23. In 2003, it was reported that an average mathematics teacher's starting salary in the United States was \$28,000. This was about \$18,000 less than the average starting salary for all mathematics majors with a bachelor's degree.

a. As a state legislator, how would you propose using these data in your state?

b. How might you be misusing the data if you were the state legislator?

24. According to the U.S. Department of Education in 2004, at least one-third of all entry level teachers will quit within 5 years of starting teaching. In urban schools, the percentage is higher.

a. As a collegiate advisor for prospective teachers, what advice would you give a student who wants to become a teacher of mathematics?

b. How might you be misusing the data in framing your advice?

25. The two possible headlines shown here were written based on exactly the same data set:

"Obesity has increased over 30%."
"On the average, Americans' weight has increased by less than 10 pounds."

a. As a reporter, which headline would you use?

b. As a reader, which headline would have the greatest impact?

c. As an educated reader, which headline might you consider the most accurate knowing that both were written based on the same data set? Explain your answer.

26. In a British study around 1950, a group of 649 men with lung cancer were surveyed. A control group of the same size was established from a set of men who did not have lung cancer. The groups were matched according to ethnicity, age, and socioeconomic status. The statistics from the survey follow.

	Lung Cancer Cases	Controls	Totals
Smokers	647	622	1269
Nonsmokers	2	27	29

a. What is the proportion of smoking in the group that has lung cancer in the study?

b. What is the proportion of smoking in the control group?

c. If one person is chosen at random from each of the two groups (smokers and nonsmokers), what is the probability that each person chosen has cancer?

d. Do you think that this data presents a strong association between lung cancer and smoking? Explain your answer.

e. Do you think that this evidence is conclusive that smoking causes lung cancer?

(** Problem taken from* Curriculum Framework)

27. What are the characteristics that you think a sample might have to have to be representative of an entire population?

28. a. Suppose you are a student doing a survey of eating habits in a high school. If you sit in the hall and ask each student who passes you in 1 hr questions about their eating habits, explain whether or not you think you have a representative sample of the school population.

b. If you are conducting the survey mentioned in part (a) and interview students in the cafeteria at noon, explain whether or not you think that you have a representative sample of the school population.

c. For this survey, explain how you could choose a sample to be reasonably sure that you got a representative sample of the population.

29. A prospective homeowner considered the dropping interest rates for house loans in the early 1990s and decided to wait until the year 2000 to buy. Explain what type of statistics might be used in making this decision and whether you consider the prospective homeowner's decision to be wise.

30. A student made 99 on a quiz and was ecstatic over the grade. What other information might you need in order to decide if the student was justified in being happy?

31. A school administrator reports to you, a school board member, that the "average" number of students in a class in a school is 32. What other information might you need in order to predict whether any single classroom was overcrowded?

32. Consider a state such as Montana that has both mountains and prairies. What numbers might you report to depict the "average" height above sea level of such a state? Why?

33. Is it possible for a state or country to have a mean sea level height that is negative? If so, what might such a region look like?

34. Describe how you might pick a random sample of adults that is representative of the members of your town.

35. A very large and successful manufacturer of computer chips released data in the early 1990s stating that approximately 65% of its chips were defective when they came off the assembly line. Give some reasons why this statistic could be accurate and yet the company could still be successful.

Review Problems

36. On the English 100 exam, the scores were as follows:

43 91 73 65
56 77 84 91
82 65 98 65

a. Find the mean.

b. Find the median.

c. Find the mode.

d. Find the variance.

e. Find the standard deviation.

f. Find the mean absolute deviation.

37. If the mean of a set of 36 scores is 27 and two more scores of 40 and 42 are added, what is the new mean?

38. On a certain exam, Tony corrected 10 papers and found the mean for his group to be 70. Alice corrected the remaining 20 papers and found that the mean for her group was 80. What is the mean of the combined group of 30 students?

39. Following are the men's gold-medal times for the 100-m run in the Olympic games from 1896 to 2000, rounded to

the nearest tenth. Construct an ordered stem-and-leaf plot for the data.

Year	Time (sec), (rounded)
1896	12.0
1900	11.0
1904	11.0
1908	10.8
1912	10.8
1920	10.8
1924	10.6
1928	10.8
1932	10.3
1936	10.3
1948	10.3
1952	10.4
1956	10.5
1960	10.2
1964	10.0
1968	10.0
1972	10.1
1976	10.1
1980	10.3
1984	10.0
1988	9.9
1992	9.7
1996	9.8
2000	9.9

40. Following are the record swimming times of the women's 100-m freestyle and 100-m butterfly in the Olympics from 1960 to 2000. Draw parallel box plots of the two sets of data to compare them.

Year	Time—100-m Freestyle (sec)	Time—100-m Butterfly (sec)
1960	61.20	69.50
1964	59.50	64.70
1968	60.00	65.50
1972	58.59	63.34
1976	55.65	60.13
1980	54.79	60.42
1984	55.92	59.26
1988	54.93	59.00
1992	54.64	58.62
1996	54.50	59.13
2000	53.83	56.61

National Assessment of Educational Progress (NAEP)

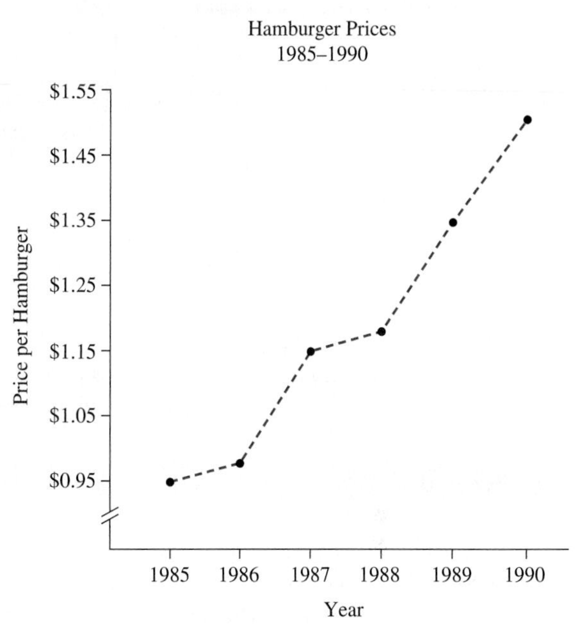

According to the graph above, how many times did the yearly increase of the price of a hamburger exceed 10 cents?

a. None **b.** One **c.** Two
d. Three **e.** Four

NAEP, Grade 8, 2003

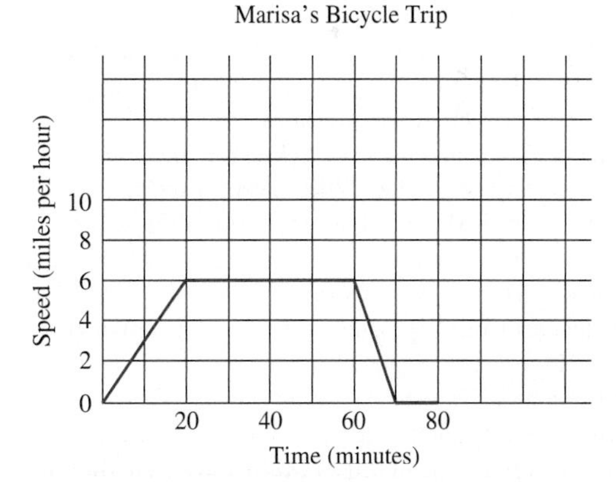

The graph above represents Marisa's riding speed throughout her 80-minute bicycle trip. Use the information in the graph to describe what could have happened on the trip, including her speed throughout the trip.

During the first 20 minutes, Marisa ______________________

From 20 minutes to 60 minutes, she ______________________

From 60 minutes to 80 minutes, she ______________________

NAEP, Grade 8, 2003

BRAIN TEASER

Ouida had test grades of 91, 89, 83, 78, 76 and 75. Her teacher uses the mean of the test scores as the final grade of the class. Ouida has the option of keeping her current test average or replacing the highest and lowest scores that she currently has with a single score on a final exam. If Ouida chooses to take the final exam, find the lowest score she can receive and still maintain at least the mean that she currently has.

HINT FOR SOLVING THE PRELIMINARY PROBLEM

To find the entry for the empty cell in the bottom row, consider the remaining percentage for Ms. Jonsey's class and how that percentage was used to determine the overall mean. For the missing entry in Mr. Ramirez's class, think algebraically about the conditions on the total percentages and how those percentages were used to obtain the overall mean for his class.

QUESTIONS FROM THE CLASSROOM

1. A student asks, "If the average income of 10 people is $10,000 and 1 person gets a raise of $10,000, is the median, the mean, or the mode changed and, if so, by how much?"
2. Jose asks, "Why can a median number of children not be 3.8?" How do you respond?
3. Suppose the class takes a test and the following averages are obtained: mean, 80; median, 90; mode, 70. Tom, who scored 80, would like to know if he did better than half the class. What is your response?
4. A student wants to know the advantages of presenting data in graphical form rather than in tabular form. What is your response? What are the disadvantages?
5. A student asks if it is possible to find the mode for data in a grouped frequency table. What is your response?
6. A student asks if she can draw any conclusions about a set of data if she knows that the mean for the data is less than the median. How do you answer?
7. A student asks if it is possible to have a standard deviation of $^{-}5$. How do you respond?
8. Mel's mean on 10 tests for the quarter was 89. She complained to the teacher that she should be given an A because she missed the cutoff of 90 by only a single point. Explain whether it is clear that she really missed an A by only a single point.
9. A student asks, "Is it always true that in any set of data, there must be at least one data point in the data that is less than or equal to the mean and at least one data point greater than or equal to the mean?"
10. A student asks if the precision with which manufacturers must calibrate their tools is at all related to statistics. How do you respond?
11. A student read in the newspaper that the pill form of a drug taken once per day is up to 92% effective in warding off the flu. She concludes that if she takes the pill 8% more times per day, she will be 100% safe in not contracting the flu. How do you respond?
12. In 2003, slightly more than 5% of the adult population in Haiti was infected with the HIV virus. Because almost 95% of the adult population was not infected at the time, Sam concluded that the virus was not a serious threat to the country. Do you agree? Why or why not?
13. Mariah read that deaths due to stampedes occurred primarily in Third World countries or at rock concerts and argued with her parents that she would be perfectly safe at the running of the bulls in Pamplona. If you were her parents, how would you reason with her?

CHAPTER OUTLINE

I. Descriptive statistics
- **A.** Types of data
  1. **Categorical data**
  2. **Numerical data**
  3. **Discrete data**
  4. **Continuous data**
- **B.** Information can be summarized in each of the following forms:
  1. **Pictographs**
  2. **Dot plots/line plots**
  3. **Stem-and-leaf plots**
  4. **Frequency tables**

5. **Histograms**
6. **Bar graphs**
7. **Line graphs/Broken line graph**
8. **Circle graphs/pie charts**
9. **Box plots/box-and-whisker plots**
10. **Scatterplots**

II. Measures of central tendency
- **A.** The **mean** of n given numbers is the sum of the numbers divided by n.
- **B.** The **median** of a set of numbers is the middle number if the numbers are arranged in numerical order; if there is no middle number, the median is the mean of the two middle numbers.
- **C.** The **mode** of a set of numbers is the number or numbers that occur most frequently in the set.

III. Measures of variation
- **A.** The **range** is the difference between the greatest and least numbers in the data.
- **B.** The **mean absolute deviation** of a data set is the mean of the absolute values of the differences of the data points and the mean.
- **C.** The **variance** of a data set is found by subtracting the mean from each value, squaring each of these differences, finding the sum of these squares, and dividing by n, where n is the number of observations.
- **D.** The **standard deviation** is equal to the square root of the variance.
- **E.** **Box plots**, or **box-and-whisker plots**, focus attention on the median, the quartiles, and the extremes and invite comparisons among them.
  1. The **lower quartile** is the median of the subset of data less than the median of all the values in the data set.
  2. The **upper quartile** is the median of the subset of data greater than the median of all the values in the data set.
  3. The **interquartile range (IQR)** is calculated as the difference between the upper quartile and the lower quartile.
  4. An **outlier** is any value more than 1.5 IQR above the upper quartile or more than 1.5 IQR below the lower quartile.
- **F.** **Scatterplots** are graphs of ordered pairs that allow us to examine the relationship (association) between two sets of data.
- **G.** A **trend line** or a line-of-best-fit is a line that approximates data in a scatterplot.
- **H.** A **normal curve** is a smooth, bell-shaped curve that depicts frequency values distributed symmetrically about the mean, as shown:

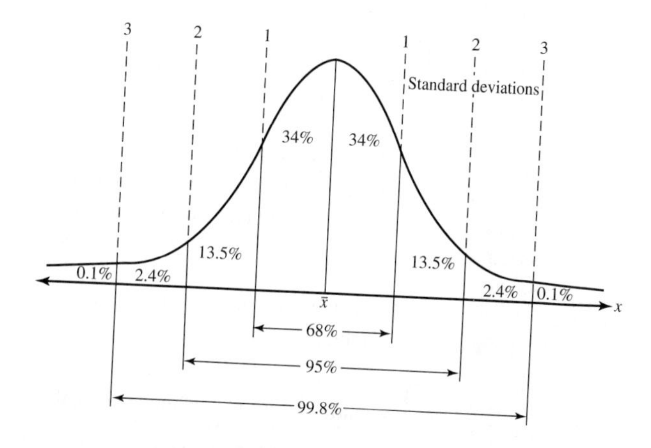

CHAPTER REVIEW

1. Suppose you read that "the average family in Rattlesnake Gulch has 2.41 children." What average is being used to describe the data? Explain your answer. Suppose the sentence had said 2.5? Then what are the possibilities?
2. At Bug's Bar-B-Q restaurant, the average weekly wage for full-time workers is \$150. There are 10 part-time employees whose average weekly salary is \$50 and the total weekly payroll is \$3950. How many full-time employees are there?
3. Find the mean, the median, and the mode for each of the following groups of data:
   a. 10, 50, 30, 40, 10, 60, 10
   b. 5, 8, 6, 3, 5, 4, 3, 6, 1, 9
4. Find the range, mean absolute deviation, interquartile range, variance, and standard deviation for each set of scores in problem 3.
5. The mass, in kilograms, of each child in Ms. Rider's class follows:

40	49	43	48	46	42	49	39	47	49
42	41	42	39	41	40	45	43	44	42

   a. Make a dot plot for the data.
   b. Make an ordered stem-and-leaf plot for the data.
   c. Make a frequency table for the data.
   d. Make a bar graph of the data.
6. The grades on a test for 30 students follow:

96	73	61	76	77	84
78	98	98	80	67	82
61	75	79	90	73	80
85	63	86	100	94	77
86	84	91	62	77	64

FIGURE 9-15

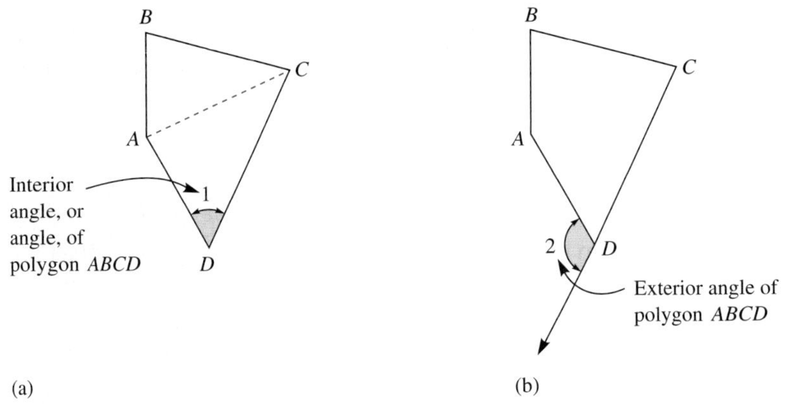

Congruent Segments and Angles

congruent parts

Most modern industries operate on the notion of creating **congruent parts**, parts that are of the same size and shape. For example, the specifications for all cars of a particular model are the same, and all parts produced for that model are basically the same. Usually congruent figures refer to figures in a plane. For example, two line **segments** are **congruent** ($\cong$) if a tracing of one line segment can be fitted exactly on top of the other line segment. If $\overline{AB}$ is congruent to $\overline{CD}$, we write $\overline{AB} \cong \overline{CD}$. Two **angles** are **congruent** if they have the same measure. Congruent segments and congruent angles are shown in Figure 9-16(a) and (b), respectively.

congruent segments

congruent angles

FIGURE 9-16

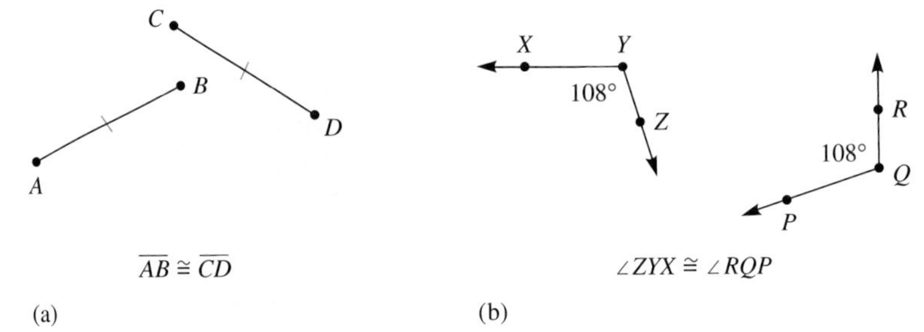

Regular Polygons

regular polygons

Polygons in which all the interior angles are congruent and all the sides are congruent are **regular polygons**. A regular polygon is both *equiangular* and *equilateral*. A regular triangle is an equilateral triangle. A regular pentagon and a regular hexagon are illustrated in Figure 9-17. The congruent sides and congruent angles are marked.

FIGURE 9-17

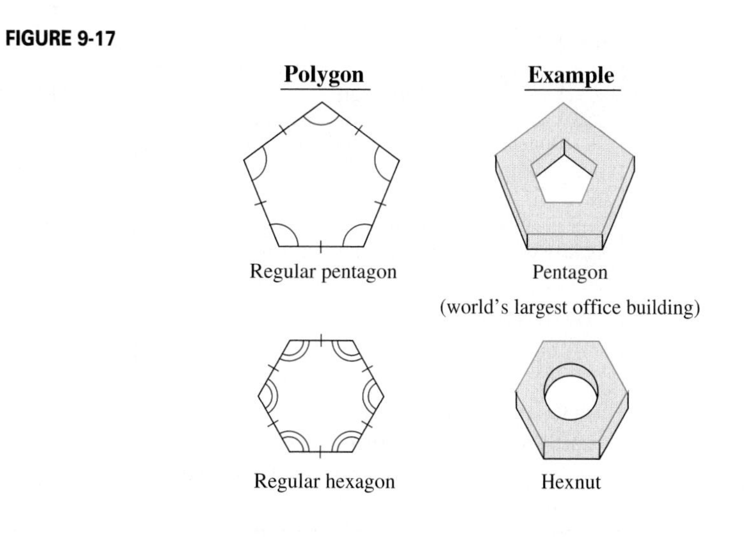

Triangles and Quadrilaterals

Triangles may be classified according to their angle measures, as shown in Table 9-6. Quadrilaterals may be classified as shown in Table 9-6.

Table 9-6

Definition	Illustration	Example
A triangle containing one right angle is a **right triangle**.	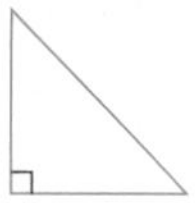	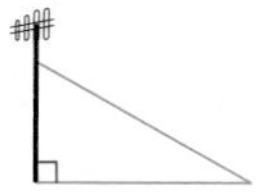
A triangle in which all the angles are acute is an **acute triangle**.	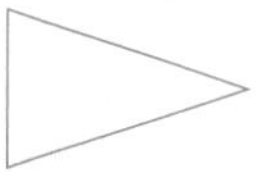	 YIELD
A triangle containing one obtuse angle is an **obtuse triangle**.	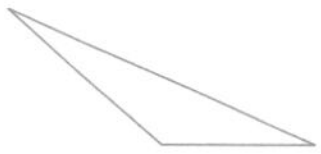	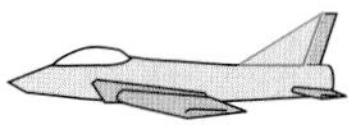
A triangle with no congruent sides is a **scalene triangle**.	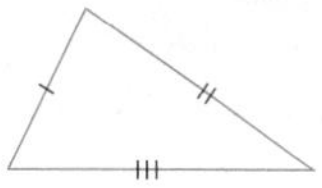	
A triangle with at least two congruent sides is an **isosceles triangle**.	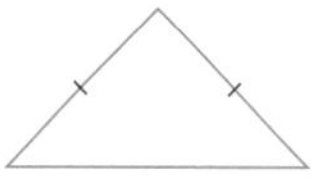	
A triangle with three congruent sides is an **equilateral triangle**.	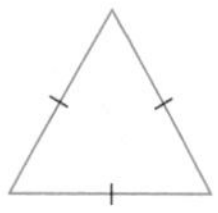	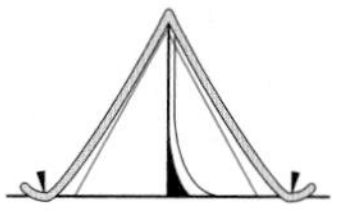

(continues)

Table 9-6 *continued*

Definition	Illustration	Example
A **trapezoid** is a quadrilateral with at least one pair of parallel sides.		
A **kite** is a quadrilateral with two adjacent sides congruent and the other two sides also congruent.		
An **isosceles trapezoid** is a trapezoid with exactly one pair of congruent sides. (Equivalently, an isosceles trapezoid is a trapezoid with two congruent base angles.)		Rax FAST FOOD WITH STYLE
A **parallelogram** is a quadrilateral in which each pair of opposite sides is parallel.		
A **rectangle** is a parallelogram with a right angle. (Equivalently, a rectangle is a quadrilateral with four right angles.)		SPEED LIMIT 40 MPH
A **rhombus** is a parallelogram with two adjacent sides congruent. (Equivalently, a rhombus is a quadrilateral with all sides congruent.)		A A
A **square** is a rectangle with two adjacent sides congruent. (Equivalently, a square is a quadrilateral with four right angles and four congruent sides.)		ROAD CONSTRUCTION AHEAD

Some texts give different definitions for trapezoids and other figures. Many elementary texts define a trapezoid as a quadrilateral with *exactly* one pair of parallel sides. Note the definition of a trapezoid on the following partial student page.

NOW TRY THIS 9-8

a. Which of the quadrilaterals on the student page are defined differently in Table 9-6?
b. Why do you think this text defines some of the quadrilaterals as in Table 9-6 and not as on the student page?

SAMPLE SCHOOL BOOK PAGE:

QUADRILATERALS

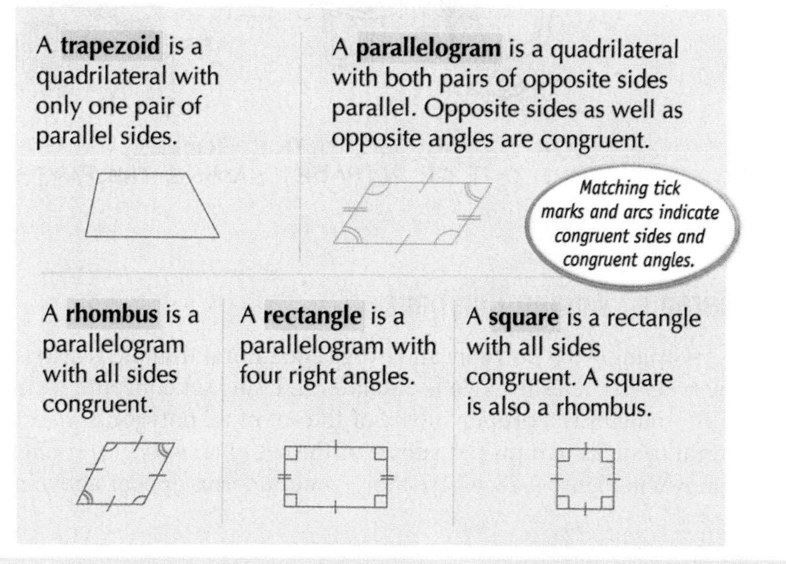
Quadrilaterals

LEARN

What are the properties of some quadrilaterals?

Quadrilaterals can be classified according to the special properties of their sides and angles.

A **trapezoid** is a quadrilateral with only one pair of parallel sides.

A **parallelogram** is a quadrilateral with both pairs of opposite sides parallel. Opposite sides as well as opposite angles are congruent.

Matching tick marks and arcs indicate congruent sides and congruent angles.

A **rhombus** is a parallelogram with all sides congruent.

A **rectangle** is a parallelogram with four right angles.

A **square** is a rectangle with all sides congruent. A square is also a rhombus.

Source: Scott Foresman–Addison Wesley Mathematics, Grade 6, 2005 (p. 500).

The many names for geometric figures can be overwhelming, as depicted in the following cartoon.

www.cartoonroom.com
WITH SO MANY GEOMETRIC FIGURES, I'M GONNA RUN OUT OF ALPHABET NAMING THE PARTS!

Hierarchy Among Polygons

Every triangle is a polygon, and every equilateral triangle is also isosceles. However, not every isosceles triangle is equilateral. Using set concepts, we can say that the set of all triangles is a proper subset of the set of all polygons. Also, the set of all equilateral triangles is a proper subset of the set of all isosceles triangles. This hierarchy is shown in Figure 9-18, where more general terms appear above more specific ones.

FIGURE 9-18

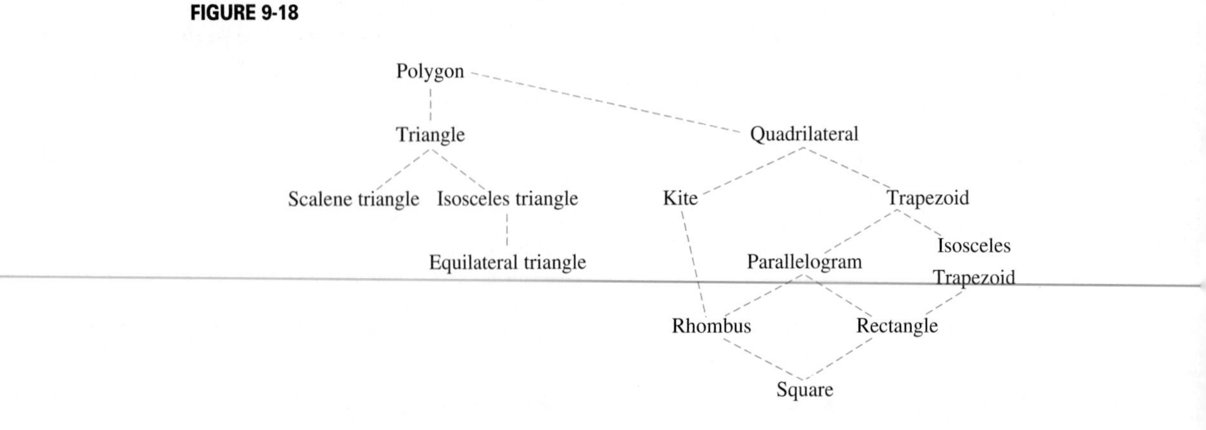

NOW TRY THIS 9-9

Use the definitions in Table 9-6 to experiment with several drawings to decide which of the following are true:

1. An equilateral triangle is isosceles.
2. A square is a regular quadrilateral.
3. If one angle of a rhombus is a right angle, then all the angles of the rhombus are right angles.
4. A square is a rhombus with a right angle.
5. All the angles of a rectangle are right angles.
6. A rectangle is an isosceles trapezoid.
7. Some isosceles trapezoids are kites.
8. If a kite has a right angle, then it must be a square.

ONGOING ASSESSMENT 9-2

1. Determine for each of the following which of the figures labeled (1) through (10) can be classified under the given term:
   a. Simple closed curve
   b. Polygon
   c. Convex polygon
   d. Concave polygon

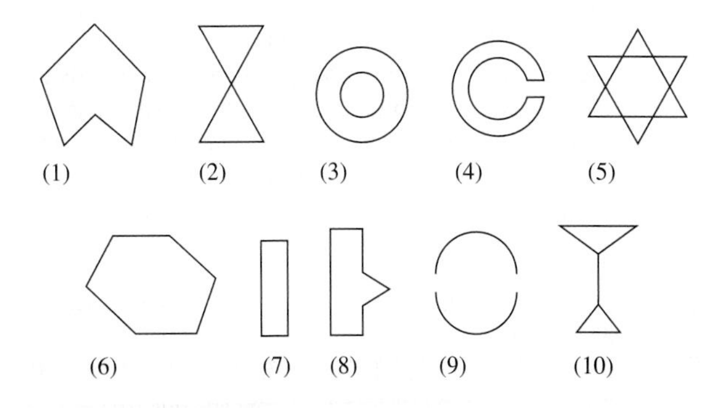

2. What is the maximum number of intersection points between a quadrilateral and a triangle (where no sides of the polygons are on the same line)?
3. What type of polygon must have a diagonal such that part of the diagonal falls outside of the polygon?
4. Which of the following figures are convex and which are concave? Why?

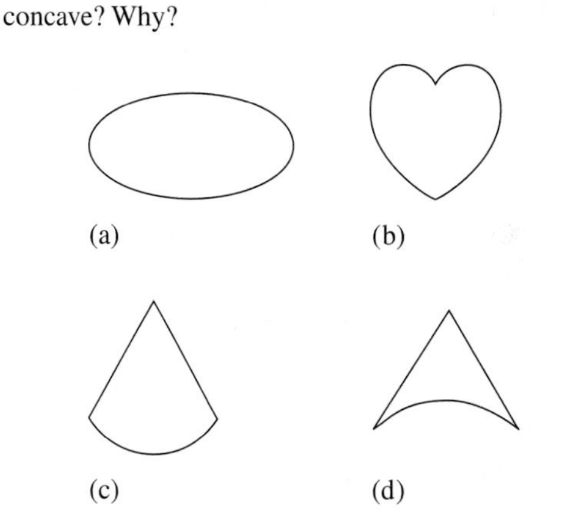

5. If possible, draw the following triangles. If it is not possible, state why.
   a. An obtuse scalene triangle
   b. An acute scalene triangle
   c. A right scalene triangle
   d. An obtuse equilateral triangle
   e. A right equilateral triangle
   f. An obtuse isosceles triangle
   g. An acute isosceles triangle
   h. A right isosceles triangle

6. Determine how many diagonals each of the following has:
   a. Decagon b. 20-gon
   c. 100-gon d. n-gon
   Justify your answer.
7. Identify each of the following triangles as scalene, isosceles, or equilateral:

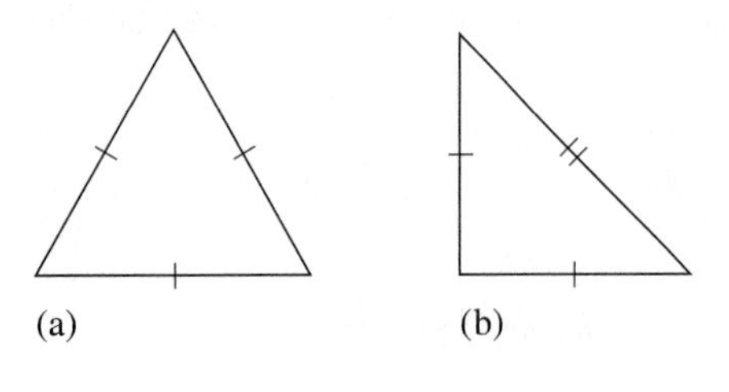

(a) (b)

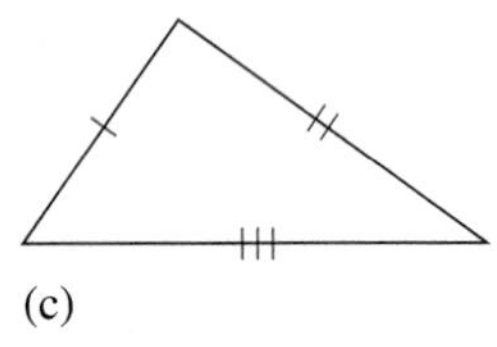

(c)

8. Describe regions (a) and (b) in the following Venn diagram, where the universal set is the set of all parallelograms:

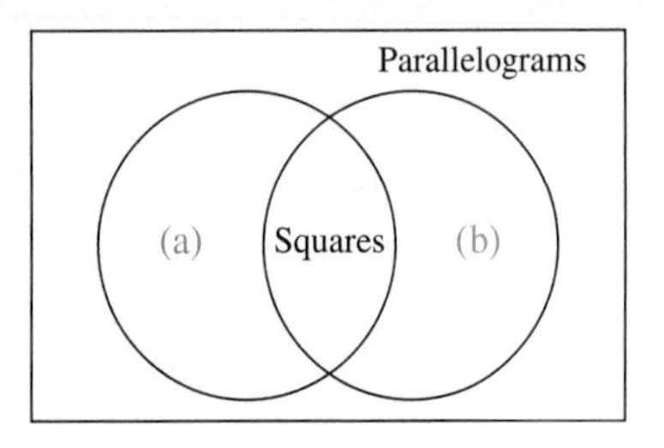

9. Use only the labeled points in the following drawing to answer the questions:

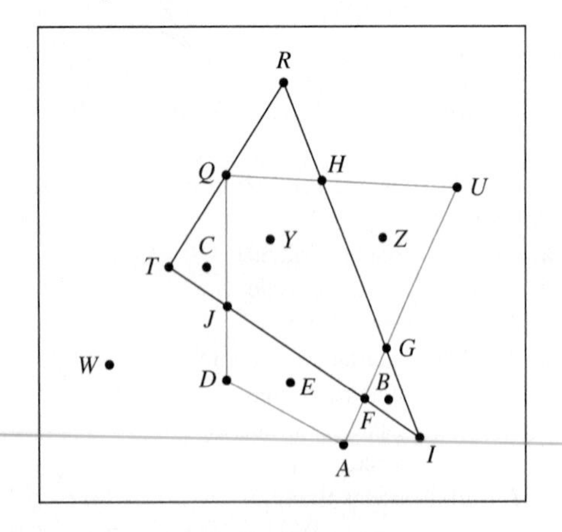

   a. Which points belong to triangle TRI?
   b. Which points belong to the interior of the quadrilateral $QUAD$?
   c. Which points belong to the exterior of triangle TRI?
   d. Which points belong to triangle TRI and quadrilateral $QUAD$?
   e. Which points belong to the intersection of the interiors of triangle TRI and quadrilateral $QUAD$?
10. Draw any angle and label its vertex A. Then construct an isosceles triangle ABC. Crease the triangle along the vertex A so that vertex B falls on C. List properties of an isosceles triangle that seem to be true.
11. Draw any triangle ABC and trace it on a second sheet of paper labeling the traced vertices A', B' and C'. Next rotate the second sheet of paper by 180° (so that the traced vertices are upside down.) Then trace the original triangle ABC on the second sheet so that A falls on B' and B on A', as shown.

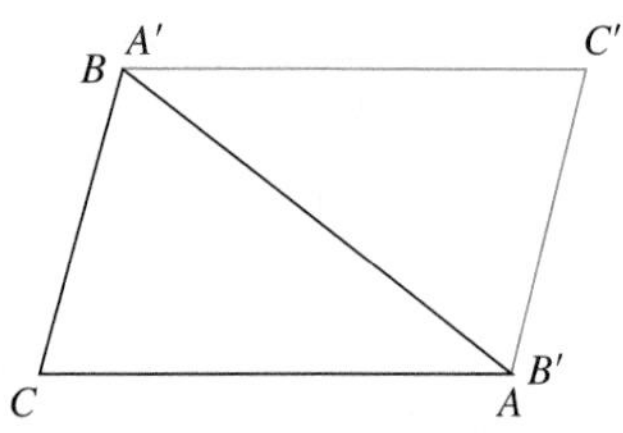

   a. What kind of figure does $ACBC'$ seem to be? List some of the properties of this figure that seem to be true.
   b. What kind of triangle must ABC in part (a) be so that the quadrilateral $ACBC'$ is
      (i) A rhombus?
      (ii) A rectangle?
      (iii) A square?

Communication

12. a. Fold a rectangular piece of paper to create a square. Describe your procedure in writing and orally with a classmate. Explain why your approach creates a square.
    b. Crease the square in (a) so that the two diagonals are shown. Use paper folding to show that the diagonals of a square are congruent and perpendicular and bisect each other. Describe your procedure and explain why it works.

Open-Ended

13. On a geoboard or dot paper, construct each of the following:
    a. A scalene triangle b. An obtuse triangle
    c. An isosceles trapezoid d. A nonisosceles trapezoid
    e. A convex hexagon f. A concave quadrilateral
    g. A parallelogram h. A rhombus which is not a square.

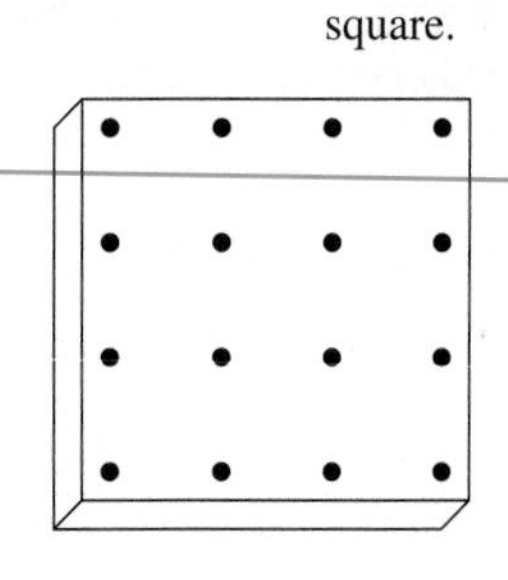

Cooperative Learning

14. Work with a partner. One of you should construct a figure on a geoboard or draw it on a piece of paper and identify it. Do not show the figure to your partner but tell your partner sufficient properties of the figure to identify it. Have your partner identify the figure you constructed. Your partner earns 1 point if the figure is correctly identified and 2 points if a figure is found that has all the required attributes but is different from the one you drew. Each of you should take the same number of turns. Try this with each of the following types of figure:
   a. Scalene triangle
   b. Isosceles triangle
   c. Square
   d. Parallelogram
   e. Trapezoid
   f. Rectangle
   g. Regular polygon
   h. Rhombus
   i. Isosceles trapezoid
   j. A kite that is not a rhombus

15. a. Compare the information in the diagram on the student page on page 595 in this section to the quadrilateral hierarchy of Figure 9-18.
   b. Work with partners to create a Venn diagram showing all triangles and isosceles, equilateral, and right triangles.

16. a. Investigate the meaning and uses of Reuleaux triangles.
   b. Explain the similarities and differences between a Reuleaux triangle and an equilateral triangle.

17. a. With partners, decide how you might define a triangle on a globe.
   b. With partners, a globe, string, and a ruler, decide if you think a square can exist on a globe.

Review Problems

18. If three distinct rays with the same vertex are drawn as shown in the following figure, then three angles are formed: $\angle AOB$, $\angle AOC$, and $\angle BOC$.

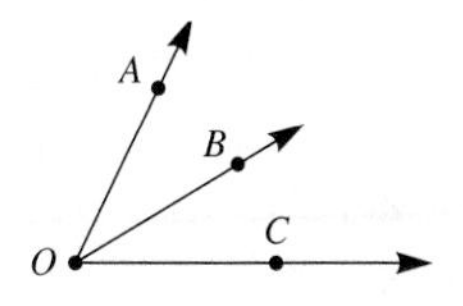

What is the maximum number of angles (measuring less than 180°) formed by using
   a. 10 distinct noncollinear rays with the same vertex?
   b. n distinct noncollinear rays with the same vertex?

19. Determine the possible intersection sets of a line and an angle.

20. Classify the following as true or false. If false, tell why.
   a. A ray has two endpoints.
   b. For any points M and N, $\overleftrightarrow{MN} = \overleftrightarrow{NM}$.
   c. Skew lines are coplanar.
   d. $\overrightarrow{MN} = \overrightarrow{NM}$
   e. A line segment contains an infinite number of points.
   f. If two distinct planes intersect, their intersection is a line segment.

Third International Mathematics and Science Study (TIMSS) Questions

a. Draw 1 straight line on this rectangle to divide it into 2 triangles.

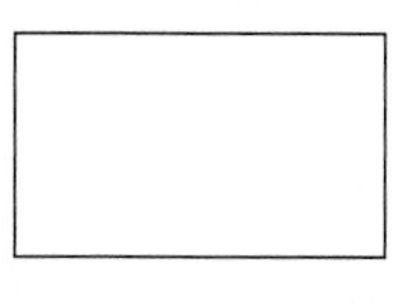

b. Draw 1 straight line on this rectangle to divide it into 2 rectangles.

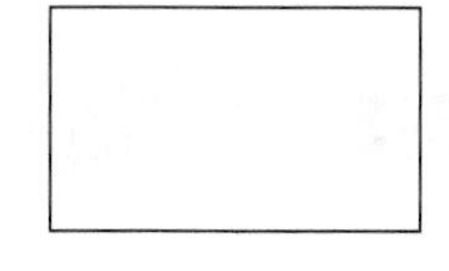

c. Draw 2 straight lines on this rectangle to divide it into 1 rectangle and 2 triangles.

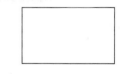

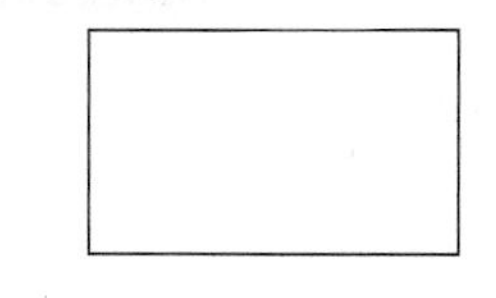

TIMSS, Grade 4, 2003

In the picture there are a number of geometric shapes, like circles, squares, rectangles, and triangles. For example, the sun looks like a circle.

Draw lines to three other different objects in the picture and write what shapes they look like.

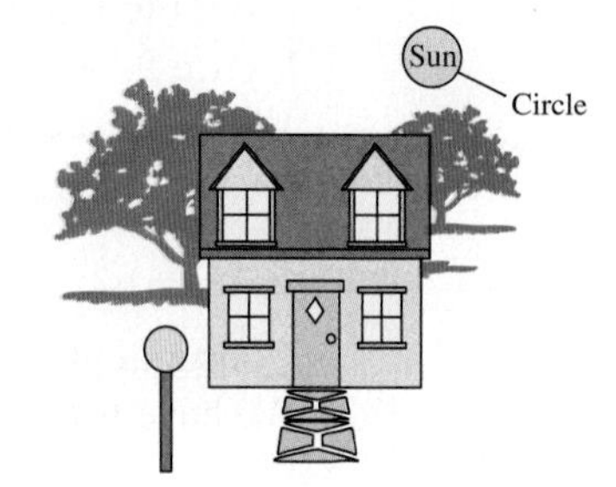

TIMSS, Grade 4, 2003

Use paper folding, tape or glue, and any other tools to create a shape larger than, but similar to, that in Figure 9-19.

FIGURE 9-19

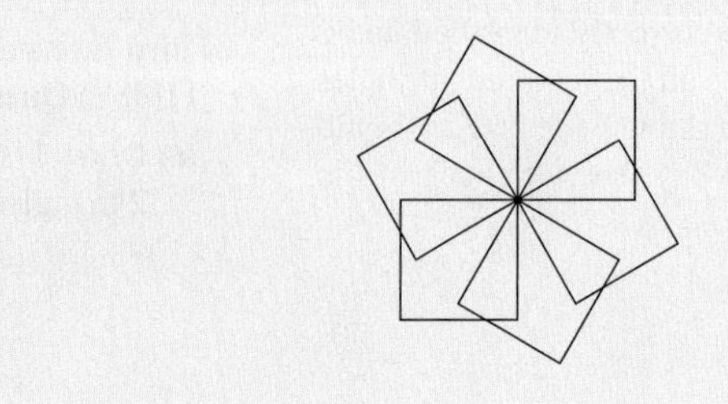

9-3 More About Angles

In Figure 9-20, two lines intersect and form the angles marked 1, 2, 3, and 4.

FIGURE 9-20

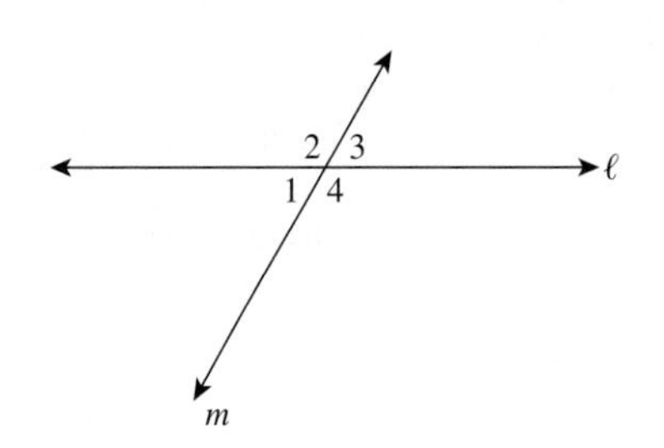

vertical angles

Vertical angles are pairs of angles such as $\angle 1$ and $\angle 3$ and appear any time two lines intersect. Another pair of vertical angles in Figure 9-20 is $\angle 2$ and $\angle 4$. You have probably noticed that vertical angles seem to be congruent. Indeed, referring to Figure 9-20, we see that $\angle 1$ and $\angle 2$ are two angles the sum of whose measure is $180°$. This is also true for $\angle 2$ and $\angle 3$. Thus,

$$m(\angle 1) = 180° - m(\angle 2)$$
$$m(\angle 3) = 180° - m(\angle 2)$$

Consequently, $m(\angle 1) = m(\angle 3)$. In a similar way we can show that $m(\angle 2) = m(\angle 3)$. We have proved therefore the following:

Theorem 9–2

Vertical angles are congruent.

Other pairs of angles appear frequently enough that it is convenient to refer to them by specific names. Table 9-7 shows several types of angle.

Table 9-7

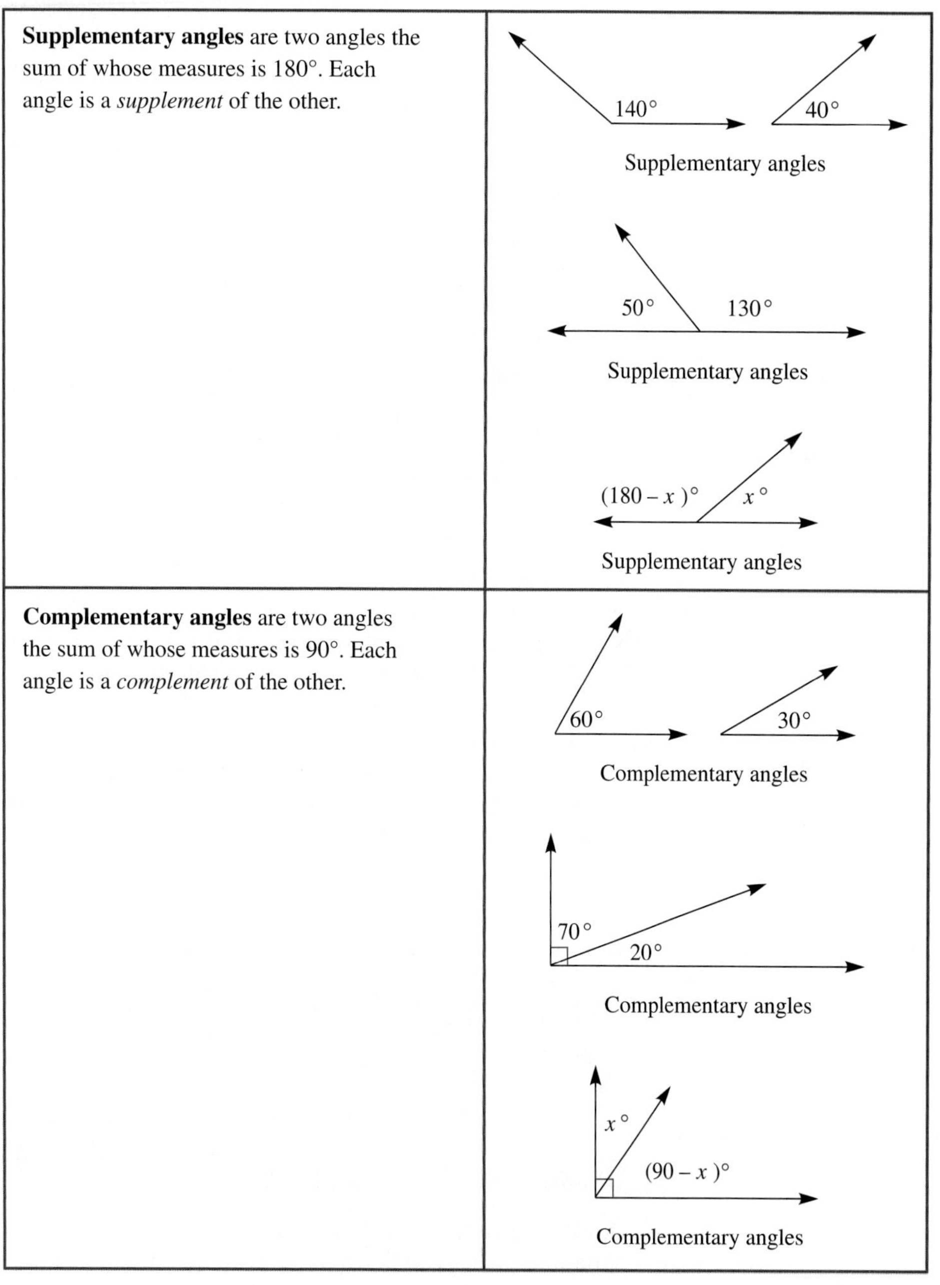

Supplementary angles are two angles the sum of whose measures is 180°. Each angle is a *supplement* of the other.	140° 40° Supplementary angles 50° 130° Supplementary angles $(180 - x)°$ $x°$ Supplementary angles
Complementary angles are two angles the sum of whose measures is 90°. Each angle is a *complement* of the other.	60° 30° Complementary angles 70° 20° Complementary angles $x°$ $(90 - x)°$ Complementary angles

Angles are also formed when a line intersects two distinct lines. Any line that intersects a pair of lines in a plane is a **transversal** of those lines. In Figure 9-21(a), line p is a transversal of lines m and n all in the same plane. Angles formed by these lines are named according to their placement in relation to the transversal and the two given lines. They are listed in Table 9-8.

transversal

FIGURE 9-21

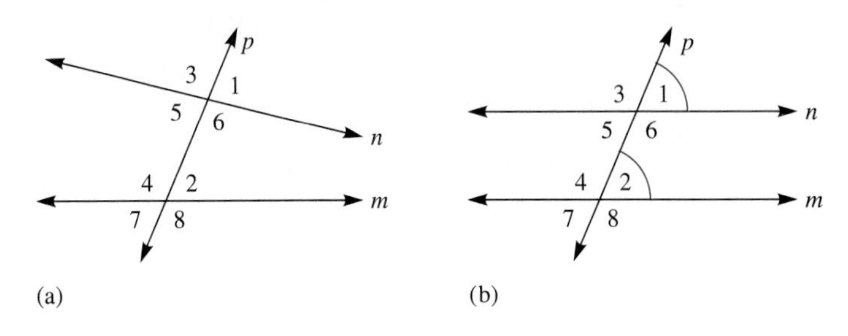

Table 9-8

Interior angles	$\angle 2, \angle 4, \angle 5, \angle 6$
Exterior angles	$\angle 3, \angle 1, \angle 7, \angle 8$
Alternate interior angles	$\angle 5$ and $\angle 2$, $\angle 4$ and $\angle 6$
Alternate exterior angles	$\angle 1$ and $\angle 7$, $\angle 3$ and $\angle 8$
Corresponding angles	$\angle 3$ and $\angle 4$, $\angle 5$ and $\angle 7$, $\angle 1$ and $\angle 2$, $\angle 6$ and $\angle 8$

Suppose corresponding angles such as $\angle 1$ and $\angle 2$ in Figure 9-21(b) are congruent. With this assumption, and because $\angle 1$ and $\angle 5$ are congruent vertical angles, we know that the pair of alternate interior angles $\angle 2$ and $\angle 5$ are also congruent. Similarly, each pair of corresponding angles, alternate interior angles, and alternate exterior angles are congruent.

If we examine Figure 9-21(b) further, we see that lines m and n appear to be parallel when $\angle 1$ is congruent to $\angle 2$. Conversely, if the lines are parallel, the sets of angles mentioned previously are congruent. This is true and is summarized in the following theorem, which we state without proof.

Theorem 9–3

If any two distinct coplanar lines are cut by a transversal, then a pair of corresponding angles, alternate interior angles, or alternate exterior angles are congruent if and only if the lines are parallel.

Constructing Parallel Lines

A method commonly used by architects to construct a line ℓ through a given point P parallel to a given line m is shown in Figure 9-22. Place the side $\overline{AB}$ of triangle ABC on line m, as shown in Figure 9-22(a). Next, place a ruler on side $\overline{AC}$. Keeping the ruler stationary, slide triangle ABC along the ruler's edge until its side $\overline{AB}$ (marked $\overline{A'B'}$) contains point P, as in Figure 9-22(b). Use the side $\overline{A'B'}$ to draw the line ℓ through P parallel to m.

To show that the construction produces parallel lines, notice that when triangle ABC slides, the measures of its angles are unchanged. The angles of triangle

ABC and triangle $A'B'C'$ in Figure 9-22(b) are correspondingly congruent angles. $\angle A$ and $\angle A'$ are corresponding angles formed by m and ℓ and the transversal $\overleftrightarrow{EF}$. Because corresponding angles are congruent, Theorem 9-3 implies that $\ell \parallel m$. In Chapter 10, we show how to construct parallel lines using only a compass and straightedge.

FIGURE 9-22

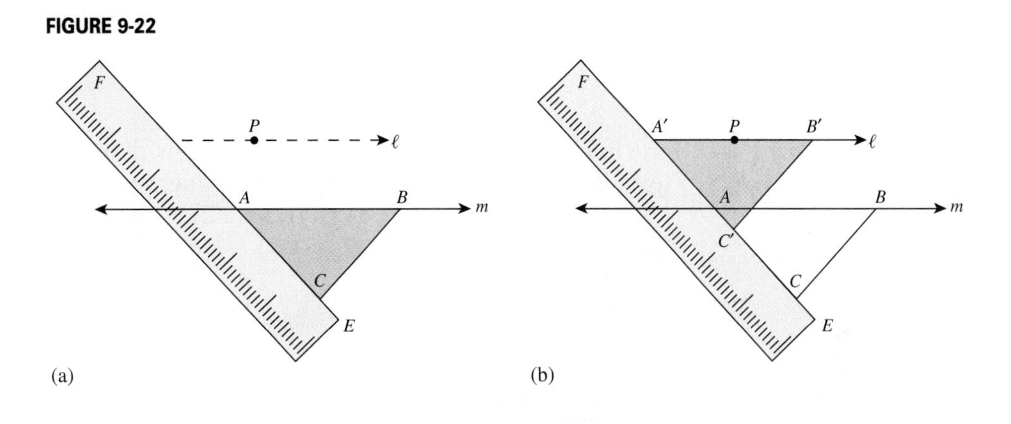

The Sum of the Measures of the Angles of a Triangle

The sum of the measures of the angles in a triangle can be observed to be 180°. We see this by using a torn triangle, as shown in Figure 9-23. Angles 1, 2, and 3 of triangle ABC in Figure 9-23(a) are torn as pictured and then replaced as shown in Figure 9-23(b). The three angles seem to lie along a single line ℓ.

FIGURE 9-23

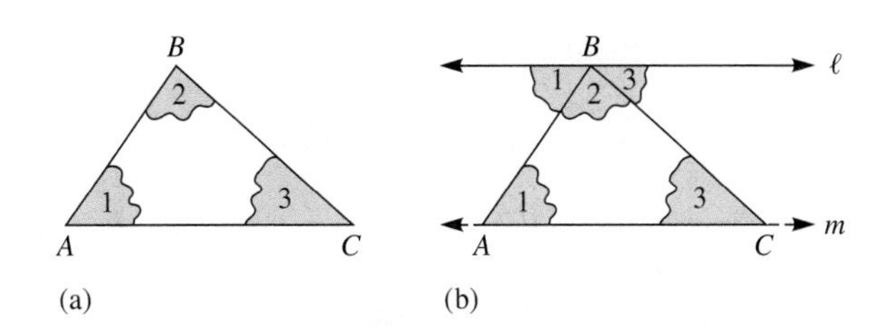

The angle measures seem to add up to a straight angle, which measures 180°. This conclusion, which is based on observation, is an example of *inductive reasoning*, discussed in Chapter 1. In contrast, *deductive reasoning* shows that a statement is true by using the given information, previously defined and undefined terms, theorems or statements assumed to be true, and logic. A conclusion based on *deductive reasoning* must be true if the hypothesis is true.

Next, we use deductive reasoning to show that the sum of the measures of the interior angles in every triangle is 180°. In Figure 9-23(b), the line ℓ appears to be parallel to m. This suggests drawing a line ℓ parallel to $\overline{AC}$ through vertex B of triangle ABC, as in Figure 9-24(b), and showing that the angles formed are congruent to the interior angles of the triangle.

FIGURE 9-24

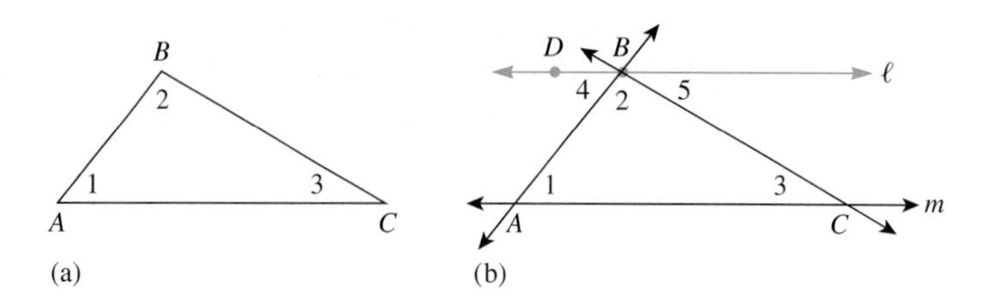

In Figure 9-24(b), ℓ and $\overline{AC}$ are parallel, with transversals $\overleftrightarrow{AB}$ and $\overleftrightarrow{BC}$, so it follows that alternate interior angles are congruent. Consequently, $m(\angle 1) = m(\angle 4)$ and $m(\angle 3) = m(\angle 5)$. Thus, $m(\angle 1) + m(\angle 2) + m(\angle 3) = m(\angle 4) + m(\angle 2) + m(\angle 5) = 180°$. So, $m(\angle 1) + m(\angle 2) + m(\angle 3) = 180°$. From this, we obtain the following theorem.

Theorem 9–4

The sum of the measures of the interior angles of a triangle is 180°.

NOW TRY THIS 9-10

An alternative way to show that the sum of the measures of the interior angles of a triangle is 180° is suggested in Figure 9-25. We start at vertex A in Figure 9-25(a) facing B, walk all the way around the triangle, and stop at the same position and facing in the same direction as when we started. This trip is shown in Figure 9-25(b).

FIGURE 9-25

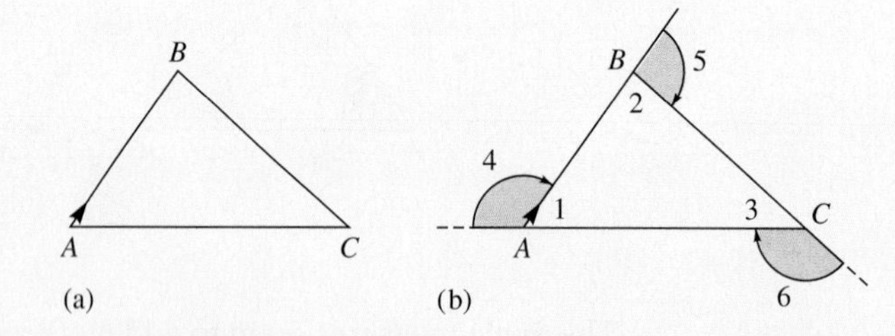

a. Write an argument for the sum of the measures of the exterior angles of the triangle to be 360°.

b. Use the fact that the sum of the measures of the exterior angles of a triangle is 360° to write an argument that the sum of the measures of the interior angles of the triangle is 180°.

c. Generalize the sum of the measures of the exterior angles of any convex polygon. Explain your reasoning.

Now Try This 9-10 can be summarized in Theorem 9-5.

Theorem 9–5

The sum of the measures of the exterior angles (one at each vertex) of a convex polygon is 360°.

EXAMPLE 9-2

In the framework for a bridge, shown in Figure 9-26(a), *ABCD* is a parallelogram. If $\angle ADC$ of the parallelogram measures 50°, what are the measures of the other angles of the parallelogram?

FIGURE 9-26

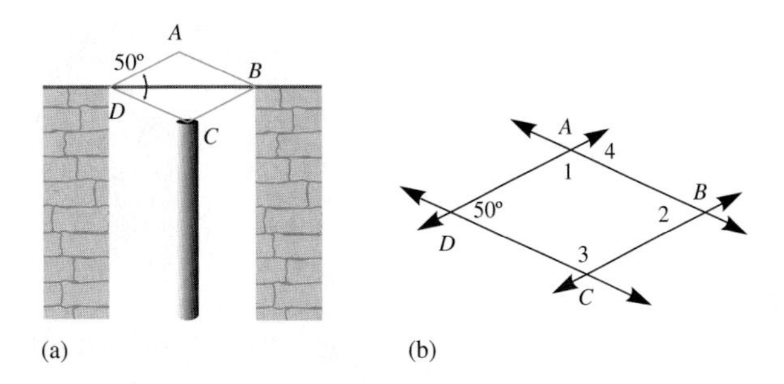

Solution Refer to Figure 9-26(b). We draw the lines containing the sides of parallelogram *ABCD*. $\angle ADC$ has a measure of 50°, and $\angle 4$ and $\angle ADC$ are corresponding angles formed by parallel lines $\overleftrightarrow{AB}$ and $\overleftrightarrow{CD}$ cut by transversal $\overleftrightarrow{AD}$. So it follows that $m(\angle 4) = 50°$. Because $\angle 1$ and $\angle 4$ are supplementary, $m(\angle 1) = 180° - 50° = 130°$. Using similar reasoning, we find that $m(\angle 2) = 50°$ and $m(\angle 3) = 130°$.

EXAMPLE 9-3

In Figure 9-27, $m \parallel n$ and k is a transversal. Explain why $m(\angle 1) + m(\angle 2) = 180°$.

FIGURE 9-27

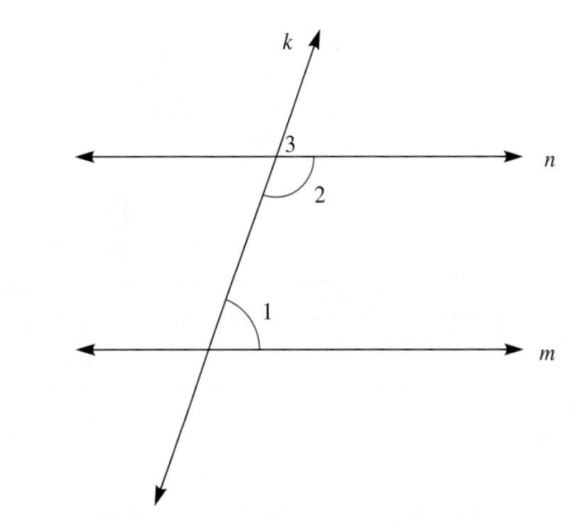

Solution Because $\angle 1$ and $\angle 3$ are corresponding angles and $m \parallel n$, $m(\angle 1) = m(\angle 3)$. Also because $\angle 2$ and $\angle 3$ are supplementary angles, $m(\angle 2) + m(\angle 3) = 180°$. Substituting $m(\angle 1)$ for $m(\angle 3)$, we have $m(\angle 2) + m(\angle 1) = 180°$.

NOW TRY THIS 9-11

When a plane flies directly from New York to London and then to Nairobi and back to New York, it flies along the sides of a spherical triangle, as shown in Figure 9-28(a), approximately following the surface of Earth. On a sphere, the shortest path between two points is along an arc of a great circle. A great circle is obtained when a plane through the center of the sphere intersects the sphere. (An example of a great circle is the equator.) To obtain a great circle through two given points on the sphere, we need consider only a plane through the two points and the center of the sphere. A great circle through points A and B is shown in Figure 9-28(b). A *spherical triangle* consists of three points on the sphere and arcs of great circles (the shortest path) connecting the points.

FIGURE 9-28

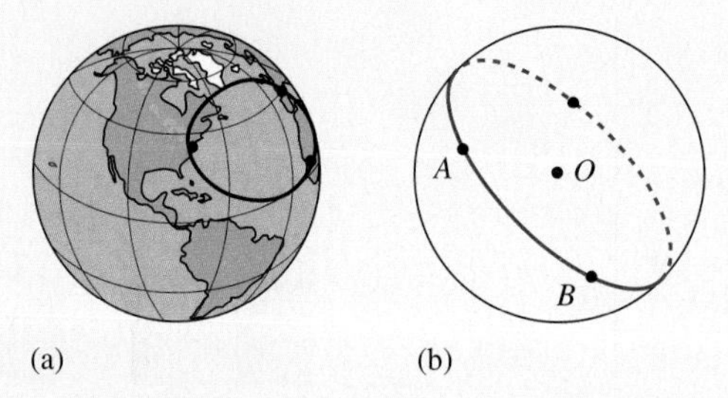

(a) (b)

a. How many great circles go through the North Pole and the South Pole on the globe?

b. Describe a spherical triangle that has one vertex at the North Pole and the other two vertices on the equator.

c. Is there a spherical triangle with three right angles? If so, find one.

The Sum of the Measures of the Interior Angles of a Convex Polygon with n Sides

We can find the sum of the measures of all the interior angles in any convex n-gon in a way different from the one in the cartoon. We first illustrate the approach for a convex pentagon. Extend each side of the pentagon as shown in Figure 9-29 so that at each vertex, an exterior angle is formed. Because at each vertex an interior angle and the corresponding exterior angle are supplementary, the sum of the

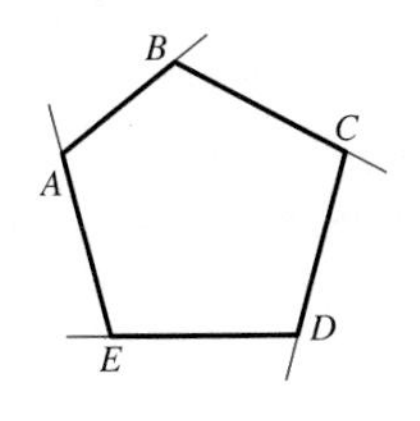

FIGURE 9-29

measures of all the exterior and interior angles of the pentagon is $5 \cdot 180°$. By Theorem 9-5, the sum of the measures of the exterior angles is 360°. Thus, the sum of the measures of the interior angles is $5 \cdot 180° - 360°$, or 540°. The same reasoning applies to any convex n-gon. The sum of the measures of an interior angle and the corresponding exterior angle at each vertex is 180°. Because there are n vertices, the sum of the measures of all the interior and exterior angles is $n \cdot 180°$. The sum of the measures of the exterior angles is 360°, so the sum of the measures of the interior angles is $n \cdot 180° - 360°$, or $(n - 2)180°$.

In a regular n-gon, all n interior angles are congruent and the sum of their measures is $180n - 360$, so the measure of a single angle is $\frac{180n - 360}{n}$, or $\frac{(n-2)180°}{n}$.

The results from this discussion are summarized by the following theorem.

Theorem 9–6

a. The sum of the measures of the interior angles of any convex polygon with n sides is $180n - 360$, or $(n - 2)180°$.

b. The measure of a single interior angle of a regular n-gon is $\frac{180n - 360}{n}$, or $\frac{(n-2)180°}{n}$.

NOW TRY THIS 9-12

We can use the fact that the sum of the measures of a triangle's interior angles is 180° to find the sum of the measures of the interior angles of a quadrilateral by dividing the quadrilateral into two triangles. Because the sum of the measures of the interior angles in each triangle is 180°, the sum of the measures of the interior angles in the quadrilateral is $2 \cdot 180°$, or 360°. Use this approach to find the sum of the measures of the interior angles for any convex n-gon. Is your result the same as in Theorem 9-6?

EXAMPLE 9-4

a. Find the measure of each interior angle of a regular decagon.

b. Find the number of sides of a regular polygon each of whose interior angles has a measure of 175°.

Solution **a.** Because a decagon has 10 sides, the sum of the measures of the angles of a decagon is $10 \cdot 180 - 360$, or 1440°. A regular decagon has 10 angles, all of which are congruent, so each one has a measure of $\frac{1440°}{10}$, or 144°. As an alternative solution, each exterior angle is $\frac{360°}{10}$, or 36°. Hence, each interior angle is $180 - 36$, or 144°.

b. Each interior angle of the regular polygon is 175°. Thus, the measure of each exterior angle of the polygon is $180° - 175°$, or 5°. Because the sum of the measures of all exterior angles of a convex polygon is 360°, the number of exterior angles is $\frac{360}{5}$, or 72. Hence, the number of sides is 72.

▲

EXAMPLE 9-5

In Figure 9-30 lines k and l are parallel. The angles at A and B are as shown. Find x, the measure of $\angle BCA$.

FIGURE 9-30

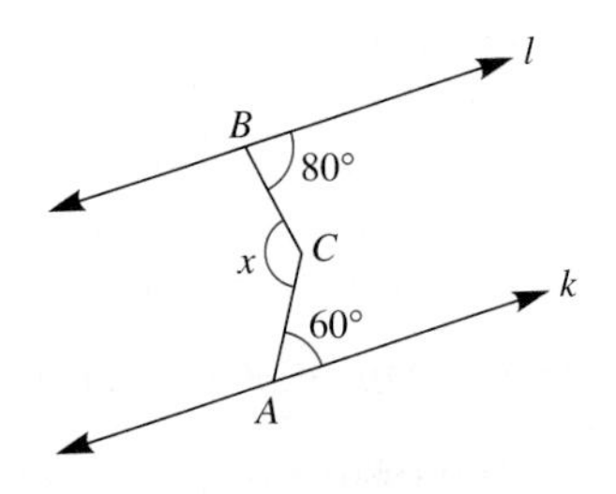

Solution We use the strategy of *examining a related problem.* If we had a transversal intersecting the parallel lines, we could use the fact that corresponding or alternate interior angles are congruent. For that purpose we need to extend either $\overline{BC}$ or $\overline{AC}$. In Figure 9-30(a), we extend $\overline{BC}$ and obtain the transversal $\overleftrightarrow{BC}$ that intersects line k at D. Because the marked angles with vertices at B and D are alternate interior angles created by the parallel lines and the transversal, they are congruent, and hence the marked angle at D measures 80°. We can now find the measure of the third angle in $\triangle ACD$; it is $180 - (60 + 80)$, or 40°. Thus, $x = 180 - 40$, or 140°.

Alternative approaches are suggested in Figure 9-31(b) where line m is constructed parallel to ℓ, and Figure 9-31(c), where a line perpendicular to k and l is drawn through point C. Solve the problem using each of these figures.

FIGURE 9-31(a) **FIGURE 9-31(b)** **FIGURE 9-31(c)**

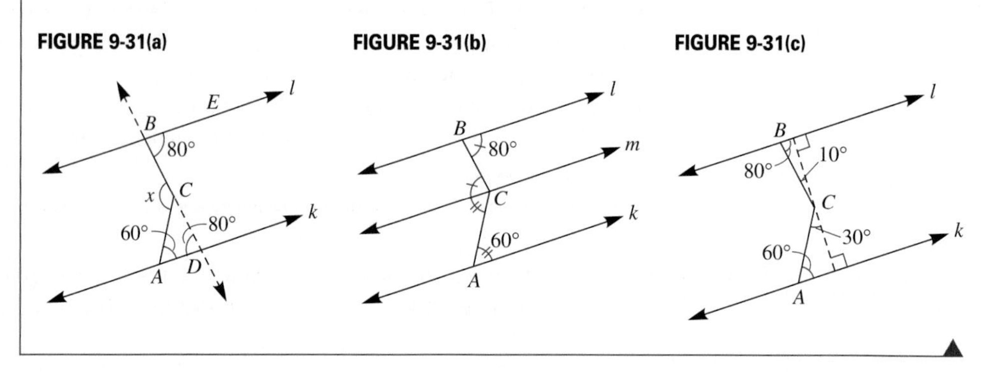

Walks Around Stars and Other Figures

FIGURE 9-32

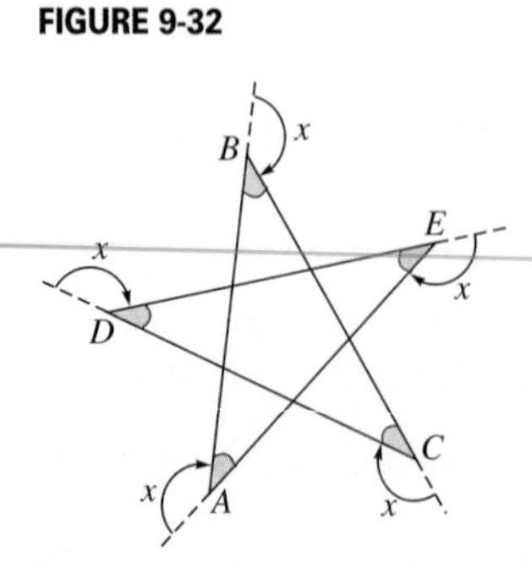

In *Now Try This 9-10* and Theorem 9-5, we have seen that the sum of the measures of the exterior angles of any convex polygon is 360°. We can use the same idea of walking around the polygon to generalize Theorem 9-5 to "polygons" that cross themselves. Consider the five-pointed star in Figure 9-32 with congruent angles at the vertices A, B, C, D, and E. We begin a trip on the star at any vertex A, walk along the line to B, and then turn by the exterior angle whose measure is x, walk to C, turn by x again, walk to D, turn by x, walk to E, turn by x, walk back to A, and finally turn again by x. In this way we not only started at A and ended there but we also face in the same direction as when we started. Our total turning is $5x$, but because we face in the same direction as we started, we must have turned a multiple of 360°. Thus, $5x = k \cdot 360°$, for some positive integer k. If $k = 1$, then $x = 72°$, which is not reasonable because it is clear from the figure that $90° < x < 180°$. If $k = 2$, then $x = 144°$, and each shaded interior angle of the five-pointed star is $180° - 144°$, or $36°$. Notice that $k = 3$ results in $x = 216°$, which is greater than $180°$ and hence impossible. For the same reason, values of k greater than 3 will yield impossible values of x.

BRAIN TEASER

Sylvia, a graphic designer, needs to know the measurement of the shaded angles at the vertices of the 12-pointed star, in which all the sides and all the marked angles are congruent. How can she find the exact measurement of the angles without using a protractor?

FIGURE 9-33

ONGOING ASSESSMENT 9-3

1. If three lines all meet in a single point, how many pairs of vertical angles are formed?
2. Find the measure of the third angle in each of the following triangles:

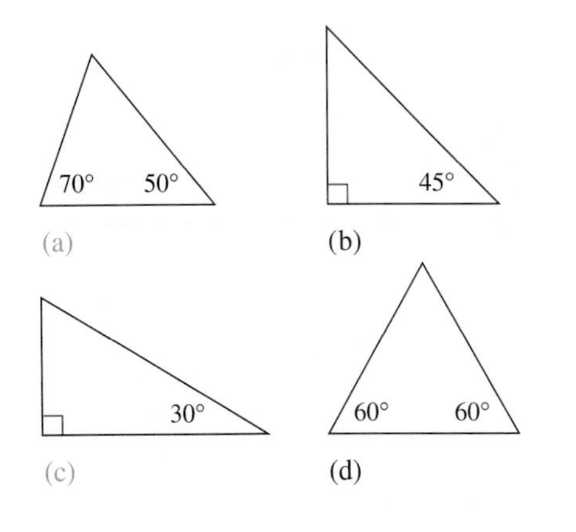

3. For each of the following figures, determine whether m and n are parallel lines. Justify your answers.

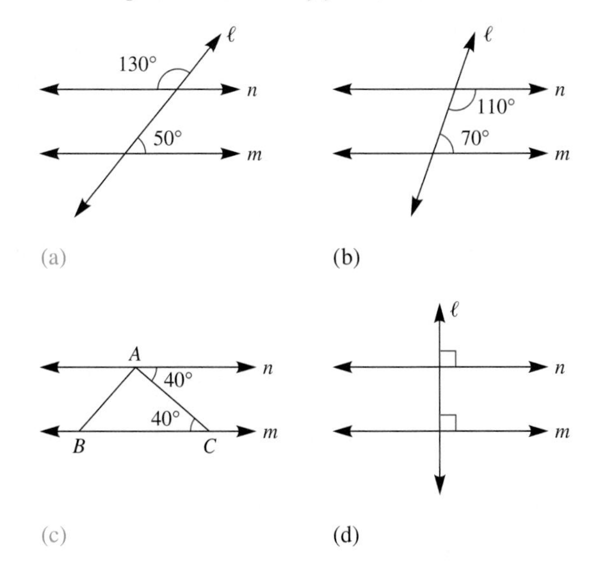

4. Two angles are complementary and the ratio of their measures is 7 : 2. What are the angle measures?

5. a. In a regular polygon, the measure of each angle is 162°. How many sides does the polygon have?

b. Find the measure of each of the angles of a regular dodecagon.

6. In the following figure, $\overleftrightarrow{DE} \parallel \overleftrightarrow{BC}$, $\overleftrightarrow{EF} \parallel \overleftrightarrow{AB}$, and $\overleftrightarrow{DF} \parallel \overleftrightarrow{AC}$, Also, $m(\angle 1) = 45°$ and $m(\angle 2) = 65°$. Find each of the following values:

a. $m(\angle 3)$
b. $m(\angle D)$
c. $m(\angle E)$
d. $m(\angle F)$

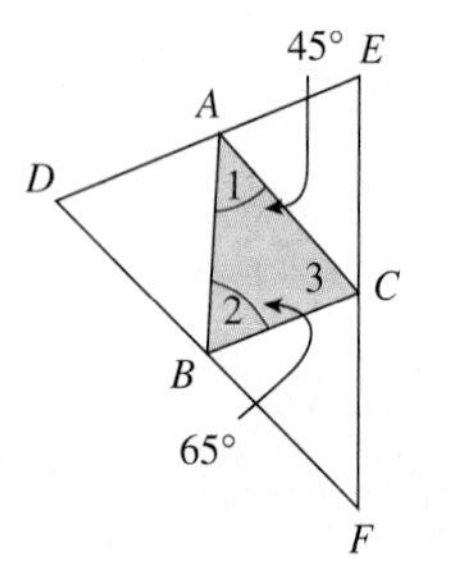

7. Lines *a, b,* and *c* are parallel. Find the measures of the numbered angles.

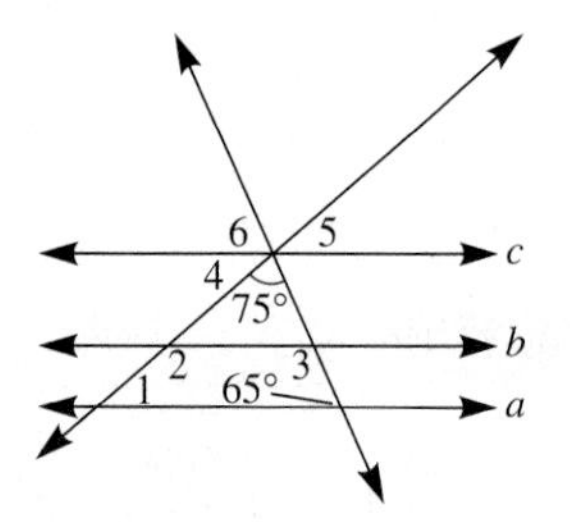

8. For each of the following, determine *x* if the lines *a* and *b* are parallel.

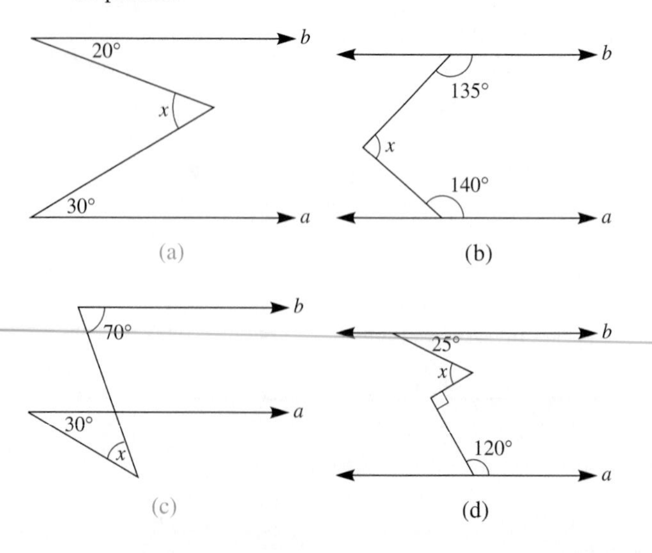

9. In the following figures, find the measures of the angles marked *x* and *y*:

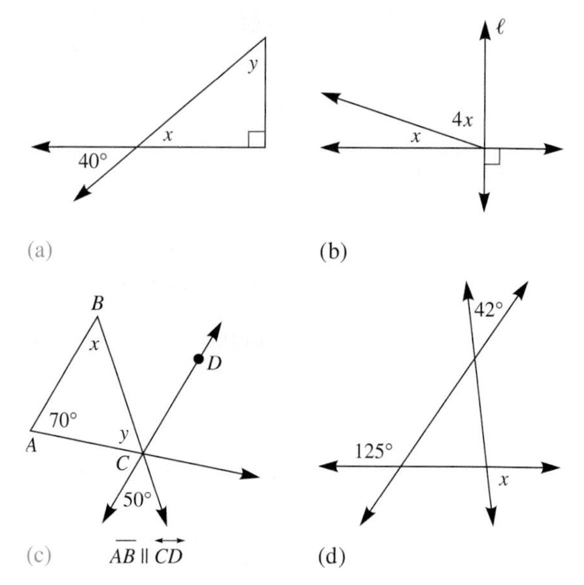

10. In each of the following, a relationship between the marked angles is given. In each case, prove that $k \parallel l$.

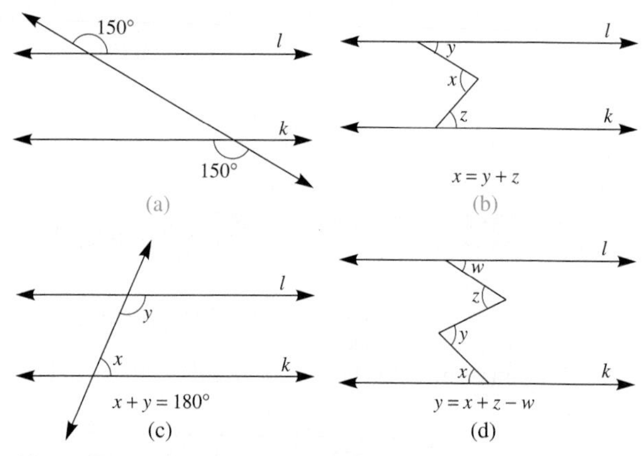

11. a. Determine the measure of an angle whose measure is twice that of its complement.

b. If two angles of a triangle are complementary, what is the measure of the third angle?

12. Find the sum of the measures of the marked angles in each of the following figures:

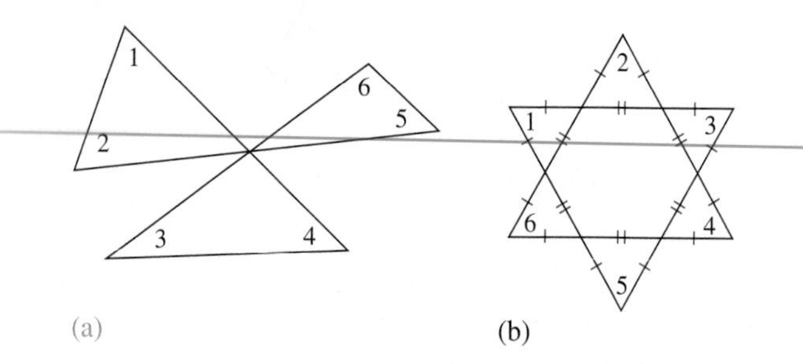

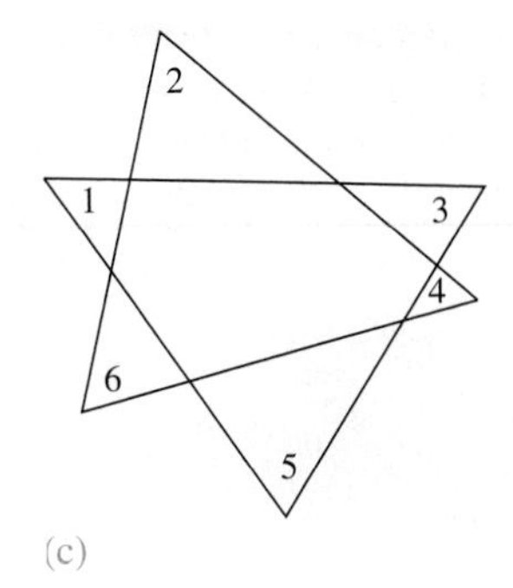

(c)

13. Calculate the measure of each angle of a pentagon, where the measures of the angles form an arithmetic sequence and the least measure is 60°.

14. Two sides of a regular octagon are extended as shown in the following figure. Find the measure of $\angle 1$.

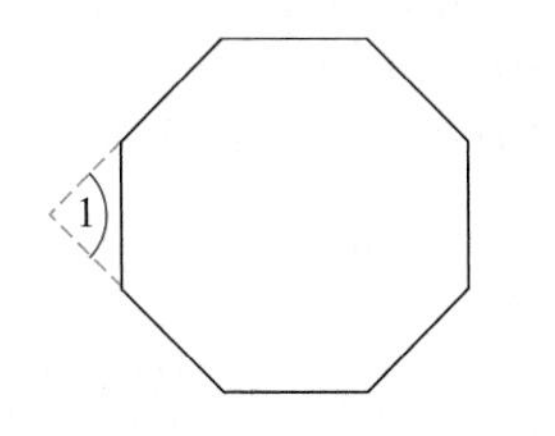

15. Find the measure of angle x in the following figure:

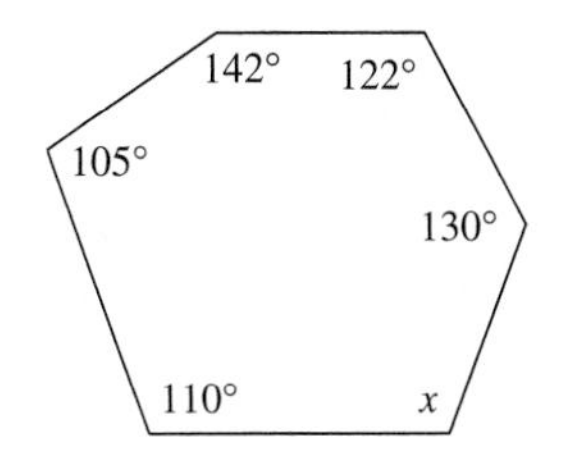

16. Find the measures of angles 1, 2, and 3 given that *TRAP* is a trapezoid with $\overline{TR} \parallel \overline{PA}$.

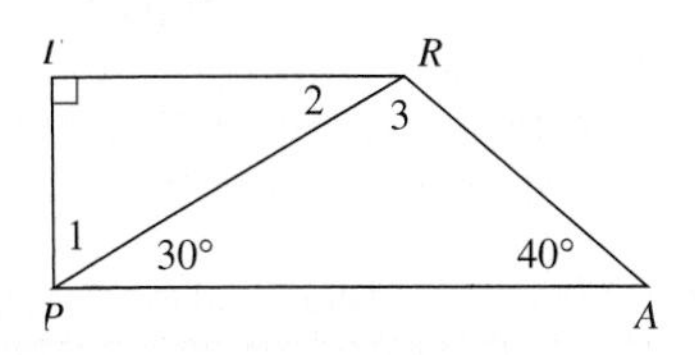

17. Home plate on a baseball field has three right angles and two other congruent angles. Refer to the following figure and find the measures of each of these two other congruent angles:

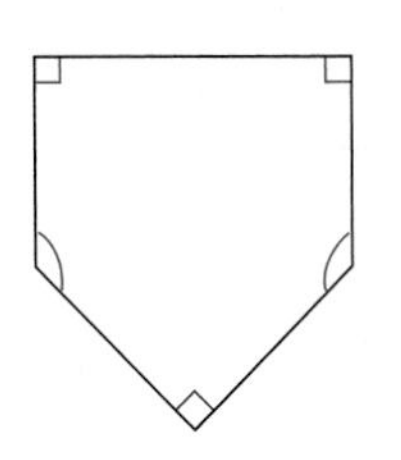

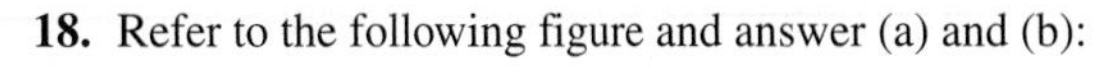

18. Refer to the following figure and answer (a) and (b):

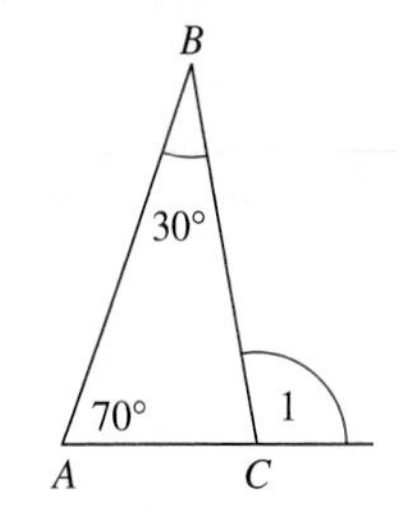

a. Find $m(\angle 1)$.

b. $\angle 1$ is an exterior angle of $\triangle ABC$. Use your answer in (a) to make a conjecture concerning the measure of an exterior angle of a triangle. Justify your conjecture.

19. Refer to the following figure and answer (a) and (b):

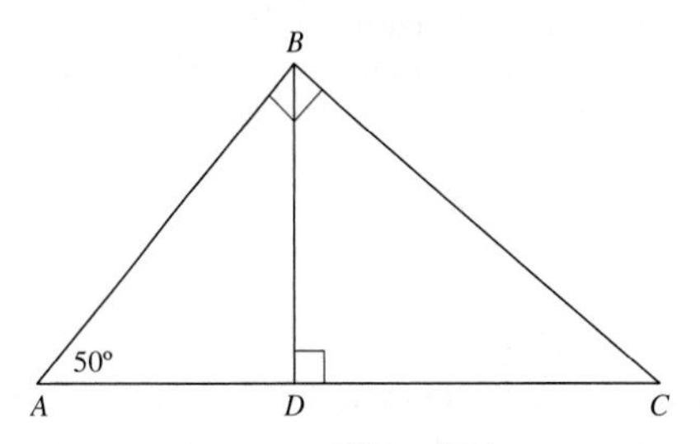

a. If $m(\angle ABC) = 90°$ and $\overline{BD} \perp \overline{AC}$ and $m(\angle A) = 50°$, find the measure of all the angles of $\triangle ABC$, $\triangle ADB$, and $\triangle CDB$.

b. If $m(\angle A) = \alpha$ in (a), find the measures of all the angles of $\triangle ABC$, $\triangle ADB$, and $\triangle CDB$ in terms of α.

20. If one interior angle of a parallelogram measures 120°, what are the measures of the other three interior angles?

21. In a quadrilateral the smallest angle has measure 49°30′. If the measures of the four angles form an arithmetic sequence, find the measures of the other angles.

22. In the figure, x, y, z, and w are measures of the angles as shown. If $y = 2x = \frac{1}{2}z = \frac{1}{3}w$, find x, y, z and w.

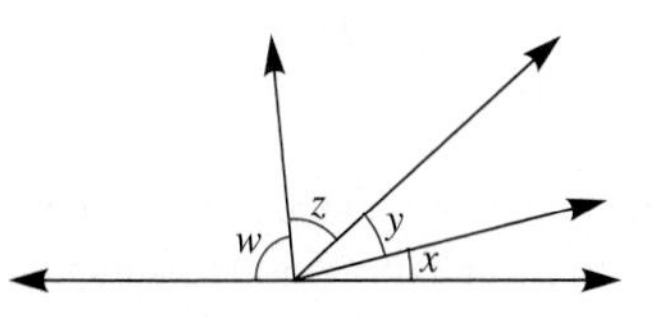

23. In each of the following figures, find the measures of the unknown marked angles.

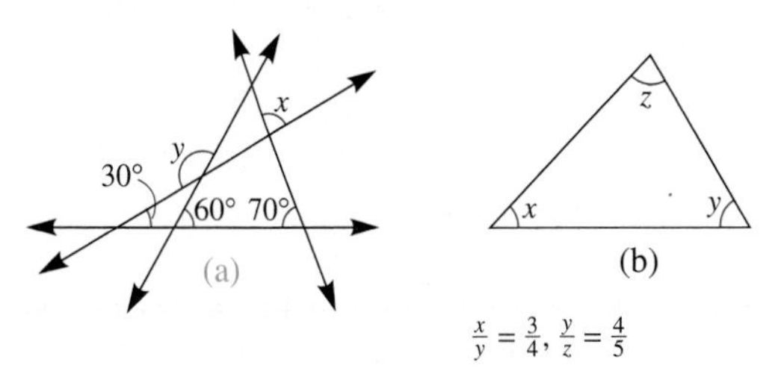

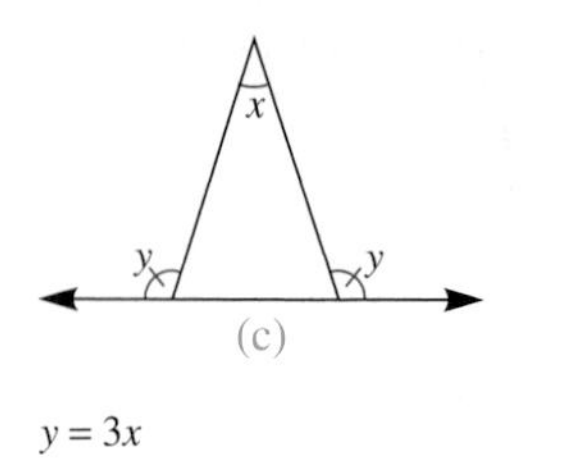

(c)

$y = 3x$

(d)

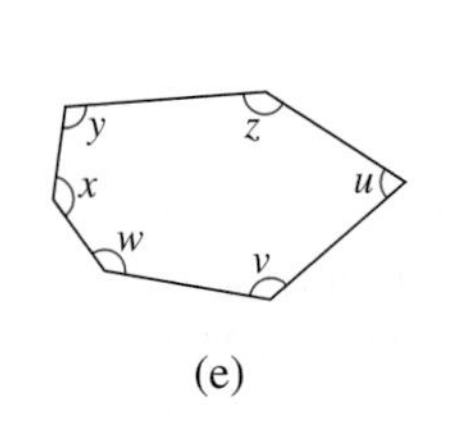

(e)

The sequence x, y, z, u, v, w is an arithmetic sequence and $v = 136°$

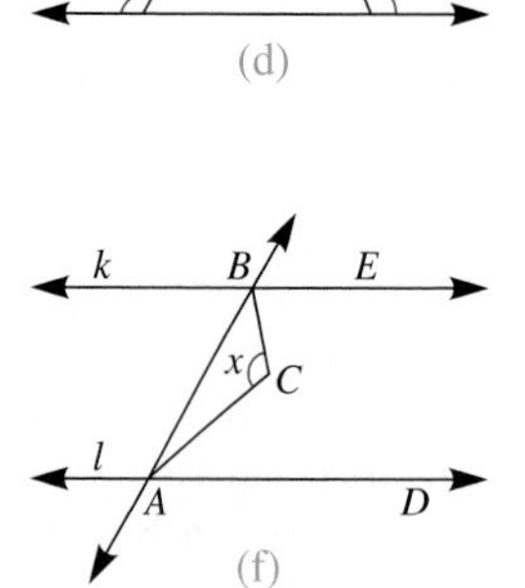

(f)

$k \parallel l$
$m(\angle CAD) = 2m(\angle BAC)$
$m(\angle CBE) = 2m(\angle ABC)$

Communication

24. Explain how you might find a measure of an angle of a staircase that will describe the staircase's steepness.

25. a. If one angle of a triangle is obtuse, can another also be obtuse? Why or why not?
b. If one angle in a triangle is acute, can the other two angles also be acute? Why or why not?
c. Can a triangle have two right angles? Why or why not?
d. If a triangle has one acute angle, is the triangle necessarily acute? Why or why not?

26. In the following figure, A is a point not on line ℓ. Discuss whether it is possible to have two distinct perpendicular segments from A to ℓ in a plane.

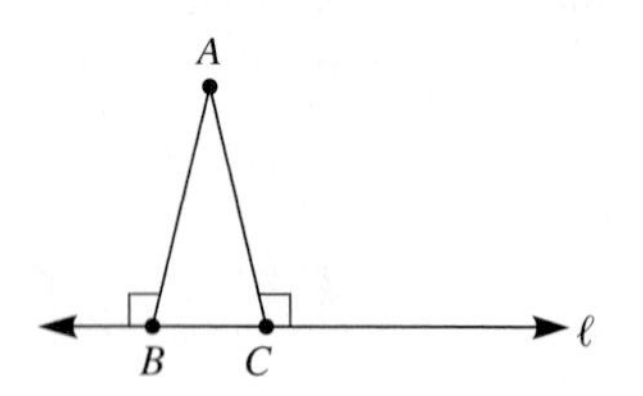

27. a. Explain how to find the sum of the measures of the interior angles of any convex pentagon by choosing any point P in the interior and constructing triangles, as shown in the following figure:

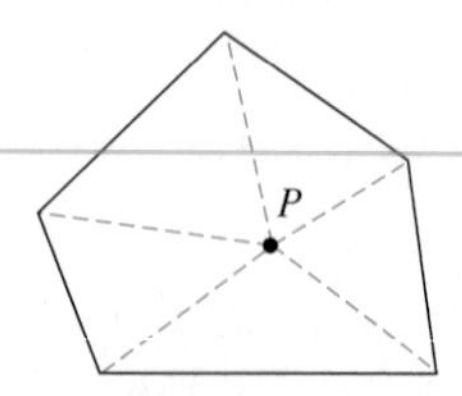

b. Using the method suggested by the diagram in (a), explain how to find the sum of the measures of the angles of any convex n-gon. Is your answer in (b) the same as the one already obtained in this section, $n \cdot 180 - 360$?

28. In the following figure, the legs of the ladder are congruent. If the ladder makes an angle of 120° with the ground, what is x? Explain your reasoning.

29. Explain how, through paper folding, you would show each of the following:
a. In an isosceles triangle, the angles opposite congruent sides are congruent.
b. If two angles of a triangle are congruent, the triangle is isosceles.
c. In an equilateral triangle, all the interior angles are congruent.
d. An isosceles trapezoid has two pairs of congruent angles.

30. a. Explain how to find the sum of the measures of the interior angles of a quadrilateral like the following:

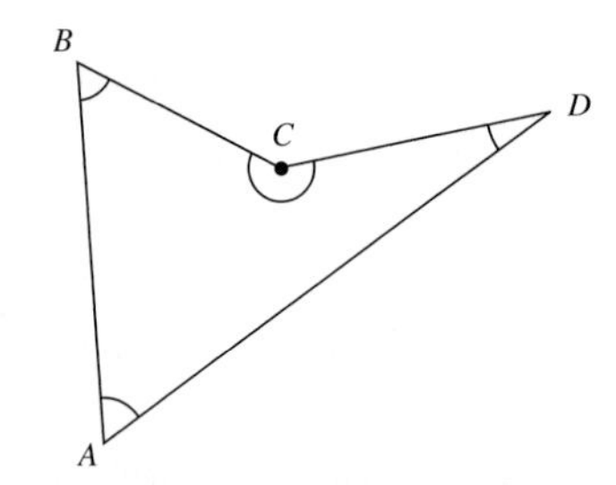

b. Conjecture whether the formula for the sum of the measures of the angles of a convex polygon is true for nonconvex polygons.
c. Justify your conjecture in (b) for pentagons and hexagons and explain why you think your conjecture is true in general.

31. Regular hexagons have been used to tile floors. Can a floor be tiled using only regular pentagons? Why or why not?

32. Lines a and b are cut by a transversal c, creating angles 1 and 2. If $m(\angle 1) + m(\angle 2) = 180°$, can you conclude that a and b are parallel? Justify your answer.

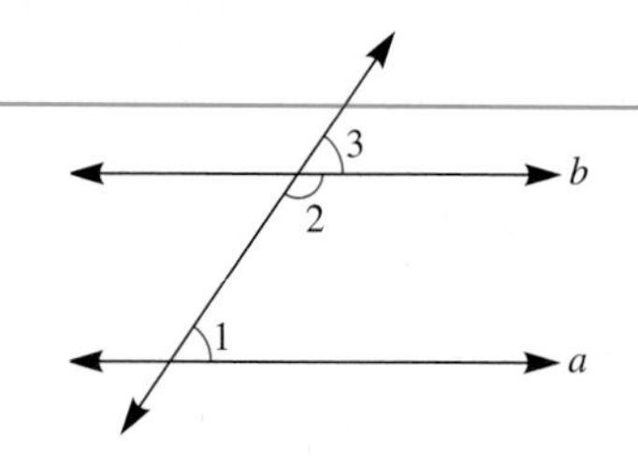

33. Prove that lines a and b are parallel.

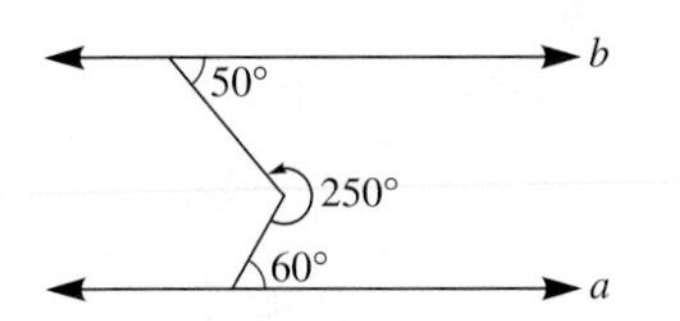

34. A beam of light from A hits the surface of a mirror at point B, is reflected, and then hits a perpendicular mirror and is reflected again. If $\angle 1 \cong \angle 2$ and $\angle 3 \cong \angle 4$, prove that the reflected beam is parallel to the incoming beam, that is, that the rays $\overrightarrow{AB}$ and $\overrightarrow{CD}$ are parallel.

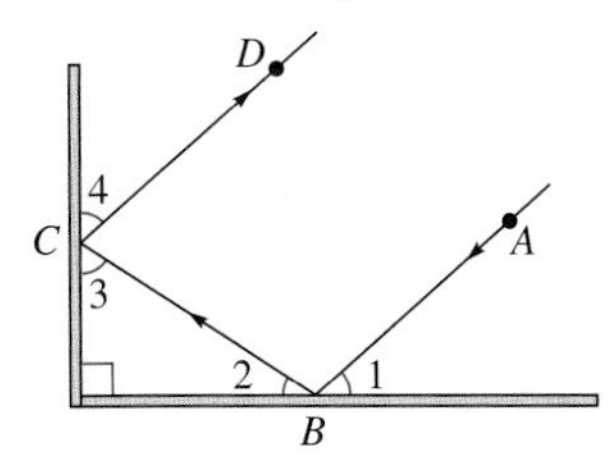

35. $ABCD$ is a quadrilateral in which opposite angles have the same measure as indicated in the following figure. What kind of quadrilateral is $ABCD$? Justify your answer.

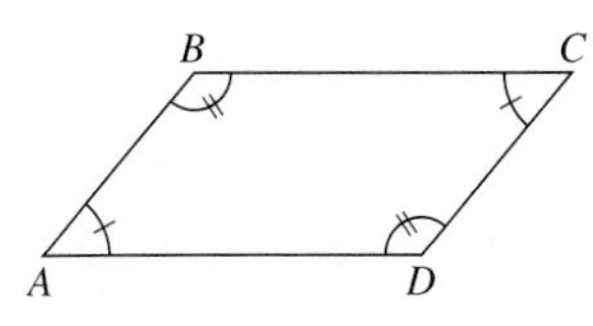

36. $ABCD$ is a parallelogram. The angle bisectors of each of the interior angles of the parallelogram intersect to form the shaded figure shown. Answer the questions that follow.

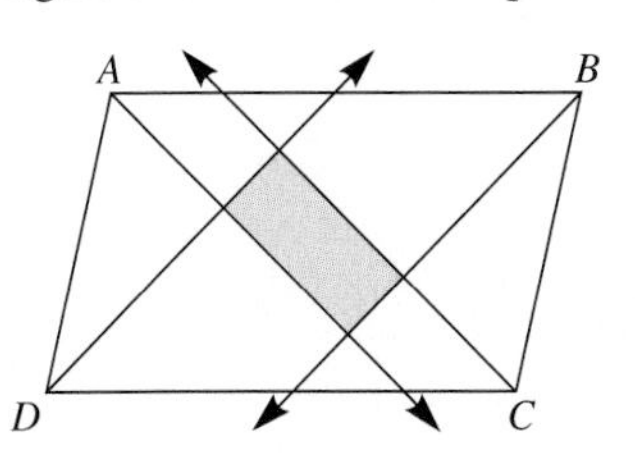

a. Draw several different parallelograms and conjecture the kind of figure the shaded quadrilateral is.

★**b.** Justify your conjecture.

37. The sides of $\triangle DEF$ are parallel to the sides of $\triangle ABC$. If the measures of two angles of $\triangle ABC$ are 60° and 70° as shown, find the measures of angles of $\triangle DEF$. Justify your answer.

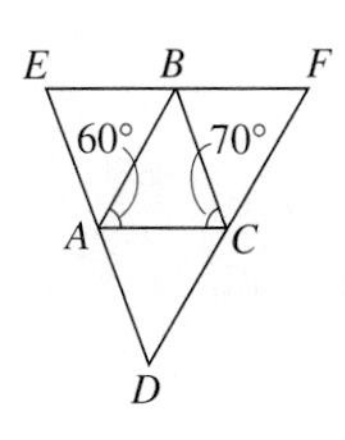

Open-Ended

38. Draw three different nonconvex polygons. When you walk around a polygon, at each vertex you need to turn either right (clockwise) or left (counterclockwise). A turn to the left is measured by a positive number of degrees and a turn to the right by a negative number of degrees. Find the sum of the measures of the turn angles of the polygons you drew. Assume you start at a vertex facing in the direction of a side, walk around the polygon, and end up at the same vertex facing in the same direction as when you started.

39. In *Now Try This 9-11*, you may have found that on a sphere a triangle with three right angles is possible. List other geometric properties of figures on a sphere that are different from corresponding properties in the plane.

Cooperative Learning

40. In $\triangle ABC$, $\overrightarrow{AD}$ and $\overrightarrow{BD}$ are *angle bisectors*; that is, they divide the angles at A and B into congruent angles.

a. If the measures of $\angle A$ and $\angle B$ are known, then $m(\angle D)$ can be found. How?

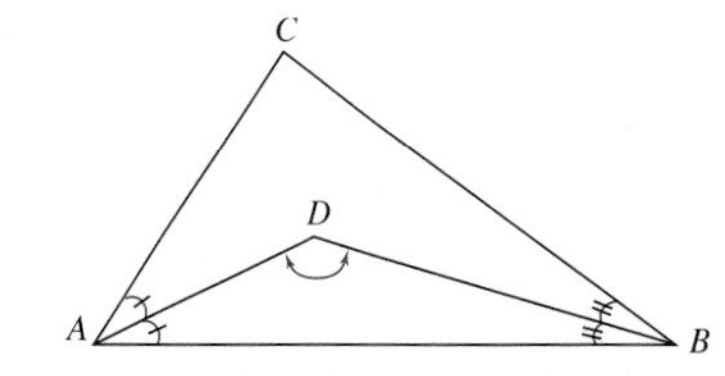

b. Suppose the measures of $\angle A$ and $\angle B$ are not known but that of $\angle C$ is. Can $m(\angle D)$ be found? To answer this question, assign each member of your group a triangle with different angles but with the same measure for $\angle C$. Each person should compute $m(\angle D)$ for his or her triangle. Use the results to make a conjecture related to the previous question.

c. Discuss a strategy for answering the question in (b) and write a solution to be distributed to the entire class.

41. Each person in your group is to draw a large triangle like $\triangle ABC$ in the following figure and cut it out. Obtain the crease $\overline{BB'}$ by folding the triangle at B so that A falls on some point A' on $\overline{AC}$. Next, unfold and fold the top B along $\overline{BB'}$ so that B falls on B'. Then fold vertices A and C along $\overline{AC}$ to match point B', as shown in the following figures:

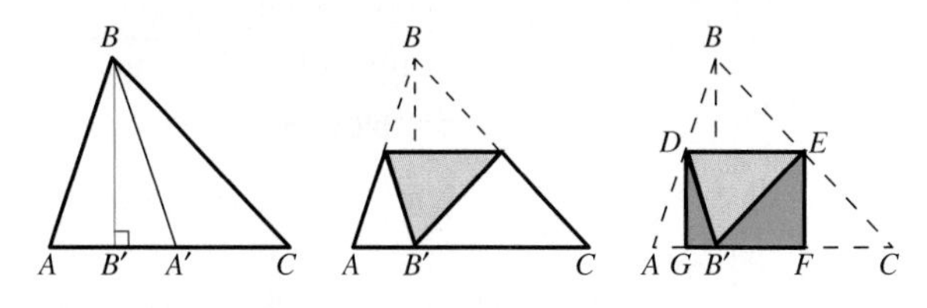

a. Why is $\overline{BB'}$ perpendicular to $\overline{AC}$?

b. What theorem does the folded figure illustrate? Why?

c. The folded figure seems to be a rectangle. Explain why.

d. What is the length of $\overline{GF}$ in terms of the length of the base $\overline{AC}$ of $\triangle ABC$? Why?

Review Problems

42. In each of the following, find the required properties. If this is not possible, explain why.
    a. Two properties that hold true for all rectangles but not for all rhombuses
    b. Two properties that hold true for all squares but not for all isosceles trapezoids
    c. Two properties that hold true for all parallelograms but not for all squares
43. Sort the shapes shown according to the following attributes:
    a. Number of parallel sides
    b. Number of right angles
    c. Number of congruent sides
    d. Polygons with congruent diagonals

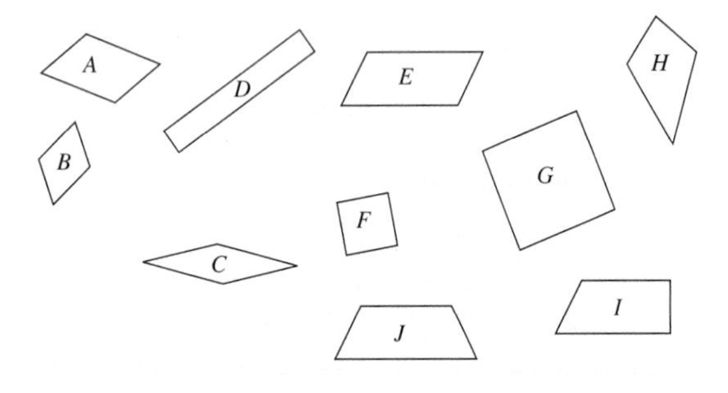

44. Use the figures in problem 43 to identify properties characteristic of different classes of figures. For example, "Congruent opposite sides describe a parallelogram."

Third International Mathematics and Science Study (TIMSS) Questions

In the figure, PQ and RS are intersecting straight lines.

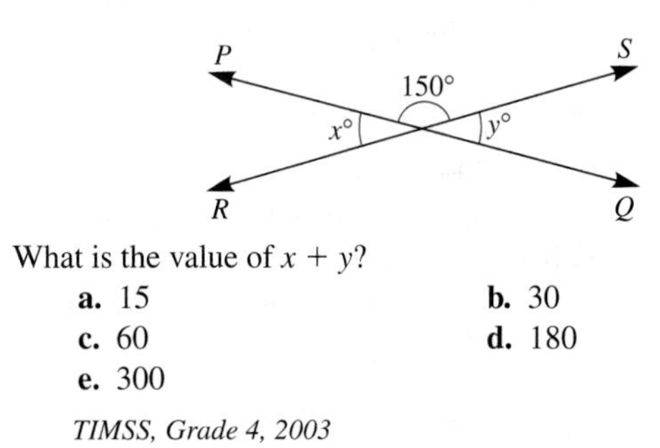

What is the value of $x + y$?

a. 15 **b.** 30
c. 60 **d.** 180
e. 300

TIMSS, Grade 4, 2003

National Assessment of Educational Progress (NAEP) Questions

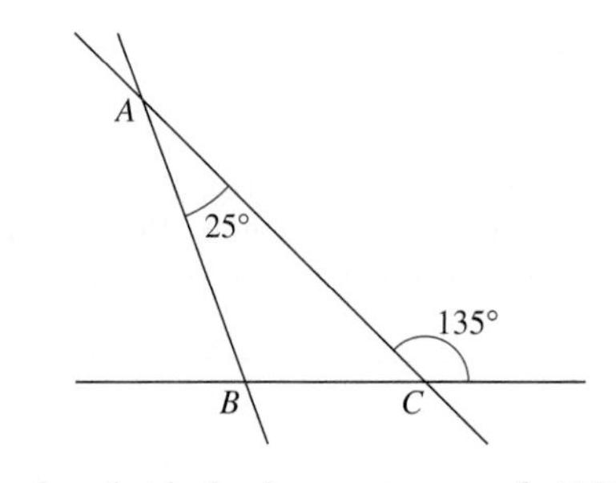

In the triangle, what is the degree measure of $\angle ABC$?

a. 45 **b.** 100
c. 110 **d.** 135
e. 160

NAEP, Grade 8, 2003

9-4 Geometry in Three Dimensions

Simple Closed Surfaces

A visit to the grocery store exposes us to many three-dimensional objects that have simple closed surfaces. Figure 9-34 shows some examples.

FIGURE 9-34

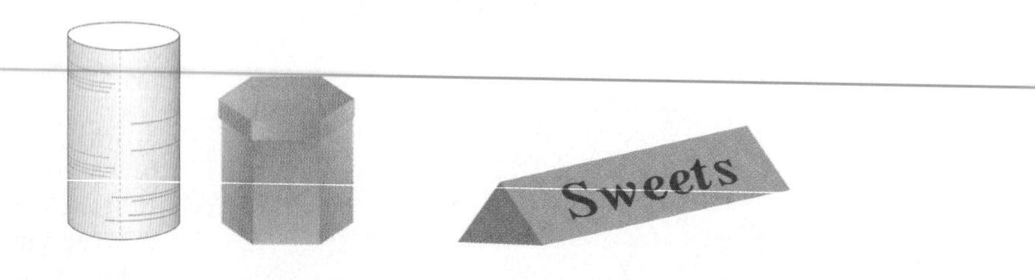

simple closed surface **sphere** **center** **solid** **polyhedron** **face** **vertices** **edges**

A **simple closed surface** has exactly one interior, has no holes, and is hollow. An example is a sphere, as shown in Figure 9-35(c). A **sphere** is defined as the set of all points at a given distance from a given point, the **center**. The set of all points on a simple closed surface with all interior points is a **solid**. Figures 9-35(a), (b), (c), and (d) are examples of simple closed surfaces; (e) and (f) are not. A **polyhedron** (*polyhedra* is the plural) is a simple closed surface made up of polygonal regions, or **faces**. The vertices of the polygonal regions are the **vertices** of the polyhedron, and the sides of each polygonal region are the **edges** of the polyhedron. Figures 9-35(a) and (b) are examples of polyhedra but (c), (d), (e), and (f) are not.

FIGURE 9-35

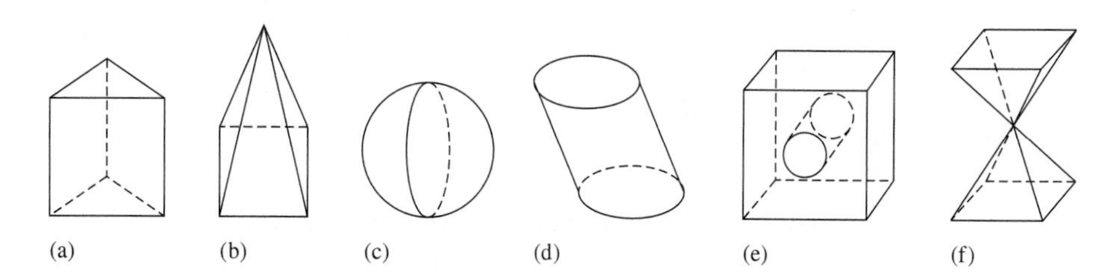

prism **bases**

A **prism** is a polyhedron in which two congruent faces lie in parallel planes and the other faces are bounded by parallelograms. Figure 9-36 shows four different prisms. The shaded parallel faces of a prism are the **bases** of the prism. A prism usually is named after its bases, as the figure suggests.

FIGURE 9-36

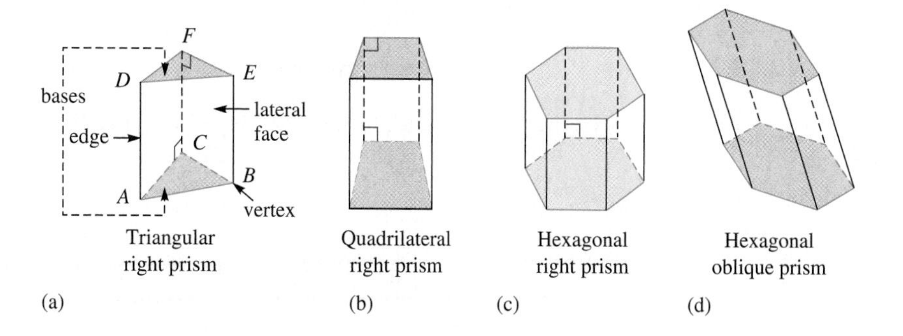

lateral face **right prism** **oblique prism**

The faces other than the bases are the **lateral faces** of a prism. If the lateral faces of a prism are all bounded by rectangles, the prism is a **right prism**, as in Figure 9-36(a)–(c). Figure 9-36(d) is an **oblique prism** because some of its lateral faces are *not* bounded by rectangles.

Students often have trouble drawing three-dimensional figures. Figure 9-37 gives an example of how to draw a right pentagonal prism.

FIGURE 9-37

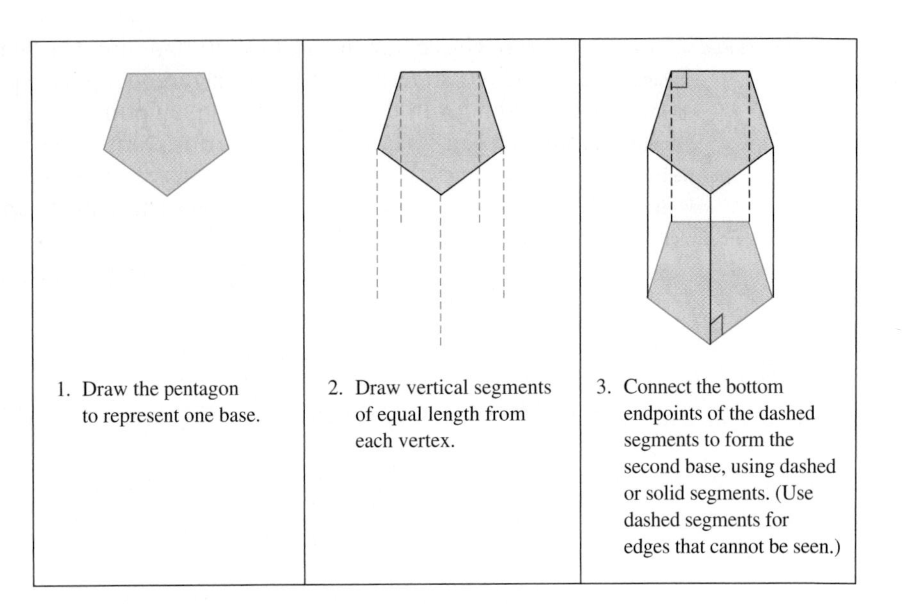

pyramid A **pyramid** is a polyhedron determined by a polygon and a point not in the plane of the polygon. The pyramid consists of the triangular regions determined by the point and each pair of consecutive vertices of the polygon and the polygonal region determined by the polygon. The polygonal region is the **base** of the pyramid, and the point is the **apex**. As with a prism, the faces other than the base are **lateral faces**. Pyramids are classified according to their bases, which are shaded in Figure 9-38. A pyramid is a **right pyramid** if all its lateral faces are congruent isosceles triangles.

base

apex

right pyramid

FIGURE 9-38

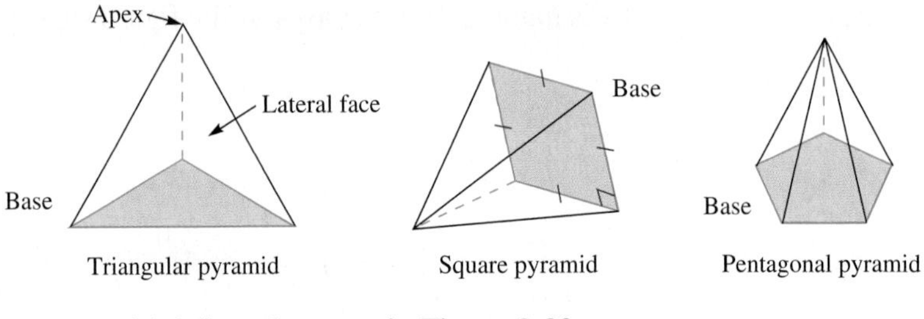

To draw a pyramid, follow the steps in Figure 9-39.

FIGURE 9-39

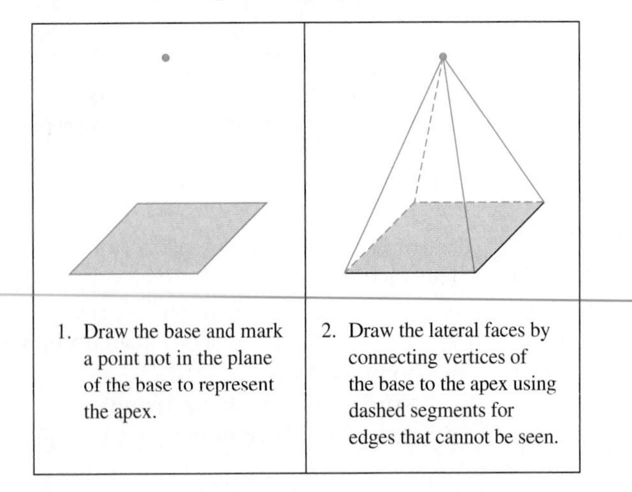

1. Draw the base and mark a point not in the plane of the base to represent the apex.
2. Draw the lateral faces by connecting vertices of the base to the apex using dashed segments for edges that cannot be seen.

Regular Polyhedra

convex polyhedron

A polyhedron is a **convex polyhedron** if and only if the segment connecting any two points in the interior of the polyhedron is itself in the interior. Figure 9-40 shows a concave polyhedron (that is, one that is caved in).

regular polyhedron

A **regular polyhedron** is a convex polyhedron whose faces are congruent regular polygonal regions such that the number of edges that meet at each vertex is the same for all the vertices of the polyhedron.

▶ HISTORICAL NOTE

The regular solid polyhedra are known as the **Platonic solids**, after the Greek philosopher Plato (ca. 350 BCE). Plato attached a mystical significance to the five regular polyhedra, associating them with what he believed were the four elements (earth, air, fire, and water) and the universe. Plato suggested that the smallest particles of earth have the form of a cube, those of air an octahedron, those of fire a tetrahedron, those of water an icosahedron, and those of the universe a dodecahedron. ◀

FIGURE 9-40

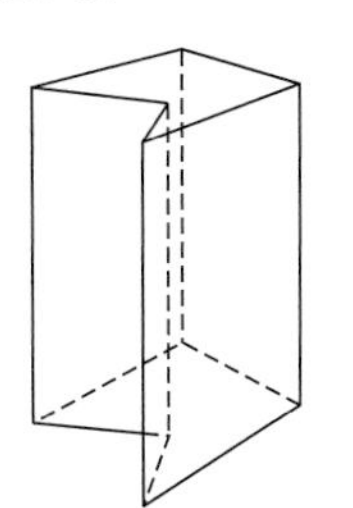

Concave polyhedron

Regular polyhedra, as in Figure 9-41, have fascinated mathematicians for centuries. At least three of them were identified by the Pythagoreans (ca. 500 BCE). Two others were known to the followers of Plato (ca. 350 BCE). Three of the five polyhedra occur in nature in the form of crystals of sodium sulphantimoniate, sodium chloride (common salt), and chrome alum. The other two do not occur in crystalline form but have been observed as skeletons of microscopic sea animals called *radiolaria*.

FIGURE 9-41

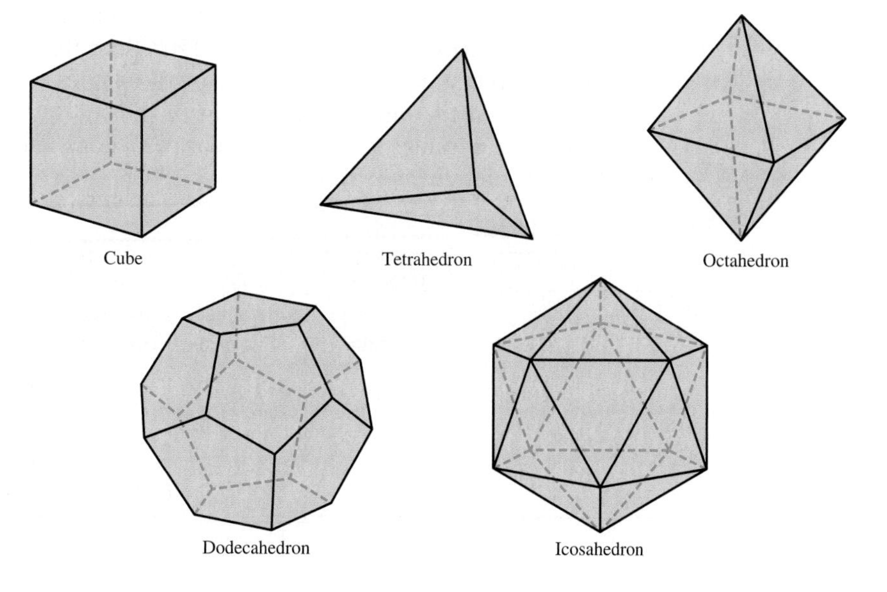

PROBLEM SOLVING **Regular Polyhedra?**

How many regular polyhedra are there?

Understanding the Problem Each face of a regular polyhedron is congruent to each of the other faces of that polyhedron, and each face is bounded by a regular polygon. We are to find the number of different regular polyhedra.

Devising a Plan The sum of the measures of all the angles at a vertex of a regular polyhedron must be less than 360°. (Do you see intuitively why this is true?) We next examine the measures of the interior angles of regular polygons to determine which of the polygons could be faces of a regular polyhedron. Then we try to determine how many types of polyhedra there are.

Carrying Out the Plan We determine the size of an angle of some regular polygons, as shown in Table 9-9. Could a regular heptagon be a face of a regular polyhedron? At least three figures must fit together at a vertex to make a polyhedron. (Why?) If three angles of a regular heptagon were together at one vertex, then the sum of the measures of these angles would be $\frac{3 \cdot 900°}{7}$, or $\frac{2700°}{7}$, which is greater than 360°. Similarly, more than three angles cannot be used at a vertex. Thus, a heptagon cannot be used to make a regular polyhedron.

Table 9-9

Polygon	Measure of an Interior Angle
Triangle	60°
Square	90°
Pentagon	108°
Hexagon	120°
Heptagon	$\left(\frac{900}{7}\right)^{\circ}$

The measure of an interior angle of a regular polygon increases as the number of sides of the polygon increases. (Why?) Thus any polygon with more than six sides has an interior angle greater than 120°. So if three angles were to fit together at a vertex, the sum of the measures of the angles would be greater than 360°. This means that the only polygons that might be used to make regular polyhedra are equilateral triangles, squares, regular pentagons, and regular hexagons. Consider the possibilities given in Table 9-10.

Table 9-10

Polygon	Measure of an Interior Angle	Number of Polygons at a Vertex	Sum of the Angles at the Vertex	Polyhedron Formed	Model
Triangle	60°	3	180°	**Tetrahedron**	

(continues)

Table 9-10 *continued*

Polygon	Measure of an Interior Angle	Number of Polygons at a Vertex	Sum of the Angles at the Vertex	Polyhedron Formed	Model
Triangle	60°	4	240°	**Octahedron**	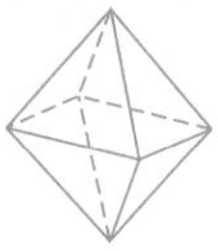
Triangle	60°	5	300°	**Icosahedron**	
Square	90°	3	270°	**Cube**	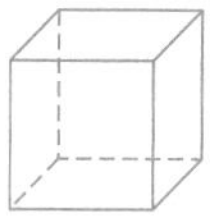
Pentagon	108°	3	324°	**Dodecahedron**	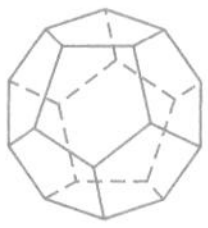

Notice that we were not able to use six equilateral triangles to make a polyhedron because $6(60°) = 360°$ and the triangles would lie in a plane. Similarly, we could not use four squares or any hexagons. We also could not use more than three pentagons because if we did, the sum of the measures of the angles would be more than 360°.

semiregular polyhedra

Looking Back Interested readers may want to investigate **semiregular polyhedra**. These are also formed by using regular polygons as faces, but the regular polygons used need not have the same number of sides. For example, a semiregular polyhedron might have squares and regular octagons as its faces.

The patterns in Figure 9-42, called *nets*, can be used to construct the five regular polyhedra. It is left as an exercise to determine other patterns for constructing the regular polyhedra.

FIGURE 9-42

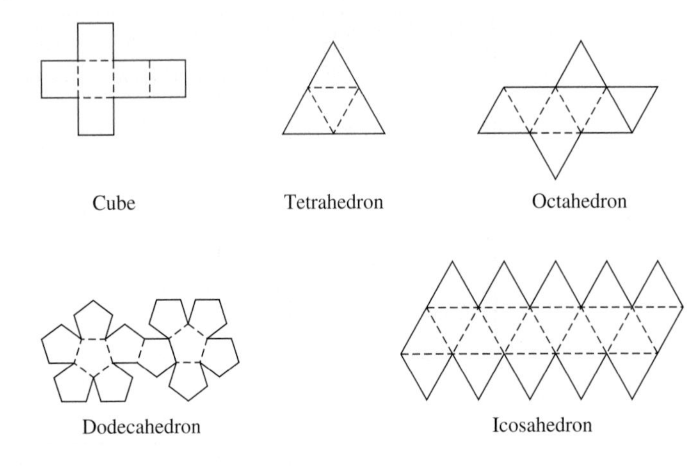

Compare these nets, with the nets introduced on the student page.

NOW TRY THIS 9-13

A simple relationship among the number of faces, the number of edges, and the number of vertices of any polyhedron was discovered by the French mathematician and philosopher René Descartes (1596–1650) and rediscovered by the Swiss mathematician Leonhard Euler (1707–1783). Table 9-11 suggests a relationship among the numbers of vertices (V), edges (E), and faces (F). This relationship is known as **Euler's formula**.

a. State a relationship suggested by the table.

b. Try your relationship on other polyhedra.

Table 9-11

Name	V	F	E
Tetrahedron	4	4	6
Cube	8	6	12
Octahedron	6	8	12
Dodecahedron	20	12	30
Icosahedron	12	20	30

▶ **HISTORICAL NOTE**

Leonhard Euler went blind in 1766 and for the remaining 17 years of his life continued to do mathematics by dictating to a secretary and by writing formulas in chalk on a slate for his secretary to copy down. He published 530 papers in his lifetime and left enough work to supply the *Proceedings of the St. Petersburg Academy* for the next 47 years. ◀

SAMPLE SCHOOL BOOK PAGE:

VIEWS OF SOLID FIGURES

Views of Solid Figures

LEARN

Activity

In what ways can you picture a solid?

A box is shown at the right. From this view, you see the top and two faces of it. You cannot see the bottom or the other faces of it.

A **net** is a plane figure which when folded gives a solid figure. Think about unfolding a box to make a net for a rectangular solid.

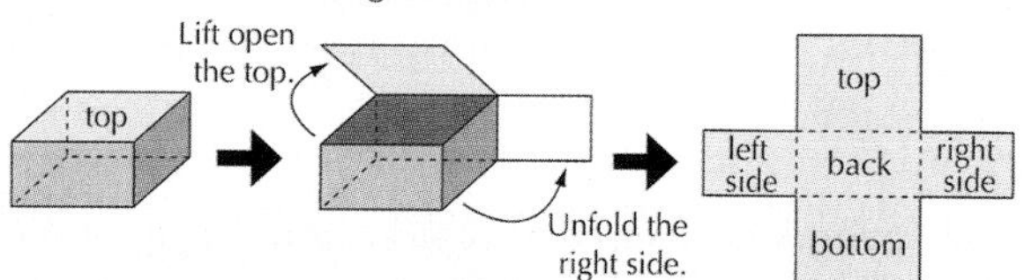

The diagram at the right is a net for a box.

Step 1 On grid paper, make a copy of this net. Be sure to copy the dotted lines, as well as the solid lines.

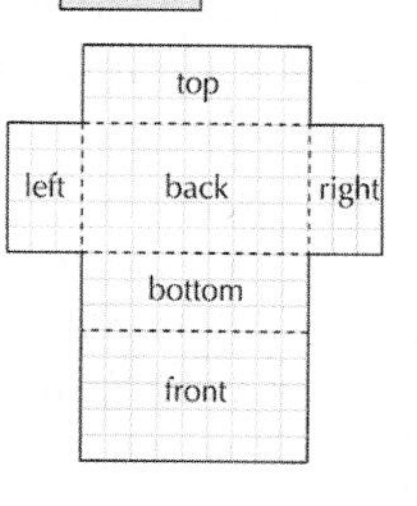

Step 2 Cut out your copy along the solid lines.

Step 3 Crease along the dotted lines.

Step 4 Fold the figure so that the six rectangles become the six faces of a box. Tape along the edges to hold the faces in position.

a. The box you made is 9 units long, 3 units wide, and 5 units high. Explain how you can find these dimensions from the drawing on the grid paper.

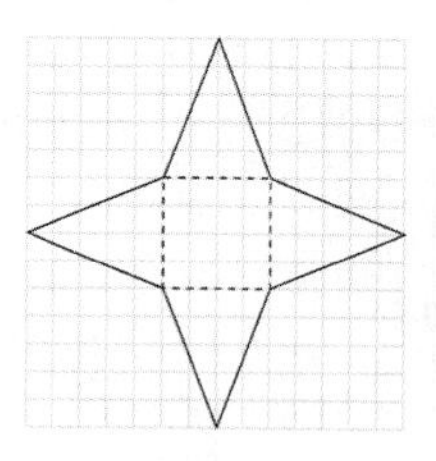

b. On another piece of grid paper, draw a net for a square pyramid. Be sure to include the dotted lines that are the fold lines. Cut out the net and make the pyramid.

c. A net has 6 squares. What solid will be formed from it?

d. Reasoning Is this figure a net for a cube? Explain.

Source: Scott Foresman–Addison Wesley Mathematics, Grade 5, 2005 (p. 598).

Cylinders and Cones

A cylinder is an example of a simple closed surface that is not a polyhedron. Consider line segment $\overline{AB}$ and a line ℓ as shown in Figure 9-43. When $\overline{AB}$ moves so that it always remains parallel to a given line ℓ and points A and B trace simple closed planar curves other than polygons, the surface generated by $\overline{AB}$, along with the simple closed curves and their interiors, form a **cylinder**. The simple closed curves traced by A and B, along with their interiors, are the **bases** of the cylinder, and the remaining points constitute the *lateral surface of the cylinder*.

cylinder
bases

FIGURE 9-43

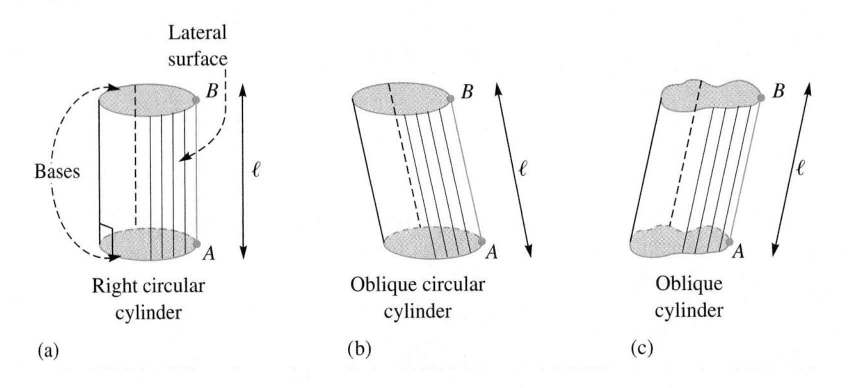

If a base of a cylinder is a circular region, as in Figure 9-43(a) and (b), the cylinder is a **circular cylinder**. If the line segment forming a cylinder is perpendicular to a base, the cylinder is a **right cylinder**. Cylinders that are not right cylinders are **oblique cylinders**. The cylinder in Figure 9-43(a) is a right cylinder; those in Figures 9-43(b) and (c) are oblique cylinders.

circular cylinder
right cylinder
oblique cylinder

Suppose we have a simple closed curve, other than a polygon, in a plane and a point P not in the plane of the curve. The union of line segments connecting point P to each point of a simple closed curve, the simple closed curve, and the interior of the curve is a **cone**. Cones are pictured in Figure 9-44. Point P is the **vertex** of the cone. The points of the cone not in the base constitute the *lateral surface of the cone*. A line segment from vertex P perpendicular to the plane of the base is the **altitude**. A **right circular cone**, such as the one in Figure 9-44(a), is a cone whose altitude intersects the base (a circular region) at the center of the circle. Figure 9-44(b) illustrates an oblique cone, and Figure 9-44(c) illustrates an **oblique circular cone**.

cone ◆ vertex
altitude
right circular cone
oblique circular cone

FIGURE 9-44

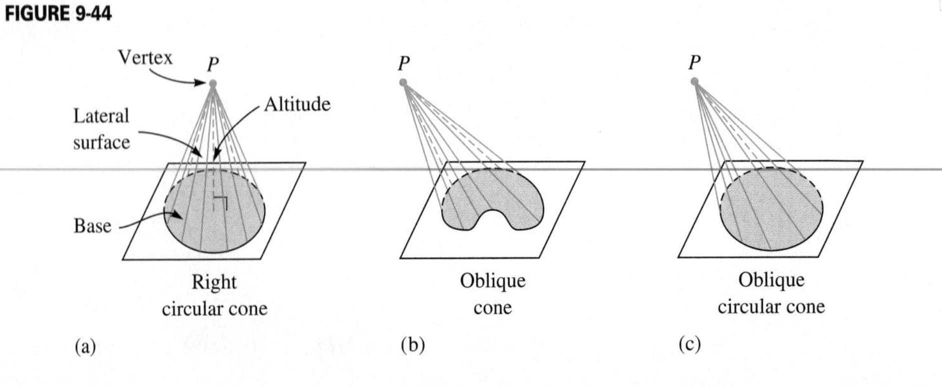

ONGOING ASSESSMENT 9-4

1. Identify each of the following polyhedra. If a polyhedron can be described in more than one way, give as many names as possible.

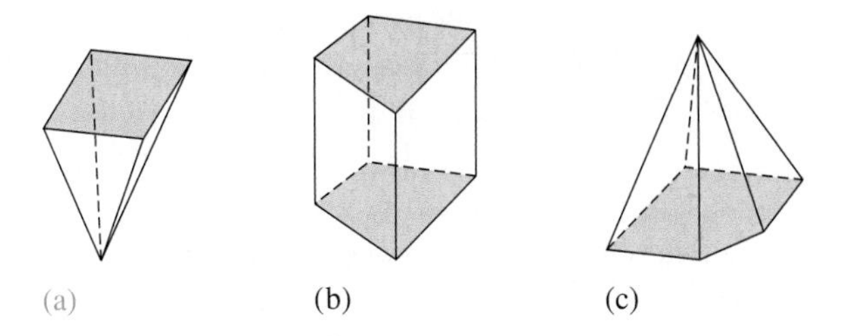

2. Given the tetrahedron shown, name the following:
   a. Vertices
   b. Edges
   c. Faces
   d. Intersection of face *DRW* and edge $\overline{RA}$
   e. Intersection of face *DRW* and face *DAW*

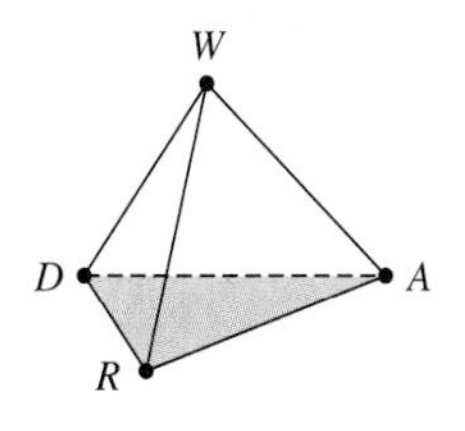

3. Identify five shapes of containers that can be found in a grocery store.
4. Determine for each of the following the minimum number of faces possible:
   a. Prism
   b. Pyramid
   c. Polyhedron
5. Classify each of the following as true or false:
   a. If the lateral faces of a prism are rectangles, it is a right prism.
   b. Every pyramid is a prism.
   c. Some pyramids are polyhedra.
   d. The bases of a prism lie in perpendicular planes.
   e. The bases of all cones are circles.
   f. A cylinder has only one base.
   g. All lateral faces of an oblique prism are rectangular regions.
   h. All regular polyhedra are convex.
6. If possible, sketch each of the following:
   a. An oblique square prism
   b. An oblique square pyramid
   c. A noncircular right cone
   d. A noncircular cone that is not right
7. For each of the following, draw a prism and a pyramid that have the given region as a base:
   a. Triangle
   b. Pentagon
   c. Regular hexagon
8. Two prisms are sketched on dot paper, as in the following figure. Complete the drawings by using dashed segments for the hidden edges.

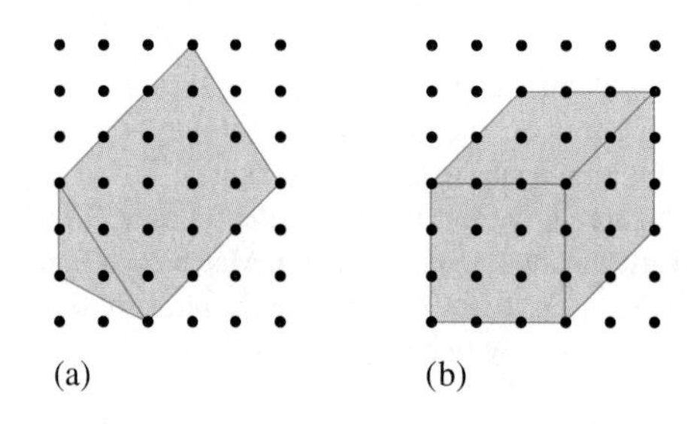

9. The following are pictures of stacks of solid cubes. In each case, determine the number of cubes in the stack and the number of faces that are glued together.

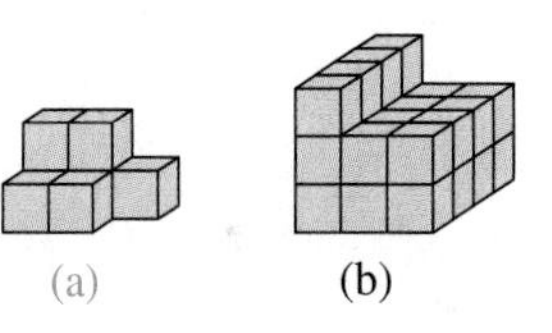

10. Name each polyhedron that can be constructed using the following flattened polyhedra:

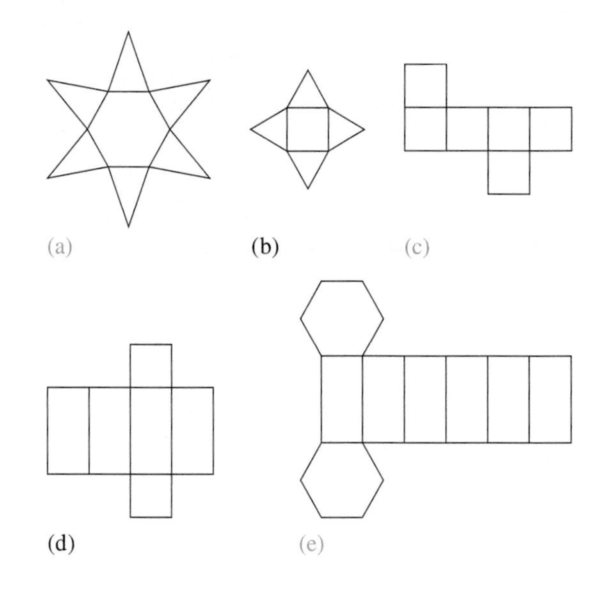

11. The following nets along with shaded tabs can be used to construct polyhedra. Copy the nets (or construct similar ones), fold them, and glue them together using the tabs. What kind of polyhedra do you obtain?

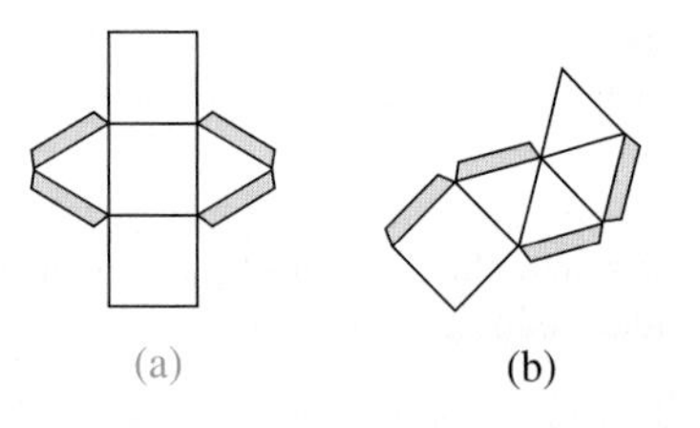

12. Draw nets like the ones in problem 11 and as few tabs as possible for each net so that when folded the nets can be glued with the tabs to create polyhedra.

13. The figure on the left in each of the following represents a card attached to a wire, as shown. Match each figure on the left with what it would look like if you were to revolve it by spinning the wire between your fingers.

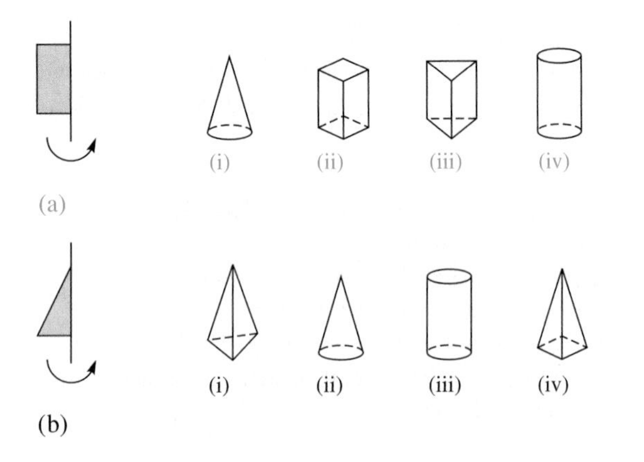

14. Which of the following three-dimensional figures could be used to make the shadow shown in (a)? In (b)?

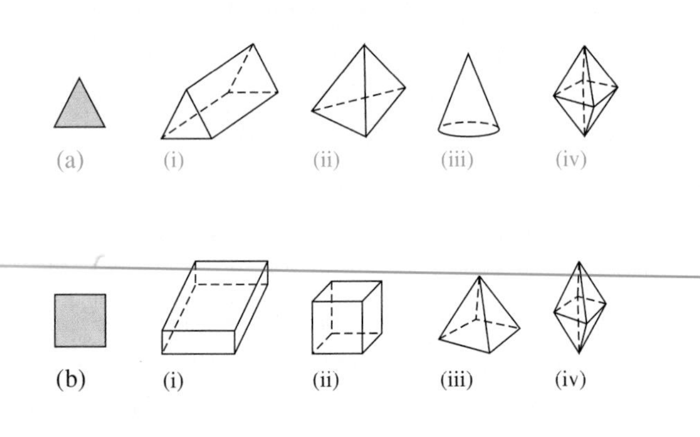

15. A diagonal of a prism is any segment determined by two vertices that do not lie in the same face, as shown in the following figure.

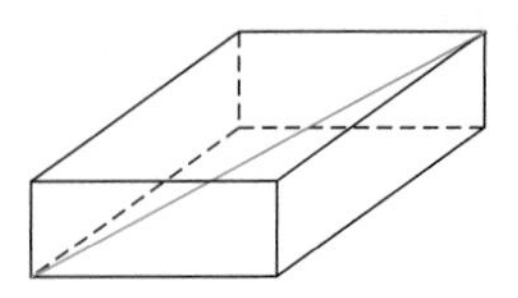

Complete the following table showing the total number of diagonals for various prisms:

Prism	Vertices per Base	Diagonals per Vertex	Total Number of Diagonals
Quadrilateral	4	1	4
Pentagonal	5		
Hexagonal			
Heptagonal			
Octagonal			
⋮			
n-gonal			

16. Consider a jar with a lid, as illustrated in the following figure. The jar is half filled with water. In drawings (a) and (b), sketch the water.

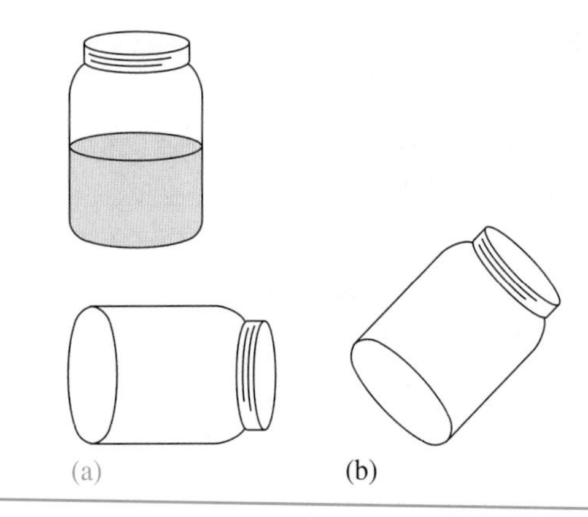

17. On the left of each of the following figures is a net for a three-dimensional object. On the right are several objects. Which object will the net fold to make?

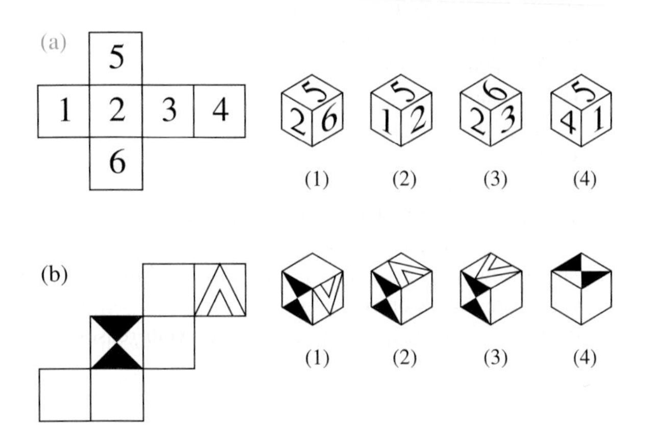

18. Sketch the intersection of each of the following with the plane shown:

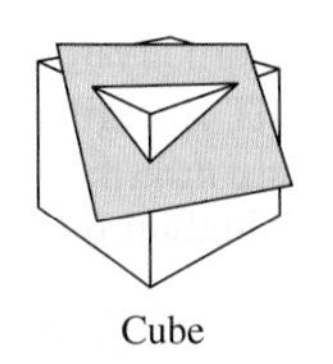
Cube

(a)

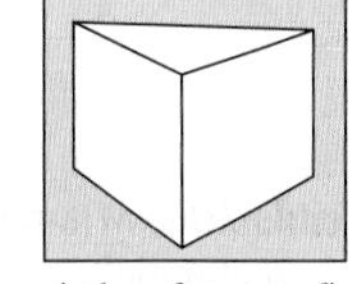

Remainder of unseen figure completes the cube

(b)

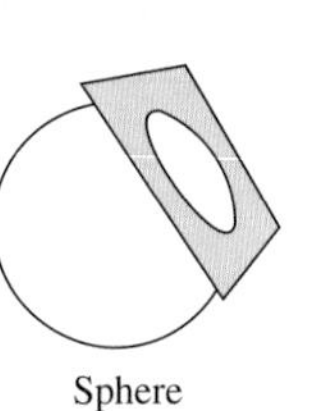
Sphere

(c)

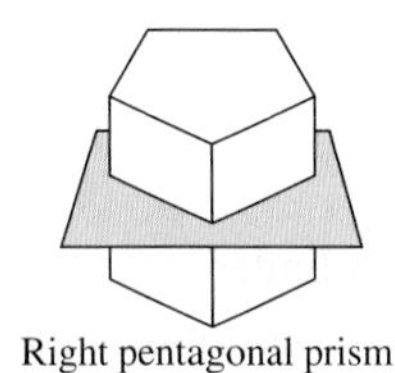
Right pentagonal prism

(d)

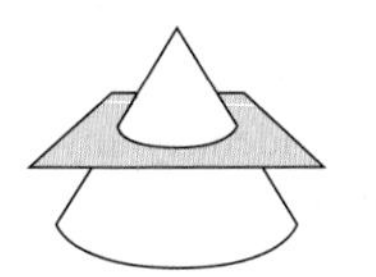
Right circular cone (plane parallel to base)

(e)

Right circular cylinder (plane not parallel to base)

(f)

19. For each of the following three-dimensional figures, draw all possible cross sections when the three-dimensional figure is sliced by a plane.
   a. Right pentagonal prism
   b. Cylinder
20. When a plane intersects a cube, different sized rectangles can be obtained. Answer the following questions and explain your reasoning.
   a. Show the largest possible rectangle that can be obtained.
   b. Is it possible to find the smallest rectangle?
21. Given a tetrahedron, show how to obtain each of the following shapes. Describe a procedure for getting each shape.
   a. An equilateral triangle
   b. A scalene triangle
   c. A rectangle
22. The following is a net that can be used to create a tetrahedron. M, N, P, Q, S are midpoints of sides of the triangles (a midpoint divides a segment into two congruent segments).
   a. Predict what kind of figure will result from the midpoints M, N, P, Q, S being the vertices when the net is folded into a tetrahedron.

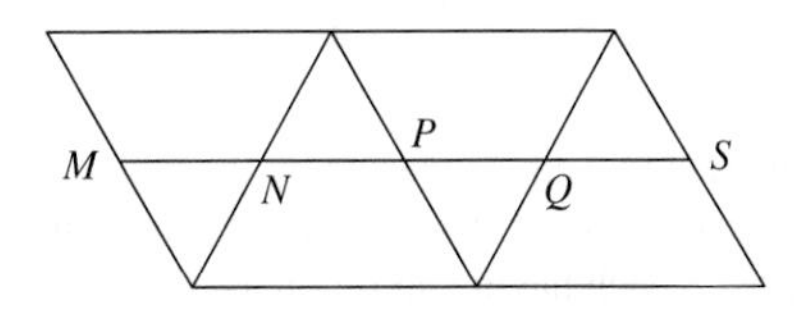

   b. Draw a larger net similar to the one shown (make angles of 60°). Mark the corresponding midpoints (use a marked ruler) and fold it into a tetrahedron. Do the midpoints create a figure you predicted in part (a)?
23. The right hexagonal prism shown has regular hexagons as bases. Answer the following questions:

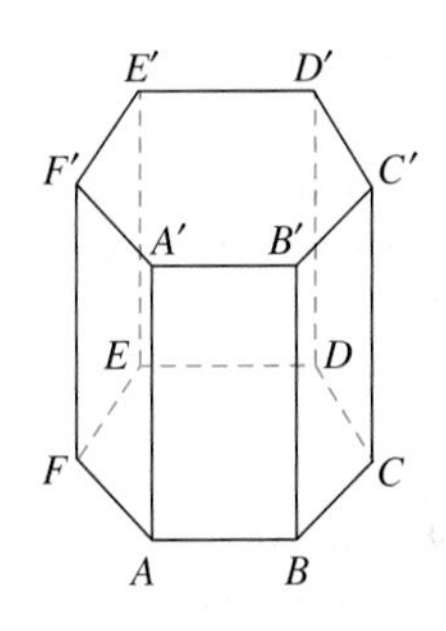

   a. Name all the pairs of parallel lateral faces. (Faces are parallel if the planes containing the faces are parallel.)
   b. What is the measure of the dihedral angle (see problem 3 in Ongoing Assessment 9-1) between two adjacent lateral faces? Why?

Hoffer, A. "Making a Better Beer Glass." *Mathematics Teacher* 75 (May 1982): 378–379.

Kennedy, J., and E. McDowell. "Geoboard Quadrilaterals." *Mathematics Teacher* 91 (April 1998): 288–290.

Kerslake, D. "Graphs." In *Children's Understanding of Mathematics*. London: John Murray, 1981, 11–16.

Lipka, J., S. Wildfeuer, N. Wahlberg, M. George, and D. Ezran. "Elastic Geometry and Storyknifing: A Yup'ik Eskimo Example." *Teaching Children Mathematics* 7 (February 2001): 337–343.

Long, V. "From Polygons to Poetry." *Mathematics Teaching in the Middle School* 6 (April 2001): 436–438.

"Menu of Problems, Desserts." *Mathematics Teaching in the Middle School* 6 (January 2001): 304.

Moyer, P. S., and J. J. Bolyard. "Classify and Capture: Using Venn Diagrams and Tangrams to Develop Abilities in Mathematical Resoning and Proof." *Mathematics Teaching in Middle School* 3 (February 2003): 325–330.

Oberdorf, C. D., and J. Taylor-Cox. "Shape Up." *Teaching Children Mathematics* 5 (February 1999): 340–345.

Olson, A. T. *Mathematics through Paper Folding*. Washington, D.C.: NCTM (1975).

Olson, J. "What Shapes Can You Make?" *Teaching Children Mathematics* 5 (February 1999): 330–331.

Olson, M. "Coloring Tetras with Two Colors." *Teaching Children Mathematics* 7 (January 2001): 294–295.

Peterson, B. "From Tessellations to Polyhedra: Big Polyhedra." *Mathematics Teaching in the Middle School* 5 (February 2000): 348–356.

Posamentier, A. "Geometry: A Remedy for the Malaise of Middle School Mathematics." *Mathematics Teacher* 82 (December 1989): 678–680.

Renne, C. G. "Is a Rectangle a Square? Developing Mathematical Vocabulary and Conceptual Understanding." *Teaching Children Mathematics* 10 (January 2004): 258–263.

Rich, W., and J. Joyner. "Using Interactive Web Sites to Enhance Mathematics Learning." *Teaching Children Mathematics* 8 (February 2002): 380–383.

Row, T. S. *Geometric Exercises in Paper Folding*. Dover Publications, Inc. (1966).

Schifter, D. "Learning Geometry: Some Insights Drawn from Teacher Writing." *Teaching Children Mathematics* 5 (February 1999): 360–366.

Sundberg, S. "A Plethora of Polyhedra." *Mathematics Teaching in the Middle School* 3 (March–April 1998): 388–391.

van Hiele, P. M. "Developing Geometric Thinking through Activities That Begin with Play." *Teaching Children Mathematics* 5 (February 1999): 310–316.

Wheatley, G. H., and A. M. Reynolds. "'Image Maker': Developing Spatial Sense." *Teaching Children Mathematics* 5 (February 1999): 374–378.

Williams, C. G. "Sorting Activities for Polygons." *Mathematics Teaching in the Middle School* 3 (March–April 1998): 444–445.

Woodward, E., and R. Brown, "Polydrons and Three-Dimensional Geometry." *Arithmetic Teacher* 41 (April 1994): 451–458.

Constructions, Congruence, and Similarity

PRELIMINARY PROBLEM

Rheba wanted a piece of red circular glass inset into her right triangular window. If the window had sides with measures of 3 ft, 4 ft, and 5 ft, and the glass was to be an inscribed circle as shown, what is the radius of the circular glass?

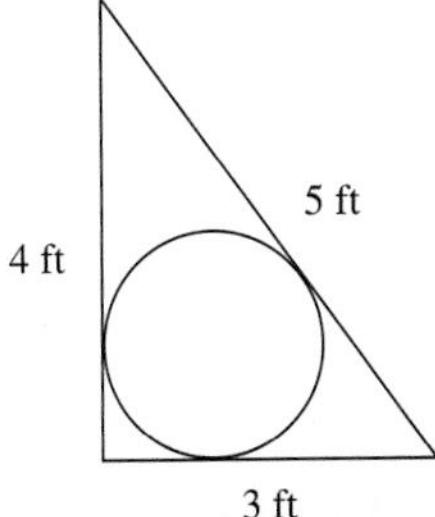

The *Principles and Standards* state that students should be able to do the following:

Pre-K–2

- *relate ideas in geometry to ideas in number and measurement.* (p. 396)

Grades 3–5

- *explore congruence and similarity;*
- *make and use coordinate systems to specify locations and to describe paths;*
- *build and draw geometric objects;*
- *use geometric models to solve problems in other areas of mathematics, such as number and measurement;*
- *recognize geometric ideas and relationships and apply them to other disciplines and to problems that arise in the classroom or in everyday life.* (p. 396)

Grades 6–8

- *use coordinate geometry to represent and examine the properties of geometric shapes;*
- *use coordinate geometry to examine special geometric shapes, such as regular polygons or those with pairs of parallel or perpendicular sides;*
- *draw geometric objects with specified properties, such as side lengths or angle measures.* (p. 397)

In this chapter, we introduce, through constructions and visualization, the concepts of congruence and similarity. We study equations of lines, investigate systems of equations both geometrically and algebraically, and consider basic trigonometry.

10-1 Congruence Through Constructions

similar
congruent

In mathematics, **similar** ($\sim$) objects have the same shape but not necessarily the same size; **congruent** ($\cong$) objects have the same size as well as the same shape. Whenever two figures are congruent, they are also similar. However, the converse is not true (why?). Examples of similar and congruent objects are seen in Figure 10-1.

FIGURE 10-1

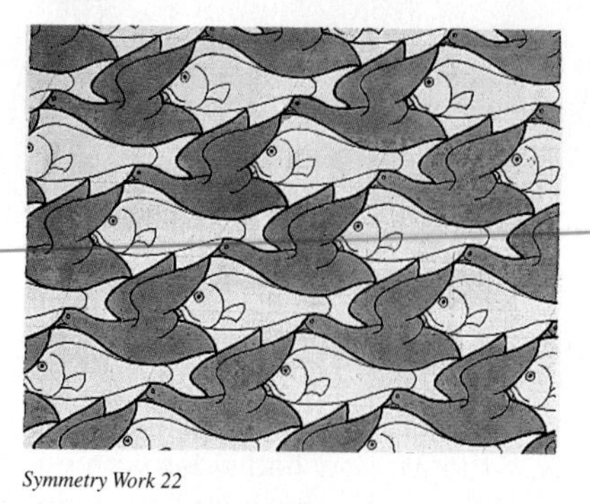

Symmetry Work 22

(a)

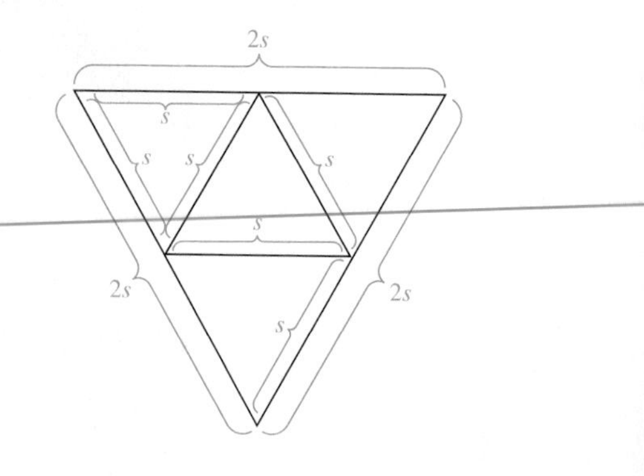

(b)

The Escher print in Figure 10-1(a) shows congruent fish and congruent birds. Figure 10-1(b) contains both congruent and similar equilateral triangles. The smaller triangles are both similar to and congruent to each other. They also are similar to the large triangle that contains them. Figure 10-1(b) is an example of a **rep-tile**, a figure that is used to construct a larger similar figure. In Figure 10-1(b), one of the smaller equilateral triangles is a rep-tile. Figure 10-2 shows two similar circles with the same shape but different sizes. The size of a circle is determined by its radius. In fact, a **circle** is the set of all points in a plane at the same distance (**radius**) from a given point, its **center**. In Figure 10-2, the larger circle (circle O) has center O and radius OA. Note that *radius* is used to describe both the length OA and the segment $\overline{OA}$.

rep-tile

circle

radius ◆ center

FIGURE 10-2

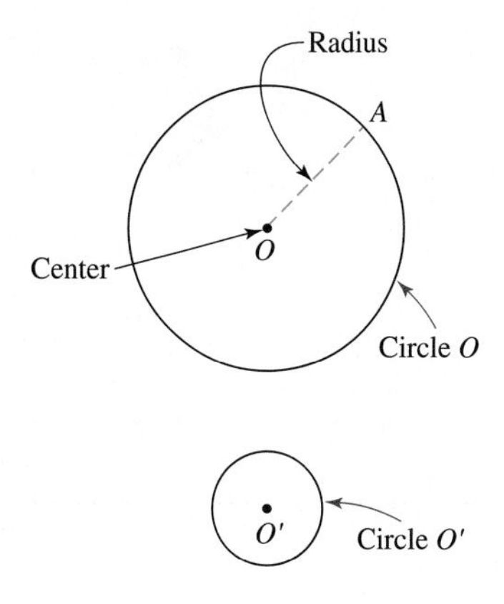

Before studying congruent and similar figures, we first consider some notation and review definitions from Chapter 9. For example, *any two line segments are congruent if they have the same length, and two angles are congruent if they have the same measure*. The length of line segment $\overline{AB}$ is denoted AB. Symbolically, we may write the following about congruent segments and angles.

Definition of Congruent Segments and Angles

$\overline{AB} \cong \overline{CD}$ if and only if $AB = CD$

$\angle ABC \cong \angle DEF$ if and only if $m(\angle ABC) = m(\angle DEF)$

▶ HISTORICAL NOTE

The straight line and circle were considered the basic geometric figures by the Greeks and the straightedge and compass are their physical analogs. It is believed that the Greek philosopher Plato (c. 427–347 BCE) rejected the use of mechanical devices other than the straightedge and compass for geometric constructions because use of other tools emphasized practicality rather than "ideas." ◀

Geometric Constructions

Ancient Greek mathematicians constructed geometric figures with a straightedge (no markings on it) and a collapsible compass. Figure 10-3 shows a modern compass that can be used to mark off and duplicate lengths and to construct circles or arcs with a radius of a given measure. To draw a circle when given the radius PQ of a circle, we follow the steps illustrated in Figure 10-3.

FIGURE 10-3 Constructing a circle given its radius

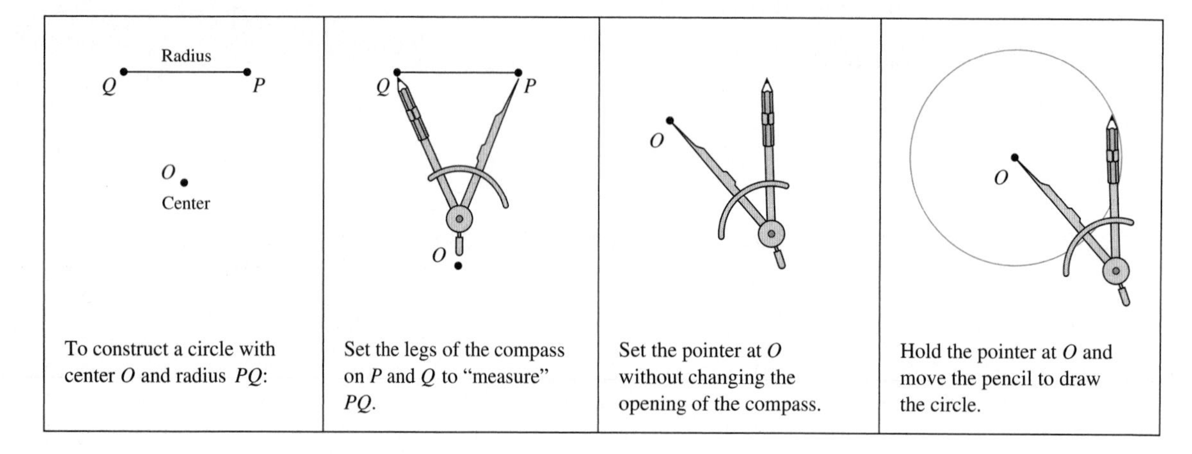

Radius
Q
P
O
Center
Q
P
O
O
O
To construct a circle with center O and radius PQ:
Set the legs of the compass on P and Q to "measure" PQ.
Set the pointer at O without changing the opening of the compass.
Hold the pointer at O and move the pencil to draw the circle.

Any circle is similar to any other circle. However, *any circle is congruent to another circle if and only if the radii of the two circles are congruent* or *the ratio of the radii is 1:1*.

arc
center of arc
minor arc ◆ major arc
semicircle

An **arc** of a circle is any part of the circle that can be drawn without lifting a pencil. (The entire circle could be considered an arc.) The **center of an arc** is the center of the circle containing the arc. If there is no danger of ambiguity in a discussion, the endpoints of the arc are used to name the arc, otherwise three points are used in the naming. In Figure 10-4, $\widehat{ACB}$ is a **minor arc** and $\widehat{ADB}$ is a **major arc**. If the major arc and the minor arc of a circle are the same size, each is a **semicircle**. We use the convention that if only two points are used in the arc name, then the minor arc is intended. Thus, in Figure 10-4, $\widehat{AB}$ is another name for $\widehat{ACB}$.

FIGURE 10-4

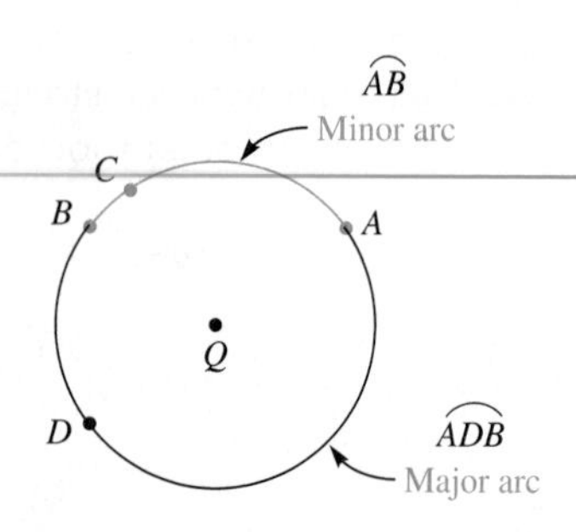

NOW TRY THIS 10-1

In the "Peanuts" cartoon, Sally draws circles in a plane. Describe all the possible ways two circles can intersect.

Constructing Segments

There are many ways to draw a segment congruent to a given segment $\overline{AB}$. A natural approach is to use a ruler, measure $\overline{AB}$, and then draw a congruent segment. A different way is to trace $\overline{AB}$ onto a piece of paper. A third method is to use a straightedge and a compass, as in Figure 10-5.

FIGURE 10-5 *Constructing a line segment congruent to a given segment*

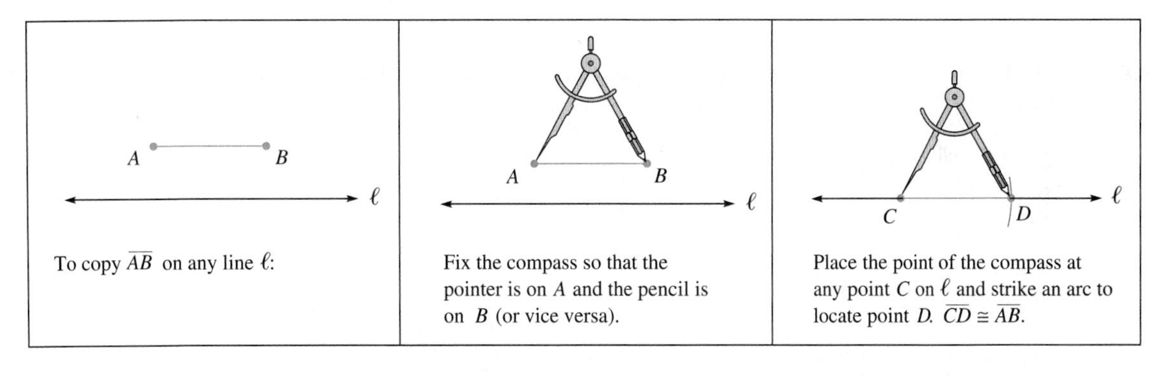

A
B
ℓ
To copy $\overline{AB}$ on any line ℓ:
Fix the compass so that the pointer is on A and the pencil is on B (or vice versa).
C
D
Place the point of the compass at any point C on ℓ and strike an arc to locate point D. $\overline{CD} \cong \overline{AB}$.

Triangle Congruence

Informally, two figures are congruent if it is possible to fit one figure onto the other so that all matching parts coincide. In Figure 10-6, $\triangle ABC$ and $\triangle A'B'C'$ have corresponding congruent parts. Tick marks are used to show congruent segments and

FIGURE 10-6

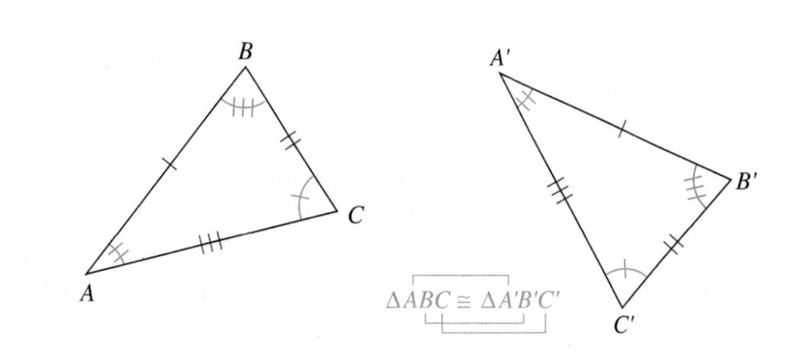

angles in the triangles. If we were to trace $\triangle ABC$ in Figure 10-6 and put the tracing over $\triangle A'B'C'$ so that the tracing of A is over A', the tracing of B is over B', and the tracing of C is over C', $\triangle ABC$ would coincide with $\triangle A'B'C'$. This suggests the following definition of congruent triangles.

Definition of Congruent Triangles

$\triangle ABC$ is congruent to $\triangle A'B'C'$, written $\triangle ABC \cong \triangle A'B'C'$, if and only if $\angle A \cong \angle A'$, $\angle B \cong \angle B'$, $\angle C \cong \angle C'$, $\overline{AB} \cong \overline{A'B'}$, $\overline{BC} \cong \overline{B'C'}$, and $\overline{AC} \cong \overline{A'C'}$.

In the definition of congruent triangles $\triangle ABC$ and $\triangle A'B'C'$, the order of the vertices is such that the listed congruent angles and congruent sides correspond, leading to the statement: "Corresponding parts of congruent triangles are congruent," abbreviated as CPCTC.

NOW TRY THIS 10-2

In Figure 10-6, $\triangle ABC \cong \triangle A'B'C'$. List all possible ways that the congruence can be symbolized.

EXAMPLE 10-1

Write an appropriate symbolic congruence for each of the pairs of congruent triangles in Figure 10-7.

FIGURE 10-7

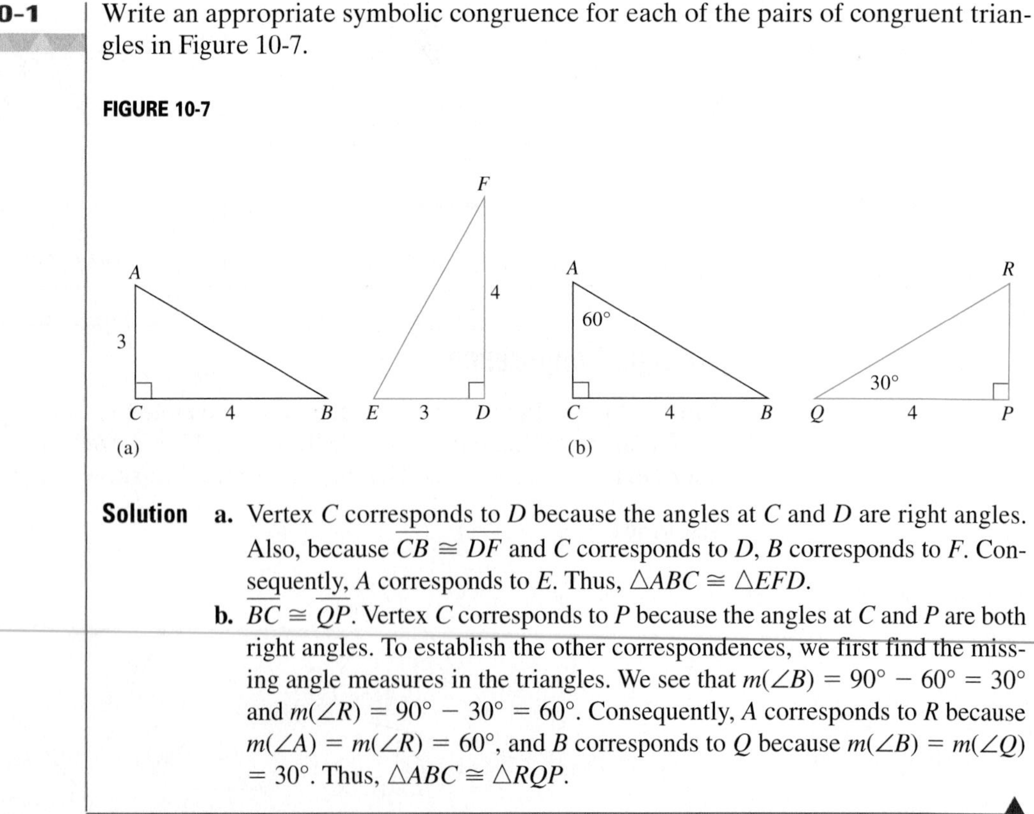

Solution **a.** Vertex C corresponds to D because the angles at C and D are right angles. Also, because $\overline{CB} \cong \overline{DF}$ and C corresponds to D, B corresponds to F. Consequently, A corresponds to E. Thus, $\triangle ABC \cong \triangle EFD$.

b. $\overline{BC} \cong \overline{QP}$. Vertex C corresponds to P because the angles at C and P are both right angles. To establish the other correspondences, we first find the missing angle measures in the triangles. We see that $m(\angle B) = 90° - 60° = 30°$ and $m(\angle R) = 90° - 30° = 60°$. Consequently, A corresponds to R because $m(\angle A) = m(\angle R) = 60°$, and B corresponds to Q because $m(\angle B) = m(\angle Q) = 30°$. Thus, $\triangle ABC \cong \triangle RQP$.

Side, Side, Side Property (SSS)

Calibration experts work to ensure that car parts on an automotive assembly line are interchangeable so that the same part fits on all basic models of the same car. For the cars to be congruent, the parts must be congruent. In design for automotive production, decisions have to be made about the minimal set of items to consider for eventual congruency. In considering congruence of figures in geometry, we consider the same process.

If three sides and three angles of one triangle are congruent to the corresponding three sides and three angles of another triangle, then the triangles are congruent. However, do we need to know that all six parts of one triangle are congruent to the corresponding parts of the second triangle in order to conclude that the triangles are congruent?

Consider the triangle formed by attaching three segments, as in Figure 10-8. Such a triangle is *rigid;* that is, its size and shape cannot be changed. Because of this property, a manufacturer can make duplicates if the lengths of the sides are known. Note that once the sides are known, all the angles are automatically determined.

FIGURE 10-8

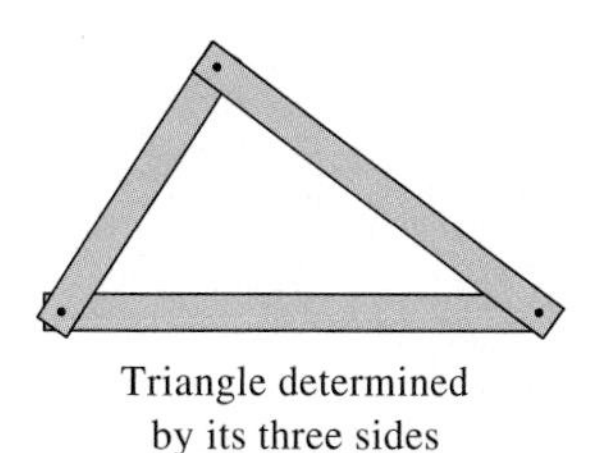

Triangle determined
by its three sides

Many bridges and other structures that have exposed frameworks demonstrate the practical use of the rigidity of triangles, as seen in Figure 10-9.

FIGURE 10-9

Because a triangle is completely determined by its three sides, we have the following property.

> **Property**
>
> **Side, Side, Side (SSS)** If the three sides of one triangle are congruent, respectively, to the three sides of a second triangle, then the triangles are congruent.

EXAMPLE 10-2

For each part in Figure 10-10, use SSS to explain why the given triangles are congruent:

FIGURE 10-10

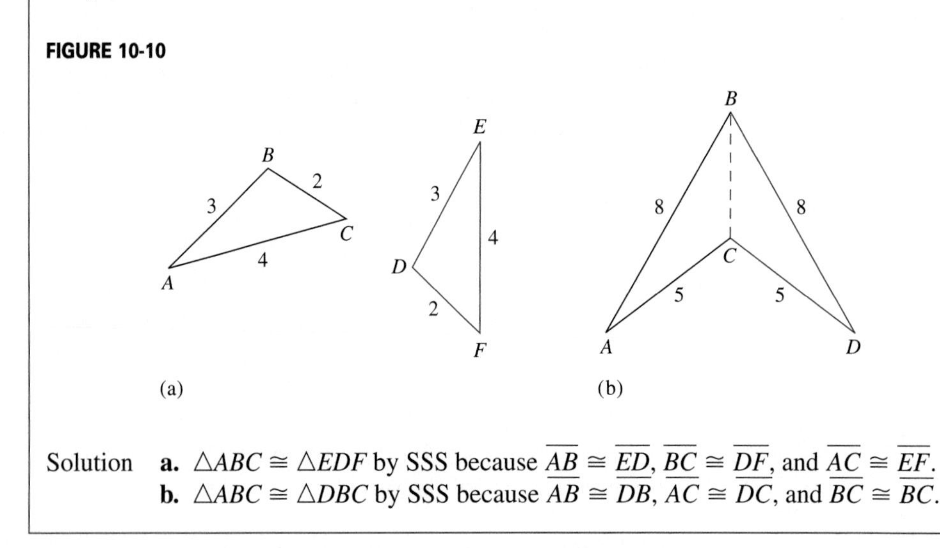

Solution **a.** $\triangle ABC \cong \triangle EDF$ by SSS because $\overline{AB} \cong \overline{ED}$, $\overline{BC} \cong \overline{DF}$, and $\overline{AC} \cong \overline{EF}$.
b. $\triangle ABC \cong \triangle DBC$ by SSS because $\overline{AB} \cong \overline{DB}$, $\overline{AC} \cong \overline{DC}$, and $\overline{BC} \cong \overline{BC}$.

Constructing a Triangle Given Three Sides

Using the SSS property, we can construct a triangle $A'B'C'$ congruent to a given triangle ABC if we know the lengths of the three sides. We illustrate this with a compass and straightedge, as in Figure 10-11.

FIGURE 10-11

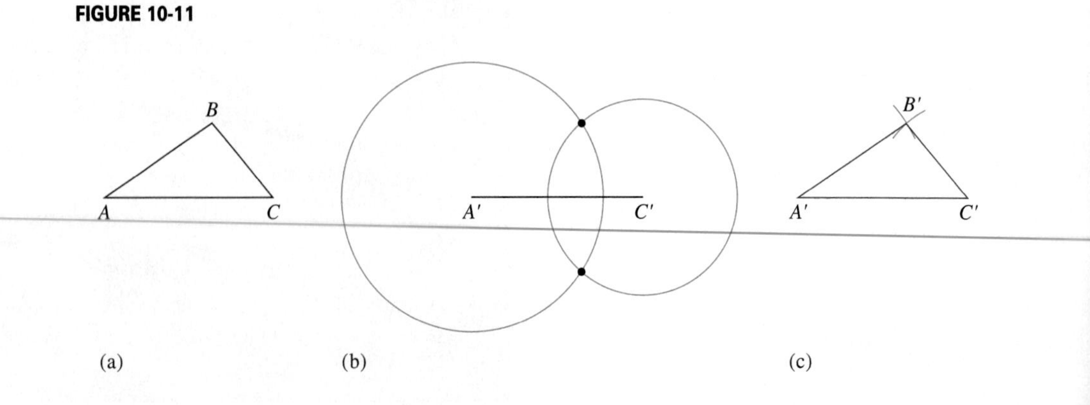

First, we construct a segment congruent to one of the three segments: for example, $\overline{A'C'}$ congruent to $\overline{AC}$. To complete the triangle construction, we must locate the other vertex, B'. The distance from A' to B' is AB. All points at a distance AB from A' are on a circle with center at A' and radius of length AB. Similarly, B' must be on a circle with center C' and radius of length BC. The only possible locations for B' are at the points where the two circles intersect. Either point is acceptable. Usually, a picture of the construction shows only one possibility and the construction uses only arcs, as seen in Figure 10-11(c).

REMARK Starting the construction with a segment $\overline{A'B'}$ congruent to $\overline{AB}$ or with $\overline{B'C'}$ congruent to $\overline{BC}$ would also result in triangles congruent to $\triangle ABC$.

NOW TRY THIS 10-3

From using the SSS construction as in Figure 10-11, we could think that given any three segments, we could construct a triangle whose sides are congruent to the given segments. To determine if this is true, cut at least ten pieces of straw of different lengths. Make all the possible triangles and answer the following questions:

a. Could a triangle be constructed from each of the three pieces of straws?

b. If three pieces were exactly the same length, what type of triangle could be constructed?

c. If the length of one piece of straw is the exact sum of the lengths of the other two pieces, can a triangle be constructed from the three pieces? Why or why not?

d. If one piece of straw is longer than the two other pieces put together, can a triangle be constructed from the three pieces? Why or why not?

TECHNOLOGY CORNER

Work through GSP Lab 4 in Appendix III to investigate congruent triangles.

Both *Now Try This 10-3* and the *Technology Corner* lend credence to the following property.

Property

Triangle inequality The sum of the measures of any two sides of a triangle must be greater than the measure of the third side.

Constructing Congruent Angles

We use the SSS notion of congruent triangles to construct an angle congruent to a given angle, $\angle B$, by making $\angle B$ a part of an isosceles triangle and then reproducing this triangle, as in Figure 10-12.

FIGURE 10-12 *Copying an angle*

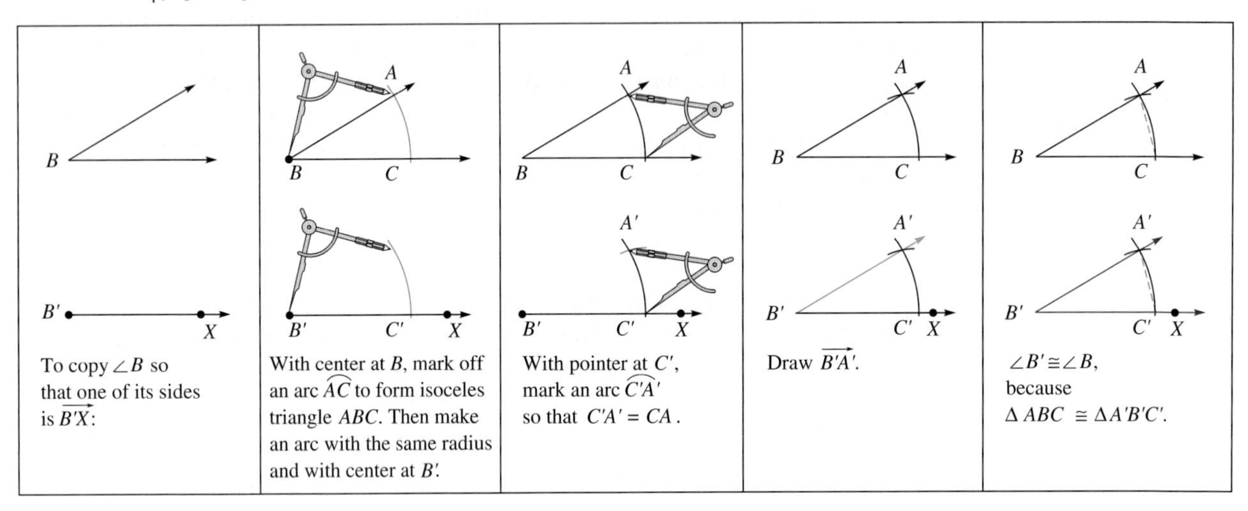

B
B'
X
To copy ∠B so that one of its sides is $\overrightarrow{B'X}$:
A
B C
B' C' X
With center at B, mark off an arc $\overset{\frown}{AC}$ to form isoceles triangle ABC. Then make an arc with the same radius and with center at B'.
A
B C
A'
B' C' X
With pointer at C', mark an arc $\overset{\frown}{C'A'}$ so that C'A' = CA.
A
B C
A'
B' C' X
Draw $\overrightarrow{B'A'}$.
A
B C
A'
B' C' X
∠B' ≅ ∠B, because Δ ABC ≅ ΔA'B'C'.

REMARK With the compass and straightedge alone, it is impossible in general to construct an angle if given only its measure. For example, an angle of measure 20° cannot be constructed with a compass and a straightedge only. Instead, a protractor or some other measuring tool must be used. A geometry drawing utility can also be used to construct angles.

▶ **HISTORICAL NOTE** Three geometry compass and straightedge construction problems that concerned mathematicians for centuries were (1) trisecting a general angle, (2) duplicating a cube (constructing a cube whose volume is twice a given cube), and (3) squaring a circle (constructing a square with exactly the same area as a given circle). These problems were proved to be impossible only in the 1800s. ◀

Side, Angle, Side Property (SAS)

We have seen that, given three segments, no more than one triangle can be constructed. Could more than one triangle be constructed when given only two segments? Consider Figure 10-13(b), which shows three different triangles with sides congruent to the segments given in Figure 10-13(a). The length of the third side depends on the measure of the **included angle** between the two given sides.

included angle

FIGURE 10-13

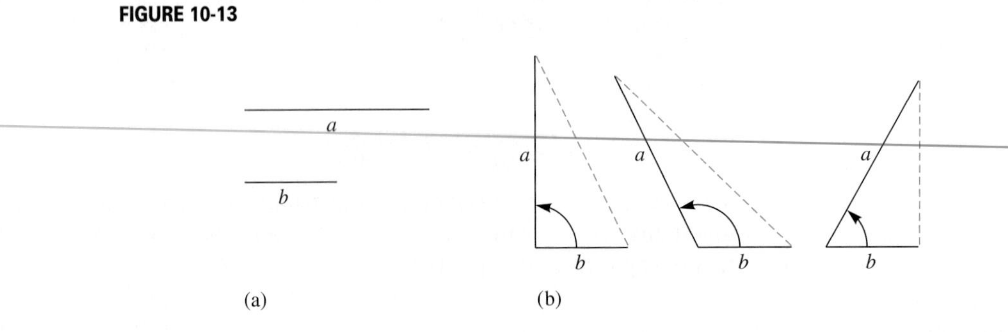

Constructions Involving Two Sides and an Included Angle of a Triangle

Figure 10-14 shows how to construct a triangle congruent to $\triangle ABC$ by using two sides $\overline{AB}$ and $\overline{AC}$ and the included angle, $\angle A$, formed by these sides. First, a ray with an arbitrary endpoint A' is drawn, and $\overline{A'C'}$ is constructed congruent to $\overline{AC}$. Then, $\angle A'$ is constructed so that $\angle A' \cong \angle A$ and B' is marked on the side of $\angle A'$ not containing C' so that $\overline{A'B'} \cong \overline{AB}$. Connecting B' and C' completes $\triangle A'B'C'$ so that $\triangle A'B'C'$ appears to be congruent to $\triangle ABC$.

FIGURE 10-14

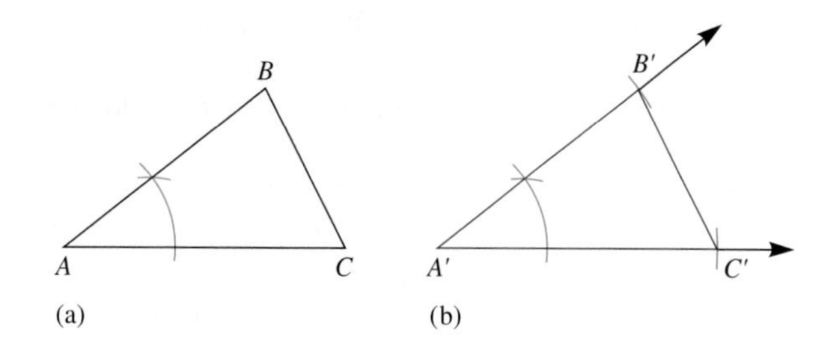

It is true that if we knew the lengths of two sides and the measure of the angle included between them, we could construct a unique triangle. We express this fact as the **Side, Angle, Side (SAS)** property.

Side, Angle, Side (SAS)

Property

Side, Angle, Side (SAS) If two sides and the included angle of one triangle are congruent to two sides and the included angle of another triangle, respectively, then the two triangles are congruent.

NOW TRY THIS 10-4

If, in two triangles, two sides and an angle not included between these sides are congruent, respectively, determine whether the triangles must be congruent.

a. If they are congruent, explain why you believe this is true.

b. If they are not congruent, state as few additional conditions as possible that can be placed on the sides or angles to make the triangles congruent.

EXAMPLE 10-3

For each part of Figure 10-15, use SAS to show that the given pair of triangles are congruent.

FIGURE 10-15

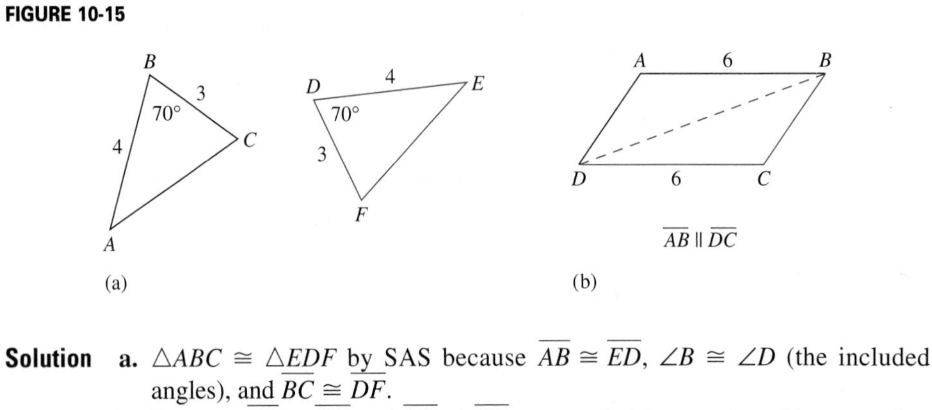

Solution **a.** $\triangle ABC \cong \triangle EDF$ by SAS because $\overline{AB} \cong \overline{ED}$, $\angle B \cong \angle D$ (the included angles), and $\overline{BC} \cong \overline{DF}$.

b. Because $\overline{AB} \cong \overline{CD}$ and $\overline{DB} \cong \overline{BD}$, we need either another side or another angle to show that the triangles are congruent. We know that $\overline{AB} \parallel \overline{DC}$. Since parallel segments $\overline{AB}$ and $\overline{DC}$ are cut by transversal $\overline{BD}$, we have alternate interior angles $\angle ABD$ and $\angle BDC$ congruent. Now $\triangle ABD \cong \triangle CDB$ by SAS.

Selected Triangle Properties

The SAS property of triangle congruence allows us to investigate a number of properties of triangles. For example, in Figure 10-16, consider isosceles triangle ABC with $\overline{AB} \cong \overline{AC}$. We use *paper folding* to construct the angle bisector of $\angle A$. For that purpose we fold a crease through vertex A so that vertex B "falls" on vertex C (this can be done because $\overline{AB} \cong \overline{AC}$). If point D is the intersection of the crease and $\overline{BC}$, then our folding assures that $\angle BAD \cong \angle CAD$. Thus, by SAS, $\triangle ABD \cong \triangle ACD$. Because corresponding parts must be congruent, we have: $\angle B \cong \angle C$, $\overline{BD} \cong \overline{CD}$, $\angle BDA \cong \angle CDA$. Because the last pair of angles are supplementary, it follows that each is a right angle. Consequently, $\overline{AD}$ is perpendicular to $\overline{BC}$ and bisects $\overline{BC}$. A line that is perpendicular to a segment and bisects it is the **perpendicular bisector** of the segment.

perpendicular bisector

FIGURE 10-16

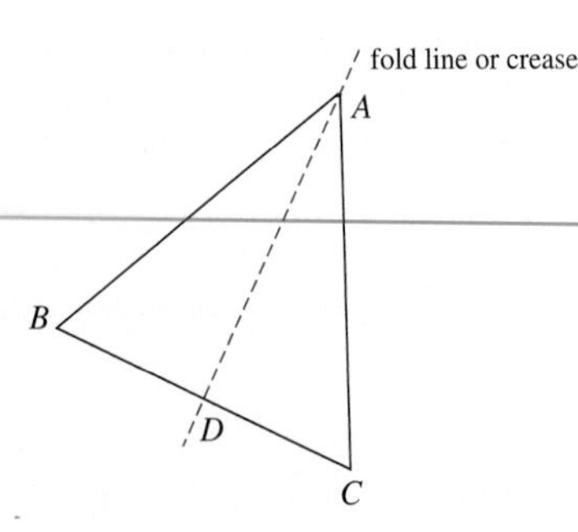

altitude An **altitude** of a triangle is the perpendicular segment from a vertex of the triangle to the line containing the opposite side of the triangle. These findings are summarized in Theorem 10–1.

> **Theorem 10–1**
>
> The following holds for every isosceles triangle:
>
> **a.** The angles opposite the congruent sides are congruent. (Base angles of an isosceles triangle are congruent.)
>
> **b.** The angle bisector of an angle formed by two congruent sides contains an altitude of the triangle and is the perpendicular bisector of the third side of the triangle.

RESEARCH NOTE Fourth graders who used Logo, a computer drawing language and tool invented by Seymour Papert, or who used traditional tools like protractors and rulers well exceeded the performance of high school students on items on the National Assessment of Educational Progress on a selected set of geometry items (Carpenter et al. 1981).

In Figure 10-16, notice that if point A is equidistant from the endpoints B and C, then A is on the perpendicular bisector of $\overline{BC}$. The converse of this statement is also true. The statement and its converse are given in Theorem 10–2.

> **Theorem 10–2**
>
> **a.** Any point equidistant from the endpoints of a segment is on the perpendicular bisector of the segment.
>
> **b.** Any point on the perpendicular bisector of a segment is equidistant from the endpoints of the segment.

Construction of the Perpendicular Bisector of a Segment

FIGURE 10-17

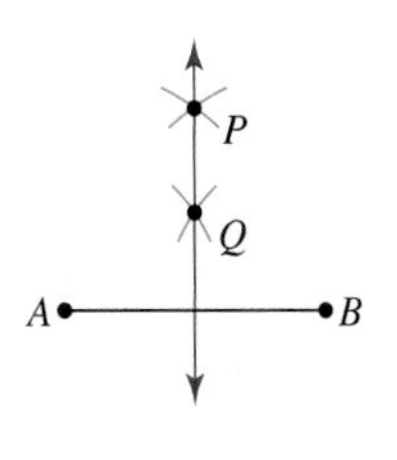

Theorem 10–2 can be used to construct the perpendicular bisector of a segment by constructing any two points equidistant from the endpoints of the segment. Each point is a vertex of an isosceles triangle, and the two points determine the perpendicular bisector of the segment. In Figure 10-17 we have constructed point P equidistant from A and B by drawing intersecting arcs with the same radius—one with center at A, and the other with center at B. Point Q is constructed similarly with two intersecting arcs. By Theorem 10-2(a), each point is on the perpendicular bisector of $\overline{AB}$. Because two points determine a unique line, $\overleftrightarrow{PQ}$ must be the perpendicular bisector of $\overline{AB}$.

NOW TRY THIS 10-5

In Figure 10-17 we constructed the perpendicular bisector of $\overline{AB}$ by drawing four arcs. An alternative construction of the perpendicular bisector of $\overline{AB}$ can be accomplished by drawing one large-enough arc (or circle) with center at A and another one with the same radius and centered at B. The two points where the arcs intersect determine the perpendicular bisector of the segment. Draw any segment and construct its perpendicular bisector using only two arcs.

Construction of a Circle Circumscribed About a Triangle

circumscribed **circumcircle**

In Figure 10-18(a), a circle is **circumscribed** about a given $\triangle ABC$; that is, every vertex of the triangle is on the circle. How can such a **circumcircle** be constructed?

FIGURE 10-18

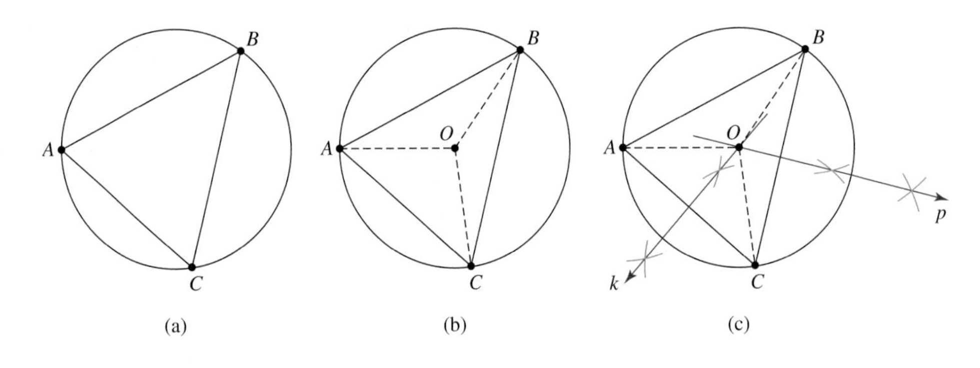

To discover an appropriate construction, imagine that we know the location of center O of the circle, as in Figure 10-18(b). The properties of the center of a circle should enable us to find its location. Because O is the center of the circle circumscribed about $\triangle ABC$, $OA = OC = OB$. The fact that $OA = OC$ implies that O is equidistant from the endpoints of segment $\overline{AC}$. Hence, by Theorem 10-2, O is on the perpendicular bisector, k, of $\overline{AC}$, as shown in Figure 10-18(c). Similarly, because $OC = OB$, O is on the perpendicular bisector, p, of $\overline{BC}$. Because O is on k and on p, it is the point of intersection of the two perpendicular bisectors. Thus, given $\triangle ABC$, we can construct the center of the circumscribed circle by constructing perpendicular bisectors of any two sides of the triangle. The point where the perpendicular bisectors intersect is the **circumcenter**. Thus, the required circle is the circle with the center at O and radius OA (the **circumradius**). The construction is shown in Figure 10-18(c).

circumcenter **circumradius**

PROBLEM SOLVING <u>Archaeological Find</u>

At the site of an ancient settlement, archaeologists found a fragment of a saucer, as shown in Figure 10-19. To restore the saucer, the archaeologists need to determine the radius of the original saucer. How can they do this?

FIGURE 10-19

Understanding the Problem The border of the shard shown in Figure 10-19 was part of a circle. To reconstruct the saucer, we are to determine the radius of the circle of which the shard is a part.

Devising a Plan We can use a *model* to determine the radius. We trace an outline of the circular edge of the three-dimensional shard on a piece of paper. The result is an arc of a circle, as shown in Figure 10-20. To determine the radius, we must find the center O. We know that by connecting O to points of the circle, we obtain congruent radii. Also, each pair of radii determines an isosceles triangle. Consider points A, B, and O. Triangle ABO is isosceles and O is on the perpendicular bisector of $\overline{AB}$.

FIGURE 10-20

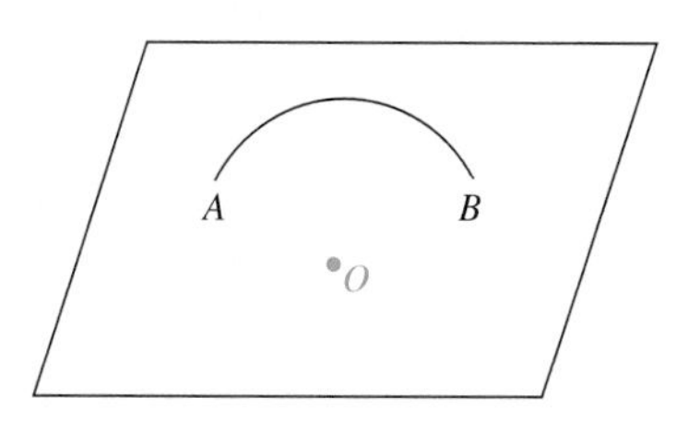

Carrying Out the Plan To find a line containing point O, construct a perpendicular bisector of $\overline{AB}$, as in Figure 10-21(a). Similarly, any other segment (for example, $\overline{AC}$) with endpoints on $\widehat{AB}$ has a perpendicular bisector containing O, as in Figure 10-21(b). The point of intersection of the two perpendicular bisectors is point O. To complete the problem, measure the length of either $\overline{OB}$ or $\overline{OA}$.

FIGURE 10-21

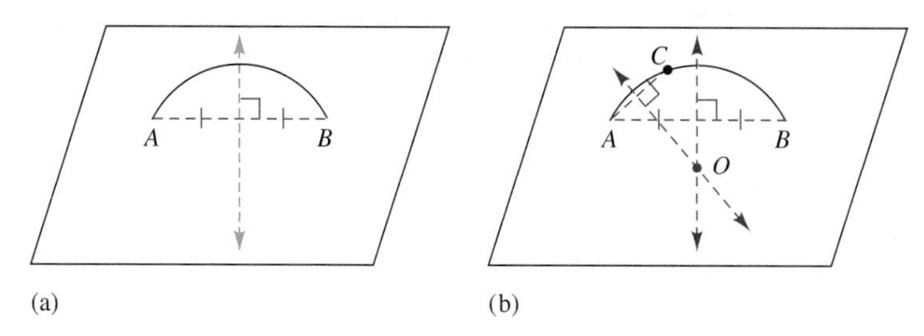

(a) (b)

Looking Back A related problem is, What would happen if the piece of pottery had been part of a sphere? Would the same ideas still work?

NOW TRY THIS 10-6

In the photograph we see congruent hexagons. Do you think that there is a Side-Side-Side-Side-Side-Side property for determining hexagons congruent? Prepare a pro or con argument.

ONGOING ASSESSMENT 10-1

1. In the definition of a circle, why are the points restricted to a plane?
2. Where do all the points in space equidistant from the endpoints of $\overline{AB}$ lie?
3. a. Use any tool to draw triangle ABC in which $\overline{BC}$ is greater than $\overline{AC}$. Measure the angles opposite $\overline{BC}$ and $\overline{AC}$. Compare the angle measures. What did you find?
   b. Based on your finding in (a), make a conjecture concerning the lengths of sides and the measures of angles of a triangle.
4. Use any tools to construct each of the following, if possible:
   a. A segment congruent to $\overline{AB}$ and an angle congruent to $\angle CAB$

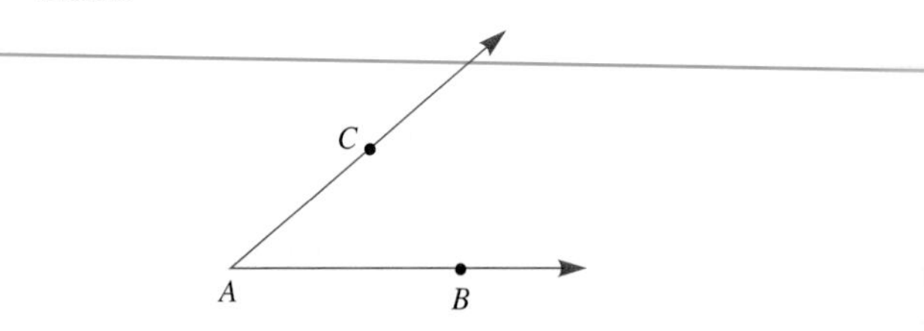

b. A triangle with sides of lengths 2 cm, 3 cm, and 4 cm
c. A triangle with sides of lengths 4 cm, 3 cm, and 5 cm (what kind of triangle is it?)
d. A triangle with sides 4 cm, 5 cm, and 10 cm
e. An equilateral triangle with sides 5 cm
f. A triangle with sides 6 cm and 7 cm and an included angle of measure 75°
g. A triangle with sides 6 cm and 7 cm and a nonincluded angle of measure 40°
h. A triangle with sides 6 cm and 6 cm and a nonincluded angle of measure 40°
i. A right triangle with legs 4 cm and 8 cm (the legs include the right angle)

5. For each of the conditions in problem 4(b) through (i), does the given information determine a unique triangle? Explain why or why not.

6. How many different triangles can be constructed with toothpicks by connecting the toothpicks only at their ends if each triangle can contain at most five toothpicks per side?

7. For each of the following, determine whether the given conditions are sufficient to prove that $\triangle PQR \cong \triangle MNO$. Justify your answers.
a. $\overline{PQ} \cong \overline{MN}$, $\overline{PR} \cong \overline{MO}$, $\angle P \cong \angle M$
b. $\overline{PQ} \cong \overline{MN}$, $\overline{PR} \cong \overline{MO}$, $\overline{QR} \cong \overline{NO}$
c. $\overline{PQ} \cong \overline{MN}$, $\overline{PR} \cong \overline{MO}$, $\angle Q \cong \angle N$

8. A rancher designed a wooden gate as illustrated in the following figure. Explain the purpose of the diagonal boards on the gate.

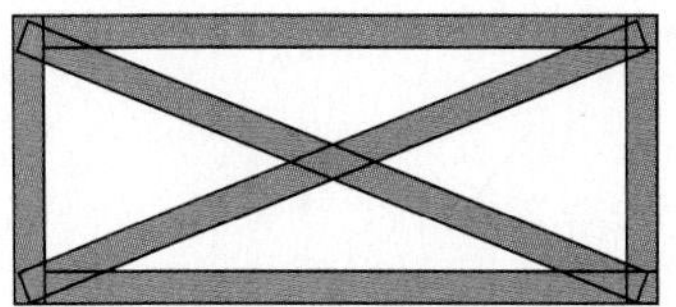

9. A rural homeowner had his television antenna held in place by three guy wires, as shown in the following figure.
a. If the stakes are on level ground and the distance from the stakes to the base of the antenna are the same, what is true about the lengths of the wires? Why?
b. If the stakes are not on level ground yet are the same distance from the base of the antenna, explain whether you can make the same conclusion regarding the lengths of the wires.

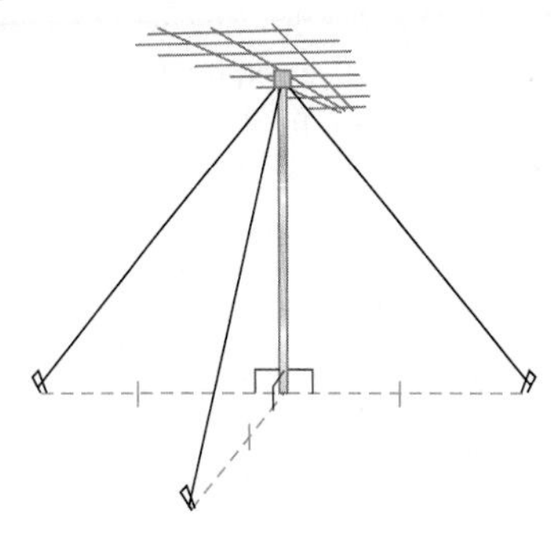

10. Is it possible to construct a right triangle with all sides congruent? Justify your answer.

11. If two students are constructing equilateral triangles with one side of length 4 in., explain whether the triangles constructed must be congruent.

12. Equilateral triangles have all angles congruent. If the sides of the respective triangles have different lengths, how can you demonstrate that the triangles are not congruent?

13. If you have a paper cutout of a circle and no other construction tools, explain how you can find the center of the circle.

14. In the following rep-tile drawing, if the circumcircles are constructed around the largest triangle and around the "middle" triangle, what can be said of the two circumcircles? Why?

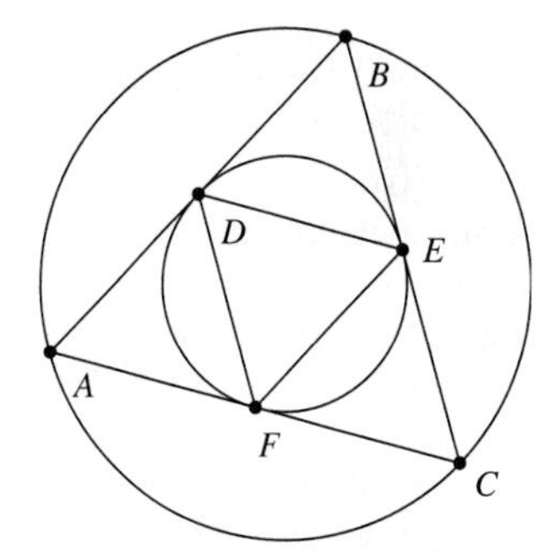

15. Given any triangle, use it to construct a rep-tile like that of problem 14.

16. What is true of the perpendicular bisectors of each of the sides of a regular hexagon? Prove your answer.

17. A group of students on a hiking trip wants to find the distance AB across a pond (see the following figure). One student suggests choosing any point C, connecting it with B, and then finding point D such that $\angle DCB \cong \angle ACB$ and $\overline{DC} \cong \overline{AC}$. How and why does this help in finding the distance AB?

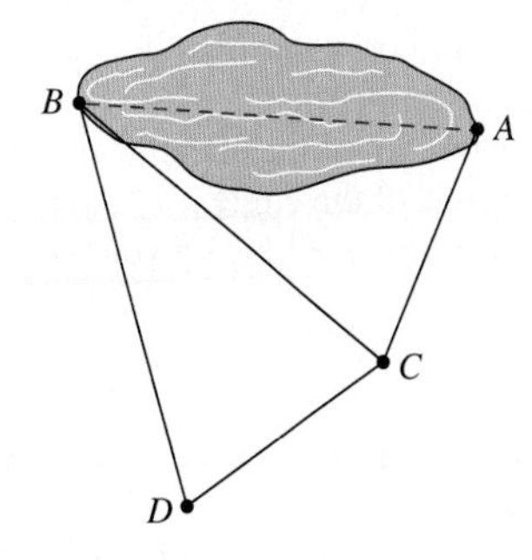

18. Using only a compass and a straightedge, perform each of the following:
a. Construct an equilateral triangle with the following side $\overline{AB}$:

A B

b. Construct a 60° angle.

c. Construct an isosceles triangle with $\angle A$ (see the following figure) as the angle included between the two congruent sides:

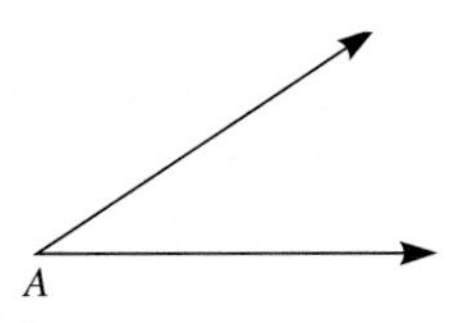

19. Refer to the figure shown and, using only a compass and a straightedge, perform each of the following:

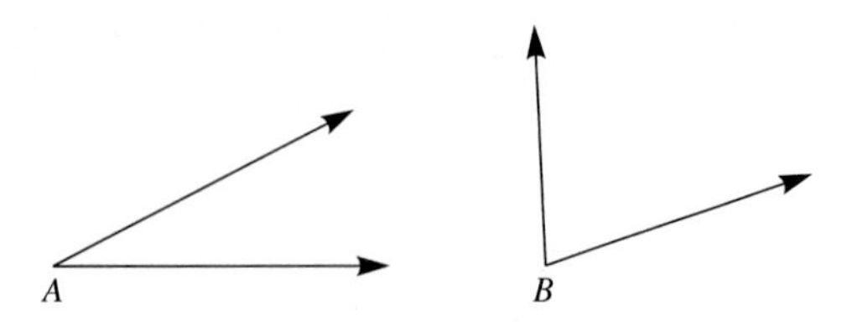

a. Construct $\angle C$ so that $m(\angle C) = m(\angle A) + m(\angle B)$.

b. Using the angles in (a), construct $\angle C$ so that $m(\angle C) = m(\angle B) - m(\angle A)$.

20. An equilateral triangle ABC is congruent to itself.

a. Write all possible true correspondences between the triangle and itself.

b. Use one of your answers in (a) to show that an equilateral triangle is also equiangular.

21. a. Draw a circle and use any method to mark off six points that are equally spaced on the circumference of the circle, as shown in the following figure:

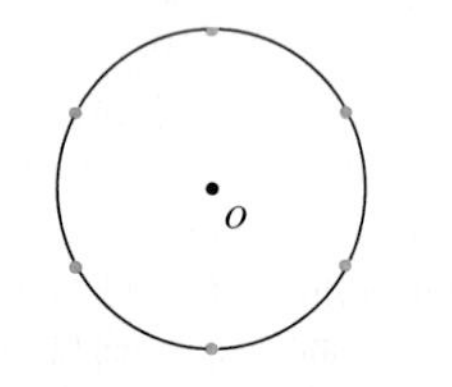

b. Connect each of the points to point O and connect the points in order around the circle to form a regular hexagon.

c. Explain why all of the triangles are congruent.

d. If the radius of the circle is r, what is the length of the side of the hexagon? Justify your answer.

e. Construct any circle, then use your answer to part (d) to inscribe a regular hexagon in the circle; that is, the sides of the hexagon are congruent and all the vertices of the hexagons are on the circle. Describe your construction in words.

f. Construct any circle. Use the construction in part (e) to construct an equilateral triangle inscribed in the circle.

★ **g.** Use the methods of this problem to construct a regular hexagon on Geometer's Sketchpad.

22. Let $ABCD$ be a square with diagonals $\overline{AC}$ and $\overline{BD}$ intersecting in point F, as shown in this figure:

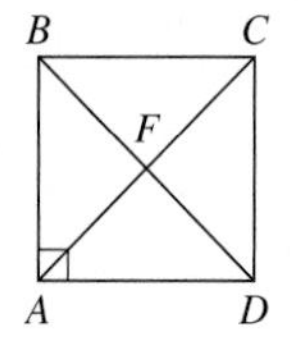

a. What is the relationship between point F and the diagonals $\overline{BD}$ and $\overline{AC}$? Why?

b. What are the measures of angles BFA and AFD? Why?

23. Write a definition for congruent arcs.

24. What minimum amount of information must you know in order to be sure that two cubes are congruent?

25. What is the least number of congruent parts that one must know in order to be sure that square pyramids are congruent?

26. How many different one-to-one correspondences could be listed between the following:

a. Vertices of two triangles

b. Vertices of two quadrilaterals

c. Vertices of two n-gons

27. In a pair of right triangles, suppose two legs of one are congruent respectively to two legs of the other. Explain whether the triangles are congruent and why.

28. If two triangles are congruent, what can be said about their perimeters? Why?

29. Construct an obtuse triangle similar to the one shown. Then, using only a compass and a straightedge, construct the circle that circumscribes the triangle.

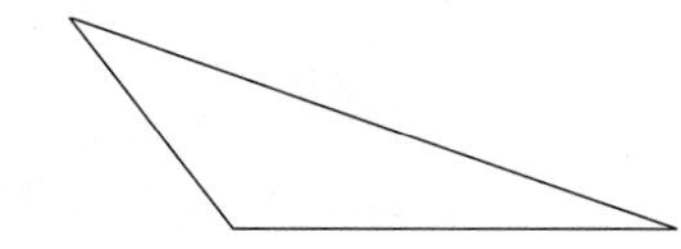

30. For which of the following figures is it possible to find a circle that circumscribes the figure? If it is possible to find such a circle, draw the figure and construct the circumscribing circle.

a. A right triangle

b. A regular hexagon

31. In the following drawing, a compass is used to draw circle O. A point P is marked on the circle. Using the compass with the same setting that was used to draw the circle and using point P as the center, draw an arc that intersects the original circle in two points A and B. Repeat the process using points A and B on the circle. Continue this process.

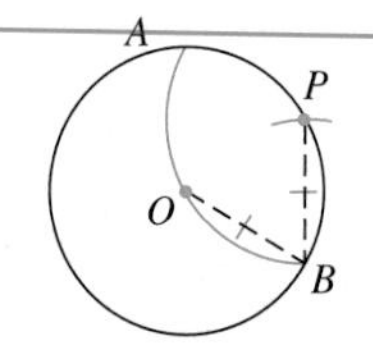

a. What do you observe?
b. If you connected the points on the circle in order with line segments, starting at P and going in a clockwise fashion around the circle, what figure would you draw? Why?

32. Consider a normal stop sign on a highway. Explain whether you think a circle could be circumscribed about this sign.
33. In a circle, consider the lines containing two perpendicular radii. The lines intersect the circle in four points. If the four points are connected.
   a. What type of figure is formed?
   b. Prove your answer in part (a).
34. a. If $\overline{MN}$ is any chord of circle O, what can you say about the perpendicular bisector of $\overline{MN}$?
   b. Prove your answer in part (a).

Communication

35. In circle A with radius AB, let P be a point of $\overline{AB}$ that is not an endpoint. Explain whether P is a part of the circle.
36. Write arguments to convince the class that Theorems 10–1 and 10–2 are true.
37. In the following figure, congruent segments are shown with tick marks.

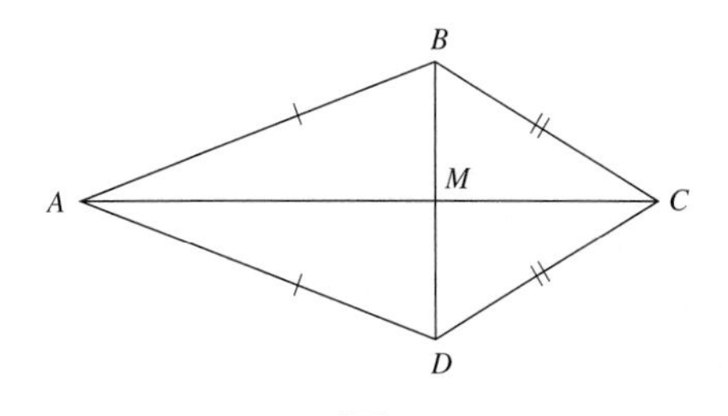

   a. Argue that the diagonal $\overline{AC}$ bisects $\angle A$ and $\angle C$.
   b. Let M be the point on which the diagonals of kite $ABCD$ intersect. Measure $\angle AMD$ and make a conjecture concerning the angle between the diagonals of a kite. Justify your conjecture.
   c. Show that $\overline{BM} \cong \overline{MD}$.
38. If the kite in Problem 37 were concave, do the same answers to parts (a) through (c) hold? Explain your answer.
39. Televisions are measured by the lengths of the diagonals across the screen. If a television is a 19-in. television, explain whether or not all televisions of this measure are congruent.
40. If given any regular polygon, describe how you can find the center of the circle that circumscribes the polygon.

Open-Ended

41. a. Find at least five examples of congruent objects.
   b. Find at least five examples of similar objects that are not congruent.
42. Design a quilt pattern that involves rep-tiles or find a pattern and describe the rep-tiles involved.
43. Use $\triangle AFB$ and $\triangle CED$ shown on the following dot paper to verify $\overline{AB}$ and $\overline{CD}$ are
   a. Congruent
   ★ **b.** Perpendicular to each other (consider the angles in $\triangle BGC$)

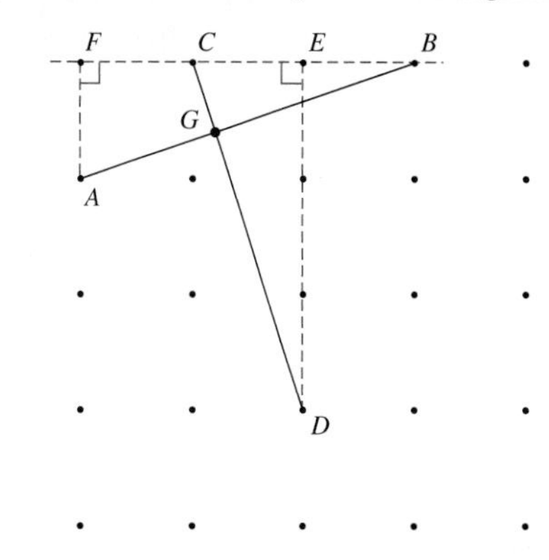

Cooperative Learning

44. Make a set of drawings of congruent figures and write an argument to convince a group of fellow students that the correspondence among vertices is necessary for determining that two triangles are congruent.

Third International Mathematics and Science Study (TIMSS) Questions

In square EFGH, which of these is FALSE?

a. $\triangle EIF$ and $\triangle EIH$ are congruent.
b. $\triangle GHI$ and $\triangle GHF$ are congruent.
c. $\triangle EFH$ and $\triangle EGH$ are congruent.
d. $\triangle EIF$ and $\triangle GIH$ are congruent.

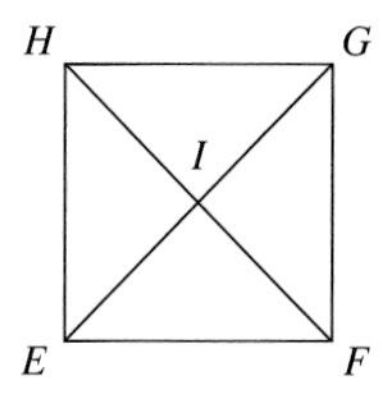

TIMSS, Grade 8, 2003

$ABCD$ is a trapezoid.

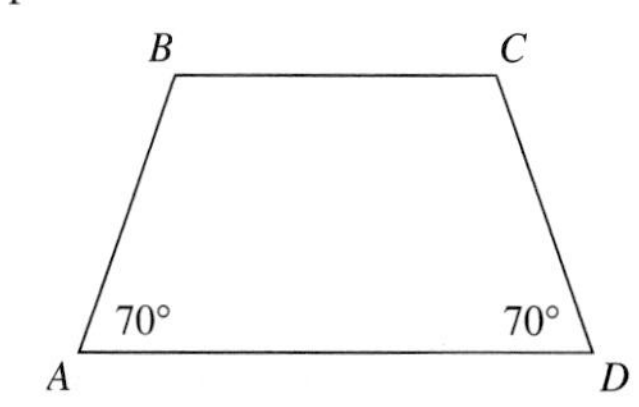

Another trapezoid, $GHIJ$ (not shown), is congruent (the same size and shape) to $ABCD$. Angles G and J each measure 70°. Which of these could be true?

a. $GH = AB$
b. Angle H is a right angle.
c. All sides of $GHIJ$ are the same length.
d. The perimeter of $GHIJ$ is 3 times the perimeter of $ABCD$.
e. The area of $GHIJ$ is less than the area of $ABCD$.

TIMSS, Grade 8, 2003

10-2 Other Congruence Properties

Angle, Side, Angle Property (ASA)

Angle, Side, Angle (ASA)

Triangles can be determined to be congruent by SSS and SAS. Can a triangle be constructed congruent to a given triangle by using two angles and a side? Figure 10-22 shows the construction of a triangle $A'B'C'$ such that $\overline{A'C'} \cong \overline{AC}$, $\angle A' \cong \angle A$, and $\angle C' \cong \angle C$. It seems that $\triangle A'B'C' \cong \triangle ABC$. This construction illustrates the **Angle, Side, Angle (ASA)** property of congruence.

FIGURE 10-22

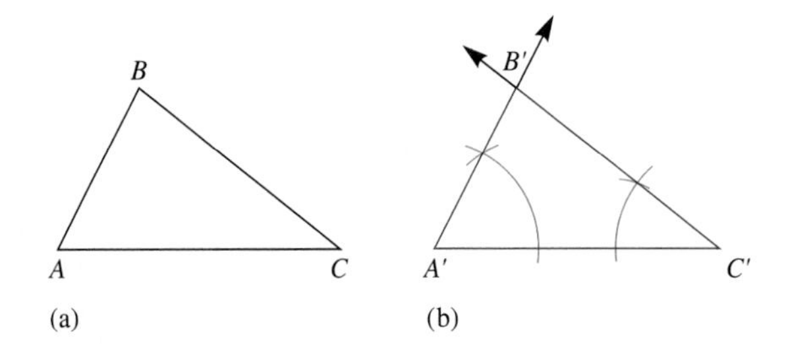

Property

Angle, Side, Angle (ASA) If two angles and the included side of one triangle are congruent to two angles and the included side of another triangle, respectively, then the triangles are congruent.

EXAMPLE 10-4

In Figure 10-23, $\triangle ABC$ and $\triangle DEF$ have two pairs of angles congruent and a pair of sides congruent. Show that the triangles are congruent.

FIGURE 10-23

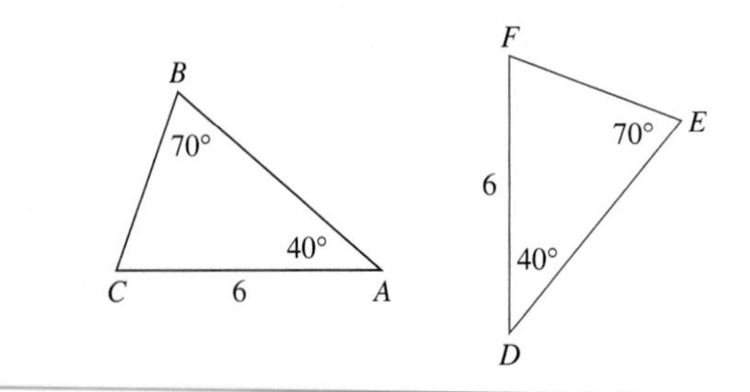

Solution Notice that $\angle A \cong \angle D$ and $\angle B \cong \angle E$, which implies that $\angle C \cong \angle F$ because the measure of each is $180° - (70 + 40)°$. Then, since $\overline{AC} \cong \overline{DF}$, we have $\triangle ABC \cong \triangle DEF$ by ASA.

Example 10-4 is a special case of the following property.

Property

Angle, Angle, Side (AAS) If two angles and a corresponding side of one triangle are congruent to two angles and a corresponding side of another triangle, respectively, then the two triangles are congruent.

EXAMPLE 10-5 Show that the triangles in each part of Figure 10-24 are congruent.

FIGURE 10-24

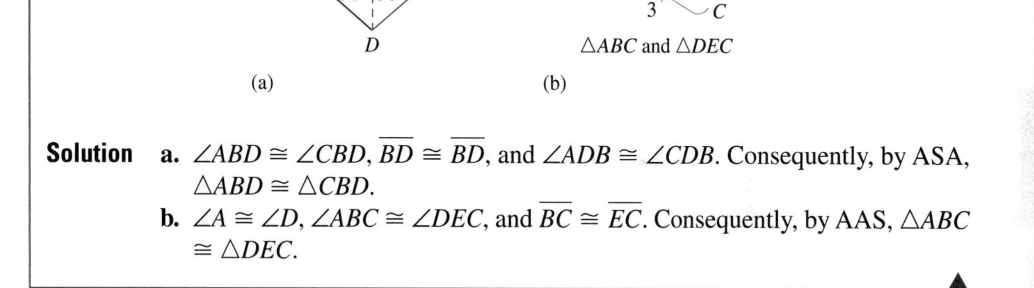

Solution **a.** $\angle ABD \cong \angle CBD$, $\overline{BD} \cong \overline{BD}$, and $\angle ADB \cong \angle CDB$. Consequently, by ASA, $\triangle ABD \cong \triangle CBD$.

b. $\angle A \cong \angle D$, $\angle ABC \cong \angle DEC$, and $\overline{BC} \cong \overline{EC}$. Consequently, by AAS, $\triangle ABC \cong \triangle DEC$.

NOW TRY THIS 10-7

Draw a parallelogram and one of its diagonals. Use the definition of a parallelogram given in Chapter 9 and the ASA congruence property to prove that opposite sides of a parallelogram are congruent.

EXAMPLE 10-6

a. Using the definition of a parallelogram and the property that opposite sides in a parallelogram are congruent (see *Now Try This 10-7*), prove that the diagonals of a parallelogram bisect each other; that is, in Figure 10-25(a), show that $AO = OC$ and $BO = OD$.

b. Draw a line through the point O where the diagonals of a parallelogram intersect, as in Figure 10-25(b). The line intersects the opposite sides of the parallelogram at points P and Q. Prove that $\overline{OP} \cong \overline{OQ}$.

c. Prove that $\overline{SQ} \cong \overline{PT}$ in Figure 10-25(c).

FIGURE 10-25

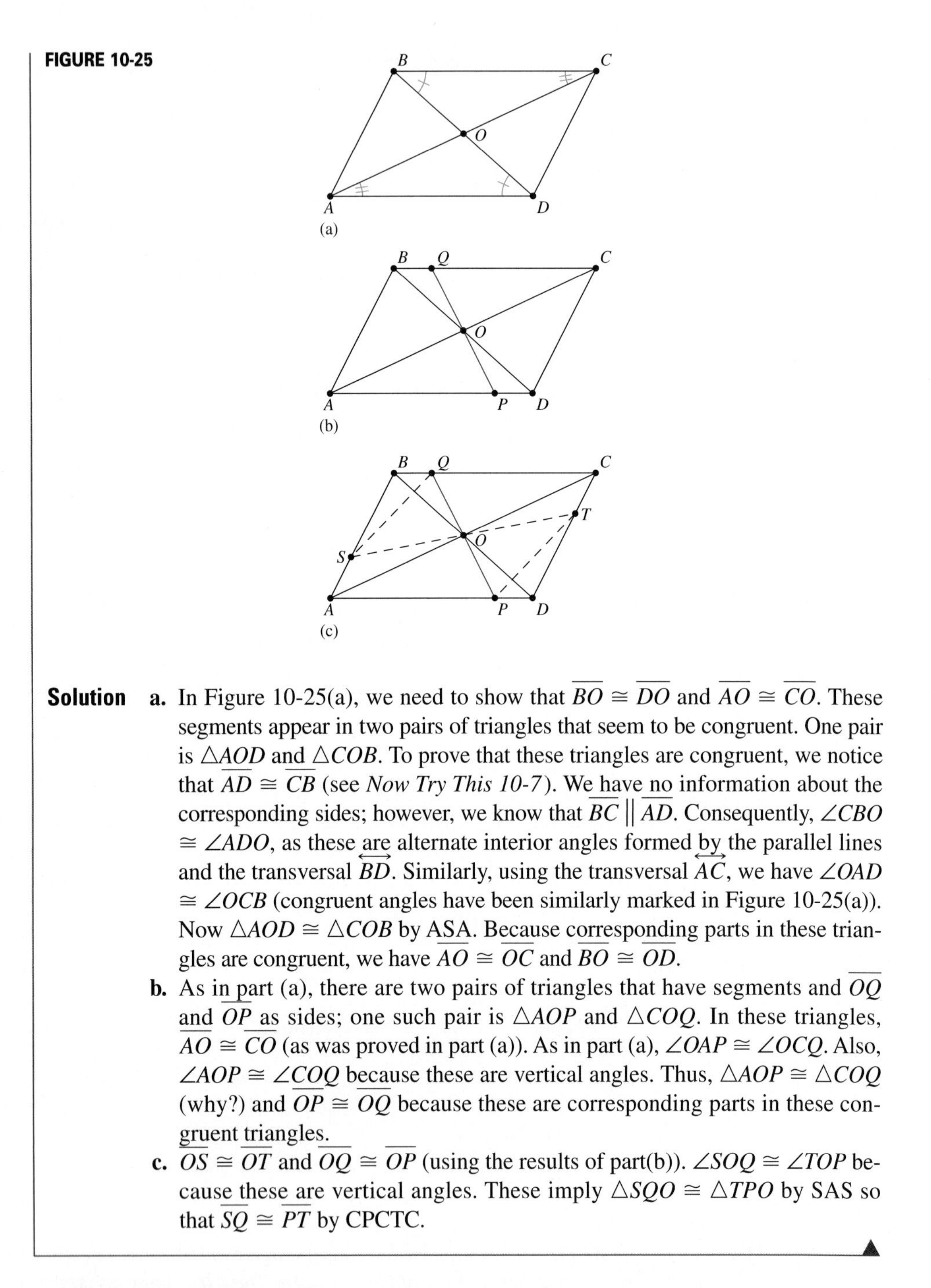

Solution **a.** In Figure 10-25(a), we need to show that $\overline{BO} \cong \overline{DO}$ and $\overline{AO} \cong \overline{CO}$. These segments appear in two pairs of triangles that seem to be congruent. One pair is $\triangle AOD$ and $\triangle COB$. To prove that these triangles are congruent, we notice that $\overline{AD} \cong \overline{CB}$ (see *Now Try This 10-7*). We have no information about the corresponding sides; however, we know that $\overline{BC} \parallel \overline{AD}$. Consequently, $\angle CBO \cong \angle ADO$, as these are alternate interior angles formed by the parallel lines and the transversal $\overleftrightarrow{BD}$. Similarly, using the transversal $\overleftrightarrow{AC}$, we have $\angle OAD \cong \angle OCB$ (congruent angles have been similarly marked in Figure 10-25(a)). Now $\triangle AOD \cong \triangle COB$ by ASA. Because corresponding parts in these triangles are congruent, we have $\overline{AO} \cong \overline{OC}$ and $\overline{BO} \cong \overline{OD}$.

b. As in part (a), there are two pairs of triangles that have segments and $\overline{OQ}$ and $\overline{OP}$ as sides; one such pair is $\triangle AOP$ and $\triangle COQ$. In these triangles, $\overline{AO} \cong \overline{CO}$ (as was proved in part (a)). As in part (a), $\angle OAP \cong \angle OCQ$. Also, $\angle AOP \cong \angle COQ$ because these are vertical angles. Thus, $\triangle AOP \cong \triangle COQ$ (why?) and $\overline{OP} \cong \overline{OQ}$ because these are corresponding parts in these congruent triangles.

c. $\overline{OS} \cong \overline{OT}$ and $\overline{OQ} \cong \overline{OP}$ (using the results of part(b)). $\angle SOQ \cong \angle TOP$ because these are vertical angles. These imply $\triangle SQO \cong \triangle TPO$ by SAS so that $\overline{SQ} \cong \overline{PT}$ by CPCTC.

▲

NOW TRY THIS 10-8

If the diagonals of a quadrilateral bisect each other, must the quadrilateral be a parallelogram? Explain why or why not.

Using properties of congruent triangles, we can deduce various properties of quadrilaterals. Table 10–1 summarizes the definitions and lists some properties of six quadrilaterals. These and other properties of quadrilaterals are further investigated in Ongoing Assessment 10-2.

Table 10-1

Quadrilateral and Its Definition	Properties of the Quadrilateral
B C A D *Trapezoid*: A quadrilateral with at least one pair of parallel sides	Consecutive angles between parallel sides are supplementary.
B D A C *Parallelogram*: A quadrilateral in which each pair of opposite sides is parallel	**a.** A parallelogram has all the properties of a trapezoid. **b.** Opposite sides are congruent. **c.** Opposite angles are congruent. **d.** Diagonals bisect each other.
B D A C *Rectangle*: A parallelogram with a right angle	**a.** A rectangle has all the properties of a parallelogram. **b.** All the angles of a rectangle are right angles. **c.** A quadrilateral in which all the angles are right angles is a rectangle. **d.** The diagonals of a rectangle are congruent and bisect each other.
D A C B *Kite*: A quadrilateral with two distinct pairs of congruent adjacent sides	**a.** Lines containing the diagonals are perpendicular to each other. **b.** A line containing one diagonal is a bisector of the other. **c.** One diagonal bisects nonconsecutive angles.
B C A D *Rhombus*: A parallelogram with all sides congruent	**a.** A rhombus has all the properties of a parallelogram and a kite. **b.** A quadrilateral in which all the sides are congruent is a rhombus. **c.** The diagonals of a rhombus are perpendicular to and bisect each other. **d.** Each diagonal bisects opposite angles.
B C A D *Square*: A rectangle with all sides congruent	A square has all the properties of a parallelogram, a rectangle, and a rhombus.

NOW TRY THIS 10-9

a. Define an isosceles trapezoid.
b. Based on your definition in part (a), is a parallelogram an isosceles trapezoid? Explain why.

RESEARCH NOTE Sixth-grade students investigating relationships in geometry often have difficulty noticing more than one role of an object. For example, if a segment is a perpendicular bisector of a side of a triangle, they might not recognize it as an altitude (Zykova 1969).

ONGOING ASSESSMENT 10-2

1. Use any tools to construct each of the following, if possible:
   a. A triangle with angles measuring 60° and 70° and an included side of 8 in.
   b. A triangle with angles measuring 60° and 70° and a nonincluded side of 8 cm on a side of the 60° angle
   c. A right triangle with one acute angle measuring 75° and a leg of 5 cm on a side of the 75° angle
   d. A triangle with angles measuring 30°, 70°, and 80°
2. For each of the conditions in problem 1(a) through (d), is it possible to construct two noncongruent triangles? Explain why or why not.
3. In both the ASA and AAS properties of congruence, if two angles of one triangle are congruent, respectively, to two angles of another triangle, what must be true about the third angles of the triangles? Justify your answer.
4. a. For two right triangles, give a minimal set of conditions based on ASA and AAS to argue that the two triangles are congruent.
   b. Using your answer in (a), write two theorems that can be used to show that right triangles are congruent.
5. For each of the following, determine whether the given conditions are sufficient to prove that $\triangle PQR \cong \triangle MNO$. Justify your answers.
   a. $\angle Q \cong \angle N$, $\angle P \cong \angle M$, $\overline{PQ} \cong \overline{MN}$
   b. $\angle R \cong \angle O$, $\angle P \cong \angle M$, $\overline{QR} \cong \overline{NO}$
   c. $\overline{PQ} \cong \overline{MN}$, $\overline{PR} \cong \overline{MO}$, $\angle N \cong \angle Q$
   d. $\angle P \cong \angle M$, $\angle Q \cong \angle N$, $\angle R \cong \angle O$
6. A parallel ruler, shown as follows, can be used to draw parallel lines. The distance between the parallel segments $\overline{AB}$ and $\overline{DC}$ can vary. The ruler is constructed so that the distance between *A* and *B* equals the distance between *D* and *C*. The distance between *A* and *C* is the same as the distance between *B* and *D*. Explain why $\overline{AB}$ and $\overline{DC}$ are always parallel.

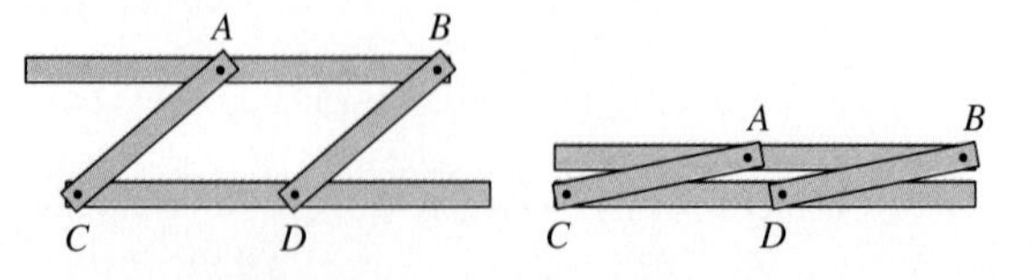

7. In each of the following, choose as many of the words *parallelogram, rectangle, rhombus, trapezoid, kite*, or *square* so that the resulting sentence is true. If none of the words makes the sentence true, answer "none" and justify your answer.
   a. A quadrilateral is a ____ if and only if its diagonals bisect each other.
   b. A quadrilateral is a ____ if and only if its diagonals are congruent.
   c. A quadrilateral is a ____ if and only if its diagonals are perpendicular.
   d. A quadrilateral is a ____ if and only if its diagonals are congruent and bisect each other.
   e. A quadrilateral is a ____ if and only if its diagonals are perpendicular and bisect each other.
   f. A quadrilateral is a ____ if and only if its diagonals are congruent and perpendicular and they bisect each other.
   g. A quadrilateral is a ____ if and only if a pair of opposite sides is parallel and congruent.
8. Create several trapezoids that have a pair of nonparallel sides congruent. Measure all angles and make a conjecture about the relationships among pairs of angles.
9. Classify each of the following statements as true or false. If the statement is false, provide a counterexample.
   a. The diagonals of a square are perpendicular bisectors of each other.
   b. If all sides of a quadrilateral are congruent, the quadrilateral is a rhombus.
   c. If a rhombus is a square, it must also be a rectangle.
   d. An isosceles trapezoid can be a rectangle.
   e. A square is a trapezoid.
   f. A trapezoid is a parallelogram.
   g. A parallelogram is a trapezoid.
   h. No rectangle is a rhombus.
   i. No trapezoid is a square.
   j. Some squares are trapezoids.
10. **a.** Construct quadrilaterals having exactly one, two, and four right angles.

b. Can a quadrilateral have exactly three right angles? Why?

c. Can a parallelogram have exactly two right angles? Why?

11. Explain whether the chevron $ABCD$ pictured is a kite.

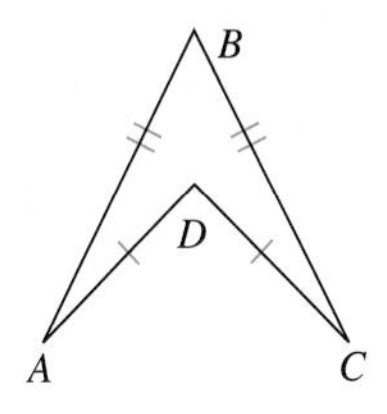

12. Classify each of the figures formed from regular hexagon $ABCDEF$.

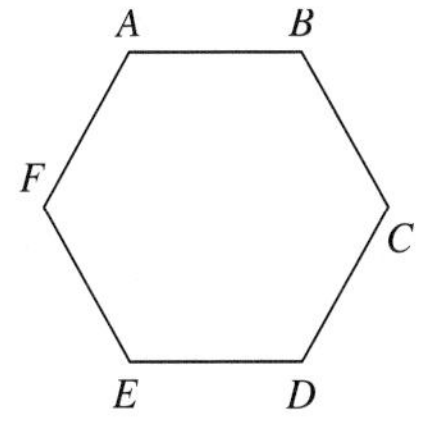

a. Quadrilateral $ABDE$

b. Quadrilateral $ABEF$

13. What is the shape that is formed by taking the inner piece of cardboard in a paper towel roll, carefully cutting along the seam, and flattening? How do you know?

14. **a.** What is the maximum number of $3'' \times 5''$ rectangular pieces of paper that can be cut from an $8\ 1/2'' \times 11''$ sheet of paper?

b. Are all $3'' \times 5''$ rectangular cards congruent? Why or why not?

15. Each fourth grader is given a protractor, two 30-cm sticks, and two 20-cm sticks and is asked to form a quadrilateral with a 75° angle. Sketch all possibilities.

16. The game of Triominoes has equilateral-triangular playing pieces with numbers at each vertex, shown as follows:

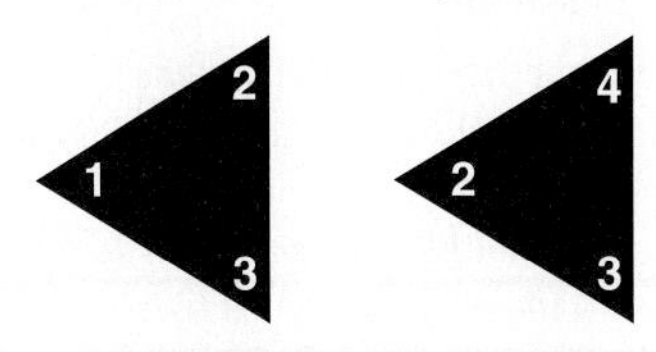

If two pieces are placed together as shown in the following figure, explain what type of quadrilateral is formed:

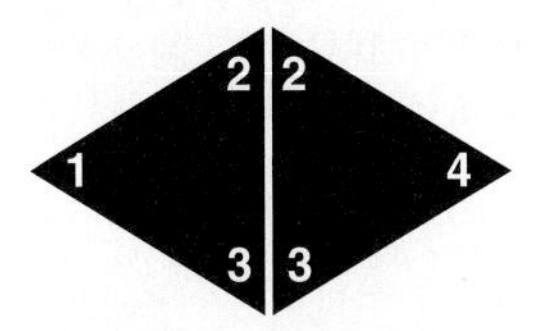

17. A **sector** of a circle is a pie-shaped section bounded by two radii and an arc. What is a minimal set of conditions for determining that two sectors of the same circle are congruent?

18. In the rectangle $ABCD$ shown, X and Y are midpoints of the given sides.

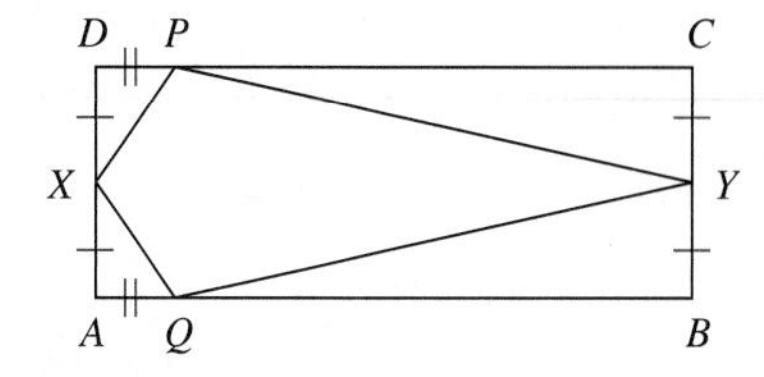

a. What type of quadrilateral is $PYQX$? Prove your answer.

b. If points P and Q are moved at a constant rate and in the same direction along $\overline{DC}$ and $\overline{AB}$, respectively, does this change your answer in part (a)? Why or why not?

19. Draw a quadrilateral so that no circle can circumscribe the quadrilateral. Explain why no such circle exists.

20. Draw any triangle and make a copy of it. Explain whether the two triangles can be "put together" to form a parallelogram.

21. a. Use a square to create a rep-tile.

b. Could your pattern from part (a) be used with any quadrilateral to create a rep-tile?

22. Prove whether a quadrilateral can have three obtuse angles.

23. In the following quadrilateral, $AB = AC = AD$. Prove that $m(\angle 1) + m(\angle 4) = m(\angle 2) + m(\angle 3)$.

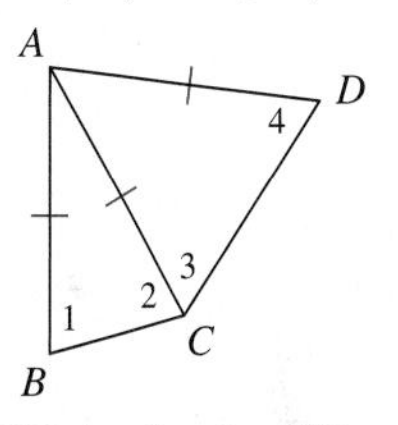

24. A compact disk (CD) is circular. Why do you think that most CD cases are square?

25. In the figure shown, $\overline{CT}$ and $\overline{CA}$ are tangent to circle O at A and T and thus perpendicular to $\overline{OT}$ and $\overline{OA}$, respectively. Prove the following:

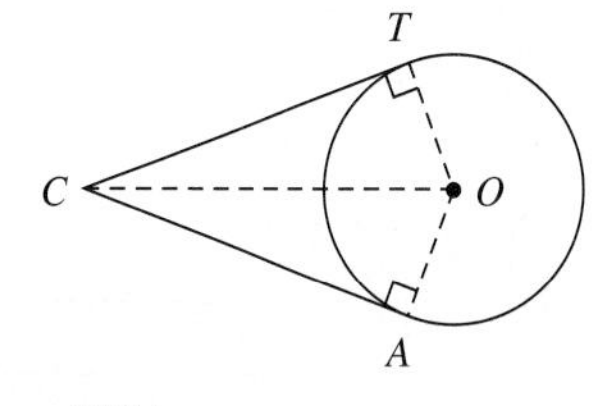

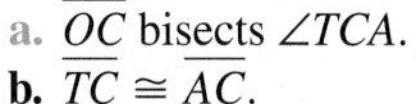
a. $\overline{OC}$ bisects $\angle TCA$.

b. $\overline{TC} \cong \overline{AC}$.

26. Draw two quadrilaterals such that two angles and an included side in one quadrilateral are congruent, respectively, to two angles and an included side in the other quadrilateral. However, the quadrilaterals are not congruent.

27. a. In an isosceles trapezoid, make a conjecture concerning the lengths of the sides that are not bases.

b. Make a conjecture concerning the diagonals of an isosceles trapezoid.

c. Justify your conjectures in (a) and (b).

28. Two angles of the trapezoid shown are 45° each. Suppose the lengths of the parallel sides of the trapezoid are a cm and b cm. Find the height of the trapezoid (the distance between the parallel sides) in terms of a and b.

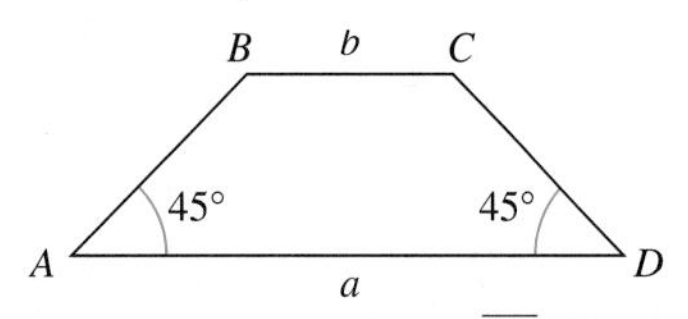

(Hint: Construct a parallel to $\overline{AB}$ through C.)

29. Using a straightedge and a compass, construct any convex kite. Then construct a second convex kite that is not congruent to the first but whose sides are congruent to the corresponding sides of the first kite.

30. **a.** What type of figure is formed by joining the midpoints of a rectangle (see the following figure)?

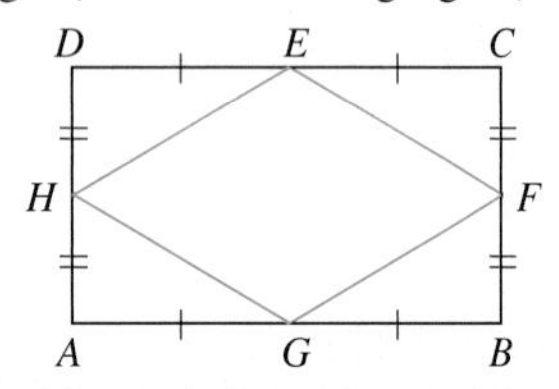

b. What type of figure is formed by joining the midpoints of the sides of a parallelogram?

★ **c.** Justify your answer to part (b).

31. What information is needed to determine congruency for each of the following?
   a. Two squares
   b. Two rectangles
   c. Two parallelograms

32. Describe a set of minimal conditions to determine if two regular polygons are congruent.

33. Suppose polygon $ABCD$ shown in the following figure is any parallelogram:

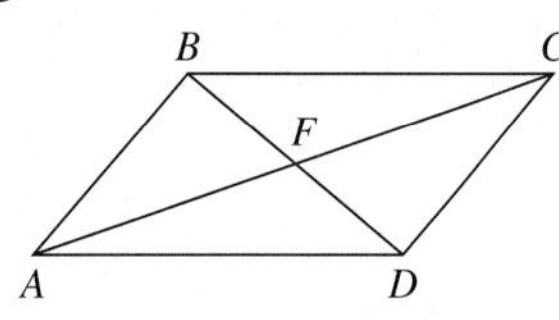

Use congruent triangles to justify each of the following:
   a. $\angle A \cong \angle C$ and $\angle B \cong \angle D$ (opposite angles are congruent).
   b. $\overline{BC} \cong \overline{AD}$ and $\overline{AB} \cong \overline{CD}$ (opposite sides are congruent).
   c. $\overline{BF} \cong \overline{DF}$ and $\overline{AF} \cong \overline{CF}$ (the diagonals bisect each other).
   d. $\angle DAB$ and $\angle ABC$ are supplementary.

Communication

34. Stan is standing on the bank of a river wearing a baseball cap. Standing erect and looking directly at the other bank, he pulls the bill of his cap down until it just obscures his vision of the opposite bank. He then turns around, being careful not to disturb the cap, and picks out a spot that is just obscured by the bill of his cap. He then paces off the distance to this spot and claims that the distance across the river is approximately equal to the distance he paced. Is Stan's claim true? Why?

35. Most ironing boards are collapsible for storage and can be adjusted to fit the height of the person using them. The surface of the board, though, remains parallel to the floor regardless of the height. Explain how to construct the legs of an ironing board to ensure the surface is always parallel to the floor.

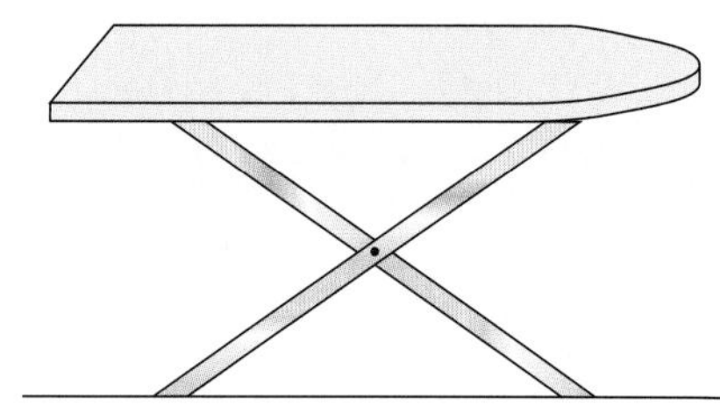

36. In the figure below, a student argues that it is a kite because two *distinct* pairs of sides are congruent. How do you respond? (The pairs are $(\overline{AB}, \overline{BC})$ and $(\overline{BC}, \overline{CD})$.)

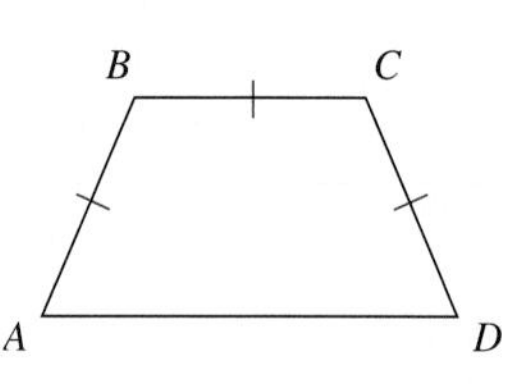

Open-Ended

37. **a.** Go to a wallpaper store and examine the pattern books. Determine whether all rolls of a specific pattern of wallpaper are congruent.
   b. On an individual roll of wallpaper, determine the length of wallpaper before a pattern is repeated.
   c. If you were wallpapering a room, explain how congruence or noncongruence could save you money.

Cooperative Learning

38. **a.** Record the definitions of *trapezoid* and *kite* given in different Grade 6–8 and secondary-school geometry textbooks.
   b. Compare the definitions found with those in this text and with those other groups found.
   c. Defend the use of one definition over another.

Review Problems

39. In the following regular pentagon, use the existing vertices to find all the triangles congruent to $\triangle ABC$. Show that the triangles actually are congruent.

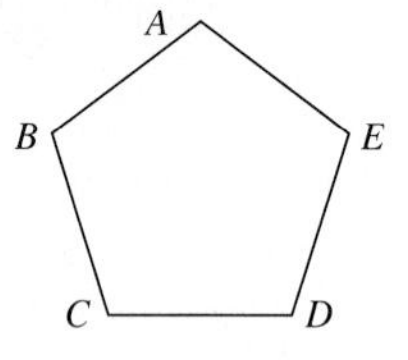

40. If possible, construct a triangle that has the three segments a, b, and c shown here as its sides; if not possible, explain why not.

41. Construct an equilateral triangle whose sides are congruent to the following segment:

42. For each of the following pairs of triangles, determine whether the given conditions are sufficient to show that the triangles are congruent. If the triangles are congruent, tell which property can be used to verify this fact.

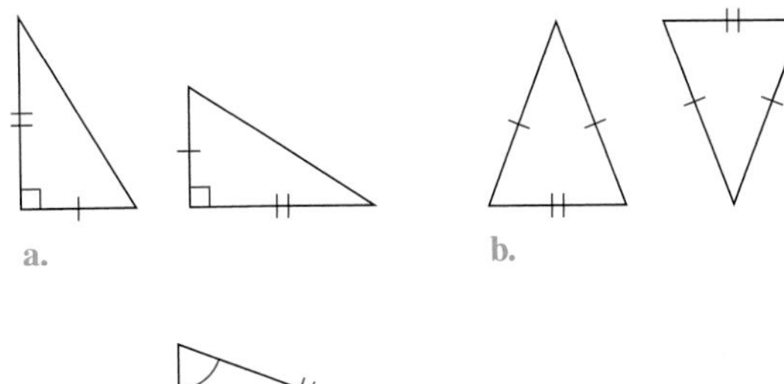

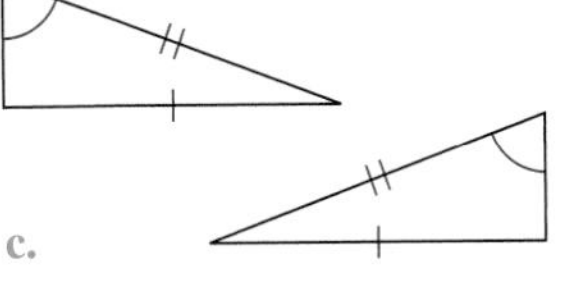

Third International Mathematics and Science Study (TIMSS) Questions

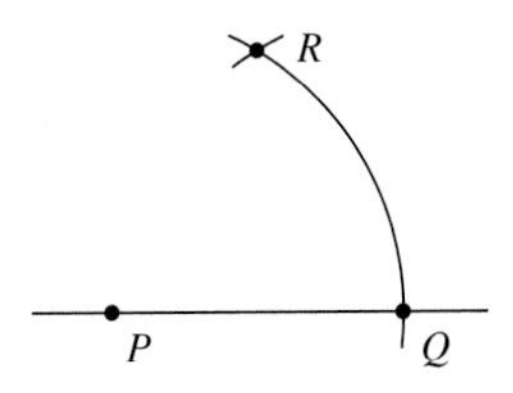

In the figure above, an arc of a circle with center P has been drawn to cut the line at Q. Then an arc with the same radius and center Q was drawn to cut the first arc at R. What would be the size of angle PRQ?

a. 30°
b. 45°
c. 60°
d. 75°

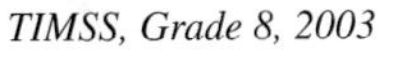

TIMSS, Grade 8, 2003

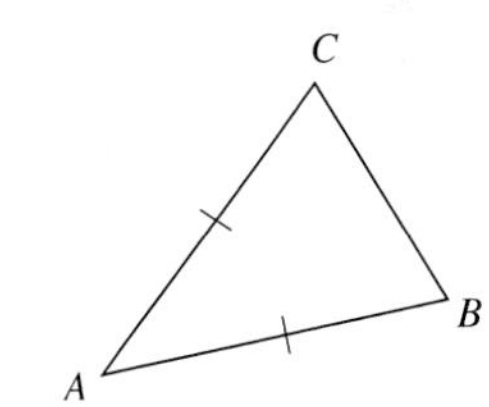

The triangle ABC has $AB = AC$.

Draw a line to divide triangle ABC into two congruent triangles.

TIMSS, Grade 8, 2003

TECHNOLOGY CORNER

Use Geometer's Sketchpad to test the truth of the following statement: "If one pair of sides of a quadrilateral is congruent and parallel, then the quadrilateral is a parallelogram."

10-3 Other Constructions

We use the definition of a rhombus and the following properties (also listed in Table 10-1) to accomplish basic compass-and-straightedge constructions:

1. A rhombus is a parallelogram in which all the sides are congruent.
2. A quadrilateral in which all the sides are congruent is a rhombus.
3. Each diagonal of a rhombus bisects the opposite angles.
4. The diagonals of a rhombus are perpendicular.
5. The diagonals of a rhombus bisect each other.

Constructing Parallel Lines

To construct a line parallel to a given line ℓ through a point P not on ℓ, as in the leftmost panel of Figure 10-26, our strategy is to construct a rhombus (using property 2 listed above) with one of its vertices at P and one of its sides on line ℓ. Because the opposite sides of a rhombus are parallel, one of the sides through P will be parallel to ℓ. This construction is shown in Figure 10-26.

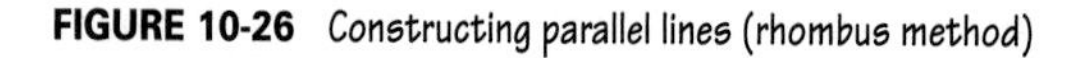

FIGURE 10-26 *Constructing parallel lines (rhombus method)*

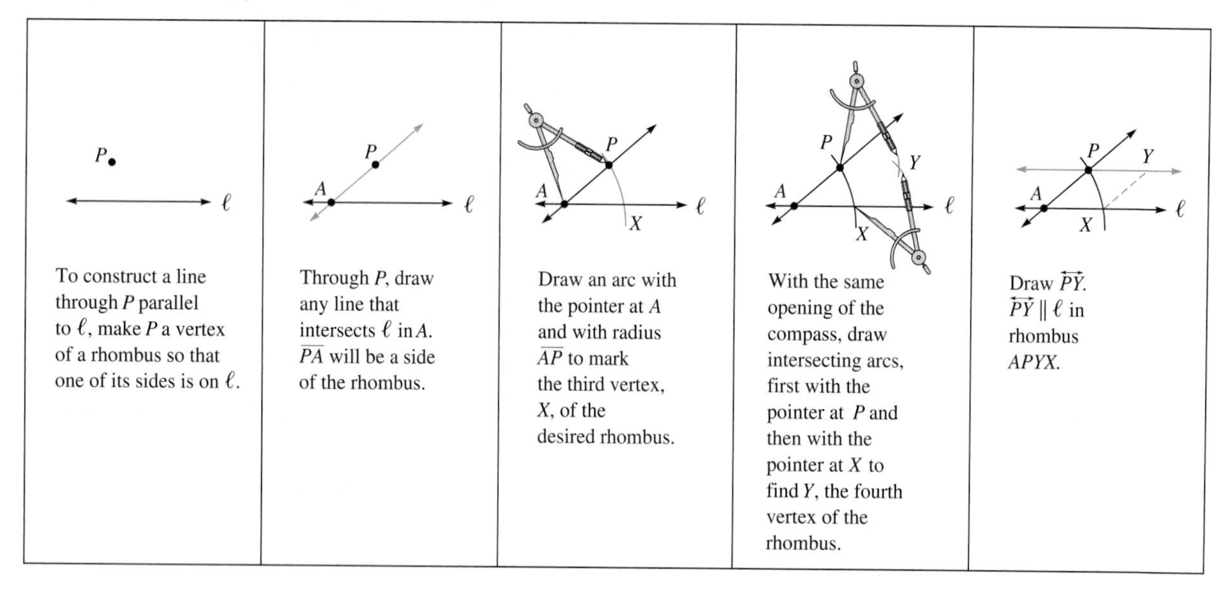

P
ℓ
To construct a line through P parallel to ℓ, make P a vertex of a rhombus so that one of its sides is on ℓ.
P
A
ℓ
Through P, draw any line that intersects ℓ in A. PA will be a side of the rhombus.
P
A
X
ℓ
Draw an arc with the pointer at A and with radius AP to mark the third vertex, X, of the desired rhombus.
P
Y
A
X
ℓ
With the same opening of the compass, draw intersecting arcs, first with the pointer at P and then with the pointer at X to find Y, the fourth vertex of the rhombus.
P
Y
A
X
ℓ
Draw PY. PY ∥ ℓ in rhombus APYX.

Figure 10-27 shows another way to do the construction of parallel lines. If congruent corresponding angles are formed by a transversal cutting two lines, then the lines are parallel. Thus, the first step is to draw a transversal through P that intersects ℓ. The angle marked α is formed by the transversal and line ℓ. By constructing an angle with a vertex at P congruent to α, we create congruent corresponding angles; therefore, $\overleftrightarrow{PQ} \parallel \ell$.

FIGURE 10-27 *Constructing parallel lines (corresponding angle method)*

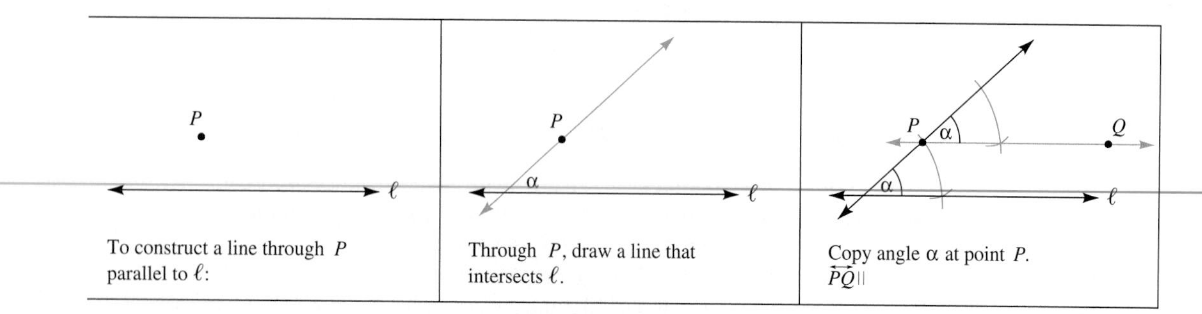

P
ℓ
To construct a line through P parallel to ℓ:
P
α
ℓ
Through P, draw a line that intersects ℓ.
P
α
Q
α
ℓ
Copy angle α at point P.
PQ ∥

NOW TRY THIS 10-10

Parallel lines are frequently constructed using either a ruler and one triangle or two triangles. If a ruler and a triangle are used, the ruler is left fixed and the triangle is slid so that one side of the triangle touches the ruler at all times. In Figure 10-28, the hypotenuses of the right triangles are all parallel (also the legs not on the ruler are all parallel). How can this method be used to construct a line through a given point parallel to a given line?

FIGURE 10-28

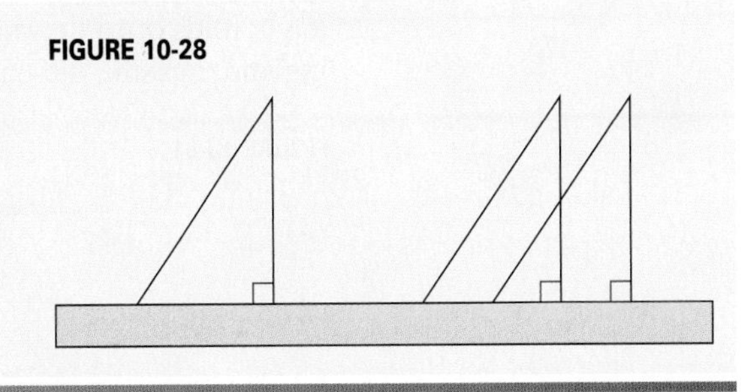

Paper folding can be used to construct parallel lines. For example, in Figure 10-29(a) to construct a line m parallel to line p through point Q, we can fold a perpendicular to line p so that the fold line does not contain point Q, as shown in Figure 10-29(b). Then by marking the image of point Q and connecting point Q and its image, Q', we have $\overleftrightarrow{QQ'}$ parallel to line p (why?).

FIGURE 10-29

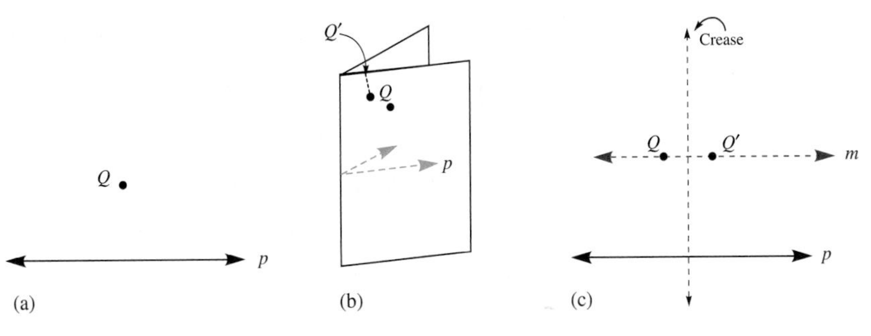

Constructing Angle Bisectors

angle bisector Another construction based on a property of a rhombus is for an **angle bisector**, a ray that separates an angle into two congruent angles. The diagonal of a rhombus with vertex A bisects $\angle A$, as shown in Figure 10-30.

FIGURE 10-30 Bisecting an angle

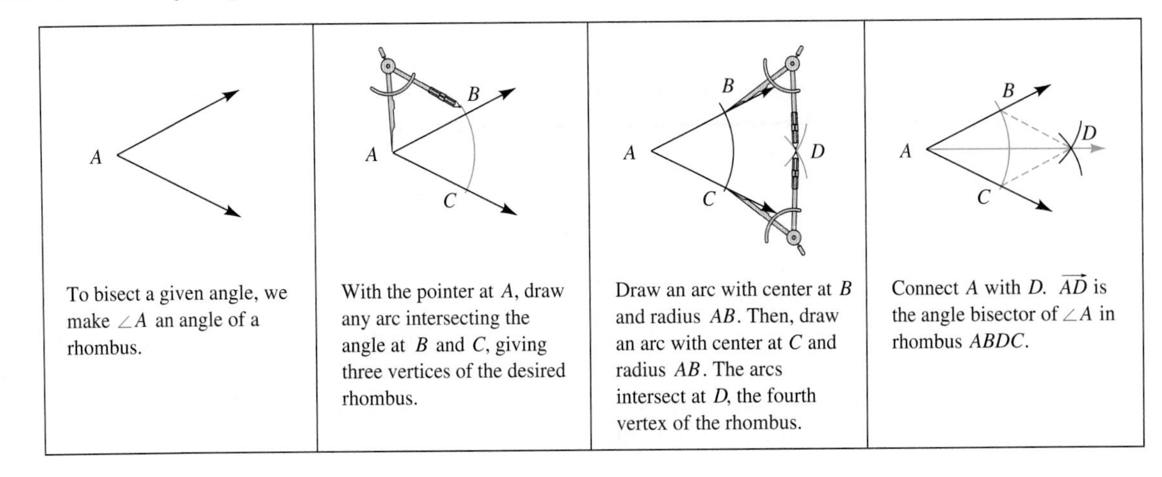

A
To bisect a given angle, we make $\angle A$ an angle of a rhombus.
A
B
C
With the pointer at A, draw any arc intersecting the angle at B and C, giving three vertices of the desired rhombus.
A
B
C
D
Draw an arc with center at B and radius AB. Then, draw an arc with center at C and radius AB. The arcs intersect at D, the fourth vertex of the rhombus.
A
B
C
D
Connect A with D. $\overrightarrow{AD}$ is the angle bisector of $\angle A$ in rhombus $ABDC$.

We can bisect an angle by folding a line through the vertex so that one side of the angle folds onto the other side. For example, in Figure 10-31 we bisect $\angle ABC$ by folding and creasing the paper through the vertex B so that $\overrightarrow{BC}$ coincides with $\overrightarrow{BA}$.

FIGURE 10-31

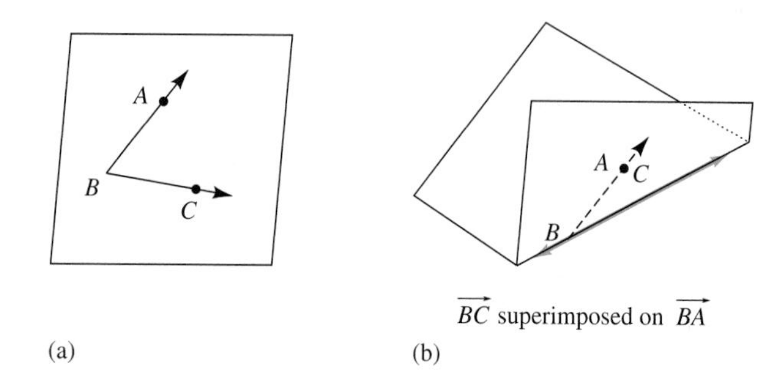

FIGURE 10-32

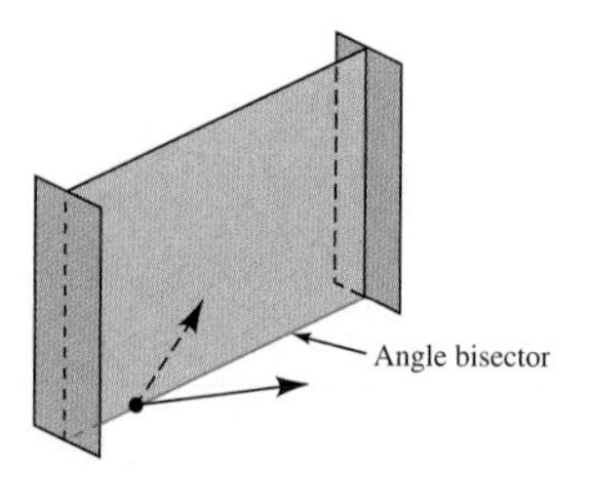

A Mira is a plastic device that acts as a reflector so that the image of an object can be seen behind the Mira. The drawing edge of the Mira acts as a folding line on paper. Any construction demonstrated in this text using paper folding can also be done with a Mira. To construct the bisector of an angle with a Mira, we place the drawing edge of the Mira on the vertex of the angle and reflect one side of the angle onto the other, as shown in Figure 10-32.

Constructing Perpendicular Lines

To construct a line through P perpendicular to line ℓ, where P is not a point on ℓ, as in Figure 10-33, recall that the diagonals of a rhombus are perpendicular to each other. If we construct a rhombus with a vertex at P and a diagonal on ℓ, as in Figure 10-33, the segment connecting the fourth vertex Q to P is perpendicular to ℓ.

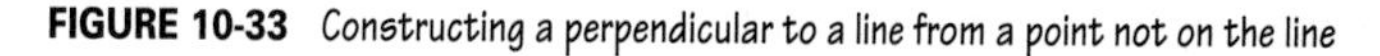
FIGURE 10-33 *Constructing a perpendicular to a line from a point not on the line*

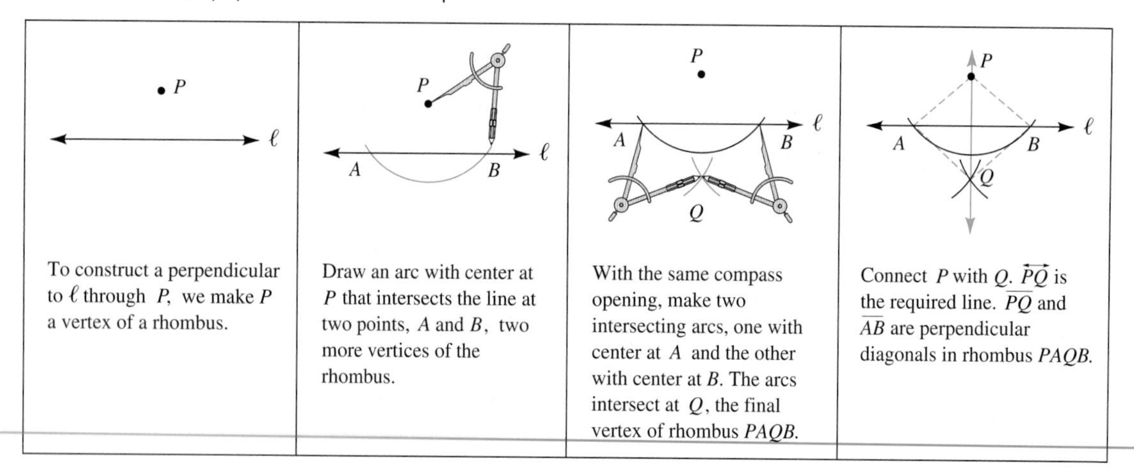

P
ℓ
To construct a perpendicular to ℓ through P, we make P a vertex of a rhombus.
P
A
B
ℓ
Draw an arc with center at P that intersects the line at two points, A and B, two more vertices of the rhombus.
P
A
B
ℓ
Q
With the same compass opening, make two intersecting arcs, one with center at A and the other with center at B. The arcs intersect at Q, the final vertex of rhombus PAQB.
P
A
B
ℓ
Q
Connect P with Q. $\overleftrightarrow{PQ}$ is the required line. $\overline{PQ}$ and $\overline{AB}$ are perpendicular diagonals in rhombus PAQB.

In Section 10-1 we saw how to construct the perpendicular bisector of a segment using a property of a perpendicular bisector stated in Theorem 10–2. Here we show how a property of a rhombus can also be used for constructing the perpendicular bisector of a segment.

To construct the perpendicular bisector of a line segment, as in Figure 10-34, we use the fact that the diagonals of a rhombus are perpendicular bisectors of each other.

FIGURE 10-34 Bisecting a line segment

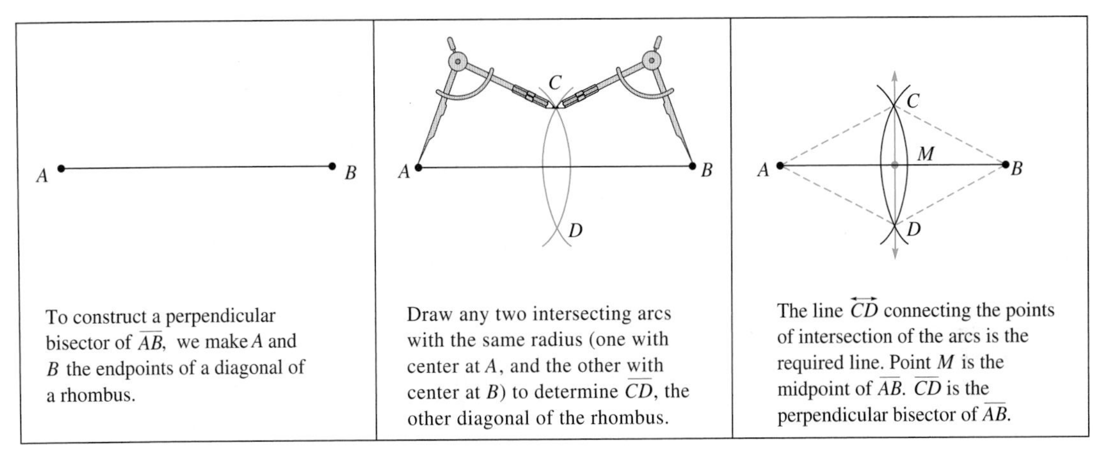

A
B
A
C
B
D
A
C
M
B
D
To construct a perpendicular bisector of $\overline{AB}$, we make A and B the endpoints of a diagonal of a rhombus.
Draw any two intersecting arcs with the same radius (one with center at A, and the other with center at B) to determine $\overline{CD}$, the other diagonal of the rhombus.
The line $\overleftrightarrow{CD}$ connecting the points of intersection of the arcs is the required line. Point M is the midpoint of $\overline{AB}$. $\overline{CD}$ is the perpendicular bisector of $\overline{AB}$.

The construction yields a rhombus such that the original segment is one of the diagonals of the rhombus and the other diagonal is the perpendicular bisector, as in Figure 10-34.

Constructing a perpendicular to a line ℓ at a point M on ℓ is based on the fact that the diagonals of a rhombus are perpendicular bisectors of each other. Observe in Figure 10-34 that $\overline{CD}$ is a perpendicular to $\overline{AB}$ through M. Thus we construct a rhombus whose diagonals intersect at point M, as in Figure 10-35.

FIGURE 10-35 Constructing a perpendicular to a line from a point on the line

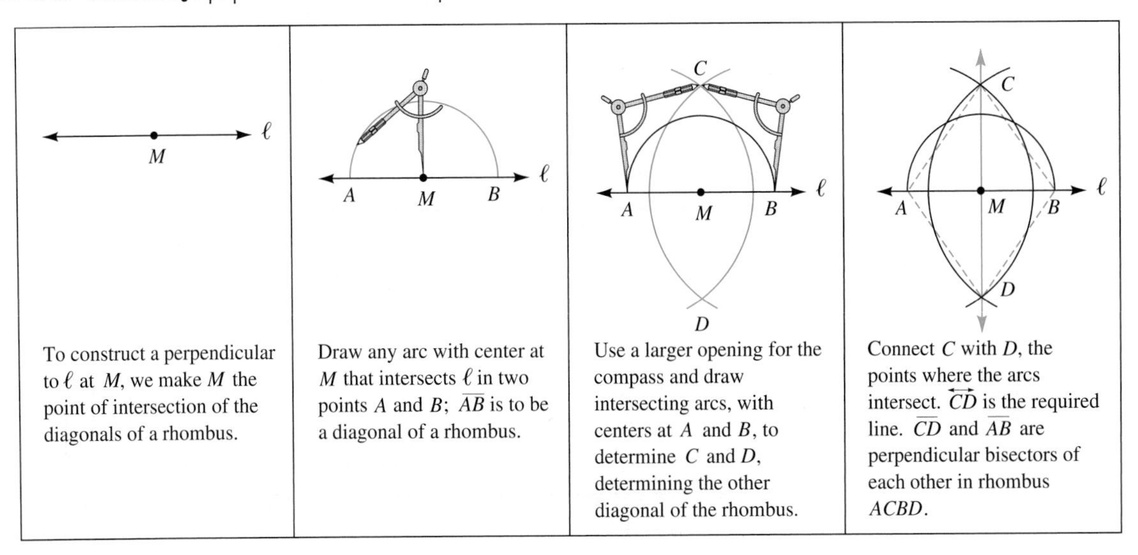

ℓ
M
ℓ
A
M
B
C
ℓ
A
M
B
D
C
ℓ
A
M
B
D
To construct a perpendicular to ℓ at M, we make M the point of intersection of the diagonals of a rhombus.
Draw any arc with center at M that intersects ℓ in two points A and B; $\overline{AB}$ is to be a diagonal of a rhombus.
Use a larger opening for the compass and draw intersecting arcs, with centers at A and B, to determine C and D, determining the other diagonal of the rhombus.
Connect C with D, the points where the arcs intersect. $\overleftrightarrow{CD}$ is the required line. $\overline{CD}$ and $\overline{AB}$ are perpendicular bisectors of each other in rhombus $ACBD$.

Perpendicularity constructions can also be completed by means of paper folding or by using a Mira. To use paper folding to construct a perpendicular to a given line ℓ at a point P on the line, we fold the line onto itself, as shown in Figure 10-36(a). The fold line is perpendicular to ℓ. To perform the construction with a Mira, we place the Mira with the drawing edge on P, as shown in Figure 10-36(b), so that ℓ is reflected onto itself. The line along the drawing edge is the required perpendicular.

FIGURE 10-36

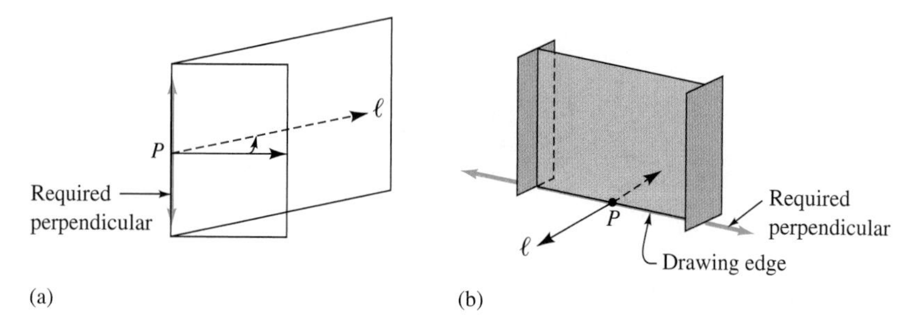

Constructing perpendiculars is useful in locating altitudes of a triangle. The construction of altitudes is described in Example 10-7.

EXAMPLE 10-7

Given triangle ABC, construct an altitude from vertex A in each part of Figure 10-37.

FIGURE 10-37

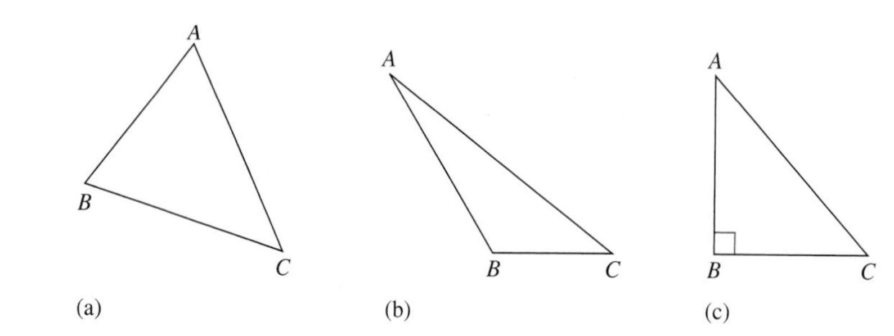

Solution **a.** An altitude is the perpendicular from a vertex to the line containing the opposite side of a triangle, so we need to construct a perpendicular from point A to the line containing $\overline{BC}$. Such a construction is shown in Figure 10-38. $\overline{AD}$ is the required altitude.

FIGURE 10-38

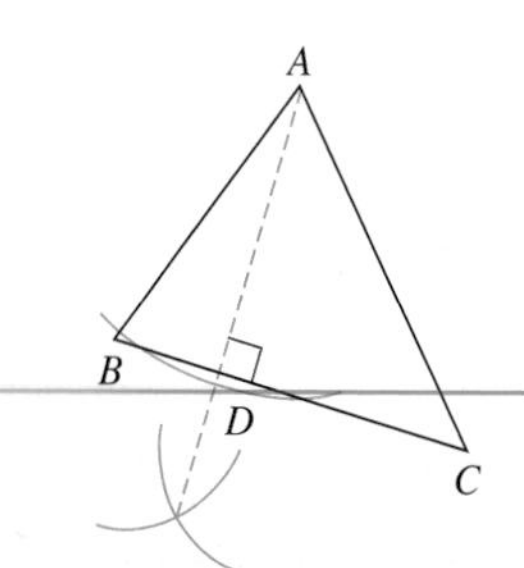

b. The construction of the altitude from vertex A is shown in Figure 10-39. Notice that the required altitude $\overline{AD}$ does not intersect the interior of $\triangle ABC$.

c. Triangle ABC is a right triangle. The altitude from vertex A is the side $\overline{AB}$. No construction is required.

FIGURE 10-39

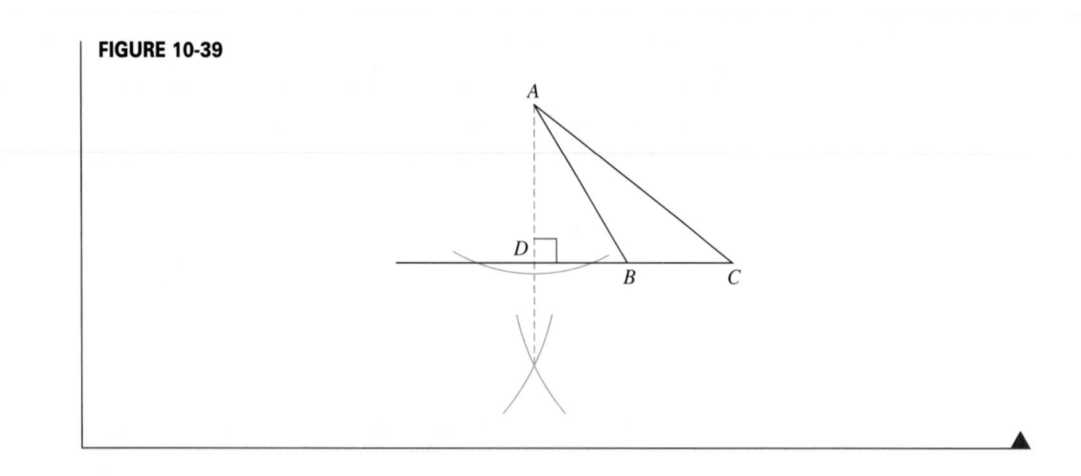

TECHNOLOGY CORNER

Use Geometer's Sketchpad to draw all the altitudes of a triangle. Make conjectures about the altitudes of each of the following types of triangles: acute, right, and obtuse.

Properties of Angle Bisectors

Consider the angle bisector in Figure 10-40. It seems that any point P on the angle bisector is equidistant from the sides of the angle; that is, $PD = PE$. (*The distance from a point to a line is the length of the perpendicular segment from the point to the line.*)

FIGURE 10-40

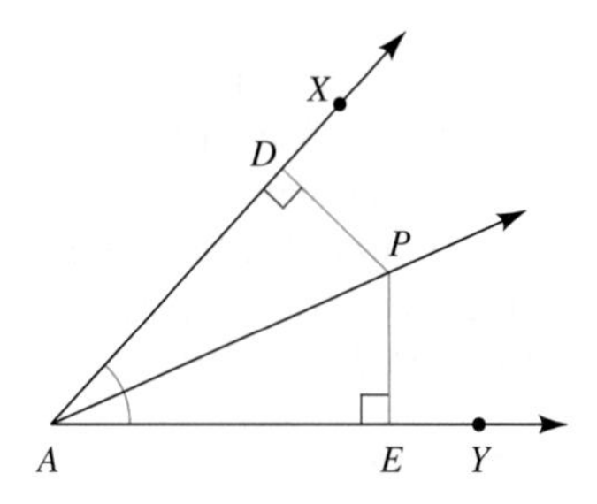

To justify this, we find two congruent triangles that have these segments as corresponding sides. The only triangles pictured are $\triangle ADP$ and $\triangle AEP$. Because $\overrightarrow{AP}$ is the angle bisector, $\angle DAP \cong \angle EAP$. Also, $\angle PDA$ and $\angle PEA$ are right angles and are thus congruent. $\overline{AP}$ is congruent to itself, so $\triangle PDA \cong \triangle PEA$ by AAS. Thus, $\overline{PD} \cong \overline{PE}$ because they are corresponding parts of congruent triangles PDA and PEA. Consequently, we have the following theorem.

Solution In Figure 10-58 the triangles are similar by AA because the tree and the stick both meet the ground at right angles and the angles formed by the Sun's rays are congruent (because the shadows are measured at the same time).

$$\frac{x}{40} = \frac{1}{2.5}$$
$$2.5x = 40$$
$$x = 16$$

The tree is 16 m tall. ▲

BRAIN TEASER

Two neighbors, Smith and Wheeler, plan to erect flagpoles in their yards. Smith wants a 10-ft pole, and Wheeler wants a 15-ft pole. To keep the poles straight while the concrete bases harden, guy wires are to be tied from the tops of the flagpoles to a fence post on the property lines and to the bases of the flagpoles, as shown in Figure 10-59. How high should the fence post be and how far apart should they erect the flagpoles for this scheme to work?

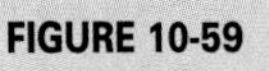

FIGURE 10-59

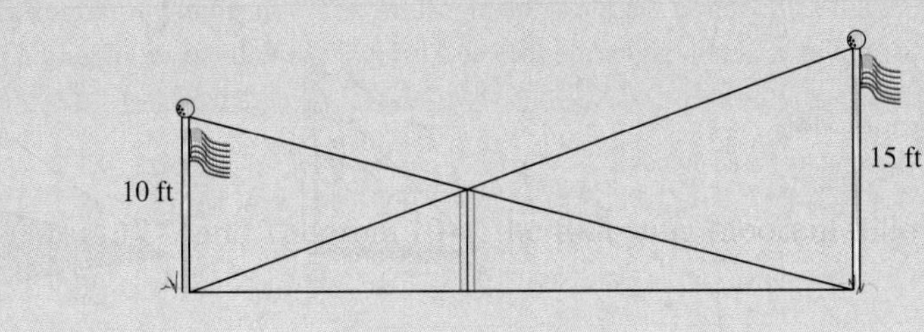

LABORATORY ACTIVITY

The device pictured in Figure 10-60 is a pantograph. It is used to draw enlarged versions of figures. In Figure 10-60 the red dots represent either brads or nuts and bolts. The strips are made of lath or cardboard and are rigid. A pointer at *D* is used to trace along an original figure, which causes the pencil at *F* to draw an enlarged version of the figure. Make or obtain a pantograph and experiment with enlarging figures. Explain how and why this works.

FIGURE 10-60

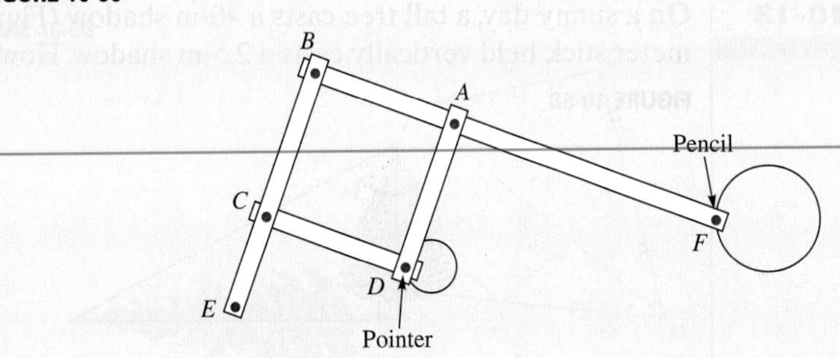

ONGOING ASSESSMENT 10-4

1. In the cartoon, the human character's size is increased. If we assume that the human figures are similar and the person's height has increased from 5′10″ to 6′11″, what is the scale factor?

2. Which of the following are always similar? Why?
   a. Any two equilateral triangles
   b. Any two squares
   c. Any two rectangles
   d. Any two rhombuses
   e. Any two circles
   f. Any two regular polygons
   g. Any two regular polygons with the same number of sides
3. Use grid paper to draw figures that have sides three times as large as the ones in the following figure and that are colored similarly.

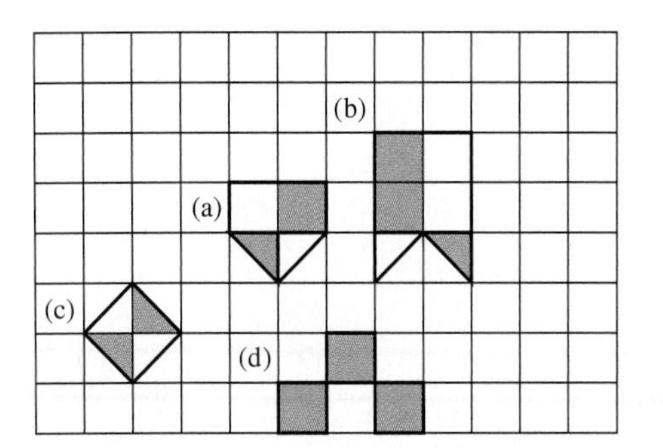

4. a. Construct a triangle with sides of lengths 4 cm, 6 cm, and 8 cm.
   b. Construct another triangle with sides of lengths 2 cm, 3 cm, and 4 cm.
   c. Make a conjecture about the similarity of triangles that have proportional sides only.
5. a. Construct a triangle with sides of lengths 4 cm and 6 cm and an included angle measuring 60°.
   b. Construct a triangle with sides of lengths 2 cm and 3 cm and an included angle measuring 60°.
   c. Make a conjecture about the similarity of triangles that have two sides proportional and congruent included angles.
6. a. Sketch two nonsimilar polygons for which corresponding angles are congruent.
   b. Sketch two nonsimilar polygons for which corresponding sides are proportional.
7. Examine several examples of similar polygons and make a conjecture concerning the ratio of their perimeters.
8. a. Which of the following pairs of triangles are similar? If they are similar, explain why.

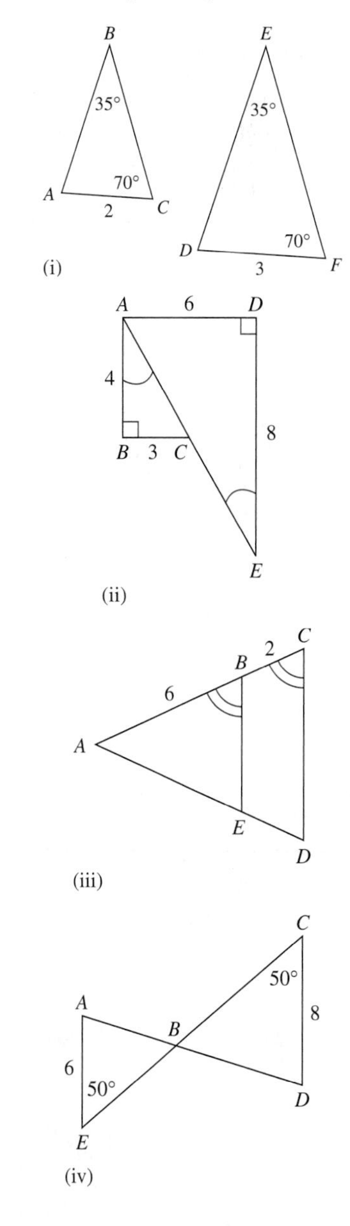

b. For each pair of similar triangles, find the scale factor of the sides of the triangles.

9. Assume that in the following figures the triangles in each part are similar and find the measures of the unknown sides.

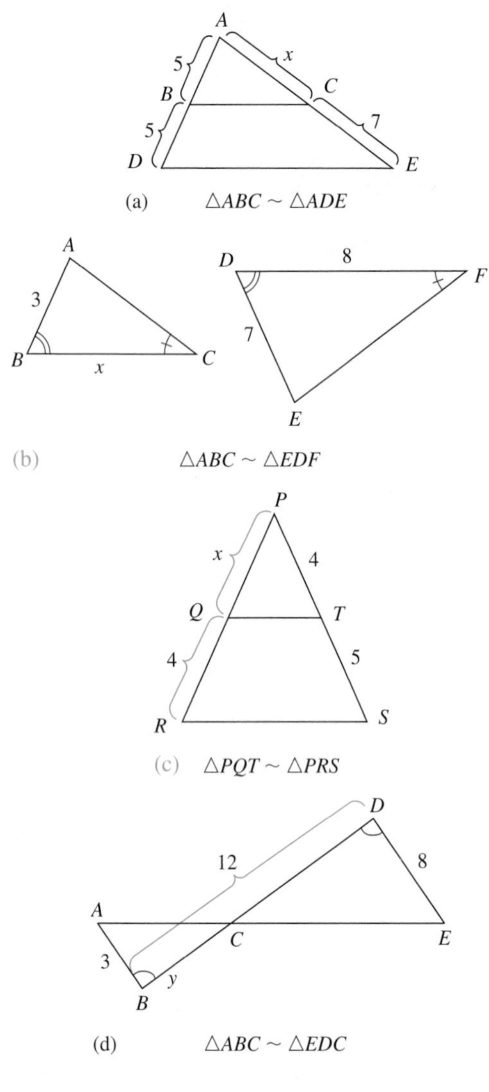

(a) $\triangle ABC \sim \triangle ADE$

(b) $\triangle ABC \sim \triangle EDF$

(c) $\triangle PQT \sim \triangle PRS$

(d) $\triangle ABC \sim \triangle EDC$

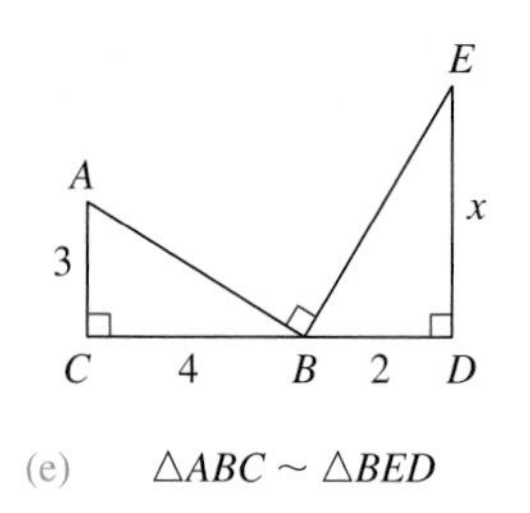

(e) $\triangle ABC \sim \triangle BED$

10. Given the following figure, use a compass and a straightedge to separate $\overline{AB}$ into five congruent pieces:

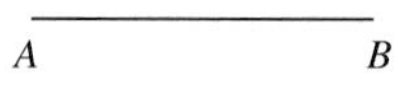

11. In the following right triangle ABC, $\overline{CD} \perp \overline{AB}$:

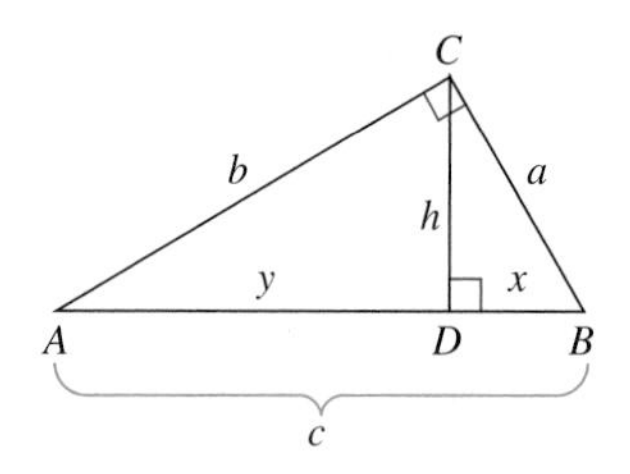

a. Find three pairs of similar triangles. Justify your answers.
b. Write the corresponding proportions for each set of similar triangles.
c. Use part (b) and the figure to show that $a^2 = xc$. Also argue that $b^2 = yc$.
d. Use part (c) to show that $a^2 + b^2 = c^2$. State this result in words using the legs and hypotenuse of a right triangle.

12. In the cartoon shown, if a smaller map were obtained and its scale were half the size of the original scale, would the distance to be traveled be different? Why?

13. Given segments of length a, b, and c as shown, construct a segment of length x so that $\frac{a}{b} = \frac{c}{x}$. (*Hint:* Use the idea shown in Figure 10-50.)

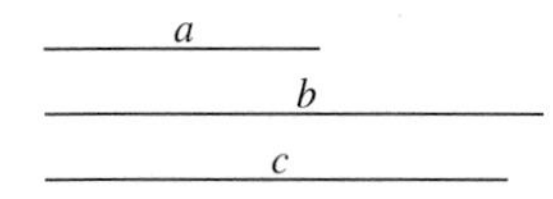

14. Triangle ABC is similar to triangle DEF with a side of triangle ABC that is 75% of its corresponding side in triangle DEF. Also, triangle GHI is similar to triangle DEF with a side of triangle GHI that is 32% of its corresponding side in triangle DEF.
   a. Are triangles ABC and GHI similar to each other? Why, or why not?
   b. What are the possible ratios of corresponding sides of triangles ABC and GHI?

15. Sketch with approximate measures two hexagons with corresponding sides proportional but so that they are not similar.

16. In the figure shown, prove that $ac = bd$.

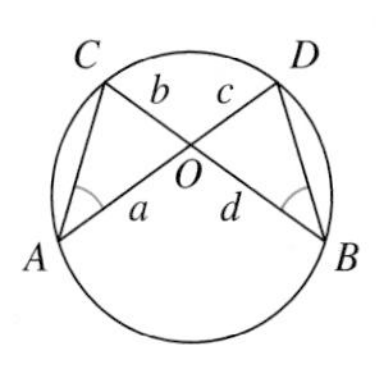

17. Sketch a pair of hexagons with corresponding angles congruent but the hexagons are not congruent.

18. Rep-tiles from Section 10-1 involved similar figures. Draw a rep-tile involving at least six parallelograms.

19. A **median** of a triangle connects a vertex to the midpoint of the opposite side. If all medians of $\triangle ABC$ shown intersect at point O, M and N are midpoints of $\overline{BC}$ and $\overline{AC}$, respectively,
   a. Prove that $\triangle ABO \sim \triangle MNO$.
   b. Find $\frac{AO}{OM}$ and $\frac{BO}{ON}$.

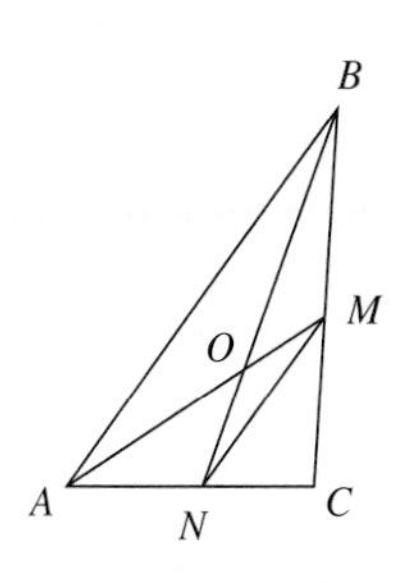

20. Explain whether any two arcs of the same circle are similar.

21. If one ruler is marked off in inches and another is marked off in centimeters, explain whether the two rulers must be similar.

22. Times is a computer font. Explain whether or not you think the Times font with sizes 16 and 36 are similar.

23. If you copy a page on a machine at 75%, you should get a similar copy of the page. What is the corresponding setting to obtain the original from the copy? Why?

24. In the following figure, find the distance AB across the pond using the similar triangles shown:

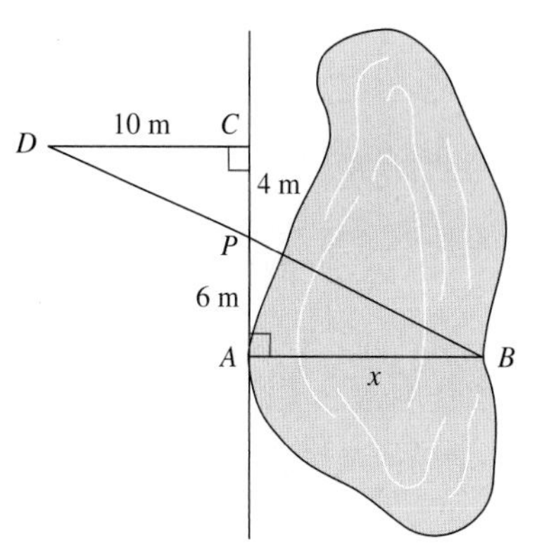

25. To find the height of a tree, a group of Girl Scouts devised the following method. A girl walks toward the tree along its shadow until the shadow of the top of her head coincides with the shadow of the top of the tree. If the girl is 150 cm tall, her distance to the foot of the tree is 15 m, and the length of her shadow is 3 m, how tall is the tree?

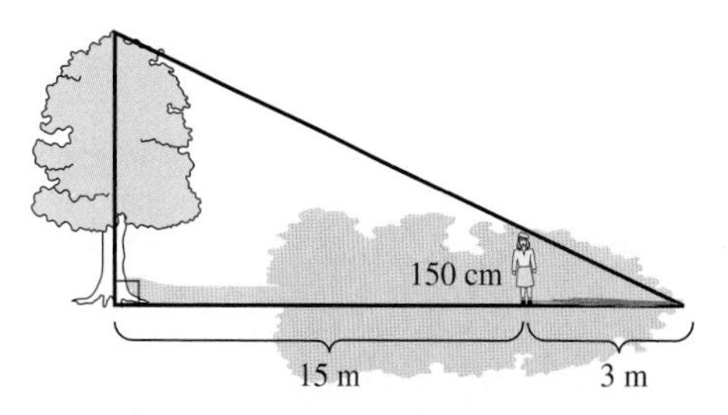

26. The angle bisector of one of the base angles in an isosceles triangle is constructed. This angle bisector partitions the original triangle into two isosceles triangles.
   a. What are the angle measures of the original triangle? (There are two possibilities.)
   b. Which, if any, triangles are congruent? Similar? Explain your answers.

27. Samantha wants to know how far above the ground the top of a leaning flagpole is. At high noon, when the Sun is directly overhead, the shadow cast by the pole is 7 ft long.

Samantha holds a plumb bob with a string 3 ft long up to the flagpole and determines that the point of the plumb bob touches the ground 13 in. from the base of the flagpole. How far above the ground is the top of the pole?

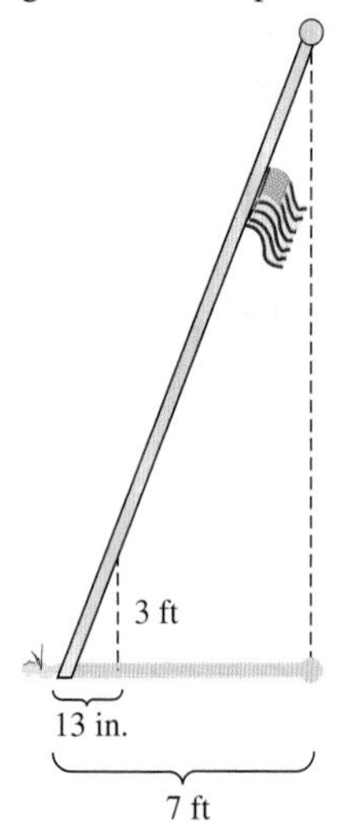

28. a. In the accompanying figure, $ABCD$ is a trapezoid. M is the midpoint of $\overline{AB}$. Through M, a line parallel to the bases has been drawn, intersecting $\overline{CD}$ at N. (i) Explain why N must be the midpoint of $\overline{CD}$ and (ii) express MN in terms of a and b, the lengths of the parallel sides of the trapezoid.

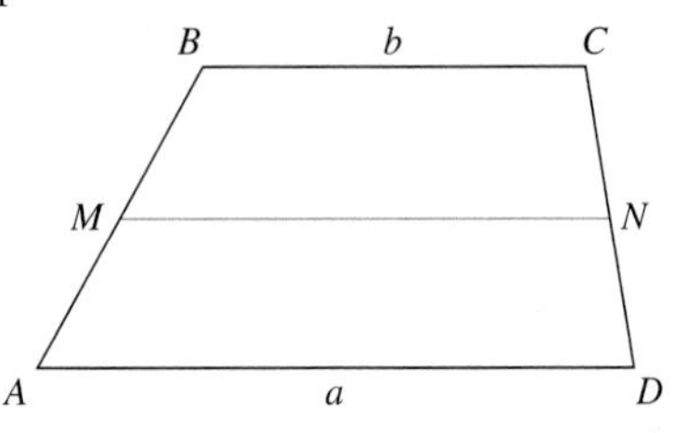

b. Denote MN by c. Use your answer to part (a) to show that b, c, a form an arithmetic sequence.

★ c. In the trapezoid $ABCD$, the lengths of the bases are a and b as shown. Side $\overline{AB}$ has been divided into 9 congruent segments. Through the endpoints of the segments, lines parallel to $\overline{AD}$ have been drawn. In this way, 8 new segments connecting the sides $\overline{AB}$ and $\overline{CD}$ have been created. Show that the sequence of 10 terms, starting with b, proceeding with the lengths of the parallel segments and ending with a, is an arithmetic sequence and find the sum of the sequence in terms of a and b.

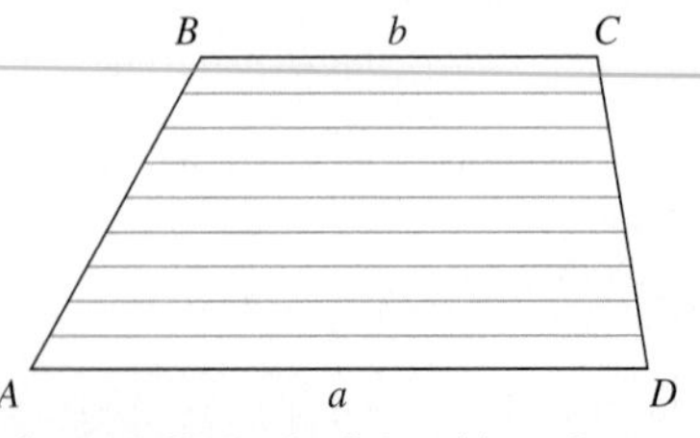

29. The midpoints M, N, P, Q of the sides of a quadrilateral $ABCD$ have been connected and an interior quadrilateral is obtained. We have shown that $MNPQ$ is a parallelogram. What is the most you can say about the kind of parallelogram $MNPQ$ is if $ABCD$ is

a. A rhombus?
b. A kite?
c. An isosceles trapezoid?
d. A quadrilateral that is neither a rhombus nor a kite but whose diagonals are perpendicular to each other?

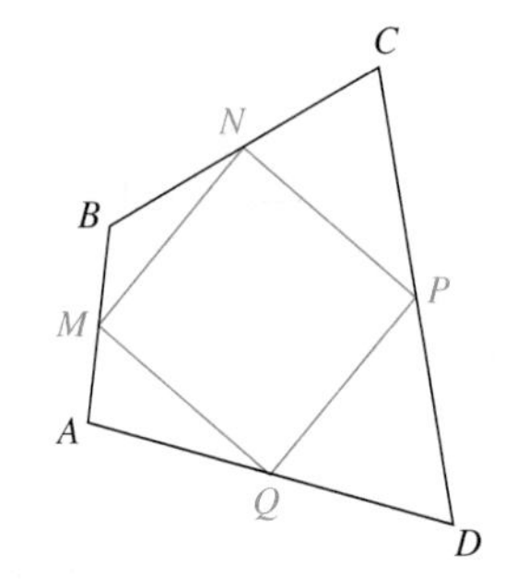

30. Explain why all pairs of regular octagons are similar.

31. If the ratio of corresponding sides of two similar triangles is $1/k$, what is the ratio of their perimeters? Why?

32. If you took cross sections of a typical ice cream cone parallel to the circular opening where the ice cream is usually placed, explain whether the cross sections would be similar.

33. Must all cross sections of a circular cylinder be similar? Why? Draw a sketch to illustrate your answer.

★ 34. A carpenter needs to string a wire between two boards $\overline{AB}$ and $\overline{BC}$ so that the colored wire is pictured as shown. How much wire is needed?

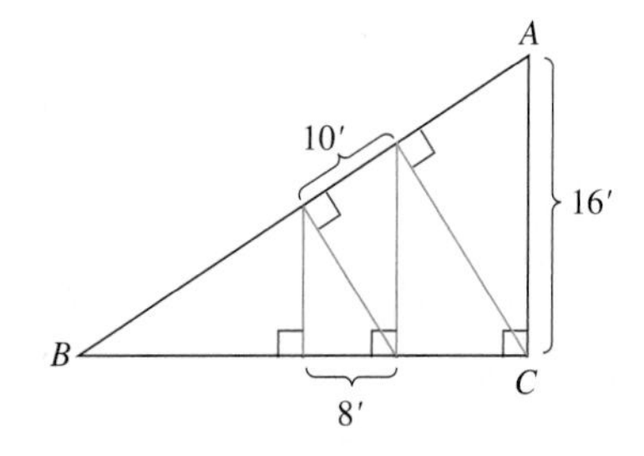

Communication

35. Do you think any two cubes are similar? Why or why not?

36. Architects frequently use scale drawings to construct models of projects. Are the models similar to the finished products? Why or why not?

37. Assuming the lines on an ordinary piece of notebook paper are parallel and equidistant, describe a method for using the paper to divide a piece of licorice evenly among 2, 3, 4, or 5 children. Explain why it works.

Open-Ended

38. Build two similar towers out of blocks.

a. What is the ratio of the heights of the towers?

b. What is the ratio of the perimeters of the bases of the towers?

39. On a sunny day, go outside and measure the heights of objects and their accompanying shadows. Use the data gathered by an entire class and plot graphs. Plot all data points on the same graph, with shadow lengths on the horizontal axis and object heights on the vertical axis. What do you observe? Why?

Cooperative Learning

40. A building was to be built on a triangular piece of property. The architect was given the approximate measurements of the angles of the triangular lot as 54°, 39°, and 87° and the lengths of two of the sides as 100 m and 80 m. When the architect began the design on drafting paper, she drew a triangle to scale with the corresponding measures and found that the lot was considerably smaller than she had been led to believe. It appeared that the proposed building would not fit. The surveyor was called. He confirmed each of the measurements and could not see a problem with the size. Neither the architect nor surveyor could understand the reason for the other's opinion.
    a. Have one person in your group play the part of the architect and explain why she felt she was correct.
    b. Have one person in the group explain the reason for the miscommunication.
    c. Have the group suggest a way to provide an accurate description of the lot.

Review Problems

41. If a person holds a mirror at arm's length and looks into it, is the image seen congruent to the original? Why or why not?
42. Given the following base of an isosceles triangle and the altitude to that base, construct the triangle:

Base

Altitude

43. Given the following length of a side of the triangle, construct an altitude of an equilateral triangle.

44. Write a paragraph describing how you could construct an isosceles right triangle when given the length of the hypotenuse of the 45°-45°-90° triangle.

45. Use a compass and a straightedge to draw a pair of obtuse vertical angles and the angle bisector of one of these angles. Extend the angle bisector. Does the extended angle bisector bisect the other vertical angle? Justify your answer.
46. Describe a minimal set of conditions that can be used to argue that two quadrilaterals are congruent.

Third International Mathematics and Science Study (TIMSS) Question

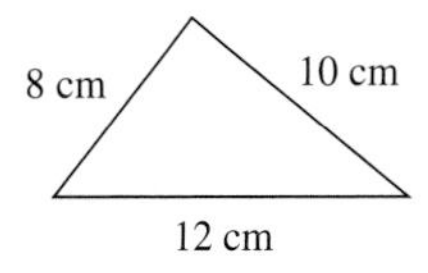

Which of the following triangles is similar to the triangle shown above?

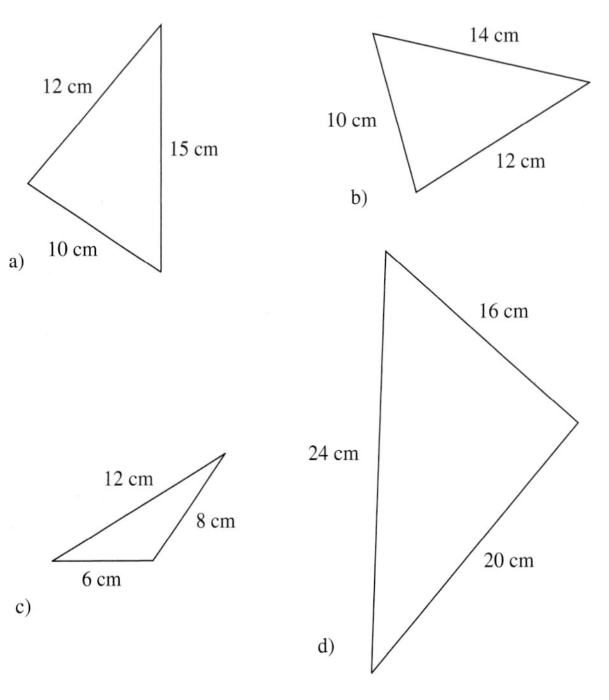

TIMSS, Grade 8, 2003

TECHNOLOGY CORNER

Work through GSP Lab 6 in Appendix III.

TECHNOLOGY CORNER

In 1975, Benoit Mandelbrot invented the word *fractal* to describe certain irregular and fragmented shapes. These shapes are such that if one looks at a small part of the shape, the small part resembles the larger shape. These shapes are somewhat like rep-tiles, although the smaller structures of fractals are not necessarily identical to the larger structure. (Recall that with rep-tiles, the smaller structures are similar and are identical in all aspects except size.) Fractal geometry was introduced for the purpose of modeling natural phenomena such as irregular coastlines, arteries and veins, the branching structure of plants, the thermal agitation of molecules in a fluid, and sponges. Figure 10-61 shows an example of a fractal—a computer-generated picture of the Mandelbrot set known as the "Tail of the Seahorse."

FIGURE 10-61

Earlier, in 1906, Helge von Koch came up with a curve that has infinite perimeter. To visualize this curve, we construct in Figure 10-62 a sequence of polygons $S_1, S_2, S_3, \ldots$ as follows:

a. S_1 is an equilateral triangle.
b. S_2 is obtained from S_1 by dividing each side of the triangle into three congruent parts and constructing on the middle part an equilateral triangle with the base removed.
c. S_3 is obtained from S_2 as S_2 was obtained from S_1.

We continue in a similar way to obtain the other polygons of the sequence in Figure 10-62(d) and (e). These polygons come closer and closer to a curve, called the *snowflake curve*, which is another example of a fractal. Use The Geometer's Sketchpad to construct Figure 10-62(a)–(c).

If S_1 has perimeter 3 units, what is the perimeter of S_2 and S_3? What do you think happens to the perimeter of the snowflakes as the number of sides of the polygons increases?

d. Suppose Figure 10-62(a) is a regular tetrahedron and Figure 10-62(b) is a regular tetrahedron with proportionately smaller regular tetrahedrons placed on the faces of the original similarly to the placement of the equilateral triangles shown. If this process continues, what do you think the resulting figure approaches?

FIGURE 10-62

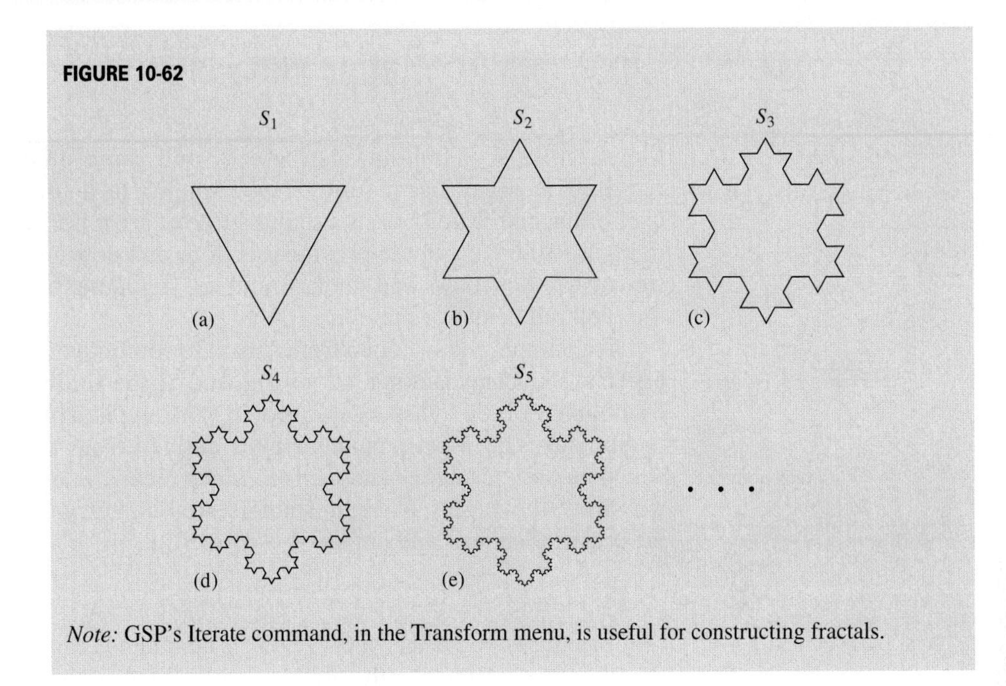

Note: GSP's Iterate command, in the Transform menu, is useful for constructing fractals.

BRAIN TEASER

A particular kaleidoscope is a right prism with an equilateral triangle as a base. A beam of light is reflected at a 60° angle from a point P on a side of the triangular base, as shown in Figure 10-63. The beam is reflected in the plane of the base to the different mirrored surfaces and continues bouncing off at 60° angles. Find the length of the path of the reflected light when it reaches the point at which it originated.

FIGURE 10-63

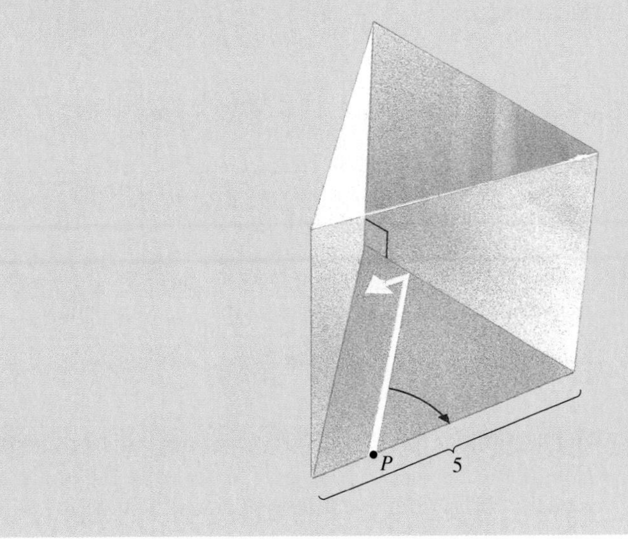

10-5 *Trigonometry Ratios via Similarity

FIGURE 10-64

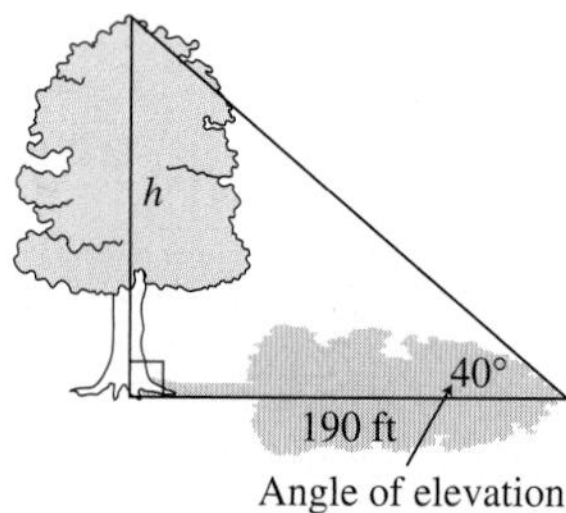

Measurements of buildings, structures, and some other objects are frequently required in many occupations. For example, in a tourist brochure advertising California, one region boasts a sequoia tree of great height, such as the one pictured in Figure 10-64. From the previous section, we know how to find the height of the tree by using shadows and similar triangles. Trigonometry might also be used to find the required height of the tree.

Trigonometry developed from a need to compute distances and angle measures, especially in map making, surveying, and range finding for artillery use. Today, trigonometry is an indispensable tool in many applied problems in both science and technology. The word **trigonometry** is derived from the Greek words *trigonom*, which means "triangle," and *metron*, which means "measurement." In this section, we study the basics of right triangle trigonometry, which has applications to measuring distances and angles.

trigonometry

▶ HISTORICAL NOTE Hipparchus of Nicea (ca. 180–125 BCE) is sometimes called the father of trigonometry because he is one of the first to try and organize a set of values associating arcs and chords of a circle. This work helped those who came later in the development of modern notions of trigonometry. ◀

The definition of trigonometric functions in a right triangle is based on properties of similar triangles. Earlier in this chapter, we saw that corresponding sides of similar triangles are proportional. Consequently, in two similar triangles the ratio of one side to another in one triangle will be the same as the ratio of the corresponding sides in the second triangle. Consider the two similar triangles in Figure 10-65. Both are right triangles, and each has an angle with measure 30°.

FIGURE 10-65

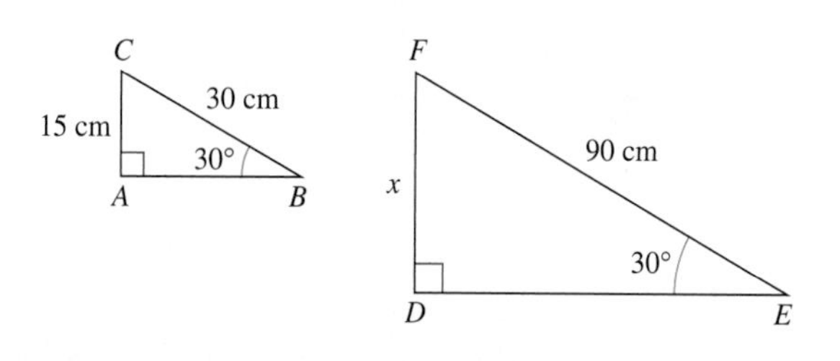

We could find x using the following proportion:

$$\frac{x}{90} = \frac{15}{30}$$

which implies that $x = 45$ cm. If we use this property, it follows that in any right triangle the ratio between the side opposite the angle with measure 30° and the side opposite the right angle, the **hypotenuse**, will be the same as that ratio in the smaller triangle in Figure 10-65, that is, $\frac{15}{30}$, or $\frac{1}{2}$.

hypotenuse

Suppose we use a protractor to construct a right triangle with an angle measuring 37° and measure both the hypotenuse and the side opposite this angle, as shown in Figure 10-66. The ratio between the side opposite the angle with measure 37° and the hypotenuse is $\frac{60}{100}$, or 0.6. Since measurements are approximate, the ratio also is only an approximation. This ratio is the "**sine** of 37°," or the **sin** 37°. Thus, in a right triangle having an angle with measure 37°,

sine ◆ sin

$$\sin 37° = \frac{\text{Length of side opposite the 37° angle}}{\text{Length of hypotenuse}}$$

FIGURE 10-66

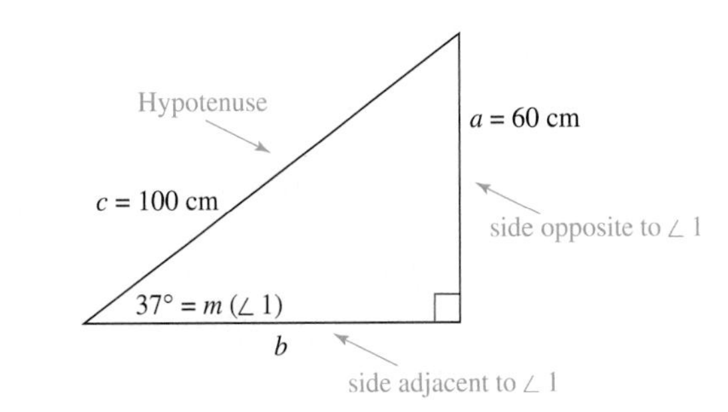

cosine (cos) ◆ tangent (tan)

Other ratios such as the **cosine (cos)** and **tangent (tan)** between lengths of sides in a right triangle are also useful and are defined in reference to Figure 10-67 as follows:

FIGURE 10-67

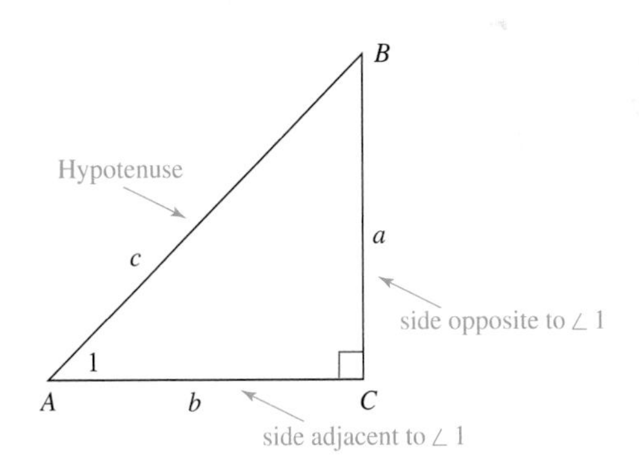

$$\sin(\angle 1) = \frac{a}{c} = \frac{\text{Length of opposite side}}{\text{Length of hypotenuse}}$$

$$\cos(\angle 1) = \frac{b}{c} = \frac{\text{Length of adjacent side}}{\text{Length of hypotenuse}}$$

$$\tan(\angle 1) = \frac{a}{b} = \frac{\text{Length of opposite side}}{\text{Length of adjacent side}}$$

REMARK Technically, we should write $\sin(m\angle 1)$, but this is typically shortened to $\sin(\angle 1)$.

TECHNOLOGY CORNER

Using The Geometer's Sketchpad, draw a right triangle ABC with right angle at C. Measure each side of the triangle and find the measures of $\angle ABC$ and $\angle BAC$. Use The Geometer's Sketchpad to find measures, calculate ratios, and build a table with headings similar to those in Table 10-2.

Table 10-2

Angle	Angle Meas.	Opp. Side Meas.	Adj. Side Meas.	Hyp. Meas.	$\frac{\text{Opp.}}{\text{Hypot.}}$	Sine of Angle	$\frac{\text{Adj.}}{\text{Hypot.}}$	Cosine of Angle	$\frac{\text{Opp.}}{\text{Adj.}}$	Tangent of Angle
$\angle ABC$										
$\angle BAC$										

1. What is true about the ratios you found using measurements and the sines, cosines, and tangents found automatically?
2. Explain whether you think it is necessary to find the measures of the sides of a triangle to determine the sine, cosine, and tangent ratios.

Students in upper middle grades use trigonometric ratios to find measures, as seen on the partial student page. Do problem 10.

Using Trigonometry Tables

In the seventeenth century, John Napier devised trigonometric tables to 14 decimal places. Today, with the help of a computer, we can find the values of the trigonometric functions to any desired degree of accuracy. Table 10-3 gives rounded trigonometric values for some angle measures.

Table 10-3

Degrees	sin	cos	tan
5	0.0872	0.9962	0.0875
10	0.1736	0.9848	0.1763
15	0.2588	0.9659	0.2679
20	0.3420	0.9397	0.3640
25	0.4226	0.9036	0.4663
30	0.5000	0.8660	0.5774
35	0.5736	0.8192	0.7002
40	0.6428	0.7660	0.8391
45	0.7071	0.7071	1.0000

SAMPLE SCHOOL BOOK PAGE:

BUILDING A RAMP

Working on the Module Project

Building a Ramp

Ramps and Angles An engineer designing a ramp often has to be concerned about the angle that the ramp makes with the ground. For example, the American National Standards Institute (ANSI) limits the possible angles for a wheelchair ramp like the one shown.

ANSI says that the measure of ∠*A* can be at most 4.76°.

h = *height*

b = *length of base*

A

9 Write an equation involving tangent that relates ∠*A*, *h*, and *b*.

10 Michelle is designing a wheelchair ramp that is supposed to reach a door 2 ft above the ground and have an angle of 4.76°. What should the length of the ramp's base be?

11 Before designing your model ramp, you may want to know some possible angles, heights, and base lengths your ramp could have.

a. Copy the table below. Replace each **?** with the length of the base of a ramp that has the given angle and height.

Base Lengths of Ramps for Different Angles and Heights

Angle \ Height	6 in.	8 in.	10 in.	12 in.
10°	?	?	?	?
20°	?	?	?	?
30°	?	?	?	?
40°	?	?	?	?

b. If you increase the height of a ramp and keep the angle constant, what happens to the base length? Explain.

c. If you increase the angle of a ramp and keep the height constant, what happens to the base length? Explain.

Source: Math Thematics, Book 3, McDougal Littell, 2005 (p. 367).

EXAMPLE 10-14 Find the height of the sequoia tree pictured in Figure 10-68.

FIGURE 10-68

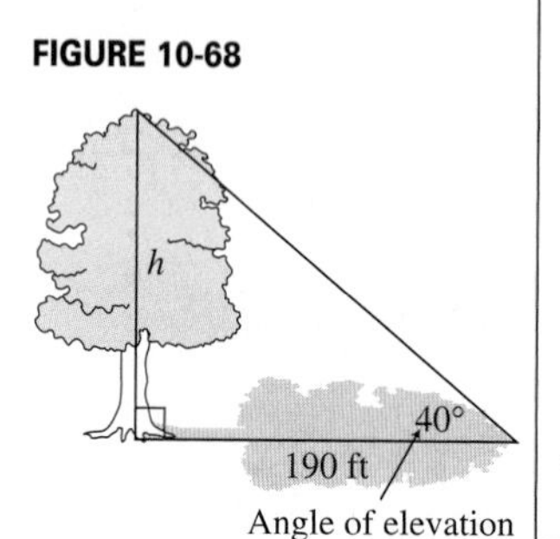

Solution To find the height of the tree, we use the following trigonometric ratio and Table 10-3:

$$\tan 40° = \frac{h}{190}$$
$$190(\tan 40°) = h$$
$$190(0.8391) \doteq h$$
$$159.4 \doteq h$$

Thus, the height of the tree is approximately 159.4 ft. ▲

Using Trigonometric Ratios on a Calculator

Tables of trigonometric ratios were essential in all parts of the world until calculators became readily available. Trigonometry in right triangles as we use it today is primarily done using a scientific calculator, graphing calculator, or computer. For example, in a given right triangle with an acute angle of 40°, we can use the [SIN], [COS], and [TAN] buttons to find the respective values of 0.643, 0.766, and 0.839 assigned to a 40° angle.

Each of the trigonometry buttons is a function button that requires an angle measure input either in degrees or radians. Because we use degrees in this text, make sure your calculator is set to use degrees and not radians. One method for telling whether your calculator is in the degree mode is to find the sine of 30°. If your answer is 0.5, or .5, then the calculator is in the proper mode. If not, consult a manual to see what you must do to put your calculator in degree mode.

REMARK Check your calculator to determine if you should use [SIN] [3] [0] [)] or [3] [0] [SIN] or [SIN] [3] [0].

EXAMPLE 10-15 Use a calculator to solve for x in Figure 10-69.

FIGURE 10-69

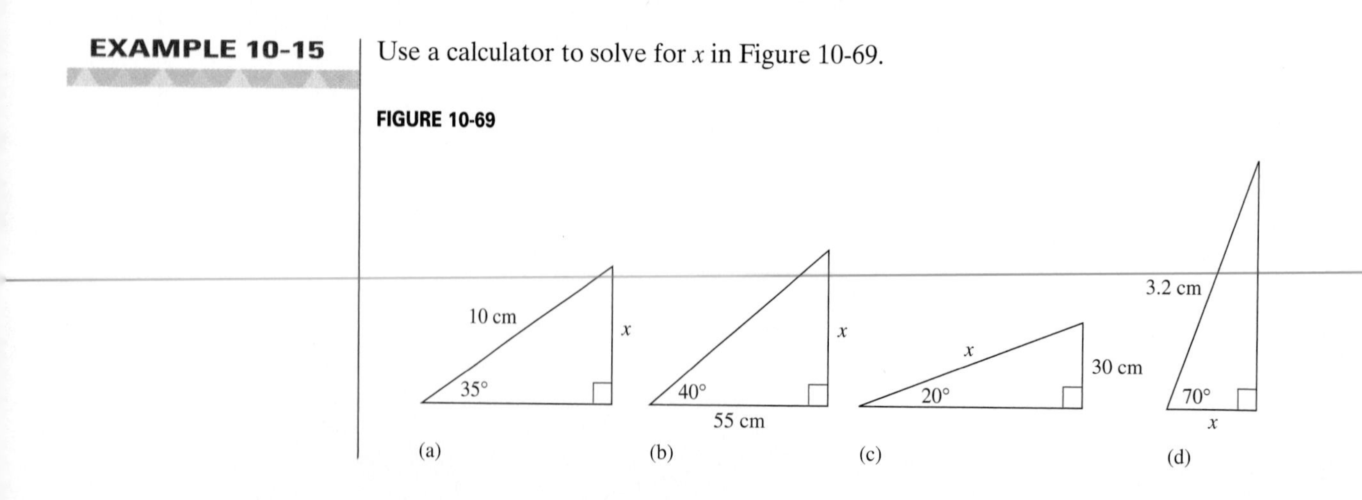

Solution **a.** $\frac{x}{10} = \sin 35°$

$x = 10 \sin 35°$

To find the answer on your calculator, you may have to press the following keys:

[1] [0] [×] [SIN] [3] [5] [)]

The resulting display is 5.7357644, or approximately 5.7 cm.

b. $\frac{x}{55} = \tan 40°$

$x = 55 \tan 40°$

$x \doteq 46.1505$, or approximately 46.2, cm

c. $\frac{30}{x} = \sin 20°$

$x = \frac{30}{\sin 20°}$

$x \doteq 87.714$, or approximately 87.7, cm

d. $\frac{x}{3.2} = \cos 70°$

$x = 3.2 \cos 70°$

$x \doteq 1.094$, or approximately 1.1, cm

▲

NOW TRY THIS 10-15

In Example 10-15, use Figure 10-69(a) to find x using the cosine function instead of the sine function.

Sometimes the lengths of the sides of a right triangle are given and we need to find the measures of the angles of the triangle. If, for example, a triangle has sides measuring 3 m, 4 m, and 5 m, as shown in Figure 10-70, we can find the measure of angle A by first finding the $\sin(\angle A)$. We have $\sin(\angle A) = \frac{4}{5}$. The [$SIN^{-1}$] key on your calculator (or possibly the [INV] key used with the [SIN] key) can be used to find the measure of the angle as follows:

[SIN^{-1}] [4] [÷] [5] [)] [=]

FIGURE 10-70

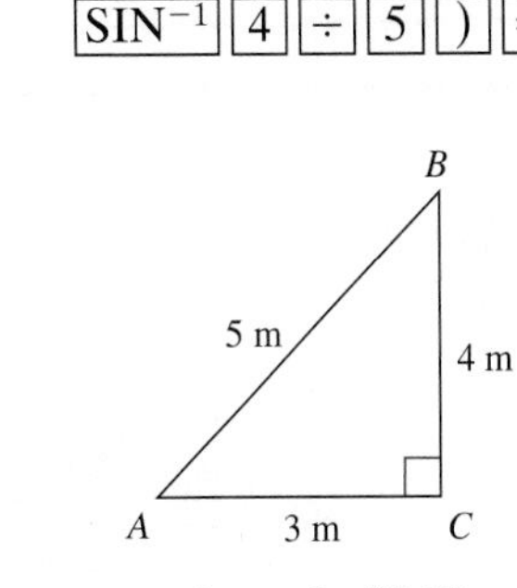

Thus, the angle has measure approximately 53.1°.

EXAMPLE 10-16

If x is an angle measure, solve for x in each part of Figure 10-71:

FIGURE 10-71

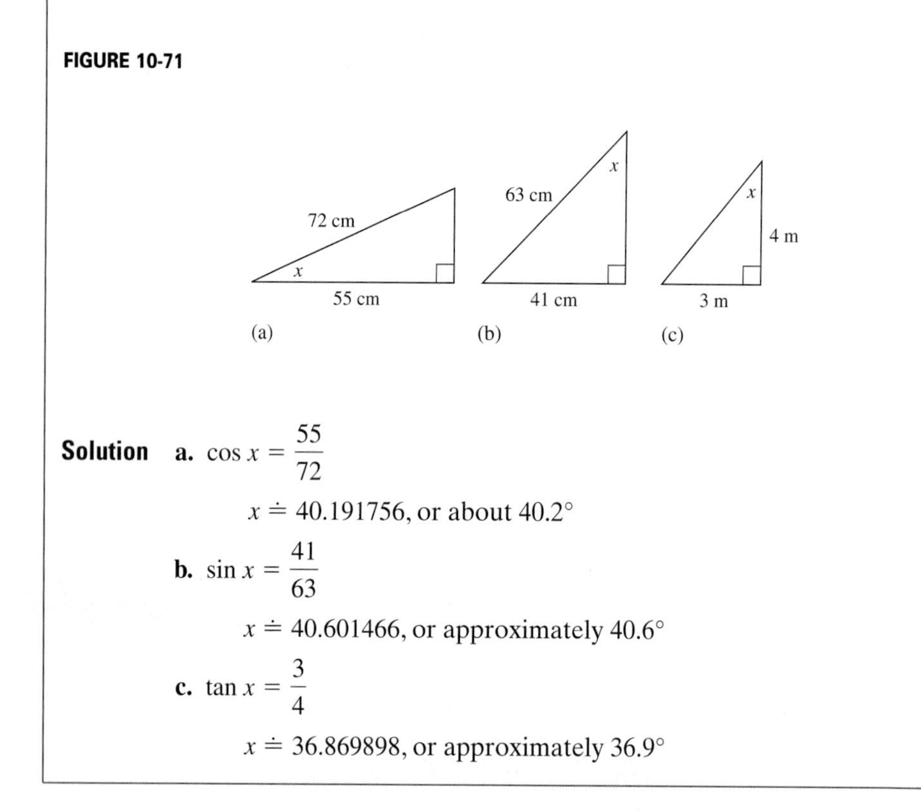

Solution **a.** $\cos x = \dfrac{55}{72}$

$x \doteq 40.191756$, or about $40.2°$

b. $\sin x = \dfrac{41}{63}$

$x \doteq 40.601466$, or approximately $40.6°$

c. $\tan x = \dfrac{3}{4}$

$x \doteq 36.869898$, or approximately $36.9°$

Finding Measurements Using Trigonometric Ratios

Just as the angle of elevation was used to find the height of the sequoia tree, we can use an *angle of depression* to find measures, as demonstrated in Example 10-17.

EXAMPLE 10-17

From the top of Mount Sentinel, the measure of the angle of depression of the administration building is 18°, as illustrated in Figure 10-72. If the top of Mount Sentinel is 1575 ft, how far through the air is the top of the mountain from the base of the building?

FIGURE 10-72

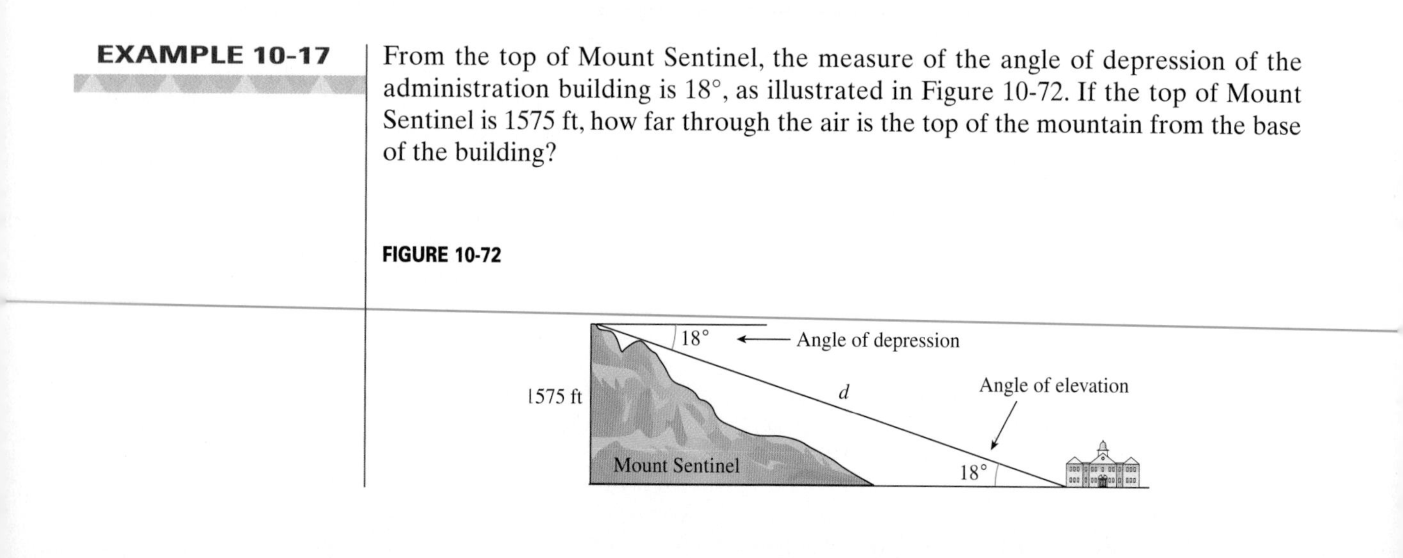

(sin A)(sin A). If sir

b are legs and the h
ric equation in term
final result.

B
a
C

Communication

18. What does your ca
Why?

origin ◆ x-axis ◆

al
x-coordinate ◆ ordin
y-coor

Solution In Figure 10-72, the measure of the angle of elevation must equal the measure of the angle of depression (why?). Using trigonometric ratios, we have the following:

$$\sin 18° = \frac{1575}{d}$$

$$d = \frac{1575}{\sin 18°}, \text{ or approximately } 5097 \text{ ft}$$

Thus, the air distance is approximately 5097 ft.

EXAMPLE 10-18

To measure the height of a flagpole on top of a building, a surveyor measures the distance to the building and the two angles from point B, as shown in Figure 10-73. Find the height of the flagpole.

FIGURE 10-73

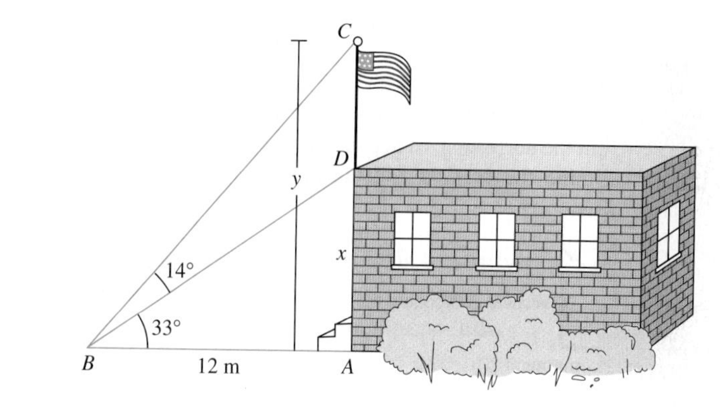

Solution We need to find CD in Figure 10-73. Because CD is not a length of a side of a right triangle, but y and x whose difference is CD are, we may find x and y first. The solution then is $CD = y - x$. In triangle BDA, we have the following:

$$\tan 33° = \frac{x}{12}$$

$$x = 12 \tan 33°$$

In triangle BCA, we have the following:

$$\tan(14 + 33)° = \frac{y}{12}$$

$$y = 12 \tan(14 + 33)°$$

Therefore, $CD = 12 \tan(14 + 33)° - 12 \tan 33°$. Using the calculator, we find 5.075 on the display. Thus, the height of the pole atop the building is about 5.1 m.

NOW TRY THIS 10-16

Find the relationships among the sin($\angle A$), cos($\angle A$), and tan($\angle A$).

1. For each of the followi
each x length to the nea

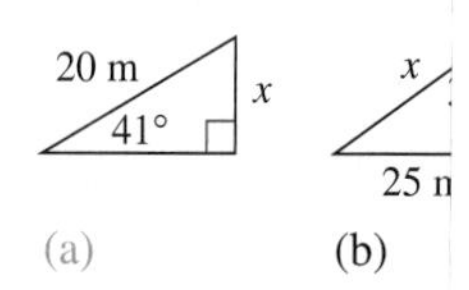

2. The angle of elevation
how far the base of the l
3. A diagonal is drawn i
the sine, cosine, and ta
diagonal and a side.
4. Determine the height
long on level ground w
Sun is 50°.
5. Vectors are used in scie
nitudes and directions.
angle measuring 38° fro
following figure, dete
components as pictured

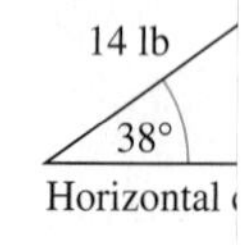

6. To find the length of a
and made the following

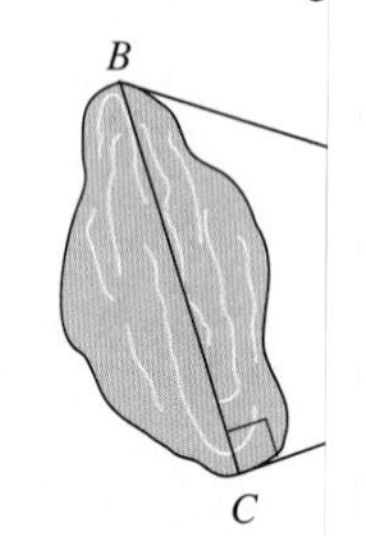

a. What is the measure
tion skills and your
b. What is the length o
7. How many feet of cable
a 30-ft vertical pole if
sume the ground is leve
8. As a plane takes off, it
and rises at an angle m
a. What is the vertical
b. After it has flown
horizontally?
9. A gutter cleaner want
ground. Find the lengt

FIGURE 10-80

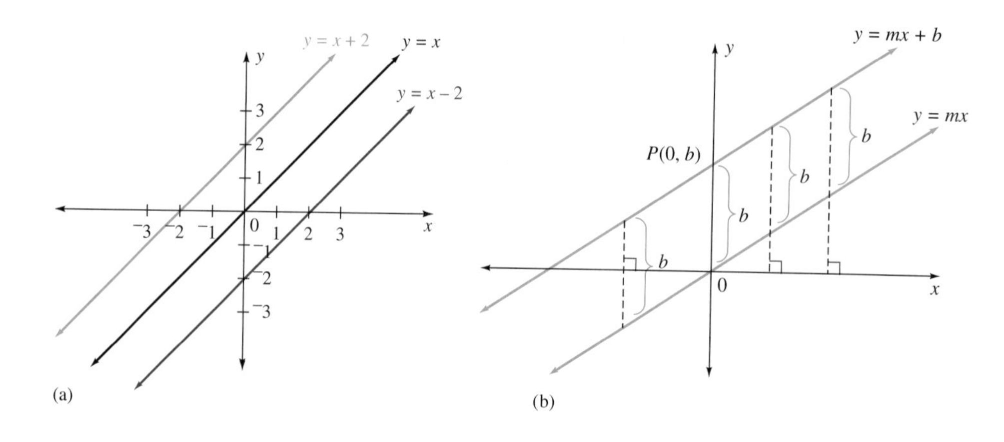

The graph of $y = mx + b$ in Figure 10-80(b) crosses the y-axis at point $P(0, b)$. The value of y at the point of intersection of any line with the y-axis is the **y-intercept**. Thus, b is the y-intercept of $y = mx + b$; this form of linear equation is the **slope-intercept form**. Similarly, the value of x at the point of intersection of a line with the x-axis is the **x-intercept**.

EXAMPLE 10-21

Given the equation $y - 3x = {}^-6$, do the following:

a. Find the slope of the line.
b. Find the y-intercept.
c. Find the x-intercept.
d. Sketch the graph of the equation.

Solution

a. To write the equation in the form $y = mx + b$, we add $3x$ to both sides of the given equation to obtain $y = 3x + ({}^-6)$. Hence, the slope is 3.

b. The form $y = 3x + ({}^-6)$ shows that $b = {}^-6$, which is the y-intercept. (The y-intercept can also be found directly by substituting $x = 0$ in the equation and finding the corresponding value of y.)

c. The x-intercept is the x-coordinate of the point where the graph intersects the x-axis. At that point, $y = 0$. Substituting 0 for y in $y = 3x - 6$ gives 2 as the x-intercept.

d. The y-intercept and the x-intercept are located at $(0, {}^-6)$ and $(2, 0)$, respectively, on the line. We may plot these points and draw the line through them to obtain the desired graph in Figure 10-81.

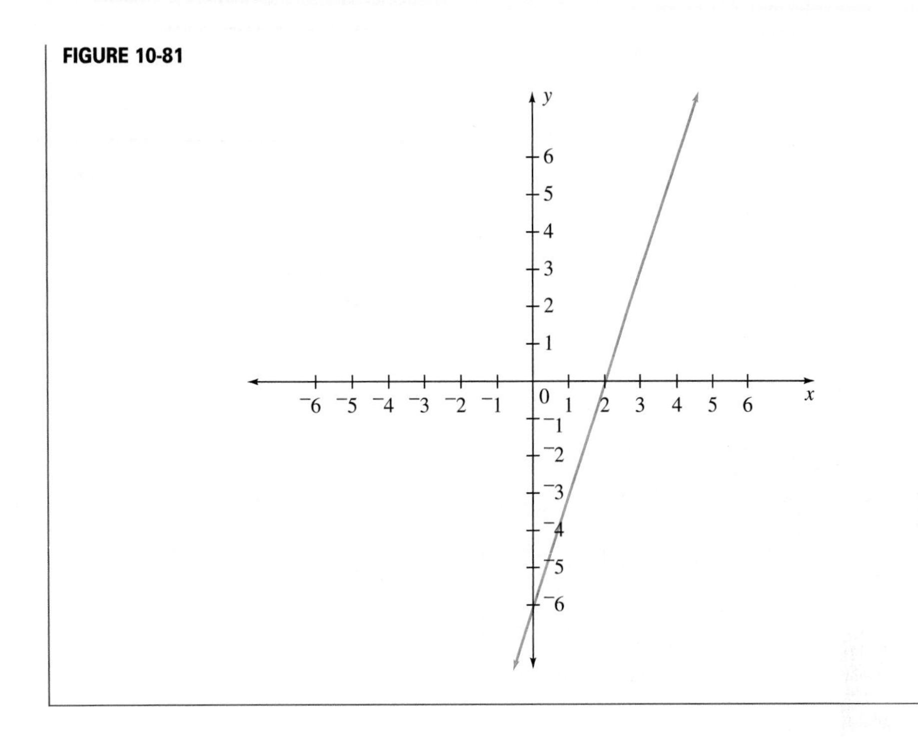

Recall that any vertical line has equation $x = a$ for some real number a. This equation cannot be written in slope-intercept form. In general, *every line has an equation of the form either* $y = mx + b$ *or* $x = a$. Any equation that can be put in one of these forms is a **linear equation**.

linear equation

> **Equation of a Line**
>
> Every line has an equation of the form either $y = mx + b$ or $x = a$, where m is the slope and b is the y-intercept.

Using Similar Triangles to Determine Slope

We have defined the slope of a line with equation $y = mx + b$ to be m. The slope is a measure of steepness of a line. A different way to discuss the steepness of a line is to consider the rate of change in y-values in relation to their corresponding x-values. In Figure 10-82, line k has a greater rate of change than line ℓ. In other words, line k rises higher than line ℓ for the same horizontal run.

FIGURE 10-82

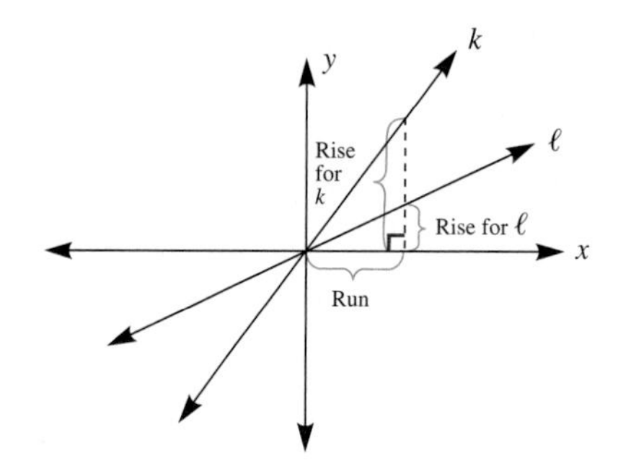

Thus, we could express the rate of change, the steepness, or the slope, as the ratio $\frac{\text{Change in } y\text{-values}}{\text{Corresponding change in } x\text{-values}}$. The rate is frequently generalized as $\frac{\text{Rise}}{\text{Run}}$. In Figure 10-83, right triangles have been constructed and shaded on several lines. In each triangle, the horizontal side is the run and the vertical side is the rise.

FIGURE 10-83

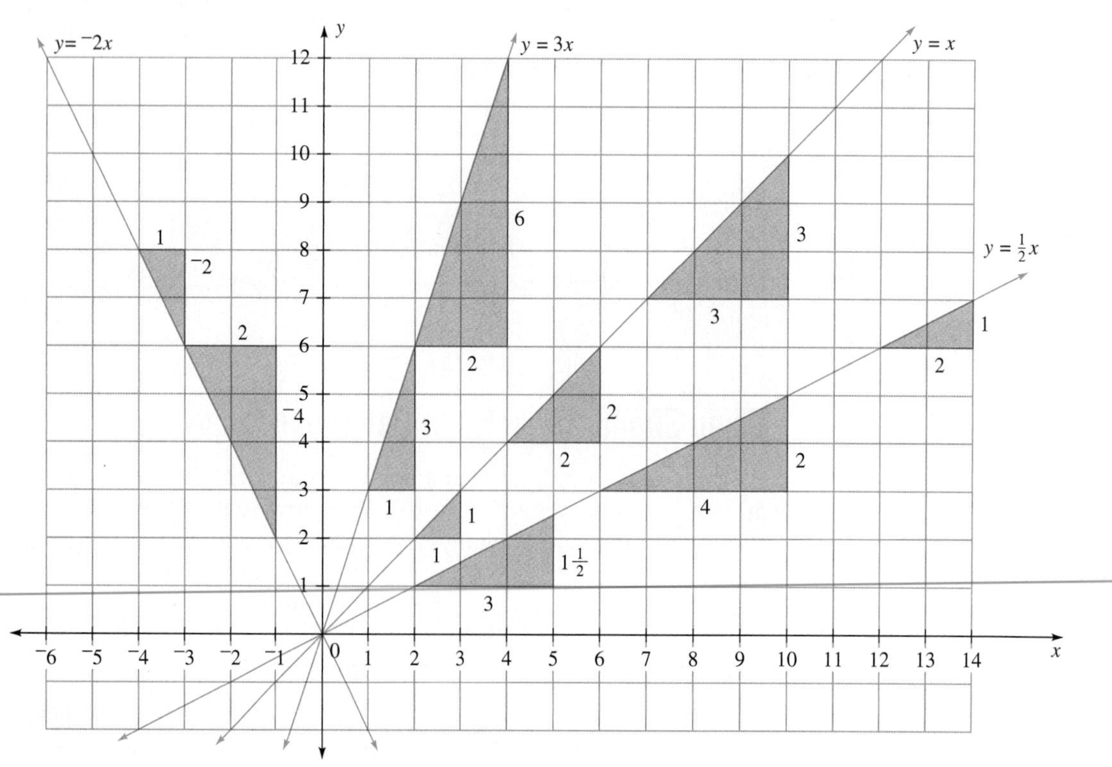

The slope of each line in Figure 10-83 can be calculated as the rise over the run in any of the shaded triangles, with hypotenuse (the side opposite the right angle) along the given line. To test this fact, notice that

$$\text{for } y = \frac{1}{2}x, \quad m = \frac{\text{Rise}}{\text{Run}} = \frac{1\frac{1}{2}}{3} = \frac{2}{4} = \frac{1}{2}$$

$$\text{for } y = x, \quad m = \frac{\text{Rise}}{\text{Run}} = \frac{1}{1} = \frac{2}{2} = \frac{3}{3} = 1$$

$$\text{for } y = 3x, \quad m = \frac{3}{1} = \frac{6}{2} = 3$$

$$\text{for } y = {}^{-}2x \quad m = \frac{{}^{-}2}{1} = {}^{-}2$$

Using the previous notion, we find that the slope of a line $\overleftrightarrow{AB}$ is the change in y-coordinates divided by the corresponding change in x-coordinates of any two points on $\overleftrightarrow{AB}$. The difference $x_2 - x_1$ is the *run*, and the difference $y_2 - y_1$ is the *rise*. The slope formula can be interpreted with coordinates, as shown in Figure 10-84. The ratio $\frac{y_2 - y_1}{x_2 - x_1}$ is always the same, regardless of which two points on a given non-vertical line are chosen.

FIGURE 10-84

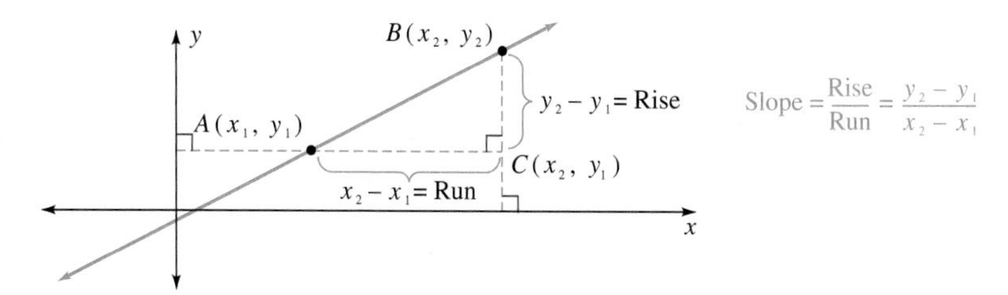

The discussion of slope is summarized in the following formula.

Slope Formula

Given two points $A(x_1, y_1)$ and $B(x_2, y_2)$ with $x_1 \neq x_2$, the slope m of the line $\overleftrightarrow{AB}$ is

$$m = \frac{y_2 - y_1}{x_2 - x_1}.$$

REMARK By multiplying both the numerator and the denominator on the right side of the slope formula by ${}^{-}1$, we obtain

$$m = \frac{y_2 - y_1}{x_2 - x_1} = \frac{(y_2 - y_1)({}^{-}1)}{(x_2 - x_1)({}^{-}1)} = \frac{y_1 - y_2}{x_1 - x_2}.$$

This shows that although it does not matter which point is named (x_1, y_1) and which is named (x_2, y_2), *the order of the coordinates in the subtraction must be the same.*

NOW TRY THIS 10-18

a. Use the slope formula to find the slope of any horizontal line.

b. What happens when we attempt to use the slope formula for a vertical line? What is your conclusion about the slope of a vertical line?

When a line is inclined downward from the left to the right, the slope is negative. This is illustrated in Figure 10-85, where the graph of the line $y = {}^{-}2x$ is shown. The slope of line $y = {}^{-}2x$ can be calculated as $\frac{\text{Rise}}{\text{Run}} = \frac{{}^{-}4}{2} = \frac{{}^{-}2}{1}$.

FIGURE 10-85

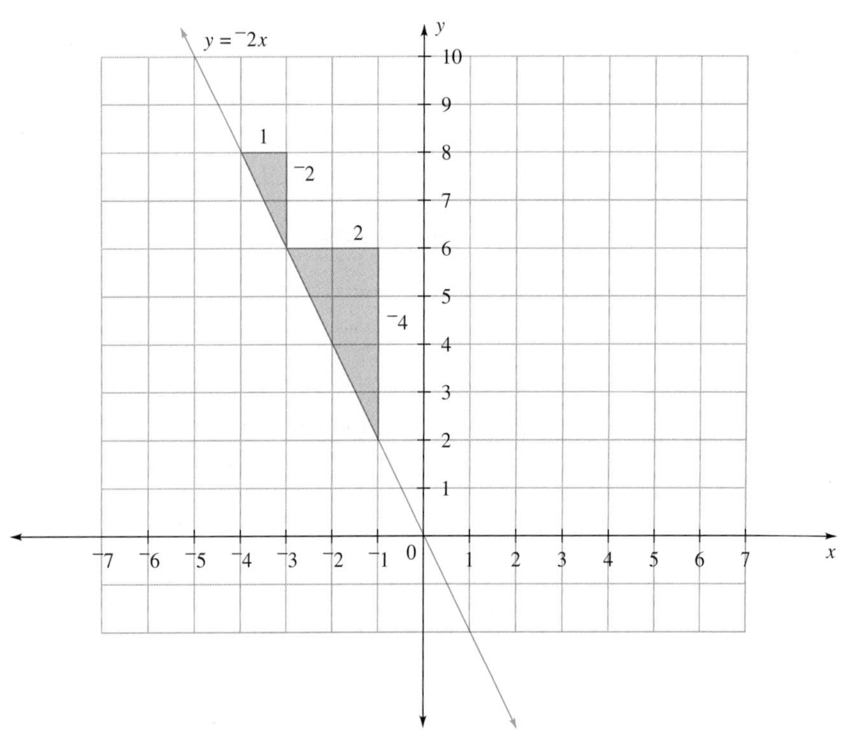

EXAMPLE 10-22

a. Given $A(3, 1)$ and $B(5, 4)$, find the slope of $\overleftrightarrow{AB}$.

b. Find the slope of the line passing through the points $A({}^{-}3, 4)$ and $B({}^{-}1, 0)$.

Solution **a.** $m = \frac{4 - 1}{5 - 3} = \frac{3}{2}$, or $\frac{1 - 4}{3 - 5} = \frac{{}^{-}3}{{}^{-}2} = \frac{3}{2}$

b. $m = \frac{4 - 0}{{}^{-}3 - ({}^{-}1)} = \frac{4}{{}^{-}2} = {}^{-}2$, or $\frac{0 - 4}{{}^{-}1 - ({}^{-}3)} = \frac{{}^{-}4}{2} = {}^{-}2$

Given two points on a nonvertical line, we can use the slope formula to find the slope of the line and its equation. This is demonstrated in the following example.

EXAMPLE 10-23 In Figure 10-86, the points $(^{-}4, 0)$ and $(1, 4)$ are on the line ℓ. Find:

FIGURE 10-86

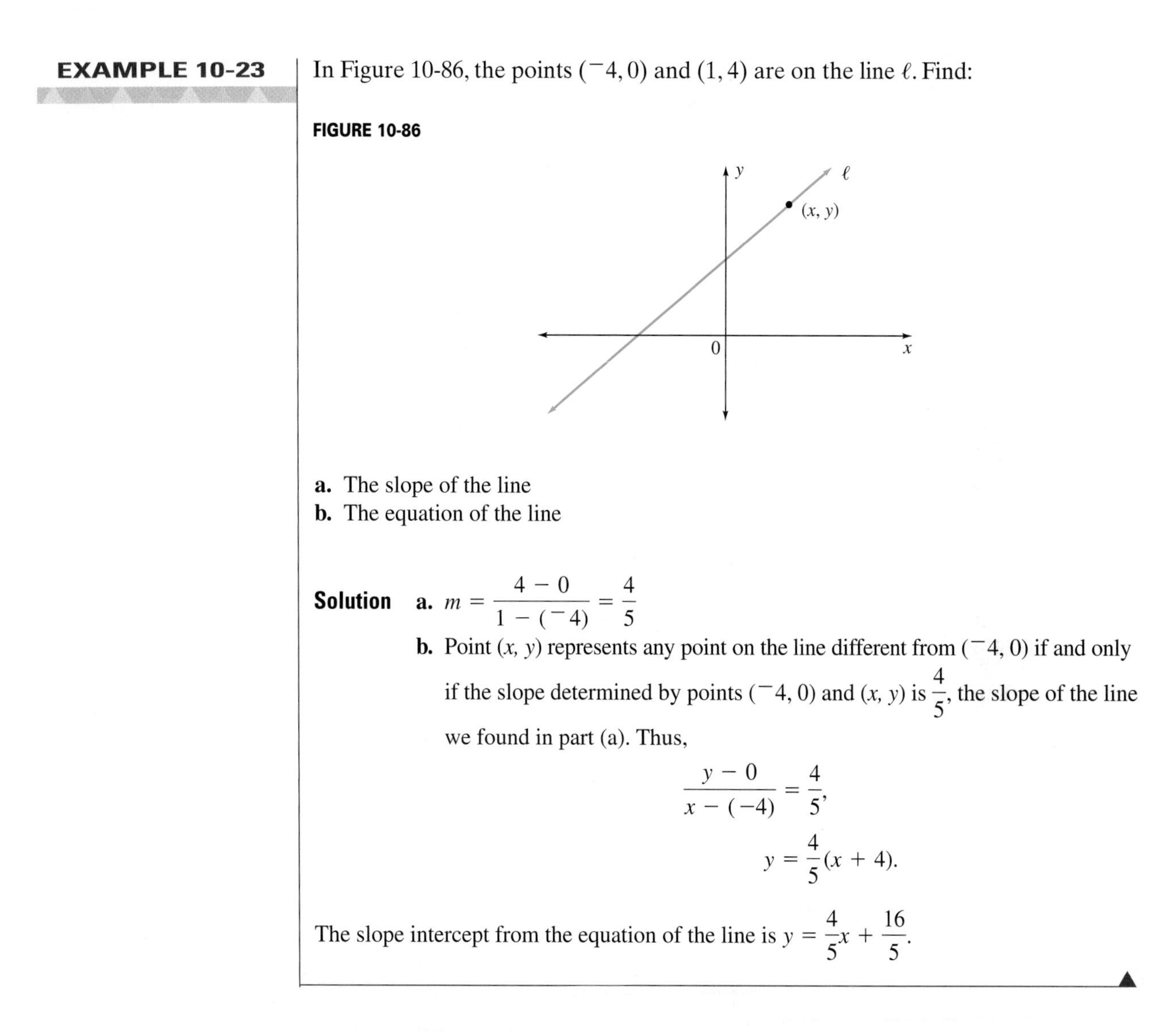

a. The slope of the line
b. The equation of the line

Solution **a.** $m = \dfrac{4 - 0}{1 - (^{-}4)} = \dfrac{4}{5}$

b. Point (x, y) represents any point on the line different from $(^{-}4, 0)$ if and only if the slope determined by points $(^{-}4, 0)$ and (x, y) is $\frac{4}{5}$, the slope of the line we found in part (a). Thus,

$$\frac{y - 0}{x - (-4)} = \frac{4}{5},$$

$$y = \frac{4}{5}(x + 4).$$

The slope intercept from the equation of the line is $y = \frac{4}{5}x + \frac{16}{5}$.

Systems of Linear Equations

The mathematical descriptions of many problems involve more than one equation, each having more than one unknown. To solve such problems, we must find a common solution to the equations, if it exists. An example is given on the following student page. Answer question 21.

SAMPLE SCHOOL BOOK PAGE:

MODELING LINEAR CHANGE

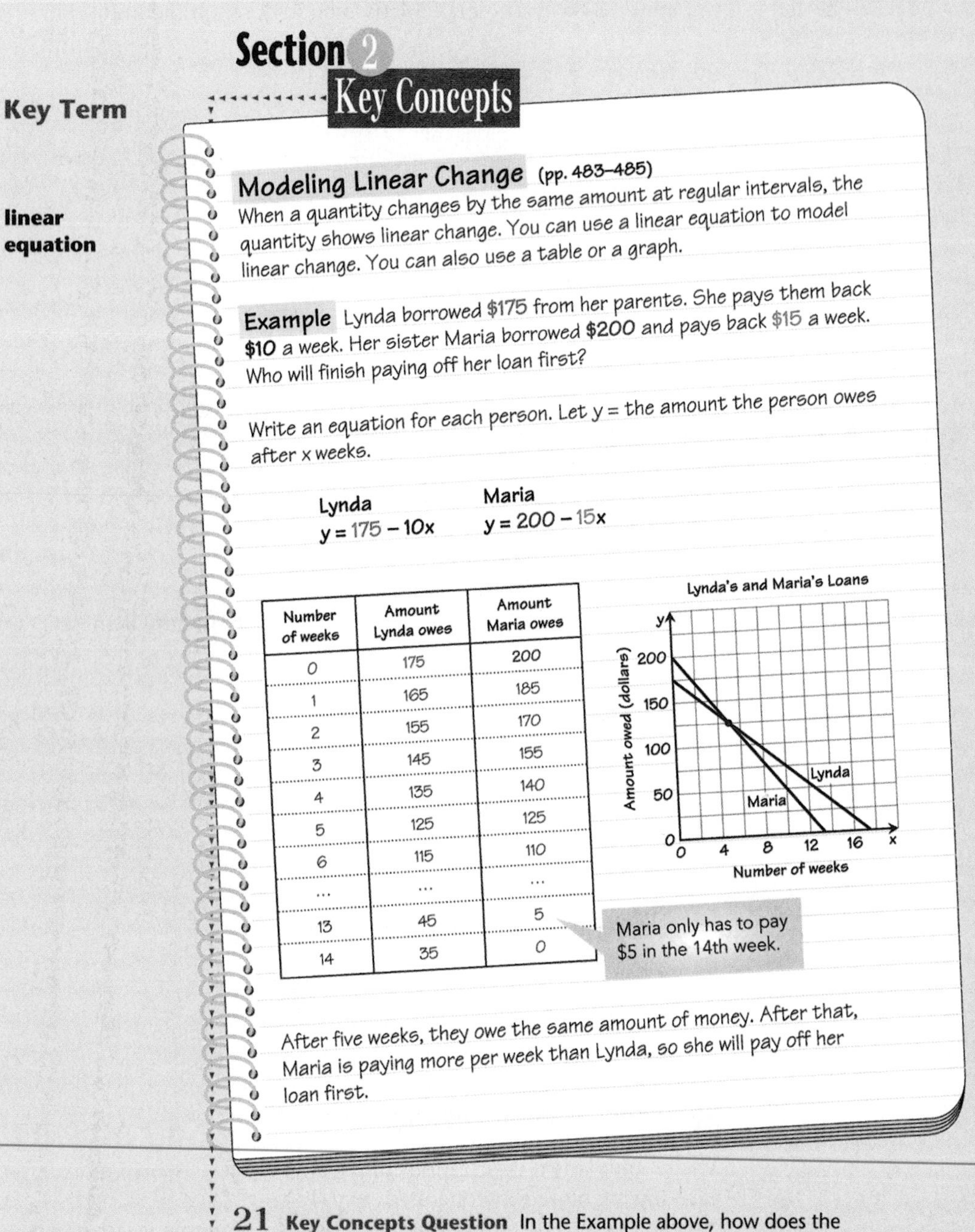

Section 2
Key Concepts
Key Term
linear equation
Modeling Linear Change (pp. 483–485)
When a quantity changes by the same amount at regular intervals, the quantity shows linear change. You can use a linear equation to model linear change. You can also use a table or a graph.
Example Lynda borrowed $175 from her parents. She pays them back $10 a week. Her sister Maria borrowed $200 and pays back $15 a week. Who will finish paying off her loan first?
Write an equation for each person. Let y = the amount the person owes after x weeks.
Lynda
y = 175 – 10x
Maria
y = 200 – 15x
Number of weeks
Amount Lynda owes
Amount Maria owes
0 175 200
1 165 185
2 155 170
3 145 155
4 135 140
5 125 125
6 115 110
... ...
13 45 5
14 35 0
Maria only has to pay $5 in the 14th week.
Lynda's and Maria's Loans
Amount owed (dollars)
Number of weeks
Lynda
Maria
After five weeks, they owe the same amount of money. After that, Maria is paying more per week than Lynda, so she will pay off her loan first.
21 Key Concepts Question In the Example above, how does the graph show when the sisters owe the same amount of money?

Source: Math Thematics, Book 3, McDougal Littell, 2005 (p. 490).

EXAMPLE 10-24

May Chin ordered lunch for herself and several friends by phone without checking prices. She paid \$6.00 for five soyburgers and two orders of fries, and another time she paid \$7.00 for four soyburgers and six orders of fries. Set up a system of equations with two unknowns representing the prices of a soyburger and an order of fries, respectively.

Solution Let x be the price in dollars of a soyburger and y be the price of an order of fries. Five soyburgers cost $5x$ dollars, and two orders of fries cost $2y$ dollars. Because May paid \$6.00 for the orders, we have $5x + 2y = 6$. Similarly, $4x + 6y = 7$.

An ordered pair satisfying two linear equations is a point that belongs to each of the lines. Figure 10-87 shows the graphs of $5x + 2y = 6$ and $x - 4y = {}^{-}1$. The two lines appear to intersect at $\left(1, \frac{1}{2}\right)$. Thus, $\left(1, \frac{1}{2}\right)$ appears to be the solution of the given system of equations. This solution can be checked by substituting 1 for x and $\frac{1}{2}$ for y in each equation. Because two distinct lines intersect in only one point, $\left(1, \frac{1}{2}\right)$ is the only solution to the system.

FIGURE 10-87

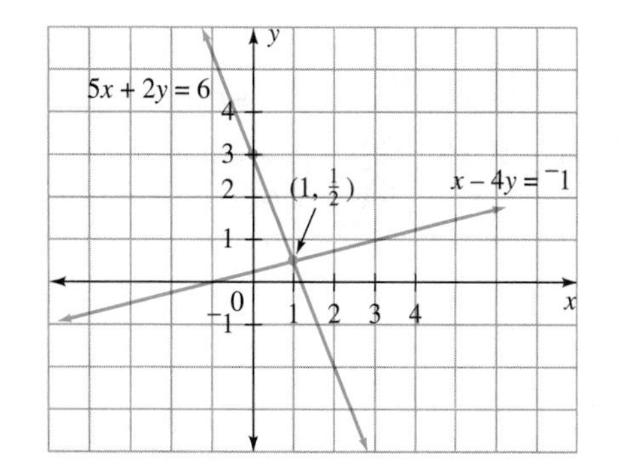

Substitution Method

Drawbacks to estimating a solution to a system of equations graphically include an inability to read some coordinates of points. However, there are algebraic methods for solving systems of linear equations. Example 10-25 demonstrates one such algebraic method: the **substitution method**.

substitution method

EXAMPLE 10-25

Solve the following system:

$$3x - 4y = 5$$
$$2x + 5y = 1$$

Solution First, rewrite each equation, expressing y in terms of x.

$$y = \frac{3x - 5}{4} \quad \text{and} \quad y = \frac{1 - 2x}{5}$$

Then equate the expressions for y and solve the resulting equation for x.

$$\frac{3x - 5}{4} = \frac{1 - 2x}{5}$$
$$5(3x - 5) = 4(1 - 2x)$$
$$15x - 25 = 4 - 8x$$
$$23x = 29$$
$$x = \frac{29}{23}$$

Substituting $\frac{29}{23}$ for x in $y = \frac{3x - 5}{4}$ gives $y = \frac{^{-}7}{23}$. Hence, $x = \frac{29}{23}$ and $y = \frac{^{-}7}{23}$. This can be checked by substituting the values for x and y in the original equations. ▲

Sometimes when using the substitution method, it is more convenient to solve a system of equations by expressing x in terms of y in one of the equations and substituting the obtained expression for x in the other equation.

Elimination Method

elimination method

The **elimination method** for solving two equations with two unknowns is based on eliminating one of the variables by adding or subtracting the original or equivalent equations. For example, consider the following system:

$$x - y = {}^{-}3$$
$$x + y = 7$$

By adding the two equations, we can eliminate the variable y. The resulting equation can then be solved for x.

$$\begin{array}{rcl} x - y & = & {}^{-}3 \\ \underline{x + y} & \underline{=} & \underline{7} \\ 2x & = & 4 \\ x & = & 2 \end{array}$$

Substituting 2 for x in the first equation (either equation may be used) gives $y = 5$. Checking this result shows that $x = 2$ and $y = 5$, or $(2, 5)$, is the solution to the system.

Often, another operation is required before equations are added so that an unknown can be eliminated. For example, consider the following system:

$$3x + 2y = 5$$
$$5x - 4y = 3$$

Adding the equations does not eliminate either unknown. However, if the first equation contained $4y$ rather than $2y$, the variable y could be eliminated by adding. To obtain $4y$ in the first equation, we multiply both sides of the equation by 2 to obtain the equivalent equation $6x + 4y = 10$. Adding the equations in the equivalent system gives the following:

$$\begin{aligned} 6x + 4y &= 10 \\ 5x - 4y &= 3 \\ \hline 11x \quad &= 13 \\ x \quad &= \frac{13}{11} \end{aligned}$$

To find the corresponding value of y, we substitute $\frac{13}{11}$ for x in either of the original equations and solve for y, or we use the elimination method again and solve for y.

An alternative method is to eliminate the x-values from the original system by multiplying the first equation by 5 and the second by $^{-}3$ (or the first by $^{-}5$ and the second by 3). Then we add the two equations and solve for y.

$$\begin{aligned} 15x + 10y &= 25 \\ ^{-}15x + 12y &= {}^{-}9 \\ \hline 22y &= 16 \\ y &= \frac{16}{22}, \text{ or } \frac{8}{11} \end{aligned}$$

Consequently, $\left(\frac{13}{11}, \frac{8}{11}\right)$ is the solution of the original system. This solution, as always, should be checked by substitution in the *original* equations.

Solutions to Systems of Linear Equations

All examples thus far have had unique solutions. However, other situations may arise. Geometrically, a system of two linear equations can be characterized as follows:

1. *The system has a unique solution if and only if the graphs of the equations intersect in a single point.*
2. *The system has no solution if and only if the equations represent parallel lines.*
3. *The system has infinitely many solutions if and only if the equations represent the same line.*

NOW TRY THIS 10-19

Find all the solutions (if any) of each of the following systems by graphing the equations in each system. Then justify your answers by an algebraic approach.

a. $x - y = 1$
$2x - y = 5$

b. $2x - y = 1$
$2y - 4x = 3$

c. $2x - 3y = 1$
$6y - 4x = 2$

Applications of Coordinate Geometry

Applications of coordinate geometry appear in many places. Examples include finding the circumcenter of a specific right triangle and in fitting data to a line. The first is seen in the following problem.

PROBLEM SOLVING **Finding the Circumcenter**

Alicia wanted to find the circumcenter of a right triangle with sides of length 3 ft, 4 ft, and 5 ft. How could this be done?

Understanding the Problem Earlier in the chapter, we discovered ways to construct the circumcenter and circumcircle of a triangle using a compass and straightedge. That same type of construction will work here. Recall that we had to find the intersection of the perpendicular bisectors of the sides of the triangle. We could do the same process here but consider a different technique: using coordinate geometry. To do this, we need to think about the right triangle in a coordinate plane.

Devising a Plan We could place the right triangle on a coordinate system with the right angle at the origin, and the two sides making the right angle on the x- and y-axes, as in Figure 10-88.

FIGURE 10-88

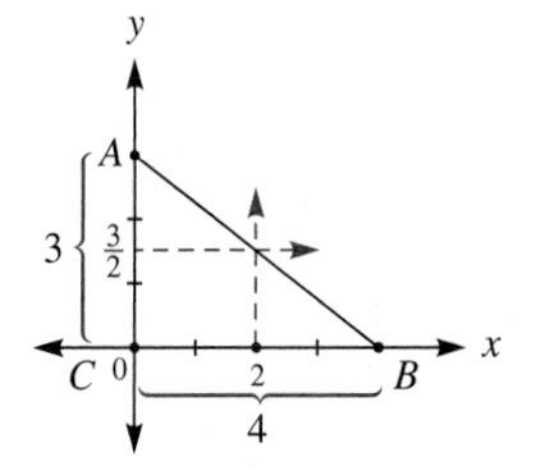

Next, if we could write equations of each of the perpendicular bisectors, we could solve the resulting equations to find the coordinates of the circumcenter.

Carrying Out the Plan We know that the perpendicular bisectors of the sides of the triangle contain the midpoints of the sides and are perpendicular to the sides. Considering Figure 10-88, we know that the perpendicular bisectors of $\overline{AC}$ and $\overline{BC}$ are, respectively, horizontal and vertical lines. Their intersection determines the circumcenter. We can check the coordinates of the circumcenter in the equation of the perpendicular bisector of $\overline{AB}$. Because the length of $\overline{AC}$ is 3, its midpoint is $\left(\frac{1}{2}\right)(3)$, or $\frac{3}{2}$. Also, the x-coordinate of the midpoint must be 0 (why?). Because the midpoint lies on the vertical axis and it must be $\frac{3}{2}$ from the origin along the line, the coordinates are $\left(0, \frac{3}{2}\right)$. Similarly, the coordinates of the midpoint of $\overline{BC}$ must be $(2, 0)$.

Now the equation of the horizontal line containing the point with coordinates $\left(0, \frac{3}{2}\right)$ is $y = \frac{3}{2}$. Similarly, the equation of the vertical line containing the point with coordinates $(2, 0)$ is $x = 2$. The intersection of the two lines is the point with coordinates $\left(2, \frac{3}{2}\right)$. Hence, the coordinates of the circumcircle are $\left(2, \frac{3}{2}\right)$.

Looking Back We should check the coordinates of the circumcircle by checking them in the equation of the perpendicular bisector of $\overline{AB}$. The midpoint of $\overline{AB}$ can be determined using the midsegment of triangle ABC that contains either the midpoint of $\overline{AC}$ or $\overline{BC}$. The parallel to $\overline{AC}$ that contains $(2, 0)$ is the perpendicular bisector of $\overline{BC}$ and has equation $x = 2$ (why?). Also, the equation of the line AB must have slope $\frac{3 - 0}{0 - 4}$, or $-\frac{3}{4}$. Because the line has y-intercept of 3, its equation then is $y = \left(-\frac{3}{4}\right)x + 3$. The intersection of $\overline{AB}$ and the line with equation $x = 2$ may be found by substituting 2 for x as follows: $y = \left(-\frac{3}{4}\right)(2) + 3 = \frac{3}{2}$. Hence, the midpoint of $\overline{AB}$ has coordinates $\left(2, \frac{3}{2}\right)$, exactly the coordinates of the circumcenter, and the coordinates must fit the equation of line $\overleftrightarrow{AB}$. Therefore, the circumcenter is a point on the hypotenuse of triangle ABC.

Will the circumcenter of a right triangle always be on the hypotenuse of the triangle? To think about this, what must be true about the measure of the intercepted arc of $\angle ACB$, an inscribed right angle? Once this has been answered, what must be the relation of the hypotenuse to the circumcircle?

Fitting a Line to Data

As noted in Chapter 8, in many practical situations, a relationship between two variables comes from collected data such as from population or business surveys. When the data are graphed, there may not be a single line that goes through all of the points, but the points may appear to approximate, or "follow," a straight line. In such cases, it is useful to find the equation of what seems to be the **best-fitting line**. Knowing the equation of such a line enables us to predict an outcome without actually performing the experiment.

best-fitting line

There are several approaches to define, and hence find, the best-fitting line. We take a graphical approach as follows:

1. Choose a line that seems to follow the given points so that there are about an equal number of points below the line as above the line.
2. Determine two convenient points on the line and approximate the x- and y-coordinates of these points.
3. Use the points in (2) to determine the equation of the line.

EXAMPLE 10-26

A shirt manufacturer noticed that the number of units sold depends on the price charged. The data in Table 10-6 shows the number of units sold for a given price per unit.

a. Find the equation of a line that seems to fit the data best.
b. Use the equation in (a) to predict the number of units that will be sold if the price per unit is $60.

Table 10-6

Price per Unit (Dollars)	Number of Units Sold (Thousands)
50	200
44	250
41	300
33	380
31	400
24.5	450
20	500
14.5	550

Solution **a.** Figure 10-89(a) shows the graph of the data displayed in Table 10-6. Figure 10-89(b) shows a line that seems to fit the data so that approximately the same number of points are below the line as above the line. We choose the points (50, 200) and (20, 500), which are on the line in Figure 10-89(b).

FIGURE 10-89

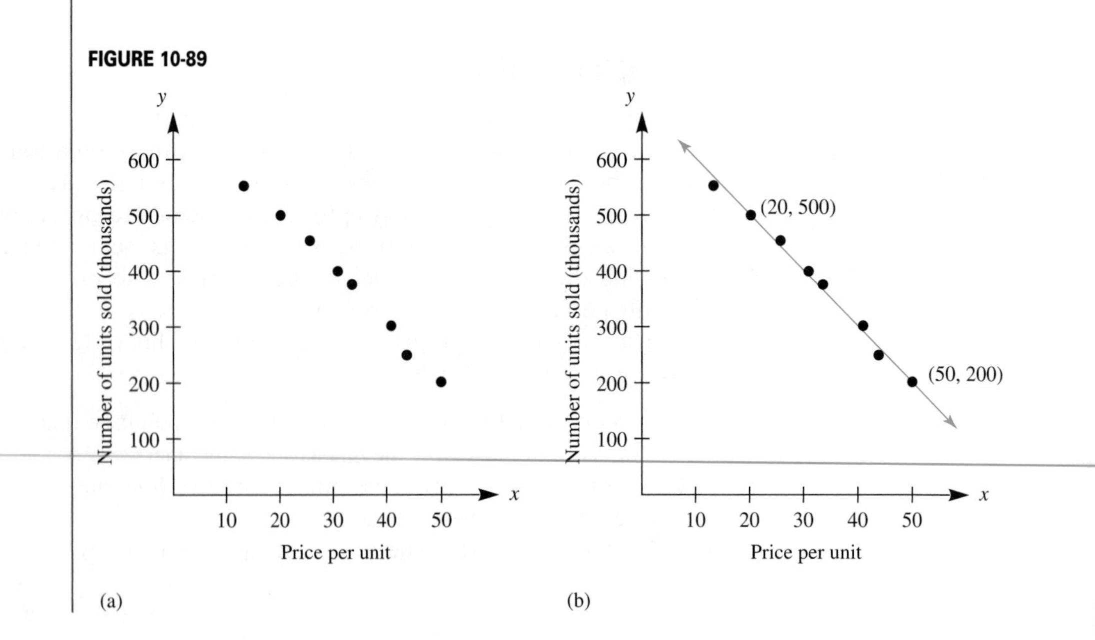

To find the equation of the line, we may find m and b in the equation $y = mx + b$. Substituting the points (50, 200) and (20, 500) into this equation, we obtain the following:

$$200 = 50m + b$$
$$500 = 20m + b$$

One way to solve the equations is to express b in terms of m for each equation:

$$b = 200 - 50m$$
$$b = 500 - 20m$$

We equate the expressions for b and solve for m:

$$200 - 50m = 500 - 20m$$
$$200 - 500 = 50m - 20m$$
$$^{-}300 = 30m$$
$$m = {}^{-}10$$

Substituting this value for m, we obtain

$$b = 200 - 50(^{-}10) = 700$$

Consequently, the equation of the fitted line is $y = {}^{-}10x + 700$.

b. Using the equation in (a), substitute $x = 60$ to obtain $y = {}^{-}10(60) + 700$, or $y = 100$. Thus, we predict that 100,000 units will be sold if the price per unit is \$60.

ONGOING ASSESSMENT 10-6

1. The graph of $y = mx$ is given in the following figure. Sketch the graphs for each of the following on the same figure. Explain your answers.
   a. $y = mx + 3$
   b. $y = mx - 3$

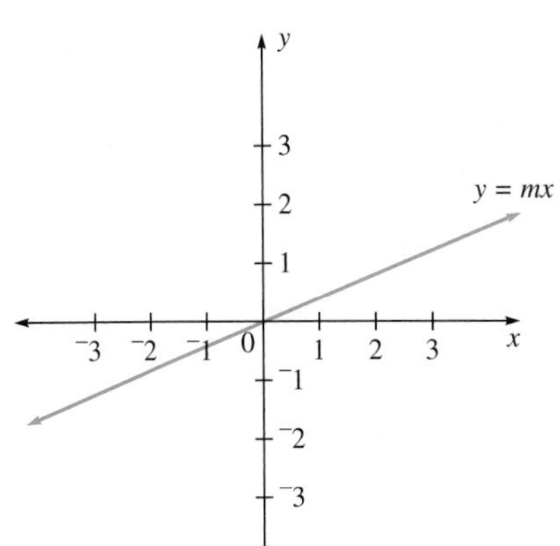

2. Sketch the graphs for each of the following equations:
   a. $y = \frac{^{-}3}{4}x + 3$
   b. $y = {}^{-}3$
   c. $y = 15x - 30$
   d. $x = {}^{-}2$
   e. $y = 3x - 1$
   f. $y = \frac{1}{20}x$
3. Find the x-intercept and y-intercept for the equations in problem 2, if they exist.

4. In the following figure, part (i) shows a dual-scale thermometer and part (ii) shows the corresponding points plotted on a graph:

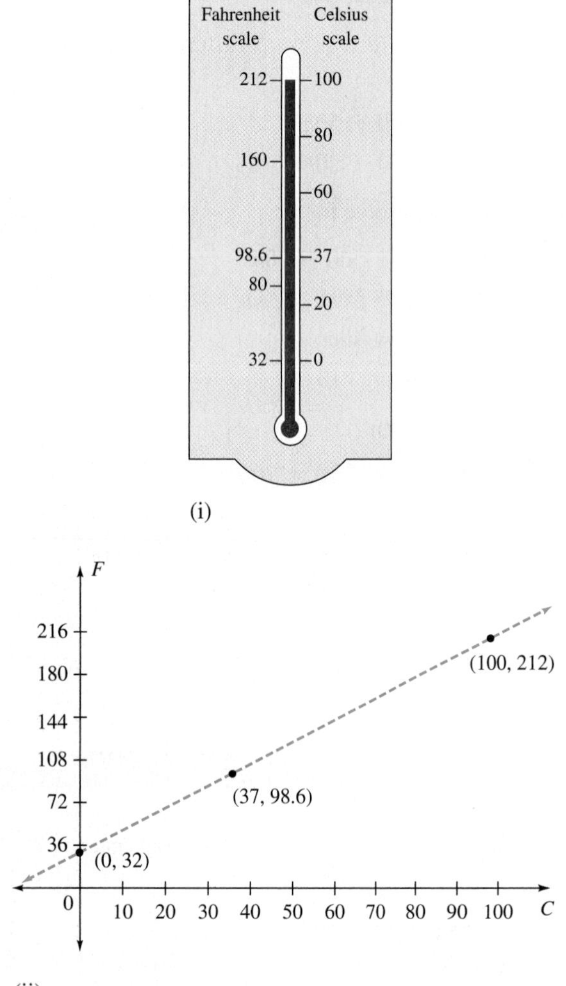

   a. Use two of the points on the graph to develop a formula for conversion from degrees Celsius (C) to degrees Fahrenheit (F).
   b. Use your answer in (a) to find a formula for converting from degrees F to degrees C.

5. Write each of the following equations in slope-intercept form and identify the slope and y-intercept:
   a. $3y - x = 0$ **b.** $x + y = 3$
   c. $\frac{x}{3} + \frac{y}{4} = 1$ **d.** $3x - 4y + 7 = 0$
   e. $x = 3y$ **f.** $x - y = 4(x - y)$

6. For each of the following, write the equation of the line determined by the given pair of points in slope-intercept form or in the form $x = a$:
   a. $(^-4, 3)$ and $(1, ^-2)$ **b.** $(0, 0)$ and $(2, 1)$
   c. $(0, 1)$ and $(2, 1)$ **d.** $(2, 1)$ and $(2, ^-1)$
   e. $\left(0, \frac{^-1}{2}\right)$ and $\left(\frac{1}{2}, 0\right)$ **f.** $(^-a, 0)$ and $(a, 0)$, $a \neq 0$

7. Find the coordinates of two other points collinear (on the same line) with each of the following pairs of given points:
   a. $P(2, 2)$, $Q(4, 2)$ **b.** $P(^-1, 0)$, $Q(^-1, 2)$
   c. $P(0, 0)$, $Q(0, 1)$ **d.** $P(0, 0)$, $Q(1, 1)$

8. For each of the following, give as much information as possible about x and y:
   a. The ordered pairs $(^-2, 0)$, $(^-2, 1)$, and (x, y) represent collinear points.
   b. The ordered pairs $(^-2, 1)$, $(0, 1)$, and (x, y) represent collinear points.
   c. The ordered pair (x, y) is in the fourth quadrant.

9. Consider the lines through $P(2, 4)$ and perpendicular to the x- and y-axes, respectively. Find the area and the perimeter of the rectangle formed by these lines and the axes.

10. Find the equations for each of the following:
    a. The line containing $P(3, 0)$ and perpendicular to the x-axis
    b. The line containing $P(0, ^-2)$ and parallel to the x-axis
    c. The line containing $P(^-4, 5)$ and parallel to the x-axis
    d. The line containing $P(^-4, 5)$ and parallel to the y-axis

11. For each of the following, find the slope, if it exists, of the line determined by the given pair of points:
    a. $(4, 3)$ and $(^-5, 0)$ **b.** $(^-4, 1)$ and $(5, 2)$
    c. $(\sqrt{5}, 2)$ and $(1, 2)$ **d.** $(^-3, 81)$ and $(^-3, 198)$
    e. $(1.0001, 12)$ and $(1, 10)$ **f.** (a, a) and (b, b)

12. Write the equation of each line in problem 11.

13. Wildlife experts found that the number of chirps a cricket makes in 15-sec intervals is related to the temperature T in degrees Fahrenheit, as shown in the following graph:

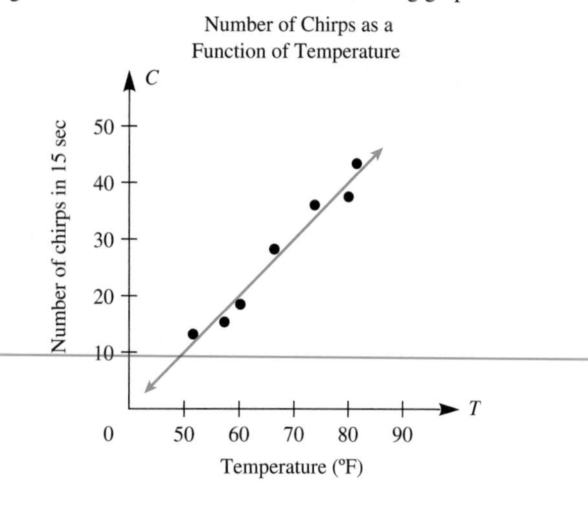

a. If C is the number of chirps in 15 sec, write a formula for C in terms of T (temperature in degrees Fahrenheit) that seems to fit the data best.

b. Use the equation in (a) to predict the number of chirps in 15 sec when the temperature is 90°.

c. If N is the number of chirps per minute, write a formula for N in terms of T.

14. An equilateral triangle whose side length is 6 is placed so that one side lies along the x-axis of a graph and one vertex is at the origin.
   a. Write the coordinates of the midpoint of the side along the x-axis.
   b. Write the equation of the perpendicular bisector of the side along the x-axis.

15. A rectangle has two vertices on the x-axis, two vertices on the y-axis, and one vertex at the point with coordinates (4, 6).
   a. Make a sketch showing that there is a rectangle with these characteristics.
   b. Write the coordinates of each of the other three vertices.
   c. Write the equations of the diagonals of the rectangle.
   d. Find the coordinates of the point of intersection of the diagonals.

16. A triangle has vertices with coordinates $(^-4, 6)$, $(2, ^-6)$, and (8, 10).
   a. Write the equations of the lines containing each of the sides of the triangle.
   b. Prove that your equations are correct by finding the coordinates of the points of intersection of each pair of sides of the triangle.

17. Write the equation of lines meeting each of the following conditions:
   a. A horizontal line containing the point with coordinates (3, 4)
   b. A vertical line containing the point with coordinates $(^-7, ^-8)$.

18. Consider a trapezoid with vertices whose coordinates are (0, 0), (4, 6), (10, 6), and (15, 0).
   a. What are the coordinates of the midsegment of the trapezoid?
   b. What is the equation of the midsegment?

19. If three vertices of a kite have coordinates (0, 0), (6, 14) and $(6, ^-14)$, write an equation of a line that contains the other vertex. Is there more than one solution to the problem?

20. Suppose a car is traveling at a constant speed of 60 mph.
   a. Draw a graph showing the relation of the distance traveled and the amount of time that it takes to travel the distance.
   b. What is the slope of the line that you drew in part (a)?

21. **a.** Graph the following data and find the equation of the best-fitting line.
   b. Use your answer in (a) to predict the value of y when $x = 10$.

x	y
1	8.1
2	9.9
3	12
4	14.1
5	15.9
6	18
7	19.9

22. Solve each of the following systems, if possible. Indicate whether the system has a unique solution, infinitely many solutions, or no solution.
   a. $y = 3x - 1$
   $y = x + 3$
   b. $2x - 6y = 7$
   $3x - 9y = 10$
   c. $3x + 4y = {}^-17$
   $2x + 3y = {}^-13$
   d. $4x - 6y = 1$
   $6x - 9y = 1.5$

23. The vertices of a triangle are given by (0, 0), (10, 0), and (6, 8). Show that the segments connecting (5, 0) and (6, 8), (10, 0) and (3, 4), and (0, 0) and (8, 4) intersect at a common point.

24. The owner of a 5000-gal oil truck loads the truck with gasoline and kerosene. The profit on each gallon of gasoline is 13¢ and on each gallon of kerosene is 12¢. How many gallons of each fuel did the owner load if the profit was $640?

25. At the end of 10 mo, the balance of an account earning simple interest is $2100.
   a. If, at the end of 18 mo, the balance is $2180, how much money was originally in the account?
   b. What is the rate of interest?

26. Josephine's bank contains 27 coins. If all the coins are either dimes or quarters and the value of the coins is $5.25, how many of each kind of coin are there?

27. a. Solve each of the following systems of equations. What do you notice about the answers?
   (i) $x + 2y = 3$
   $4x + 5y = 6$
   (ii) $2x + 3y = 4$
   $5x + 6y = 7$
   (iii) $31x + 32y = 33$
   $34x + 35y = 36$
   b. Write another system similar to those in (a). What solution did you expect? Check your guess.
   c. Write a general system similar to those in (a). What solution does this system have? Why?

28. If a triangle has vertices with coordinates (0, 32), (20, 0), and (100, 212) and a similar triangle has vertices (0, 16) and (10, 0), find a possible set of coordinates for the third vertex.
29. If a line ℓ forms a 45° angle with the x-axis, what is the slope of
   a. Line ℓ?
   b. Line m, which is perpendicular to line ℓ?
30. If two sets of data have a negative association, what can be said about the slope of the line of best fit for the data?

Communication

31. An arithmetic sequence having a linear graph was depicted in this chapter. Explain whether a geometric sequence has a linear graph.
32. Dahlia wanted to find what temperature has the same measure when measured in degrees Fahrenheit or degrees Celsius. To do that, she graphed the equation of the conversion formula $C = \frac{5}{9}(F - 32)$. On the same coordinate system, she also graphed the equation $C = F$ and found the intersection point ($^{-}40$, $^{-}40$). Explain why this procedure answers the question Dahlia asked.

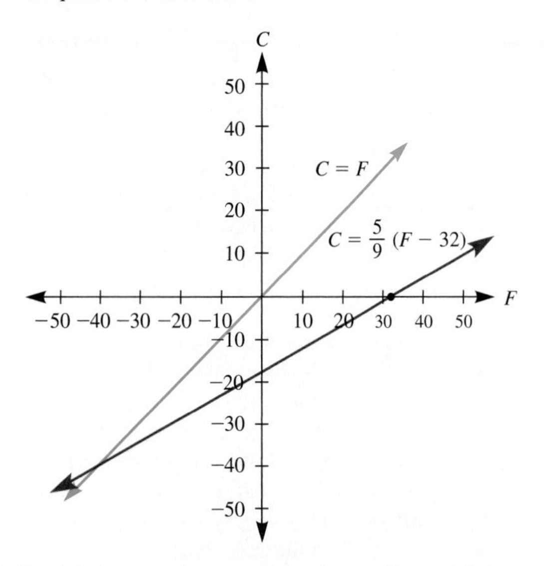

33. Explain in more than one way why two lines with the same slope are parallel.
34. Jonah tried to solve the equation $^{-}5x + y = 20$ by adding 5 to both sides. He wrote $5 - 5x + y = 5 + 20$ or $0 \cdot x + y = 25$ and finally $y = 25$. How would you help Jonah?
35. Jill would like to know why two lines with an undefined slope are parallel. How would you respond?

Open-Ended

36. Look for data in newspapers, magazines, or books whose graphs appear to be close to linear and find the equations of the lines that you think best fit the data.
37. a. Write equations of two lines that intersect but when graphed look parallel.
   b. At what point do those two lines intersect?

Cooperative Learning

38. Play the following game between your group and another group. Each group makes up four linear equations that have a common property and presents the equations to the other group. For example, one group could present the equations $2x - y = 0$, $4x - 2y = 3$, $y - 2x = 3$, and $3y - 6x = 5$. If the second group discovers a common property that the equations share, such as the graphs of the equations are four parallel lines, they get 1 point. Each group takes a specified number of turns.
39. In the following drawing, show that $a + b = c + d$:

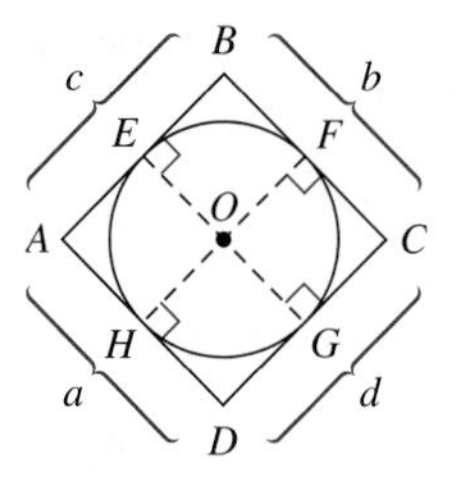

40. On a circular clock face consider a triangle formed by connecting the center of the face, the point marking 1, and the point marking 9. Find a different congruent triangle and prove the triangles are congruent.

Review Problems

41. $ABCD$ is a trapezoid with $\overline{BC} \parallel \overline{AD}$. Solve for x.

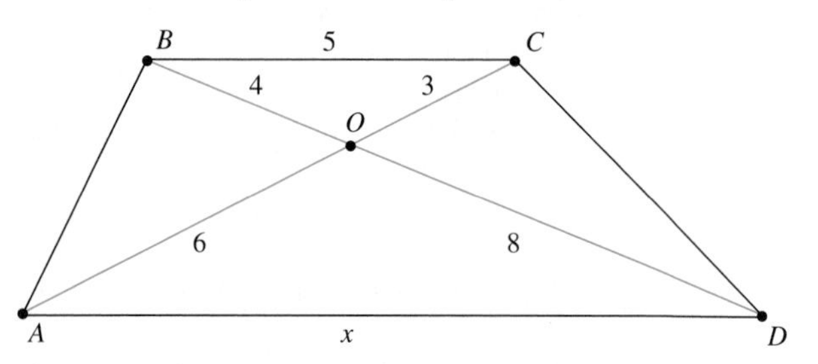

42. Draw an arbitrary triangle ABC. Let M, N, and P be the corresponding midpoints of the sides $\overline{AB}$, $\overline{BC}$, and $\overline{AC}$. Connect the points M, N, and P. Four triangles are created. Is it true that all of the smaller triangles are congruent and each is similar to $\triangle ABC$? Why or why not?
43. In triangle ABC, a square has been inscribed as shown. The lengths of the sides of $\triangle ABC$ are 3, 4, and 5 as shown. Find the length of a side of the square. (*Hint:* First show that $\triangle AED \sim \triangle ABC$.)

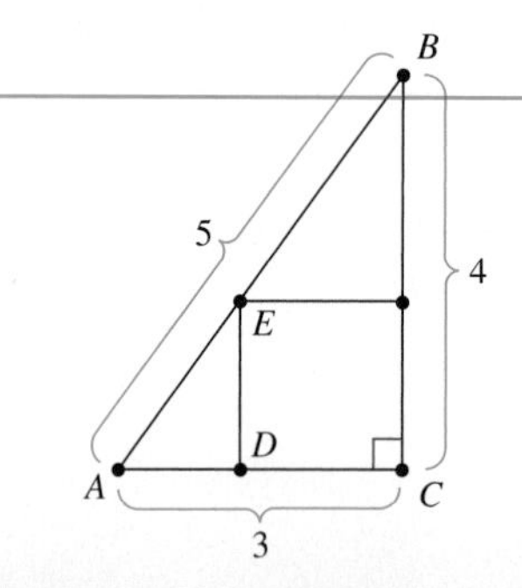

44. In the figure shown, find BE.

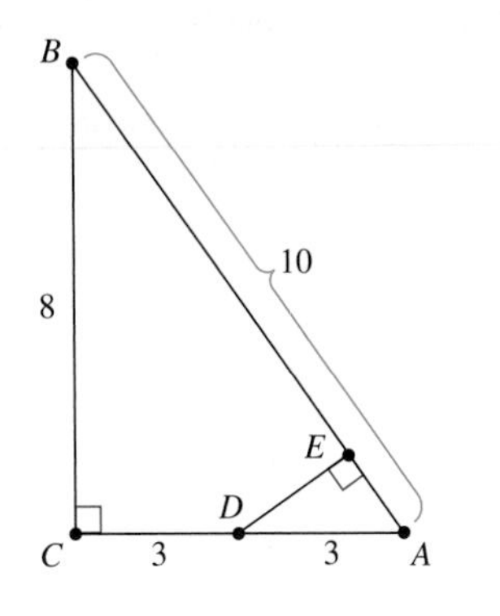

Third International Mathematics and Science Study (TIMSS) Question

A straight line passes through the points (2, 3) and (4, 7). Which of these points is also on the line?

a. (0, 2)
b. (1, 2)
c. (2, 4)
d. (3, 5)
e. (4, 5)

TIMSS, Grade 8, 2003

HINT FOR SOLVING THE PRELIMINARY PROBLEM

Recall that any point on an angle bisector is equidistant from the sides of the angle. Use this information, put the measures that you know in a diagram, sketch the figure as if completed, label the radii of the incircle as r and label one unknown part of a side of the right triangle as x, use congruent triangles to find congruent parts, and write all unknown parts with expressions involving x. Finally, solve for x.

QUESTIONS FROM THE CLASSROOM

1. On a test, a student wrote $AB \cong CD$ instead of $\overline{AB} \cong \overline{CD}$. Is this answer correct? Why?

2. A student asks if there are any constructions that cannot be done with a compass and a straightedge. How do you answer?

3. A student asks for a mathematical definition of congruence that holds for all figures. How do you respond? Is your response the same for similarity?

4. One student claims that by trisecting $\overline{AB}$ and drawing $\overrightarrow{CD}$ and $\overrightarrow{CE}$, as shown in the following figure, she has trisected $\angle ACB$:

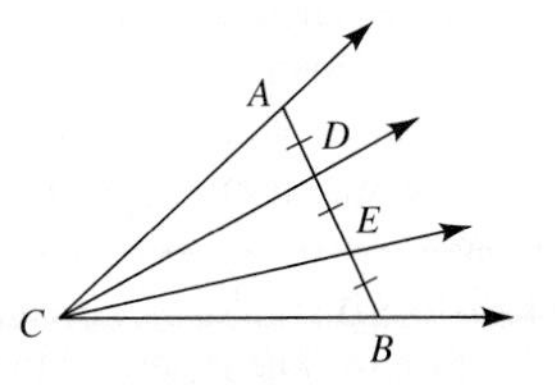

How do you convince her that her construction is wrong?

5. A student claims that polygon $ABCD$ in the following drawing is a parallelogram if $\angle 1 \cong \angle 2$. Is he correct? Why?

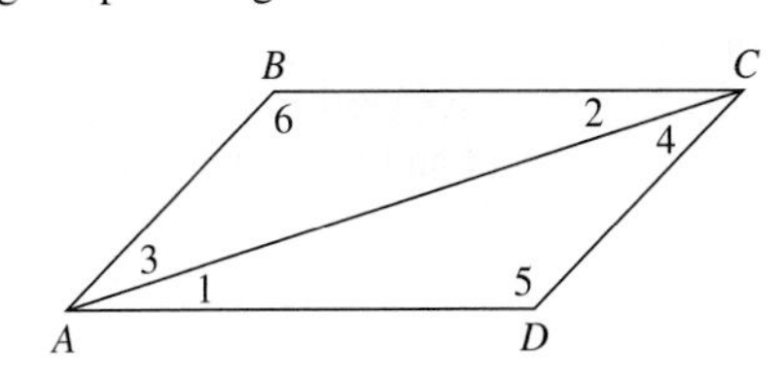

6. A student claims that when the midpoints of the sides of any polygon are connected, a polygon similar to the original results. Is this true? Why?

7. A student asks why "congruent" rather than "equal" is used to discuss triangles that have the same size and shape. What do you say?

8. A student draws the following figure and claims that because every triangle is congruent to itself, we can write $\triangle ABC \cong \triangle BCA$. What is your response?

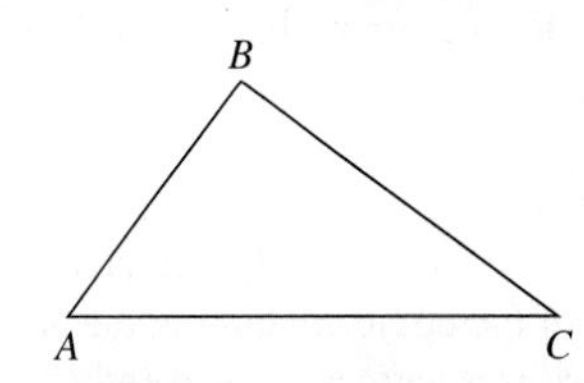

9. A student asks whether there is an AAAA similarity condition for quadrilaterals. How do you respond?

10. A student asks if, for the same n, all regular n-gons are similar. How do you respond?

11. A student says she thinks all circles are similar but would like to know why. How do you answer her?

12. A student says that a vertical line has an infinite slope. How do you respond?

13. A student claims that the lines whose equations are $y = 3x + 5$ and $y = 3x + 4$ are parallel because when she tried to solve the equations, she got the false statement $4 = 5$. How do you respond?

14. A student claims that because one can use slope to move horizontally and then vertically to find other points on a line as well as doing the reverse, then slope could be defined as run/rise. How do you respond?

15. Sue asks if coordinate geometry is "real" geometry. What do you say?

16. A student argues that in the following figure triangles ABC and DBE are similar because $\overline{DE}$ divides the side proportionally. One side is divided in the ratio $\frac{2}{4}$, while the other is divided in the ratio $\frac{3}{6}$. How do you respond?

17. A student argues that there is no line of best fit for a set of data. Could the student be correct? Give an example to make your point to the student.

***18.** One student says that a tangent is another expression for a slope. Is the student correct?

CHAPTER OUTLINE

I. Congruence
- **A.** Two geometric figures are **congruent** if and only if they have the same size and shape.
- **B.** Two triangles are congruent if they satisfy any of the following properties:
  1. **Side, Side, Side (SSS)**
  2. **Side, Angle, Side (SAS)**
  3. **Angle, Side, Angle (ASA)**
  4. **Angle, Angle, Side (AAS)**
- **C.** **Triangle inequality:** The sum of the measure of any two sides of a triangle must be greater than the measure of the third side.
- **D.** Corresponding parts of congruent figures are congruent.
- **E.** A **rep-tile** is a figure that is used in the construction of a similar figure.

II. Circles
- **A.** An **arc** of a circle is part of the circle that can be drawn without lifting a pencil. The **center of an arc** is the center of the circle containing the arc.
- **B.** A **chord** is a segment whose endpoints lie on a circle.

III. Proportion
- **A.** If a line parallel to one side of a triangle intersects the other sides, it divides those sides into proportional segments.
- **B.** If parallel lines cut off congruent segments on one transversal, they cut off congruent segments on any transversal.

IV. Constructions that can be accomplished using a compass and a straightedge
- **A.** Copy a line segment.
- **B.** Copy a circle.
- **C.** Copy an angle.
- **D.** Bisect a segment.
- **E.** Bisect an angle.
- **F.** Construct a perpendicular from a point to a line.
- **G.** Construct a perpendicular bisector of a segment.
- **H.** Construct a perpendicular to a line through a point on the line.
- **I.** Construct a parallel to a line through a point not on the line.
- **J.** Divide a segment into congruent parts.
- **K.** Inscribe some regular polygons in a circle.
- **L.** Circumscribe a circle about a triangle.
- **M.** Inscribe a circle in a triangle.

V. Similarity and proportion
- **A.** Two polygons are **similar** if and only if their corresponding angles are congruent and their corresponding sides are proportional.
- **B.** **AA:** If two angles of one triangle are congruent to two angles of a second triangle, respectively, the triangles are similar.
- **C.** If a line divides two sides of a triangle into proportional segments, then the line is parallel to the third side.

VI. Lines in a Cartesian coordinate system
- **A.** Every nonvertical line has an equation of the form $y = mx + b$, where m is the slope and b is the y-intercept.
- **B.** The equation of any vertical line can be written in the form $x = a$, and any horizontal line in the form of $y = b$.

C. Slope formula: Given two points $A(x_1, y_1)$ and $B(x_2, y_2)$, the slope m of line AB is given by the following:

$$m = \frac{y_2 - y_1}{x_2 - x_1} = \frac{y_1 - y_2}{x_1 - x_2} = \frac{\text{Rise}}{\text{Run}}$$

D. The equation of a line with slope m through a given point with coordinates (x_1, y_1) is $y - y_1 = m(x - x_1)$.

E. Parallel nonvertical lines have the same slope.

F. If two lines have the same slope, they are parallel.

G. The slope of a horizontal line is 0 and it is impossible to define the slope of a vertical line.

H. A system of **linear equations** can be solved graphically by drawing the graphs of the equations.

I. A system of linear equations can be solved algebraically by either the **substitution method** or the **elimination method**.

J. The **best-fitting line** is a straight line that approximates the data.

*__VII.__ Trigonometric functions

In a right triangle, as shown, we have the following:

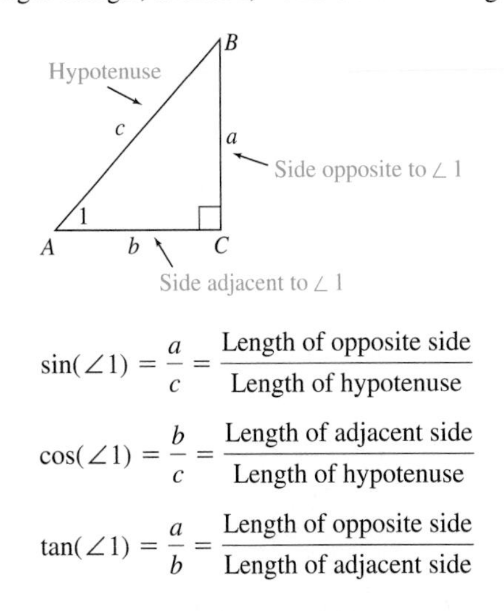

$$\sin(\angle 1) = \frac{a}{c} = \frac{\text{Length of opposite side}}{\text{Length of hypotenuse}}$$

$$\cos(\angle 1) = \frac{b}{c} = \frac{\text{Length of adjacent side}}{\text{Length of hypotenuse}}$$

$$\tan(\angle 1) = \frac{a}{b} = \frac{\text{Length of opposite side}}{\text{Length of adjacent side}}$$

CHAPTER REVIEW

1. Each of the following figures contains at least one pair of congruent triangles. Identify them and tell why they are congruent.

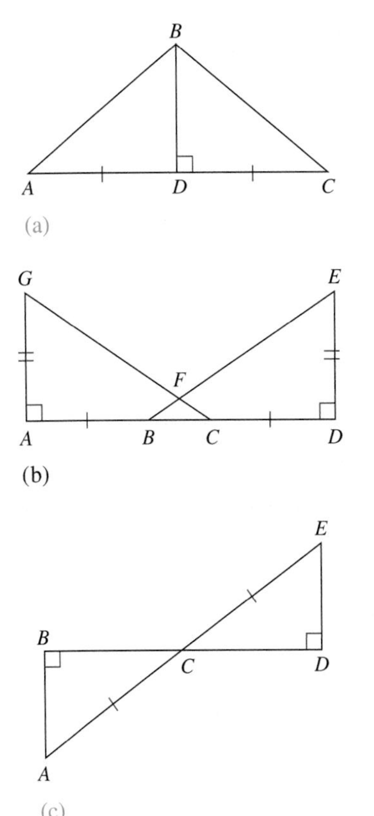

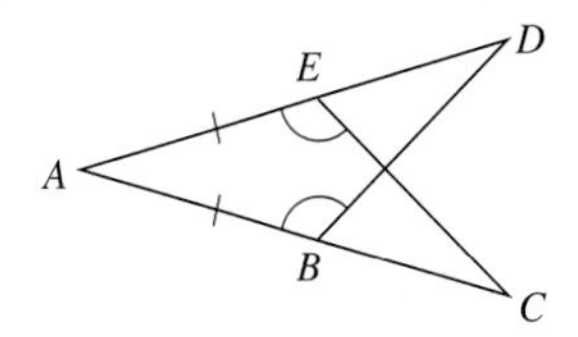

(d)

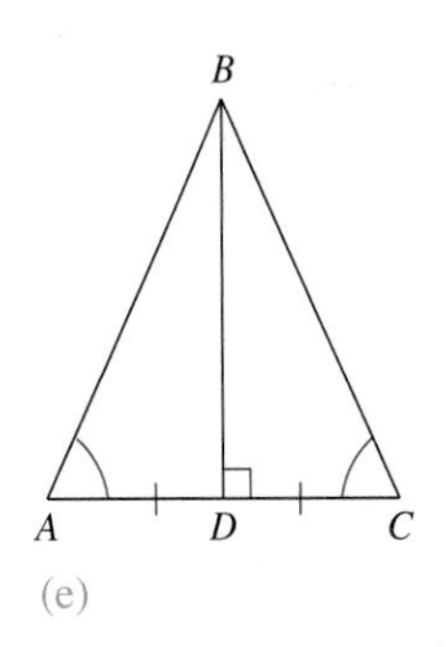

(e)

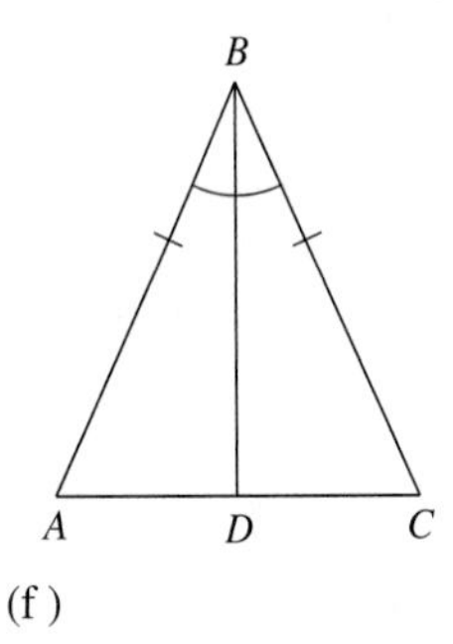

(f)

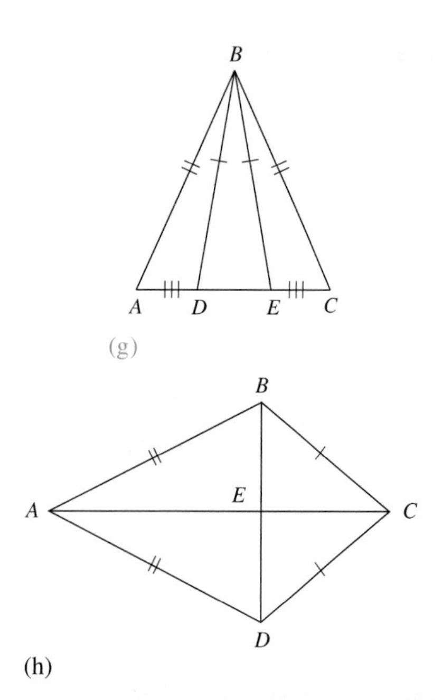

2. In the following figure, $ABCD$ is a square and $\overline{DE} \cong \overline{BF}$. What kind of figure is $AECF$? Justify your answer.

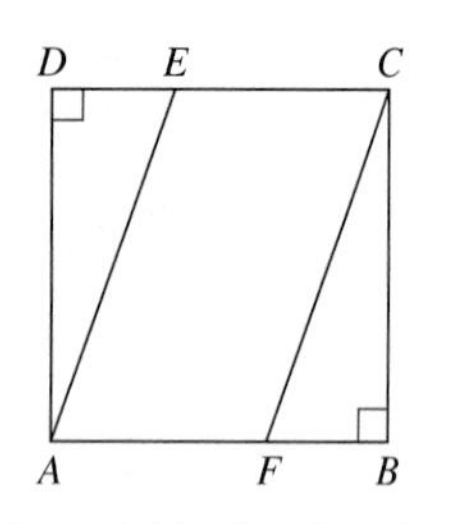

3. Construct each of the following by (1) using a compass and straightedge and (2) paper folding:

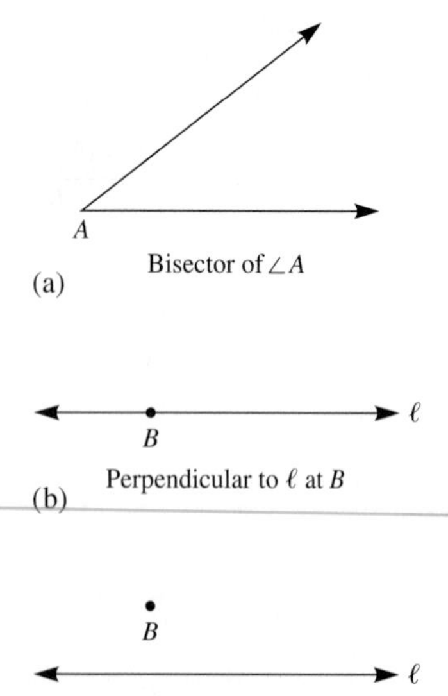

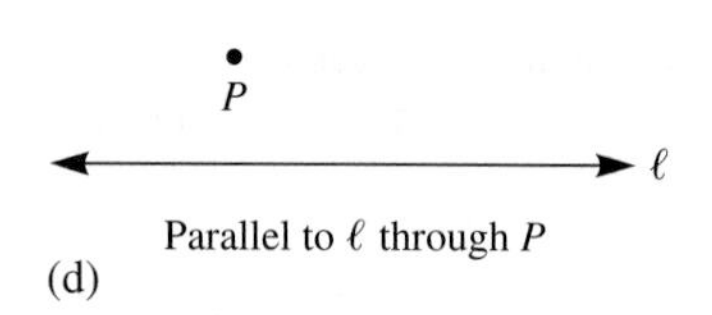

4. For each of the following pairs of similar triangles, find the missing measures:

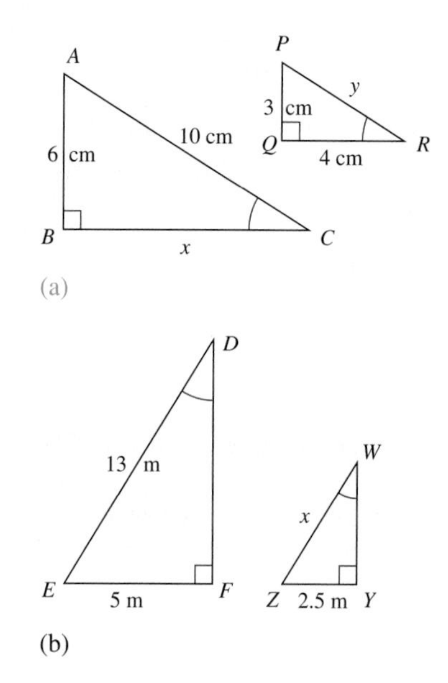

5. Divide the following segment into five congruent parts:

6. If $ABCD$ is a trapezoid, $\overline{EF} \parallel \overline{AD}$ and $\overline{AC}$ is a diagonal. What is the relationship between $\frac{a}{b}$ and $\frac{c}{d}$? Why?

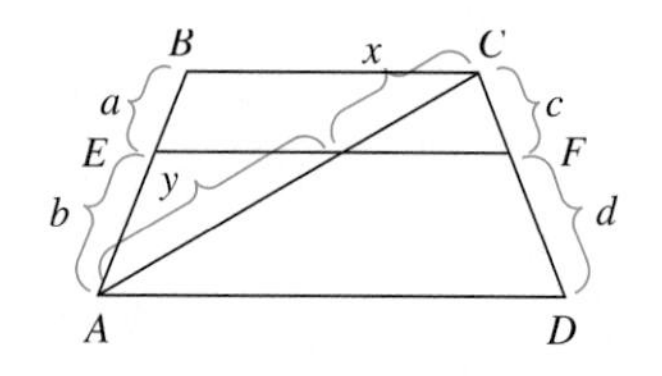

7. Given the following figure, construct a circle that contains A and B and has its center on ℓ:

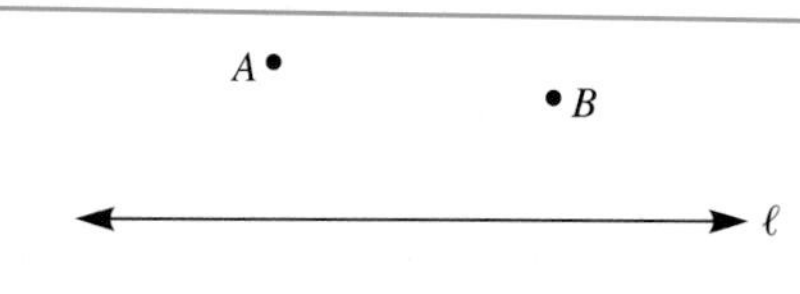

8. For each of the following figures, show that appropriate triangles are similar and find x and y:

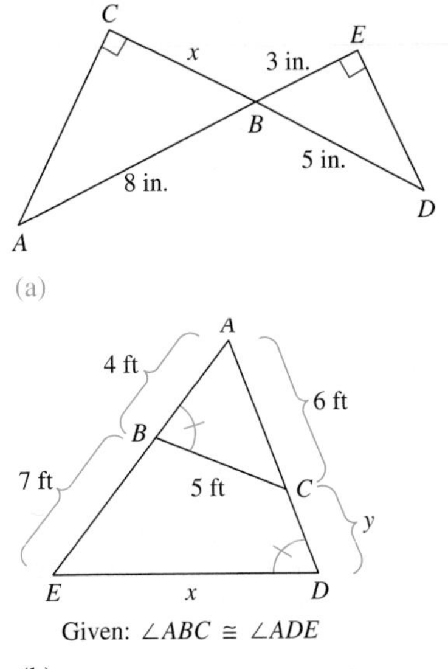

9. Determine if each of the following is true or false. If false, explain why.
   a. A radius of a circle is a chord of the circle.
   b. If a radius bisects a chord of a circle, then it is perpendicular to the chord.

10. A person 2 m tall casts a shadow 1 m long when a building has a 6-m shadow. How high is the building?

11. a. Which of the following polygons can be inscribed in a circle? Assume that all sides of each polygon are congruent and that all the angles of polygons (ii) and (iii) are congruent.

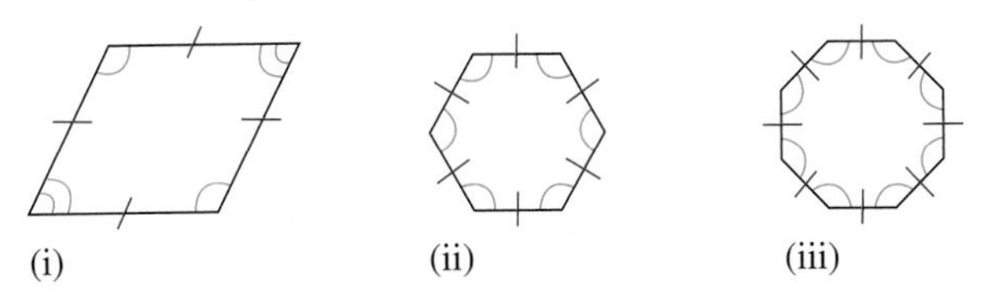

   b. Based on your answer in (a), make a conjecture about what kinds of polygons can be inscribed in a circle.

12. Determine the vertical height of the playground slide shown in the following figure:

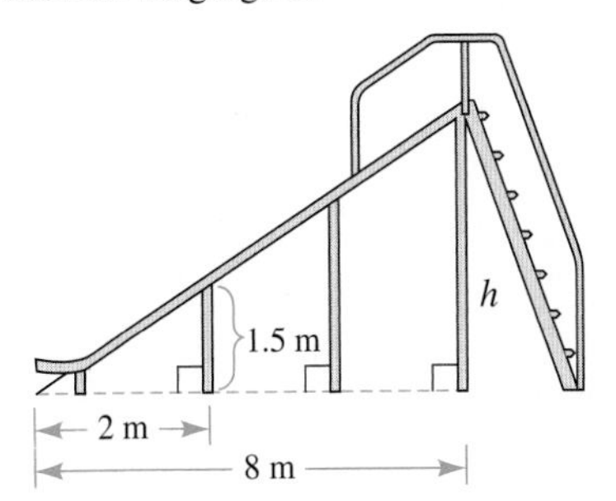

13. Find the distance d across the river sketched.

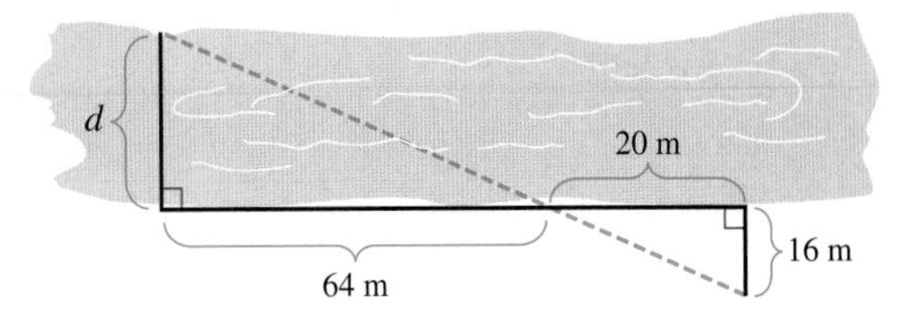

14. Describe how any parallelogram can be dissected and reassembled to form a rectangle.

15. A regular hexagon $ABCDEF$ can be divided into two congruent trapezoids by drawing $\overline{AD}$.
   a. What is the relation of $\overline{AD}$ to the circumcircle of the regular hexagon?
   b. Prove that $\angle ABD$ is a right angle.
   *c. Use trigonometry to find BD in terms of the length of a side of the regular hexagon.

16. If two lines on a coordinate plane are parallel, justify whether or not their slopes must always be equal.

17. In an equilateral triangle, what must be true about all midsegments?

18. In any triangle, a midsegment is drawn. The triangle is divided into two parts, a triangle and a quadrilateral.
   a. Prove that the triangle is similar to the original.
   b. Prove that the quadrilateral is a trapezoid.

19. Is the following statement always true, always false, or true in some cases and false in others? Explain your answer.
   A quadrilateral whose diagonals are congruent and perpendicular is a square.

20. For each of the following, write the equation of the line determined by the given pair of points.
   a. $(2, {}^-3)$ and $({}^-1, 1)$
   b. $({}^-3, 0)$ and $(3, 2)$

21. Use slope to determine if there is a single line through the points with coordinates $(4, 2)$, $(0, {}^-1)$, and $(7, {}^-5)$. Explain your reasoning.

22. Solve each of the following systems, if possible. If the system does not have a unique solution, explain why not.
   a. $x + 2y = 3$
      $2x - y = 9$
   b. $\dfrac{x}{2} + \dfrac{y}{3} = 1$
      $4y - 3x = 2$
   c. $x - 2y = 1$
      $4y - 2x = 0$

23. Prove that two congruent triangles ABC and $A'B'C'$ are similar.

*24. Prove that $\sin 60° \neq 2 \sin 30°$.

* 25. Solve for x in the following:

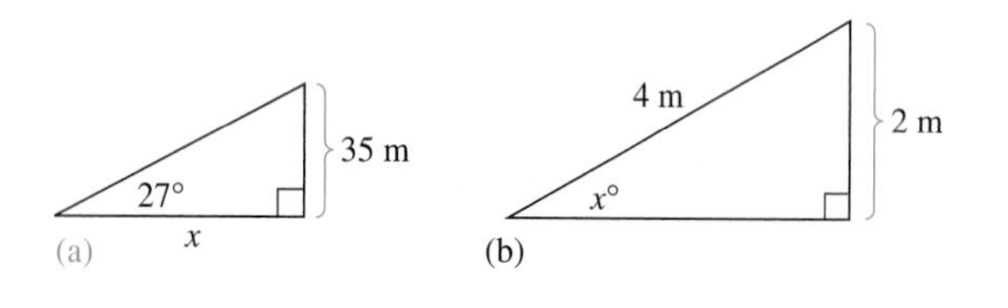

* 26. a. Argue that $\tan 30° = \dfrac{\sin 30°}{\cos 30°}$.

b. Argue that in general $\tan A = \dfrac{\sin A}{\cos A}$.

SELECTED BIBLIOGRAPHY

Bartels, B. "Truss(t)ing Triangles." *Mathematics Teaching in the Middle School* 3 (March–April 1998): 394–396.

Beckmann, C., and K. Rozanski. "Graphs in Real Time." *Mathematics Teaching in the Middle School* 5 (October 1999): 92–93.

Beigie, D. "Coordinate Plane Set Detective," *Mathematics Teaching in the Middle School* 9 (January 2004): 251–255.

Beigie, D. "Computer-Generated Fractal Art," *Mathematics Teaching in the Middle School* 10 (February 2005): 262–269.

Carpenter, T., M. Corbitt, H. Kepner, M. Lindquist, and R. Reys. *Results from the Second Mathematics Assessment of the National Assessment of Educational Progress*. Reston, VA.: National Council of Teachers of Mathematics, 1981.

Drum, R. "Miniature Toys Introduce Ratio and Proportion with a Real-world Connection." *Mathematics Teaching in the Middle School* 7 (September 2001): 50–54.

Dwyer, N., B. Causey-Lee, and N. Irby. "Conceptualizing Ratios with Look-alike Polygons," *Mathematics Teaching in the Middle School* 8 (April 2005): 426–431.

Greer, G., and L. Wantuck. "Menu of Problems." *Mathematics Teaching in the Middle School* 1 (March–April 1996): 729.

Kastberg, S. "Euclidean Tools and the Creation of Euclidean Geometry." *Mathematics Teaching in the Middle School* 7 (January 2002): 294–296.

Kennedy, J., and E. McDowell. "Geoboard Quadrilaterals." *Mathematics Teacher* 91 (April 1998): 288–290.

Manouchehri, A., M. Enderson, and L. Pagnucco. "Exploring Geometry with Technology." *Mathematics Teaching in the Middle School* 3 (March–April 1998): 436–442.

Nieuwoudt, H., and R. van Niekerk. "The Spatial Competence of Young Children through the Development of Solids." Paper presented at the meeting of the American Educational Research Association, Chicago, March, 1997.

Sanders, C. "Geometric Constructions: Visualizing and Understanding Geometry." *Mathematics Teacher* 91 (October 1998): 554–556.

Sharp, J., and K. Hoiberg. "And Then There Was Luke: The Geometric Thinking of a Young Mathematician." *Teaching Children Mathematics* 7 (March 2001): 432–439.

Slavit, D. "Above and beyond AAA: The Similarity and Congruence of Polygons." *Mathematics Teaching in the Middle School* 3 (January 1998): 276–280.

Slovin, H. "Moving to Proportional Reasoning." *Mathematics Teaching in the Middle School* 6 (September 2000): 58–60.

Van Dyke, F., and J. Tomback. "Collaborating to Introduce Algebra," *Mathematics Teaching in the Middle School* 10 (December/January 2005): 236–242.

Vennebush, P. "Menu of Problems." *Mathematics Teaching in the Middle School* 3 (February 1998): 352–353.

Welchman, R., and J. Urso. "Midpoint Shapes." *Teaching Children Mathematics* 6 (April 2000): 506–509.

Zykova, V. "The Psychology of Sixth-Grade Pupils' Mastery of Geometric Concepts," In *The Learning of Mathematical Concepts*, edited by J. Kilpatrick and I. Wirzup. Soviet Studies in the Psychology of Learning and Teaching Mathematics (vol. 1). Palo Alto, Ca.: SMSG, 1969.

Concepts of Measurement

PRELIMINARY PROBLEM

Chuck used a dynamic geometry utility and kept drawing bow tie shapes. To do this, he used two parallel lines where *A* and *B* are on one line, *C* and *D* are on the other. He moved only point *D* to change the shape. Elizabeth noted that no matter where he moved *D*, one thing remained true about the triangles making the bow tie. What did she observe about triangles *ACE* and *DBE*?

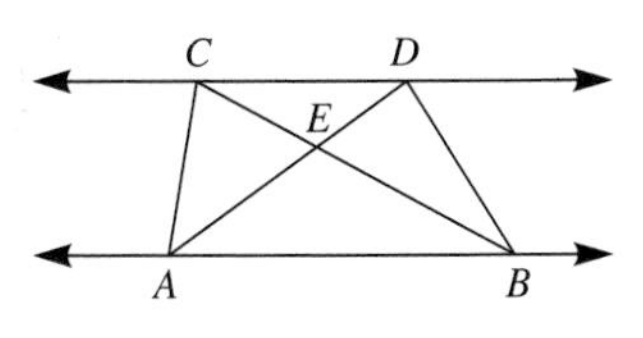

Principles and Standards discusses the measurable attributes of objects as follows:

> *A measurable attribute is a characteristic of an object that can be quantified. Line segments have length, plane regions have area, and physical objects have mass. As students progress through the curriculum from preschool through high school, the set of attributes they can measure should expand. Recognizing that objects have attributes that are measurable is the first step in the study of measurement. Children in prekindergarten through grade 2 begin by comparing and ordering objects using language such as longer and shorter. Length should be the focus in this grade band, but weight, time, area, and volume should also be explored. In grades 3–5, students should learn about area more thoroughly, as well as perimeter, volume, temperature, and angle measure. In these grades, they learn that measurements can be computed using formulas and need not always be taken directly with a measuring tool. Middle-grade students build on these earlier measurement experiences by continuing their study of perimeter, area, and volume and by beginning to explore derived measurements, such as speed.* (p. 44)

In the United States, two measurement systems are used regularly: the English system and the metric system. In *Principles and Standards*, we find the following concerning the use of the two measurement systems:

> *Since the customary English system of measurement is still prevalent in the United States, students should learn both customary and metric systems and should know some rough equivalences between the metric and customary systems—for example, that a two-liter bottle of soda is a little more than half a gallon. The study of these systems begins in elementary school, and students at this level should be able to carry out simple conversions within both systems. Students should develop proficiency in these conversions in the middle grades and should learn some useful benchmarks for converting between the two systems.* (pp. 45–46)

Many concepts of measurement are more confusing for children than for adults because students lack everyday measurement experiences. In this chapter, we use both systems of measurement for length, area, volume, mass, and temperature with the philosophy that students should learn to think within a system. We also develop formulas for the area of plane figures and for surface areas and volumes of solids. We use the concept of area in discussing the Pythagorean Theorem.

▶ HISTORICAL NOTE Howard Eves, a noted mathematics writer and historian (1911–2004) described the construction of the metric system as one of the most important accomplishments of the eighteenth century. He wrote that the system replaced the "world's vast welter of chaotic and unscientific systems of weights and measures by one that is orderly, uniform, scientific, exact, and simple" (p. 342). ◀

11-1 Linear Measure

To measure a segment, we must decide on a unit of measure. Early attempts at measurement lacked a standard unit and used fingers, hands, arms, and feet as units of measure. "Hands" are still used to measure the heights of horses. These early crude measurements were refined eventually and standardized by the English.

NOW TRY THIS 11-1

The *Principles and Standards* state that in grades PreK–2, students should begin their study of measurement using nonstandard units such as paper clips (p. 45).

a. Estimate the length of this textbook page in terms of paper clips.
b. Measure the textbook in paper clips.
c. How close was your estimate to the measure?
d. Repeat parts (a–c) using *hands*.

The English System

Originally, in the English system, a yard was the distance from the tip of the nose to the end of an outstretched arm of an adult person and a foot was the length of a human foot. In 1893, the United States defined the yard and other units in terms of metric units.

HISTORICAL NOTE

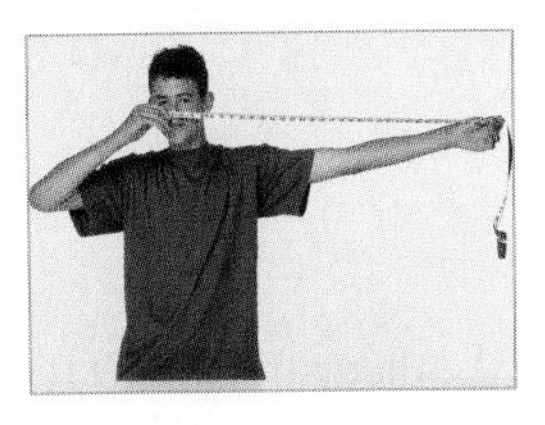

The yard has been defined in many different ways in the United States over the history of the country. In 1832, it was defined as the distance between the 27th and 63rd inches on a certain brass bar made by Troughton of London. In 1856, the yard was redefined in terms of the British Bonze Yard No. 11 (which was 0.00087 in. longer than the Troughton yard). In 1893, the yard was redefined in terms of the international meter as 3600/3937 of a meter through use of an 1866 U.S. law making the use of the metric system permissible in the United States. In 1960, the meter was redefined in terms of the wavelength of light from krypton-86 and still later in terms of the distance light travels in $\frac{1}{299,792,458}$ sec. Effective July 1, 1959, the yard is defined in terms of the international yard, which in turn was based on the international definition of a meter. ◀

Some units of length in the English system and relationships among them are summarized in Table 11-1.

Table 11-1

Unit	Equivalent in Other Units
yard (yd)	3 ft
foot (ft)	12 in.
mile (mi)	1760 yd, or 5280 ft

Dimensional Analysis (Unit Analysis)

dimensional analysis

To convert from one unit of measure to another different processes are used by different people, and one such process is known as **dimensional analysis**. This process works with *unit ratios* (ratios equivalent to 1). For example, $\frac{1 \text{ yd}}{3 \text{ ft}}$ and $\frac{5280 \text{ ft}}{1 \text{ mi}}$ are unit ratios. Therefore, to convert 5.25 mi to yards, we have the following:

$$5.25 \text{ mi} = 5.25 \text{ mi} \cdot \frac{5280 \text{ ft}}{1 \text{ mi}} \cdot \frac{1 \text{ yd}}{3 \text{ ft}} = 9240 \text{ yd}$$

EXAMPLE 11-1

If a cheetah is clocked at 60 miles per hour (mph), what is its speed in feet per second?

Solution

$$60\frac{\text{mi}}{\text{hr}} = 60\frac{\text{mi}}{\text{hr}} \cdot \frac{5280 \text{ ft}}{1 \text{ mi}} \cdot \frac{1 \text{ hr}}{60 \text{ min}} \cdot \frac{1 \text{ min}}{60 \text{ sec}} = 88\frac{\text{ft}}{\text{sec}}$$

▲

REMARK The use of *dimensional analysis* will be seen in the solution to many parts of the examples in this text that involve conversions. It will not be the only technique used.

EXAMPLE 11-2

Convert each of the following:

a. 219 ft = _____ yd
b. 8432 yd = _____ mi
c. 0.2 mi = _____ ft
d. 64 in. = _____ yd

Solution **a.** Because $1 \text{ ft} = \frac{1}{3} \text{ yd}$, $219 \text{ ft} = 219 \cdot \frac{1}{3} \text{ yd} = 73 \text{ yd}$. Alternatively, $219 \text{ ft} = 219 \text{ ft} \cdot \frac{1 \text{ yd}}{3 \text{ ft}} = 73 \text{ yd}$.

b. Because $1 \text{ yd} = \frac{1}{1760} \text{ mi}$, $8432 \text{ yd} = 8432 \cdot \frac{1}{1760} \text{ mi} \doteq 4.79 \text{ mi}$. Alternatively, $8432 \text{ yd} = 8432 \text{ yd} \cdot \frac{3 \text{ ft}}{1 \text{ yd}} \cdot \frac{1 \text{ mi}}{5280 \text{ ft}} \doteq 4.79 \text{ mi}$.

c. $1 \text{ mi} = 5280 \text{ ft}$. Hence, $0.2 \text{ mi} = 0.2 \cdot 5280 \text{ ft} = 1056 \text{ ft}$. Alternatively, $0.2 \text{ mi} = 0.2 \text{ mi} \cdot \frac{5280 \text{ ft}}{1 \text{ mi}} = 1056 \text{ ft}$.

d. We first find a connection between yards and inches. We have $1 \text{ yd} = 3 \text{ ft}$ and $1 \text{ ft} = 12 \text{ in.}$ Hence, $1 \text{ yd} = 3 \text{ ft} = 3 \cdot 12 \text{ in.} = 36 \text{ in.}$ Hence, $1 \text{ in.} = \frac{1}{36} \text{ yd}$; therefore $64 \text{ in.} = 64 \cdot \frac{1}{36} \text{ yd} \doteq 1.78 \text{ yd}$. Alternatively, $64 \text{ in.} = 64 \text{ in.} \cdot \frac{1 \text{ ft}}{12 \text{ in}} \cdot \frac{1 \text{ yd}}{3 \text{ ft}} \doteq 1.78 \text{ yd}$.

▲

The Metric System

metric system

At this time, the United States is the only major industrial nation in the world that continues to use the English system. However, the use of the **metric system** in the United States has been increasing, particularly in the scientific community and in industry.

▶ HISTORICAL NOTE

The metric system, a decimal system, was proposed in France in 1670 by Gabriel Mouton. However, not until the French Revolution in 1790 did the French Academy of Sciences bring various groups together to develop the system. The Academy recognized the need for a standard base unit of linear measurement. The members chose $\frac{1}{10{,}000{,}000}$ of the distance from the equator to the North Pole on a meridian through Paris as the base unit of length and called it the **meter (m)**.

meter

The U.S. Congress included encouragement for U.S. industrial metrication in the Omnibus Trade and Competitiveness Act of 1988 by designating the metric system as the preferred system of weights and measures for U.S. trade and commerce and by requiring each federal agency to be metric by the end of fiscal year 1992. By December 31, 2009, all products sold in Europe (with limited exceptions) will be required to have only metric units on their labels. Dual labeling will not be permitted. ◀

Different units of length in the metric system are obtained by multiplying a power of 10 times the base unit. Table 11-2 gives the prefixes for these units, their symbols, and the multiplication factors.

Table 11-2

Prefix	Symbol	Factor
kilo	k	1000 (one thousand)
*hecto	h	100 (one hundred)
*deka	da	10 (ten)
*deci	d	0.1 (one tenth)
centi	c	0.01 (one hundredth)
milli	m	0.001 (one thousandth)

*Not commonly used

The metric prefixes, combined with the base unit meter, name the different units of length. Table 11-3 gives these units, the symbol for each, and their relationship to the meter.

Table 11-3

Unit	Symbol	Relationship to Base Unit
kilometer	km	1000 m
*hectometer	hm	100 m
*dekameter	dam	10 m
meter	**m**	**base unit**
*decimeter	dm	0.1 m
centimeter	cm	0.01 m
millimeter	mm	0.001 m

*Not commonly used

REMARK Two other prefixes, mega (1,000,000) and micro (0.000001), are used for very large and very small units, respectively. The symbols for mega and micro are M and μ, respectively.

Benchmarks that can be used for estimations for a meter, a decimeter, a centimeter, and a millimeter are shown in Figure 11-1. The kilometer is commonly used for measuring longer distances: 1 km = 1000 m. Nine football fields, including end zones, laid end to end are approximately 1 km long.

FIGURE 11-1

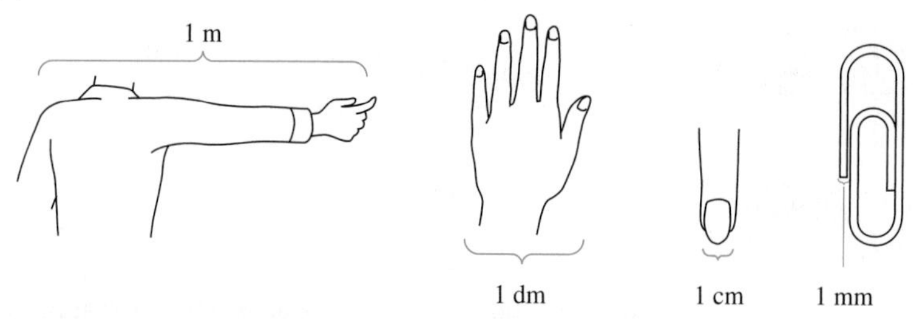

Conversions among metric lengths are accomplished by multiplying or dividing by powers of 10. As with money, we simply move the decimal point to the left or right, depending on the units. For example,

$$0.123 \text{ km} = 1.23 \text{ hm} = 12.3 \text{ dam} = 123 \text{ m} = 1230 \text{ dm} = 12{,}300 \text{ cm} = 123{,}000 \text{ mm}$$

It is possible to convert units by using the chart in Figure 11-2. We count the number of steps from one unit to the other and move the decimal point that many steps in the same direction.

FIGURE 11-2

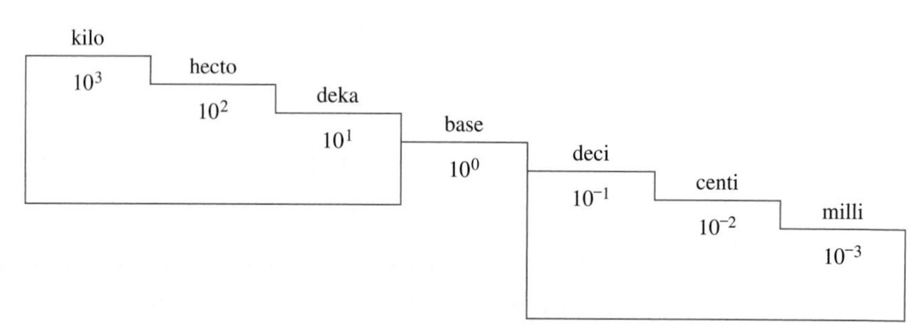

▶ **RESEARCH NOTE**

Research is inconclusive about prerequisite relationships between conservation of measure and a child's ability to measure attributes. Conservation is, however, a prerequisite for understanding the relationship between the size of a unit and the number of units involved in a measurement. For example, the number of inches in a measurement will be greater than the number of yards in the same measurement (Hiebert 1981). ◀

NOW TRY THIS 11-2

If our money system used metric prefixes and the base unit was a dollar, give metric names to each of the following:

a. dime **b.** penny **c.** $10 bill **d.** $100 bill **e.** $1000 bill

EXAMPLE 11-3

Convert each of the following:

a. 1.4 km = _____ m
b. 285 mm = _____ m
c. 0.03 km = _____ cm

Solution **a.** Because 1 km = 1000 m, to change kilometers to meters, we multiply by 1000. Hence, 1.4 km = 1.4 (1000 m) = 1400 m. Alternatively,

$$1.4 \text{ km} = 1.4 \text{ km} \cdot \frac{1000 \text{ m}}{1 \text{ km}} = 1400 \text{ m}.$$

b. Because 1 mm = 0.001 m, to change from millimeters to meters, we multiply by 0.001. Thus, 285 mm = 285 (0.001 m) = 0.285 m. Alternatively,

$$285 \text{ mm} = 285 \text{ mm} \cdot \frac{1 \text{ m}}{1000 \text{ mm}} = 0.285 \text{ m}.$$

c. To change kilometers to centimeters, we first multiply by 1000 to convert kilometers to meters and then multiply by 100 to convert meters to centimeters. Therefore, we move the decimal five places to the right to obtain 0.03 km = 3000 cm. An alternative approach is to use Figure 11-2. To go from kilo to centi on the steps, we move five places to the right, so we need to move the decimal point in 0.03 five places to the right, which is 3000.

Linear units of length are commonly measured with rulers. Figure 11-3 shows part of a centimeter ruler. Rulers can be used to measure distance.

FIGURE 11-3

0 1 2 3
Centimeters

Measuring distances in the real world frequently results in errors. Because of this, many industrial plants using parts from a variety of sources rely on portable calibration units that are taken from plant to plant to test measuring instruments used in constructing the parts. This is done so that the final assembly plant can fit all the parts together to make the product. To calibrate the measuring instruments, technicians must establish the greatest possible error (GPE) allowable in order to obtain the final fit. The **greatest possible error (GPE)** of a measurement is one-half the unit used. For example, if the width of a piece of board was measured to the nearest centimeter as 5 cm, the actual width must be between 4.5 cm and 5.5 cm. Therefore, the GPE for this measurement is 0.5 cm. If the width of a button is measured as 1.2 cm, then the actual width is between 1.15 cm and 1.25 cm, and so the GPE is 0.05 cm or 0.5 mm.

greatest possible error (GPE)

When drawings are given, we assume that the listed measurements are accurate. When actually measuring objects in the real world, we find that such accuracy is usually impossible.

Distance Properties

A person using the expression "the shortest distance between two points is a straight line" may have good intentions, but the expression is actually false. (Why?) The shortest among all the polygonal paths in a plane connecting two points A and B is the segment $\overline{AB}$. (The length of $\overline{AB}$ is denoted by AB.) This fact is one of the basic properties of distance listed next.

Properties

1. The distance between any two points A and B is greater than or equal to 0, written $AB \geq 0$.
2. The distance between any two points A and B is the same as the distance between B and A, written $AB = BA$.
3. For any three points A, B, and C, the distance between A and B plus the distance between B and C is greater than or equal to the distance between A and C, written $AB + BC \geq AC$.

In the special case where A, B, and C are collinear and B is between A and C, as in Figure 11-4(a), we have $AB + BC = AC$. Otherwise, if A, B, and C are not collinear, as in Figure 11-4(b), then they form the vertices of a triangle and $AB + BC > AC$. This inequality, $AB + BC > AC$, is the **Triangle Inequality**.

Triangle Inequality

Triangle Inequality

The sum of the lengths of any two sides of a triangle is greater than the length of the third side.

FIGURE 11-4

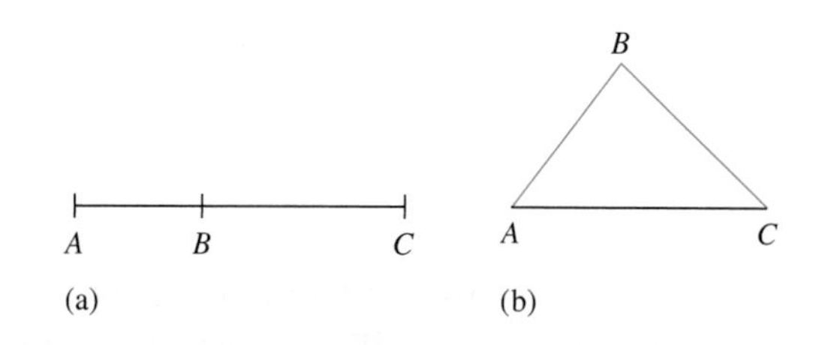

NOW TRY THIS 11-3

If two sides of a triangle are 31 cm and 85 cm long and the measure of the third side must be a whole number of centimeters,

a. what is the longest the third side can be?
b. what is the shortest the third side can be?

REMARK Notice the difference between AB, $\overline{AB}$, and $\overleftrightarrow{AB}$. AB is the distance between two points A and B and therefore a nonnegative real number, $\overline{AB}$ is the segment connecting points A and B and therefore a set of points, $\overleftrightarrow{AB}$ is the line through points A and B.

Distance Around a Plane Figure

perimeter The **perimeter** of a simple closed curve is the length of the curve. If a figure is a polygon, its perimeter is the sum of the lengths of its sides. A perimeter has linear measure.

EXAMPLE 11-4 Find the perimeter of each of the shapes in Figure 11-5.

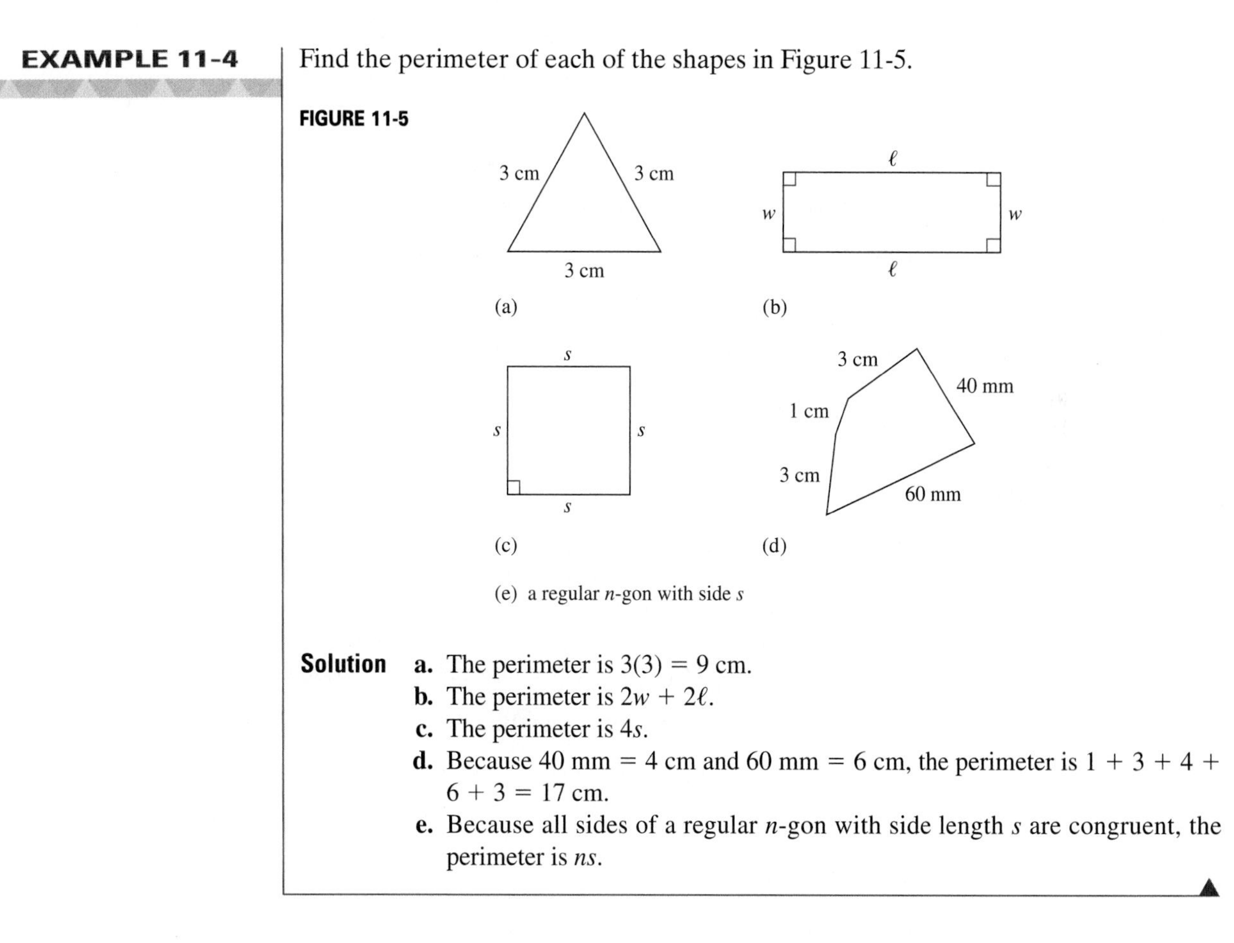

FIGURE 11-5

Solution
- **a.** The perimeter is $3(3) = 9$ cm.
- **b.** The perimeter is $2w + 2\ell$.
- **c.** The perimeter is $4s$.
- **d.** Because 40 mm = 4 cm and 60 mm = 6 cm, the perimeter is $1 + 3 + 4 + 6 + 3 = 17$ cm.
- **e.** Because all sides of a regular n-gon with side length s are congruent, the perimeter is ns.

REMARK When letters are used for distances, as in Example 11-5(b), (c), and (e), we do not usually attach dimensions to an answer. If the perimeter is $4s$, we actually mean that it is $4s$ units of length.

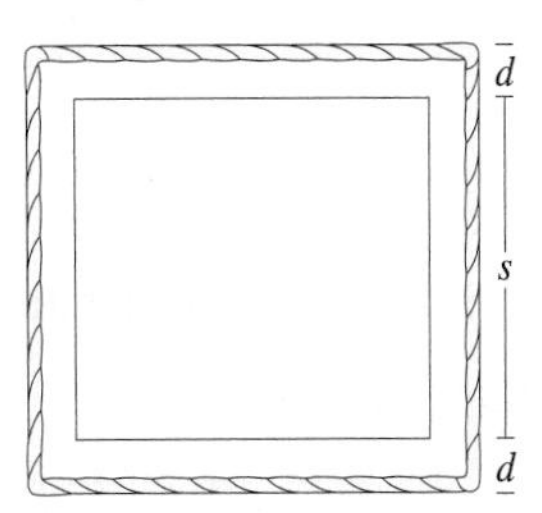

FIGURE 11-6

PROBLEM SOLVING **Roping a Square**

Given a square of any size, stretch a rope tightly around it. Now take the rope off, add 100 ft to it, and put the extended rope back around the square so that the new rope makes a square around the original square, as in Figure 11-6. Find d, the distance between the squares.

Understanding the Problem We are to determine the distance between a square and a new square formed by adding 100 ft of rope to a rope that was stretched around the original square. Figure 11-6 shows the situation if the length of the side of the original square is s and the unknown distance is d.

Devising a Plan If we use variables to represent the unknowns, we can *write an equation* to model the problem. The perimeter of the new square is $4s + 100$. Another way to represent this perimeter is $4(s + 2d)$. Therefore, we have $4s + 100 = 4(s + 2d)$. We must solve this equation for d.

Carrying Out the Plan We solve the equation as shown next.

$$4s + 100 = 4(s + 2d)$$
$$4s + 100 = 4s + 8d$$
$$100 = 8d$$
$$12.5 = d$$

FIGURE 11-7

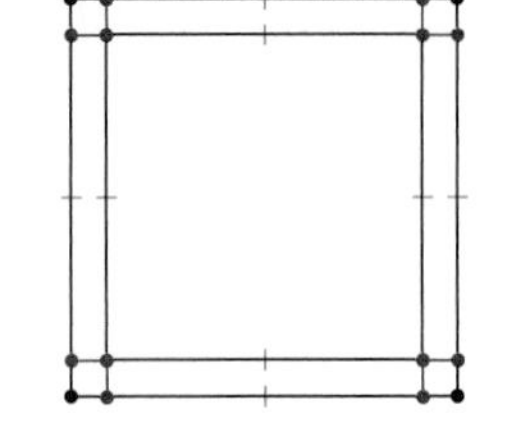

Therefore, the distance between the squares is 12.5 ft.

Looking Back A different way to think about the problem is to consider the individual parts that must sum to 100. Note that the perimeter of the original square (the sum of the marked congruent segments) does not change in Figure 11-7 but the eight red segments sum to 100 ft so one red segment has length 100/8, or 12.5 ft. This problem can be extended to figures other than squares. Try the Brainteaser on page 745.

Circumference of a Circle

circle center circumference pi

A **circle**, as shown in Figure 11-8, is defined as the set of all points in a plane that are the same distance from a given point, the **center**. The perimeter of a circle is its **circumference**. The ancient Greeks discovered that if they divided the circumference of any circle by the length of its diameter, they always obtained approximately the same number. The ratio of circumference C to diameter d is symbolized π (**pi**). In the late eighteenth century, mathematicians proved that the ratio $\frac{C}{d}$, or π, is neither a terminating nor a repeating decimal. Rather, it is an irrational number. For most practical purposes, π is approximated by $\frac{22}{7}$, $3\frac{1}{7}$, or 3.14. These values are not exact values of π.

FIGURE 11-8

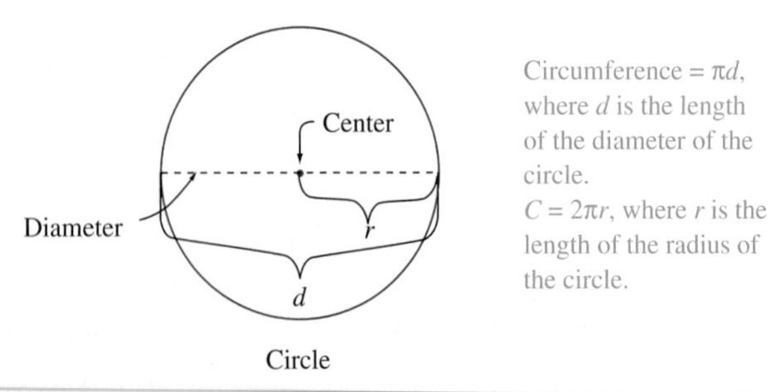
Center
Diameter
r
d
Circle
Circumference = πd, where d is the length of the diameter of the circle.
C = 2πr, where r is the length of the radius of the circle.

The relationship $\frac{C}{d} = \pi$ is used for finding the circumference of a circle and normally is written $C = \pi d$ or $C = 2\pi r$ because the length of the diameter d is twice the radius (r) of the circle. The exact circumference of a circle with diameter 6 cm is 6π cm, and an approximation might be $6(3.14) = 18.84$ cm.

LABORATORY
ACTIVITY

To approximate the value of π, you need string, a marked ruler, and several different-sized round tin cans or jars. Pick a can and wrap the string tightly around the can. Use a pen to mark a point on the string where the beginning of the string meets the string again. Unwrap the string and measure its length. Next, determine the diameter of the can by tracing the bottom of the can on a piece of paper. Fold the circle onto itself to find a line of symmetry. The chord determined by the line is a diameter of the circle. Measure the diameter and determine the ratio of the circumference to the diameter. (Use the same units in all of your measurements.) Repeat the experiment with at least three cans and find the average of the corresponding ratios.

Do this activity with a class and record diameters and corresponding circumferences in columns A and B. Plot the graph of this data with circumference on the vertical axis and the diameter on the horizontal axis. What do you observe? Find a line of best fit for the data? What is the slope of this line?

▶ HISTORICAL NOTE

π = 3.14159 26535 89793 23846 26433 83279 50288 41971 69399 37510 58209 . . . Archimedes (b. 287 BCE) found an approximation for π given by the inequality $3\frac{10}{71} < \pi < 3\frac{10}{70}$. A Chinese astronomer thought that $\pi = \frac{355}{113}$. Ludolph van Ceulen (1540–1610), a German mathematician, calculated π to 35 decimal places. The approximation was engraved on his tombstone. Leonhard Euler adopted the symbol π in 1737. In 1761, Johann Lambert, an Alsatian mathematician, proved that π is an irrational number. In 1989, Columbia University mathematicians and Soviet émigré brothers David and Gregory Chudnovsky used computers to establish 480 million digits of π. If these digits were printed along a line, the line would extend 600 miles. Since 1989 better approximations have been obtained. ◀

BRAIN
TEASER

Suppose a wire is stretched tightly around Earth. (The radius of Earth is approximately 6400 km.) Then suppose the wire is cut and its length is increased by 20 m. It is then placed back around the planet so that it is the same distance from Earth at every point. Could you walk under the wire?

NOW TRY THIS 11-4

Explain the following cartoon.

Arc Length

The length of an arc depends on the radius of the circle and the central angle determining the arc. If the central angle has a measure of 180°, as in Figure 11-9(a), the arc is a **semicircle**. The length of a semicircle is $\frac{1}{2} \cdot 2\pi r$, or πr. The length of an arc whose central angle is $\theta°$ can be developed as in Figure 11-9(b) by using proportional reasoning. Since a circle has 360°, an angle of $\theta°$ determines $\theta/360$ of a circle. Because the circumference of a circle is $2\pi r$, an arc of $\theta°$ has length $\frac{\theta}{360} \cdot 2\pi r$, or $\frac{\pi r \theta}{180}$.

semicircle

FIGURE 11-9

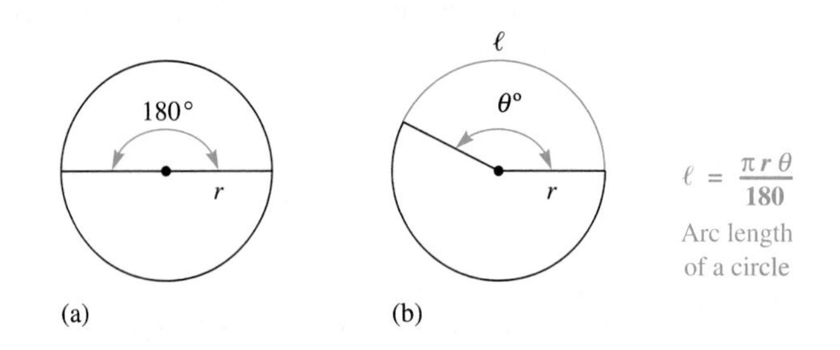

$\ell = \frac{\pi r \theta}{180}$
Arc length of a circle

EXAMPLE 11-5

Find each of the following:

a. The circumference of a circle if the radius is 2 m.
b. The radius of a circle if the circumference is 15π m.
c. The length of a 25° arc of a circle of radius 10 cm.
d. The radius of an arc whose central angle is 87° and whose arc length is 154 cm.

Solution **a.** $C = 2\pi(2) = 4\pi$; thus the circumference is 4π m.

b. $C = 2\pi r$ implies $15\pi = 2\pi r$. Hence, $r = \frac{15}{2}$ and the radius is 7.5 m.

c. The arc length is $\frac{\pi r \theta}{180} = \frac{\pi \cdot 10 \cdot 25}{180}$ cm, or $\frac{25\pi}{18}$ cm, or approximately 4.36 cm.

d. The arc length ℓ is $\frac{\pi r \theta}{180}$, so that $154 = \frac{\pi r \cdot 87}{180}$.

Thus, $r \doteq 101.4$ cm.

ONGOING ASSESSMENT 11-1

1. Use the following picture of a ruler to find each of the lengths in centimeters:

a. AB b. DE c. CJ d. EF
e. IJ f. AF g. IC h. GB

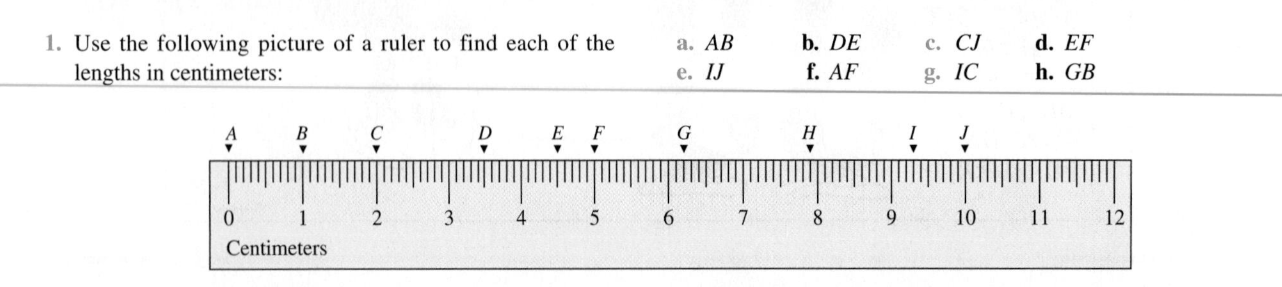

2. Estimate and then measure each of the following in terms of the units listed:

a. The length of your desktop in cubits (elbow to outstretched fingers)

b. The length of this page in pencil lengths

c. The width of this book in pencil widths

3. The United Kingdom used the following money system in the past: 1 pound = 20 shillings, 1 shilling = 12 pence, 1 penny = 2 half-pence, (pence is the plural of penny), 1 penny = 4 farthings. Use dimensional analysis to convert each of the following:

a. 1 pound = ____ farthings

b. 30 shillings = ____ pound

c. 12 pounds = ____ pence

4. Is the American money system an example of a "metric" money system? Why or why not?

5. On a 4×4 geoboard as shown, find examples of the following:

a. The square with greatest perimeter

b. A square with least perimeter

c. A triangle with least perimeter

6. Convert each of the following:

a. 100 in. = ____ yd

b. 400 yd = ____ in.

c. 300 ft = ____ yd

d. 372 in. = ____ ft

7. Draw segments that you estimate to be of the following lengths. Use a metric ruler to check the estimates.

a. 10 mm

b. 100 mm

c. 1 cm

d. 10 cm

e. 0.01 m

f. 15 cm

8. Estimate the length of the following segment and then measure it:

Express the measurement in each of the following units:

a. Millimeters

b. Centimeters

9. Choose an appropriate metric unit and estimate each of the following measures (check your estimates):

a. The length of a pencil

b. The diameter of a nickel

c. The width of the top of a desk

d. The thickness of the top of a desk

e. The length of this sheet of paper

f. The height of a door

10. Redo problem 9 using English measures.

11. Complete the following table:

Item	m	cm	mm
a. Length of a piece of paper		35	
b. Height of a woman	1.63		
c. Width of a filmstrip			35
d. Length of a cigarette			100
e. Length of two meter sticks laid end to end	2		

12. For each of the following, place a decimal point in the number to make the sentence reasonable:

a. A stack of 10 dimes is 1000 mm high.

b. The desk is 770 m high.

c. The distance from one side of a street to the other is 100 m.

d. A dollar bill is 155 cm long.

e. The basketball player is 1950 cm tall.

f. A new piece of chalk is about 8100 cm long.

g. The speed limit in town is 400 km/hr.

13. List the following in decreasing order: 8 cm, 5218 mm, 245 cm, 91 mm, 6 m, 700 mm.

14. Draw each of the following as accurately as possible:

a. A regular polygon whose perimeter is 12 cm

b. A circle whose circumference is 4 in.

c. A triangle whose perimeter is 4 in.

d. A nonconvex quadrilateral whose perimeter is 8 cm

15. a. What is the length of a semicircle of a circle whose radius is 1 unit?

b. What is the length of a semicircle of a circle whose radius is 1/2 unit?

c. Are the two semicircles in parts (a) and (b) similar? If so, what is the ratio of their lengths? If not, why not?

16. Guess the perimeter in centimeters of each of the following figures and then check the estimates using a ruler:

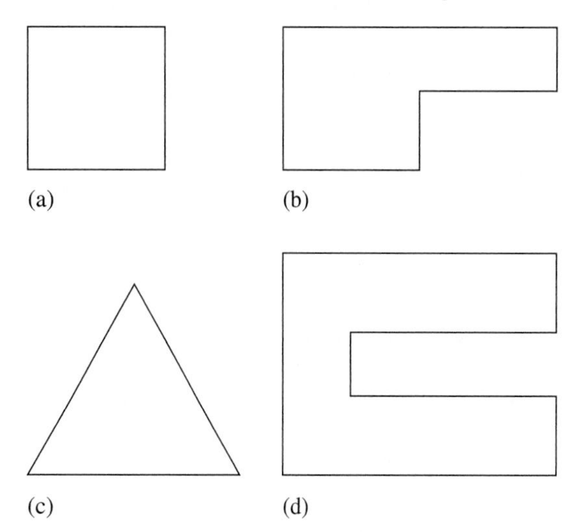

17. Complete each of the following:
    a. 10 mm = ____ cm
    b. 262 m = ____ km
    c. 3 km = ____ m
    d. 30 mm = ____ m
    e. 35 m = ____ cm
    f. 359 mm = ____ m
    g. 647 mm = ____ cm
    h. 0.1 cm = ____ mm
18. Draw a triangle ABC. Measure the length of each of its sides in millimeters. For each of the following, tell which is greater and by how much:
    a. $AB + BC$ or AC
    b. $BC + CA$ or AB
    c. $AB + CA$ or BC
19. Can the following be the lengths of the sides of a triangle? Why or why not?
    a. 23 cm, 50 cm, 60 cm
    b. 10 cm, 40 cm, 50 cm
    c. 410 mm, 260 mm, 14 cm
20. a. How do the ratios of the perimeters of similar triangles compare to the ratios of the corresponding sides? Why?
    b. Does the relationship in part (a) hold for all similar polygons?
21. Take an $8\frac{1}{2}$- × 11-in. piece of typing paper, fold it as shown in the following figure, and then cut the folded paper along the diagonal as shown.

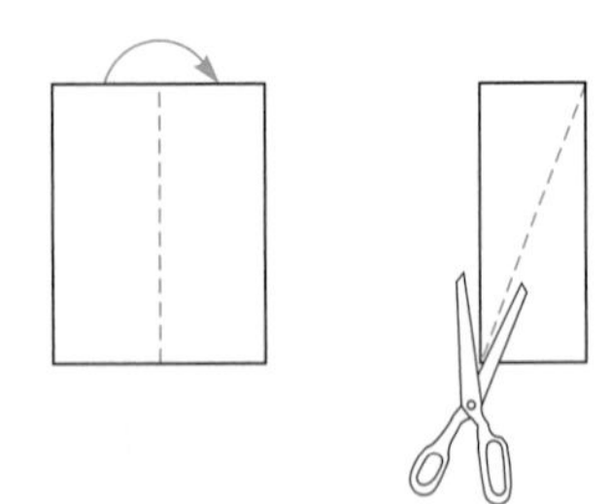

    a. Rearrange the two smaller pieces to find a triangle with the minimum perimeter.
    b. Arrange the two smaller pieces to form a triangle with the maximum perimeter.
22. The following figure made of 6 unit squares has a perimeter of 12 units. The figure is made in such a way that each square must share at least one complete side with another square.

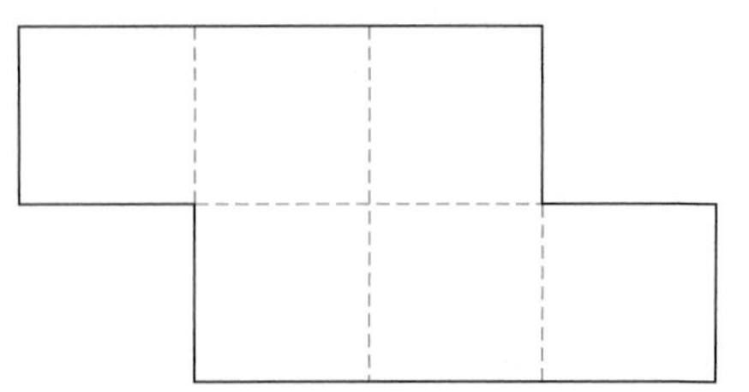

    a. Add more squares to the figure so that the perimeter of the new figure is 18.
    b. What is the minimum number of squares required to make a new figure of perimeter 18?
    c. What is the maximum number of squares that can be used to make a new figure of perimeter 18?
23. For each of the following circumferences, find the radius of the circle:
    a. 12π cm
    b. 6 m
    c. 0.67 m
    d. 92π cm
24. In the "Frank and Ernest" cartoon, how would the uncrossed formula be simplified?

Frank and Ernest

©1990 Thaves. Reprinted with permission. Newspaper dist. by NEA, Inc.

25. For each of the following, if a circle has the dimensions given, determine its circumference:
    a. 6 cm diameter
    b. 3 cm radius
    c. $\frac{2}{\pi}$ cm radius
    d. 6π cm diameter
26. What happens to the circumference of a circle if the length of the radius is doubled?
27. The following figure is a circle whose radius is r units. The diameters of the two semicircular regions inside the large circle are also r units long. Compute the length of the curve that separates the shaded and white regions.

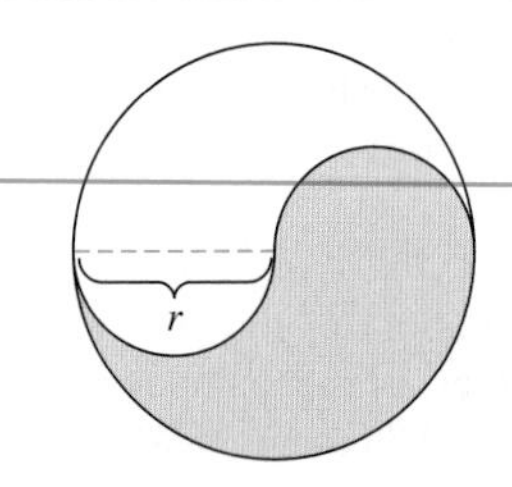

28. Astronomers use a light-year to measure distance. A light-year is the distance light travels in 1 yr. The speed of light is 300,000 km/sec.
    a. How long is 1 light-year in kilometers?
    b. The nearest star (other than the sun) is Alpha Centauri. It is 4.34 light-years from Earth. How far is that in kilometers?
    c. How long will it take a rocket traveling 60,000 km/hr to reach Alpha Centauri?
    d. How long will it take the rocket in (c) to travel to the Sun if it takes approximately 8 min 19 sec for light from the sun to reach Earth?
29. Jet planes can exceed the speed of sound, so a new measurement called *Mach number* was invented to measure the speed of such planes. Mach 2 is twice the speed of sound. (Mach number is a number that indicates the ratio of the speed of an object through a medium to the speed of sound in the medium.) The speed of sound in air is approximately 344 m/sec.
    a. Express Mach 2.5 in kilometers per hour.
    b. Express Mach 3 in meters per second.
    c. Express the speed of 5000 km/hr as a Mach number.
30. Refer to the following figure and determine the perimeter of the paint lane (shaded area) and the semicircle determined by the free-throw line on a basketball court:

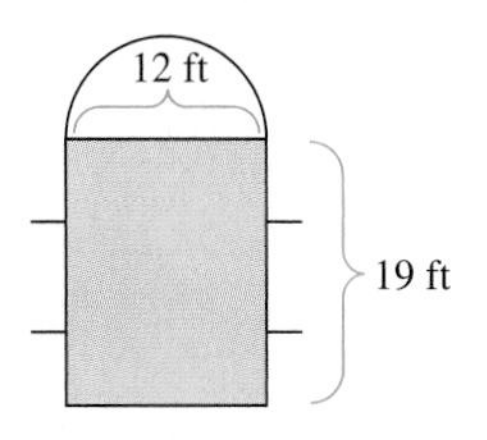

31. Find the length of the side of a square that has the same perimeter as a rectangle that is 66 cm by 32 cm.
32. Give the greatest possible error for each of the following measurements:
    a. 23 m
    b. 3.6 cm
    c. 3.12 m
33. On a circular merry-go-round, one horse is 3 m from the center and another is 6 m from the center. The merry-go-round makes 3 revolutions per minute. Are the bases of the merry-go-round that the two horses are standing on traveling at the same speed? If not, how fast is each traveling?

Communication

34. Howard Eves (1911–2004) described the creation of the metric system as one of the greatest accomplishments of the eighteenth century. Do you agree or disagree? Explain.
35. How could you define the perimeter of a nonsimple closed curve? Give an example in your explanation.
36. A student claims that the circumference of a semicircle is the same as that of a circle from which the semicircle was taken because one has to measure the "outer part" and the "inner part." How do you respond?
37. What are some of the ramifications of changing the standard for the length of a yard over the history of the United States?
38. Explain whether or not you think that using an unsharpened pencil as a measuring tool is an example of a standard or nonstandard measuring.
39. A contractor claimed to sell concrete by the yard. Is that claim accurate? Explain your reasoning.
40. There has been considerable debate about whether the United States should change to the metric system. (a) Based on your experiences with linear measure, what do you see as the advantages of changing? (b) Which system do you think would be easier for children to learn? Why? (c) What things will probably not change if the United States adopts the metric system?
41. A student has a tennis ball can with a flat top and bottom containing three tennis balls. To the student's surprise, the perimeter of the top of the can is longer than the height of the can. The student wants to know if this fact can be explained without performing any measurements. Can you help?
42. In track, the second lane from the inside of the track is longer than the inside lane. Use this information to explain why, in running events that require a complete lap of the track, runners are lined up at the starting blocks as shown in the following figure:

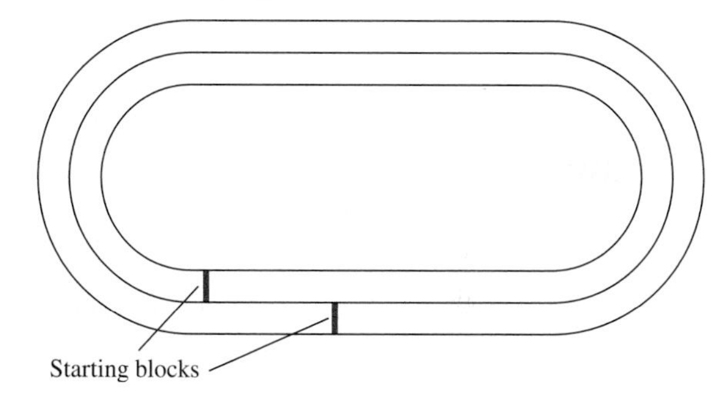

Open-Ended

43. Observe that it is possible to build a triangle with toothpicks that has sides of 3, 4, and 5 toothpicks, as shown, and answer the questions that follow.

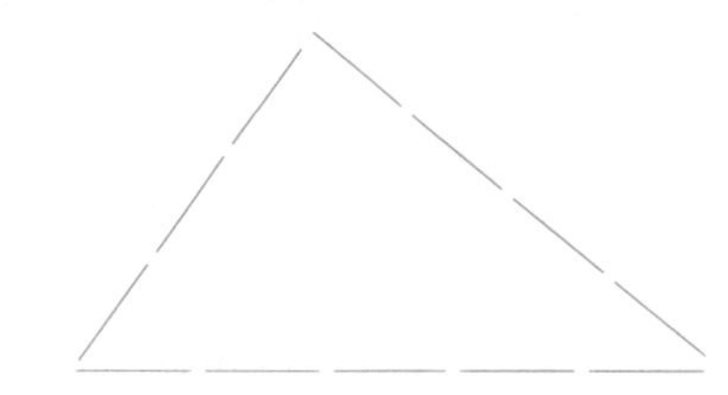

    a. Find two other triples of toothpicks that can be used as sides of a triangle and two other triples that cannot be used to create a triangle.
    b. Describe how to tell whether a given triple of numbers a, b, c can be used to construct a triangle with sides of $a, b,$ and c toothpicks. Explain why your rule is valid.

Cooperative Learning

44. **a.** Help each person in the group find his or her height in centimeters.

b. Help each person in the group find the length of his or her outstretched arms (horizontal) from fingertip to fingertip.

c. Compare the difference between the two measurements in parts (a) and (b). Compare the results of the group members and make a conjecture about the lengths of the two measurements.

d. Compare your group's results with other groups to determine if they have similar findings.

45. Jerry wants to design a gold chain 60 cm long made of thin gold wire circles, each of which is the same size. He wants to use the least amount of wire and wonders what the radius of each circle should be.

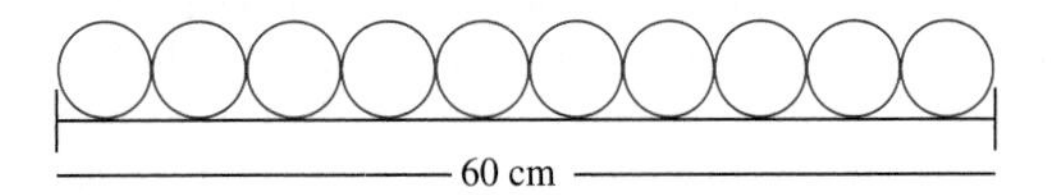

a. Each member of the group should choose a specific number of circles and find the length of wire needed to make a 60-cm chain with the chosen number of circles.

b. Compare your results and make a conjecture based on the results.

c. Justify your conjecture.

Third International Mathematics and Science Study (TIMSS) Questions

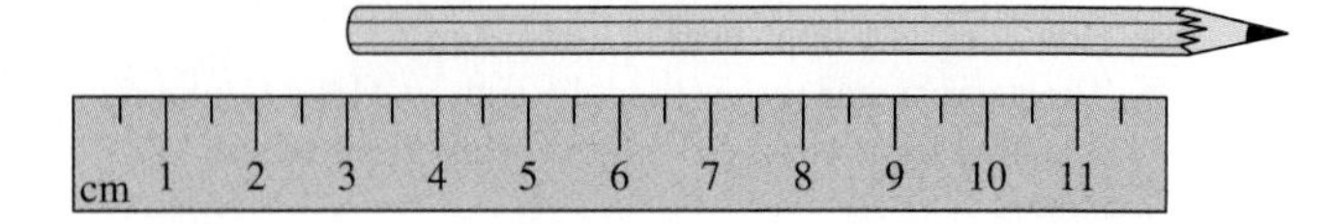

Which of these is closest to the length of the pencil in the figure?

a. 9 cm

b. 10.5 cm

c. 12 cm

d. 13.5 cm

TIMSS, Grade 8, 1994

What is the ratio of the length of a side of a square to its perimeter?

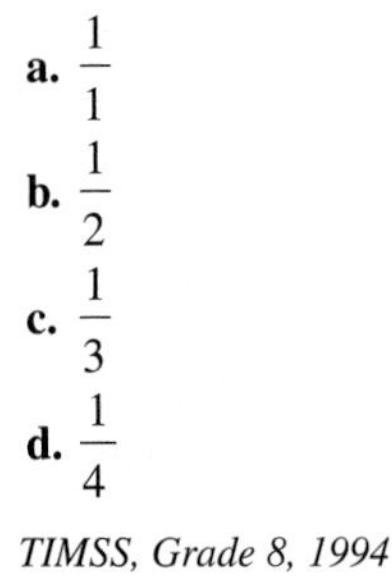

a. $\frac{1}{1}$

b. $\frac{1}{2}$

c. $\frac{1}{3}$

d. $\frac{1}{4}$

TIMSS, Grade 8, 1994

LABORATORY ACTIVITY

Use an ordinary 8½ by 11-in. typing paper to answer the following:

a. What is its perimeter?

b. Cut a shape from the page that you can walk through *without* cutting any of the edges of the original page. What is the approximate perimeter of the new shape?

11-2 Areas of Polygons and Circles

Area is measured using square units and the area of a region is the number of nonoverlapping square units that covers the region. A square measuring 1 ft on a side has an area of 1 square foot, denoted 1 ft^2. A square measuring 1 cm on a side has an area of 1 square centimeter, denoted 1 cm^2.

REMARK Students sometimes confuse the area of 5 cm^2 with the area of a square 5 cm on each side. The area of a square 5 cm on each side is $(5\text{ cm})^2$, or 25 cm^2. Five squares each 1 cm by 1 cm have the area of 5 cm^2. Thus, 5 $cm^2 \neq (5\text{ cm})^2$.

Areas on a Geoboard

In teaching the concept of area, intuitive activities should precede the development of formulas. Many such activities can be accomplished using a *geoboard* or *dot paper*. Notice that the square unit is defined in the upper left corner of the dot paper in Figure 11-10(a). The area of the pentagon can be found by finding the sum of the areas of smaller pieces. Finding the area in this way is the *addition method*. The region in Figure 11-10(b) has been divided into smaller pieces in Figure 11-10(c). What is the area of this shape?

FIGURE 11-10

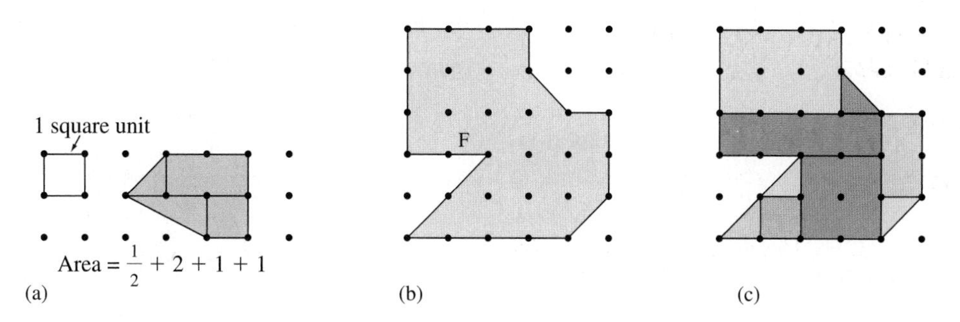

FIGURE 11-11

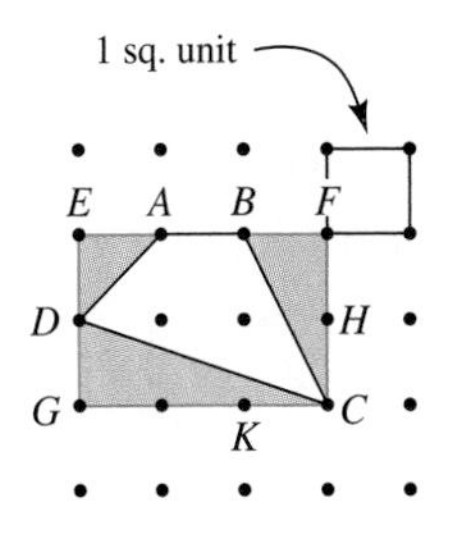

Another method of finding the area of shapes on the dot paper is the *rectangle method*. To find the area of quadrilateral *ABCD* in Figure 11-11, we construct the rectangle *EFCG* around the quadrilateral and then subtract the areas of the shaded triangles *EAD, BFC,* and *DGC*. The area of the rectangle *EFCG* can be counted to be 6 square units. The area of $\triangle EAD$ is $\frac{1}{2}$ square unit, and the area of $\triangle BFC$ is half the area of rectangle *BFCK*, or $\frac{1}{2}$ of 2, or 1 square unit. Similarly, the area of $\triangle DGC$ is half the area of rectangle *DHCG*, that is, $\frac{1}{2} \cdot 3$, or $\frac{3}{2}$ square units. Consequently, the area of *ABCD* is $6 - \left(\frac{1}{2} + 1 + \frac{3}{2}\right)$, or 3 square units.

EXAMPLE 11-6 Using a geoboard, find the area of each of the shaded parts of Figure 11-12.

FIGURE 11-12

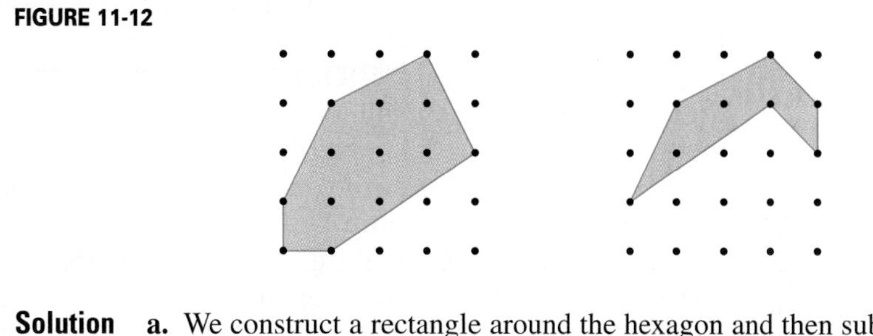

Solution **a.** We construct a rectangle around the hexagon and then subtract the areas of regions *a, b, c, d,* and *e* from the area of this rectangle, as shown in Figure 11-13. Therefore, the area of the hexagon is $16 - (3 + 1 + 1 + 1 + 1)$, or 9 square units. The addition method could also be used in this problem.

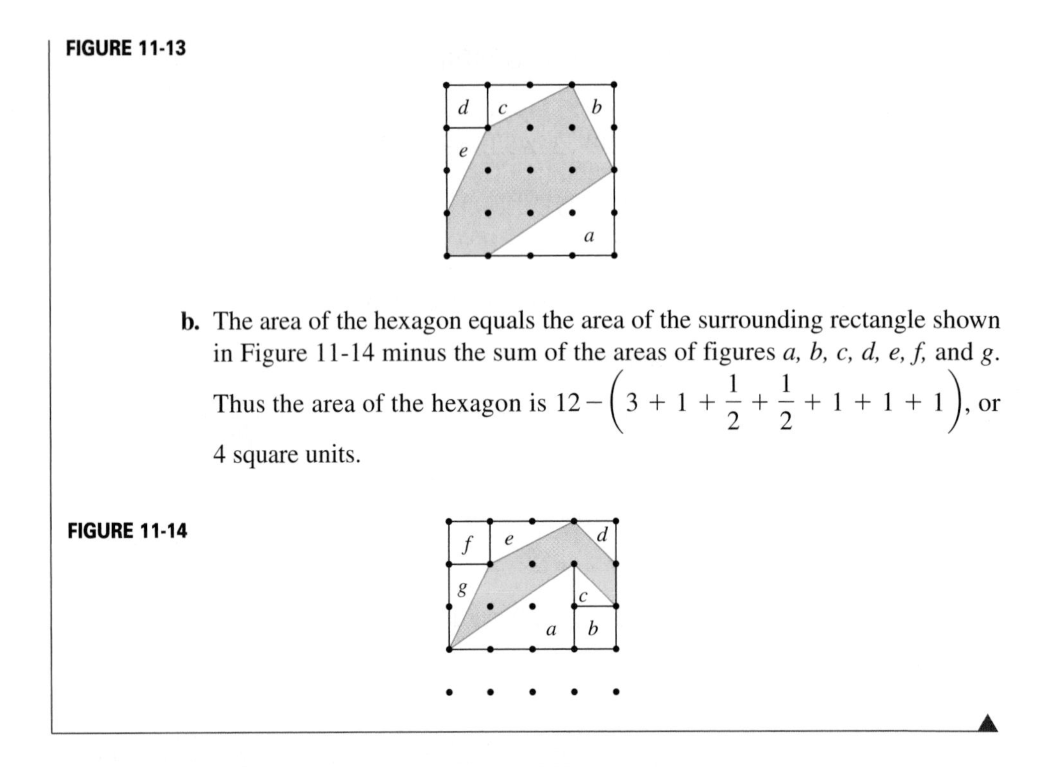

FIGURE 11-13

b. The area of the hexagon equals the area of the surrounding rectangle shown in Figure 11-14 minus the sum of the areas of figures *a, b, c, d, e, f,* and *g*. Thus the area of the hexagon is $12 - \left(3 + 1 + \frac{1}{2} + \frac{1}{2} + 1 + 1 + 1\right)$, or 4 square units.

FIGURE 11-14

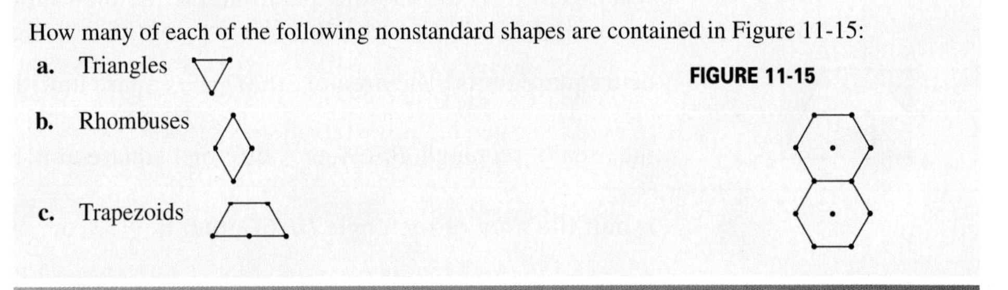

NOW TRY THIS 11-5

How many of each of the following nonstandard shapes are contained in Figure 11-15:

a. Triangles

b. Rhombuses

c. Trapezoids

FIGURE 11-15

Converting Units of Area

The price of new carpet for a house is quoted in terms of dollars per square yard, for example, $12.50/yd^2. The most commonly used units of area in the English system are the square inch (in.^{2}), the square foot (ft^2), the square yard (yd^2), the square mile (mi^2), and, for land measure, the acre (A). In the metric system, the most commonly used units are the square millimeter (mm^2), the square centimeter (cm^2), the square meter (m^2), the square kilometer (km^2), and, for land measure, the hectare (ha). It is often necessary to convert from one area measure to another within a system.

To determine how many 1-cm squares are in a square meter, look at Figure 11-16(a). There are 100 cm in 1 m, so each side of the square meter has a measure of 100 cm. Thus, it takes 100 rows of 100 1-cm squares each to fill a square meter, that is, $100 \cdot 100$, or 10,000 1-cm squares. Because the area of each centimeter square is $1 \text{ cm} \cdot 1 \text{ cm}$, or 1 cm^2, there are 10,000 cm^2 in 1 m^2. In general, the area A of a square that is s units on a side is s^2, as shown in Figure 11-16(b).

FIGURE 11-16

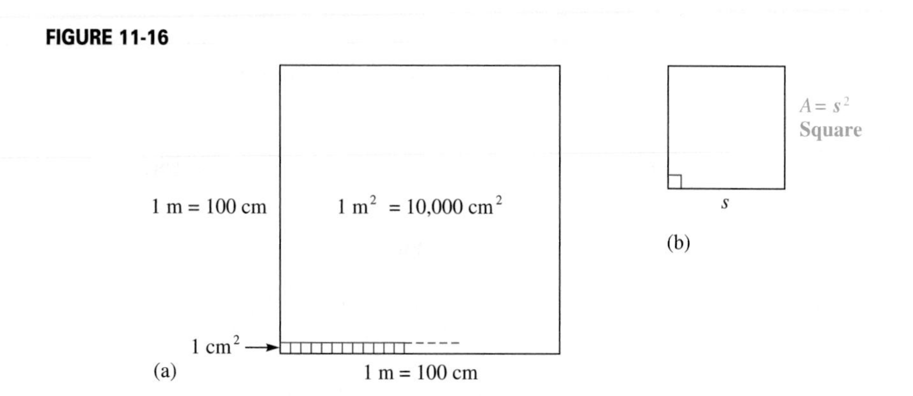

Other metric conversions of area measure can be developed similarly. For example, Figure 11-17(a) shows that $1 \text{ m}^2 = 10{,}000 \text{ cm}^2 = 1{,}000{,}000 \text{ mm}^2$. Likewise, Figure 11-17(b) shows that $1 \text{ m}^2 = 0.000001 \text{ km}^2$. Similarly, $1 \text{ cm}^2 = 100 \text{ mm}^2$ and $1 \text{ km}^2 = 1{,}000{,}000 \text{ m}^2$.

FIGURE 11-17

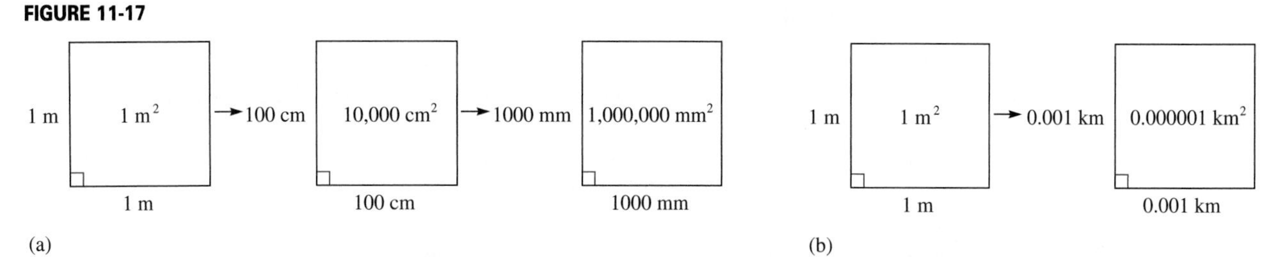

Table 11-4 shows the symbols for metric units of area and their relationship to the square meter.

Table 11-4

Unit	Symbol	Relationship to Square Meter
square kilometer	km^2	1,000,000 m^2
*square hectometer	hm^2	10,000 m^2
*square dekameter	dam^2	100 m^2
square meter	$\mathbf{m^2}$	**1** $\mathbf{m^2}$
*square decimeter	dm^2	0.01 m^2
square centimeter	cm^2	0.0001 m^2
square millimeter	mm^2	0.000001 m^2

*Not commonly used

EXAMPLE 11-7

Convert each of the following:

a. $5 \text{ cm}^2 = \_\_\_ \text{ mm}^2$

b. $124{,}000{,}000 \text{ m}^2 = \_\_\_ \text{ km}^2$

Solution **a.** $1 \text{ cm}^2 = 100 \text{ mm}^2$ implies $5 \text{ cm}^2 = 5 \cdot 1 \text{ cm}^2 = 5 \cdot 100 \text{ mm}^2 = 500 \text{ mm}^2$.

b. $1 \text{ m}^2 = 0.000001 \text{ km}^2$ implies $124{,}000{,}000 \text{ m}^2 = 124{,}000{,}000 \cdot 1 \text{ m}^2 = 124{,}000{,}000 \cdot 0.000001 \text{ km}^2 = 124 \text{ km}^2$.

An alternate solution follows:

$$124{,}000{,}000 \text{ m}^2 \cdot \frac{0.000001 \text{ km}^2}{1 \text{ m}^2} = 124 \text{ km}^2$$

Based on the relationship among units of length in the English system, it is possible to convert among English units of area. For example, because 1 yd = 3 ft, it follows that $(1 \text{ yd})^2 = 1 \text{ yd} \cdot 1 \text{ yd} = 3 \text{ ft} \cdot 3 \text{ ft} = 9 \text{ ft}^2$. Similarly, because 1 ft = 12 in., $(1 \text{ ft})^2 = 1 \text{ ft} \cdot 1 \text{ ft} = 12 \text{ in.} \cdot 12 \text{ in.} = 144 \text{ in.}^2$ Table 11-5 summarizes various relationships among units of area in the English system.

Table 11-5

Unit of Area	Equivalent of Other Units
1 ft^2	$\frac{1}{9} \text{ yd}^2$, or 144 in.^2
1 yd^2	9 ft^2
1 mi^2	$3{,}097{,}600 \text{ yd}^2$, or $27{,}878{,}400 \text{ ft}^2$

Land Measure

acre

square mile

One application of area today is in land measure. The common unit of land measure in the English system is the **acre**. Historically, an acre was the amount of land a man with one horse could plow in one day. There are 4840 yd^2 in 1 acre. For very large land measures in the English system, the **square mile** (mi^2), or 640 acres, is used.

are

hectare

square kilometer

In the metric system, small land areas are measured in terms of a square unit 10 m on a side, called an **are** (pronounced "air") and denoted by **a**. Thus, $1 \text{ a} = 10 \text{ m} \cdot 10 \text{ m}$, or 100 m^2. Larger land areas are measured in **hectares**. A hectare is 100 a. A hectare, denoted by **ha**, is the amount of land whose area is $10{,}000 \text{ m}^2$. It follows that 1 ha is the area of a square that is 100 m on a side. Therefore, $1 \text{ ha} = 1 \text{ hm}^2$. For very large land measures, the **square kilometer**, denoted by km^2, is used. One square kilometer is the area of a square with a side 1 km, or 1000 m, long. Land area measures are summarized in Table 11-6.

Table 11-6

Unit of Area	Equivalent of Other Units
1 a	100 m^2
1 ha	100 a, or $10{,}000 \text{ m}^2$, or 1 hm^2
1 km^2	$1{,}000{,}000 \text{ m}^2$
1 acre	4840 yd^2
1 mi^2	640 acres

EXAMPLE 11-8

a. A square field has a side of 400 m. Find the area of the field in hectares.
b. A square field has a side of 400 yd. Find the area of the field in acres.

Solution **a.** $A = (400 \text{ m})^2 = 160{,}000 \text{ m}^2 = \frac{160{,}000}{10{,}000} \text{ ha} = 16 \text{ ha}$

b. $A = (400 \text{ yd})^2 = 160{,}000 \text{ yd}^2 = \frac{160{,}000}{4840} \text{ acre} \doteq 33.1 \text{ acre}$

▲

TECHNOLOGY CORNER

GSP Lab 8 in Appendix III can be used at this point to motivate the formulas for finding the areas of a rectangle, parallelogram, and trapezoid.

Area of a Rectangle

To measure area, we may count the number of units of area contained in any given region. For example, suppose the square in Figure 11-18(a) represents 1 square unit. Then, the rectangle *ABCD* in Figure 11-18(b) contains $3 \cdot 4$, or 12 square units.

FIGURE 11-18

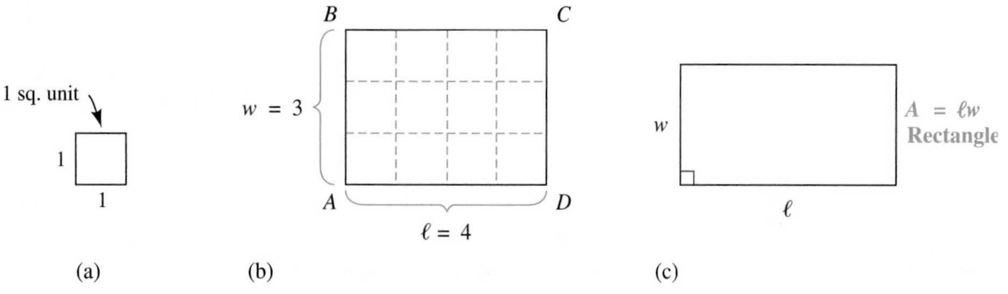

If the unit in Figure 11-18(a) is 1 cm^2, then the area of rectangle *ABCD* is 12 cm^2. In general, the area *A* of any rectangle can be found by multiplying the lengths of two adjacent sides ℓ and w, or $A = \ell w$, as given in Figure 11-18(c).

EXAMPLE 11-9 Find the area of each rectangle in Figure 11-19.

FIGURE 11-19

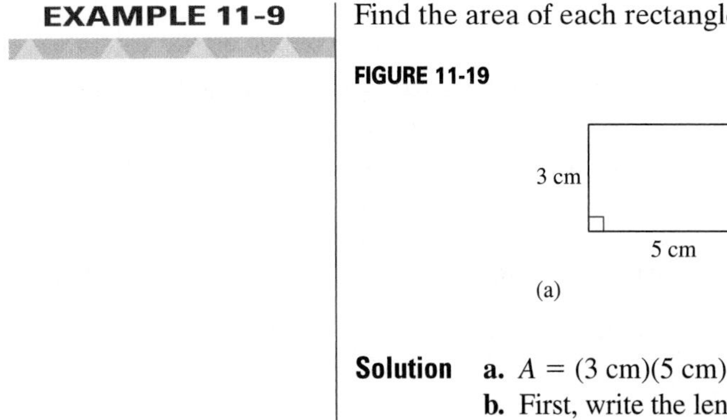
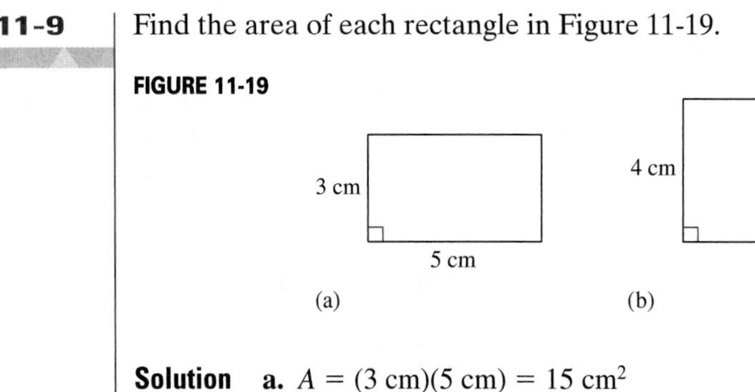

Solution **a.** $A = (3\text{ cm})(5\text{ cm}) = 15\text{ cm}^2$

b. First, write the lengths of the sides in the same unit of length. Because 0.2 m = 20 cm, $A = (4\text{ cm})(20\text{ cm}) = 80\text{ cm}^2$. Alternatively, 4 cm = 0.04 m, so $A = (0.04\text{ m})(0.2\text{ m}) = 0.008\text{ m}^2$.

NOW TRY THIS 11-6

Estimate the area in square centimeters of a dollar bill. Measure and calculate how close your estimate is to the actual area.

Area of a Parallelogram

The area of a parallelogram can be found by *reducing the problem to one that we already know how to solve*, in this case, finding the area of a rectangle. To develop the area formula for a parallelogram, complete Now Try This 11-7.

NOW TRY THIS 11-7

Cut out a parallelogram *ABCD* similar to the one in Figure 11-20(a). Now cut off a shaded triangle as shown and move it to the right to obtain a rectangle as in Figure 11-20(b).

FIGURE 11-20

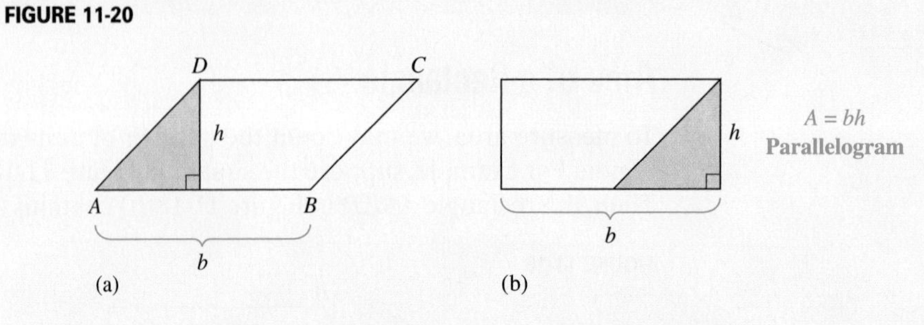

a. How do the areas of the parallelogram and the rectangle compare? Why?
b. How does this experiment lead to the formula for finding the area of a parallelogram?

base ◆ height

In general, any side of a parallelogram can be designated as a **base**. The **height** (h) is the distance between the bases and is always the length of a segment perpendicular to the lines containing the bases. Now Try This 11-7 shows that the area of parallelogram *ABCD* is given by $A = bh$, that is, the length of the base times the corresponding height. Similarly, in Figure 11-21, *EB*, or g, is the height that corresponds to the bases $\overline{AD}$ and $\overline{BC}$, each of which has measure a. Consequently, the area of the parallelogram *ABCD* is ag. Similarly, its area can be expressed as bh. Therefore, $A = ag = bh$.

FIGURE 11-21

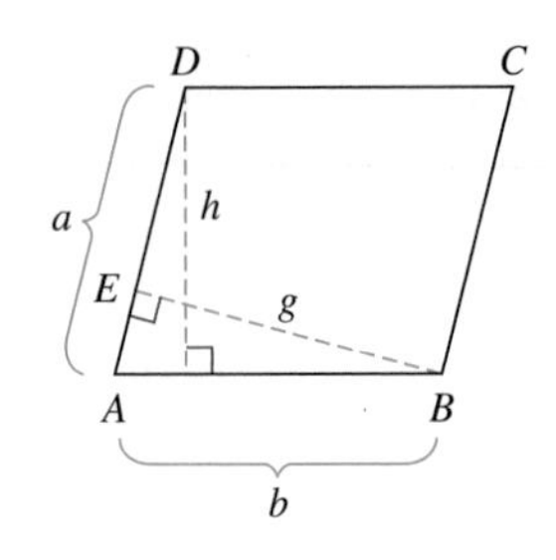

Area of a Triangle

The formula for the area of a triangle can be derived from the formula for the area of a parallelogram. To explore this, suppose $\triangle BAC$ in Figure 11-22(a) has base of length b and height h. Let $\triangle BAC'$ be the image of $\triangle BAC$ when $\triangle BAC$ is rotated 180° about M, the midpoint of $\overline{AB}$, as in Figure 11-22(b). Proving that quadrilateral $BCAC'$ is a parallelogram is left as an exercise. Parallelogram $BCAC'$ has area bh and is constructed of congruent triangles BAC and BAC'. So the area of $\triangle ABC$ is $\frac{1}{2}bh$. In general, the area of a triangle is equal to half the product of the length of a side and the altitude to that side.

FIGURE 11-22

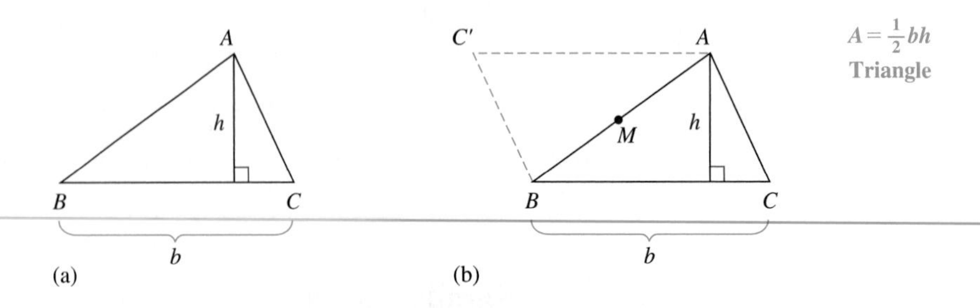

In Figure 11-23, $\overline{BC}$ is a base of $\triangle ABC$, and the corresponding height h_1, or $\overline{AE}$, is the distance from the opposite vertex A to the line containing $\overline{BC}$. Similarly, $\overline{AC}$ can be chosen as a base. Then h_2, or BG, the distance from the opposite vertex B to the line containing $\overline{AC}$, is the corresponding height. If $\overline{AB}$ is chosen as a base, then the corresponding height is h_3, or FC. Thus, the area of $\triangle ABC$ is

$$\text{Area}(\triangle ABC) = \frac{bh_1}{2} = \frac{ah_2}{2} = \frac{ch_3}{2}.$$

FIGURE 11-23

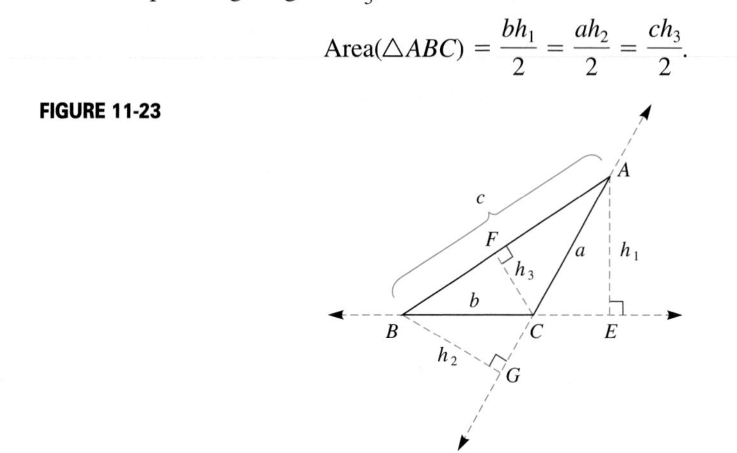

NOW TRY THIS 11-8

To see how the formula for the area of a triangle can be derived from the formula for the area of a rectangle, cut out any triangle as in Figure 11-24(a); fold to find the attitude h; and fold the altitude in half as shown in Figure 11-24(b). Next fold along the colored segments in the trapezoid in Figure 11-24(c) to obtain the rectangle in Figure 11-24(d).

FIGURE 11-24

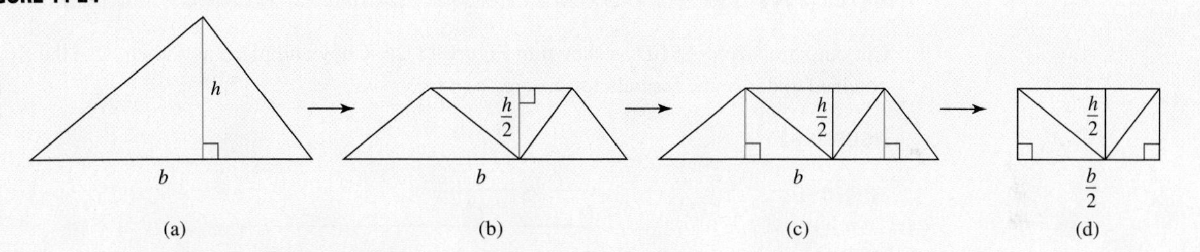

a. What is the area of the rectangle in Figure 11-24(d)?

b. How can the formula for the area of a triangle be developed from your answer in part (a)?

EXAMPLE 11-10

Find the areas in Figure 11-25. Assume the quadrilaterals *ABCD* in (a) and (b) are parallelograms.

FIGURE 11-25

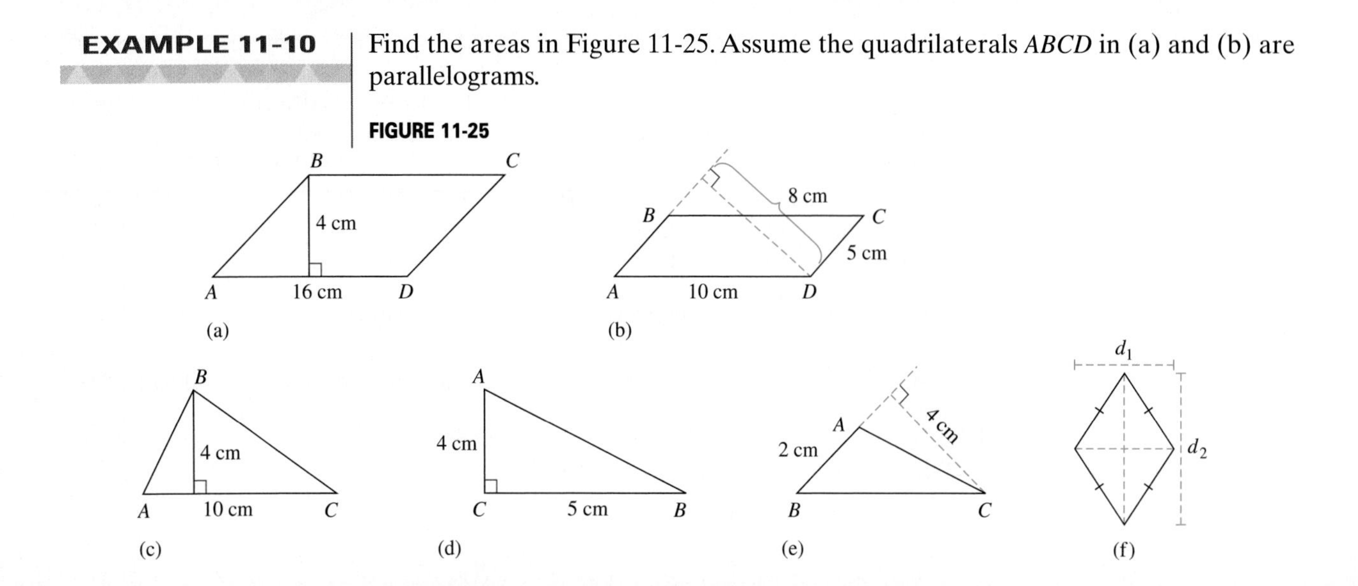

Solution **a.** $A = bh = (16 \text{ cm})(4 \text{ cm}) = 64 \text{ cm}^2$

b. $A = bh = (5 \text{ cm})(8 \text{ cm}) = 40 \text{ cm}^2$

c. $A = \frac{1}{2}bh = \frac{1}{2}(10 \text{ cm})(4 \text{ cm}) = 20 \text{ cm}^2$

d. $A = \frac{1}{2}bh = \frac{1}{2}(5 \text{ cm})(4 \text{ cm}) = 10 \text{ cm}^2$

e. $A = \frac{1}{2}bh = \frac{1}{2}(2 \text{ cm})(4 \text{ cm}) = 4 \text{ cm}^2$

f. $A = 2\left(\frac{1}{2} \cdot \frac{1}{2}d_1d_2\right) = \frac{d_1d_2}{2}$

Note: In Example 11-10(f), *the area of a rhombus is one half the product of its diagonals.*

Area of a Trapezoid

The formula for the area of a trapezoid can be developed using the formula for the area of a parallelogram, as shown in Now Try This 11-9.

NOW TRY THIS 11-9

Cut out a trapezoid *ABCD* as shown in Figure 11-26. Copy and place as shown. Use the figure obtained to derive the formula for the area of a trapezoid.

FIGURE 11-26

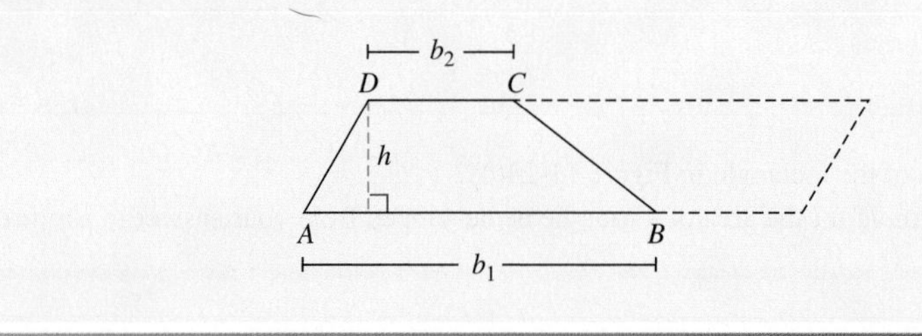

Areas of general polygons can also be found by partitioning the polygons into triangles, finding the areas of the triangles, and summing those areas. In Figure 11-27(a), trapezoid *ABCD* has bases b_1 and b_2 and height h. By connecting points *B* and *D*, as in Figure 11-27(b), we create two triangles: one with base of length *AB* and height *DE* and the other with base of length *CD* and height *BF*. Because $\overline{DE} \cong \overline{BF}$, each has length h. Thus, the areas of triangles *ADB* and *DCB* are $\frac{1}{2}(b_1h)$ and $\frac{1}{2}(b_2h)$, respectively. Hence, the area of trapezoid *ABCD* is $\frac{1}{2}(b_1h) + \frac{1}{2}(b_2h), = \frac{1}{2}h(b_1 + b_2)$. That is, the area of a trapezoid is equal to half the height times the sum of the lengths of the bases, which should be the same as the formula you derived in Now Try This 11-9.

FIGURE 11-27

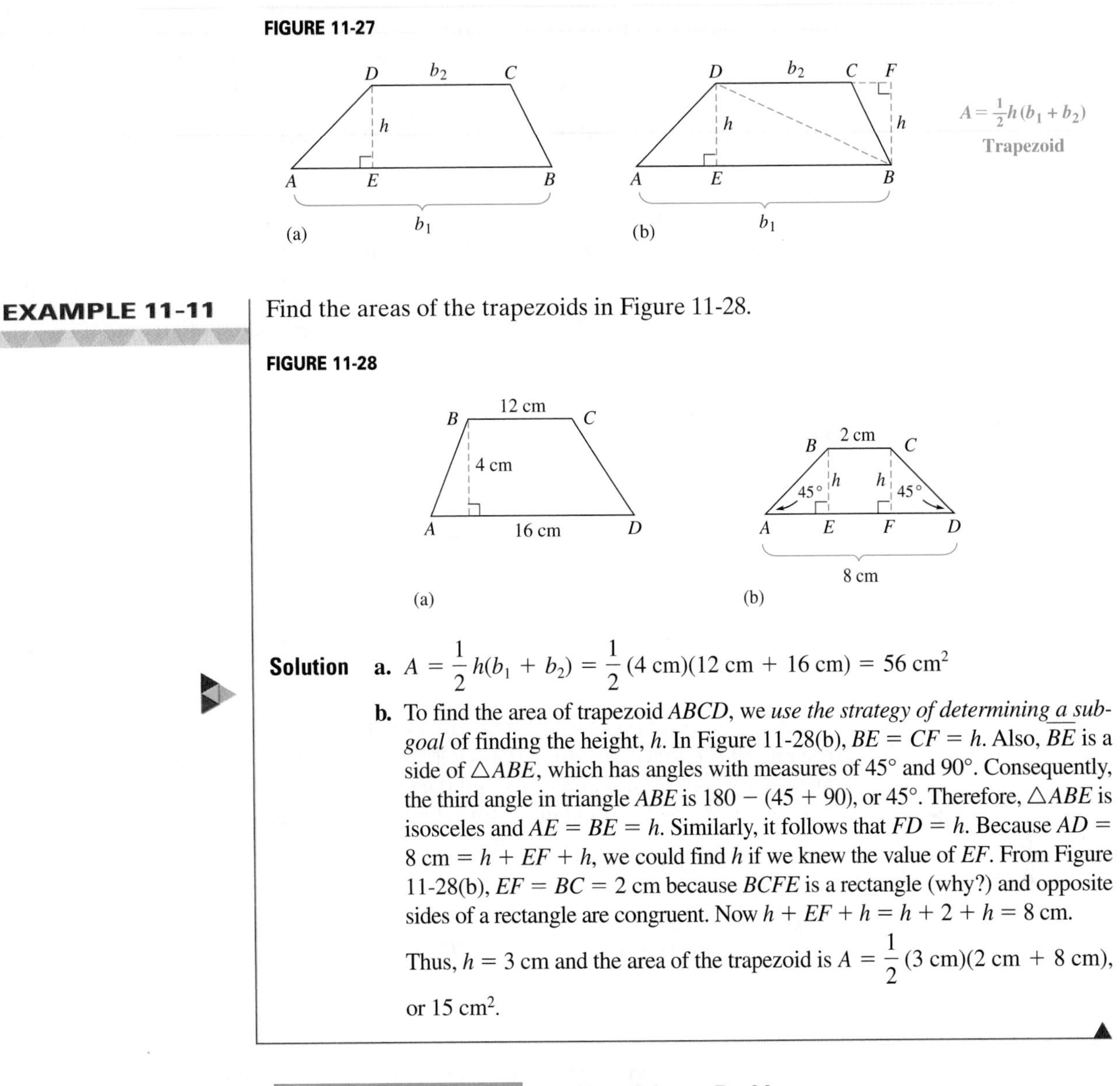

EXAMPLE 11-11 Find the areas of the trapezoids in Figure 11-28.

FIGURE 11-28

Solution **a.** $A = \frac{1}{2}h(b_1 + b_2) = \frac{1}{2}(4 \text{ cm})(12 \text{ cm} + 16 \text{ cm}) = 56 \text{ cm}^2$

b. To find the area of trapezoid $ABCD$, we *use the strategy of determining a subgoal* of finding the height, h. In Figure 11-28(b), $BE = CF = h$. Also, $\overline{BE}$ is a side of $\triangle ABE$, which has angles with measures of 45° and 90°. Consequently, the third angle in triangle ABE is $180 - (45 + 90)$, or 45°. Therefore, $\triangle ABE$ is isosceles and $AE = BE = h$. Similarly, it follows that $FD = h$. Because $AD = 8 \text{ cm} = h + EF + h$, we could find h if we knew the value of EF. From Figure 11-28(b), $EF = BC = 2$ cm because $BCFE$ is a rectangle (why?) and opposite sides of a rectangle are congruent. Now $h + EF + h = h + 2 + h = 8$ cm. Thus, $h = 3$ cm and the area of the trapezoid is $A = \frac{1}{2}(3 \text{ cm})(2 \text{ cm} + 8 \text{ cm})$, or 15 cm^2.

PROBLEM SOLVING **Equal Areas Problem**

Larry purchased a plot of land surrounded by a fence. The former owner had subdivided the land into 13 equal-sized square plots, as shown in Figure 11-29. To reapportion the property into two plots of equal area, Larry wishes to build a single, straight fence beginning at the far left corner (point P on the drawing). Is such a fence possible? If so, where should the other end be?

FIGURE 11-29

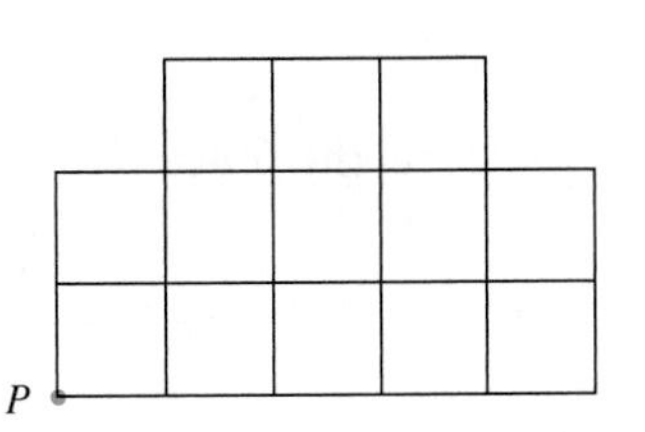

Understanding the Problem We want to divide the land in Figure 11-29 into two plots of equal area by means of a straight fence starting at point P. A *subgoal* is to find the other endpoint. Because the area of the entire plot is 13 square units, the area of each part formed by the fence must be $\frac{1}{2} \cdot 13$, or $6\frac{1}{2}$ square units.

Devising a Plan To find an approximate location for the fence, consider a fence connecting P with point A, as shown in Figure 11-30. The area of the land below $\overline{PA}$ is the sum of the areas of $\triangle APD$ and the rectangle $DAFE$. The area of $\triangle APD$ is 4 and the area of rectangle $DAFE$ is 2, so the area below $\overline{PA}$ is $4 + 2$, or 6, square units. We want an area of $6\frac{1}{2}$ square units. Consequently, the other end of the fence should be above point A.

FIGURE 11-30

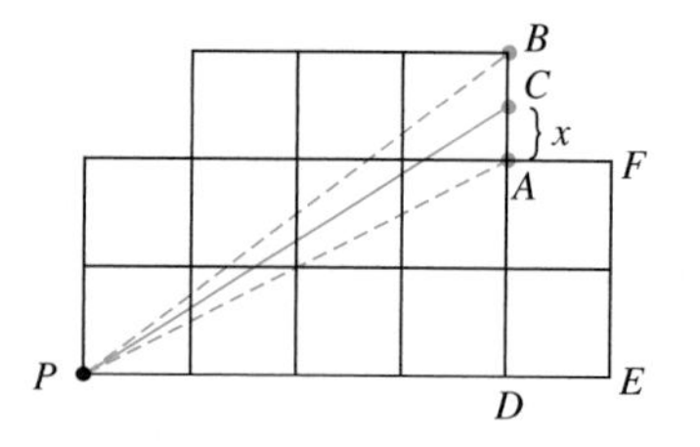

A similar argument shows that the area below $\overline{PB}$ is 8 square units and hence the end of the fence should be below B. Thus, the other end of the fence should be at a point C between A and B. To find the exact location of point C, we designate $\overline{CA}$ by x. We then *write an equation* for x by finding the area below $\overline{PC}$ in terms of x, make the area equal to $6\frac{1}{2}$, and solve for x.

Carrying Out the Plan The area below $\overline{PC}$ equals the area of $\triangle PCD$ plus the area of the rectangle $DAFE$. The area of $\triangle PCD$ is

$$\frac{PD \cdot DC}{2} = \frac{4(2 + x)}{2} = 2(2 + x) = 4 + 2x.$$

The area of rectangle $DAFE$ is 2, so $(4 + 2x) + 2$ should equal half the area of the plot. Consequently, we have the following:

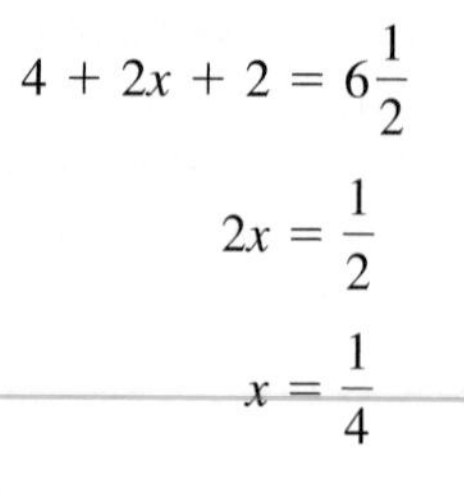

$$4 + 2x + 2 = 6\frac{1}{2}$$
$$2x = \frac{1}{2}$$
$$x = \frac{1}{4}$$

Therefore, the fence should be built along the line connecting point P to point C, which is $\frac{1}{4}$ unit directly above point A.

Looking Back We check that the solution is correct by finding the area above $\overline{PC}$. The problem can be varied by changing the shape of the plot. Another variation is to ask if Larry could divide the plot into thirds, fourths, and so on.

Area of a Regular Polygon

The area of a triangle can be used to find the area of any regular polygon, as illustrated *using a simpler case strategy* involving a regular hexagon in Figure 11-31(a). The hexagon can be separated into six congruent triangles, each with a vertex at the center (the center of the circle containing each vertex), with side s and height a. The height of such a triangle of a regular polygon is the *apothem* and is denoted a. The area of each triangle is $\frac{1}{2}as$. Because six triangles make up the hexagon, the area of the hexagon is $6(\frac{1}{2}as)$, or $\frac{1}{2}a(6s)$. However, $6s$ is the perimeter p of the hexagon, so the area of the hexagon is $\frac{1}{2}ap$. The same process can be used to develop the formula for the area of any regular polygon. That is, the area of any regular polygon is $\frac{1}{2}ap$, where a is the height of one of the triangles involved and p is the perimeter of the polygon, as shown in Figure 11-31(b).

FIGURE 11-31

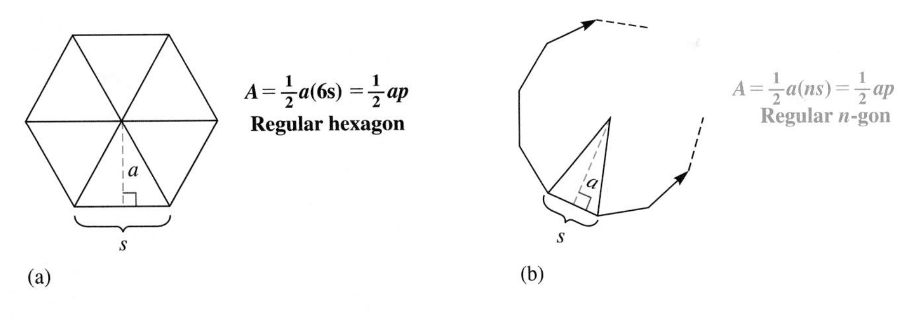

Area of a Circle

We use the strategy of *examining a related problem* to find the area of a circle. The area of a regular polygon inscribed in a circle, as in Figure 11-32, approximates the area of the circle, and we know that the area of any regular n-gon is $\frac{1}{2}ap$, where a is the height of a triangle of the n-gon and p is the perimeter. If the number of sides n is made very large, then the perimeter and the area of the n-gon are close to those of the circle.

FIGURE 11-32

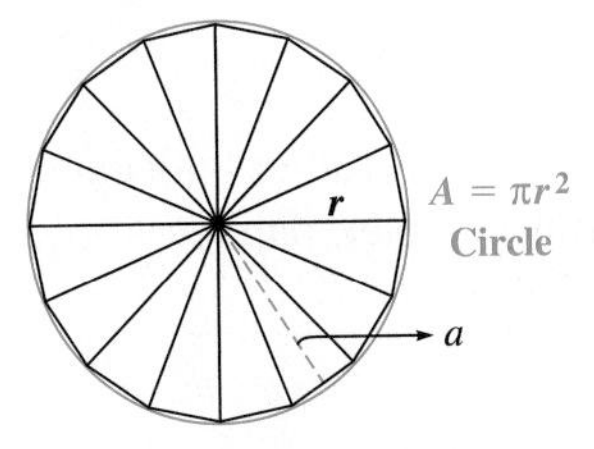

Also, the apothem a is approximately equal to the radius r of the circle, and the perimeter p approximates the circumference $2\pi r$. Because the area of the circle is approximately equal to the area of the n-gon, $\frac{1}{2}ap = \frac{1}{2}r \cdot 2\pi r = \pi r^2$. In fact, the area of the circle is precisely πr^2.

Another approach for leading students to discover the formula for finding the area of a circle is given in Now Try This 11-10.

NOW TRY THIS 11-10

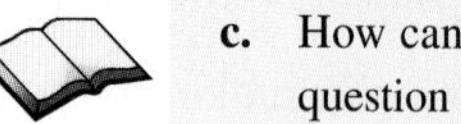

a. Draw a circle of radius 4 cm and divide it into eight equal-sized sections.
b. Using the circle you constructed, answer questions 24–28 on the student page.
c. How can you obtain the formula for the area of a circle by following the instructions in question 28 on the student page?

Area of a Sector

sector
central angle

A **sector** of a circle is a pie-shaped region of the circle determined by an angle whose vertex is the center of the circle. This angle is a **central angle**. The area of a sector depends on the radius of the circle and the measure of the central angle determining the sector. If the angle has a measure of 90°, as in Figure 11-33(a), the area of the sector is one-fourth the area of the circle, or $\frac{90}{360}\pi r^2$. The area of a sector with central angle of 1° is $\frac{1}{360}$ of the area of a circle, and a sector with central angle $\theta°$ has area $\frac{\theta}{360}$ of the area of the circle, or $\frac{\theta}{360}(\pi r^2)$, as shown in Figure 11-33(b).

FIGURE 11-33

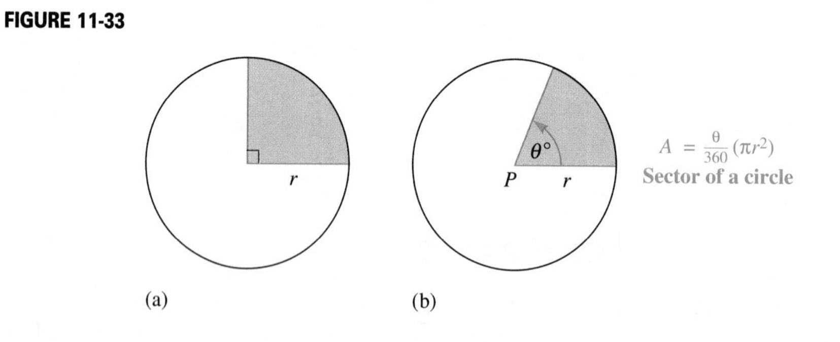

▶ RESEARCH NOTE

Middle-school students find areas and volumes by counting visual units rather than using formulas learned in the past, even if counting is tedious and complex (Figueras and Waldegg 1984). ◀

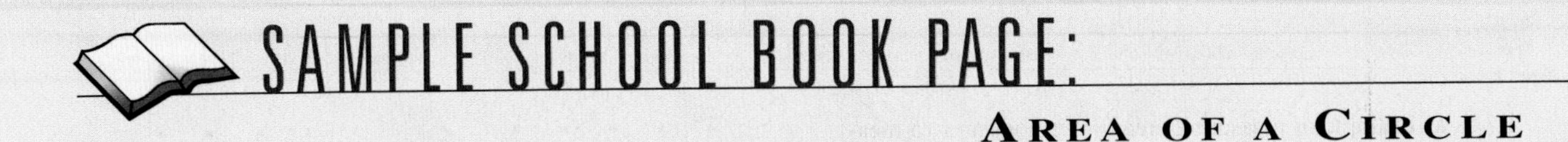

AREA OF A CIRCLE

24 **Use Labsheet 2A.** Follow the directions below.

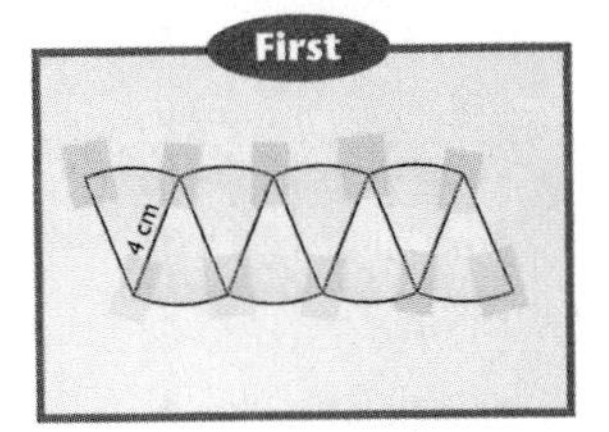

Then

Cut out the *Circle*. Then cut apart the eight sectors and arrange them to form the figure shown. Tape the figure to a sheet of paper.

Use a ruler to draw segments across the top and bottom of your figure. Extend the sides of the figure to meet the bottom segment.

25 What kind of polygon is the new figure you drew in Question 24?

26 **Try This as a Class** Use the figure you made in Question 24.

a. Estimation Explain how you could use the new figure you drew to estimate the area of the circle. Then estimate the area. Do you think this is a good estimate? Why or why not?

b. How is the length of the base of the figure related to the circumference of the circle?

c. How is the height of the figure related to the radius of the circle?

27 **Discussion** Examine the drawings shown.

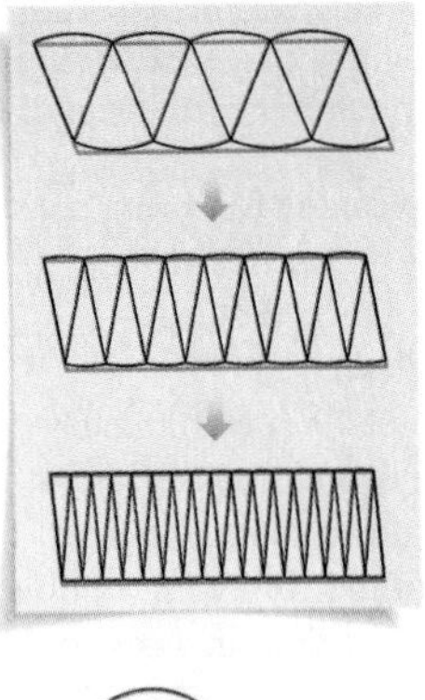

a. As a circle is cut into more and more sectors and put back together as shown, what begins to happen to its shape?

b. The area A of a parallelogram is found by multiplying the length of its base b by the height h, or $A = bh$. Use your figure to explain why this formula can be written as $A = \frac{1}{2}Cr$ to find the area of a circle, where C is the circumference and r is the length of the radius of the circle.

28 The circumference C of a circle is equal to $2\pi r$, where r is the length of the radius. Rewrite the formula $A = \frac{1}{2}Cr$ by substituting $2\pi r$ for C.

Source: Math Thematics, Book 2, McDougal Littell, 2005 (p. 402).

ONGOING ASSESSMENT 11-2

1. As an example of measuring area with a nonstandard measure, estimate the area of your desktop in terms of a piece of notebook paper as your unit of area. Then measure the area of your desktop with the paper and compare it to your estimate.
2. Choose the most appropriate metric units (cm^2, m^2, or km^2) and English units ($in.^2$, yd^2, or mi^2) for measuring each of the following:
   a. Area of a sheet of notebook paper
   b. Area of a quarter
   c. Area of a desktop
   d. Area of a classroom floor
   e. Area of a parallel parking space
   f. Area of an airport runway
3. Estimate and then measure each of the following using cm^2, m^2, or km^2:
   a. Area of a door
   b. Area of a chair seat
   c. Area of a desktop
   d. Area of a whiteboard
4. Complete the following conversion table:

Item	m^2	cm^2	mm^2
a. Area of a sheet of paper		588	
b. Area of a cross section of a crayon			192
c. Area of a desktop	1.5		
d. Area of a dollar bill		100	
e. Area of a postage stamp		5	

5. Using a calculator, complete the following conversions:
   a. $4000\ ft^2 =$ _____ yd^2 **b.** $10^6\ yd^2 =$ _____ mi^2
   c. $10\ mi^2 =$ _____ acre **d.** 3 acres $=$ _____ ft^2
6. Complete each of the following:
   a. A football field is about $49\ m \times 100\ m$ or _____ m^2.
   b. About _____ a are in two football fields.
   c. About _____ ha are in two football fields.
7. Find the areas of each of the following figures if the distance between two adjacent dots in a row or a column is one unit:

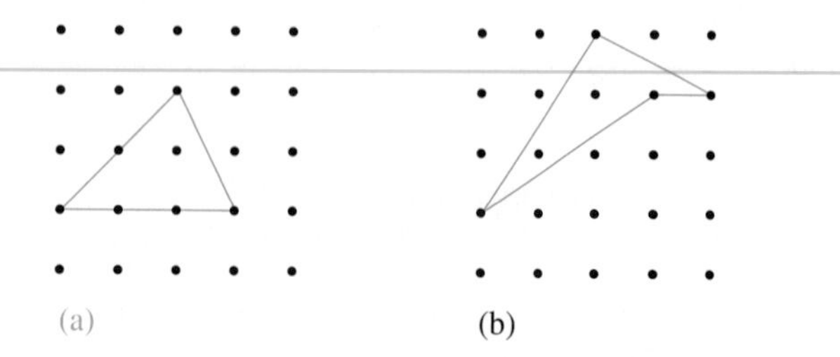

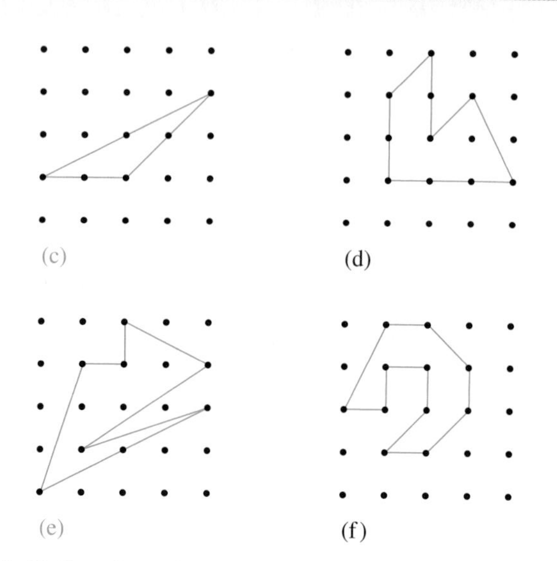

8. If all vertices of a polygon are points on square-dot paper, the polygon is a **lattice polygon**. In 1899, G. Pick discovered a surprising theorem involving I, the number of dots *inside* the polygon, and B, the number of dots that lie *on* the polygon. The theorem states that the area of any lattice polygon is $I + \frac{1}{2}B - 1$. Check that this is true for the polygons in problem 7.
9. Find the area of $\triangle ABC$ in each of the following triangles:

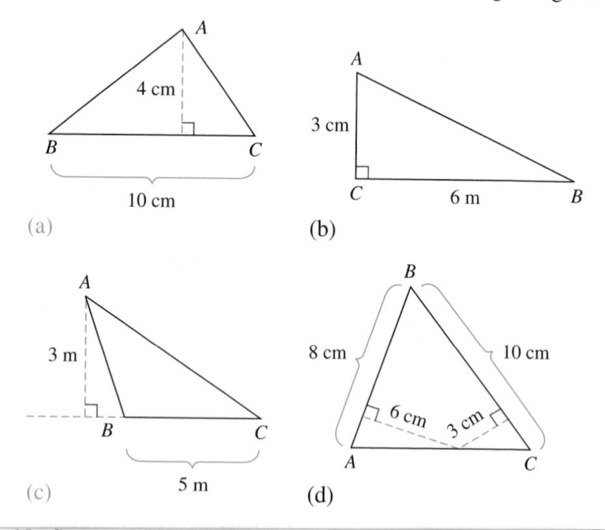

10. Suppose a triangle has sides of lengths 6 in., 11 in., and 13 in. Explain how you can find the area of this triangle when the height is not given.
11. a. If a triangle is inscribed in a square so that the base of the triangle corresponds to the base of the square and the third vertex of the triangle is on the side of the square parallel to this base, how do the areas compare?

b. If the triangle in part (a) has its third vertex at the midpoint of one of the adjacent sides of the square, how do the areas compare?

12. If a triangle is inscribed in a circle so that one of the triangle's sides is a diameter of the circle, what is the greatest area that the triangle can have in terms of the radius, r, of the circle?

13. If a triangle and a square have one side in common, where could the third vertex of the triangle lie if the area of the triangle is exactly equal to the area of the square?

14. a. If triangle ABC is similar to triangle DEF and $\frac{AB}{DE} = \frac{2}{3}$, what is the ratio of the heights of the triangles?

b. What is the ratio of the areas of the two triangles? Why?

15. If two squares have sides in the ratio $\frac{a}{b}$,

a. Are the squares similar? Why?

b. What is the ratio of the areas of the squares?

16. Find the area of each of the following quadrilaterals:

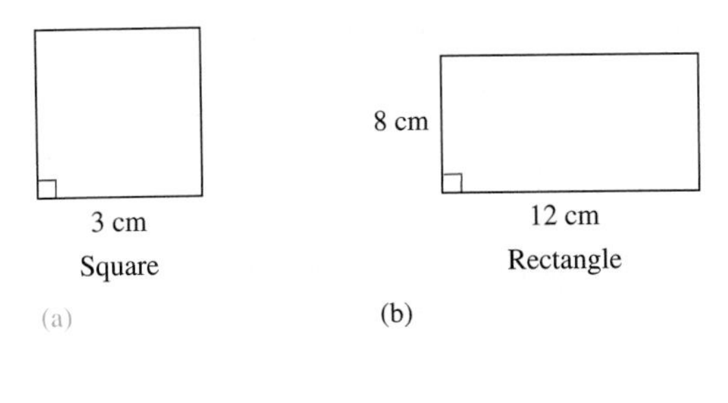

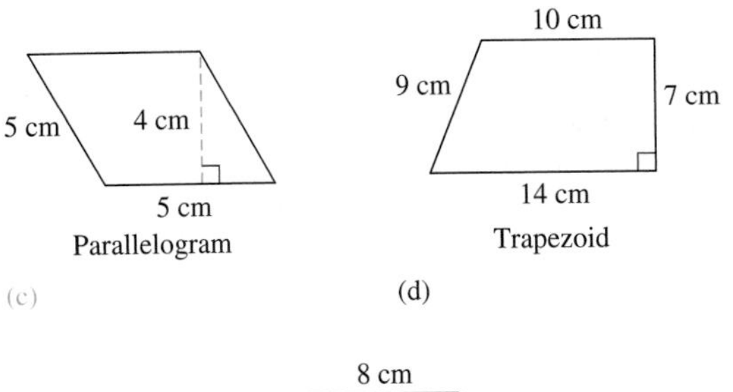

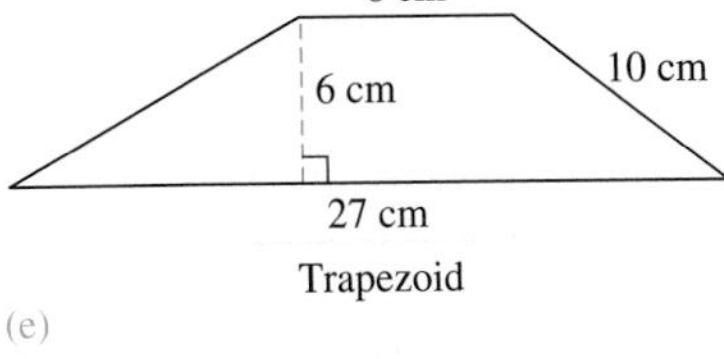

17. a. A rectangular piece of land is 1300 m × 1500 m.
   (i) What is the area in square kilometers?
   (ii) What is the area in hectares?

b. A rectangular piece of land is 1300 yd × 1500 yd.
   (i) What is the area in square miles?
   (ii) What is the area in acres?

c. Explain which measuring system you would rather use to solve problems like those in (a) and (b).

18. For a parallelogram whose sides are 6 cm and 10 cm, which of the following is true?

a. The data are insufficient to enable us to determine the area.

b. The area equals 60 cm^2.

c. The area is greater than 60 cm^2.

d. The area is less than 60 cm^2.

19. If the diagonals of a rhombus are a and b units long, find the area of the rhombus in terms of a and b.

20. Find the cost of carpeting the following rectangular rooms:

a. Dimensions: 6.5 m × 4.5 m; cost = \$13.85/m^2

b. Dimensions: 15 ft × 11 ft; cost = \$30/yd^2

21. Find the area of each of the following. Leave your answers in terms of π.

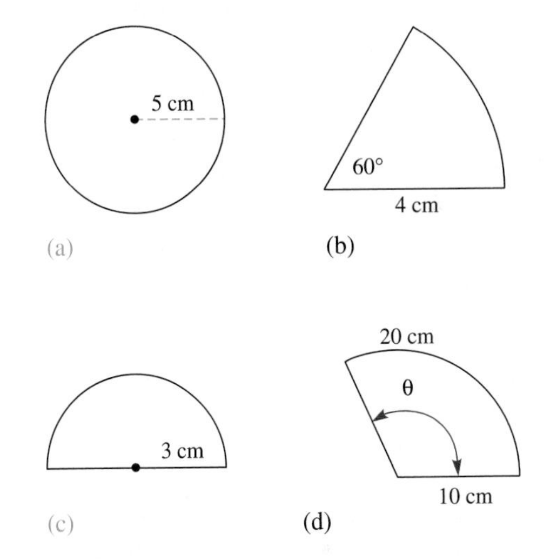

22. Joe uses stick-on square carpet tiles to cover his 3 m × 4 m bathroom. If each tile is 10 cm on a side, how many tiles does he need?

23. To tile a rectangular room, the manufacturer of square tiles recommends that the midsegments of the room be drawn to find the "center" of the room. The center is the starting point.

a. What is true about the midsegments?

b. The first tile is laid with one vertex at the center and two of the sides of the tile on the midsegments. The second tile is laid directly to the "north" of the first tile; the third tile is laid directly to the "west" of the second; the fourth directly to the east of the third; and so on, forming a spiral as the tiles are laid. If the room is 10.5 ft by 11.4 ft and the tiles are 1-ft squares, how many tiles will have to be cut to completely cover the room?

24. Describe how to cut a square and reassemble the parts to form each of the following noncongruent figures that have the same area but different shapes:

a. a rectangle

b. a trapezoid

c. a parallelogram

25. A rectangular plot of land is to be seeded with grass. If the plot is 22 m × 28 m and a 1-kg bag of seed is needed for 85 m^2 of land, how many bags of seed are needed?

26. Find the area of each of the following regular polygons:

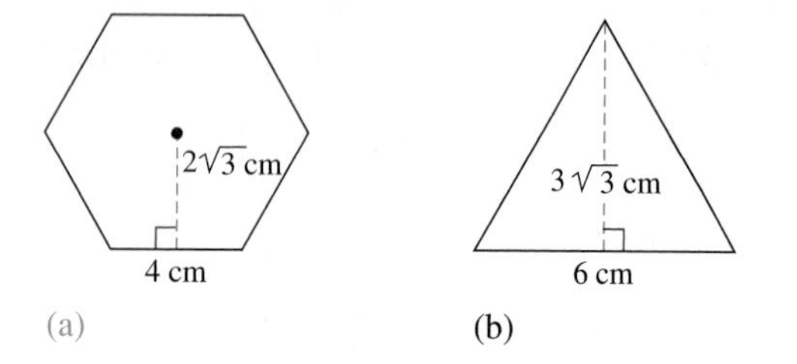

27. Suppose the largest square peg possible is placed in a circular hole as shown in the following figure and that the largest circular peg possible is placed in a square hole. In which case is there a smaller percentage of space wasted?

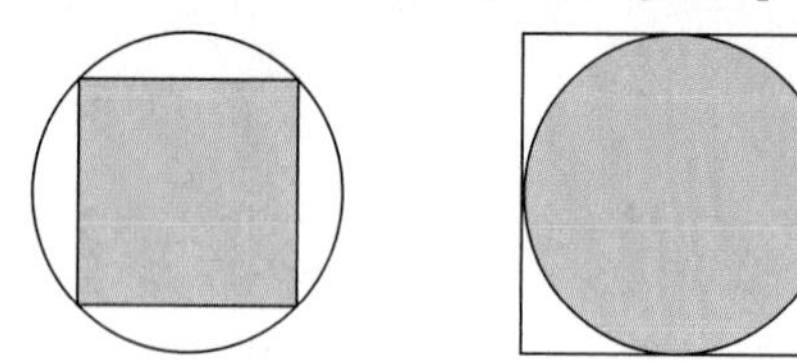

28. a. If a circle has a circumference of 8π cm, what is its area?
    b. If a circle of radius r and a square with a side of length s have equal areas, express r in terms of s.

29. Find the area of each of the following shaded parts. Assume all arcs are circular.

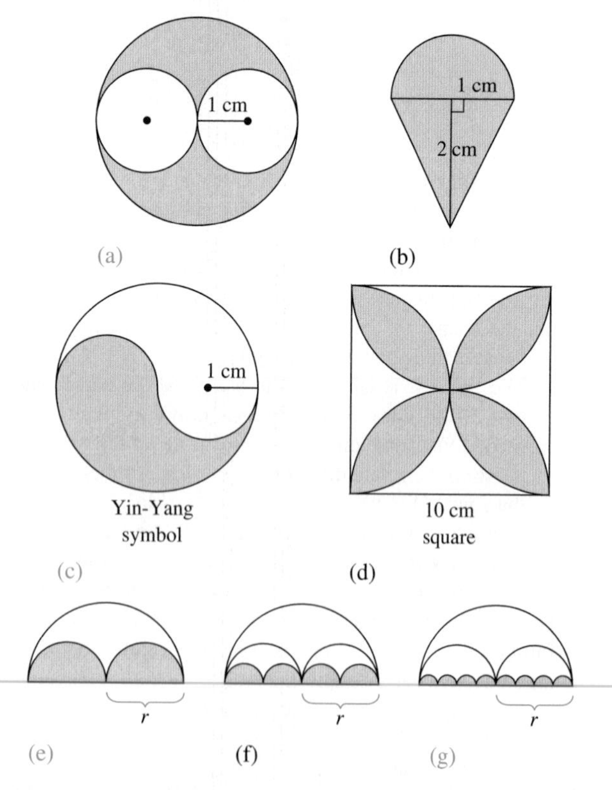

30. A circular flower bed is 6 m in diameter and has a circular sidewalk around it 1 m wide. Find the area of the sidewalk in square meters.

31. a. If the area of a square is 144 cm^2, what is its perimeter?
    b. If the perimeter of a square is 32 cm, what is its area?

32. a. What happens to the area of a square when the length of each side is doubled?
    b. If the ratio of the sides of two squares is 1 to 5, what is the ratio of their areas?

33. a. What happens to the area of a circle if its diameter is doubled?
    b. What happens to the area of a circle if its radius is increased by 10%?
    c. What happens to the area of a circle if its circumference is tripled?

34. A rectangular field is 64 m $\times$ 25 m. Shawn wants to fence a square field that has the same area as the rectangular field. How long are the sides of the square field?

35. A store has wrapping paper on sale. One package is 3 rolls of $2\frac{1}{2}$ ft $\times$ 8 ft for \$6.00. Another package is 5 rolls of $2\frac{1}{2}$ ft $\times$ 6 ft for \$8.00. Which is the better buy?

36. Find the shaded area in the following figure:

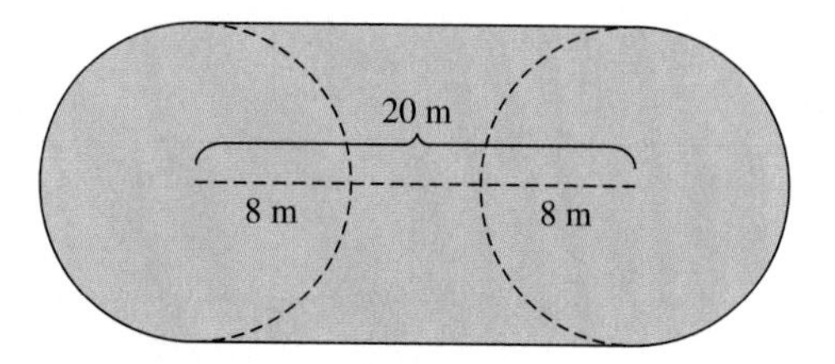

37. An aircraft company starts with a square piece of metal measuring 10 in. $\times$ 10 in. and wants to remove a strip x in. wide from all sides to form another square with an area of 64 $in.^2$. Find x.

38. The following figure consists of five congruent squares. Find a segment through point P that divides the figure into two parts of equal area.

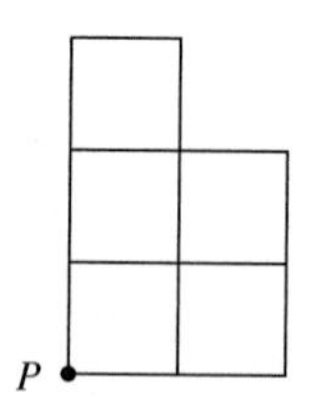

39. a. Sketch a graph showing the relationship between the length and width of all rectangles with perimeters of 12 cm.
    b. Sketch a graph showing the relationship between the length and width of all rectangles with areas of 12 cm^2.

40. For a dartboard (see the following figure), Joan is trying to determine how the area of the outside shaded region compares with the area of the inside shaded region so that she can determine payoffs. How do they compare?

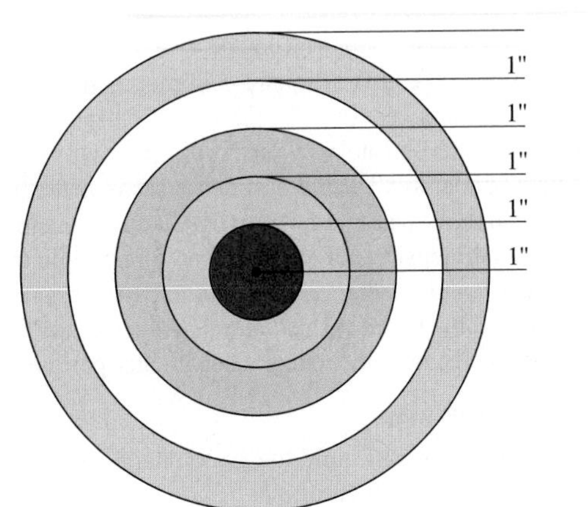

41. Quadrilateral *MATH* has been dissected into squares. The area of the red square is 64 square units and the area of the blue square is 81 square units. Determine the dimensions and the area of quadrilateral *MATH*.

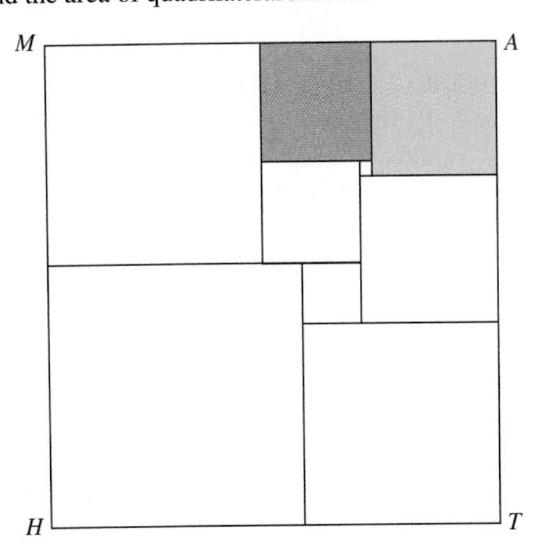

42. Complete and explain how to use geometric shapes to find an equivalent algebraic expression not involving parentheses for each of the following:

a. $a(b + c)$

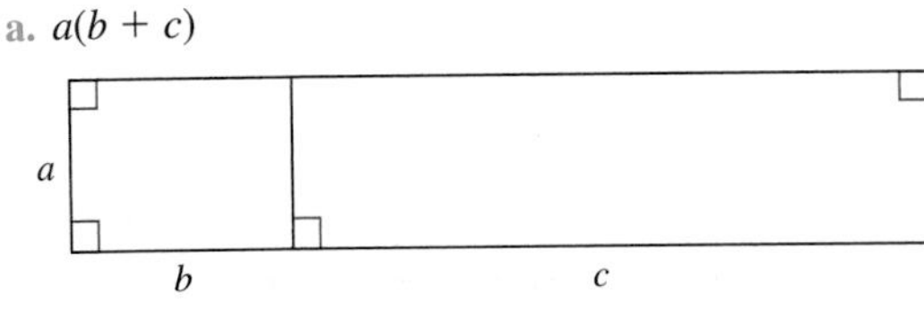

b. $(a + b)(c + d)$

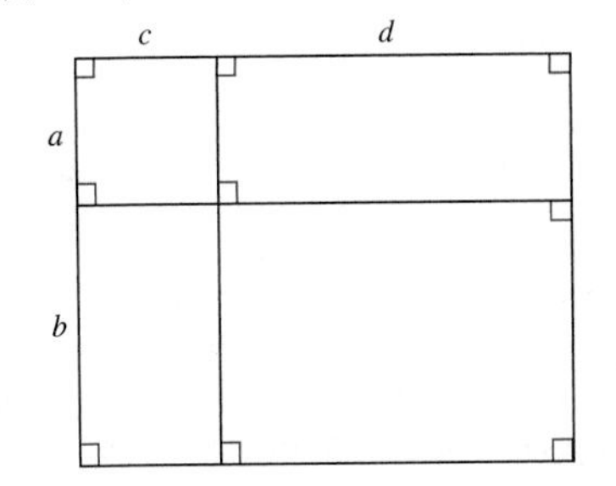

43. Draw two rectangles that have the same perimeter but different areas.

44. a. If the area of a rectangle remains constant but its perimeter has increased, how has the shape of the rectangle changed?

b. If the perimeter of a rectangle remains constant but its area has increased, how has the shape of the rectangle changed?

45. In the following figure, $\ell \parallel \overleftrightarrow{AB}$. If the area of $\triangle ABP$ is 10 cm^2, what are the areas of $\triangle ABQ$, $\triangle ABR$, $\triangle ABS$, $\triangle ABT$, and $\triangle ABU$? Explain your answers.

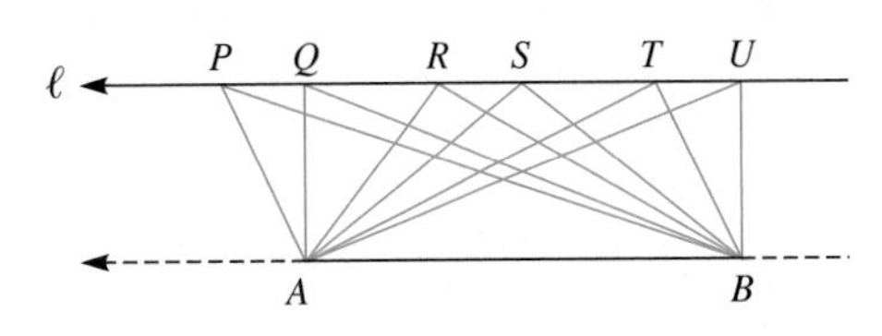

46. Heron's formula can be used to find the area of a triangle if the lengths of the three sides are known. If the lengths of the three sides are a, b, and c units and the semiperimeter $s = \dfrac{a + b + c}{2}$, then the area of the triangle is given by $\sqrt{s(s - a)(s - b)(s - c)}$. Use Heron's formula to find the areas of the right triangles with the sides given.

a. 3 cm, 4 cm, 5 cm

b. 5 cm, 12 cm, 13 cm

47. Given $\triangle ABC$ with parallel lines dividing $\overline{AB}$ into three congruent segments as shown, how does the area of $\triangle BDE$ compare with the area of $\triangle ABC$?

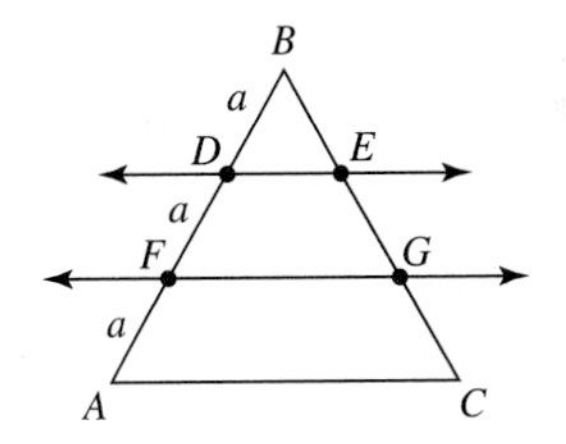

48. Use problem 47 and compare the following areas:

a. Triangle *DBE* and trapezoid *DEGF*

b. Triangle *DBE* and trapezoid *FGCA*

c. Trapezoids *DEGF* and *FGCA*

d. Triangle *DBE* and triangle *ABC*

e. Trapezoid *DEGF* and triangle *ABC*

f. Trapezoid *FGAC* and triangle *ABC*

g. Triangle *ABC* and trapezoid *DECA*

49. In parallelogram *ABCD*, compare the areas of triangles *BCX* and *ADX*.

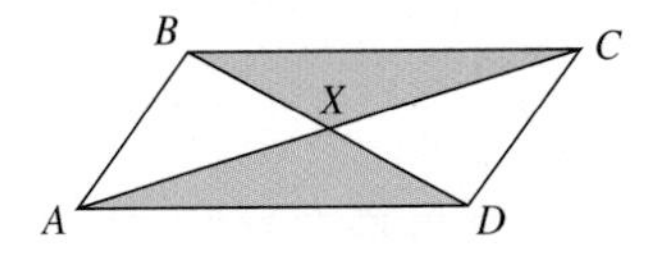

50. In the following figure, quadrilateral $ABCD$ is a parallelogram and P is any point on $\overline{AC}$. Prove that the area of $\triangle BCP$ is equal to the area of $\triangle DPC$.

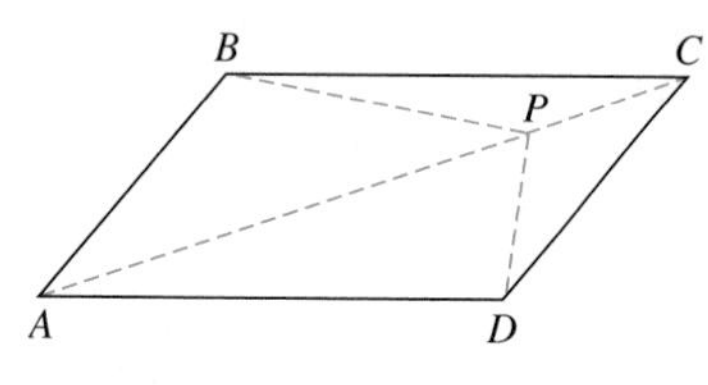

Communication

51. On a field trip, Glenda, a sixth-grade student was looking at a huge dinosaur footprint and wondering about its area. Glenda said all you have to do is place a string around the border of the print and then take the string off and form it into a square and compute the area of the square. How would you help her?
52. John claimed he had a garden twice as large as Al's rectangular-shaped garden that measured 15 ft by 30 ft. When they visited John's rectangular-shaped garden, they found it measured 18 ft by 50 ft. Al claimed that it could not be twice as large since neither the length nor the width were twice as large. Who was correct and why?
53. **a.** If a 10-in. (diameter) pizza costs $10, how much should a 20-in. pizza cost? Explain the assumptions you made in your answer.
    b. If the ratio between the diameters of two pizzas is $1 : k$, what should the ratio be between the prices? Explain the assumptions you made in your answer.
54. **a.** Explain how the following drawing can be used to determine a formula for the area of $\triangle ABC$:

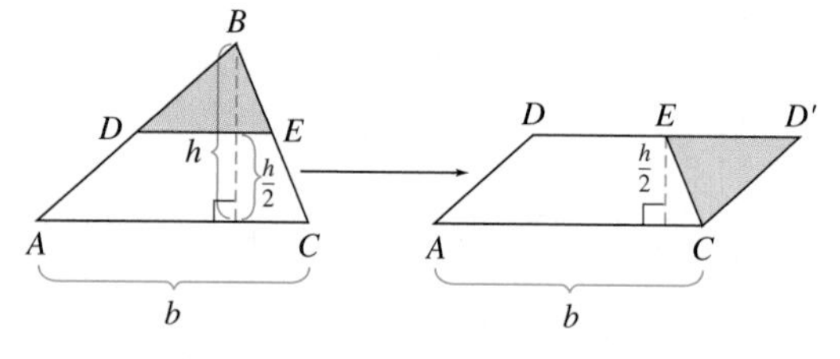

    b. Use paper cutting to reassemble $\triangle ABC$ in (a) into parallelogram $ADD'C$.
55. The area of a parallelogram can be found by using the concept of a half-turn (a turn by 180°). Consider the parallelogram $ABCD$ and let M and N be the midpoints of $\overline{AB}$ and $\overline{CD}$, respectively. Rotate the shaded triangle with vertex M about M by 180° clockwise and rotate the shaded triangle with vertex N about N by 180° counterclockwise. What kind of figure do you obtain? Now complete the argument to find the area of the parallelogram.

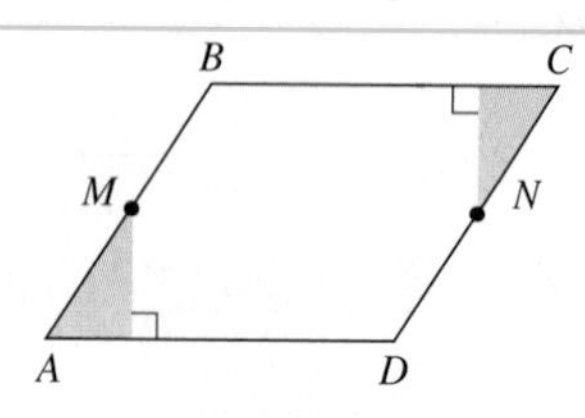

Open-Ended

56. **a.** Estimate the area in square centimeters that your handprint will cover.
    b. Trace the outline of your hand on square-centimeter grid paper and use the outline to obtain an estimate for the area. Explain how you arrived at your estimate.
57. **a.** Give dimensions of a square and a rectangle that have the same perimeter but the square has the greater area.
    b. Give dimensions of a square and a rectangle that have equal area but the rectangle has greater perimeter.

Cooperative Learning

58. Use five squares to build the cross shape shown and discuss the questions that follow.

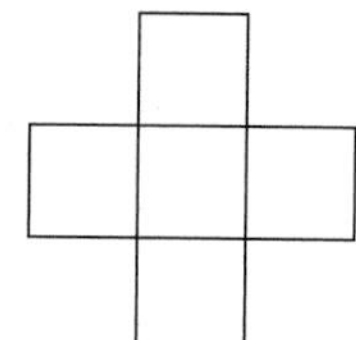

    a. What is the area of this shape?
    b. What is the perimeter of this shape?
    c. Add squares to the shape so that each square added touches a complete edge with at least one other square.
    (i) What is the minimum number of squares that can be added so that the shape has a perimeter of 18?
    (ii) What is the maximum number?
    (iii) What is the maximum area the new shape could have and still have a perimeter of 18?
    d. Using the five squares, have members of the group start with shapes different from the original cross shape and answer the questions in (c). Discuss your results.
    e. Explore shapes that are made up of more than five squares.

Review Problems

59. A glass coffee table top is essentially a square but has rounded circular corners. Find its perimeter.

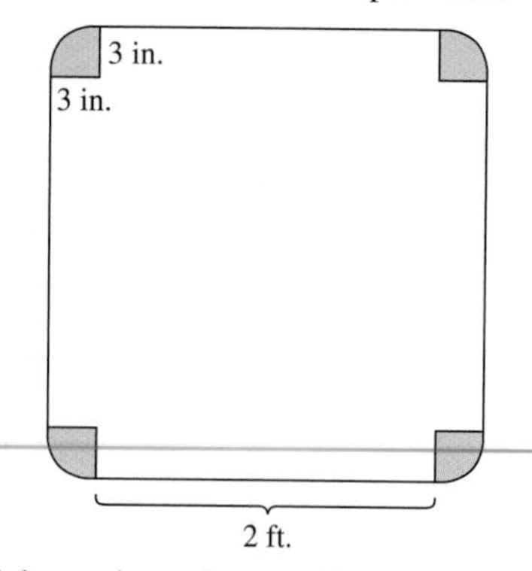

60. **a.** Earth has a circumference of approximately 39,750 km. With this circumference, what is its radius?
    b. Use the measurement in part (a) to determine the length of the arc from the North Pole to the equator.

61. Compare the perimeter of a regular hexagon to the circumference of its circumscribed circle.

Third International Mathematics and Science Study (TIMSS) Questions

A rectangular picture is pasted to a sheet of white paper as shown.

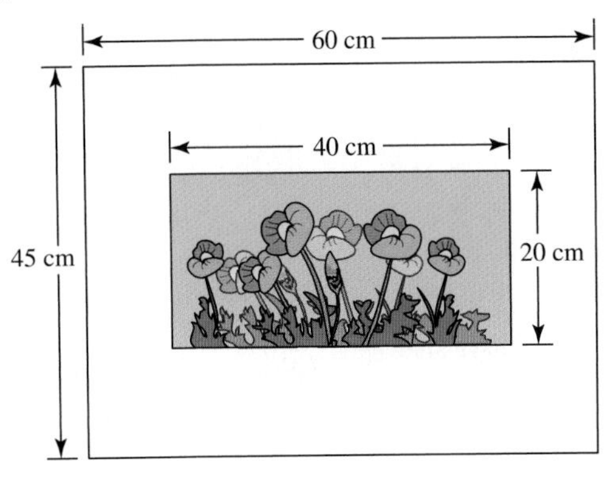

What is the area of the white paper not covered by the picture?

a. 165 cm^2
b. 500 cm^2
c. 1900 cm^2
d. 2700 cm^2

TIMSS, Grade 8, 1994

The length of a rectangle is 6 cm, and its perimeter is 16 cm. What is the area of the rectangle in square centimeters?

TIMSS, Grade 8, 1994

TECHNOLOGY CORNER

Use The Geometer's Sketchpad (GSP) to explore some area concepts:

1. Complete each of the following:
   a. Construct a segment and label the endpoints A and B.
   b. Choose a point not on $\overline{AB}$ and label it C.
   c. Select $\overline{AB}$ and the point C and construct a line through C parallel to $\overline{AB}$.
   d. Choose a point on the new line and label it D.
   e. Select points A and D and construct a segment connecting points A and D.
   f. Select $\overline{AD}$ and point B and construct a line through B parallel to $\overline{AD}$.
   g. Select the two constructed lines. Construct the point at their intersection and label it E.
   h. Select the vertices (E, D, A, B) and construct the interior of the polygon.
   i. Measure the area of the polygon.
   j. Select $\angle DAB$ and measure the angle.
   k. Move point D along $\overleftrightarrow{DC}$ and find the area of all the new parallelograms formed by D, C, B, and A. How are the areas related?
   l. How does this activity lead to the formula for finding the area of a parallelogram if you know how to find the area of a rectangle?

2. Complete each of the following:
   a. Repeat Steps (a) through (d) from part 1.
   b. Select points A, B, and D and construct segments connecting these points to form a triangle.
   c. Select the vertices A, D, and B and construct the interior of the polygon.
   d. Measure the area of $\triangle ADB$.
   e. Move point D along $\overleftrightarrow{DC}$ and find the area of all the triangles that are formed by A, D, and B.
   f. How do these areas compare with the area of the original triangle?
   g. How can this activity be used to motivate the formula for finding the area of a triangle?

3. Take your triangle and line in 2(b) and add a line through B parallel to $\overline{AD}$ to form a parallelogram. Find the area of the parallelogram and compare it to the area of the triangle. How can this activity be used to motivate the formula for finding the area of a triangle?

4. Devise a way to motivate the formula for finding the area of a trapezoid using the geometry utility.

LABORATORY ACTIVITY

1. On a 5×5 geoboard, make $\triangle DEF$ as shown in Figure 11-34. Keep the rubber band around D and E fixed and move the vertex F to all the possible locations so that the triangles formed will have the same area as the area of $\triangle DEF$. How do the locations for the third vertex relate to D and E?

FIGURE 11-34

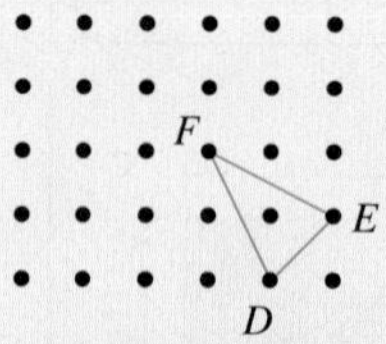

2. On a 5×5 geoboard, construct, if possible, squares of areas 1, 2, 3, 4, 5, 6, and 7 square units.
3. On a 5×5 geoboard, construct triangles that have areas $\frac{1}{2}$, 1, $1\frac{1}{2}$, 2, . . . , until the maximum-sized triangle is reached.

BRAIN TEASER

The rectangle in Figure 11-35(b) was apparently formed by cutting the square in Figure 11-35(a) along the dotted lines and reassembling the pieces as pictured.

FIGURE 11-35

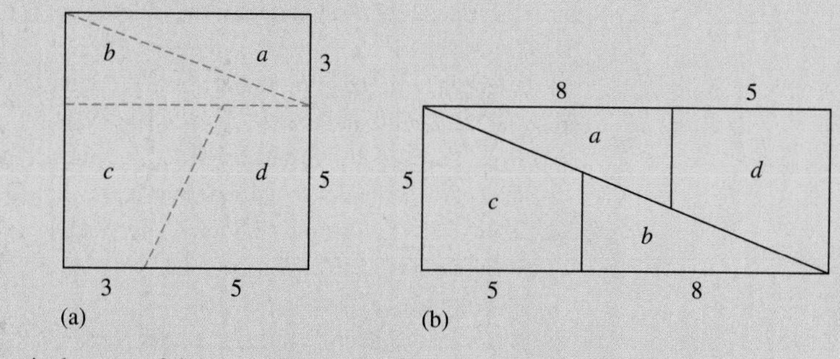

1. What is the area of the square in 11-35(a)?
2. What is the area of the rectangle in 11-35(b)?
3. How do you explain the discrepancy between the areas?

11-3 The Pythagorean Theorem and the Distance Formula

Surveyors often have to calculate distances that cannot be measured directly, such as distances across water, as illustrated in Figure 11-36.

FIGURE 11-36

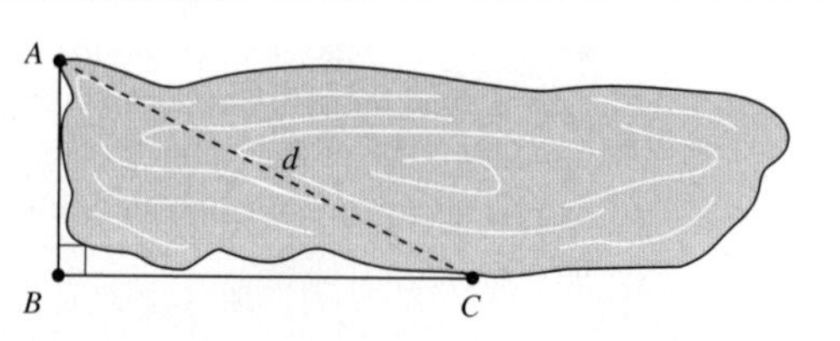

To measure the distance from point A to point C, they could use one of the most remarkable and useful theorems in geometry: the Pythagorean Theorem. This theorem was illustrated on a Greek stamp in 1955, as shown in Figure 11-37(a), to honor the 2500th anniversary of the founding of the Pythagorean School.

FIGURE 11-37

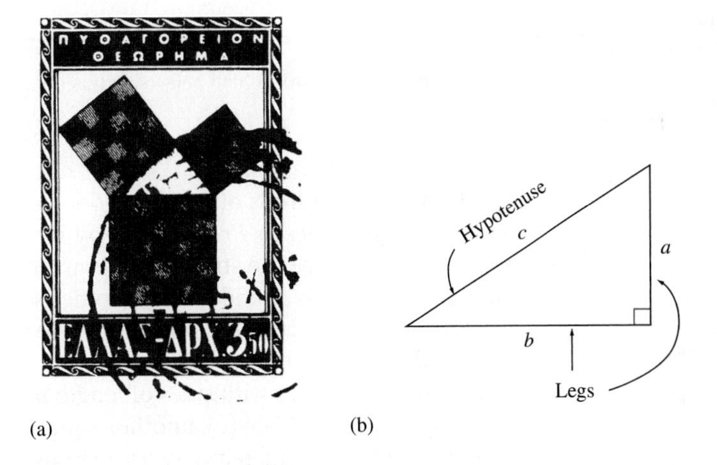

(a) (b)

hypotenuse ◆ legs

In the triangle on the stamp and in triangle ABC in Figure 11-37(b), the side opposite the right angle is the **hypotenuse**. The other two sides are **legs**. Interpreted in terms of area, the Pythagorean Theorem states that the area of a square with the hypotenuse of a right triangle as a side is equal to the sum of the areas of the squares with the legs as sides.

▶ **HISTORICAL NOTE**

Pythagoras (ca. 582–507 BCE), a Greek philosopher and mathematician, was head of a group known as the Pythagoreans. Members of the group regarded Pythagoras as a demigod and attributed all their discoveries to him. One of Pythagoras's most unusual discoveries was the dependence of the musical intervals on the ratio of the length of strings at the same tension, with the ratio 2 : 1 giving the octave, 3 : 2 the fifth, and 4 : 3 the fourth. ◀

FIGURE 11-38

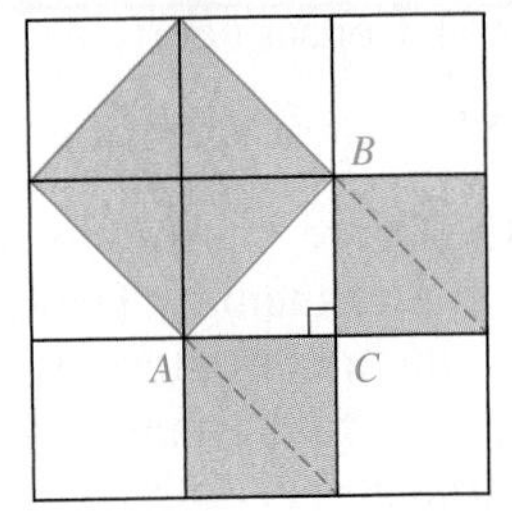

Because the Pythagoreans affirmed geometric results on the basis of special cases, mathematical historians believe it is possible they may have discovered the theorem by looking at a floor tiling like the one illustrated in Figure 11-38. Each square can be divided by its diagonal into two congruent isosceles right triangles, so we see that the shaded square constructed with $\overline{AB}$ as a side consists of four triangles, each congruent to $\triangle ABC$. Similarly, each of the shaded squares with legs $\overline{BC}$ and $\overline{AC}$ as sides consists of two triangles congruent to $\triangle ABC$. Thus, the area of the larger square is equal to the sum of the areas of the two smaller squares. The theorem is true in general and is stated as follows using Figure 11-37(b).

Theorem 11-1

Pythagorean Theorem If a right triangle has legs of lengths a and b and hypotenuse of length c, then $c^2 = a^2 + b^2$.

TECHNOLOGY CORNER

THE GEOMETER'S SKETCHPAD

GSP Lab 9 in Appendix III can be used at this point to investigate the Pythagorean Theorem.

There are hundreds of known proofs for the Pythagorean Theorem. The classic book *The Pythagorean Proposition*, by E. Loomis, contains many of these proofs. Some proofs involve the strategy of *drawing diagrams* with a square area c^2 equal to the sum of the areas a^2 and b^2 of two other squares. One such proof is given in Figure 11-39; others are discussed in Ongoing Assessment 11-3. In Figure 11-39(a), the measures of the legs of a right triangle ABC are a and b and the measure of the hypotenuse is c. We draw a square with sides of length $a + b$ and subdivide it, as shown in Figure 11-39(b). In Figure 11-39(c), another square with side of length $a + b$ is drawn and each of its sides is divided into two segments of length a and b, as shown.

FIGURE 11-39

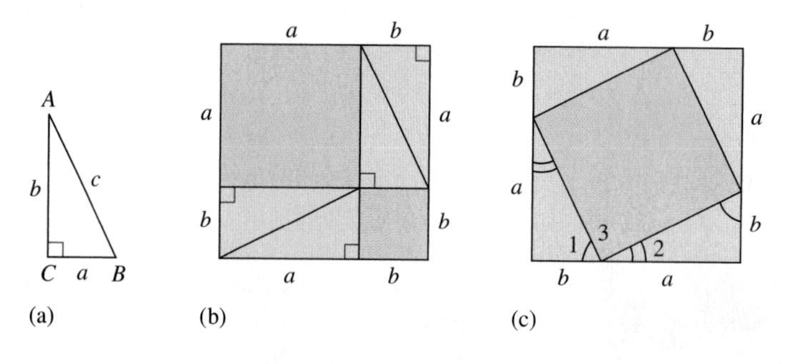

REMARK One version of the proof shown is found in the 1904 proceedings of the Royal Society of Canada.

Each cream-colored triangle is congruent to $\triangle ABC$ (why?). Consequently, each triangle has hypotenuse c and the same area, $\frac{1}{2}ab$. Thus, the length of each side of the inside blue quadrilateral in Figure 11-39(c) is c and so the figure is a rhombus. Because the triangles are right triangles, their acute angles are complementary. Hence $m(\angle 1) + m(\angle 2) = 90°$ so that $m(\angle 3) = 90°$. Therefore, the blue quadrilateral is a square whose area is c^2. To complete the proof, we consider the four triangles in Figure 11-39(b) and (c). Because the areas of the sets of four triangles in both Figure 11-39(b) and (c) are equal, the sum of the areas of the two shaded squares in Figure 11-39(b) equals the area of the shaded square in Figure 11-39(c), that is, $a^2 + b^2 = c^2$.

NOW TRY THIS 11-11

In 1873, Henry Perigal, a London stockbroker, printed what has been called the "paper and scissors" proof of the Pythagorean Theorem. It is illustrated in Figure 11-40. Explain how this figure could be used to justify the theorem.

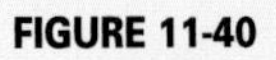
FIGURE 11-40

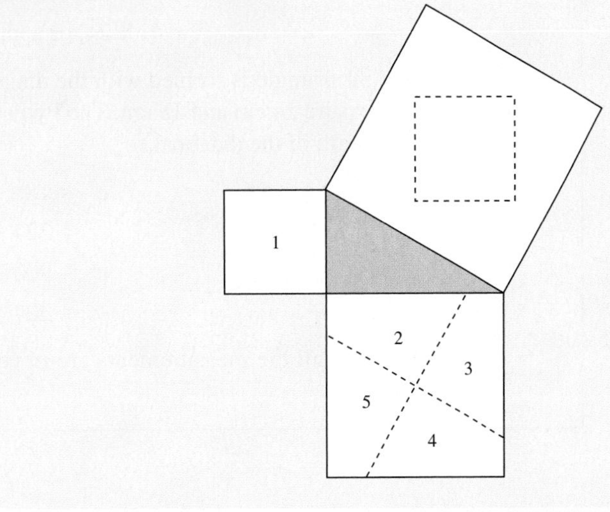

EXAMPLE 11-12

a. For the drawing in Figure 11-41, find the value of x.

FIGURE 11-41

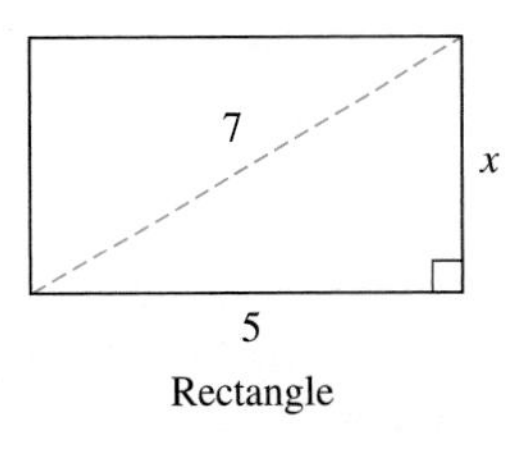

Rectangle

b. The size of a rectangular television screen is given as the length of the diagonal of the screen. If the length of the screen is 24 cm and the width is 18 cm, as shown in Figure 11-42, what is the diagonal length?

FIGURE 11-42

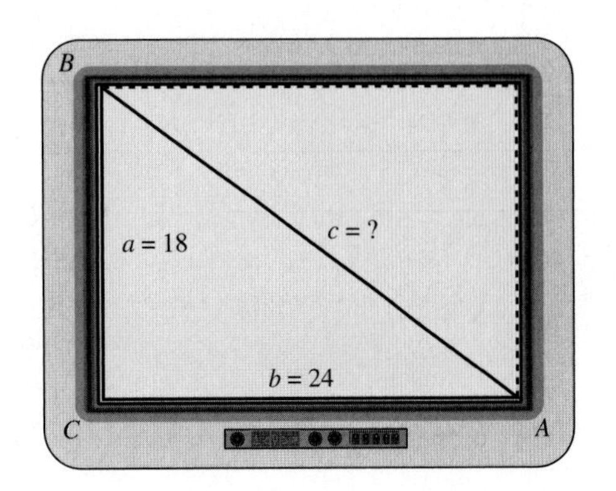

Solution **a.** In the rectangle, the diagonal partitions the rectangle into two right triangles, each with length 5 units and width x units. Thus we have the following:

$$5^2 + x^2 = 7^2$$
$$25 + x^2 = 49$$
$$x^2 = 24$$
$$x = \sqrt{24}\text{, or approximately 4.9 units.}$$

b. A right triangle is formed with the diagonal as the hypotenuse and the legs of measure 24 cm and 18 cm. The Pythagorean Theorem can be used to find the length of the diagonal.

$$c^2 = 18^2 + 24^2$$
$$c^2 = 324 + 576$$
$$c^2 = 900$$
$$c = 30$$

Because all the measurements are in centimeters, the diagonal has length 30 cm.

EXAMPLE 11-13 A pole $\overline{BD}$, 28 ft high, is perpendicular to the ground. Two wires $\overline{BC}$ and $\overline{BA}$, each 35 ft long, are attached to the top of the pole and to stakes A and C on the ground, as shown in Figure 11-43. If points A, D, and C are collinear, how far are the stakes A and C from each other?

FIGURE 11-43

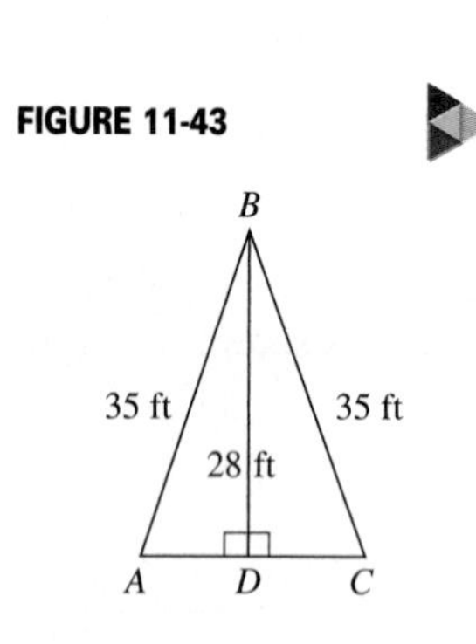

Solution $\overline{AC}$ is not a side in any known right triangle, but we want to find AC. Because a point equidistant from the endpoints of a segment must be on a perpendicular bisector of the segment, then $AD = DC$. Therefore, AC is twice as long as DC. Our *subgoal* is to find DC. We may find DC by applying the Pythagorean Theorem in triangle BDC. This results in the following:

$$28^2 + (DC)^2 = 35^2$$
$$(DC)^2 = 35^2 - 28^2$$
$$DC = \sqrt{441}\text{, or 21 ft}$$
$$AC = 2 \cdot DC = 42\text{ ft}$$

EXAMPLE 11-14 How tall is the Great Pyramid of Cheops, a right regular square pyramid, if the base has a side 771 ft and the slant height is 620 ft?

Solution In Figure 11-44, $\overline{EF}$ is a leg of a right triangle formed by $\overline{FD}$, $\overline{EF}$, and $\overline{ED}$. Because the pyramid is a right regular pyramid, EF intersects the base at its center. Thus,

FIGURE 11-44

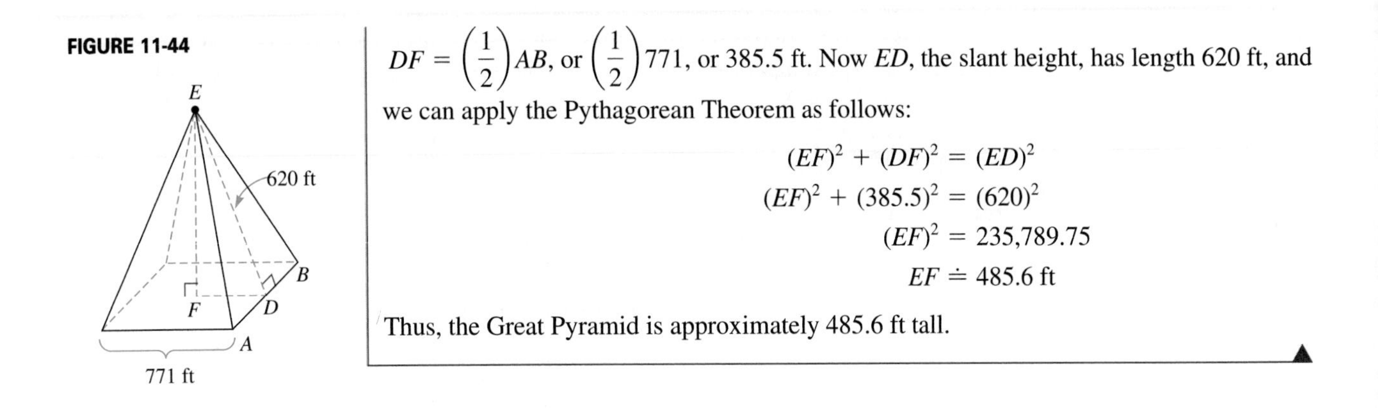

$DF = \left(\frac{1}{2}\right)AB$, or $\left(\frac{1}{2}\right)771$, or 385.5 ft. Now ED, the slant height, has length 620 ft, and we can apply the Pythagorean Theorem as follows:

$$(EF)^2 + (DF)^2 = (ED)^2$$
$$(EF)^2 + (385.5)^2 = (620)^2$$
$$(EF)^2 = 235{,}789.75$$
$$EF \doteq 485.6 \text{ ft}$$

Thus, the Great Pyramid is approximately 485.6 ft tall. ▲

NOW TRY THIS 11-12

In the following cartoon, Jason runs a different pattern than he was told. Explain how he arrived at this pattern.

Special Right Triangles

45°-45°-90° right triangle

An isosceles right triangle has two legs of equal length and two 45° angles. Any such triangle is a **45°-45°-90° right triangle**. Drawing a diagonal of a square forms two of these triangles, as shown in Figure 11-45.

FIGURE 11-45

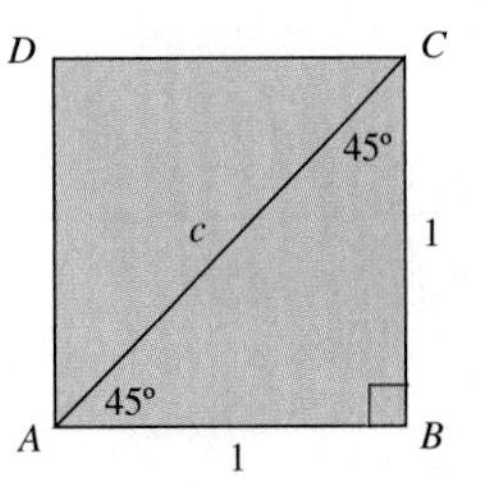

In Figure 11-46, we see several 45°-45°-90° triangles. Each side of the shaded square is a hypotenuse of a 45°-45°-90° triangle. The area of the shaded square is 2 square units (why?). Therefore, $c^2 = 2$ and $c = \sqrt{2}$. Another way to see that $c = \sqrt{2}$ is to apply the Pythagorean Theorem to one of the nonshaded triangles. Because $c^2 = 1^2 + 1^2 = 2$, then $c = \sqrt{2}$.

FIGURE 11-46

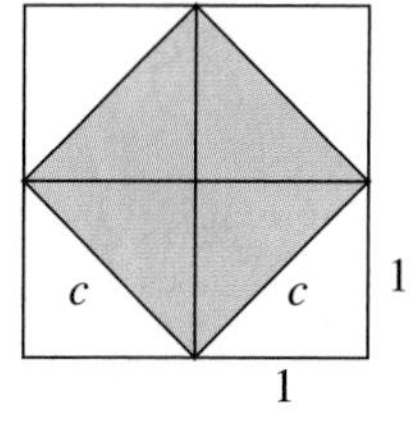

In one isosceles right triangle in Figure 11-46, each leg is 1 unit long and the hypotenuse is $\sqrt{2}$ units long. This property is generalized when the isosceles right triangle has a leg of length a, as follows.

Property

Property of 45°-45°-90° triangle In an isosceles right triangle, if the length of each leg is a, then the hypotenuse has length $a\sqrt{2}$.

Similarly, Figure 11-47(a), shows that a 30°-60°-90° triangle is half of an equilateral triangle. When the equilateral triangle has side 2 units long, then in the 30°-60°-90° triangle, the leg opposite the 30° angle is 1 unit long and the leg opposite the 60° angle has a length of $\sqrt{3}$ units.

FIGURE 11-47

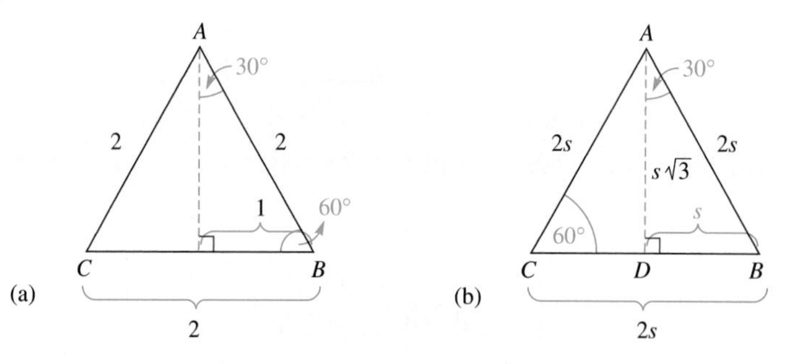

This example may also be generalized using the triangle in Figure 11-47(b). When the side of the equilateral triangle ABC is $2s$, then in triangle ABD, the side opposite the 30° angle, $\overline{BD}$, is s units long, and AD may be found using the Pythagorean Theorem to have a length of $s\sqrt{3}$ units. This discussion is summarized in the following property.

Property

Property of 30°-60°-90° triangle In a 30°-60°-90° triangle, the length of the hypotenuse is two times as long as the leg opposite the 30° angle and the leg opposite the 60° angle is $\sqrt{3}$ times the shorter leg.

Converse of the Pythagorean Theorem

The converse of the Pythagorean Theorem is also true. It provided a useful way for early surveyors—in particular, the Egyptian rope stretchers—to determine right angles. Figure 11-48(a) shows a knotted rope with 12 equally spaced knots. Figure 11-48(b) shows how the rope might be held to form a triangle with sides of lengths 3, 4, and 5. The triangle formed is a right triangle and contains a 90° angle. Note that $5^2 = 3^2 + 4^2$.

FIGURE 11-48

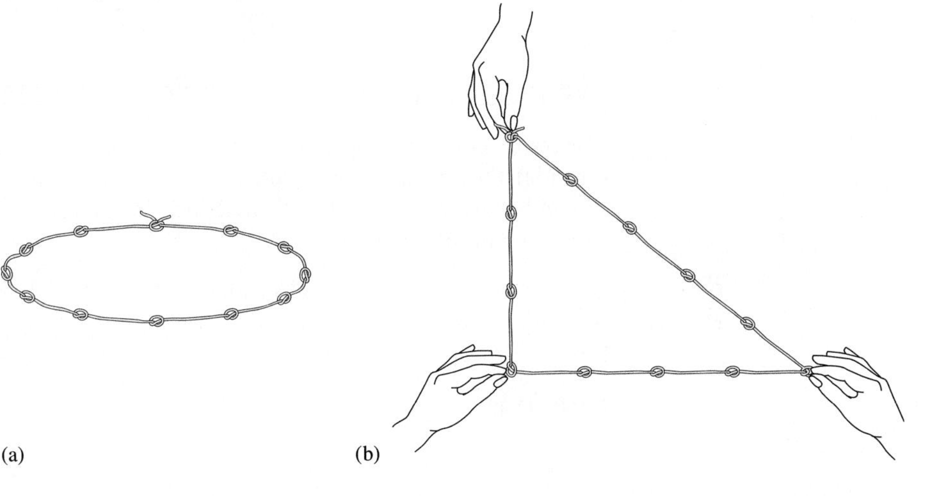

Given a triangle with sides of lengths a, b, and c such that $a^2 + b^2 = c^2$, must the triangle be a right triangle? The answer is yes, and we state the following theorem without proof.

Theorem 11-2

Converse of the Pythagorean Theorem If $\triangle ABC$ is a triangle with sides of lengths a, b, and c such that $a^2 + b^2 = c^2$, then $\triangle ABC$ is a right triangle with the right angle opposite the side of length c.

EXAMPLE 11-15

Determine if the following can be the lengths of the sides of a right triangle:

a. 51, 68, 85
b. 2, 3, $\sqrt{13}$
c. 3, 4, 7

Solution **a.** $51^2 + 68^2 = 7225 = 85^2$, so 51, 68, and 85 can be the lengths of the sides of a right triangle.

b. $2^2 + 3^2 = 4 + 9 = 13 = (\sqrt{13})^2$, so 2, 3, and $\sqrt{13}$ can be the lengths of the sides of a right triangle.

c. $3^2 + 4^2 \neq 7^2$, so the measures cannot be the lengths of the sides of a right triangle. In fact, since $3 + 4 = 7$, these segments do not form a triangle.

NOW TRY THIS 11-13

a. Draw three segments that could be used to form the sides of a right triangle and discuss how you would show that these three lengths determine a right triangle.
b. Multiply the lengths of the three segments in (a) by a fixed number and determine if the resulting three lengths could be sides of a right triangle.
c. Using three new numbers, repeat the experiment in (a) and (b). Form a conjecture based on your experiments.

The Distance Formula: An Application of the Pythagorean Theorem

Given the coordinates of two points A and B, we can find the distance AB. We first consider the special case in which the two points are on one of the axes. For example, in Figure 11-49(a), $A(2, 0)$ and $B(5, 0)$ are on the x-axis. The distance between these two points is 3 units:

$$AB = OB - OA = 5 - 2 = 3$$

FIGURE 11-49

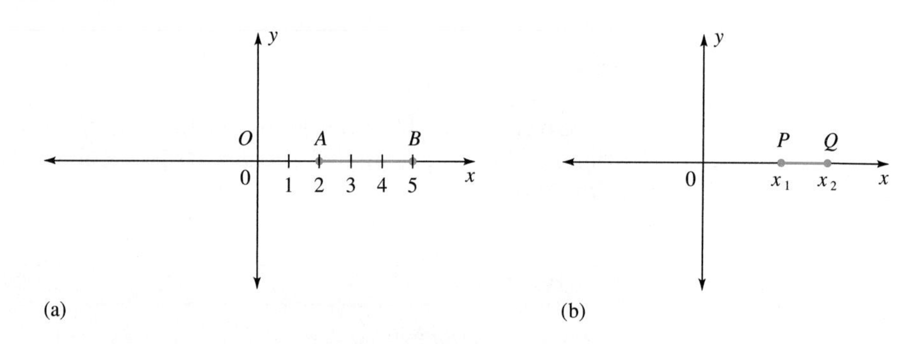

In general, if two points P and Q are on the x-axis, as in Figure 11-49(b), with x-coordinates x_1 and x_2, respectively, and $x_2 > x_1$, then $PQ = x_2 - x_1$. In fact, *the distance between two points on the x-axis is always the absolute value of the difference between the x-coordinates of the points* (why?). A similar result holds for any two points on the y-axis.

Figure 11-50 shows two points in the plane: $C(2, 5)$ and $D(6, 8)$. The distance between C and D can be found by using the strategy of *looking at a related problem.* We know how to find the length of a segment if the segment is a side or the hypotenuse of a right triangle. We obtain a right triangle by drawing perpendiculars from the points to the x-axis and to the y-axis, as shown in Figure 11-50, thus defining triangle CDE. The lengths of the legs of triangle CDE are found by using horizontal and vertical distances and properties of rectangles.

$$CE = |6 - 2| = 4$$
$$DE = |8 - 5| = 3$$

FIGURE 11-50

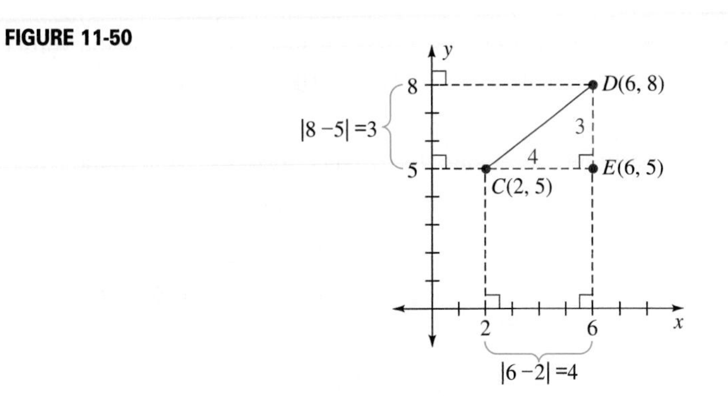

The distance between C and D can be found by applying the Pythagorean Theorem to the triangle.

$$\begin{aligned} CD^2 &= DE^2 + CE^2 \\ &= 3^2 + 4^2 \\ &= 25 \\ CD &= \sqrt{25}, \text{ or } 5 \end{aligned}$$

The method can be generalized to find a formula for the distance between any two points $A(x_1, y_1)$ and $B(x_2, y_2)$. Construct a right triangle with $\overline{AB}$ as one of its sides by drawing a segment through A parallel to the x-axis and a segment through B parallel to the y-axis, as shown in Figure 11-51. The lines containing the segments intersect at point C, forming right triangle ABC. Now, apply the Pythagorean Theorem.

FIGURE 11-51

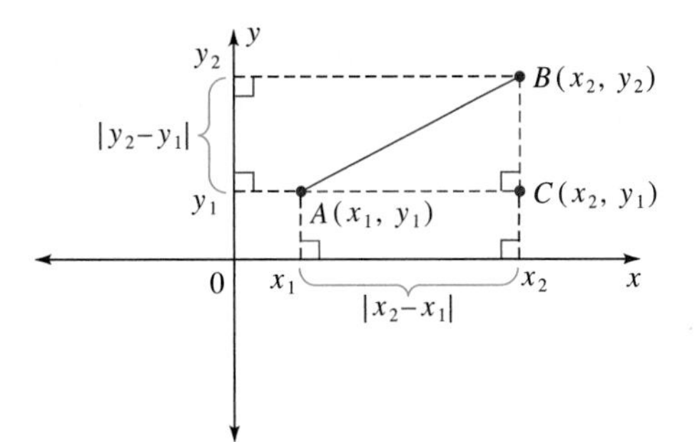

In Figure 11-51, we see that $AC = |x_2 - x_1|$ and $BC = |y_2 - y_1|$. By the Pythagorean Theorem, $(AB)^2 = |x_2 - x_1|^2 + |y_2 - y_1|^2$, and consequently $AB = \sqrt{|x_2 - x_1|^2 + |y_2 - y_1|^2}$. Because $|x_2 - x_1|^2 = (x_2 - x_1)^2$ and $|y_2 - y_1|^2 = (y_2 - y_1)^2$, $AB = \sqrt{(x_2 - x_1)^2 + (y_2 - y_1)^2}$. This result is known as the **distance formula**.

distance formula

Distance Formula

The distance between the points $A(x_1, y_1)$ and $B(x_2, y_2)$ is given by

$$AB = \sqrt{(x_2 - x_1)^2 + (y_2 - y_1)^2}$$

NOW TRY THIS 11-14

Investigate whether it makes any difference in the distance formula if $(x_1 - x_2)$ and $(y_1 - y_2)$ are used instead of $(x_2 - x_1)$ and $(y_2 - y_1)$, respectively.

EXAMPLE 11-16

a. Show that $A(7, 4)$, $B(^-2, 1)$, and $C(10, ^-5)$ are the vertices of an isosceles triangle.
b. Show that $\triangle ABC$ in (a) is a right triangle.

Solution **a.** Using the distance formula, we find the lengths of the sides.

$$AB = \sqrt{(^-2 - 7)^2 + (1 - 4)^2} = \sqrt{(^-9)^2 + (^-3)^2} = \sqrt{90}$$
$$BC = \sqrt{[10 - (^-2)]^2 + (^-5 - 1)^2} = \sqrt{12^2 + (^-6)^2} = \sqrt{180}$$
$$AC = \sqrt{(10 - 7)^2 + (^-5 - 4)^2} = \sqrt{3^2 + (^-9)^2} = \sqrt{90}$$

Thus, $AB = AC$, and so the triangle is isosceles.

b. Because $(\sqrt{90})^2 + (\sqrt{90})^2 = (\sqrt{180})^2$, $\triangle ABC$, is a right triangle with $\overline{BC}$ as hypotenuse and $\overline{AB}$ and $\overline{AC}$ as legs.

EXAMPLE 11-17

Determine whether the points $A(0, 5)$, $B(1, 2)$, and $C(2, {}^-1)$ are collinear.

Solution It is hard to tell by graphing the points whether they are collinear (on the same line). If they are not collinear, they would be the vertices of a triangle, and hence $AB + BC$ would be greater than AC (triangle inequality). If $AB + BC = AC$, a triangle could not be formed and the points would be collinear. Using the distance formula, we find the lengths of the sides:

$$AB = \sqrt{(0-1)^2 + (5-2)^2} = \sqrt{1 + 9} = \sqrt{10}$$
$$BC = \sqrt{(2-1)^2 + ({}^-1-2)^2} = \sqrt{1 + 9} = \sqrt{10}$$
$$AC = \sqrt{(0-2)^2 + (5-({}^-1))^2} = \sqrt{4 + 36} = \sqrt{40}$$

Thus, $AB + BC = 2\sqrt{10}$. Is this sum equal to $\sqrt{40}$? Because $\sqrt{40} = \sqrt{4 \cdot 10} = \sqrt{4} \cdot \sqrt{10} = 2\sqrt{10}$, we have $AB + BC = AC$, and consequently, A, B, and C are collinear.

REMARK An alternative solution to Example 11-17 is to show that the slopes of each of this segments that can be formed are the same.

BRAIN TEASER

A farmer has a square plot of land. An irrigation system can be installed with the option of one large circular sprinkler, as in Figure 11-52(a), or nine small sprinklers, as in Figure 11-52(b). The farmer wants to know which plan will provide water to the greatest percentage of land in the field, regardless of the cost and the watering pattern. What advice would you give?

FIGURE 11-52

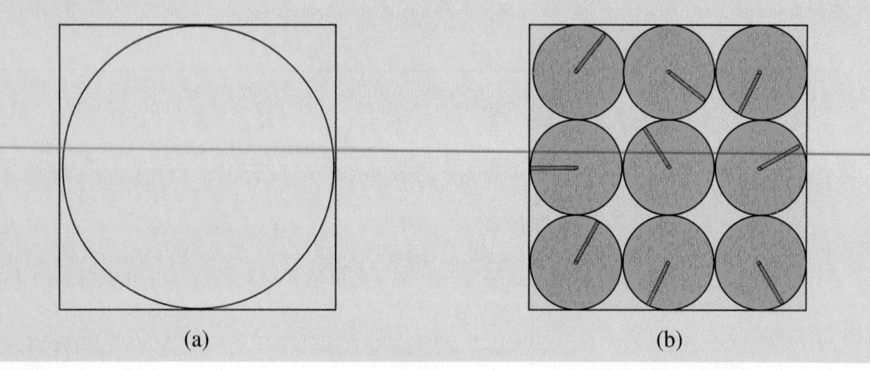

(a) (b)

ONGOING ASSESSMENT 11-3

1. Find the length of the following segments. Assume that the horizontal and vertical distances between neighboring dots is 1 unit.

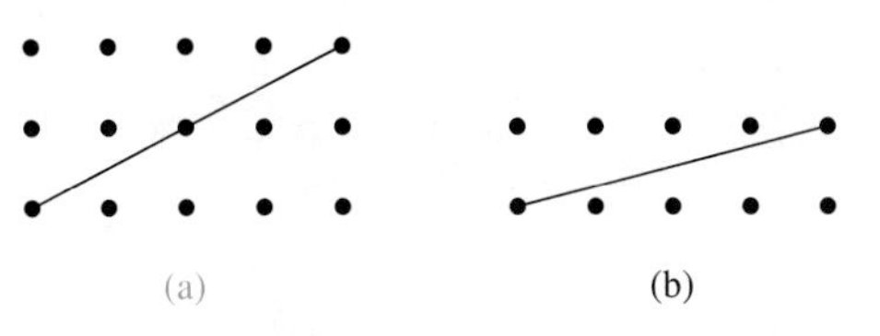

2. Use the Pythagorean Theorem to find x in each of the following:

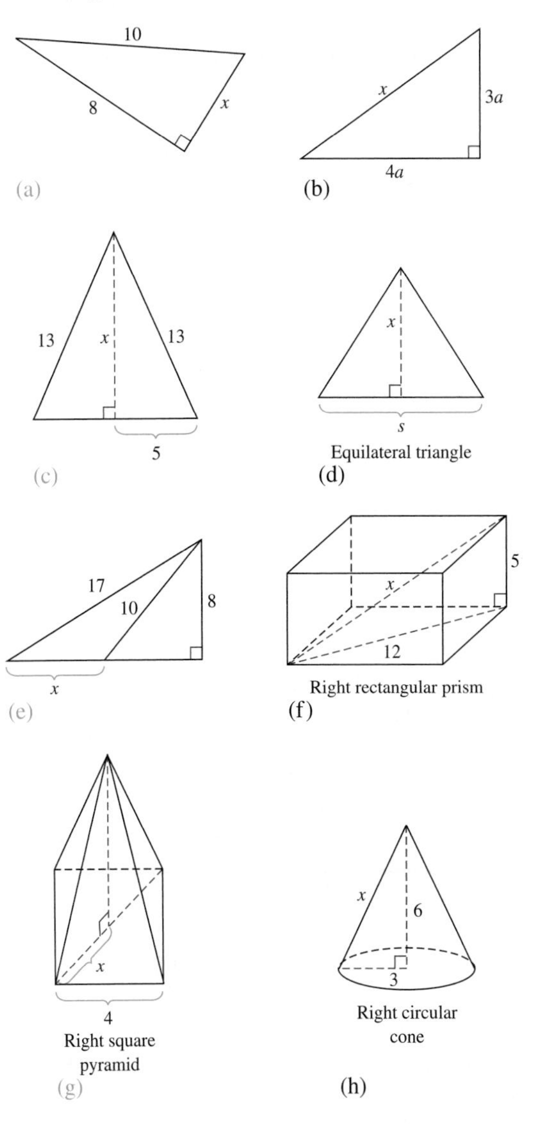

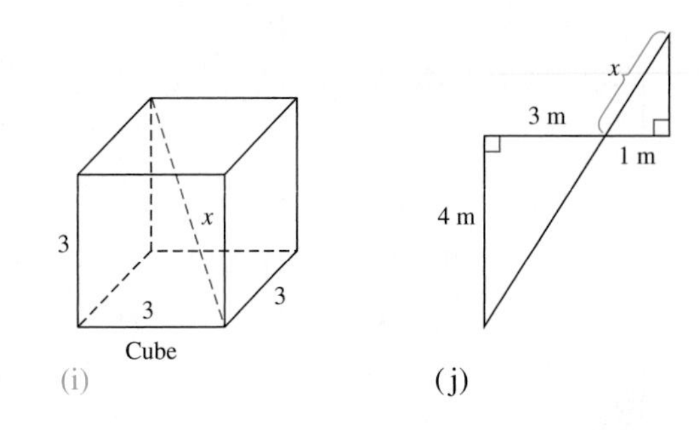

3. On a geoboard, construct segments of the following lengths.
   a. $\sqrt{13}$
   b. $\sqrt{5}$
   c. $\sqrt{10}$
4. Find a square on a 4×4 geoboard that has an area of 2 square units and a perimeter of $4\sqrt{2}$ units.
5. On a 4×4 geoboard the greatest square has a perimeter of 12 units. Find a polygon on the geoboard with a greater perimeter. Prove your answer.
6. What is the perimeter of the largest isosceles triangle on a 5×5 geoboard?
7. Prove that an equilateral triangle either can or cannot be constructed on a 4×4 geoboard.
8. If the hypotenuse of a right triangle is 30 cm long and one leg is twice as long as the other, how long are the legs of the triangle?
9. For each of the following, determine whether the given numbers represent lengths of sides of a right triangle:
   a. 10, 24, 16
   b. 16, 34, 30
   c. $\sqrt{2}, \sqrt{2}, 2$
   d. $\frac{3}{2}, \frac{4}{2}, \frac{5}{2}$
10. What is the longest line segment that can be drawn in a right rectangular prism that is 12 cm wide, 15 cm long, and 9 cm high?
11. Two airplanes depart from the same place at 2:00 P.M. One plane flies south at a speed of 376 km/hr, and the other flies west at a speed of 648 km/hr. How far apart are the airplanes at 5:30 P.M.?
12. Starting from point A, a boat sails due south for 6 mi, then due east for 5 mi, and then due south for 4 mi. How far is the boat from A?

Calvin and Hobbes by Bill Watterson

13. a. In the cartoon above, is Hobbes's square really a square? Why?
    b. If Hobbes's figure in the second panel is a rectangle that is 6 units by 3 units, find y.
    c. If Hobbes draws a rectangle with each dimension doubled, how does the length of the diagonal change?
    d. Was Calvin's measure of the diagonal correct?

14. A 15-ft ladder is leaning against a wall. The base of the ladder is 3 ft from the wall. How high above the ground is the top of the ladder?

15. In the following figure, two poles are 25 m and 15 m high. A cable 14 m long joins the tops of the poles. Find the distance between the poles.

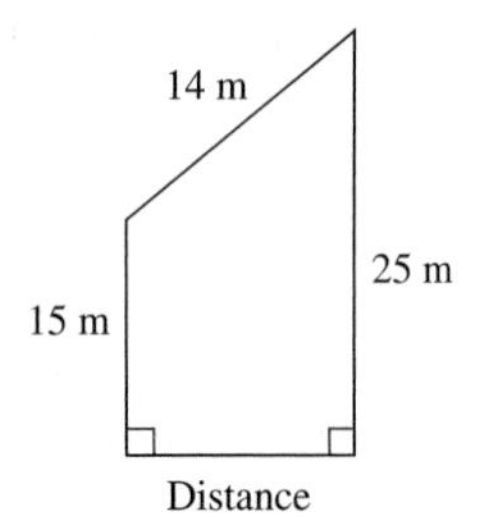

16. Find the area of each of the following:

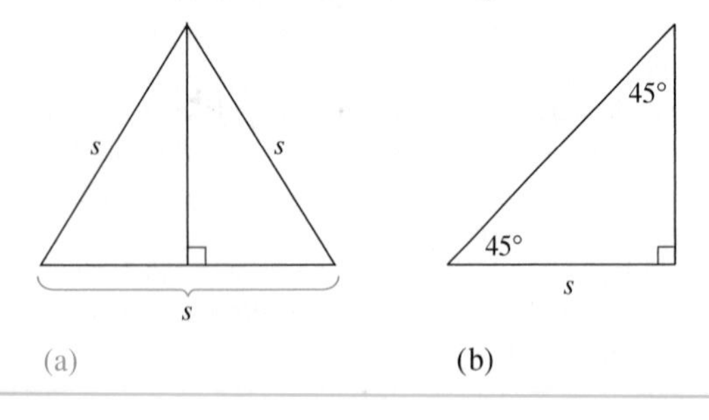

17. For each of the following, solve for the unknowns:

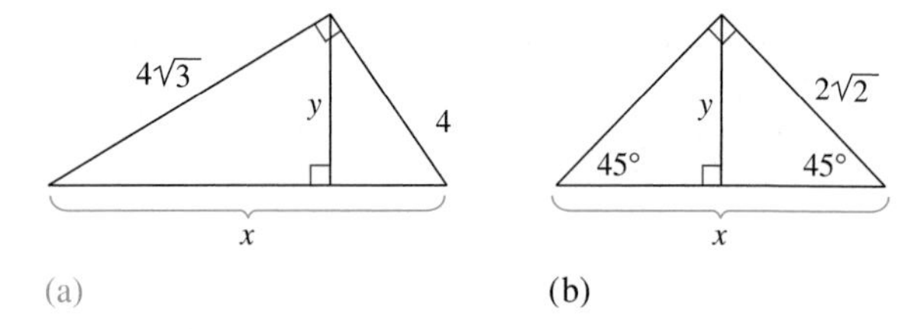

18. A builder needs to calculate the dimensions of a regular hexagonal window. Assuming the height CD of the window is 1.3 m, find the width AB (O is the midpoint of $\overline{AB}$) in the following figure:

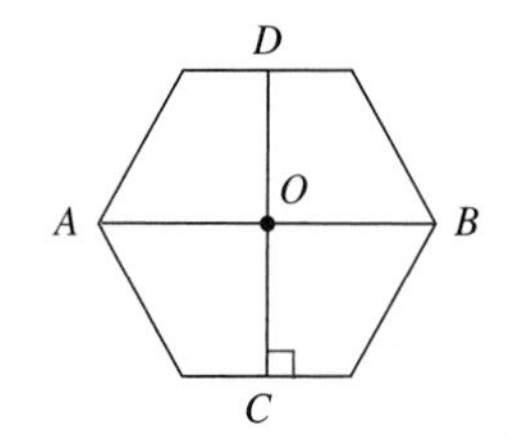

19. The following two rhombuses have perimeters that are equal. Use the properties of a rhombus to find the area of each rhombus.

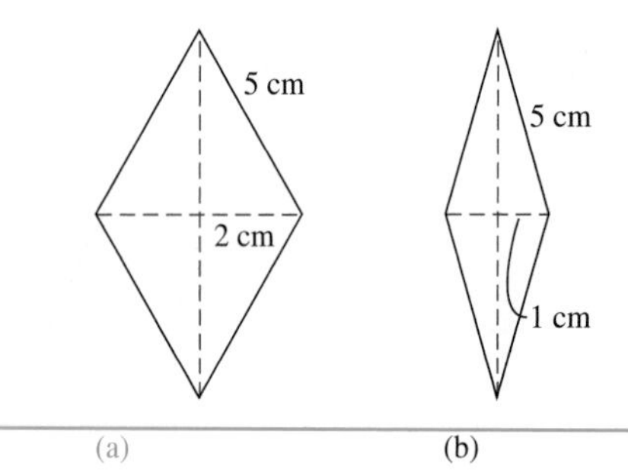

20. The length of the diagonal $\overline{AC}$ of a rhombus $ABCD$ is 20 cm. The distance between $\overline{AB}$ and $\overline{DC}$ is 12 cm. Find the length of the sides of the rhombus and the length of the other diagonal.

21. If $\overline{AB}$, a diameter of circle O, has length 10 cm, point C is on circle O, and AC is 6 cm, how long is $\overline{BC}$?
22. Georgette wants to put a diagonal brace on a rectangular gate that is 3 ft wide and 5 ft high. If she uses a board that is 6 in. wide and 8 ft long for the diagonal, how much will she have left?
23. If a third baseman on the base throws to first base, how far is the ball thrown? (*Hint:* The distance from home plate to first base is 90 ft.)
24. What is the longest piece of straight dry spaghetti that will fit in a cylindrical can that has a radius of 2 in. and height of 10 in.?
25. If possible, draw a square with the given number of square units on a geoboard grid. (You will have to draw your own geoboard grid.)
a. 5 **b.** 7 c. 8 **d.** 14 e. 15
26. Use the following drawing to prove the Pythagorean Theorem by using corresponding parts of similar triangles $\triangle ACD$, $\triangle CBD$, and $\triangle ABC$. Lengths of sides are indicated by a, b, c, x, and y. (*Hint:* Show that $b^2 = cx$ and $a^2 = cy$.)

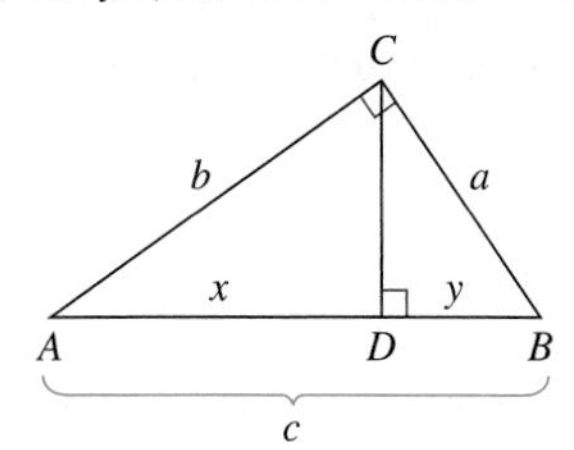

27. An access ramp enters a building 1 m above ground level and starts 3 m from the building. How long is the ramp?
28. To make a homeplate for a neighborhood baseball park, we can cut the plate from a square, as shown in the following figure. If A, B, and C are midpoints of the sides of the square, what are the dimensions of the square to the nearest tenth of an inch?

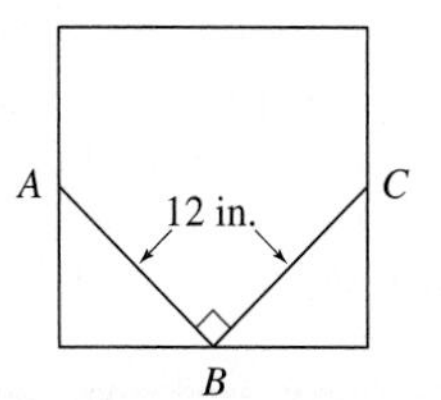

29. A company wants to lay a string of buoys across a lake. To find the length of the lake, they made the following measurements. What is the length of the lake?

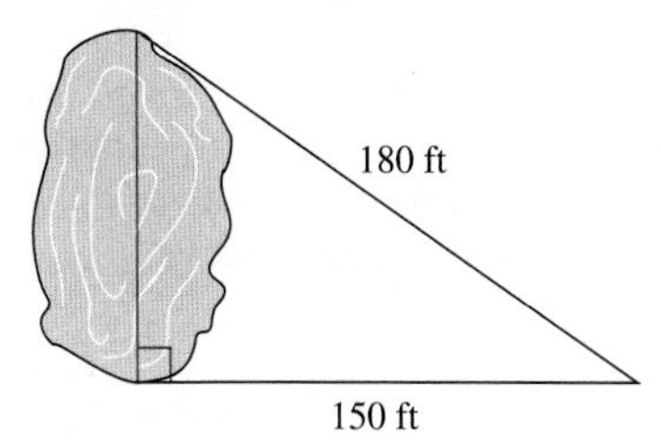

30. A CB radio station C is located 3 mi from the interstate highway h. The station has a range of 6.1 mi in all directions from the station. If the interstate is along a straight line, how many miles of highway are in the range of this station?

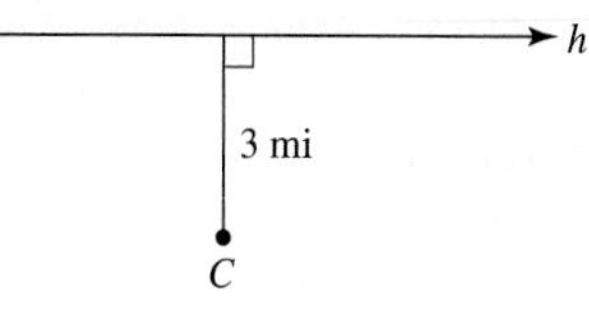

31. Before James Garfield was elected President of the United States, he discovered a proof of the Pythagorean Theorem. He formed a trapezoid like the one that follows and found the area of the trapezoid in two ways. Can you discover his proof?

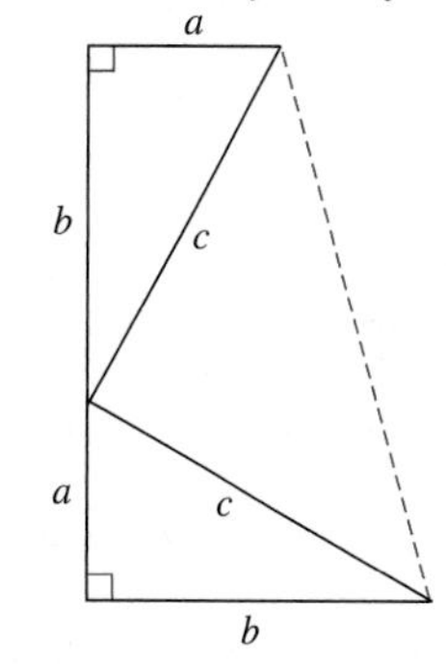

32. Use the following figure to prove the Pythagorean Theorem by first proving that the quadrilateral with side c is a square. Then, compute the area of the square with side $a + b$ in two ways: (a) as $(a + b)^2$ and (b) as the sum of the areas of the four triangles and the square with side c.

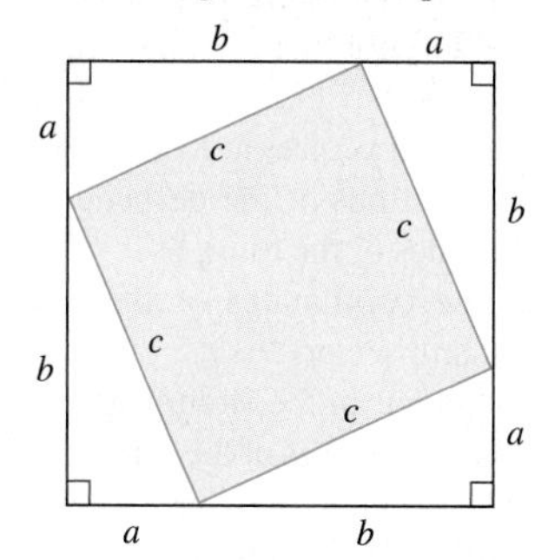

33. Show how the following figure could be used to prove the Pythagorean Theorem:

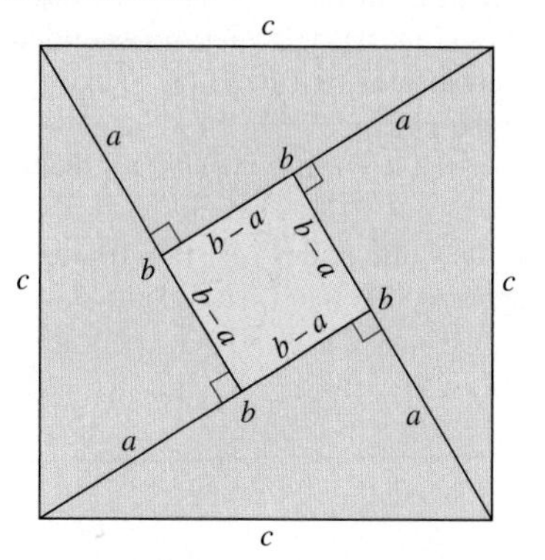

34. Construct semicircles on right triangle ABC with $\overline{AB}$, $\overline{BC}$, and $\overline{AC}$ as diameters. Is the area of the semicircle on the hypotenuse equal to the sum of the areas of the semicircles on the legs? Why?

35. On each side of a right triangle, construct an equilateral triangle. Is the area of the triangle constructed on the hypotenuse always equal to the sum of the areas of the triangles constructed on the legs? Why?

36. For each of the following, find the length of $\overline{AB}$:
   a. $A(0, 3), B(0, 7)$
   b. $A(0, 3), B(4, 0)$
   c. $A(^-1, 2), B(3, ^-4)$
   d. $A(4, ^-5), B\left(\frac{1}{2}, \frac{^-7}{4}\right)$

37. Find the perimeter of the triangle with vertices at $A(0, 0)$, $B(^-4, ^-3)$, and $C(^-5, 0)$.

38. Show that $(0, 6)$, $(^-3, 0)$, and $(9, ^-6)$ are the vertices of a right triangle.

39. Show that the triangle whose vertices are $A(^-2, ^-5)$, $B(1, ^-1)$, and $C(5, 2)$ is isosceles.

40. a. Find x if the distance between $P(1, 3)$ and $Q(x, 9)$ is 10 units.
   ★ **b.** Find the set of all points (x, y) in the plane that are equidistant from $A(^-2, 0)$ and $B(0, 1)$.

41. If the hypotenuse in a 30°-60°-90° triangle is $c/2$ units, what is the length of the side opposite the 60° angle? Explain your answer.

42. In triangle ABC with vertices at $A(0, 0)$, $B(6, 0)$, and $C(0, 8)$,
   a. Write the equations of each of the altitudes.
   b. What are the coordinates of the intersection of the altitudes?

43. In triangle ABC with vertices at $A(0, 0)$, $B(6, 0)$, and $C(0, 8)$,
   a. Write the equations of the perpendicular bisectors of each of the sides of the triangle.
   b. What are the coordinates of the intersection of the perpendicular bisectors?
   c. What is the radius of the circumcircle for the triangle?
   d. What are the coordinates of the center of the circumcircle?

44. In an equilateral triangle with one vertex at the origin and one with coordinates $(8, 0)$, find the possible coordinates of the third vertex.

45. a. Find the possible coordinates of the third vertex of an isosceles right triangle that has endpoints of one leg with coordinates $(0, 0)$ and $(8, 8)$ and the hypotenuse lies on the x-axis.
   b. What are the lengths of the sides of the triangle in part (a)?
   c. Show that the sides of the triangle satisfy the Pythagorean Theorem.

Communication

46. Given the following square, describe how to use a compass and a straightedge to construct a square whose area is as follows:

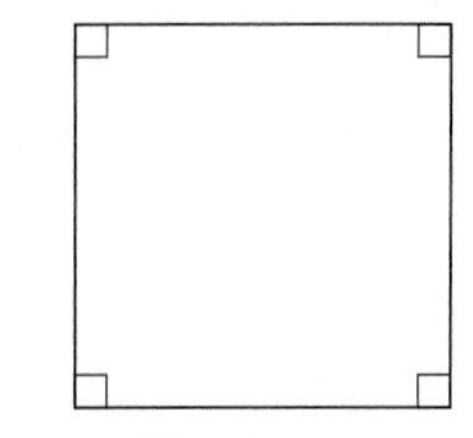

   a. Twice the area of the given square
   b. Half the area of the given square

47. If the hypotenuse and a leg of one right triangle are congruent to the hypotenuse and a leg of another right triangle, respectively, must the triangles be congruent? Explain.

48. Gail tried the Egyptian method of using a knotted rope to determine a right angle so that she could build a shed. She placed her knots so that each was 1 ft from the next. She stretched out her rope in the form of a triangle whose sides were of lengths 5, 12, and 13 ft. Did she have a right angle? Explain why or why not.

Open-Ended

49. Draw several kinds of triangles including a right triangle. Draw a square on each of the sides of the triangles. Compute the areas of the squares and use this information to investigate whether the Pythagorean Theorem works for only right triangles. Use a geometry utility if available.

50. Find an application from real life in which knowing the Pythagorean Theorem would be useful. Write a problem about the application to share with the class.

51. **Pythagorean triples** are three natural numbers a, b, and c that satisfy the relationship $a^2 + b^2 = c^2$. The least three numbers that are Pythagorean triples are 3-4-5. Another triple is 5-12-13 because $5^2 + 12^2 = 13^2$.
   a. Find two other Pythagorean triples.
   b. Does doubling each number in a Pythagorean triple result in a new Pythagorean triple? Why or why not?
   c. Does adding a fixed number to each number in a Pythagorean triple result in a new Pythagorean triple? Why or why not?
   d. Suppose $a = 2uv$, $b = u^2 - v^2$, and $c = u^2 + v^2$, where u and v are whole numbers. Determine whether a-b-c is a Pythagorean triple.

Cooperative Learning

52. Have each person in the group use a 1-m string to make a different right triangle. Measure each side to the nearest centimeter. Use these measurements to see if the Pythagorean Theorem holds for your measurements.

Review Problems

53. Arrange the following in decreasing order: 3.2 m, 322 cm, 0.032 km, 3.020 mm.
54. Find the area of each of the following figures:

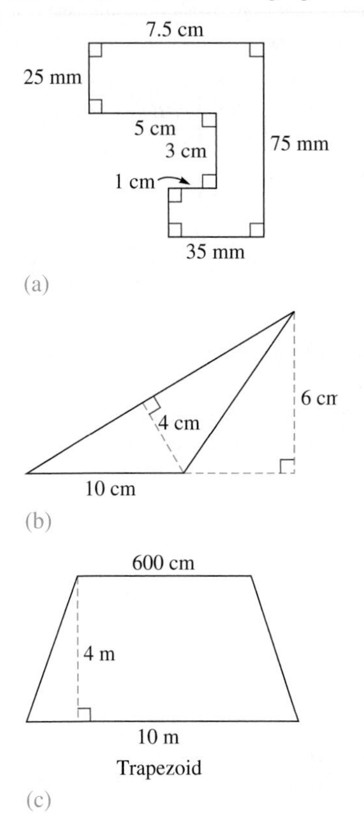

55. Complete the following table, which concerns circles:

	Radius	Diameter	Circumference	Area
a.	5 cm			
b.		24 cm		
c.				$17\pi m^2$
d.			20π cm	

56. A 10-m wire is wrapped around a circular region. If the wire fits exactly, what is the area of the region?

Third International Mathematics and Science Study (TIMSS) Questions

A string is wound symmetrically around a circular rod. The string goes exactly 4 times around the rod. The circumference of the rod is 4 cm and its length is 12 cm.

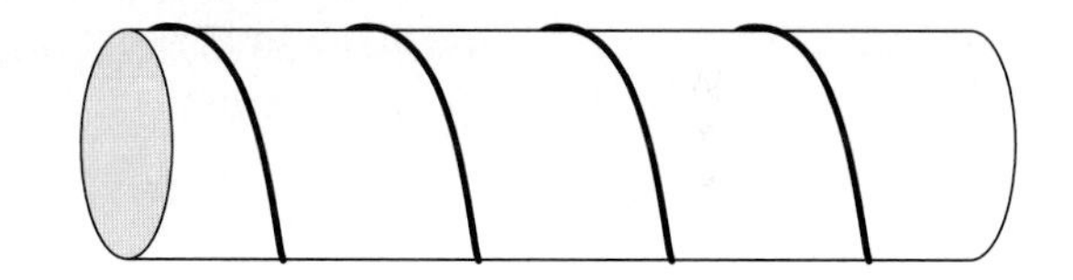

Find the length of the string. Show all your work.

TIMSS, Grade 12, 1994

Which of these is closest to $11^2 + 9^2$?

a. 20 + 20
b. 20 + 80
c. 120 + 20
d. 120 + 80

TIMSS, Grade 8, 2003

BRAIN TEASER

A spider is sitting at A, the midpoint of the edge of the ceiling in the room shown in Figure 11-53. It spies a fly on the floor at C, the midpoint of the edge of the floor. If the spider must walk along the wall, ceiling, or floor, what is the length of the shortest path the spider can travel to reach the fly?

FIGURE 11-53

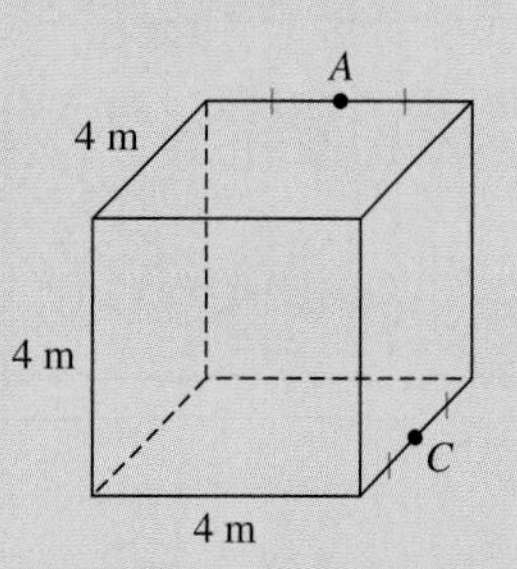

TECHNOLOGY CORNER

THE GEOMETER'S SKETCHPAD

Use GSP to determine the relationship between the length of the hypotenuse of a 45°-45°-90° triangle and the length of a leg.

a. Construct a 45°-45°-90° triangle, label the vertices as in Figure 11-54, and measure the lengths of the sides. Record the data for triangle 1 in the following table and compute the ratio:

FIGURE 11-54

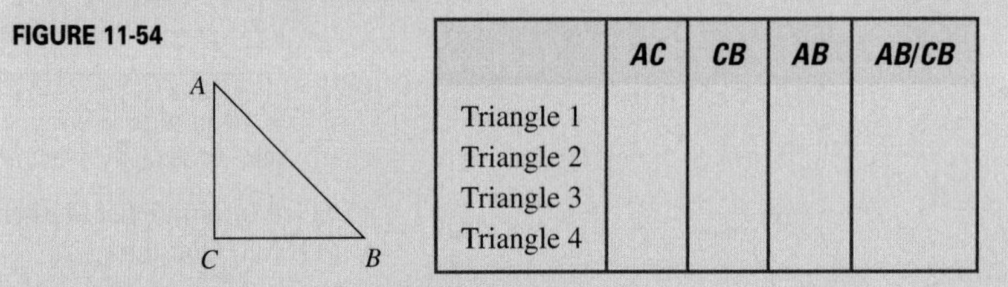

	AC	*CB*	*AB*	*AB/CB*
Triangle 1				
Triangle 2				
Triangle 3				
Triangle 4				

b. Repeat (a) for three other triangles.

c. Make a conjecture about the relationship between the length of the hypotenuse and the length of a leg for these triangles.

d. Given a 30°-60°-90° triangle, determine the relationship between the lengths of the hypotenuse and the shorter leg and the relationships between the lengths of the longer and shorter legs.

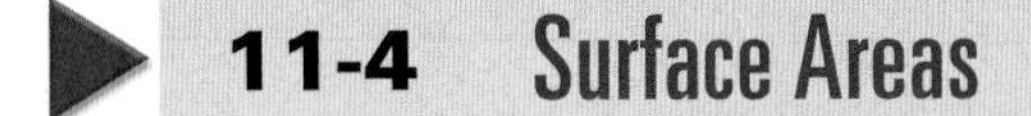

11-4 Surface Areas

Painting houses, buying roofing, seal-coating driveways, and buying carpet are among the common applications that involve computing areas. In many real-world problems, we must find the surface areas of such three-dimensional figures as prisms, cylinders, pyramids, cones, and spheres. Formulas for finding these areas are usually based on finding the area of two-dimensional pieces of the three-dimensional figures.

Surface Area of Right Prisms

Consider the cereal box shown in Figure 11-55(a). If we ignore the flaps for gluing the box together, to find the amount of cardboard necessary to make the box, we cut the box along the edges and make it lie flat, as shown in Figure 11-55(b). When we do this we obtain a *net* for the box. The box is composed of a series of rectangles. We find the area of each rectangle and sum those areas to find the surface area of the box.

FIGURE 11-55

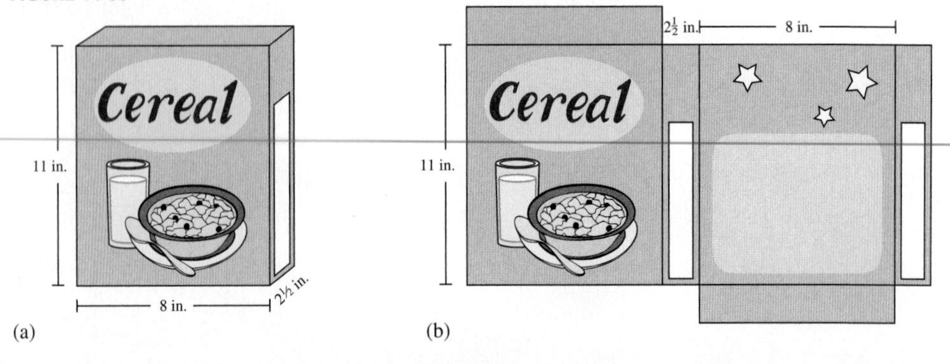

NOW TRY THIS 11-15

a. Find the surface area of the box in Figure 11-55.

b. Could the box be made from a rectangular piece of cardboard 21 in. by 15 in.? If not, what size rectangle could you use and how would you do it?

A similar process can be used for many three-dimensional figures. For example, the surface area of the cube in Figure 11-56(a) is the sum of the areas of the faces of the cube. A net for the cube is shown in Figure 11-56(c). Because each of the six faces is a square of area 16 cm^2, the surface area is $6 \cdot (16\text{ cm}^2)$, or 96 cm^2, or in general, for a cube whose edge is e units as in Figure 11-56(b), the surface area is $6e^2$.

FIGURE 11-56

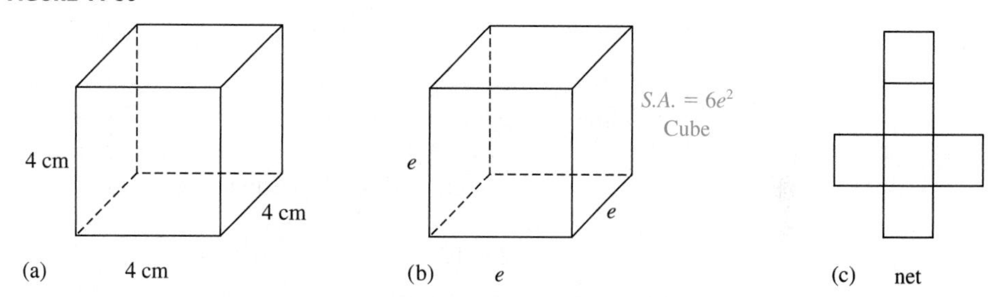

To find the surface area of a right prism, like the cereal box in Figure 11-55(a), we find the sum of the areas of the rectangles that make up the lateral faces and the areas of the top and bottom. The sum of the areas of the lateral faces is the **lateral surface area**. The **surface area** (*S.A.*) is the sum of the lateral surface area and the area of the bases.

lateral surface area ◆ surface area

NOW TRY THIS 11-16

Figure 11-57 shows a right pentagonal prism with a net for the prism. If B stands for the area of each of the prism's bases, show that the surface area of the prism could be computed as $S.A. = ph + 2B$, where p is the perimeter of the base of the prism and h is the height. Does this formula hold for all right prisms? Why or why not?

FIGURE 11-57

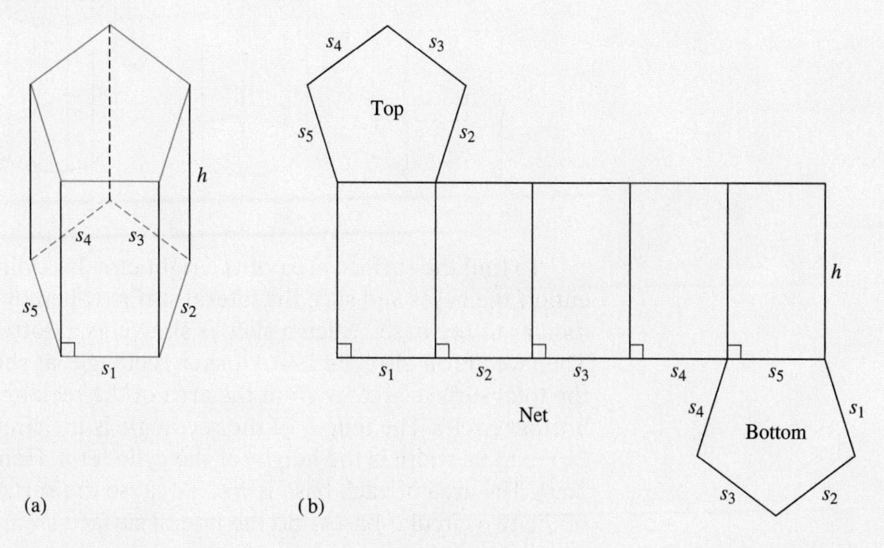

EXAMPLE 11-18 Find the surface area of each of the right prisms in Figure 11-58.

FIGURE 11-58

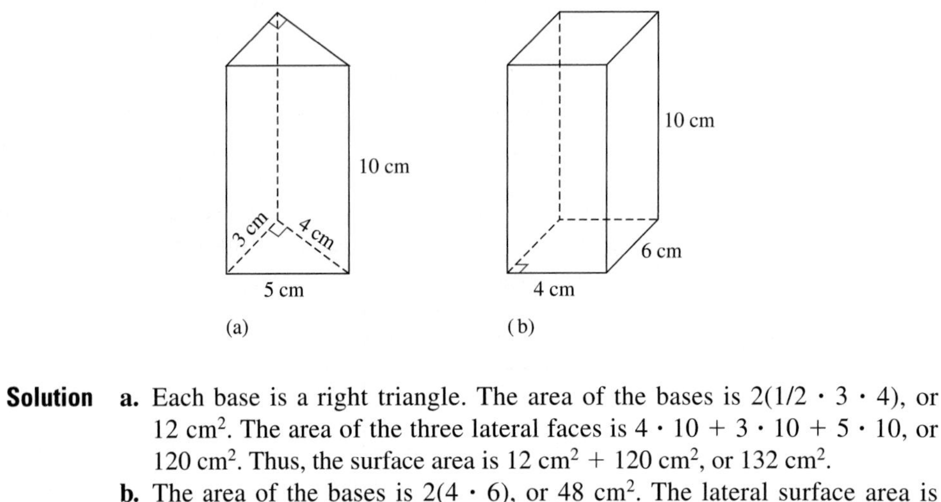

Solution **a.** Each base is a right triangle. The area of the bases is $2(1/2 \cdot 3 \cdot 4)$, or 12 cm^2. The area of the three lateral faces is $4 \cdot 10 + 3 \cdot 10 + 5 \cdot 10$, or 120 cm^2. Thus, the surface area is $12\text{ cm}^2 + 120\text{ cm}^2$, or 132 cm^2.

b. The area of the bases is $2(4 \cdot 6)$, or 48 cm^2. The lateral surface area is $2 \cdot (10 \cdot 6) + 2 \cdot (4 \cdot 10)$, or 200 cm^2. Thus, the surface area is 248 cm^2.

Surface Area of a Cylinder

As the number of sides of a right regular prism increases, as shown in Figure 11-59, then the figure approaches the shape of a right regular cylinder.

FIGURE 11-59

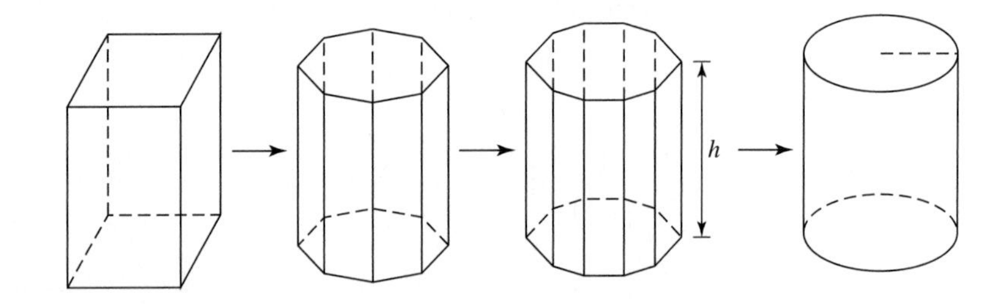

To find the surface area of the right circular cylinder shown in Figure 11-60, we cut off the bases and slice the lateral surface open by cutting along any line perpendicular to the bases. Such a slice is shown as a dotted segment in Figure 11-60(a). Then we unroll the cylinder to form a rectangle, as shown in Figure 11-60(b). To find the total surface area, we find the area of the rectangle and the areas of the top and bottom circles. The length of the rectangle is the circumference of the circular base $2\pi r$, and its width is the height of the cylinder h. Hence, the area of the rectangle is $2\pi rh$. The area of each base is πr^2. Because the surface area is the sum of the areas of the two circular bases and the lateral surface area, we have $S.A. = 2\pi r^2 + 2\pi rh$.

FIGURE 11-60

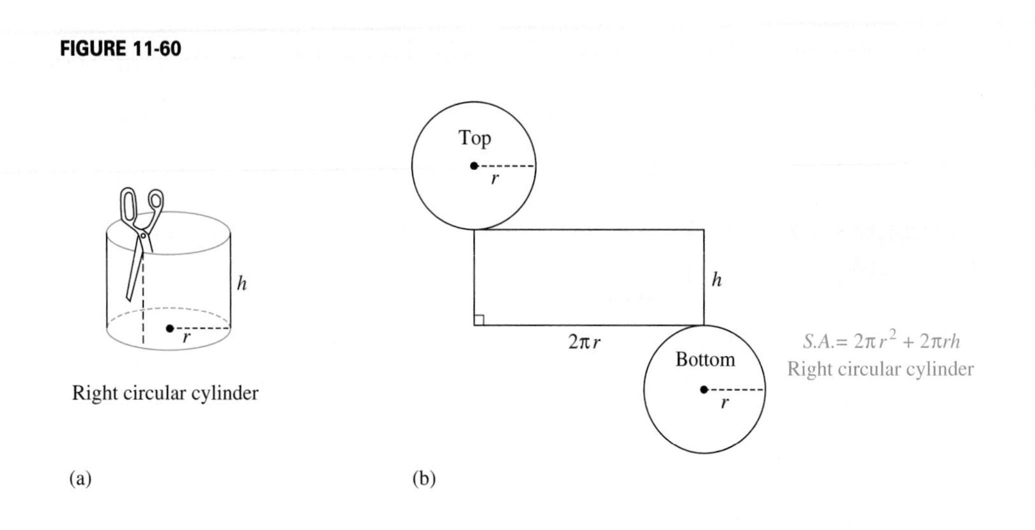

Surface Area of a Pyramid

The surface area of a pyramid is the sum of the lateral surface area of the pyramid and the area of the base. A right regular pyramid is a pyramid such that the segments connecting the apex to each vertex of the base are congruent and the base is a regular polygon. The lateral faces of the right regular pyramid pictured in Figure 11-61 are congruent triangles. Each triangular face has an altitude of length ℓ, the *slant height*. Because the pyramid is right regular, each side of the base has the same length b. To find the lateral surface area of a right regular pyramid, we need to find the area of one face, $\frac{1}{2}b\ell$, and multiply it by n, the number of faces. Adding the lateral surface area $n\left(\frac{1}{2}b\ell\right)$ to the area of the base B gives the surface area.

FIGURE 11-61

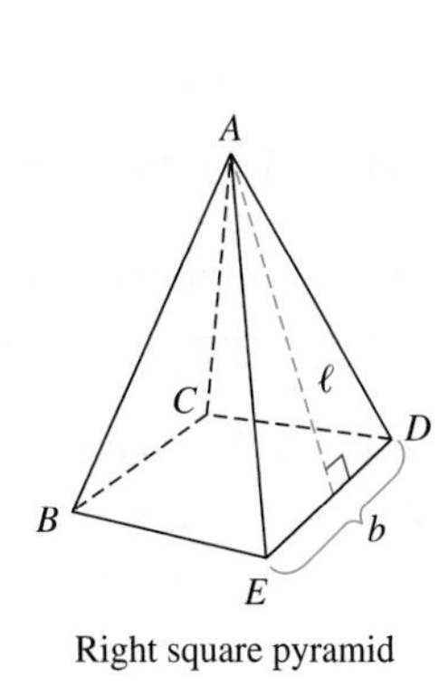

Right square pyramid

(a)

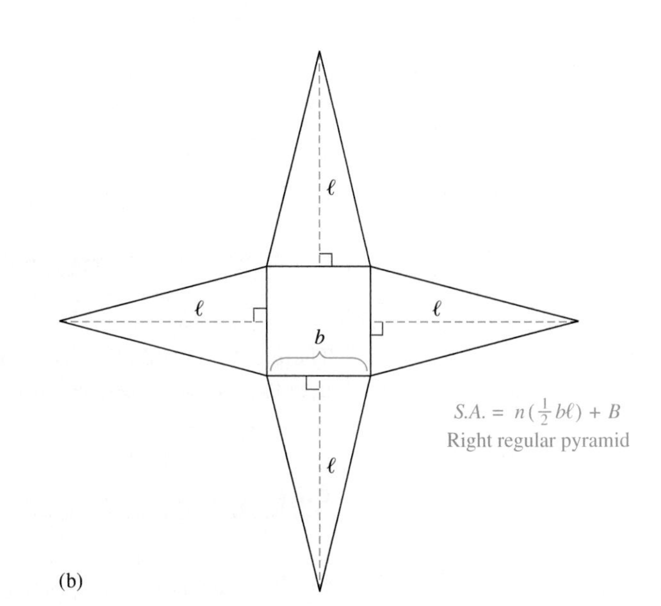

(b)

REMARK The formula for the surface area of a right regular pyramid can be simplified because nb is the perimeter, p, of the base. Thus, $S.A. = \frac{1}{2}p\ell + B$.

EXAMPLE 11-19

Find the surface area of the right regular pyramid in Figure 11-62.

FIGURE 11-62

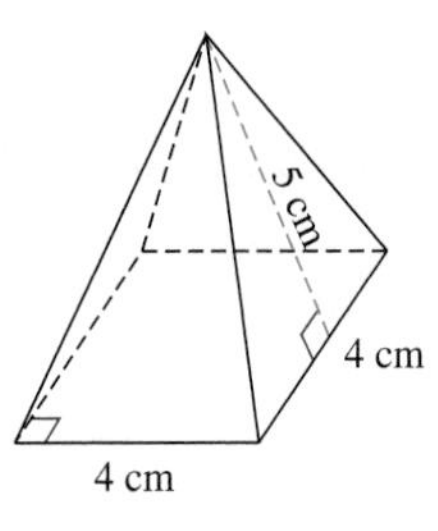

Solution The surface area consists of the area of the square base plus the area of the four triangular faces. Hence, the surface area is

$$4\text{ cm} \cdot 4\text{ cm} + 4 \cdot \left(\frac{1}{2} \cdot 4\text{ cm} \cdot 5\text{ cm}\right) = 16\text{ cm}^2 + 40\text{ cm}^2$$
$$= 56\text{ cm}^2$$

EXAMPLE 11-20

The Great Pyramid of Cheops is a right square pyramid with a height of 148 m and a square base with perimeter of 940 m. The altitude of each triangular face is 189 m. The basic shape of the Transamerica Building in San Francisco is a right square pyramid that has a height of 260 m and a square base with a perimeter of 140 m. The altitude of each triangular face is 261 m. How do the lateral surface areas of the two structures compare?

Solution The length of one side of the square base of the Great Pyramid is $\frac{940}{4}$, or 235, m. Likewise, the length of one side of the square base of the Transamerica Building is $\frac{140}{4}$, or 35 m. The lateral surface area (*L.S.A.*) of the two are computed as follows:

$$\text{(Great Pyramid) } L.S.A. = 4 \cdot (1/2 \cdot 235 \cdot 189) = 88{,}830\text{ m}^2$$
$$\text{(Transamerica) } L.S.A. = 4 \cdot (1/2 \cdot 35 \cdot 261) = 18{,}270\text{ m}^2$$

Therefore, the lateral surface area of the Great Pyramid is approximately 4.9 times greater than that of the Transamerica Building.

Surface Area of a Cone

As the number of sides of a right regular pyramid increases, as shown in Figure 11-63, then the figure approaches the shape of a right circular cone.

FIGURE 11-63

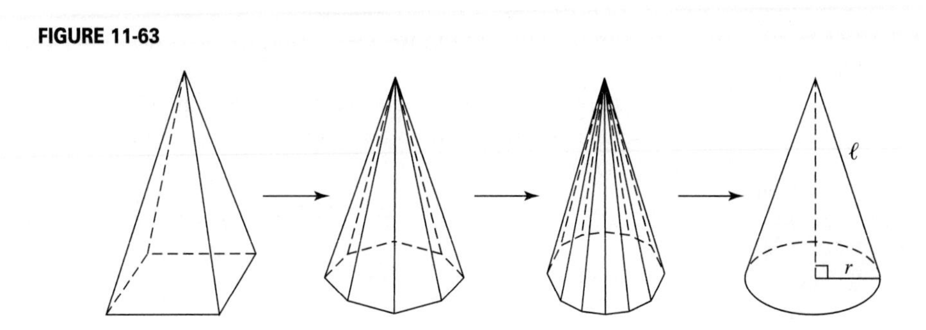

It is possible to find a formula for the surface area of a right circular cone by approximating the cone with a pyramid with many sides. We could inscribe in the circular base of the cone a regular polygon with many sides. The polygon can be used as the base of a right regular pyramid. The lateral surface area of the pyramid is close to the lateral surface area of the cone. The greater the number of faces of the pyramid, the closer the surface area of the pyramid is to that of the cone. In Figure 11-64(a), the lateral surface of the pyramid is $\frac{1}{2}p \cdot h$, where p is the perimeter of the base and h is the height of each triangle. With many sides in the pyramid, the perimeter of its base is close to the perimeter of the circle, $2\pi r$. The height of each triangle of the pyramid is close to the slant height ℓ, a segment that connects the vertex of the cone with a point on the circular base. Consequently, it is reasonable that the lateral surface area of the cone becomes $\frac{1}{2} \cdot 2\pi r \cdot \ell$, or $\pi r\ell$. To find the total surface area of the cone, we add πr^2, the area of the base. Thus, $S.A. = \pi r^2 + \pi r\ell$, as given in Figure 11-64(b).

FIGURE 11-64

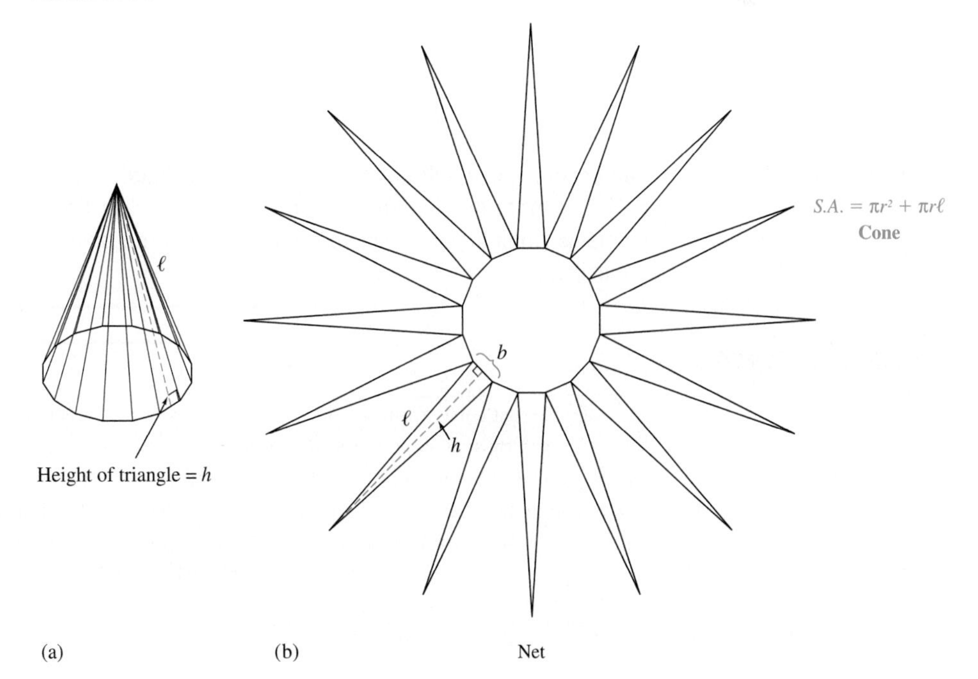

EXAMPLE 11-21

Given the cone in Figure 11-65, find the surface area of that cone.

FIGURE 11-65

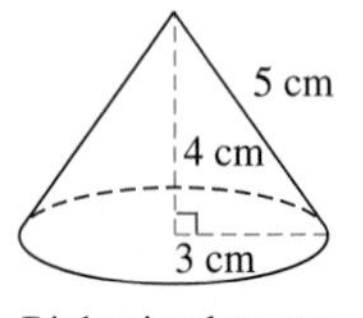

Right circular cone

Solution The base of the cone is a circle with radius 3 cm and area $\pi(3 \text{ cm})^2$, or $9\pi \text{ cm}^2$. The lateral surface has area $\pi(3 \text{ cm})(5 \text{ cm})$, or $15\pi \text{ cm}^2$. Thus we have the following surface area:

$$\begin{aligned} S.A. &= \pi(3 \text{ cm})^2 + \pi(3 \text{ cm})(5 \text{ cm}) \\ &= 9\pi \text{ cm}^2 + 15\pi \text{ cm}^2 \\ &= 24\pi \text{ cm}^2 \end{aligned}$$

Surface Area of a Sphere

great circle

A **great circle** of a sphere is a circle on the sphere whose radius is equal to the radius of the sphere. A great circle is obtained whenever a plane through the center of the sphere intersects the sphere. There are infinitely many great circles on a sphere. However, they are all congruent. Finding a formula for the surface area of a sphere is a simple task using calculus, but it is not easy in elementary mathematics. The surface area of a sphere is four times the area of a great circle of the sphere. Therefore, the formula is $S.A. = 4\pi r^2$, as pictured in Figure 11-66.

FIGURE 11-66

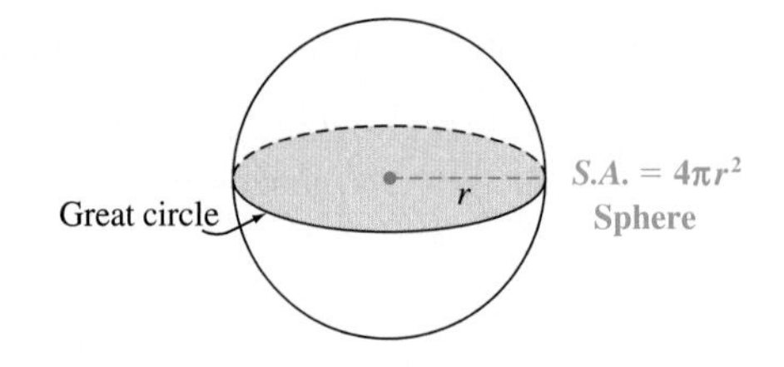

RESEARCH NOTE

Adolescents have a difficult time translating visual information from a three-dimensional environment to two dimensions and vice versa (Ben-Chaim et al. 1989). This may cause problems when trying to find surface areas of three-dimensional figures using two-dimensional images.

LABORATORY ACTIVITY

Create different cones from sectors of a circle. Use a compass to draw a sector of a circle whose diameter is almost as large as the width of a piece of paper. Draw two such sectors that have the same radii but different central angles. In one sector, make the central angle measure smaller than 180°, and in the other, make it measure greater than 180°. Then make a cone from each sector by gluing the edges of each sector together. Can you predict which cone will be taller? Without performing the experiment, can you explain why that cone will be taller?

ONGOING ASSESSMENT 11-4

1. Find the surface area of each of the following:

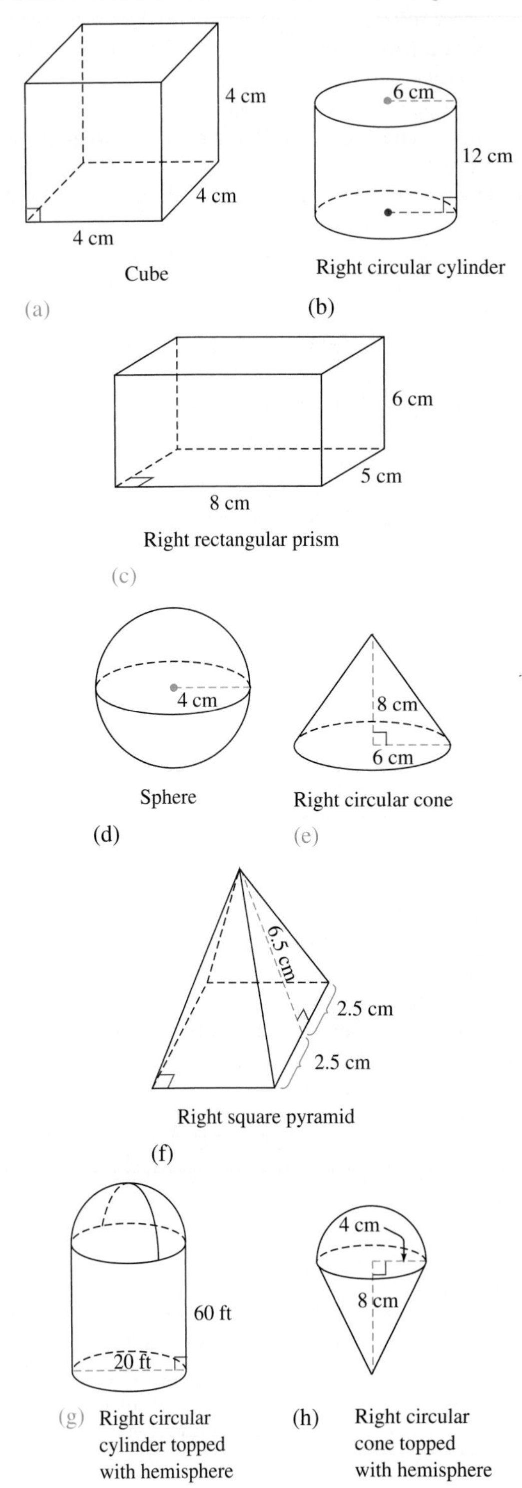

2. How many liters of paint are needed to paint the walls of a rectangular prism-shaped room that is $6 \text{ m} \times 4 \text{ m} \times 2.5 \text{ m}$ if 1 L of paint covers 20 m^2? (Assume there are no doors or windows.)
3. The napkin ring pictured in the following figure is to be re-silvered. How many square millimeters must be covered?

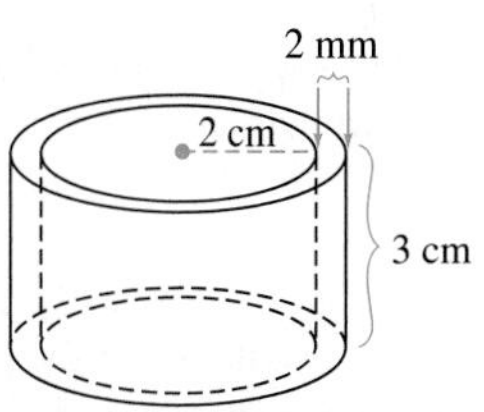

4. Assume the radius of Earth is 6370 km and Earth is a sphere. What is its surface area?
5. Two cubes have sides of length 4 cm and 6 cm, respectively. What is the ratio of their surface areas?
6. Suppose one right circular cylinder has radius 2 m and height 6 m and another has radius 6 m and height 2 m.
   a. Which cylinder has the greater lateral surface area?
   b. Which cylinder has the greater total surface area?
7. The base of a right pyramid is a regular hexagon with sides of length 12 m. The altitude of the pyramid is 9 m. Find the total surface area of the pyramid.
8. A soup can has a $2\frac{5}{8}$-in. diameter and is 4 in. tall. What is the area of the paper that will be used to make the label for the can if the paper covers the entire lateral surface area?
9. A square piece of paper 10 cm on a side is rolled to form the lateral surface area of a right circular cylinder and then a top and bottom are added. What is the surface area of the cylinder?
10. Approximately how much material is needed to make the tent illustrated in the following figure (both ends and the bottom should be included)?

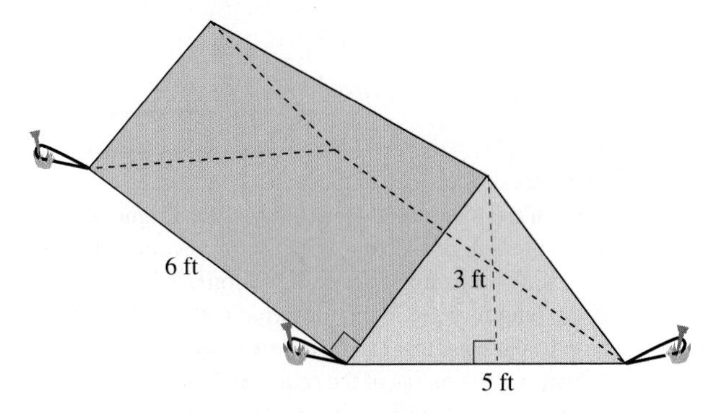

11. The top of a rectangular box has an area of 88 cm^2. The sides have areas 32 cm^2 and 44 cm^2. What are the dimensions of the box?

12. How does the surface area of a box (including top and bottom) change if
    a. Each dimension is doubled?
    b. Each dimension is tripled?
    c. Each dimension is multiplied by a factor of k?
13. How does the lateral surface area of a cone change if
    a. The slant height is tripled but the radius of the base remains the same?
    b. The radius of the base is tripled but the slant height remains the same?
    c. The slant height and the radius of the base are tripled?
14. What happens to the surface area of a sphere if the radius is
    a. Doubled?
    b. Tripled?
15. Find the surface area of a square pyramid if the area of the base is 100 cm^2 and the height of the pyramid is 20 cm.
16. Suppose a structure is composed of unit cubes with at least one face of each cube connected to the face of another cube, as shown in the following figure:

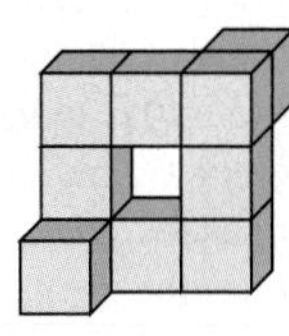

    a. If one cube is added, what is the maximum surface area the structure can have?
    b. If one cube is added, what is the minimum surface area the structure can have?
    c. Is it possible to design a structure so that one can add a cube and yet add nothing to the surface area of the structure? (*Hint:* Cubes might have to be glued together.) Explain your answer.
17. The sector shown in the following figure is rolled into a cone so that the dotted edges just touch. Find the following:

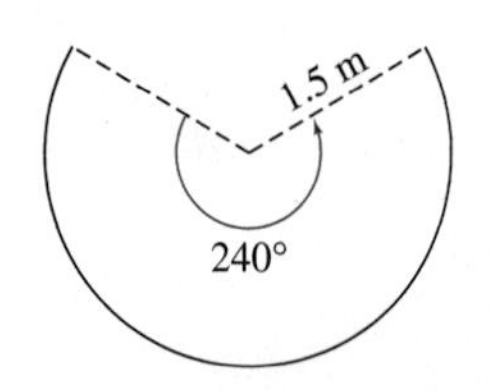

    a. The lateral surface area of the cone
    b. The total surface area of the cone
18. A sector of a circle can be used to construct a right circular cone. The length of the arc of the sector becomes the circumference of the circular base of the cone.
    a. If the length of the arc is 6π, what is the radius of the base of the cone that can be constructed?
    b. In part (a), the radius of the sector is 5 units, what is the slant height of the cone that can be constructed?
    c. Using the information in parts (a) and (b), what is the height of the cone that can be constructed?
    d. Using the information in parts (a)–(c), what is the angle measure for the original sector?
19. If the cardboard tube of a toilet paper roll has diameter of 2.5 in. and is 4 in. tall, what is the lateral surface area of the cardboard roll?
20. Suppose a paper cup is a frustum of a cone (that is, the cone is truncated).
    a. If the paper cup is rolled in a plane, with the paths of the top and bottom of the cup traced, what figure is formed?
    b. Describe the measures of the figure formed in relation to the original cup.
21. If two right circular cones are similar with radii of the bases in the ratio $1:2$, what is the ratio of their surface areas?
22. Must every two right quadrilateral prisms be similar? Why?
23. If two cubes have total surface areas of 64 in.^{2} and 36 in.^{2}, what is the ratio of their edges?
24. Each region in the following figure revolves about the indicated axis. For each case, sketch the three-dimensional figure obtained and find its surface area.

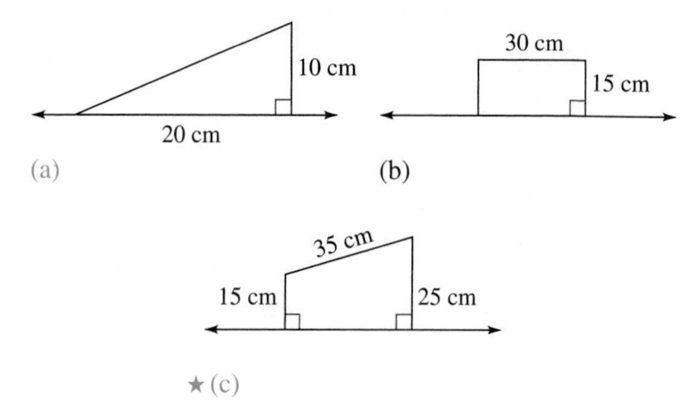

25. The total surface area of a cube is 10,648 cm^2. What is the length of each of the following?
    a. One of the sides
    b. A diagonal that is not a diagonal of a face
26. A square pyramid and a right circular cone are inscribed in a cube as shown. Find each of the following:
    a. Surface area of the pyramid
    b. Surface area of the cone

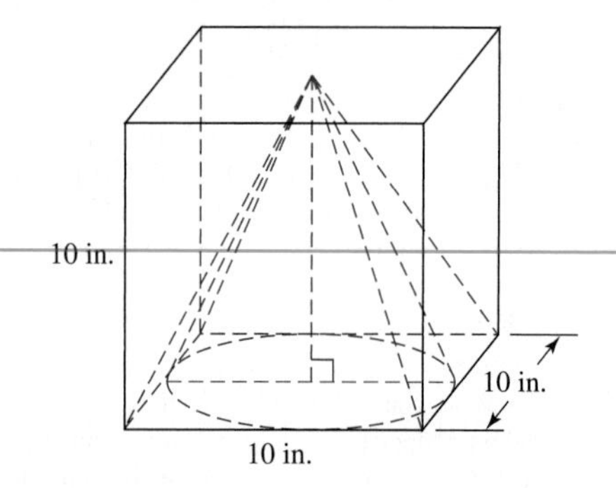

★ 27. Find the total surface area of the following stand, which was cut from a right circular cone:

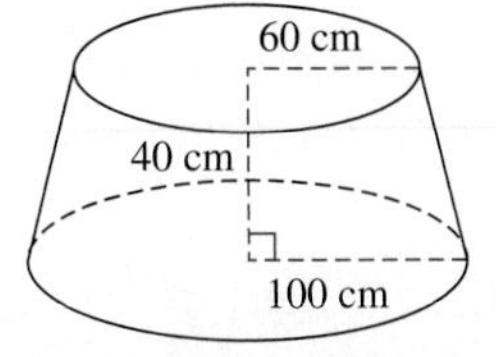

★ 28. In the following figure, a right circular cylinder is inscribed in a right circular cone. Find the lateral surface area of the cylinder if the height of the cone is 40 cm, the height of the cylinder 30 cm, and the radius of the base of the cone is 25 cm.

Communication

29. Which do you think would melt faster in an ice chest: a block of ice made from a gallon of water or ice cubes made from a gallon of water? Explain why.
30. A student wonders if she doubles each measurement of a cereal box, will she need twice as much cardboard to make the new box. How would you help her decide?
31. Tennis balls are packed tightly three to a can that is shaped like a cylinder.
   a. Estimate how the surface area of the balls compares to the lateral surface area of the can. Explain how you arrived at your estimate.
   b. See how close your estimate in (a) was by actually computing the surface area of the balls and the lateral surface area of the can.
32. Consider a pie in a normal circular 9-in. pie plate. Mathematically describe the pie.
33. A water hose that is 50 ft long is described as a $\frac{1}{2}$-in. hose.
   a. What is the meaning of the description?
   b. What is the surface area of the exterior of the hose?
   c. How does the exterior surface area of the $\frac{1}{2}$-in. hose compare to the exterior surface area of a 1-in. hose that is 50 ft long?

Open-Ended

34. One method of estimating body surface area in burn victims uses the fact that 100 handprints will approximately cover the whole body.
   a. What percentage of the body surface area is the surface area of two handprints?
   b. Estimate the percentage of the body surface area of one arm. Explain how you arrived at your estimate.
   c. Estimate your body surface area in square centimeters. Explain how you arrived at your estimate.
   d. Find the area of the flat part of your desk. How does the area of the desk compare with the surface area of your body?
35. Design a net for a polyhedron in such a way that the surface area of the polyhedron is 10 cm². Explain what polyhedron the net will form and why its surface area is 10 cm².

Cooperative Learning

36. a. Shawn used small cubes to build a bigger cube that was solid and was three cubes long on each side. He then painted all the sides of the new, large cube red. He dropped the newly painted cube and all the little cubes came apart. He noticed that some cubes had only one side painted, some had two sides painted, and so on. Describe the number of cubes with 0, 1, 2, 3, 4, 5, or 6 sides painted. Have each member of the group choose a different number of sides and then combine your data to see if it makes sense. Look for any patterns that occur.
   b. What would the answers be if the large cube was four small cubes long on a side?
   c. Make a conjecture about how to count the cubes if the large cube were n small cubes long on a side.
37. In groups, work through the Package Design Contest as described on the student page on page 796.

Review Problems

38. Complete each of the following:
   a. $10 \text{ m}^2 = \_\_\_\_ \text{ cm}^2$
   b. $13{,}680 \text{ cm}^2 = \_\_\_\_ \text{ m}^2$
   c. $5 \text{ cm}^2 = \_\_\_\_ \text{ mm}^2$
   d. $2 \text{ km}^2 = \_\_\_\_ \text{ m}^2$
   e. $10^6 \text{ m}^2 = \_\_\_\_ \text{ km}^2$
   f. $10^{12} \text{ mm}^2 = \_\_\_\_ \text{ m}^2$
39. The sides of a rectangle are 10 cm and 20 cm long. Find the length of a diagonal of the rectangle.
40. The length of the side of a rhombus is 30 cm. If the length of one diagonal is 40 cm, find the length of the other diagonal.

SAMPLE SCHOOL BOOK PAGE:

PACKAGE DESIGN CONTEST

The Unit Project

Package Design Contest

The Worldwide Sporting Company wants new package designs for its table-tennis balls (Ping-Pong balls). The company's table-tennis balls are about 3.8 cm in diameter. There are three main requirements for the packages:

- The board of directors wants to have three different size packages: small, medium, and large.
- The president of the company wants the cost of the packages to be a primary consideration.
- The sales manager wants the packages to be appealing to customers, to stack easily, and to look good on store shelves.

The company holds a package design contest, and you decide to enter.

- You must design three different packages for the table-tennis balls.
- You must submit your designs and a written proposal to WSC.
- You must try, in your written proposal, to convince WSC to use your designs.

Include the following things in your proposal:

1. A description of the shape or shapes of the packages and an explanation for why you selected these shapes.
2. Patterns for each package that, when they are cut out, folded, and taped together, will make models of your packages. Use centimeter grid paper to make your patterns.
3. Cost estimates to construct your designs. The packaging material costs \$0.005 per square centimeter.
4. An explanation of how you have addressed WSC's requirements.

Remember, you are trying to convince WSC that your designs are the best and that they meet the requirements. Your written proposal should be neat, well organized, and easy to read so that the company officials can follow your work and ideas easily.

Source: Prentice Hall, Connected Mathematics: Filling and Wrapping, 2002 (p. 73).

41. Find the perimeters and the areas of the following figures:

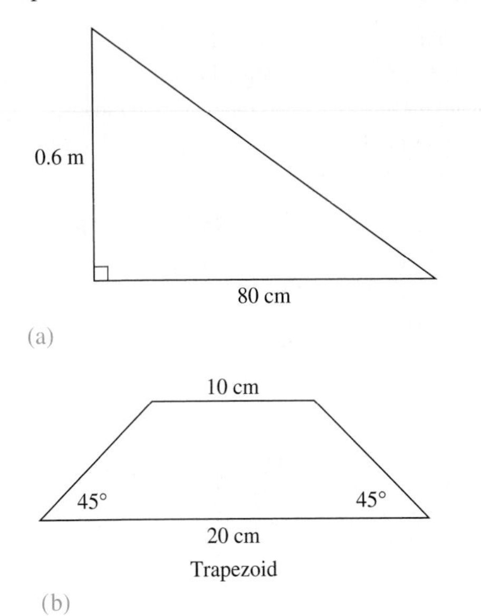

42. In the following, the length of the diagonal $\overline{AC}$ of rhombus $ABCD$ is 40 cm; $AE = 24$ cm. Find the length of a side of the rhombus and the length of the diagonal $\overline{BD}$.

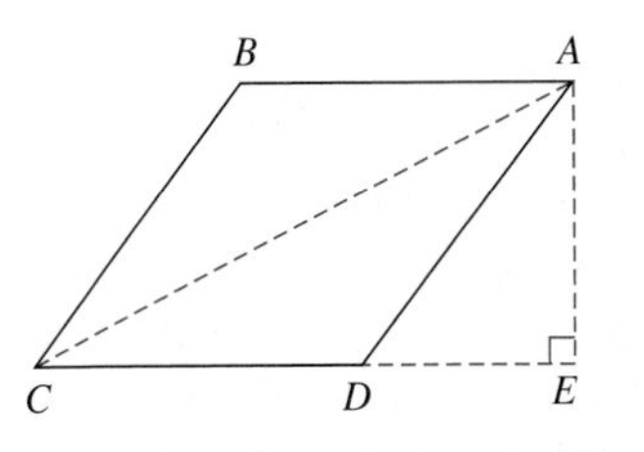

43. Choose any square. Inscribe a circle in it. Circumscribe a circle about it.
   a. What is the ratio of the circumference of the inscribed circle to the perimeter of the square?
   b. What is the ratio of the circumference of the circumcircle to the perimeter of the square?
   c. What is the ratio of the circumference of the incircle to the circumference of the circumscribed circle?

44. a. In the following drawing, where is point C on line so that $AC + CB$ is the shortest?

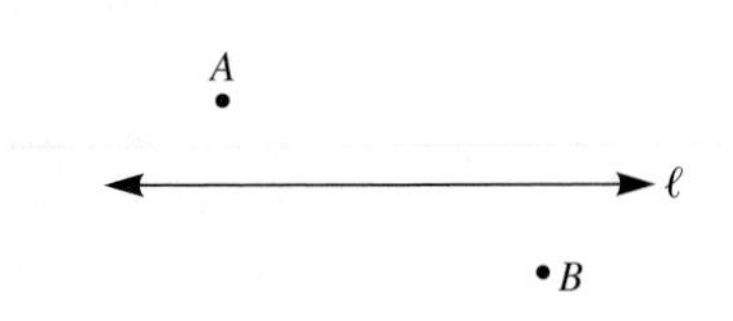

   b. Prove your answer in part (a).

Third International Mathematics and Science Study (TIMSS) Questions

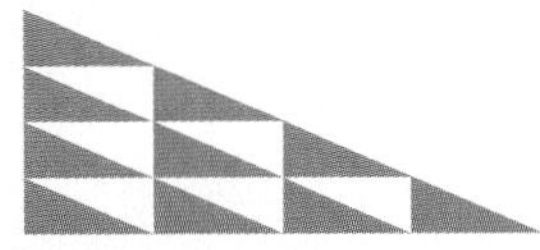

In the figure above, each of the smaller triangles has the same area. What is the ratio of the shaded area to the unshaded area?

a. 5 : 3
b. 8 : 5
c. 5 : 8
d. 3 : 5

TIMSS, Grade 8, 2003

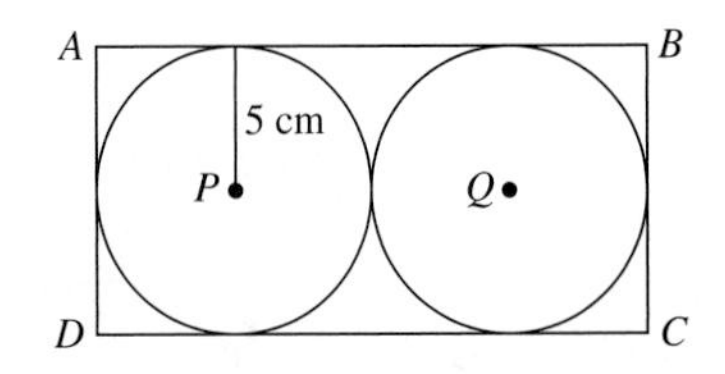

In the figure above, $ABCD$ is a rectangle, and circles P and Q each have a radius of 5 cm. What is the area of the rectangle?

a. 50 cm^2
b. 60 cm^2
c. 100 cm^2
d. 200 cm^2

TIMSS, Grade 8, 2003

TECHNOLOGY CORNER

Find various-shaped cardboard containers and cut them apart to form nets. Make sure the nets are smaller than your computer screen.

a. Estimate the surface area of the nets.
b. Trace the perimeter of each net on a transparency sheet and hang the sheet on the monitor of your computer. Use GSP to trace the outline from the transparency and compute the area.
c. Compare your estimate in (a) with the answer in (b).

BRAIN TEASER

A manufacturer of paper cups wants to produce paper cups in the form of truncated cones 16 cm high, with one circular base of radius 11 cm and the other of radius 7 cm, as shown in Figure 11-67. When the base of such a cup is removed and the cup is slit and flattened, the flattened region looks like a part of a circular ring. To design a pattern to make the cup, the manufacturer needs the data required to construct the flattened region. Find these data.

FIGURE 11-67

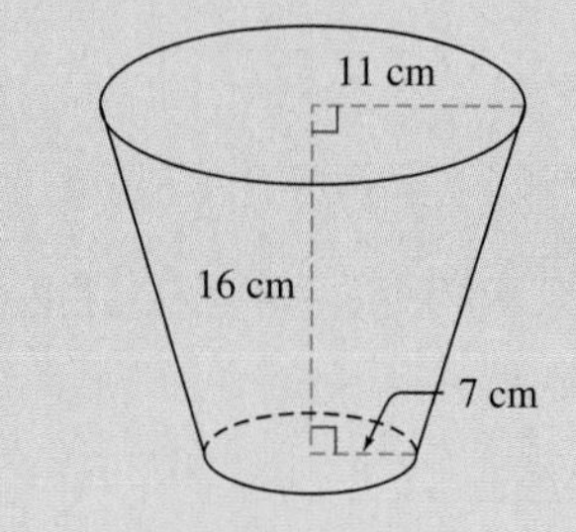

11-5 Volume, Mass, and Temperature

In Section 11-4, we investigated surface areas of various-shaped containers. In this section, we explore how much the containers will hold. This distinction is sometimes confused by elementary school students. Whereas the surface area is the number of square units covering a three-dimensional figure, volume describes how much space a three-dimensional figure contains. The unit of measure for volume must be a shape that tessellates space. Cubes tessellate space; that is, they can be stacked so that they leave no gaps and fill space. Standard units of volume are based on cubes and are *cubic units*. A cubic unit is the amount of space enclosed within a cube that measures 1 unit on a side. The distinction between surface area and volume is demonstrated in Figure 11-68.

FIGURE 11-68

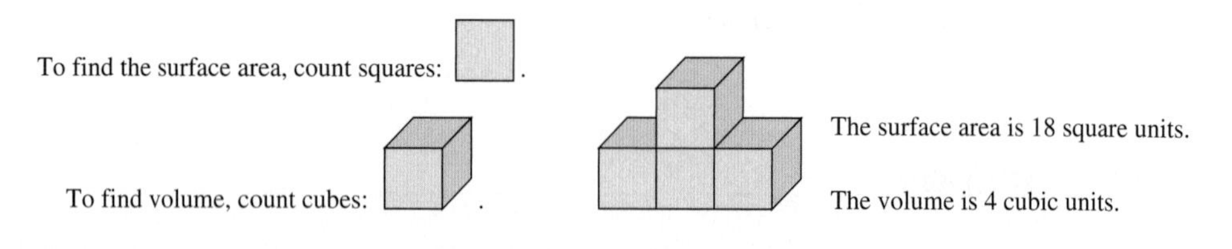

To find the surface area, count squares: .
To find volume, count cubes: .
The surface area is 18 square units.
The volume is 4 cubic units.

NOW TRY THIS 11-17

In Figure 11-69, the purple block is moved from one position to another.

FIGURE 11-69

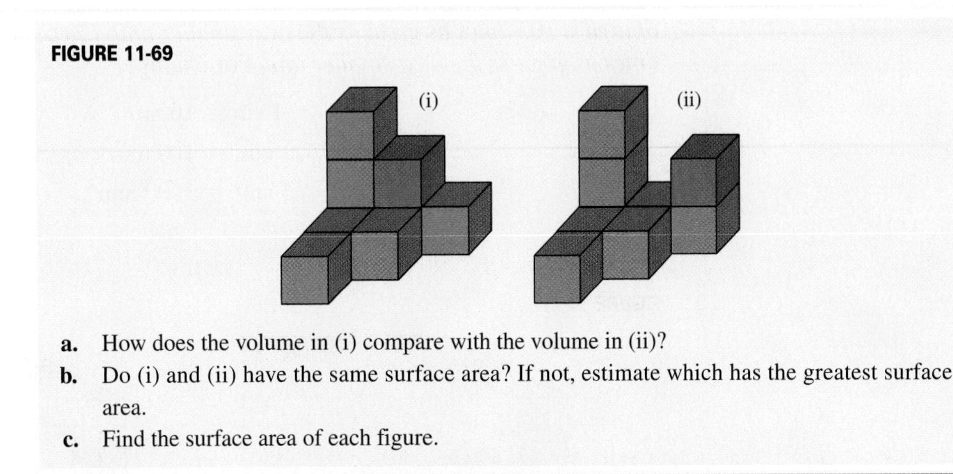

a. How does the volume in (i) compare with the volume in (ii)?

b. Do (i) and (ii) have the same surface area? If not, estimate which has the greatest surface area.

c. Find the surface area of each figure.

Volume of Right Rectangular Prisms

The volume of a right rectangular prism can be measured by determining how many cubes are needed to build it as a solid. To find the volume, count how many cubes cover the base and then how many layers of these cubes are used to fill the prism. As shown in Figure 11-70(a), there are $8 \cdot 4$, or 32, cubes required to cover the base and there are five such layers. The volume of the rectangular prism is $(8 \cdot 4) \cdot 5$, or 160 cubic units. For any right rectangular prism with dimensions ℓ, w, and h measured in the same linear units, the volume of the prism is given by the area of the base, ℓw, times the height, h, or $V = \ell wh$, as shown in Figure 11-70(b).

FIGURE 11-70

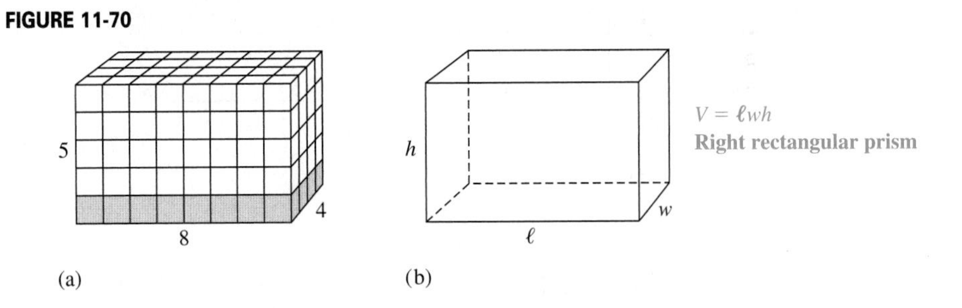

Converting Metric Measures of Volume

cubic centimeter
cubic meter

The most commonly used metric units of volume are the **cubic centimeter** and the **cubic meter**. A cubic centimeter is the volume of a cube whose length, width, and height are each 1 cm. One cubic centimeter is denoted 1 cm^3. Similarly, a cubic meter is the volume of a cube whose length, width, and height are each 1 m. One cubic meter is denoted 1 m^3. Other metric units of volume are symbolized similarly.

Figure 11-71 shows that since 1 dm = 10 cm, 1 dm^3 = (10 cm) · (10 cm) · (10 cm) = 1000 cm^3. Figure 11-72 shows that 1 m^3 = 1,000,000 cm^3 and that 1 dm^3 = 0.001 m^3. *Each metric unit of length is* 10 *times as great as the next smaller unit. Each metric unit of area is* 100 *times as great as the next smaller unit. Each metric unit of volume is* 1000 *times as great as the next smaller unit.* For example:

$$1 \text{ cm} = 10 \text{ mm}$$
$$1 \text{ cm}^2 = 100 \text{ mm}^2$$
$$1 \text{ cm}^3 = 1000 \text{ mm}^3$$

FIGURE 11-71

FIGURE 11-72

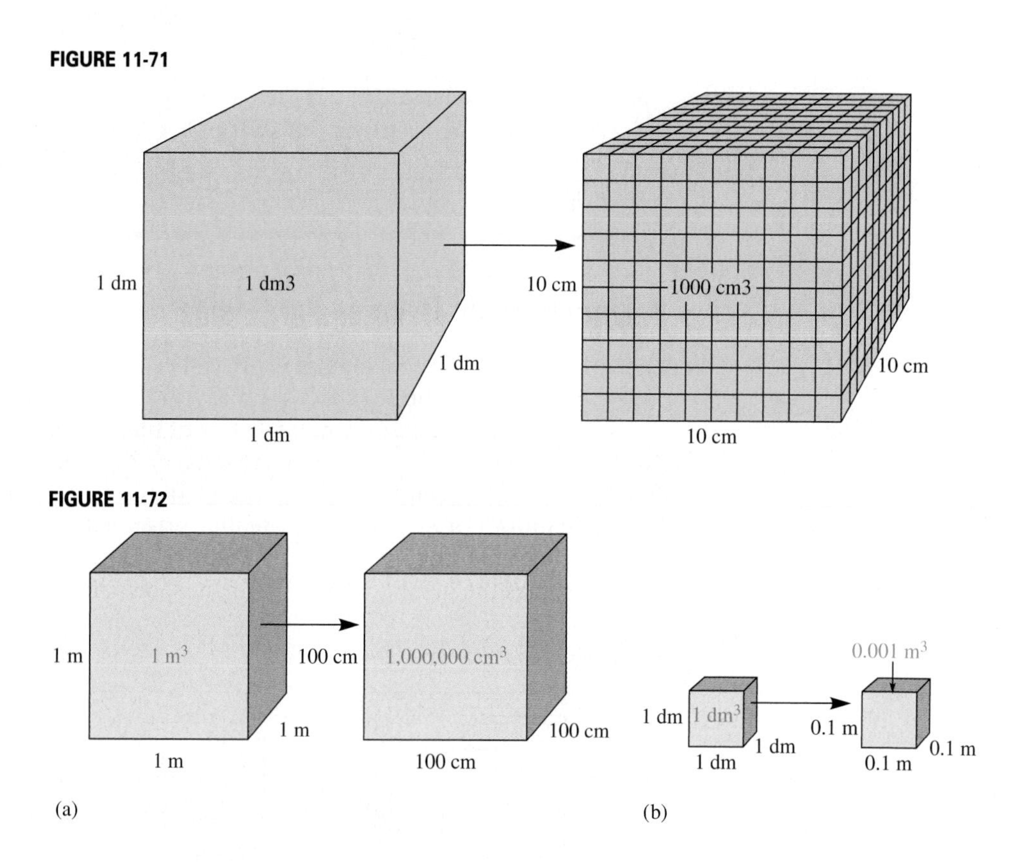

Because 1 cm = 0.01 m, then $1 \text{ cm}^3 = (0.01 \cdot 0.01 \cdot 0.01) \text{ m}^3$, or 0.000001 m^3. To convert from cubic centimeters to cubic meters, we move the decimal point six places to the left.

NOW TRY THIS 11-18

a. Describe how to determine how many places to move the decimal point in a metric area conversion if you know how many places and the direction to move in the corresponding length conversion.

b. Describe how to determine how many places to move the decimal point in a metric volume conversion if you know how many places and the direction to move in the corresponding length conversion.

EXAMPLE 11-22 Convert each of the following:

a. $5 \text{ m}^3 = \_\_\_\_ \text{ cm}^3$ **b.** $12{,}300 \text{ mm}^3 = \_\_\_\_ \text{ cm}^3$

Solution **a.** $1 \text{ m} = 100 \text{ cm}$, so $1 \text{ m}^3 = (100 \text{ cm})(100 \text{ cm})(100 \text{ cm})$, or $1{,}000{,}000 \text{ cm}^3$. Thus, $5 \text{ m}^3 = (5)(1{,}000{,}000 \text{ cm}^3) = 5{,}000{,}000 \text{ cm}^3$.

b. $1 \text{ mm} = 0.1 \text{ cm}$, so $1 \text{ mm}^3 = (0.1 \text{ cm})(0.1 \text{ cm})(0.1 \text{ cm})$, or 0.001 cm^3. Thus, $12{,}300 \text{ mm}^3 = 12{,}300(0.001 \text{ cm}^3) = 12.3 \text{ cm}^3$.

In the metric system, cubic units may be used for either dry or liquid measure, although units such as liters and milliliters are usually used for liquid measures. By definition, a **liter**, symbolized L, equals, or is the capacity of, a cubic decimeter; that is, $1 \text{ L} = 1 \text{ dm}^3$. (In the United States, L is the symbol for liter, but this is not universally accepted.)

liter

Because $1 \text{ L} = 1 \text{ dm}^3$ and $1 \text{ dm}^3 = 1000 \text{ cm}^3$, it follows that $1 \text{ L} = 1000 \text{ cm}^3$ and $1 \text{ cm}^3 = 0.001 \text{ L}$. Also, $0.001 \text{ L} = 1$ milliliter $= 1 \text{ mL}$. Hence, $1 \text{ cm}^3 = 1 \text{ mL}$. These relationships are summarized in Figure 11-73 and Table 11-7.

FIGURE 11-73

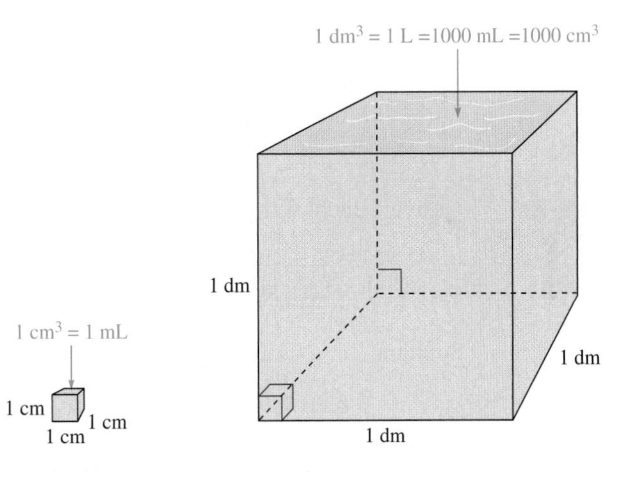

Table 11-7

Unit	Symbol	Relation to Liter
kiloliter	kL	1000 L
*hectoliter	hL	100 L
*dekaliter	daL	10 L
liter	**L**	**1 L**
*deciliter	dL	0.1 L
centiliter	cL	0.01 L
milliliter	mL	0.001 L

*Not commonly used

EXAMPLE 11-23 Convert each of the following as indicated:

a. 27 L = ____ mL **b.** 362 mL = ____ L

c. 3 mL = ____ cm^3 **d.** 3 m^3 = ____ L

Solution **a.** 1 L = 1000 mL, so 27 L = 27 · 1000 mL = 27,000 mL. Alternatively,

$$27 \text{ L} = 27 \text{ L} \cdot \frac{1000 \text{ mL}}{1 \text{ L}} = 27{,}000 \text{ mL}.$$

b. 1 mL = 0.001 L, so 362 mL = 362(0.001 L) = 0.362 L.

c. 1 mL = 1 cm^3, so 3 mL = 3 cm^3.

d. 1 m^3 = 1000 dm^3 and 1 dm^3 = 1 L, so 1 m^3 = 1000 L and 3 m^3 = 3000 L.

$$\text{Alternatively, } 3 \text{ m}^3 = 3 \text{ m}^3 \cdot \frac{1000 \text{ dm}^3}{1 \text{ m}^3} \cdot \frac{1 \text{ L}}{1 \text{ dm}^3} = 3000 \text{ L}.$$

Converting English Measures of Volume

Basic units of volume in the English system are the cubic foot (1 ft^3), the cubic yard (1 yd^3), and the cubic inch (1 $in.^3$). In the United States, 1 gal = 231 $in.^3$, which is about 3.8 L, and 1 qt = $\frac{1}{4}$ gal, or about 58 $in.^3$

Relationships among the one-dimensional units enable us to convert from one unit of volume to another, as shown in the following example.

EXAMPLE 11-24 Convert each of the following, as indicated:

a. 45 yd^3 = ____ ft^3 **b.** 4320 $in.^3$ = ____ yd^3

c. 10 gal = ____ ft^3 **d.** 3 ft^3 = ____ yd^3

Solution **a.** Because 1 yd^3 = $(3 \text{ ft})^3$ = 27 ft^3, 45 yd^3 = 45 · 27 ft^3, or 1215 ft^3.

b. Because 1 in. = $\frac{1}{36}$ yd, 1 $in.^3$ = $\left(\frac{1}{36}\right)^3$ yd^3. Consequently, 4320 $in.^3$ = 4320 · $\left(\frac{1}{36}\right)^3$ $yd^3 \doteq 0.0926$ yd^3, or approximately 0.1 yd^3.

c. Because 1 gal = 231 $in.^3$ and 1 $in.^3$ = $\left(\frac{1}{12}\right)^3$ ft^3, 10 gal = 2310 $in.^3$ = 2310$\left(\frac{1}{12}\right)^3$ $ft^3 \doteq 1.337$ ft^3, or approximately 1.3 ft^3.

$$\text{Alternatively, } 10 \text{ gal} = 10 \text{ gal} \cdot \frac{231 \text{ in.}^3}{1 \text{ gal}} \cdot \frac{\left(\frac{1}{12}\right)^3 \text{ ft}^3}{1 \text{ in.}^3} \doteq 1.3 \text{ ft}^3$$

d. From (a), 1 ft^3 = $\frac{1}{27}$ yd^3. Hence, 3 ft^3 = 3 · $\frac{1}{27}$ yd^3 = $\frac{1}{9}$ $yd^3 \doteq 0.1$ yd^3.

Volumes of Prisms and Cylinders

We have shown that the volume of a right rectangular prism, as shown in Figure 11-74, involves multiplying the area of the base times the height. If we denote the area of the base by B and the height by h, then $V = Bh$.

FIGURE 11-74

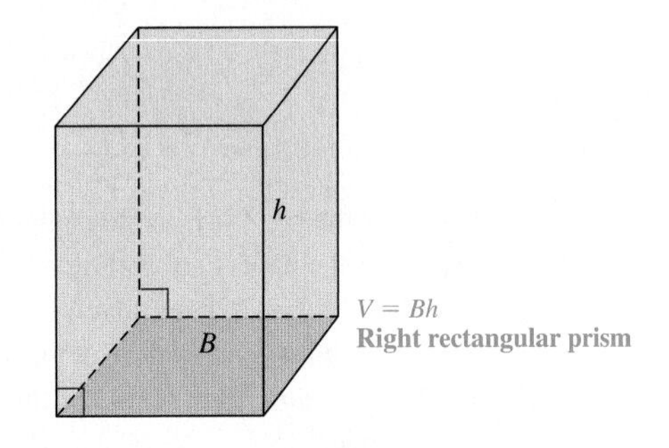

Formulas for the volumes of many three-dimensional figures can be derived using the volume of a right prism. In Figure 11-75(a), a rectangular solid box has been sliced into thin layers. If the layers are shifted to form the solids in Figure 11-75(b) and (c), the volume of each of the three solids is the same as the volume of the original solid. This idea is the basis for **Cavalieri's Principle**.

Cavalieri's Principle

FIGURE 11-75

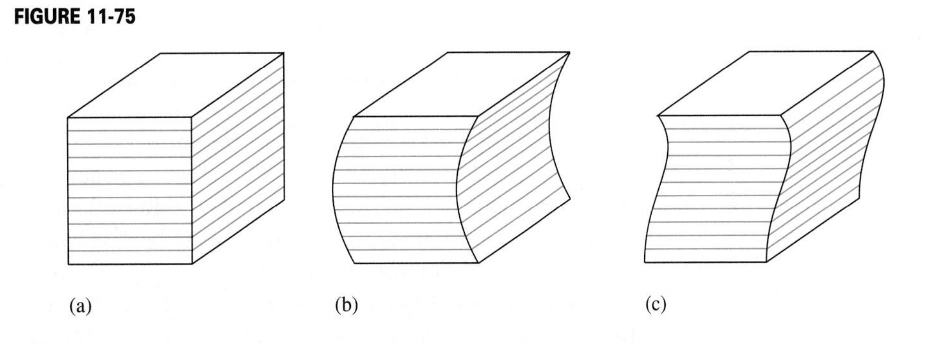

Cavalieri's Principle

Two solids each with a base in the same plane have equal volumes if every plane parallel to the bases intersects the solids in cross sections of equal area.

▶ **HISTORICAL NOTE**

Bonaventura Cavalieri (1598–1647), an Italian mathematician and disciple of Galileo, contributed to the development of geometry, trigonometry, and algebra in the Renaissance. He became a Jesuit at an early age and later, after reading Euclid's *Elements*, was inspired to study mathematics. In 1629, Cavalieri became a professor at Bologna and held that post until his death. Cavalieri is best known for his principle concerning the volumes of solids. ◀

NOW TRY THIS 11-19

a. The two right prisms in Figure 11-76 have the same height. How do their volumes compare? Explain why.

FIGURE 11-76

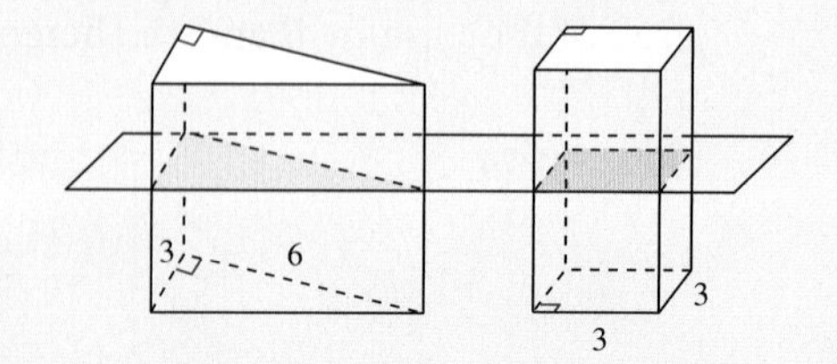

b. Consider the right prism and right circular cylinder in Figure 11-77(a) and (c) as stacks of papers. If the papers are shifted as shown in Figure 11-77(b) and (d), an oblique prism and an oblique cylinder, respectively, are formed.

(i) Explain how the volume of the oblique prism is related to the volume of the right prism.

(ii) Explain how the volume of the oblique cylinder is related to the volume of the right cylinder.

FIGURE 11-77

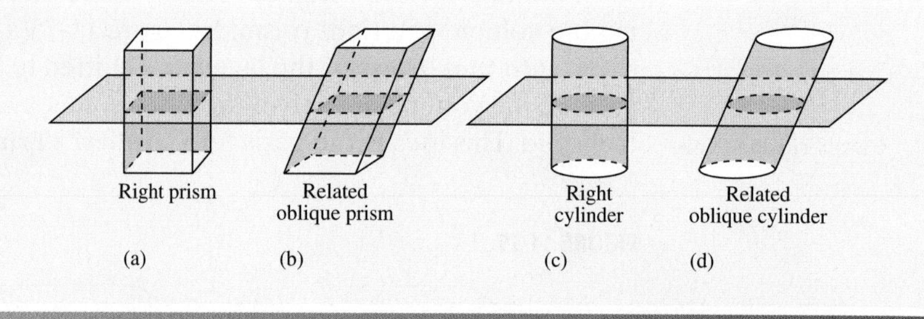

The volume of a cylinder can be approximated using prisms with increasing numbers of sides in their bases. The volume of each prism is the product of the area of the base and the height. Similarly, the **volume V of a cylinder** is the product of the area of the base B and the height h. If the base is a circle of radius r, and the height of the cylinder is h, then $V = Bh = \pi r^2 h$.

volume of a cylinder

Another way to conceptualize the fact that the volume of a cylinder is equal to the area of the base times the height is shown on the student page on page 806. Answer the questions in the Problem 4.1 Follow-Up.

EXAMPLE 11-25

Find the volume of each figure in Figure 11-78.

FIGURE 11-78

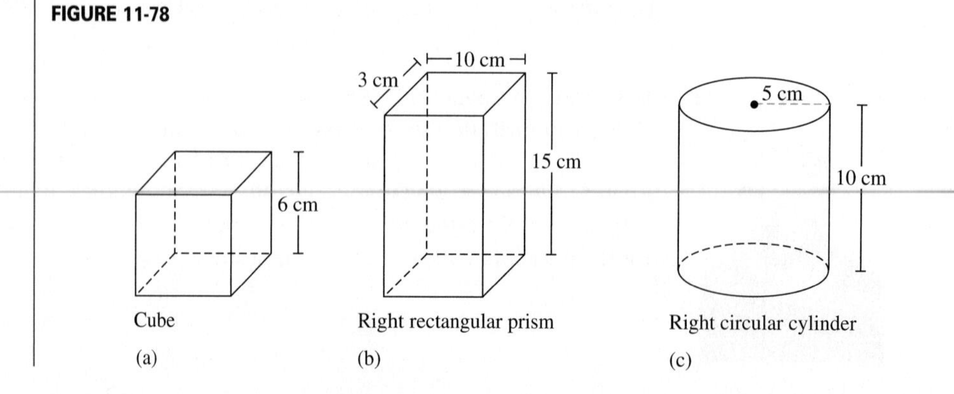

Solution **a.** $V = Bh = (6\text{ cm} \cdot 6\text{ cm})6\text{ cm} = 216\text{ cm}^3$
b. $V = Bh = (10\text{ cm} \cdot 3\text{ cm})15\text{ cm} = 450\text{ cm}^3$
c. $V = \pi r^2 h = \pi(5\text{ cm})^2\, 10\text{ cm} = 250\pi\text{ cm}^3$

Volumes of Pyramids and Cones

Figure 11-79(a) shows a right prism and a right pyramid with congruent bases and equal heights.

FIGURE 11-79

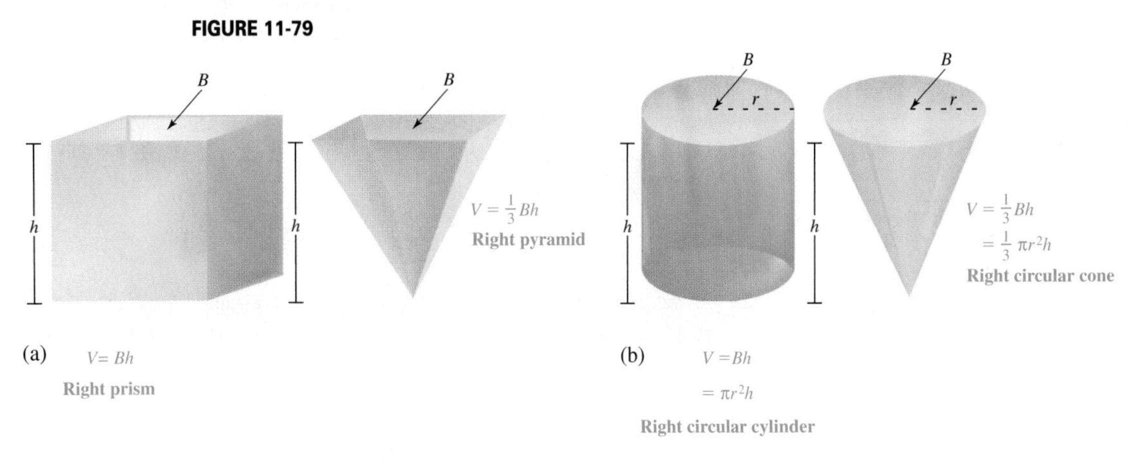

How do we know how the volumes of these containers are related? Students may explore the relationship by filling the pyramid with water, sand, or rice and pouring the contents into the prism. They should find that it takes three full pyramids to fill the prism. Therefore the volume of the pyramid is equal to one-third the volume of the prism. This relationship between prisms and pyramids with congruent bases and heights, respectively, is true in general; that is, for a pyramid $V = (1/3)Bh$, where B is the area of the base and h is the height. The same relationship holds between the volume of a cone and the volume of a cylinder, where they have congruent bases and equal heights, as shown in Figure 11-79(b). Therefore, the volume of a right circular cone is given by $V = (1/3)Bh$, or $V = (1/3)\pi r^2 h$.

This relationship between the volume of a cone and the volume of a cylinder is explored in Problem 5.2 on the student page on page 807.

Another way to determine the area of a pyramid in terms of a prism is to start with a cube and three diagonals from one vertex drawn to other vertices of the "opposite face," as shown in Figure 11-80(a). We can see that there are three pyramids formed inside the cube, as shown in Figure 11-80(b), (c), and (d).

FIGURE 11-80

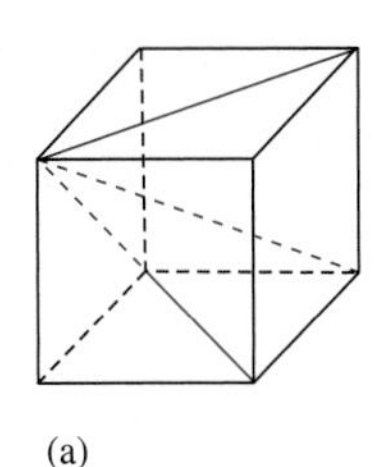

(a)

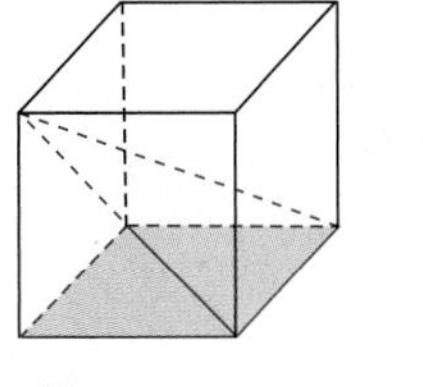

(b)

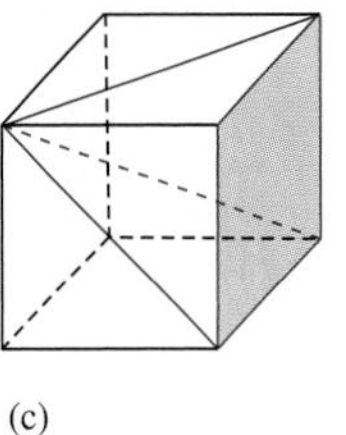

(c)

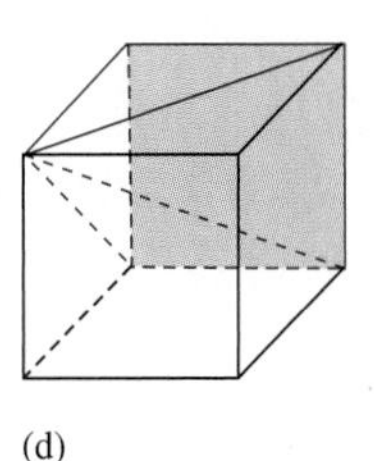

(d)

SAMPLE SCHOOL BOOK PAGE:

VOLUME OF A CYLINDER

4.1 Filling a Cylinder

The *volume* of a container is the number of unit cubes it will hold. In the last investigation, you saw that you could find the volume of a prism-shaped box by figuring out how many unit cubes will fit in a single layer at the bottom of the box and then multiplying by the total number of layers needed to fill the box. In this problem, you will develop a method for determining how many cubes will fit inside a cylinder.

Problem 4.1

Make a cylinder by taping together the ends of a sheet of paper. Use the same size paper you used to make the prism shapes in Problem 3.3.

A. Set the cylinder on its base on a sheet of centimeter grid paper. Trace the cylinder's base. Look at the centimeter squares inside your tracing. How many cubes would fit in one layer at the bottom of the cylinder? Consider whole cubes and parts of cubes.

B. How many layers of cubes would it take to fill the cylinder?

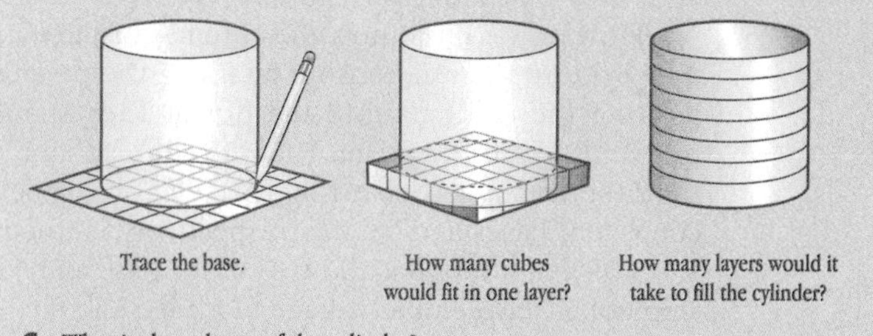

Trace the base. How many cubes would fit in one layer? How many layers would it take to fill the cylinder?

C. What is the volume of the cylinder?

Problem 4.1 Follow-Up

1. How can you use the dimensions of the cylinder to help you estimate its volume more accurately? Explain.
2. How does the volume of the cylinder compare to the volumes of the prisms you made in Problem 3.3?

Source: Prentice Hall, Connected Mathematics: Filling and Wrapping, 2004 (p. 38).

SAMPLE SCHOOL BOOK PAGE:

CONES AND CYLINDERS

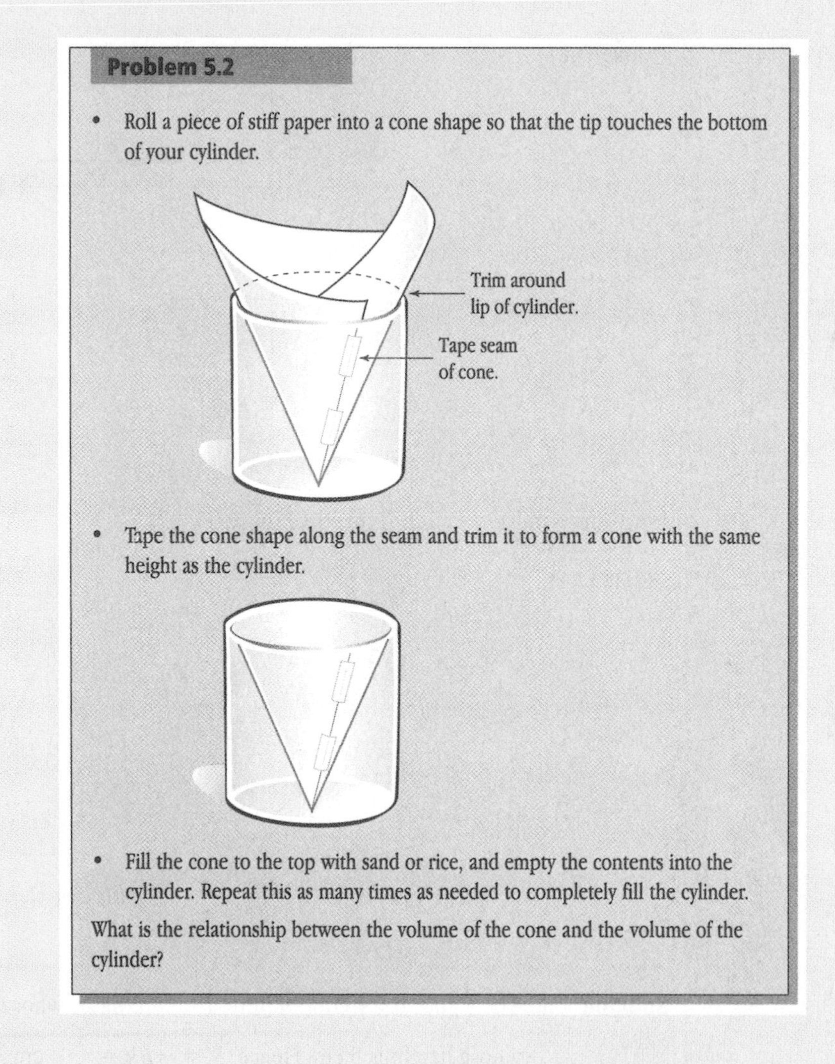
Problem 5.2

- Roll a piece of stiff paper into a cone shape so that the tip touches the bottom of your cylinder.

Trim around
lip of cylinder.
Tape seam
of cone.

- Tape the cone shape along the seam and trim it to form a cone with the same height as the cylinder.
- Fill the cone to the top with sand or rice, and empty the contents into the cylinder. Repeat this as many times as needed to completely fill the cylinder.

What is the relationship between the volume of the cone and the volume of the cylinder?

Source: Prentice Hall, Connected Mathematics: Filling and Wrapping, 2004 (p. 49).

The three pyramids are identical in size and shape, do not overlap, and their union is the whole cube. Therefore, each pyramid has a volume one-third that of the cube. This result is true in general, and once again we see that for a pyramid $V = (1/3)Bh$, where B is the area of the base and h is the height. This can be demonstrated by building three paper models of the pyramids and fitting them together into a prism. A net that can be enlarged and used for the construction is given in Figure 11-81.

FIGURE 11-81

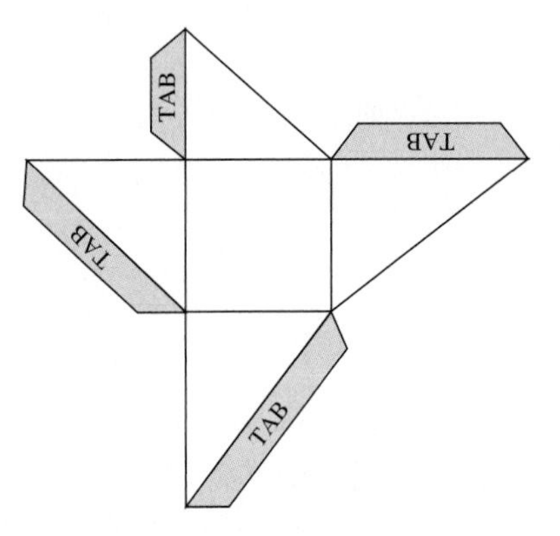

EXAMPLE 11-26

Find the volume of each figure in Figure 11-82.

FIGURE 11-82

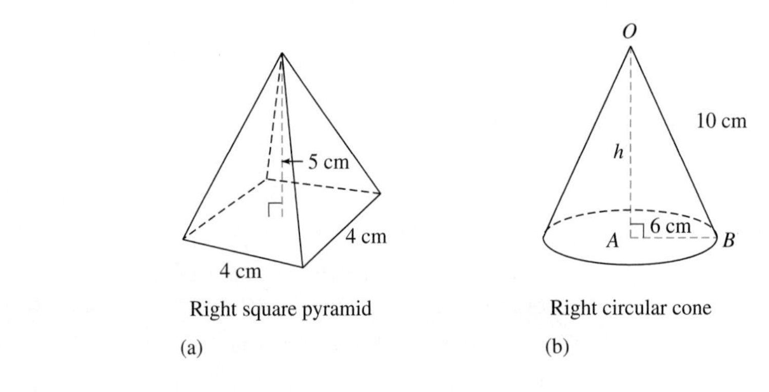

Solution

a. The figure is a pyramid with a square base, whose area is 4 cm · 4 cm and whose height is 5 cm. Hence, $V = \frac{1}{3}Bh = \frac{1}{3}(4 \text{ cm} \cdot 4 \text{ cm})(5 \text{ cm}) = \frac{80}{3}\text{cm}^3$.

b. The base of the cone is a circle of radius 6 cm. Because the volume of the cone is given by $V = \frac{1}{3}\pi r^2 h$, we need to know the height. In the right triangle OAB, $OA = h$ and by the Pythagorean Theorem, $h^2 + 6^2 = 10^2$. Hence, $h^2 = 100 - 36$, or 64, and $h = 8$ cm. Thus, $V = \frac{1}{3}\pi r^2 h = \frac{1}{3}\pi(6 \text{ cm})^2(8 \text{ cm}) = 96\pi \text{ cm}^3$.

▲

EXAMPLE 11-27 Figure 11-83 is a net for a pyramid. If each triangle is equilateral, find the volume of the pyramid.

FIGURE 11-83

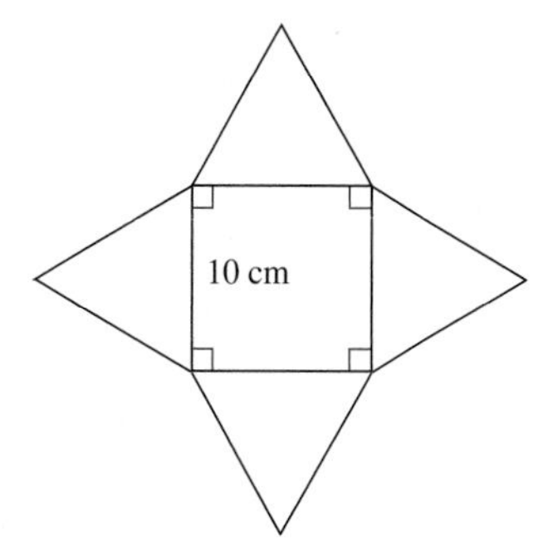

Solution The pyramid obtained from the folded model is shown in Figure 11-84. The volume of the pyramid is $V = \frac{1}{3}Bh = \frac{1}{3} \cdot 10^2 h$. We must find h. Notice that h is a leg in the right triangle EOF, where F is the midpoint of $\overline{CB}$. We know that $OF = 5$ cm. If we knew EF, we could find h by applying the Pythagorean Theorem to $\triangle EOF$. To find the length of $\overline{EF}$, notice that $\overline{EF}$ is a leg in the right triangle EBF. ($\overline{EF}$ is the perpendicular bisector of $\overline{BC}$ in the equilateral triangle BEC.) In the right triangle EBF, we have $(EB)^2 = (BF)^2 + (EF)^2$. Because $EB = 10$ cm and $BF = 5$ cm, it follows that $10^2 = 5^2 + (EF)^2$, or $EF = \sqrt{75}\text{ cm} \doteq 8.66\text{ cm}$. In $\triangle EOF$, we have $h^2 + 5^2 = (EF)^2$, or $h^2 + 25 = 75$. Thus, $h = \sqrt{50}\text{ cm} \doteq 7.07\text{ cm}$, and $V \doteq \frac{1}{3} \cdot 10^2 \cdot 7.07 \doteq 235.7\text{ cm}^3$.

FIGURE 11-84

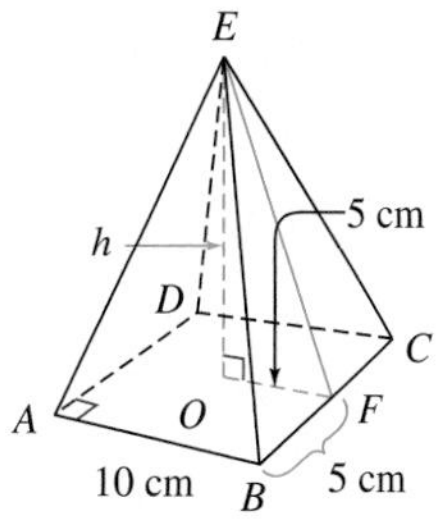

Volume of a Sphere

To find the volume of a sphere, imagine that a sphere is composed of a great number of congruent pyramids with apexes at the center of the sphere and that the vertices of the base touch the sphere, as shown in Figure 11-85. If the pyramids have very small bases, then the height of each pyramid is nearly the radius r. Hence, the volume of each pyramid is $\frac{1}{3}Bh$ or $\frac{1}{3}Br$, where B is the area of the base. If there are n pyramids each with base area B, then the total volume of the pyramids is $V = \frac{1}{3}nBr$. Because nB is the total surface area of all the bases of the pyramids and because the sum of the areas of all the bases of the pyramids is very close to the surface area of the sphere, $4\pi r^2$, the volume of the sphere is given by $V = \frac{1}{3}(4\pi r^2)r = \frac{4}{3}\pi r^3$.

FIGURE 11-85

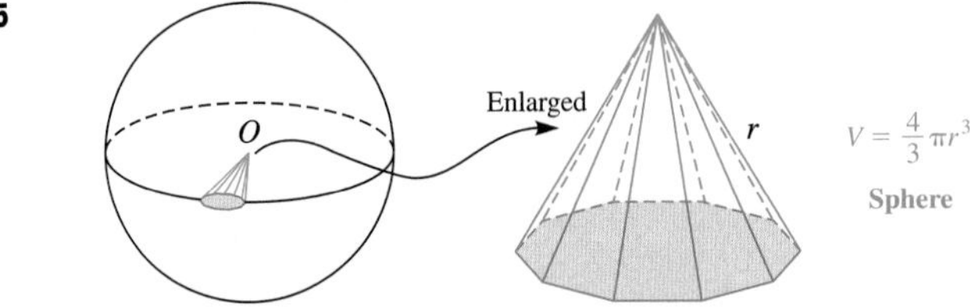

EXAMPLE 11-28 Find the volume of a sphere whose radius is 6 cm.

Solution $V = \frac{4}{3}\pi(6 \text{ cm})^3 = \frac{4}{3}\pi(216 \text{ cm}^3) = 288\pi \text{ cm}^3$

NOW TRY THIS 11-20

A cylinder, a cone, and a sphere have the same radius and same height, as shown in Figure 11-86.

FIGURE 11-86

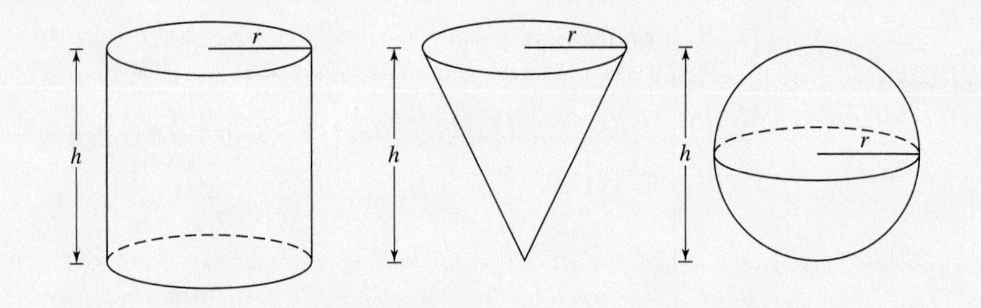

a. Find the volume of each figure in terms of r and h.
b. Use your answer to part (a) to show that the volumes are in the ratio $3:1:2$.

PROBLEM SOLVING Volume Comparisons: Cylinders and Boxes

A metal can manufacturer has a large quantity of rectangular metal sheets 20 cm × 30 cm. Without cutting the sheets, the manufacturer wants to make cylindrical pipes with circular cross sections from some of the sheets and box-shaped pipes with square cross sections from the other sheets. The volume of the box-shaped pipes is to be greater than the volume of the cylindrical pipes. Is this possible? If so, how would the pipes be made and what are their volumes?

Understanding the Problem We are to use 20 cm × 30 cm rectangular sheets of metal to make some cylindrical pipes as well as some box-shaped pipes with square cross sections that have a greater volume than do the cylindrical pipes. Is this possible, and if so, how should the pipes be designed and what are their volumes?

Figure 11-87 shows a sheet of metal and two sections of pipe made from it, one cylindrical and the other box-shaped. A model for such pipes can be designed from a piece of paper by bending it into a cylinder or by folding it into a right rectangular prism, as shown in the figure.

FIGURE 11-87

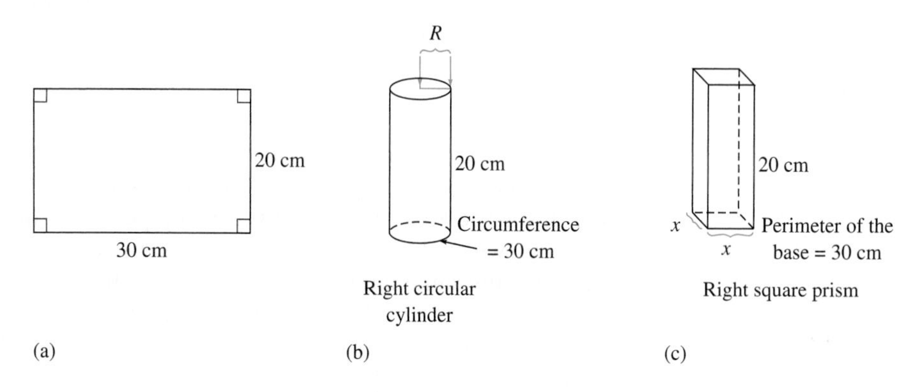

Devising a Plan If we compute the volume of the cylinder in Figure 11-87(b) and the volume of the prism in Figure 11-87(c), we can determine which is greater. If the prism has a greater volume, the solution of the problem will be complete. Otherwise, we look for other ways to design the pipes before concluding that a solution is impossible.

To compute the volume of the cylinder, we find the area of the base. The area of the circular base is πr^2. To find r, we note that the circumference of the circle $2\pi r$ is 30 cm. Thus, $r = \frac{30}{2\pi} \doteq 4.77$ cm, and the area of the circle is $\pi r^2 \doteq (4.77)^2 \doteq 71.48\text{ cm}^2$.

With the given information, we can also find the area of the base of the rectangular box. Because the perimeter of the base of the prism is $4x$, we have $4x = 30$, or $x = 7.5$ cm. Thus, the area of the square base is $x^2 = (7.5)^2$, or 56.25 cm^2.

Carrying Out the Plan Denoting the volume of the cylindrical pipe by V_1 and the volume of the box-shaped pipe by V_2, we have $V_1 \doteq 71.48 \cdot 20$, or approximately 1429.6 cm^3. For the volume of the box-shaped pipe, we have $V_2 \doteq 56.25 \cdot 20$, or 1125 cm^3. We see that in the first design for the pipes, the volume of the cylindrical pipe is greater than the volume of the box-shaped pipe. This is not the required outcome.

Rather than bend the rectangular sheet of metal along the 30-cm side, we could bend it along the 20-cm side to obtain either pipe, as shown in Figure 11-88. Denoting the radius of the cylindrical pipe by r, the side of the box-shaped pipe by y, and their volumes by V_3 and V_4, respectively, we have $V_3 = \pi r^2 \cdot 30 = \pi(20/2\pi)^2 \cdot 30 = (10^2 \cdot 30)/\pi$, or approximately 954.9 cm^3. Also, $V_4 = y^2 \cdot 30 = \left(\frac{20}{4}\right)^2 \cdot 30 = 25 \cdot 30$, or 750 cm^3. Because $V_2 = 1125\text{ cm}^3$ and $V_3 = 945.9\text{ cm}^3$, we see that the volume of the box-shaped pipe with an altitude of 20 cm is greater than the volume of the cylindrical pipe with an altitude of 30 cm. Determine the order of volumes from greatest to least for the four shapes.

20. If 50 steel marbles that are 1 cm in diameter are melted down, will enough steel result to build a marble that is 4 cm in diameter? Explain.

21. A heavy metal sphere with radius 10 cm is dropped into a right circular cylinder with base radius of 10 cm. If the original cylinder has water in it that is 20 cm high, how high is the water after the sphere is placed in it?

22. A cone-shaped paper water cup has a height of 8 cm and a radius of 4 cm. If the cup is filled with water to half its height, what portion of the volume of the cup is filled with water?

23. If each edge of a cube is increased by 30%, by what percent does the volume increase?

24. One freezer measures 1.5 ft × 1.5 ft × 5 ft and sells for \$350. Another freezer measures 2 ft × 2 ft × 4 ft and sells for \$400. Which freezer is the better buy in terms of dollars per cubic foot?

25. A tennis ball can in the shape of a cylinder holds three tennis balls snugly. If the radius of a tennis ball is 3.5 cm, what percentage of the tennis ball can is occupied by air?

26. A box is packed with six soda cans, as in the following figure. What percentage of the volume of the interior of the box is not occupied by the cans?

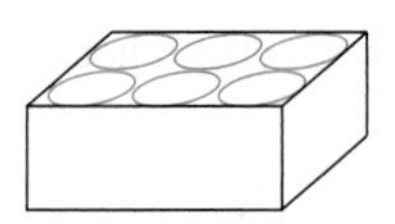

27. a. If a room filled with ping-pong balls were a cube that is 10 ft on a side, and a ping-pong ball has a diameter of 1.5 in., approximately how many ping-pong balls could be placed in the room?

b. A pop can holds 12 fluid oz. About how much liquid would the cubic room in part(a) hold? (*Hint*: There are 16 fluid oz in 1 pint.)

28. A professor chose to move offices because the new office was 40 cm wider than his old office. If all other dimensions were the same and the office was a right rectangular prism, write a formula to determine the percentage of gained space in terms of the width w in centimeters.

29. If a storage space is described as having 200 ft^3, explain whether it is possible that a car would not fit in the space.

30. If coins followed geometric principles of volume, what would be the size of a dime in relation to a penny? Why is this not feasible?

31. What unit would you use to describe the volume of Earth? Justify your reasoning.

32. a. If two cubes have sides in the ratio $2:3$, what is the ratio of their volumes?

b. If two similar cones have heights in the ratio $a:b$, what is the ratio of their volumes?

33. A right rectangular prism with base $ABCD$ as the bottom is shown in the following figure:

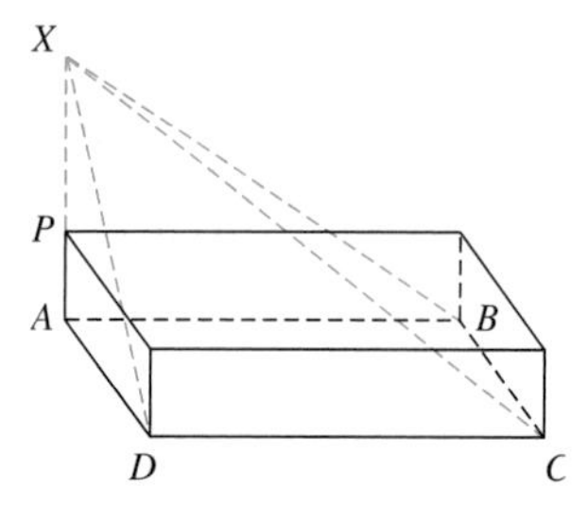

Suppose X is drawn so that $AX = 3 \cdot AP$, where AP is the height of the prism and X is connected to A, B, C, and D to form a pyramid. How do the volumes of the pyramid and the prism compare?

34. A right cylindrical can is to hold approximately 1 L of water. What should be the height of the can if the radius is 12 cm?

35. A theater decides to change the shape of its popcorn container from a regular box to a right regular pyramid, as shown in the following figure, and charge only half as much.

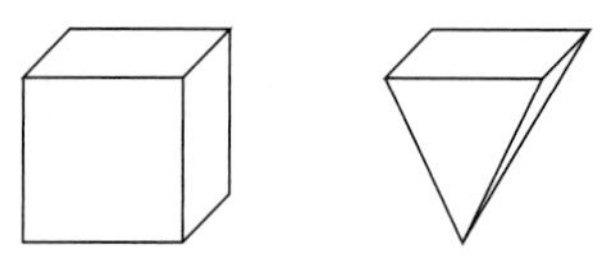

If the containers are the same height and the tops are the same size, is this a bargain for the customer? Explain.

36. Which is the better buy: a grapefruit 5 cm in radius that costs 22¢ or a grapefruit 6 cm in radius that costs 31¢? Explain.

37. Two spherical cantaloupes of the same kind are sold at a fruit and vegetable stand. The circumference of one is 60 cm and that of the other is 50 cm. The larger melon is $1\frac{1}{2}$ times as expensive as the smaller. Which melon is the better buy and why?

38. An engineer is to design a square-based pyramid whose volume is to be 100 m^3.

a. Find the dimensions (the length of a side of the square and the altitude) of one such pyramid.

b. How many (noncongruent) such pyramids are possible? Why?

39. A square sheet of cardboard measuring y cm on a side is to be used to produce an open-top box when the maker cuts off a small square x cm × x cm from each corner and bends up the sides. Find the volume of the box if $y = 200$ cm and $x = 20$ cm.

40. a. Suppose a scoop of ice cream that is a sphere with radius 5 cm fits exactly along one of its great circles in a sugar cone, as shown in the following figure:

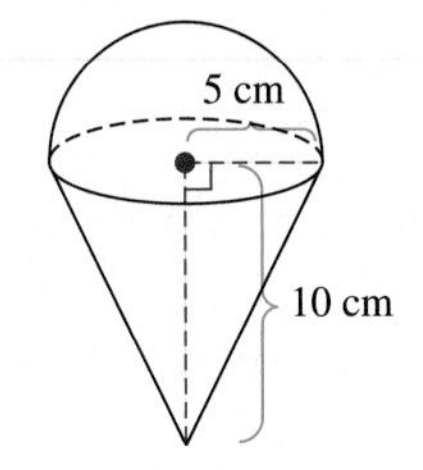

Suppose the ice cream melts and the cone does not absorb any of it. If the cone is 10 cm tall, will it hold the melted ice cream? If not, how tall would the cone have to be to hold the melted ice cream?

b. Suppose the scoop of ice cream in part (a) fits exactly along the top of a right circular cylinder of height 10 cm. Will the cylinder hold the melted ice cream? If yes, what height of the cylinder will it reach?

★ **41.** Half of the air is let out of a spherical balloon. If the balloon remains in the shape of a sphere, how does the radius of the smaller balloon compare to the original radius?

42. For each of the following, select the appropriate metric unit of measure (gram, kilogram, or metric ton):

a. Car **b.** Adult
c. Can of frozen orange juice **d.** Elephant
e. Jar of mustard **f.** Bag of peanuts
g. Army tank **h.** Cat

43. For each of the following, choose the correct unit (milligram, gram, or kilogram) to make each sentence reasonable:

a. A staple has a mass of about 340 _____.
b. A professional football player has a mass of about 110 _____.
c. A vitamin tablet has a mass of about 1100 _____.
d. A dime has a mass of 2 _____.
e. The recipe said to add 4 _____ of salt.
f. One strand of hair has a mass of 2 _____.

44. Complete each of the following:

a. 15,000 g = _____ kg **b.** 8000 kg = _____ t
c. 0.036 kg = _____ g **d.** 72 g = _____ kg
e. 4320 mg = _____ g **f.** 5 kg 750 g = _____ g
g. 0.03 t = _____ kg **h.** 2.6 lb = _____ oz
i. 25 oz = _____ lb **j.** 3.8 lb = _____ oz

45. A paper dollar has a mass of approximately 1 g. Is it possible to lift $1,000,000 in the following denominations:

a. $1 bills **b.** $10 bills
c. $100 bills **d.** $1000 bills
e. $10,000 bills

46. A fish tank, which is a right rectangular prism, is 40 cm × 20 cm × 20 cm. If it is filled with water, what is the mass of the water?

47. Convert each of the following from degrees Fahrenheit to the nearest integer degree Celsius:

a. 10°F **b.** 0°F
c. 30°F **d.** 100°F
e. 212°F **f.** ⁻40°F

48. Answer each of the following:

a. The thermometer reads 20°C. Can you go snow skiing?
b. The thermometer reads 26°C. Will the outdoor ice rink be open?
c. Your body temperature is 39°C. Are you ill?
d. It is 40°C. Will you need a sweater at the outdoor concert?
e. The temperature reads 35°C. Should you go water skiing?
f. Your bath water is 16°C. Will you have a hot, warm, or chilly bath?
g. It's 30°C in the room. Are you comfortable, hot, or cold?

49. a. Rainfall is usually measured in linear measure. Suppose St. Louis received 2 cm of rain on a given day. If a certain lot in St. Louis has measure 1 ha, how many liters of rainfall fell on the lot?
b. What is the mass of the water that fell on the lot?

Communication

50. a. Which will increase the volume of a circular cylinder more: doubling its height or doubling its radius? Explain.
b. Is your answer the same for a circular cone? Why?

51. Write a one-page paper to a sixth-grade student explaining the difference between surface area and volume.

52. Explain how you would find the volume of an irregular shape.

53. Read the following problems (i) and (ii):

(i) A tank in the shape of a cube 5 ft 3 in. on a side is filled with water. Find the volume in cubic feet, the capacity in gallons, and the weight of the water in pounds.
(ii) A tank in the shape of a cube 2 m on a side is filled with water. Find the volume in cubic meters, the capacity in liters, and the mass of the water in kilograms.

Discuss which problem is easier to work and why.

54. A statue in front of the mathematics building is described as being half of a right circular cone. Sketch two possible shapes that fit the given description and tell why the shapes you drew fit the description.

55. A furniture company gives an estimate for moving based upon the size of the rooms in an apartment. Write a rationale for why this is feasible. What assumptions are being made?

56. Why is using the Kelvin scale not practical in the daily world?

57. Which measure, pounds or kilograms, would you rather use to describe your weight? Why?

58. Styrofoam spheres are sold in craft stores. Describe how you might stack the spheres to conserve space. On what did you base your decision?

Open-Ended

59. A right circular cylinder has a 4-in. diameter, is 6 in. high, and is completely full of water. Design a right rectangular prism that will hold the water as exactly as possible.

60. Circular-shaped cookies are to be packaged 48 to a box. Each cookie is approximately 1 cm thick and has a diameter of 6 cm. Design a box that will hold this volume of cookies and has the least amount of surface area.

61. Design a cylinder that will hold 1 L of juice. Give the dimensions of your cylinder and tell why you designed the shape as you did.

Cooperative Learning

62. **a.** Find many different types of cans that are in the shape of a cylinder. Measure the height and diameter of each can.
b. Find the surface area of each can.
c. Find the volume of each can.
d. Compute the ratio of surface area to volume for each can.
e. Compare your results with those of other groups.
f. Based on the information collected, write recommendations to the manufacturers of the cans about an ideal surface-area-to-volume ratio.

Review Problems

63. Find the perimeter and the area of the following figures:

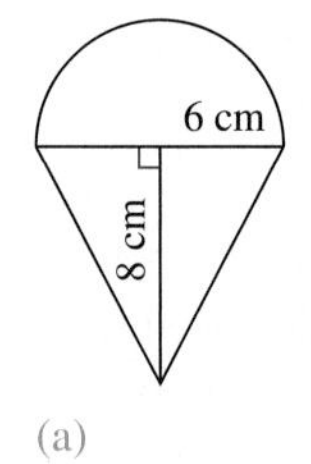

(a)

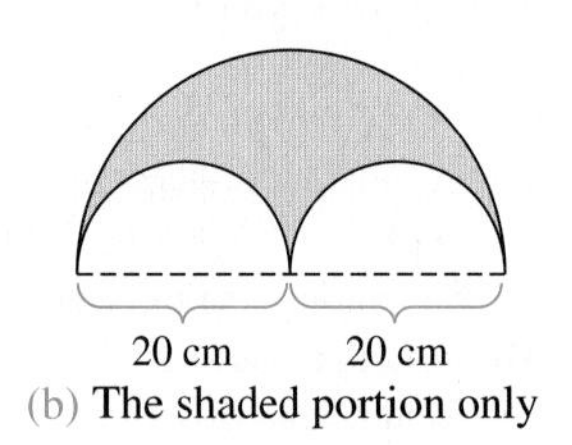

(b) The shaded portion only

64. Complete the following:
a. 350 mm = ____ cm
b. 1600 cm^2 = ____ m^2
c. 0.4 m^2 = ____ mm^2
d. 5.2 cm^2 = ____ m^2

65. Determine whether each of the following is a right triangle:

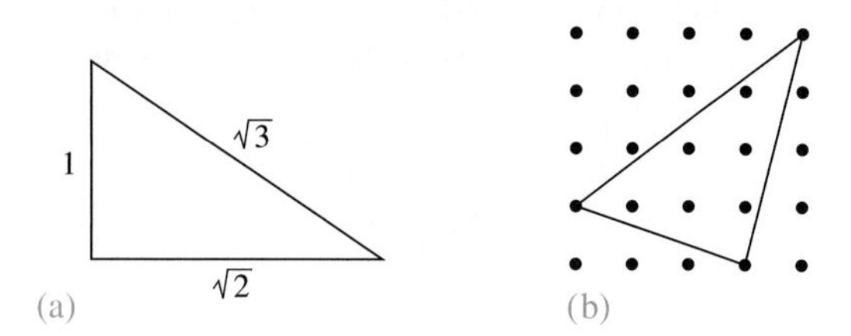

66. Find the surface area of each of the following:

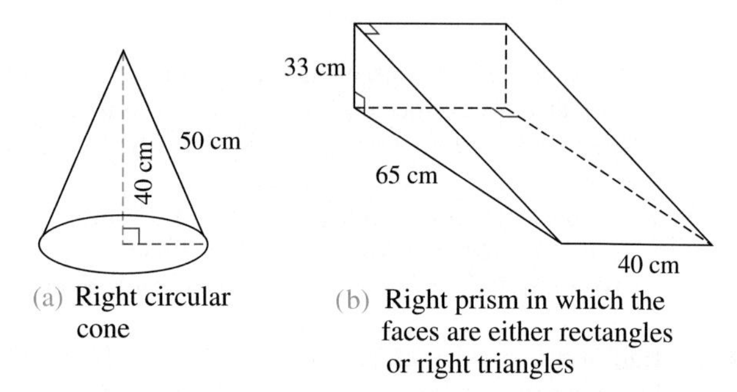

(a) Right circular cone
(b) Right prism in which the faces are either rectangles or right triangles

Third International Mathematics and Science Study (TIMSS) Questions

All the small blocks are the same size. Which stack of blocks has a different volume from the others?

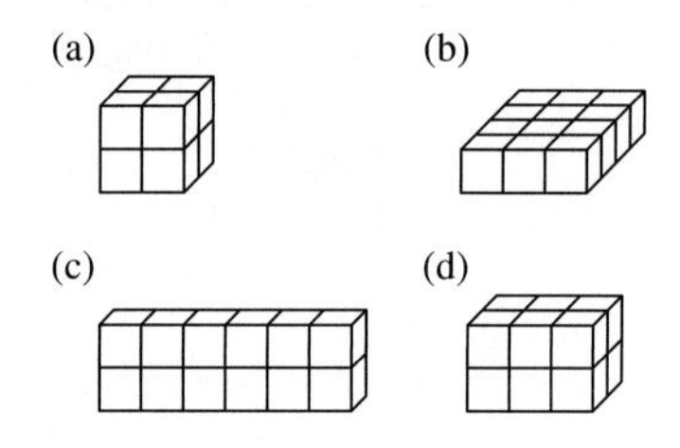

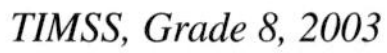

TIMSS, Grade 8, 2003

Oranges are packed in boxes. The average diameter of the oranges is 6 cm, and the boxes are 60 cm long, 36 cm wide, and 24 cm deep.

Which of these is the BEST approximation of the number of oranges that can be packed in a box?
a. 30
b. 240
c. 360
d. 1920

TIMSS, Grade 8, 2003

LABORATORY ACTIVITY

Record the mass of each U.S. coin. Which coin has the greatest mass? Which of the following have the same mass?
a. A half-dollar vs. two quarters
b. A quarter vs. two dimes and a nickel
c. A dime vs. two nickels
d. A dime vs. ten pennies
e. A nickel vs. five pennies

HINT FOR SOLVING THE PRELIMINARY PROBLEM

Consider the areas of triangles *ACE* and *BED*. Each is a part of a larger triangle that includes triangle *AEB*.

(*Note:* The authors wish to thank Charles Vonder Embse for suggesting this problem.)

QUESTIONS FROM THE CLASSROOM

1. A student asks if the units of measure must be the same for each term in order to use the formulas for volumes. How do you respond?
2. As part of the discussion of the Pythagorean Theorem, squares were constructed on each side of a right triangle. A student asks, "If different similar figures are constructed on each side of the triangle, does the same type of relationship still hold?" How do you reply?
3. A student asks, "Can I find the area of an angle?" How do you respond?
4. A student argues that a square has no area because its interior can be thought of as the union of infinitely many points, each of which has no area. How do you react?
5. A student asks whether the volume of a prism can ever be the same number as its surface area. How do you answer?
6. A student asks, "Should the United States switch to the metric system?" How do you reply?
7. A student claims that in a triangle with 20° and 40° angles, the side opposite the 40° angle is twice as long as the side opposite the 20° angle. How do you reply?
8. A student interpreted 5 cm^3 as shown below. What is wrong with this interpretation?

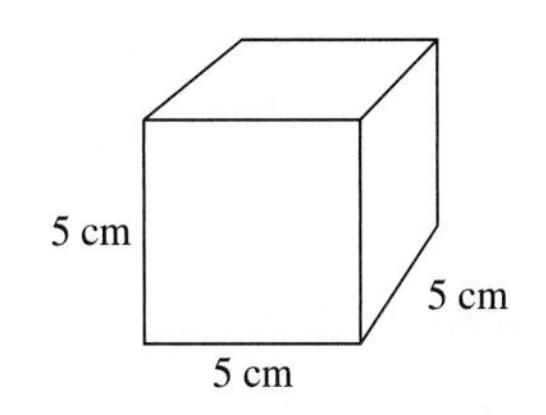

9. A student claims that because *are* and *hectare* are measures of area, we should say "square are" and "square hectare." How do you respond?
10. A student claims that the area of his hand does not exist because it cannot be found by any formula. How do you respond?
11. A student claims that since a circular cylinder has a curved surface, its lateral surface area should not be expressed in square units. How do you respond?
12. A student claims that it does not make any difference if his temperature is 2 degrees above normal Fahrenheit or 2 degrees above normal Celsius because in either case he is only 2 degrees above normal. How do you respond?
13. Larry and Gary are discussing whose garden has the most area to plant flowers. Larry claims that all they have to do is walk around the two gardens to get the perimeter and the one with the greatest perimeter has the greatest area. How would you help these students?
14. Andrea claims that if she doubles the length and width of the base of a rectangular prism and triples the height, she has increased the volume by a factor $2 \cdot 2 \cdot 3 = 12$. What would you tell her?
15. Jimmy claims that to find the area of a parallelogram he just has to multiply length times width. In the figure he multiplies $25 \times 20 = 500$ in.^{2}. What would you tell him?

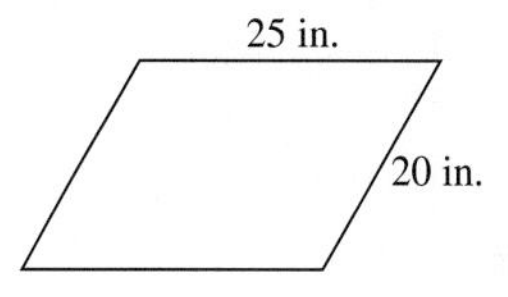

16. A student asked whether a pop can, reputed to hold 591 ml, actually held that much liquid. How do you respond?
17. A student asks why necklaces are sold in only common lengths. How might you frame an answer that ties to data analysis?
18. A well-known mathematician stated that there is no general relation between perimeter and area. If he were your student, would you agree or disagree? Why?
19. A student read about Volkswagen-packing in the 1960s. She asked about the maximum number of students that might have fit into a Volkswagen. How would you help her estimate an answer in a reasonable way?
20. A student claimed that pop packaging in plastic was more economical than the same number of cans (a six-pack) packaged in cardboard boxes. What type of research might you suggest to help her determine an answer?

CHAPTER OUTLINE

I. The English system of measure

A. Linear measure
1 ft = 12 in.
1 yd = 3 ft
1 mi = 5280 ft = 1760 yd

B. Area measure
1. Units commonly used are the **square inch** (in.^{2}), **square yard** (yd^2), and **square foot** (ft^2).
2. Land can be measured in **acres.**

C. Volume measure
Units commonly used are the **cubic inch** (in.^{3}), **cubic foot** (ft^3), **cubic yard** (yd^3), and **gallon.**

D. Mass
Units of mass commonly used are **pound** (lb), **ounce** (oz) (1 oz = $\frac{1}{16}$ lb), and **ton** (t) (1 ton = 2000 lb).

II. The metric system of measure

A. A summary of relationships among prefixes and the base unit of linear measure follows:

Prefix	Unit	Relationship to Base Unit	Symbol
kilo	kilometer	1000 m	km
*hecto	hectometer	100 m	hm
*deka	dekameter	10 m	dam
	meter	**1 m**	**m**
*deci	decimeter	0.1 m	dm
centi	centimeter	0.01 m	cm
milli	millimeter	0.001 m	mm

*Not commonly used

B. Area measure
1. Units commonly used are the **square kilometer** (km^2), **square meter** (m^2), **square centimeter** (cm^2), and **square millimeter** (mm^2).
2. Land can be measured using the **are** (a) (100 m^2) and the **hectare** (ha) (10,000 m^2).

C. Volume measure
1. Units commonly used are the **cubic meter** (m^3), **cubic decimeter** (dm^3), and **cubic centimeter** (cm^3).
2. 1 dm^3 = 1 L and cm^3 = 1 mL.

D. Mass
1. Units of mass commonly used are the **milligram** (mg), **gram** (g), **kilogram** (kg), and **metric ton** (t).
2. 1 L and 1 mL of water have masses of 1 kg and 1 g, respectively.

E. Temperature
1. In the metric system the unit commonly used is the **degree Celsius** (°C). In the English system, the unit of temperature is the **degree Fahrenheit** (°F). The scientific unit used is **degree Kelvin** (K).
2. Basic temperature reference points are the following:
100°C—boiling point of water
37°C—normal body temperature
20°C—comfortable room temperature
0°C—freezing point of water
3. $C = \frac{5}{9}(F - 32)$ and $F = \frac{9}{5}C + 32$

III. Distance

A. Distance properties. Given points A, B, and C,
a. $AB \geq O$
b. $AB = BA$
c. $AB + BC \geq AC$

B. The distance around a two-dimensional figure is the **perimeter**. The distance C around a circle is the **circumference.** $C = 2\pi r = \pi d$, where r is the radius of the circle and d is the **diameter.**

C. Distance formula. The distance between the points A (x_1, y_1) and B (x_2, y_2) is given by
$AB = \sqrt{(x_2 - x_1)^2 + (y_2 - y_1)^2}$

IV. Areas

A. Formulas for areas
1. **Square:** $A = s^2$, where s is the length of a side.
2. **Rectangle:** $A = \ell w$, where ℓ is the length and w is the width.
3. **Parallelogram:** $A = bh$, where b is the length of the base and h is the height.
4. **Triangle:** $A = \frac{1}{2}bh$, where b is the length of the base and h is the altitude to that base.
5. **Trapezoid:** $A = \frac{1}{2}h(b_1 + b_2)$, where b_1 and b_2 are the lengths of the bases and h is the height.
6. **Regular polygon:** $A = \frac{1}{2}ap$, where a is the apothem and p is the perimeter.
7. **Circle:** $A = \pi r^2$, where r is the radius.
8. **Sector:** $A = \theta\pi r^2/360$, where θ is the measure of the central angle forming the sector and r is the radius of the circle containing the sector.

B. The Pythagorean Theorem: In any right triangle, the square of the length of the hypotenuse is equal to the sum of the squares of the lengths of the legs.

C. Triangle relations
  1. **Property of 30°-60°-90° triangle:** The length of the hypotenuse in a 30°-60°-90° triangle is 2 times the length of the leg opposite the 30° angle, and the length of the leg opposite the 60° angle is $\sqrt{3}$ times the length of the short leg.
  2. **Property of 45°-45°-90° triangle:** The length of the hypotenuse of a 45°-45°-90° triangle is $\sqrt{2}$ times the length of a leg.

D. **Converse of the Pythagorean Theorem:** In any triangle ABC with sides of lengths a, b, and c such that $a^2 + b^2 = c^2$, $\triangle ABC$ is a right triangle with the right angle opposite the side of length c.

V. Surface areas and volumes
  A. Formulas for areas
    1. **Right prism:** $S.A. = 2B + ph$, where B is the area of a base, p is the perimeter of the base, and h is the height of the prism.
    2. **Right circular cylinder:** $S.A. = 2\pi r^2 + 2\pi rh$, where r is the radius of the circular base and h is the height of the cylinder.
    3. **Right circular cone:** $S.A. = \pi r^2 + \pi r\ell$, where r is the radius of the circular base and ℓ is the slant height.
    4. **Right regular pyramid:** $S.A. = B + \frac{1}{2}p\ell$, where B is the area of the base, p is the perimeter of the base, and ℓ is the slant height.
    5. **Sphere:** $S.A. = 4\pi r^2$, where r is the radius of the sphere.
  B. Formulas for volumes
    1. **Right prism:** $V = Bh$, where B is the area of the base and h is the height.
      a. **Right rectangular prism:** $V = \ell wh$, where ℓ is the length, w is the width, and h is the height.
      b. **Cube:** $V = e^3$, where e is an edge.
    2. **Right circular cylinder:** $V = \pi r^2 h$, where r is the radius of the base and h is the height of the cylinder.
    3. **Pyramid:** $V = \frac{1}{3}Bh$, where B is the area of the base and h is the height of the pyramid.
    4. **Circular cone:** $V = \frac{1}{3}\pi r^2 h$, where r is the radius of the circular base and h is the height.
    5. **Sphere:** $V = \frac{4}{3}\pi r^3$, where r is the radius of the sphere.

CHAPTER REVIEW

1. Complete the following.
   a. 50 ft = _____ yd
   b. 947 yd = _____ mi
   c. 0.75 mi = _____ ft
   d. 349 in. = _____ yd
   e. 5 km = _____ m
   f. 165 cm = _____ m
   g. 52 cm = _____ mm
   h. 125 m = _____ km
2. Given three segments of length p, q, and r, where $p > q$, determine if it is possible to construct a triangle with sides of length p, q, and r in each of the following cases. Justify your answers.
   a. $p - q > r$ **b.** $p - q = r$
3. Determine the area of the shaded region on each of the following geoboards if the unit of measure is 1 cm²:

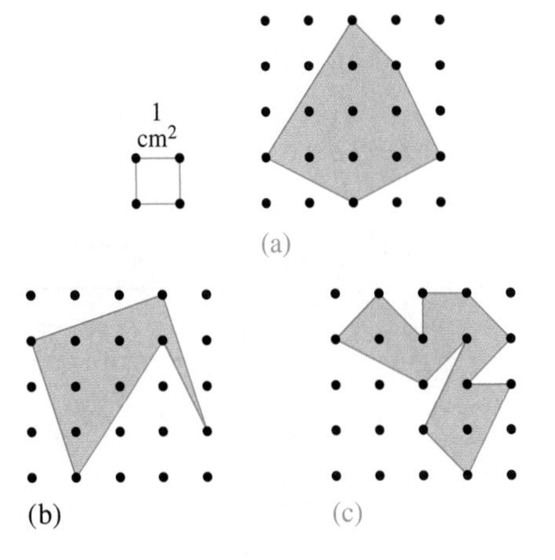

4. Explain how the formula for the area of a trapezoid can be found by using the following figures:

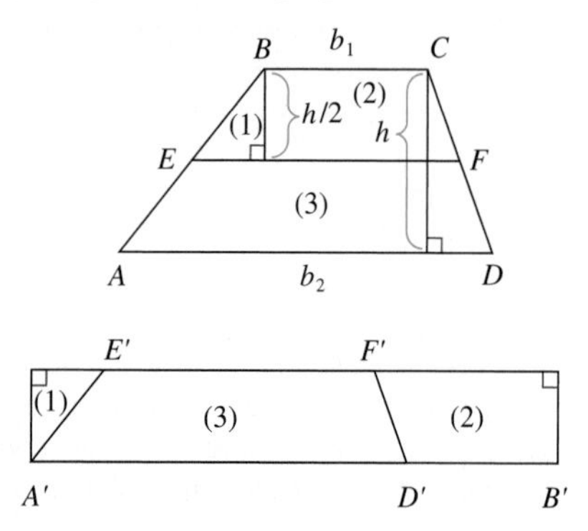

5. Lines a, b, and c are parallel to the line containing side AB of the triangles shown.

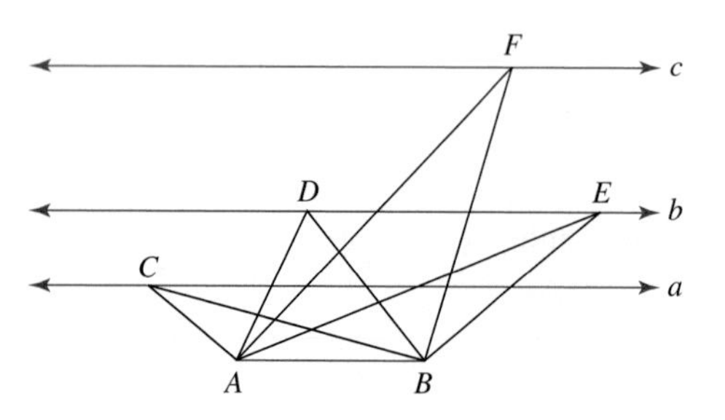

List the triangles in order of size of their areas from least to greatest. Explain why your order is correct.

6. Use the figure shown to find each of the following areas:
   a. The area of the regular hexagon
   b. The area of the circle

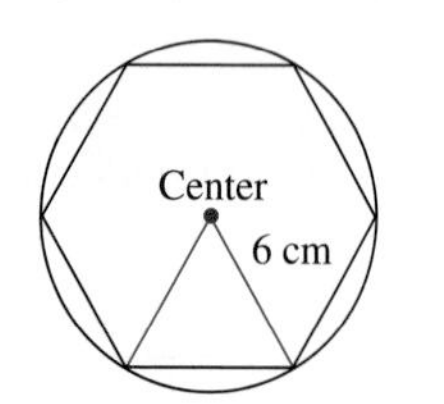

7. Find the area of each shaded region in the following figures:

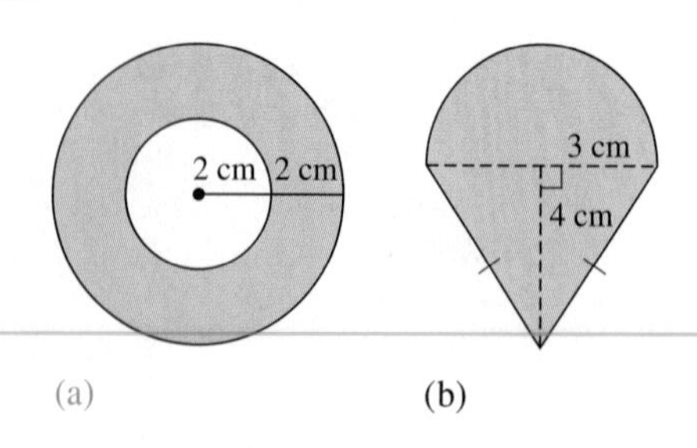

(a) (b)

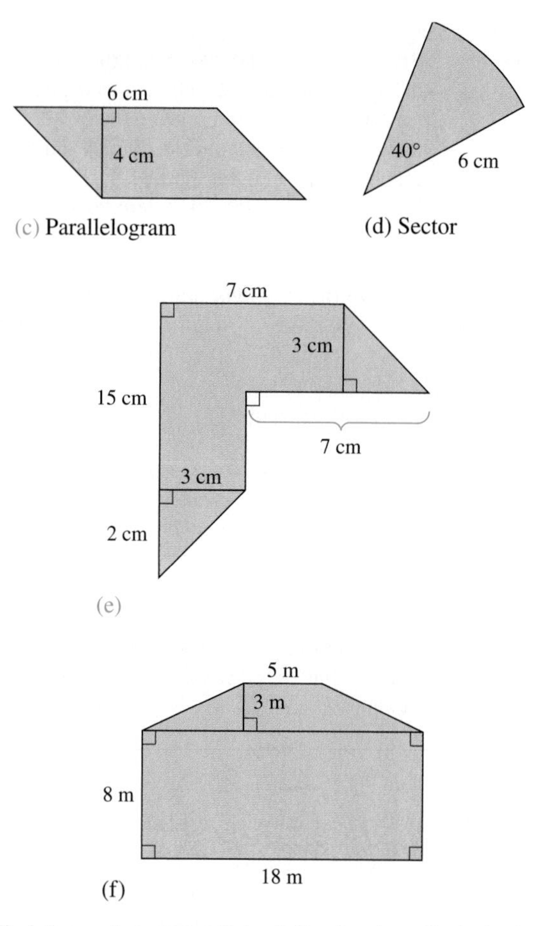

(c) Parallelogram (d) Sector

(e)

(f)

8. Find the surface area of the following box (include the bottom):

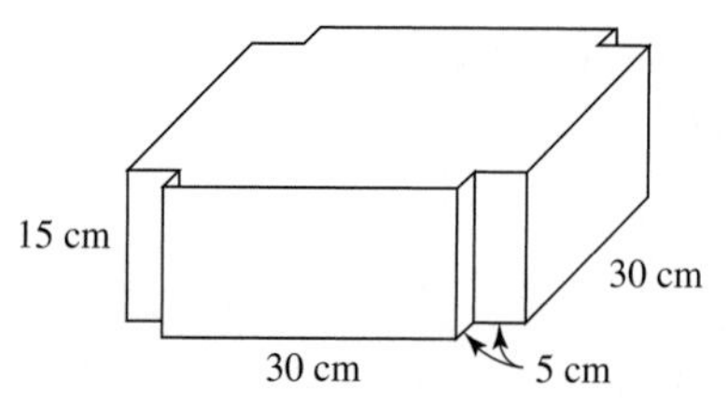

9. A baseball diamond is actually a square 90 ft on a side. What is the distance a catcher must throw from home plate to second base?

10. Find the length of segment AG in the spiral shown.

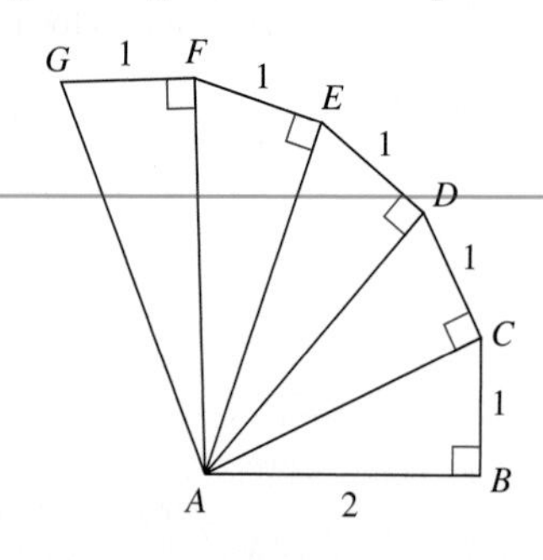

11. For each of the following, determine whether the measures represent sides of a right triangle. Explain your answers.
    a. 5 cm, 12 cm, 13 cm
    b. 40 cm, 60 cm, 104 cm
12. Find the surface area and volume of each of the following figures:

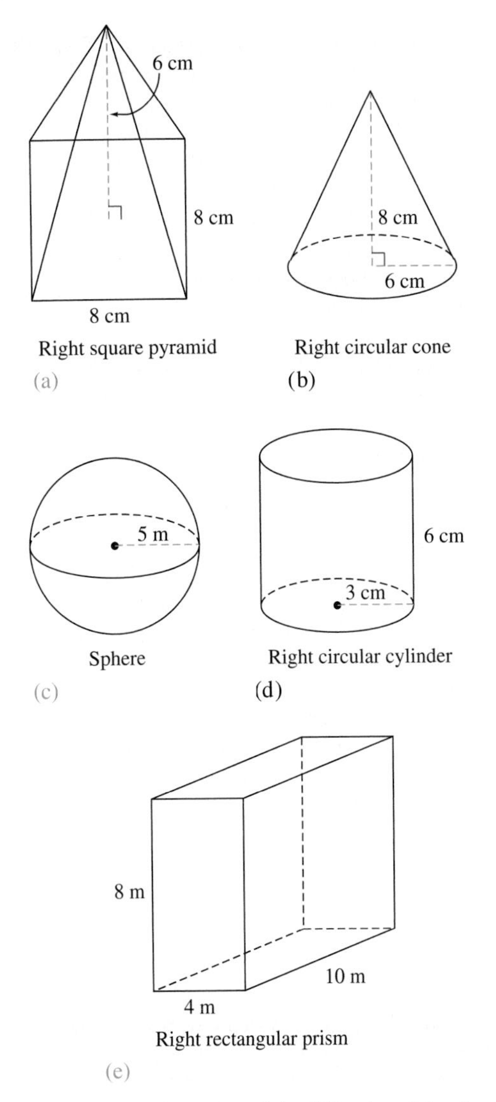

13. Find the lateral surface area of the following right circular cone:

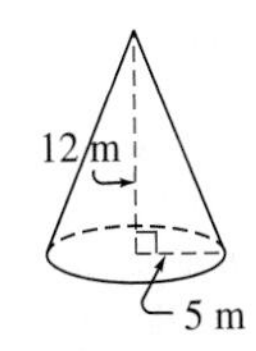

14. Doug's Dog Food Company wants to impress the public with the magnitude of the company's growth. Sales of Doug's Dog Food doubled from 2000 to 2003, so the company is displaying the following graph, which shows the radius of the base and the height of the 2003 can to be double those of the 2000 dog food can. What does the graph really show with respect to the company's growth? Explain your answer.

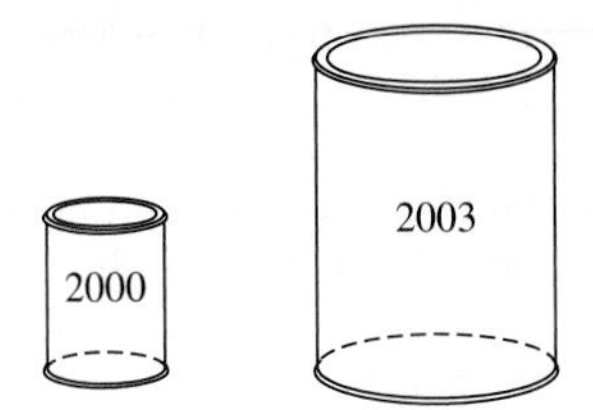

15. Find the area of the kite shown in the following figure:

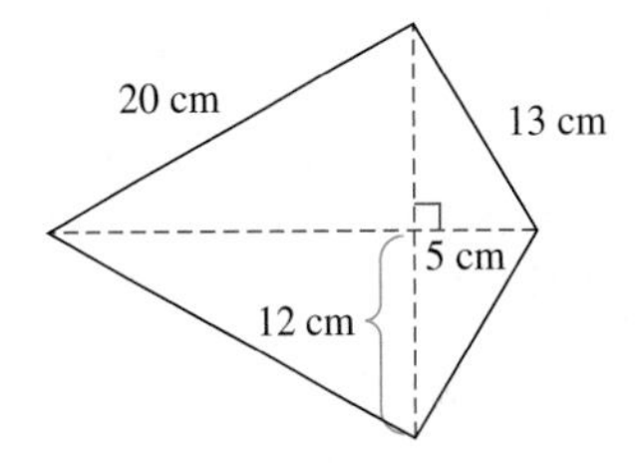

16. The diagonal of a rectangle has measure 1.3 m, and a side of the rectangle has measure 120 cm. Find the following:
    a. Perimeter of the rectangle
    b. Area of the rectangle
17. Find the area of a triangle that has sides of 3 m, 3 m, and 2 m.
18. A poster is to contain 0.25 m^2 of printed matter, with margins of 12 cm at top and bottom and 6 cm at each side. Find the width of the poster if its height is 74 cm.
19. A right circular cylinder and a right circular cone share a circular base and have the same volume. What is the height of the cone?

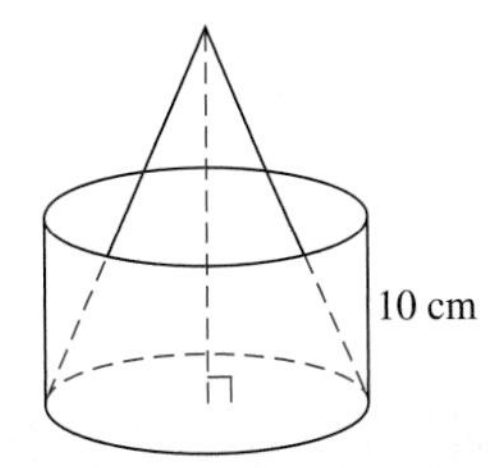

20. Use a coordinate system to locate triangle ABC with vertices having coordinates $A(0, 0)$, B(8, 0), and $C(6, 10)$.
    a. Write equations for each of the sides of the triangle.
    b. Find the midpoints of the sides of the triangle. (*Hint*: Use similar triangles and midsegments if you need to do so.)
    c. Write the equations of each of the medians of the triangle. A **median** connects a vertex with the midpoint of the opposite side.
    d. Find the point of intersection (the **centroid**) of the medians.
    e. The centroid of the triangle divides the medians in some given ratio. What is that ratio? Prove your answer.

21. On a 5 × 5 geoboard,
   a. Find a polygon whose perimeter is greater than 16.
   b. Find a polygon with least perimeter. What is the perimeter?
   c. Find the polygon with greatest area.
22. What is the length of the diagonal of a 8½ × 11-in. page of typing paper?
23. What is the circumference of the circumcircle of a triangle whose sides have length 6, 8, and 10 in.?
24. What is the radius of a circumcircle for an equilateral triangle of side 6 in.?
25. Parallel lines could be defined using the notion of distance. How might that be done?
26. Consider a three-dimensional "geospace board." If it is 5 × 5 × 5,
   a. What is the length of the diagonal?
   b. What is the polyhedron with least volume on the board?
   c. What is the polyhedron with greatest volume on the board?
27. Complete each of the following:
   a. A very heavy object has mass that is measured in ____.
   b. A cube whose length, width, and height are each 1 cm has a volume of _____.
   c. If the cube in (b) is filled with water, the mass of the water is _____.
   d. Which has a larger volume: 1 L or 1 dm^3?
   e. If a car uses 1 L of gas to go 12 km, the amount of gas needed to go 300 km is _____ L.
   f. 20 ha = _____ a
   g. 51.8 L = _____ cm^3
   h. 10 km^2 = _____ m^2
   i. 50 L = _____ mL
   j. 5830 mL = _____ L
   k. 25 m^3 = _____ dm^3
   l. 75 dm^3 = _____ mL
   m. 52,813 g = _____ kg
   n. 4800 kg = _____ t
28. Two cones are defined to be similar if the ratio between their heights equals the ratio between their radii. If two similar cones have heights h_1 and h_2, find the ratio between their volumes in terms of h_1 and h_2.
29. a. A tank that is a right rectangular prism is 1 m × 2 m × 3 m. If the tank is filled with water, what is the mass of the water?
   b. Suppose the tank is exactly half full of water and then a heavy metal sphere of radius 30 cm is put into the tank. How high is the water now if the height of the tank is 3 m?
30. For each of the following, fill in the correct unit to make the sentence reasonable:
   a. Anna filled the gas tank with 80 _____.
   b. A man has a mass of about 82 _____.
   c. The textbook has a mass of 978 _____.
   d. A nickel has a mass of 5 _____ .
   e. A typical adult cat has a mass of about 4 _____.
   f. A compact car has a mass of about 1.5 _____.
   g. The amount of coffee in the cup is 180 _____.
31. For each of the following, decide if the situation is likely or unlikely:
   a. Carrie's bath water has a temperature of 15°C.
   b. Anne found 26°C too warm and so lowered the thermostat to 21°C.
   c. Jim is drinking water that has a temperature of ⁻5°C.
   d. The water in the teakettle has a temperature of 120°C.
   e. The outside temperature dropped to 5°C, and ice appeared on the lake.
32. Complete each of the following:
   a. 2 dm^3 of water has a mass of _____ g.
   b. 1 L of water has a mass of _____ g.
   c. 3 cm^3 of water has a mass of _____ g.
   d. 4.2 ml of water has a mass of _____ kg.
   e. 0.2 L of water has a volume of _____ m^3.

SELECTED BIBLIOGRAPHY

Ameis, J. "Developing an Area Formula for a Circle with Goldilocks and the Three Bears." *Mathematics Teaching in the Middle School* 7 (November 2001): 140–142.

Bellasanta, B., B. Hunter, K. Irwin, M. Sheldon, C. Thompson, and C. Vistro-Yu. "By the Unit or Square Unit." *Mathematics Teaching in the Middle School* 7 (November 2001): 132–137.

Ben-Chaim, D., G. Lappan, and R. Houang. "Adolescents' Ability to Communicate Spatial Information: Analyzing and Affecting Students' Performance." *Educational Studies in Mathematics* 15 (5, 1984): 323–341.

Berry, R., and J. Wiggins. "Measurement in the Middle Grades." *Mathematics Teaching in the Middle School* 7 (November 2001): 154–156.

Chappell, M. "Geometry in the Middle Grades: From Past to the Present." *Mathematics Teaching in the Middle School* 9 (May 2001): 516–519.

Eves, H. *An Introduction to the History of Mathematics.* Philadelphia, PA: Saunders College Publishing, 1983.

Figueras, O., and G. Waldegg. "A First Approach to Measuring (Children between 11 and 13 Years Old)." *Proceedings of the Sixth Annual Meeting of the North American Branch of the International Group for the Psychology of Mathematics Education*, edited by J. Moser. Madison, WI: University of Wisconsin, 1984.

Hart, K. "Measurement," In *Children's Understanding of Mathematics: 11–16*. London: John Murray, 1981.

Hartzler, S. "Ratios of Linear, Area, and Volume Measures in Similar Solids." *Mathematics Teaching in the Middle School* 8 (January 2003): 228–232.

Hayes, N. "Cardboard Tubes Bring Geometry from Home." *Mathematics Teaching in the Middle School* 4 (October 1998): 120–122.

Hiebert, J. "Children's Thinking," In *Mathematics Education Research: Implications for the '80s*, edited by E. Fennema. Alexandria, VA: ASCD, 1981.

LaSaracina, B., and S. White. "The Restless Rectangle and the Transforming Trapezoid." *Teaching Children Mathematics* 5 (February 1999): 336–337, 366.

Loomis, E. *The Pythagorean Proposition*. National Council of Teachers of Mathematics. Reston, VA: NCTM, 1976.

Merz, A. "Hurry Up and Weight." *Teaching Children Mathematics* 10 (September 2003): 8–14.

Moore, D. "Some Like It Hot: Promoting Measurement and Graphical Thinking by Using Temperature." *Teaching Children Mathematics* 5 (May 1999): 538–543.

Moyer, P., and E. Mailley. "*Inchworm and a Half:* Developing Fraction and Measurement Concepts Using Mathematical Representations." *Teaching Children Mathematics* 10 (January 2004): 244–252.

Pumala, V., and D. Klabunde. "Learning Measurement through Practice." *Mathematics Teaching in the Middle School* 10 (May 2005): 452–460.

Robertson, S. "Getting Students Actively Involved in Geometry." *Teaching Children Mathematics* 5 (May 1999): 526–529.

Rozanski, K., C. Beckmann, and D. Thompson. "Exploring Size with the *Grouchy Ladybug*." *Teaching Children Mathematics* 10 (October 2003): 84–89.

Scanlon, G. "Sweet-Tooth Geometry." *Mathematics Teaching in the Middle School* 8 (May 2003): 466–469.

Scott, M. "Using *Measuring Penny* to Introduce the Unit." *Teaching Children Mathematics* 9 (October 2002): 70–74.

Smith, L. "Using Dragon Curves to Learn about Length and Area." *Mathematics Teaching in the Middle School* 5 (December 1999): 222–223.

Taylor, M. "Do Your Students Measure Up Metrically?" *Teaching Children Mathematics* 7 (January 2001): 282–287.

Tent, M. "Circles and the Number II." *Mathematics Teaching in the Middle School* 6 (April 2001): 452–457.

Weinberg, S. "How Big Is Your Foot?" *Mathematics Teaching in the Middle School* 6 (April 2001): 476–481.

Weinberg, S., P. Hammrich, and M. Bruce. "The Giants Project." *Mathematics Teaching in the Middle School* 8 (April 2003): 406–413.

Young, S., and R. O'Leary. "Creating Numerical Scales for Measuring Tools." *Teaching Children Mathematics* 8 (March 2002): 400–405.

Motion Geometry and Tessellations

PRELIMINARY PROBLEM

An architect needs to determine the location of an airport T serving three cities A, B and C so that the sum of the distances from the airport to the three cities is minimal and the airport is in the interior of triangle ABC. The architect's friend, a mathematician, suggests investigating the problem by picking any point for T and rotating $\triangle ATB$ clockwise about B by 60°. Assuming $\triangle ABC$ is acute, can you help the architect to find the true location of the airport?

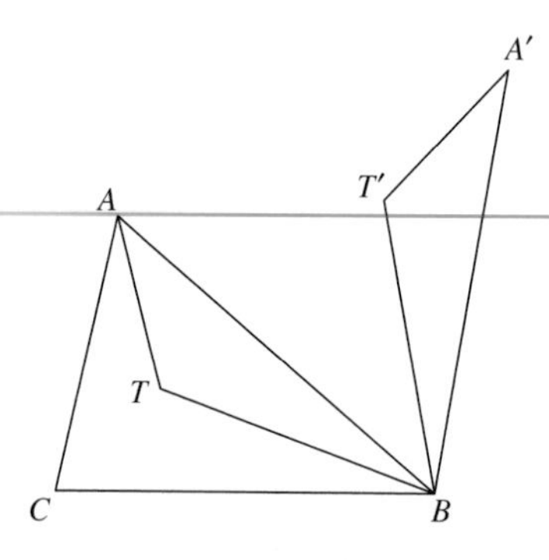

Euclid envisioned moving one geometric figure in a plane and placing it on top of another to determine if the two figures were congruent. Intuitively, we know this can be done by making a tracing of one figure, then sliding, turning, or flipping the tracing, and finally placing it back down atop the other figure.

Symmetries are fundamental in the study of geometry, nature, and shapes. Many symmetries are the results of sliding, flipping, and turning shapes. A study of these "motions" and symmetries leads to tessellations of the plane or space. A **tessellation** is the filling of a plane or space with repetitions of congruent figures in such a way that none overlap and there are no gaps. This chapter contains sections on motions, symmetries, and tessellations.

tessellation

In the *Principles and Standards*, we find the following:

Young children come to school with intuitions about how shapes can be moved. Students can explore motions such as slides, flips, and turns by using mirrors, paper folding, and tracing. Later, their knowledge about transformations should become more formal and systematic. In grades 3–5 students can investigate the effects of transformations and begin to describe them in mathematical terms. Using dynamic geometry software, they can begin to learn the attributes needed to define a transformation. In the middle grades, students should learn to understand what it means for a transformation to preserve distance, as translations, rotations, and reflections do. . . . At all grade levels, appropriate consideration of symmetry provides insights into mathematics and into art and aesthetics. (p. 43)

The *Principles and Standards* also lists the following expectations for different grade levels.

In prekindergarten through Grade 2, all students should

- *recognize and apply slides, flips, and turns;*
- *recognize and create shapes that have symmetry.* (p. 96)

In Grades 3–5 all students should

- *predict and describe the results of sliding, flipping, and turning two-dimensional shapes;*
- *describe a motion or a series of motions that will show that two shapes are congruent;*
- *identify and describe line and rotational symmetry in two- and three-dimensional shapes and designs.* (p. 164)

In Grades 6–8 all students should

- *describe sizes, positions, and orientations of shapes under informal transformations such as flips, turns, slides, and scaling;*
- *examine the congruence, similarity, and line or rotational symmetry of objects using transformations.* (p. 232)

▶ **HISTORICAL NOTE**

In 1872, at age 23, Felix Klein (1849–1925) was appointed to a chair at the University of Erlangen, Germany. His inaugural address, referred to as the *Erlanger Programm*, described geometry as the study of properties of figures that do not change under a particular set of transformations. Specifically, Euclidean geometry was described as the study of such properties of figures as area and length, which remain unchanged under a set of transformations called *isometries*.

One of the first attempts to introduce geometric transformations into the elementary school in the United States was by the University of Illinois Committee on School Mathematics (UICSM). UICSM published *Motion Geometry* in four volumes in 1969. ◀

12-1 Translations and Rotations

Translations

translation/slide
preimage
slide line ◆ image
slide arrow ◆ vector

Figure 12-1 shows a two-dimensional representation of a child moving down a slide without twisting or turning. This type of motion is a **translation**, or **slide**. In Figure 12-1, the child (**preimage**) at the top of the slide moves a certain distance in a certain direction along a **slide line** to obtain the **image** at the bottom of the slide. Figure 12-1 shows a translation that takes the preimage to its image. The translation is determined by the **slide arrow**, or **vector**, from M to N. The vector determines the image of any point in a plane in the following way: The image of a point A in the plane is the point A' obtained by sliding A along a line parallel to $\overleftrightarrow{MN}$ in the direction from M to N by the distance MN. (MN is also denoted by d in Figure 12-1.) Notice that $AA' = MN = d$ and $\overleftrightarrow{MN}$ is parallel to $\overleftrightarrow{AA'}$. It appears that under the translation, figures change neither their shapes nor their sizes. In fact, a translation preserves both length and angle size, and thus congruence of figures.

FIGURE 12-1

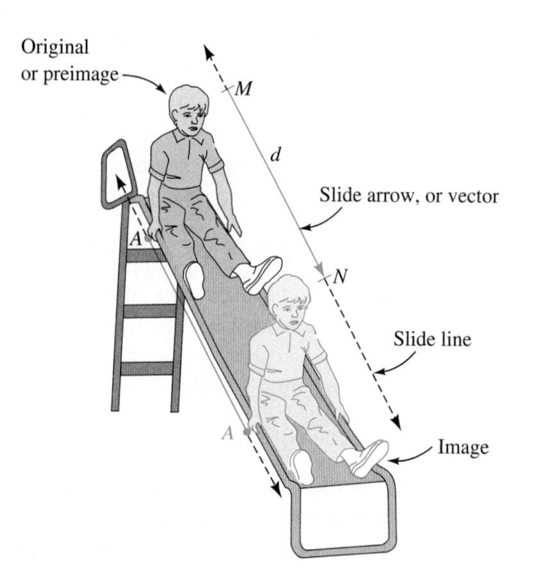

rigid motion ◆ isometry

Any **rigid motion** that preserves length or distance is an **isometry** (derived from Greek and meaning "equal measure"). For example, a translation is an isometry. A translation is a function whose domain and range both are the set of all the points in the plane.

transformation

Notice that a translation of a plane is a one-to-one correspondence between the plane and itself. Any function from a plane to itself that is a one-to-one correspondence between a plane and itself is a **transformation** of the plane.

We assume the following properties of a translation from a plane to a plane.

Definition of a Translation

A **translation** is a motion of a plane that moves every point of the plane a specified distance in a specified direction along a straight line.

TECHNOLOGY
CORNER

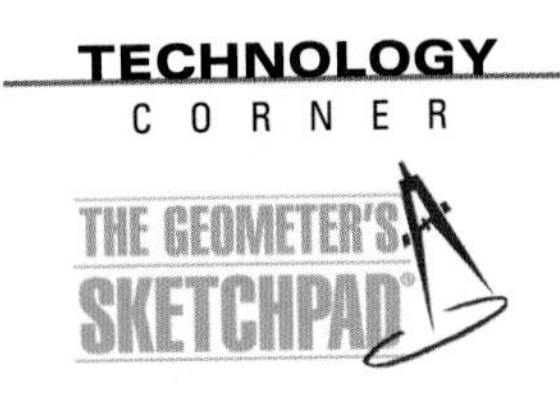

GSP Lab 10 activities 6–8 can be used to investigate properties of translations.

Properties of Translations

- A figure and its image are congruent.
- The image of a line is a line parallel to it.

Constructions of Translations

The image of a figure under a translation can be constructed easily with tracing paper or by using only a compass and straightedge. Using tracing paper may be the more natural way to construct a "motion." However, since traditionally most constructions in geometry are accomplished with a compass and a straightedge, we consider first how that might be done and then leave the actual construction as an activity in "Now Try This 12-1."

To construct the image of an object under a translation, we first need to know how to construct the image A' of a single point A. In Figure 12-1, by definition of translation $MN = AA'$ and $\overleftrightarrow{MN}$ is parallel to $\overleftrightarrow{AA'}$. Thus, $MAA'N$ will be a parallelogram. Why? Consequently, to construct the image A' of a point A with a compass and straightedge, we need only to construct a parallelogram $MAA'N$ so that $\overrightarrow{AA'}$ is in the same direction as $\overrightarrow{MN}$.

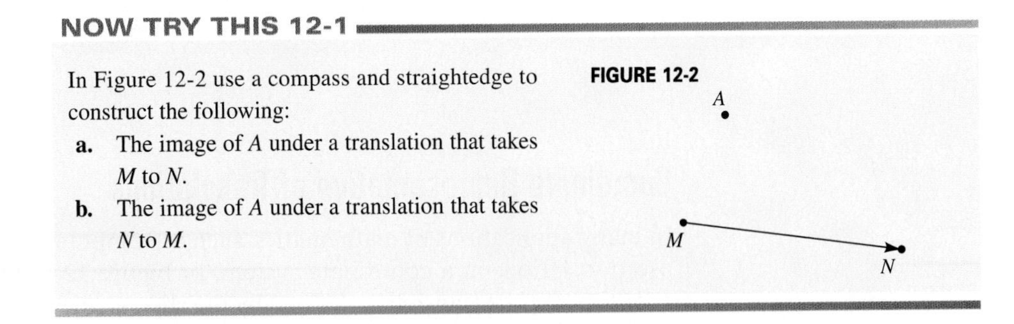

NOW TRY THIS 12-1

In Figure 12-2 use a compass and straightedge to construct the following:

a. The image of A under a translation that takes M to N.

b. The image of A under a translation that takes N to M.

FIGURE 12-2

To find the image of a triangle under a translation, it suffices to find the images of the three vertices of the triangle by using a process similar to that used in Figure 12-1 or Figure 12-2 and connect these images with segments to form the triangle's image.

It is possible to use a geoboard or a grid to find an image of a segment, as the following example shows.

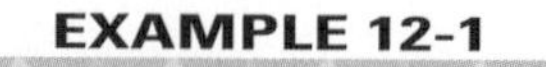

EXAMPLE 12-1

Find the image of $\overline{AB}$ under the translation from X to X' pictured on the dot paper in Figure 12-3.

FIGURE 12-3

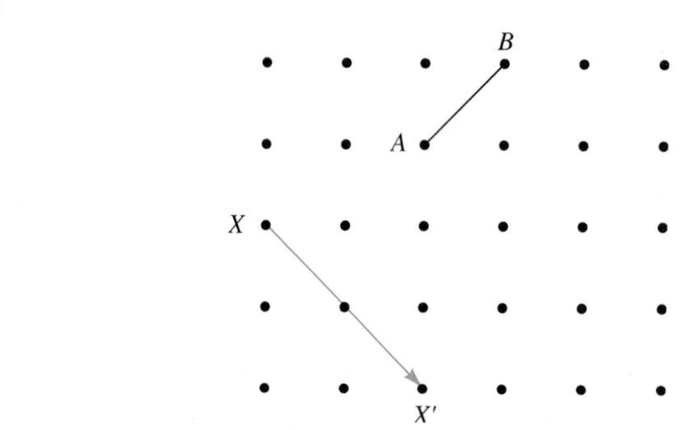

Solution X' is the image of X under the translation. X' could be obtained from X by shifting X 2 units vertically down and then 2 units horizontally to the right, as shown in Figure 12-4. This shifting determines the slide arrow from X to X'. The image of each point on the dot paper can be obtained by first shifting it 2 units down and then 2 units to the right. The image of $\overline{AB}$ is found in this way in Figure 12-4.

FIGURE 12-4

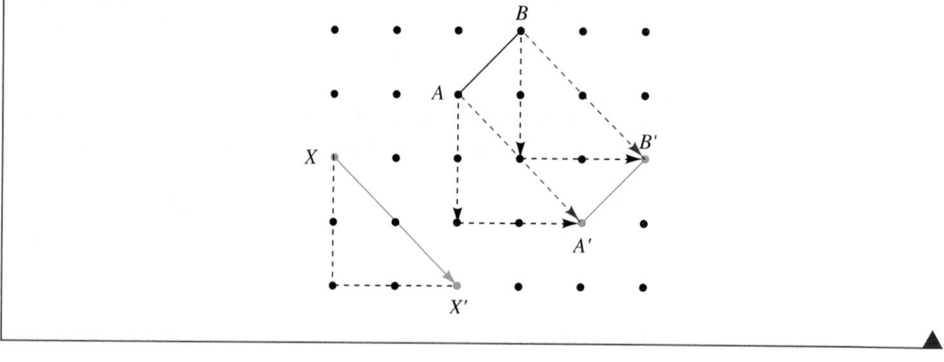

Coordinate Representation of Translations

In many applications of mathematics, such as computer graphics, it is necessary to use translations in a coordinate system. In Figure 12-5, $\triangle A'B'C'$ is the image of $\triangle ABC$ under the translation defined by the slide arrow from O to O', where O is the origin and O' has coordinates $(5, {}^{-}2)$. Point O' is the image of point O under the given translation. The point $O'(5, {}^{-}2)$ can be obtained by moving O horizontally to the right 5 units and then 2 units down. As each point in the triangle is translated in the direction from O to O' by the distance OO', we can obtain the image of any point by moving horizontally to the right 5 units and then vertically 2 units down. This is shown in Figure 12-5 for points A, B, C and their corresponding images A', B', and C'. Table 12-1 shows how the coordinates of the image vertices A', B', and C' in Figure 12-5 are obtained from the coordinates of A, B, and C.

FIGURE 12-5

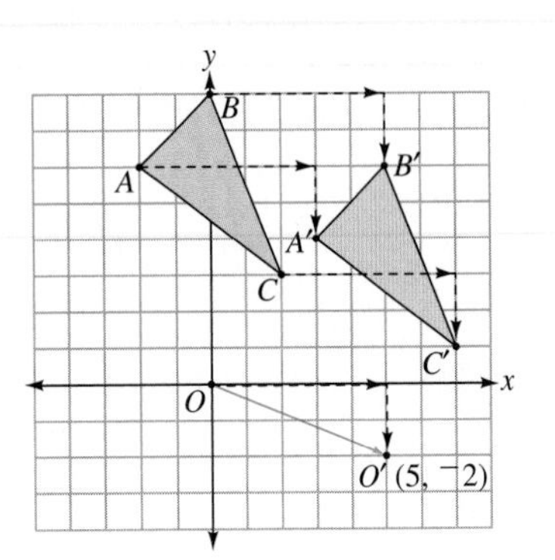

Table 12-1

Point (x, y)	Image Point $(x + 5, y - 2)$
$A\ (^-2, 6)$	$A'\ (3, 4)$
$B\ (0, 8)$	$B'\ (5, 6)$
$C\ (2, 3)$	$C'\ (7, 1)$

This discussion suggests that we could describe a translation by showing how the coordinates of any point (x, y) are changed. The translation described in Table 12-1 can be written symbolically as $(x, y) \rightarrow (x + 5, y - 2)$, where "$\rightarrow$" denotes "moves." In this notation, $(x + 5, y - 2)$ is the image of (x, y).

Property of a Translation in a Coordinate System

A translation is a function from the plane to the plane such that to every point (x, y) corresponds the point $(x + a, y + b)$, where a and b are real numbers.

In general, a translation is symbolized as $(x, y) \rightarrow (x + a, y + b)$, where the point $(x + a, y + b)$ is the image of the point (x, y).

EXAMPLE 12-2

Find the coordinates of the image of the vertices of quadrilateral $ABCD$ in Figure 12-6 under the translations in parts (a) through (c). Draw the image in each case.

FIGURE 12-6

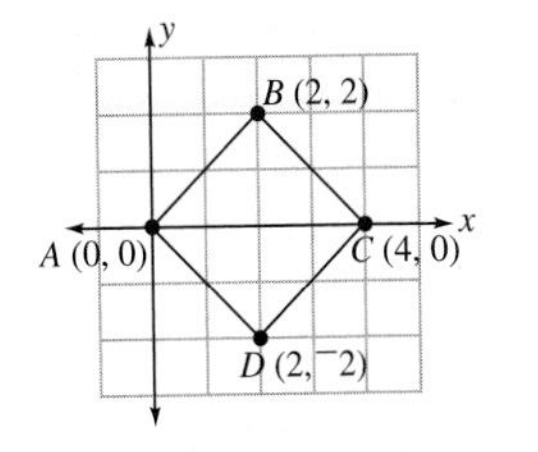

a. $(x, y) \rightarrow (x - 2, y + 4)$
b. A translation determined by the slide arrow from $A(0, 0)$ to $A'(^-2, 4)$
c. A translation determined by the slide arrow from $S(4, ^-3)$ to $S'(2, 1)$

Solution **a.** Because $(x, y) \rightarrow (x - 2, y + 4)$, the images A', B', C', and D' of the corresponding points A, B, C, and D can be found as follows:

$$A(0, 0) \rightarrow A'(0 - 2, 0 + 4), \text{ or } A'(^-2, 4)$$
$$B(2, 2) \rightarrow B'(2 - 2, 2 + 4), \text{ or } B'(0, 6)$$
$$C(4, 0) \rightarrow C'(4 - 2, 0 + 4), \text{ or } C'(2, 4)$$
$$D(2, ^-2) \rightarrow D'(2 - 2, ^-2 + 4), \text{ or } D'(0, 2)$$

The square $ABCD$ and its image $A'B'C'D'$ are shown in Figure 12-7.

FIGURE 12-7

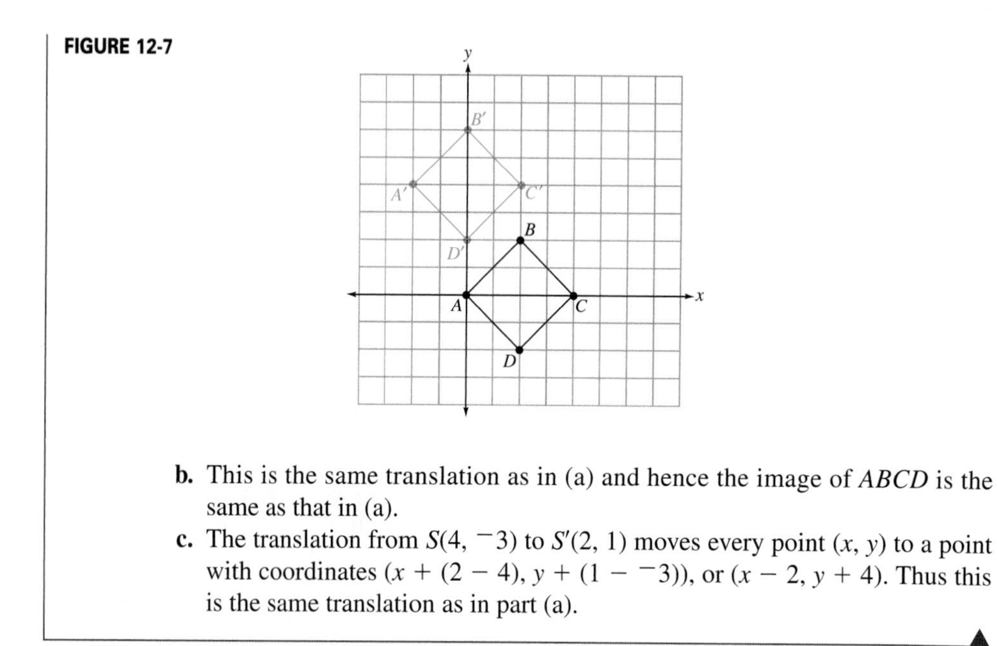

b. This is the same translation as in (a) and hence the image of $ABCD$ is the same as that in (a).

c. The translation from $S(4, {}^{-}3)$ to $S'(2, 1)$ moves every point (x, y) to a point with coordinates $(x + (2 - 4), y + (1 - {}^{-}3))$, or $(x - 2, y + 4)$. Thus this is the same translation as in part (a).

Translations are useful in determining frieze patterns such as those appearing in wallpaper designs. Such patterns are explored in problems 18 and 19 of the Ongoing Assessment 12-4.

Rotations

rotation ◆ turn

A **rotation**, or **turn**, is another kind of isometry. Figure 12-8 illustrates congruent figures that resulted from a rotation about point O. The image of the letter **F** is shown in green.

FIGURE 12-8

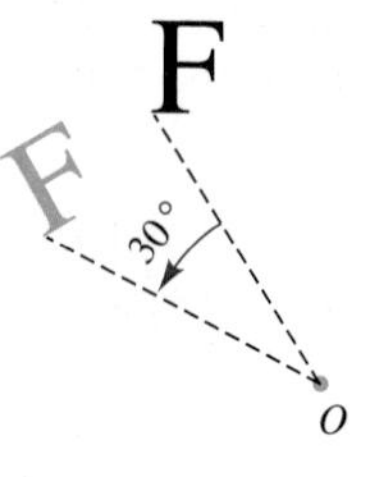

▶ RESEARCH NOTE

Students have some informal understanding of geometric transformations such as reflections and rotations but have a difficult time operating on shapes using these transformations (Kuchemann 1981). ◀

A rotation can be constructed by using tracing paper, as in Figure 12-9. In Figure 12-9(a), $\triangle ABC$ and point O are traced on tracing paper. Holding point O fixed, we turn the tracing paper to obtain the image $\triangle A'B'C'$, as shown in Figure 12-9(b). Point O is the **turn center**, and $\angle COC'$ is the **turn angle**.

turn center ◆ turn angle

FIGURE 12-9 Construction of a rotation using tracing paper

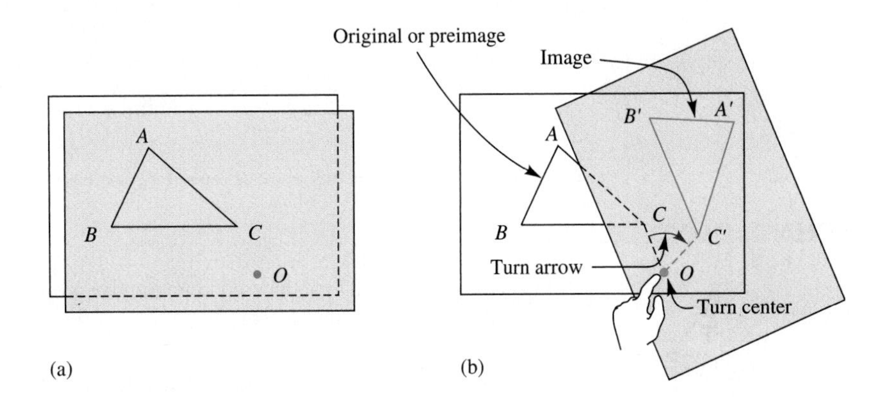

To determine a rotation, we must know three pieces of data: the turn center; the direction of the turn, either clockwise or counterclockwise; and the amount of the turn. The amount and the direction of the turn can be illustrated by a **turn arrow**, or they can be specified in numbers of degrees.

turn arrow

This discussion leads to the following definition.

Definition of Rotation

A **rotation** is a transformation of the plane determined by holding one point—the center—fixed and rotating the plane about this point by a certain amount in a certain direction.

REMARK In higher mathematics and its applications, it is customary to assign a positive measure to a counterclockwise turn and a negative measure to a clockwise turn.

To construct an image of a figure under a rotation, observe that every point on the tracing-paper construction moves along a circle. Also the angle formed by any point, the center of the rotation O, and the image of the point is the angle of the turn. Why? With this in mind, we construct the image of a point under a rotation in "Now Try This 12-2."

NOW TRY THIS 12-2

Use a compass and a straightedge to construct the image of point P under a rotation with center O through the angle and in the direction given in Figure 12-10. (*Hint*: Construct an isosceles triangle BAC with B on one side of the given angle and C on the other side so that $AB = AC = OP$. Then construct $\triangle POP'$ congruent to $\triangle BAC$.)

FIGURE 12-10

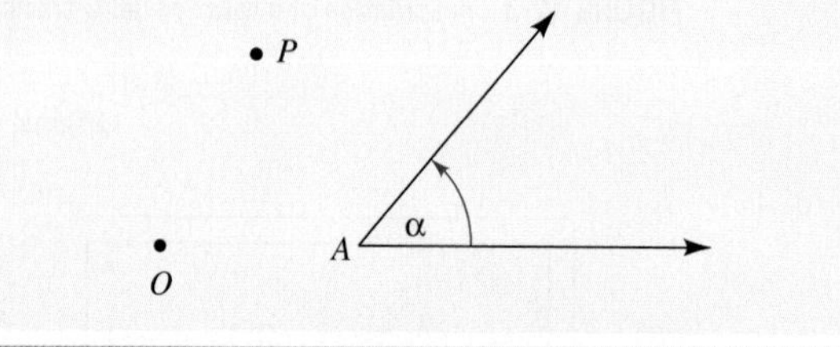

TECHNOLOGY CORNER

GSP Lab 10 activities 9–11 can be used to investigate properties of rotations.

Because a rotation is an isometry, the image of a figure under a rotation is congruent to the original figure. It can be shown that *under any isometry, the image of a line is a line, the image of a circle is a circle, and the images of parallel lines are parallel lines*.

For certain angles like 90°, rotations may be constructed on a geoboard or dot paper, as demonstrated in Example 12-3.

EXAMPLE 12-3

In Figure 12-11, find the image of $\triangle ABC$ under the rotation with center O.

FIGURE 12-11

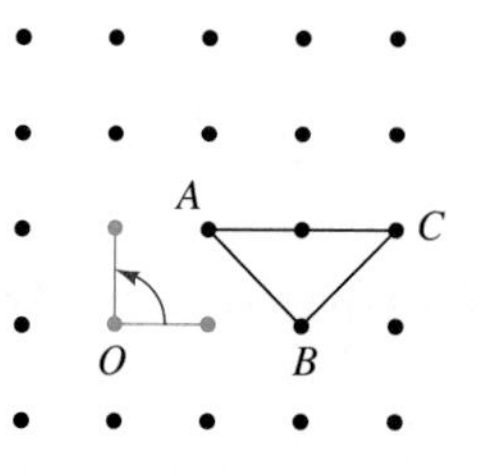

Solution $\triangle A'B'C'$, the image of $\triangle ABC$, is shown in Figure 12-12. The image of A is A' because $\angle A'OA$ is a right angle (why?) and $OA = OA'$. Similarly, B' is the image of B. To find the location of C', we use the fact that rotation is an isometry and hence $\triangle A'B'C' \cong \triangle ABC$. Thus, $\angle B \cong \angle B'$. The location of point C' shown makes $\angle B \cong \angle B'$, $C'B' = CB$ (why?), and the direction of the rotation is counterclockwise as specified.

FIGURE 12-12

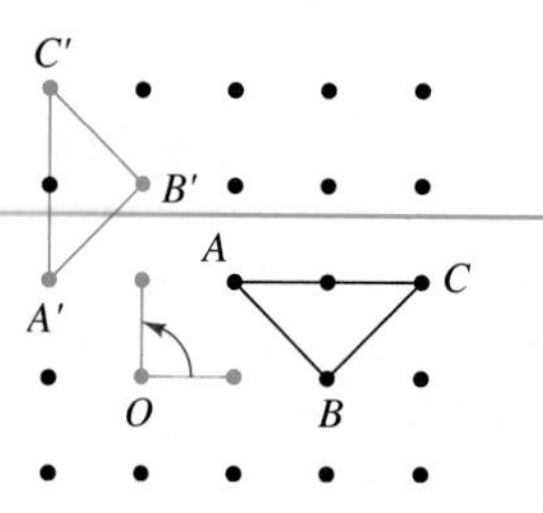

A rotation of 360° about a point will move any point and hence any figure onto itself. Such a transformation is an **identity transformation**. It plays an important role in more advanced studies of transformations.

identity transformation

A rotation of 180° about a point is also of particular interest. Such a rotation is a **half-turn**. Because a half-turn is a rotation, it has all the properties of rotations. Figure 12-13 shows some shapes and their images under a half-turn about point O.

half-turn

FIGURE 12-13

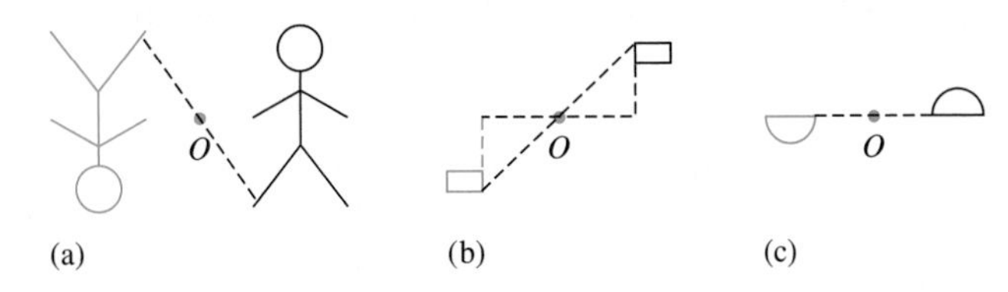

Rotations are useful in determining turn symmetries, which we investigate later. A different application of rotations appears when we consider the slopes of perpendicular lines, as seen in the following section.

Slopes of Perpendicular Lines

Transformations can be used to investigate various mathematical relationships. For example, they can be used to determine the relationship between the slopes of two perpendicular lines, neither of which is vertical.

We first consider a special case in which the lines go through the origin. Suppose the slopes of the lines ℓ_1 and ℓ_2, shown in Figure 12-14, are m_1 and m_2, respectively. Because the slope of a line is equal to rise over run, the slope of ℓ_1 can also be determined from $\triangle OBA$, in which we choose $OA = 1$. We have $m_1 = \frac{\text{rise}}{\text{run}} = \frac{BA}{1} = BA$.

FIGURE 12-14

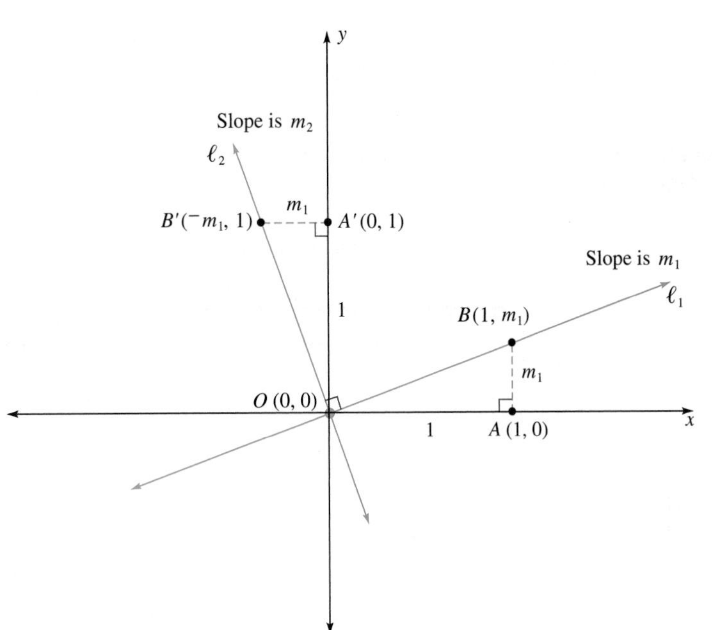

We then rotate the plane 90° counterclockwise about center O. The image of ℓ_1 is ℓ_2. To determine the image of $\triangle OBA$, we need only determine the images of each of the vertices of the triangle. The image of O is O itself. The image of A is A' on the y-axis. Why? The image of B is B' on ℓ_2. Why?. Because rotation preserves congruence, $\triangle OB'A' \cong \triangle OBA$; consequently, $\angle B'A'O$ is a right angle, $A'B' = m_1$, and $OA' = 1$. Thus, as shown in Figure 12-14, point B' is at $(^-m_1, 1)$. We can use the slope formula to find the slope of ℓ_2 as follows:

$$m_2 = \frac{1-0}{^-m_1 - 0} = \frac{1}{^-m_1} = \frac{^-1}{m_1}$$

Thus, $m_2 = {}^-1/m_1$, or $m_1m_2 = {}^-1$.

The relationship between the slopes m_1 and m_2 of two perpendicular lines (neither of which is vertical) that do not intersect at the origin can always be found using two lines parallel to the original lines but that pass through the origin. Because parallel lines have equal slopes, the relationship between the slopes of the perpendicular lines is the same as the relationship between the slopes of the perpendicular lines through the origin; that is, $m_1m_2 = {}^-1$.

REMARK The relationship described between slopes of perpendicular lines is only true if each line has a slope.

It is also possible to prove the converse statement; that is, if the slopes of two lines satisfy the condition $m_1m_2 = {}^-1$, then the lines are perpendicular. We summarize these results in the following property.

Property of Slopes of Perpendicular Lines

Two lines, neither of which is vertical, are perpendicular if and only if their slopes m_1 and m_2 satisfy the condition $m_1m_2 = {}^-1$. Every vertical line has no slope but is perpendicular to a line with slope 0.

EXAMPLE 12-4

Find the equation of line ℓ through point $(^-1, 2)$ and perpendicular to the line $y = 3x + 5$.

Solution If m is the slope of ℓ as in Figure 12-15, then ℓ can be written as $y = mx + b$.

FIGURE 12-15

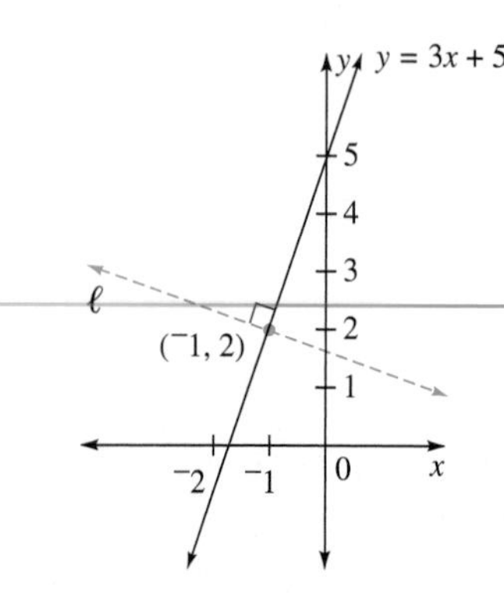

Because the line $y = 3x + 5$ has slope 3 and is perpendicular to ℓ, we have $m \cdot 3 = {}^{-}1$; therefore, $m = -\frac{1}{3}$. Consequently, the equation of ℓ is

$$y = -\frac{1}{3}x + b.$$

Because the point $({}^{-}1, 2)$ is on ℓ, we can substitute $x = {}^{-}1, y = 2$ in $y = -\frac{1}{3}x + b$ and solve for b as follows:

$$2 = -\frac{1}{3} \cdot ({}^{-}1) + b$$

$$\frac{5}{3} = b.$$

Consequently, the equation of ℓ is

$$y = -\frac{1}{3}x + \frac{5}{3}.$$

ONGOING ASSESSMENT 12-1

1. For each of the following, find the image of the given quadrilateral under a translation from A to B:

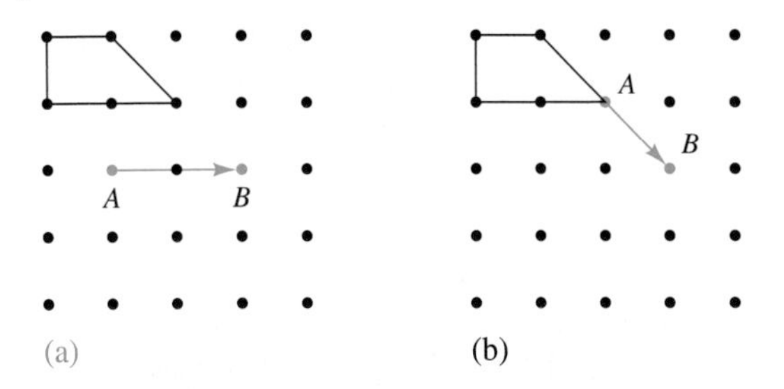

2. Find the figure whose image is given in each of the following under a translation from X to X':

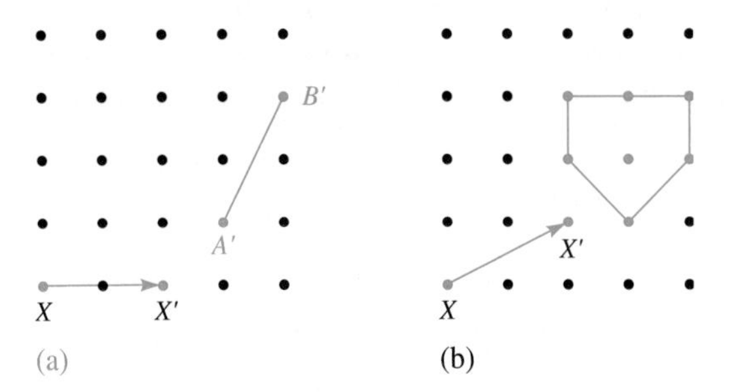

3. Construct the image of $\overline{BC}$ under the translation pictured in the figure by using the following:
   a. Tracing paper
   b. Compass and straightedge

B
C

4. Find the coordinates of the image for each of the following points under the translation defined by $(x, y) \rightarrow (x + 3, y - 4)$:
   a. $(0, 0)$
   b. $({}^{-}3, 4)$
   c. $({}^{-}6, {}^{-}9)$
   d. $(7, 14)$
   e. (h, k)
5. Find the coordinates of the points whose images under the translation $(x, y) \rightarrow (x - 3, y + 4)$ are the following:
   a. $(0, 0)$
   b. $({}^{-}3, 4)$
   c. $({}^{-}6, {}^{-}9)$
   d. $(7, 14)$
   e. (h, k)
6. Consider the translation $(x, y) \rightarrow (x + 3, y - 4)$. In each of the following, draw the image of the figure under the trans-

lation and find the coordinates of the images of the labeled points.

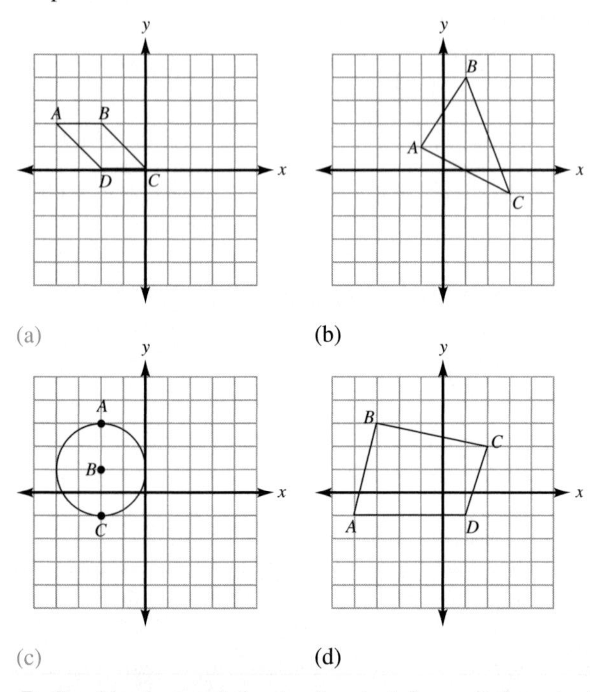

7. Consider the translation $(x, y) \to (x + 3, y - 4)$. In each of the following, draw the figure whose image is shown:

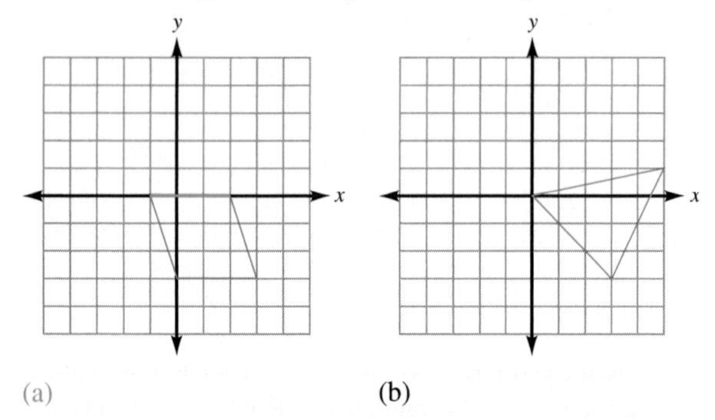

8. Find the image of the following quadrilateral in a 90° counterclockwise rotation about O:

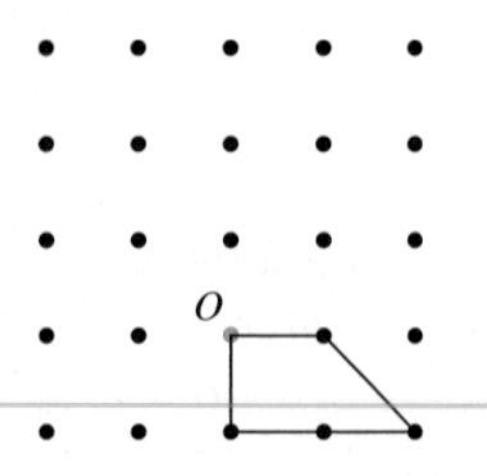

9. If ℓ is a line whose equation is $y = 2x - 1$, find the equation of the image of ℓ under each of the following translations:
   a. $(x, y) \to (x, y - 2)$
   b. $(x, y) \to (x + 3, y)$
   c. $(x, y) \to (x - 3, y + 2)$

10. If $y = {}^-2x + 3$ is the image of line k under the translation $(x, y) \to (x + 3, y - 2)$, find the equation of k.
11. If P' is the image of point P under a half-turn about its center O, what can be said about points P', P, and O? Why?
12. Use a compass and a straightedge to find the image of line ℓ under a half-turn about point O as shown.

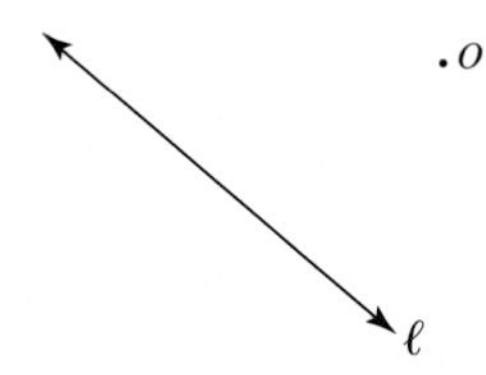

13. The images of $\overline{AB}$ under various rotations are given in the following figures. Find $\overline{AB}$ in each case.

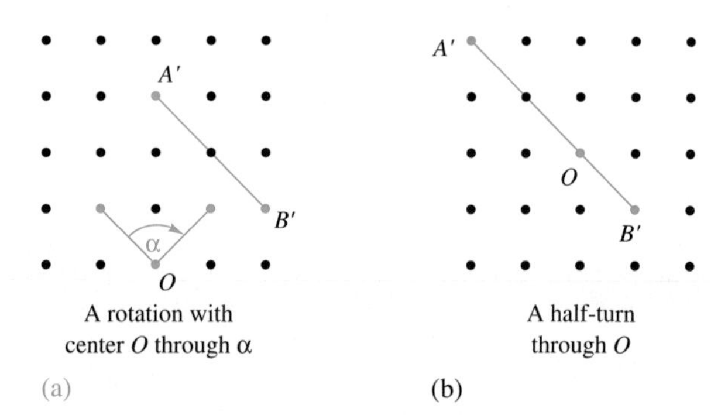

14. The image of NOON is still NOON after a special half-turn. List some other words that have the same property. What letters can such words contain?
15. Answer each of the following:
    a. Draw a line ℓ and any two points A and B so that $\overline{AB}$ is parallel to the line. Find the image of ℓ under a translation from A to B.
    b. Draw a line ℓ and any two points A and B so that $\overline{AB}$ is not parallel to ℓ. Construct ℓ', the image of ℓ under the translation from A to B.
    c. How are ℓ and ℓ' in part (b) related? Why?
    d. What is the image of $\angle ABC$ under 10 successive rotations about B if $m(\angle ABC) = 36°$.
16. a. Refer to the following figure and use paper folding or any other method to show that if P' is the image of P under rotation about point O by a given angle, then O is on the perpendicular bisector of $\overline{PP'}$.

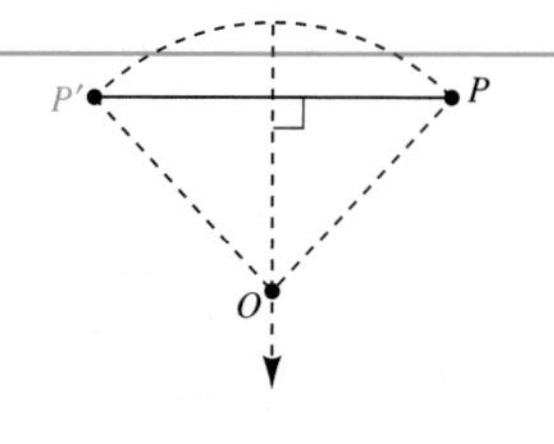

b. $\triangle A'B'C'$ shown in the following figure was obtained by rotating $\triangle ABC$ about a certain point O. Explain how to find the point O and the angle of rotation.

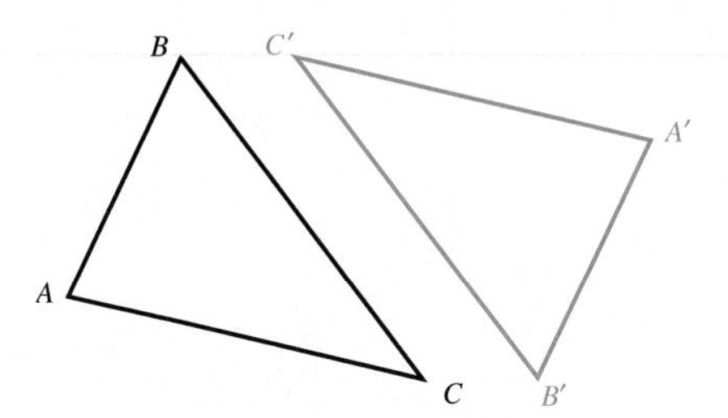

c. Triangles ABC and $A'B'C'$ shown are congruent. Trace them and explain why it is impossible to find a rotation under which $\triangle A'B'C'$ is the image of $\triangle ABC$.

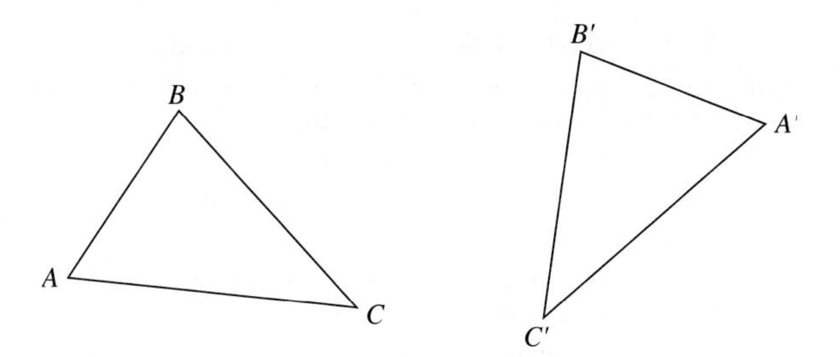

17. The images of any point under a rotation by certain angles can be found with only a compass and straightedge (without the use of a protractor). Construct P', the image of P when it is rotated about O, as shown in the figure, for angles with the following measures and direction:
   a. 45° counterclockwise
   b. 60° clockwise
   c. 30° counterclockwise

• P

• O

18. For each of the following points, find the coordinates of the image point under a half-turn about the origin:
   a. (4, 0) **b.** (0, 3)
   c. (2, 4) **d.** ($^-$2, 5)
   e. ($^-$2, $^-$4) **f.** (a, b)
   g. (^-a, ^-b)

19. In each of the following figures, find the image of the figure under a half-turn about O:

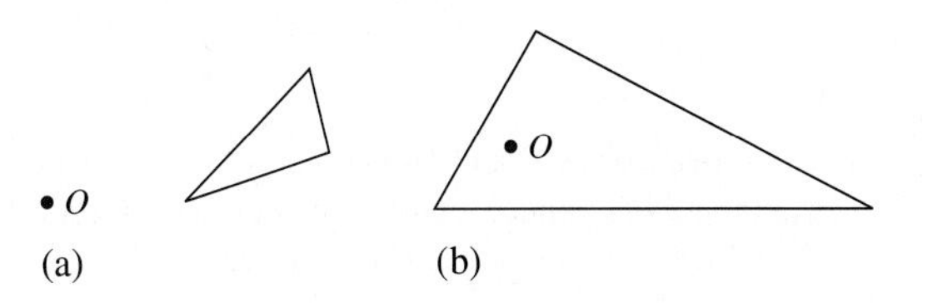

20. Draw any line and label it ℓ. Use tracing paper to find ℓ', the image of ℓ under each of the following rotations. In each case, describe in words how ℓ' is related to ℓ.
   a. Half-turn about point O on ℓ
   b. Half-turn about point O, not on ℓ
   c. A 90° turn counterclockwise about point O, not on ℓ
   d. A 60° turn counterclockwise about point O, not on ℓ

21. a. Find the final image of $\triangle ABC$ by performing two rotations in succession each with center O, one by angle α and one by angle β, in the directions shown in the figure.

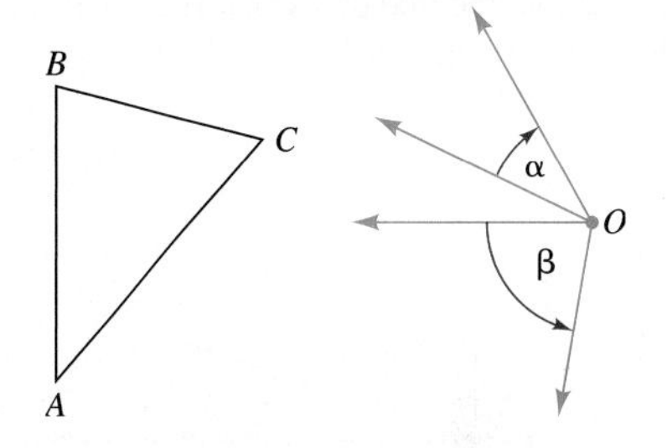

   b. Is the order of the rotations important?
   c. Could the result have been accomplished in one rotation?

22. When $\triangle ABC$ in the following figure is rotated about a point O by 360°, each of the vertices traces a path:

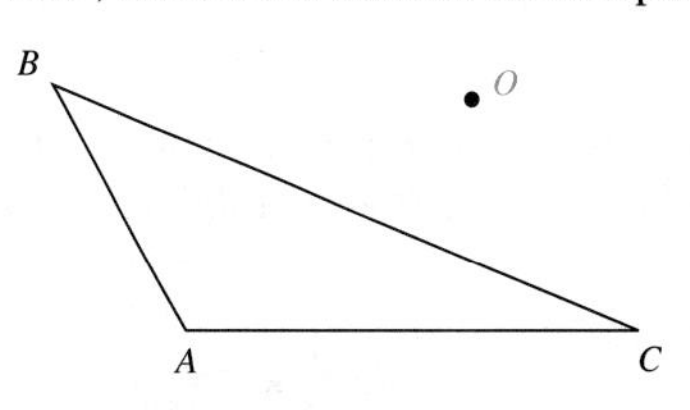

   a. What geometric figure does each vertex trace?
   b. Identify all points O for which two vertices trace an identical path. Justify your answer.
   c. Given any $\triangle ABC$, is there a point O such that the three vertices trace an identical path? If so, describe how to find such a point.

23. Use a drawing similar to Figure 12-14 to find:
   a. The images of the following points under rotation counterclockwise about the origin:
   (i) (2, 3) (ii) ($^-$1, 2) (iii) $P(a, b)$ in terms of a and b
   b. Use what you found in part (a) to get the image of $P(a, b)$ under rotation clockwise by 90° about the origin, (*hint*: Rotate first as in part (a), then apply a half-turn about the origin).

24. Find the equation of the image of the line $y = 3x - 1$ under the following transformations:
   a. Half-turn about the origin
   b. A 90° rotation about the origin counterclockwise

25. Three vertices of a rhombus are at $O(0,0)$, $A(3,4)$, $C(a,0)$. Find the following:
   a. All possible coordinates of point C
   b. All possible coordinates for the fourth vertex B
   c. Verify that the diagonals of each of the rhombi you found in part (b) are perpendicular to each other.

26. Find the equation of the line through $(^{-}1, {}^{-}3)$ and perpendicular to
   a. $y = 2x + 1$
   b. $y = {}^{-}2x + 3$
   c. The line $x = {}^{-}4$
   d. The line $y = {}^{-}3$
27. The following figure is a rhombus with sides of length a and one of the vertices at (h, k):
   a. Explain why the coordinates of B are $(h + a, k)$.
   b. Explain why $h^2 + k^2 = a^2$.
   c. Prove that the diagonals of a rhombus are perpendicular to each other.

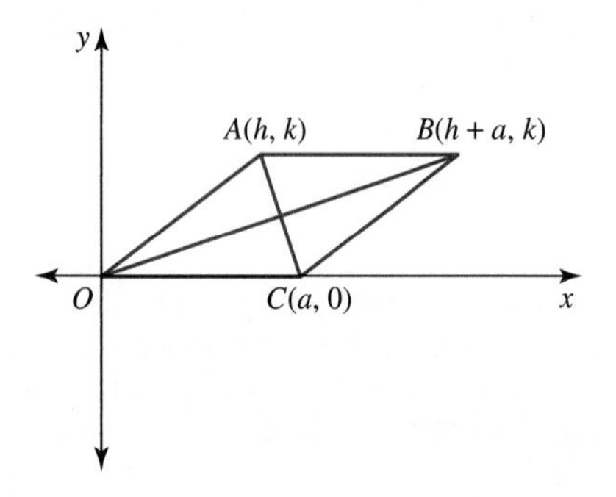

28. Use any tools to construct a regular octagon (8-gon) inscribed in a circle and answer the following questions:

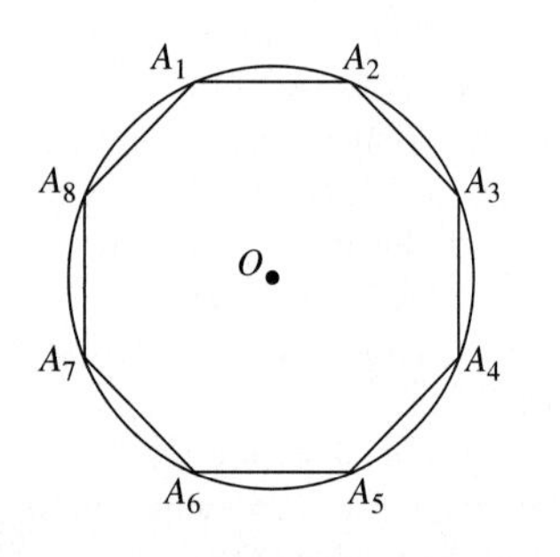

   a. $\triangle A_5OA_6$ is the image of $\triangle A_1OA_2$ under a rotation about O clockwise by an angle whose measure is α. Find α, if $0 < \alpha \leq 180°$.
   b. Find the smallest angle by which the octagon needs to be rotated so that its image is itself.
   c. What is the image of each vertex under a half-turn in O? Why?
   d. What is the image of the octagon under a half-turn in O? Why?

★ 29. In problem 17, you have explored finding the image of a point (x, y) under a half-turn about the origin. To find the image of a point $P(x, y)$ under a half-turn about an arbitrary point $M(h, k)$, Sophie has the following idea: "I move M to the origin $O(0, 0)$ under the translation that takes M to O. Under this translation, P moves to P'. Then I find P'', the image of P' under the half-turn about O, and finally I translate that image by a vector that takes O to M, to obtain P''."

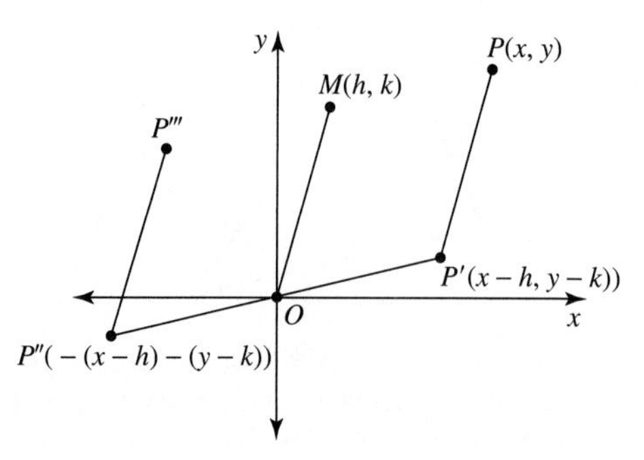

   a. Why do you think Sophie's approach is valid?
   b. If $M(1, 2)$ and $P(4, 5)$, use Sophie's approach to find the image of P under the half-turn in M.
   c. If $M(h, k)$ and $P(x, y)$, find the image of P under the half-turn in M. (Your answer should be in terms of x, y, h, and k).

30. a. A translation from A to B is followed by a translation from B to C, as shown. Show that the same result can be accomplished by a single translation. What is that translation?
   b. Find a single translation equivalent to a translation from A to B followed by a translation from D to E.
   c. Suppose the order of the translations in part (b) is reversed; that is, a translation from D to E is followed by a translation from A to B. Is the result different from the one in part (b)?

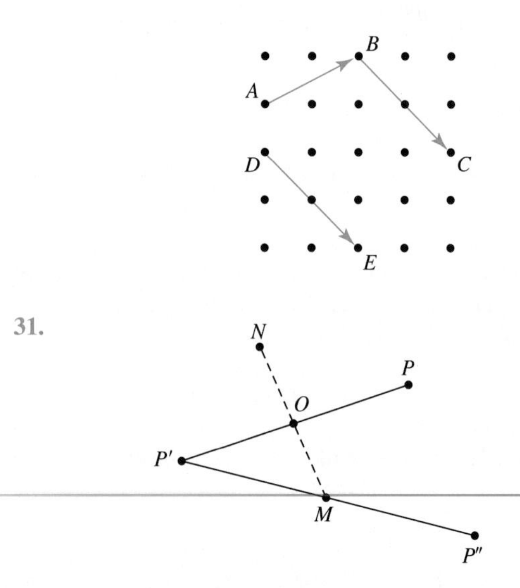

31.
   a. Draw two points O and M and a point P. Construct P', the image of P under a half-turn about O and then P''; the image of P' under a half-turn in M.

b. Find the image of P under a single translation that takes N to M, where O is the midpoint of $\overline{MN}$. How does the image of P under this translation compare with P''?

c. Repeat parts (a) and (b) for a new point Q (first find Q', the image of Q under a half-turn about O and then Q'', the image of Q' under half-turn in M).

d. Conjecture what single transformation will have the same effect on any point in the plane as a half-turn in O followed by a half-turn in M.

★ **e.** Justify your conjecture in part (d).

Communication

32. If we are given two congruent nonparallel segments, is it always possible to find a rotation so that the image of one segment will be the other segment? Explain why or why not.

33. a. If you rotate an object 180° clockwise or counterclockwise, using the same center, is the image the same in both cases? Explain.

b. Answer part (a) if you rotate the object 360°.

34. For each of the following figures, trace the figure on tracing paper, rotate the tracing by 180° about the given point O, sketch the image, and then make a conjecture about the kind of figure that is formed by the union of the original figure and its image. In each case, explain why you think your conjecture is true.

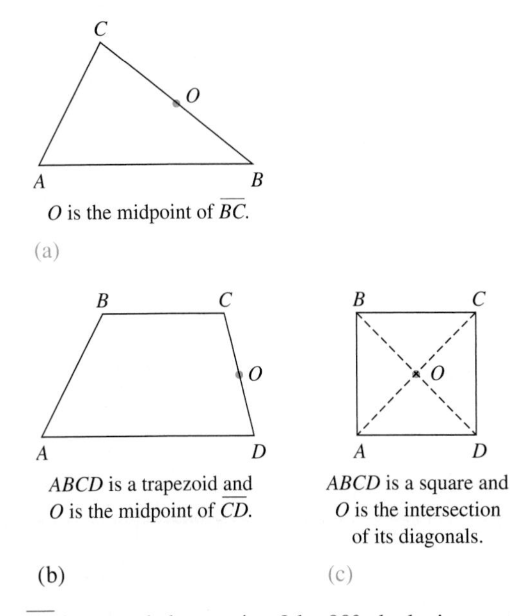

O is the midpoint of $\overline{BC}$.

(a)

$ABCD$ is a trapezoid and O is the midpoint of $\overline{CD}$.

(b)

$ABCD$ is a square and O is the intersection of its diagonals.

(c)

35. If $\overline{AB}$ is rotated about point O by 90° clockwise, explain whether its image is perpendicular to $\overline{AB}$, regardless of the location of point O.

Open-Ended

36. A drawing of a cube, shown in the following figure, can be created by drawing a square $ABCD$, finding its image under translation defined by the slide arrow from A to A' so that $AA' = AB$, and connecting the points A, B, C, and D with their corresponding images.

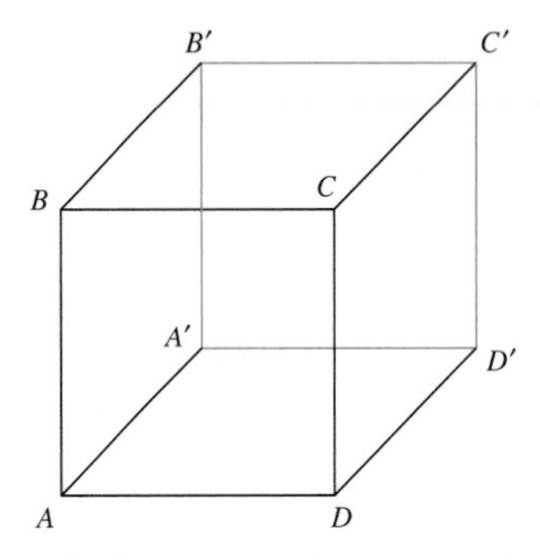

Draw several other perspective geometric figures using translations. In each case, name the figures and indicate the slide arrow that defines the translation.

37. Wall stenciling has been used to obtain an effect similar to that of wallpapering. The stencil pattern in the following figure can be used to create a border on a wall:

Measure the length of a wall of a room and design your own stencil pattern to create a border. Cut the pattern from a sheet of plastic or cardboard. Define the translation that will accomplish creating an appropriate border for the wall.

38. The following pattern can be created by rotating figure A about O by the indicated angle, then rotating the image B about O by the same angle, and then rotating the image C about O by the same angle, and so on until one of the images coincides with the original figure A:

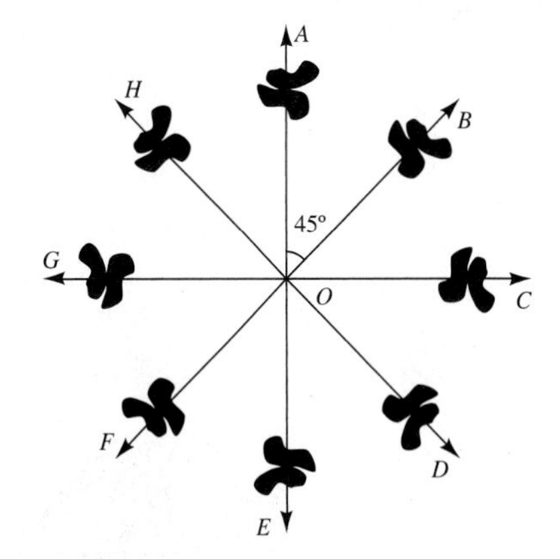

Make several designs with different numbers of congruent figures around a circle in which the image of each figure under the same rotation is the next figure and so that one of the images coincides with the original figure.

Cooperative Learning

39. Mark a point A on a sheet of paper and set a straightedge through A as shown in the following figure. Find a circular shape (a jar lid is a good choice) and mark point P on the edge of the shape. Place the shape on the straightedge so that P coincides with A. Consider the path traced by P as the circle rolls so that its edge stays in contact with the straightedge all the time and until P comes in contact with the straightedge again at point B.

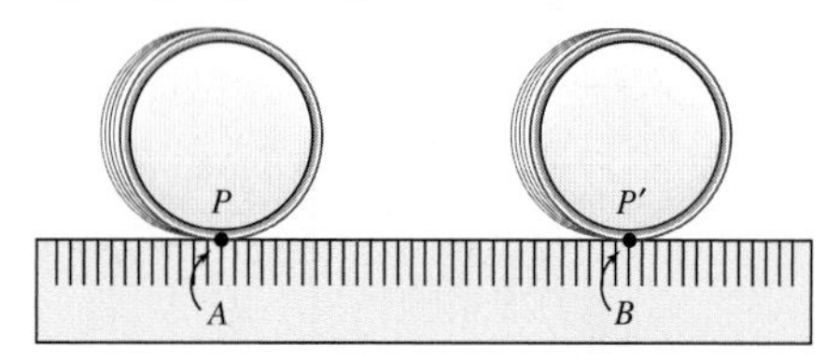

a. Have one member of your group roll the circular shape and another draw the path traced by P as accurately as possible.
b. Discuss how to check if the path traced by point P is an arc of a circle.
c. Find the length of $\overline{AB}$.

Third International Mathematics and Science Study (TIMSS) Questions

This figure will be turned to a different position.

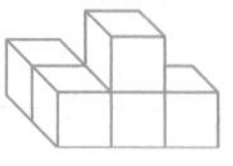

Which of these could be the figure after it is turned?

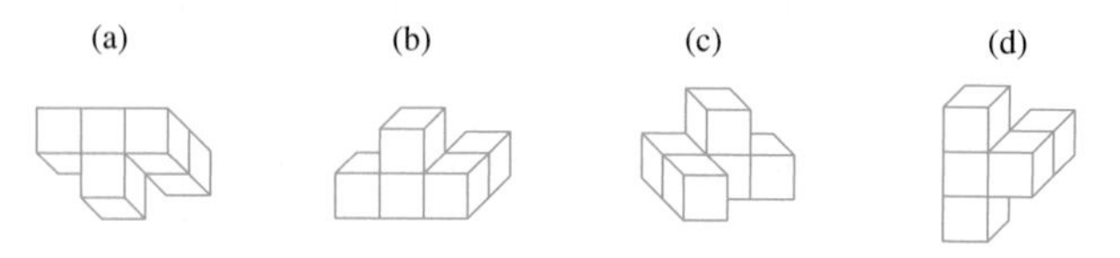

TIMSS, Grade 4, 2003

Rectangle $PQRS$ can be rotated (turned) onto rectangle $UVST$.

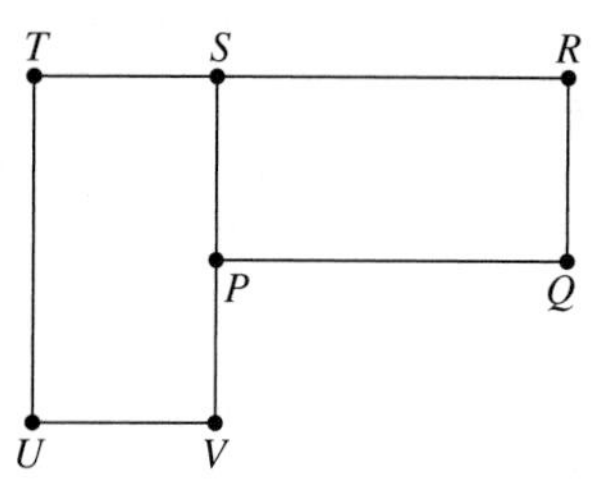

What point is the center of rotation?

a. P
b. R
c. S
d. T
e. V

TIMSS, Grade 8, 2003

TECHNOLOGY CORNER

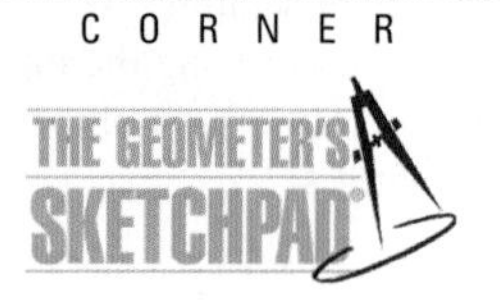

Use GSP to draw an equilateral triangle and two altitudes of the triangle. Let O be the point at which the altitudes intersect. Rotate the triangle by 120° about O in any direction. Make a conjecture based on this experiment. Do you think your conjecture may be true for some triangles that are not equilateral? Why?

BRAIN TEASER

In Figure 12-16, a coin is shown above and touching another coin. Suppose the top coin is rotated around the circumference of the bottom coin until it rests directly below the bottom coin. Will the head be straight up or upside down?

FIGURE 12-16

12-2 Reflections and Glide Reflections

Reflections

reflection ◆ flip Another isometry is a **reflection**, or **flip**. One example of a reflection often encountered in our daily lives is a mirror image. Figure 12-17 shows a figure with its mirror image.

FIGURE 12-17

Another reflection is shown in the following "B.C." cartoon.

reflecting line
mirror image

We can obtain reflections in a line in various ways. Consider the half tree shown in Figure 12-18(a). Folding the paper along the **reflecting line** and drawing the image gives the **mirror image**, or *image*, of the half tree. In Figure 12-18(b), the paper is shown unfolded. Another way to simulate a reflection in a line involves using a Mira, as illustrated in Figure 12-18(c).

FIGURE 12-18

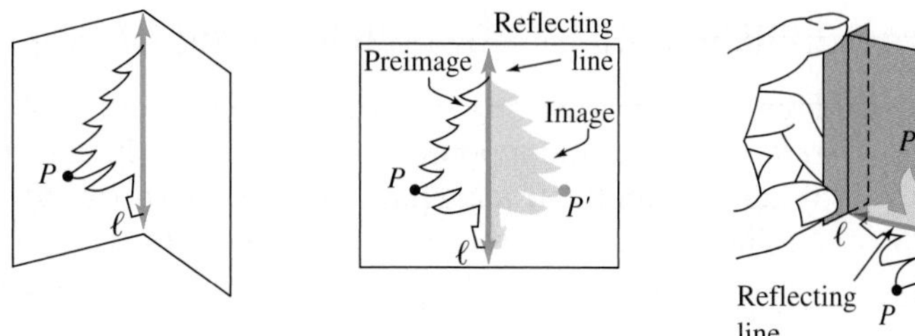

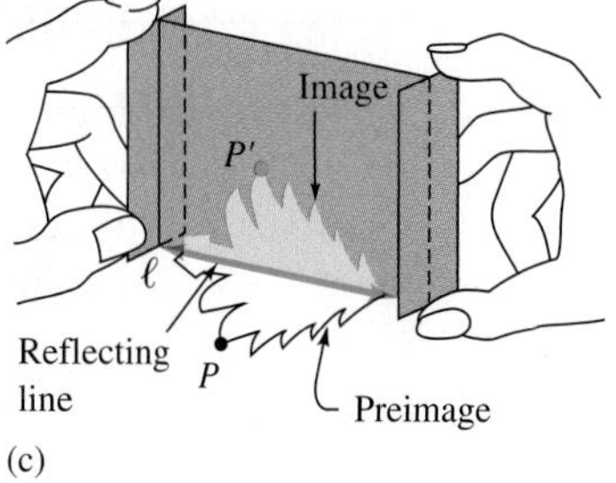

In Figure 12-19(a), the image of P under a reflection in line ℓ is P'. $\overline{PP'}$ is both perpendicular to and bisected by ℓ, or equivalently, ℓ is the perpendicular bisector of $\overline{PP'}$. In Figure 12-19(b), P is its own image under the reflection in line ℓ. If ℓ were a mirror, then P' would be the mirror image of P. This leads us to the following definition of a reflection.

FIGURE 12-19

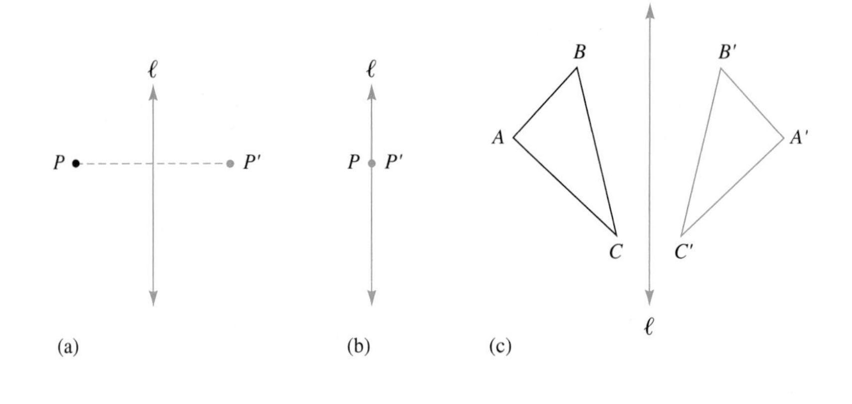

Definition of Reflection

A **reflection** in a line ℓ is a transformation of a plane that pairs each point P of the plane with a point P' in such a way that ℓ is the perpendicular bisector of $\overline{PP'}$, as long as P is not on ℓ. If P is on ℓ, then $P = P'$.

In Figure 12-19(c), we see another property of a reflection. In the original triangle ABC, if we walk clockwise around the vertices, starting at vertex A, we see the vertices in the order A-B-C. However, in the reflection image of triangle ABC, if we start at A' (the image of A) and walk clockwise, we see the vertices in the following order: A'-C'-B'. Thus a reflection does something that neither a translation nor a rotation does; it reverses the **orientation** of the original figure.

orientation

There are many methods of constructing a reflection image. We already illustrated such constructions with paper folding and a Mira. Next, we illustrate the construction of the image of a figure under a reflection in a line with tracing paper.

Constructing a Reflection by Using Tracing Paper

Figure 12-20(a) shows the use of tracing paper. We trace the original figure, the *reflecting line*, and a point on the reflecting line, which we use as a *reference point*. When we flip the tracing paper over to perform the reflection, we align the reflecting line and the reference point, as in Figure 12-20(b). Aligning the reference point ensures that no translating occurs along the reflecting line when the reflection is performed. If we wish the image to be on the paper with the original, we may indent the tracing paper or acetate sheet to mark the images of the original vertices.

FIGURE 12-20

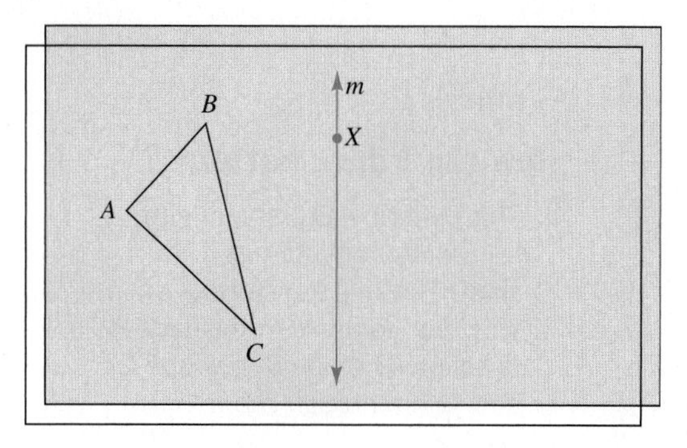

(a)

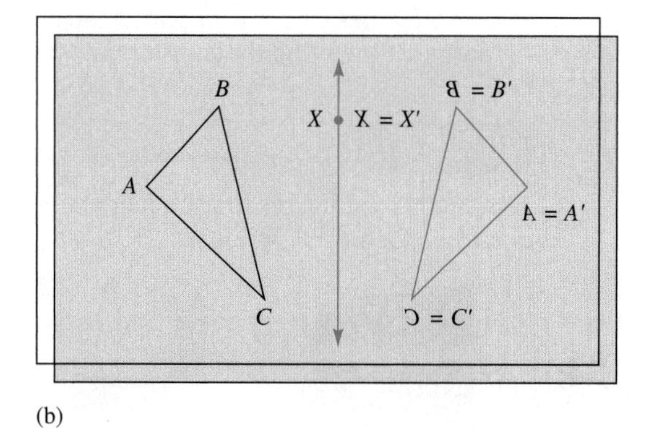

(b)

NOW TRY THIS 12-3

Use the definition of a reflection in a line and properties of a rhombus to construct the image P' of point P in Figure 12-21 under reflection in line m using only a compass and straightedge.

FIGURE 12-21

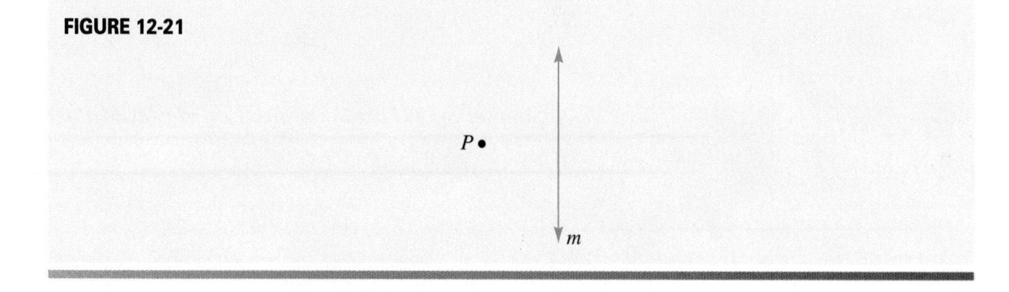

Reflections, rotations, and translations appear in the elementary curriculum, as shown in the following partial student page. Work through the activity and questions 1 through 5 on the page.

SAMPLE SCHOOL BOOK PAGE:

Flip, Turn, and Slide

Activity

How can I draw patterns?

a. This pattern was created with transformations starting with shape A. What transformation was used to move from shape A to shape B? shape B to shape C? and so on through shape I?

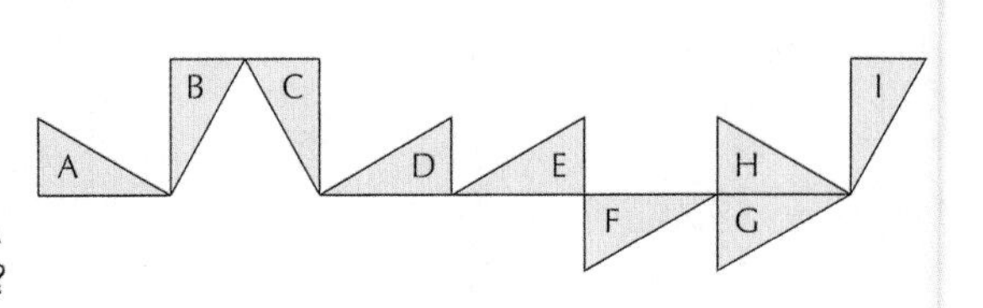

b. Work with a partner. Choose a polygon. Trace around it. Apply a transformation and trace around it again. Continue at least 8 more times. Show your partner your pattern and ask your partner to describe the transformations you used.

CHECK *For another example, see Set 6-10 on p. 387.*

Tell whether the figures in each pair are related by a slide, a flip, or a turn. If a turn, describe it.

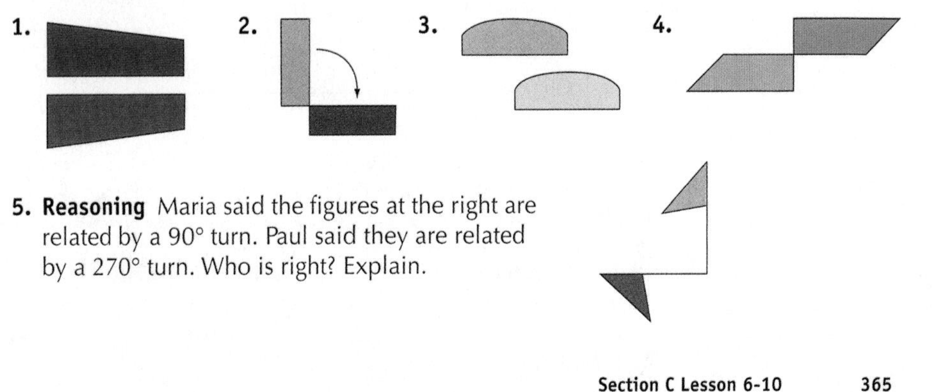

5. Reasoning Maria said the figures at the right are related by a 90° turn. Paul said they are related by a 270° turn. Who is right? Explain.

Section C Lesson 6-10 365

Source: Scott Foresman–Addison Wesley Mathematics, Grade 5, 2005 (p. 365)

EXAMPLE 12-5

Describe how to construct the image of $\overleftrightarrow{AB}$ under a reflection in line m in Figure 12-22.

FIGURE 12-22

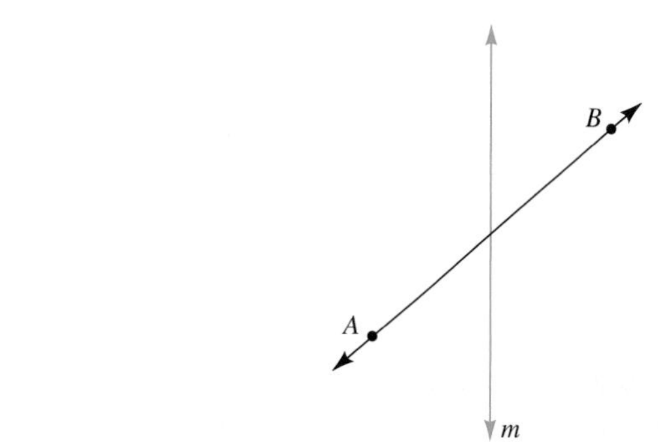

Solution Under a reflection, the image of a line is a line. Thus, to find the image of $\overleftrightarrow{AB}$, it is sufficient to choose any two points on the line and find their images. The images determine the line that is the image of $\overleftrightarrow{AB}$. We choose two points whose images are easy to find. Point X, the intersection of $\overleftrightarrow{AB}$ and m, is its own image. If we choose point A and use a compass and straightedge, we produce the construction shown in Figure 12-23.

FIGURE 12-23

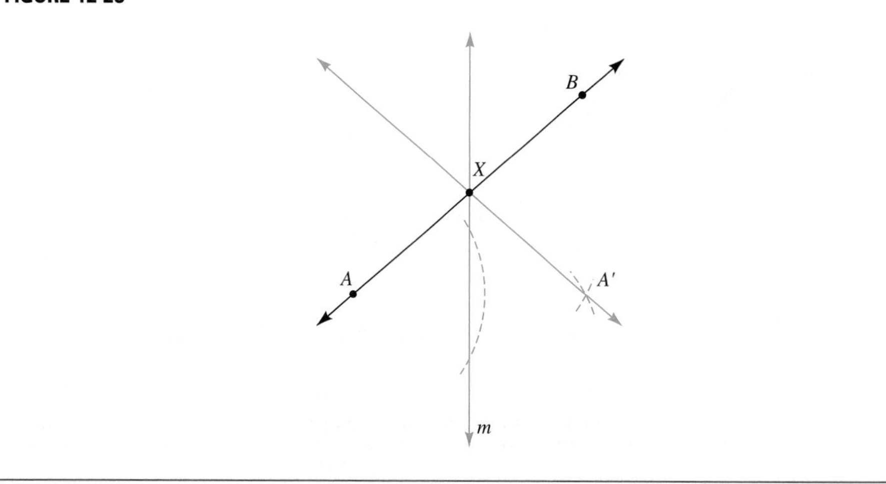

Constructing a Reflection on Dot Paper or a Geoboard

On dot paper or a geoboard, the images of figures under a reflection can sometimes be found by inspection, as seen in Example 12–6.

EXAMPLE 12-6

Find the image of $\triangle ABC$ under a reflection in line m, as in Figure 12-24.

FIGURE 12-24

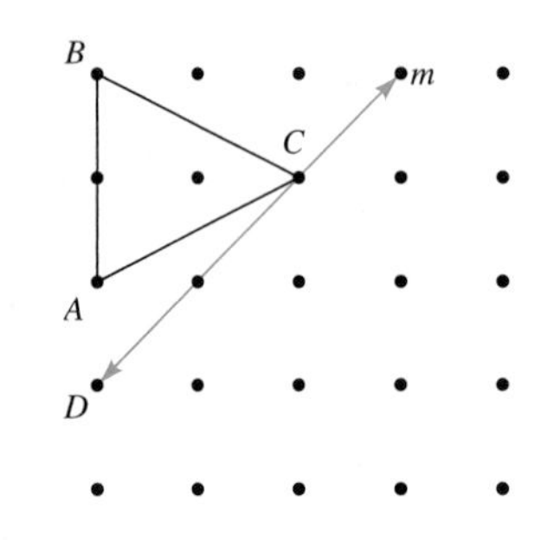

Solution The image $A'B'C'$ is given in Figure 12-25. Note that C is the image of itself and the images of the vertices A and B are A' and B' such that m is the perpendicular bisector of $\overline{AA'}$ and $\overline{BB'}$.

FIGURE 12-25

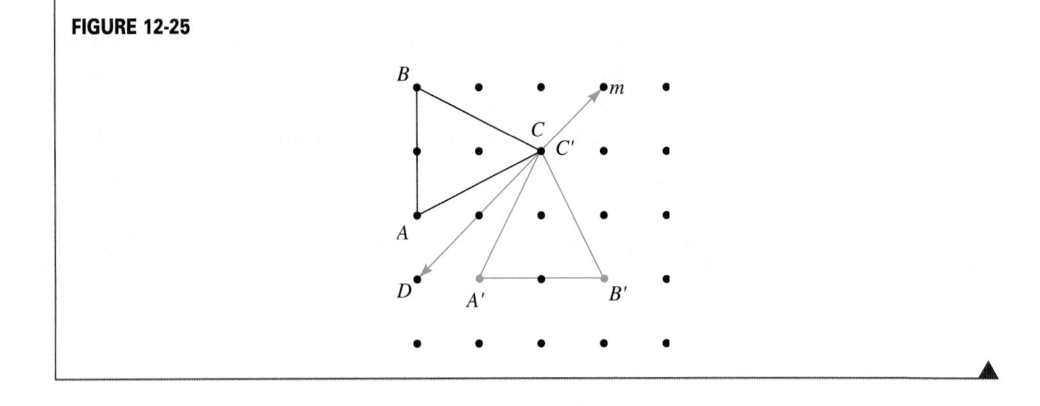

It is possible to find the reflecting line, if we are given an original figure and its reflection image. This is left as an exercise.

NOW TRY THIS 12-4

Trace $\triangle ABC$ and $\triangle A'B'C'$ from Figure 12-25 on a sheet of paper (without tracing line m). Find the line of reflection using paper folding.

Reflections in a Coordinate System

For some reflecting lines like the x-axis and y-axis and the line $y = x$, it is quite easy to find the coordinates of the image, given the coordinates of the point. In Figure 12-26, the line $y = x$ bisects the angle between the x-axis and y-axis. The image of $A(1, 4)$ is the point $A'(4, 1)$. Also the image of $B(^{-}3, 0)$ is $B'(0, ^{-}3)$.

FIGURE 12-26

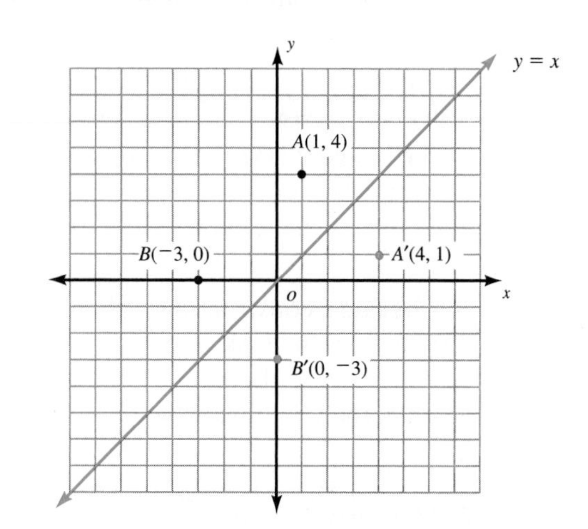

NOW TRY THIS 12-5

Show that in general the image of $P(a, b)$ under reflection in line $y = x$ is the point $P'(b, a)$ and consequently that the reflection in the line $y = x$ exchanges the coordinates of the point. Is this still true if the point is not above line $y = x$?

TECHNOLOGY CORNER

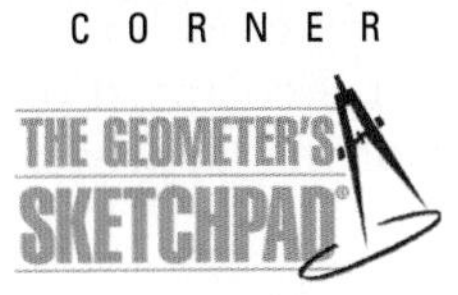

GSP Lab 10 activities 1–5 can be used to investigate properties of reflections.

Glide Reflections

glide reflection

Another basic isometry is a **glide reflection.** An example of a glide reflection is shown in the footprints of Figure 12-27. We consider the footprint labeled F_1 to have been translated to footprint F_2 and then reflected over line m (parallel to $\overline{F_1F_2}$) to yield F_3, the image of F_2. F_3 is the final image of F_1.

FIGURE 12-27

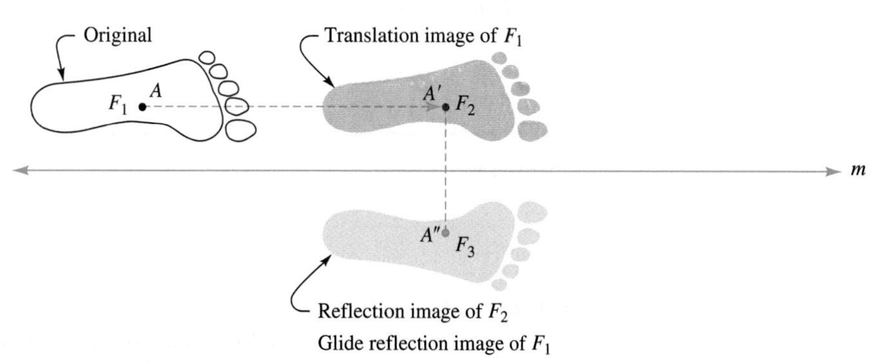

Note that point A is translated to find point A' and then point A' is reflected over line m (parallel to $\overline{F_1F_2}$) to obtain point A''. Thus, A'' is the image of A in the glide reflection.

The illustration in Figure 12-27 leads us to the following definition.

Definition of Glide Reflection

A **glide reflection** is a transformation consisting of a translation followed by a reflection in a line parallel to the slide arrow.

Because constructing a glide reflection involves constructing a translation and a reflection, the task of constructing a glide reflection is not a new problem. Exercises involving the construction of images of figures under glide reflections are given in the Ongoing Assessment 12-2.

RESEARCH NOTE Children first discover that figures can be superimposed by motion when one glides the transparent sheet on the plane without lifting it, or by reversals when it is necessary to lift the transparent sheet, flip it in space to change sides, and then glide it (Michel Demal 2004).

Congruence via Isometries

We have seen that under an isometry, the image of a figure is a congruent figure. Also, given two congruent polygons, it is possible to show that one can be transformed to the other by isometries. In fact, we can define two figures as *congruent* if one is an image of the other under an isometry or under a composition of isometries. The following example shows one illustration of this approach to congruence.

EXAMPLE 12-7 *ABCD* in Figure 12-28 is a rectangle. Describe a sequence of isometries to show

a. $\triangle ADC \cong \triangle CBA$
b. $\triangle ADC \cong \triangle BCD$
c. $\triangle ADC \cong \triangle DAB$

FIGURE 12-28

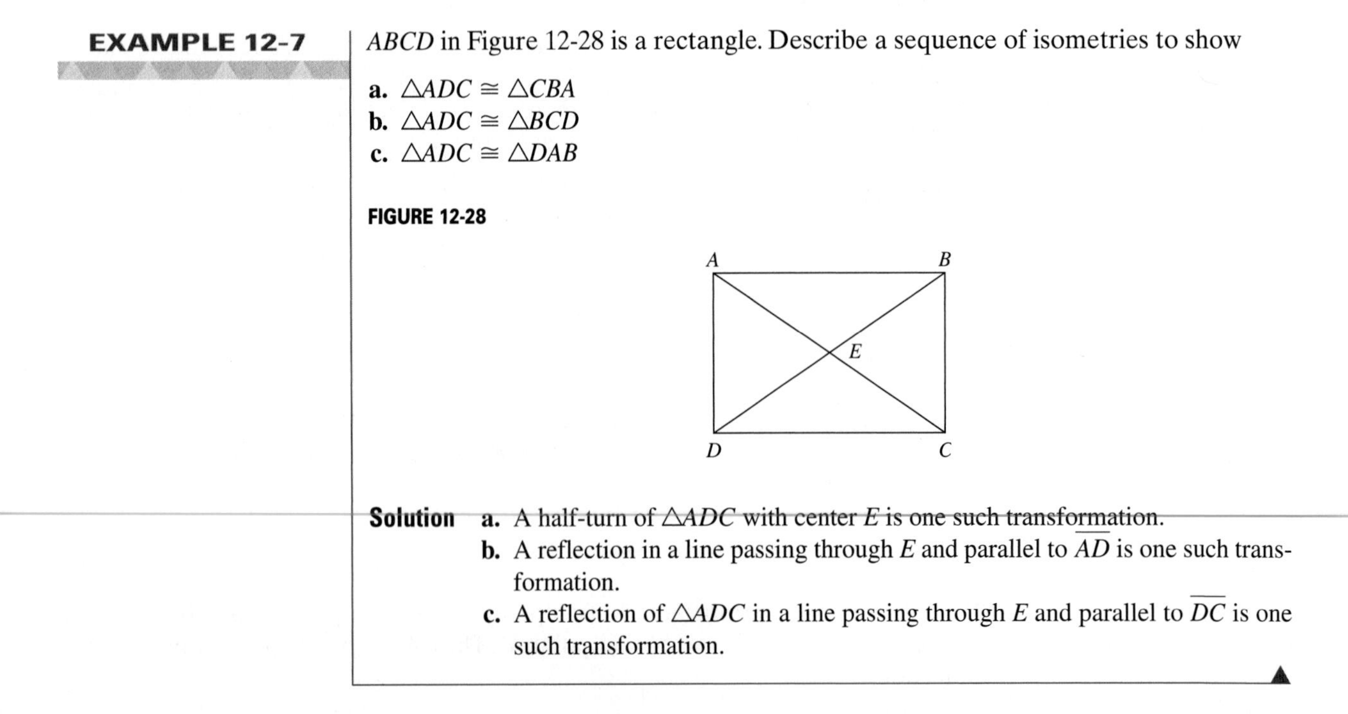

Solution **a.** A half-turn of $\triangle ADC$ with center E is one such transformation.
b. A reflection in a line passing through E and parallel to $\overline{AD}$ is one such transformation.
c. A reflection of $\triangle ADC$ in a line passing through E and parallel to $\overline{DC}$ is one such transformation.

Light Reflecting from a Surface

angle of incidence
angle of reflection

When a ray of light bounces off a mirror or when a billiard ball bounces off the rail of a billiards table, the **angle of incidence**, the angle formed by the incoming ray in Figure 12-29 and a line perpendicular to the mirror, is congruent to the **angle of reflection**, the angle between the reflected ray and the line perpendicular to the mirror.

FIGURE 12-29

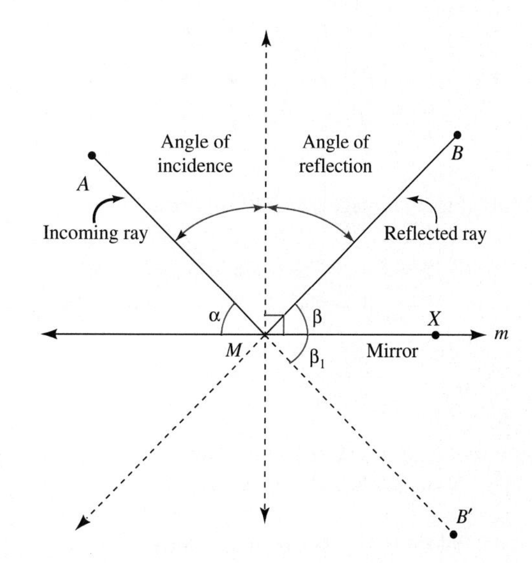

Because the angle of incidence is congruent to the angle of reflection, the respective complements of these angles must be congruent. If the measures of these complements are α and β, as indicated in Figure 12-29, then $\alpha = \beta$. Figure 12-29 shows B', the image of B under reflection in m, and hence $\angle XMB'$ is the image of $\angle XMB$. Notice that β, the measure of $\angle XMB$ must equal β_1, the measure of $\angle XMB'$, because reflection preserves angle measurement. Because $\alpha = \beta$ and $\beta = \beta_1$, we get $\alpha = \beta_1$. For that reason, points A, M, B' are collinear (why?). We can show that these facts imply *Fermat's Principal: Light follows the path of shortest distance*; that is, the path A-M-B that light travels is the shortest among all the paths connecting A with a point in the mirror to B. The proof is explored in Now Try This 12-6.

NOW TRY THIS 12-6

Trace Figure 12-29 and mark on the mirror m a point P other than M. Prove that $AM + MB < AP + PB$ by showing that $MB = MB'$ and $BP = B'P$, and then using the triangular inequality applied to $\triangle APB'$ and the fact that A, M, B' are collinear.

ONGOING ASSESSMENT 12-2

1. For each of the following figures, find the image of the given quadrilateral under a reflection in ℓ:

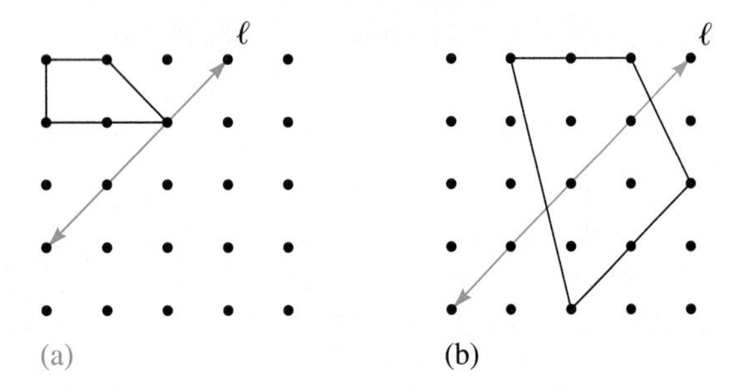

2. In the following figure, find the image of $\triangle ABC$ under a reflection in line ℓ:

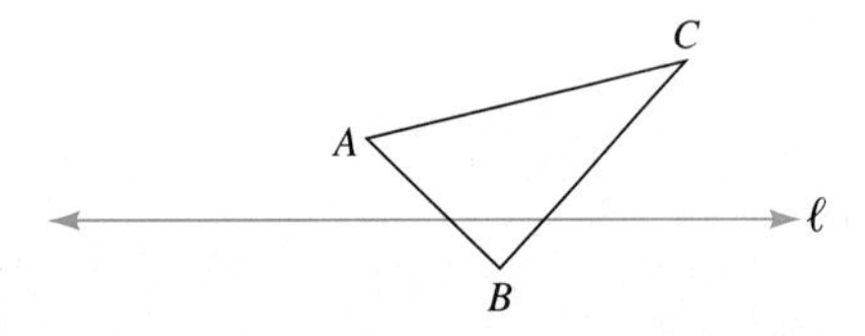

3. Draw a line and then draw a circle whose center is not on the line. Find the image of the circle under a reflection in the line.
4. Determine which of the following figures have a reflecting line such that the image of the figure under the reflecting line is the figure itself. In each case, find as many such reflecting lines as possible, sketching appropriate drawings.
   a. Circle
   b. Segment
   c. Ray
   d. Line
   e. Square
   f. Rectangle
   g. Scalene triangle
   h. Isosceles triangle
   i. Equilateral triangle
   j. Trapezoid whose base angles are not congruent
   k. Isosceles trapezoid
   l. Arc
   m. Kite
   n. Rhombus
   o. Regular hexagon
   p. Regular n-gon
5. Determine the final result when $\triangle ABC$ is reflected in line ℓ and then its image is reflected again in ℓ.

6. a. Refer to the following figure and suppose lines ℓ and m are parallel and $\triangle ABC$ is reflected in ℓ to obtain $\triangle A'B'C'$ and then $\triangle A'B'C'$ is reflected in m to obtain $\triangle A''B''C''$. Determine whether the same final image is obtained if $\triangle ABC$ is reflected first in m and then its image is reflected in ℓ.

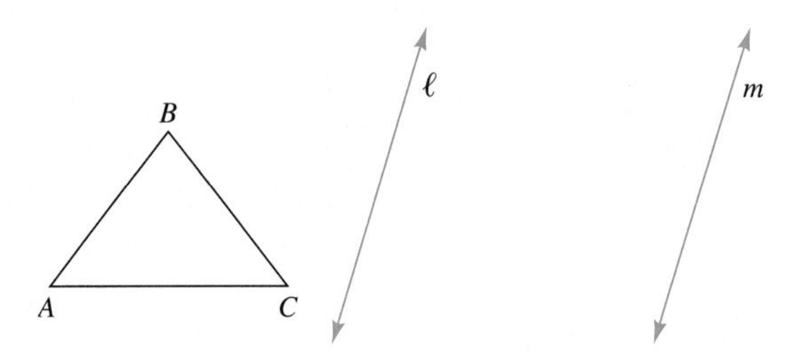

   b. Conjecture what single transformation will take $\triangle ABC$ directly to $\triangle A''B''C''$. Check your conjecture using tracing paper.
7. a. For the following figure, use any construction method to find the image of $\triangle ABC$ if $\triangle ABC$ is reflected in ℓ to obtain $\triangle A'B'C'$ and then $\triangle A'B'C'$ is reflected in m to obtain $\triangle A''\ B''\ C''$ (ℓ and m intersect at O):

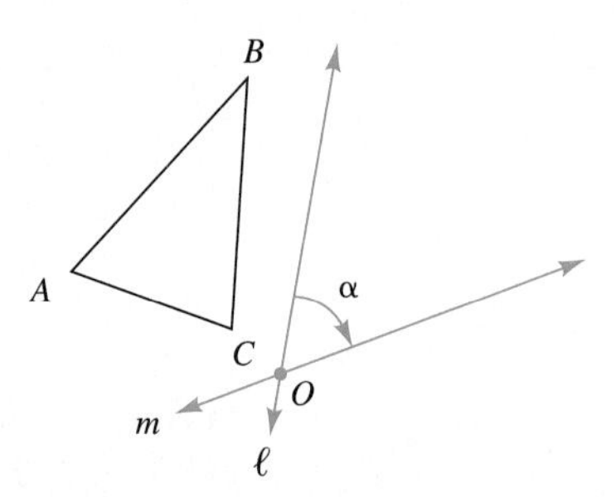

   b. Conjecture what single transformation will take $\triangle ABC$ directly to $\triangle A''B''C''$. Check your conjecture using tracing paper.
   c. Answer the questions in (a) and in (b) for the case in which ℓ and m are perpendicular.
8. Use a Mira, if available, to investigate problems 6 and 7.
9. Given $\triangle ABC$ and its reflection image $\triangle A'B'C'$, find the line of reflection.

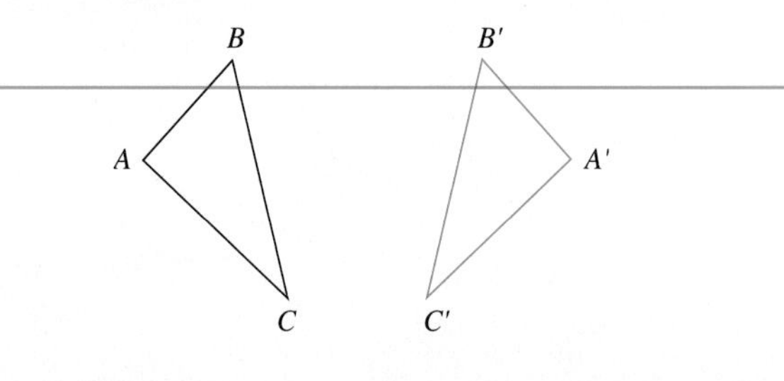

10. a. The word TOT is its own image when it is reflected through a vertical line through O, as shown in the following figure. List some other words that are their own images when reflected similarly.

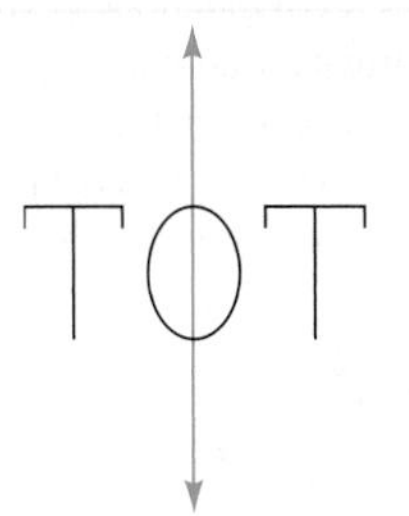

b. The image of BOOK is still BOOK when it is reflected through a horizontal line. List some other words that have the same property. Which uppercase letters can you use?

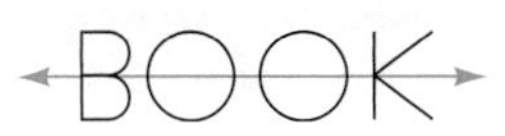

c. With an appropriate font, the image of 1881 is 1881 after reflection in either a horizontal or vertical line, as shown in the following figure. What other natural numbers less than 2000 have this property?

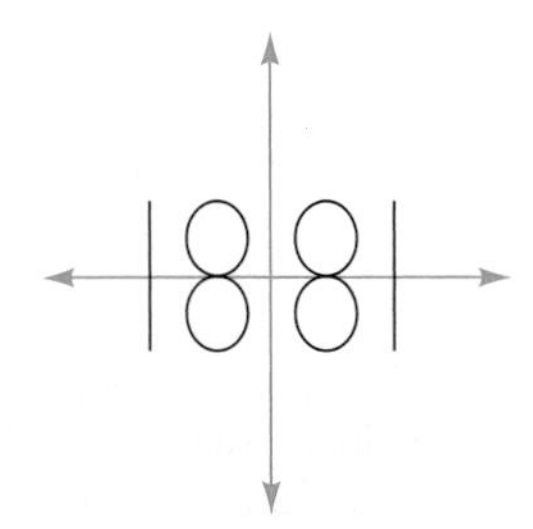

11. A glide reflection is determined by a translation parallel to a given line followed by a reflection in that line.
   a. Determine whether the same final image is obtained if the reflection is followed by the translation.
   b. Use your answer in (a) to determine whether the reflection and translation involved in the glide reflection are commutative.
12. For the following figure numbered 1, decide whether a reflection, a translation, a rotation, or a glide reflection will transform the figure into each of the other numbered figures (there may be more than one answer).

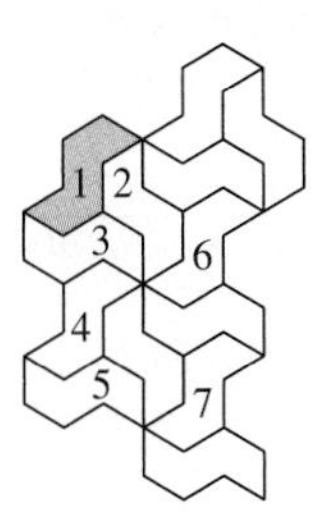

13. Given points $A(3, 4)$, $B(2, {}^{-}6)$, and $C({}^{-}2, 5)$, find the coordinates of the images of these points under each of the following transformations:
   a. Reflection in the x-axis
   b. Reflection in the y-axis
   c. Reflection in the line $y = x$
   d. Reflection in the line $y = {}^{-}x$
14. a. Conjecture what the image of a point with coordinates (x, y) will be under each of the transformations in problem 13.
   b. Suppose a point P with coordinates (x, y) is reflected in the x-axis and then its image P' is reflected in the y-axis to obtain P''. What are the coordinates of P'' in terms of x and y? Justify your answer.
   c. What single isometry will take point P in part (b) to P''?
15. Find the equations of the images of the following lines when reflected in the x-axis:
   a. $y = 3x$
   b. $y = {}^{-}x$
   c. $y = {}^{-}x + 3$
   d. $x = 0$
   e. $y = 0$
16. Find the equation of the images of the lines in problem 15 when the reflection line is the y-axis.
17. A quadrilateral has vertices at $(a, 0)$, $(0, a)$, $({}^{-}a, 0)$, and $(0, {}^{-}a)$, where $a > 0$.
   a. What is the most you can say about the quadrilateral?
   b. Find the image of each vertex under the reflection in the line $y = x$.
18. In which line will the two intersecting circles reflect onto themselves; that is, the image of the circles will be the same two circles?

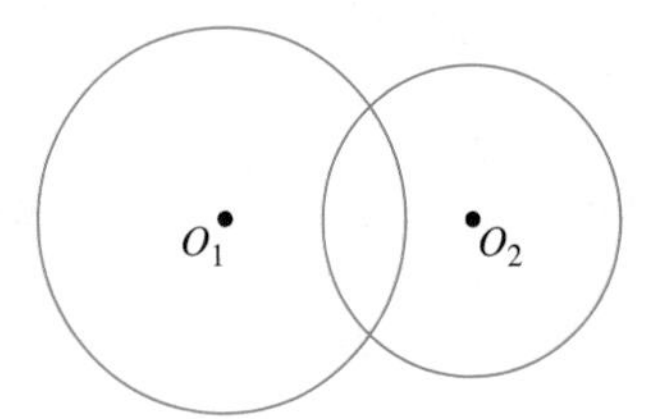

19. The two circles "touch" each other at point P; that is, the tangent to one circle at P is also the tangent to the other circle. Find the line of reflection such that the image of the smaller circle will be in the interior of the larger circle, but still "touching" the larger circle.

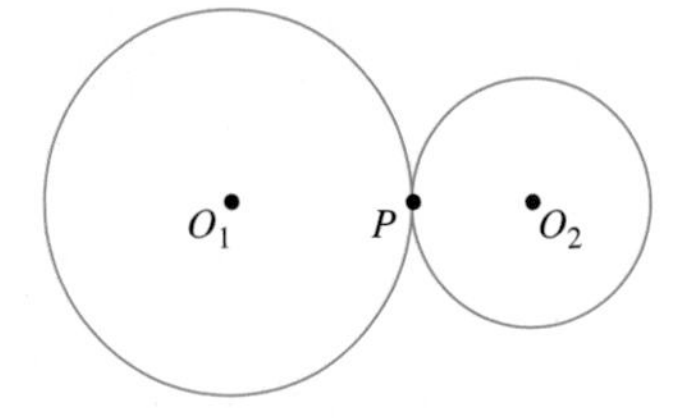

20. Two congruent circles with centers O_1 and O_2 intersect at points A and B.

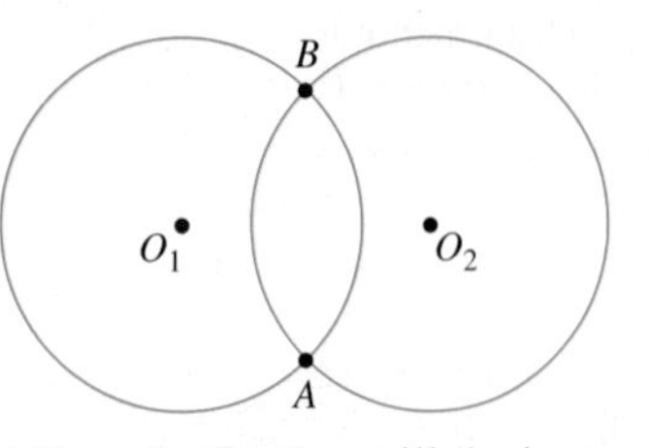

a. In what line of reflection will the image of the circle with center O_1 be the circle with center O_2?
b. Can the circle with center O_1 be transformed into the circle with center O_2 by a translation? If so, describe the translation.

21. Construct a square and a circle of an appropriate size so that it will be possible to transform the circle by a reflection into the square in a way that the image of the circle will be tangent to all sides of the square. Identify the line of reflection.

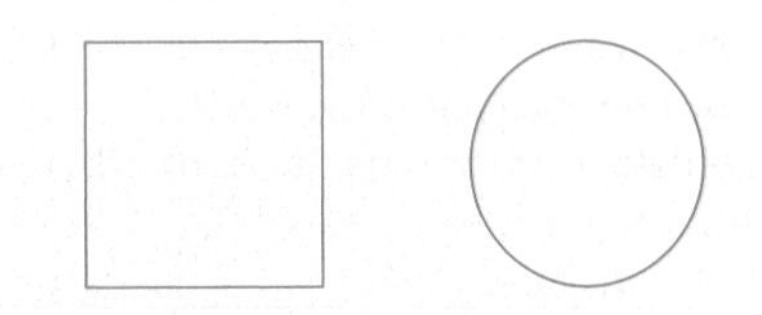

22. Graph each pair of the following lines and for each construct the corresponding line of reflection so that the image of one line in the pair will be the second line. For each, identify the line of reflection.
a. $y = {}^{-}x$ and $y = x$
b. $y = 2x$ and $y = {}^{-}2x$
c. $x = 0$ and $y = 0$
d. $x = {}^{-}2$ and $x = 3$

23. a. Construct any three lines k, l and m that intersect at a single point O. Draw any point P and reflect it in line k, then reflect its image P' in l, and finally reflect P'', the image of P' in line m to obtain P'''.
b. Next construct line n as shown and so that the marked angles are congruent. Reflect point P directly in line n. How does the image compare to P''' found earlier?

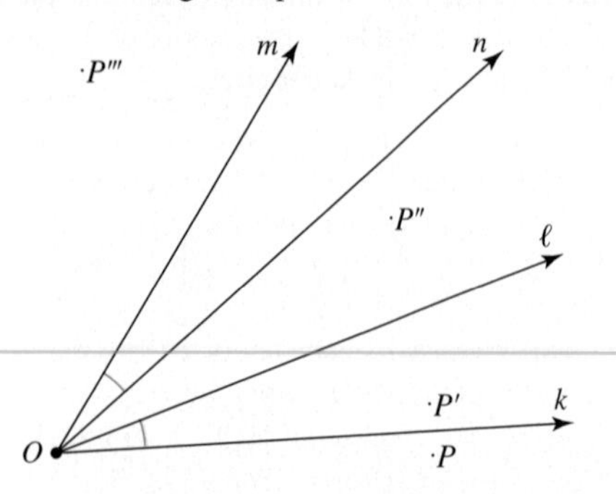

c. Draw any point Q on line n and reflect it successively in lines k, l and then m. What do you find about its final image?
d. Can three successive reflections in three lines intersecting in a single point be equivalent (have the same effect) to a reflection in a single line? Explain your reasoning.

24. Two farm houses are located away from a road, as shown. A telephone company wants to construct a telephone pole at the edge of the road so that the telephone cable connecting the houses to the pole is as short as possible. Where should the pole be located?

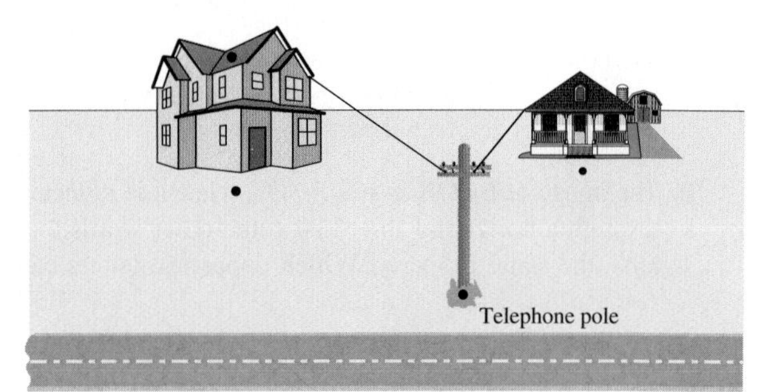

25. A *fixed point* of a transformation is a point whose image is the point itself. List all the fixed points of any of the following isometries:
a. Reflection in line ℓ
b. Rotations by a given angle and direction about point O
c. Translation with slide arrow from A to B
d. Glide reflection determined by a translation with slide arrow $\overrightarrow{AB}$ followed by a reflection in line ℓ parallel to line AB

Communication

26. If you look into a hinged mirror (as depicted) and see a polygon with six sides, what is the minimum number of sides that the original could have? Explain your reasoning.

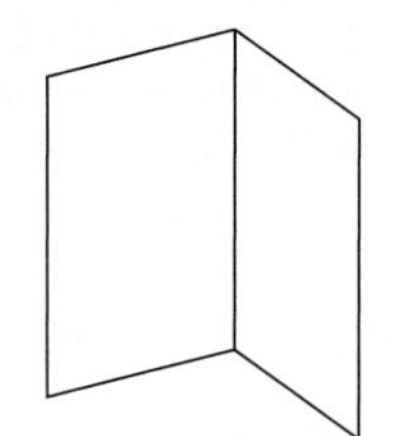

27. When a billiard ball bounces off a side of a pool table, the angle of incidence is usually congruent to the angle of reflection. In the following figure showing a scale drawing of a pool table, a cue ball is at point A. Show how a player should aim to hit two sides of the table and then the ball at B. Justify your solution.

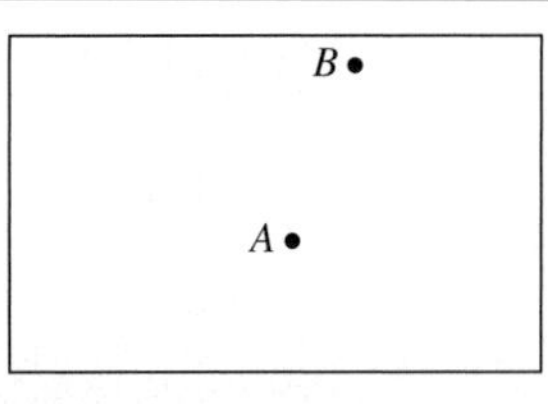

28. a. Draw an isosceles triangle ABC and then construct a line such that the image of $\triangle ABC$ when reflected in the line is $\triangle ABC$ though every point is not necessarily its own image. Explain why the line you constructed has the required property.
   b. For what kind of triangles is it possible to find more than one line with the property in (a)? Justify your answer.
   c. Given a scalene triangle ABC, is it possible to find a line ℓ such that when $\triangle ABC$ is reflected in ℓ, its image is itself? Explain your answer.
   d. Draw a circle with center O and a line with the property that the image of the circle, when reflected in the line, is the original circle. Identify all such lines. Justify your answer.

29. Use the following drawing to explain how a periscope works:

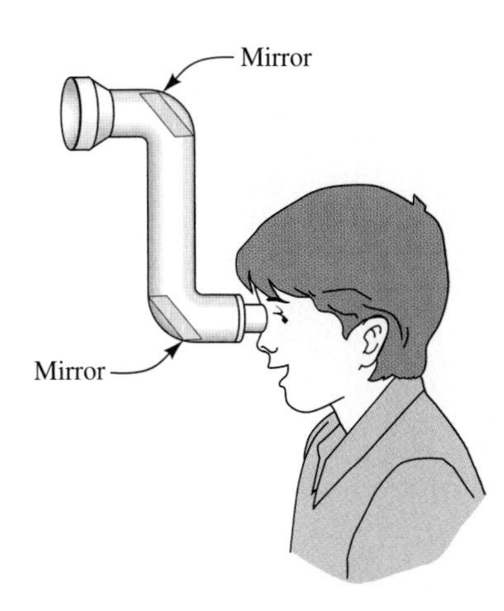

Open-Ended

30. In the following figure representing a miniature golf course, explain and justify the procedure showing how to aim the ball so that it gets in the hole if it is to bounce off
   a. One wall only
   b. Two walls

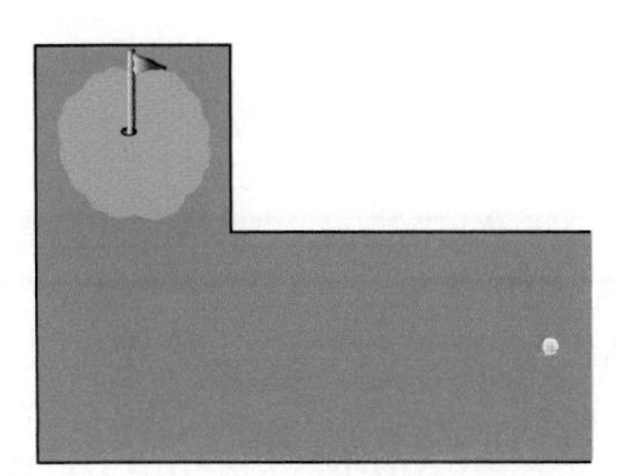

31. Design several wall stencil patterns using a reflection. In each case, explain how you would use the stencil in practice.

32. Design wall stencil patterns using a glide reflection.

33. If a right triangle $\triangle ABC$ is reflected in one of its legs as shown in the following triangle, the triangle and its image form an isosceles triangle $\triangle ABA'$:

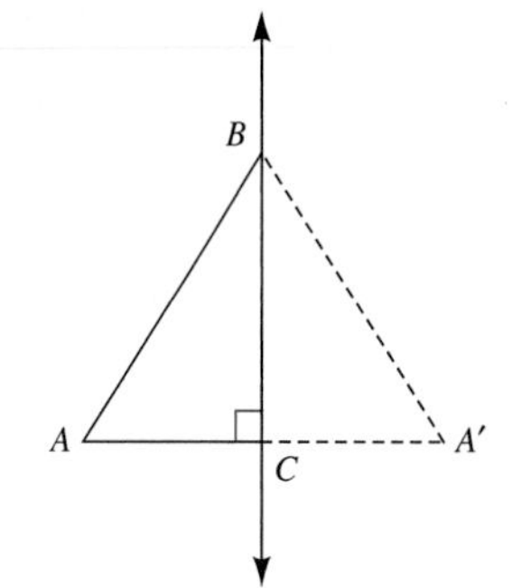

From the properties of reflection, we can deduce that the base angles in an isosceles triangle are congruent and that the altitude to the base bisects $\angle ABA'$ as well as $\overline{AA'}$.

Apply the concept of reflection to deduce properties of other geometric figures by reflecting **(a)** a scalene triangle in one of its sides, **(b)** a right angle trapezoid in one of its sides, and **(c)** other figures. In each case, define the reflection, list the geometric properties of the figure obtained from the union of your original figure and its image, and justify the properties.

Cooperative Learning

34. In the following figure representing a pool table, ball B is sent on a path that makes a 45° angle with the table wall, as shown. It bounces off the wall 5 times and returns to its original position.

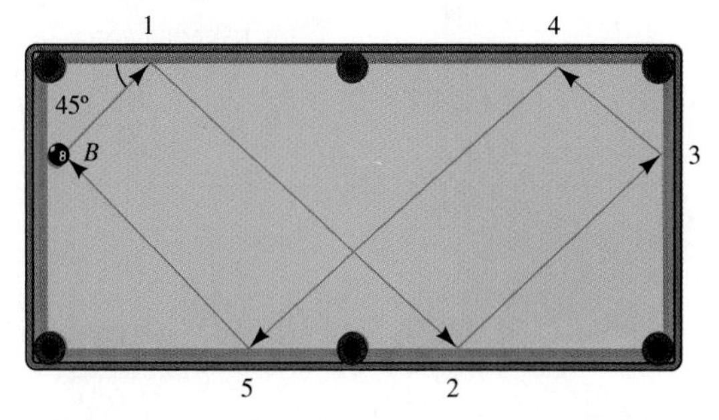

   a. Have each member of your group use graph paper to construct rectangular models of different-sized pool tables. Simulate the experiment using any tools (such as a straightedge, compass, and protractor) by choosing different positions for ball B.
   b. Share the results of your experiments with the rest of the group and together conjecture for which dimensions of the pool table and for what positions of B the experiment described in the problem will work.

Review Exercises

35. Which single digits are their own images under a rotation by an angle whose measure is less than 360°?
36. What is the image of a point (a, b) under a half-turn in the origin?
37. a. Find all possible rotations that transform a circle onto itself.
    b. By what other kinds of transformations can a circle be transformed onto itself?
38. Explain how a translation can be used to construct a rectangle whose area is equal to that of the parallelogram *ABCD* in the following figure:

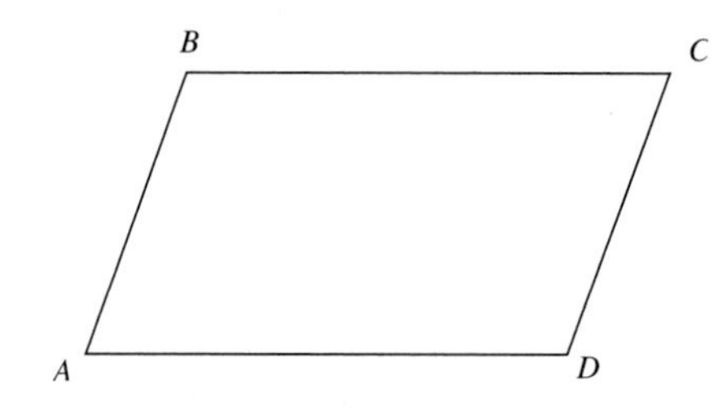

BRAIN TEASER

Two cities are on opposite sides of a river, as shown in Figure 12-30.

FIGURE 12-30

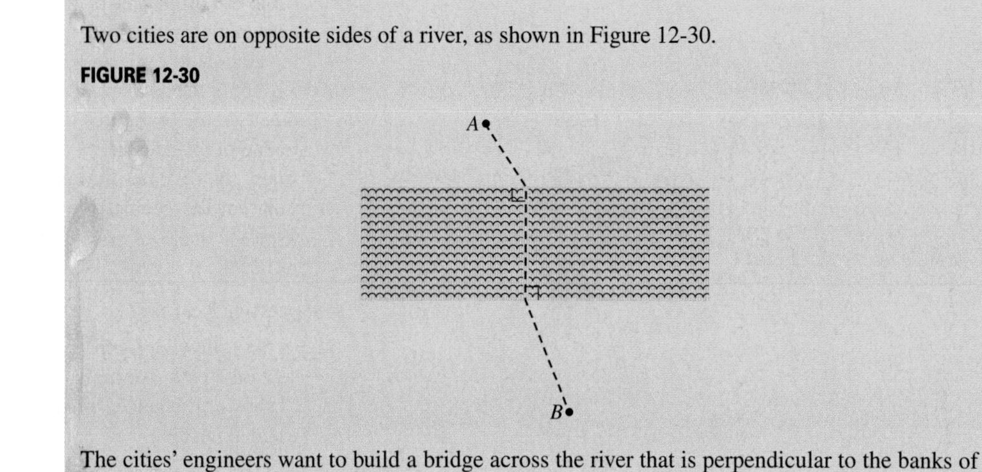

The cities' engineers want to build a bridge across the river that is perpendicular to the banks of the river and access roads to the bridge so that the distance between the cities is as short as possible. Where should the bridge and the roads be built?

12-3 Size Transformations

Isometries preserve distance. Consequently, the image of a figure under an isometry is a figure congruent to the original. A different type of transformation happens when a slide is projected on a screen. All objects on the slide are often enlarged on the screen by the same factor. Figure 12-31 is another example of such a transformation. The point O is the *center* of the *size transformation* and 2 is the *scale factor*. Points O, A, and A' are collinear and $OA' = 2 \cdot OA$; also, O, C, and C' are collinear and $OC' = 2 \cdot OC$. Similarly, O, B, and B' are collinear and $OB' = 2 \cdot OB$. It can be shown that $\triangle A'B'C'$ is similar to $\triangle ABC$ and hence that each side of $\triangle A'B'C'$ is twice as long as the corresponding sides of $\triangle ABC$.

FIGURE 12-31

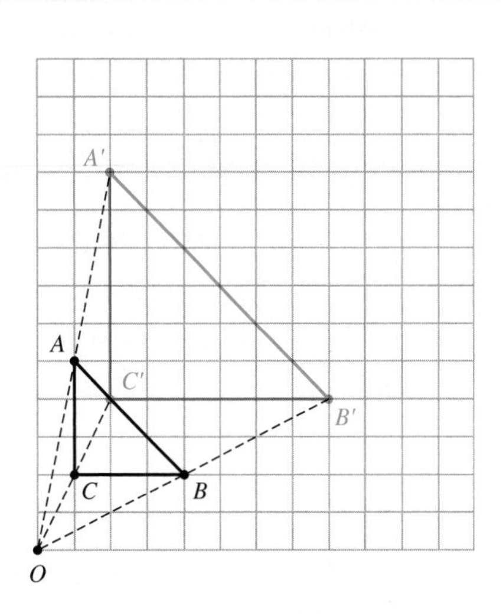

In general, we have the following definition.

Definition of Size Transformation

A size transformation from the plane to the plane with center O and scale factor r ($r > 0$) is a transformation that assigns to each point A in the plane a point A' such that O, A, and A' are collinear and $OA' = r \cdot OA$ and so that O is not between A and A'.

Notice that a size transformation is a function from the plane to the plane. What is the range of this function?

REMARK It is possible to define a size transformation when the scale factor is negative in the same way as in the preceding definition for positive scale factor, except that O *must be between A and A'*.

EXAMPLE 12-8

a. In Figure 12-32(a), find the image of point P under a size transformation with center O and scale factor $\frac{2}{3}$.

FIGURE 12-32

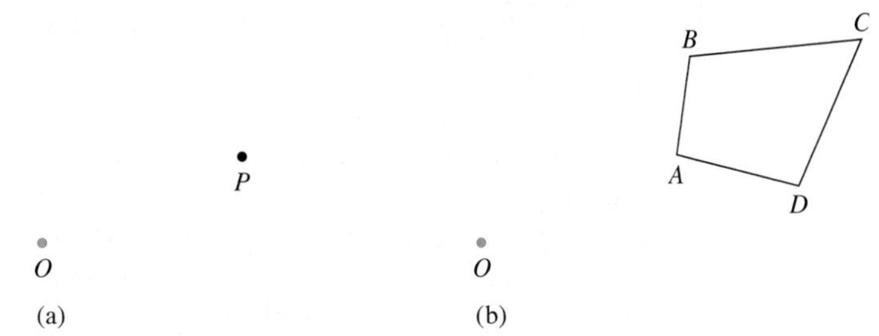

P

O

(a)

B C A D

O

(b)

b. Find the image of the quadrilateral $ABCD$ in Figure 12-32(b) under the size transformation with center O and scale factor $\frac{2}{3}$.

Solution **a.** In Figure 12-33(a), we connect O with P and divide $\overline{OP}$ into three congruent parts. The point P' is the image of P because $OP' = \frac{2}{3}OP$.

b. We find the image of each of the vertices and connect the images to obtain the quadrilateral $A'B'C'D'$, shown in Figure 12-33(b).

FIGURE 12-33

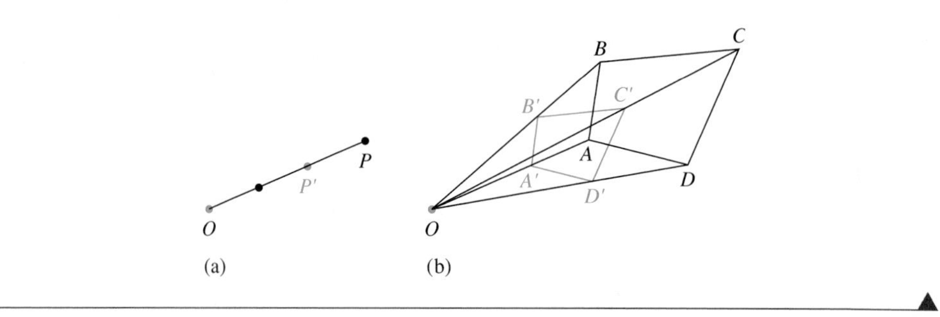

In Figure 12-33(b), the sides of the quadrilateral $A'B'C'D'$ are all parallel to the corresponding sides of the original quadrilateral, and the angles of the quadrilateral $A'B'C'D'$ are congruent to the corresponding angles of quadrilateral $ABCD$. Also, each side in the quadrilateral $A'B'C'D'$ is $\frac{2}{3}$ as long as the corresponding side of quadrilateral $ABCD$.

These properties are true for any size transformation and are summarized in the following theorem.

Theorem 12-1

A size transformation with center O and scale factor r $(r > 0)$ has the following properties:

1. The image of a line segment is a line segment parallel to the original segment and r times as long.
2. The image of an angle is an angle congruent to the original angle.

From Theorem 12-1, it follows that the image of a polygon under a size transformation is a similar polygon. Why? However, for any two similar polygons it is not always possible to find a size transformation so that the image of one polygon under the transformation is the other polygon. But, given two similar polygons, we can "move" one polygon to a place so that it will be the image of the other under a size transformation. The following examples show such instances.

EXAMPLE 12-9

Show that $\triangle ABC$ in Figure 12-34 is the image of $\triangle ADE$ under a size transformation. Identify the center of the size transformation and the scale factor.

FIGURE 12-34

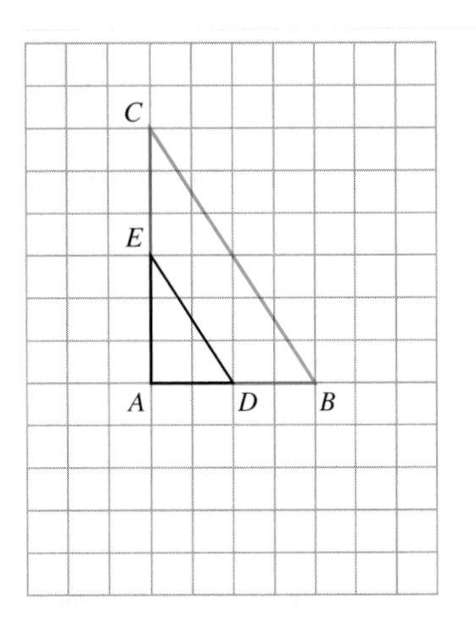

Solution Because $\frac{AB}{AD} = \frac{AC}{AE} = 2$, we choose A as the center of the size transformation and 2 as the scale factor. Notice that under this transformation, the image of A is A itself. The image of D is B, and the image of E is C.

EXAMPLE 12-10

Show that $\triangle ABC$ in Figure 12-35 is the image of $\triangle APQ$ under a succession of isometries with a size transformation.

FIGURE 12-35

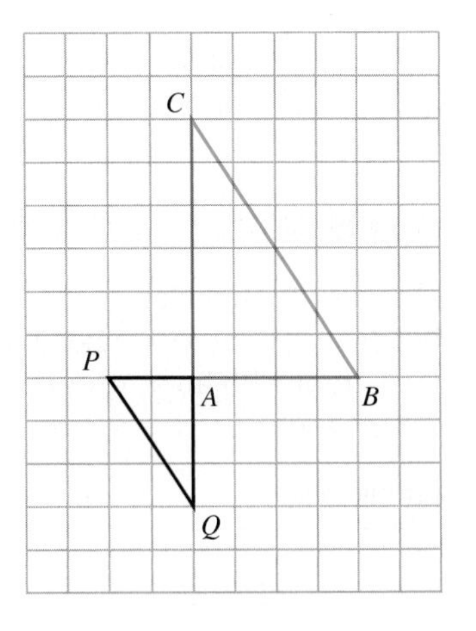

Solution We use the strategy of *looking at a related problem*. In Example 12–9, the common vertex served as the center of the size transformation. This was possible because the corresponding sides of the triangles were parallel. To achieve a similar situation, we first transform $\triangle APQ$ by a half-turn in A and obtain $\triangle AP'Q'$, as shown in Figure 12-36.

FIGURE 12-36

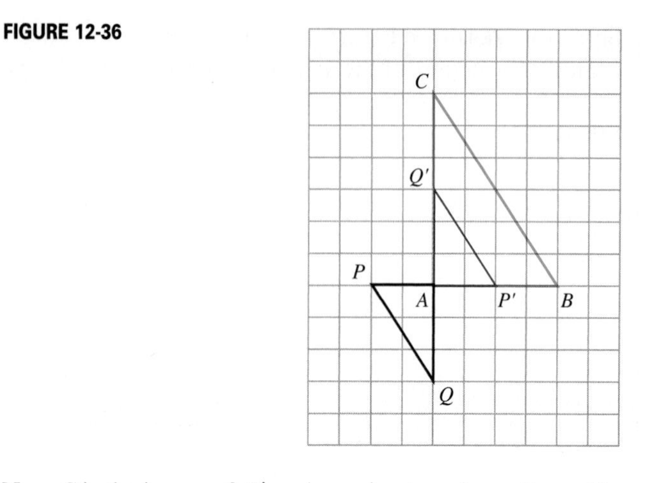

Now C is the image of Q' under a size transformation with center at A and scale factor 2. B is the image of P', and A is the image of itself under this transformation. Thus, $\triangle ABC$ can be obtained from $\triangle APQ$ by first finding the image of $\triangle APQ$ under a half-turn in A and then applying a size transformation with center A and a scale factor 2 to that image.

Examples 12-9 and 12-10 are a basis for an alternative definition of similar figures.

Definition of Similar Figures

Two figures are similar if it is possible to transform one onto the other by a sequence of isometries followed by a size transformation.

Size transformations and similarity in the coordinate plane are discussed on the following student page. Answer the questions on the student page. Conjecture the coordinates of the image of the point (x, y) under a size transformation with center at the origin and scale factor r.

Applications of Size Transformations

perspective drawing

One way to make an object appear three-dimensional is to use a **perspective drawing**. For example, to make a letter appear three-dimensional we can use a size transformation with an appropriate center O and a scale factor, as shown in Figure 12-37 for the letter L.

FIGURE 12-37

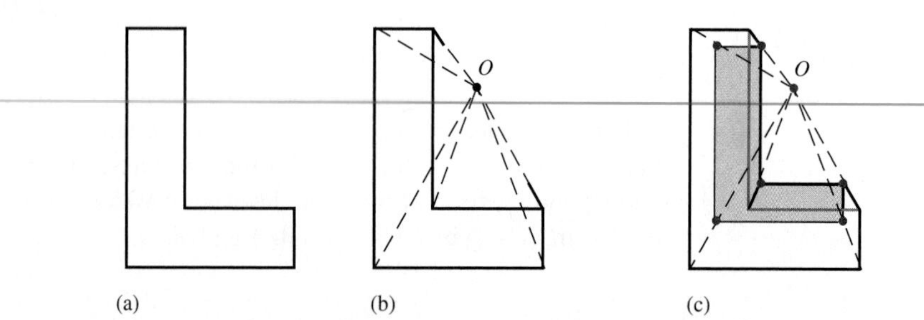

SAMPLE SCHOOL BOOK PAGE:

TRANSFORMATIONS AND SIMILARITY

15. You can use an algorithm to model a change that involves stretching. The transformation shown is modeled by the algorithm below.

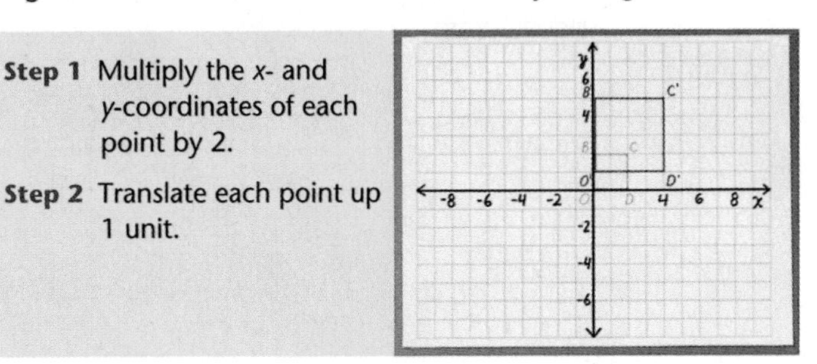

a. Draw *OBCD* with coordinates *O*(0, 0), *B*(0, 2), *C*(2, 2) and *D*(2, 0). Translate the figure up 1 unit. Then multiply each coordinate by 2. Draw the final image.

b. Compare your results from part (a) with the transformation shown. Does the order in which you perform the steps matter?

16. Copy the figure below. Then use the algorithm to transform the figure.

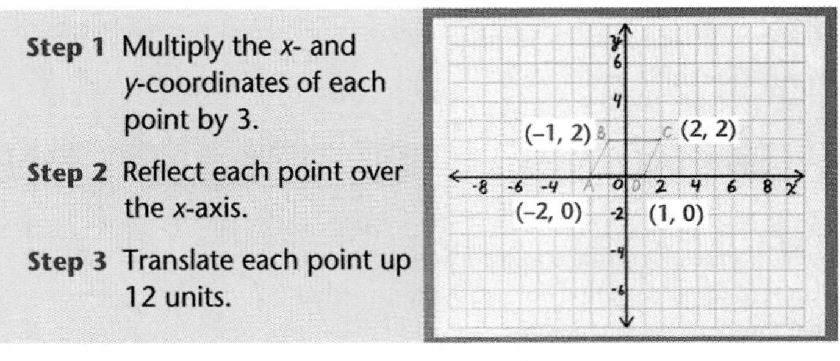

Step 1 Multiply the x- and y-coordinates of each point by 3.
Step 2 Reflect each point over the x-axis.
Step 3 Translate each point up 12 units.
(−1, 2)
(2, 2)
(−2, 0)
(1, 0)
-8 -6 -4 O 2 4 6 8 x
-2
-4
-6

17. You already know how to stretch a figure using multiplication. You can also use multiplication to shrink a figure. Draw the figure below. Then use the algorithm to transform the figure.

Step 1 Multiply the *x*- and *y*-coordinates of each point by $\frac{1}{2}$.

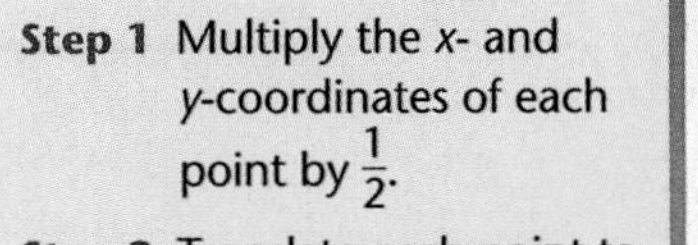

Step 2 Translate each point to the left 1 unit.

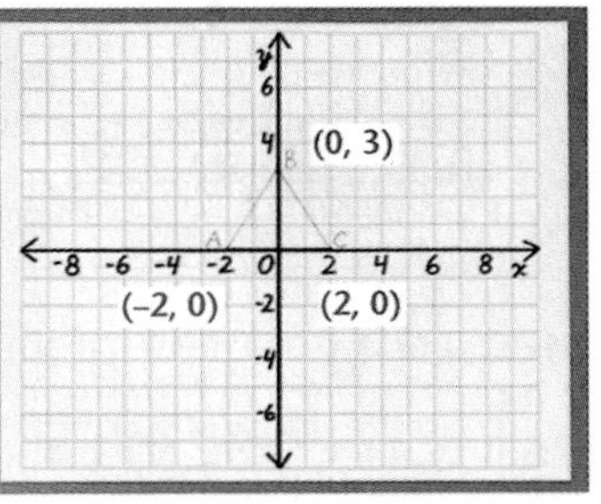

Source: Math Thematics, Book 3, McDougal Littell, 2005 (p. 517).

When a picture of an object is taken, the object appears upside down on the negative. The picture of the object on the negative can be interpreted as an image under composition of a half-turn and a size transformation. Figure 12-38(a) illustrates the image of an arrow from A to B under a composition of a half-turn followed by a size transformation with scale factor $\frac{1}{2}$.

FIGURE 12-38

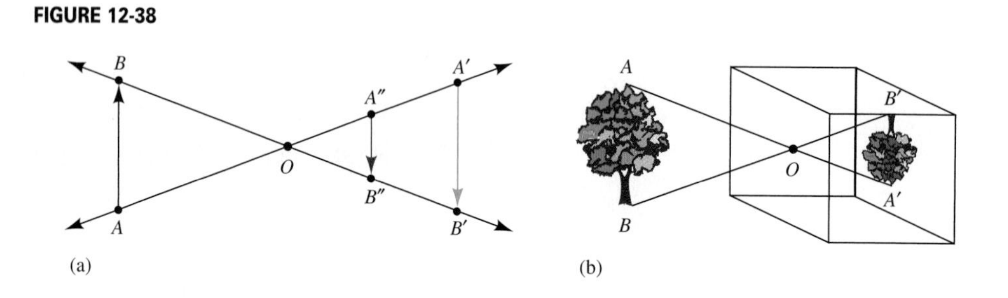

The image A' of A under the half-turn with center O is found on the ray opposite $\overrightarrow{OA}$ so that $OA' = OA$. The point B', the image of B under the half-turn, is found similarly on the ray opposite $\overrightarrow{OB}$. The images of A' and B' under the size transformation are A'' and B'', respectively. Consequently, the image of the arrow from A to B under the composition of the half-turn followed by the size transformation is the arrow from A'' to B''. Figure 12-38(b) illustrates another composition of a half-turn and a size transformation in a simple box camera.

ONGOING ASSESSMENT 12-3

1. In the following figures, describe a sequence of isometries followed by a size transformation so that the larger triangle is the final image of the smaller one:

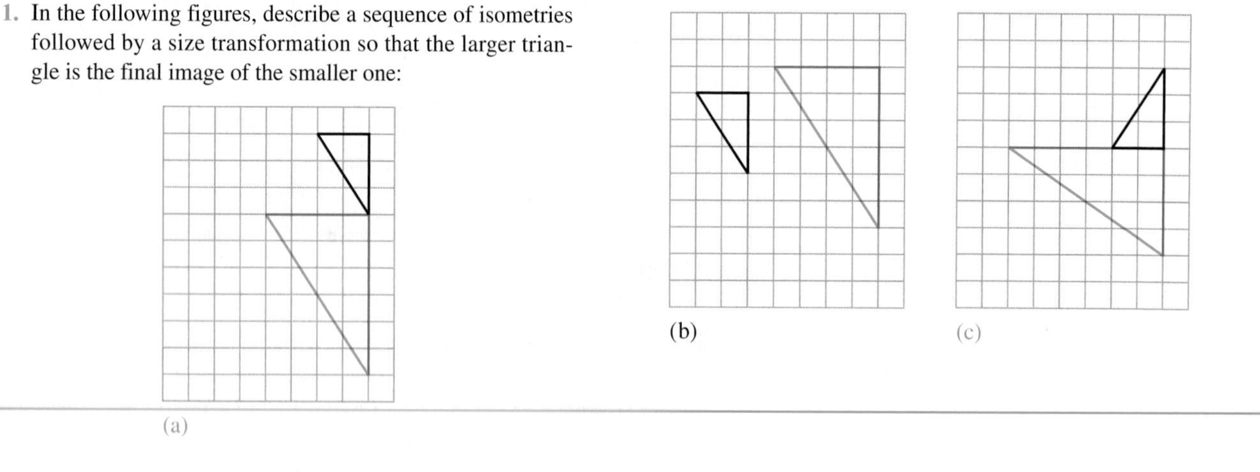

2. In the following drawing, find the image of $\triangle ABC$ under the size transformation with center O and scale factor $\frac{1}{2}$:

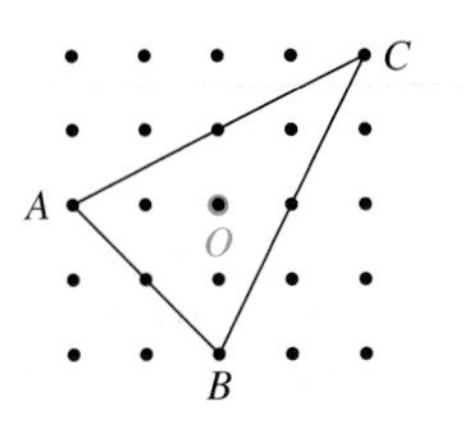

3. In each of the following drawings, find transformations that will take $\triangle ABC$ to its image, $\triangle A'B'C'$, which is similar:

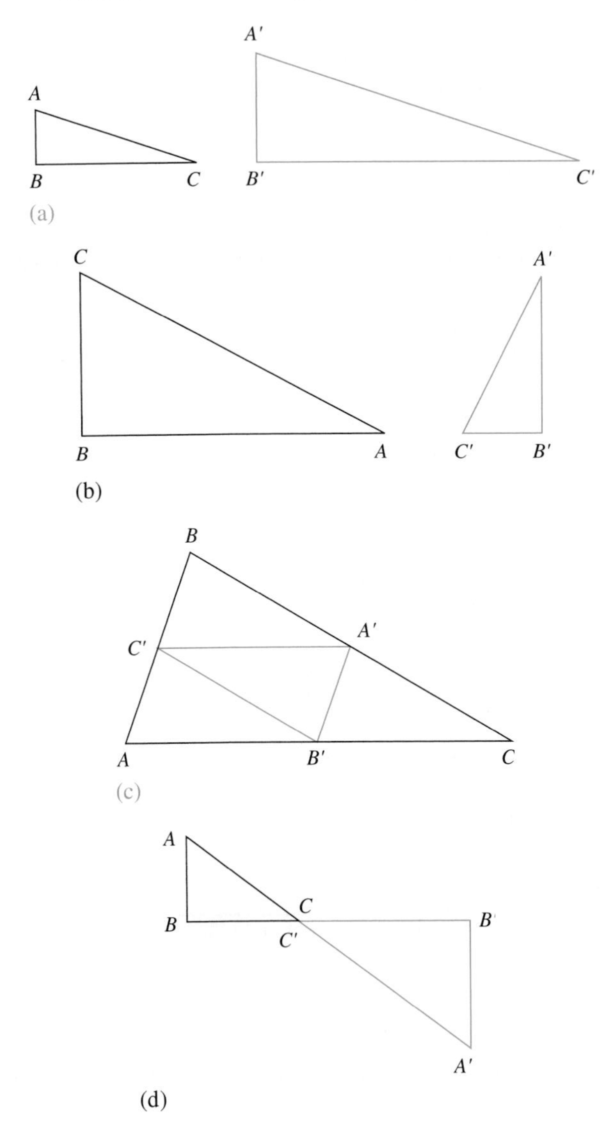

4. In each of the following figures, the smaller one is the image of the larger under a size transformation centered at point O. In each case, find the scale factor and the length of x and y as pictured.

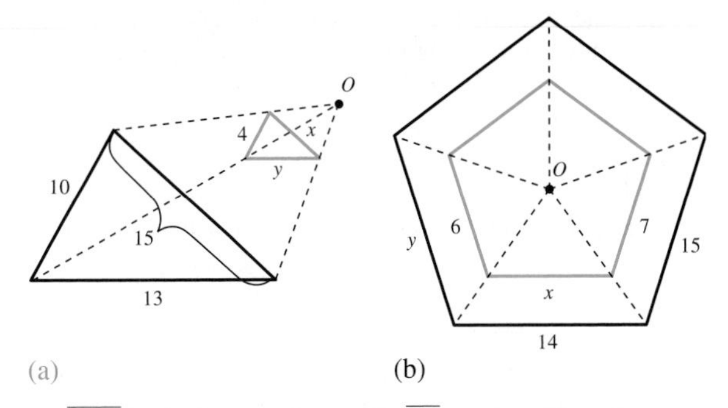

5. $\overline{A'B'}$ is the image of a candle $\overline{BA}$ produced by a box camera. Given the measurement of the figure, find the height of the candle.

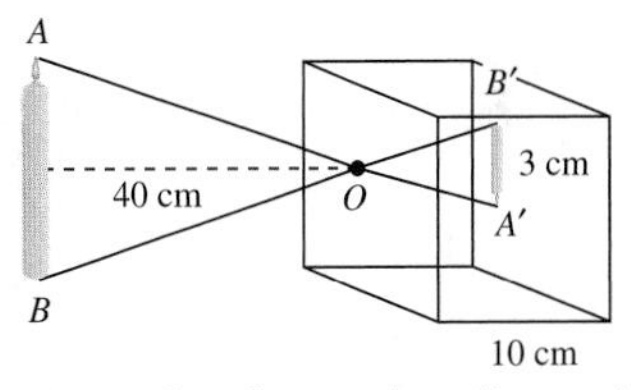

6. Each of the following figures describes a size transformation with center O and images in blue. Find the scale factor and the lengths designated by x and y.

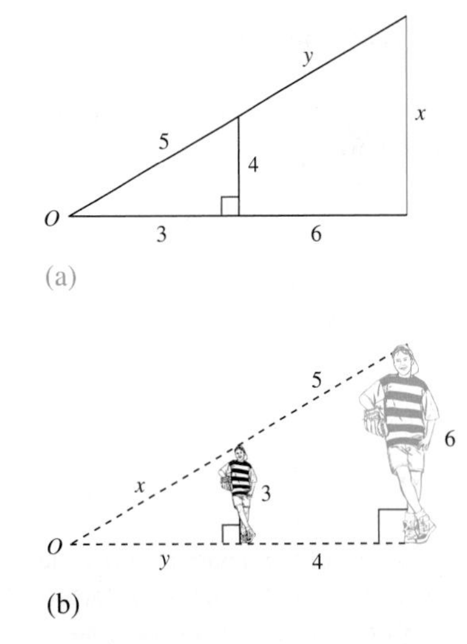

7. Find the coordinates of the images of $A(2, 3)$, and $B(^{-}2, 3)$ under the following transformations. Assume that all size transformations are centered at the origin.
   a. A size transformation with a scale factor 3 followed by a size transformation with a scale factor 2
   b. A size transformation with scale factor 2 followed by a size transformation with a scale factor 3

c. A size transformation with scale factor 2 followed by a translation with a slide arrow from (2, 1) to (3, 4)

d. A translation with a slide arrow from (2, 1) to (3, 4) followed by a size transformation with scale factor 2

e. What can you conclude from your answers to parts (c) and (d) concerning the order in which the transformations are performed?

8. If a size transformation with center O and scale factor r takes a quadrilateral $ABCD$ to $A'B'C'D'$, what size transformation will take $A'B'C'D'$ back to $ABCD$?

9. a. Explain why in a coordinate system a size transformation with center at the origin and scale factor r $(r > 0)$ is given by $(x, y) \rightarrow (rx, ry)$.

b. The transformation $(x, y) \rightarrow (^{-}2x, {}^{-}2y)$ can be achieved by a size transformation followed by an isometry. Find that size transformation and the isometry.

c. Find the equations of the images of each of the following lines under size transformation in part (a) with the given scale factor r:

(i) $y = 2x, r = \frac{1}{2}$

(ii) $y = 2x, r = 2$

(iii) $y = 2x + 1, r = \frac{1}{3}$

(iv) $y = {}^{-}x - 1, r = 3$

10. What size transformation will transform a circle with center at the origin and radius 4 onto a circle with the same center and radius 3?

11. What sequence of transformations will transform a circle with center O_1 and radius 2 onto a circle with center O_2 and radius 3?

12. Construct any triangle ABC and a square $DEFG$ so that D is on $\overline{AB}$ and the vertices F and G are on $\overline{AC}$, as shown, and answer the following:

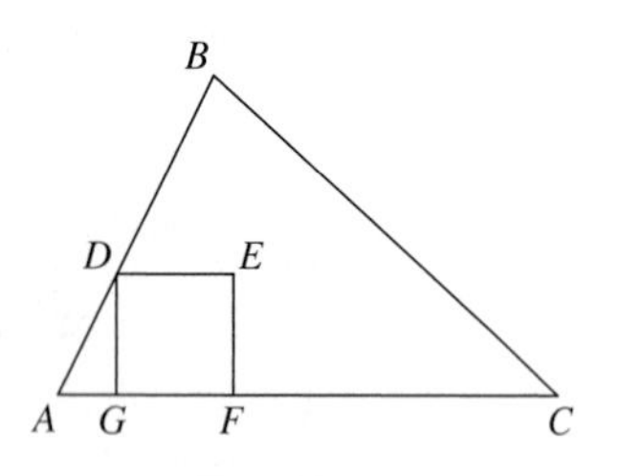

a. Find the images of the square $DEFG$ under several size transformations with different scale factors. What do all the images of vertex E have in common?

b. Explain why the intersection of $\overleftrightarrow{AE}$ and $\overline{BC}$ is a vertex of a square. Construct that square.

Communication

13. Which of the following properties do not change under a size transformation? Explain how you can be sure of your answers.

a. Distance between points

b. Angle measure

c. Parallelism; that is, if two lines are parallel to each other, then their images are parallel to each other.

14. Given two similar figures, explain how to tell if there is a size transformation that transforms one of the figures onto the other.

15. a. Consider two consecutive size transformations, each with center O and corresponding scale factors $\frac{1}{2}$ and $\frac{1}{3}$, respectively. Suppose the image of figure F under the first transformation is F' and the image of F' under the second transformation is F''. What single transformation will map F directly onto F''? Explain why.

b. What would be the answer to (a) if the scale factors were r_1 and r_2?

16. Copy the following figure onto grid paper and determine the center and the scale factor of the size transformation. Explain why there is only one possibility for the center.

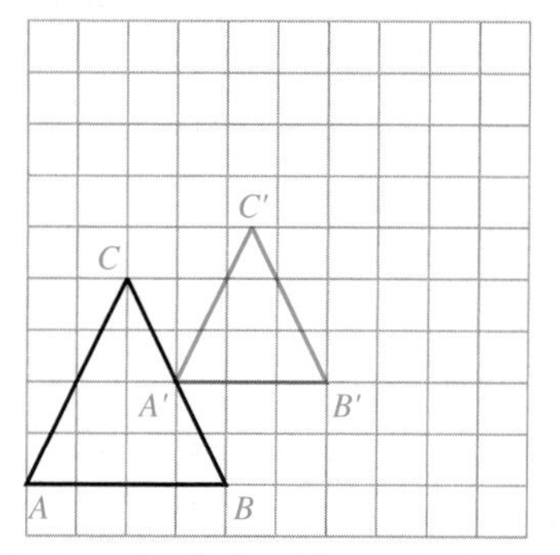

17. a. Is the image of a circle with center O under a size transformation with center O always a circle? Explain why or why not.

b. Assume that under a size transformation with scale factor r, the image of a segment of length a is a segment of length ra and answer part (a) of this question in case the center of the circle is not at the origin.

Open-Ended

18. Describe several real-life situations other than the ones discussed in this section in which size transformations occur.

19. Use a sheet of graph paper with a coordinate system. Locate the origin as the center of a transformation with a scale factor of $^{-}1$ and find the image of some triangle located in the first quadrant. What other transformation that you've studied has the same image as this transformation?

Cooperative Learning

20. Have members of your group draw several figures and find their images under a size transformation with a scale factor of 3 and center of your choice.
    a. How does the perimeter of each image compare to the perimeter of the original figure? Compare your answers.
    b. How does the area of each image compare to the area of the original figure? Compare your answers.
    c. Make a conjecture concerning the relationship between the perimeter of each image and the perimeter of the original figure under a size transformation with a scale factor r.
    d. Repeat (c) for the area of each figure.
    e. Discuss your findings and come up with a group conjecture.

Review Problems

21. Describe a transformation that would "undo" each of the following:
    a. A translation determined by slide arrow from M to N
    b. A rotation of 75° with center O in a clockwise direction
    c. A rotation of 45° with center A in a counterclockwise direction
    d. A glide reflection that is the composition of a reflection in line m and a translation that takes A to B
    e. A reflection in line n

22. In the following coordinate plane, find the images of each of the given points in the transformation that is the composition of a reflection in line m followed by a reflection in line n.

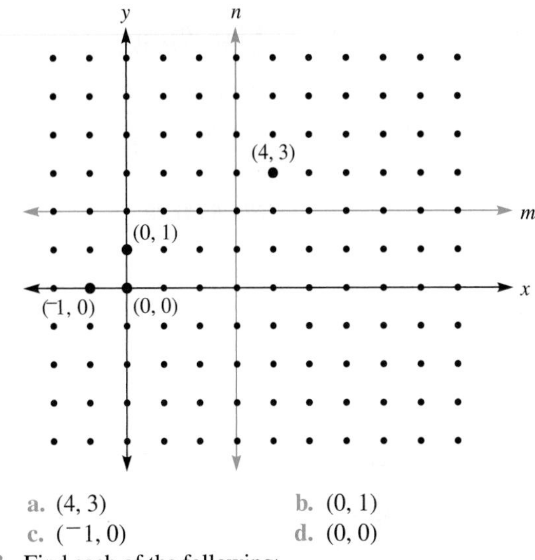

a. (4, 3) b. (0, 1)
c. $(^-1, 0)$ d. (0, 0)

23. Find each of the following:
    a. Reflection image of an angle in its angle bisector
    b. Reflection image of a square in one of its diagonals

TECHNOLOGY CORNER

Use GSP to draw $\triangle ABC$ and three lines m, n, and p that intersect in a single point as shown in Figure 12-39. Reflect $\triangle ABC$ in line m to obtain its image $\triangle A'B'C'$. Then reflect $\triangle A'B'C'$ in line n to obtain $\triangle A''B''C''$. Finally, reflect $\triangle A''B''C''$ in line p to obtain the final image $\triangle A'''B'''C'''$.

FIGURE 12-39

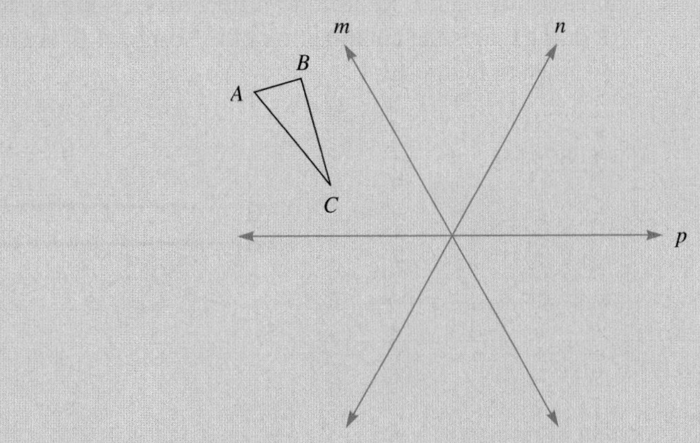

Find a single line q that could be used to reflect the original $\triangle ABC$ onto the final image.

12-4 Symmetries

Line Symmetries

The concept of a reflection can be used to identify line symmetries of a figure. All the drawings in Figure 12-40 have symmetries about the dashed lines.

FIGURE 12-40

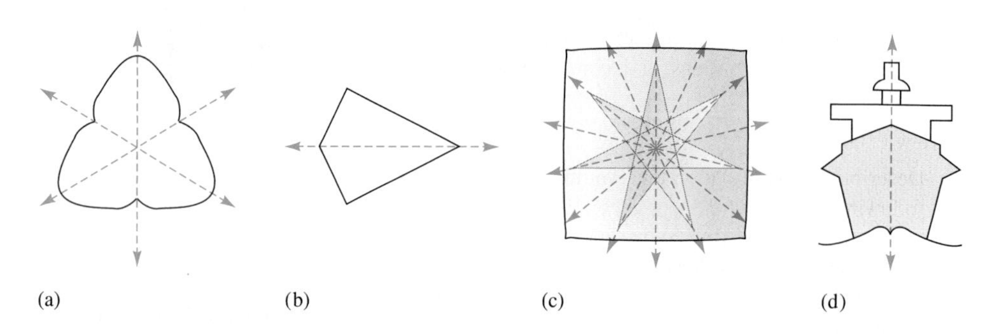

line of symmetry

Mathematically, a geometric figure has a **line of symmetry** ℓ if it is its own image under a reflection in ℓ. A method of creating a symmetrical figure is seen in Example 12-11.

EXAMPLE 12-11

In Figure 12-41, we are given a figure and a line m. Do the minimum amount of drawing to create a figure from the given figure so that the result has line m as its line of symmetry.

FIGURE 12-41

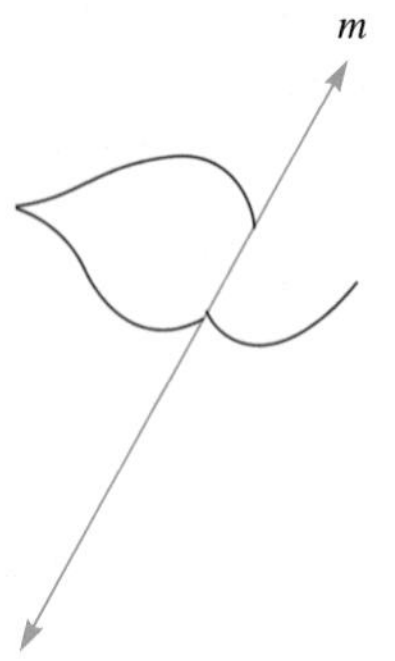

m

Solution For the resulting figure both to be symmetric about line m and to incorporate the existing figure, we need to reflect the existing figure about line m. The desired result of doing that is the combination of the original figure and the image. The resulting figure is shown in Figure 12-42.

FIGURE 12-42

m

EXAMPLE 12-12

How many lines of symmetry does each drawing in Figure 12-43 have?

FIGURE 12-43

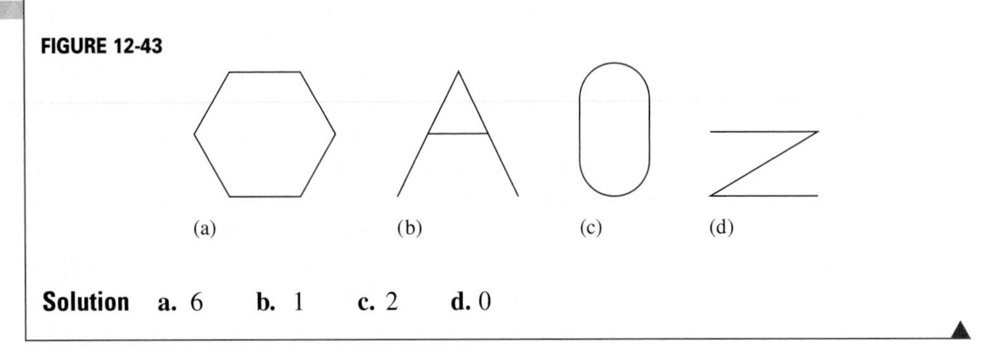

Solution **a.** 6 **b.** 1 **c.** 2 **d.** 0

EXAMPLE 12-13

Given an arc AB in Figure 12-44, use symmetry and paper folding to find the center and the radius of the circle to which the arc belongs.

FIGURE 12-44

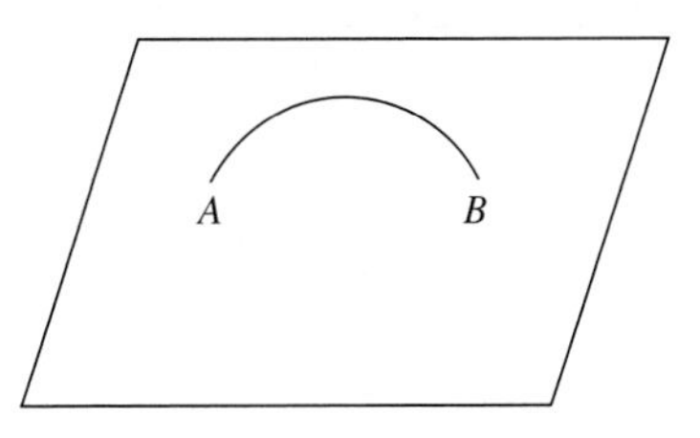

Solution The arc is part of a circle. The center of the circle is equidistant from the endpoints of any chord of the circle and hence is on the perpendicular bisector of the chord. That perpendicular bisector is a line of symmetry of the circle. A circle has infinitely many lines of symmetry and each line passes through the center of the circle, where all the lines of symmetry intersect. To find a line of symmetry, fold the paper containing AB so that a portion of the arc is folded onto itself. Then unfold the paper and draw the line of symmetry on the fold mark, as shown in Figure 12-45(a). By refolding the paper in Figure 12-45(a) so that a different portion of the arc AB is folded onto itself, we can determine a second line of symmetry, as shown in Figure 12-45(b). The two dotted lines of symmetry intersect at O, the center of the circle of which AB is an arc. Either $\overline{OB}$, $\overline{OA}$, or any segment connecting O to a point on the arc represent a radius of the circle.

FIGURE 12-45

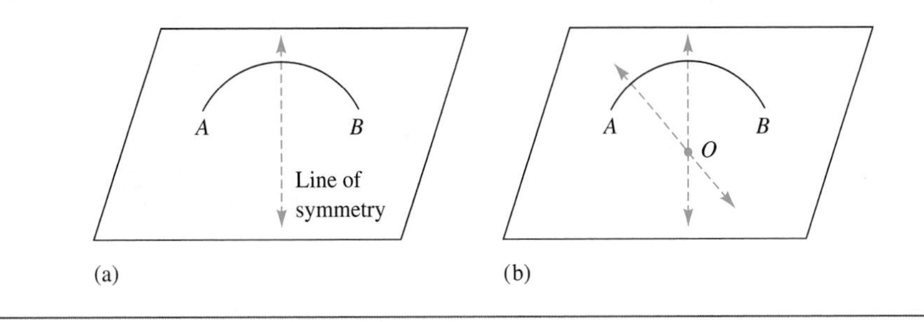

Rotational (Turn) Symmetries

rotational symmetry ◆ turn symmetry

FIGURE 12-46

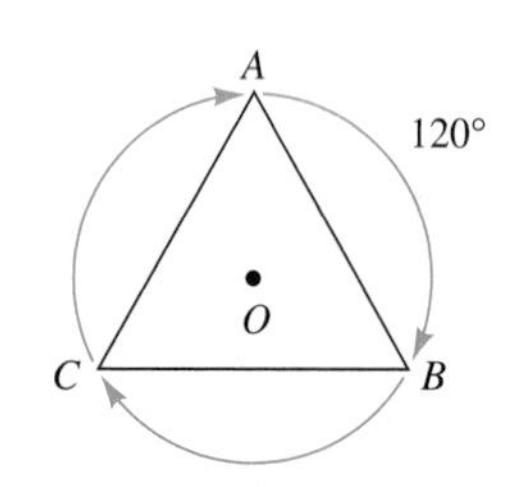

A figure has **rotational symmetry**, or **turn symmetry**, when the traced figure can be rotated less than 360° about some point so that it matches the original figure. Note that the condition "less than 360°" is necessary because any figure will coincide with itself if it is rotated 360° about any point. In Figure 12-46, the equilateral triangle coincides with itself after a rotation of 120° about point O. Hence, we say that the triangle has 120° rotational symmetry. Also in Figure 12-46, if we were to rotate the triangle another 120°, we would find again that it matches the original. So we can say that the triangle also has 240° rotational symmetry.

In general, if a figure has $\alpha°$ rotational symmetry, it also will coincide with itself when rotated by $n\alpha°$ for any nonzero integer n. For this reason, in rotational symmetry it is sufficient to report the smallest possible positive angle measure that turns the figure onto itself. Notice that a circle has a rotational symmetry by any turn around its center. Thus, a circle has infinitely many rotational symmetries.

Other examples of figures that have rotational symmetry are shown in Figure 12-47. Figures 12-47(a), (b), (c), and (d) have 72°, 90°, 180°, and 180° rotational symmetries, respectively [(a) and (b) also have other rotational symmetries].

FIGURE 12-47

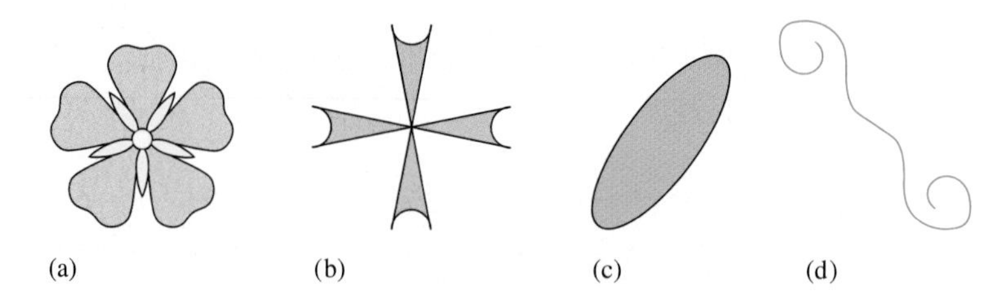

In general, we can determine whether a figure has rotational symmetry by tracing it and turning the tracing about a point (the center of the figure) to see if it aligns on the figure before the tracing has turned in a complete circle, or 360°. The amount of the rotation can be determined by measuring the angle $\angle POP'$ through which a point P is rotated around a point O to match another point P' when the figures align. Such an angle, $\angle POP'$, is labeled with points P, O, and P' in Figure 12-48 and has measure 120°. Point O, the point held fixed when the tracing is turned, is the *turn center*.

FIGURE 12-48

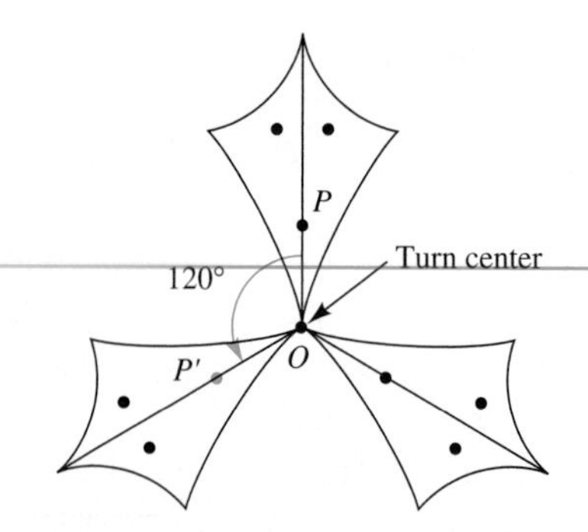

EXAMPLE 12-14 Determine the amount of the turn for the rotational symmetries of each part of Figure 12-49.

FIGURE 12-49

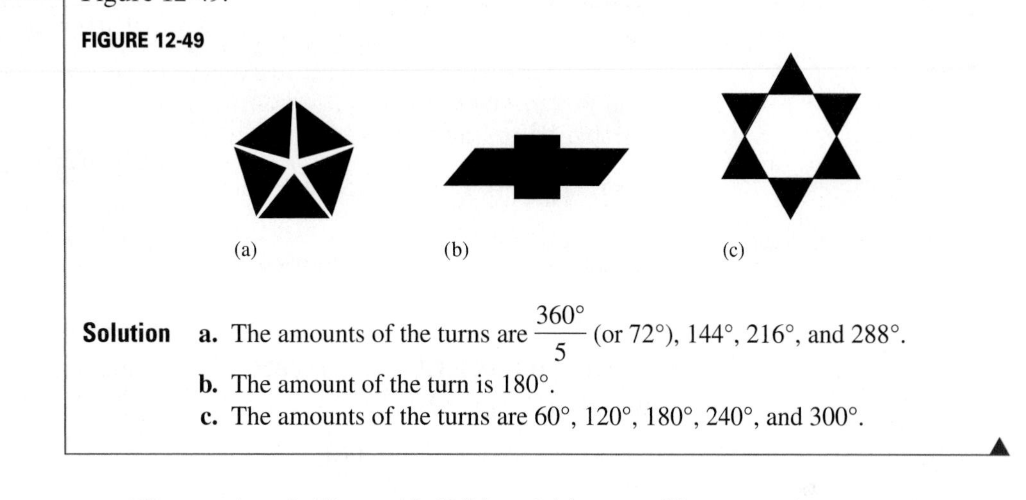

Solution **a.** The amounts of the turns are $\frac{360°}{5}$ (or 72°), 144°, 216°, and 288°.

b. The amount of the turn is 180°.

c. The amounts of the turns are 60°, 120°, 180°, 240°, and 300°.

The rotations in Figure 12-49(b) and (c) exemplify yet another type of symmetry, namely, point symmetry.

Point Symmetry

point symmetry

Any figure that has 180° rotational symmetry is said to have **point symmetry** about the turn center. Some figures with point symmetry are shown in Figure 12-50. As illustrated in Figure 12-50, any figure with point symmetry is its own image under a half-turn. This makes the center of the half-turn the midpoint of a segment connecting a point and its image.

FIGURE 12-50

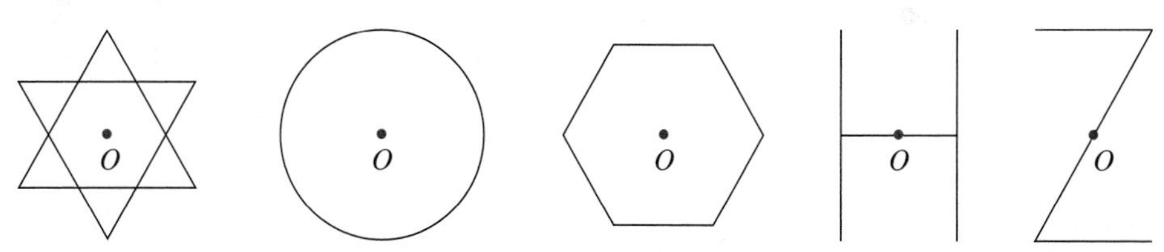

Tracing and paper folding are commonly used in elementary school to investigate symmetries of figures, as shown in the following student page. Work through the questions on the student page.

FIGURE 12-51

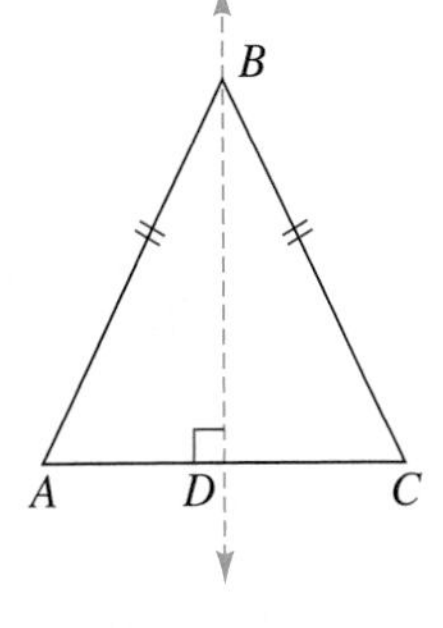

Classification of Figures by Their Symmetries

Geometric figures in a plane can be classified according to the number of symmetries they have. Consider a triangle described as having exactly one line of symmetry and no turn symmetries. What could the triangle look like? The only possibility is a triangle in which two sides are congruent; that is, an isosceles triangle that is not equilateral. The line of symmetry passes through a vertex, as shown in Figure 12-51. We can describe equilateral and scalene triangles in terms of the number of lines of symmetry they have. This is left as an exercise.

SAMPLE SCHOOL BOOK PAGE:

INVESTIGATING SYMMETRY

Symmetry

LEARN

WARM UP

Draw an example of each figure. Then draw a flip.

1. rectangle
2. trapezoid
3. right triangle
4. obtuse triangle

Activity

How can you describe and create symmetric figures?

An artist designed the trademark at the right for a sporting goods company. Many trademarks are **symmetric figures.** This means they can be folded into two congruent parts that fit on top of each other. The fold line is a **line of symmetry.**

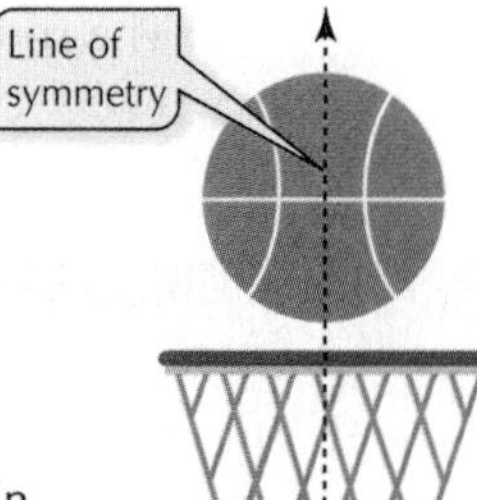

You can follow the steps below to create a design with two lines of symmetry.

a. Fold a sheet of paper in half. Then fold it in half again the other way (so the second crease is perpendicular to the first).

b. Draw a path that starts on one folded edge and ends at the other folded edge, as shown below.

c. Cut along the curve. Then open up the folded paper.

The figure you made should be symmetric with two lines of symmetry.

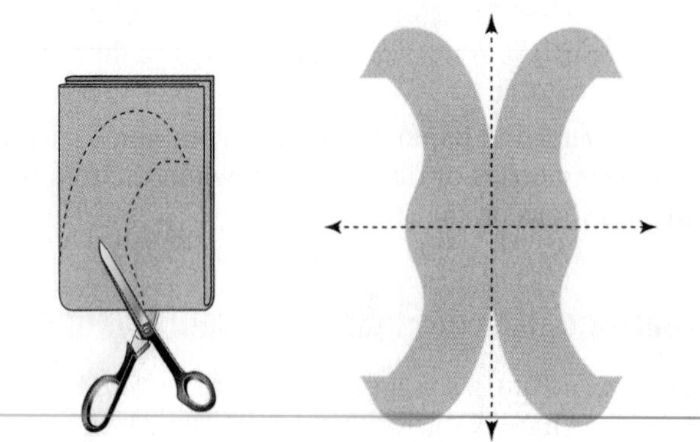

d. How many congruent parts are there in the figure you made?

e. Are the congruent parts related by slides, reflections, or turns?

Source: Scott Foresman–Addison Wesley Mathematics, Grade 5, 2005 (p. 368)

A square, as in Figure 12-52, can be described as a four-sided polygon with four lines of symmetry—d_1, d_2, h, and v—and three turn symmetries about point O. In fact, we can use lines of symmetry and turn symmetries to define various types of quadrilaterals normally used in geometry. It is left as an exercise to see how these definitions differ from those in Chapter 9.

FIGURE 12-52

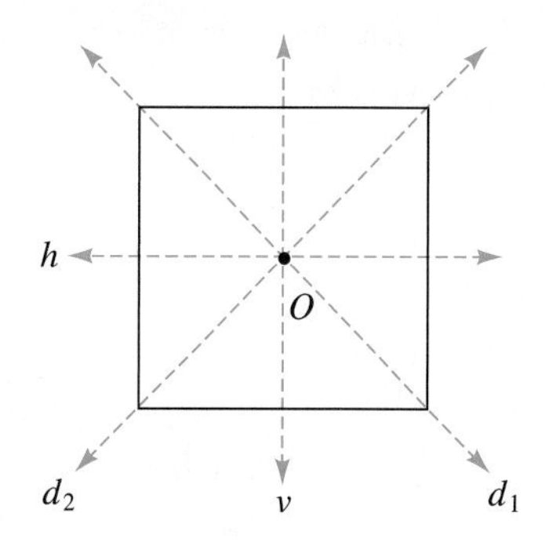

Symmetries of Three-Dimensional Figures

plane of symmetry

axis of rotation

A three-dimensional figure has a **plane of symmetry** when every point of the figure on one side of the plane has a mirror image on the other side of the plane. Examples of figures with plane symmetry are shown in Figure 12-53. Solids can also have point symmetry, line symmetry, and turn symmetry. Notice that the right circular cylinder in Figure 12-53(a) has turn symmetries about the line n, which goes through the centers of the upper and lower bases and is perpendicular to them. The line n is the **axis of rotation**. If the cylinder is rotated by any number of degrees about its axis of rotation, its image is the cylinder itself. Hence, the cylinder has infinitely many rotational symmetries. The cube in Figure 12-53(b) also has rotational symmetries. For the cube, if n is a line connecting the centers of two opposite square faces, then a rotation of 90° in any direction will result in an image that coincides with the original cube. Because we can create an axis of rotation by connecting the centers of any two opposite faces, the cube has three axes of rotation, each with a 90° rotational symmetry. These three-dimensional symmetries are analogous to the two-dimensional symmetries and are further investigated in Ongoing Assessment 12-4.

FIGURE 12-53

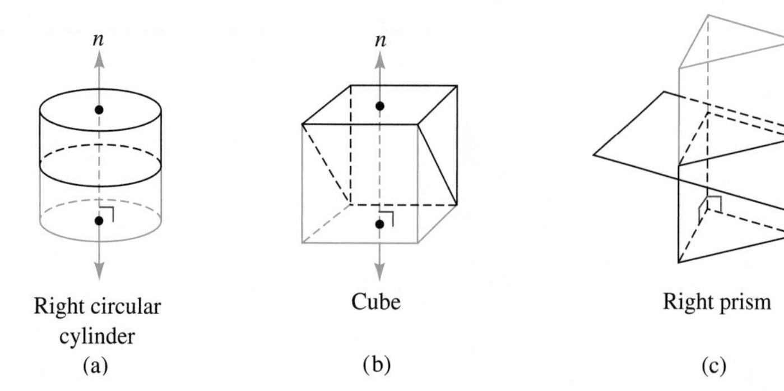

Right circular cylinder
(a)

Cube
(b)

Right prism
(c)

NOW TRY THIS 12-7

a. If we connect two opposite vertices of a cube that are farthest apart, like the pair in Figure 12-54, we obtain an axis of rotation. (i) What is the smallest angle by which the cube needs to be rotated about this axis so that its image will coincide with the original cube? (ii) How many such axes are there?

b. Figure 12-54(b) shows another axis of rotation for the cube. (i) By what angle does the cube need to be rotated about this axis for the image to be the same as the original cube? (ii) How many such axes of rotation are there for a cube?

FIGURE 12-54

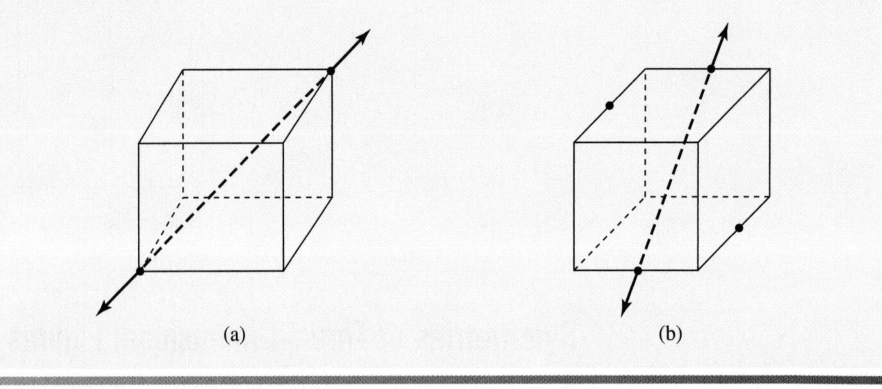

ONGOING ASSESSMENT 12-4

1. What geometric properties make "SOS" a good choice for the international distress symbol?
2. Various international signs have symmetries. Determine which of the following have (i) line symmetry, (ii) rotational symmetry, and/or (iii) point symmetry:

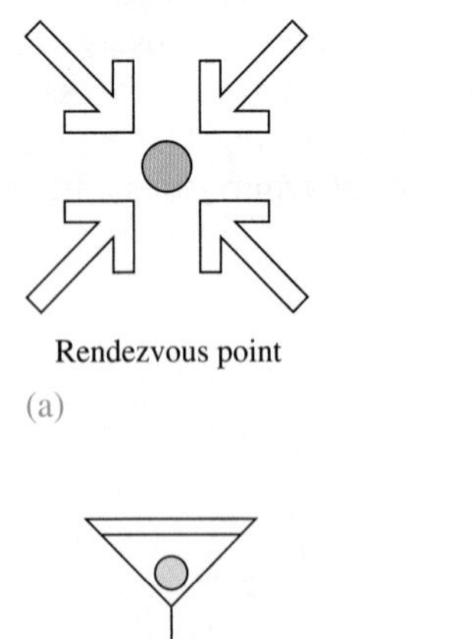

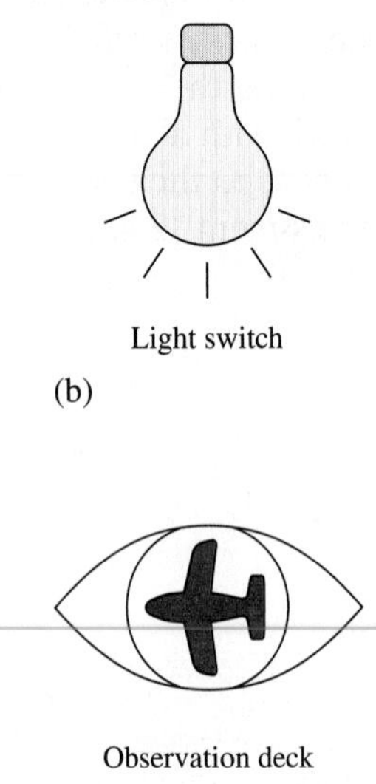

3. Design symbols that have each of the following symmetries, if possible:
   a. Line symmetry but not rotational symmetry
   b. Rotational symmetry but not point symmetry
   c. Rotational symmetry but not line symmetry
4. In each of the following figures, complete the sketches so that they have line symmetry about ℓ:

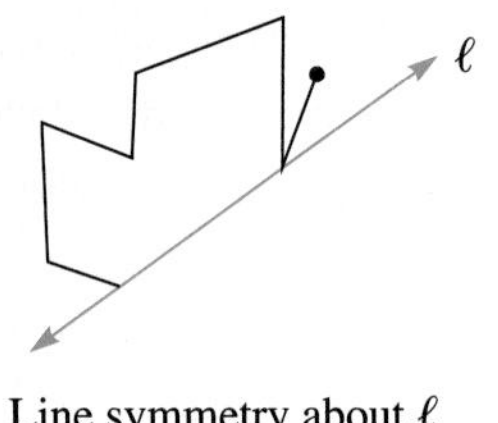

(a) Line symmetry about ℓ

(b) Line symmetry about ℓ

5. **a.** Determine the number of lines of symmetry in each of the following flags.
   b. Sketch the lines of symmetry for each.

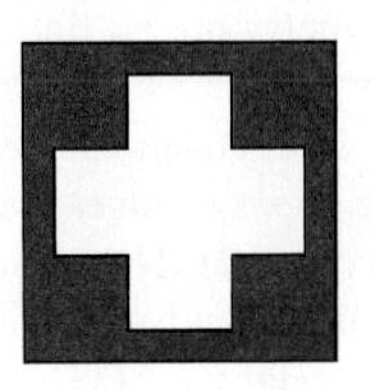

Switzerland

(i)

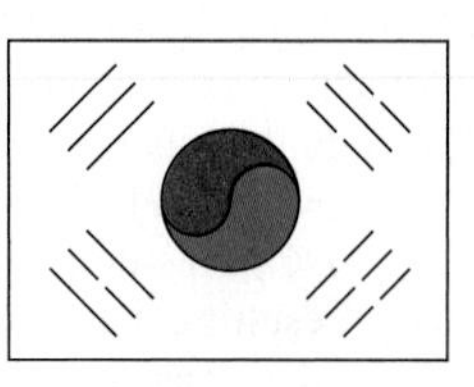

South Korea

(ii)

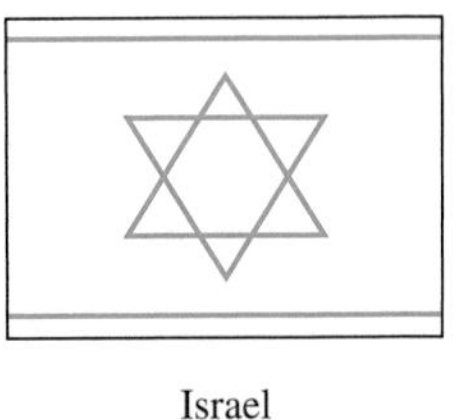

Israel

(iii)

Barbados

(iv)

6. **a.** Determine how many lines of symmetry the following figure has:

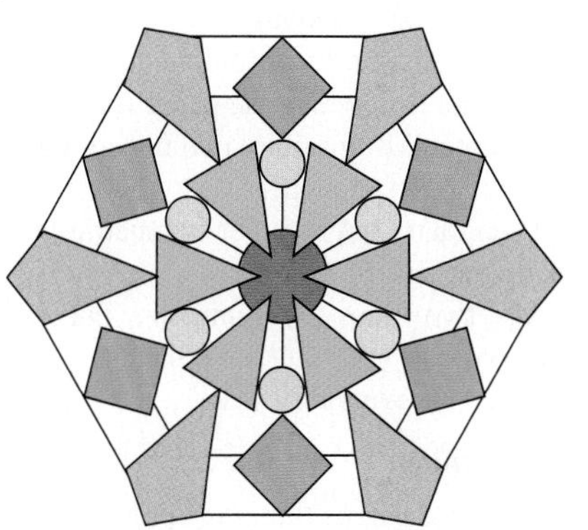

   b. Does the figure have rotational symmetry?

7. Find the lines of symmetry, if any, for each of the following trademarks:

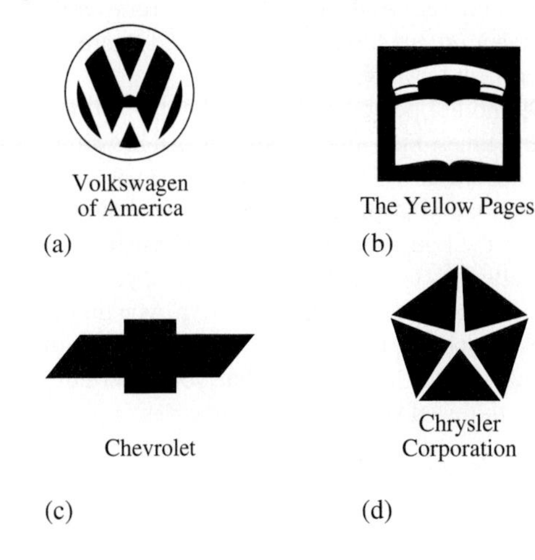

(a) Volkswagen of America

(b) The Yellow Pages

(c) Chevrolet

(d) Chrysler Corporation

8. In each of the following figures, complete the sketches so that they have the indicated symmetry:

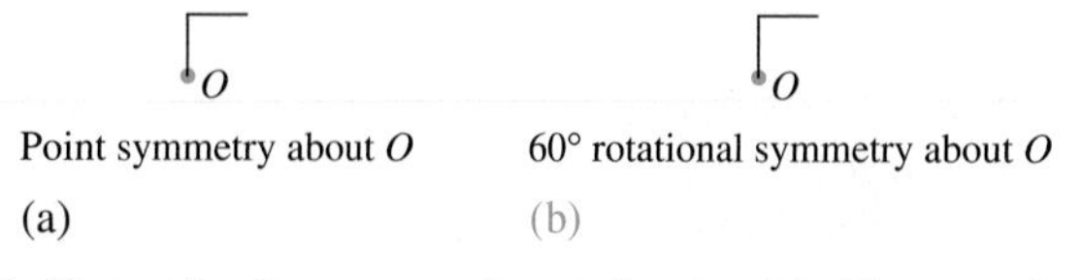

(a) Point symmetry about O

(b) 60° rotational symmetry about O

9. Determine how many planes of symmetry, if any, each of the following three-dimensional vehicle controls has:

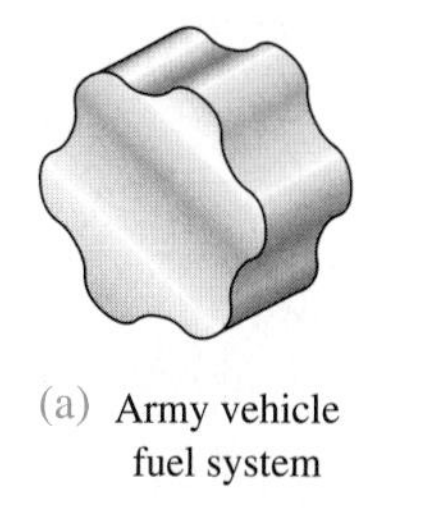

(a) Army vehicle fuel system

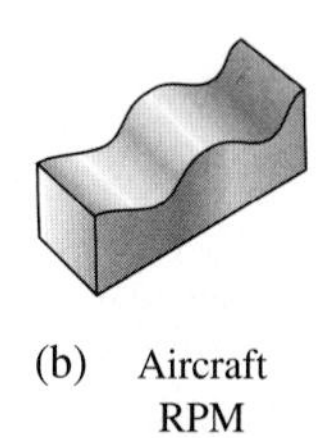

(b) Aircraft RPM

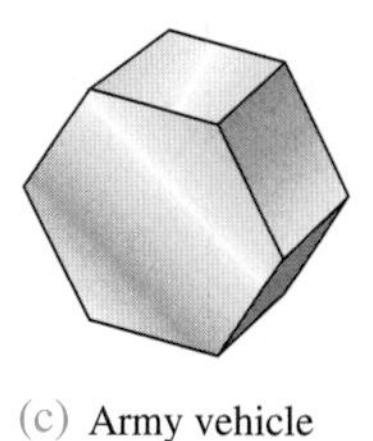

(c) Army vehicle special-purpose equipment

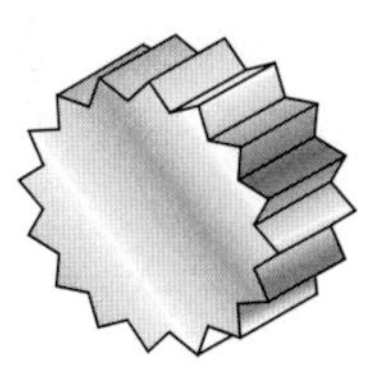

(d) Automotive finger-operated continuous multiturn

10. The following figures were drawn using the computer language Logo. Determine what kinds of symmetries each figure has.

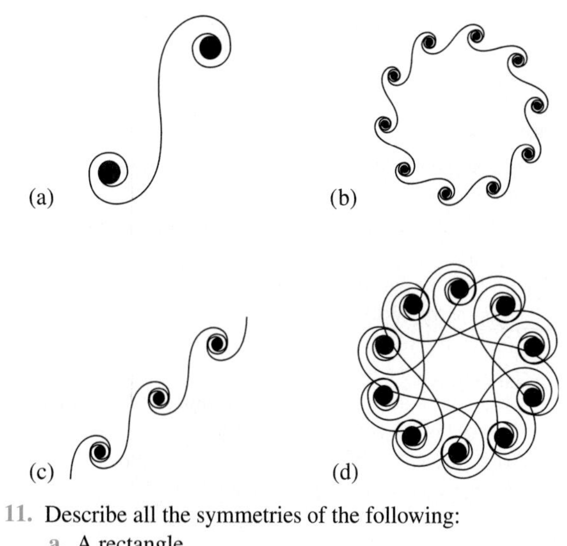

(a)

(b)

(c)

(d)

11. Describe all the symmetries of the following:
    a. A rectangle
    b. An isosceles triangle
    c. An isosceles trapezoid
    d. A parallelogram
    e. A rhombus

f. A kite
g. An angle
h. An equilateral triangle
i. A square
j. A regular pentagon
k. A regular hexagon
★**l.** A regular n-gon
m. Two intersecting circles with centers O_1 and O_2 and intersection points A and B

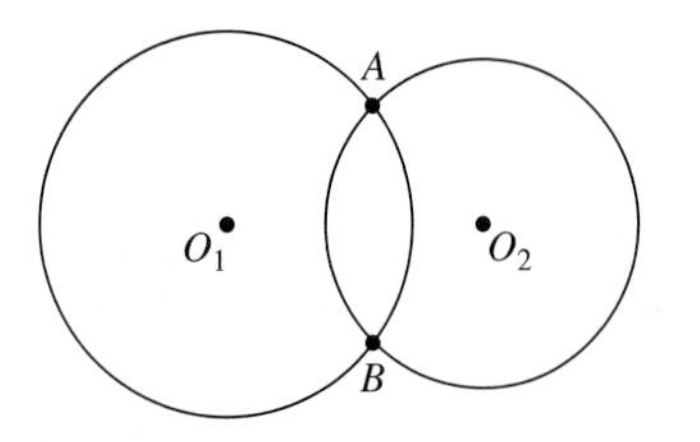

n. Two touching circles with centers O_1 and O_2

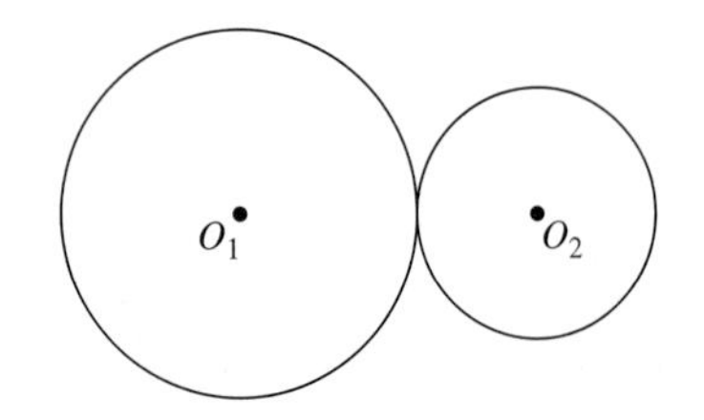

o. The colored ring between two circles with the same center O

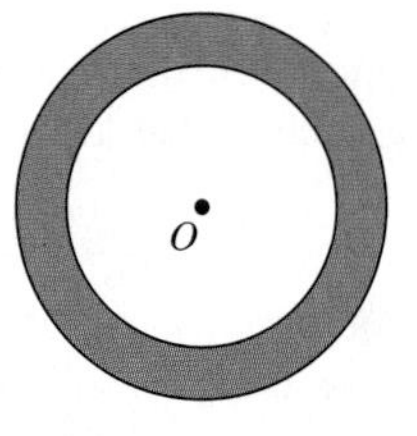

Communication

12. Answer each of the following. If your answer is no, provide a counterexample.
a. If a figure has point symmetry, must it have rotational symmetry? Why?
b. If a figure has rotational symmetry, must it have point symmetry? Why?
c. Can a figure have point, line, and rotational symmetry? If so, sketch a figure that has these properties.
d. If a figure has point symmetry, must it have line symmetry? Is the converse true? Why?
e. If a figure has both point and line symmetry, must it have rotational symmetry? Why?

Open-Ended

13. If possible, sketch a triangle that satisfies each of the following:
a. It has no lines of symmetry.
b. It has exactly one line of symmetry.
c. It has exactly two lines of symmetry.
d. It has exactly three lines of symmetry.

14. Sketch a figure that has point symmetry but no line symmetry.

15. a. In the following figure, $ABCD$ is a rectangular sheet of paper. Fold the paper so that the opposite edges $\overline{AB}$ and $\overline{DC}$ coincide (the crease $\overline{EF}$ is created). Then fold the resulting rectangle $ABEF$ so that $\overline{BE}$ and $\overline{AF}$ coincide (the crease $\overline{HG}$ is created). The rectangle $AFGH$ is obtained. Now cut a curved piece of paper out of the corner G as shown and unfold the paper. Describe all the symmetries that the unfolded figure has.

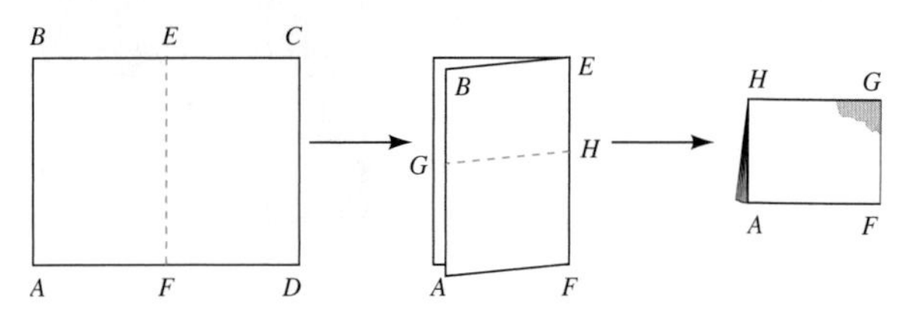

b. Repeat the experiment in (a) by successively folding a new sheet of paper 3 times and cutting out a curved piece containing a corner that resulted from the three folds. Predict all the symmetries of the unfolded figure. Check your answer by unfolding the paper.

16. For each of the following, use paper folding or any other method to create, if possible, two nonsimilar figures that have the specified symmetries:
a. Point symmetry but no other symmetries
b. Exactly one line of symmetry
c. Exactly two lines of symmetry but no rotational symmetry
d. 45° rotational symmetry

Cooperative Learning

17. With a partner, design a three-mirror kaleidoscope by fastening three mirrors together, each perpendicular to a flat surface so that they form an equilateral triangular prism. Place colored paper with a pattern (or design your own pattern) in the base of the kaleidoscope. Peer over the edge of the kaleidoscope to view the generated figure. Repeat the experiment for different patterns.

18. With another person or group, play the following game several times. First, draw a polygon that has different kinds of symmetries. Without revealing your polygon to your partner, tell your partner all you know about the symmetries of the figure. Next, from this information, your opponent attempts to draw the type of figure you drew. If your opponent produces a figure of the type you drew, he or she earns two points. If your opponent produces a different figure having the symmetries that you reported or if you failed to reveal some of the symmetries of your polygon, your opponent gets three points. Alternate roles several times.

19. A *frieze pattern* is a pattern that extends indefinitely in both directions and the image of the pattern under a translation is the original pattern. Pictured here are seven frieze patterns, each of which is the image of itself under a translation and also by the other isometries indicated. Every frieze pattern can be classified into one of these seven categories.

a. Have each member of your group draw one or two of the patterns (depending on the size of the group) on tracing paper or a transparency with a corresponding label using the letters *T, R, H, V,* or *G*, as in the figure. Exchange the drawings so that each member of your group checks how each of the patterns can be transformed onto itself by each of the isometries. Compare your answers with those of your group members.

b. Create other, more elaborate patterns for each of the seven categories and present each pattern to a partner for identification. (Label the patterns with numbers 1 through 7, keeping to yourself the corresponding *T, R, H, V, G* labeling.) Compare your partner's answers with yours.

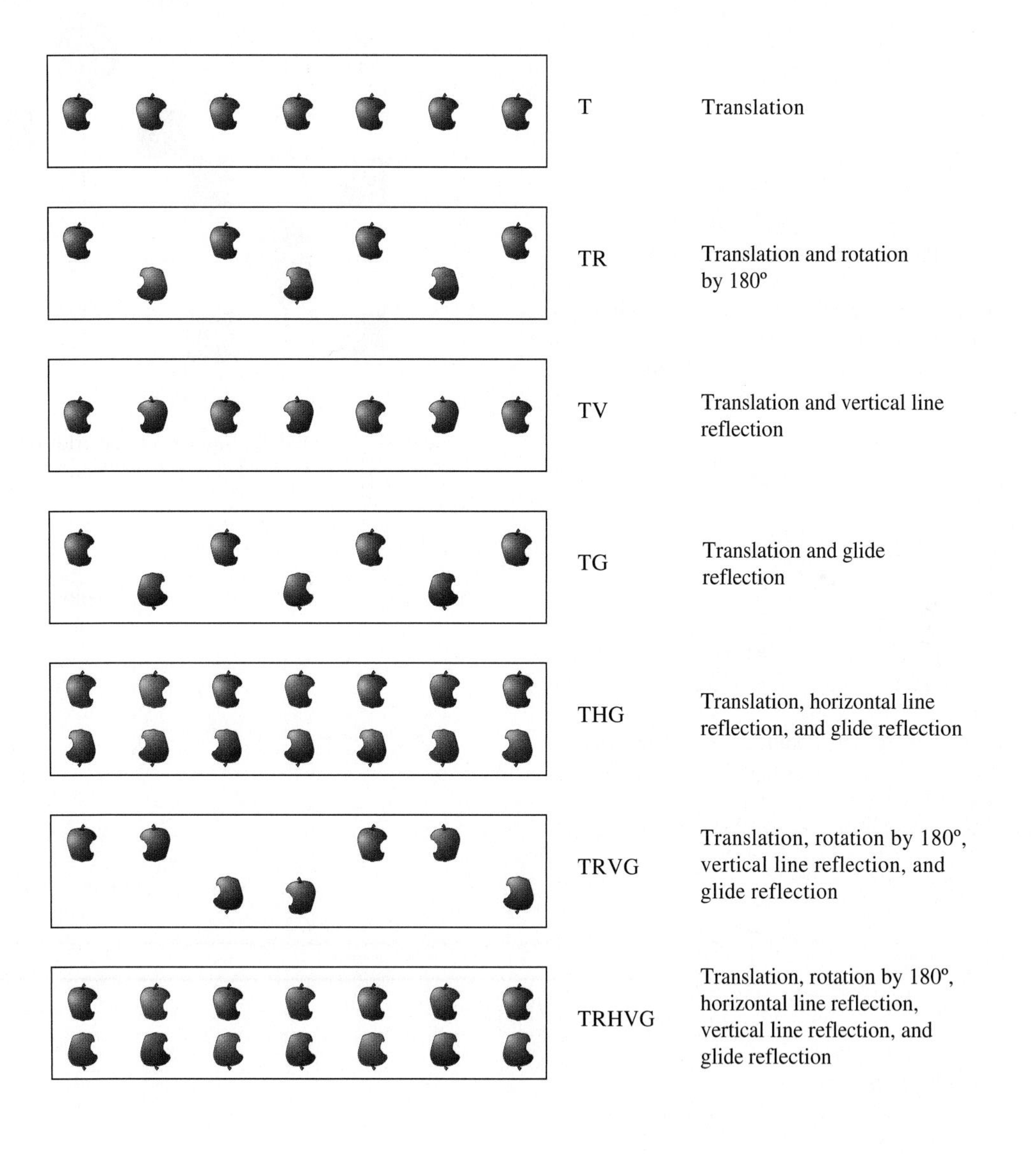

Review Problems

20. For each of the following cases, find the image of the given figure using paper folding:

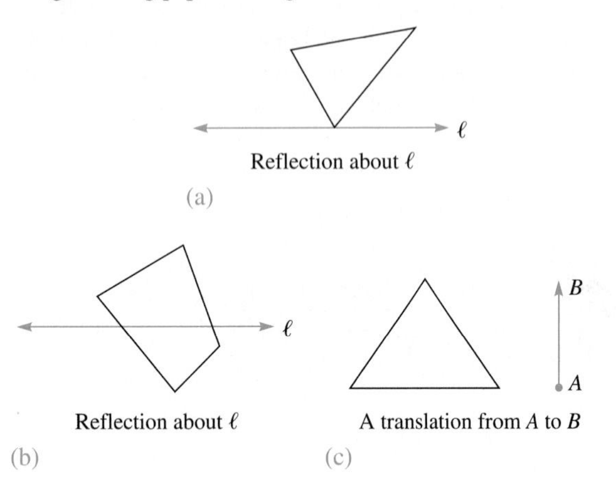

21. Construct each image in problem 20 using a compass and a straightedge.

22. Given two circles with centers O_1 and O_2 and point P as shown, follow the steps below to determine points X on the smaller circle and Y on the larger circle so that P is the midpoint of $\overline{XY}$.

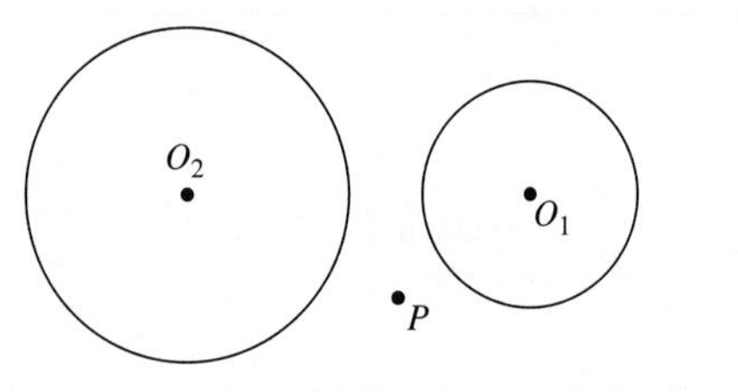

a. Find the image of the smaller circle under half-turn in P.
b. Find the intersection points of the image circle with the larger circle.
c. Check if each intersection point in part (b) can serve as point Y by drawing a line through Y and P and finding where the lines intersect the smaller circle.
d. Draw a point P for which the problem has no solution.

Third International Mathematics and Science Study (TIMSS) Questions

There are several ways of arranging the tiles so that they form patterns. The grid below has been shaded to show how tiles can be placed on some of the squares. The pattern can be continued so that AB and CD are lines of symmetry.

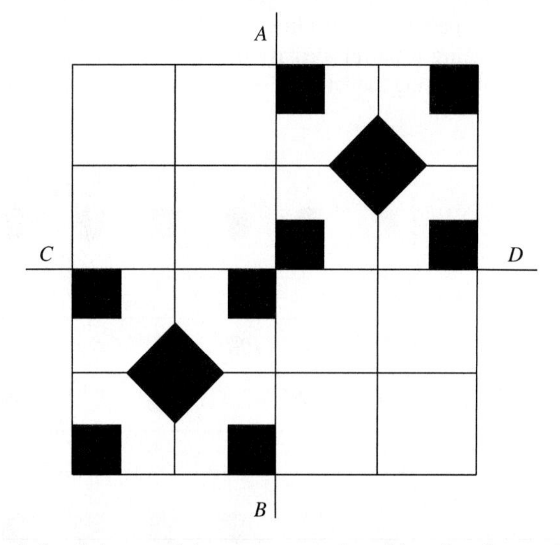

Shade in all the remaining squares on the grid so that the resulting pattern is symmetrical about line AB, and also is symmetrical about line CD.

TIMSS, Grade 8, 2003

Draw a line of symmetry on the triangle below.

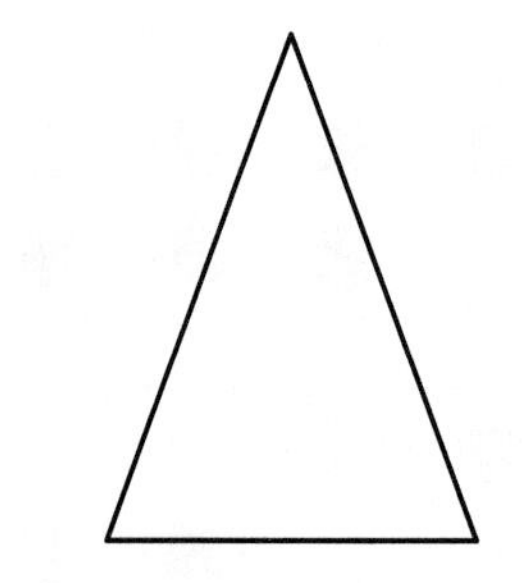

TIMSS, Grade 4, 2003

BRAIN TEASER

Suppose you are looking in a mirror hung flat on a wall and your image goes from the top of the mirror to the bottom. How does the length of the part of your body that you see compare with the length of the mirror?

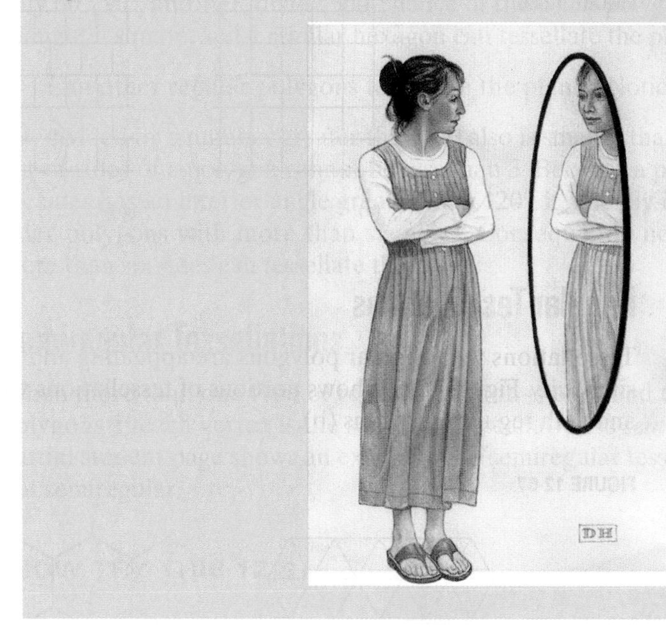

12-5 *Tessellations of the Plane

In this section we use concepts from motion geometry to study *tessellations* of the plane. A **tessellation** of a plane is the filling of the plane with repetitions of figures in such a way that no figures overlap and there are no gaps. (Similarly, one can tessellate space.) The tiling of a floor and various mosaics are examples of tessellations. Maurits C. Escher, born in the Netherlands in 1902, was a master of tessellations. Many of his drawings have fascinated mathematicians for decades. An example from his work *Study of Regular Division of the Plane with Reptiles* (pen, ink, and watercolor), 1939, contains an exhibit of a tessellation of the plane by a lizardlike shape, as shown in Figure 12-55.

tessellation

FIGURE 12-55

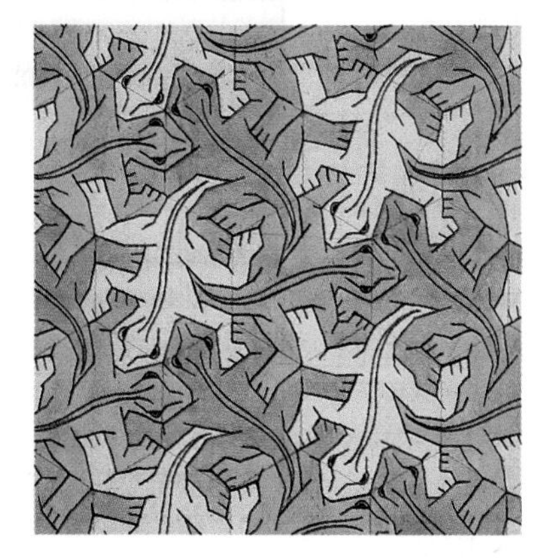

SAMPLE SCHOOL BOOK PAGE: SEMIREGULAR TESSELLATIONS

Extension ▶▶

Semi-Regular Tessellations

In Exercise 6 on page 573, you saw that one regular polygon can be used to create a tessellation. You can also create tessellations using more than one regular polygon. For example, an equilateral triangle and a regular hexagon were used to create the tessellation below.

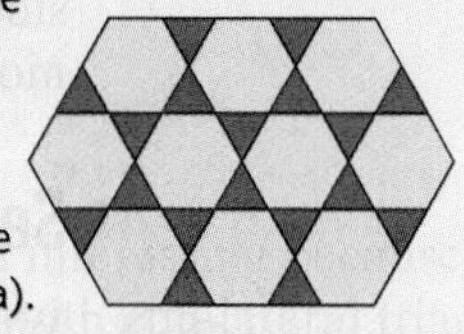

17. a. Which pattern blocks are regular polygons?

b. Create two or three different tessellations that include more than one type of the pattern blocks you identified in part (a).

A *semi-regular tessellation* uses more than one kind of regular polygon, and has the same arrangement of polygons at every vertex. The tessellation shown above is an example of a semi-regular tessellation.

18. Two different arrangements of regular polygons appear in the tessellation below, so it is not a semi-regular tessellation.

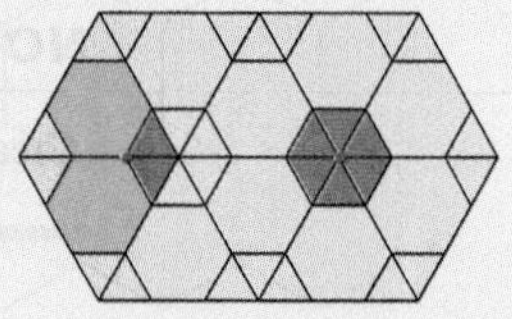

a. Identify which of your tessellations from Exercise 17(b) are semi-regular tessellations.

b. Is the tessellation shown on page 565 a semi-regular tessellation? Explain.

Section 4 Tessellations and Volumes of Cones 575

Source: Math Thematics, Book 2, McDougal Littell, 2005 (p. 575).

As we saw earlier in this section, a regular pentagon does not tessellate the plane. However, some nonregular pentagons do. One is shown in Figure 12-59, along with a tessellation of the plane by the pentagon.

FIGURE 12-59

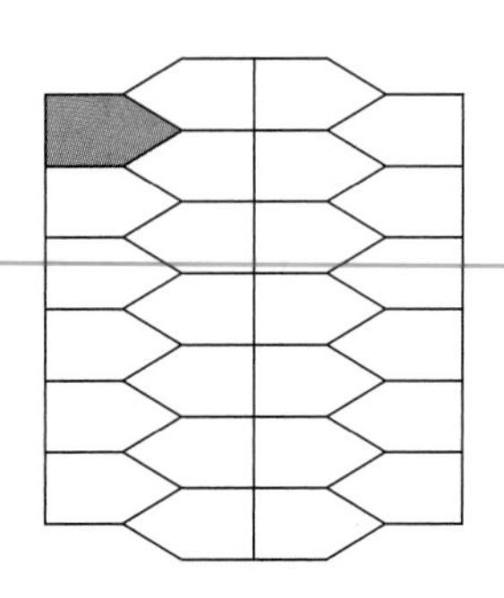

▶ **HISTORICAL NOTE** Determining which irregular pentagons tessellate is a surprisingly rich problem. Mathematicians thought they had solved it when they had classified eight types of pentagons that would tessellate. They believed they had all of them. But then in 1975, Marjorie Rice, a woman with no formal training in mathematics, discovered a ninth type of tessellating pentagon. She went on to discover four more by 1977. Her interest was piqued by reading an article in *Scientific American* by Martin Gardner. Two of the pentagons she found are shown in Figure 12-60. The problem of how many types of pentagons tessellate remains unsolved.

FIGURE 12-60

Type 9 discovered in February 1976

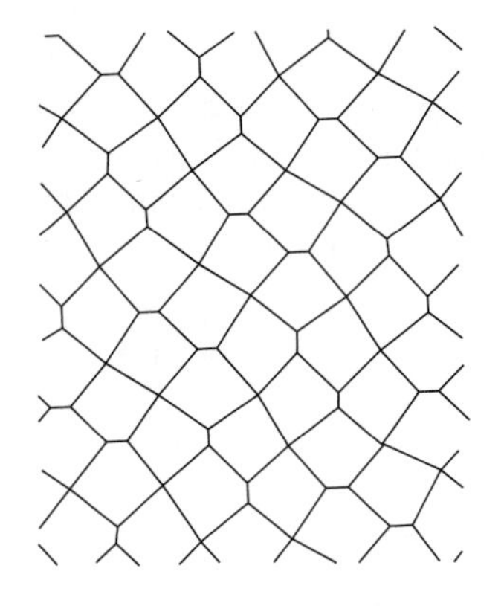

Type 13 discovered in December 1977

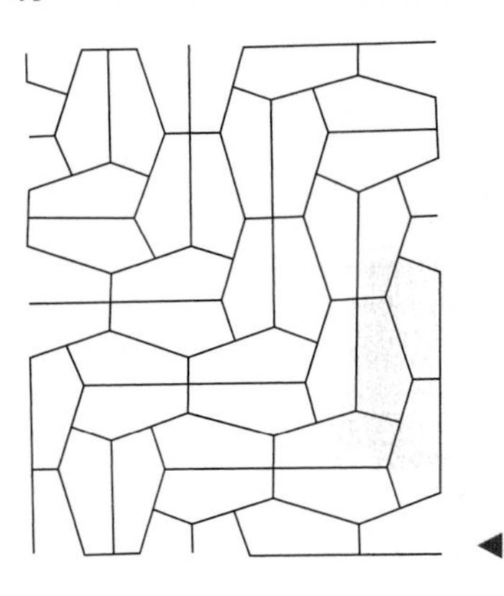

What other types of designs can be made that tessellate a plane? The plane geometry and motions studied earlier give some clues on how to design shapes that work. Consider one of the methods used in Chapter 10 to determine the area of a parallelogram. In Figure 12-61(a), triangle *ABE* was removed from the left of parallelogram *ABCD* and slid to the right, forming the rectangle *BB'E'E* of Figure 12-61(b). This same notion can be used to create a tessellating shape.

FIGURE 12-61

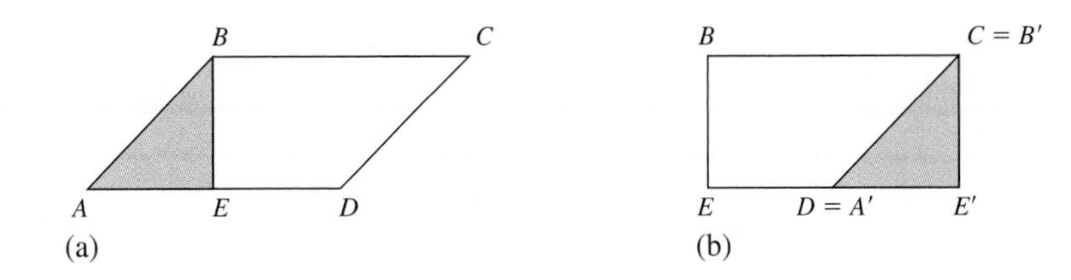

Consider any polygon known to tessellate a plane, such as rectangle *ABCD* in Figure 12-62(a). On the left side of the figure draw any shape in the interior of the rectangle, as in Figure 12-62(b). Cut this shape from the rectangle and slide it to the right by the slide that takes *A* to *B*, as shown in Figure 12-62(c). The resulting shape will tessellate the plane. Why?

FIGURE 12-62

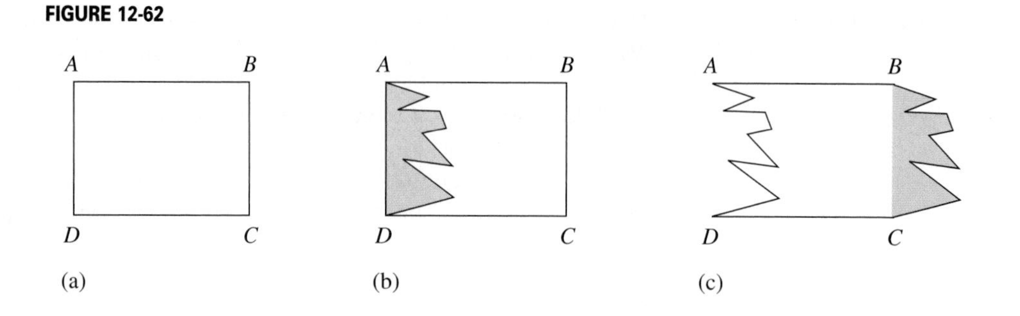

A second method of forming a tessellation involves a series of rotations of parts of a figure. In Figure 12-63(a), we start with an equilateral triangle *ABC*, choose the midpoint *O* of one side of the triangle, and cut out a shape, as in Figure 12-63(b), being careful not to cut away more than half of angle *B*, and then rotate the shape 180° clockwise around point *O*. If we continue this process on the other two sides, then we obtain a shape that can be rotated around point *A* to tessellate the plane. Complete the tessellating shape and tessellate the plane with it "Now Try This 12-9."

FIGURE 12-63

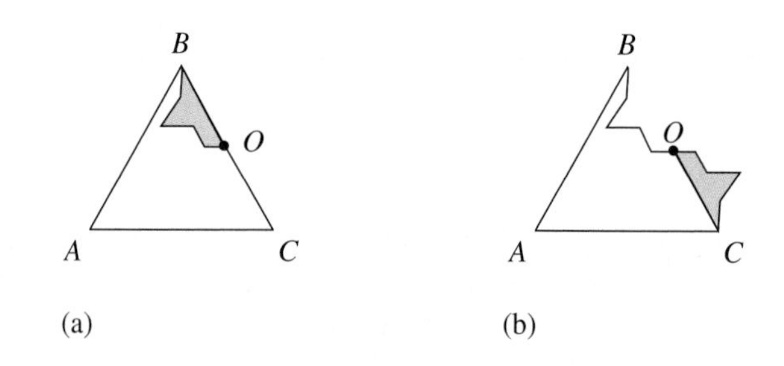

NOW TRY THIS 12-9

Continue the drawing of the tessellating shape in Figure 12-63. Cut out the shape and use it to draw a tessellation of the plane.

ONGOING ASSESSMENT 12-5

1. On dot paper, draw a tessellation of the plane using the following figures:

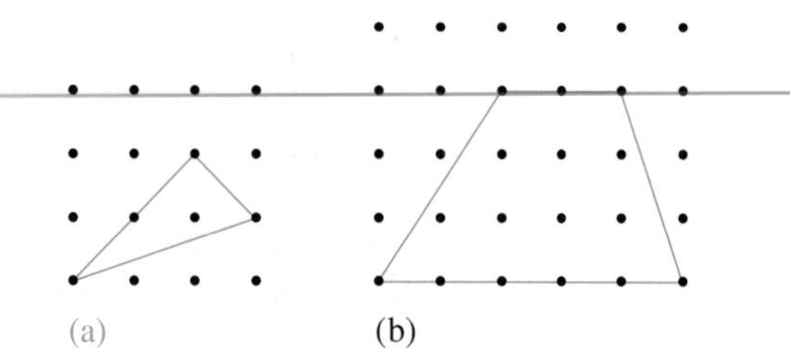

2. **a.** Tessellate the plane with the following quadrilateral:

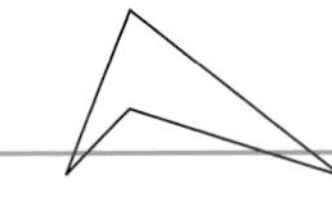

   b. Is it possible to tessellate the plane with any quadrilateral? Why or why not?
3. On square-dot paper, use each of the following four pentominoes, one at a time, to make a tessellation of the plane, if possible. (A *pentomino* is a polygon composed of five

congruent, nonoverlapping squares.) Which of the pentominoes tessellate the plane?

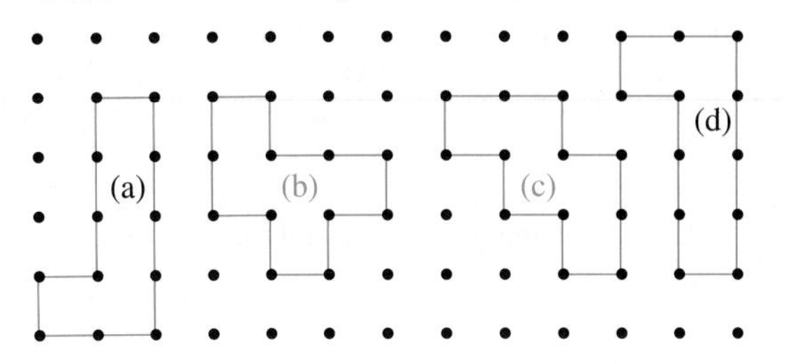

4. We have seen that equilateral triangles, squares, and regular hexagons are the only regular polygons that will tessellate the plane by themselves. However, there are many ways to tessellate the plane by using combinations of these and other regular polygons, as shown in the following figure:

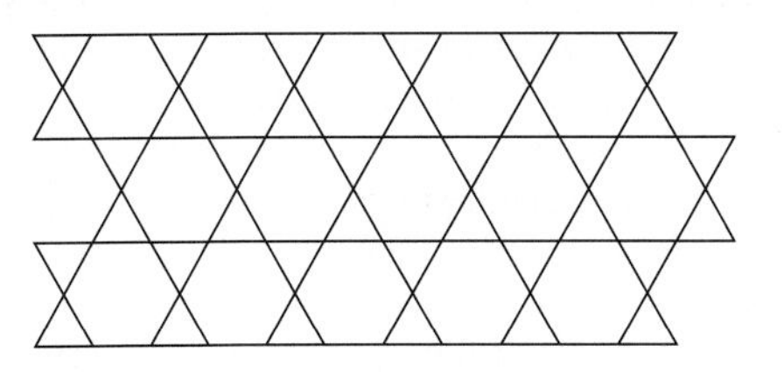

Try to produce other such tessellations by using the following:

a. Only equilateral triangles, squares, and regular hexagons
b. Regular octagons (8-gons) and squares

5. To determine if a shape created using a glide reflection will tessellate the plane, complete the following:

a. Start with a rectangle. Determine some shape that you might use with a slide to form a tessellating shape. Slide it as shown. Determine the horizontal line of symmetry of the rectangle, and reflect as shown.

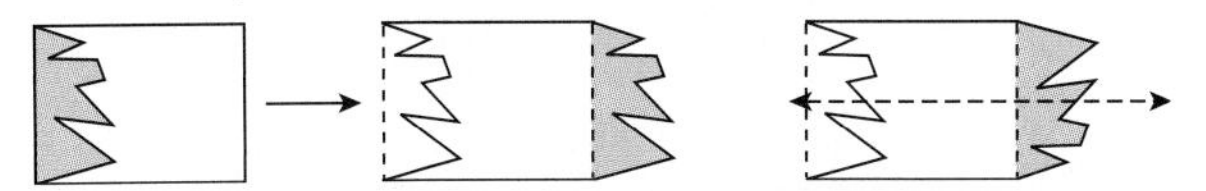

b. Explain why the described series of motions is a glide reflection.
c. Determine whether the final shape will tessellate the plane.

6. The **dual of a regular tessellation** is the tessellation obtained by connecting the centers of the polygons in the original tessellation that share a common side. The dual of the tessellation of equilateral triangles is the tessellation of regular hexagons, shown in color in the following figure:

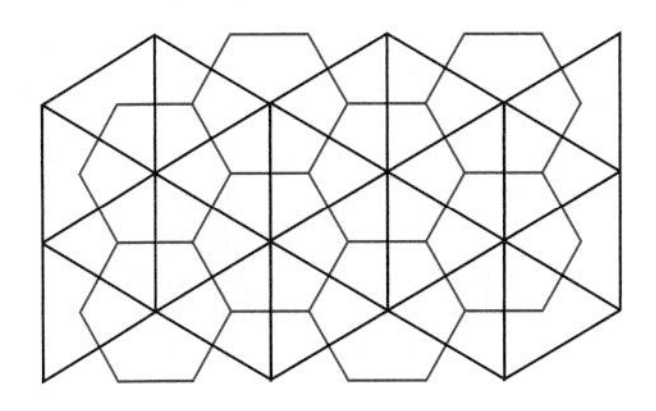

Describe and show the dual of each of the following:

a. The regular tessellation of squares shown in Figure 12-56(a)
b. The tessellation of squares in Figure 12-56(b)
c. A tessellation of regular hexagons

7. A sidewalk is made of tiles of the type shown in the following figure:

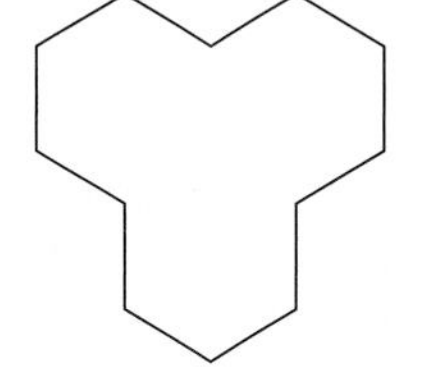

Each tile is made of three regular hexagons from which three sides have been removed. Draw a partial tessellation composed of seven such figures.

8. Which of the following tessellations are semiregular and which are not? Justify your answers.

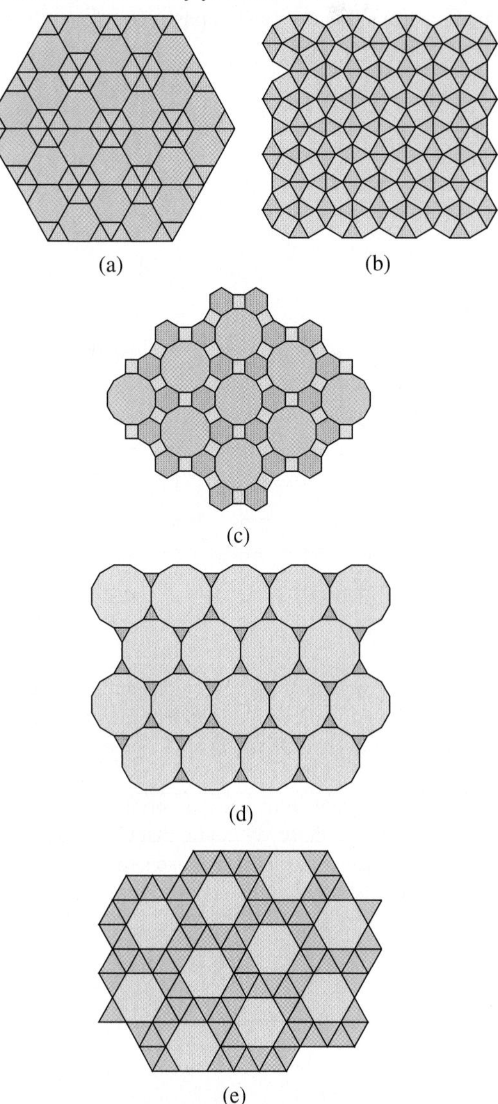

Communication

9. The following figure is a partial tessellation of the plane with the trapezoid $ABCD$:

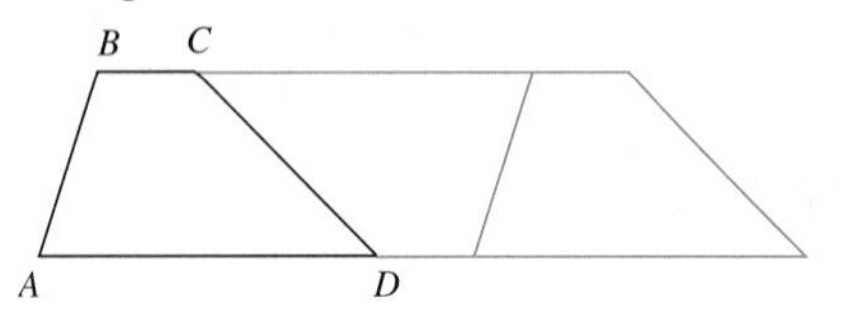

   a. Explain how the tessellation can be used to find a formula for the area of the trapezoid.
   b. Tessellate the plane with a triangle and show how the tessellation can be used to find the relationship between the length of the segment connecting the midpoints of the two sides of a triangle and the length of the third side.

10. Explain in your own words why only three types of regular polygons tessellate the plane.

11. The following figure shows how to tessellate the plane with irregular pentagons. Explain how the pentagons can be constructed and actually construct at least three such pentagons.

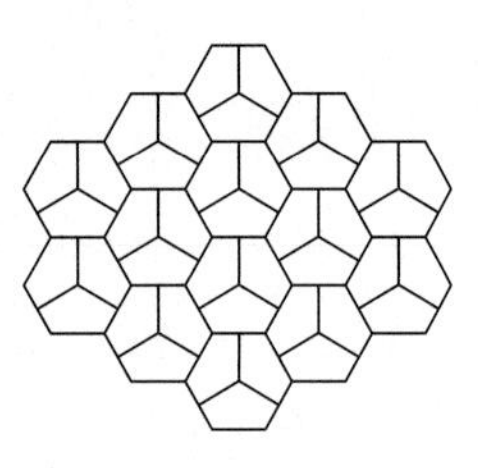

Open-Ended

12. There are endless numbers of figures that tessellate a plane. In the following drawing, the shaded figure is shown to tessellate the plane:

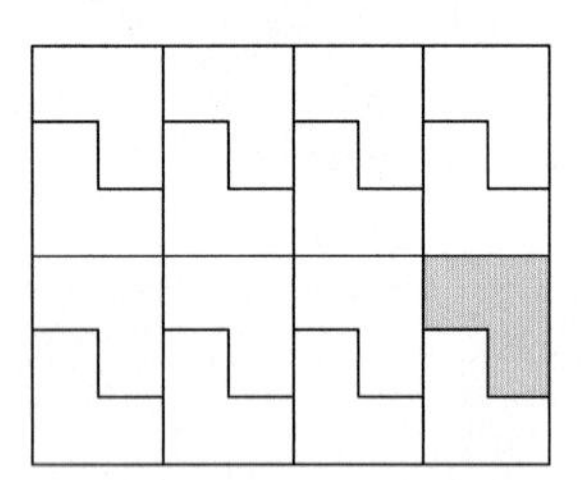

Design several different polygons and show how each can tessellate the plane. What transformations are used in each of your designs? Explain how they are used to tessellate the plane.

13. Examine different quilt patterns or floor coverings and make a sketch of those you found that tessellate a plane.

14. A cube will tessellate space but a sphere will not. List several other solids that will tessellate space and several that will not.

Cooperative Learning

15. Each member of a small group is to find a drawing by M. C. Escher that does not appear in this text and in which the concept of tessellation is used. Each then shows the other members of the group, in detail, how he or she thinks Escher created the tessellation in a drawing.

16. a. Convince the members of your group that the following figure containing six equilateral triangles tessellates the plane:

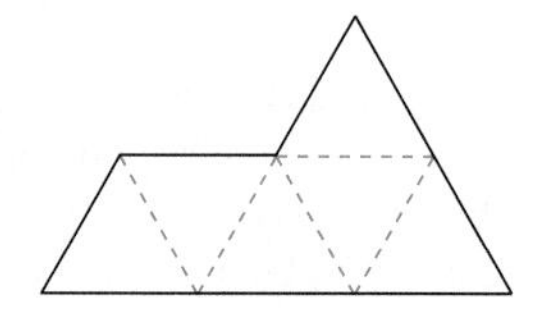

   b. As a group, find different figures that contain six equilateral triangles. How many such figures can you find? Discuss the meaning of "different."
   c. Find some of the figures in (b) that are *rep-tiles*. (A rep-tile is a figure whose copies can be used to form a larger figure similar to itself.) Convince other members of your group that your figures are rep-tiles and that they tessellate the plane.

17. Trace each of the following shapes and cut out several copies of each. Each member of a small group should pick a shape, decide if it tessellates the plane, and convince other members of the group if it does or does not tessellate. The group should report the answers with figures or an argument why a shape does not tessellate.

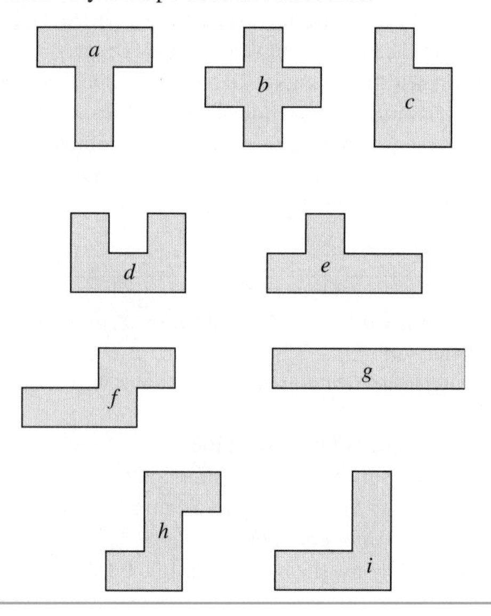

LABORATORY ACTIVITY

Use pattern blocks to construct tessellations using each of the following types of pieces:

1. Squares
2. Equilateral triangles
3. Octagons and squares
4. Rhombuses

HINT FOR SOLVING THE PRELIMINARY PROBLEM

Show that $\triangle TBT'$ and $\triangle BAA'$ are equilateral. Then prove that $AT + BT + CT = A'T' + TT' + CT'$. To find the location of T' that yields the minimal sum of required distances, connect C to A'.

QUESTIONS FROM THE CLASSROOM

1. A student asks, "If I have a point and its image, is that enough to determine whether the image was found using a translation, reflection, rotation, or glide reflection?" How do you respond?
2. Another student asks a question similar to question 1 but is concerned about a segment and its image. How do you respond to this student?
3. A student claims that a kite has no lines of symmetry. How do you respond?
4. A student says that every three-dimensional figure that has plane symmetry automatically has line symmetry. Do you agree? Why or why not?
5. A student says that in a size transformation with the scale factor 0, we do not have a transformation. Is that true?
6. A student asks if every translation on a grid can be accomplished by a translation along a vertical direction followed by a translation along a horizontal direction. How do you respond?
7. A student asks why the images of two perpendicular lines will also be perpendicular under a size transformation. How do you respond?
8. When asked if symmetries are used in occupations other than mathematics, how do you respond?
9. A student asks why some wheel covers have symmetry. How do you respond?
10. A student asks, "If I have two triangles that are not similar, is it possible to transform one onto the other by a sequence of isometries followed by a size transformation?" How do you respond?
11. A student wants to know if it is true that a line has infinitely many lines of symmetry. How do you respond?
12. A student wants to know all the symmetries of two intersecting lines. How do you respond?
13. A student asks if a semicircle (a circular arc consisting of half a circle) has half as many symmetries as the entire circle. The same student also wants to know how to find all the symmetries of any arc of a circle. How do you respond?
14. A student heard that the graph of $y = x^2$ as well as the graph of $y = |x|$ has a line symmetry but the graph of $y = x^3$ has point symmetry. The student wants to know why. How do you respond?

CHAPTER OUTLINE

I. Motions of the plane
   A. A **transformation** of the plane is a function from the plane to itself that is a one-to-one correspondence. **Isometries** are transformations that preserve distance.
      1. A **translation** is a transformation of the plane that moves every point a specified distance in a specified direction along a straight line.
         In terms of coordinates, a translation of the plane is a transformation of the plane such that to every point (x, y) corresponds the point $(x + a, y + b)$ for real numbers a and b.
      2. A **rotation** is a transformation of the plane determined by holding one point (the center) fixed and rotating the plane about this point by a certain amount in a certain direction.
      3. A **half-turn** is a rotation of 180°.
      4. Properties of a rotation can be used to show that two lines, neither of which is vertical, are perpendicular if and only if their slopes m_1 and m_2 satisfy the condition $m_1m_2 = {}^{-}1$.
      5. A **reflection** in a line m is a transformation among points of the plane that pairs each point P of the plane with a point P' in such a way that m is the perpendicular bisector of $\overline{PP'}$, as long as P is not on m. If P is on m, then $P = P'$.

6. A **glide reflection** is the composition of a translation and a reflection in a line parallel to the slide arrow of the translation.

B. A **size transformation** S from the plane to the plane is defined as follows: Some point O, the center of the size transformation, is its own image. For any other point Q of the plane, its image Q' is such that $OQ'/OQ = r$, where r is a positive real number and O, Q, and Q' are collinear.

C. Two figures are **similar** if it is possible to transform one onto the other by a sequence of isometries followed by a size transformation.

II. Symmetries

A. A figure has **line symmetry** if it is its own image under a reflection.

B. A figure has **rotational symmetry** if it is its own image under a rotation of less than 360° about a turn center.

C. A figure has **point symmetry** if it has 180° rotational symmetry.

D. A three-dimensional figure has a **plane of symmetry** when every point of the figure on one side of the plane has a mirror image on the other side of the plane.

E. A three-dimensional figure has **rotational symmetry** if when rotated about an axis of rotation it coincides with itself.

***III.** Tessellations

A **tessellation** of a plane is the filling of the plane with repetitions of figures in such a way that no figures overlap and there are no gaps.

CHAPTER REVIEW

1. Complete each of the following motions:

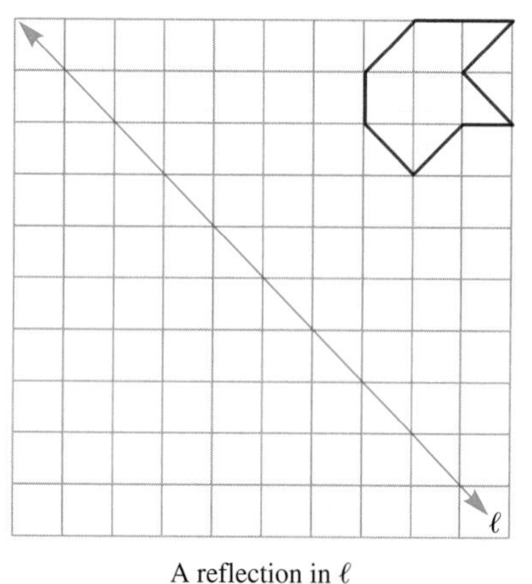

A reflection in ℓ

(a)

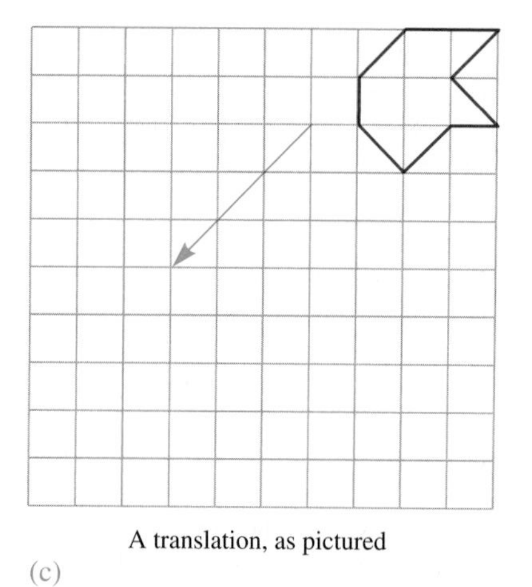

A translation, as pictured

(c)

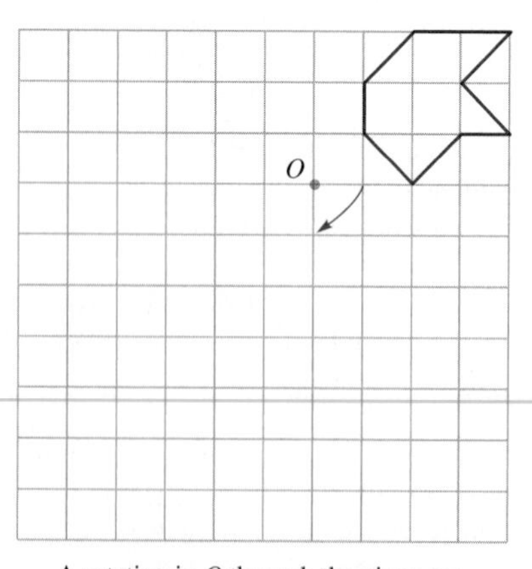

A rotation in O through the given arc

(b)

2. For each of the following figures, construct the image of $\triangle ABC$.

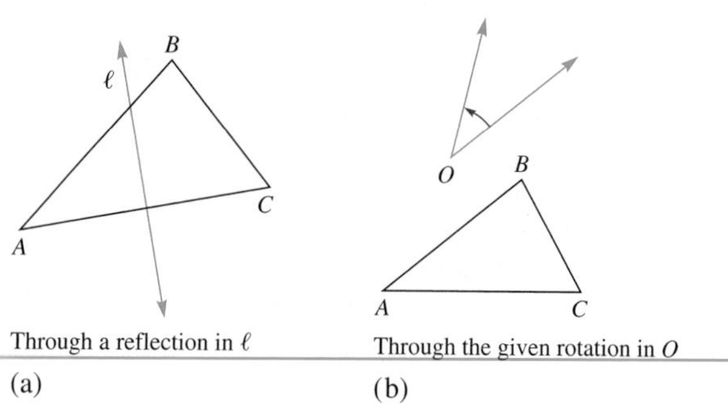

Through a reflection in ℓ

(a)

Through the given rotation in O

(b)

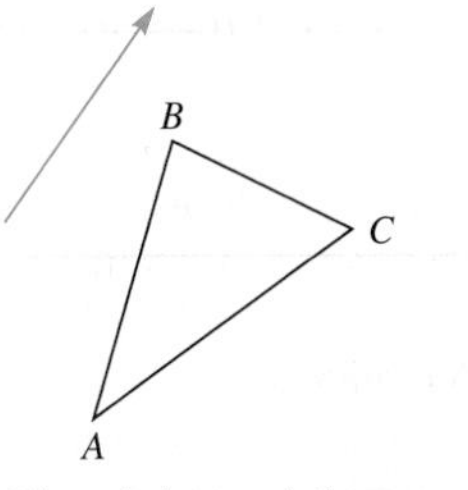

Through the translation arrow pictured

(c)

3. Determine how many lines of symmetry, if any, each of the following figures has:

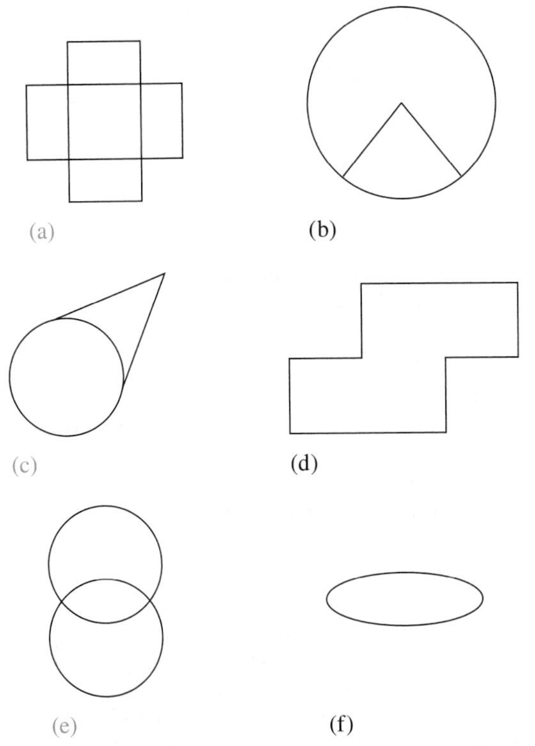

4. For each of the following figures, identify the types of symmetry (line, rotational, or point) it possesses:

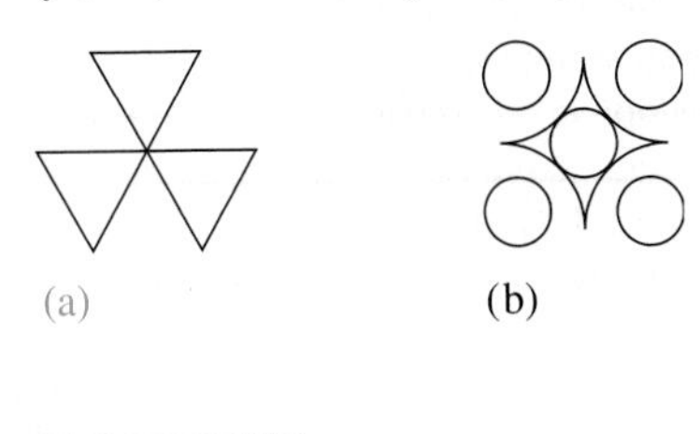

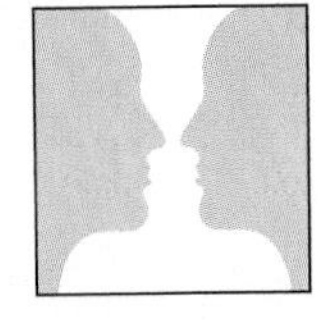

(c)

5. Determine how many planes of symmetry each of the following has:
   a. A ball
   b. A right cylindrical water pipe
   c. A box that is a right rectangular prism but not a cube
   d. A cube
6. What type of symmetry (line, rotational, or point) does each of the lowercase letters of the printed English alphabet have?
7. In the following figure, $\triangle A'B'C'$ is the image of $\triangle ABC$ under a size transformation:

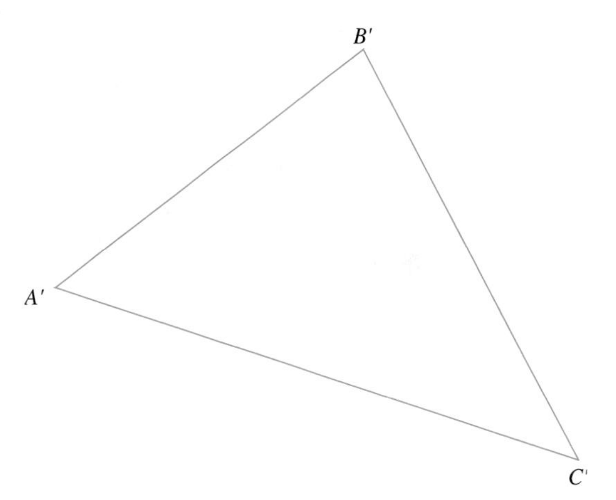

Locate points A, B, and C such that A' is the center of the size transformation and $BC = \frac{1}{2}B'C'$.

8. Given that $STAR$ in the figure shown is a parallelogram, describe a sequence of isometries to show the following:
   a. $\triangle STA \cong \triangle ARS$
   b. $\triangle TSR \cong \triangle RAT$

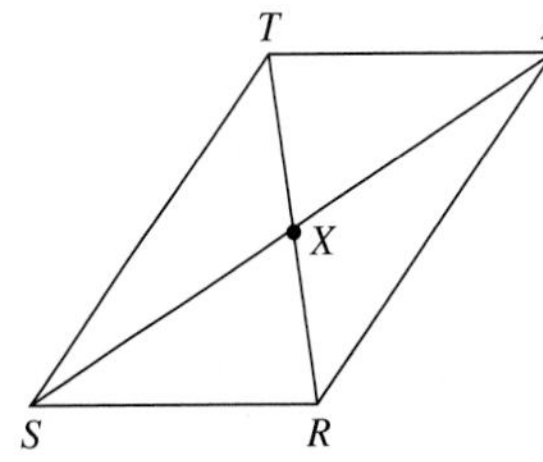

9. Given that $BEAUTY$ in the figure shown is a regular hexagon, describe a sequence of isometries that will transform the following:
   a. $BEAU$ into $AUTY$
   b. $BEAU$ into $YTUA$

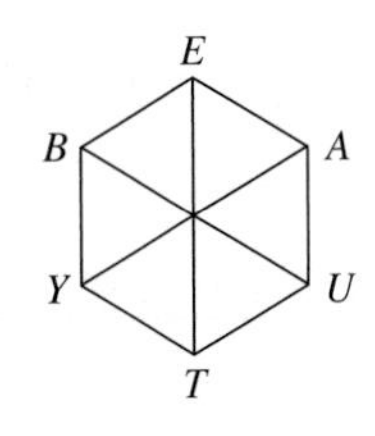

10. Given that $\triangle SNO \cong \triangle SWO$ in the following figure, describe one or more isometries that will transform $\triangle SNO$ into $\triangle SWO$:

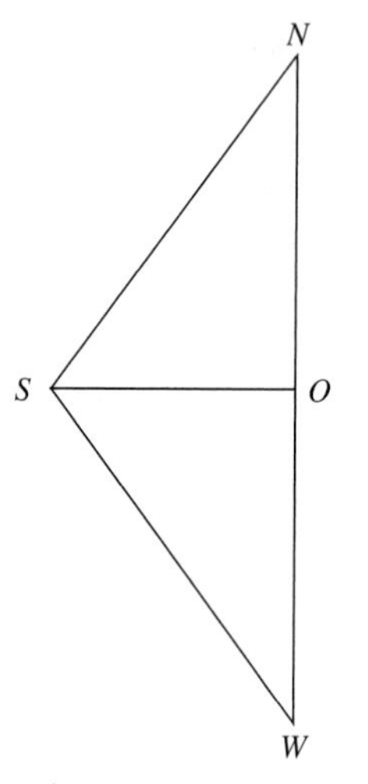

11. Show that $\triangle SER$ in the following figure is the image of $\triangle HOR$ under a succession of isometries with a size transformation:

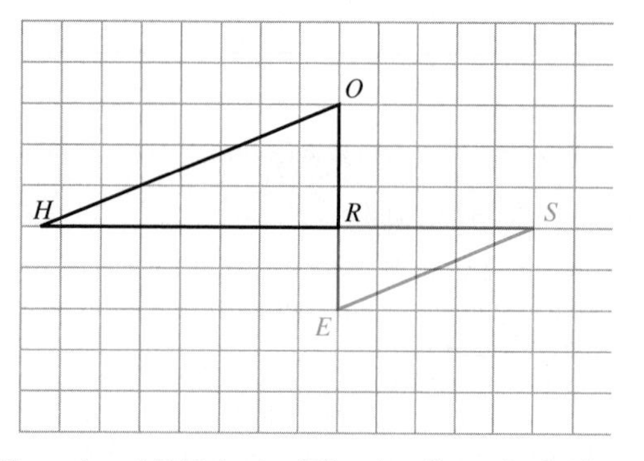

12. Show that $\triangle TAB$ in the following figure is the image of $\triangle PIG$ under a succession of isometries with a size transformation:

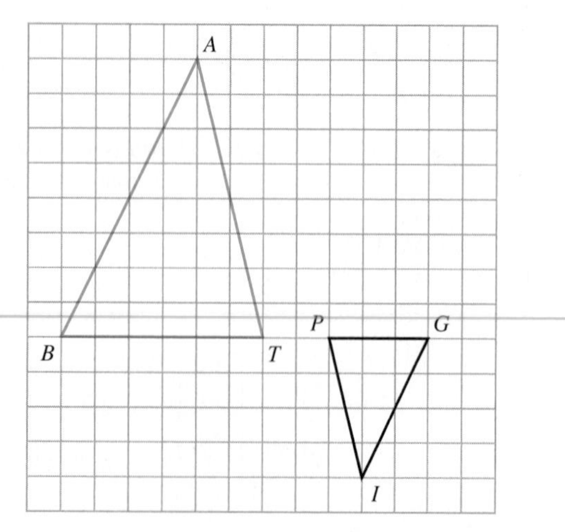

13. The triangle $A'B'C'$ with $A'(0, 7.91)$, $B'(^-5, ^-4.93)$, $C'(4.83, 0)$ is the image of triangle ABC under the translation $(x, y) \to (x + 3, y - 5)$.
    a. Find the coordinates of A, B, and C.
    b. Under what translation will the image of $\triangle A'B'C'$ be $\triangle ABC$?

14. If a translation is determined by $(x, y) \to (x + 3, y - 2)$ is followed by another translation determined by $(x, y) \to (x - 3, y + 2)$, describe a transformation that would achieve the same thing.

15. Suppose $\triangle A''B''C''$ with $A(0, 0)$, $B(1, 5)$, $C(^-2, 7)$ is the image of $\triangle ABC$ under the translation $(x, y) \to (x + 2, y - 1)$ followed by the translation $(x, y) \to (x + 1, y + 3)$.
    a. Find the coordinates of A, B, and C.
    b. Under what single transformation will the image of $\triangle ABC$ be $\triangle A''B''C''$?

16. Write each of the following as a single transformation.
    a. (i) A translation from A to B followed by a translation from B to C.
    (ii) A translation from B to C followed by a translation from A to B.
    b. A rotation about O by 90° counterclockwise, followed by a rotation about O by 30° clockwise.
    c. **(i)** A size transformation with center O and scale factor 3 followed by a size transformation with center O and scale factor 2.
    (ii) The size transformations in part (i) in reverse order.

17. Which of the following figures has more symmetries? Justify your answer.
    a. A rhombus or a square
    b. An isosceles trapezoid or a parallelogram
    c. A cube or a box; that is, a rectangular right prism

18. Given the line $y = 2x + 3$ and the point $P(^-1,3)$, find the equation of each of the following:
    a. The line through P parallel to the given line
    b. The line through P intersecting the given line at the point whose y-coordinate is 1
    c. The line through P perpendicular to the given line

19. Find the equation of the image of the line $y = -x + 3$ under each of the following transformations:
    a. The translation $(x, y) \to (x + 2, y - 3)$
    b. Reflection in the x-axis
    c. Reflection in the y-axis
    d. Reflection in the line $y = x$
    e. Half-turn about the origin
    f. Size transformation with center at the origin and scale factor 2

★ **20.** A line has the property that its reflection in the line $y = x$ is itself. Find the equations of all such lines. Explain your reasoning.

21. The square in the first diagram rolls around a regular hexagon until it is at the bottom in figure (4). Use this information to draw the position of the solid colored triangle and the segment when the square is in diagram (4).

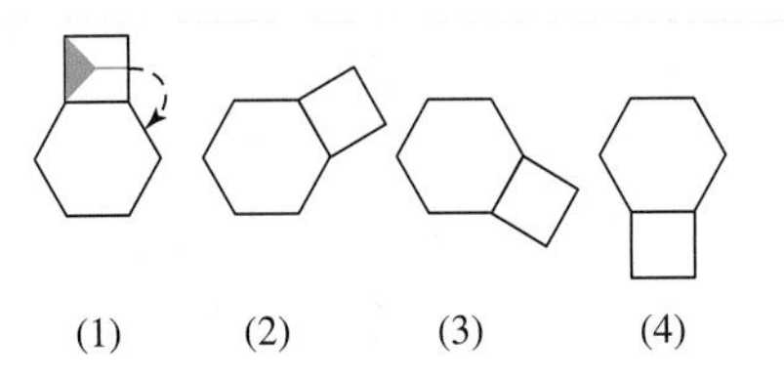

22. Explain why a regular octagon cannot tessellate the plane.

23. A square will tessellate a plane; can we always find a square that will tessellate a rectangle?

★ **24.** Can any tessellating shape be used as a unit for measuring area? If so, explain why.

SELECTED BIBLIOGRAPHY

Clason, R., D. Ericksen, and C. Ericksen. "Cross-and-Turn Tile Patterns." *Mathematics Teaching in the Middle Schools* 2 (May 1997): 430–437.

Clements, D., and J. Sarama. "The Earliest Geometry." *Teaching Children Mathematics* 7 (October 2000): 82–86.

Dayoub, I., and J. Lott. *Geometry: Constructions and Transformations*. White Plains, NY: Dale Seymour Publishing Company, 1977.

Demal, M., and D. Popeler. "*L'enseignement de la géométrie*." *Bulletin de l'apmep* Num 454 (2004), APMEP, Paris, pp. 717–728.

Fossnaugh, L., and M. Harrell. "Covering the Plane with Rep-Tiles." *Mathematics Teaching in the Middle Schools* 1 (January–February 1996): 666–670.

Harrell, M., and L. Fossnaugh. "Allium to Zircon: Mathematics & Nature." *Mathematics Teaching in the Middle Schools* 2 (May 1997): 380–389.

Harris, J. "Using Literature to Investigate Transformations." *Teaching Children Mathematics* 4 (May 1998): 510–513.

Kuchemann, D. "Reflections and Rotations." In *Children's Understanding of Mathematics: 11–16*. London: John Murray, 1981b.

Peterson, B. "From Tessellations to Polyhedra." *Mathematics Teaching in the Middle Schools* 5 (February 2000): 348–356.

Phillips, J. M., and R. E. Zwoyer. *Motion Geometry Book 2: Congruence*. New York: Harper & Row, Publishers, 1969.

——— *Motion Geometry Book 1: Slides, Flips, and Turns*. New York: Harper & Row, Publishers, 1969.

——— *Motion Geometry Book 3: Symmetry*. New York: Harper & Row, Publishers, 1969.

——— *Motion Geometry Book 4: Area, Similarity, and Constructions*. New York: Harper & Row, Publishers, 1969.

Sakshaug, L. "Patterns in Squares." *Teaching Children Mathematics* 7 (November 2000): 172–173.

Seidel, J. "Symmetry in Season." *Teaching Children Mathematics* 4 (January 1998): 244–249.

Sellke, D. "Geometric Flips via the Arts." *Teaching Children Mathematics* 4 (February 1999): 379–383.

Speer, W., and J. Dixon. "Reflections of Mathematics." *Teaching Children Mathematics* 2 (May 1996): 537–543.

Walter, M. *The Mirror Puzzle Book*. New York: Parkwest Publications (1985).

Wesslen M., and S. Fernandez. "Transformational Geometry." *Mathematics Teaching* 191 (June 2005).

White, D. "Kenta, Kilts, and Kiminos: Exploring Cultures and Mathematics through Fabrics." *Teaching Children Mathematics* 7 (February 2001): 354–359.

Zaslavsky, C. "Symmetry in American Folk Art." *Arithmetic Teacher* 37 (January 1990): 6–12.

Zilliox, J., and S. Lowrey. "Many Faces Have I." *Mathematics Teaching in the Middle Schools* 3 (November–December 1997): 180–183.

APPENDIX I

Using a Spreadsheet

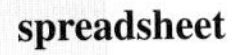

Introduction

spreadsheet

An electronic **spreadsheet** is a table of rows and columns in which each cell may contain a value that can be operated on and changed at any time. A spreadsheet has the capacity to allow a change in one cell to be reflected in any other cell that relies on the data in that original cell.

In *Principles and Standards*, we find the following:

Students can learn more mathematics more deeply with the appropriate use of technology. . . .

Students' learning is assisted by feedback, which technology can supply: . . ., change the defining rules for a spreadsheet, and watch as dependent values are modified.

Spreadsheets, dynamic geometry software, and computer microworlds are also useful tools for posing worthwhile problems. (p. 26)

Table AI-1

x	$f(x)$
1	1
2	4
3	9
4	16
5	25

Table AI-1 depicts certain values for the function $f(x) = x^2$. In Table AI-1, each item in the $f(x)$ column depends on the value in the x column. A spreadsheet, as it might appear on a computer, is shown in Figure AI-1, depicting the information in Table AI-1. (The labels x and $f(x) = x$^2 are for headings only.)

FIGURE AI-1

Workbook1

	A	B	C	D	E	F
1	x	f(x)=x^2				
2	1	1				
3	2	4				
4	3	9				
5	4	16				
6	5	25				
7						

In Figure AI-1, an index column down the left side numbers the rows of the table. Across the top, an index row of letters identifies the columns. In the body of the table, the value 3 is in column A and row 4. As a result, the value 3 could be identified by label A4, where A is the column heading and 4 is the row number. The labels can be used to write formulas for the values of column B in terms of column A. For example, the value in B2 could be found by multiplying the A2 value times itself, or in computer language as = A2^2, depicting $(A2)^2$. Similarly, the value in B5 could be written = A5^2. In Figure AI-1, the items in row 1 are used as headings to indicate what the table is depicting.

The user bars on an opening of the *Microsoft Excel* spreadsheet may be shown as in Figure AI-2. (The bars may be customized by the user.)

FIGURE AI-2

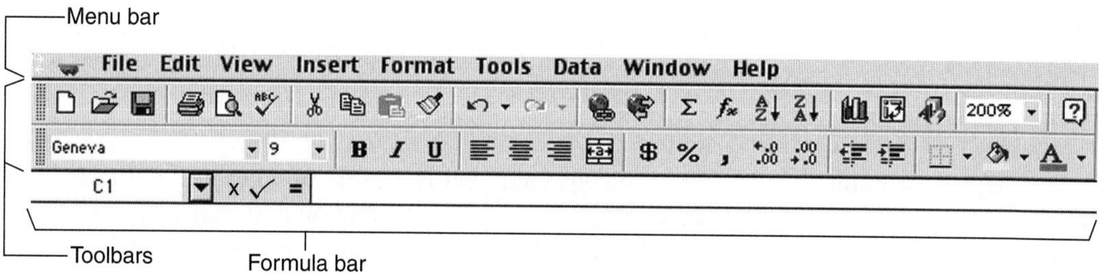

Menu bar

File Edit View Insert Format Tools Data Window Help

Toolbars

Formula bar

Figure AI-2 depicts the **Menu** bar, the **Standard** toolbars, and a **Formula** bar. The **Menu** bar offers such options as **File, Edit, Format,** and so on. For example, as a beginning user you might choose to click **File** and choose *New* to obtain a new worksheet on the screen. The standard toolbars show different options available to the user, including file folder, disk, and printing options on the left along with other types of options, such as **B** for bold type and the *I* for italics. The **Formula** bar is for entry of items in a worksheet and is discussed later in this appendix.

Developing a Spreadsheet

To create the spreadsheet in Figure AI-1, we first open a new worksheet and then type the entries in cells A1 and B1 by typing x and $f(x) = x$^2, respectively, in the entry line (also known as the **Formula** bar) and pressing RETURN after each. These are simply headings for our reference.

1. To create the column of values listed under x in the A column, we enter the first item as 1 in A2 and press RETURN. Then we use a formula to create the value for A3 by using the value for A2 in the formula. We tell the spreadsheet that a formula is being used by highlighting the cell to be filled and typing = in the entry line followed by the formula we wish to use. In this case, we want A3 to have the value A2 +1. We signal this by highlighting cell A3, typing = A2 +1, and pressing RETURN. That done, the value 2 appears in A3, as shown in Figure AI-3. (Near the **Formula** bar are two boxes, one containing an × and one containing a tick mark. The × is used to cancel any changes you have made in a formula. The tick is used in the same way as the RETURN key.)

FIGURE AI-3

Workbook1

	A	B	C	D	E	F
1	x	f(x)=x^2				
2	1					
3	2					
4						

2. To complete the column under A, we use the *Fill Down* command by first highlighting A3 and all the cells to be filled and then using the *Fill Down* command in the **Menu** bar normally found under the **Edit** menu. The *Fill Down* command fills successive cells in column A by adapting the created formula to accommodate the cell number. For example, the entry in A4 is automatically created as the value of A3 + 1, and so on. The *Fill Down* command will fill all highlighted cells. If the cells A3 through A6 are highlighted, the result is seen in Figure AI-4. (If there is no data in a cell to create other cells, the program assumes that the value in the cell is 0.)

FIGURE AI-4

Workbook1

	A	B	C	D	E	F
1	x	f(x)=x^2				
2	1					
3	2					
4	3					
5	4					
6	5					

3. To complete the column under the heading B and $f(x) = x$^2, we highlight cell B2, type = A2^2, and press RETURN. This should cause a 1 to be placed in cell B2. We then highlight cell B2 and the rest of the column through row 6 and use the *Fill Down* command to complete column B.

REMARK Most spreadsheets also have a *Fill Right* command to fill in cells in rows as well as columns.

To clear a table or a set of values in *Excel*, highlight the desired values to be deleted and pull down the **Edit** menu. If we choose *Clear*, then we are asked about clearing *All, Formats, Formulas,* or *Notes*. To clear all highlighted values, choose *All* and press RETURN.

Most spreadsheets allow the use of various functions, including $+, -, \div, \cdot, ^\wedge$, trigonometric functions, square roots, and absolute values. To determine what your spreadsheet can do, consult the software manual or the **Help** menu. (Note that many graphing calculators have the capability of acting like a spreadsheet.)

Graphing with a Spreadsheet

In addition to allowing the use of arithmetic operations, most spreadsheets can create graphs of the data presented in a table. For example, to create a graph of the data in Figure AI-1, we follow these steps:

1. Highlight the information to be graphed. In the graph in Figure AI-5, the data from rows A2 through A6 and columns B2 through B6 were highlighted. After the data is highlighted, then we choose the *Graph* icon from the **Toolbar**. (On some spreadsheets, when you choose the *Graph* icon, you can size the graph as you want; other spreadsheets will do this automatically.) In *Excel*, clicking the icon once causes the highlighted information to be placed in a flashing dashed rectangle. If we move the mouse pointer on the worksheet, we see a + symbol. Click the mouse, and drag it while holding down the mouse button to size the graph. Once the graph is sized, most spreadsheets ask for a variety of information, such as what you want for the x-values, what you want for the y-values, what legends you want, and what maximum and minimum values you want. Because each spreadsheet is different, we suggest that you consult your user's manual to see what features are available. For example, in *Excel* the first question is about the range of data. With the data we are graphing, the range is written $A$2:$B$6 automatically. We can change or accept this range. The $A$2 indicates that the first values from our table to be used in the graph start at A2 and this will not change in the graph; $B$6 tells the computer that cell B6 contains the last value of the graph.

FIGURE AI-5

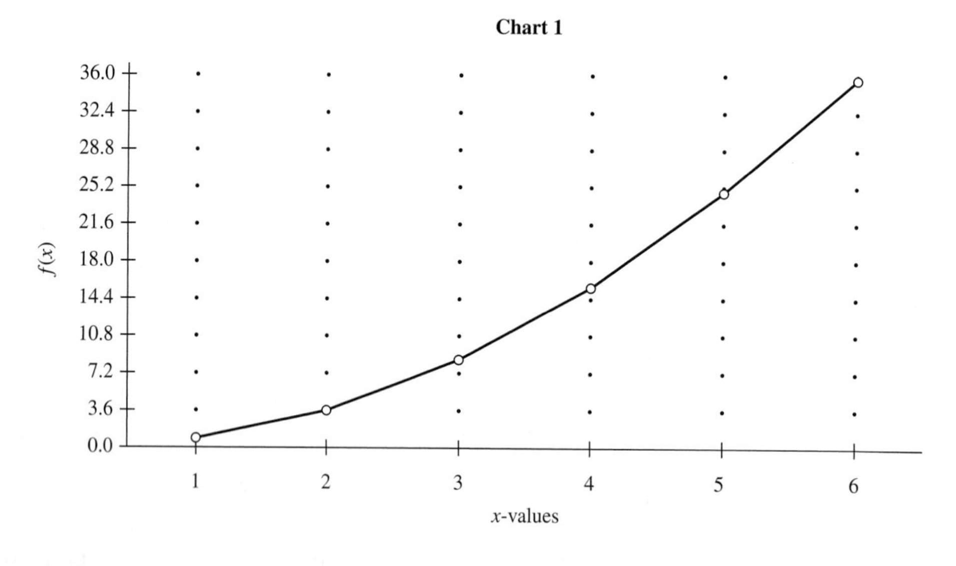

REMARK The $ is used before and after a variable to indicate that the variable is fixed and does not change in this application. This feature is useful particularly when we want to use a specific value of a variable over and over in an application. An example is seen in Ongoing Assessment problem 4, where we use the calculated mean of a set of data over and over to find the standard deviation of the set of data.

2. Because the range from A2 through B6 contains the values we intend to graph, we move to the next screen, where we must select a chart type, usually from several options. *Excel* presents these options in icon form. Figure AI-5 was created as a scatterplot. Once we choose the scatterplot style, by double-clicking an icon, the next screen asks about the type of scatterplot desired. Again, in *Excel*, the choices are presented as icons. We chose the connected scatterplot (option 2), after which a graph is drawn. But there are other options to consider. To use the data as ordered pairs with the x data as the first column, we are using the data series in columns. Finally, we are asked if we want to use the first row as a legend or as data. Because we started with A2, we want this as the first value of x data. Next, we are asked if we want to add a legend to the graph. Figure AI-5 shows the data from Figure AI-1 as a connected scatterplot. Using a title of Chart 1, we see a finished product with the $f(x)$ values and x-values labeled along the axes.

The same information could be depicted as a bar graph, as shown in Figure AI-6, by using other graph icons.

FIGURE AI-6

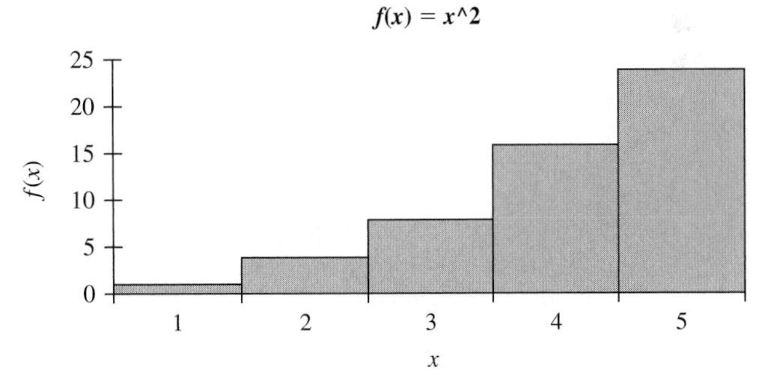

Many spreadsheets allow the options of using scatterplots, line plots, bar graphs, stem and leaf plots, and three-dimensional plots. Again, check the user's manual to see what is available for your spreadsheet.

Explicit and Recursive Formulas

When using a spreadsheet as in the previous paragraphs, most users discover they need to know the difference between explicit and recursive formulas. In Figure AI-4, a **recursive formula** was used—a value in a cell was determined by the value in the cell immediately above it. For example, the value in cell A5 is 1 plus the value in cell A4. Whenever a value in any cell is expressed in terms of a value in a previous cell (or cells), we say that the expression is a *recursive expression* or *formula*. Recursive formulas are typically used by most young students to describe patterns. For example, most students would describe the following pattern as adding 3, meaning that to find the next item, a person would add 3 to the previous term:

recursive formula

$$3, 6, 9, 12, 15, 18, 21, 24, \ldots$$

On a spreadsheet, the pattern could be depicted using two columns. Row A may serve to count the numbers of the terms, as seen in Figure AI-7. (Note that column A serves the same purpose as the index column, but we can use the numbers in the A cells to create the terms in the B column.)

FIGURE AI-7

	A	B
1	1	3
2	2	6
3	3	9
4	4	12
5	5	
6	6	

To fill in additional rows, we can highlight cell B5 and type = B4 + 3 in the entry line and then use the *Fill Down* command by highlighting the number of entries needed. Clearly, the value of a cell is determined by the cell immediately above it. In this example, the values in column B are multiples of 3. If n is the term number, then the corresponding value is $3n$. The expression $3n$ is an *explicit formula* for finding the value of an expression as a function of the number of the term n.

To use the explicit formula $3n$ to create the entries in column B, we use the formula = 3*A:A to fill in the entry in B1 and then use the *Fill Down* command to fill all highlighted cells of column B where there is a corresponding A entry. The use of :A at the end of the formula tells the spreadsheet to use column A to fill column B entries.

Other Features of a Spreadsheet

Spreadsheets can act as simple calculators. For example, determine the number of handshakes at a party that 25 people attended. Each person shook hands with everyone else, and no one shook hands with himself or herself. This problem can be solved using a spreadsheet. We consider a *simpler case* whereby there is first only one person in the room, in which case there are no handshakes. Then another person enters for a total of two persons, and so one handshake occurs. When a third person enters the room, two more handshakes take place, and so on. This could be recorded as in Figure AI-8, in which column A records the number of people in the room and is developed by using the *Fill Down* command and the formula = A2 + 1. Column B depicts the number of additional handshakes when another person enters the room. Column B is 1 less than the number in column A and can be set up using the formula = A2 − 1.

FIGURE AI-8

	A	B
1	No. People	No. of Additional Handshakes
2	1	0
3	2	1
4	3	2
5	4	3
...	...	...
26	25	24
27	Total	300

To find the total number of handshakes at the party, we need to find the total number of handshakes in column B. The spreadsheet does this when we highlight the cell in which we want the total, in this case B27, and then click the Σ button. (On some spreadsheets, you may need to use the *SUM* feature.) When we click Σ, the spreadsheet surrounds the numbers immediately above the highlighted cell with a flashing rectangle to show what will be added. (In some cases, we may have to click and drag a rectangle around the numbers that are to be added. This is true especially if there is an empty cell among the other cells to be added.) When the RETURN or ENTER key is pressed, the total appears in the highlighted cell. As Figure AI-8 shows, the total is 300 handshakes for the 25 people.

Many other features available on most spreadsheets are not discussed here. We suggest that you consult the user's manual that comes with your spreadsheet program and try the following problems.

ONGOING ASSESSMENT A-I

1. a. Write any arithmetic sequence.
   b. In column A of your spreadsheet, list the number of the term of the arithmetic sequence.
   c. What formula did you use to fill down the column?
   d. List the first 25 terms of the arithmetic sequence in column B.
   e. What formula could be used to fill in the terms in column B?
   f. Find the sum of the first 25 terms in the arithmetic sequence. Describe how that was done with a spreadsheet.
   g. Plot the number of the terms of the arithmetic sequence versus the actual terms using the graphing or chart option of your spreadsheet. Describe the graph of the arithmetic sequence.
2. **a.** Write any geometric sequence.
   b. In column A of your spreadsheet, list the number of the term of the geometric sequence.
   c. List the first 25 terms of the geometric sequence in column B.
   d. What formula could be used to fill in the terms of column B?
   e. Find the sum of the first 25 terms of the geometric sequence. Describe how that was done by using a spreadsheet.
   f. Plot the number of the terms of the geometric sequence versus the actual terms using the graphing or chart option of your spreadsheet. Describe the graph of the geometric sequence.
3. Given the following set of data, use a spreadsheet to find the arithmetic mean:

   23, 45, 67, 78, 98, 54, 36, 76, 75, 24, 43, 54, 100, 99

4. Use the data in problem 3 to develop a spreadsheet to find the standard deviation of the data. Use the columns of the spreadsheet to represent the number of the term, the term, and the difference of the mean and the term. Then find the square of each of the differences, the sum of those squares, and the quotient of the sum and n, where n is the number of terms. Finally, find the square root of the quotient.
5. Use the data in problem 3 to develop a spreadsheet to find the mean absolute deviation of the data. Use the columns of the spreadsheet to represent the number of the term, the term, and the absolute value of the difference of the mean and the term. (To make use of the absolute feature of the spreadsheet, we pull down the Insert feature of the Menu Bar and choose Function. . . . Under that menu, choose ABS and enter the spreadsheet formula that allows finding the difference of the mean and the term.). Sum the absolute values and divide the sum by the number of terms to find the MAD.
6. Businesspeople use spreadsheets for the calculation of interest on loans or outstanding bills. Consider a debt of \$1000 with payments of \$40 per month, which includes 1.5% interest per month on the unpaid balance. Develop a spreadsheet that shows the number of the month the payment was made, the amount of payment in each month, and the outstanding balance. If no other debts accrue, how many months will it take to pay off the debt?
7. Use a spreadsheet to show the first 100 multiples of 13. Explain all steps in developing this spreadsheet.
8. Develop a spreadsheet for finding your college grade-point average. Explain all steps used in developing the spreadsheet.

9. The sequence of Fibonacci numbers is 1, 1, 2, 3, 5, 8, . . ., where each successive term after the first two is the sum of the two preceding terms.

a. Develop a spreadsheet to find the first 25 Fibonacci numbers.

b. Extend your spreadsheet to find the square of each term of the Fibonacci sequence and the sum of those squares. Make a conjecture about the sum of the squares of the first n terms of the sequence.

c. To examine the updating feature of your spreadsheet, change the first two terms of the Fibonacci sequence and observe how each cell that was written based on these terms is changed. Does the conjecture in (b) still hold?

10. Develop a spreadsheet for finding $n!$. Explain the steps used to develop the spreadsheet.

Open-ended

11. In the chapter by Caulfield et al., in the references, find the section entitled "What Are the Odds? Simulating the Lottery."

a. Read that section and simulate a lottery.

b. Use the information gained in part (a) to discuss the reality of winning the lottery.

REFERENCES

Caulfield, R., P. Smith, and K. McCormick. "The Spreadsheet: A Vehicle for Connecting Proportional Reasoning to the Real World in a Middle School Classroom," in *Technology-Supported Mathematics Learning Environments, 67th Yearbook*, edited by W. Masalski and P. Elliott. Reston, VA: National Council of Teachers of Mathematics, 2005.

Parke, C. "Using Spreadsheet Software to Explore Measures of Central Tendency and Variability," in *Technology-Supported Mathematics Learning Environments, 67th Yearbook*, edited by W. Masalski and P. Elliott. Reston, VA: National Council of Teachers of Mathematics, 2005.

APPENDIX II

Graphing Calculators

Introduction

According to the *Principles and Standards of School Mathematics*, we have the following:

. . . graphing utilities facilitate the exploration of characteristics of classes of functions. Because of technology, many topics in discrete mathematics take on new importance in the contemporary mathematics classroom; the boundaries of the mathematical landscape are being transformed. (p. 27)

A graphing calculator can be a useful tool in problem solving. By working through problems we show how a graphing calculator might be used. One advantage of the graphing calculator is the display, not only of graphs but also of text. With this feature, computations and answers can be viewed, various methods can be compared, and changes can be made easily.

In this appendix, we assume your calculator has at least the capabilities of the Texas Instruments TI-73 middle-school graphing calculator. If you use other calculators, you will have to determine whether the discussed features are available on them.

PROBLEM SOLVING **Molly's CD Problem**

Molly belongs to a club that sells CDs by mail. The CDs sell for $12.95 each, and she can order as many as she wants each month. Each order has a shipping charge of $5.00 no matter how many CDs are ordered. She would like to develop a table that tells her how much money she owes when she orders from 1 to 10 CDs. Develop such a table.

This problem is worked using a variety of techniques to show some of the capabilities of the graphing calculator.

Using the Replay Feature Suppose we want to solve Molly's CD Problem using paper and pencil to build a table and using the calculator only to perform the computations. First, we must realize that the cost for any number of CDs is given by multiplying the number of CDs by $12.95 and adding $5.00. To find the cost of one CD using this process, enter the following:

1 × 1 2 . 9 5 + 5

and press ENTER. The answer, 17.95, is displayed. This number can then be recorded in a table. It is important for elementary school students to see the whole problem as well as the answer. This allows students to check visually whether they have entered their numbers and operations correctly.

replay feature

To find the cost for other quantities of CDs, we use the **replay feature** of the calculator. To activate this feature, press 2nd ENTRY; the previous entry is displayed. We then use the left arrow to move the cursor over the first 1 and replace it with a 2. When ENTER is pressed, the next answer, 30.90, is displayed. We could shorten the left arrow strokes even further by pressing 2nd followed by the left arrow; the cursor will go to the beginning of the line. If we continue in this manner, as shown in Figure AII-1, we could find the cost of any number of CDs and enter these costs into a table.

FIGURE AII-1

```
1*12.95+5
                17.95
2*12.95+5
                 30.9
3*12.95+5
                43.85
4*12.95+5
```

Using Data Lists Many graphing calculators can display sequences of numbers as

CLRLIST

lists. To clear existing lists, press STAT and choose the **CLRLIST** (under OPS on the TI-73) feature with the name of the list as input. For example, to clear list 1 (2nd L1), we enter CLRLIST L1 and press ENTER. A message is displayed telling us that the list is cleared. Other lists are cleared in the same way.

We use L1 to represent the number of CDs and then use L2 to represent the total cost for the CDs. To enter the values for L1, we press STAT and choose the

EDIT

EDIT feature. We then press ENTER, and the lists are shown in columns as in Figure AII-2(a). (On a TI-73 press LIST.) On the bottom edit line where L1(1) = is displayed, type the numeral 1 followed by ENTER. A 1 will appear as the first entry in L1, as shown in Figure AII-2(a).

FIGURE AII-2

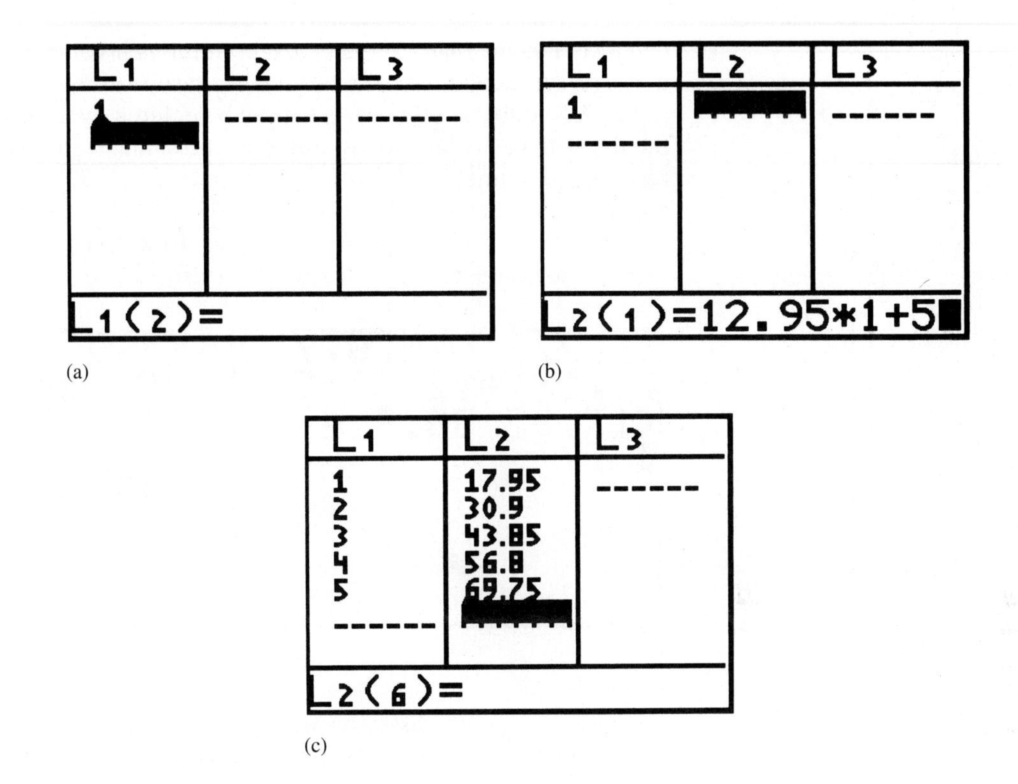

(a) (b) (c)

By pressing the right arrow, we move to list L2. The first entry in L2 can be entered directly or it can be computed by entering the expression L2(1) = 12.95 × 1 + 5 in the edit line at the bottom of the screen, as shown in Figure AII-2(b). When ENTER is pressed, 17.95 appears as the first entry in L2. We next use the left and down arrows to return to the second entry in L1 and enter 2. The remainder of the table as shown in Figure AII-2(c) can be completed.

An easier method for finding the various costs is to define a general pattern and have the calculator do all the computations at once. To do this, enter the numbers 1 through 10 in L1. Move the cursor to the top of L3 so that L3 is highlighted. When L3 is highlighted, anything that is done in the edit line on the bottom will happen to every element in L3. In the edit line, enter L3 = 12.95L1 + 5, as shown in Figure AII-3(a), and press ENTER. The entire list in L3 is then computed based on the entries in L1 and the rule given in the edit line (see Figure AII-3b). Lists L2 and L3 should be the same even though they were obtained in two different ways.

FIGURE AII-3

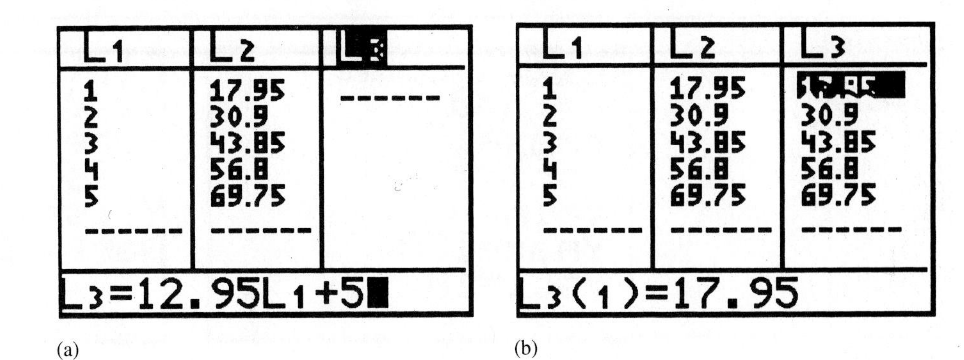

(a) (b)

Using Lists on the Home Screen Computations involving lists can be done on the home screen using the braces { and } located above the parentheses keys (or in the TEXT editor on a TI-73). A list can be entered by using a left brace, followed by the list, with each element separated by a comma. Figure AII-4(a) shows a list representing the number of possible CDs ordered: {1, 2, 3, 4, 5, 6, 7, 8, 9, 10}. We then multiply each element in this list by 12.95 and add 5 to the result; the answers are given on the screens by pressing ENTER. To see the entire list of answers, scroll to the right using the right arrow key.

FIGURE AII-4

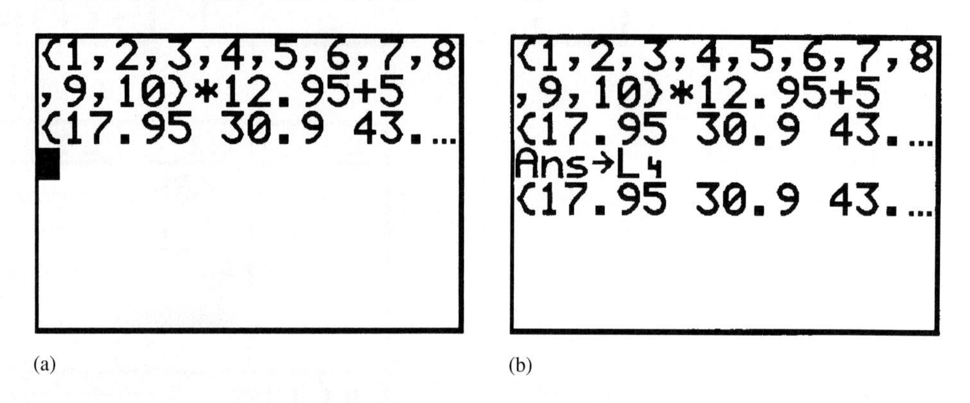

(a) (b)

You can store the list of answers in one of the list memories by using STO and the list number you choose. By pressing STO L4 ENTER, the answers are stored in L4. The list of answers still appears on the home screen, as shown in Figure AII-4(b), even though the answers are stored in L4. If we press STAT, select EDIT, and use the right arrow to move to the L4 column, we can see that the list of answers has been stored in L4. (On a TI-73, press LIST.) Notice that this list matches the ones in L2 and L3.

Using STAT Plot Graphing Capabilities Next we investigate how Molly's CD Problem could be solved using a graphical representation. First, we need to set an appropriate-size graphing window. If we plot the number of CDs ordered on the *x*-axis and the total costs on the *y*-axis, we need only look at our lists to determine appropriate values. The number of CDs ranges from 1 to 10, and the cost ranges from \$12.95 to \$134.50. We have several choices. If we press WINDOW, then the choices can be entered, as shown in Figure AII-5(a). After selecting WINDOW, we can enter the minimum and maximum values for the *x*- and *y*-axes and the scale that gives the distance between the marks on each axis. (On certain calculators you can have the calculator choose the appropriate window.)

FIGURE AII-5

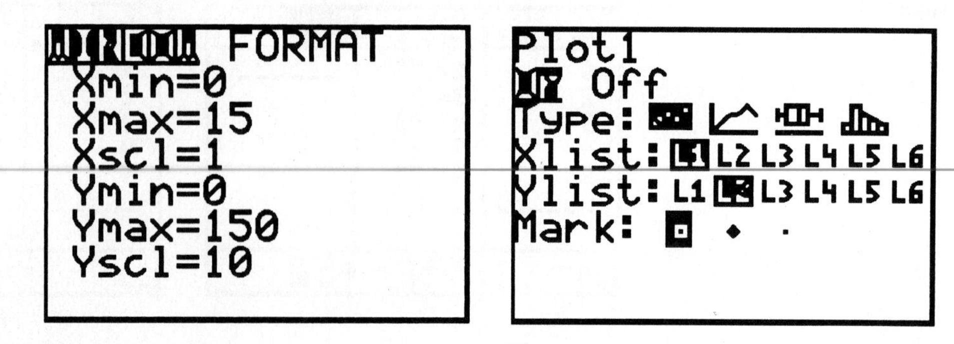

(a) (b)

We can plot these points by using 2nd to choose STAT PLOT (or PLOT) and then select Plot1, as shown in Figure AII-5(b). We see in Figure AII-5(b) that Plot1 is on, the first choice of a scatterplot is chosen, the x-values come from L1, and the y-values come from L2. We can choose five plots and each plot can be a scatterplot, a line graph, a box-and-whisker plot, or a histogram.

By pressing GRAPH, we obtain the graph shown in Figure AII-6(a). If we then press TRACE and use the right and left arrow keys, we can see the trace cursor move along the dots, displaying the x- and y-values at each point, as shown in Figure AII-6(b).

FIGURE AII-6

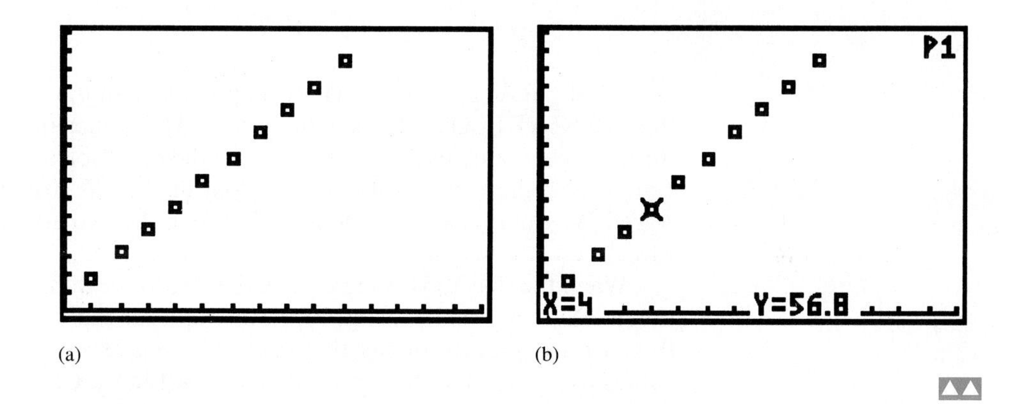

(a) (b)

NOW TRY THIS AII-1

a. How are the values that are displayed using the TRACE feature related to L1 and L2?
b. Why are just the points shown rather than a line containing the points?
c. What pattern is shown in the graphical representation that is not shown when working with only numbers or lists?
d. Use the graph to approximate the cost of 12 CDs.

RESEARCH NOTE Students can learn more mathematics more deeply with the appropriate use of technology (Dunham and Dick 1994).

Connecting Graphing and Algebra

Earlier in this appendix, we generated a mathematical rule for computing the total cost for any number of CDs in Molly's CD Problem. The rule was to multiply the number of CDs by 12.95 and add 5. We can enter this rule as a function by pressing Y =. We can enter the rule for Y_1 using x to represent the number of calculators. The variable x has a separate key—X, X,T, or X,T,θ. Figure AII-7(a) shows the entered function, and Figure AII-7(b) shows the graph obtained when GRAPH is pressed.

FIGURE AII-7

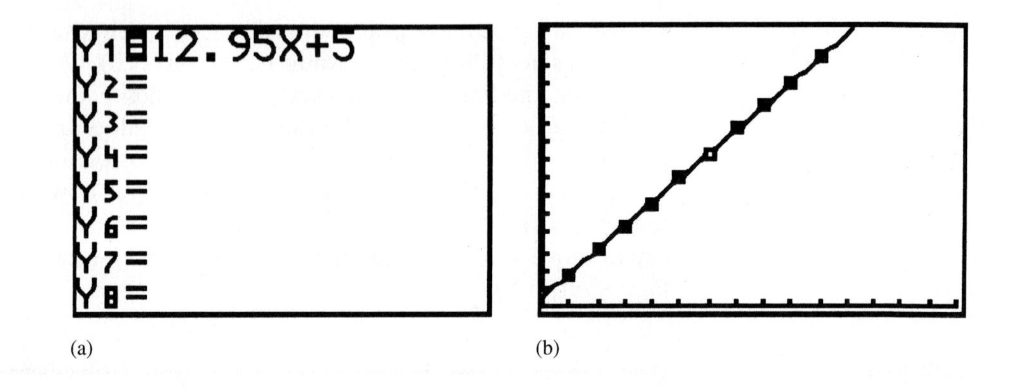

(a) (b)

Press TRACE and a P1 (Plot1) appears in an upper corner. The P1 indicates that the STAT PLOT (P1) is being traced. The values that appear are the same as those we generated earlier when we built the cost tables. If we press the up arrow, a 1 appears in the corner. This indicates that the line Y_1 from the Y = menu is being traced. We can toggle back and forth between the two graphs by pressing the up arrow.

When the line Y_1 is being traced, the x-values can be nonintegers. The graph of the line Y_1 is not really an appropriate graph for this problem because the data in this case are not continuous; that is, the y-values exist only for nonnegative integer values of x. There is no such thing as a cost for 1.5 CDs as implied by the graph. However, the graph can be used to find values of y (costs) when x is a nonnegative integer.

Using Tables

We can build a table based on the algebraic rule developed for Molly's CD Problem. To do this, we press 2nd to choose TBLSET and set the table menu, as shown in Figure AII-8(a). The TblMin, or TblStart on some calculators, set at 1 starts the table at 1 and the ΔTbl setting of 1 sets the increment between x-values at 1. If we next press 2nd to select TABLE, we see the table of values generated by the rule $y = 12.95x + 5$. Notice that the x-values are incremented by 1 in each case because this is what we selected in the TblSet menu. Also notice that if we use the up and down arrow keys, we can continue to obtain y-values for any integer x-value, as shown in Figure AII-8(b). From the result in Figure AII-8(b), we see that the cost of 15 CDs is 199.25. Use the down arrow to determine the cost of 17 CDs.

FIGURE AII-8

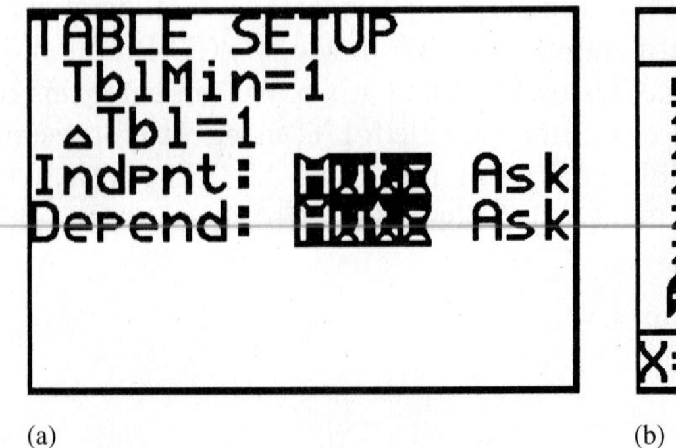

(a)

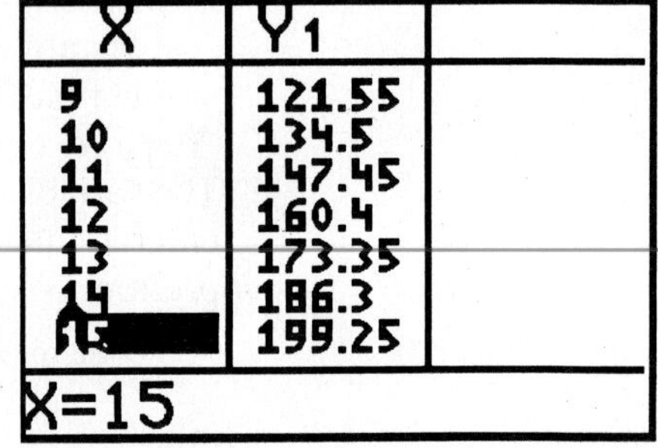

(b)

PROBLEM SOLVING **Choosing a CD Company**

Another company is advertising CDs for \$7.95 per CD with no shipping charge. However, the company charges a processing fee of \$20 for each order. Molly is considering switching companies. Compare the prices of the two companies and find which one has the better price.

The rule for computing the cost under the second plan is $Y = 7.95x + 20$. If we enter this rule as Y_2 in the Y = menu, as shown in Figure AII-9(a), we can compare the two plans using tables. If we keep the same table setup as in the earlier problem, then Figure AII-9(b) shows a comparison of the two rules. It follows that the first plan is better if we order either one or two CDs. If we order three CDs, it makes no difference which plan we use. If we order more than three CDs, then the second plan is better.

FIGURE AII-9

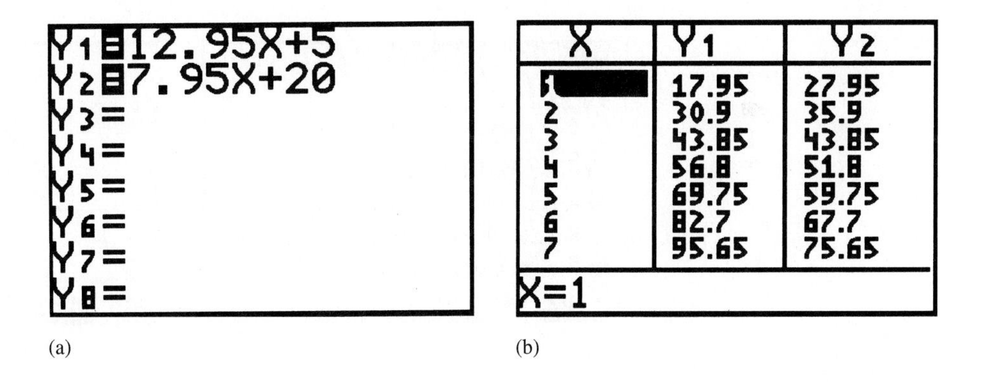

(a) (b)

We can compare the two plans graphically by pressing GRAPH. The two graphs are shown in Figure AII-10(a). The same comments that were made about continuous data are also true in this case. If TRACE and the arrow keys are used, then we see in Figure AII-10(b) that the two lines appear to intersect at point (3, 43.85), which is the point at which the two plans match. After $x = 3$, we see that the second line illustrates the less expensive cost.

FIGURE AII-10

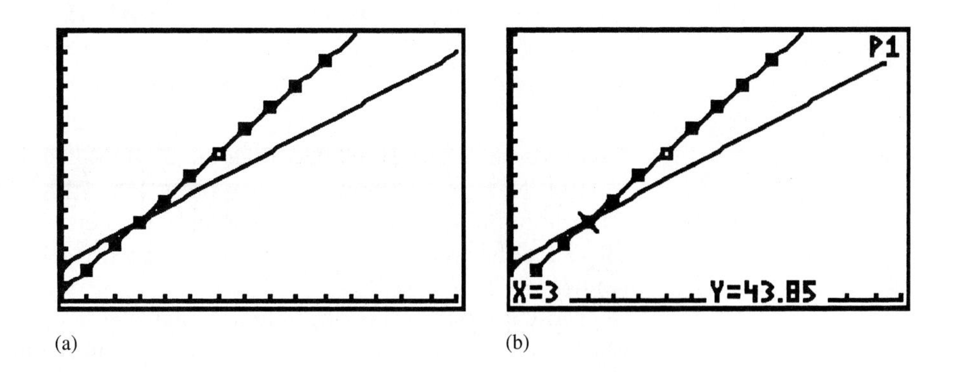

(a) (b)

Using ZOOM Various zoom features can be used to adjust the viewing window to see more of the features of the graph. Press ZOOM to see the menu shown in Figure AII-11(a). The first item on the menu, ZBox, allows the cursor to select opposite corners of a box to define a new viewing window. To use this feature, we move the zoom cursor to any point on the screen on which we want to locate a corner of the box and then press ENTER. As we move the cursor away from the selected point, we see a small square dot indicating the selected corner. We can then move the cursor to the diagonal corner of the box that we want to define. As we use the arrow keys to move the cursor, we see the box change on the screen. When we get the box located where we want it, as shown in Figure AII-11(b), we press ENTER to replot the graph as shown in Figure AII-11(c). The TRACE feature can now be used to find values of various points. We can cancel ZBox at any time by pressing CLEAR before pressing ENTER.

FIGURE AII-11

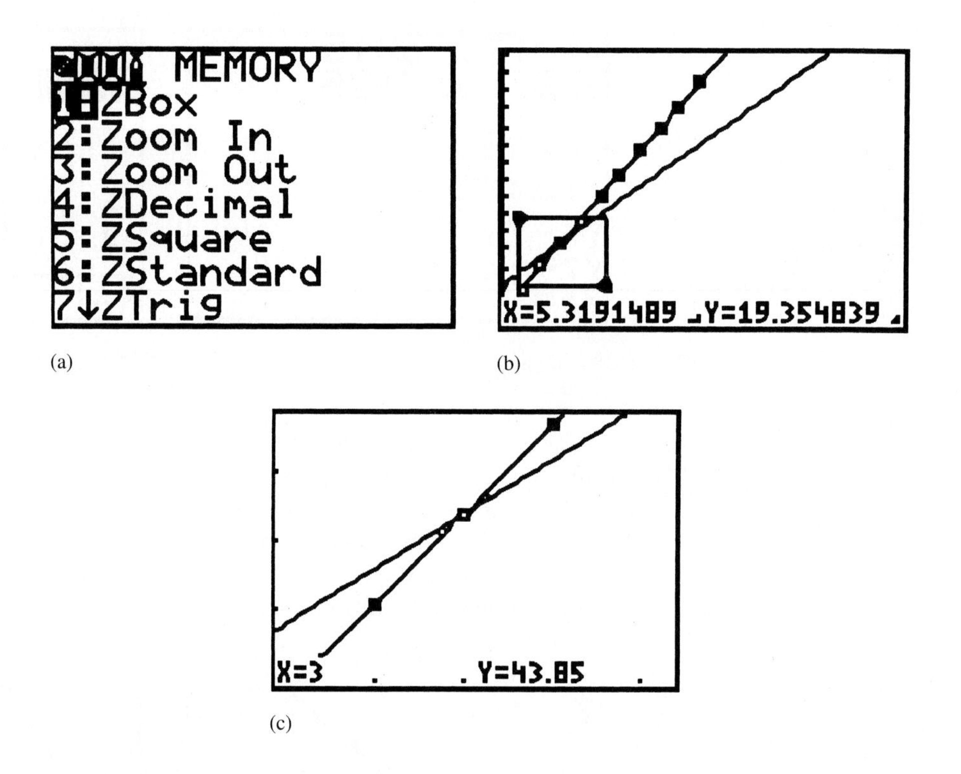

(a) (b) (c)

There are other zoom features in the ZOOM menu. For example, we can select ZOOM IN from the menu and then press ENTER. We then can move the cursor to the point on which we want the center of the new viewing window to be and press ENTER. The graph will be replotted. We can also select ZOOM OUT in a similar manner.

These are only a few of the features available on a graphing calculator. Many other important features are contained in menus under keys such as MATH. Consult your user's manual to see how you can use these other tools to solve mathematical problems.

ONGOING ASSESSMENT A-II

In the following problems, the section in the book dealing with this topic is listed in parentheses.

1. Use trial and error to determine when $1000 invested at 6% annual compounded interest will double in value. (*Hint:* The value after 10 yr is given by 1000 × 1.06^10.) (Section 6-6)
2. Given the sequence 3, 6, 9, 12, 15, 18, . . . , use the TABLE feature to find the 32nd term by entering the appropriate equation using the Y = menu. (Section 1-2)
3. Enter y = $3x^2 + 5x + 7$ in the Y = menu and use the TABLE feature to evaluate the function at 10 and $^-10$. (Section 4-2)
4. Set the WINDOW to graph on each axis from $^-10$ to 10 with increments of 1 unit, then graph each of the following, and answer the questions that follow:
   - **i.** $y = x + 3$
   - **ii.** $y = 2x + 3$
   - **iii.** $y = 3x + 3$
   - **iv.** $y = 4x + 3$
   - **a.** What do the graphs have in common?
   - **b.** How do they differ?
   - **c.** How does changing the slope, m, change the shape of the graph in the equation $y = mx + b$? (Section 10-5)
5. Leave the WINDOW the same as it was in problem 4, graph each of the following, and answer the questions that follow:
   - **i.** $y = x + 3$
   - **ii.** $y = x + 4$
   - **iii.** $y = x + 5$
   - **iv.** $y = x + 6$
   - **a.** What do the graphs have in common?
   - **b.** How do they differ?
   - **c.** How does changing the value, b, in the equation $y = mx + b$ change the shape of the graph? (Section 10-5)
6. Graph the following system. Make sure to choose an appropriate window. Use ZOOM and TRACE features to find the point of intersection to two decimal places. (Section 10-5)

$$y = 2x - 3$$
$$y = -7x + 8$$

7. The distance from Missoula to Billings is 350 mi. To investigate how long it takes to drive this distance, Joan entered the following equations for Y =. (Section 10-5)

$$y_1 = 50X, y_2 = 60X, y_3 = 70X, y_4 = 350.$$

   She set the window for Xmin = 0, Xmax = 10, Xscl = 1, Ymin = $^-100$, Ymax = 600, and Yscl = 20 and then graphed the functions.
   - **a.** How could she tell the time it takes to make the trip traveling at 50, 60, and 70 mph?
   - **b.** How much time is saved traveling at 70 mph rather than 50 mph? (Section 6-6)
8. **a.** Evaluate the function $y = x^2 - 5x + 4$ at each integer value between $^-10$ and 10. How many sign changes are there in this range, and where do they occur?
   - **b.** Set the window from -10 to 10 on the x-axis and -10 to 10 on the y-axis with increments of 1 and graph the equation.
   - **c.** Use the TRACE feature to examine where the function crosses the x-axis.
9. Linda has 33 coins in dimes and quarters. The value of the coins is $5.55. Determine how many of each coin she has. (Section 1-3 or 10-5)

REFERENCES

Burrill, G. *Handheld Graphing Technology in Secondary Mathematics: Research Findings and Implications for Classroom Practice.* Grant to Michigan State University by Texas Instruments, 2002.

Dunham, P., and T. Dick. "Research on Graphing Calculators." *Mathematics Teacher* 87 (September, 1994): 440–445.

Using a Geometry Drawing Utility

Introduction

In the time of Plato, a compass and straightedge were the norm for constructions. Today many other tools are available, including technology. *Principles and Standards of School Mathematics* (NCPM 2000) asserts that

> *Technology is essential in teaching mathematics, it influences the mathematics that is taught and enhances student's learning. (p. 24).*

In this appendix, we present a series of labs appropriate for a computer geometry utility. Most geometry utilities allow both drawing and construction. In a drawing, an "eyeballing" approach is used to draw a figure to look as it should, but the figure may not be constrained by elements of its geometric properties. For example, a segment may be drawn in a circle to look like a diameter, as in Figure AIII-1(a). However, if it is not constructed to pass through the center of the circle and one moves the circle as in Figure AIII-1(b), the segment may no longer move with the circle and may not appear to be a diameter.

FIGURE AIII-1

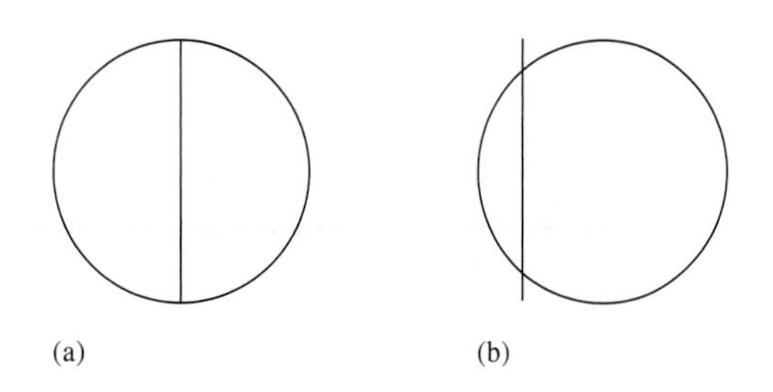

The constraint of having the diameter pass through the center was not used. If the segment is constructed as a chord to contain the center as a part of the construction, then when the circle is moved, the segment moves accordingly (as a diameter should). For most purposes in this appendix, we are using constructions with all the geometric constraints applied.

The Lab Activities in this appendix are arranged in the order they might be used with the geometry chapters of the book. The activities may be used independently if the teacher provides the language and definitions as needed. Some of the materials in this appendix are based on the work of Jim Williamson and Terry Souhrada at the University of Montana.

About a Geometry Utility

Figure AIII-2 is an example of a drawing window with the Toolbox and Menu Bar shown from *The Geometer's Sketchpad (GSP)*. Other utilities have different windows, but we use *GSP* for illustrations in this appendix.

The Menu bar of Figure AIII-2 shows the following menu options:

- **File**, for opening, closing, saving, and printing documents.
- **Edit**, for selecting objects, editing object properties, and setting preferences.
- **Display**, for changing the appearance of drawings.
- **Construct**, for constructing figures.
- **Transform**, for translating, rotating, dilating, reflecting, and iterating [optimal] figures.
- **Measure**, for displaying measurements and making calculations.
- **Graph**, for options with coordinate axes and function plots.
- **Help**, for accessing Sketchpad's help system.

FIGURE AIII-2

File Edit Display Construct Transform Measure Graph Help

Untitled 1

Selection Arrow (**Translate**) tool: Click on objects in sketch to select them. Drag objects to move (translate) them. (See **Translate**, **Rotate**, and **Dilate** tools below.)

Point tool: Click in blank sketch area to create an independent point. Or click on object to create a point on object.

Compass (**Circle**) tool: Press mouse button to create center point, then press elsewhere to create radius point (or press, drag, and release). Center and radius point can be independent points or points on objects.

Straightedge (**Segment**) tool: Press to create first endpoint, then press elsewhere to create second endpoint (or press, drag, and release). (See **Segment**, **Ray**, and **Line** tools below.)

Text tool: Drag in blank area to create caption. Click on object to display or hide label. Drag label to reposition. Double-click on label, measure, or caption to edit or change style.

Custom Tools tool: Press to open menu of available tools or to create a new tool.

Translate, Rotate, and Dilate Tools

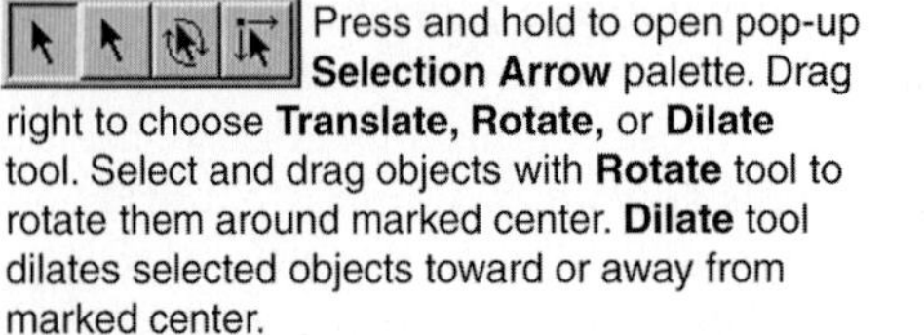

Press and hold to open pop-up **Selection Arrow** palette. Drag right to choose **Translate, Rotate,** or **Dilate** tool. Select and drag objects with **Rotate** tool to rotate them around marked center. **Dilate** tool dilates selected objects toward or away from marked center.

Segment, Ray, and Line Tools

Press and hold to open pop-up Straightedge tool palette. Drag right to choose **Segment, Ray,** or **Line** tool. Construct segments, rays, and lines by clicking twice in sketch.

The Laboratory Activities that follow contain a reference to a section in the book where they are appropriate to use. In many cases, the mathematical content of a section can be explored by working through the appropriate activity. GSP Lab 1 provides an introduction to GSP and can be used to acquaint users with the main features of the program. Additional features will be introduced as needed. More activities and problems are given in Ongoing Assessment A-III. Many of the Technology Corners in Chapters 9–12 also contain GSP activities or problems.

The Geometer's Sketchpad

GSP Lab 1 (Section 9-1)

What Is GSP?

The Geometer's Sketchpad (GSP) is a graphics program that allows us to

- Simulate straightedge and compass constructions.
- Save the steps in a construction so we can do them over and over again.
- Explore and discover mathematical, especially geometric, properties.

Accessing GSP

Double-click on the GSP icon. In a second, the GSP title box appears on the screen. Then the GSP window appears in the background. Click anywhere inside the title box to activate the GSP window. (See Figure AIII-3)

FIGURE AIII-3

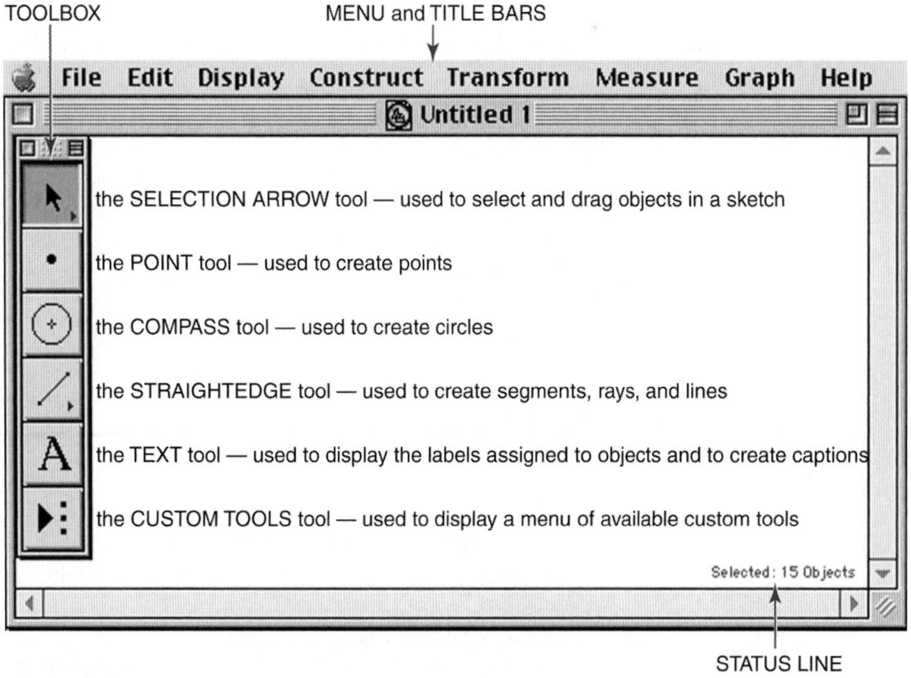

TOOLBOX
MENU and TITLE BARS
File Edit Display Construct Transform Measure Graph Help
Untitled 1
the SELECTION ARROW tool — used to select and drag objects in a sketch
the POINT tool — used to create points
the COMPASS tool — used to create circles
the STRAIGHTEDGE tool — used to create segments, rays, and lines
the TEXT tool — used to display the labels assigned to objects and to create captions
the CUSTOM TOOLS tool — used to display a menu of available custom tools
Selected: 15 Objects
STATUS LINE

Think of the window as a "blank sheet of paper" on which we can draw a sketch. On the left side of the window is the **Toolbox**. Each icon stands for a different tool. In this lab, we will experiment with some of these tools to learn how GSP operates.

Working with GSP

As we work through the exercises, keep an eye on the **Status Line** in the lower left (Windows) or lower right (Mac) corner of the sketch. This box explains exactly what GSP is doing. It can be a real help if we are unsure about what is going on!

1. **a.** Click on the **POINT** tool. How does GSP indicate that the **POINT** tool is active?
b. Notice how the cursor changes when we move it onto the sketch. Click to create several points.
c. Choose the **TEXT** tool. How does the cursor change?
d. Click on each point created. What happens?

GSP automatically labels points with capital letters, starting with A, then B, and so on. We can change a label using the **TEXT** tool; just double-click on the label. A box will appear that allows us to type any label wanted. Try it!

2. With the **TEXT** tool active, double-click in an empty part of the sketch. A text box is created that can be used to type captions in the sketch. After typing a caption, use the **ARROW** tool to "grab" the caption (point with the mouse and hold the button down) and move it around the sketch ("drop" it by releasing the mouse button). Go back to the **TEXT** tool to edit the caption.

3. Place several points in the sketch. Then choose the **ARROW** tool. Notice the cursor is now an arrow. Click on a point. How does GSP tell that the point has been selected? Click anywhere else in the sketch to deselect it.

With the **ARROW** tool active, we can select several points at once by dragging a selection rectangle around them. Start in one corner and hold the button on the mouse down while the cursor is dragged down and over. A dashed box is drawn. When the button is released, everything inside the box is selected. Notice that if we click the mouse button in blank space, everything gets deselected.

Another way to select several objects at the same time is simply by selecting them with the **ARROW** tool one by one. Click on an unselected object to select it, click on a selected object to deselect it, click in blank space to deselect all objects. Experiment with it.

We can get rid of created objects by selecting them and pressing the DELETE or BACKSPACE key. Experiment with that key.

Finally, all the objects on the screen can be selected by using the **Select All** command found in the **Edit** menu at the top of the screen while the **ARROW** tool is active. Select the remaining points and delete them. *Note:* If the **POINT** is active, only points will be selected using this method. Likewise, if the **COMPASS** tool, one of the **STRAIGHTEDGE** tools, or the **TEXT** tool is active, only objects of that type will be selected.

4. Choose the **COMPASS** tool.
a. To create circles in a sketch, click the mouse button once to create the center point, then click elsewhere to create the radius point (or click, drag, and release).
b. Use the **TEXT** tool to label the two points.
c. Choose the **ARROW** tool. Drag either of the points. What happens?
d. Point at the circle and select it. How does GSP indicate that the circle has been selected?

5. Clear any objects from the sketch.
a. Choose the **SEGMENT** tool. (If the **SEGMENT** tool isn't showing, press and hold down the button on the current **STRAIGHTEDGE** tool to see the three available options: SEGMENT, RAY, and LINE.)

b. To create a segment in the sketch, click at two locations where you'd like the two endpoints to be. The segment is selected. Use the **TEXT** tool to see what kinds of labels are assigned to segments.

c. Go back to the **Toolbox** and choose the **LINE** option of the **STRAIGHTEDGE** tool. Return to the sketch and see how lines are drawn and labeled. Do the same with the **RAY** option.

The Construct Menu

The **Construct** menu on the menu bar across the top of the sketch contains some very powerful tools. A key idea for GSP success is:

Whenever possible, use a command from the **Construct** menu to define objects.

6. Here is an example of how to use the **Construct** menu.

a. Use the **POINT** tool to create two points. Use the **SELECT ARROW** tool to select both points. (If you hold down the **SHIFT** key while creating the points, they remain selected.)

b. Choose **Segment** from the **Construct** menu.

7. a. Clear the objects and then use the **POINT** tool to create three points. Select the three points, then use the segment construction. What happens?

b. Can we move the points around after the construction?

c. Can we move the segments around?

8. a. What happens when we use the segment construction with four points?

b. Does the order in which we select the four points matter?

9. There are two ways to find the point(s) where two objects intersect.

a. Clear any objects from the screen. Draw two circles that intersect. Then draw a ray and a segment that intersect.

b. Method 1: Select the two circles, then choose **Intersections** from the **Construct** menu. What happens?

c. Method 2: Choose the **ARROW** or **POINT** tool from the **Toolbox**. Move the cursor so that it points at the intersection of the segment and the ray. Be careful—look at the status line to be sure that GSP recognizes that the intersection point is desired. Then click the mouse button once.

If we want a new window, we choose **New Sketch** from the **File** menu. You can have several sketches available at one time. *Note:* There is a Window menu in Windows GSP, but not in all versions of Mac GSP.

The Measure Menu

10. a. In a new sketch, draw a segment and select it. Choose **Point On Segment** from the **Construct** menu. What happens?

b. Select the segment. Pull down the **Measure** menu and choose **Length**. What happens? (*Note:* We can select the unit of measurement, inches, centimeters, or pixels, by going to **Preferences** in the **Edit** menu.)

c. Select an endpoint of the segment and the point on the segment. Choose **Distance** from the **Measure** menu. What happens?

11. We can use **Calculate** in the **Measure** menu to do calculations with numbers, measurements, and functions.

a. Choose **Calculate** from the **Measure** menu. A calculator should appear. Click on one of the measurements on the screen. Next, click on the minus sign (−) on the key pad. Finally, click on the second measurement. Click on **OK** and the difference of the measurements should appear on the screen.

b. Select the point on the segment and move it back and forth. What happens to the measurements and the difference?

c. Change the length and position of the segment by dragging one of the endpoints. What happens to the measurements and the difference?

GSP Lab 2: Measuring Angles of Polygons (Sections 9-2 and 9-3)

1. a. Open a new sketch.

b. Construct a triangle using three points from the **POINT** tool and **Segment** from the **Construct** menu and label its vertices.

c. Measure each interior angle of the triangle.

Measuring an Angle

To measure an angle:

- Select three points that determine the angle. Be sure the vertex is the second point selected.
- From the **Measure** menu, choose **Angle**. The measure of the angle will appear on the sketch.

Calculating a Sum

d. Calculate the sum of the interior angles of the triangle.

To calculate a sum:

- Open the Calculator by choosing **Calculate** from the **Measure** menu.
- Enter a measurement to be added by clicking on the desired measurement. *Note:* It is possible to enter numbers on the calculator keypad as well, but these numbers will not be updated when the sketch is changed.
- Click on the desired operation on the calculator keypad.
- Continue to alternate between selecting measurements from the sketch and operations on the calculator until all measurements and operations needed are entered. When you are done, click on **OK**. The operation and result will appear on the sketch.

e. Move the vertices of the triangle to change its size and shape. What happens to the sum of the measures of the interior angles?

Making a Conjecture

f. Repeat part (e) several times, then make a *conjecture* (that is, an educated guess based on your observations) about the sum of the measures of the interior angles of a triangle.

2. a. Repeat exercise 1(b–f) for the following convex polygons: a quadrilateral, a pentagon, and a hexagon. Use the results to make a conjecture about the sum of the interior angles of a convex n-gon (a polygon with n sides).

Testing Your Conjecture

b. Test the conjecture by repeating exercise 1(b–f) for a convex octagon.

3. a. Open a new sketch.

b. Construct a triangle. Moving clockwise, label the vertices A, B, and C. Form the exterior angles of the triangle by selecting two points and then choosing **Ray** from the **Construct** menu. Construct rays $\overrightarrow{AB}$, $\overrightarrow{BC}$, and $\overrightarrow{CA}$. Place a point on each ray so that you can measure each angle. Calculate the sum of the measures of the exterior angles.

c. Change the size and shape of the triangle several times. What happens to the sum of the measures of the exterior angles?

d. What seems to be true about the sum of the measures of the exterior angles of a triangle? Record your conjecture in a caption in the sketch.

e. Repeat parts (b–d) for the following convex polygons: a quadrilateral, a pentagon, and a hexagon. Use the results to make a conjecture about the sum of the measures of the exterior angles of a convex n-gon.

f. Test the conjecture in part (e) by repeating parts (b–d) for a convex octagon.

GSP Lab 3: Regular Polygons (Section 9-2)

In this section, we construct regular polygons by using the **Rotate** command from the **Transform** menu. Transformations are discussed in detail in Chapter 12.

The Transform Menu

FIGURE AIII-4

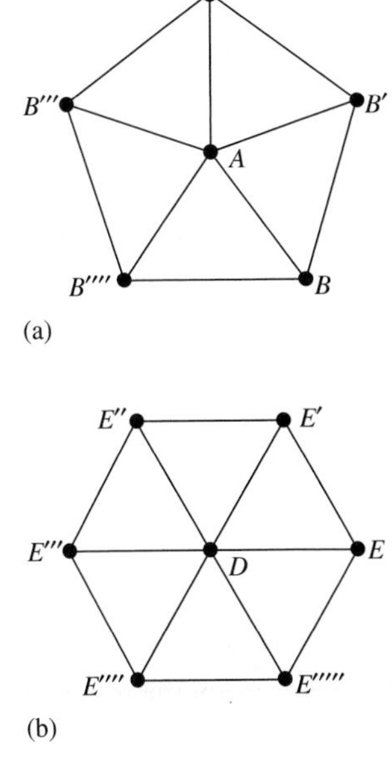

1. a. In a new sketch, draw a segment $\overline{AB}$ and select point A.

b. Go to the **Transform** menu and choose **Mark Center**.

c. Select segment AB and point B. Go back to the **Transform** menu and choose **Rotate** . . . A dialog box allows us to tell GSP how many degrees to rotate the selected objects. Because the central angle of a regular pentagon is 72°, put that number in the box and click on OK. What happens?

d. Continue to rotate the segment and point around point A until there are five segments, each making a 72° angle with the next one at the center A. What are the labels on the points? (Depending on the Preferences setting, labels may not be showing.)

e. Now connect the vertices to make a regular pentagon as in Figure AIII-4(a). Why does a rotation of 72° result in the vertices of a pentagon?

2. a. In a new sketch, create a regular hexagon as in Figure AIII-4(b) using the method in exercise 1 but adjusting the angle measure.

b. By how many degrees did you rotate the segment around the center each time to create a regular hexagon?

c. By how many degrees would you rotate the segment each time to create a regular n-gon?

There are two ways to create a circle using the **Construct** menu.

3. The first way is to use a center point and a radius.

a. Create a segment and a point not on the segment. Select them both.

b. From the **Construct** menu, choose **Circle By Center+Radius** Notice that the segment (the radius) does not have to have the center of the circle as one of its endpoints.

c. What happens to the circle if the center is moved?

d. What happens if the length controlling the radius is changed?

4. The second way is to use two points.

a. In a new sketch, create and select two points.

b. From the **Construct** menu, choose **Circle By Center+Point**. Select the points in the opposite order. Does the order of selecting points matter?

5. a. In a new sketch, draw a segment.

b. Use two circles to create an equilateral triangle that has the segment in part (a) as one of its sides. The size and position of the triangle can be changed by dragging a vertex, but it should always remain an equilateral triangle.

c. Hide everything except the triangle and its vertices.

Hiding Objects

To hide objects:

- Select all the objects to be hidden.
- From the **Display** menu, select **Hide Objects** (or **Points, Lines**, etc.).

GSP Lab 4: Congruent or Not Congruent? (Section 10-1)

Side-Side-Side

1. In a new sketch, draw $\triangle ABC$. Measure its sides and angles.

In part 2 below, we construct a triangle, $\triangle XYZ$, whose sides are congruent to the sides of $\triangle ABC$.

2. a. Create a point X. Next, we construct $\overline{XY}$, a copy of $\overline{AB}$. Select X and $\overline{AB}$. Use the **Circle By Center+Radius** command from the **Construct** menu to construct a circle.
What is the center of the circle? What is the radius of the circle? What does the circle represent?

b. Select the circle and use **Point On Circle** to construct a point on the circle; label it Y. Construct $\overline{XY}$. What is true about the measures of segments $\overline{XY}$ and $\overline{AB}$? Hide the circle.

c. Side $\overline{XZ}$ in $\triangle XYZ$ will be congruent to side $\overline{AC}$ in $\triangle ABC$. To create $\overline{XZ}$, we first find the possible locations of point Z from X. Use the **Circle By Center+Radius** command to construct a circle with radius $\overline{AC}$ and center X.
What does this circle represent?

d. Side $\overline{YZ}$ in $\triangle XYZ$ will be congruent to $\overline{BC}$. To create $\overline{YZ}$, we find the possible locations of point Z from Y. Use the **Circle By Center+Radius** command to construct a circle with radius $\overline{BC}$ and center Y.

e. There are two possible locations for point Z. Construct one of the points and label it Z. Construct $\overline{XZ}$ and $\overline{YZ}$ to complete $\triangle XYZ$. Then hide the two circles.

3. a. Measure the sides and angles of $\triangle XYZ$. How are $\triangle ABC$ and $\triangle XYZ$ related? Support your answer.

b. Change the shape and size of $\triangle ABC$. Does the relationship between the triangles change?

c. When the sides of one triangle are congruent respectively to the sides of another triangle (SSS), what appears to be true about the triangles?

Side-Angle-Side

4. Draw a new triangle, $\triangle PQR$. Measure its sides and angles.

In exercise 5, we construct a triangle, $\triangle LMN$, that has two sides and the angle between them congruent, respectively, to two sides and the included angle in $\triangle PQR$.

5. **a.** Use the method from exercises 2a and 2b to construct $\overline{LM}$ with the same measure as $\overline{PQ}$. Use the **Line** command from the **Construct** menu to construct ray *LM* as well. Hide the circle. Locate and label point *M*.
   b. Next construct a copy of $\angle QPR$ with its vertex at *L*. To do this, first select $\angle QPR$ by selecting points *Q, P,* and *R* in that order. Then, choose **Mark Angle** under the **Transform** menu. What happens?
   Select point *L*. Choose **Mark Center** from the **Transform** menu. Create a copy of $\angle QPR$ by selecting $\overleftrightarrow{LM}$ and rotating it using the **Rotate** command from the **Transform** menu. When the dialogue box appears, choose **By Marked Angle**. What happens?
   Move point *R* to help identify which angle is congruent to $\angle QPR$. Place a point *U* on the rotated line to create $\angle MLU$ so it appears to be congruent to $\angle QPR$. Measure $\angle MLU$ and $\angle QPR$ to be sure they are congruent.
   c. Construct $\overline{LN}$ congruent to side $\overline{PR}$ in $\triangle PQR$ by constructing a circle with center *L* and side $\overline{PR}$ as a radius. Locate and label point *N* so that $\triangle LMN$ contains $\angle MLU$. Construct point *N* at the intersection of the circle and the rotated line so that $\angle MLN \cong \angle QPR$. (If it's not clear which intersection to use, try dragging point *R*.)
   d. Complete the triangle by constructing segments $\overline{LN}$ and $\overline{MN}$. Hide the two lines and the circle.
6. **a.** How are $\triangle LMN$ and $\triangle PQR$ related? Justify your answer.
   b. Change the shape and size of $\triangle PQR$ Does the relationship between the triangles change? Drag $\triangle LMN$ on top of $\triangle PQR$ so that corresponding parts of $\triangle LMN$ fall on top of $\triangle PQR$ How do they compare?
   c. When two sides and the included angle in one triangle are congruent to two sides and the included angle in another triangle (SAS), what appears to be true about the triangles?

Side-Side-Angle

7. **a.** In a new window, draw a new scalene triangle, $\triangle TUV$. Measure its sides and angles. We will attempt to construct a triangle, $\triangle DEF$, that is congruent to $\triangle TUV$ using two sides and a nonincluded angle. Adjust the triangle so that $TU > UV$.
   b. Construct $\overline{DE}$ with the same length as side $\overline{TU}$. Use points *D* and *E* to construct ray *DE*.
   c. Construct the set of possible locations for *F* so that $\overline{EF}$ will have the same length as side $\overline{UV}$.
   d. Construct a copy of $\angle UTV$ with its vertex at *D*. Remember to first select $\angle UTV$. Then, under the **Transform** menu, choose **Mark Angle**.
   Select point *D*. Under the **Transform** menu, choose **Mark Center**. Finally, create the copy of $\angle UTV$ by selecting $\overleftrightarrow{DE}$ and rotating it using the **Rotate** command.
   e. There should be two possible places where *F* could be located. Label one of them *F* and the other one *G*. Draw $\overline{EF}$ and $\overline{EG}$. Hide any remaining circles.

   Are $\triangle DEF$ and $\triangle TUV$ congruent? Explain.
   Are $\triangle DEG$ and $\triangle TUV$ congruent? Explain
   Are $\triangle DEF$ and $\triangle DEG$ congruent? Explain.

   f. If we know that two sides and a nonincluded angle in one triangle are congruent to two sides and a nonincluded angle in another triangle (SSA), is that enough information to determine that the triangles are congruent? Explain.

GSP Lab 5: Circles and Chords (Section 10-1)

Drawing and Bisecting Chords

In this lab we explore some relationships among circles, radii, and chords.

1. a. Construct a large circle by drawing a segment and a point, then using the **Circle By Center+Radius** command. Then place two points on the circle by selecting the circle and using **Point On Circle** from the **Construct** menu.

b. Construct the segment joining the two points. A segment joining two points on a circle is a **chord** of the circle. Move one end of the chord so the chord is clearly visible.

c. Construct a **perpendicular bisector** of the chord, that is, a line that bisects the chord and is perpendicular to it, by performing the following steps:

- Select the segment. Choose **Midpoint** from the **Construct** menu.
- Select the segment and the midpoint. Choose **Perpendicular Line** from the **Construct** menu. What happens?

d. What happens if you move one of the points on the circle? What happens if you change the radius of the circle?

e. Create another chord on the circle and construct its perpendicular bisector. What do you notice about the intersection of the two perpendicular bisectors?

f. Change the location of the chords and the size of the circle. How does this affect the intersection of the two perpendicular bisectors?

Constructing a Circle Circumscribed about a Triangle

2. a. Draw a triangle. Construct the perpendicular bisectors of two sides of the triangle.

b. Construct a point at the intersection of the two perpendicular bisectors.

c. Draw a segment from the intersection point to one of the vertices of the triangle.

d. Construct a circle that has the point of intersection of the perpendicular bisectors as its center and the segment created in part (c) as a radius, as shown in Figure AIII-5.

e. Describe the relationship between the triangle and the circle.

f. Move the vertices of the triangle to change its shape and size. How does this affect the relationship between the circle and the triangle?

g. If we were to construct the perpendicular bisector of the third side of the triangle, would it intersect the other two? If so, where? If not, why not?

3. a. Create three noncollinear points. Construct a single circle that contains them.

b. Describe how the circle was found and how this relates to exercise 2.

FIGURE AIII-5

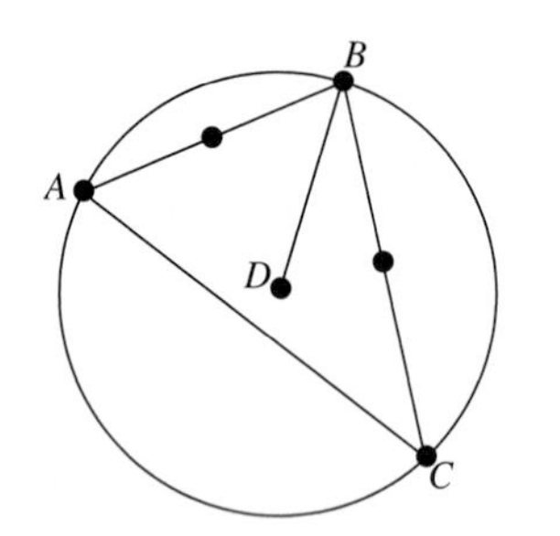

GSP Lab 6: Circles Circumscribing Rectangles

1. a. Construct two congruent segments $\overline{AB}$ and $\overline{CD}$ that bisect each other at O as in Figure AIII-6. (This can be accomplished by constructing a segment AB, finding its midpoint, and rotating the segment by any angle or, alternatively, by constructing two nonparallel congruent segments and their midpoints and then translating one segment by a translation that takes its midpoint to the other midpoint.) Construct the quadrilateral $ACBD$. Drag one of the vertices to different locations. What kind of quadrilateral is $ACBD$? Why?

FIGURE AIII-6

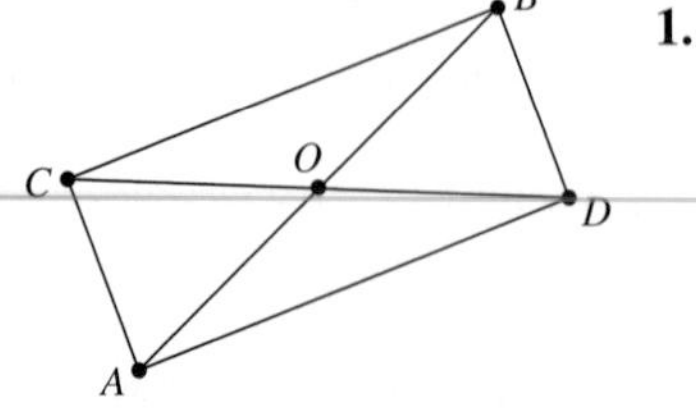

FIGURE AIII-7

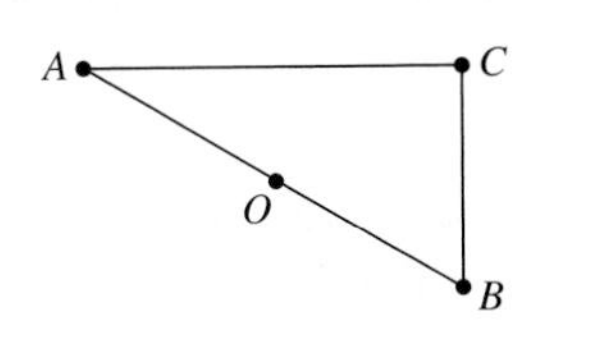

b. Use the construction in part (a) and construct a circle with center O and radius OD. Why are the vertices of the quadrilateral on the circle?

2. a. Draw a circle and any diameter AB of the circle, as in Figure AIII-7. Choose any point C on the circle. Measure $\angle ACB$. Drag point C to different locations on the circle. What happens to $m\ (\angle ACB)$?

b. Use parts (a) or (b) of exercise 1 to explain what you observed about $m(\angle ACB)$.

GSP Lab 7: Split the Sides (Section 10-4)

1. a. Start with a new sketch. Draw $\triangle ABC$. Construct a point M on $\overline{AB}$.

b. Construct a segment $\overline{MN}$ such that $\overline{MN}$ is parallel to $\overline{AC}$ and N is on $\overline{BC}$, as in Figure AIII-8.

FIGURE AIII-8

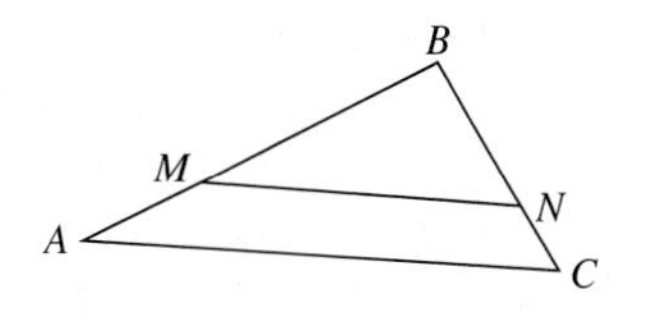

c. Measure $\overline{BM}$, $\overline{BN}$, $\overline{BA}$, and $\overline{BC}$ What is true about the ratios $\frac{BM}{BA}$ and $\frac{BN}{BC}$?

d. Drag point M to a new position. What is true about the ratios $\frac{BM}{BA}$ and $\frac{BN}{BC}$?

2. What are the ratios $\frac{BM}{BA}$ and $\frac{BN}{BC}$ when M is the midpoint of $\overline{AB}$?

3. a. Drag B to form a new triangle. What is true about the ratios $\frac{BM}{BA}$ and $\frac{BN}{BC}$?

b. Drag point M to a new position. What is true about the ratios?

4. a. Measure $\overline{MN}$ and $\overline{AC}$. How does the ratio $\frac{MN}{AC}$ compare to the ratios $\frac{BM}{BA}$ and $\frac{BN}{BC}$?

b. Drag B to form a new triangle. Now how do the ratios in part (a) compare?

5. What seems to be true when a segment intersects two sides of a triangle and is parallel to the third side?

6. What can you conclude about $\triangle ABC$ and $\triangle MBN$? Justify your answer.

FIGURE AIII-9

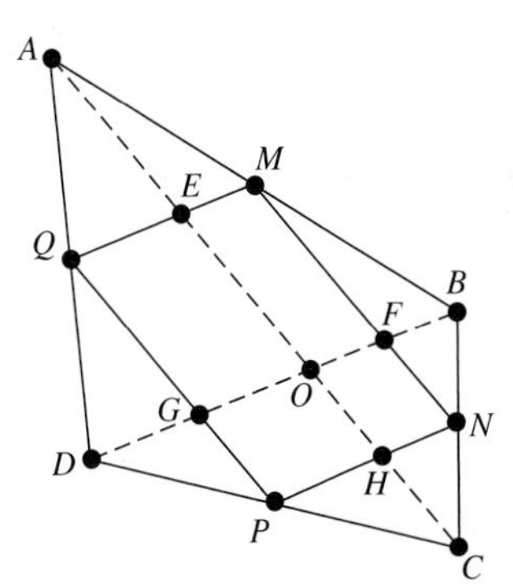

GSP Lab 8: Mysterious Midpoints (Section 10-9)

1. a. Start with four points that will form the vertices of a convex quadrilateral. Draw the quadrilateral.

b. Construct the midpoints of the sides of the quadrilateral and connect the midpoints to form a new polygon, as shown in Figure AIII-9.

c. What kind of quadrilateral is the new polygon? Why?

d. What happens when you change the shape of the original quadrilateral?

Repeat exercise 1 starting with four points that are the vertices of

2. A rectangle.

3. A square.

4. A rhombus.

5. A kite.

6. Label the midpoints of the sides of the original quadrilateral $ABCD$ by M, N, P, Q as shown in Figure AIII-9. Construct the diagonals $\overline{AC}$ and $\overline{BD}$ and label some of the points of intersection of the various segments as shown. Conjecture the kind of quadrilateral $EMFO$ is.

7. a. Follow the steps below:

i. Construct two congruent intersecting segments that are neither perpendicular to nor bisect each other.

ii. Construct the quadrilateral whose diagonals are the constructed segments in part (i).

iii. Construct the midpoint-quadrilateral of the quadrilateral in part (ii).

b. Conjecture what kind of quadrilateral the midpoint quadrilateral in part (iii) is.

c. Complete the following statement and justify it.

The midpoint quadrilateral is _____ if and only if the diagonals of the original quadrilateral are congruent.

8. a. Construct two segments that are perpendicular to each other but not congruent. Make these segments the diagonals of quadrilateral $ABCD$. (You may choose any segment and rotate it by an appropriate angle about a point not on the segment.) Next construct the midpoint quadrilateral. Check, by dragging, what kind of quadrilateral it always is.

b. Complete the following statement and justify it: The midpoints quadrilateral is a rectangle if and only if the diagonals of the original quadrilateral are _____.

c. Drag one of the vertices of the quadrilateral $ABCD$ in part (a) so that the midpoint quadrilateral becomes a square. Complete the following: The midpoint quadrilateral is a square if and only if the diagonals of the original quadrilateral are _____.

d. Based on your answer to part (c) construct two segments to serve as diagonals of the original quadrilateral for which the midpoint quadrilateral is a square. Check, by dragging, that it always is a square.

9. Which of the following statements are true? Explain why. If a statement is false, change part of the "if and only if" phrase to make the statement true.

a. The midpoint quadrilateral is a rectangle if and only if the original quadrilateral is a rhombus or a kite.

b. The midpoint quadrilateral is a square if and only if the original quadrilateral is a square.

GSP Lab 9: Areas of Polygons (Section 11-2)

REMEMBER Whenever possible, use a command from the **Construct** menu to define objects.

1. Start with a new sketch. Choose **Preferences** from the **Edit** menu and make sure that the *Distance Unit* is cm and the *Precision* is set to hundredths for both distance and scalars.

2. Complete the following steps to construct a rectangle with a line containing one side shown in Figure AIII-10.

FIGURE AIII-10

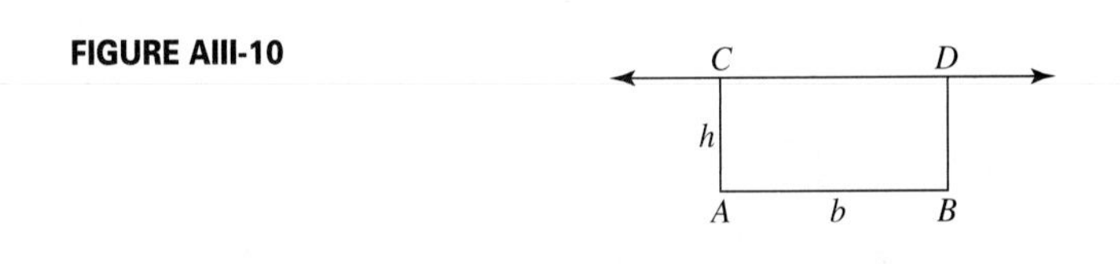

a. Draw a horizontal segment $\overline{AB}$.
b. Construct lines through A and B perpendicular to $\overline{AB}$.
c. Construct a point C on the line through A and construct a line through C parallel to $\overline{AB}$. Construct the point of intersection of this line and the vertical line through B. Label the point D.
d. Hide the two vertical lines and complete the rectangle by constructing segments $\overline{AC}$, $\overline{BD}$, and $\overline{CD}$.
e. Show the label for $\overline{AB}$ and change it to b for *base*. Show the label for $\overline{AC}$ and change it to h for *height*. Measure the lengths of b and h.

3. In this exercise, we review how to find the area of a rectangle.
a. Choose **Show Grid** from the **Graph** menu. Hide the two axes and two new points. Choose **Snap Points** from the **Graph** menu. Drag A and B so that they snap to points on the grid, making $\overline{AB}$ horizontal. The length b should be a whole number.
b. Drag C so that $\overleftrightarrow{CD}$ passes through grid points. C won't snap to the grid, but h should be very close to a whole number.
c. Count the number of squares in the rectangle. How does the number of squares appear to be related to b and h?
d. Use **Calculate** from the **Measure** menu to create an algebraic expression in terms of b and h that gives the area of the rectangle. Check that the expression works by dragging A, B, and C to form a rectangle with a different height and base and counting the number of square units as in part (c).

4. Now we investigate the area of a parallelogram using the sketch from part (3).
a. Use **Point On Parallel Line** from the **Construct** menu to place a point G on $\overleftrightarrow{CD}$. Then construct $\overline{AG}$.
b. Construct a line through B parallel to $\overline{AG}$. Construct the point where this line intersects $\overleftrightarrow{CD}$ and label it H. Hide $\overleftrightarrow{BH}$ and construct $\overline{BH}$. Measure AG and BH.
c. Construct the polygon interior $AGHB$ by selecting the vertices of the parallelogram in order and then choosing **Quadrilateral Interior** from the **Construct** menu.
d. As you move point G, what measures in the sketch change? What measures stay the same?
e. How are the heights of the parallelogram and the rectangle related? How are the lengths of their bases related?
f. Measure the area of the parallelogram by selecting its interior and then choosing **Area** from the **Measure** menu.
g. How does the area of the parallelogram change as we drag G? How does the area of the parallelogram compare to the area of the rectangle?
h. Write a rule for finding the area of a parallelogram. Move points A and C to check that the rule works for parallelograms with different heights and bases.

5. Now we investigate the area of a triangle.
a. In a new sketch, draw two segments $\overline{IJ}$ and $\overline{JK}$. Construct a line through points I and K. Construct the altitude $\overline{JL}$ from J to $\overleftrightarrow{IK}$, as shown in Figure AIII-11(a).

(c)

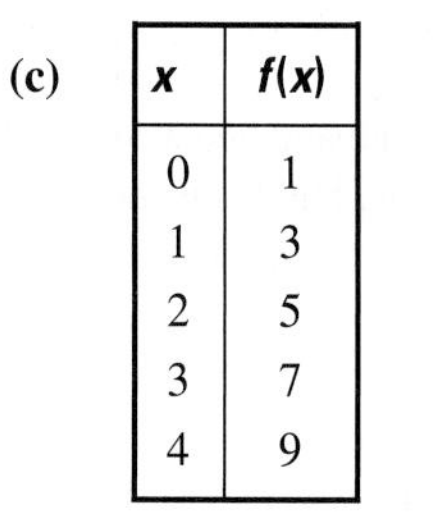

x	$f(x)$
0	1
1	3
2	5
3	7
4	9

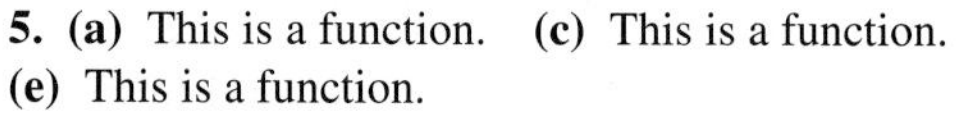

5. (a) This is a function. **(c)** This is a function.
(e) This is a function.

6. (a) **(i)**

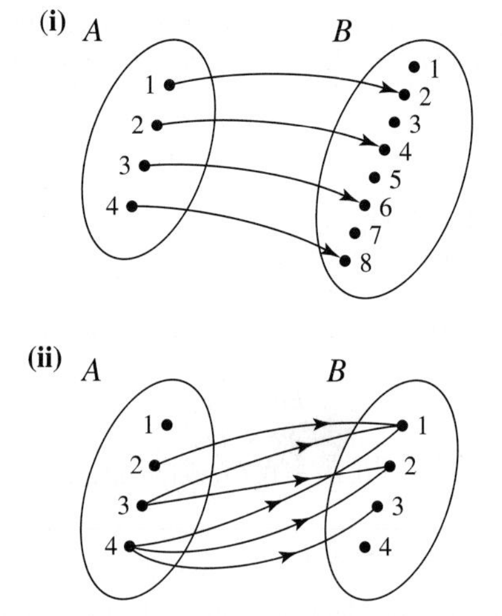

(b) Part (i) is a function from A to B. For each element in A, there is a unique element in B. The range of the function is $\{2, 4, 6, 8\}$.

7. (a) $f(x) = 2x + 1$

8.

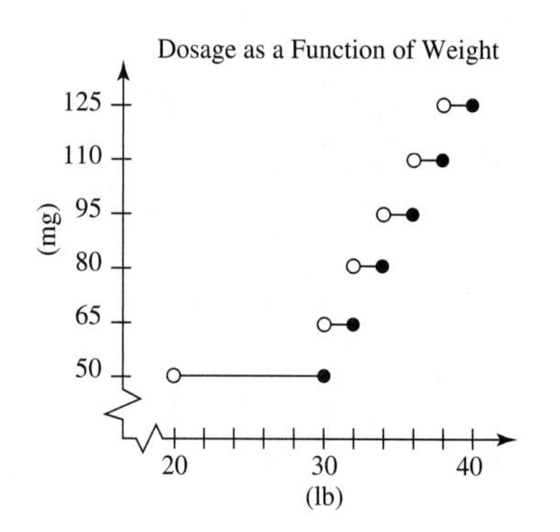

9. (a) 2 chirps per sec

10. (a) 8 dollars

11. (a) $L(n) = 2n + (n - 1)$, or $3n - 1$

12. (a) 7 **(b)** 55

13. (a) Cost per minute in the first and the total cost for a 6-minute call in the second

14. (b) That the company charges for each part of a minute at the rate of \$0.50 per minute **(d)** If $C(t) = 0.50t$ if $O \le t \le 60$, $C(t) = 30 + 0.10(t - 60)$ if $t > 60$.

15. (a) $5n - 2$

16. (a) 30 **(c)** 65

17. (a) 1, 385, 389 **(c)** 2, 12, 2550

18. (a) $2 \cdot 1 + 2 \cdot 7 = 16; 2 \cdot 2 + 2 \cdot 6 = 16; 2 \cdot 6 + 2 \cdot 2 = 16; 2 \cdot 5 + 2 \cdot 5 = 20$ **(c)** The domain is $N \times N$, and the range is the set of all even numbers greater than or equal to 4.

19. (a) 50 cars **(c)** 0 **(e)** Segments are used because the data are continuous rather than discrete. For example, there are a number of cars at 5:20 A.M.

20. (a) $C(x) = 100 + 40x$ **(d)** 20 months

21. (a) $H(2) = 192$; $H(6) = 192$; $H(3) = 240$; $H(5) = 240$. Some of the heights correspond to the ball going up, some to the ball coming down. **(c)** 8 sec **(e)** $0 \le H(t) \le 256$

22. (a) $A(x) = x \cdot \frac{1}{2}(900 - x)$

(c) Length = 450, width = 225

23. (a) **(i)** 40
(ii) 38
(b) **(i)** $S(n) = 2n(n + 1)$
(ii) $S(1) = 4$, $S(n) = 2n(n + 1) - 2$ when $n \ge 2$

24. (a) $1 + 2 + 3 + \ldots + n$, or $\frac{n(n + 1)}{2}$

(c) $n + 2 + 2n$, or $3n + 2$

25. The converse is false. For example, the set of ordered pairs $\{(1, 2), (1, 3)\}$ does not represent a function, since the element 1 is paired with two different second components.

26. (a) Boys: *B, J*; Girls: *A, C, D, F, G, I*

27. (a) Function **(c)** Relation, but not a function (assuming there is a mother in Birmingham who has more than one child) **(e)** Relation, but not a function.

28. (a) Yes **(b)** No

29. (a) None **(c)** Reflexive, symmetric, and transitive (and so an equivalence relation) **(e)** Symmetric

30. (a) Reflexive, symmetric, and transitive; equivalence relation **(c)** Symmetric

Communication

31. Yes, since each element of A is paired with exactly one element of B.

33. (a) This is not a function, since a faculty member may teach more than one class.

34. (b) $f(d) = 50 - d$

Review Problems

42. (a) $109 = 98 + 11$ **(b)** $x = z + y$ **(c)** $60 = 4 \cdot 15$ **(d)** $x = y \cdot 3$ **(e)** $x = 10 \cdot 5 + 3$

43. No, because if $A \cap B \neq \emptyset$ then we would be counting those elements twice.

44.

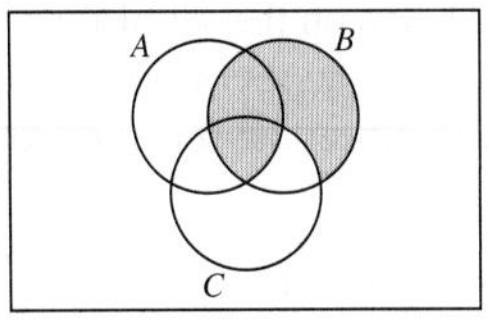

$(B - C) \cup (A \cap B)$

Chapter Review

1. $\emptyset$, $\{m\}$, $\{a\}$, $\{t\}$, $\{h\}$, $\{m, a\}$, $\{m, t\}$, $\{m, h\}$, $\{a, t\}$, $\{a, h\}$, $\{t, h\}$, $\{m, a, t\}$, $\{m, a, h\}$, $\{m, t, h\}$, $\{a, t, h\}$, $\{m, a, t, h\}$
2. **(a)** $A \cup B = A$ **(c)** $\overline{D} = \{u, n, i, v, r\}$
(e) $\overline{B \cup C} = \{s, v, u\}$ **(g)** $\{i, n\}$ **(i)** 5

3.

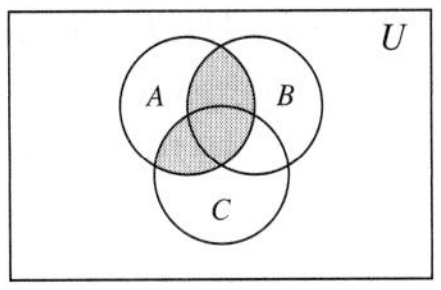

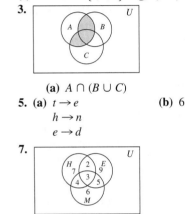

(a) $A \cap (B \cup C)$

5. **(a)** $t \to e$
$h \to n$
$e \to d$
(b) 6

7.

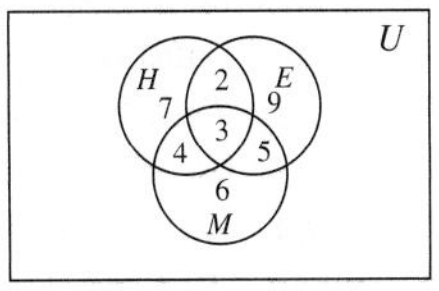

(a) 36 **(b)** 6
8. Answers vary. **(a)** $B \cup (C \cap A)$
9. **(a)** False. Consider the sets $\{a\}$ and $\{2\}$. **(c)** False. Consider the sets $\{t, h, e\}$ and $\{e, n, d\}$. They have the same number of elements, but they are not equal. **(e)** False. The set $\{5, 10, 15, 20, \ldots\}$ is a proper subset of the natural numbers and is equivalent to the natural numbers, since there is a one-to-one correspondence between the two sets.
(g) True
10. **(a)** Because $A \cup B$ is the union of disjoint sets, $A - B$, $B - A$, and $A \cap B$, the equation is true.
11. **(a)** 17 **(c)** 0
14. **(a)** Distributive property of multiplication over addition **(c)** Identity property of multiplication for whole numbers **(e)** Commutative property of multiplication
15. **(a)** Let $A = \{1, 2, 3, \ldots, 13\}$ and $B = \{1, 2, 3\}$, then B is a proper subset of A. Therefore, B has fewer elements than A, so $n(B)$ is less then $n(A)$; thus, $3 < 13$.
16. **(a)** 15, 14, 13, 12, 11, or 10 **(c)** Any whole number
17. **(a)** $15a$ **(b)** $5x^2$ **(c)** $xa + xb + xy$
18. 40 cans
20. $w = 26$
22. \$214
24. **(a)** If n is the original number, then each of the following lines shows the result of performing the instruction:

$$\begin{array}{c} n \\ n + 17 \\ 2(n + 17) = 2n + 34 \\ 2n + 30 \\ 4n + 60 \\ 4n + 80 \\ n + 20 \\ n \end{array}$$

(c) Answers vary.
25. **(a)** Function **(b)** Not a function **(c)** Function
26. **(a)** Range $= \{3, 4, 5, 6\}$ **(c)** Range $= \{0, 1, 4, 9, 16\}$
27. **(a)** This is not a function, since one student can have two majors. **(c)** This is a function. The range is $\{6, 8, 10, 12, \ldots\}$.
(e) This is a function. The range is N.
28. **(a)** $C(x) = 200 + 55(x - 1)$
(c) After the ninth month, the cost exceeds \$600.
29. **(a)** 4 **(c)** 40 **(e)** 13
31. **(a)** Yes **(c)** No, for $x = 2$, there are two values for y.

Answers to Now Try This

2-1. **(a)** and **(b)**: Number the swimming lanes 1, 2, 3, 4 and name the people A, B, C, D. Then we represent the correspondence:

$$\begin{array}{ccc} 1 & \leftrightarrow & A \\ 2 & \leftrightarrow & B \\ 3 & \leftrightarrow & C \\ 4 & \leftrightarrow & D \end{array} \quad \text{as} \quad \begin{array}{cccc} 1 & 2 & 3 & 4 \\ A & B & C & D \end{array}$$

The 24 one-to-one correspondences are

1 2 3 4	1 2 3 4	1 2 3 4	1 2 3 4
A B C D	B A C D	C A B D	D A B C
A B D C	B A D C	C A D B	D A C B
A C B D	B C A D	C B A D	D B A C
A C D B	B C D A	D B D A	D B C A
A D B C	B D A C	C D A B	D C A B
A D C B	B D C A	C D B A	D C B A

(c) We notice that $24 = 4 \cdot 3 \cdot 2 = 4 \cdot 3 \cdot 2 \cdot 1$. We also notice that we had four choices for people to swim in lane 1. After making a choice, we see that we had three choices to swim in lane 2, leaving us with two choices for lane 3 and, finally, one choice for lane 4. Extrapolating from this, we conjecture that there are

$$5 \cdot 4 \cdot 3 \cdot 2 \cdot 1 = 120$$

distinct one-to-one correspondences between a pair of five-element sets.
2-2. If event M_1 can occur in m_1 ways and after it has occurred, event M_2 can occur in m_2 ways, and after it has occurred event M_3 can occur in m_3 ways, and so on, where events $M_1, M_2, M_3, \ldots$ can occur correspondingly in $m_1, m_2, m_3, \ldots$ ways, then event M_1 followed by event M_2 followed

The graphs intersect when $x = 2$ hr **(c)** When the bowling time is more than 2 hr **(d)** When the time is 2 hr

2-22. Answers vary; for example, no they should not be connected since this would imply you could have a price for $1\frac{1}{3}$ videos.

Chapter 3

Ongoing Assessment 3-1

1. (a) $\overline{\overline{\text{M}}}$CDXXIV; The double bar over M represents $1000 \cdot 1000 \cdot 1000$. **(b)** 46,032; the 4 in 46,032 represents 40,000 while the 4 in 4632 represents only 4000. **(c)** < ▼▼ : the space in the latter number indicates < is multiplied by $10 \cdot 60$ rather than by 10. **(d)** The 𓆼 represents 1000 while 𓍢 represents only 100. **(e)** 𝋣𝋠 represents three groups of 20 plus zero 1s and 𝋲 represents three 5s and three 1s.
2. (a) MCML; MCMXLVIII **(c)** M; CMXCVIII
(e) 𓆼𓍢𓍢𓏺 ; 𓆼𓍢𓎆𓎆𓎆𓎆𓎆𓎆𓎆𓎆𓎆𓏺𓏺𓏺𓏺𓏺𓏺𓏺𓏺
3. 1922
4. (a) CXXI **(c)** LXXXIX.
5. (a) ▼ < ▼▼ ; 𓎆𓎆𓎆𓎆𓎆𓎆𓎆𓏺𓏺 LXXII; 𝋣𝋬
(c) 1223; << <<▼▼▼ MCCXXIII; 𝋣𝋧𝋣
6. (a) Hundreds **(c)** Thousands
7. (a) 3,004,005 **(c)** 3560
8. 811 or 910
9. (a) 86
11. (a) $(1, 10, 11, 100, 101, 110, 111, 1000, 1001, 1010, 1011, 1100, 1101, 1110, 1111)_{\text{two}}$ **(c)** $(1, 2, 3, 10, 11, 12, 13, 20, 21, 22, 23, 30, 31, 32, 33)_{\text{four}}$
12. 20
14. (a) 111_{two} **(c)** 999_{ten}
15. (a) ETE_{twelve}; $EE1_{\text{twelve}}$ **(c)** 554_{six}; 1000_{six}; **(e)** 444_{five}; 1001_{five}
16. (a) There is no numeral 4 in base four. **(c)** There is no numeral T in base three.
17. 3 blocks, 1 flat, 1 long, 2 units
18. (a)

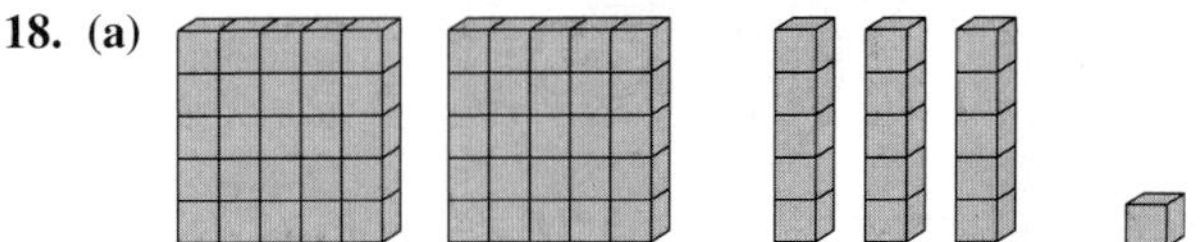

19. (b) Assume that you have 73 cents in any possible combination, for example, 19 nickels and 23 pennies. Because 23 pennies can be traded for 4 nickels and 3 pennies, we have 14 nickels and 3 pennies. 14 nickels can be traded for 2 quarters and 4 nickels. After the second trade, we should have 2 quarters, 4 nickels, and 3 pennies. We obtain: $73 = 243_{\text{five}}$.
20. (a) $3TE9E_{\text{twelve}} < 3E000_{\text{twelve}}$ because the second digit from the left in $3TE9E_{\text{twelve}}$ is less than the corresponding digit in $3E000_{\text{twelve}}$.
22. By $10{,}000_{\text{five}}$ or 5^4 or 625.
24. (a) 3212_{five} **(c)** 12110_{four}
26. (a) 117 **(c)** 1331 **(e)** 157
27. 1 prize of \$625, 2 prizes of \$125, and 1 of \$25
28. (a) 8 weeks, 2 days **(c)** 1 day, 5 hr
30. (a) 6 **(c)** 9
33. (b) 9, 004, 005, 645 **(d)** 100, 000, 000, 000, 000
34. Above the bar are depicted 5s, 50s, 500s, and 5000s. Below the bar are 1s, 10s, 100s, and 1000s. Thus there are $1 \cdot 5000$, $1 \cdot 500$, $3 \cdot 100$, $1 \cdot 50$, $1 \cdot 5$, and $2 \cdot 1$ depicted for a total of 5857. The number 4869 could be depicted as follows:

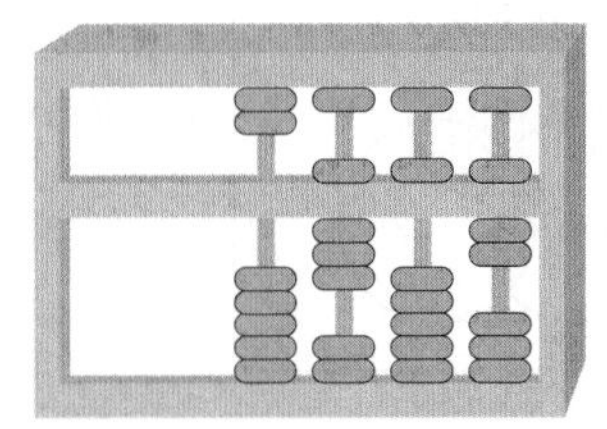

35. Assume an eight-digit display without scientific notation. **(a)** 98,765,432 **(c)** 99,999,999

Communication

37. Answers will vary. Ben is incorrect. Zero is a place holder in the Hindu-Arabic system. It is used to differentiate between numbers like 54 and 504. If zero were nothing, then we could eliminate it without changing our number system. Zero is the cardinal number of the empty set.
39. (a) This is primarily for readability. It has been proposed with the metric system to drop the commas and simply use spaces instead.

$q \cdot q_2$. This means that q (which was $a \div b$) equals the quotient of q_1 (which was $a \div c$) and q_2 (which was $b \div c$).
(b) i. $34000 \div 17000 - (34000 \div 10000) \div (17000 \div 10000) = 34 \div 17 = 2$ **ii.** $(176 \cdot 97) \div (44 \div 97) = ((176 \cdot 97) \div 97) \div (44 \div 97) \div 97) = 176 \div 44 = 4$
iii. $(a \cdot d) \div (b \cdot d) = ((a \cdot d) \div d) \div ((b \cdot d) \div d) = a \div b$
(c) Let q be $a \div b$. Then $a = b \cdot q + r$, and recall that $0 < r < b$. Then $a - r = b \cdot q$. From here,

$$(a - r) \cdot d = b \cdot q \cdot d$$

This means that since $a \cdot d = b \cdot d \cdot q + r \cdot d$, the remainder when $a \cdot d$ is divided by $b \cdot d$ is $r \cdot d$.
15. (a) $(19 \cdot 194) \cdot 10 = 36{,}860$ **(b)** $(379 \cdot 193) \cdot 100 = 7{,}314{,}700$ **(c)** $481 \cdot 73 \cdot (8 \cdot 125) = (481 \cdot 73) \cdot 1000 = 35{,}113{,}000$ **(d)** $374 \cdot 893 \cdot (200 \cdot 50) = (374 \cdot 893) \cdot 10{,}000 = 3{,}339{,}820{,}000$
16. \$395
18. 2600 cases
20. 40 cans
21. 26
23. \$6000
25. \$594
26. \$400
27. Selling pencils by the units, dozens, and gross is an example of the use of base 12.
28. (a) The Egyptian system had seven symbols. It was a *tally* system, a *grouping* system, and it used the *additive property*. It did not have a symbol for zero but this was not very important because they did not use place value.
29. (a) 1003_{five} **(c)** $T8_{\text{twelve}}$
30. (a) 212_{five} **(c)** 1442_{five}
31. (a) 4210014_{five}
32. (a) For example, $(26 + 24) + (37 - 7) = 50 + 30 = 80$
34. Answers vary. For example, **(a)** 2300 + 300 (adjustment) = 2600. **(b)** 2600
35. $2400 \cdot 4 = 9600$
37. (a) $999 \cdot 47 + 47 = 47\,(999 + 1) = 47 \cdot 1000 = 47{,}000$
(b) $43 \cdot 59 + 41 \cdot 43 = 43 \cdot (59 + 41) = 43 \cdot 100 = 4300$
(c) $1003 \cdot 70 - 3 \cdot 79 = 79 \cdot (1003 - 3) = 79 \cdot 1000 = 79{,}000$
(d) $1001 \cdot 113 - 113 = 113 \cdot (1001 - 1) = 113 \cdot 1000 = 113{,}000$ **(e)** $101 \cdot 35 = (100 + 1) \cdot 35 = 100 \cdot 35 + 1 \cdot 35 = 3500 + 35 = 3535$ **(f)** $98 \cdot 35 = (100 - 2) \cdot 35 = 100 \cdot 35 - 2 \cdot 35 = 3500 - 70 = 3430$
38. (a) $27^2 = 26^2 + 2 \cdot 26 + 1 = 676 + 52 + 1 = 729$
(c) Let a be an integer. Then,

$$(a-1)^2 = (a-1) \cdot (a-1) = a \cdot (a-1) - (a-1)$$
$$= a^2 - a - a + 1 = a^2 - 2a + 1.$$

In order to find the previous perfect square, one would have to subtract twice the number a that was squared and add 1.
39. (a) $3x^3 + 4x^2 + 7x + 8 + (5x^2 + 2x + 1) = 3x^3 + 9x^2 + 9x + 9$ **(b)** Answers vary. **(c)** Answers vary; for example, $34 \cdot 10^2 = (3 \cdot 10 + 4) \cdot 10^2 = 3 \cdot 10^3 + 4 \cdot 10^2 + 0 \cdot 10 + 0 = 3400$ and $(3x + 4)x^2 = 3x^3 + 4x^2$.

Answers to
3-1. 3 block
= 3 blocks
= 4 blocks
= 4 blocks
= 4327
3-2. (a)
(b) 203,034
(c) Answers
cumbersome
value. Perfor
multiplicatio
numbers are
3-3. (a) ▼▼
(b) $2 \cdot 60^2$ +
(c) The Hinc
very importa
uses base six
many symbol
3-4. (a) The
5's. This show
remainder. N
remainder. Ea
in 824 with 4
continues unt
2 25's; 4 5's;
(b) 5 | 728
5 | 145 3
5 | 29 0
5 | 5 4
1 → 0
3-5. (0001, 0
1010, 1011, 1
10011, 10100
3-6. (a) 182
1 blc
1 blo
or 1 blo
or 2 blo
or 243
(b) (g) 5 tens
15 te
(h) (i) 2 hu
1 hu
1 hu
1 h

9.
(d)
23fi
mul
12.
14.
16.
17.
18.
19.
21.
100
99a
22.
23.
25.
27.
exp
mul
345
resu
29.
ove
of a
30.
33.
34.
36.
38.
39.
41.
42.
43.
fact

Ongoing Assessment 3-2

1. (a)
```
  981
+ 421
 1402
```
(c)
```
  383
- 159
  224
```
3. (a) One possibility:
```
  863
+ 752
 1615
```
4. Only if positive numbers are used:
(a)
```
  876
- 235
  641
```
5. No, not all at dinner. He can have either the fish or the salad.
6. Yes, \$124
7.
```
  3428
+ 5631
  9059
```
8. (a)
```
  93        93 + 3        96
- 37  →   - 37 + 3  →   - 40
                          56
```
9. (a) (i) 1236 (ii) 1032
10. Answers may vary; for example: **(a)** The student did not regroup from the ones place to the tens place. **(c)** The units minuend is subtracted from the subtrahend.
11. 1 hr 34 min 15 sec
13.
(b) $174 + 285 = (1 \cdot 100 + 7 \cdot 10 + 4) + (2 \cdot 100 + 8 \cdot 10 + 5)$
$= (1 \cdot 100 + 2 \cdot 100) + (7 \cdot 10 + 8 \cdot 10) + (4 + 5)$
$= (1 + 2) \cdot 100 + (7 + 8) \cdot 10 + (4 + 5)$
$= 3 \cdot 100 + 15 \cdot 10 + 9$
$= 4 \cdot 100 + 5 \cdot 10 + 9$
$= 459$
14. (a)
```
   4  3  5  8
 + 3  8  6  4
 |0/7|1/1|1/1|1/2|
   8  2  2  2
```
(b)
```
     4  9  2  3
 +   9  8  9  7
   |1/3|1/7|1/1|1/0|
  1  4  8  2  0
```
(c)
```
     2  3  4  5
 +   8  8  8  8
   |1/0|1/1|1/2|1/3|
  1  1  2  3  3
```
15. Answers vary. This approach emphasizes the meaning of place value of the digits and may be easier for young children than the standard algorithm. It can also serve as a transition to the standard algorithm.
16. Answers vary. As in problem 15, this approach may serve as a transition from a concrete base-ten blocks approach to the standard algorithm.

17. (a) 121_{five} **(c)** 1010_{five} **(e)** 1001_{two}
18.

+	0	1	2	3	4	5	6	7
0	0	1	2	3	4	5	6	7
1	1	2	3	4	5	6	7	10
2	2	3	4	5	6	7	10	11
3	3	4	5	6	7	10	11	12
4	4	5	6	7	10	11	12	13
5	5	6	7	10	11	12	13	14
6	6	7	10	11	12	13	14	15
7	7	10	11	12	13	14	15	16

Base eight

19. (b) 1 hr 39 min 40 sec
20. (a) 2 qt, 1 pt. 0 c or $\frac{1}{2}$, 1 pt, 0 c **(c)** 2 qt, 1 pt, 1 c
21. It is doubling the second number in the operation.
22. (a) 34; 34; 34 **(c)** 34
23. (b) $^{3}\not{3}_{1}2$
$1\,\not{3}_{0}$
$2\,2$
$\not{4}_{3}\,\not{3}_{0}$
$\not{2}_{0}\,3$
$1\,\not{2}_{0}$
310_{five}
24. (a) 3 gross 10 doz 9 ones
26. (a)
230_{five}
$-\,22_{\text{five}}$
203_{five}
27. (a) 1241_{five} **(c)** TET_{twelve}
29. (a) The method produces a palindrome in each case: (i) 363 (ii) 9339 (iii) 5005.
30. (a)
32, 31, 28, 30, 24, 27, 29, 26, 25
24, 25, 29, 26, 28, 30, 27, 31, 32
24, 25, 28, 26, 32, 29, 27, 30, 31
(b) Three, (24, 28, 32)

Commur

33. For
reflect th
subtracti
mechani
37. Eve
is not co
which is

Review

40. This
with the
1 dekam
1 kilom
41. 528
42. 141
43. (a)
44. (a)
example
(if leap
for exan
beat 37,

Ongoi
1. (a)

2. (a)

5. (a)
6. (a)

16. Answers vary; for example, $3300 - 100 - 300 - 400 - 500 = 2000$. The estimate is high because the amounts that were taken away were rounded up and resulted in more being taken away than the $13 that was added to the $3287.
18. (a) Different answers since the estimates of 800 and 220 are way off.
20. The clustering strategy gives $6 \cdot 70{,}000$, or 420,000.
21. (a) Since $40 + 30 + 60 + 250 = 380$, he will spend less than $380 and has enough in his checking account.
(c) Alberto; $3473 + 5615 = 3472 + 5616 > 3463 + 5616$
22. (c) Low; $6{,}000 \div 299 > 6000 \div 300$
(d) Low; $10{,}000 \div 999 > 10{,}000 \div 1000$
(f) High; $1{,}999 \div 201 < 2{,}000 \div 200$
23. One possibility is that to find $x5^2$ we could write $x(x + 1)$ and append the digits 25. For example, in 65^2, we take $6 \cdot 7 = 42$ and append 25 to obtain 4225.

Communication

25. Mental mathematics is the process of producing an exact answer to a computation without using external aids. Computational estimation is the process of forming an approximate answer to a numerical problem.

Review Problems

33. We will exhibit the property for a three-digit number, abc:

$$\begin{aligned} abc \cdot 10 &= (a \cdot 10^2 + b \cdot 10 + c) \cdot 10 \\ &= a \cdot 10^3 + b \cdot 10^2 + c \cdot 10 = abc0 \end{aligned}$$

34. (a)

$$\begin{array}{rll} 18\overline{)623} & & \\ -180 & \leftarrow & 10 \\ 443 & & \\ -180 & \leftarrow & 10 \\ 263 & & \\ -180 & \leftarrow & 10 \\ 83 & & \\ -18 & \leftarrow & 1 \\ 65 & & \\ -18 & \leftarrow & 1 \\ 47 & & \\ -18 & \leftarrow & 1 \\ 29 & & \\ -18 & \leftarrow & \underline{1} \\ 11 & & 34 \end{array} \qquad \begin{array}{rll} 18\overline{)623} & & \\ -540 & \leftarrow & 30 \\ 83 & & \\ -72 & \leftarrow & \underline{4} \\ 11 & & 34 \end{array}$$

(b)

$$\begin{array}{rl} 21\overline{)493} & \\ -210 & \leftarrow \\ 283 & \\ -210 & \leftarrow \\ 73 & \\ -21 & \leftarrow \\ 52 & \\ -21 & \leftarrow \\ 31 & \\ -21 & \leftarrow \\ 10 & \end{array}$$

(c)

$$\begin{array}{rl} 97\;\;1000 & \\ -970 & \leftarrow \\ 30 & \end{array}$$

35. (a) 623 = 3
(c) 1000 = 10 ·
36. 34 · 79 = (3
= 3
= 21
= 21
= 21
= 26

Chapter Review

1. (a) tens **(c)**
2. (a) 400,044
3. (a) CMXCIX
4. (a) 3^{17} **(c)**
5. 2020_{three}
6. 1 block, 2 flats
7. (a)
8. (a) 10^{10} **(b)**
9. (a) 10,000,00
(c) 1,000,000,00
(e) 1,111,111,11
10. 1119
11. 60,074
12. (a) 5 remain
(c) 120_{six} remain
13. (a) $5 \cdot 912 +$
(c) $23_{five} \cdot 120_{five}$
$= 1011_{two}$
14. (a) Let $q_1 =$
and $b = c \cdot q_2$. Als
$a =$
But $a = q_1 \cdot c$, wh
$(q \cdot q_2 - q_1)\, c = 0$

(ii)

$$\begin{array}{r} 1 \text{ hundred} + 4 \text{ tens} + 5 \\ -\; 1 \text{ hundred} + 2 \text{ tens} + 6 \end{array} \Big\},$$

$$\begin{array}{r} 1 \text{ hundred} + 3 \text{ tens} + 15 \\ -\; 1 \text{ hundred} + 2 \text{ tens} + 6 \\ \hline 1 \text{ ten} + 9 \;=\; 19 \end{array}$$

(iii)

$$\begin{array}{r} 2 \text{ hundred} + 2 \text{ tens} + 3 \\ -\,1 \text{ hundred} + 5 \text{ tens} + 6 \end{array} \Big\}$$

$$\begin{array}{r} 1 \text{ hundred} + 12 \text{ tens} + 3 \\ -\,1 \text{ hundred} + 5 \text{ tens} + 6 \end{array} \Big\}$$

$$\begin{array}{r} 1 \text{ hundred} + 11 \text{ tens} + 13 \\ -\,1 \text{ hundred} + 5 \text{ tens} + 6 \\ \hline 6 \text{ tens} + 7 \;=\; 67 \end{array}$$

3-7. (i) The method is valid because subtracting and adding the same number does not change the original sum.
(ii)

$$\begin{aligned} 97 + 69 &= (97 + 3) + (69 - 3) \\ &= 100 + 66 \\ &= 166 \end{aligned}$$

3-8. (a) 1000_{five} **(b)** 31_{five}
3-9. (a)

+	0_{two}	1_{two}
0_{two}	1_{two}	1_{two}
1_{two}	1_{two}	10_{two}

(b) (i)

$$\begin{array}{r} 1101_{two} \\ -\,111_{two} \\ \hline 110_{two} \end{array}$$

(ii) 10110_{two}

3-10. The only time the expressions are equal for all non-zero whole numbers m and n is if $a = 0$. If $n = m = 1$ and $a = 2$ the expressions are also equal.
3-11.

$$\begin{aligned} 7 \cdot 4589 &= 7(4 \cdot 10^3 + 5 \cdot 10^2 + 8 \cdot 10 + 9) \\ &= (7 \cdot 4) \cdot 10^3 + (7 \cdot 5) \cdot 10^2 + (7 \cdot 8) \\ &\quad \cdot 10 + 7 \cdot 9 \\ &= 28000 + 3500 + 560 + 63 \\ &= 32123 \end{aligned}$$

3-12. (a) (i) 40 **(ii)** 6 **(iii)** 6 **(iv)** 0 **(b)** 12
3-13. Answers vary, for example,
(a) $40 + 160 = 200$ and $29 + 31 = 60$ so the sum is 260.
(b) $3679 - 400 = 3279$ and $3279 - 74 = 3205$.
(c) $75 + 25 = 100$ and $100 + 3 = 103$.
(d) $2500 - 500 = 2000$ and $2000 - 200 = 1800$.
3-14. Answers vary, for example,
(a) $4 \cdot 25 = 100$ and $32 \cdot 100 = 3200$.
(b) $123 \cdot 3 = 100 \cdot 3 + 23 \cdot 3 = 300 + 69 = 369$.
(c) $25 \cdot 35 = (30 - 5)(30 + 5) = 30^2 - 5^2 = 900 - 25 = 875$.
(d) $5075/25 = 5000/25 + 75/25 = 200 + 3 = 203$.
3-15. Answers vary, for example,
(a) To estimate $4525 \cdot 9$ we know $4525 \cdot 10 = 45{,}250$ and since we have only 9 sets of 4525 we can take away approximately 5000 from our estimate and we have 40,250.
(b) To estimate 3625/42 we know the answer will be close to 3600/40 or 90.

Chapter 4

Ongoing Assessment 4-1

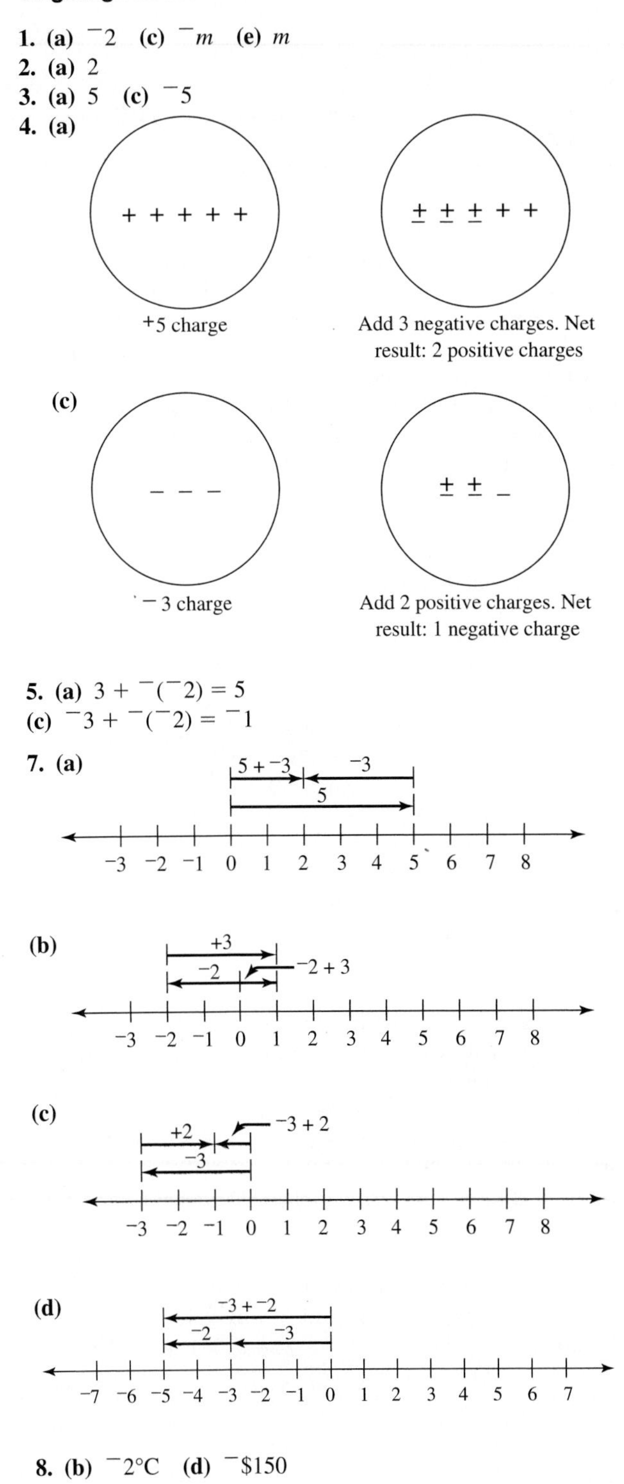

1. (a) $^{-}2$ **(c)** ^{-}m **(e)** m

2. (a) 2

3. (a) 5 **(c)** $^{-}5$

4. (a)

(c)

5. (a) $3 + {}^{-}({}^{-}2) = 5$
(c) ${}^{-}3 + {}^{-}({}^{-}2) = {}^{-}1$

7. (a)

(b)

(c)

(d)

8. (b) $^{-}2$°C **(d)** $^{-}$\$150

9. (a) $({}^{-}45) + ({}^{-}55) + ({}^{-}165) + ({}^{-}35) + ({}^{-}100) + 75 + 25 + 400$

10. (a)

11. (a) ${}^{-}4 - 2 = {}^{-}6$; ${}^{-}4 - 1 = {}^{-}5$; ${}^{-}4 - 0 = {}^{-}4$; ${}^{-}4 - {}^{-}1 = {}^{-}3$

12. (a) $^{-}9$ **(c)** 1 **(e)** $^{-}13$

13. (a) (i) $55 - 60$ **(ii)** $55 + ({}^{-}60)$ **(iii)** $^{-}5$

14. (a) 59 **(c)** 192°F

15. (a) 10W–40 or 10W–30 **(c)** 10W–40, 5W–30 or 10W–30 **(e)** 10W–30 or 10W–40

16. (a) $1 + 4x$ **(c)** $x - 2$

17. The equation holds if and only if $c = 0$ (a and b can be any integers). Justification: It can be shown that $a - (b - c) = a - b + c$ for all integers a, b, and c. Thus, the original equation holds if and only if $(a - b) + {}^{-}c = (a - b) + c$, which in turn holds if and only if ${}^{-}c = c$. This last equation is true if and only if $c = 0$.

18. (a) I **(c)** $I - \{0\}$ **(e)** $\varnothing$ **(g)** $\{0\}$ **(i)** I

19. (a) Let ${}^{-}x - y = n$, where n is the unique integer so that ${}^{-}x = y + n$. Note that

$$\begin{aligned} y + ({}^{-}y - x) &= y + ({}^{-}y + {}^{-}x) \\ &= (y + {}^{-}y) + {}^{-}x \\ &= 0 + {}^{-}x \\ &= {}^{-}x \end{aligned}$$

Because n was the unique integer so that ${}^{-}x = y + n$ and ${}^{-}x = y + ({}^{-}y - x)$, then $n = {}^{-}y - x$, or ${}^{-}x - y = {}^{-}y - x$.
(b) No. Because ${}^{-}x - y = {}^{-}x + {}^{-}y$ and because ${}^{-}y - x = {}^{-}y + {}^{-}x$, the property in part (a) is just the commutativity property of addition.

21. (a) 0 **(c)** 1 **(e)** $^{-}4$

22. (a) All negative integers
(c) All integers less than $^{-}1$ **(e)** There are none.
(g) There are none.

23. (a) 9 **(c)** 0 or 2

24. (a) x

25. (b) -8 and 8

26. $f(x) = \begin{cases} x - 6, & \text{if } x \geq 6; \\ -x + 6, & \text{if } x < 6. \end{cases}$

27. (a) 89 **(c)** 19

29. (a) $d = {}^{-}3$, next terms: $^{-}12$, $^{-}15$ **(c)** $d = {}^{-}y$, next terms: $x - 2y$, $x - 3y$

30. (a) 0 **(c)** 2538

31. $^{-}1$

32. (a) True **(c)** True **(e)** False; let $x = {}^{-}1$.

33. (a) -4 **(c)** -5 **(e)** 14

34. The smaller gear rotates 28 times in the opposite direction of the larger gear.

35. **(a)** $^{-}18$ **(c)** 22 **(e)** 23
36. **(a)** $^{-}101$ **(c)** 10,894

Communication

37. He could have driven 12 mi in either direction from milepost 68. Therefore, his location could be either at the $68 - 12 = 56$ milepost or at the $68 + 12 = 80$ milepost.
38. **(a)** By definition of subtraction $(a - b) + b = a$. Also, using the associative property for addition and the definition of an additive inverse:

$$\begin{aligned}[a + (^{-}b)] + b &= a + [(^{-}b) + b] \\ &= a + 0 \\ &= a\end{aligned}$$

Hence, $(a - b) + b = [a + (^{-}b)] + b$. Adding ^{-}b to both sides of the last equation, we get the required result.
40. Two numbers are opposite if and only if their sum is 0. We have

$$\begin{aligned}(b - a) + (a - b) &= b + (^{-}a) + a + (^{-}b) \\ &= b + 0 + {}^{-}b \\ &= b + {}^{-}b \\ &= 0.\end{aligned}$$

42.

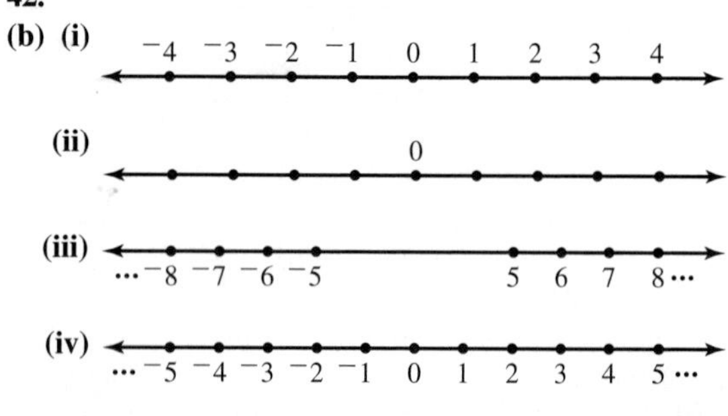

Open-Ended

45. Answers vary; for example, the floors aboveground could be numbered as usual 1, 2, 3, 4, . . . , *n*, the zero or ground floor could be called *G*, and the floors belowground could be called 1*B*, 2*B*, 3*B*, 4*B*, . . . , *nB*. The system could be modeled on a vertical number line with *G* replacing 0 and 1*B*, 2*B*, 3*B*, . . . , *nB* replacing the negative integers. We could have computations like $3 + G = 3$, or $5B + 3B = 8B$.
46. **(a)** 4
47. **(a)** Answers vary. For example, $f(x) = -|x| - 1$.

Ongoing Assessment 4-2

1. $3(^{-}1) = {}^{-}3$; $2(^{-}1) = {}^{-}2$; $1(^{-}1) = {}^{-}1$; $0(^{-}1) = 0$; $(^{-}1)(^{-}1) = 1$, by continuing the pattern of an arithmetic sequence with fixed difference 1.

3.

−8 −7 −6 −5 −4 −3 −2 −1 0 1 2 3 4 5 6 7 8

4. **(a)** $(^{-}3)(^{-}3) = 9$ **(b)** $(^{-}5)2 = {}^{-}10$
5. **(a)** $^{-}20 \cdot 4$ **(c)** $^{-}20n$
6. **(a)** 5 **(c)** $^{-}11$ **(e)** Impossible; division by 0 is not defined.
7. **(a)** $^{-}10$ **(c)** $^{-}10$ **(e)** Not defined **(g)** Not defined
8. **(a)** $^{-}30$, $^{-}30 \div {}^{-}6 = 5$, $^{-}30 \div 5 = {}^{-}6$ **(c)** 0, $0 \div {}^{-}3 = 0$; division by 0 is not defined.
9. **(a)** $ac \div bc = x$ if and only if $ac = xbc$ if and only if $ac - xbc = 0$ if and only if $(a - xb)c = 0$. Because $c \neq 0$, we have $a - xb = 0$ or $a = xb$, which implies $a \div b = x$.
(b) No; if $c = 0$, the relation in (a) is not true anymore. Then notice that $ac = bc = 0$ and $ac \div bc$ is no longer unique.
If $a \div b = 0$, the relation holds because in this case $a = b \cdot 0 = 0$. Hence, $ac = 0$ or $ac = bc \cdot 0$, which means $ac \div bc = 0 = a \div b$.
10. **(a)** $(4x) \div 4 = a$ if and only if $4x = 4a$ if and only if $a = x$ **(b)** If $x \neq 0$, $(^{-}4x) \div x = a$ if and only if $^{-}4x = ax$ if and only if $a = {}^{-}4$. **(c)** $(^{-}xy) \div y = a$ if and only if $^{-}xy = ay$ if and only if $a = {}^{-}x$. **(d)** $(^{-}10x + 5) \div 5 = a$ if and only if $^{-}10x + 5 = 5a$ if and only if $5(^{-}2x + 1) = 5a$ if and only if $a = {}^{-}2x + 1$.
11. All answers are in °C **(a)** $32 + (^{-}3)\,30$ or $32 - 3 \cdot 30$ **(c)** $^{-}20 + (^{-}4)(^{-}30)$ or $^{-}20 + 4 \cdot 30$ **(e)** $^{-}5 + dm$ **(f)** dm **(g)** $20 - md$ degrees C
12. **(a)** $4(^{-}11) = {}^{-}44$. Thus, 44 yd were lost.
14. **(a)** $^{-}1(^{-}5 + {}^{-}2) = {}^{-}1(^{-}7) = 7$; $(^{-}1)(^{-}5) + (^{-}1)(^{-}2) = 5 + 2 = 7$ **(c)** $^{-}5(2 + {}^{-}6) = {}^{-}5(^{-}4) = 20$; $(^{-}5)(2) + (^{-}5)(^{-}6) = {}^{-}10 + 30 = 20$
15. **(a)** $^{-}8$ **(c)** $^{-}1000$ **(e)** 1 **(g)** 12 **(m)** $^{-}13$ **(o)** $^{-}32$
16. **(a)** 0 **(b)** 2^{63} **(c)** 0 **(d)** $^{-}159$ **(e)** 9
17. (b), (c), (g), (h) are always positive; (a), (f) are always negative.
18. (b) = (c); (d) = (e); (g) = (h)
19. **(a)** Commutative property of multiplication **(b)** Closure property of multiplication **(d)** Distributive property of multiplication over addition
20. **(a)** xy **(c)** $3x - y$ **(g)** $3y - 2x$
21. **(a)** $^{-}2x + 2y$ **(c)** $^{-}x^2 + xy$ **(e)** $^{-}x^2 + xy + 3x$ **(g)** $x^2 - y^2 - 1 - 2y$
22. **(a)** $^{-}2$ **(e)** $^{-}36$ **(i)** No solution is possible. **(k)** No solution is possible. **(m)** All integers except 0 **(o)** All integers
23. **(a)** $^{-}5$ **(c)** No solutions **(e)** $^{-}2$ and 2 **(g)** $x = {}^{-}1$ **(h)** 1 and $^{-}3$ **(k)** $x < {}^{-}4$ **(m)** $3 - 5x < 13$
24. **(a)** $(50 + 2)(50 - 2) = 50^2 - 2^2 = 2500 - 4 = 2496$ **(c)** $x^2 - y^2$ **(e)** $x^2 - 1$

25. **(a)** $8x$ **(c)** $x(y + 1)$ **(e)** $x(x + y)$ **(g)** $x(3y + 2 - z)$
(i) $a(b(c + 1) - 1)$ **(k)** $(4 + a)(4 - a)$
(m) $(2x + 5y)(2x - 5y)$
26. **(a)**

$$\begin{aligned}(a - b)^2 &= (a + {}^{-}b)(a + {}^{-}b) = a(a + {}^{-}b) + {}^{-}b(a + {}^{-}b)\\ &= a^2 + a({}^{-}b) + ({}^{-}b)a + ({}^{-}b)({}^{-}b)\\ &= a^2 - 2ab + b^2.\end{aligned}$$

27. **(a)** False. Let $x = 3$, $y = {}^{-}5$, then $|x + y| = 2$ but $|x| + |y| = 8$. **(c)** True. Because $x^2 \geq 0$, $|x^2| = x^2$.
28. **(a)** The sums are 9 times the middle number.
29. **(a)** 8, 11, $d = 3$, nth term is $3n - 13$. **(c)** $^{-}128$, $^{-}256$, $r = 2$, nth term $^{-}(2)^n$
31. **(a)** $^{-}9$, $^{-}6$, $^{-}1$, 6, 15 **(c)** $^{-}3$, 3, $^{-}9$, 15, $^{-}33$
(e) $^{-}1$, 4, $^{-}9$, 16, $^{-}25$
33. **(a)** No **(b)** yes; the order is 10.
34. After 17 min; the temperature was $^{-}108$°C.
35. **(a)** Sometimes true. If $x \geq 0$ and $y \leq 0$, then $^{-}|x| \cdot |y| = {}^{-}x({}^{-}y) = xy$. Similarly, the statement is true for $x \leq 0$ and $y \geq 0$. **(c)** Sometimes true. True if and only if $x = 0$. If $x > 0$, $^{-}x^2$ is negative and x^2 is positive and hence the expressions cannot be equal. **(e)** Sometimes true. If x and y are positive or zero, the statement is true. It can be shown that the statement is true if and only if $x + y > 0$.
(f) False. If $a = 0$ and $x < y$, then $-x > -y$.
37. Answers vary; for example, each of the following is correct: **(i)** $|x| < |y|$ **(ii)** $-|y| < x < |y|$ **(iii)** If $y > 0$, then $-y < x < y$, and if $y < 0$, then $y < x < -y$. Justification for **(i)** and **(ii)**: $x^2 < y^2$ is equivalent to each of the following: $|x|^2 < |y|^2$, $|x| < |y|$, $^{-}|y| < x < |y|$.

Communication

38. No; it is not of the form $(a - b)(a + b)$.
40. **(a)**

$$\begin{aligned}({}^{-}1)a + a &= ({}^{-}1)a + 1 \cdot a\\ &= ({}^{-}1 + 1)a\\ &= 0 \cdot a\\ &= 0\end{aligned}$$

41.

$({}^{-}a)b = [({}^{-}1)a]b$	
$= {}^{-}1(ab)$	By the associative property of multiplication
$= -(ab)$	Problem 40(b)

46. **(b)** No, there is no least integer. Because x is an integer the inequality is equivalent to $x \leq {}^{-}5$ and there is no such smallest integer.
47. Denote the number that the classmate chooses by x. The operations that Jill performs on these numbers are described by

$$\begin{aligned}[({}^{-}2)({}^{-}3x + 2) - 14] \div 6 &= (6x - 4 - 14) \div 6\\ &= 6(x - 3) \div 6 = x - 3\end{aligned}$$

By adding 3 to the result, we obtain the original number.

Open-Ended

49. Answers vary; for example, if the student answered only one problem and missed it, he or she would score $^{-}1$. If the student answered 5 correct and missed more than 20, he or she would receive a negative score. Any coordinate falling above the line $4x - y = 0$ would result in a negative score, where x is the number of correct and y is the number of incorrect answers.

Review Problems

54.

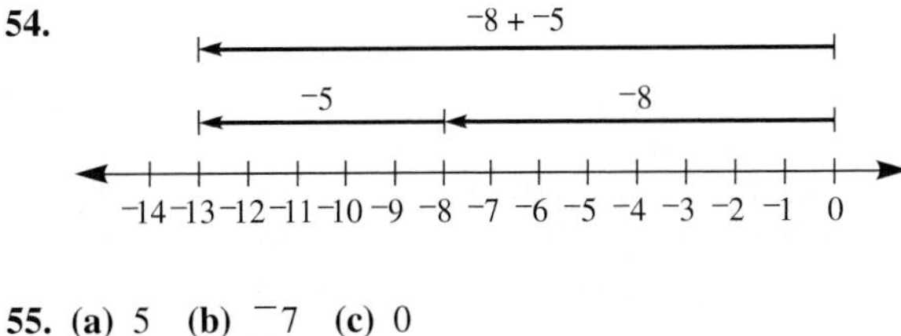

55. **(a)** 5 **(b)** $^{-}7$ **(c)** 0
56. **(a)** 14 **(b)** 21 **(c)** $^{-}4$ **(d)** 22
57. 400 lb
58. If $x \geq 0$, then $f(x) = x$; if $x < 0$, then $f(x) = (x + {}^{-}x) \div 2 = 0$.
59. **(a)** $x = 3$ or $^{-}3$ **(b)** No x possible **(c)** $x \geq 0$ **(d)** $x \leq 0$

Ongoing Assessment 4-3

1. **(a)** True **(c)** True **(e)** True
2. **(a)** Yes **(c)** Yes **(e)** Yes
3. **(a)** 2, 3, 4, 6, 11 **(c)** 2, 3, 5, 6, 10 **(e)** 11
4. **(b)** Yes, $17|34{,}000$ and $17|51$, so $17|34{,}051$. **(d)** Yes, 5 is a factor of $2 \cdot 3 \cdot 5 \cdot 7$.
5. **(a)** $a^3|a^4$ because $a^4 = a^3 \cdot a$. **(c)** $a^n|a^m$ because $a^m = a^n \cdot a^{m-n}$ and a^{m-n} is an integer because $m - n \geq 0$.
6. **(b)** True by Theorem 4-2(b) **(d)** True by Theorem 4-2(b)
7. **(a)** Sometimes true; for example, if $c = 0$ **(c)** Always true. Because $b|a$, there is c such that $a = bc$. Then $^{-}a = {}^{-}(bc) = b({}^{-}c)$, which implies that $b|{}^{-}a$. Also, $^{-}a = {}^{-}(bc) = ({}^{-}b)c$, which implies $^{-}a|{}^{-}b$.
8. **(a)** $210 = 7 \cdot 30$ **(b)** $19|1900$ and $19|38$ **(d)** $7|4200$ but $7 \nmid 22$ **(e)** $13^4 \cdot 100 = (13 \cdot 2) \cdot 13^3 \cdot 50 = 26 \cdot (13^3 \cdot 50)$
9. **(a)** True **(c)** False **(e)** True **(g)** True
10. **(a)** Always **(b)** Sometimes **(c)** Never **(d)** Always **(e)** Always **(f)** Sometimes **(g)** Always
11. **(a)** A number is divisible by 16 if and only if the last four digits form a number divisible by 16.
13. **(a)** 7
14. **(a)** Any digit 0 through 9 **(c)** 1, 3, 5, 7, 9 **(e)** 7
15. **(a)** True, because $7 | 2800$ and $7 | 21$ **(d)** False, because $23 | 460{,}000$ and $23 | 46$, thus $23 | 460{,}000 + 46 = 460{,}046$
16. 17
18. **(b)** No **(d)** Yes

19. **(a)** 1, 2, 4, 5, 8, 11 **(c)** 5 field goals
21. **(b)** $987 + 456 + 8765 = 10{,}208$; $6 + 6 + 8 = 20$ has a remainder of 2 when divided by 9, as does $1 + 0 + 2 + 0 + 8 = 11$. **(d)** Answers vary. **(f)** $345 \cdot 56 = 19{,}320$. 345 has a remainder of 3 when divided by 9; 56 has a remainder of 2 when divided by 9; $3 \cdot 2 = 6$ and $345 \cdot 56 = 19{,}320$ when divided by 9.
22. **(b)** 1, 2, 4, and 8 divide n. 1 divides every number. Also, because $16|n$, then $n = 16 \cdot d$, $d \in I$. Therefore, $n = (2 \cdot 8)d = 2(8d) = 8(2d) = 4(4d)$. Hence, n is divisible by 2, 4, and 8.
23. **(a)** All of the numbers are divisible by 11. **(c)** No, for example, consider 12,321, which is a five-digit palindrome and it is not divisible by 11.
24. In a palindromic pair, the sum of the digits in one number is the same as the sum of the digits in the other number. Therefore, if 3 divides one number, then 3 must divide the other number.
25. **(a)** Yes, if $5|x$ and $5|y$, then $5|(x + y)$. **(c)** Yes, if $5|x$, then 5 divides all the multiples of x and in particular $5|xy$.
27. **(a)** False; $2|4$, but $2 \nmid 1$ and $2 \nmid 3$ **(c)** False; $12|72$, but $12 \nmid 8$ and $12 \nmid 9$ **(e)** False; if $a = 5$ and $b = -5$, then $a|b$ and $b|a$, but $a \neq b$. **(g)** False; $2 \nmid 3$ and $2 \nmid 9$, but $2|(3 + 9)$ **(i)** False; $50 \nmid 10$, but $50|100$.
30. **(a)** The result is always 9. **(c)** Let the number be $a \cdot 10 + b$. The number with the digits reversed is $b \cdot 10 + a$.

$$\begin{aligned}\text{Now, } a \cdot 10 + b - (b \cdot 10 + a) &= a \cdot 10 + b - b \cdot 10 - a \\ &= a \cdot 10 - a + b - b \cdot 10 \\ &= 9a - 9b \\ &= 9(a - b)\end{aligned}$$

Thus, the difference is a multiple of 9.

Communication

32. **(a)** Yes, $d|n$ because a display of 32 implies that $n = 32 \cdot d$ and this implies that $d|n$.
33. **(b)** Yes, 2 is a factor of 52,834 and 2 is a factor of 324,514, so $2 \cdot 2$ or 4 is a factor of $52{,}834 \cdot 324{,}514$, which is the area.
35. **(a)** No. If $5 \nmid d$ for any integer d, then there is no integer m such that $5m = d$. If we assume $10|d$, this means there exists n such that $10n = d$ or $5(2n) = d$. This contradicts the original assumption that d is not divisible by 5.
39. **(a)** Let $abcd$ be a number. Subtracting the last digit gives $abc0$. Thus, the result is divisible by 2, 5, and 10. **(b)** If one subtracts cd from $abcd$, one obtains $ab00$. This number is divisible by 2, 4, 5, 10, 20, 25, 50, and 100.
(c) $abcd - (a + b + c + d) = a \cdot 10^3 + b \cdot 10^2 + c \cdot 10 + d - a - b - c - d = a \cdot 999 + b \cdot 99 + c \cdot 9 = 9(a \cdot 111 + b \cdot 11 + c)$. Thus, the result is divisible by 3 and 9.
40. **(a)** Divisibility by 2 will depend only on the last digit in even bases. Consider, for example, $abcd_x$, where x is even. Then,

$$abcd_x = ax^3 + bx^2 + cx + d$$

Because x is even, then $2|x^3$ and $2|x^2$ and $2|x$, which means $2|ax^3 + bx^2 + cx$. Then, $2|abcd_x$ if and only if $2|d$.

Open-Ended

41. **(a)** Answers vary; for example, upon inspection of the given numbers we notice that all the numbers are multiples of 3. Since 3 divides each number, then 3 divides the sum of any of the numbers. Because 100 is not divisible by 3, there is no winning combination of the given numbers that will sum to 100. **(b)** Answers vary; for example, since many of these multiples of 3 sum to 99 (33 + 66, 45 + 51 + 3, etc.), the company could place at most 1000 cards with the number 1 on the card. This would ensure that there are at most 1000 winners. Other numbers could also be used.
42. **(a)** If some objects can be rearranged in two rows so that there is a one-to-one correspondence between the objects in one and the other, then the number of objects is even. If we can divide the objects in two rows so that in one row there is one more object than in the other, the number of objects is odd.

Cooperative Learning

43. Answers vary.

Review Problems

44. **(a)** -2 **(b)** No such integers exist. **(c)** No such integers exist. **(d)** All integers **(e)** No such integers exist. **(f)** All positive integers
45. **(a)** $5x - 1$ **(b)** x^2 **(c)** $x - y$ **(d)** 1
46. **(a)** $f(^-5) = 10 - 3 = 7$ **(b)** $^-10$ **(c)** No. If $-2x - 3 = 2$, then $^-2x = 5$, which has no solution in integers. **(d)** The set of all odd integers
47. **(a)** Profit of $\$^-400$, that is, a loss of \$400 **(b)** $P(x) = 2x - 1800$ **(c)** 900

Ongoing Assessment 4-4

2. **(a)** Prime **(c)** Not prime **(e)** Prime
3. **(a)**

504
2 252
2 126
2 63
3 21
3 7

$504 = 2^3 \cdot 3^2 \cdot 7$

(b)

2475
3 825
3 275
5 55
5 11

$2475 = 3^2 \cdot 5^2 \cdot 11$

4. (a)

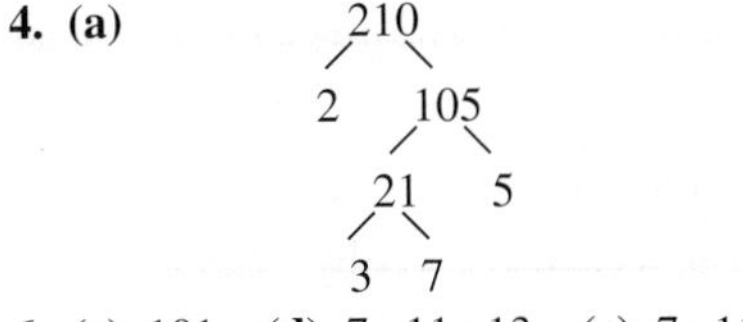

6. (c) 101 **(d)** $7 \cdot 11 \cdot 13$ **(e)** $7 \cdot 11 \cdot 13 \cdot 10001$
(f) $7^2 \cdot 11^2 \cdot 13^2$ **(g)** $3^{30} \cdot 37^{10}$
7. (a) $1 \times 48, 2 \times 24, 3 \times 16, 4 \times 12$
8. Yes; 177 flotillas of 1 ship each, 1 flotilla of 177 ships; 59 flotillas of 3 ships each; 3 flotillas of 59 ships.
9. (a) 3, 5, 15, 29 people **(b)** 145 committees of 3; 87 committees of 5; 29 committees of 15; 15 committees of 29.
10. (a) 1, 2, 3, 4, 6, 9, 12, 18, or 36 **(c)** 1 or 17
11. 27,720
13. (a) The Fundamental Theorem of Arithmetic says that n can be written as a product of primes in one and only one way. Since $2|n$ and $3|n$ and 2 and 3 are both prime, they must be included in the unique factorization.

That is, $2 \cdot 3 \cdot p_1 \cdot p_2 \cdot \ldots \cdot p_m = n$.
Therefore, $(2 \cdot 3)(p_1 \cdot p_2 \cdot \ldots \cdot p_m) = n$.
Thus, $6|n$.

15. 101, 103, 107, 109, 113, 127, 131, 137, 139, 149, 151, 157, 163, 167, 173, 179, 181, 191, 193, 197, 199
17. Every number would have its "usual" factorization
with infinitely many other
$= 1$; n may be any natural

1000. The other 15 factors are:
00, 125, 200, 250, and 500.
45, . . . , which are in arithmetic
belong to the original sequence

$2^3) = 3^4 \cdot 2^7$
posite because it is divisible by
$7 \cdot 11 \cdot 13) + 5 = 5((3 \cdot 7 \cdot 11$
.
31, 37, 53, 59, 71, 73, 79
$\cdot 5^{110}$

e numbers, there is one
at least one of the other two
re, the product will be
divisible by 6.
Using 3 and 4 is correct
isors. But 2 and 6 have a
st will ensure only that the

$\cdot 13 \cdot 17 \cdot 19$. Then
9 appears in the prime
both. If p is in the prime
and hence $p \nmid a + b$.
rime factorization of b.

37. Suppose n is composite and d is its least positive divisor other than 1. We need to show that d is prime. Suppose that some prime p less than d will divide d and hence will divide n, which contradicts the fact that d is the smallest divisor of n greater than 1.
38. (a) No, only perfect squares have an odd number of divisors.
39. (a) Each prime in the list divides $2 \cdot 3 \cdot 5 \cdot 7 \cdot \ldots \cdot p$ but does not divide the sum $2 \cdot 3 \cdot 5 \cdot 7 \cdot \ldots \cdot p + 1$. **(c)** Make p the discovered prime. The argument assures us that there is a prime greater than p. **(e)** Apply the argument for $p = 19$.
40. (a) Each number is written as a sum or difference of two numbers having the following property: whenever one number is divisible by one of the primes 2, 3, 5 or 7, then the other is not divisible by that prime.

Open-Ended

43. (a) (i) $1 + 2 + 3 + 4 + 6 = 16$, so 12 is abundant. **(ii)** $1 + 2 + 4 + 7 + 14 = 28$, so 28 is perfect. **(iii)** $1 + 5 + 7 = 13$, so 35 is deficient.

Review Problems

45. (a) False **(b)** True **(c)** True **(d)** True
46. (a) 2, 3, 6 **(b)** 2, 3, 5, 6, 9, 10
47. If $12|n$, there exists an integer a such that $12a = n$.
$(3 \cdot 4)\, a = n$
$3(4a) = n$
Thus, $3|n$.
48. Yes, among eight people, each would get $422.

Ongoing Assessment 4-5

1. (a) $D_{18} = \{1, 2, 3, 6, 9, 18\}$
$D_{10} = \{1, 2, 5, 10\}$
$\text{GCD}(18, 10) = 2$
$M_{18} = \{18, 36, 54, 72, 90, \ldots\}$
$M_{10} = \{10, 20, 30, 40, 50, 60, 70, 80, 90, \ldots\}$
$\text{LCM}(18, 10) = 90$
(c) $D_8 = \{1, 2, 4, 8\}$
$D_{24} = \{1, 2, 3, 4, 6, 8, 12, 24\}$
$D_{52} = \{1, 2, 4, 13, 26, 52\}$
$\text{GCD}(8, 24, 52) = 4$
$M_8 = \{8, 16, 24, 32, 40, 48, 56, 64, 72, 80, 88, 96, \ldots\}$
$M_{24} = (24, 48, 72, 96, 120, 144, 168, 192, 216, 240, 264, 288, 312, \ldots)$
$M_{52} = \{52, 104, 156, 208, 260, 312, \ldots\}$
$\text{LCM}(8, 24, 52) = 312$
2. (a) $132 = 2^2 \cdot 3 \cdot 11$
$504 = 2^3 \cdot 3^2 \cdot 7$
$\text{GCD}(132, 504) = 2^2 \cdot 3 = 12$
$\text{LCM}(132, 504) = 2^3 \cdot 3^2 \cdot 7 \cdot 11 = 5544$

(c) $96 = 2^5 \cdot 3$
$900 = 2^2 \cdot 3^2 \cdot 5^2$
$630 = 2 \cdot 3^2 \cdot 5 \cdot 7$
$GCD(96, 900, 630) = 2 \cdot 3 = 6$
$LCM(96, 900, 630) = 2^5 \cdot 3^2 \cdot 5^2 \cdot 7 = 50{,}400$
(e) GCD is 1, LCM is $11 \cdot 19$
3. (a) GCD(2924, 220) = GCD(220, 64) = GCD(64, 28) = GCD(28, 8) = GCD(8, 4) = GCD(4, 0) = 4
(c) GCD(123, 152, 122, 368) = GCD(122, 368, 784) = GCD(784, 64) = GCD(64, 16) = GCD(16, 0) = 16
4. (a) 72 **(c)** 630
5. (a) $220 \cdot 2924/4$, or 160,820 **(c)** $123{,}152 \cdot 122{,}368/16$, or 941,866,496
7. (a) LCM(15, 40, 60) = 120 min = 2 hr, so the clocks alarm again together at 8:00 A.M.
9. 24 nights
11. After 7 ½ hr, or 2:30 A.M.
14. (a) 25 ft by 25 ft; 75 squares needed
16. (a) ab **(c)** $GCD(a^2, a) = a$, $LCM(a^2, a) = a^2$
(e) $GCD(a, b) = 1$; $LCM(a, b) = ab$ **(g)** $b|a$
17. (a) True. If a and b are even, then $GCD(a, b) \geq 2$.
(c) False. The GCD could be a multiple of 2; for example, GCD(8, 12) = 4. **(e)** True, by Theorem 4-8
18. (a) 15
19. (a) $4 = 2^2$. Since 97,219,988,751 is odd, it has no prime factors of 2. Consequently, 1 is their only common divisor and they are relatively prime.
21. $48
23. Two packages of plates, four packages of cups, and three packages of napkins
25. 3 for gear 1, 5 for gear 2, and 2 for gear 3
28. {1, 2, 3, 4, 6, 7, 8, 9, 11, 12, 13, 14, 16, 17, 18, 19, 21, 22, 23, 24}
29. (a) $6x^3y^3(2x + 3y)$ **(c)** $54(x - y)^2$
(d) $6(x - y)(x + y - 2)$
30. (a) Always true. The common divisors of a and b are the same to the common divisors of $|a|$ and b and the same as the common divisors of $|a|$ and $|b|$.
(d) Always true. Since $LCM(a, b, c) = abc$, then a, b, c have no primes in common; otherwise, the LCM will be smaller than abc, so $GCD(a, b, c) = 1$
31. (a) GCD(10!, 11!) = 10! and LCM(10!, 11!) = 11!
(c) $GCD(pqr, qrs) = qr$ and $LCM(pqr, qrs) = pqrs$

Communication

34. No; for example, consider GCD(2, 4, 10) = 2. LCM(2, 4, 10) = 20, and the GCD · LCM = (2 · 20) = 40, while $abc = 2 \cdot 4 \cdot 10 = 80$.
36. No. Let $a = 2 \cdot 3$, $b = 3 \cdot 5$, $c = 5 \cdot 7$. Then $GCD(a,b,c) = 1$, but $GCD(a,b) = 3$ and $GCD(b,c) = 5$.
38. The LCM equals the product of the numbers if and only if the numbers have no prime factors in common.
Because $GCD(a, b) \cdot LCM(a, b) = ab$, $LCM(a, b) = ab$ if and only if $GCD(a, b) = 1$; that is, a and b have no prime factors in common.
40. (a) Every divisor of $2^{20} + 1$ and $2^{18} - 1$ is also a divisor of $2^{20} + 1 - (2^{18} - 1)$ and $2^{18} - 1$ by Theorem 4-2(c). Also, every divisor of $2^{20} + 1 - (2^{18} - 1)$ and $2^{18} - 1$ is a divisor of $2^{20} + 1$ and $2^{18} - 1$. Repeating this argument, we have $GCD(2^{20} + 1, 2^{18} - 1)$ $GCD(2^{20} + 1 - 2^2(2^{18} - 1), (2^{18} - 1)$ **(b)** The last digit of 2^{18} is 4. Therefore, the last digit of $2^{18} - 1$ is $4 - 1 = 3$, and so $2^{18} - 1$ cannot be divisible by 5. Therefore, $GCD(5, 2^{18} - 1) = 1$.

Open-Ended

45. Answers vary. If a and b are relatively prime, then **(a)** $GCD(2a, 2b) = 2$.

Review Problems

47. $x = 15{,}625$; $y = 64$
48. (a) 83,151; 83,451; 83,751 **(b)** 86,691 **(c)** 10,396
49. No, 3 divides the sum of the digits and therefore 3 divides 3111.
50. Answers may vary. $30{,}030 = 2 \cdot 3 \cdot 5 \cdot 7 \cdot 11 \cdot 13$
51. 27,720
52. 43

Ongoing Assessment 4-6

1. 2:00 P.M.
2. (a) 3 **(c)** 6 **(e)** 3 **(g)** Does not exist
3. (a) 2 **(c)** 2 **(e)** 1 **(g)** 2
4. (b) $6 = 5 \ominus 6$; $4 = 2 \ominus 5$;
5. (a)

⊗	1	2	3	4	5	6	7
1	1	2	3	4	5	6	7
2	2	4	6	1	3	5	7
3	3	6	2	5	1	4	7
4	4	1	5	2	6	3	7
5	5	3	1	6	4	2	7
6	6	5	4	3	2	1	7
7	7	7	7	7	7	7	7

6. (a) 10 **(c)** 7 **(e)** 1
7. (a) 2, 9, 16, 30
8. (b) 0 **(d)** 7
9. (b) $x - 1 = 2k$ implies $x = 2k + 1$, where k is an integer.

Communication

11. (b) 5 **(c)** Let $abcd$ be a four-digit number, for example. Then $abcd = a \cdot 10^3 + b \cdot 10^2 + c \cdot 10 + d = 100(10 \cdot a \mid b) \equiv cd$. Note that $100(10a + b) \equiv 0 \pmod{100}$; therefore, $abcd + cd \pmod{100}$.

12. **(a)** $10 \equiv 1 \pmod 9$ and hence $10^n \equiv 1^n = 1 \pmod 9$. Thus $abcd = a \cdot 10^3 + b \cdot 10^2 + c \cdot 10 + d \equiv a \cdot 1 + b \cdot 1 + c \cdot 1 + d \pmod 9$.

Chapter Review

1. **(a)** $^-3$ **(c)** 0 **(e)** $x - y$
2. **(a)** $^-7$ **(c)** 8 **(e)** 8
3. **(a)** 3 **(c)** Any integer except 0 **(e)** $^-41$
5. **(a)** $10 - 5 = 5$ **(b)** $1 - (^-2) = 3$
6. **(a)** ^-x **(c)** $3x - 1$ **(e)** 0
7. **(a)** ^-2x **(c)** $(x - 6)(x + 6)$ **(e)** $5(1 + x)$
8. **(a)** False; it is not positive for $x = 0$. **(c)** False, if $b < 0$
9. **(a)** $2/1 \neq 1/2$ **(c)** 1/2 is not an integer.
10. **(a)** $^-10$ **(b)** $^-2^{99}$ **(d)** 0 **(f)** $x \leq 0$; that is, $0, ^-1, ^-2, ^-3, \ldots$ **(g)** $x \geq 4$ or $x \leq {^-4}$; that is $\{\ldots ^-6, ^-5, ^-4\} \cup \{4, 5, 6, 7, \ldots\}$
11. **(a)** $^-1, 1, ^-1, 1, ^-1, 1$
(d) $^-5, ^-8, ^-11, ^-14, ^-17, ^-20$
13. **(a)** False **(c)** True **(e)** False; 9, for example
14. **(a)** False; $7|7$ and $7 \nmid 3$, yet $7|3 \cdot 7$ **(b)** False; $3 \nmid (3 + 4)$, but $3|3$ and $3 \nmid 4$ **(d)** True **(f)** False; $4 \nmid 2$ and $4 \nmid 22$, but $4|44$
15. **(a)** Divisible by 2, 3, 4, 5, 6, 8, 9, 11
16. If 10,007 is prime, $17 \nmid 10{,}007$. We know $17|17$, so $17 \nmid (10{,}007 + 17)$ by Theorem 4-2(b).
17. **(a)** $87\underline{2}4$; $87\underline{5}4$; $87\underline{8}4$ **(c)** $87{,}\underline{1}74$; $87{,}\underline{4}64$; $87{,}\underline{7}54$
18. **(a)** The student's claim is true. Examples vary.
19. **(a)** Composite
20. Check for divisibility by 3 and 8, $24|4152$.
21. No, they are the same if the numbers are equal.
23. The number is not divisible by 2, 3, 5, 7, 11, and 13 because each of these primes divides one product in the sum $2 \cdot 3 \cdot 5 \cdot 7 + 11 \cdot 13$ but not the other. The student checked that $17 \nmid 353$ and because $19^2 = 361 > 353$, no other primes need to be tested.
24. **(a)** 4
25. **(b)** 77,562
29. The LCM of all positive integers less than or equal to 10 is $2^3 \cdot 3^2 \cdot 5 \cdot 7$, or 2520.
31. 9:30 A.M.
33. 5 packages
36. We first show that among any three consecutive odd integers, there is always one that is divisible by 3. For that purpose, suppose that the first integer in the triplet is not divisible by 3. Then by the division algorithm that integer can be written in the form $3n + 1$ or $3n + 2$ for some integer n. Then the three consecutive odd integers are $3n + 1, 3n + 3, 3n + 5$ or $3n + 2, 3n + 4, 3n + 6$. In the first triplet, $3n + 3$ is divisible by 3, and in the second, $3n + 6$ is divisible by 3. This implies that if the first odd integer is greater than 3 and not divisible by 3, then the second or the third must be divisible by 3, and hence cannot be prime.
38. Friday
40. **(a)** $2^{10} \cdot 3^{10}$ **(c)** 97^4 since 97 is prime.
(e) $2^3 \cdot 3^2 (1 + 2 \cdot 3 \cdot 7) = 2^3 \cdot 3^2 \cdot 43$

Answers to Now Try This

4-1. **(a)** Since $x \leq 0$, $|x| = {^-x}$ and $|x| + x = {^-x} + x = 0$
(b) Since $x \leq 0$, $^-|x| + x = {^-(^-x)} + x = 2x$
(c) Since $x \geq 0$, $^-|x| + x = {^-x} + x = 0$
4-2. **(a)** Answers will vary. For example, a mail carrier brings you three letters, one with a check for \$23 and the other two with bills for \$13 and \$12, respectively. Are you richer or poorer and by how much? **(b)** Answers will vary. For example, a mail carrier brings you one letter with a check for \$18 and takes away a bill for \$37 that was intended for someone else. Are you richer or poorer and by how much?
4-3. **(a)** Yes. Because $a - b = a + {^-b}$ and the sum of two integers is an integer. **(b)** None of the properties holds for integers because:

$$a - b \neq b - a \text{ (if } a \neq b)$$
$$(a - b) - c \neq a - (b - c) \text{ (if } c \neq 0)$$

There is no single integer i such that for all integers $a - i = a$ and $i - a = a$ (the first equation implies that $i = 0$ but i does not satisfy the second equation).
4-4. **(a)** $101 \cdot 99 = (100 + 1)(100 - 1) = 100^2 - 1^2 = 9999$
(b) $22 \cdot 18 = (20 + 2)(20 - 2) = 20^2 - 2^2 = 400 - 4 = 396$
(c) $24 \cdot 36 = (30 - 6)(30 + 6) = 30^2 - 6^2 = 900 - 36 = 864$ **(d)** $998 \cdot 1002 = (1000 - 2)(1000 + 2) = 1000^2 - 2^2 = 1{,}000{,}000 - 4 = 999{,}996$
4-5. $a \div 0 = x$ if and only if $0 \cdot x = a$ and x is unique. Because $0 \cdot x = 0$ for all integers x, the equation has no solution if $a > 0$. If $a = 0$ then for all integers x, $0 \cdot x = 0$. Because the solution is not unique, $0 \div 0$ is not defined.
4-6.

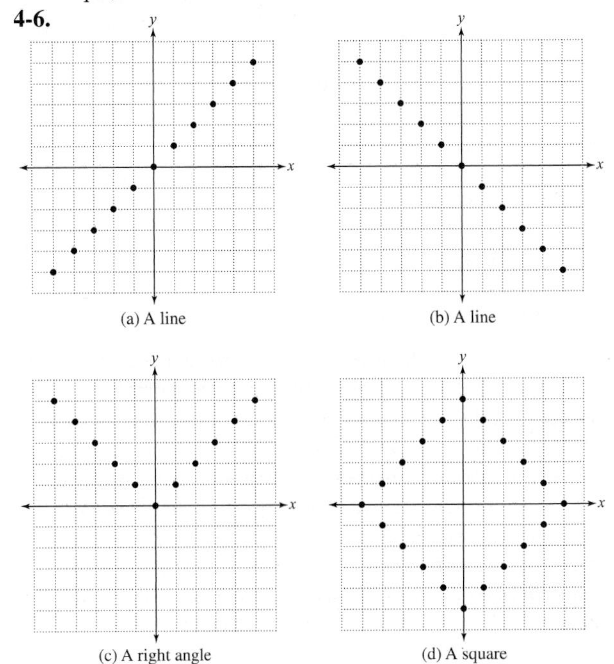

(a) A line (b) A line

(c) A right angle (d) A square

4-7. If $5 \nmid a$ and $5 \nmid b$ nothing can be concluded with only this information. For some a and b, $5 \nmid a + b$ and for some $5 \mid a + b$. For example $5 \nmid 8$ and $5 \nmid 12$. But $5 \mid 8 + 12$. On the other hand $5 \nmid 7$ and $5 \nmid 12$ and $5 \nmid 7 + 12$.
4-8. Yes, it is true. If $3 \mid x$, then 3 divides any numbers times x and in particular $3 \mid xy$.
4-9. $1 + 2 + 5 + 0 + 6 + 5 = 19$, so we must find numbers x and y such that $9 \mid [19 + (x + y)]$. Any two numbers that sum to 8 or 17 will satisfy this. Therefore the blanks could be filled with 8 and 9, or 9 and 8, or 0 and 8, or 8 and 0.
4-10. **(a)** Answers vary. For example, only square numbers are listed in Column 3; 2 is the only even number that will ever be in Column 2; and Column 2 contains prime numbers. The powers of 2 appear in successive columns. **(b)** There will never be other entries in Column 1 because 1 is the only number with one factor. Other numbers have at least the number itself and 1. **(c)** 49, 121, 169 **(d)** 64 **(e)** The square numbers have an odd number of factors. Factors occur in pairs; for example, for 16 we have 1 and 16, 2 and 8, and 4 and 4. When we list the factors, we list only the distinct factors, so 4 is not listed twice, thereby making the number of factors of 16 an odd number. Similar reasoning holds for all square numbers.
4-11. **(a)** 1, 2, 3, 6, 9 **(b)** 1, 2, 3, 4, 6, 8 **(c)** Only white rods can be used to form one-color trains for prime numbers if two or more rods must be used. **(d)** The number must have at least 8 factors: 1, 2, 3, 5, 6, 10, 15, 30.
4-12. **(a)** No, because the multiples of 2 have 2 as a factor. **(b)** The multiples of 3: {3, 6, 9, 12, 15, . . . } **(c)** The multiples of 5: {5, 10, 15, 20, . . . } **(d)** The multiples of 7: {7, 14, 21, . . . } **(e)** We have to check only divisibility by 2, 3, 5, and 7.
4-13. The 1, 2, 3, and 6 rods can all be used to build both the 24 and 30 train. The greatest of these is 6, so *GCD* (24, 30) = 6.
4-14. **(a)** Yes
(b) Yes, 12 is the identity
(c) Yes

Answers to Technology Corner

Section 4-1

1. The entries in column *A* stay 4 while the entries in column *B* start with 3 and decrease by 1. The sum of columns *A* and *B* is entered in column *C* starting with 7. The entries in column *C* are the integers in decreasing order starting with 7. The patterns show that the sum of two positive numbers is positive. The sum of a positive and a negative number is positive if the absolute value of the positive number is greater than the absolute value of the negative number. The sum is 0 if both numbers have the same absolute value. The sum is negative otherwise. Similar results can be obtained if column *A* is changed to $^{-}4$.

2. **(a)** The graph should appear as shown below.

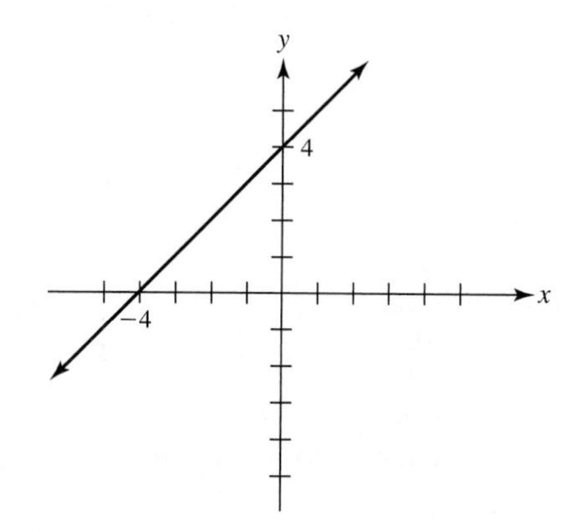

(b) When x is less than $^{-}4$, the y-values are negative; when $x = {}^{-}4$, $y = 0$; when x is greater than $^{-}4$, the y-values are positive.

Section 4-2

The product of two negative numbers is a positive number. The numbers in the *C* column are perfect squares.

Section 4-5

1. The intersection is the first twelve multiples of 12.
2. You only need to fill down to 47.
3. **(a)** 180
(b) You need to use one of the techniques in the section to find LCM (6, 9, 12, 15).

Chapter 5

Ongoing Assessment 5-1

1. (a) The solution to $8x = 7$ is $\frac{7}{8}$. **(c)** The ratio of boys to girls is 7 to 8.

2. (a) $\frac{1}{6}$ **(e)** $\frac{5}{16}$

3. (a) $\frac{2}{3}$

4. (a)

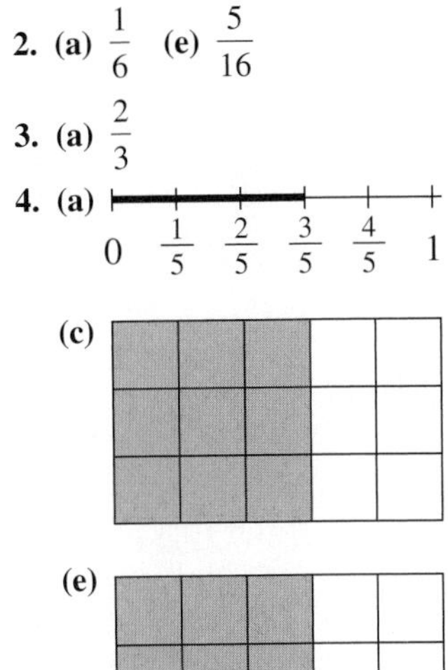

(c)

(e)

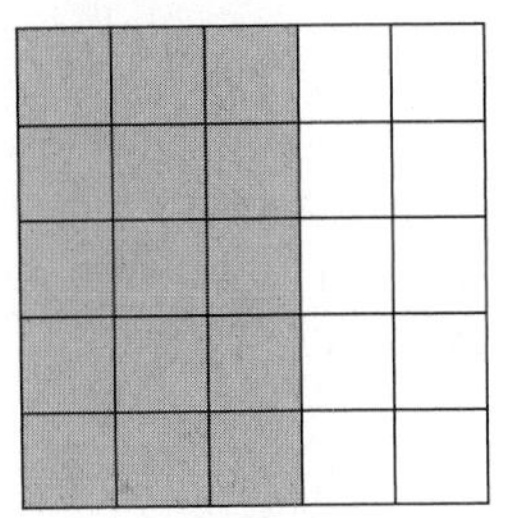

5. (a) No, not equal parts **(c)** Yes

6. (a)

(c)

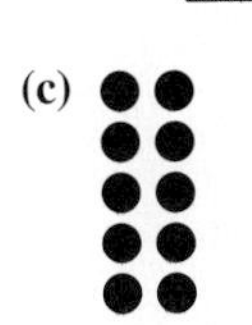

7. (a) **(c)**

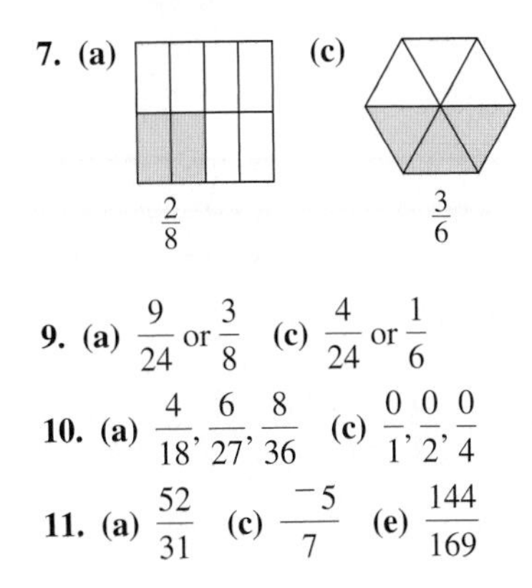

9. (a) $\frac{9}{24}$ or $\frac{3}{8}$ **(c)** $\frac{4}{24}$ or $\frac{1}{6}$

10. (a) $\frac{4}{18}, \frac{6}{27}, \frac{8}{36}$ **(c)** $\frac{0}{1}, \frac{0}{2}, \frac{0}{4}$

11. (a) $\frac{52}{31}$ **(c)** $\frac{^-5}{7}$ **(e)** $\frac{144}{169}$

13. (a) Undefined **(c)** 0 **(e)** Cannot be simplified

14. (a) $\frac{a-b}{3}$ **(c)** $\frac{a}{1}$

15. (a) Equal **(c)** Equal

16. (a) Not equal **(c)** Equal

17. Yes, $\frac{1}{32}$ in.

19. $\frac{36}{48}$

21.

22. (a) $2\frac{7}{8}$ in **(c)** $1\frac{3}{8}$ in.

23. (a) False, $\frac{ab + c}{b} = \frac{ab}{b} + \frac{c}{b} = a + \frac{c}{b}$

(c) True, $\frac{ab + ac}{ac} = \frac{a(b + c)}{ac} = \frac{b + c}{c}$

24. (a) $\frac{32}{3}$ **(c)** x is any rational number except 0.

25. (a) $a = b, c \neq 0$

26. (a) $N \subseteq W \subseteq I \subseteq Q$

28. (a) $>$ **(c)** $<$ **(e)** $=$

29. (a) $\frac{11}{13}, \frac{11}{16}, \frac{11}{22}$

30. $\frac{a}{b} < 1$ and $\frac{c}{d} > 0$ imply $\frac{a}{b} \cdot \frac{c}{d} < 1 \cdot \frac{c}{d}$ or $\frac{a}{b} \cdot \frac{c}{d} < \frac{c}{d}$.

32. Answers may vary. The following are possible answers:
(a) $\frac{10}{21}, \frac{11}{21}$ **(c)** $\frac{997}{1200}, \frac{998}{1200}$

33. (a) 1

34. 456 mi

35. (a) $\frac{6}{16}$, or $\frac{3}{8}$ of a pound; $\frac{6}{32{,}000}$, or $\frac{3}{16{,}000}$ of a ton

(b) $\frac{10}{100}$, or $\frac{1}{10}$ **(c)** $\frac{8}{24}$, or $\frac{1}{3}$

Communication

39. Each class does not have to have the same number of females. One class could have 8 females in a class of 24 and another could have 12 females in a class of 36.

40. The new fraction is equal to $\frac{1}{2}$. The principle can be generalized as follows:

$$\frac{a}{b} = \frac{ar_1}{br_1} = \frac{ar_2}{br_2} = \frac{ar_3}{br_3} = \ldots = \frac{ar_n}{br_n} = \frac{a(r_1 + r_2 + r_3 + \ldots + r_n)}{b(r_1 + r_2 + r_3 + \ldots + r_n)} = \frac{a}{b}$$

42. **(a)** Suppose the rational numbers are $\frac{2}{16}$ and $\frac{1}{4}$. In this case, $\frac{1}{4} > \frac{2}{16}$ and Iris is incorrect. **(b)** Suppose the rational numbers are $\frac{5}{1}$ and $\frac{1}{2}$. In this case, $\frac{5}{1} > \frac{1}{2}$. Shirley is incorrect.
44. Because 36 in. = 1 yd, then to convert x in. to yards, we could consider how to write equivalent fractions $\frac{36}{1}$ and $\frac{x}{1}$ unknown yards. By seeing how many 36s are in x, we could determine the unknown number of yards. Because 1 in. $= \frac{1}{36}$ yd, the process to convert yards to inches could be accomplished using similar methods.

Open-Ended

45. Frequently in recipes, measurements are found in fractional parts of cups, teaspoons, tablespoons. For example, a recipe might call for $\frac{1}{2}$ TSP of salt and $\frac{2}{3}$ c of flour.

Cooperative Learning

47. **(a)** Answers vary; students must work together to determine the heights and to order the people according to height. For example, if the heights are 60″, 52″, 56″, and 58″, the fractions could be $\frac{60}{60}, \frac{52}{60}, \frac{56}{60}$, and $\frac{58}{60}$.
(b) Answers vary depending on part (a).

Ongoing Assessment 5-2

1. **(a)**

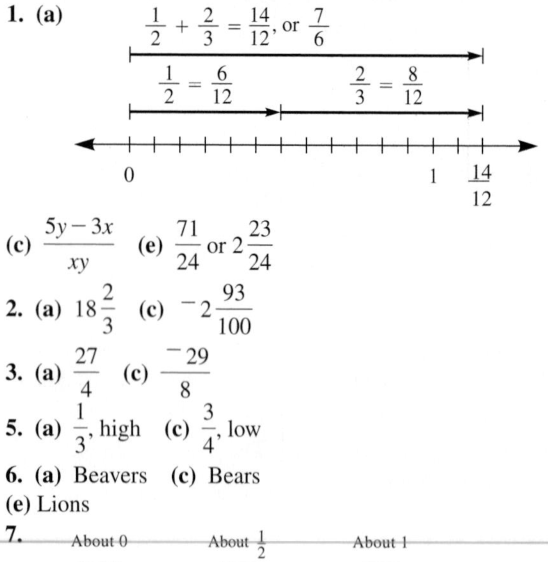

(c) $\frac{5y - 3x}{xy}$ **(e)** $\frac{71}{24}$ or $2\frac{23}{24}$
2. **(a)** $18\frac{2}{3}$ **(c)** $^{-}2\frac{93}{100}$
3. **(a)** $\frac{27}{4}$ **(c)** $\frac{^{-}29}{8}$
5. **(a)** $\frac{1}{3}$, high **(c)** $\frac{3}{4}$, low
6. **(a)** Beavers **(c)** Bears
(e) Lions
7.

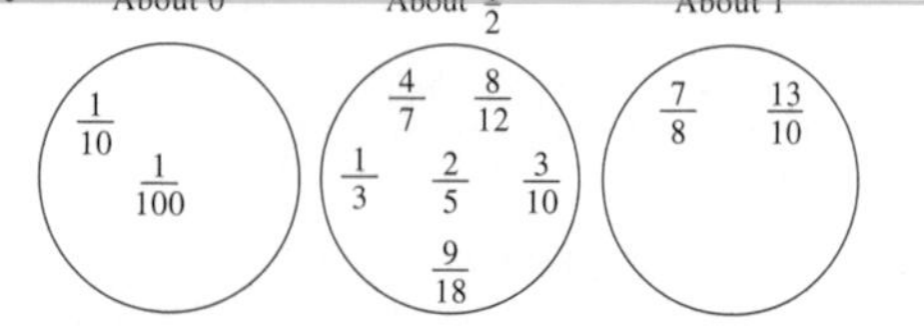

8. **(a)** $\frac{1}{2}$, high **(c)** $\frac{3}{4}$, high **(e)** 1, low **(g)** $\frac{3}{4}$, low
9. **(a)** 2 **(c)** 0
10. **(a)** $\frac{1}{4}$ **(c)** 0
11. **(a)** A **(c)** T
14. **(a)** $\frac{1}{30}$ **(c)** $\frac{1}{60}$
15. $6\frac{7}{12}$ yd
17. $2\frac{5}{6}$ yd
19. **(a)** Team 4, $76\frac{11}{16}$ lb
20. **(a)** $\frac{1}{2} + \frac{3}{4} \in Q$ **(c)** $\left(\frac{1}{2} + \frac{1}{3}\right) + \frac{1}{4} = \frac{1}{2} + \left(\frac{1}{3} + \frac{1}{4}\right)$
21. **(a)** $\frac{3}{2}, \frac{7}{4}$, 2; arithmetic,
$\frac{1}{2} - \frac{1}{4} = \frac{3}{4} - \frac{1}{2} = 1 - \frac{3}{4} = \frac{5}{4} - 1$
(c) $\frac{17}{3}, \frac{20}{3}, \frac{23}{3}$, arithmetic;
$\frac{5}{3} - \frac{2}{3} = \frac{8}{3} - \frac{5}{3} = \frac{11}{3} - \frac{8}{3} = \frac{14}{3} - \frac{11}{3}$
23. $1, \frac{7}{6}, \frac{8}{6}, \frac{9}{6}, \frac{10}{6}, \frac{11}{6}, 2$
24. **(a)** **(i)** $\frac{3}{4}$ **(ii)** $\frac{25}{12}$, or $2\frac{1}{12}$ **(iii)** 0
25. **(a)** $f(0) = {}^{-}2$ **(c)** $f({}^{-}5) = \frac{1}{2}$
26. **(b)** $\frac{1}{n} = \frac{1}{n + 1} + \frac{1}{n(n + 1)}$
27. **(a)** $\frac{9}{6}$ or $\frac{3}{2}$ or $1\frac{1}{2}$
(c) $\frac{17}{2^3 3^2}$ or $\frac{17}{72}$
28. **(a)** $\frac{3x^2 + y^3}{x^2y^2}$

Communication

29. Answers vary; for example, $\frac{1}{3} + \frac{1}{4}$ or $\frac{7}{12}$ does not represent the amount received since the fractions did not come from the same size "whole."
32. Answers vary; for example, $5\frac{3}{4} = \frac{5}{1} + \frac{3}{4} = \frac{5 \cdot 4 + 1 \cdot 3}{1 \cdot 4}$
$= \frac{5 \cdot 4 + 3}{4}$

34. No, for example, Kara spent $\frac{1}{3}$ of $\frac{1}{2}$ of her allowance on Sunday, so she has $\frac{1}{2} - \frac{1}{3} \cdot \frac{1}{2} = \frac{3}{6} - \frac{1}{6} = \frac{2}{6}$, or $\frac{1}{3}$, of her allowance left.
36. It might be easier but she would be incorrect. Think of the numerator as the number of pieces of pie cut into the denominator's value of slices. Then to add the pieces of pie, we would add the numerators and have that number of pieces.
37. **(a)** Yes. If a, b, c, and d are integers, $b \neq 0$, $d \neq 0$, then $\frac{a}{b} - \frac{c}{d} = \frac{ad - bc}{bd}$ is a rational number.
(c) No. For example, $\frac{1}{2} - \left(\frac{1}{4} - \frac{1}{8}\right) \neq \left(\frac{1}{2} - \frac{1}{4}\right) - \frac{1}{8}$.
38. **(a)** Like digits are being canceled. **(c)** Numerators and denominators of fractional portions are both being subtracted.

Cooperative Learning

41. Depending on the people interviewed, students may hear an answer like the following from a teacher. I use fractions in determining total grades for my classes. For example, if a paper is $\frac{1}{2}$ of the grade and a test is another $\frac{1}{3}$ of the grade, I need to know what fractional part of the grade is yet to be determined.

Review Problems

42. **(a)** $\frac{2}{3}$ **(b)** $\frac{13}{17}$ **(c)** $\frac{25}{49}$ **(d)** $\frac{a}{1}$, or a **(e)** Simplified
43. **(a)** Equal **(b)** Not equal **(c)** Equal **(d)** Not equal
44. **(a)** February **(b)** The answer depends upon whether the year is a leap year. If it is a leap year, the answer is $\frac{185}{366}$; if not, the answer is 184/365. **(c)** Most people consider there to be $365\frac{1}{4}$ days in a year. As an improper fraction, this number is $\frac{1461}{4}$.
45. Greater than. Consider $\frac{a}{b}$ and $\frac{a + x}{b + x}$ when $a < b$. $\frac{a}{b} < \frac{a + x}{b + x}$ because $ab + ax < ab + bx$. This is true because $a < b$ and $x > 0$ imply $ax < bx$.

Ongoing Assessment 5–3

1. **(a)** $\frac{1}{4} \cdot \frac{1}{3} = \frac{1}{12}$
2. **(a)**

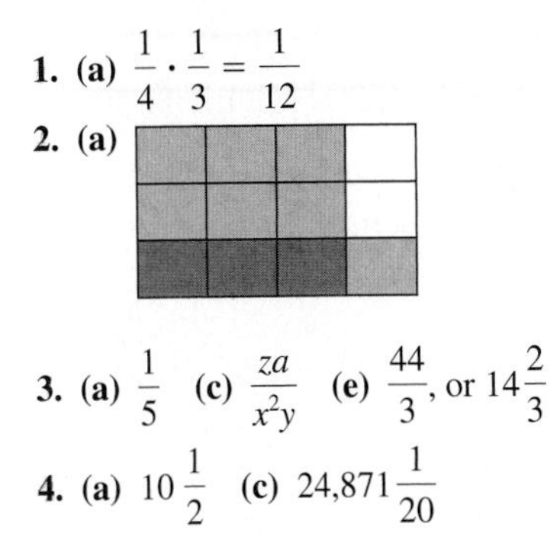

3. **(a)** $\frac{1}{5}$ **(c)** $\frac{za}{x^2y}$ **(e)** $\frac{44}{3}$, or $14\frac{2}{3}$
4. **(a)** $10\frac{1}{2}$ **(c)** $24{,}871\frac{1}{20}$
5. **(a)** $^{-}3$ **(c)** $\frac{y}{x}$
6. **(a)** $\frac{21}{12}$ or $\frac{7}{4}$ **(c)** $\frac{^{-}1}{8}$
7. $\frac{1}{2}$
8. Answers vary; for example, **(a)** $6 \div 2 \neq 2 \div 6$ **(c)** There is no rational number a such that $2 \div a = a \div 2 = 2$.
9. **(a)** 26 **(c)** 92 **(e)** 6 **(g)** 9
10. **(a)** 20 **(c)** 2
11. **(a)** 18 **(c)** 7
12. **(a)** Less than 1 **(c)** Greater than 2 **(e)** Greater than 4
13. (c)
14. $\frac{29}{36}$
16. 400 members
18. **(a)** 39 uniforms
19. \$240
20. **(a)** \$121,000 **(c)** \$300,000
21. 246 pages
22. \$225
23. **(a)** Peter, 30 min.; Paul, 25 min.: Mary, 20 min.
24. **(a)** $89\frac{3}{5}$°F
26. $120\frac{1}{4}$ lb
27. **(a)** $1/3^{13}$ **(c)** 5^{11} **(e)** $1/(^{-}5)^2$, or $1/5^2$ **(g)** a^2
28. **(a)** $\left(\frac{1}{2}\right)^{10}$ **(c)** $\left(\frac{2}{3}\right)^{9}$ **(e)** $\left(\frac{3}{5}\right)^{11}$
29. **(a)** False. $2^3 \cdot 2^4 \neq (2 \cdot 2)^{3+4}$ **(c)** False. $2^3 \cdot 2^3 \neq (2 \cdot 2)^{2 \cdot 3}$ **(e)** False. $(2 + 3)^2 \neq 2^2 + 3^2$ **(g)** False. $2^{2 \cdot 3} \neq 2^2 \cdot 2^3$
30. **(a)** 5 **(c)** $^{-}2$ **(e)** 0
31. **(a)** $1 \cdot 10^{11}$
32. **(a)** $x \leq 4$ **(c)** $x \geq 2$
33. **(a)** $\left(\frac{1}{2}\right)^{3}$ **(c)** $\left(\frac{4}{3}\right)^{10}$ **(e)** $\left(\frac{4}{3}\right)^{10}$
34. **(a)** 10^{10}
35. **(a)**
$$2S = 2\left(\frac{1}{2} + \frac{1}{2^2} + \cdots + \frac{1}{2^{64}}\right) = 1 + \frac{1}{2} + \frac{1}{2^2} + \cdots + \frac{1}{2^{63}}$$
36. **(a)** $1\frac{49}{99}$
37. **(a)** $\frac{3}{4}$ **(d)** $^{-}9$
38. **(a)** $\frac{3}{2}, \frac{3}{4}, \frac{3}{8}, \frac{3}{16}, \frac{3}{32}$ **(c)** $\frac{3}{1024}$
39. **(a)** 32^{50}, since $32^{50} = (2^5)^{50} = 2^{250}$ and $4^{100} = (2^2)^{100} = 2^{200}$
40. **(a)** $n(n + 1) + \left(\frac{1}{2}\right)^2$

41. (a) First 3, second 4, third 5, Guess 6. The guess is correct since $\left(1+\frac{1}{1}\right)\left(1+\frac{1}{2}\right)\left(1+\frac{1}{3}\right)\left(1+\frac{1}{4}\right)\left(1+\frac{1}{5}\right) = 5\left(1+\frac{1}{5}\right) = 6.$

Communication

44. Never less than n. $0 < \frac{a}{b} < 1$ implies $0 < 1 < \frac{b}{a}$. The last inequality implies $0 < n < n \cdot \left(\frac{b}{a}\right)$. Also $n \div \left(\frac{a}{b}\right) = n \cdot \frac{b}{a} > n.$

49. Answers vary; for example, yes she is correct, since $\frac{3}{4} \cdot \frac{2}{5} = \frac{3 \cdot 2}{4 \cdot 5} = \frac{3 \cdot 2}{5 \cdot 4} = \frac{3}{5} \cdot \frac{2}{4} = \frac{3}{5} \cdot \frac{1}{2} = \frac{3 \cdot 1}{5 \cdot 2} = \frac{3}{10}.$

Open-Ended

50. (a) Answers vary; for example, if you have a board $1\frac{3}{4}$ yd long, how many $\frac{1}{2}$-yd lengths can you divide it into?
(b) Answers vary.
51. Answers will vary about class use. 7 oz

Cooperative Learning

52. The answers depend on the size of the bricks and the size of the joints. In all likelihood, the measurements will be made in fractions of inches for the size of the joints. The size of the bricks may be done in inches. An alternative is to measure in centimeters. Thus, all measurement is approximate and some rounding or estimation may occur.

Review Problems

53. (a) $\frac{25}{16}$, or $1\frac{9}{16}$ **(b)** $\frac{25}{18}$, or $1\frac{7}{18}$ **(c)** $\frac{5}{216}$
(d) $\frac{259}{30}$, or $8\frac{19}{30}$ **(e)** $\frac{37}{24}$, or $1\frac{13}{24}$ **(f)** $\frac{^-39}{4}$, or $^-9\frac{3}{4}$
54. 120 students

Ongoing Assessment 5-4

1. (a) 5:21 **(c)** 21:26
2. (a) 30 **(c)** $23\frac{1}{3}$
3. 36 lb
5. \$1.19
7. 64 pages
9. (a) 42, 56
11. \$14,909.09; \$29,818.18; \$37,272.73
13. 135
14. (a) 1:6 **(c)** 7:12
16. (a) 5/7
17. 120 ft
19. (a) 27
21. 312 lb
22. (a) 2:5. Because the ratio is 2:3, there are $2x$ boys and $3x$ girls; hence, the ratio of boys to all students is $2x/(2x + 3x) = \frac{2}{5}$.
23. (a) $\frac{2}{3}$ tsp mustard seeds, 1 c scallions, $2\frac{1}{6}$ c beans
(c) $\frac{7}{13}$ tsp mustard seeds, $1\frac{8}{13}$ c tomato sauce, $\frac{21}{26}$ c scallions
26. 35 ft
29. (a) \$320
30. (a)

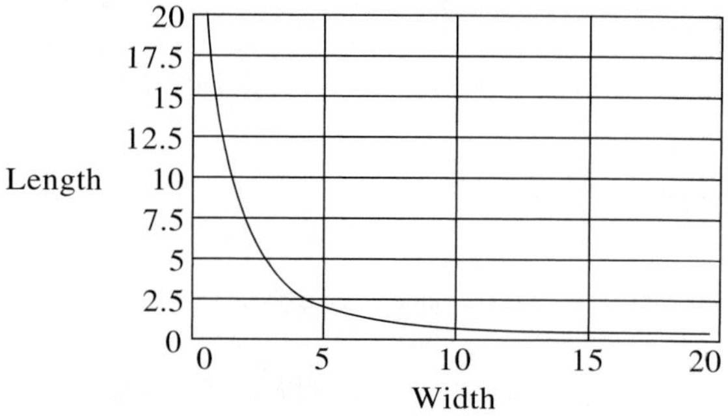

Length
Width

(c) $W = 10/L$
31. (b) Let $\frac{a}{b} = \frac{c}{d} = \frac{e}{f} = r.$

Then, $a = br$
$c = dr$
$e = fr.$
So, $a + c + e = br + dr + fr$
$a + c + e = r(b + d + f)$
$\frac{a + c + e}{b + d + f} = r.$

32. (a) $\frac{a}{b} = \frac{c}{d}$ implies $\frac{a}{b} + 1 = \frac{c}{d} + 1$, which implies $\frac{a + b}{b} = \frac{c + d}{d}$.

Communication

33. (a) 40/700, or 4/70, or 2/35 **(c)** For the first set, $\frac{\text{footprint length}}{\text{thighbone length}} = \frac{40}{100} = \frac{20}{50}$; that is, a 50-cm thighbone would correspond to a 20-cm footprint. Thus, it is not likely that the 50-cm thighbone is from the animal that left the 30-cm footprint. (Notice that $\frac{20}{50} \neq \frac{30}{50}$.)
34. No, the ratio of the prices is proportional to the ratio of the areas and not to the ratio of the diameters.

Open-Ended

40. (a) 57.6 lb per sq in.

Review Problems

44. **(a)** 0 **(b)** 5 **(c)** $\frac{588}{13}$, or $45\frac{3}{13}$ **(d)** $x \leq \frac{588}{13}$ or $45\frac{3}{13}$

45. All rational numbers except 0 have a multiplicative inverse.

46. $2, 2\frac{1}{4}, 2\frac{1}{2}, 2\frac{3}{4}, 3$

47. $\frac{256}{9}$, or $28\frac{4}{9}$

48. **(a)** $\frac{3+3}{3} \neq \frac{3}{3} + 3$ **(b)** $\frac{4}{2+2} \neq \frac{4}{2} + \frac{4}{2}$

(c) $\frac{\cancel{a}b + c}{\cancel{a}} \neq \frac{ab+c}{a}$ **(d)** $\frac{a \cdot a - b \cdot b}{a - b} \neq \frac{a \cdot \cancel{a} - b \cdot \cancel{b}}{\cancel{a} - \cancel{b}}$

(e) $\frac{a+c}{b+c} \neq \frac{a+\cancel{c}}{b+\cancel{c}}$

Chapter Review

1. **(a)** **(c)**

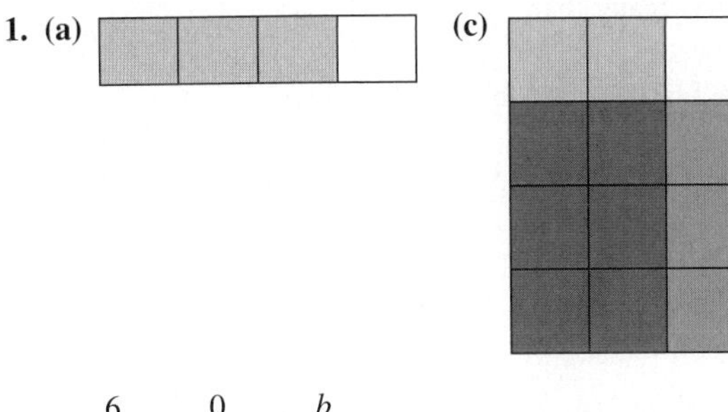

3. **(a)** $\frac{6}{7}$ **(c)** $\frac{0}{1}$ **(e)** $\frac{b}{1}$

4. **(a)** $=$ **(c)** $>$

5. **(a)** $^{-}3, \frac{1}{3}$ **(c)** $\frac{^{-}5}{6}, \frac{6}{5}$

7. Yes. By the definition of multiplication and the commutative and associative laws of addition, we can do the following:

$$\frac{4}{5} \cdot \frac{7}{8} \cdot \frac{5}{14} = \frac{4 \cdot 7 \cdot 5}{5 \cdot 8 \cdot 14} = \frac{4 \cdot 7 \cdot 5}{8 \cdot 14 \cdot 5} = \frac{4}{8} \cdot \frac{7}{14} \cdot \frac{5}{5}.$$

9. 17 pieces, $\frac{11}{6}$ yd

10. **(a)** 15 **(b)** 15 **(c)** 4

11. $\frac{a}{b} \div \frac{c}{d} = x$ if and only if $\frac{a}{b} = \frac{c}{d} \cdot x$, $x = \frac{d}{c} \cdot \frac{a}{b}$ is the solution of the equation because $\frac{c}{d} \cdot \left(\frac{d}{c} \cdot \frac{a}{b}\right) = \frac{a}{b}$.

13. $\frac{76}{100}, \frac{78}{100}$, but answers may vary.

15. $333\frac{1}{3}$ calories

17. $\frac{240}{1000} = \frac{6}{25}$

19. The numerators of the rational numbers are integers and follow the properties of integers: the same is true of the denominators. Thus, both the numerator and denominator of the answer are integers, and we can apply another property of integers to determine the sign of the answer.

21. 112 bags

22. $\frac{4}{15}$

23. $\frac{^{-}12}{10}$ is greater than $\frac{^{-}11}{9}$ because $\frac{^{-}12}{10} - \left(\frac{^{-}11}{9}\right)$ is a positive number.

24. **(a)** 3

25. The ratio of hydrogen to the total is 1:9. Therefore, $\frac{1}{9} = \frac{x}{16}$ implies $x = 1\frac{7}{9}$ oz.

26. 560 fish

28. $1 : r^2$

29. **(a)** $\frac{5}{4}$, or $1\frac{1}{4}$ **(c)** $^{-}100$

30. **(a)** $\frac{a^3}{x^7}$

31. **(a)** $\frac{3ax + b}{x^2y^2}$ **(c)** $\frac{a - bx^2y}{x^3y^2z}$

32. **(a)** 18 : 7

33. **(a)** 1 : 5

34. Answers vary; for example, the problem is to find how many $\frac{1}{2}$-yd pieces of ribbon there are in $1\frac{3}{4}$ yd. There are 3 pieces of length $\frac{1}{2}$ yd with $\frac{1}{4}$ yd left over. This $\frac{1}{4}$ yd is $\frac{1}{2}$ of a $\frac{1}{2}$-yd piece. Therefore, there are 3 pieces of $\frac{1}{2}$-yd ribbon and 1 piece that is $\frac{1}{2}$ of the $\frac{1}{2}$-yd piece or $3\frac{1}{2}$ of the $\frac{1}{2}$-yd pieces. This agrees with the answer obtained by direct computation.

Answers to Now Try This

5-1. Fractions are defined in relation to a whole. In this case, there are 2 wholes. In Figure 5-2(a), $\frac{1}{3}$ of a large circular whole is shaded. In Figure 5-2(b), $\frac{1}{2}$ of a smaller square whole is shaded. To show $\frac{1}{3} > \frac{1}{2}$, the fractions would have to be associated with the same whole, for example, by comparing $\frac{1}{3}$ of the circle with $\frac{1}{2}$ of the circle. It is true that the area shaded in the circle is larger than the area shaded in the square, but this does not show $\frac{1}{3} > \frac{1}{2}$.

5-2. (a)

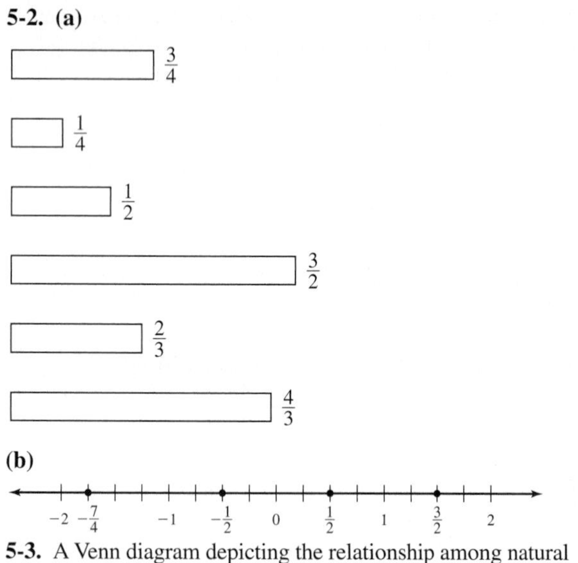

5-3. A Venn diagram depicting the relationship among natural numbers, whole numbers, integers and rational numbers follows:

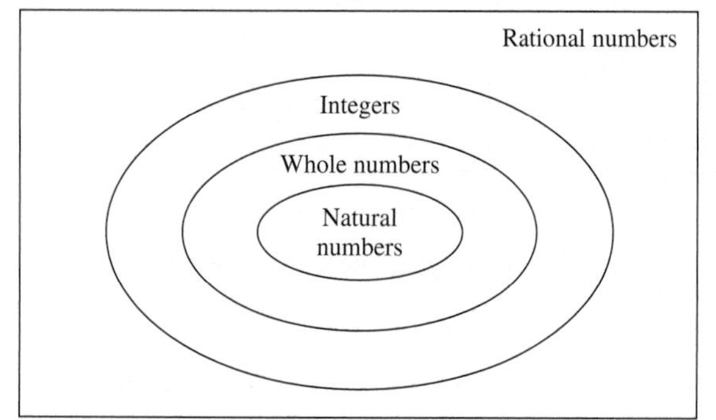

5-4. If n were zero, then bn would be zero and division by zero is undefined.

5-5. Theorem 5-1 states that if a, b, and c are integers and $b > 0$, then $\frac{a}{b} > \frac{c}{b}$ if and only if $a > c$. The question is to investigate if the theorem is true if $b < 0$. Consider $2 > 1$ and $^-1 < 0$. Now $2/(^-1) < 1/(^-1)$, which contradicts an expanded theorem when $b < 0$.

5-6. $\frac{9}{16}, \frac{5}{8}, \frac{2}{3}$, and $\frac{3}{4}$

5-7. Consider two rational numbers $\frac{a}{b}$ and $\frac{c}{d}$. Find the difference between them and add half the difference to the smaller rational number to find $\frac{e}{f}$. Now repeat the process with $\frac{a}{b}$ and $\frac{e}{f}$. This process can be repeated an infinite number of times because there are infinitely many rational numbers between any two rational numbers.

5-8. $\frac{a}{b}$ and $\frac{c}{d}$ are rational numbers.

$$\begin{array}{rcl} \frac{a}{b} < \frac{c}{d} & \text{similarly,} & \frac{a}{b} < \frac{c}{d} \\ ad < bc & & ad < bc \\ ad + ab < bc + ab & & ad + cd < bc + cd \\ a(d + b) < b(c + a) & & d(a + c) < c(b + d) \\ \frac{a}{b} < \frac{a + c}{b + d} & & \frac{a + c}{b + d} < \frac{c}{d} \end{array}$$

Therefore, $\frac{a}{b} < \frac{a + c}{b + d} < \frac{c}{d}$.

5-9. $\frac{a}{b} + \frac{c}{b} = \frac{ab + bc}{bb} = \frac{b(a + c)}{b \cdot b} = \frac{a + c}{b}$

5-10. $A\frac{b}{c} = A + \frac{b}{c} = \frac{A}{1} + \frac{b}{c} = \frac{A \cdot c + 1 \cdot b}{1 \cdot c} = \frac{A \cdot c + b}{c}$

5-11. $\frac{3}{4}$ is greater than $\frac{1}{2}$; $\frac{1}{2} + \frac{1}{2} = 1$, so $\frac{3}{4} + \frac{1}{2} > 1$. $\frac{4}{6}$ is less than 1, so it cannot be the correct answer for $\frac{3}{4} + \frac{1}{2}$.

5-12. $$\frac{a}{b} \div \frac{c}{d} = \frac{\frac{a}{b}}{\frac{c}{d}} = \frac{\frac{a}{b} \cdot \frac{d}{c}}{\frac{c}{d} \cdot \frac{d}{c}} = \frac{\frac{a}{b} \cdot \frac{d}{c}}{\frac{c}{d} \cdot \frac{d}{c}} = \frac{\frac{a}{b} \cdot \frac{d}{c}}{1} = \frac{a}{b} \cdot \frac{d}{c}$$

5-13. $$\frac{a \div c}{b \div d} = \frac{\frac{a}{c}}{\frac{b}{d}} = \frac{\frac{a}{c} \cdot \frac{d}{b}}{\frac{b}{d} \cdot \frac{d}{b}} = \frac{\frac{a}{c} \cdot \frac{d}{b}}{\frac{b}{d} \cdot \frac{d}{b}} = \frac{\frac{a}{c} \cdot \frac{d}{b}}{1} = \frac{ad}{bc} = \frac{a}{b} \div \frac{c}{d}$$

5-14.

$$\begin{aligned} \frac{a}{b} &= \frac{c}{d} \\ \frac{a}{b} \cdot \frac{bd}{1} &= \frac{c}{d} \cdot \frac{bd}{1} \\ \frac{ad}{1} &= \frac{cb}{1} \\ ad &= cb \\ ad &= bc \end{aligned}$$

Chapter 6

Ongoing Assessment 6-1

1. (a) $0 \cdot 10^0 + 0 \cdot 10^{^-1} + 2 \cdot 10^{^-2} + 3 \cdot 10^{^-3}$
(c) $3 \cdot 10^2 + 1 \cdot 10^1 + 2 \cdot 10^0 + 0 \cdot 10^{^-1} + 1 \cdot 10^{^-2} + 0 \cdot 10^{^-3} + 3 \cdot 10^{^-4}$
2. (a) 4356.78 **(c)** 40,000.03
3. (a) 536.0076 **(c)** 0.000436 **(d)** 5,000,000.2
4. (a) Thirty-four hundredths **(c)** Two and thirty-four thousandths
5. (a) $\frac{436}{1000} = \frac{109}{250}$ **(c)** $\frac{^-316{,}027}{1000}$ **(e)** $\frac{^-43}{10}$

6. (a) Yes **(c)** Yes **(e)** Yes **(g)** No

8. a. Nonterminating; the fraction $\frac{7}{60}$ does not terminate when written as a decimal. The denominator has a factor of 3.
9. 3.56
10. Answers may vary. Some of the numbers that are composed of whole-number powers of 2 and 5 are 1, 5, 10, 25, and 50. These all divide 100 and can be written as a two-digit decimal between 0 and 1, but there are others.
11. (a) 13.492, 13.49199, 13.4919, 13.49183
(b) $^-1.4053$, $^-1.45$, $^-1.453$, $^-1.493$
12. (a) 0.014 in.
14. 0.84
15. For example, 8.3401
16. (a) One method for doing this is to find the first place value where the two numbers differ. If that place value is 10^n, then add $1 \cdot 10^{n+1}$ to the lesser of the two numbers.
17. (a) If the block is used as 1 unit, the 0.613 can be represented by 6 flats, 1 long, and 3 units.
18. Mathematically, the meaning is 0.05 of 1¢ or 0.0005 of a dollar.
21. $100{,}000^3 = 1000^5$ and these are the greatest.
23. (a) A meaning could be as follows: $3 \cdot 6^0 + 1 \cdot 6^{^-1} + 4 \cdot 6^{^-2} + 5 \cdot 6^{^-3}$
24. Rhonda, Martha, Kathy, Molly, Emily

Communication

25. (a) 3① 2② 5③ 6④ **(b)** 0① 0② 3③ 2④
27. One day cannot be expressed as a terminating decimal because $\frac{1}{365}\left(\text{or } \frac{1}{365\frac{1}{4}}\right)$ cannot be written as a terminating decimal.
28. There is no greatest terminating decimal less than 1 because the unit segment with endpoints 0 and 1 can be divided into smaller and smaller segments and the smaller segments can be counted from the left, always letting another number be between any supposedly "closest" number and 1.
30. Answers vary; for example, you would use 2 blocks to represent the units, 3 flats to represent the tenths, 4 longs to represent the hundredths, and 5 cubes to represent the thousandths. In this way, each base-ten block represents $\frac{1}{10}$ of the base-ten block to its left and there are enough base-ten blocks for all the numbers used.
31. Answers vary; for example, a fraction can be written as a terminating decimal if it can be written as a fraction with a denominator that is a power of 10. The denominator can be written as a power of 10 if it contains only factors of 2 and 5. Other factors may appear in the denominator if the fraction is not in simplest form. For example, in $\frac{28}{35}$ the denominator of 35 has a factor of 7, but in its simplest form, $\frac{4}{5}$, there is no factor of 7.

Open-Ended

33. Answers vary, but most countries will be similar to ours.

Cooperative Learning

35. Answers vary depending on the objects chosen. For example, if a 5×5 flat is used to represent 1, then a 1×5 long would represent $\frac{1}{5}$, and a 1×1 unit would represent $\frac{1}{25}$.

Ongoing Assessment 6-2

1. \$231.24
2. 35.83
3. (a)

8.2	1.9	6.4
3.7	5.5	7.3
4.6	9.1	2.8

(c) Yes; 8.25
4. \$8.00/lb
6. (a) Approx. 6391 cm^3 (rounded to nearest cm^3)
8. (a) 5.4, 6.3, 7.2 **(c)** 0.0625, 0.03125, 0.015625
9. 1.12464
11. 0.6, 0.018, 0.0054, 0.00162
14. (a) 0.0000000032 **(c)** 0.42
15. (a) $1.27 \cdot 10^7$ m **(c)** $5 \cdot 10^7$ cans
16. (a) 0.0000044 sec **(c)** 3,000,000,000 yr
17. (a) $4.8 \cdot 10^{28}$ **(c)** $2 \cdot 10^2$
18. (a) 200 **(c)** 204 **(e)** 203.65
19. 19 mpg
22. Profit of \$2098 (rounded to nearest dollar)
23. (a) $2.3 \cdot 1 = 2.3$; $8.7 \cdot 9 = 78.3$
28. (a) 181.56
29. (a) Yes **(b)** No

31. **(a)** $1.5^2, 2.5^2; 3 \cdot 4 + 0.25 = 3.5^2; 4 \cdot 5 + 0.25 = 4.5^2$
(c) $n \cdot (n + 1) + 0.25 = (n + 0.5)^2$

Communication

32. Suppose a deposit is recorded with $10.00 over the true amount and a check is recorded with $10.00 less than the true amount. The ultimate balance is correct.
33. Answers vary; for example, the multiplication of decimals is exactly like the multiplication of whole numbers if the decimal points are ignored. The difference occurs when the decimal point is placed to obtain the final answer.
34. Answers vary; for example, many of the estimation techniques that work for whole-number division also work for decimal division. The long division algorithm is more efficient when good estimates are used. Also, estimates are important to determine whether an answer obtained by long division is reasonable. Estimation techniques can also be used to place the decimal point in the quotient when decimals are divided.

Open-Ended

38. Answers vary; for example, the calculator could be used to explore the result of placing the decimal point when multiplying by a power of 10 or of placing the decimal point when multiplying two decimals.
39. If a calculator displays 0.3 as a result when computing $0.6 \cdot 0.5$, misinterpretations can be made.

Review Problems

43. $14.0479 = 1 \cdot 10^1 + 4 \cdot 10^0 + 0 \cdot 10^{^-1} + 4 \cdot 10^{^-2} + 7 \cdot 10^{^-3} + 9 \cdot 10^{^-4}$
44. **(a)** Not a terminating decimal **(b)** $\frac{35}{56}$ is a terminating decimal.
45. Yes; for example, $\frac{13}{26} = \frac{1}{2} = 0.5$
46. 0.625. The decimal terminates because $\frac{35}{56} = \frac{7}{8}$ in simplest form and $8 = 2^3$.

Ongoing Assessment 6-3

1. **(a)** $0.\overline{4}$ **(c)** $0.\overline{27}$ **(e)** $0.02\overline{6}$ **(g)** $0.8\overline{3}$ **(i)** $0.\overline{047619}$
2. **(a)** $\frac{4}{9}$ **(c)** $\frac{7}{5}$ **(e)** $\frac{^-211}{90}$
4. $^-1.454 > {}^-1.45\overline{4} > {}^-1.\overline{454} > {}^-1.4\overline{54} = {}^-1.\overline{45}$
5. **(a)** $1.\overline{6}, 2, 2.\overline{3}$
6. $0.\overline{3}_{\text{five}}$
8. **(a)** Six digits
9. Yes. Zeros can be repeated.
10. **(a)** Answers vary, for example, 3.221, 3.2211, 3.22111.
11. **(a)** $0.47\overline{2}$
12. **(a)** Answers vary, for example, 0.751, 0.752, 0.753.
13. **(a)** 8
14. **(b)** $\frac{1}{9999}$
15. **(a)** $\frac{2}{9}$ **(c)** $\frac{5}{9}$ **(e)** 10
16. **(a)** $\frac{5}{99}$ **(c)** $\frac{322}{99}$
17. 0.775
18. **(a)** $\frac{3}{10}$ **(c)** $\frac{2009}{990}$
19. $1.\overline{0}$; the zero repeats in this representation but adds no value.
20. **(a)** No; for example, $0.\overline{3} + 0.\overline{6} = 1$. The answer could be yes if one considered $1.\overline{0} = 1$ as a repeating decimal.
(c) See part **(a)** if 0 is considered a repeating block.

Communication

21. **(a)** Mathematically the cost is $33\frac{1}{3}$¢ but realistically, the cost is 34¢. **(c)** The cost is rounded up.
22. Yes. Zeros can be appended to any finite decimal to make it an infinite decimal.

Open-Ended

24. **(a)** They each have six-digit repetends using each of the digits 1, 2, 4, 5, 7, 8 exactly once. **(c)** For example, $\frac{n}{14}$
25. **(a)** Any integer n can be written as a decimal by appending .0 to the right. **(b)** **(i)** $0.\overline{6}$ **(iii)** $1.\overline{06}$ **(c)** It would have to be adapted to allow multiplication from the left.
26. **(a)** Answers will vary, but in general, the answer is no.
27. Most would prefer $\frac{7}{3}$ as a solution.

Cooperative Learning

29. **(a)** $\frac{8}{3} = 2.\overline{6}$ **(c)** $3 = 3.0$ **(e)** $\frac{^-7}{3} = -2.\overline{3}$

Review Problems

30. $22,761.95
31. $2.35 \cdot 10^{13}$ mi
32. 0.077. The rule says that the placement should be four places or 0.0770. Because 0.077 and 0.0770 are equivalent, the rule still works.
33. Answers may vary. **(a)** 3.024 **(b)** $^-3.024$ **(c)** 134.8 **(d)** 0.00713

Ongoing Assessment 6-4

2. **(a)** $\sqrt{6}$ **(c)** 3
3. $0.\overline{9}, 0.9\overline{98}, \sqrt{0.98}, 0.\overline{98}, 0.9\overline{88}, 0.9, 0.\overline{898}$
4. **(a)** Yes **(c)** No **(e)** Yes
5. **(a)** 15 **(c)** Impossible
6. **(a)** 2.65 **(c)** 4.51

7. **(b)** False; $-\sqrt{2} + \sqrt{2}$ **(d)** False; $\sqrt{2} - \sqrt{2}$
9. Answers vary. For example, assume the following pattern continues: 0.54544544454444 . . .
10. Answers vary, for example, 0.5155155515555 . . . .
11. Answers vary, for example, 0.56166166616666 . . . .
13. There are infinitely many positive rational numbers. Add $\sqrt{2}$ to each of those and infinitely many irrational numbers result.
14. **(a)** R **(c)** Q **(e)** R
15. **(a)** Q, R **(c)** R, S **(e)** Q, R
16. **(a)** N, I, Q, R **(c)** R, S **(e)** None
17. **(a)** 64 **(c)** $^{-}64$ **(e)** All real numbers greater than zero
18. 6.4 ft
19. **(a)** $\sqrt{180} = 6\sqrt{5}$ **(c)** $\sqrt{252} = 6\sqrt{7}$
20. **(a)** $^{-}3\sqrt[3]{2}$ **(c)** $5\sqrt[3]{2}$
21. **(a)** $5, 5\sqrt[3]{2}, 5\sqrt[3]{4}, 10$
22. **(a)** 2^{10} **(c)** 2^{12}
23. **(a)** 4 **(c)** $\dfrac{^{-}4}{7}$
24. **(a)** Rational **(c)** Irrational
25. **(a)** Yes when $a \leq 0$
27. Between 0 and 0.8
29. $Guess2 = \dfrac{\dfrac{n}{Guess1} - Guess1}{2}$
If $Guess1$ is accurate, then let $x = Guess1 = Guess2$.

$$x = \frac{\frac{n}{x} + x}{2}$$
$$2x = \frac{n}{x} + x$$
$$2x^2 = n + x^2$$
$$x^2 = n$$
$$x = \sqrt{n}$$

Communication

31. $^{-}2$ is a valid solution, as seen by substituting $^{-}2$ for x. The solution is seen from the following:

$$x^2 = 4$$
$$x^2 - 4 = 0$$
$$(x-2)(x+2) = 0$$
$$x = 2 \text{ or } {}^{-}2$$

32. Answers vary; for example, to be rational $\sqrt{2}$ must be able to be written in the form $\frac{a}{b}$, where a and b are integers. When $\sqrt{2}$ is written as $\frac{\sqrt{2}}{1}$, it still fails to have a numerator that is an integer.
34. False: $\sqrt{64 + 36} \neq \sqrt{64} + \sqrt{36}$
35. No; $\frac{22}{7}$ is a rational number that can be represented by the repeating decimal $3.\overline{142857}$.
36. No; $\sqrt{13}$ is an irrational number. So when it is expressed as a decimal, it is nonterminating and nonrepeating.
38. Notice that $\left(\frac{4}{25}\right)^{-1/3} = \left(\frac{25}{4}\right)^{1/3}$ and $\left(\frac{4}{25}\right)^{-1/4} = \left(\frac{25}{4}\right)^{1/4}$. Because $\left(\frac{25}{4}\right)^{1/4} < \left(\frac{25}{4}\right)^{1/3}$, we have $\left(\frac{4}{25}\right)^{-1/4} < \left(\frac{4}{25}\right)^{1/3} = \left(\frac{25}{4}\right)^{1/3}$.

Open-Ended

40. Answers vary. For example:
(a) $\sqrt{1/2}, \pi/6, \sqrt[2]{2/5}, 0.505005000500005\ldots, \sqrt{3/10}$
41. **(a)** When a number between 0 and 1 is raised to larger and larger exponents, the results approach 0. **(b)** Answers vary. **(c)** Answers vary.
43. There are more irrational numbers than rational numbers. Search the Internet for Cantor or transfinite numbers.

Review Problems

44. **(a)** 21.6 lb **(b)** 48 lb
45. **(a)** $\frac{418}{25}$ **(b)** $\frac{3}{1000}$ **(c)** $\frac{^{-}507}{100}$ **(d)** $\frac{123}{1000}$
46. **(a)** $4.\overline{9}$ **(b)** $5.0\overline{9}$ **(c)** $0.4\overline{9}$
47. $\frac{3}{12{,}500}$
48. $\frac{8}{33}$
49. **(a)** 208,000 **(b)** 0.00038

Ongoing Assessment 6-5

1. **(a)** 789% **(c)** 19,310% **(e)** $83\frac{1}{3}\%$ **(g)** 12.5% **(i)** 62.5% **(k)** 80%
2. **(a)** 0.16 **(c)** 0.002 **(e)** $0.13\overline{6}$ **(g)** $0.00\overline{3}$
3. **(a)** 4 **(c)** 25 **(e)** 12.5
5. **(a)** 2.04 **(c)** 60 **(e)** 300%
6. **(a)** $\frac{5x}{100}$
7. 63 boxes
9. $25,500
11. 20%
13. Approximately 18.4%
15. 100%
17. $5.10
19. $336
21. $3200
23. $\frac{325}{500}; \frac{325}{500} = \frac{650}{1000} = \frac{65}{100} = 65\%$, while $\frac{600}{1000} = \frac{60}{100} = 60\%$.

25. It is less than 30; 30 is the number plus 50% of the number.
27. $16.\overline{6}\%$ or $16\frac{2}{3}\%$
29. $11.\overline{1}\%$ (approximately)
31. $33\frac{1}{3}\%$
33. **(a)** \$3.30 **(c)** \$1.90
34. **(a)** 4% **(c)** 64%
36. Apprentice makes \$700. Journeyman makes \$1400. Master makes \$2100.
37. **(a)** 4%
38. \$82,644,63
39. **(a)** The answer is essentially true. Spending of that magnitude for 815 years is a bit more.
40. **(a)** Cherokee 700,000
Navajo 300,000
Latin 200,000
Choctaw 200,000
Sioux 200,000
Chippewa 100,000
Apache 100,000
(c) Respectively, $38.\overline{8}\%$; $16.\overline{6}\%$; $11.\overline{1}\%$; $11.\overline{1}\%$; $11.\overline{1}\%$; $5.\overline{5}\%$; $5.\overline{5}\%$
41. No; the respective salaries would be approximately \$2000 and \$1696.43.
42. Yes, if the class has twice as many boys as girls
44. **(a)** 3
45. **(a)** \$4.50 **(c)** 100%

Communication

47. Yes. 40% of 30 $= \left(\frac{40}{100}\right)\left(\frac{30}{1}\right) = \frac{40 \cdot 30}{100} = \frac{30 \cdot 40}{100} = \left(\frac{30}{100}\right)\left(\frac{40}{1}\right) = 30\%$ of 40

48. Answers vary; for example, a price can go up 150%. If an item is bought for \$100, it can be sold for \$150. The price can't go down 150%, because 100% is all there is and it can't go lower. A price cannot be less than 0.
50. Let x be the amount invested. The first stock option will yield $(1.15x) \cdot 0.85$ after 2 yr. The second stock will yield $(0.85x) \cdot 1.15$. Because each yield equals $(1.15 \cdot 0.85)x$, the investments are equally good.

Open-Ended

53. Answers may vary. **(a)** 115 is 37% of what number?

Review Problems

55. **(a)** 2.5 **(b)** 2.5
56. $\frac{6544}{900} = \frac{1636}{225}$
57. Answers vary, for example, $0.2\overline{1}$.
58. $\frac{5}{3}, 2.0\overline{5}, 2.\overline{15}, \frac{7}{3}, 2.5$
59. Answers vary; for example, $\frac{2}{3} = 0.\overline{6}$ and $\frac{1}{3} = 0.\overline{3}$, so

$0.\overline{9} = 0.\overline{6} + 0.\overline{3} = \frac{2}{3} + \frac{1}{3} = 1$; $n = 0.\overline{9}$, so $10n = 9.\overline{9}$;

$10n - n = 9.\overline{9} - 0.\overline{9}$, so $9n = 9$ and $n = 1$; $0.\overline{4} = \frac{4}{9}$, $0.\overline{5} = \frac{5}{9}$,

$0.\overline{6} = \frac{6}{9}$, $0.\overline{7} = \frac{7}{9}$, $0.\overline{8} = \frac{8}{9}$ so $0.\overline{9} = \frac{9}{9}$ or 1.

Ongoing Assessment 6-6

1.	Int. Rate per Period	No. of Periods	Amt. of Int. Paid
(a)	3%	4	\$125.51
(c)	$\frac{10}{12}\%$	60	\$645.31

2. \$3,675.00
3. \$24.45
6. \$64,800
8. \$1015.20
9. **(c)** at 13.2%
10. Approximately \$2.53
12. Approximately \$4044.98
14. Approximately $7.026762 \cdot 10^8$ or $7.03 \cdot 10^8$
16. \$81,628.81
18. \$10,935
19. A geometric sequence with a ratio of 1.04
21. 8 editions
22. 25,000 students

Communication

24. Let a be the original value of the house. Because it depreciates 10% each year for the first 3 yr, using compound depreciation the price after 3 yr will be $a(1 - 0.10)^3$ or $a \cdot 0.9^3$. Because of compound appreciation, after another 3 yr the value of the house will be $a(0.9^3) \cdot 1.1^3$ or $a(0.9^3 \cdot 1.1^3)$, which equals approximately $a \cdot 0.9703$. Because $a \cdot 0.9703 < a$, the value of the house decreased after 6 yr. The value of the house decreases by approximately 3%.
26. No. The percentages cannot be added because each time the percent is of a different quantity. The 15% is a savings of the original price of the fuel. The second savings of 35% is of the new price after the first savings. We could find the percentage of savings as follows. For each \$100 of the cost of fuel, the new cost after a 15% savings will be \$85. When the second device is installed, an additional 35% is saved, and the new cost is $\left(\frac{65}{100}\right) \cdot 85$. After the third

device is installed, an additional 50% is saved, and the new cost is $\left(\frac{50}{100}\right) \cdot \left(\frac{65}{100}\right) \cdot 85$. The savings on a \$100 initial cost is $100 - \left(\frac{50}{100}\right) \cdot \left(\frac{65}{100}\right) \cdot 85$, or \$72.375. Because the savings were based on a \$100 initial cost, the percentage of total savings with all three devices is 72.375%.

Open-Ended

28. (b) Let a be an initial cost of an item that depreciates at a rate of $p\%$ each period of time. If n is the number of periods and $C(n)$ is the cost (as a function of n) after n periods, then $C(n) = a(1 - p/100)^n$.

Chapter Review

1. (a) $A = 0.02, B = 0.05, C = 0.11$
2. (a) $\frac{32012}{100}$
4. 8 shelves
5. (a) $0.\overline{571428}$ **(c)** $0.\overline{6}$
6. (a) $\frac{7}{25}$ **(c)** $\frac{1}{3}$
7. (a) 307.63 **(c)** 308
8. (a) $4.26 \cdot 10^5$ **(c)** $2.37 \cdot 10^{-6}$
9. $1.451\overline{9}, 1.4\overline{519}, 1.45\overline{19}, 1.4519, 0.13\overline{401}, {}^{-}0.134, {}^{-}0.13\overline{401}$.
10. (a) $5\sqrt{2}$
11. (a) $1.78341156 \cdot 10^6$ **(c)** $4.93 \cdot 10^9$ **(e)** $4.7 \cdot 10^{35}$
13. (a) Irrational **(c)** Rational **(e)** Irrational
14. (a) $11\sqrt{2}$ **(c)** $6\sqrt{10}$
15. (a) No; $\sqrt{2} + ({}^{-}\sqrt{2})$ is rational.
(c) No; $\sqrt{2} \cdot \sqrt{2}$ is rational.
17. (a) 25% **(c)** $56.\overline{6}$
18. (a) 12.5% **(c)** 627% **(e)** 150%
19. (a) 0.60 **(c)** 1
20. \$9280
21. $3.\overline{3}\%$
23. \$5750
25. \$80
26. 25%
27. (a) \$26,600
29. Answers vary. If the dress originally cost \$100, then with 60% off, the cost is \$40. With a 40% off coupon, the cost should be \$40 − (0.40)40 = \$24.00.
30. All are mathematically meaningful.
31. $\frac{1}{\pi}$
33. No. π is irrational so ${}^{-}\pi$ is irrational, but $\pi + ({}^{-}\pi) = 0$, a rational number.
34. \$15,000
35. \$15,110.69

Answers to Now Try This

6-1. The cartoon reports the cost of prime rib is .75¢, which is not one-half of \$1.50. If the waiter intended the price of the prime rib to be one-half the price of the steak, then the amount should be written as \$0.75 or 75¢. The notation .75¢ indicates 0.75 of 1 cent. This mistake is commonly made in stores when the money amount is less than 1 dollar; for example, an item is marked for .50¢ when it really should be \$0.50 or 50¢.

6-2. $3.6 \cdot 1000 = 3.6 \cdot 10^3 = \left(3 + \frac{6}{10}\right) \cdot 10^3 = 3 \cdot 10^3 + \frac{6}{10} \cdot 10^3 = 3 \cdot 10^3 + 6 \cdot 10^2 = 3 \cdot 10^3 + 6 \cdot 10^2 + 0 \cdot 10^1 + 0 \cdot 1 = 3600$. Thus, we see that multiplication by 1000 results in moving the decimal point three places to the right. In general, multiplication by 10^n, where n is a positive integer, results in moving the decimal point n places to the right.
6-3. \$1.19/32 oz is about \$0.037/oz, while \$1.43/48 oz is about \$0.030/oz, so the 48-oz jar is a better buy.
6-4. Answers vary; for example, using the front digits the first estimate is \$2 + \$0 + \$6 + \$4 + \$5 = \$17. Next, we adjust the estimate. Because \$0.89 + \$0.13 is about \$1.00 and \$0.75 + \$0.05 is \$0.80 and \$0.80 + \$0.39 is about \$1.20, the adjustment is \$2.20 and the estimate is \$19.20.
6-5. (a) The answer is rounded to 0.0000014.
(b) The answers is 0.00000136.
6-6. (a) $\frac{1}{9} = 0.\overline{1}$ **(b) (i)** $\frac{2}{9} = 2(0.\overline{1}) = 0.\overline{2}$
(ii) $\frac{3}{9} = 3(0.\overline{1}) = 0.\overline{3}$ **(iii)** $\frac{5}{9} = 5(0.\overline{1}) = 0.\overline{5}$
(iv) $\frac{8}{9} = 8(0.\overline{1}) = 0.\overline{8}$
6-7. (a) If $r = 1$, the denominator of $\frac{a}{1-r}$ is 0. **(b)** As n becomes greater, so does r^{n+1} when $r > 1$.
(c) Let $S = a + ar + ar^2 + \cdots + ar^n$

$$-rS = {}^{-}(ar + ar^2 + \cdots + ar^n + ar^{n+1})$$
$$S - rS = a - ar^{n+1}$$
$$S = \frac{a(1 - r^{n+1})}{1 - r}$$

6-8. (a) Approximately 42.47753 . . . . The decimal is nonterminating because $\left(\frac{207}{365}\right) \cdot \frac{749}{10} = \frac{(207 \cdot 749)}{3650} = \frac{(3 \cdot 3 \cdot 23 \cdot 7 \cdot 107)}{2 \cdot 5 \cdot 5 \cdot 73}$ and the fraction in simplest form has factors other than 2s and 5s in the denominator. **(b)** Arlo's age in years and months is 42 yr and approximately 5.7 mo.
(c) Answers will vary.
6-9. (a) $0.3\overline{55}$; $0.36\overline{5}$ **(b)** With each repeating decimal (like 0.36) one can move halfway between the number and the two found to write more repeating decimals—for example, $0.3\overline{55} < 0.36 < 0.36\overline{5}$—and continue the process indefinitely.

6-10. **(a)** $\sqrt{13} \approx 3.6056$ **(b)** $Guess2 = \dfrac{\dfrac{n}{Guess1} - Guess1}{2}$

6-11. **(a)** The approach works because

$$\sqrt{\sqrt{\sqrt{a}}} = \left(\left(a^{\frac{1}{2}}\right)^{\frac{1}{2}}\right)^{\frac{1}{2}} = \left(a^{\frac{1}{4}}\right)^{\frac{1}{2}}$$
$$= a^{1/8}$$
$$= \sqrt[8]{a}$$

(b) For $n = 2^k$, where k is a positive integer. As shown in part (a), by repeatedly applying the square root function to a we get

$\overbrace{a^{\frac{1}{2} \cdot \frac{1}{2} \cdot \frac{1}{2} \cdot \frac{1}{2}}}^{k \text{ times}}$ or $a^{\frac{1}{2^k}} = \sqrt[2^k]{a}$.

6-12. **(a)** Answers vary, for example, most calculators will convert the decimal form of a number to a percent by moving the decimal point two places to the left. Other calculators actually place a % symbol in the display when the [%] key is pushed. **(b)** $33.\overline{3}\%$

6-13. The tip is 30%.

Answers to Technology Corner

(a) Column d shows that the 30% mixture is reached before 5 L of lemon juice and 12 L of water are mixed.

(b) $\dfrac{5}{5 + x} \cdot 100$ changes the ratio of lemon juice to water to a percentage, where x is the number of liters of water added and $5 + x$ is the number of total liters in the mixture.

Chapter 7

Ongoing Assessment 7-1

2. **(a)** No **(c)** Yes

3. **(a)** $\{0, 1, 2, 3, 4, 5, 6, 7, 8, 9\}$ **(c)** $\{1, 3, 5, 7, 9\}$

(e) $\frac{5}{10}$, or $\frac{1}{2}$; $\frac{5}{10}$, or $\frac{1}{2}$; $\frac{9}{10}$

4. **(a)** $\frac{5}{26}$ **(c)** $\frac{11}{26}$

5. **(a)** $\frac{3}{8}$ **(c)** $\frac{4}{8}$, or $\frac{1}{2}$ **(e)** 0 **(g)** $\frac{1}{8}$

6. **(a)** $\frac{26}{52}$, or $\frac{1}{2}$ **(c)** $\frac{28}{52}$, or $\frac{7}{13}$ **(e)** $\frac{48}{52}$, or $\frac{12}{13}$ **(g)** $\frac{3}{52}$

7. **(a)** $\frac{4}{12}$, or $\frac{1}{3}$ **(c)** 0 **(e)** 1

8. 25 cards

10. **(a)** $\frac{8}{36}$, or $\frac{2}{9}$ **(c)** $\frac{24}{36}$, or $\frac{2}{3}$ **(e)** 0 **(g)** 10 times

12. **(a)** $\frac{18}{38}$, or $\frac{9}{19}$ **(c)** $\frac{26}{38}$, or $\frac{13}{19}$

14. **(a)** No **(c)** Yes **(e)** No **(g)** No

15. **(a)** $\frac{2}{4}$, or $\frac{1}{2}$ **(c)** $\frac{3}{4}$

18. Assuming the dealt cards are not put back into the deck, the answers are: **(a)** $\frac{20}{50}$, or $\frac{2}{5}$ **(c)** 0

19. The answers may vary depending on how the 6, 7, and 9 are formed. The following answers are based on a Casio digital watch. **(a)** $\frac{8}{10}$, or $\frac{4}{5}$ **(c)** $\frac{7}{10}$

20. **(a)** The probability of students taking algebra or chemistry **(c)** This represents 1 minus the probability of a student taking chemistry or the probability of a student not taking chemistry.

22. **(a)** $\frac{45}{80}$, or $\frac{9}{16}$ **(c)** $\frac{60}{80}$, or $\frac{3}{4}$

23. 0.4

24. **(a)** 22

Open-Ended

32. **(a)** $\frac{1}{4}$

Cooperative Learning

36. **(a)** Most often, 1 and 2 and then 0 and 3; least often, 6 and then 5

Ongoing Assessment 7-2

1. (a)

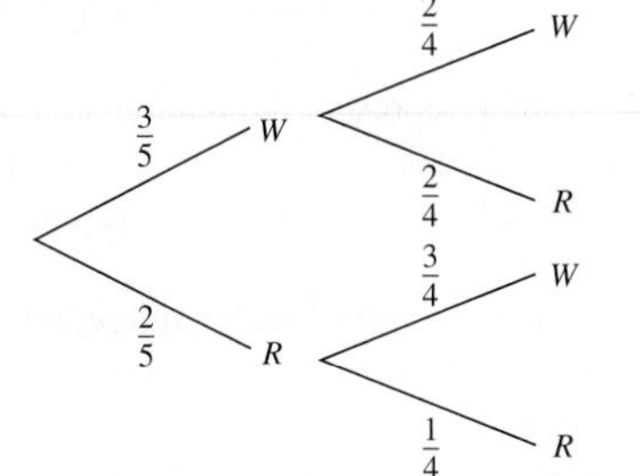

The probability that two balls are of different colors is $\frac{3}{5} \cdot \frac{2}{4} + \frac{2}{5} \cdot \frac{3}{4} = \frac{3}{5}$

2. (a) {(1,1), (1, 2), (1, 3), (2, 1), (2, 2), (2, 3)}
(c) {(1, 2), (2, 1), (2, 2), (2, 3)}

3. (a) $\frac{1}{216}$

4. (a) $\frac{1}{24}$ **(c)** $\frac{1}{84}$

6. (a) Box 1, with probability $\frac{1}{3}$ (Box 2 has probability $\frac{1}{5}$.)

7. (a) $\frac{64}{75}$

9. (a) $\frac{1}{5}$ **(c)** $\frac{11}{15}$

11. $\frac{1}{16}$

12. (a) $\frac{1}{4}$ **(b)** $\frac{5}{8}$ **(c)** $\frac{1}{8}$

13. (a) $\frac{4}{10}$, or $\frac{2}{5}$

14. $\frac{2}{16}$, or $\frac{1}{8}$

15. (a) $\frac{1}{320}$ **(c)** 0

16. (a) The first spinner. If you choose the first spinner, you win if and only if the spinning combinations are as follows:

Outcome on spinner *A*	Outcome on spinner *B*
4	3
6	3 or 5
8	3 or 5

The probability of this happening is

$\frac{1}{3} \cdot \frac{1}{3} + \frac{1}{3} \cdot \frac{2}{3} + \frac{1}{3} \cdot \frac{2}{3} = \frac{5}{9}$.

If you choose spinner *B*, the probability of winning is only $\frac{4}{9}$.

17. (a) $\frac{5}{8}$

18. $\frac{1}{32}$

22. (a) $\frac{1}{25}$ **(c)** $\frac{16}{25}$

23. (a) 100 square units
(c) $\frac{1}{625}$

24. 0.7

27. $\frac{2}{5}$

28. $\frac{25}{30}$, or $\frac{5}{6}$

30. $\frac{69}{3000}$, or $\frac{23}{1000}$

31. (a)

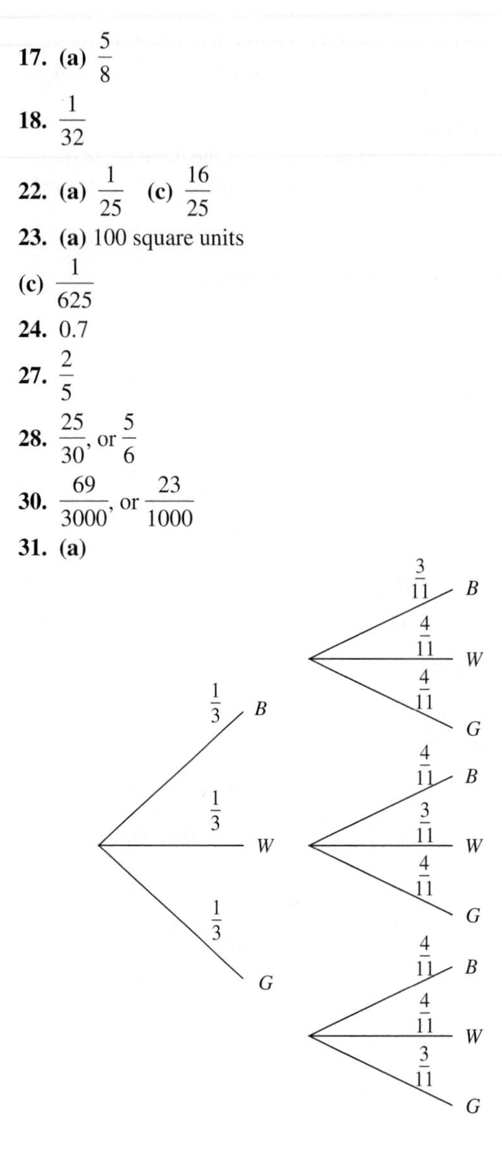

(c) $\frac{1}{3} \cdot \frac{3}{11} = \frac{1}{11}$

32. Each room has probability $\frac{1}{2}$ of being chosen. Hence it does not matter where the car is placed.

33. a. $\frac{11}{16}$

35. Billie-Bobby-Billie because the probability of winning two in a row is greater this way. Note, it does not say win two out of three.

Communication

38. Finding a witness with blond hair and blue eyes has a greater probability. The probability of a red car introduces another number that is less than 1 to the product. When you multiply a, the probability that you have for blond hair and blue eyes, by a number less than 1 and greater than 0, the probability becomes smaller.

39. The red ball in one of the boxes and the three white balls in the other box. In this case the probability of getting the red ball is $\frac{1}{2} \cdot 1 + \frac{1}{2} \cdot 0$, or $\frac{1}{2}$. In the three other arrangements, the probability of getting the red ball is smaller.

Cooperative Learning

43. c.

x	1	2	3	4
1	1	2	3	4
2	2	4	6	8
3	3	6	9	12
4	4	8	12	16

d. Yes, each player has probability $\frac{1}{2}$ of winning.

44. If the first player chooses *TT* and the second *HT*, then the first player wins the game if and only if tails appears on the first and second flips. The probability of this happening is $\frac{1}{4}$. The probabilities of winning the game for other choices are summarized below:

		First Player's Choice			
		HH	*HT*	*TH*	*TT*
Second	*HH*	—	0.50	0.75	0.50
Player's	*HT*	0.50	—	0.50	0.25
Choice	*TH*	0.25	0.50	—	0.50
	TT	0.50	0.75	0.50	—

Of the 12 possible games, only 8 result in choices with equally likely probabilities. Therefore, the game is not fair.

45. (a) The game is not fair. You should choose spinner *A*.

Review Problems

46. (a) v **(b)** iii **(c)** ii **(d)** i **(e)** iv

47. (a) $\frac{1}{30}$ **(b)** 0 **(c)** $\frac{19}{30}$

Ongoing Assessment 7-3

1. Answers vary; for example, a black card might represent the birth of a boy and a red card might represent the birth of a girl. Choose a card to represent a birth.

2. Answers vary; for example, let the digits 1 and 2 represent Diamonds, digits 3 and 4 represent Hearts, digits 5 and 6 represent Spades, and 7 and 8 represent Clubs. If 0 or 9 appear, then disregard these digits.

3. (a) Let 1, 2, 3, 4, 5, and 6 represent the numbers on the die and ignore the numbers 0, 7, 8, 9. **(c)** Represent red by the numbers 0, 1, 2, 3, 4; green by the numbers 5, 6, 7; yellow by the number 8; and white by the number 9.

5. To simulate Monday, let the digits 1 through 8 represent rain and 0 and 9 represent no rain. If rain occurred on Monday, repeat the same process for Tuesday. If it did not rain on Monday, let the digits 1 through 7 represent rain and 0, 8, and 9 represent dry. Repeat a similar process for the rest of the week.

6. Answers vary; for example, mark off blocks of two digits and let the digits 00, 01, 02, . . . , 13, 14 represent contracting the disease and 15 to 99 represent no disease. Mark off blocks of six digits to represent the three children. If at least one of the numbers is in the range 00 to 14, then this represents a child in the three-child family having strep.

8. 1200 fish

9. (a) 7

Communication

12. Answers vary; one way is to turn left when an even number comes up, turn right when the number 1 comes up, and move straight when 3 or 5 come up.

Open-Ended

13. Answers vary; one could use a random-digit table with blocks of two digits, or a spinner could be designed with 12 sections representing the different months. The spinner could be spun 5 times to represent the birthdays of five people. We could then keep track of how many times at least two people have the same birthday.

14. Answers vary; for example, use a random-digit table. Let the digits 1–8 represent a win and the digits 0 and 9 represent losses. Mark off blocks of three. If only the digits 1–8 appear, then this represents three wins in a row.

15. Let the 10 ducks be represented by the digits 0, 1, 2, 3, . . . , 8, 9. Then pick a starting point in the table and mark off 10 digits to simulate which ducks the hunters shoot at. Count how many of the digits 0 through 9 are not in the 10 digits; this represents the ducks that escaped. Do this experiment many times and take the average to determine an answer. See how close your simulation comes to 3.49 ducks.

Cooperative Learning

16. (d) The probability of two boys and two girls is $\frac{6}{16}$, or $\frac{3}{8}$.

17. (d) $\left(\frac{1}{2}\right)^{10} = \frac{1}{1024}$

Review Problems

20. No, you will win about $\frac{6}{16}$ or $\frac{3}{8}$ of the games.

21. (a) $\frac{1}{4}$ **(b)** $\frac{1}{52}$ **(c)** $\frac{48}{52}$, or $\frac{12}{13}$ **(d)** $\frac{3}{4}$
(e) $\frac{1}{2}$ **(f)** $\frac{1}{52}$ **(g)** $\frac{16}{52}$, or $\frac{4}{13}$ **(h)** 1

22. (a) $\frac{15}{19}$ **(b)** $\frac{56}{361}$ **(c)** $\frac{28}{171}$

Ongoing Assessment 7-4

1. (a) 12 to 40, or 3 to 10
3. 15 to 1
4. (a) $\frac{1}{2}$ **(c)** 1023 to 1
5. $\frac{5}{8}$
6. 1 to 1
7. 4 to 6, or 2 to 3
9. $\frac{1}{27}$
11. 73 : 27
13. 4 : 7
14. (a) $\frac{1}{2}$
15. (a) 0 **(c)** $\frac{1}{5}$
17. $E = \frac{1}{6}(10) + \frac{5}{6}(^-2) = \frac{10}{6} + \left(\frac{^-10}{6}\right) = 0$

Therefore, you should come out about even if you play a long time.

18. (a) $\frac{1}{38}$ + **(c)** $-\frac{2}{38}$ or $\frac{^-1}{19}$ dollars
20. \$10,000
21. \$3.50
23. (a) \$7 **(b)** No

Communication

24. Odds are determined from probabilities. The *odds in favor* of an event. E, are determined by $\frac{P(E)}{P(\overline{E})}$. The *odds against* an event are determined by $\frac{P(\overline{E})}{P(E)}$.

Review Problems

30. (a) {1, 2, 3, 4} **(b)** {Red, Blue} **(c)** {(1, Red), (1, Blue), (2, Red), (2, Blue), (3, Red), (3, Blue), (4, Red), (4, Blue)}
(d) {(Blue, 1), (Blue, 2), (Blue, 3), (Blue, 4), (Blue, 5), (Blue, 6), (Red, 1), (Red, 2), (Red, 3), (Red, 4), (Red, 5), (Red, 6)} **(e)** {(1, 1), (1, 2), (1, 3), (1, 4), (2, 1), (2, 2), (2, 3), (2, 4), (3, 1), (3, 2), (3, 3), (3, 4), (4, 1), (4, 2), (4, 3), (4, 4)}
(f) {(Red, Red), (Red, Blue), (Blue, Red), (Blue, Blue)}
31. The blue section must have 300 degrees; the red has 60 degrees.
32. $\frac{25}{676}$

Ongoing Assessment 7-5

1. 224
2. 32
6. (a) True **(c)** False **(e)** False **(g)** True
8. 15
9. (a) 12 **(c)** 3360 **(e)** 3780
11. (a) 24,360
12. 792
13. $\frac{1}{120}$
15. 1260
16. (a) 6
18. (a) $\frac{1}{13}$
19. 2,598,960 different 5-card hands (Order within the hand is not important.)
21. $\frac{1}{25,827,165}$
23. $\frac{13,440}{59,049}$, or approx. 0.228
24. (a) $\frac{1}{6^8}$ **(c)** $1-\left(\frac{5}{6}\right)^8 \approx 0.767$
25. (a) $\left(\frac{6}{36}\right)^5$, or $\frac{1}{6^5} \approx 0.00013$
26. (a) $\frac{{}_{12}C_4}{{}_{22}C_4} = \frac{9}{133}$ **(c)** $\frac{{}_{10}C_4}{{}_{22}C_4}$, or $\frac{6}{209}$
27. (a) $({}_{20}C_2 \cdot {}_{21}C_4 \cdot {}_4C_2) \div {}_{45}C_8 \approx 0.032$
(c) $1 - \frac{{}_{25}C_8}{{}_{45}C_8} \approx 0.995$
30. a. $\frac{1}{120}$ **c.** 0
31. a. 10^4, or 10,000 **c.** $\frac{625}{10,000}$, or $\frac{1}{16}$

Communication

32. Answers vary. For example, the Fundamental Counting Principle (FCP) says that to find the number of ways of making several decisions in a row, multiply the number of choices that can be made for each decision. The FCP can be used to find the number of permutations. A permutation is an arrangement of things in a definite order. A combination is a selection of things in which the order is not important. We could find the number of combinations by using the FCP and then divide by the number of ways in which the things can be arranged.

33. (a) There are 10 choices for the number on the first reel, 10 choices for the number on the second reel, and 10 choices for the number on the third reel. By the Fundamental Counting Principle, there are $10 \cdot 10 \cdot 10$, or 1000, choices for the combination of the lock.

34. (a) $8! \cdot 3!$. If the family is considered a unit and each of the remaining people also a unit, we have 8 units. There are 8! ways to arrange the 8 units. For each of the 8! ways, the family unit can be arranged in 3! ways and hence the number of seating arrangements is $8! \cdot 3!$.

35. 3840. Consider each couple as a unit. The five units can be arranged in 5! ways. For each of the 5! arrangements, each of the five couples can be arranged in 2 ways. Consequently there are $5! \cdot 2^5$ arrangements.

Open-Ended

36. (a) 10^6, or 1,000,000
(c) Answers vary. For example, you would first find the population of California and then experiment with using letters in the license plates. This would help because the choice is for 26 letters in a slot rather than 10 numbers.

Cooperative Learning

37. (c) The sums of the numbers in the rows are 1, 2, 4, 8, 16, 32, 64. The sum in the 10th row is $2^{10} = 1024$. **(d)** Yes, a similar relationship holds in all the rows for the entries in Pascal's triangle.

Review Problems

38. (a) $\frac{396}{2652}$, or $\frac{33}{221}$ **(b)** $\frac{1352}{2652}$, or $\frac{26}{51}$

39. $\frac{3}{36}$, or $\frac{1}{12}$

40. $E = \$0$, so the game is fair.

41. (a) The game is not fair. **(b)** The expected payoff is $\frac{1}{38} \cdot \$36$, or approximately 95¢. Therefore, you can expect to lose on the average about 5¢ a game if you play it a large number of times.

Chapter Review

2. (a) {Monday, Tuesday, Wednesday, Thursday, Friday, Saturday, Sunday} **(c)** $\frac{2}{7}$

3. There are 800 blue ones, 125 red ones, and 75 that are neither blue nor red.

4. (a) Approximately 0.501
(c) 34,226,731 to 34,108,157 $\doteq$ 1.003:1

5. (a) $\frac{5}{12}$ **(c)** $\frac{5}{12}$ **(e)** 0

6. (a) $\frac{13}{52}$, or $\frac{1}{4}$ **(c)** $\frac{22}{52}$, or $\frac{11}{26}$

7. (a) $\frac{64}{729}$

8. $\frac{6}{25}$

9. $\frac{14}{80}$, or $\frac{7}{40}$

10. $\frac{7}{45}$

11. 4 to 48, or 1 to 12

13. $\frac{3}{8}$

14. \$0.30

16. The expected value is \$1.50. The expected average earnings are $^{-}$\$0.50.

17. 900

19. 5040

21. (a) $5 \cdot 4 \cdot 3$, or 60 **(c)** $\frac{1}{60}$

23. $\frac{2}{5}$

30. (a) $\frac{1}{8}$ **(c)** $\frac{1}{16}$

Answers to Now Try This

7-1 (a) 1
(b) 1
(c) Yes, they always sum to 1. 1 is the sum of the probabilities of all the different elements in any sample space.

7-2 1. (a) $P(\text{both balls are red}) = \frac{1}{9}$
(b) $P(\text{no ball is red}) = \frac{4}{9}$
(c) $P(\text{at least one, ball is red}) = \frac{5}{9}$
(d) $P(\text{at most one ball is red}) = \frac{8}{9}$
(e) $P(\text{both balls are the same color}) = \frac{3}{9} = \frac{1}{3}$
2. The answers are the same as in part 1.

7-3 (a) $P(\text{rain both days}) = (0.3)(0.6) = 0.18$
(b) $P(\text{no rain either day}) = (0.7)(0.4) = 0.28$
(c) $P(\text{rain exactly one day}) = (0.3)(0.4) + (0.7)(0.6) = 0.12 + 0.42 = 0.54$
(d) $P(\text{rain at least one day}) = 1 - 0.28 = 0.72$
(e) Yes. If conditions are favorable for rain tonight, they are more likely to be favorable for rain tomorrow.

7-4 **(a)** P(the last letter drawn is a B given that the first letter drawn was an A) $= \frac{2}{3}$

(b) P(the last letter drawn is an A given that the first letter drawn was an A) $= \frac{1}{3}$

(c) P(the last two letters drawn will match) $= \frac{8}{15}$

7-5 **(a)** Yes

(b) With replacement: any game with the same number of white and colored marbles. Without replacement: 3 colored and 6 white or 3 white and 6 colored, 6 colored and 10 white or 6 white and 10 colored marbles.

(c) Without replacement: the white and colored marbles need to be two consecutive triangular numbers, that is, $1 + 2 + 3 + \ldots n$ and $1 + 2 + 3 + \ldots n + n + 1$ (this is easier to discover and verify using combinations introduced in Section 7–5).

7-6 **(a)** Answers will vary. Earth is approximately 30% land.

(b) 0.7

(c) $500(0.3) = 150$

7-7 **(a)** Answers vary.

(b) $\frac{3}{8}$

(c) No, simulations will not always result in the same probability as the theoretical probability. However, if the experiment is repeated a great number of times, the simulated probability should approach the theoretical probability

7-8 **(b)** There is one way to toss a head and one way of not tossing a head, so the odds in favor are $1:1$.

(c) There are four ways to draw an ace and 48 ways of not drawing an ace, so the odds in favor are $4:48$, or $1:12$.

(d) There are 13 ways of drawing a heart and 39 ways of not drawing a heart, so the odds in favor are $13:39$, or $1:3$.

7-9 **(a)** $n(n-1), n(n-1)(n-2), n(n-1)(n-1)(n-3)$

(b) $n(n-1)(n-2)\ldots(n-r+1)$

(c) $4 \cdot 3 \cdot 2 = 24$ ways to choose a president, vice president, and secretary

7-10 **(a)** We get an error message because 100! and 98! are too large for the calculator to handle.

(b) $\frac{100!}{98!} = \frac{98! \cdot 99 \cdot 100}{98!} = 9900$

7-11 $\frac{1}{9}$

Answers to Technology Corner

Section 7-3

1. **(a)** Use **randint**(1, 6) to randomly generate a number from 1 to 6. **randint**(1, 6, 5) will randomly generate five numbers from 1 to 6.

(b) Use **randint**(1, 6) + **randint**(1, 6) to find the sum of the two randomly generated numbers. Or use the **dice** function to simulate the sum of two dice. Using **dice**(5, 2), for example, gives the sum of the numbers on two dice on five rolls of the dice.

2. **(a)** Combine the integer function with the rand function. **ipart**(6*__rand__ + 1) will randomly generate integers from 1 to 6.

(b) **ipart**(6*__rand__ + 1) + **ipart**(6*__rand__ + 1) will give the sum of two randomly generated numbers.

3. Answers will vary.

Chapter 8

Ongoing Assessment 8-1

2. (a) 225 million **(c)** 550 million

3. Answers may vary. An example follows:

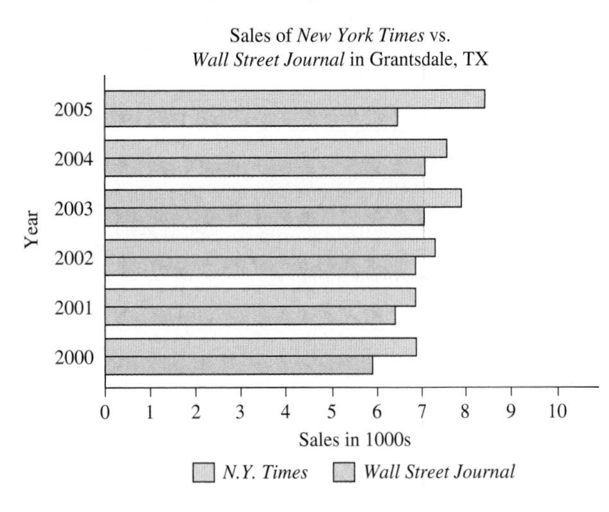

4.

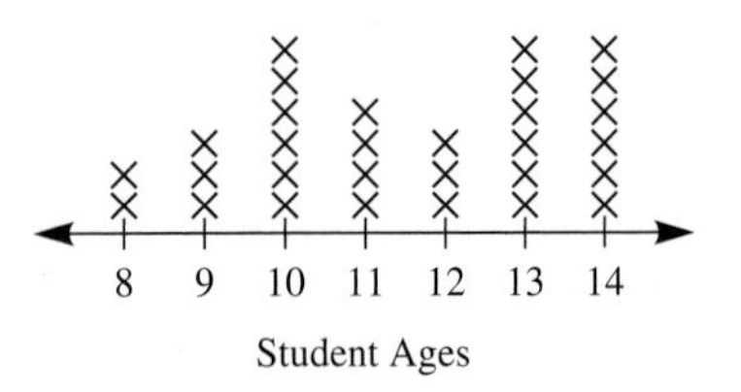

5. Answer may vary. An example follows:

(a)

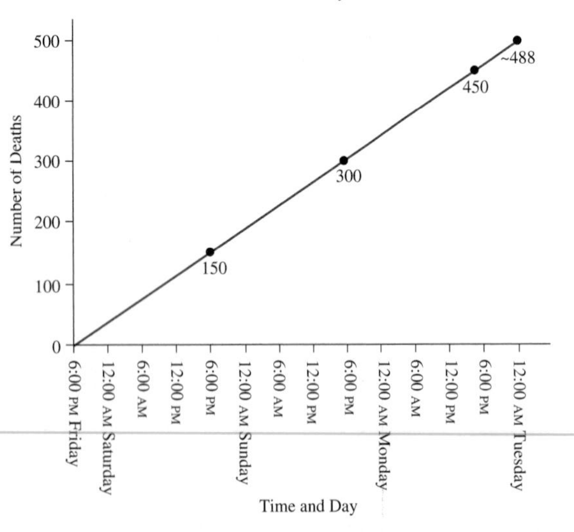

7. (a) 72, 74, 81, 81, 82, 85, 87, 88, 92, 94, 97, 98, 103, 123, 125 **(c)** 125 lbs

8.

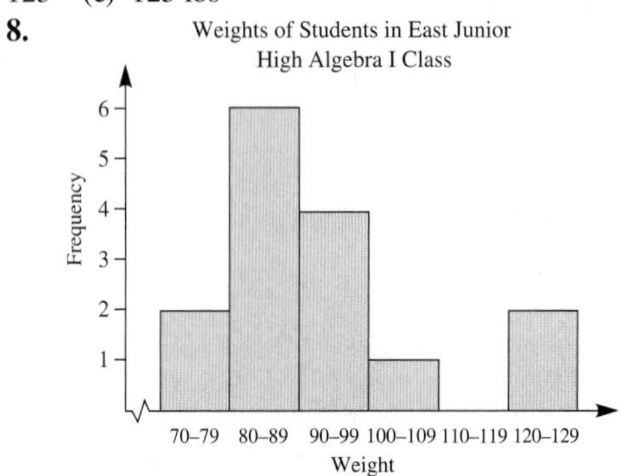

10. (a) November, 30 cm

11. (a)

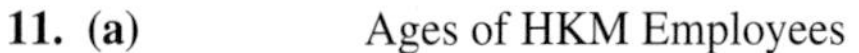

Ages of HKM Employees

Stem	Leaves
1	889
2	01113333334566679
3	224447
4	1115568
5	2248
6	233

3 | 4 represents 34 years old

(c) 20

12. (a) Approximately 3800 km

14.

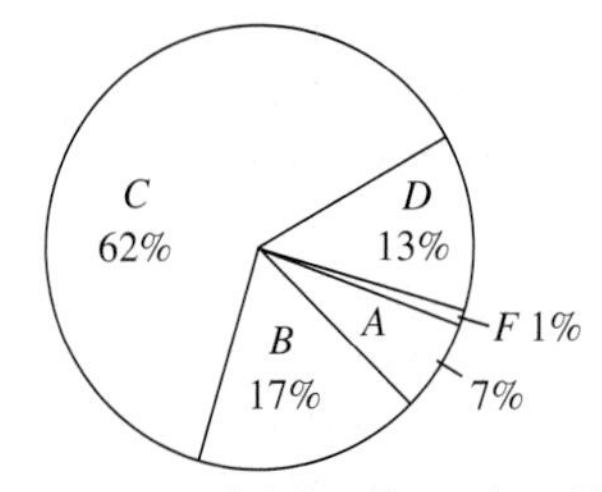

Percentages are rounded. Part F rounds to 2%.

16. (a) Answers may vary. For example, histogram might be used to depict real or estimated census data over the 200-year period.

18. (a) 19% of 21 people is approximately 4 people
14% of 21 people ≐ 3 people
52% of 21 people ≐ 11 people

Each of the 5% sectors represents approximately 1 person.

19. Answers will vary. A graph of this type might be suspect. To have a population where there are more people in the 70s–90s might mean the people are in a home for the aging. Also to have more people in their 90s than 80s reading might make one suspicious. Without more information, one cannot tell.

20. (a) In this particular population, one would suspect that people in their 70s read the most, **but** the graph only depicts the frequency of choice for a book type, not the number of readers.

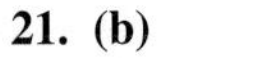

21. (b) Fall Textbook Costs

Classes	Tally	Frequency
\$15–19	I	1
\$20–24	II	2
\$25–29		0
\$30–34	II	2
\$35–39	~~IIII~~	5
\$40–44	IIII	4
\$45–49	III	3
\$50–54	IIII	4
\$55–59	I	1
\$60–64	III	3
		25

(c)

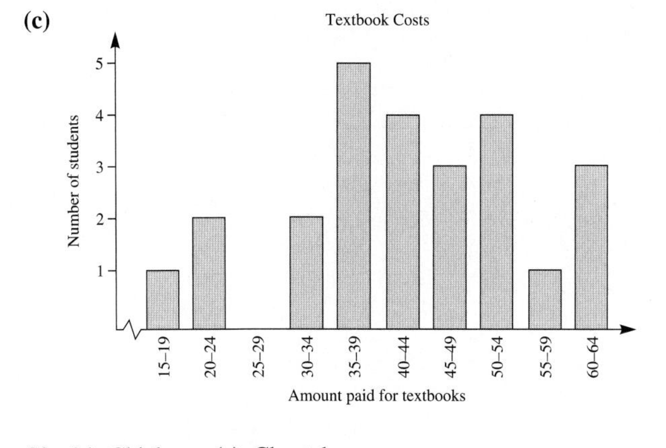

22. (a) Chicken **(c)** Cheetah
23. (a) Women **(c)** About 5.5 yr
25. (a) Asia **(c)** It is about 2/3 as large. **(e)** 5 : 16
26. (a)

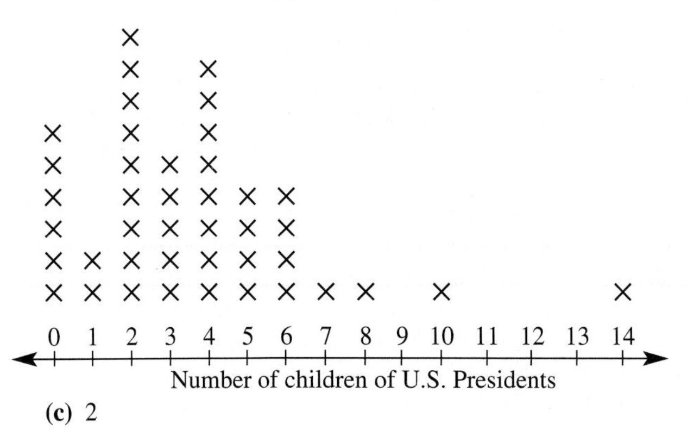

(c) 2

27. (a)

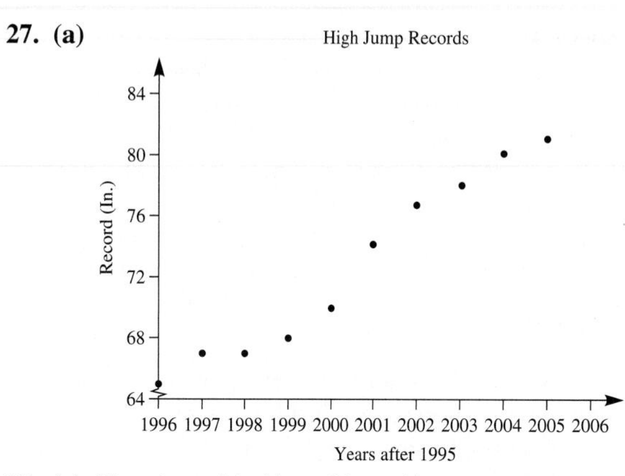

28. (a) Negative **(c)** About 22 yr old

Communication

31. Answers vary; for example, if only percentages are shown on the circle graph, then the circle graph would not change. If data are shown, the numbers in the graph change but not the shape of the graph.
33. Answers vary; for example, the budget for 1 yr is usually represented with a circle graph because the circle graph looks somewhat like a dollar and a circle graph allows for a visual comparison of the relative sizes of fractional parts, which is what the graph of the federal budget is attempting to show.
36. Answers vary; for example, the line graph is more useful because we can approximate the point midway between 8:00 A.M. and 12:00 noon and then draw a vertical line upward until it hits the line graph. An approximation for the 10:00 A.M. temperature can be obtained from the vertical line.

Open-Ended

37. Answers may vary. **(a)** If the faces were circular, we would expect a positive association.
40. (a)

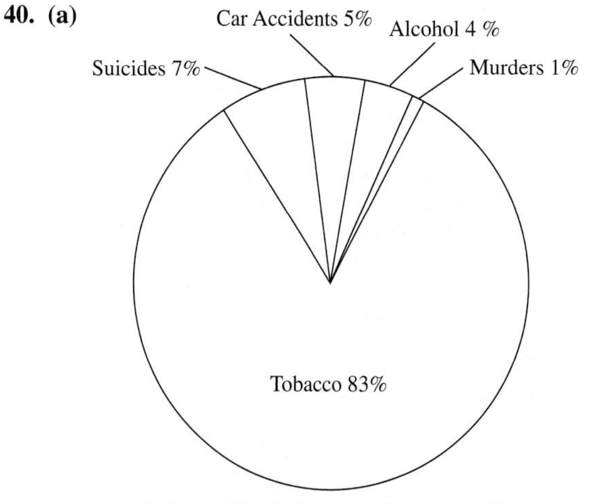

Estimated Death Causes in Canada in 1996

41. Answers vary depending on data chosen.
42. Answers vary; for example, death rates have been decreasing from cardiovascular diseases since around 1965. Deaths from cancer have been on the rise since 1900. Deaths from influenza and pneumonia have been decreasing since 1900 except for a very serious outbreak in the years around 1918. Tuberculosis deaths continued to drop from 1900 to about 1960 and since then there have been few or no deaths.

Cooperative Learning

43. Answers vary; for example, students may not be able to make definitive decisions other than for their class. You may want to check in department stores to see if they can determine the desired size.
44. Answers will vary. For example, in the 2004 elections for the U.S. Senate, the candidate with the most money in the campaign fund won. For the Senate races, 85% of the candidates who spent the most money were successful at the polls. In all, 28 Senate winners outspent their opponents. Only 5 won without being outspent. Candidates who used personal fortunes in election bids received mixed results on election day. Jon Corzine of New Jersey spent an estimated $60 million of his personal fortune to win—twice as much as anyone had spent previously to win a Senate seat. Mark Dayton in Minnesota spent $8.6 million of his own money to win. Herb Kohl of Wisconsin spent just over $4 million of his own money to win re-election to the Senate. Winning Senate campaigns cost an average of $5.6 million ($6.07 per vote). That figure is skewed by the Corzine race in New Jersey. All spending figures are based on the October 18, 2004, campaign spending reporting.

Based on the data that candidates spending the most money won campaigns, the first implication is that you stand a better chance to win if you spend more money; the issue is not as clear for personal spending. However, wherever it comes from, the candidate needs a vast amount of money to campaign.
47. Answers will vary. Expectations of word length on future pages depend on the page chosen.

Ongoing Assessment 8-2

1. **(a)** Mean = 6.625, median = 7.5, mode = 8.
(c) Mean $\doteq$ 19.9, median = 18, modes = 18 and 22.
(e) Mean = $5.8\overline{3}$, median = 5, mode = 5.
3. **(a)** The mean, median, and mode are all 80.
4. 1500
6. $78.\overline{3}$
7. **(a)** $\overline{x}$ = 18.4 yr **(c)** 28.4 yr
9. Answers will vary. **(a)** Suppose the scores were 100, 100, 50, 50. The mean is 75.

11. $\dfrac{100m + 50n}{m + n} = \text{mean}$

13. **(a)** 84
14. Approximately 2.59
16. $1880
17. **(a)** $41,275 **(c)** $38,000
18. **(a)** $132,000 **(c)** Approximately $21,756.60
19. **(a)** Answers may vary. The median of $38,000 reported with the interquartile range of $13,500 would be one appropriate set of choices.
20. **(a)** Balance beam—Olga (9.575); uneven bars—Lisa (9.85); floor—Lisa (9.925)
22. 58 years old
23. **(a)** *A* **(c)** *C*
25. 96, 90, and 90
27. **(a)** **(i)** $\overline{x}$ = 5, median = 5 **(ii)** $\overline{x}$ = 100, median = 100 **(iii)** $\overline{x}$ = 307, median = 307.
28.

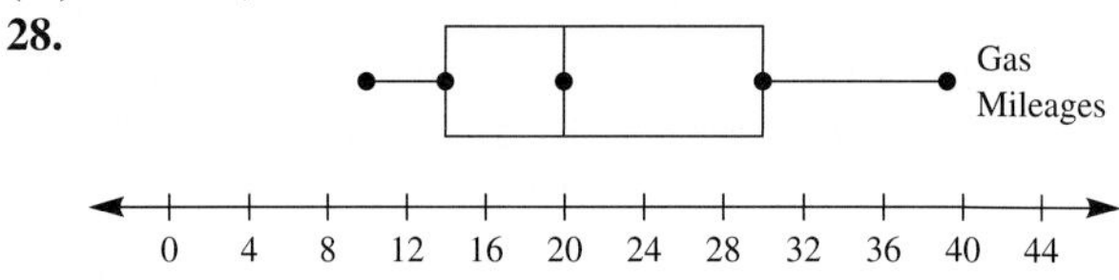

29. **(a)** A—$25, B—$50 **(c)** $80 at B
31. **(a)**

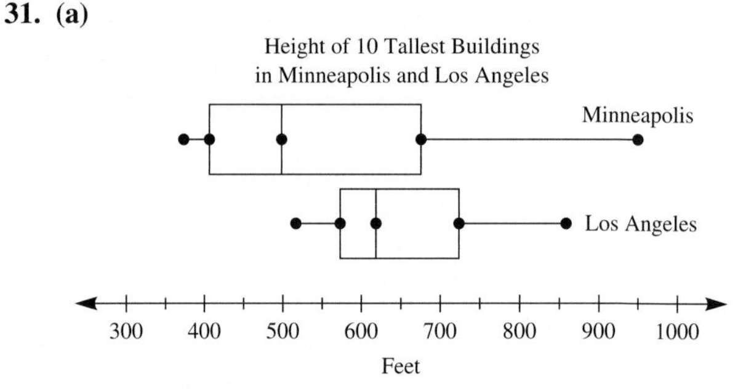

(b) There are no outliers.
34. **(a)** $s = 0$ **(b)** Yes
35. **(a)** Approximately 76.8 **(c)** 71
(e) Approximately 156.8 **(g)** Approximately 9.04
36. **(a)** **(i)** Increase by $1000 **(ii)** Increase by $1000 **(iii)** Increase by $1000 **(iv)** Increase by $1000 **(v)** Stays the same **(vi)** Stays the same **(b)** **(i)** Increase by 5% **(ii)** Increase by 5%
37. **(a)** 1020 **(c)** 1.5, so 1 or 2 people
39. 0.68
40. 16%
42. **(a)** 47.5%
43. 0.84
45. Between 60.5 in and 70.5 in.
47. 1600 students

Communication

49. One might use the mode. If you collect the data of all states and consider the most common age at which one could get a license, that would be the mode. Both the mean and median might be decimals, and no state would worry about a decimal age for a driver's license.

51. The government probably uses the mean of data collected over a period of time.

52. (a) Mean = 90, median = 90, mode = 90. **(c)** Mean

54. No. To find the average speed we divide the distance traveled by the time it takes to drive it. The first part of the trip took $\frac{5}{30}$, or $\frac{1}{6}$, of an hour. The second part of the trip took $\frac{5}{50}$, or $\frac{1}{10}$, of an hour. Therefore, to find the average speed we compute $\frac{10}{\frac{1}{6}+\frac{1}{10}}$ to obtain 37.5 mph.

55. Answers may vary. The mode could be an extreme value in a data set (for example, the least value) and not represent the center at all.

57. Answers may vary. A set of data that contains wait time might be appropriate *if* the times were determined during peak people-period events. In that case, the mean with absolute mean deviation and standard deviation would be appropriate for this numerical data.

58. (b) In an arithmetic sequence we have a common difference, d, so that the sequence might appear as follows:

$$\begin{aligned}&a_1\\&a_2 = a_1 + d\\&a_3 = a_1 + 2d\\&a_4 = a_1 + 3d\\&\vdots\\&a_n = a_1 + (n-1)d\end{aligned}$$

The mean is $\frac{a_1 + (a_1 + d) + \ldots + (a_1 + (n-1)d)}{n}$

$$= \frac{na_1 + d(1 + 2 + 3 + \ldots + n - 1)}{n}$$

$$= \frac{na_1 + d\left(\frac{(1 + n - 1)(n - 1)}{2}\right)}{n}$$

$$= \frac{na_1 + \frac{d(n(n-1))}{2}}{n} = a_1 + \frac{d(n-1)}{2} = \frac{2a_1 + d(n-1)}{2}$$

$$= \frac{a_1 + [a_1 + d(n-1)]}{2} = \frac{a_1 + a_n}{2}$$

Cooperative Learning

61. Answers will vary. To do this problem, the class will have to be divided in such a way that the choice of newspapers does not overlap. A teacher may want to discuss how one might do a random sample of newspapers.

Review Problems

62. Answers vary; for example, the number of men in the workforce has increased by a factor of about 4 over the last hundred years, while the number of women in the workforce has increased by about a factor of 17. As late as 1960, there were more than 200% more men in the workforce than there were women. In 1970, this was no longer true, and in 2000 there were only about 15% more men.

63. Approximately, midsize—173°, large—25°, luxury—61°, small—101°

64. (a) Everest, approximately 8500 m **(b)** Aconcagua. Everest, McKinley

65. (b) History Test Scores

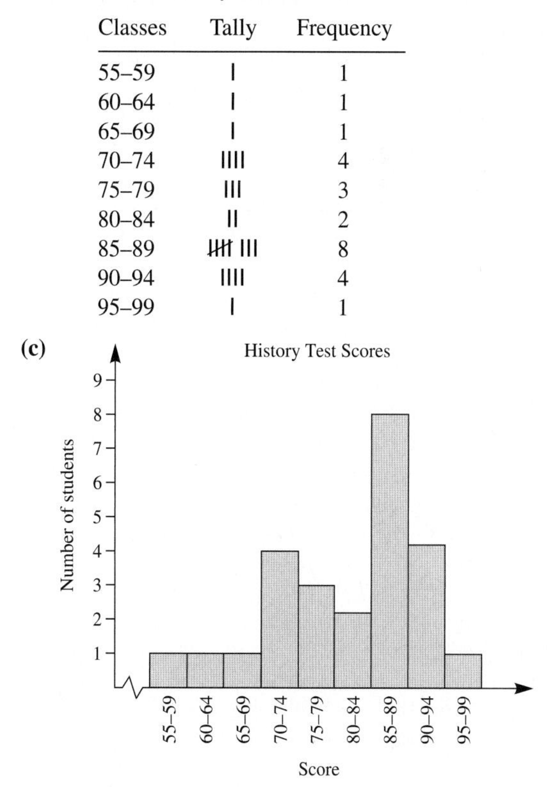

Classes	Tally	Frequency
55–59	|	1
60–64	|	1
65–69	|	1
70–74	||||	4
75–79	|||	3
80–84	||	2
85–89	~~||||~~ |||	8
90–94	||||	4
95–99	|	1

(c)

(d) Approximately 115°

Ongoing Assessment 8-3

1. Answers may vary in all parts of this question.
(a) A question to ask is whether the car is running. If it is not running, then there is no sound and the car would be quieter.
(c) 11% more fruit solids than 10% is very little.
(e) "Up to" is very indefinite, as are the conditions under which the 30 was obtained. 6 mpg is "up to" 30 mpg.
(g) How many dentists responded? **(i)** A question to ask is whether another airline flies to the city.

3. She could have taken a different number of quizzes during the first part of the quarter than the second part.
4. When the radius of a circle is doubled, the area is quadrupled, which is misleading since the population has only doubled.
5. The horizontal axis does not have uniformly sized intervals and both the horizontal axis and the graph are not labeled.
7. It could very well be that most of the pickups sold in the last 10 years were actually sold during the last 2 years. In such a case, most of the pickups have been on the road for only 2 years, and therefore the given information might imply, but would not substantiate, that the average life of a pickup is around 10 years.
9. Only part (b) might not be helpful.
10. Answers may vary. It is expected that the 892 respondents represent a large number of those surveyed and further that the 892 respondents are representative of the entire population of Americans.
11. The three-dimensional drawing distorts the graph. The result of doubling the radius and the height of the can is to increase the volume by a factor of 8.
12. No labels for vertical axis, so we cannot compare actual sales. Also, there is no scale on the vertical axis.
14. **(a)** False; prices vary only by $30. **(c)** True
15. **(a)** This bar graph could have perhaps 20 accidents at the point where the scale starts. Then 38 in 2005 would appear to be almost double the 24 of 2001, when in fact it is only 58% higher.
16. Answers may vary, but one such would be 5, 5, 5, 5, 5, 5, 100, 100. The mean would be 28.75 and the median 5. Neither are representative.
17. You could not automatically conclude correctly that the population of the West Coast has increased since 1790. However, based on the westward movement of the mean center of population, there would be strong suspicion that that was the case.
19. **(a)** Answers may vary. The survey depicts that teachers are generally happy with their mathematics textbooks. The articles (given without data) lead one to infer that national experts in mathematics are unhappy with textbooks that teachers use.
20. Answers may vary. **(a)** At grades K–4, homework of less than 1 hr (60 min) could be justified because 75% of the teachers surveyed assign homework that takes this long on the average. Perhaps 0–30 minutes for homework would be more appropriate at this grade level because allmost 50% of the surveyed teachers assign this amount. It is less easy to justify a specific amount for grades 5–8, but 71% of teachers surveyed use the range from 31–120 min of homework.
21. Answers may vary. **(a)** As a teacher at either level, if I were not using calculators/computers, I would question why more than 40% and 50% at the two respective levels were doing something I was not. Not knowing about the number of teachers surveyed and the population allows for misuse at all levels by all types of people. **(c)** As a parent, I would need to know how my child's teachers would have answered the survey before reading any conclusions pro or con on calculator/computer usage.
22. Answers may vary. **(b)** The data could be misused to justify actions if the state's middle-school teachers did not match the survey results.
23. Answers may vary. **(b)** A knowledge of the mean, absolute deviation, and standard deviation for a particular state's comparison would help avoid misuse.
24. Answers may vary. **(b)** There are many factors that affect teachers who quit. An informed counselor might seek specific data about the area or region where an advisee plans to teach.
25. Answers may vary, and the problem's information is true. **(c)** If the headlines are based on the same data set as indicated, the second headline is the most informative. The first reports a percentage with no base.
26. **(a)** $\frac{647}{649}$, or about 99.7% **(b)** $\frac{622}{649}$, or about 95.8%
(c) For the lung cancer group, the probability is $\frac{647}{1269}$, or about 51%. For the control group, it is $\frac{2}{29}$, or about 7%.
(d) There is a strong association because of the high probabilities. **(e)** The evidence is not conclusive.
28. Answers may vary. **(a)** You probably would not have a representative sample. This is a *convenience sample.*
(c) Assign students numbers from 1 to n, where there are n students. Choose your sample using random-number selection methods. Make sure your sample is big enough.
30. A student would need to know the highest possible score that a person could make. Also the scores of other students would be important.
32. You could report the mode of a selected number of spots if enough spots were chosen at random. It is also possible that the mode would not exist. A median might be misleading, depending on the number of data points given. Also the mean would not be sufficient. A report of the mean, median, and standard deviation would be the most helpful of all "averages" studied.
34. Answers may vary. A sample size is needed. Then, one way to pick a random sample of adults in the town is to use the telephone book or a voter registration list. These methods will not list all adults in the town, but these are probably the most accessible sets of data. To pick the sample, one might roll a die and consider the number n that appears on it. Then starting at some point in the adult list, choose every nth person after the start on the list.

Review Problems

36. (a) About 74.17 **(b)** 75 **(c)** 65 **(d)** About 237.97
(e) About 15.43 **(f)** 13
37. About 27.74
38. $76.\overline{6}$
39. Men's Olympic 100-m Run Times 1896–2000

9	7899
10	00011233334568888
11	00
12	0

10 | 0 represents 10.0 sec

40. Women's Olympic 100-m Swim Times 1960–2000

100-m Butterfly
100-m Freestyle
52 54 56 58 60 62 64 66 68 70
Time (sec)

From examining the box plot. we can see that the times on the 100-m butterfly are much greater (relatively speaking) than the times on the 100-m freestyle.

Chapter Review

1. If the average is 2.41 children, then the mean is being used. If the average is 2.5, then the mean or the median might have been used.
3. (a) Mean = 30, median = 30, mode = 10.
5. (a) Miss Rider's Class Masses in Kilograms

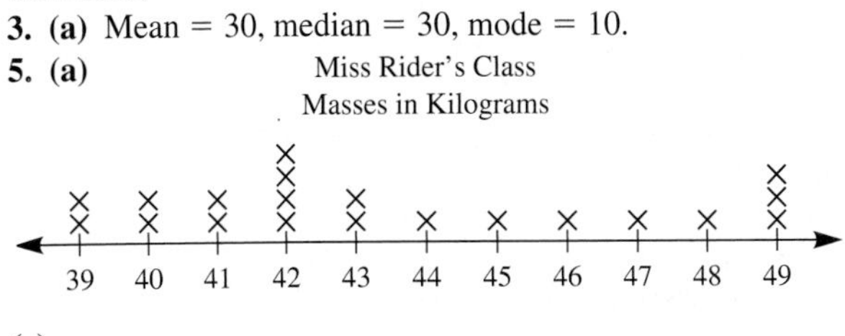

(c) Miss Rider's Class Masses in Kilograms

Mass	Tally	Frequency
39	II	2
40	II	2
41	II	2
42	IIII	4
43	II	2
44	I	1
45	I	1
46	I	1
47	I	1
48	I	1
49	III	3
		20

6. (a) Test Grades

Classes	Tally	Frequency
61–70	~~IIII~~ I	6
71–80	~~IIII~~ ~~IIII~~ I	11
81–90	~~IIII~~ II	7
91–100	~~IIII~~ I	6
		30

(b) Grade Distribution

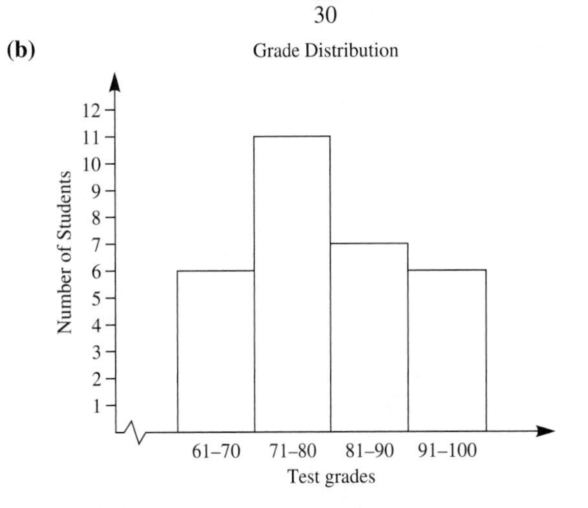

8. The widths of the bars are not uniform.
9. \$2840
11. (a) Life Expectancies of Males and Females

Females		Males
	67	1446
	68	28
	69	156
	70	0049
	71	02235578
	72	01145
	73	1689
7	74	147
9310	75	
86	76	
88532	77	
9999854332211	78	
98554210	79	

7 | 74 | represents 74.7 years old

| 67 | 1 represents 67.1 years old

(c) Because the interquartile range (as well as all data but one point) is significantly above that of men, women are expected to live longer.
12. Larry was correct because his average was $3.2\overline{6}$, while Marc's was $2.7\overline{3}$.
13. (a) 360 **(c)** 350
14. (a) 67 mph

(c)

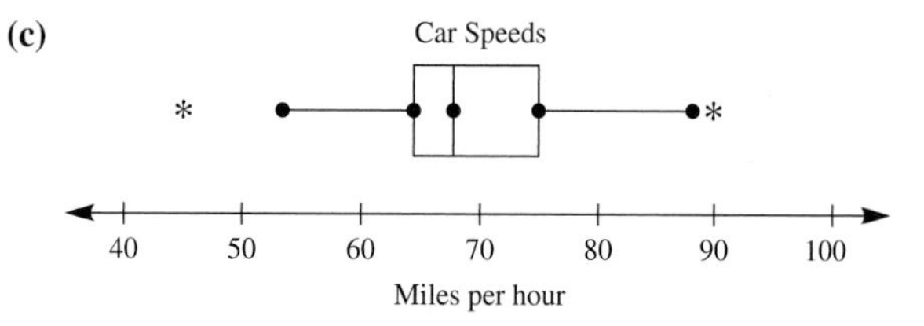

(e) 30% **(g)** about 50% of people checked drove between 65 and 75 mph. At least 25% sped.

15. (a) Positive **(c)** 67 in. **(e)** 50 lb

16. (a) Collette is more consistent. There is less deviation from the mean in her scoring.

(c) Collette scored more than might have been expected.

17. 2.5%

18. (a) The score for each **(c)** Same as (a) **(e)** 0

19. The bars would be the same height in a vertical bar graph.

20. (a) 19 lb **(c)** No. The first statement should be modified.

22. No reportable single number would be meaningful for such a pool.

23. Answers may vary. A reasonable way to make the claim is to pull boxes at random after they are packed and weigh the contents. If more than 98% of the boxes' contents have weight within 1g, the claim is reasonable.

24. (a) Answers will vary. Perhaps to be with those of comparable age a typical senior might choose the first.

25. Answers may vary. The validity is suspect, though kids may reflect parents' voting habits.

28. The advertisement is not reasonable. One would expect the snow to be okay, but with reports like those given, there is little information about the snow in the middle of the mountain. If it is cold enough to have 23 in. at the bottom of a hill, one would expect there to be snow all the way to the top. However, if the bottom and the top are shaded, then there could be much variation if a part of the hill were sunny.

29. (b) Between 83–84 and 84–85 **(d)** Answers vary; for example, you could change the scale on the y-axis to go up by 5s or 10s rather than by 20s.

30. (a)

U.S. Infant Mortality Rates 1980–99
(rates are per 1000 live births)

Females		Males
8655421	6	
8642	7	5678
9981	8	0038
631	9	34
720	10	038
2	11	02599
	12	38
6 | 6 | represents 6.6	13	19 | 13 | 1 represents 13.1

(c) Answers vary; for example, a double-bar graph showing male and female deaths for each year or a box plot showing male and female deaths would be appropriate.

31. (a) 25 **(c)** 16%

32. 475

33. The circle graph depicts a sector with $^{-}5\%$ indicated. The positive percentages sum to 105%. While the $^{-}5\%$ may indicate a loss in investments, the graph is inappropriate.

Answers to Now Try This

8-1 Ages of Presidents at Death

4	69
5	36778
6	003344567778
7	0112347889
8	01358
9	003

4 | 9 represents 49 years old

8-2 It appears that the fifth-period class did better. We can see that there are more scores grouped toward the bottom, which is where the higher grades are located. We can easily determine that the median for the second-period class is 78.5 and the median for the fifth-period class is 85, and this again shows that the fifth-period class did better.

8-3 (a) A histogram is more appropriate to compare numbers of data that are grouped into numerical intervals. With the line graph there appears to be a frequency for *Times per month* such as 4.5, etc.

(b) Connecting the dots is visually okay, but is mathematically meaningless.

(c) (i) A circle graph is appropriate for showing the division of a whole into parts.

(ii) A line graph is appropriate for showing how data values change over time.

8-4 (a) Answers vary, for example, all the puppies could weigh 7 lb. Then $(7 \cdot 7)/7 = 7$ lb. Another possibility is 4, 5, 6, 7, 8, 9, 10. The mean for these scores is also 7.

(b) If all the scores were equal, then the mean is equal to the high score and also the low score.

8-5 Answers may vary. For example, one set of data is 92, 94, 94, 94, 96 in which the mean, median, and mode are all 94.

8-6 (a) The average of 2.58 could only be a mean. To be a mode it would have to be a whole number. To be a median it would have to be a whole number or end in 5.

(b) The average of 2.5 could be a mean or a median. It could not be a mode since the mode for the number of children must be a whole number.

8-7 One possible set is 0, 0, 20, 20, 20, 70, 70, 90, 90, 90, 100, 100.

8-8 One probably would not expect any outliers because the lower end of the salary scale was dropped. One would expect all the five-number summary points to increase. There will probably be little noticeable difference in the story the data tell.

8-9 Because there is no base for the percentages given, arguments are difficult to make. Assuming the base is the set of all parents, consider the following:

(a) The argument cannot be made because we do not know even if the other 82% answered this part of the survey.

(b) The argument cannot be answered. The statement may be the result of considering test/real learning to be two equal parts that make 18%. Hence, half of 18% is 9%.

(c) The argument cannot be made; there is no cause-effect relationship shown.

(d) This argument cannot be made. Possibly it could be made for 11%.

(e) The argument cannot be made. The indicated 9% does not tell us about the other 91% of parents.

Answer to Technology Corner

Section 8-2

The mean is 46. The standard deviation is approximaely 8.85. The variance is approximately 78.32. The mean absolute deviation is 7.6.

Chapter 9

Ongoing Assessment 9-1

1. (a) In 12 ways, if different order implies different name
2. (a) For example: $\overleftrightarrow{BC}$ and $\overleftrightarrow{DH}$ or $\overleftrightarrow{AE}$ and $\overleftrightarrow{BD}$
(c) No **(e)** Point H **(g)** For example: $\overleftrightarrow{BF}$ and $\overleftrightarrow{DH}$
3. (a) $\angle EFB$ **(c)** Two planes are perpendicular if the measure of any of the four dihedral angles created by the planes is 90°.
4. (a) $\varnothing$ **(c)** A **(e)** $\overleftrightarrow{AC}$ and $\overleftrightarrow{DE}$ or $\overleftrightarrow{AD}$ and $\overleftrightarrow{CE}$
5. 14
7. (a) 110° **(c)** 20°
9. (a) (i) 41°31′10″ **(ii)** 79°48′47″ **(b) (i)** 54′
(ii) 15°7′48″
10. (a) (i) 90° **(ii)** 12°30′ **(iii)** 205°
(c) Approximately 32 min and 43.6 sec after 12 noon
12. (a) Let $m(\angle AOB) = x$. Then $x + 90° + 3x = 180°$, $x = 22°30'$. Thus, $m(\angle AOB) = 22°30'$, $m(\angle COD) = 67°30'$.
(c) $m(\angle AOB) = x$, $x + 2x + 4x + 20 = 180$, $x \doteq 22.857°$, $m(\angle AOB) \doteq 22.857°$, $m(\angle BOC) \doteq 45.714°$,
$m(\angle DOC) \doteq 91.428°$
13. $x \doteq 23.89°$, $y \doteq 31.86°$, $z \doteq 47.79°$, $w \doteq 76.46°$
14. (a) 90° **(b)** Always 90°.
15. (a) 6 **(c)** $2(n - 1)$
16. (a)

Number of Intersection Points

Number of Lines	0	1	2	3	4	5
2			Not possible	Not possible	Not possible	Not possible
3					Not possible	Not possible
4			Not possible			
5			Not possible	Not possible		
6			Not possible	Not possible	Not possible	

19. (a) No

20. (a) 1 **(c)** ${}_nC_3$, or $\dfrac{n(n-1)(n-2)}{3 \cdot 2 \cdot 1}$, or $\dfrac{n(n-1)(n-2)}{6}$

Communication

22. No; if all four points are collinear, then one line is determined; if only three are collinear, then four lines are determined; if no three are collinear, then six lines are determined.

Cooperative Learning

24. (a) An angle of 20° can be drawn by tracing a 50° angle and a 30° angle, as shown in the following figure. Another 20° angle adjacent to the first 20° angle can be drawn in a similar way, thus creating a 40° angle.

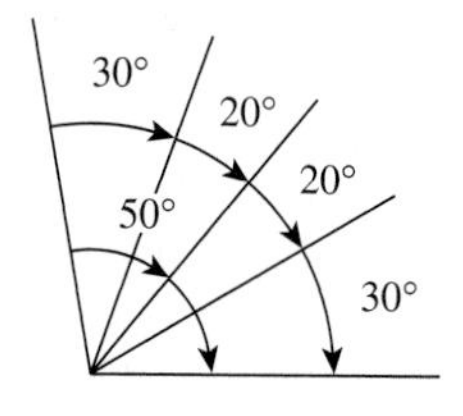

(c) All multiples of 10°

Ongoing Assessment 9-2

1. (a) 1, 4, 6, 7, 8 **(c)** 6, 7
3. A concave polygon
6. (a) 35 **(c)** 4850
8. (a) and (b) represent rhombuses and rectangles.
9. (a) T, Q, R, H, G, I, F, J **(c)** W, D, A, Z, U, E **(e)** Y
11. (a) A parallelogram. Answers vary; for example, opposite angles and opposite sides are congruent. **(b) (i)** Isosceles **(ii)** A right triangle **(iii)** An isosceles right triangle

Cooperative Learning

15. (a) The trapezoid definition is different.
(b)

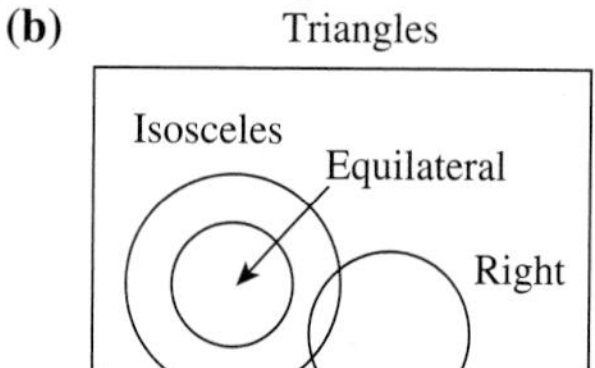

17. (a) Answers will vary. Typically the "sides" of a triangle on a globe are arcs of great circles.

Review Problems

18. (a) 45 **(b)** $n(n - 1)/2$
19. $\varnothing$, 1 point, 2 points, ray
20. (a) False. A ray has only one endpoint. **(b)** True
(c) False. Skew lines cannot be contained in the single plane.
(d) False. $\overrightarrow{MN}$ has endpoint M and extends in the direction of point N; $\overrightarrow{NM}$ has endpoint N and extends in the direction of point M. **(e)** True **(f)** False. Their intersection is a line.

Ongoing Assessment 9-3

1. 6
2. (a) 60° **(c)** 60°
3. (a) Yes. A pair of corresponding angles are 50° each. **(c)** Yes. A pair of alternate interior angles are 40° each.
5. (a) 20
6. (a) 70° **(c)** 65°
8. (a) 50° **(c)** 40°
9. (a) $x = 40°$ and $y = 50°$ **(c)** $x = 50°$ and $y = 60°$
10. (a) The marked angles are congruent corresponding angles. By Theorem 9–3, $k \parallel \ell$.

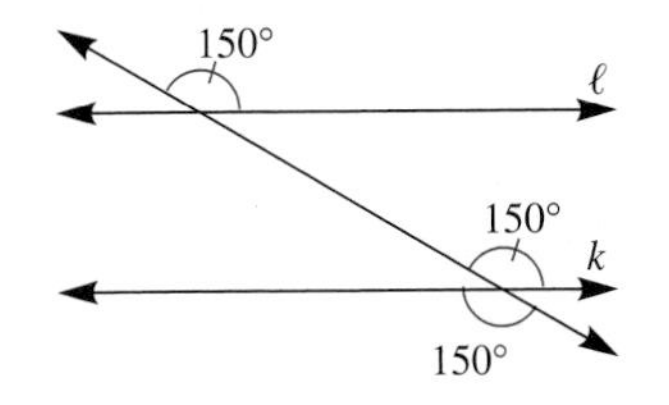

(b) One way is to extend $\overline{AC}$ to intersect k at D. Because x is the measure of an exterior angle in triangle BCD, $x = z + w$. But it is given that $x = y + z$. Hence, $z + w = z + y$, $w = y$. Because w and y are measures of alternate interior angles, $k \parallel \ell$.

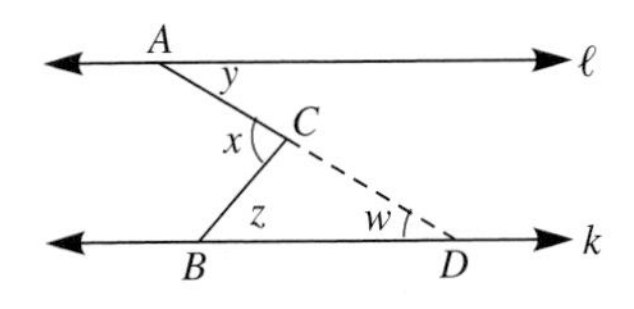

11. (a) 60°
12. (a) 360° **(c)** 360°
13. 60°, 84°, 108°, 132°, 156°
15. 111°
17. 135°
19. (a) $m(\angle ACB) = 40°$, $m(\angle ABD) = 40°$, $m(\angle DBC) = 50°$
21. 76°30′, 103°30′, 103°30′
23. (a) $x = 80°$, $y = 150°$ **(c)** $x = 36°$, $y = 108°$ **(d)** $x = 120°$ **(f)** $x = 120°$

Communication

25. (a) No. Two or more obtuse angles will produce a sum of more than 180°. **(c)** No. The sum of the measures of the three angles would be more than 180°.
27. (a) Five triangles will be constructed in which the sum of the angles of each triangle is 180°. The sum of the measures of the angles of all the triangles equals 5(180°), from which we subtract 360° (the sum of all the measures of the angles of the triangles with vertex P). Thus, $5(180°) - 360° = 540°$.
28. $x = 30°$. Possible explanation: The ladder makes an isosceles triangle with base angles of 60° (since the angle supplementary to 120° is 60°, as shown in the following figure). The bar across the ladder is parallel to the ground and hence creates corresponding angles formed by the bar, the ground, and the side of the ladder. Consequently, the angle formed by the bar and the side of the ladder is 60°. Since the line from the apple is perpendicular to the bar crossing the ladder, we have $x = 90° - 60° = 30°$. Hence, $x = 30°$.

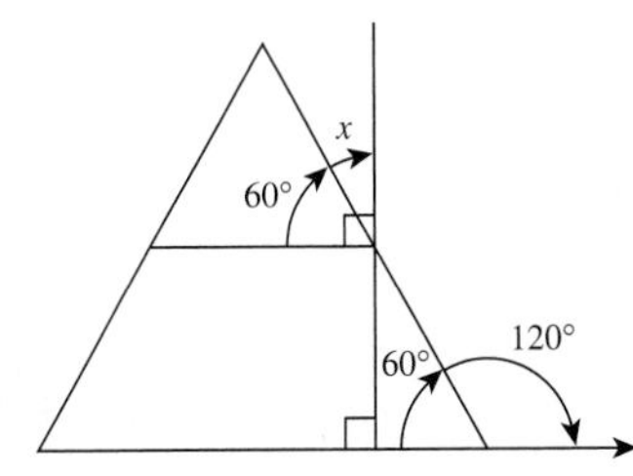

29. (a) Fold the isosceles triangle so the congruent sides match up. The two angles will fall on top of each other. **(c)** Choose a vertex and fold the triangle at that vertex so that the other two vertices fall on top of each other. Two angles of the triangle should fall on top of each other. Repeat at a second vertex.
30. (a) Divide the quadrilateral into two triangles: $\triangle ABC$ and $\triangle ACD$. The sum of the measures of the interior angles of each triangle is 180°, so the sum of the measures of the interior angles of the quadrilateral is 2(180°), or 360°. **(c)** Any concave pentagon can be divided into three triangles and a concave hexagon can be divided into four triangles. In general, it seems that any concave n-gon can be divided into $n - 2$ triangles so that the sum of the measures of the angles of the triangles is the sum of the measures of the interior angles of the n-gon.
32. Yes. From the hypothesis, we see that $\angle 1$ is supplementary to $\angle 2$. Because $\angle 3$ is also supplementary to $\angle 2$, we have $\angle 1 \cong \angle 3$. Thus, corresponding angles created by lines a and b and the transversal c are congruent. Consequently, a and b are parallel.
34.

$$\begin{aligned} m(\angle DCB) + m(\angle ABC) &= 180° - (m(\angle 3) + m(\angle 4)) \\ &\quad + 180° - (m(\angle 2) + m(\angle 1)) \\ &= 360° - (2m(\angle 3) + 2m(\angle 2)) \\ &= 360° - 2(m(\angle 3) + m(\angle 2)) \\ &= 360° - 2 \cdot 90° \\ &= 180° \end{aligned}$$

Thus by the result in problem 32, the lines CD and AB are parallel.

Open-Ended

39. Answers vary. The following are some properties of figures in a plane and corresponding properties on a sphere:

Property in a Plane	Property on a Sphere
Two lines in a plane may intersect in one point or may be parallel.	Two great circles (lines on a sphere) always intersect in two points.
Two points determine a unique line.	Two points do not always determine a unique great circle.
A line has no length, and any point on the line separates it into two parts.	A great circle has a finite length ($2\pi r$, where r is the radius of the sphere) and is not separated into two parts by a point on the circle.

Cooperative Learning

40. (a) If $m(\angle A) = \alpha$ and $m(\angle B) = \beta$, then $m(\angle D) = 180 - \left(\frac{\alpha}{2} + \frac{\beta}{2}\right)$.
41. (a) Because congruent supplementary angles are formed **(c)** When A and C are folded, congruent supplementary angles are formed and, hence, each is a right angle. When B is folded along $\overline{BB'}$, the crease $\overline{DE}$ formed is perpendicular to $\overline{BB'}$ (again because congruent supplementary angles are formed); consequently, $\overline{DE} \parallel \overline{GF}$. Thus, D and E are also right angles.

Review Problems

42. (a) All angles must be right angles, and all diagonals are the same length. **(b)** All sides are the same length, and all angles are right angles. **(c)** This is impossible because all squares are parallelograms.
43. (a) Two sets of parallel sides—B, C, D, E, F, G; one set of parallel sides—A, I and J (exactly one); No parallel lines: H
(b) Four right angles—D, F, G; exactly two right angles—I No right angles: A, B, C, E, H, J **(c)** Four congruent sides—B, C, F, G; Two pairs of congruent sides only—E, H, D; One pair of congruent sides only—J; No sides congruent—A, I **(d)** D, F, G, J
44. Answers will vary but should include the following:
- Congruent opposite sides describe a parallelogram.
- All sides congruent describe a rhombus (and square).
- Four right angles describe a rectangle (and square).
- Two pairs of consecutive sides congruent describe a kite.

Ongoing Assessment 9-4

1. (a) Quadrilateral pyramid
2. (a) A, D, R, W **(c)** $\triangle ARD$, $\triangle AWD$, $\triangle AWR$, $\triangle WDR$
(e) $\overline{DW}$
4. (a) 5 **(c)** 4
5. (a) True **(c)** True **(e)** False **(g)** False **(h)** True
6. All are possible.
9. (a) 7 cubes; 10 faces.
10. (a) Right hexagonal pyramid **(c)** Cube
(e) Right hexagonal prism
11. (a) Triangular right prism
13. (a) iv
14. (a) i, ii, and iii
16. (a)

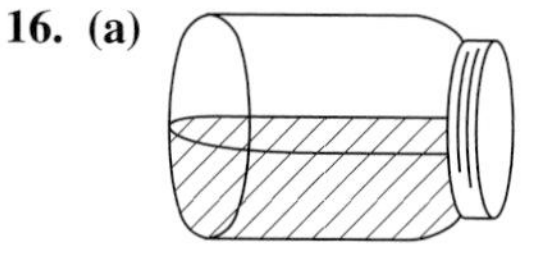

17. (a) (2)
18. (a) **(c)** **(e)**

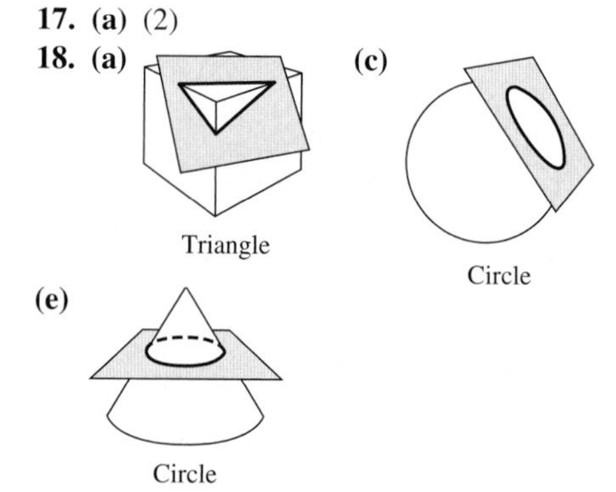

19. (a) Parallelogram, rectangle, square, triangle
20. (a) A rectangle whose opposite sides are diagonals of two opposite faces and the other two sides, edges of the cube
(b) No. Answers vary.
23. (a) Three pairs of faces determined by three pairs of opposite sides of a hexagonal base
24. (a) $10 + 7 - 15 = 2$
26.

Pyramid	Prism
(a) $n + 1$	$n + 2$
(c) $2n$	$3n$

Communication

28. Answers vary; for example, a cone is a many-sided pyramid and a cylinder is a many-sided prism.

29. Both could be drawings of a quadrilateral pyramid. In **(a)**, we are directly above the pyramid, and in **(b)**, we are directly below the pyramid.

31. **(a)** A rectangle since $\angle HMN$ is a right angle

Cooperative Learning

35. **(a)** Parallelogram, rectangle, square, scalene triangle, isosceles triangle, equilateral triangle

Review Problems

36. Yes because the sum of the angles of a triangle is 180°.

37. $m(\angle BCD) = 60°$.

38. 140°

39. **(a)** True **(b)** True **(c)** False (e.g., equilateral triangle)

40. Lines in the same plane and perpendicular to the same line are parallel. This is because corresponding angles are right angles and hence are congruent.

Ongoing Assessment 9-5

1. All except (d), (f), and (i) are traversable; (a) and (j) are Euler circuits (starting and stopping points are the same)

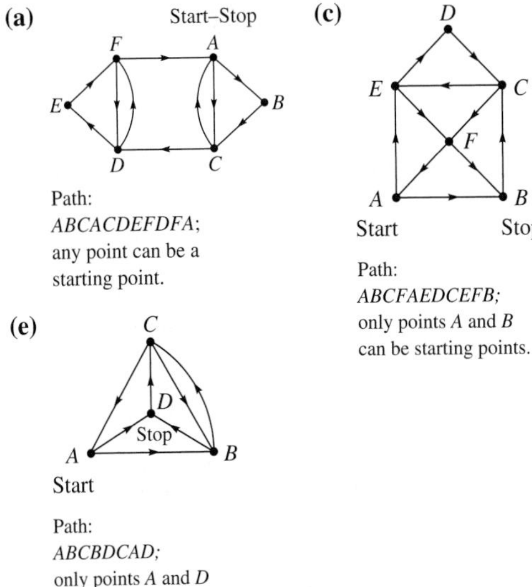

(i) Not traversable; has more than two odd vertices

3.

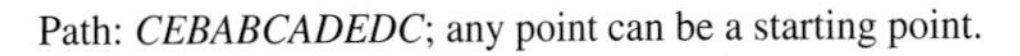
Path: *CEBABCADEDC*; any point can be a starting point.

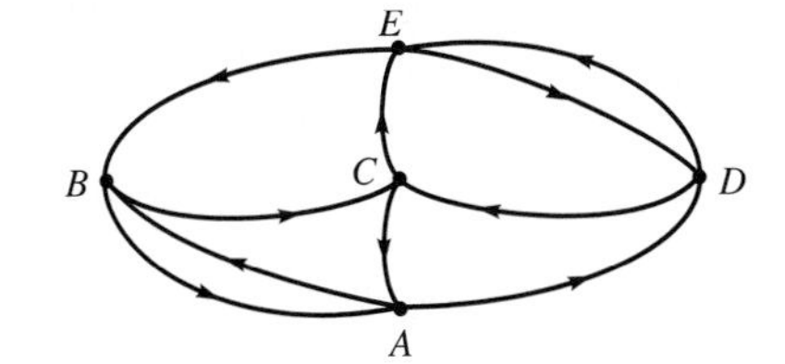

4. **(a)**

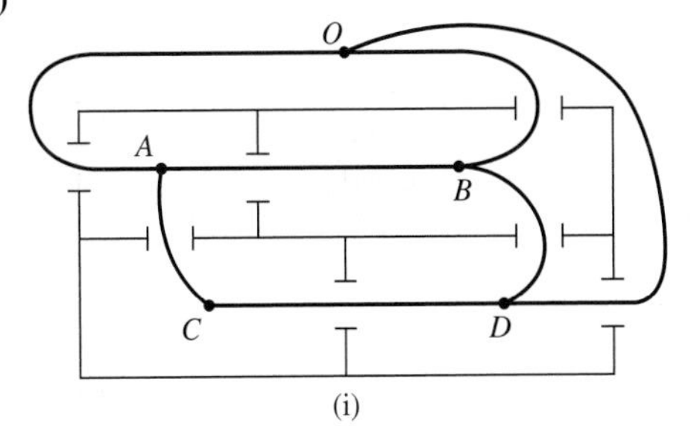

(i)

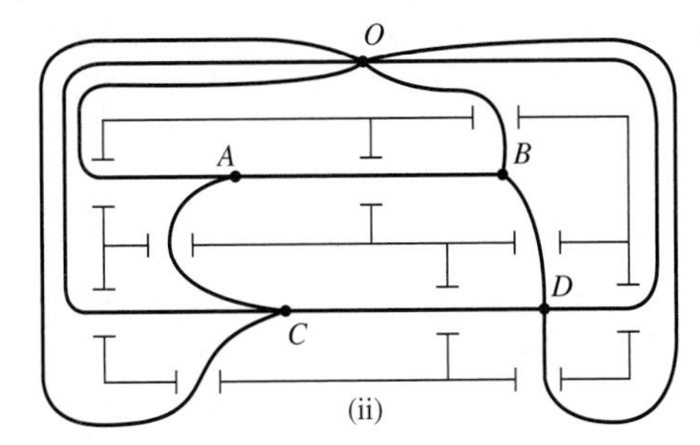

(ii)

6. It is not possible.

8.

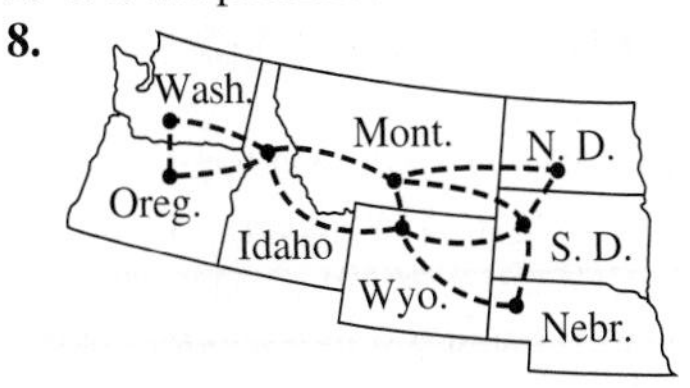

All vertices are even, so the trip is possible. It makes no difference where she starts.

Communication

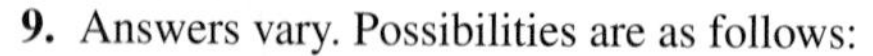

9. Answers vary. Possibilities are as follows:

(a)

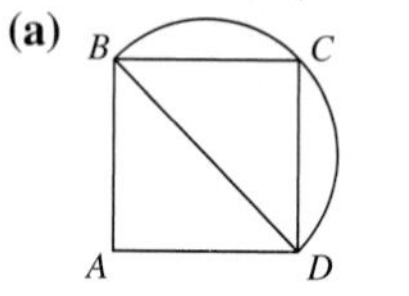

10. The bridge could be built anywhere and we would have two odd vertices. Thus, the network would be traversable.

Open-Ended

12. (a) A postal worker would have to travel each route twice (both sides). A traveling salesperson would only be interested in vertices (cities or stores). A highway inspector would travel each arc only once.

Chapter Review

1. (a) $\overleftrightarrow{AB}$, $\overleftrightarrow{BC}$, and $\overleftrightarrow{AC}$ **(c)** $\overline{AB}$
2. (a) Answers may vary. **(c)** $\overleftrightarrow{AQ}$
4. (a) No. The sum of the measures of two obtuse angles is greater than 180°, which is the sum of the measures of the angles of any triangle.
5. (a) 132°11′ **(c)** 132°21′
6. 18°, 36°, 126°
7. (a) Given any convex n-gon, pick any vertex and draw all possible diagonals from this vertex. This will determine $n - 2$ triangles. Because the sum of the measures of the angles in each triangle is 180°, the sum of the measures of the angles in the n-gon is $(n - 2)180°$.
10. $m(\angle 3) = m(\angle 4) = 45°$
12. (a) 60° **(c)** 120°
13. (a) Alternate interior angles are congruent.
14. (a) $5 \cdot 180° - 360°$, or $3 \cdot 180$ **(c)** Answer vary. One way is to connect B with E and A with F. We get two quadrilaterals and a triangle. Thus, the sum of all the measures of the interior angles of the polygon is $2 \cdot 360° + 180°$, or 900°.
15. $m(\angle 1) = m(\angle 2) = 65°$, $m(\angle 3) = m(\angle 4) = 115°$, $m(\angle 5) = 70°$.
17. (a) $\angle 3$ and $\angle 4$ are supplements of congruent angles.
18. (a) The measure of one of the dihedral angles formed by planes α and γ is $m(\angle APS) = 90°$. The measure of one of the dihedral angles formed by planes β and γ is $m(\angle BPQ) = 90°$ because line PQ is perpendicular to line AB (given).
21. (a) ${}_{10}C_3 = \dfrac{10 \cdot 9 \cdot 8}{3!} = 120$. The number of triangles equals the number of ways of choosing 3 points out of 10 points where order is not important.
22. (a) $\dfrac{360}{20}$, or 18 **(b)** Does not exist, $25 \nmid 360$.
(c) Does not exist, the sum is always 360°.
(d) Does not exist, the equation $\dfrac{n(n-3)}{2} = 4860$, or $n(n - 3) = 9720$, has no solution in integers because if $n = 100$ then $100 \cdot 97 = 9700$ and if $n = 101$ then $101 \cdot 98 = 9898 > 9700$.
23. (a) 188° **(b)** Approximately 169.8°
26. 48°
27. Answers may vary.
(a) **(c)** **(e)**

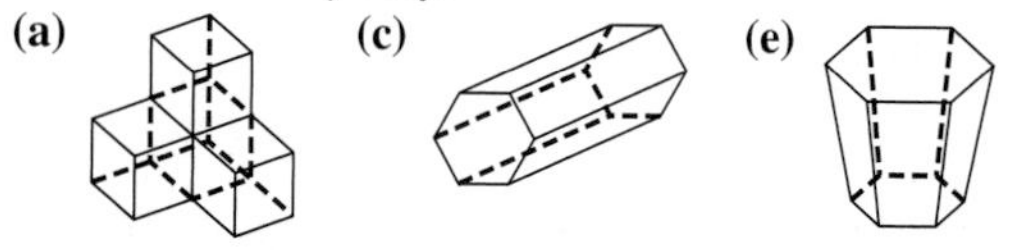

31. (a) (i), (ii), and (iv)

Answers to Now Try This

9-1 (a) Exactly one line can be drawn through any two points.
(b) Skew lines cannot be parallel. By definition, parallel lines are in the same planes. Skew lines are not.
(c) Let O be the center of the globe. First consider two points A and B on the globe which are the endpoints of a diameter, that is, A, O, and B are collinear. There are infinitely many planes containing A and B and hence the center of the globe, O. Each of these planes intersects the globe in a "great" circle. Consequently if A, O, B are collinear, there is no unique "line" through A and B.

9-2 Statement 5 follows from statements 1 and 3. The two points on a line and a third noncollinear point determine a plane. Statement 6 also follows from statements 1 and 3. Two points are from one line and the other from the line parallel to it.
9-3 The problem on page 501 gives the number of lines that can be drawn through n points. This problem is a model for the number of handshakes that take place if everyone shakes hands with everyone else. Everyone shakes hands with everyone except himself. Because we don't want to count one hand shake twice, we need to divide by 2. The number of handshakes is $\dfrac{n(n-1)}{2}$.
9-4 $8.42° = 8° + 0.42(60)' = 8°25.2' = 8°25' + (.2)(60)'' = 8°25'12''$
9-5 (a) It is possible for a line intersecting a plane to be perpendicular to only one line in the plane. Drawings will vary.
(b) It is not possible for a line intersecting a plane to be perpendicular to two distinct lines and not be perpendicular to the plane.
(c) Yes. If a line intersects a plane in point P and is perpendicular to two lines in the plane through P, then it is perpendicular to every line in the plane through P.

9-6 Answers will vary

9-7 One approach to this problem is to start shading the area surrounding point X. If we stay between the lines, we should be able to decide whether the shaded area is inside our outside the curve.

The shaded part of the figure below indicates that point X is located outside the curve.

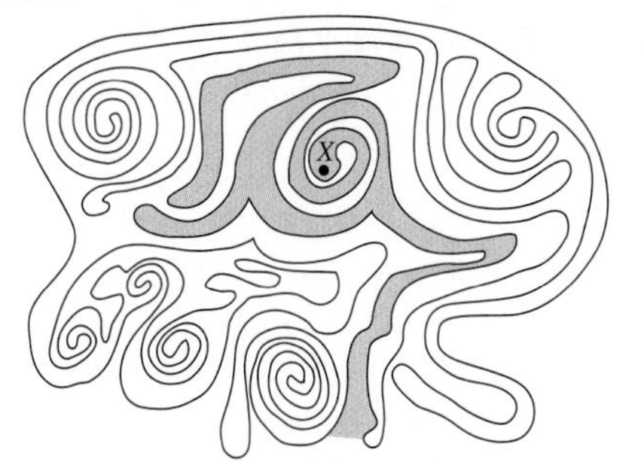

9-8 (a) 1. A square is a regular rhombus because the sides and angles are congruent.
2. Answers may vary
3. No. Triangles don't have "opposite sides."

(b) The trapezoid is defined differently.

(c) Answers will vary.

9-9 (1) True. From the definition (Table 9-6) we know that an equilateral triangle has three congruent angles. Therefore it has at least two congruent sides and hence it is isosceles.

(2) True. By definition, a regular quadrilateral is a four-sided figure that is both, equilateral and equiangular. A square has four sides of equal length and four right angles—so a square is a regular quadrilateral.

(3) True. Because a rhombus is a parallelogram and a rectangle is a parallelogram with a right angle, if one angle of a rhombus is a right angle then all of its angles are right.

(4) True; follows from part (3).

(5) True, because a rectangle is a parallelogram and if one angle of a parallelogram is a right angle all its angles are right angles.

(6) True. A rectangle is a trapezoid since it is a quadrilateral with at least one pair of parallel sides.

(7) True. A square is a rectangle that is an isosceles trapezoid. Also a square is a kite. Hence there are isosceles trapezoids which are kites.

(8) False as shown in the figure below.

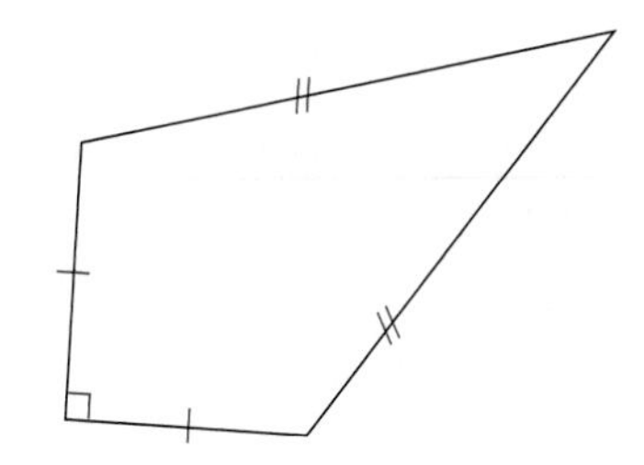

9-10 (a) The amount of turning that takes place when one "walks around the triangle and ends up pointed in the same direction in which you started" is 360°. Thus the sum of the measure of the exterior angles of the triangle is 360°.

(b) With both the exterior and interior angles of a triangle, there are three sets of supplementary angles having a total sum of 540° for their measures. Subtracting the sum of the measures of the exterior angles from the total sum, we have $540° - 360°$, or 180° for the sum of the measure of the interior angles of the triangle.

(c) The sum of the measures of the exterior angles for any convex polygon is 360° for the same reason as given in part (a).

9-11 (a) There are infinitely many. Consider the number of lines of longitude on the globe.

(b) A spherical triangle could have as vertices the North Pole, Quito, Ecuador, and Kisumu, Kenya.

(c) Consider any triangle with the North Pole as one vertex and the other two vertices on the equator such that the angle formed at the North Pole is a right angle.

9-12 Any convex polygon of n sides can be separated into $n - 2$ triangles by drawing all the diagonals from one vertex of the polygon. The sum of the measures of these triangles is $180(n - 2)$. The result is the same as that of Theorem 9-6.

9-13 (a) $V + F - E = 2$

(b) The relationship is true for any polyhedron.

9-14 (a) There can be no traversable network with more than two odd vertices. An odd vertex must be either a starting point of a stopping point of the path.

(b) This is not possible.

Chapter 10

Ongoing Assessment 10-1

1. Otherwise the definition applies to three-dimensional space and thus a sphere.
3. **(a)** $m(\angle A) > m(\angle B)$ **(b)** The side of greater length is opposite the angle of greater measure.
4. **(c)** A scalene right triangle **(d)** Not possible
6. 22
7. **(a)** Yes; SAS **(c)** No
9. **(a)** The lengths must be the same because they are corresponding parts of congruent triangles.
10. No; there could be no right angle if the triangle in equilateral.
12. The ratio of corresponding sides of congruent figures must be 1:1. That is not the case here.
14. The circumcircles must be contentric. The perpendicular bisectors of the sides of both the large and small triangles intersect at the same point, the circumcenter.
16. They must meet at the center of the circumcircle. The radius of that circle is the length of a side of the hexagon. It is the only point equidistant from all vertices.
20. **(b)** $\triangle ABC \cong \triangle BCA$, so $\angle A \cong \angle B$ and $\angle B \cong \angle C$, implying $\angle A \cong \angle B \cong \angle C$.
21. **(a)** One method is to start at any point on the circle and with a compass opened to the length of the radius of the circle mark off equally spaced points; there will be six equal arcs. **(c)** Congruent by SSS **(e)** Anywhere on the circle construct a chord $\overline{AB}$ congruent to the radius of the circle, then construct $\overline{BC}$, $\overline{CD}$, $\overline{DE}$, and $\overline{EF}$ all congruent to the radius of the circle; the hexagon $ABCDEF$ is a regular hexagon.
22. **(a)** F is the midpoint of both diagonals ($\triangle ABC \cong \triangle ADC$ by SSS and $\angle CAB \cong \angle CAD$ by CPCTC) and $\triangle BAF \cong \triangle DAF$ by SAS $\Rightarrow \overline{BF} \cong \overline{DF}$, so F is a midpoint; a similar argument will show $\overline{AF} \cong \overline{CF}$.
24. The side of one cube is congruent to a side of the other cube.
26. **(a)** 6 **(c)** $n(n-1)(n-2) \cdot \ldots \cdot 3 \cdot 2 \cdot 1$, or $n!$
28. Perimeters are equal. If sides are congruent the sums of their lengths must be the same.
30. **(a)** Yes **(b)** Yes
31. **(a)** Only six points are determined.
32. A circle can be circumscribed about the sign..
33. **(a)** A square is formed.
34. **(a)** The perpendicular bisector of $\overline{MN}$ contains O.

Communication

35. P is not on the circle because it doesn't fit the definition.
37. **(a)** $\triangle ABC \cong \triangle ADC$ by SSS. Hence, $\angle BAC \cong \angle DAC$ and $\angle BCM \cong \angle DCM$ by CPCTC. Therefore, $\overleftrightarrow{AC}$ bisects $\angle A$ and $\angle C$. **(c)** By part (b) $\overline{BM} \cong \overline{MD}$; CPCTC.
38. **(a)** This depends upon where the concavity is. If point C lies in the interior of $\triangle ABD$, part (a) is consistent. **(b)** For part (b) to hold, the diagonals must be extended in order to meet. By extending and allowing one diagonal to be outside the figure, the result holds. **(c)** See parts (b) and (a). With these conditions, the result is true.
40. Construct perpendicular bisectors of the sides. They meet at the center of the circumcircle.

Open-Ended

41. **(a)** Answers may vary. Some objects are tins of food and floor tiles. **(b)** Answers may vary. Photographs and their enlargements, an original and its projected image in an overhead projector, and a slide and its image are examples.
43. **(a)** $\triangle AFB \cong \triangle CED$ by SAS because $\overline{AF} \cong \overline{CE}$, $\overline{FB} \cong \overline{ED}$, and $\angle F \cong \angle E$ (each is a right angle). Hence, $\overline{AB} \cong \overline{CD}$ (corresponding parts in congruent triangles).

Ongoing Assessment 10-2

1. **(d)** Infinitely many are possible.
2. **(a)** No; by ASA, the triangle is unique. **(b)** No; by AAS, the triangle is unique. **(c)** No; by ASA, the triangle is unique. **(d)** Yes; AAA determines a unique shape, but not size.
4. **(a)** **(i)** ASA: If one leg and an acute angle of one right triangle are congruent, respectively, to a leg and an acute angle of another right triangle, the triangles are congruent. **(ii)** AAS: If the hypotenuse and an acute angle of one right triangle are congruent, respectively, to the hypotenuse and an acute angle of another right triangle, the triangles are congruent.
5. **(a)** Yes; ASA **(c)** No; SSA does not assure congruence.
7. **(a)** Parallelogram **(c)** None **(e)** Rhombus **(g)** Parallelogram
9. **(a)** True **(c)** True **(e)** True **(g)** True **(i)** False. A square satisfies all conditions of a trapezoid, so some trapezoids must be squares.
10. **(c)** No; any parallelogram with a pair of right angles must have right angles as its other pair and hence it must also be a rectangle.
11. The sides fit the definition. Also, $\triangle ABD \cong \triangle CBD$ by SSS. So $\angle BDA \cong \angle BDC$ by CPCTC. This implies $\angle ADE \cong \angle CDE$ because they are supplements of congruent angles. Now $\triangle ADE \cong \triangle CDE$ by SAS. Hence, $\angle DEA \cong \angle DEC$.

Because these congruent angles are supplementary, they have measure 90° each. Therefore, $\overline{BE} \perp \overline{AC}$, so the diagonals of the chevron are congruent. Thus $ABCD$ is a kite.

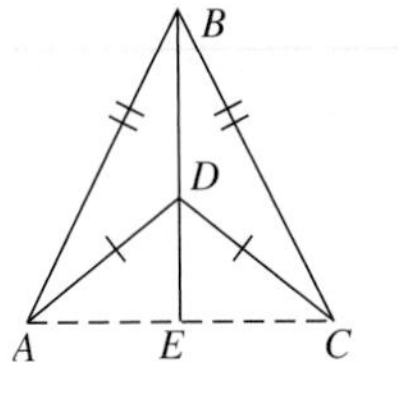

12. (a) Rectangle
14. (b) Yes; they are congruent rectangles.
15. There are five possibilities: one parallelogram and four kites.
16. A rhombus, because all the sides are congruent
18. (a) Kite. $\overline{DX} \cong \overline{AX}$; $\angle D$ and $\angle A$ are right angles; $\overline{DP} \cong \overline{AQ}$. Thus, $\triangle PDX \cong \triangle QAX$ by SAS. Hence, $\overline{PX} \cong \overline{QX}$ by CPCTC. $\overline{DC} \cong \overline{BA}$ as opposite sides of a rectangle. $\overline{PD} \cong \overline{QA}$. Hence, $\overline{CP} \cong \overline{BQ}$. $\overline{CY} \cong \overline{BY}$, given; and $\angle C$ and $\angle B$ are right angles. Therefore, $\triangle CYP \cong \triangle BYQ$ by SAS so that $\overline{QY} \cong \overline{PY}$ and $PXQY$ is a kite.
19. *Hint:* Make the quadrilateral concave.
21. (a) Answers may vary. For example,

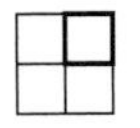

22. The answer is yes. If a quadrilateral had three obtuse angles, it could appear as shown.

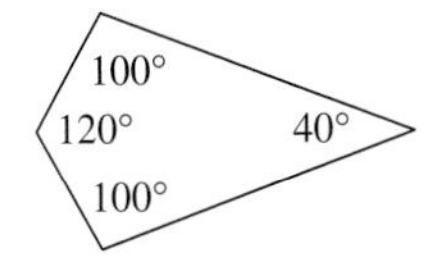

25. (a) $\overline{TO} \cong \overline{AO}$, radii of the same circle, are congruent. $\overline{CO} \cong \overline{CO}$. $\triangle CTO \cong \triangle CAO$ by hypotenuse-leg. $\angle TCO \cong \angle ACO$, CPCTC. Therefore, $\overline{OC}$ bisects $\angle TCA$.
27. (a) The sides which are not bases in an isosceles trapezoid are congruent. **(b)** The diagonals are congruent.
28. $h = \dfrac{a - b}{2}$
30. (b) Parallelogram **(c)** Suppose $ADCB$ in part (a) is a parallelogram; by SAS $\triangle EDH \cong \triangle GBF$, so $\overline{EH} \cong \overline{GF}$. Similarly, $\triangle ECF \cong \triangle GAH$, so $\overline{EF} \cong \overline{GH}$; by SSS $\triangle EFG \cong \triangle GHE$. Now $\angle GEH \cong \angle EGF$ and $\overline{FG} \| \overline{EH}$. Similarly, $\overline{EF} \| \overline{HG}$.
31. (b) The lengths of the sides of two perpendicular sides of the rectangles must be equal. **(c)** Answers may vary; one solution is that two adjacent sides must have equal lengths and the included angle of one must be congruent to the other.
33. (a) Use the definition of a parallelogram and ASA to prove that $\triangle ABD \cong \triangle CDB$ and $\triangle ADC \cong \triangle CBA$.
(c) *Hint:* Prove that $\triangle ABF \cong \triangle CDF$.

Communication

34. The triangle formed by Stan's head, Stan's feet, and the opposite bank is congruent to the triangle formed by Stan's head, Stan's feet, and the spot just obscured by the bill of his cap. These triangles are congruent by ASA since the angle at Stan's feet is 90° in both triangles, Stan's height is the same in both triangles, and the angle formed by the bill of his cap is the same in both triangles. The distance across the river is approximately equal to the distance he paced off, since these distances are corresponding parts of congruent triangles.
36. The student is incorrect *unless* the figure is also a rhombus. In general, the pairs must be distinct (have distinct components).

Open-Ended

37. (a) All rolls of wallpaper are not congruent. They come in different widths and lengths. Typically with the same pattern, they are congruent. **(c)** If rolls were congruent and there was a zero drop length in the pattern, then there would be less waste in cutting and matching patterns.

Review Problems

39. The triangles that are congruent to triangle ABC are triangles AED, CDE, EAB, and BCD. They are all congruent by SAS. Students may need to cut these triangles out and compare shapes.
40. Constructions
41. Constructions
42. (a) Yes; SAS **(b)** Yes; SSS **(c)** No

Ongoing Assessment 10-3

2. Constructions. Paper folding is a tactile approach to the problem. The Mira is easy to use when the paper on which the constructions are to be performed may not be altered. Using a compass and straightedge is the classical way to do constructions. The Geometer's Sketchpad demands that exact measurements be used on the screen.
3. (b) The altitude is the extension of the cable from vertex A to the ground.
4. (a) The perpendicular bisectors of the sides of an acute triangle meet inside the triangle. **(c)** The perpendicular bisectors of the sides of an obtuse triangle meet outside the triangle.
5. (a) This point is equidistant from all vertices because it is on all three perpendicular bisectors. Being at the intersection

of two of the perpendicular bisectors forces the point to be equidistant from all three vertices.

7. (a) If the rectangle is not a square, it is impossible to construct an inscribed circle. The angle bisectors of a rectangle do not intersect in a single point. **(c)** Possible. The angle bisectors intersect at a single point.

9. Answers may vary; for example:
(i) Draw a line segment (10¢).
(ii) Draw two intersecting arcs (20¢) to construct a perpendicular segment (10¢).
(iii) With compass point at the intersection of the two segments, sweep a wide arc (10¢) intersecting both segments.
(iv) Maintain the same compass setting and measure an arc from each of these points to determine the fourth point (20¢).
(v) Draw the two segments to complete the square (20¢).
The total cost is 90¢.

12. 4″

14. r

15. (a) If the parallelogram is not a rectangle, cut along any altitude. If the paralellogram is a rectangle, cut along any line through the point where the diagonals meet (where the line is not a diagonal and is not parallel to any side).

16. (a) Reassemble as shown.

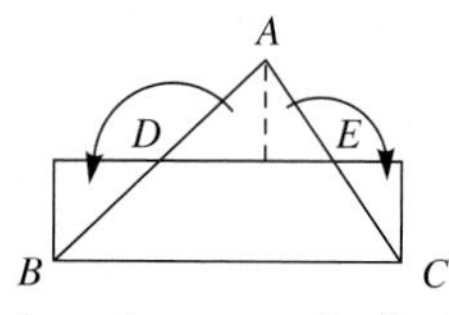

17. The center is where the perpendicular bisectors of the sides meet (the incenter). The radius is the distance from the incenter to a side.

18. $x = 3''$

19. (a) $\overrightarrow{PQ}$ is the perpendicular bisector of $\overline{AB}$.
(c) $\overrightarrow{PQ}$ is the angle bisector of $\angle APB$; $\overrightarrow{QC}$ is the angle bisector of $\angle AQB$.

21. Trapezoid because $\overline{XT} \parallel \overline{IO}$

22. Rhombus; see problem 31.

23. (b) Possible. Construct two perpendicular segments bisecting each other and congruent to the given diagonal.
(d) There is no unique parallelogram. Without the angle between the sides, there are infinitely many parallelograms that could be constructed from the two given sides. **(f)** Not possible because the sum of the measurements of the angles would be greater than 180°.

24. (a) Construct an equilateral triangle and bisect one of its angles. **(c)** Add 30° and 15° angles or bisect a right angle.
(e) Add 60°, 30°, and 15° angles.

26. (a) Because the triangles are congruent, the acute angles formed by the hypotenuse and the line are congruent; the corresponding angles are congruent, so the hypotenuses are parallel (the line formed by the top of the ruler is the transversal).

28. (a) The point is determined by the intersection of the angle bisector of $\angle A$ and the perpendicular bisector of $\overline{BC}$. Because the point is on the angle bisector of $\angle BAD$, it is equidistant from its sides. Because it is on the perpendicular bisector of $\overline{BC}$, it is equidistant from B and C. **(c)** The point is determined by the intersection of the perpendicular bisectors of $\overline{AB}$ and $\overline{BC}$.

29. Construct two perpendicular diameters. Their endpoints will determine a square.

Communication

32. Answers may vary. Most students will probably lobby to have other tools included as construction tools.

33. Answers may vary. Miras may lead to more precision, but the cost is more.

Open-Ended

36. (a) Concentric circles are circles with the same center.

37. Most students will probably say that there are more perpendiculars from a point to a line using the North Pole as a point and the equator as a line. All lines of longitude intersect at the North Pole and all are perpendicular to the equator.

Review Problems

41. $\triangle ABC \cong \triangle DEC$ by ASA ($\overline{BC} \cong \overline{CE}$: $\angle ACB \cong \angle ECD$ as vertical angles, and $\angle B \cong \angle E$ as alternate interior angles formed by the parallels AB and ED and the transversal $\overleftrightarrow{EB}$). $\overline{AB} \cong \overline{DE}$ by CPCTC.

42. Construction

43. AB in the trapezoid shown equals CD:

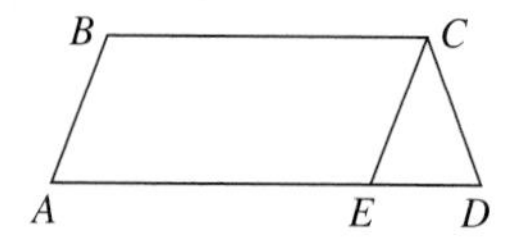

(a) Draw a line through C parallel to $\overline{AB}$; let E be the point where the line intersects $\overline{AD}$. $ABCE$ is a parallelogram ($\overline{BC} \parallel \overline{AE}$ and $\overline{AB} \parallel \overline{EC}$): in a parallelogram opposite sides are congruent (properties of a parallelogram). $\overline{AB} \cong \overline{CE}$, then, along with the hypothesis $\overline{AB} \cong \overline{CD}$, $\overline{CE} \cong \overline{CD}$. Thus, $\triangle ECD$ is an isosceles triangle. $\angle E \cong \angle D$; $\angle E$ and $\angle A$ are corresponding angles formed by the parallel segments $\overline{AB}$ and $\overline{CE}$ and the transversal $\overline{AD}$. Thus, $\angle A \cong \angle E$, so $\angle A \cong \angle D$.
(b) $\triangle ABD \cong \triangle DCA$ by SAS since $\overline{AB} \cong \overline{DC}$, $\overline{AD} \cong \overline{DA}$, and $\angle BAD \cong \angle CDA$, so $\overline{BD} \cong \overline{CA}$ by CPCTC.

44. The triangles are not necessarily congruent.

Ongoing Assessment 10-4

2. (a) Yes; AAA **(e)** Yes; radii are proportional. **(f)** No
(g) Yes; sides are proportional and angles are congruent.

4. (c) The triangles are similar when their corresponding sides are proportional.

5. **(c)** Thetriangles are similar if in $\triangle ABC$ and $\triangle DEF$, $\dfrac{AB}{DE} = \dfrac{AC}{DF}$ and $\angle A \cong \angle D$.

6. Answers may vary. **(a)** Two rectangles, one of which is a square and the other is not

8. **(b)** **(i)** $\dfrac{2}{3}$

(ii) $\dfrac{1}{2}$

(iii) $\dfrac{3}{4}$

(iv) $\dfrac{3}{4}$

9. **(b)** $\dfrac{24}{7}$ **(c)** $\dfrac{16}{5}$ **(e)** $\dfrac{8}{3}$

11. **(a)** **(i)** $\triangle ABC \sim \triangle ACD$ by AA because $\angle ADC$ and $\angle ACB$ are right angles and $\angle A$ is common to both.
(ii) $\triangle ABC \sim \triangle CBD$ by AA because $\angle CDB$ and $\angle ACB$ are right angles and $\angle B$ is common to both.
(iii) Using **(i)** and **(ii)**, $\triangle ACD \sim \triangle CBD$ by the transitive property.

14. **(a)** The triangle can be proved similar by AA.

16. $\triangle ACO \sim \triangle BDO$ by AA. Thus, $\dfrac{a}{b} = \dfrac{d}{c}$, which implies $ac = bd$.

18. Answers may vary. For example,

19. **(a)** Because M and N are midpoints of $\overline{BC}$ and $\overline{AC}$, respectively, $\overline{MN} \parallel \overline{BA}$ and $MN = \dfrac{1}{2}BA$. Also, $\angle BAM \cong \angle AMN$ and $\angle ABN \cong \angle BNM$. Thus, $\triangle ABO \sim \triangle MNO$ by AA.

21. Without knowing the lengths of the rulers, one cannot tell. However, 1″ corresponds to 2.54 cm, so it is possible to have them similar with ratio $\dfrac{1}{2.54}$.

23. $133\frac{1}{3}\%$, but most copy machines will not allow this setting

24. 15 m

25. 9 m

27. About 232.6 in., or 19.4 feet

28. **(a)** **(i)** Connect B with D and then apply Theorem 10–8 to triangles ABD and BCD. Let P be the midpoint of $\overline{MN}$.
(ii) Theorems 10–7 and 10–8 imply that $MP = \dfrac{1}{2}a$ and $PN = \dfrac{1}{2}b$. Thus, $MN = MP + PN = \dfrac{1}{2}a + \dfrac{1}{2}b = \dfrac{1}{2}(a + b)$.

29. **(c)** Rhombus

31. $\dfrac{1}{k}$; their sum (i.e., the perimeter) must be in the same proportion.

33. No; cross sections would be circular (and so similar) if cut parallel to the bases but could also be rectangular or elliptical if cut perpendicularly or at an angle, respectively, to the bases.

Communication

35. Any two cubes are similar because they have the same shape.

37. Lay the licorice diagonally on the paper so it spans a number of spaces equal to the number of children. (See the figure.) Cut on the lines. Equidistant parallel lines will divide any transversal into congruent segments.

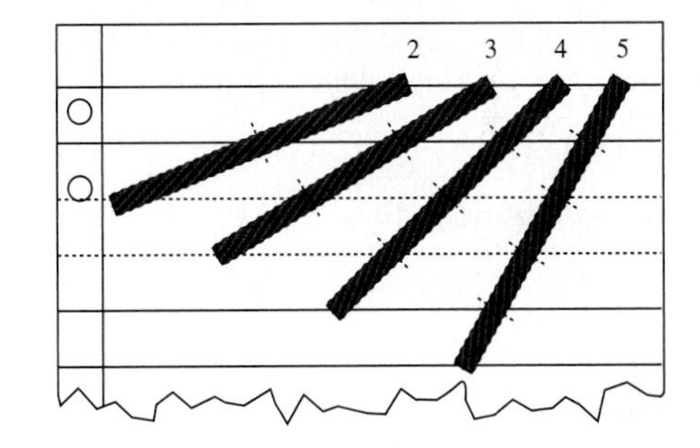

Open-Ended

39. Points lie along the same line because similar triangles are formed.

Cooperative Learning

40. **(b)** The following are two different-size triangles with the given data (the given angle measures are approximate). The triangles are approximately similar but not congruent. (The ratio of the corresponding sides is $\dfrac{80}{100}$, or $\dfrac{4}{5}$.) Hence, the surveyor and the architect could both have been correct in their conclusions.

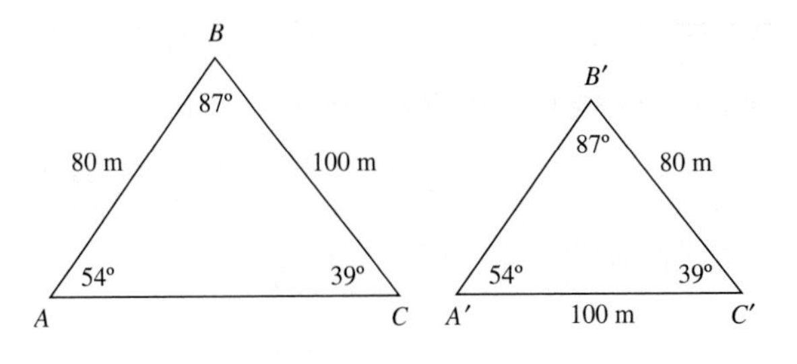

Review Problems

41. No; the image is two-dimensional while the original person is three-dimensional.
42. Construction
43. Start with the given base and construct the perpendicular bisector of the base. The vertex of the required triangle must be on that perpendicular bisector. Starting at the point where the perpendicular bisector intersects the base, mark on the perpendicular bisector a segment congruent to the given altitude. The endpoint of the segment not on the base is the vertex of the required isosceles triangle.
44. Construct an equilateral triangle with the given side, then construct the perpendicular from any vertex to the opposite side.
45. Answers may vary. Students may suggest that angles of measure 45° be constructed with the endpoints of the hypotenuse as vertices of the 45° angles and the hypotenuse as one of the sides of the angles. Both angles need to be constructed on the same side of the hypotenuse.
46. SASAS or ASASA

*Ongoing Assessment 10-5

1. (a) About 13.1 m **(c)** About 9.5 m
2. About 5.1′
4. About 8.3 m
6. (a) About 35.7°
7. About 63.9′
8. (a) About 561.9′
9. A 42′ ladder is needed.
10. About 8.4 mi
11. (b) The sum of the squares of the sine and cosine is 1 for any angle.
12. About 297′
14. Tangent; we measure the ratio of rise (change along the y-axis) to run (change along the x-axis).
16. (a) About 121.8° **(c)** About 1.7 m
17. $a^2 + b^2 = c^2$

Open-Ended

19. Answers may vary, but the answer is no.

Cooperative Learning

20. Answers may vary. **(d)** The values should be comparable.

Ongoing Assessment 10-6

1. (c) The graph of $y = mx + 3$ contains the point (0, 3) and is parallel to the line $y = mx$. Similarly, the graph of $y = mx - 3$ contains the point $(0, {}^-3)$ and is parallel to $y = mx$.

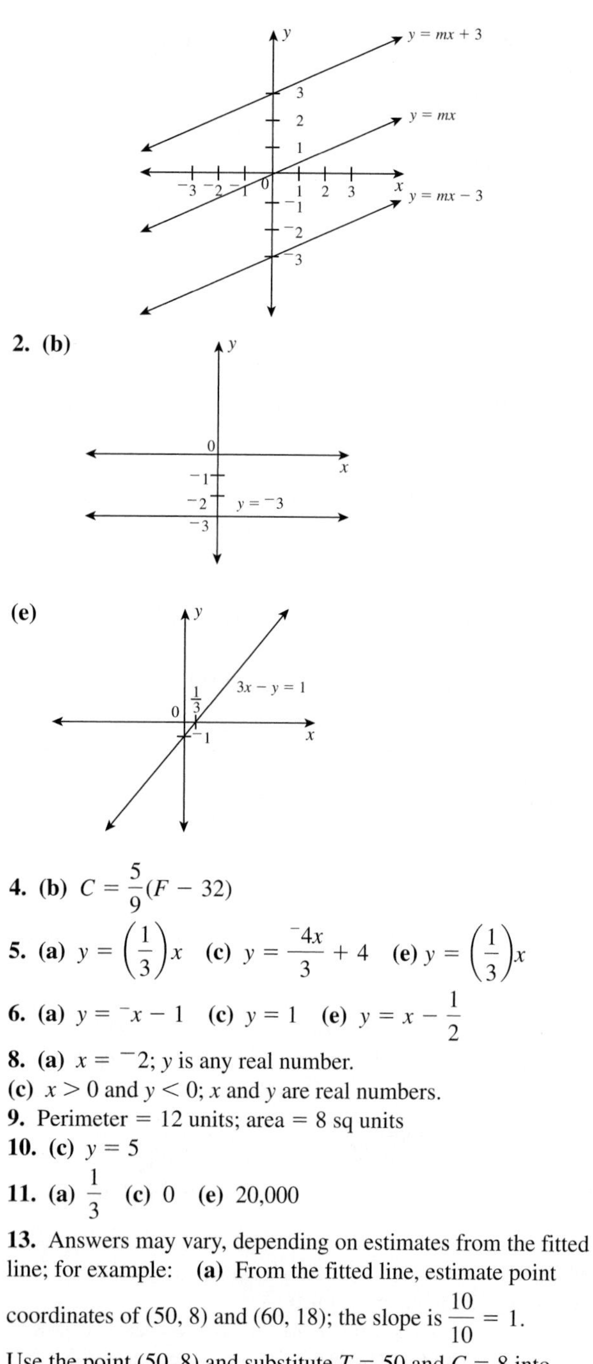

2. (b)

(e)

4. (b) $C = \frac{5}{9}(F - 32)$

5. (a) $y = \left(\frac{1}{3}\right)x$ **(c)** $y = \frac{{}^-4x}{3} + 4$ **(e)** $y = \left(\frac{1}{3}\right)x$

6. (a) $y = {}^-x - 1$ **(c)** $y = 1$ **(e)** $y = x - \frac{1}{2}$

8. (a) $x = {}^-2$; y is any real number.
(c) $x > 0$ and $y < 0$; x and y are real numbers.
9. Perimeter = 12 units; area = 8 sq units
10. (c) $y = 5$

11. (a) $\frac{1}{3}$ **(c)** 0 **(e)** 20,000

13. Answers may vary, depending on estimates from the fitted line; for example: **(a)** From the fitted line, estimate point coordinates of (50, 8) and (60, 18); the slope is $\frac{10}{10} = 1$.
Use the point (50, 8) and substitute $T = 50$ and $C = 8$ into $C = 1T + b$ (i.e., an equation of the form $y = mx + b$), so $8 = 1(50) + b \Rightarrow b = {}^-42$; the equation is then $C = T - 42$.
(c) $N = 4T - 168$
14. (a) (3, 0) or $({}^-3, 0)$

15. (a)

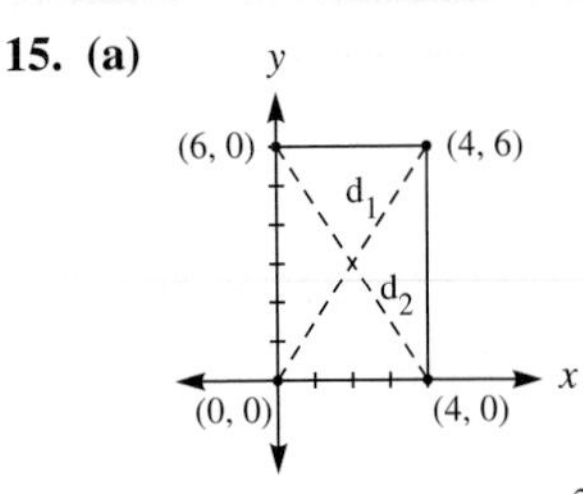

(c) The equation of d_1 is $y = \frac{3}{2}x$.

The equation of d_2 is $y = \frac{^-3}{2}x + 6$.

16. (a) $y = \frac{1}{3}x + \frac{22}{3}; y = \frac{8}{3}x - \frac{34}{3}; y = -2x - 2$

17. (a) $y = 4$

18. (a) $(2, 3)$ and $\left(12\frac{1}{2}, 3\right)$

19. $y = 0$; it is possible to have other solutions.

20. (b) 60

22. (a) Unique solution of $(2, 5)$ **(d)** Infinite number of solutions; x may be any real number $\Rightarrow y = \frac{4x - 1}{6}$.

25. (a) $2000

26. 17 quarters, 10 dimes

27. (a) All solutions are $(^-1, 2)$. **(c)** (i) These equations are of the form $nx + (n + 1)y = (n + 2)$ and $(n + 3)x + (n + 4)y = (n + 5)$. (ii) All solutions will be $(^-1, 2)$.

28. Answers vary; (50, 106).

29. (a) 1 **(b)** $^-1$

Communication

31. No; it is curved.

32. The intersection point gives the solution of the system.

$$\left.\begin{aligned} C^\circ &= F^\circ \\ C^\circ &= \frac{5}{9}(F^\circ - 32^\circ) \end{aligned}\right\}$$ The solution of this system tells when $C^\circ = F^\circ$, that is, when the temperature measured in degrees Celsius equals the temperature measured in degrees Fahrenheit.

33. Answers will vary.

Explanation 1: If two distinct lines have the same slope m, then the equations are $y = mx + b$ and $y = mx + c$ for some real numbers b and c (with $b \neq c$). To show that the lines are parallel, it is sufficient to show that the lines do not intersect; that is, that the system of equations has no solution. Indeed, if we try to solve the equations, we get $mx + b = mx + c$. Because $b \neq c$, this equation has no solution. Explanation 2: With $y = mx + b$ and $y = mx + c$, each is a vertical shift of the same line, $y = mx$.

34. Tell Jonah that $5 - 5x$ is not equal to 0 (his order of operations is incorrect). We have $5 - 5x = 5 \cdot 1 - 5x = 5(1 - x)$, which is not 0.

Cooperative Learning

39. Use the result of Ongoing Assessment 10-2, problem 25(b) that proves tangents to a circle from some outside point are congruent. Thus, $AH = AE$; $BE = BF$; $CF = CG$; $DG = DH$. Hence, $AE + EB + CG + DG = BF + FC + DH + HA$ or $c + d = a + b$.

Review Problems

41. 10

42. Yes. Parallel segments are created. Use the midsegments theorem.

43. $\frac{12}{7}$

44. 8.2

Chapter Review

1. (a) $\triangle ADB \cong \triangle CDB$ by SAS
(c) $\triangle ABC \cong \triangle EDC$ by AAS
(e) $\triangle ABD \cong \triangle CBD$ by ASA or SAS
(g) $\triangle ABD \cong \triangle CBE$ by SSS; $\triangle ABE \cong \triangle CBD$ by SSS

2. Parallelogram; $\triangle ADE \cong \triangle CBF$ by SAS, so $\angle DEA \cong \angle CFB$; $\angle DEA \cong \angle EAF$ (alternate interior angles between the parallels $\overleftrightarrow{DC}$ and $\overleftrightarrow{AB}$ and the transversal $\overleftrightarrow{AE}$), and $\angle EAF \cong \angle CFB$; and therefore $\overline{AE} \parallel \overline{FC}$.
$\overline{EC} \parallel \overline{AF}$ (parallel sides of the square); two pairs of parallel opposite sides implies a parallelogram.

4. (a) (i) $x = 8$ cm (ii) $y = 5$ cm

6. $\frac{a}{b} = \frac{c}{d}$ (by the transitive property)

8. (a) (i) $\triangle ACB \sim \triangle DEB$ by AA **(ii)** $x = \frac{24}{5}$ in.

9. (a) False: A chord has its endpoints on the circle.

11. (a) Polygons (ii) and (iii)

12. $h = 6$ m

13. $d = \frac{256}{5}$ m

14. Cut as shown. Slide shaded triangle to the right.

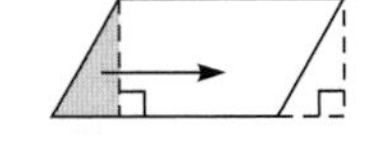

15. (a) AD is a diameter.

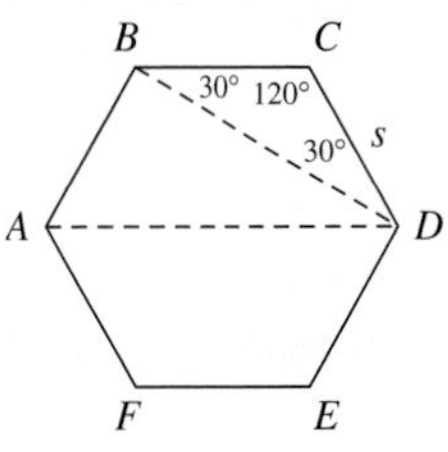

(c) $S\sqrt{3}$, where S is the length of a side.

16. They are equal unless vertical when the slopes do not exist.

17. All midsegments are congruent.
18. **(a)** Use AA.
20. **(b)** $y = \frac{1}{3}x + 1$
21. No; if a line connects all three points, the slope between any two must be the same.
Slope between (4, 2) and (0, $^{-}1$) $= \frac{^{-}1 - 2}{0 - 4} = \frac{3}{4}$; slope between (0, $^{-}1$) and (7, $^{-}5$) $= \frac{^{-}5 - {}^{-}1}{7 - 0} = \frac{^{-}4}{7}$.
Slopes are unequal, so there is no line through the points.
22. **(a)** Unique solution of $\left(4\frac{1}{5}, \frac{^{-}3}{5}\right)$ **(b)** Unique solution of $\left(\frac{10}{9}, \frac{4}{3}\right)$ **(c)** No solution; parallel lines
23. Congruent triangles have two corresponding angles that are congruent. This makes them similar by AA.
★ **24.** sin 60 is about 0.87; 2 sin 30 is 1.
25. **(a)** $x \doteq 68.7$ m

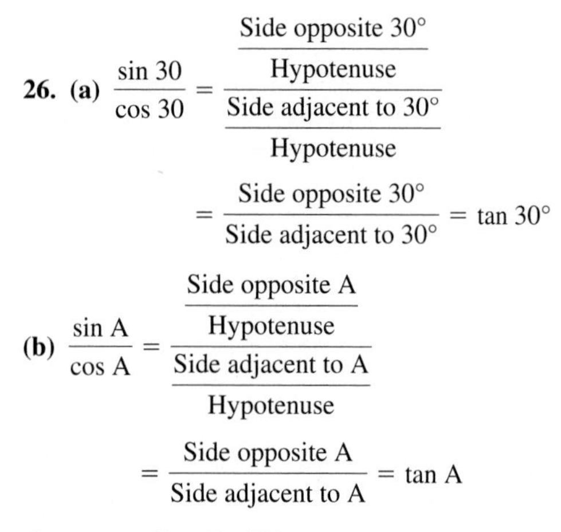

26. **(a)** $$\frac{\sin 30}{\cos 30} = \frac{\frac{\text{Side opposite } 30°}{\text{Hypotenuse}}}{\frac{\text{Side adjacent to } 30°}{\text{Hypotenuse}}} = \frac{\text{Side opposite } 30°}{\text{Side adjacent to } 30°} = \tan 30°$$

(b) $$\frac{\sin A}{\cos A} = \frac{\frac{\text{Side opposite A}}{\text{Hypotenuse}}}{\frac{\text{Side adjacent to A}}{\text{Hypotenuse}}} = \frac{\text{Side opposite A}}{\text{Side adjacent to A}} = \tan A$$

Answers to Now Try This

10-1

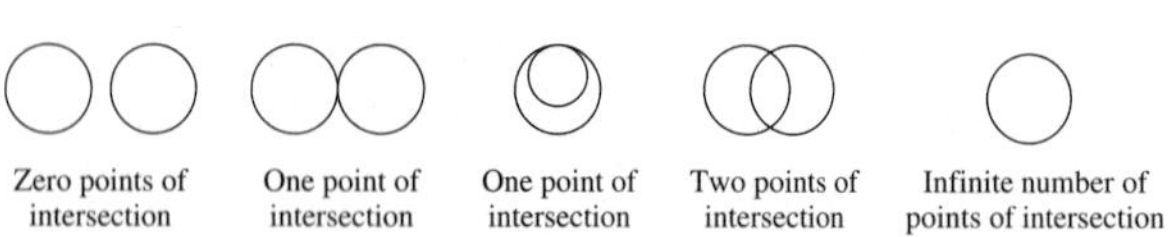

10-2 There are six possible ways to write the congruence: *ABC* paired with *A′B′C′*, *ACB* paired with *A′C′B′*, *BAC* paired with *B′A′C′*, *BCA* paired with *B′C′A′*, *CAB* paired with *C′A′B′*, and *CBA* paired with *C′B′A′*.
10-3 **(a)** In order to construct a triangle from three lengths of straws, the length of one piece must be less than the sum of the lengths of the other two pieces.
(b) The triangle would be equilateral and equiangular.
(c) There can be no triangle constructed. The three pieces would lie along a segment.
(d) No. See the answer to part (a).
10-4 **(a)** The triangles do not have to be congruent. An example is seen in the following drawing.

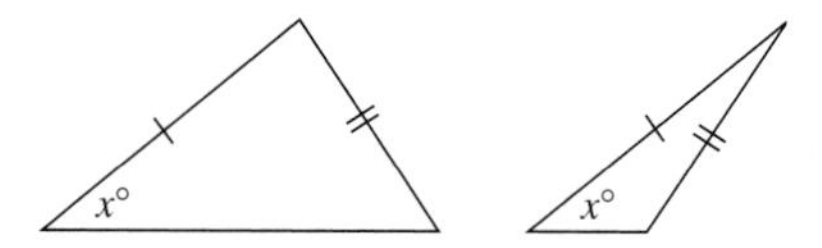

(b) If the nonincluded angle is a right angle, the triangles are congruent. More generally, if the nonincluded angle is opposite the larger of the two sides, the triangles are congruent.
10-5 Draw two intersecting arcs with the same radius, one with center at *A* and the other with center at *B*. The two points of intersection are opposite vertices of a rhombus. The line connecting the points is the perpendicular bisector of the segment.
10-6 The answer is no. Consider the following drawing:

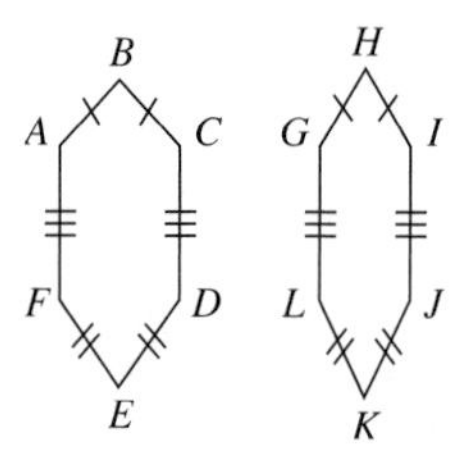

10-7 In a parallelogram *ABCD* opposite sides are parallel. This implies that $\angle BAC \cong \angle ACD$ as well as $\angle BCA \cong \angle DAC$. Hence, $\triangle ABC \cong \triangle CDA$ by *ASA* since $\overline{AC}$ is a common side. By CPCTC, $\overline{AB} \cong \overline{CD}$ and $\overline{BC} \cong \overline{DA}$.
10-8 Yes. If *ABCD* is a quadrilateral whose diagonals intersect at *M*, then congruent vertical angles with vertex at *M* are formed. Because the diagonals bisect each other, it follows from *SAS* that $\triangle AMD \cong \triangle CMB$. Hence, $\angle MAD \cong \angle MCB$ and therefore $\overline{BC}$ is parallel to $\overline{AD}$. In a similar way, we can show that $\overline{AB}$ is parallel to $\overline{DC}$ and hence *ABCD* is a parallelogram.
10-9 **(a)** An isosceles trapezoid is a trapezoid with one pair of congruent base angles.
(b) Using this definition, a general parallelogram would not be considered an isosceles trapezoid.
10-10 One way to accomplish the construction is to place the hypotenuse of the triangle on the given line and to place the ruler so that one of the legs of the triangle will be on the ruler. Now, keeping the ruler fixed, slide the triangle so that the side of the triangle on the ruler touches the ruler all the time. Slide the triangle on the ruler until the given point is on the hypotenuse. The line containing the hypotenuse is parallel to the given line.

10-11 One way is to consider the equivalent contrapositive statement: If a point is not on the angle bisector of an angle, then it is not equidistant from the sides of the angle. Choose points not on the angle bisector of an angle and use the measurement tool to see that for each point the distances to the sides of the angle are not the same. It is possible to prove the original statement by using the Hypotenuse-Leg Congruency Theorem (if the hypotenuse and one leg of one right triangle are correspondingly congruent to the hypotenuse and a leg of another right triangle, then the triangles are congruent).
10-12 Construct the diagonals of the square. Use the intersection of the diagonals as the center of the inscribed circle.
10-13 The statement is not true for polygons. For example, a rectangle and a square have congruent angles but do not have to be similar.
10-14 The triangles formed are similar and one would expect the segments cut by the parallel segments to have proportional lengths. They do not. Thus, there is the possibility for a misleading graph.

10-15 $\cos \angle 1 = \dfrac{\text{Length of adjacent side}}{\text{Length of hypotenuse}}$ so we need to use the angle adjacent to x, which measures 55°.

$$\frac{x}{10} = \cos 55°$$
$$x = 10 \cos 55°$$
$$x \doteq 5.7$$

10-16 Answers may vary. For example, the sine of an angle is equal to the cosine of the complement of an angle.
10-17 **(a)** $y = 0$
(b) As m increases, the slope is positive and the graph of the line becomes steeper as m becomes larger.
(c) As m decreases, the slope is negative and the graph of the line becomes steeper as the absolute value of m becomes larger.
10-18 **(a)** All the points on a horizontal line have the same y-coordinate. Thus, two points on a horizontal line have the form (x_1, y_1) and (x_2, y_2), where $y_2 = y_1$.
The slope is $\dfrac{y_2 - y_1}{x_2 - x_1} = \dfrac{0}{x_2 - x_1} = 0$.
(b) For any vertical line, $x_2 = x_1$ and therefore $x_2 - x_1 = 0$. If we attempted to find the slope, we would have to divide by 0, which is impossible. Hence, the slope of a vertical line is not defined.
10-19 **(a)** The graphs are lines that intersect at $x = 4$ and $y = 3$.
(b) The lines are parallel and hence the system has no solution. Algebraic approach: Assuming that there is a solution x and y, we multiply the first equation by 2 and add it side-by-side to the second. Therefore, x and y must satisfy $0 \cdot x + 0 \cdot y = 5$. However, no x and y satisfies this equation (a solution would imply $0 = 5$). This contradicts our assumption that the original system has a solution.
(c) The lines are parallel. The algebraic approach is similar to part (b).

Answers to Technology Corners

Section 10-1
2. (a) X is the center of the circle. The radius is $\overline{XY}$. The center is the set of all points that are the same distance, the radius, from X.
(b) The measures of $\overline{XY}$ and $\overline{AB}$ are equal.
(c) The circle represents all the points that are the same distance, AC from X.
3. (a) The sides and angles of $\triangle ABC$ and $\triangle XYZ$ arc congruent. Therefore, the triangles are congruent.
(b) The relationship between the triangles does not change when the size and shape of $\triangle ABC$ changes.
(c) When the sides of one triangle are congruent, respectively, to the sides of another triangle, the triangles are congruent.
5. (b) When **Mark Angle** and **Transform** are selected, an arc appears briefly that looks like it is measuring the angle. When **Rotate By Marked Angle** is chosen, another line appears making an angle congruent to $\angle PQR$.
6. (a) They are congruent.
(b) The relationship does not change.
(c) They are congruent.
7. (e) Answers may vary.
(f) If we know that two sides and a nonincluded angle in one triangle are congruent to two sides and a nonincluded angle in another triangle, it is not enough information to determine that the two triangles are congruent.

Section 10-2
It is true that if one pair of sides of a quadrilateral is congruent and parallel, then the quadrilateral is a parallelogram.

Section 10-3
The altitudes of an acute triangle intersect in the interior of the triangle. The altitudes of a right triangle intersect on the hypotenuse of the triangle. The altitudes of an obtuse triangle intersect in the exterior of the triangle.

1. $m(\angle BOC) = 2m(\angle BPC)$
2. The measure is 90°.

All of the ratios are equal. You can conclude that the triangles are similar because the ratios of their sides are equal.

The perimeter of the S_1 is 3; the perimeter of S_2 is 4; and the perimeter of S_3 is $\dfrac{48}{9}$. Students may have trouble thinking about the perimeters of the snowflake curves.
(d) A cube

Section 10-5

1. The ratios found corresponded to the sines, cosines, and tangents.
2. It is not necessary to find the measures of the sides of a triangle to determine the sine, cosine, and tangent ratios.

Chapter 11

Ongoing Assessment 11-1

1. (a) 1 cm **(c)** 8 cm **(e)** 0.7 cm **(g)** 7.3 cm
3. (a) 960 farthings **(c)** 2880 pence
4. It is not a true metric system because of nickels, quarters, half-dollars, and $2 coins and several bills.
5. (a) Perimeter is 12 units.

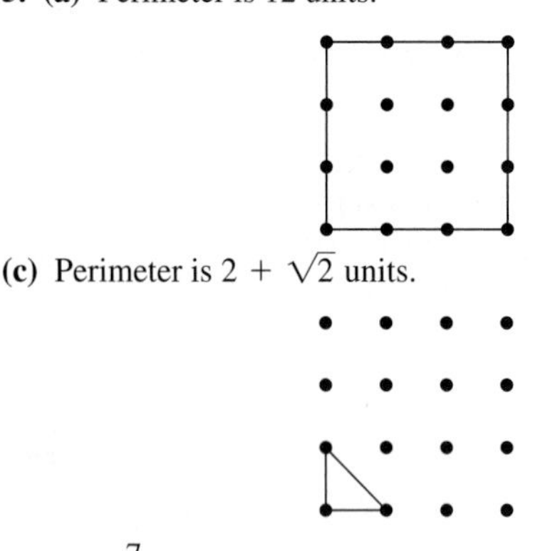

(c) Perimeter is $2 + \sqrt{2}$ units.

6. (a) $2\frac{7}{9}$ **(c)** 100
11. (a) 0.35, 350 **(c)** 0.035, 3.5 **(e)** 200, 2000
12. (a) 10.00 **(c)** 10.0 **(e)** 195.0 **(g)** 40.0
13. 6 m, 5218 mm, 245 cm, 700 mm, 91 mm, 8 cm
15. (a) π **(c)** yes; $\frac{1}{2}$
17. (a) 1 cm **(c)** 3000 m **(e)** 3500 cm **(g)** 64.7 cm
19. (a) Can be **(c)** Cannot be. In (b) and (c) the numbers cannot be the lengths of the sides of a triangle because in (b) $10 + 40 = 50$ and in (c) 260 mm + 14 cm < 410 cm, each of which contradicts the triangle inequality.
20. (a) The ratios of the perimeters are the same as the ratio of the sides.
21. (a) The minimum perimeter is attained when the second longest side is not part of the perimeter.

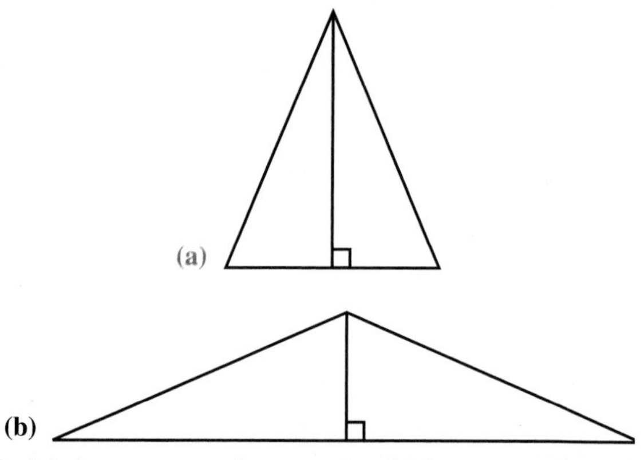

22. (a) Answers vary; for example, add 4 squares to the end of row one. **(c)** 20 squares.
23. (a) 6 cm **(c)** $0.335/\pi$ m
24. $A = at^2r^3$
25. (a) 6π cm **(c)** 4 cm
26. The circumference will double.
27. πr
28. (a) About $9.5 \cdot 10^{12}$ km **(c)** About $6.8 \cdot 10^{8}$ hr or about 78,000 years
29. (a) 3096 km/hr **(c)** Approximately Mach 4.04
31. 49 cm
32. (a) 0.5 m **(c)** 0.005 m or 5 mm

Communication

35. Answers may vary. A practical approach might be to tape string to the curve to mark it and then measure the string needed.
36. Answers may vary. Though the question may appear confusing, some students may think that the perimeter measures both the "outside" and "inside" of the curve instead of the curve itself.
37. The standard actually changed the length of a yard. Check the Historical Note in the section. Do research on the web if needed.
39. The contractor's claim is not accurate. Concrete is sold by the cubic yard.
41. The height is $3d$, where d is the diameter of a tennis ball. The perimeter is πd, or about $3.14d$, which is greater than $3d$.

42. The outer curve has a greater radius and a correspondingly greater distance (that is, arc length) to run. To compensate for the extra distance, the runner in the outer lane is given an apparent head start.

Open-Ended

43. (a) Answers vary. For example: 1-1-1 and 2-2-1 work, 2-3-6 and 1-2-3 do not

Cooperative Learning

45. (a) Answers vary. **(b)** If the circles are the same size, then regardless of the radius of each circle the amount of wire needed is the same. **(c)** Let the radius of each circle be r cm. Then the number of circles needed for each chain is $\frac{60}{2r}$ (why?). The circumference of each circle is $2\pi r$ and hence the amount of wire, which is the total circumference of all the circles, is $\frac{60}{2r} \cdot 2\pi r$, or 60π cm. Because this number does not depend on the value of r, the amount of wire needed for the chain is the same regardless of the radius of each circle.

Ongoing Assessment 11-2

2. (a) cm^2, in.^2 **(c)** m^2 or cm^2, yd^2 or in.^2 **(e)** m^2, yd^2
4. (a) 0.0588 m^2, 58,800 mm^2 **(c)** 15,000 cm^2, 1,500,000 mm^2 **(e)** 0.0005 m^2, 500 mm^2
5. (a) 444 4/9 yd^2 **(c)** 6400 acres
6. (a) 4900 m^2 **(c)** 0.98 ha
7. (a) 3 sq. units **(c)** 2 sq. units **(e)** 6 sq. units
9. (a) 20 cm^2 **(c)** 7.5 m^2
10. Answers may vary. The triangle may be drawn on graph paper and estimated. Students may look up Heron of Alexandria to find the formula $A = \sqrt{s(s-a)(s-b)(s-c)}$, where s is one-half of the perimeter of the triangle with sides a, b, and c. In this case, the area is $\sqrt{15(15-6)(15-11)(15-13)}$, or $\sqrt{1080}$, approximately 32.9 in.^2.
11. (a) The ratio is $\frac{1}{2}$.
12. r^2
14. (a) $\frac{2}{3}$
15. (a) Yes. All squares are similar.
16. (a) 9 cm^2 **(c)** 20 cm^2 **(e)** 105 cm^2
17. (a) (i) 1.95 km^2 **(ii)** 195 ha **(c)** Answers vary; for example, the metric system is easier because you only have to move the decimal point to convert units within the system.
18. (a) True **(c)** It could be 60 cm^2, so false.
19. $(1/2)ab$
20. (a) \$405.11
21. (a) 25π cm^2 **(c)** $(9/2)\pi$ cm^2
22. 1200 tiles
23. (a) The midsegments intersect at the center of the room (the point where the diagonals intersect).
24. Answers may vary.
(a) Cut as shown:

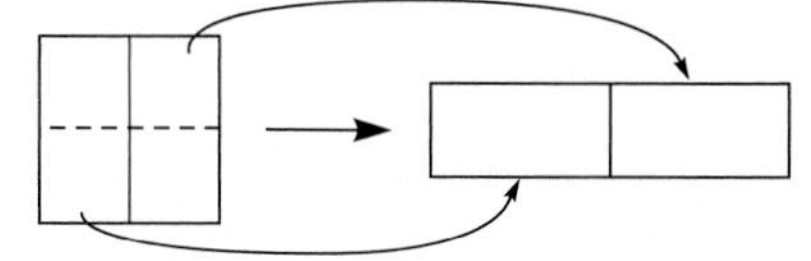

(c) Cut as shown:

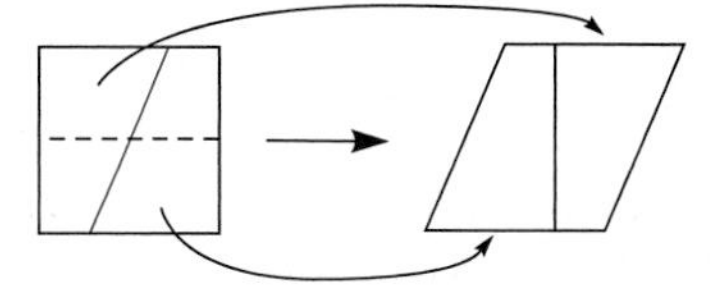

25. 8 bags of seed
26. (a) $24\sqrt{3}$ cm^2
28. (a) 16π cm^2
29. (a) 2π cm^2 **(c)** 2π cm^2 **(e)** $(1/4)\pi r^2$ **(g)** $(1/16)\pi r^2$
30. 7π m^2
31. (a) 48 cm
32. (a) The area is quadrupled.
33. (a) The area is quadrupled. **(c)** The area is increased by a factor of 9.
35. The first is a better buy at 10 ft^2 per dollar versus 9.375 ft^2 per dollar.
36. $(320 + 64\pi)$ m^2
37. 1 in.
39. (a)

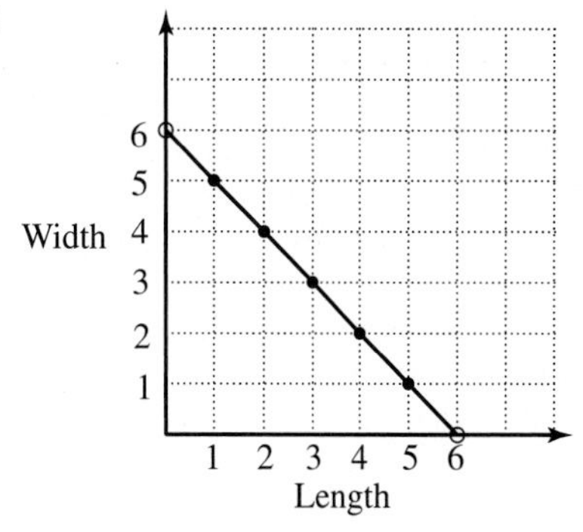

(b)

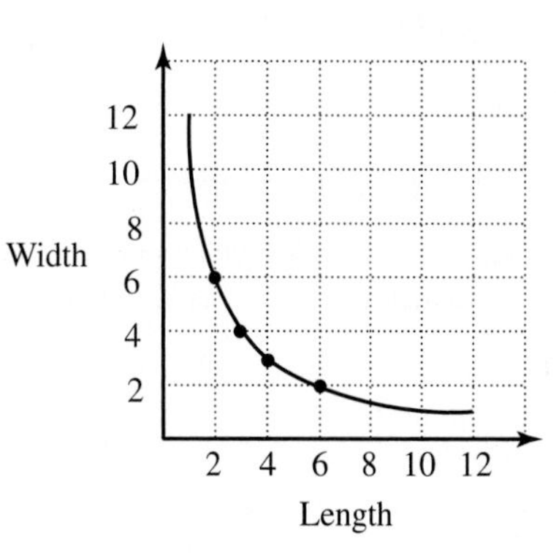

40. The area of the outside shaded portion is the same as the area of the inside shaded region, 9π in.^{2}.

42. (a)

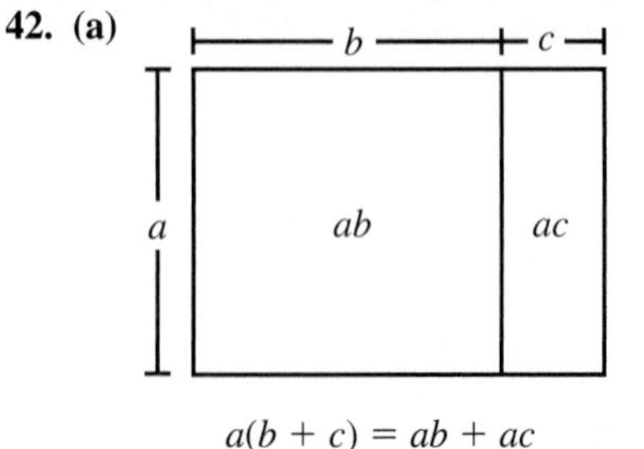

$a(b + c) = ab + ac$

44. (a) The shape of the rectangle becomes less like a square; that is, it becomes elongated.

45. The area of each triangle is 10 cm^2 because the base of each triangle is $\overline{AB}$ and the height of each triangle is the perpendicular distance between the two lines. Because each triangle has the same base and height, the areas of the triangles are the same.

46. (a) 6 cm^2

47. $\triangle BDE$ is $\frac{1}{9}$ the area of $\triangle BAC$.

48. (a) $\frac{1}{3}$ **(c)** $\frac{3}{5}$ **(e)** $\frac{3}{9}$, or $\frac{1}{3}$ **(g)** $\frac{9}{8}$

49. The areas are equal.

50. Draw altitudes $\overline{BE}$ and $\overline{DF}$ of triangles BCP and DCP, respectively. $\triangle ABE \cong \triangle CDF$ by AAS. Thus, $\overline{BE} \cong \overline{DF}$. Because $\overline{CP}$ is a base of $\triangle BCP$ and $\triangle DCP$, and because the heights are the same, the areas must be equal.

Communication

53. (a) The area of the 10-in. pizza is 25π in.^{2}. The area of the 20-in. pizza is 100π in.^{2}. Because the area of the 20-in. pizza is 4 times as great, this pizza might cost 4 times as much, or \$40. However, this is not the case because other factors are considered rather than just the area of the pizza.

54. (b) Use paper cutting and try it.

55. After the rotations, a rectangle is formed. The area of the rectangle is length times width, which in the case of the parallelogram is the same as the base times the height.

Cooperative Learning

58. (a) 5 sq. units **(b)** 12 units

Review Problems

59. $\left(8 + \frac{\pi}{2}\right)$ ft, or approximately 9.57 ft

60. (a) Approximately 6330 km **(b)** 9937.5 km

61. $(2\pi - 6)r$, or approximately $0.28r$ is the difference.

Ongoing Assessment 11-3

1. (a) $\sqrt{20} \doteq 4.5$

2. (a) 6 **(c)** 12 **(e)** 9 **(g)** $2\sqrt{2}$ **(i)** $3\sqrt{3}$

3. (a) Make a right triangle with sides 2 and 3, and then $h = \sqrt{13}$. **(c)** Make a right triangle with sides 1 and 3, and then $h = \sqrt{10}$.

5. Answers may vary. The perimeter of the figure shown is $13 + 2\sqrt{2} + \sqrt{5}$, which is certainly greater than 12.

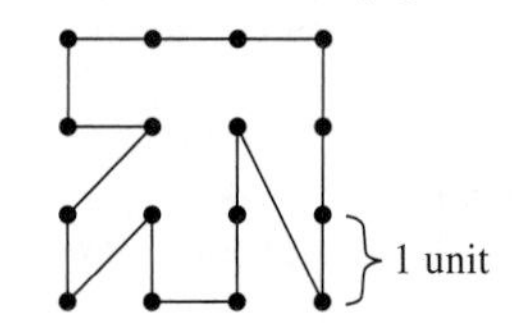

7. The lengths that can be constructed are $1, \sqrt{2}, 2, \sqrt{5}, 3, \sqrt{10}, \sqrt{13}, 2\sqrt{2}$, and $3\sqrt{2}$. The only sides acceptable for a 30°-60°-90° triangle are $\sqrt{2}$ for the short leg and $2\sqrt{2}$ for the hypotenuse. There is no way to construct such as triangle with these specifications.

9. (a) No **(c)** Yes

11. About 2622 km

13. (a) No, in order for it to be a square the sides must be the same length and the angles must be right angles. The sides are of different lengths. **(c)** The length of the diagonal is doubled.

15. 9.8 m approx.

16. (a) $(s^2\sqrt{3})/4$

17. (a) $x = 8, y = 2\sqrt{3}$

19. (a) $4\sqrt{21}$ cm^2

20. 12.5 cm; 15 cm

21. 8 cm

23. $90\sqrt{2}$, or about 127.28 ft

25. (a)

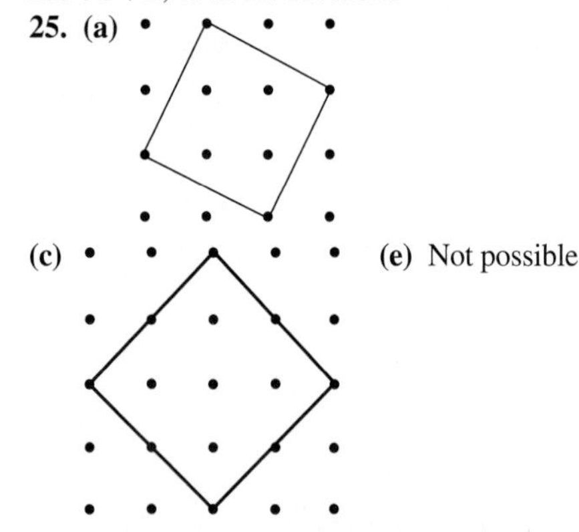

(c) **(e)** Not possible

26. $\triangle ACD \sim \triangle ABC$; $AC/AB = AD/AC$ implies $b/c = x/b$, which implies $b^2 = cx$; $\triangle CBD \sim \triangle ABC$; $AB/CB = CB/DB$ implies $c/a = a/y$, which implies $a^2 = cy$; $a^2 + b^2 = cx + cy = c(x + y) = cc = c^2$

27. $\sqrt{10}$, or approximately 3.16 m
29. Approximately 99.5 ft
30. Approximately 10.6 mi
31. The area of the trapezoid is equal to the sum of the areas of the three triangles. Thus,

$$1/2(a + b)(a + b) = (1/2)ab + (1/2)ab + (1/2)c^2$$
$$1/2(a^2 + 2ab + b^2) = ab + (1/2)c^2$$
$$a^2/2 + ab + b^2/2 = ab + c^2/2$$

Subtracting ab from both sides and multiplying both sides by 2, we have $a^2 + b^2 = c^2$. The reader should also verify that the angle formed by the two sides of the length c has measure 90°.
34. Yes

36. **(a)** 4 **(c)** $2\sqrt{13}$

37. $10 + \sqrt{10}$
39. The side lengths are 5, $7\sqrt{2}$, and 5 and so the triangle is isosceles.
40. **(a)** $x = 9$ or $^-7$

42. **(a)** $x = 0; y = 0; y = \frac{3}{4}x$

43. **(a)** $x = 3; y = 4; y = \frac{3}{4}x + \frac{7}{4}$ **(c)** 5

44. $(4, 4\sqrt{3})$ or $(4, {}^-4\sqrt{3})$
45. **(a)** (16, 0)
(c) $(8\sqrt{2})^2 + (8\sqrt{2})^2 = 128 + 128 = 256 = 16^2$

Communication

46. **(a)** Let the length of the side of the square be s. Draw the diagonal of the square and make the new square have side lengths equal to the diagonal. Then the area is $(\sqrt{2}\,s)^2 = 2s^2$.
48. Yes, she had a right triangle because the converse of the Pythagorean Theorem implies that if $13^2 = 12^2 + 5^2$, then the triangle is a right triangle. Because this is true, she has a right triangle and therefore a right angle.

Open-Ended

51. **(b)** Yes, we know that if a-b-c is a Pythagorean triple, then $a^2 + b^2 = c^2$. This implies that $4(a^2 + b^2) = 4c^2$, and that $(2a)^2 + (2b)^2 = (2c)^2$.
(d)
$$\begin{aligned} a^2 + b^2 &= (2uv)^2 + (u^2 - v^2)^2 \\ &= 4u^2v^2 + u^4 - 2u^2v^2 + v^4 \\ &= u^4 + 2u^2v^2 + v^4 \\ &= (u^2 + v^2)^2 \\ &= c^2 \end{aligned}$$

Review Problems

53. 0.032 km, 322 cm, 3.2 m, 3.020 mm.
54. **(a)** 33.25 cm^2 **(b)** 30 cm^2 **(c)** 32 m^2
55. **(a)** 10 cm, 10π cm, 25π cm^2 **(b)** 12 cm, 24π cm, 144π cm^2 **(c)** $\sqrt{17}$ m, $2\sqrt{17}$ m, $2\pi\sqrt{17}$ m **(d)** 10 cm, 20 cm, 100π cm^2
56. $25/\pi$ m^2

Ongoing Assessment 11-4

1. **(a)** 96 cm^2 **(c)** 236 cm^2 **(e)** 96π cm^2 **(g)** 1500π ft^2
3. 2688π mm^2
5. 4 : 9
6. **(a)** They have equal lateral surface areas.
8. Approx. 32.99 $in.^2$
10. Approx. 91.86 ft^2
11. $\ell = 11$ cm, $w = 8$ cm, $h = 4$ cm
12. **(a)** The surface area is multiplied by 4. **(c)** The surface area is multiplied by k^2.
13. **(a)** The lateral area is multiplied by 3. **(c)** The lateral area is multiplied by 9.
14. **(a)** The surface area is multiplied by 4.
16. **(a)** 44 **(c)** Yes, for example, place five cubes in the shape of a C. Then adding a cube to the center of the *C* would add no surface area.
17. **(a)** 1.5π m^2
18. **(a)** 3 units **(c)** 4 units
19. 10π $in.^2$
20. **(a)** A ring or an annulus as shown

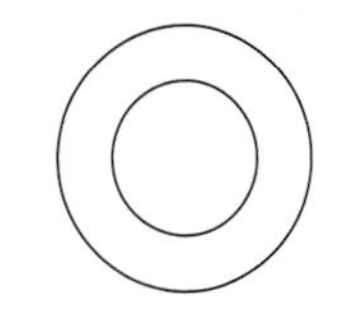

21. 1/4
23. 4/3 is the ratio of the edges.
24. **(a)** $100\pi(1 + \sqrt{5})$ cm^2 **(c)** 2250π cm^2
25. **(a)** Approx. 42 cm
26. **(a)** $S.A. = 100 + 20\sqrt{125} = 100 + 100\sqrt{5}$ in., or about 323.6 $in.^2$
★ **27.** $(6400\pi\sqrt{2} + 13{,}600\pi)$ cm^2
★ **28.** 375π cm^2

Communication

29. The ice cubes would melt faster because they have a greater surface area that is exposed to the air.
30. She would need 4 times as much cardboard. If each measurement is doubled, then the area of each face is increased by a factor of 4; that is, $A_1 = \ell w$ and $A_2 = (2\ell)(2w) = 4\ell w = 4A_1$. Because this is true for all faces, the surface area is multiplied by 4.
31. **(b)** The height of the can is $3d$ or $6r$, where d is the diameter of the ball. The circumference of the can is $2\pi r$. Therefore, the $L.S.A.$ is $(2\pi r)6r = 12\pi r^2$. The surface area of

three balls is $3(4\pi r^2) = 12\pi r^2$. Therefore the surface area of the three balls is the same as the lateral surface area of the can.
33. **(a)** The ½ in. is the diameter of the hose. **(c)** The 1-in. hose has exterior surface area 600π in.^{2}, or has twice the exterior surface area as the ½-in. hose.

Open-Ended

34. **(a)** 2% **(c)** Estimates vary depending on the size of the person.

Review Problems

38. **(a)** 100,000 **(b)** 1.3680 **(c)** 500 **(d)** 2,000,000
(e) 1 **(f)** 1,000,000
39. $10\sqrt{5}$ cm
40. $20\sqrt{5}$ cm
41. **(a)** 240 cm; 2400 cm^2 **(b)** $(10\sqrt{2} + 30)$ cm , 75 cm^2
42. Length of side is 25 cm. Diagonal $\overline{BD}$ is 30 cm.
43. **(a)** $\frac{\pi}{4}$ **(b)** $\frac{\pi\sqrt{2}}{4}$ **(c)** $\frac{1}{\sqrt{2}}$
44. **(a)** C is the intersection of $\overline{AB}$ and line ℓ. **(b)** *Hint*: Use the Triangle Inequality.

Ongoing Assessment 11-5

1. **(a)** 8000 **(c)** 0.000675 **(e)** 7 **(g)** 0.00857 approx.
(i) 345.6
2. 32.4 L
4. 2000 ft^3
5. **(a)** $\left(\frac{256}{3}\right)\pi$ cm^3 **(c)** 216 cm^3 **(e)** 15π cm^3
(g) $(4000/3)\pi$ cm^3 **(i)** $(20{,}000/3)\pi$ ft^3
6. **(a)** 2000, 2, 2000 **(b)** 0.5, 0.5, 500 **(c)** 1500, 1.5, 1500
(d) 5000, 5, 5 **(e)** 0.75, 0.75, 750 **(f)** 4.8, 4.8, 4800
7. **(a)** 200.0 **(c)** 1.0
8. 1680π mm^3
10. It is multiplied by 8.
11. **(a)** 2000, 2, 2 **(b)** 6000, 6, 6 **(c)** 2 dm, 4000, 4
(d) 2.5 dm, 7500, 7.5
12. $253{,}500\pi$ L
14. 2,500,000 L
15. π mL
16. **(a)** It is multiplied by 8. **(b)** It is multiplied by 27.
17. The Great Pyramid has the greater volume. It is approx. 25.66 times greater.
18. More than 17,196.3 or 17,197 apartments
22. 1/8 of the cone is filled.
24. The $2 \times 2 \times 4$ ft freezer is a better buy at \$25/ft^3.
26. About 21.5%
27. **(a)** Approximately 512,000 ping-pong balls
29. It is very possible that a car would not fit. Suppose the space were only 3 ft tall.
31. Answer may vary. One unit might be mi^3.
32. **(a)** $\frac{8}{27}$
33. They are equal.
35. No, it is only 1/3 of the volume for 1/2 the price.
37. The larger is the better buy. The volume of the larger melon is 1.728 times the volume of the smaller but is only 1.5 times as expensive.
39. 512,000 cm^3
40. **(a)** It won't hold the ice cream at 10 cm tall: it would have to be 20 cm tall.
42. **(a)** Kilograms or metric tons **(c)** Grams **(e)** Grams
(g) Metric tons
43. **(a)** Milligrams **(c)** Milligrams **(e)** Grams
44. **(a)** 15 **(c)** 36 **(e)** 4.320 **(g)** 30
(i) 1.5625
45. **(a)** No **(c)** Yes **(e)** Yes
46. 16 kg
47. **(a)** $^-12$°C **(c)** $^-1$°C **(e)** 100°C
48. **(a)** No **(c)** Yes **(e)** Yes **(g)** Hot
49. **(a)** 200,000 L

Communication

50. **(a)** Doubling the height will only double the volume. Doubling the radius will multiply the volume by 4. This happens because the value of the radius is squared after it is doubled.
54. Answers may vary. Possible shapes are shown with one based on surface area and the other on volume.

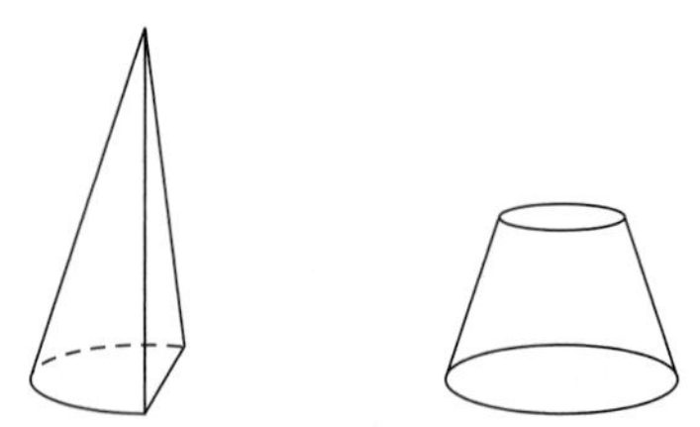

55. Answers may vary. An assumption is that comparable-sized rooms hold a comparable amount of furniture.
57. The numerical value of kilograms would be a smaller number.

Open-Ended

60. If the cookies are stood up so the 6-cm diameter is toward the front of the box, then the minimum surface area is obtained by having two cookies across (12 cm), two cookies high (12 cm), and 12 cookies deep (12 cm).

Review Problems

63. (a) $(20 + 6\pi)$ cm; $(48 + 18\pi)$ cm^2 **(b)** 40π cm; 100π cm^2
64. (a) 35 **(b)** 0.16 **(c)** 400,000 **(d)** 0.00052
65. (a) Yes **(b)** No
66. (a) 2400π cm^2 **(b)** $(6065 + 40\sqrt{5314})$ cm^2

Chapter Review

1. (a) $16\frac{2}{3}$ yd **(c)** 3960 ft **(e)** 5000 m **(g)** 520 mm
2. (a) Not possible, $p > q + r$
3. (a) $8\frac{1}{2}$ cm^2 **(c)** 7 cm^2
4. The pieces of the trapezoid are rearranged to form a rectangle with width $h/2$ and length $(b_2 + b_1)$. The area is $A = h/2(b_2 + b_1)$, which is the area of the initial trapezoid.
5. $\triangle ABC$, $\triangle ABD$, and $\triangle ABE$ (same) and $\triangle ABF$. All the triangles have the same base, so the ordering is just by height and $\triangle ABD$ and $\triangle ABE$ have the same height.
6. (a) $54\sqrt{3}$ cm^2
7. (a) 12π cm^2 **(c)** 24 cm^2 **(e)** 64.5 cm^2
8. 5400 cm^2
9. $\sqrt{16{,}200} \doteq 127.3$ ft
10. 3
11. (a) Yes
12. (a) $S.A. = 32(2 + \sqrt{13})$ cm^2; $V = 128$ cm^3 **(c)** $S.A. = 100\pi$ m^2; $V = (500\pi)/3$ m^3 **(e)** $S.A. = 304$ m^2; $V = 320$ m^3
13. 65π m^2
14. The graph on the right has 8 times the volume of the figure on the left, rather than double as it should be.
15. 252 cm^2
16. (a) 340 cm **(b)** 6000 cm^2
17. $2\sqrt{2}$ m^2
18. 62 cm
19. 30 cm
20. (a) $y = 0$; $y = {}^{-}5x + 40$; $y = 5/3x$
(c) $y = \frac{5}{7}x$; $y = {}^{-}x + 8$; $y = 5x - 20$ **(d)** $\left(\frac{14}{3}, \frac{10}{3}\right)$

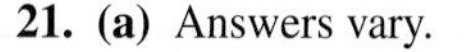
21. (a) Answers vary.

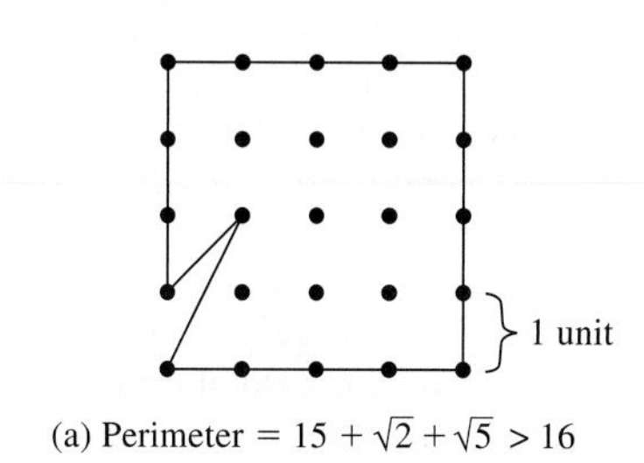

(a) Perimeter = $15 + \sqrt{2} + \sqrt{5} > 16$

(b) Answers may vary.

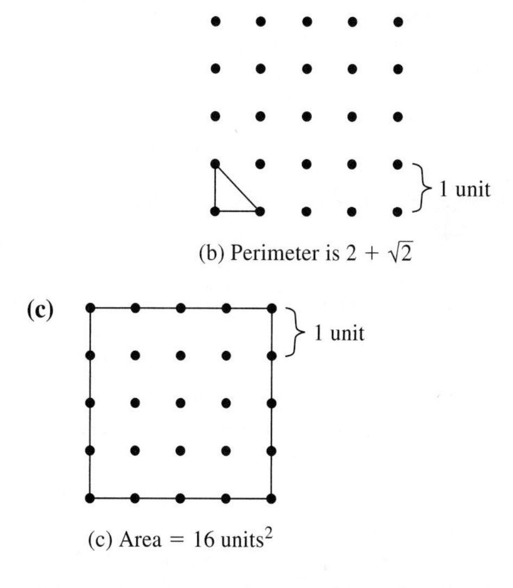

(b) Perimeter is $2 + \sqrt{2}$

(c) Area = 16 units^2

22. Approximately 13.9 in.
24. $2\sqrt{3}$ in.
25. Find the length of a common perpendicular segment. The lines have that common distance between them everywhere.
26. (a) $4\sqrt{3}$ units **(c)** The polyhedron is the cube determined by the outside dots. The volume is 64 units^3.
27. (a) Metric tons **(c)** 1 g **(e)** 25 **(g)** 51,800 **(i)** 50,000 **(k)** 25,000 **(m)** 52.813
28. $h_1^3/h_2^3 = V_1/V_2$
29. (a) 6000 kg
30. (a) L **(c)** g **(e)** kg **(g)** mL
31. (a) Unlikely **(c)** Unlikely **(e)** Unlikely
32. (a) 2000 **(c)** 3 **(e)** 0.0002

Answers to Now Try This

11-1 (a) Answers will vary.
(b) Answers may vary depending upon the size of paper clips used. The length is approximately 8 paper clips.
(c) Answers will vary.
(d) Answers will vary.

11-2 **(a)** decidollar **(b)** centidollar **(c)** dekadollar **(d)** hectodollar **(e)** kilodollar
11-3 **(a)** 115 cm **(b)** 55 cm
11-4 π is an irrational number. The decimal representation never terminates and never repeats.
11-5 **(a)** 12 **(b)** 6 **(c)** 4
11-6 Approximately 100 cm^2
11-7 **(a)** The areas of the parallelogram and the rectangle are the same because they are composed of the same two pieces and the area of the figures is the sum of the areas of the two pieces.
(b) The formula for the area of a rectangle is $A = \ell w$. The length of the new rectangle is just base b of the parallelogram and the width of the rectangle is just the height of the parallelogram. Therefore, the formula for the area of the parallelogram is $A = bh$.
11-8 **(a)** $\frac{b}{2} \cdot \frac{h}{2} = \frac{bh}{4}$
(b) The triangle was twice as large as the folded rectangle, so multiply by 2. $2 \cdot \frac{bh}{4} = \frac{1}{2}bh$
11-9 The new figure is a parallelogram with base $(b_1 + b_2)$ and height h, where b_1 and b_2 are the bases of the original trapezoid. The area of the parallelogram is $A = h(b_1 + b_2)$. Because this is twice the area of the original trapezoid, we divide by 2 to obtain $A = h/2(b_1 + b_2)$, which is the formula for the area of the original trapezoid.
11-10 **(a)** Answers will vary.
(b) 25. Parallelogram
26. The length is approximately equal to $\pi \cdot$ radius, half of the circumference. The width is approximately equal to the radius. The area is approximately equal to the length $\cdot$ width.
The length of the base is 1/2 of the circumference.
The height of the figure is the radius of the circle.
27. As the circle is cut into more and more sectors and put back together, the shape becomes more and more like a parallelogram.
28. $A = \frac{1}{2}Cr = \frac{1}{2}2\pi r \cdot r = \pi r^2$
11-11 The square on one leg that is labeled 1 could be cut off and placed in the dashed space on the square of the hypotenuse. Then pieces 2, 3, 4, and 5 could be cut off and placed around piece 1 so that the square on the hypotenuse is filled exactly with the five pieces. This shows that the sum of the areas of the squares on the two legs of a right triangle is equal to the area on the square of the hypotenuse.
11-12 Jason ran the hypotenuse of a right triangle with legs each 10 yd long.
11-13 **(a)** You could build the triangle and then measure the angles to see if there was a right angle. If the angle is a right angle, then the triangle is a right triangle. You could measure the three sides and use the converse of the Pythagorean Theorem to see if a right triangle is formed.
(b) If the three lengths of a right triangle are multiplied by a fixed number, then the resulting lengths determine a right triangle; for example, if the right triangle lengths are 3-4-5, and the fixed number is 5, then 15-20-25 is a right triangle.
(c) If the three lengths of a right triangle are multiplied by a fixed number, then the resulting numbers determine a right triangle.
11-14 It makes no difference in the distance formula if $(x_1 - x_2)$ and $(y_1 - y_2)$ are used instead of $(x_2 - x_1)$ and $(y_2 - y_1)$, respectively. Because both quantities in the formula are squared, the result is the same whether the difference is positive or negative.
11-15 **(a)** $S.A. = 2 \cdot (5/2) \cdot 8 + 2 \cdot (8 \cdot 11) + 2 \cdot (5/2) \cdot 11 = 271 \text{ in.}^2$
(b) No, the rectangle would have to be 21 in. by 16 in.
11-16 Because we want the surface area of a right prism, we must include the top and bottom so we need $2B$ (where B is the area of the base, which is the same as the area of the top) in the formula $S.A. = ph + 2B$. From the net, we see that the lateral surface area opens up into a rectangle that has width equal to the height, h. The length of the rectangle is equal to the sum of the lengths of the sides of the base, which is the perimeter of the base. Therefore, the area of the rectangle (lateral surface area of the prism) is $A = \ell w = ph$. Hence, the surface area for any right prism is given by $S.A. = ph + 2B$.
11-17 **(a)** Both figures have a volume of 9 cubic units.
(b) No, the second figure has the greater surface area.
(c) The first figure has surface area 34 square units and the second figure has area 36 square units.
11-18 **(a)** Move the decimal point twice as many places as in a linear equation. For example, the area of a square that is 1 m on each side is $1 \cdot 1 = 1 \text{ m}^2$. $1 \text{ m} = 10 \text{ dm}$. $10 \cdot 10 = 100 \text{ dm}^2$. $1 \text{ m}^2 = 100 \text{ dm}^2$.
(b) Move the decimal point 3 times as many places as in a linear equation. $1 \text{ m}^3 = 1000 \text{ dm}^3$.
11-19 **(a)** The two figures have bases in the same plane and the figures have the same height. Figure 11-76 shows that if a plane parallel to the base is passed through the figures, then equal areas are obtained. By Cavalieri's Principle, these two figures have equal volumes.
(b) **(i)** By Cavalieri's Principle, the volumes are the same.
(ii) By Cavalieri's Principle, the volumes are the same.
11-20 **(a)** We know that the height of each figure is $2r$ because the height of the sphere is $2r$. The volume of the cylinder is $\pi r^2 2r = 2\pi r^3$. The volume of the cone is $\frac{1}{3}\pi r^2 \cdot 2r = \frac{2}{3}\pi r^3$. The volume of the sphere is $\frac{4}{3}\pi r^3$.

(b) Using a common denominator of 3, the three formulas are $\frac{6}{3}\pi r^3$, $\frac{2}{3}\pi r^3$, and $\frac{4}{3}\pi r^3$. The ratio is 6 : 2 : 4, which simplifies to 3 : 1 : 2.

11-21 **(a)** 1 g **(b)** 1 kg **(c)** 1 dm^3 **(d)** 1 mL **(e)** 1 g **(f)** 1 kL **(g)** 1 t

11-22 Yes, when it is $^{-}40$°C it is $^{-}40$°F.

Answers to Technology Corner

Section 11-2

Geometer's Sketchpad is used to derive the formulas for the area of a parallelogram, a triangle, and a trapezoid.

Geometer's Sketchpad is used to explore the Pythagorean Theorem.

Section 11-3

4c In a right triangle, if a and b are the lengths of the legs and c is the length of the hypotenuse, then, $a^2 + b^2 = c^2$.

6f If the square on the length of the longest side of a triangle is **greater than** the sum of the squares of the lengths of the two shorter sides, then the triangle is an obtuse triangle.
If the square on the length of the longest side of a triangle is **less than** the sum of the squares of the lengths of the two shorter sides, then the triangle is an acute triangle.
If the square on the length of the longest side of a triangle is **equal to** the sum of the squares of the lengths of the two shorter sides, then the triangle is a right triangle.

7c The relationships that are true for squares also are true for other regular polygons. The area of the hexagon on side a + the area of the hexagon on side b is equal to the area of the hexagon on side c.
(a) Answers vary depending on the lengths of the sides.
(b) Answers vary depending on the lengths of the sides.
(c) The length of the hypotenuse of a 45-45-90 degree triangle is $\sqrt{2}$ times the length of a leg.
(d) The length of the hypotenuse in a 30-60-90 degree triangle is 2 times the length of the leg opposite the 30-degree angle, and the length of the leg opposite the 60-degree angle is $\sqrt{3}$ times the length of the short leg.

The answers to all the parts will depend on the nets that are chosen.

Chapter 12

Ongoing Assessment 12-1

1. (a)

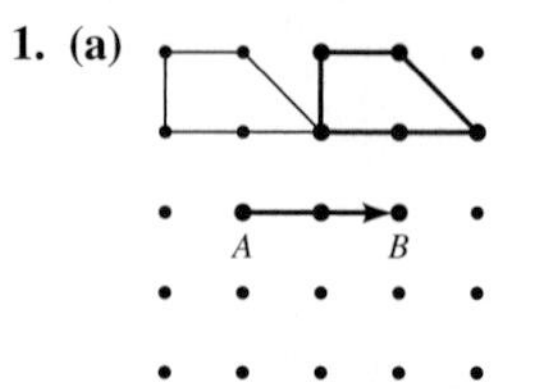

2. Reverse the translation so that the image completes a slide from X' to X (to its preimage). Then check by carrying out the given motion in the "forward" direction; that is, see if $\overline{AB}$ goes to $\overline{A'B'}$.

(a)

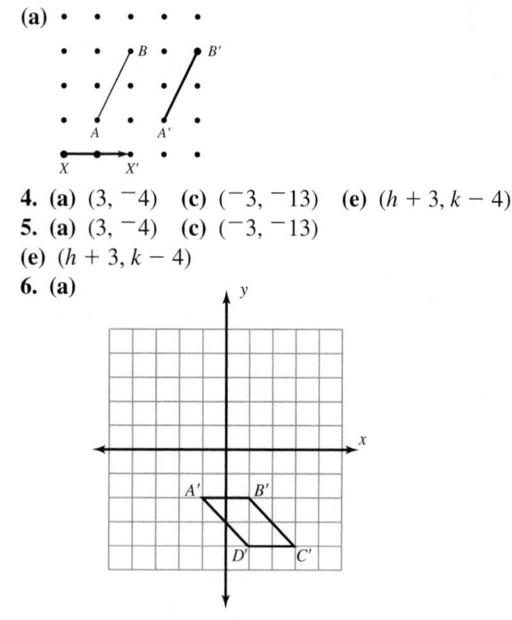

4. (a) $(3, ^-4)$ **(c)** $(^-3, ^-13)$ **(e)** $(h + 3, k - 4)$

5. (a) $(3, ^-4)$ **(c)** $(^-3, ^-13)$

(e) $(h + 3, k - 4)$

6. (a)

(c)

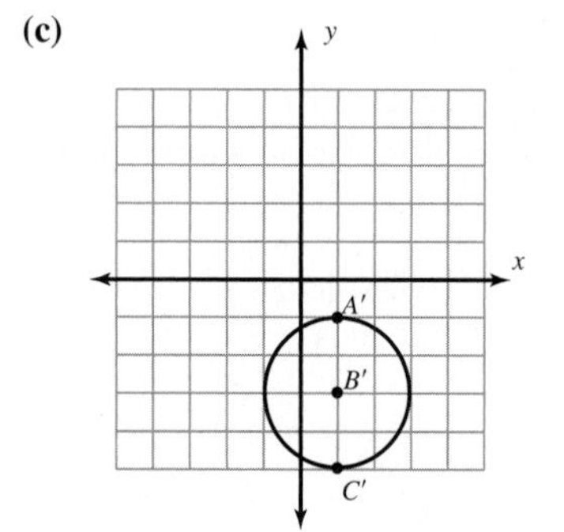

7. (a)

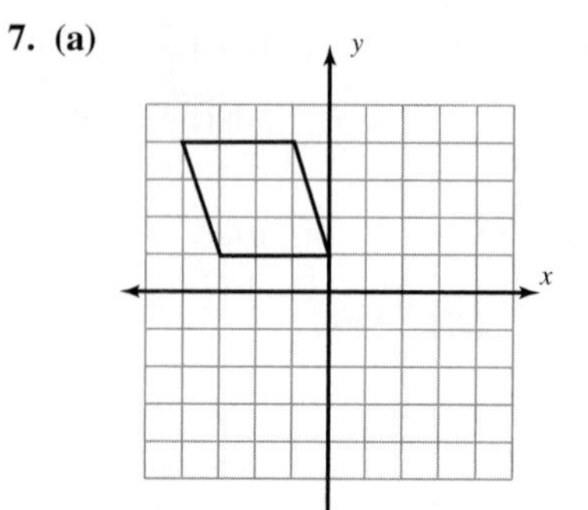

9. (a) $y = 2x - 3$ **(c)** $y = 2x + 7$

11. P', P and O are collinear because the measure of angle POP' must be 180°.

13. Reverse the rotation (to the counterclockwise direction) to locate $\overline{AB}$, that is, the preimage.

(a)

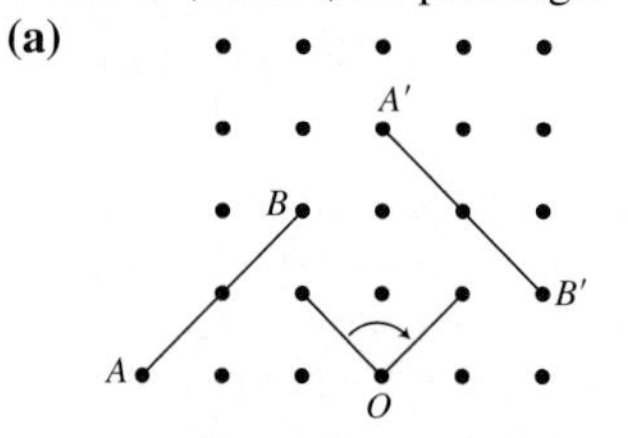

14. Answers may vary, but *H, I, N, O, S, X*, or *Z* could appear in such rotational words. Examples include SOS. Variations could use *M* and *W* in rotational images, for example, MOW.

15. (a) The image is the line ℓ itself.

17. *Hints*: An angle whose measure is 45° can be constructed by bisecting a right angle. An angle whose measure is 60° can be constructed by first constructing an equilateral triangle.

18. (a) $(^-4, 0)$ **(c)** $(^-2, ^-4)$ **(e)** $(2, 4)$ **(g)** (a, b)

19. *Hint*: Find the images of the vertices.

21. (a) First rotate $\triangle ABC$ by angle α to obtain $\triangle A'B'C'$, and then rotate $\triangle A'B'C'$ by angle β to obtain $\triangle A'B'C'$.

22. (a) A circle

23. (a) (i) $(^-3, 2)$ **(ii)** $(^-2, ^-1)$ **(iii)** $(^-b, a)$

24. (a) $y = 3x + 1$

25. (a) $C(5, 0)$ or $(^-5, 0)$

26. (a) $y = -\frac{1}{2}x - \frac{7}{2}$

(c) $y = ^-3$

27. (a) The image of $A(h, k)$ under the translation from O to C is $B(h + a, k + 0)$ or $B(h + a, k)$

(b) Using the distance formula, we get $OA^2 = h^2 + k^2$. Since $OA = OC$, $h^2 + k^2 = a^2$.

28. **(a)** $\alpha = 180°$
(b) 45°
★ **29.** **(a)** Answers vary. One possible explanation is that $PP'P''P'''$ is a parallelogram and $\overline{OM}$ is a midsegment of the parallelogram. Thus, taking the path from P to P' to P'' and finally to P''' via a translation, a half-turn about O, and again by a translation leads to the same point, P''', as via a half-turn in M. **(b)** $P'''(^-2, ^-1)$
30. **(a)** A translation from A to C
(c) No; it is the same.
31. **(b)** It is the same.
(c) The images of Q are the same.

Communication

34. **(a)** A parallelogram. Under a half-turn, the image of a line is parallel to the line. Thus, $\overline{AB} \parallel \overline{CD}$ and $\overline{AC} \parallel \overline{DB}$; therefore, $ABCD$ is a parallelogram.

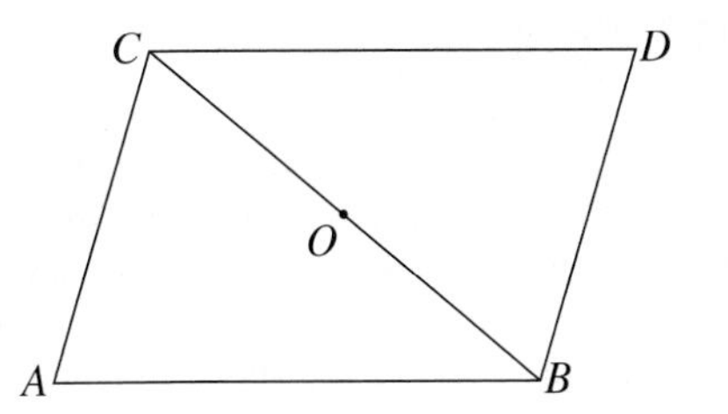

(c) A square. Because the image of A is C and the image of B is D, the image of $ABCD$ is $CDAB$.

Open-Ended

38. Answers vary. If rotation is by 30°, there will be 360/30, or 12, images.

Cooperative Learning

39. **(a)** The path will look like the one shown in the following figure. Such a path traced by P on the circle is called a *cycloid*.

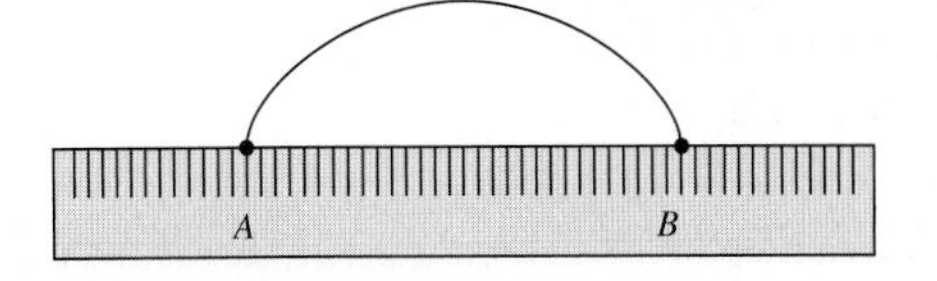

Ongoing Assessment 12-2

1. Locate the image of vertices directly across (perpendicular to) ℓ on the geoboard.

(a)

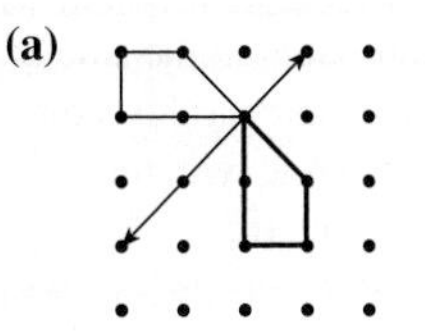

4. Reflecting lines are described for each. **(a)** All diameters (infinitely many) **(c)** The line containing the ray **(e)** Perpendicular bisectors of the sides and lines containing the diagonals **(g)** None **(i)** Perpendicular bisectors of each side **(k)** Perpendicular bisector of parallel sides **(m)** The line containing the diagonal determined by vertices of the noncongruent angles **(o)** Perpendicular bisectors of parallel sides and three diameters determined by vertices on the circumscribed circle
6. **(a)** The final images are congruent but in different locations, and hence not the same.

7. **(a)**

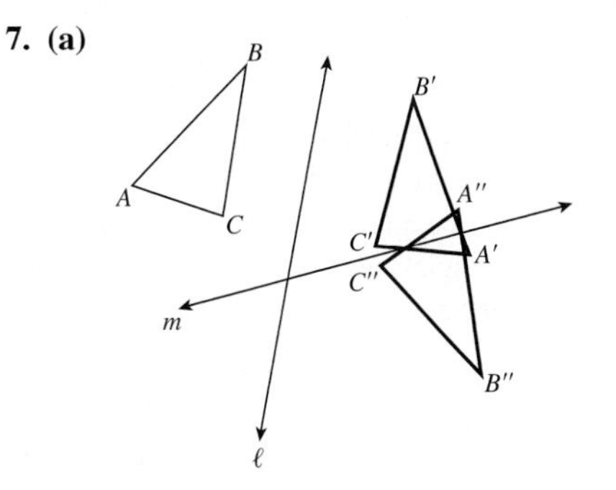

(c) A half-turn about O
9. The line of reflection is the perpendicular bisector of $\overline{BB'}$.
10. **(a)** Examples include MOM, WOW, TOOT, and HAH.
11. **(a)** The images are the same.
12. None of the images has a reverse orientation, so there are no reflections or glide reflections involved. Thus,

1 to 2 is a counterclockwise rotation.
1 to 3 is a clockwise rotation.
1 to 4 is a translation down.
1 to 5 is a rotation (with an exterior point as the center of rotation).
1 to 6 is a translation (sides are parallel to 1).
1 to 7 is a translation (sides are parallel).

13. **(a)** $A'(3, ^-4)$, $B'(2, 6)$, $C'(^-2, ^-5)$ **(c)** $A'(4, 3)$, $B'(^-6, 2)$, $C'(5, ^-2)$
14. **(a)** **(i)** $(x, ^-y)$ **(ii)** $(^-x, y)$ **(iii)** (y, x) **(iv)** $(^-y, ^-x)$
15. **(a)** $y = ^-3x$ **(e)** $y = 0$
17. **(a)** It is a square.
(b) The respective images are $(0, a)$, $(a, 0)$, $(0, ^-a)$, $(^-a, 0)$.
20. **(a)** $\overleftrightarrow{O_1O_2}$

21. The diameter of the circle must be congruent to the side of the square. The line of reflection is the perpendicular bisector of the segment connecting the center of the circle with the point of intersection of the diagonals of the square.
22. **(b)** $x = 0$ or $y = 0$
23. **(b)** They are the same.
(d) The successive reflections in lines k, ℓ, and m are equivalent to a single reflection in line n. A possible reasoning is: The successive reflections preserve distance and hence the resulting transformation must be an isometry. Because O is a fixed point—that is, its image after the three reflections is itself—the single transformation can only be a rotation or a reflection. However, the three reflections reverse orientation of any figure, so the transformation cannot be a rotation, and therefore must be a reflection. Since all points on line n are fixed points, line n is the line of reflection.
25. **(a)** All the points on ℓ

Communication

28. **(a)** If $\overline{AB} \cong \overline{BC}$, then the perpendicular bisector of $\overline{AC}$ is the required line. The image of B is B, the image of A is C, and the image of C is A. Hence, the image of $\triangle ABC$ is $\triangle CBA$.
(c) No. Since no sides (or angles) are congruent, bisecting any side or angle will leave noncongruent portions of the triangle on opposite sides of the bisector.

Open-Ended

30. **(a)** Reflect the location P of the ball in $\overline{AB}$. The intersection of $\overline{P'Q}$ with $\overline{AB}$ is the point at which the ball should be aimed. The justification is similar to the one given in problem 27.

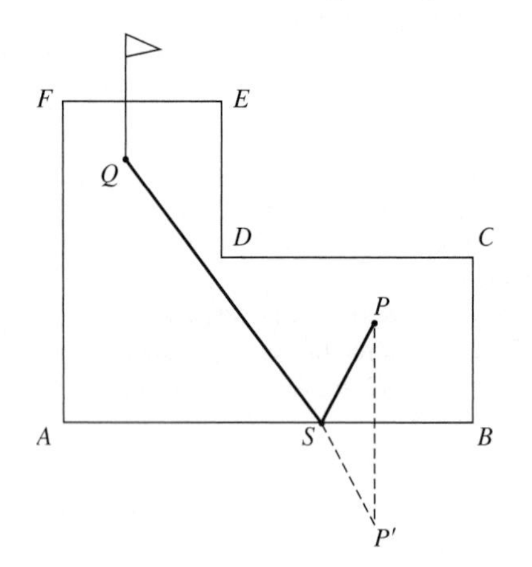

Review Problems

35. Depending on how they are drawn: 0, 1, and 8
36. $(^-a, {}^-b)$
37. **(a)** A rotation by any angle about the center of the circle will result in the same circle. **(b)** Reflections about lines containing diameters
38. Construct $\overline{BE}$ perpendicular to $\overline{AD}$, as shown. Next translate $\triangle ABE$ by the slide arrow from B to C. The image of $\triangle ABE$ is $\triangle DCE'$. The rectangle $BCE'E$ is the required rectangle.

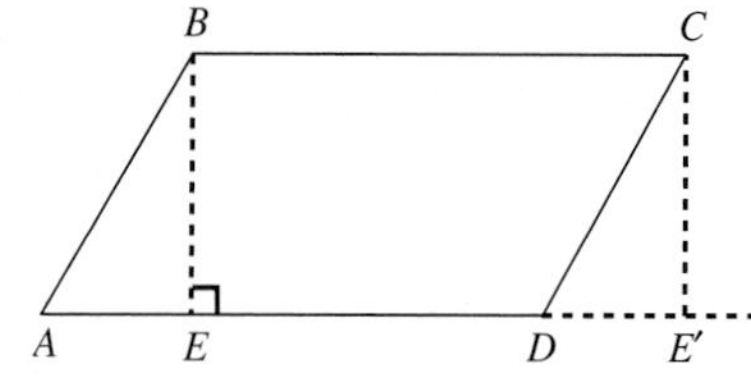

Ongoing Assessment 12-3

1. **(a)** Slide the smaller triangle down 3 units (translation). Then complete a size transformation with scale factor 2 using the top-right vertex as the center. **(c)** Rotate 90° counterclockwise with the lower-right vertex of the smaller triangle as the center of rotation. Then complete a size transformation with scale factor 2 using the same point as center.
2.

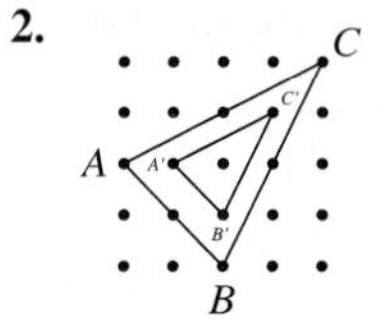

3. **(a)** Translation taking B to B' followed by a size transformation with center B' (and scale factor 2)
(c) Size transformation with scale factor 1/2 and center A followed by a half-turn with the midpoint of $\overline{AA'}$ as center
4. **(a)** $x = 6$, $y = 5.2$, scale factor 2/5
6. **(a)** Scale factor 3, $x = 12$, $y = 10$
7. **(a)** (12, 18), ($^-12$, 18)
(c) (5, 9), ($^-3$, 9)
(e) Answers depend on the order.
8. The size transformation with center O and scale factor $1/r$

9. (a) Answers vary. A possible explanation is:

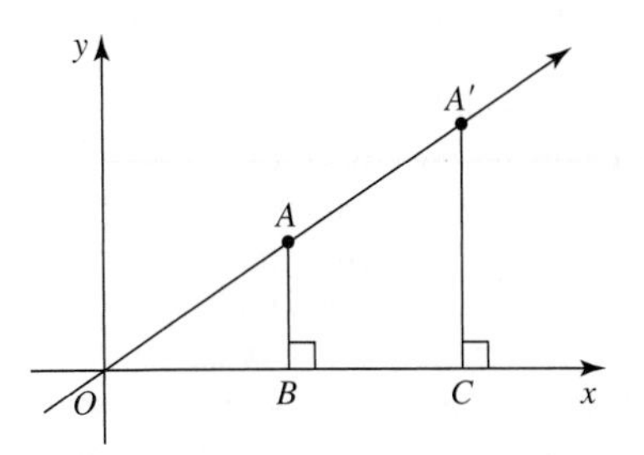

Let $A'(x', y')$ be the image of A under the size transformation. From the definition of a size transformation, we have $\frac{OA'}{OA} = r$. Since $\triangle OA'C \cong \triangle OAB$, $\frac{OC}{OB} = r$, which implies that $\frac{x'}{x} = r$ and hence $x' = rx$. Similarly, $y' = ry$.

10. The size transformation with center at the origin and scale factor $\frac{3}{4}$
11. A translation taking O_1 to O_2 followed by a size transformation with center at O_2 and scale factor $\frac{3}{2}$
12. (a) They all are on $\overrightarrow{AE}$.

Communication

13. (a) It does change. For example, consider the segment whose endpoints are (0, 0) and (1, 1) and that has length $\sqrt{2}$. Under the size transformation with center at (0, 0) and scale factor 2, the image of the segment is a segment whose endpoints are (0, 0) and (2, 2). That segment has length $2\sqrt{2}$.
(b) It does not change. Under a size transformation, the image of a triangle is a similar triangle and the angles of corresponding angles of two similar triangles are congruent.
(c) It does not change. Given two parallel lines, draw a transversal that intersects each line. Because the lines are parallel, the corresponding angles are congruent. From (b), the images of the angles will also be congruent and hence the image lines will be parallel.
15. (a) A single size transformation with center O and scale factor $\frac{1}{2} \cdot \frac{1}{3}$ or $\frac{1}{6}$. Let P be any point and P' its image under the first size transformation and P'' the image of P' under the second size transformation. Then $\frac{OP'}{OP} = \frac{1}{2}$ and $\frac{OP''}{OP'} = \frac{1}{3}$. Consequently, $\left(\frac{OP'}{OP}\right) \cdot \left(\frac{OP''}{OP'}\right) = \frac{1}{2} \cdot \frac{1}{3}$, or $\frac{OP''}{OP} = \frac{1}{6}$. Thus, P' can be obtained from P by a size transformation with center O and scale factor $\frac{1}{2} \cdot \frac{1}{3}$ or $\frac{1}{6}$.
16. The scale factor $= \frac{A'B'}{AB} = \frac{3}{4}$. The center must be the intersection of $\overleftrightarrow{BB'}$ and $\overleftrightarrow{CC'}$.
17. (a) Yes. Suppose the size transformation with center O has a scale factor r. The image of any point P on the circle with radius d is P' such that $\frac{OP'}{OP} = r$. Thus, $OP' = r(OP)$ or $OP' = rd$. This means that the image of every point on the circle is at the same distance rd from O and hence on a circle with radius rd.

Open-Ended

19. The result is equivalent to a half-turn through the origin.

Cooperative Learning

20. (a) If p is the perimeter of the figure and p' the perimeter of the image, then $p' = 3p$. **(c)** $p' = rp$

Review Problems

21. (a) The translation given by slide arrow from N to M
(b) A counterclockwise rotation of 75° about O **(c)** A clockwise rotation of 45° about A **(d)** A reflection in m and translation from B to A **(e)** A second reflection in n
22. (a) (4, 3) reflects about m to (4, 1); (4, 1) reflects about n to (2, 1). **(b)** $(0, 1) \rightarrow (0, 3) \rightarrow (6, 3)$
(c) $(^-1, 0) \rightarrow (^-1, 4) \rightarrow (7, 4)$ (d) $(0, 0) \rightarrow (0, 4) \rightarrow (6, 4)$
23. (a) The angle itself
(b) The square itself

Ongoing Assessment 12-4

2. (a) (i) Yes. A line may be drawn through the center of the circle, either horizontally or vertically. A line may also be drawn through any of the sets of arrows. **(ii)** Yes. The figure will match the original figure after rotations of 90°, 180°, or 270° about the center of the circle. **(iii)** Yes. Any figure having 180° rotational symmetry has point symmetry about the turn center. **(d) (i)** Yes. A horizontal line through the middle of the plane is a line of symmetry.
(ii) No **(iii)** No

4. Reflect the given portions about ℓ.
(b)

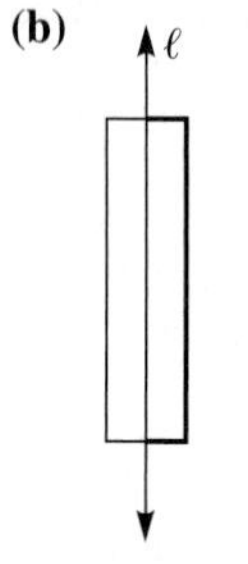

5. (a) (i) Four lines of symmetry; the diagonals and horizontal or vertical lines through the center **(ii)** No lines of symmetry **(iii)** Two lines of symmetry; horizontally and vertically through the center **(iv)** One line of symmetry; vertically through the center
6. (a) 6 **(b)** Yes. The figure has 60° rotational symmetry.

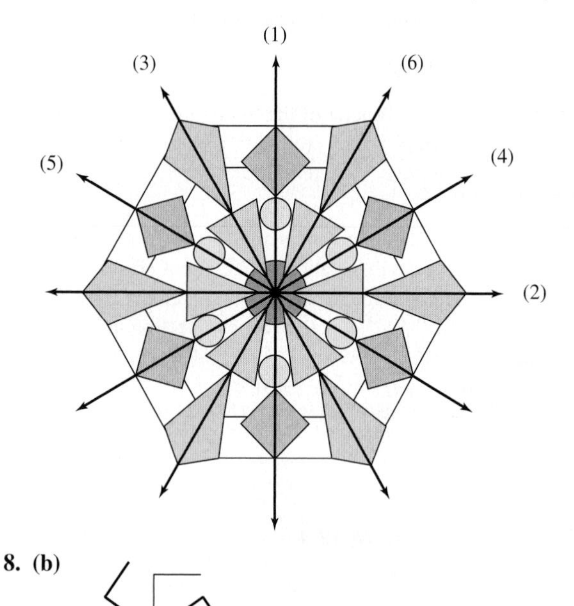

8. (b)

9. (a) Seven; three through the "peaks," three through the "valleys," and one perpendicular to the others through the width of the figure **(c)** Seven; three through the vertices, three through the faces, and one perpendicular to the others through the width of the figure

11. (a) Reflections in lines that bisect opposite sides of the rectangle, a half-turn in O the point of intersection of the diagonals
(d) A half-turn about the point of intersection of the diagonals
(g) Reflection in the angle bisector
(h) Reflections in the perpendicular bisectors of the sides, rotations about the center (where the perpendicular bisectors intersect) by 120°, 240° (the rotation by 360° is the identity transformation, technically also a symmetry)
(k) Six reflection symmetries, three in perpendicular bisectors of each pair of opposite sides and three in lines determined by the three longest diagonals (segments connecting a pair of opposite vertices). Also, six rotations about the center of the hexagon by 60°, $2 \cdot 60°$, $3 \cdot 60°$, $4 \cdot 60°$, $5 \cdot 60°$, and $6 \cdot 60°$.
(m) If the circles have equal radii, then there are two symmetries—one a reflection in $\overleftrightarrow{O_1O_2}$ and the other a reflection in $\overleftrightarrow{AB}$. If the radii are different; only one symmetry, determined by the line of reflection $\overleftrightarrow{O_1O_2}$.
(n) Reflection in $\overleftrightarrow{O_1O_2}$ if the radii are different. Two reflections—one in $\overleftrightarrow{O_1O_2}$ the other in the perpendicular bisector of $\overleftrightarrow{O_1O_2}$ if the radii are the same.

Communication

12. (a) Yes. The definition of point symmetry is that it is rotational symmetry of 180°. **(d)** No. Consider the letter Z. (Nor is the converse true.)

Open-Ended

13. (a) A scalene triangle **(c)** Not possible
15. (b) All figures will consist of a union of disjoint curves and hence will have only two line symmetries.

Cooperative Learning

18. Answers vary; for example, consider a rectangle. You would report that your figure has two lines of symmetry and a rotational symmetry of 180°.

Review Problems

20. One method is to trace the figure. Then fold at ℓ and trace along the figure as seen through the paper.
21. Find the images of the vertices.
22. (a) The image O_1' of O_1 under half turn in P will be the center of the image circle, its radius will be the same.
(b) The intersection points are X and X_1
(c) The line intersects the smaller circle in Y and Y'. The segments XY and X_1Y_1 are the required segments

(d) For the point P_1 in the figure the problem has no solution as the image circle with center Q does not intersect the larger circle with center O_2.

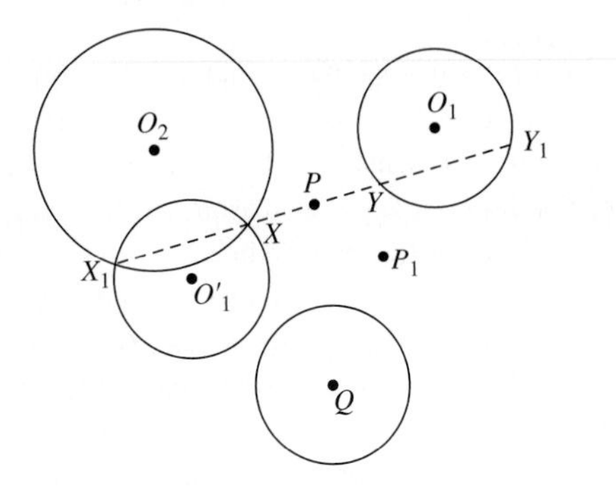

Ongoing Assessment 12-5

1. (a)

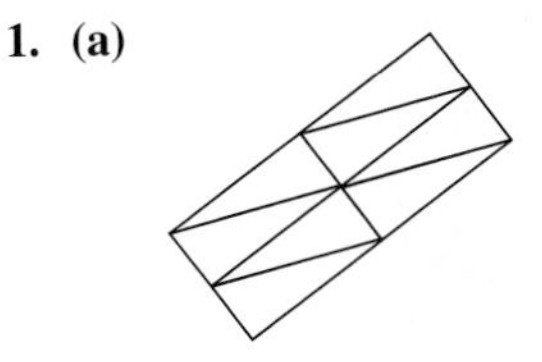

2. (b) Yes. If a polygon tessellates the plane, the sum of the angles around every vertex must be 360°. Successive 180° turns of a quadrilateral about the midpoints of its sides will produce four congruent quadrilaterals around a common vertex, with each of the quadrilateral's angles being represented at each vertex. These angles must add up to 360°, as angles of any quadrilateral do.
3. Experimentation by cutting out shapes and moving them about is one way to learn about these types of problems.
(b) Cannot be tessellated
(c)

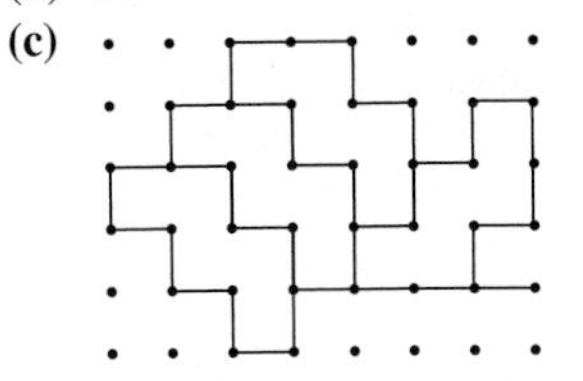

5. (b) It combines a translation and a reflection.
6. (a) The dual is another tessellation of squares (congruent to those given).
8. (a) is not semiregular; the rest are semiregular.

Communication

9. (a) The image $ABCD$ under a half-turn in M (the midpoint of CD) is the trapezoid $FEDC$. Because the trapezoids are congruent, $ABFE$ is a parallelogram. The area of the parallelogram is $AE \cdot h$ or $(a + b) \cdot h$. The parallelogram is the union of two nonoverlapping congruent trapezoids. The area of each trapezoid is $(a + b)(h/2)$.

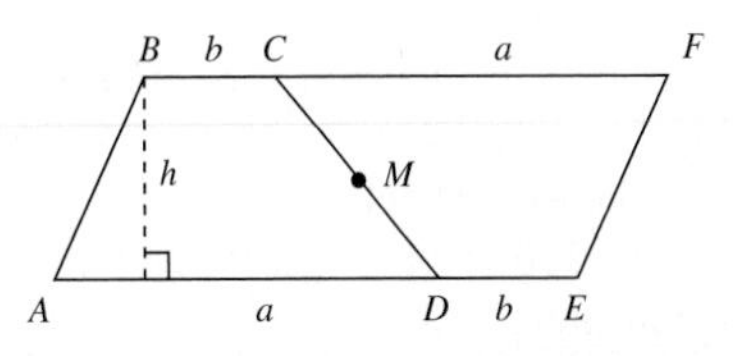

11. First tessellate the plane with regular hexagons and then divide each hexagon into congruent pentagons by three segments connecting the center of a hexagon to the midpoint of every second side. The segments inside one hexagon should not have points in common with segments inside a neighboring hexagon. Actually, one could draw the Y-shape segments in one hexagon and then translate the shape in all directions by a translation, taking the center of one hexagon to the center of a neighbouring one.

Cooperative Learning

16. (a) Two such figures can be put together to form a parallelogram. Because parallelograms tessellate the plane, the original figure tessellates the plane. **(c)** Answers vary. For example the following figure is a rep-tile.

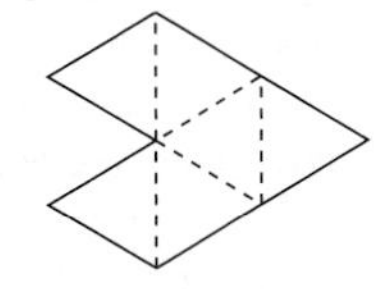

Chapter Review

1. (a)

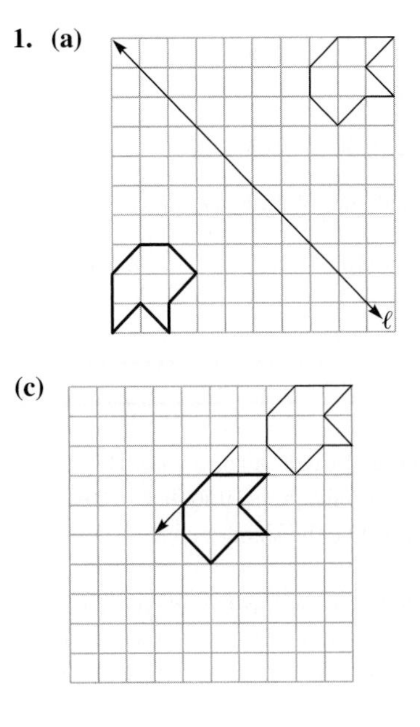

(c)

3. **(a)** 4 **(c)** 1 **(e)** 2
4. **(a)** Line and rotational **(c)** Line
5. **(a)** Infinitely many **(c)** 3
7. $A = A'$, B is the midpoint of $\overline{A'B'}$, and C is the midpoint of $\overline{A'C'}$.
8. In each case, half-turn about X.
9. **(a)** Rotation by 120° about the center of the hexagon
10. Reflection in $\overleftrightarrow{SO}$
11. Let $\triangle H'O'R'$ be the image of $\triangle HOR$ under a half-turn about R. Then $\triangle SER$ is the image of $\triangle H'O'R'$ under a size transformation with center R and scale factor $\frac{2}{3}$. Thus, $\triangle SER$ is the image of $\triangle HOR$ under the half-turn about R followed by the size transformation described above.
12. Rotate $\triangle PIG$ 180° (half-turn) about the midpoint of $\overline{PT}$, then perform a size transformation with scale factor 2 and center $P'(= T)$.
13. **(a)** $A(^-3, 12.91)$, $B(^-8, 0.07)$, $C(1.83, 5)$
(b) Under the translation $(x, y) \rightarrow (x - 3, y + 5)$
15. **(a)** $A(^-3, ^-2)$, $B(^-2, 3)$, $C(^-5, 5)$
16. **(a)** **(i)** A translation from A to C **(ii)** Same as (i)
17. **(a)** Square. A square has four lines of symmetry, a 90° rotational symmetry about its center, which implies that it has a 180° rotational symmetry (i.e., point symmetry) as well as 270° rotational symmetry. A rhombus has only two lines of symmetry and a 180° rotational symmetry (point symmetry).
(c) A cube. All the symmetries of a rectangular right box are also symmetries of a cube but not vice versa. For example, a cube has a plane symmetry determined by the diagonals of opposite faces (there are two such planes), but a rectangular box that is not a cube does not have such planes of symmetries.
18. **(a)** $y = 2x + 5$ **(c)** $y = -\frac{1}{2}x + \frac{5}{2}$
19. **(a)** $y = {}^-x + 2$ **(b)** $y = x - 3$
(d) $y = {}^-x + 3$
22. The measure of each exterior angle of a regular octagon is $\frac{360°}{8}$, or 45°. Hence the measure of each interior angle is $180° - 45°$, or 135°. Because $135 \nmid 360$, a regular octagon does not tessellate the plane.
23. No

Answers to Now Try This

1. **(a)** One way to do this is to connect A and M. Draw a line parallel to line MN through A; find a point A' so that $MN = AA'$.
(b) The direction of $\overline{AA'}$ must be the same as that of the direction of $\overline{MN}$.
2. Draw a circle with center O and radius OP. Next draw the same size circle with center the angle given. Use your compass to measure the arc of the circle cut off by the angle. On circle O, mark the image P' of P.
3. Construct a perpendicular from P to line m. Find P' so that line m is the perpendicular bisector of $\overline{PP'}$. Points P and P' are the endpoints of a diagonal of a rhombus. The other diagonal lies along line m.
4. Find the line so that the figure folds onto the image. The fold line is the reflecting line.
5. The answer is the same for A below the line $y = x$. The accompanied figure suggest the reason why the image of $P(a, b)$ is $P'(b, a)$. Notice that the line $y = x$ bisects the right angle in the first quadrant formed by the x and y axes. Because $y = x$ is the perpendicular bisector of $\overline{AA'}$, $OA = OA'$. Thus $\triangle OAM \cong \triangle OA'M$. Consequently the marked angles with vertex O are congruent. Now it is easy to show that $\angle BOA \cong \angle B'OA'$ and consequently that $\triangle OBA \cong \triangle OB'A'$, which in turn implies that $OB' = b$ and $A'B' = a$.

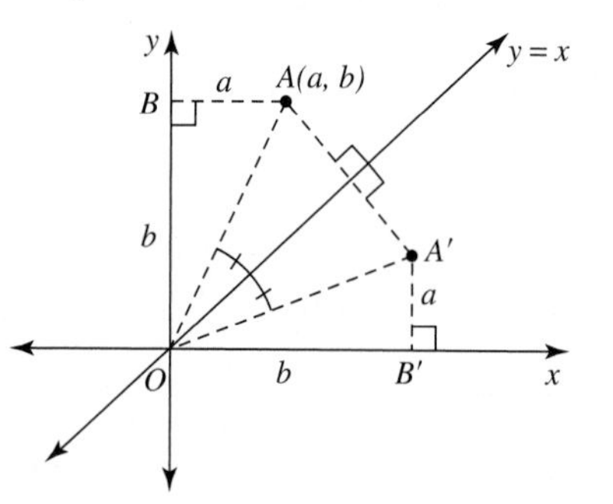

6. $MB = MB'$ and $BP = B'P$. Points A, P, and B' make a triangle. Therefore they are not all on the same line and by the triangle inequality we know that $AP + PB'$ must be longer than AB'. A, M, and B' are collinear. Therefore $AM + MB < AP + PB$.
7. **(a)** 120 degrees. 4 axes of rotation
(b) 180 degrees. 6 axes of rotation
8. Answers will vary.
9. Continue to create the shapes and rotate them to form the tessellation.

Answers to Technology Corner

Section 12-1

(a) The segments joining each vertex to its image are congruent.
(c) The object and its translation image have the same orientation.
The angles are congruent to each other and to $\angle STU$. The object and its image do not have the same orientation.
An equilateral triangle has 120° rotational symmetry. Other triangles do not have this property.

Section 12-2

5. **(a)** The measure of an angle and the measure of its reflected angle are equal.

(b) The length of a segment and the length of its reflected segment are equal.
(c) A polygon and its reflection are congruent.
(d) The orientations of an object and its reflection are opposite.
(e) The reflecting line contains the midpoint of each segment joining a point and its reflection.

Section 12-3
It is always possible to find such a line. The line, x, needed is such that the angle formed by lines m and n in that order is the same as the angle formed by the lines x and p in that order.

Appendix I

Ongoing Assessment A-I

1. Answers will vary. An example follows:
(a) 5, 9, 13, 17, 21, . . . **(b)** Column A should contain the numbers 1, 2, 3, 4, 5, 6, ... **(c)** The formula = A1 + 1 should be used to find A2 and then to fill down the rest of the column. **(d)** Column B for this example will start with 5, and each successive term is 4 more than the previous one.
(e) The formula = B1 + 4 could be used here. **(f)** The Σ key should be used when the cell where the sum is wanted is highlighted.
3. If all the data are listed in the first 14 cells of column A, the Σ key can be used to find the sum of all the data, and the sum can be placed in cell A15. The formula = A15/14 can be used to place the mean in cell A16.

5.

Number of the Term	The Term	Absolute Value of Difference of Mean and Term
1	23	39.2857
2	45	17.2857
3	67	4.7143
4	78	15.7143
5	98	35.7143
6	54	8.2857
7	36	26.2857
8	76	13.7143
9	75	12.7143
10	24	38.2857
11	43	19.2857
12	54	8.2857
13	100	37.7143
14	99	36.7143
Sum	872	314
Mean	62.2857	Sum of Absolute Value of Differences MAD 22.42857143

The MAD is approximately 22.4.

7. The first step is to create a column A that is the number of the term by placing 1 in cell A1 and using the formula = A1 + 1 to fill down for a total of 100 cells in column A. Next we could use the formula = 13*A:A and fill down for a total of 100 cells in column B.
8. Answers may vary depending upon the method used for computing grade points at your college. In general, you may use 4 quality points for each hour of A, 3 quality points for each hour of B, 2 for each hour of C, 1 for each hour of D, and 0 for each hour of F. After multiplying each hour by the respective number of quality points, find the total of all quality points, then divide by the total number of hours.

9. (a) A sample spreadsheet follows where the A column is created by entering 1 in cell A1 and using the formula = A1 + 1 and Fill Down to create a total of 25 cells. The B column is found by entering 1 in cell B1 and 1 in cell B2. The remainder of the cells are created using the formula = B1 + B2. Column C is created using the formula = B1^2. Column D is created by filling cell D1 with = C1, and using the formula = D1 + C2 to create the remainder of the cells.

	A	B	C	D
1	1	1	1	1
2	2	1	1	2
3	3	2	4	6
4	4	3	9	15
5	5	5	25	40
6	6	8	64	104
7	7	13	169	273
8	8	21	441	714
9	9	34	1156	1870
10	10	55	3025	4895
11	11	89	7921	12816
12	12	144	20736	33552
13	13	233	54289	87841

	A	B	C	D
14	14	377	142129	229970
15	15	610	372100	602070
16	16	987	974169	1576239
17	17	1597	2550409	4126648
18	18	2584	6677056	10803704
19	19	4181	17480761	28284465
20	20	6765	45765225	74049690
21	21	10946	119814916	193864606
22	22	17711	313679521	507544127
23	23	28657	821223649	1328767776
24	24	46368	2149991424	3478759200
25	25	75025	5628750625	9107509825

10. The A column can be created by entering 1 in cell A1 and using the formula = A1 + 1 and Fill Down to fill as many cells as wanted. The B column can be created by entering = A1 in cell B1 and = B1*A2 with the Fill Down feature to complete the wanted cells.

11. Answers will vary.

(b) The odds of winning are not good. The reality is that it costs more to play than one can expect to win.

Appendix II

Ongoing Assessment A-II

1. Approximately 11.896, or 11.9 yr
2. 96
4. (b) They differ in steepness
5. (a) The graphs are all linear and are parallel. They all have the same slope.
6. Approximately (1.22, $^{-}0.55$)
7. (b) 2 hr
8. (a) Sign changes occur at $x = 1$ and $x = 4$. **(c)** The graph crosses at $x = 1$ and $x = 4$.

Answers to Now Try This

Page 905
AII-1 (a) The values displayed using TRACE are the exact values in L1 and L2.
(b) Only points are displayed because this is the graph of the ordered pairs where the x-value comes from L1 and the y-value is the corresponding value from L2.
(c) A linear pattern is shown on the graph; that is, all the points fall along a line.
(d) Depending on the window, the cost of 12 CDs can be obtained from the graph of $y = 12.95x + 5$ by using the TRACE feature.

INDEX

A

B

C

D

E

F

Q

R

S

T

U

V

W

X

Y

Z

p. 1 IndexStock Imagery, **p. 4** AP Wide World Photos, **p. 5** Library of Congress, **pp. 6 and 30** © 1997 Carolina Biological Supply Company. Burlington, NC. Used by permission. **p. 31** The Granger Collection, **p. 48** Photo Researchers, Inc., **p. 55** The Image Works, **p. 69** Lucy Nicholson/Reuters/Corbis, **p. 70** Corbis/Bettmann, **p. 77** © United Feature Syndicate October 5, 1965, **p. 103** © King Features Syndicate September 12, 1990, **p. 104** © 2005 Creators Syndicate. Used by permission of John L. Hart FLP, and Creators Syndicate, Inc. January 21, 2005, **p. 112** St. Andrews University MacTutor Archive, **p. 122** © 1997 Carolina Biological Supply Company. Burlington, NC. Used by permission, **p. 129** Library of Congress, **p. 152** PhotoDisc, **p. 155** Dorling Kindersley Media Library, **p. 157** © 1992 Creators Syndicate. Used by permission of John L. Hart FLP, and Creators Syndicate, Inc. February 8, 1992, **p. 199** Calvin and Hobbes © 1990 Bill Watterson, September 15, 1990. Distributed by Universal Press Syndicate. Reprinted with permission. All rights reserved., **p. 214** PhotoDisc, **p. 216** University of St. Andrews, MacTutor Archive, **p. 234** Bryn Mawr College, **p. 247** Photo Researchers, Inc., **p. 255** Constance Reid, **p. 265** © Sidney Harris/ScienceCartoonsPlus.com, **p. 266** Culver Pictures, Inc., **p. 268** The Granger Collection, **p. 269** Picture Desk, Inc./Kobal Collection, **p. 297** PhotoDisc, **p. 353** © King Features Syndicate. February 13, 1988, **p. 364** Digital Vision, **p. 365** Corbis/Bettmann, **p. 399** The Granger Collection, **p. 427** © Scott Adams (March 17, 1998)/Distributed by United Feature Syndicate, Inc., **p. 431** Getty Images, Inc., **p. 432** Photo Researchers, Inc., **p. 444** © Newspaper Enterprise Association. September 26, 1998, **p. 454** © Creators Syndicate. Used by permission of John L. Hart FLP, and Creators Syndicate, Inc. March 17, 1992, **p. 477** © 1988 Carole Cable; April 6, 1988. Used with permission., **p. 501** Comstock RF, **p. 502** Hulton Archive, **p. 503** © 1981 LaughingStock Licensing, Inc., October 20, 1981. All rights reserved. HERMAN® is a registered trademark of LaughingStock Licensing, Inc., **p. 544** Kobal Collection **p. 570** PhotoDisc, **p. 572** (illustration) The Granger Collection, (cartoon) © 1985 Creators Syndicate. Used by permission of John L. Hart FLP and Creators Syndicate, Inc. December 1, 1985, **p. 596** Education Cartoons by Merv Magus, http://www.CartoonRoom.com, **p. 606** FOXTROT © 2005 Bill Amend. May 3, 2005. Reprinted with permission of Universal Press Syndicate. All rights reserved. **p. 617** Picture Desk, Inc./Kobal Collection, **p. 620** Library of Congress, **p. 643** © Beth Anderson, **p. 644** © M.C. Escher Company B.V. Used with permission. **p. 645** Picture Desk, Inc./Kobal Collection, **p. 649** © Peter M. Bagdigian. **p. 658** Digital Vision, **p. 692** © King Features Syndicate. November 4, 1988, **p. 696** © Benoit Mandelbrot. Used with permission. **p. 735** Digital Vision, **p. 737** Dorling Kindersley, **p. 739** Photo Researchers, Inc. **p. 745** Photo Researchers, Inc., **p. 771** Photo Researchers, Inc., **p. 781** Calvin and Hobbes © 1990 Bill Watterson. September 10, 1990. Distributed by Universal Press Syndicate. Reprinted with permission. All rights reserved **p. 803** University of St. Andrews, MacTutor Archive **p. 828** Digital Vision, **p. 829** University of St. Andrews, MacTutor Archive **p. 845** © 1983 Creators Syndicate. Used by permission of John L. Hart FLP, and Creators Syndicate, Inc. December 15, 1983, **pp. 871 and 875** Chrysler logo courtesy of Chrysler Corporation, Chevrolet logo courtesy of Chevrolet Motor Division, **p. 875** Volkswagen of America logo courtesy of Volkswagen of America, **p. 879** © M.C. Escher Company B.V. Used with permission.

PreK–2 Standards

DATA ANALYSIS AND PROBABILITY

In prekindergarten through grade 2 all students should—

Formulate questions that can be addressed with data and collect, organize, and display relevant data to answer them and should—

- Pose questions and gather data about themselves and their surroundings;
- Sort and classify objects according to their attributes and organize data about the objects;
- Represent data using concrete objects, pictures, and graphs.

Select and use appropriate statistical methods to analyze data and should—

- Describe parts of the data and the set of data as a whole to determine what the data show.

Develop and evaluate inferences and predictions that are based on data and should—

- Discuss events related to students' experiences as likely or unlikely.

Grades 3–5

NUMBER AND OPERATIONS

In grades 3–5 all students should—

Understand numbers, ways of representing numbers, relationships among numbers, and number systems and should—

- Understand the place-value structure of the base-ten number system and be able to represent and compare whole numbers and decimals;
- Recognize equivalent representations for the same number and generate them by decomposing and composing numbers;
- Develop understanding of fractions as parts of unit wholes, as parts of a collection, as locations on number lines, and as divisions of whole numbers;
- Use models, benchmarks, and equivalent forms to judge the size of fractions;
- Recognize and generate equivalent forms of commonly used fractions, decimals, and percents;
- Explore numbers less than 0 by extending the number line and through familiar applications;
- Describe classes of numbers according to characteristics such as the nature of their factors.

Understand meanings of operations and how they relate to one another and should—

- Understand various meanings of multiplication and division;
- Understand the effects of multiplying and dividing whole numbers;
- Identify and use relationships between operations, such as division as the inverse of multiplication, to solve problems;
- Understand and use properties of operations, such as the distributivity of multiplication over addition.

Compute fluently and make reasonable estimates and should—

- Develop fluency with basic number combinations for multiplication and division and use these combinations to mentally compute related problems, such as 30 × 50;
- Develop fluency in adding, subtracting, multiplying, and dividing whole numbers;
- Develop and use strategies to estimate the results of whole-number computations and to judge the reasonableness of such results;

Grades 3–5

NUMBER AND OPERATIONS (continued)

In grades 3–5 all students should—

- Develop and use strategies to estimate computations involving fractions and decimals in situations relevant to students' experience;
- Use visual models, benchmarks, and equivalent forms to add and subtract commonly used fractions and decimals;
- Select appropriate methods and tools for computing with whole numbers from among mental computation, estimation, calculators, and paper and pencil according to the context and nature of the computation and use the selected method or tool.

ALGEBRA

In grades 3–5 all students should—

Understand patterns, relations, and functions and should—

- Describe, extend, and make generalizations about geometric and numeric patterns;
- Represent and analyze patterns and functions, using words, tables, and graphs.

Represent and analyze mathematical situations and structures using algebraic symbols and should—

- Identify such properties as commutativity, associativity, and distributivity and use them to compute with whole numbers;
- Represent the idea of a variable as an unknown quantity using a letter or a symbol;
- Express mathematical relationships using equations.

Use mathematical models to represent and understand quantitative relationships and should—

- Model problem situations with objects and use representations such as graphs, tables, and equations to draw conclusions.

Analyze change in various contexts and should—

- Investigate how a change in one variable relates to a change in a second variable;
- Identify and describe situations with constant or varying rates of change and compare them.

GEOMETRY

In grades 3–5 all students should—

Analyze characteristics and properties of two- and three-dimensional geometric shapes and develop mathematical arguments about geometric relationships and should—

- Identify, compare, and analyze attributes of two- and three-dimensional shapes and develop vocabulary to describe the attributes;
- Classify two- and three-dimensional shapes according to their properties and develop definitions of classes of shapes such as triangles and pyramids;
- Investigate, describe, and reason about the results of subdividing, combining, and transforming shapes;
- Explore congruence and similarity;
- Make and test conjectures about geometric properties and relationships and develop logical arguments to justify conclusions.

Specify locations and describe spatial relationships using coordinate geometry and other representational systems and should—

- Describe location and movement using common language and geometric vocabulary;
- Make and use coordinate systems to specify locations and to describe paths;
- Find the distance between points along horizontal and vertical lines of a coordinate system.

Grades 6–8

NUMBER AND OPERATIONS (continued)

Understand meanings of operations and how they relate to one another and should—

- Understand the meaning and effects of arithmetic operations with fractions, decimals, and integers;
- Use the associative and commutative properties of addition and multiplication and the distributive property of multiplication over addition to simplify computations with integers, fractions, and decimals;
- Understand and use the inverse relationships of addition and subtraction, multiplication and division, and squaring and find square roots to simplify computations and solve problems

Compute fluently and make reasonable estimates and should—

- Select appropriate methods and tools for computing with fractions and decimals from among mental computation, estimation, calculators or computers, and paper and pencil, depending on the situation, and apply the selected methods;
- Develop and analyze algorithms for computing with fractions, decimals, and integers and develop fluency in their use;
- Develop and use strategies to estimate the results of rational-number computations and judge the reasonableness of the results;
- Develop, analyze, and explain methods for solving problems involving proportions, such as scaling and finding equivalent ratios

ALGEBRA

In grades 6–8 all students should—

Understand patterns, relations, and functions and should—

- Represent, analyze, and generalize a variety of patterns with tables, graphs, words, and, when possible, symbolic rules;
- Relate and compare different forms of a representation for a relationship;
- Identify functions as linear or nonlinear and contrast their properties from tables, graphs, or equations

Represent and analyze mathematical situations and structures using algebraic symbols and should—

- Develop an initial conceptual understanding of different uses of variables;
- Explore relationships between symbolic expressions and graphs of lines, paying particular attention to the meaning of intercept and slope;
- Use symbolic algebra to represent situations and to solve problems, especially those that involve linear relationships;
- Recognize and generate equivalent forms for simple algebraic expressions and solve linear equations

Use mathematical models to represent and understand quantitative relationships and should—

- Model and solve contextualized problems using various representations, such as graphs, tables, and equations

Analyze change in various contexts and should—

- Use graphs to analyze the nature of changes in quantities in linear relationships

Grades 6–8

GEOMETRY

In grades 6–8 all students should—

Analyze characteristics and properties of two- and three-dimensional geometric shapes and develop mathematical arguments about geometric relationships and should—

- Precisely describe, classify, and understand relationships among types of two- and three-dimensional objects using their defining properties;
- Understand relationships among the angles, side lengths, perimeters, areas, and volumes of similar objects;
- Create and critique inductive and deductive arguments concerning geometric ideas and relationships, such as congruence, similarity, and the Pythagorean relationship

Specify locations and describe spatial relationships using coordinate geometry and other representational systems and should—

- Use coordinate geometry to represent and examine the properties of geometric shapes;
- Use coordinate geometry to examine special geometric shapes, such as regular polygons or those with pairs of parallel or perpendicular sides

Apply transformations and use symmetry to analyze mathematical situations and should—

- Describe sizes, positions, and orientations of shapes under informal transformations such as flips, turns, slides, and scaling;
- Examine the congruence, similarity, and line or rotational symmetry of objects using transformations

Use visualization, spatial reasoning, and geometric modeling to solve problems and should—

- Draw geometric objects with specified properties, such as side lengths or angle measures;
- Use two-dimensional representations of three-dimensional objects to visualize and solve problems such as those involving surface area and volume;
- Use visual tools such as networks to represent and solve problems;
- Use geometric models to represent and explain numerical and algebraic relationships;
- Recognize and apply geometric ideas and relationships in areas outside the mathematics classroom, such as art, science, and everyday life